LIT

Active Leadership in Education
Enterprise and Engagement

Pearson's
Composition and Analysis of Foods

Pearson's Composition and Analysis of Foods

Ronald S. Kirk
MSc (Lond), CChem, FRSC
Consultant.
Previously Head of Food Section, Laboratory of the Government Chemist, Teddington, Middlesex

Ronald Sawyer
BSc (Shef), CChem, FRSC
Consultant.
Previously Superintendent of the Food and Nutrition Division, Laboratory of the Government Chemist, London

NINTH EDITION

Longman Scientific & Technical
Longman Group UK Limited
Longman House, Burnt Mill, Harlow
Essex CM20 2JE, England
and Associated Companies thoughout the world.

Copublished in the United States with
John Wiley & Sons Inc., 605 Third Avenue, New York, NY 10158

First Edition 1926
Second Edition 1938
Third Edition 1946
Fourth Edition 1950
Fifth Edition 1962
Sixth Edition 1970
Seventh Edition 1976
Eighth Edition first published by Churchill Livingstone 1981 under the title *Pearson's Chemical Analysis of Foods*
Reprinted by Longman Scientific & Technical 1987
Ninth Edition 1991

British Library Cataloguing in Publication Data
Pearson's composition and analysis of foods.
1. Food. Chemical analysis
I. Kirk, Ronald S. II. Sawyer, Ronald III. Pearson, David *1919–1977. Pearson's chemical analysis of foods*
664.07

ISBN 0–582–40910–1

Library of Congress Cataloging-in-Publication Data

Kirk, Ronald S.
Pearson's composition and analysis of foods.—9th ed. / Ronald S. Kirk, Ronald Sawyer.
p. cm.
Rev. ed. of: Pearson's chemical analysis of foods. 8th ed. / Harold Egan, Ronald S. Kirk, Ronald Sawyer. 1981.
Includes bibliographical references and index.
1. Food—Analysis. 2. Food adulteration and inspection.
I. Sawyer, Ronald. II. Egan, Harold, 1922– Pearson's chemical analysis of foods. III. Title. IV. Title: Composition and analysis of foods.
TX545.K53 1991
664'.07—dc20

ISBN 0–470–21693–X (USA only)

Set in 10/12 Linotron 202 Plantin Roman
Produced by Longman Singapore Publishers (Pte) Ltd.
Printed in Singapore

Preface to the Ninth Edition

In this edition we have tried to strike a balance in content to suit the needs of readers who are interested in differing fields in food analysis. This has called for the inclusion of some methods in reasonable detail, with an indication of the main reactions involved and references to alternative procedures, together with some indication of how the results are to be interpreted, especially from the legal viewpoint. As with the eighth edition, the choice of material has been increasingly influenced by the effect of international organisations, particularly the European Community, on methods and standards.

Analysis plays an important role in the assessment and maintenance of food quality and safety, both in industry and for enforcement authorities at the national and international levels. The phrase 'chemical analysis' has been used in previous editions in the restricted sense of only those methods of analysis which use chemical processes; in the eighth edition it was used in the broader sense to encompass physical techniques used to characterise and measure chemical substances, and in this edition the trend has been continued. However, for the purpose of this new edition we have come to the conclusion that the long established title 'Chemical Analysis of Foods' is no longer suitable. In view of the extended inclusion of information on food standards and on food composition we believe that the new title 'Composition and Analysis of Foods' more accurately reflects the contents. We have felt no need to exclude the name of David Pearson from the new title since we believe that the spirit of his long association with the book should be maintained.

Opportunity has been taken to consolidate the changes made in the eighth edition, not only in relation to changes on account of newer EC Directives and other international requirements such as those of the Codex Alimentarius Commission but also by taking into account substantive changes in food legislation which have taken place in the 1980s, to a lesser extent revisions of BSI and ISO standards and methods published and recommended by the Analytical Methods Committee of the Analytical Division of the Royal Society of Chemistry. In all respects we have also tried to retain the benefit of experience which the previous authors have incorporated in the earlier editions, particularly in relation to the readership. This can range from the student of food science to the professional laboratory manager requiring technical information and available methods of analysis, but we hope that with the inclusion in this edition of more compositional data, this book will also be found useful by such workers as dietitians, home economists, etc. We have included a short table of nutritional data in Appendix 13. The text has been further amplified by references for more detailed or more specialised interests. Although comprehensiveness is impossible in so vast a field we have tried to provide a broad cross-section of methods of analysis of interest to school, college, university, industrial and enforcement laboratories. Minor revisions of the distribution of material between chapters have been made. The reader should note that the use of proprietary names in this book does not imply endorsement of the product or Company.

Harold Egan's untimely death in June 1984 was a sad occasion. His activities in the wider sphere of food science and analysis at the national and international level were such that the scientific community lost a wealth of experience. It was perhaps fortunate that he was able to contribute some of his knowledge to the content of the eighth edition of this book. Although before his retirement he had conveyed to the present authors his intention to withdraw from active participation in the production of a ninth edition, his exceptional energy and drive to complete on target will be missed by both of us.

We would also like to thank all those friends and colleagues both within and without LGC who have troubled to provide comment and criticism on the earlier edition. Many of their suggestions have been incorporated into the present text. We would like to acknowledge the official associations, societies, industrial organisations and others who have given permission to use published and other information in this and earlier editions.

1989 RSK, RS

Contents

Contents

Abbreviations

used in this book or commonly met in analytical, standards, trade and official documentation dealing with food. See also Appendix 1.

AACC	American Association of Cereal Chemists
AAS	atomic absorption spectroscopy
AMC	Analytical Methods Committee
AOAC	Association of Official Analytical Chemists
AOCS	American Oil Chemists' Society
APA	Association of Public Analysts
BFMIRA	British Food Manufacturing Industries Research Association (now Leatherhead Food RA)
BP	British Pharmacopoeia
BPC	British Pharmaceutical Codex
BS	British Standard
BSI	British Standards Institution
COT	Committee on Toxicology
EC (EEC)	European Community (European Economic Community)
ECD	electron capture detector
FAC (FdAC)	Food Advisory Committee
FACC	Food Additives and Contaminants Committee
FAO	Food and Agriculture Organization
FDA	US Food and Drug Administration
FDF	Food and Drink Federation
FID	flame ionisation detector
FMF	Food Manufacturers' Federation
FSC	Food Standards Committee
FTNMR	Fourier transform NMR
GC/MS	combined gas chromatography and mass spectrometry
GLC	gas–liquid chromatography
GMP	Good manufacturing practice
GSC	gas solid chromatography
HPLC	high performance liquid chromatography
HPTLC	high performance TLC
ICP-OES	inductively coupled plasma-optical emission spectroscopy
ICUMSA	International Commission for Uniform Methods of Sugar Analysis
IDF	International Dairy Federation
IOB	Institute of Brewing
ISO	International Organization for Standardization
IUPAC	International Union for Pure and Applied Chemistry
LACOTS	Local Authorities' Coordinating Body on Trading Standards
LAJAC	Local Authorities' Joint Advisory Committee
LGC	Laboratory of the Government Chemist
MAFF	Ministry of Agriculture, Fisheries and Food
NMR	nuclear magnetic resonance (spectroscopy)
OIV	International Office of Wine

p.p.m.	parts per million (equivalent to mg/kg or mg/l)
RSC	Royal Society of Chemistry
SAC	Society for Analytical Chemistry
SI	Statutory Instrument, also Système International d'Unités
SPA	Society of Public Analysts (now the Analytical Division of the Royal Society of Chemistry)
SR & O	Statutory Rules and Orders
TLC	thin-layer chromatography
USDA	United States Department of Agriculture
WCOT	wall coated open tubular column
WHO	World Health Organization

1

Introduction. Legislation, Standards and Nutrition

Food analysis

In earlier times the food analyst was mainly concerned with gross adulteration. Nowadays there is an increasing tendency to examine food from a more positive viewpoint. Processed foods are produced within the limits of prescribed manufacturing formulations, set also to comply with legal or other recognised requirements. This is attained by standardising, as far as possible, the process at each of the following stages: at the farm, the 'raw' material, the process itself and finally the product and its storage. Enhanced manufacturing capacity and the complexity of modern products has necessitated the development of techniques which are suitable for rapid assessment and control. Concurrently, attempts have been made to replace subjective methods of assessing various organoleptic qualities by more precise objective procedures. Our knowledge of even minor food constituents has improved markedly due particularly to the application of newer techniques of separation, identification and measurement.

In many food laboratories, most of the routine work comprises methods of proximate analysis and the study of additives and contaminants. The main compositional components of interest are moisture, fat, protein, ash and available and unavailable carbohydrate. In practice the methods used may vary according to the food under examination: they may also be empirical. For example, the results obtained for moisture and fat depend on the procedure adopted and it is often only the free moisture or free fat that is determined. Moisture (or more correctly, loss on drying) values obtained by oven drying may include other volatile matter such as essential oils, traces of volatile acids and amines. Much attention must be paid to the preparation of the sample and the particle size. Sugars from natural sources such as fruits may include a mixture of several compounds but it is often convenient to express these together as the total soluble solids (uncorrected), as measured by a simple refractometric determination. Protein can be calculated from the total nitrogen as determined by the Kjeldahl method using an arbitrary factor which, because of the differing proportions of amino acid groups present, varies according to the food. 'Fibre' and 'ash' are essentially analytical terms. Neither represent a precise component or group of components in the original food but as long as the same standard procedure is applied to the same food each time, the results provide an adequate basis for interpretation. In many estimations allowance has to be made for interference arising from the food itself or from contamination of the reagents and blank determinations must also be performed. Allowance may sometimes have to be made for changes which occur during storage. Many different alternative methods may be available for some analyses and it is obviously not possible to quote all of these in working detail. We have therefore had to be selective, sometimes on an arbitrary basis, but in doing this we have tried also to indicate where other methods (or reviews of these) may be found. In particular, preference has been given to methods which have been studied collaboratively. Such preference recognises the need to establish tolerance limits which are appropriate not only to the method but also to the matrix analysed. The importance of these factors and the associated sampling problems must be taken into account in the application of analytical procedures to the enforcement of standards of composition and to limits of contamination. In this respect recognition must be given to the need to obtain the performance characteristics of methods in the region of specification or legal limits. Horwitz and Albert (1987) have discussed methods for assessment of analytical procedures; the AMC (1987a, b) has presented recommendations on the estimation, definition and use of detection limits and also on the conduct of trials of methods. Validation of methods of analysis in respect of quality control of the performance of laboratories has also been discussed by Wood (1986).

Internationally recognised reference procedures have been published by bodies such as ISO, IUPAC, Codex Alimentarius and the AOAC. In the UK, standard methods for some foods have been published by the British Standards Institution and the Analytical Division of the Chemical Society (1974), which receives reports from its Analytical Methods Committee (Chirnside and Hamence, 1974).

Frequent reference is made in the text to the AOAC methods book which is the Official Methods of Analysis of the Association of Official Analytical Chemists (editor S. Williams), published by the Association of Official Analytical Chemists, 1111 N 19th Street, Arlington VA, USA. The AOAC methods book is now in its fifteenth edition, published in 1990, in a new format.

Some organisations which serve a particular industry are concerned with the development of analytical methods suitable for use within that industry, e.g. IFJU (fruit juice), ICUMSA (sugar). A list of such organisations is given in Appendix 1.

Some of the problems in food analysis arise from the fact that many of the ingredients used in the manufacture of a composite food are themselves foods of biological origin and vary widely in chemical composition. For example, flour used in the manufacture of bread, cakes and biscuits may be prepared by various milling procedures from different varieties of wheat. The composition of the manufactured article may therefore vary widely, depending on the composition of the natural or prepared ingredients. Difficulties can thus arise when analysis of the final product is contemplated in order to assess the amount of the natural ingredients used in its manufacture, e.g. tomato content of a ketchup, fruit juice content of a soft drink or cocoa content of chocolate. This can only be achieved by the analysis of the product for a major component of the natural ingredient which is not provided significantly by other ingredients used. However, the natural variation in the concentration of the selected indicator component in the ingredient makes it difficult to choose a suitable mean or minimum figure to use in the calculation.

As in other fields of analysis, the availability of good methods is essential if accurate results are to be obtained. The overall importance of cleanliness at the bench and meticulous attention to detail at each step cannot be over-emphasised. Good analytical results also depend on careful sample preparation and accurate analyses are of little use unless the results can be interpreted meaningfully. It is therefore important that laboratory workers acquire a realistic appreciation of the fundamentals of sampling, statistics and assessment of quality criteria, together with an understanding of the significance of the data obtained. The development in recent years of a wide range of convenience foods has been due both to advances in food science and technology and to changes in the style of food distribution and marketing. Convenience foods include ready meals (dehydrated, frozen or preserved), low calorie dishes and products, instant soups and desserts, sauces and garnishes, bread and cake mixes, pie fillings, fruit drink powders, dietetic and baby foods, snack foods, processed meat products and analogues. It is not possible to cover in detail the composition and analysis of this wide range of food products. Many can be analysed by methods used for examination of the traditional classes of foodstuffs. Certain convenience foods contain technological additives such as emulsifiers, stabilisers, colours and flavours and they may also contain modified ingredients derived from starches, carbohydrates, lipids and proteins.

Furthermore, as a result of the growth in food science and technology, consumer awareness and the ever increasing range of food products, there has been a change in emphasis in food legislation in recent years, both in the EC and in the UK, away from compositional regulation and more toward informative labelling. This has led to new technical challenges to the food analyst, who will be required to confirm the correctness of food label declarations, e.g. to determine which species of meat are present in the sample which may be a cooked product, or whether a pasta product is prepared solely from durum wheat, or whether a so-called non-dairy product contains milk ingredients.

The application of standardised methods of analysis to products which contain substantial amounts of modified ingredients may yield results which are significantly in error; the analyst is advised to proceed with caution in such circumstances and where possible to cross check method performance against products of known or similar composition. Typically, the presence of emulsifiers, stabilisers, gelling agents, modified starches and similar ingredients will significantly affect the performance of extraction processes and in some circumstances the recommended method may prove to be totally inoperable.

Food legislation

Food has been liable to adulteration to a greater or lesser degree since very early times. The first recorded instances of adulteration in Britain date back to the Middle Ages when organised commerce in food began. At later dates, individual statutes for foods such as beer, wine, tea and coffee were introduced by the Excise Authorities in order to protect the revenue. Scientific methods for the detection of the more subtle forms of sophistication were then almost non-existent and it was almost impossible to give the ordinary consumer any real protection.

The turning-point came in the nineteenth century. Between 1820 and 1860, attention was drawn to the prevalence of food adulteration by a few medical practitioners, Members of Parliament and microscopists. In 1850 Wakley, the editor of *The Lancet*, established a Sanitary Commission to review the types of fraud which were being practised in relation to the food supply. The Commission's reports, published in *The Lancet* from 1851 to 1854, were largely responsible for the Government's decision to set up a Select Committee on the Adulteration of Food in 1855. Subsequently, an Act for Preventing the Adulteration of Articles of Food or Drink became law in 1860. This established the appointment of Public Analysts, who included at that time a number of medically qualified analysts. In the decade after 1860 much attention was paid to the need for a legally acceptable and workable definition of 'adulteration'. One of the early proposals formulated for inclusion in the new Sale of Food Act 1875, was to the effect that a food or drink shall be deemed to be adulterated if:

1. It contains any ingredients which may render it injurious to the health of the consumer.
2. It contains any substance that sensibly increases its weight, bulk, or strength, unless the presence of such substance be due to circumstances necessarily appertaining to its collection or manufacture, or be necessary for its preservation, or be acknowledged at its time of sale.
3. Any important constituent has been wholly or in part abstracted, without acknowledgement being made at the time of sale.
4. It be a colourable imitation of, or be sold under the name of, another article.

The 1875 Sale of Food Act established two principal offences, the mixing of injurious ingredients and selling to the prejudice of the purchaser a food not of the nature, substance or quality demanded. Subsequent Acts in 1928 and 1938 were eventually succeeded by the 1955 Food and Drugs Act; which in turn was replaced by the 1968 Medicines Act and the 1984 Food Act. The two principal offences however have stood the test of time and are the basis of present-day legislation. The first of these requires proof of injuriousness, the second evidence of composition: both are matters of opinion though in some areas Ministers have made regulations which in effect leave room only for analytical opinion. This is especially the case in respect of compositional matters and in the use of permitted additives, which in effect control the commercial adulteration. Contamination may be regarded as the harmful (accidental or deliberate) aspect of adulteration and as such, and with the notable exceptions of lead, arsenic and (more recently) vinyl chloride monomer for which specific regulations apply, are controlled only by general provisions of the present Food Acts.

The main provisions of the Foods Acts in the UK may be summarised as follows. Section 1 of the Food Act 1984 makes it an offence to sell for human consumption any food to which substances have been added or abstracted or which have been processed to render it injurious to health. Section 2 prohibits the sale to the prejudice of the consumer of food not of the nature, substance or quality demanded. Sections 6 and 8 prevent the use of false or misleading descriptions and the sale of unfit food. Powers to make regulations to control composition (including use of additives and extent of contamination) are conferred under Section 4 whilst Section 7 confers similar powers in respect of labelling. New regulations are developed and old regulations are revised taking into account the advice of the Food Advisory Committee.

As this edition goes to print, the Food Safety Bill, successor to the Food Act 1984, has been published and is expected to become law later in 1990. It is essentially an enabling instrument which, through regulations, will strengthen the enforcement of food law in the UK, particularly with regard to food safety. It will continue to encompass the main features of the 1984 Act relating to food analysis, i.e. nature, substance and quality, composition, labelling, additives and contaminants and will absorb EC obligations into regulations.

The UK joined the European Economic Community (EC) on 1 January 1973. The Treaty of Rome gives the Community institutions the powers and procedures necessary to achieve the objective of establishing a common market by harmonising economic activities and the laws governing these. The Laws of the Community are expressed in the form of Regulations and Directives. The essential difference is that the Regulations must be incorporated into national legislation, whilst requirements of the Directives are built into the framework of individual legislative systems. In general, Regulations are concerned with primary agricultural produce whilst Directives are concerned with detailed compositional aspects of manufactured foods. A list of Directives and Regulations is given in Appendix 3. A comparative study of food law enforcement in the EC Member States has been published by Roberts (1977) and the implications for the analyst have been described by Crosby (1982), Sawyer (1976) and Shenton (1976). Various aspects of the development of EC legislation in relation to the food industry and the impact on UK legislation have been described by Turner (1978) and Goodall (1977); Haigh (1978) has outlined the aims, objectives and philosophies of the European Commission.

Nutritional evaluation

Analysis is widely accepted as a basis for the nutritional evaluation of food, whether natural or processed, and is comprehensively treated as such in the widely accepted publication on the composition of food by Paul and Southgate (1978). Various conversion factors have been used for the calculation of food energy values. The SI unit of energy is the Joule and recommendations to adopt this in place of the calorie as the unit of food energy were made by the Food Standards Committee; these recommendations have been implemented in part by the 1984 Labelling Regulations which require that the expression for energy calculated in kilojoules shall predominate. The Labelling Regulations also include factors for use when calculating food energy values in Joules: 16 kJ/g (3·75 kcal/g) for carbohydrate, expressed as monosaccharide, 17 kJ/g (4·0 kcal/g) for protein, 37 kJ/g (9.0 kcal/g) for fat and 29 kJ/g (7·0 kcal/g) for alcohol. Paul and Southgate list food energy values in both kilocalories (kcal) and kilojoules (kJ) and use the same factors.

The Food Standards Committee's second report on Claims and Misleading Descriptions (1980) supported a recommendation of its first (1966) Report that the labelling of foods, in respect of claims for nutritive value due to vitamin and mineral content, should be controlled by a Code of Practice and that this should be incorporated in future labelling legislation. For a claim to be made, at least one-sixth of the recommended daily amount of a vitamin or mineral should be provided by the amount of the food normally consumed in one day. For the food to be described as a 'rich' or 'excellent' source the amount of vitamin or mineral present must be at least one-half of the recommended daily amount. These proposals, along with others in respect of claims on polyunsaturated fats, cholesterol, slimming properties, energy content, diabetic and other medical qualities, were all included in the 1984 Food Labelling Regulations. The schedule set out in Table 1.1 gives detailed information regarding the levels of vitamins and minerals which are allowed in food in relation to a specific claim. Each nutrient specified shall include its biologically active equivalent or derivative.

Table 1.1 Recommended daily allowances (RDA) of vitamins and minerals

Recommended name	RDA	To be calculated as
Vitamin A	750 µg	µg of retinol or µg of retinol equivalents on the basis that 6 of beta-carotene or 12 µg of other biologically active carotenoids equals 1 µg of retinol equivalent
Thiamin (vitamin B_1)*	1·2 mg	mg of thiamin hydrochloride
Riboflavin (vitamin B_2)*	1·6 mg	mg of riboflavin
Niacin	18 mg	mg of nicotinic acid or mg of nicotinamide or tryptophan on the basis that 60 mg equals 1 mg of niacin
Folic acid	300 µg	µg of total folic acid
Vitamin B_{12}	2 µg	mg of cobalamins
Vitamin C (ascorbic acid)*	30 mg	mg of ascorbic acid
Vitamin D	2·5 µg	µg of ergocalciferol (vitamin D_2) or µg of cholecalciferol (vitamin D_3)
Calcium	500 mg	mg of calcium
Iodine	140 µg	µg of iodine
Iron	12 mg	mg of iron

* This name may also be added in parenthesis

So far as polyunsaturated fatty acid content is concerned the 1984 Regulations stipulate that the food must contain at least 35 per cent fat by weight and at least 45 per cent of the fatty acid content be polyunsaturated and not more than 25 per cent of the fatty acid content be saturated. The claim regarding cholesterol levels require that the food contains no more than 0·005 per cent cholesterol and that it also satisfies the provisions of the polyunsaturated fatty acid claim.

However, a recent Commission proposal for a Council Directive on compulsory nutrition labelling of foodstuffs (89/C296/04) (OJ No. C296, 24.11.89, p. 3) lists the following energy conversions: carbohydrate 17 kJ/g, 4 kcal/g; sugar alcohols 10, 2·4; protein 17, 4; fat 37, 9; ethanol 29, 7; organic acid 13, 3 respectively. The Directive will also deal with nutritional claims. Protein is defined as N × 6·25 for all foods except where the protein is totally from milk when 6·38 applies. Fat means total lipids including phospholipids. Percentages of the Recommended Daily Allowances (see Table 1.2) may be declared numerically or graphically. Fifteen per cent of the RDA per 100 g or per a less amount should be regarded as a significant amount.

International standards

The United Nations, through the Food and Agriculture Organization, the World Health Organization and the joint Codex Alimentarius Commission, aims to promote international trade through the development of internationally accepted food standards. There are recommended international standards published for the individual foods such as sugars, edible oils, canned fruits and vegetables, fruit juices, quick frozen foods, processed meats, fish products and cocoa products. A list of

Table 1.2 Vitamins and minerals which may be declared and their Recommended Daily Allowances (RDAs). Annex to proposal for a Council Directive on nutrition labelling, see text

Vitamin A (μg)	1000	Vitamin B_{12} (μg)	3
Vitamin D (μg)	5	Biotin (mg)	0·15
Vitamin E (μg)	10	Pantothenic acid (mg)	6
Vitamin C (μg)	60	Calcium (mg)	800
Thiamin (mg)	1·4	Phosphorus (mg)	800
Riboflavin (mg)	1·6	Iron (mg)	12
Niacin (mg)	18	Magnesium (mg)	300
Vitamin B_6 (μg)	2	Zinc (mg)	15
Folacin (μg)	400	Iodine (μg)	150

Codex Standards is given in Appendix 2. All of the standards are detailed specifications most of which require analytical methods for their realisation and enforcement. Consultations have been established between the various international organisations with interests in development of analytical methods. Other bodies concerned include the AOAC, IUPAC and ISO, which also has interests in the development of standards and methods of sampling and analysis especially for primary agricultural produce, together with the various international bodies representing individual commodity interest indicated in Appendix 1. Attention has also been given to the development of a common sampling terminology by international standards organisations; this is in recognition of the need to obtain international agreement on enforcement standards.

REFERENCES

AMC (1987a) *Analyst*, **112**, 199.
AMC (1987b) *Analyst*, **112**, 679.
Chirnside, R. C. & Hamence, J. H. (1974) *The Practising Chemists*. London: The Society for Analytical Chemistry.
Crosby, N. T. (1982) *Trends in Analytical Chemistry*, **1**, iv.
Goodall, H. M. (1977) *Journal of the Royal Society of Health*, **6**, 263.
Haigh, R. (1978) *Journal of Food Technology*, **13**, 255.
Horwitz, W. & Albert, R. (1987) *Analytical Proceedings*, **24**, 49.
Paul, A. A. & Southgate, D. A. T. (1978) *The Composition of Foods*. London: HMSO.
Roberts, W. (1977) *Journal of the Royal Society of Health*, **6**, 272.
Sawyer, R. (1976) *Proceedings of the Analytical Division of the Chemical Society*, **13**, 238.
Shenton, F. C. (1976) *Proceedings of the Analytical Division of the Chemical Society*, **13**, 241.
Turner, A. (1978) *Environmental Health*, **86**, 252.
Wood, R. (1986) *Analytical Proceedings*, **23**, 329.

2

General chemical methods

SAMPLING AND SAMPLE PREPARATION

Sampling

The value of the result of a chemical analysis on a well prepared laboratory sample will depend on how representative the sample is of the lot, batch, package or consignment of the particular food from which it was taken and on the kind of chemical information that is required. Foodstuffs and food ingredients are relatively heterogeneous materials, so it is difficult to obtain a single absolutely representative sample for laboratory analysis. The problem may be minimised by selecting several samples from the lot either randomly or according to a plan. These samples may be analysed separately to yield results from which the average composition of the lot may be computed, or in certain cases the samples may be thoroughly mixed to give a single large representative bulk sample from which a sample may be taken for laboratory analysis.

The process of sampling is one facet of statistics and most books on statistics include chapters describing the elementary mathematical principles involved. Kratochvil and Taylor (1981) give a useful short introduction and the same authors (1982) present a survey of recent literature on sampling. The Codex Alimentarius Commission has published sampling plans for prepackaged foods (CAC/RM 42–1969) and the Codex Committee on Methods of Analysis and Sampling has produced a document giving guidance notes for commodity committees on the general principles for Codex sampling procedures (Wood, 1985). Other useful compilations have been published by the Ministry of Defence (1965), Kramer and Twigg (1966) and Herschdoerfer (1967). BS 5309: Part I: 1976 gives an introduction and describes the general principles of methods for sampling chemical products. There are ISO standards for the sampling of various foodstuffs. Because of the practical difficulties and economic aspects of full statistical sampling, and the natural variation in the composition of foodstuffs, food analysis is often carried out on randomly chosen single samples.

Preparation of laboratory samples

In order to obtain precise analytical results, the laboratory sample must be made as homogeneous as possible so that, within the limits of the analytical method used, the replicate analyses agree closely. The method of homogenisation will depend on the type of food being analysed. A number of very efficient electrical mechanical devices are available to reduce the size of food particles and to mix food products thoroughly. Mincers, graters, blenders and homogenisers for dry, moist and wet food and various types of powder mills or grinders are all essential pieces of equipment in a food laboratory. All mehanical devices produce heat, so that care must be taken not to alter the composition of the sample by the loss of moisture due to over-running the equipment. Dry foods require to be brought to at least a coarse powder by means of a mechanical grinder and then mixed thoroughly with a spoon or spatula. Bulk samples of dry or powdery foods can be reduced in size by the process known as quartering. This involves tipping the bulk sample into a uniform pile on a large sheet of glazed paper, glass or the surface of a clean laminated bench or table top. The pile is divided into four roughly equal parts by separating quarter segments. A pair of opposite segments are combined; the other pair are rejected. The combined segments are thoroughly mixed and the process of quartering is repeated until a laboratory sample of about 200–400 g is obtained by the combination of opposite quarter segments.

Moist solid foods such as meat products are best homogenised by chopping rather than mincing, using for example a modern domestic food processor. Fluid foods are best emulsified by top-drive or bottom-drive blenders. Oils and fats are easily prepared by gentle warming and mixing. Fatty mixed-phase poducts such as cheese, butter, margarine and chocolate are difficult. Cheese and chocolate are best grated followed by hand mixing of the grated material; butter and margarine may be re-emulsified by shaking by hand in a glass jar after warming to 35 °C to melt the fat.

The preparation of dry foods for analyses should be carried out rapidly because of possible moisture loss or gain from the laboratory atmosphere. All prepared food samples should be rapidly transferred to dry glass or rigid plastic containers with well-fitting closures and be clearly labelled and stored at a suitable cool or cold temperature.

MOISTURE

The moisture content of foods is of great importance for many scientific, technical and economic reasons (Food Standards Committee, 1979) but its accurate determination is very difficult. Water occurs in foods essentially in two forms, as bound water and as available or free water. Bound water includes water molecules chemically or hydrogen bonded to ionic and polar groups whereas free water is that which is not physically linked to the food matrix and which is freezable and easily lost by evaporation or drying. Since most foods are heterogeneous mixtures of substances, they may contain varying proportions of the two types.

There are many methods for the determination of the moisture content of foods, varying in their involvement with the three types of water and so often resulting in poor correlations of the results obtained. Most of the methods give reproducible results, however, if the empirical instructions are followed closely, and so can be satisfactory for practical use.

Mitchell and Smith (1977) have published a treatise on the determination of water. Harris (1972) and Pyper (1985) have reviewed the subject and Karmas (1980) has classified and listed methods. Practical texts for food analysis have been written by

Pearson (1973), Lees (1975) and Osborne and Voogt (1978). Methods may be classified as drying, distillation, chemical and instrumental methods.

Drying methods

These involve the measurement of the weight lost due to the evaporation of water at or near the boiling point. Although such methods are frequently used, since when considered on a comparative basis they can give accurate results, it should be borne in mind that the value obtained may not be a true measure of the water content of the sample. For example, volatile oil may be lost at drying temperatures such as 100 °C. With some foods (e.g. cereals) only a proportion of the water present is lost at the drying temperature. The remainder (mainly bound water) is difficult to remove and appears to be associated with the proteins present. The proportion of water lost increases as the temperature is raised, so it is especially important only to compare the results obtained using the same conditions of drying. Furthermore, if decomposition is likely to occur, as with foods which contain an appreciable proportion of sugars, it is advisable to use a lower drying temperature, such as 70 °C and to apply vacuum. Baking powder has to be dried at room temperature in a vacuum desiccator for a considerable period as heat causes appreciable losses of carbon dioxide.

The loss in weight may also depend on other factors, including particle size and weight of sample used, type of dish and temperature variations in the oven from shelf to shelf. Ovens which are mechanically ventilated by means of an internal fan tend to give more consistent results and an increased rate of drying.

Rapid approximate moisture determinations for use during food manufacture can be carried out by using special drying ovens which operate at elevated temperatures. Others contain radiant infra-red drying lamps and incorporate a direct reading crude balance. Minor *et al.* (1984) have reported favourable comparisons of rapid infra-red oven drying with AOAC methods for moisture in a variety of foods. Microwave ovens can be used for rapid laboratory moisture determinations, for example Davis and Lai (1984) for flour and Eustace and Jones (1984) for meat and meat products. Bostian *et al.* (1985) have compared the use of a commercial microwave instrument against AOAC methods for the determination of moisture (and fat) in meat and poultry products.

Distillation methods

These involve the distillation of the food with an immiscible solvent having a higher boiling point and a lower specific gravity than water, e.g. toluene, heptane and xylene. The distilled water falls below the condensed solvent in a graduated receiver in which the aqueous phase is measured by volume. A long wire or 'policeman' should be pushed into the condenser tube near the end of the distillation in order to coax any adhering water into the graduated tube. Although low results are not uncommon with the distillation method, it has the advantages that it needs little attention once the apparatus has been set up and any volatile oils in the sample which distil over mix with the solvent and are not measured. This method is therefore the preferred one for herbs and spices (see Chapter 11).

Chemical methods

The sensitive titration method of determining water developed originally by Karl Fischer is described in a British Standard (BS 2511:1970). This is based on the nonstoichiometric reaction of water with iodine and sulphur dioxide in pyridine–methanol solution. Although the end-point of the titration can be detected visually, most workers use one of the commercially available electrometric instruments which incorporate semi-automatic microprocessor controlled titration with digital read-out. The reagent is standardised against a standard water in methanol solution or a pure salt hydrate such as sodium tartrate dihydrate.

In the modern instruments the prepared food sample where appropriate is weighed into the titration vessel and the moisure extracted by stirring with a suitable solvent, e.g. anhydrous methanol, or chloroform/methanol. If the moisture is extracted away from the filtrator, e.g. in a flask or in a refluxing extractor, great care needs to be exercised to avoid pick-up of moisture from the air. The Karl Fischer method is employed mainly for the determination in materials of fairly low moisture content, e.g. sugar confectionery (Sandell, 1960), chocolate products (AOAC Methods), molasses (Kviesitis, 1975) and dried vegetables (Thung, 1964). Instead of using methanol, some workers have employed other solvents such as formamide, dimethylformamide, ethanol, acetic acid and dioxan.

Zurcher and Hadorn (1981a) have determined the moisture content of nearly 100 foodstuffs using the Karl Fischer method and an alternative method appropriate to the food, and have also reported (1981b) that interferences from sugars and solvents are negligible provided that the titrations are carried out at below 25 °C. Hadorn (1980) describes a satisfactory collaborative evaluation of three Karl Fischer methods: direct cold titration, direct hot titration and titration after high frequency comminution on maize starch, wheat flour and sultanas, respectively. There are now available pyridine-free Karl Fischer reagents. Scholz (1983) found that analysis time using these reagents could be significantly reduced by titrating at 50 °C.

Instrumental methods

A very wide range of instrumental methods based on physical or physico-chemical principles have been applied to the determination of moisture. Many of these have been devised for rapid results on a large number of samples of the same type, such as may be required for production on-line checking for quality control in food manufacture (Kropf, 1984). Instruments based on electrical resistance, conductance and capacitance are widely used; more recent applications have involved NMR (Hester and Quine, 1976), near infra-red reflectance (Osborne, 1981; Davies, 1985) and microwaves (Okabe *et al.*, 1973). Measurement of density and refractive index are used for wet and liquid foods. Kropf (1984) has reviewed new rapid instrumental methods for moisture incorporating microwaves, NMR, electro-optical scanners, capacitance, and near infra-red transmittance, reflectance and absorbance. Thermal gravimetric analysis (Mackenzie, 1974) is also useful since it gives information on the types of water present.

Water activity

The regulation of water as an ingredient of foods was the earliest form of food preservation. The availability of water for microbiological growth and biochemical reaction can be controlled by dehydration, freezing or the addition of solutes such as salt and sugar. The role of water in foods has been reviewed by Hardman (1978), Duckworth (1975) and by the Food Standards Committee (1979). Water activity

(a_w) is a measure of the availability of liquid water and is defined as the ratio of the equilibrium vapour pressure of the sample (P) to the equilibrium vapour pressure of pure water (P_0) at the same temperature, $a_w = \frac{P}{P_0}$, and can have values varying from zero to unity (Scott, 1957; Troller and Christian, 1978). Equilibrium relative humidity (ERH) refers to the atmosphere surrounding food and is equal to $100 \times a_w$; whereas a_w refers to the activity of water in solid and liquid foods.

Dried foods ($a_w < 0{\cdot}6$) are not generally at risk of spoilage by the growth of microorganisms. Approximate threshold values for microbiological growth in foods are as follows:

bacteria	a_w 0·91
yeasts	a_w 0·85
moulds and fungi	a_w 0·75
xerophilic moulds and fungi	a_w 0·65
osmophilic yeasts	a_w 0·60

Spores have much greater tolerance to low a_w.

Water activity is usually determined by measurement of ERH. This can be carried out either statically by the measurement of ERH of the food in an enclosed space once moisture equilibrium has been attained, or dynamically by passing a gas stream of known relative humidity over the food and ascertaining the moisture lost or gained by the sample.

Simple laboratory procedures for the determination of ERH include the use of salt impregnated filter paper (Kvaale and Dalhoff, 1963) and the non-equilibrium sorption rate method of Landrock and Proctor (1951) in which samples of the food are placed in atmospheres of known ERH, such as those above selected saturated salt solutions. The food, depending on its a_w, will either absorb or lose moisture and from the plot of weight change after a fixed period of time against relative humidity, the interpolated ERH at which no weight change occurs gives the figure for a_w. This method is best suited to relatively dry and solid foods with a_w below 0·90. Several instrumental methods are available which find widespread use in the food industry. The simplest is the hair or polyamide thread dial type hygrometer, e.g. the LUFFT hygrometer, which depends on the change in length of the fibre with change in relative humidity. It can be used to measure the a_w of meat products at levels higher than 0·85. Psychrometers, vapour pressure manometers and dew point apparatus are also employed (Prior, 1979). Because of their convenience, accuracy and precision, electronic hygrometers are now widely used in food manufacture. These instruments incorporate sensors to detect the equilibrium humidity above the sample contained in a small chamber maintained at a selected temperature. Sensors depend on substances such as lithium chloride which show variations in conductivity with humidity (e.g. Sinascope) or on capacitance changes in a polymer thin film capacitor (e.g. Vaisala Humicap). The electric signal created by the sensor is electronically amplified and a_w or ERH is presented as a recorded sorption isotherm or digital display. With different sensors a_w can be measured over the full range on all types of food using only a few grams of sample.

Troller (1977) has statistically analysed a_w measurements obtained using the Sinascope and Favetto *et al.* (1983) have measured the accuracy and precision of the Vaisala Humicap instrument and compared it with the LUFFT meter.

ASH AND MINERAL MATTER

Total ash. The ash of a foodstuff is the inorganic residue remaining after the organic matter has been burnt away. The ash obtained is not necessarily of exactly the same composition as the mineral matter present in the original food as there may be losses due to volatisation or some interaction between constituents. Conditions of ignition are specified for various materials in a British Standard (BS 4603:1970). The ash figure can be regarded as a general measure of quality or grade (e.g. in tea, flour and edible gelatine) and often is a useful criterion in identifying the authenticity of a food. When a high ash figure suggests the presence of an inorganic adulterant, it is often advisable to also determine the acid-insoluble ash.

The general method for determining the total ash involves weighing out 5 g of the prepared sample into a silica dish (about 7 cm diam.) that has previously been ignited and cooled before weighing. Then the dish and contents are ignited, first gently over a low flame or under an infra-red lamp until charred and then in a muffle furnace at 500–550 °C. Alternatively, if the muffle furnace is fitted with a temperature controller, the sample may be ignited and ashed by placing the dish in the cold furnace which is then switched on and left overnight. With some foods, e.g. cereals such as barley and oats, it is difficult to burn away all the organic matter, but the ashing can sometimes be completed if the particles are broken up with a stout platinum wire or by moistening the cold carbonaceous residue with water, drying and then gently re-igniting. Alternatively, higher ignition temperatures of 550–600 °C can be employed if the sample is heated in the presence of a measured volume of an ashing aid, e.g. ISO 936:1978 for meat and meat products specifies 1 ml of magnesium acetate solution, 150 g/l. The residue due to the ash-aid is allowed for by evaporating and igniting the same volume in another dish. Care must be taken when moving dishes containing fluffy ashes (e.g. of gelatine), which tend to blow away easily. Such ashes should be covered with a Petri dish or watch glass after placing in the desiccator prior to weighing.

Methods of determining the ash based on electrical conductivity measurements have been recommended for sugar by ICUMSA. The conductivity of the sugar solution is measured under standardised conditions and after multiplying by an empirical factor (C-ratio) the conductivity ash (equivalent to the sulphated ash) is obtained.

Water-soluble ash. The ash is boiled with 25 ml water and the liquid filtered through an ashless filter paper and thoroughly washed with hot water. The filter paper is then ignited in the original dish, cooled and the *water-insoluble ash* weighed:

$$\text{Water-soluble ash (\%)} = \text{Total ash (\%)} - \text{Water-insoluble ash (\%)}.$$

If the filtrate is cooled and titrated with 0·05 M (0·1 N) sulphuric or 0·1 M hydrochloric acids using methyl orange, the *alkalinity of the soluble ash* can be obtained. The alkalinity is best expressed as ml M acid/100 g sample. This value multiplied by 0·0691, 0·053 or 0·0471 gives the alkalinity calculated as K_2CO_3, Na_2CO_3 or K_2O, respectively. This is often useful when considered in conjunction with the ash figure for confirming other results in relation to the original composition. In some foods, such as ginger and tea, a low water-soluble ash and low alkalinity of the ash indicates previous abstraction of important constituents with consequent lowering of quality.

Acid-insoluble ash. The ash or water-insoluble ash is boiled with 25 ml of dilute hydrochloric acid, 2 + 5, (10 per cent m/m HCl) for 5 min, the liquid filtered through an ashless filter paper and thoroughly washed with hot water. The filter paper is then ignited in the original dish, cooled and weighed. In some instances it is advisable to commence by evaporating the ash to dryness with concentrated hydrochloric acid to render the silica insoluble before repeated treatment with hot dilute acid. The acid-insoluble ash is a measure of the sandy matter and maxima are prescribed for herbs and spices in US regulations.

Sulphated ash. This involves moistening the ash with concentrated sulphuric acid and igniting gently to constant weight. The sulphated ash gives a more reliable ash figure for samples containing varying amounts of volatile inorganic substances which may be lost at the ignition temperature used.

SODIUM CHLORIDE

For preservation and taste reasons, a knowledge of the concentration of salt (sodium chloride) in foodstuffs is very often required. Normally, it is sufficient to determine total chloride and express this in terms of sodium chloride. Minor components of foods can also contribute chloride ions but foods which require an analysis for salt are normally those in which salt is an added or significant ingredient. Titrimetric methods are the most commonly used. Salt must first be extracted from the food either by careful ashing at 500–550 °C (alkali chlorides are relatively volatile at higher temperatures), followed by dissolution of the ash, or by boiling the foodstuff in diluted nitric acid. In the absence of acid the chloride ion can be determined by the Mohr procedure involving direct titration with 0·1 M silver nitrate; or in the presence of acid by the Volhard procedure involving the addition of excess silver nitrate and back titration with potassium thiocyanate. The latter method is generally more accurate and precise.

Mohr method. Wash the ash into a conical flask or white porcelain dish with minimal water. Add 1 ml of 5 per cent potassium chromate solution and titrate with 0·1 M silver nitrate solution to the first appearance of an orange colour. 1 ml 0·1 M $AgNO_3 \equiv$ 0·005 844 g NaCl.

Volhard method. Accurately weigh about 5 g of the prepared sample into a 250 ml conical flask, add 10 ml water and 25·0 ml of 0·1 M silver nitrate solution (or sufficient excess to precipitate all the chloride in the weighed sample). Add 10 ml of concentrated nitric acid and some boiling granules. Alternatively, dissolve the ash from a sample in the nitric acid, and add water and silver nitrate. Boil gently with a small funnel in the neck of the flask for about 10 minutes until the solution is pale yellow. Cool. Add 50 ml water, 5 ml of saturated ammonium ferric sulphate solution and a few drops of nitrobenzene (see Note). Shake the flask to coat the precipitated silver chloride with nitrobenzene and titrate the excess silver nitrate with 0·1 M ammonium or potassium thiocyanate solution until a permanent reddish colour persists for 15 s. 0·5 g urea may be added to the hot solution to remove the yellow nitrous fumes. Carry out a blank determination on the reagents alone. The difference between the titrations of the blank determination and the test is equivalent to the chloride concentration.

1 ml 0·1 M $AgNO_3 \equiv$ 0·005 844 g NaCl.

NOTE

Since nitrobenzene is regarded as toxic, its use is not now recommended. It can be replaced as a coating agent by triacetin (glycerol triacetate) which has no smell and is considerably cheaper. Some workers however consider that the use of a coating agent is unnecessary.

The potentiometric titration of sodium chloride in foods without ashing, using a pH meter and a silver electrode has been collaboratively studied (Brammell, 1974) and reported to be accurate and rapid. However, more recently Beljaars and Horwitz (1985) have compared the Volhard method and the Codex potentiometric method which is described in the AOAC methods book. With meat products the potentiometric method gave consistently high results, whereas the Volhard method was imprecise below 1 per cent NaCl.

Sekerka and Lechner (1978) have reported a method using the chloride ion selective electrode in which the common interferences from other halides and thio compounds have been eliminated. A dry sample addition method using a solid state ion selective electrode has been described by Chapman and Goldsmith (1982a). Instruments based on electrode reactions, e.g. Corning Chloride Analyser, can also be applied to the determination of salt in foods such as cheese (Johnson and Olson, 1985). A semi-quantitative dip-stick device based on the colour change from brown silver chromate to white silver chloride is described in the AOAC methods book.

Kindstedt and Kosikowski (1984) have reported a close correlation for salt in cheeses when determined as chloride by the Volhard method, and as sodium by an ion selective electrode.

TITRATABLE ACIDITY AND pH

Introduction

Acidity can be measured by titration with alkali to an end-point depending on the selected indicator, and the result may be expressed in terms of a particular acid. The titration value does not indicate whether the acids present are strong or weak. However, if the titration is followed potentiometrically, the plotted titration curve can give information on the relative strengths of acids present. In many instances, a knowledge of the hydrogen ion activity is more useful than the titratable acidity. During food storage and in food spoilage, changes may occur which are due to enzymic action and microbiological growth. The extent of these changes is strongly influenced by hydrogen ion concentration rather than the titratable acidity present. Protein stability is also influenced by hydrogen ion activity. Hence the measurement of pH is important in assessing the effectiveness of preservatives, and also in monitoring the performance of food manufacturing operations.

The pH value may be defined as the common logarithm of the number of litres of solution that contain 1 gram-equivalent of hydrogen ion.

$$pH = -\log [H^+]$$

Pure water is neutral, with a pH of 7·00. As the acid strength or hydrogen ion concentration becomes greater, the pH value reduces towards zero. Thus 0·1 M solution hydrochloric acid is pH 1·0 whereas 0·1 M acetic acid is pH 2·9. In solutions where the hydrogen ion concentration is less than that in water, alkaline solutions, the pH value exceeds 7. A 0·1 M solution of sodium hydroxide is pH 13·0. Practically all foods have pH values lower than 7·0, but tap water can be alkaline due to dissolved salts.

Measurement of pH

The pH of a food can be measured either by the use of colour indicators or electrometrically. In acid–base titrations, indicators are used which change colour at selected pH values. Thus phenolphthalein changes from purple to colourless at around pH 9.0. The pH values of foodstuffs which are not too highly coloured can be readily determined by the use of pH indicator papers. These are now available in wide and narrow ranges of pH, enabling values to be measured within 0·5 or less units of pH.

Electrometric measurement using pH meters is now simple and accurate. Microelectronic components have made possible small portable high quality instruments with digital displays, some with built-in electrodes in dip-stick form. These meters measure the potential difference between a glass electrode and a standard calomel electrode, both of which can form parts of a combination electrode, and are calibrated by the use of prepared or purchased buffer solutions of accurately known pH. Two suitable buffers are:

1. 0·0496 M potassium hydrogen phthalate. Dissolve 10·12 g of pure $KHC_8H_4O_4$ (which has been dried at 105 °C) in 1 litre of distilled or demineralised water. This solution has a pH of 4·002 at 20 °C.
2. 0·009 97 M borax. Dissolve 3·80 g of pure $Na_2B_4O_7 . 10H_2O$ in 1 litre of distilled or demineralised water. This solution has a pH of 9·22 at 20 °C.

There are also available several other types of electrodes designed for special purposes, such as probe electrodes for the examination of carcass meat. Very accurate pH measurement, though seldom required in food analysis, is very susceptible to the temperature of the test solution. This is due principally to the temperature dependence of electrode and liquid junction potentials. For measurements of pH to within ±0·01 units, the temperature of the test solution and calibrating buffer solutions should be known within ±2 °C. Temperature compensation devices are incorporated in pH meters to correct for known temperature deviations.

For immediate use and long life, glass electrodes must be kept moist, usually by immersion in water or a buffer solution. Unless the electrodes are used very frequently, open beakers are unsuitable because of unobserved water evaporation. The measurement of pH of aqueous foods using a glass electrode is straightforward. BS 4401: Part 9: 1975 describes the measurement of pH in meat and meat products.

Moist and semi-moist foods can be examined afer making a slurry with water. This is possible because foods in general contain sufficient natural pH buffering salts to allow some dilution without affecting pH. However, this relatively high electrolyte content in the aqueous phase of foods is not present in the pH meter calibrating buffer solutions. Hence the measured pH cannot be equated precisely to the theoretical value.

NITROGEN AND CRUDE PROTEIN

Until comparatively recent times, the total protein content of foods was invariably estimated from the organic nitrogen content determined by the Kjeldahl procedure. Nowadays there are several alternative chemical and physical methods available, some of which have been automated or semi-automated. General accounts of methods for the determination of nitrogen in food materials are given by Lillevik (1970) and Pomeranz and Meloan (1978). Lakin (1978) has reviewed modern developments.

The Kjeldahl procedure

Although it has been subject to modification over the years (Hatfull, 1983; Morries, 1983), the basic Kjeldahl procedure still maintains its position as the most reliable technique for the determination of organic nitrogen. In consequence it is included in official and statutory methods and is approved by international organisations. Furthermore, the results obtained by Kjeldahl are used to calibrate physical and automatic methods. The Kjeldahl method is based on the wet combustion of the sample by heating with concentrated sulphuric acid in the presence of metallic and other catalysts to effect the reduction of organic nitrogen in the sample to ammonia, which is retained in solution as ammonium sulphate. The digest, having been made alkaline, is distilled or steam distilled to release the ammonia which is trapped and titrated.

Many catalysts have been employed. Mercury, as mercuric oxide has been thought to be the most effective, with selenium almost as effective, but both have toxic hazards and waste disposal problems. Moreover, mercury forms ammonia complexes in the digester requiring the addition of sodium thiosulphate to break the complex and to release the ammonia. Williams (1976) and Croll *et al.* (1985) recommend the use of a mixture of copper (II) sulphate and titanium dioxide and in a large study involving 22 laboratories, Kane (1984) reported comparable results on animal feeds when using mercuric oxide or copper sulphate.

Reduction in the digestion time has also been achieved by the addition of potassium or sodium sulphate which raises the digestion temperature. Metal catalysts are conveniently available in tablet form compounded in a potassium sulphate base. Concon and Soltness (1973) and Koops *et al.* (1975) have reported that the addition of hydrogen peroxide significantly accelerates digestion and decreases foaming. Singh *et al.* (1984) and Florence and Milner (1979) have shown that satisfactory digestion can be achieved in the presence of hydrogen peroxide without catalysts. Traditionally, the ammonia liberated from the digest having been made alkaline, is distilled into a standard quantity of dilute acid which is finally titrated with standard alkali to give the organic nitrogen content of the sample. More popular nowadays is to distil into 4 per cent boric acid solution and to titrate the ammonia direct with standard sulphuric acid.

The following procedure for macro-Kjeldahl is basically that of the AOAC.

Weigh out (note 1) a portion of the prepared sample containing 0·03–0·04 g N and transfer to a Kjeldahl digestion flask. Add 0·7 g mercuric oxide, 15 g powdered potassium sulphate and 40 ml concentrated sulphuric acid. Heat the flask gently in an inclined position until frothing ceases (note 2) then boil briskly for 2 hours. Allow to cool (note 3). Add approximately 200 ml water and 25 ml sodium thiosulphate solution (80 g/l) and mix. Add a piece of granulated zinc or anti-bump granules and carefully pour down the side of the flask sufficient sodium hydroxide solution (450 g/l) to make the contents strongly alkaline (about 110 ml). Before mixing the acid and alkaline layers, connect the flask to a distillation apparatus (note 4) incorporating an efficient splash head and condenser. To the condenser fit a delivery tube which dips just below the surface of a pipetted volume of standard acid contained in a conical flask receiver. Mix the contents of the digestion flask, then boil until at least 150 ml have distilled into the receiver. Add 5 drops of methyl red indicator solution (0·5 g/100 ml ethanol) and titrate with 0·1 M sodium hydroxide. Carry out a blank titration. 1 ml of 0·1 M hydrochloric acid or 0·05 M sulphuric acid ≡ 0·0014 g N.

NOTES

1. Semi-solid materials, such as meat products, can be conveniently weighed on to a small filter paper and the filter paper, wrapped around the sample, dropped to the bottom of the digestion flask.

2. Anti-foaming preparations can be added.
3. The amount of sample, sulphuric acid and potassium sulphate employed must be such that the digest does not solidify on cooling.
4. Purpose designed distillation apparatus is available. Figure 2.1 shows an example.

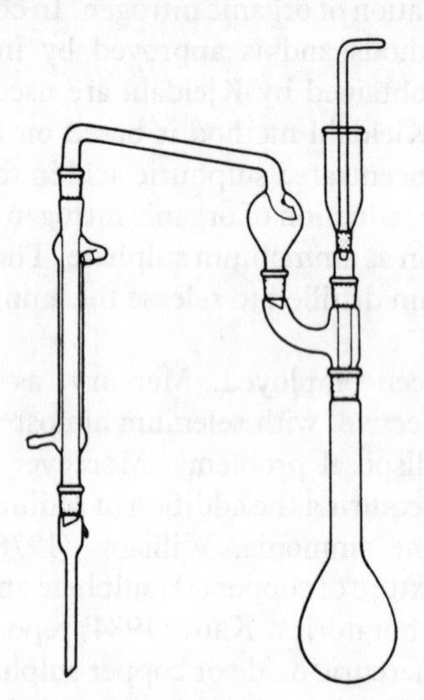

Fig. 2.1 Macro-Kjeldahl distillation assembly. (Courtesy of Corning Medical and Scientific (England))

Other official or recommended procedures have been published by the SAC (Hanson, 1973), by the American Association of Cereal Chemists (1983), in ISO 3188: 1978 for starch and its derivatives, in BS 4401: Part 2: 1980 for meat and meat products, in EC Directive 72:199 for animal feedingstuffs, and also in the Feeding Stuffs (Sampling and Analysis) Regulations 1982, SI No. 1140, as amended.

Adaptations of the Kjeldahl procedure

Apparatus for the distillation of ammonia on a semi-micro scale was described by Markham (1942). A still bearing his name and an alternative semi-micro distillation apparatus, both available commercially, are shown in Figs 2.2. and 2.3. A semi-micro Kjeldahl method is described in the AOAC methods book and micro-diffusion of milk digests using modified Conway micro-diffusion chambers has been described by Lin and Randolph (1978) and appears as a method for volatile nitrogen bases in feeding stuffs in Directive 71/393/EEC (OJ No. 279, 20.12.1971, p. 7). The Kjeldahl procedure has been mechanised and automated. Wall *et al.* (1975) and McGill (1981) have reported on the accuracy and precision of the Kjel-Foss fully automatic version of the traditional procedure. Tecator, Büchi and Gerhardt instrumental Kjeldahl systems employ digestion in tubes in an electrically heated aluminium block digestor, followed by rapid steam distillation of ammonia from the digestion tubes into acid or boric acid for manual titration. Some models incorporate fully automatic titration of the distilled ammonia, detected colorimetrically (Tecator) or potentiometrically (Büchi, Gerhardt). At the Laboratory of the Government Chemist, the following procedure has been applied when using such equipment:

Pre-heat the block digestor 30 min before it is required to 435 °C. Transfer weighed homogenised

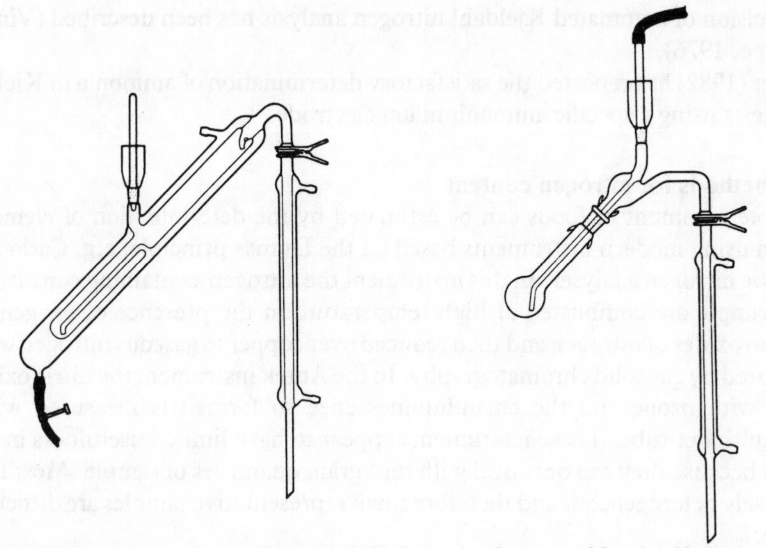

Fig. 2.2 Markham type semi-micro Kjeldahl distillation unit. (Courtesy of Corning Medical and Scientific (England))
Fig. 2.3 Quickfit semi-micro Kjeldahl apparatus. (Courtesy of Corning Medical and Scientific (Europe))

sample, on a filter paper, to a digestion tube, taking an amount calculated to provide a titration of about 25 ml of 0·05 M sulphuric acid, or a maximum weight of 2 g. Add two Kjeltab TCT tablets (Thompson and Capper Ltd, Liverpool)—each tablet contains 3·5 g K_2SO_4, 0·105 g $CuSO_4$ and 0·105 g TiO_2. Add 10 ml concentrated sulphuric acid, sp. gr. 1·84, 10 ml hydrogen peroxide (100 volume) and mix. Leave for a few minutes. Place the tube in the heated digestor and attach the venting tube which removes the acid vapours. Allow digestion to proceed for 45 min. Remove the tube from the block digestor and allow to cool for 5 min on the metal rack. Add 75 ml demineralised water and mix. For instruments employing manual titration, place 25 ml boric acid solution (4 per cent) in a titration receiver flask, and place it on the distillation unit. Attach the tube containing the digested sample to the distillation unit and press the start button to effect a metered addition of sodium hydroxide solution and to initiate the steam distillation. When the receiver platform falls and the distillation stops, remove the flask, add 5 drops of screened methyl red indicator solution (0·125 g methyl red and 0·082 g methylene blue per 100 ml ethanol) and titrate with 0·05 M sulphuric acid solution to a grey end-point. Carry out a blank determination. For fully automatic titration, place the tube in the distillation unit and follow the manufacturer's instructions.

Suhre *et al.* (1982) have described a collaborative study of the Kjel-Foss, block digestion/steam distillation system and the AOAC reference method.

The Kjeldahl procedure has also been adapted for automatic analysis and micro-analysis of ammonia by a colorimetric finish. The reaction between ammonia and Nessler reagent is utilised in ISO 5378: 1978 (BS 5697: Part 2: 1979) for the determination of nitrogen in starch and its derivatives. Berthelot's reaction of ammonia with an alkaline mixture of phenol and hypochlorite, and the bright green colour formed by the reaction of ammonia with an alkaline solution containing sodium salicylate, nitroprusside and dichloroisocyanurate have also been employed (Scheiner, 1976; Havilah *et al.*, 1977; Nkonge and Ballance, 1982). The last two reactions are used in Technicon Auto Analyser systems (Crooke and Simpson, 1971; Isaac and Johnson, 1976). The importance of calibration procedures on the accuracy and precision of automated Kjeldahl nitrogen analysis has been described (Vincent and Shipe, 1976).

Pailler (1982) has reported the satisfactory determination of ammonia in Kjeldahl acid digests using a specific ammonium ion electrode.

Other methods for nitrogen content

The protein content of foods can be estimated by the determination of elemental nitrogen using modern instruments based on the Dumas principle, e.g. Carlo Erba automatic nitrogen analyser. In this instrument the nitrogen-containing constituents of the sample are combusted at high temperature in the presence of oxygen and helium to oxides of nitrogen and then reduced over copper to gaseous nitrogen which is measured by gas solid chromatography. In the Antek instrument the nitric oxide is reacted with ozone and the chemiluminescence so formed is measured with a photomultiplier tube. These instruments appear to have limited usefulness in food analysis because they can only deal with milligram quantities of sample. Most foods are coarsely heterogeneous and therefore small representative samples are difficult to prepare.

Arneth (1984) proposes an automatic method for protein nitrogen based on the colorimetric reaction between trinitrobenzene sulphonic acid and the amino acids in the sample hydrolysate prepared for hydroxyproline analysis. Polarography was used by Fosdick and Pike (1982) to determine hexamethylenetetramine produced by the reaction of formaldehyde and the ammonia in Kjeldahl digests.

Rapid, simple and safe automatic techniques have been described, based on neutron activation analysis (Doty *et al.*, 1970) and proton activation analysis (Dohan *et al.*, 1976). Comparative studies using Kjeldahl, Kjel-Foss, Kjeltec, infra-red reflectance, neutron activation, proton activation and thermal-decomposition analysis have been reported by Williams *et al.* (1978). The authors claim that for the determination of nitrogen in wheat, all the methods were satisfactory and could take the place of Kjeldahl as a standard method. Neutron activation analysis is claimed to be the most accurate, precise and economical.

Conversion factors for nitrogen to crude protein

The determination of total nitrogen by normal Kjeldahl procedures will not include inorganic nitrogen from, for example, nitrates and nitrites. Radiochemical and elemental analyser methods, however, will detect and measure nitrogen in all forms of combination. Non-protein nitrogen is high in certain foods (fish, fruit and vegetables) but the factors commonly used to convert nitrogen to crude protein are based on the average nitrogen content of proteins found in particular foods (Jones, 1931). FAO/WHO (1973) recommended factors include the following:

Wheat—wholemeal	5·83		Soya	5·71
flours, except wholemeal	5·70		Nuts—peanuts, brazil nuts	5·41
macaroni	5·70		almonds	5·18
bran	6·31		other nuts	5·30
Rice	5·95		Milk and milk products	6·38
Barley, oats, rye	5·83		Gelatin and collagen	5·55
Maize	6·25		All other foods	6·25

These factors, based on early protein analyses, have been criticised by Heidelbaugh *et al.* (1975), who have published factors for a wide range of foods based on detailed amino acid analysis. De Rham (1982) has suggested that the range of protein nitrogen factors for foods depends heavily on the proportion of arginine and the amides of glutamic and aspartic acids. He has proposed factors for a range of food based on the FAO/WHO amino acid profiles and his assumptions for the proportions of the amides present.

It is good practice for food analysts to quote the factor used when reporting protein contents.

THE DIRECT DETERMINATION OF PROTEIN

Food proteins consist of amino acids possessing various functional groups and hence show a wide variation of chemical reactions. Because foods contain mixtures of proteins, methods for the direct estimation of protein need to be calibrated against a reference standard method for nitrogen, e.g. the Kjeldahl procedure.

Formol titration

When formalin is added to neutralised aqueous solution containing protein, the $-NH_2$ group reacts to form the methylene-amino group $-N=CH_2$ with the release of protein which may be titrated. The procedure is described by Taylor (1957) and has been used for the determination of milk solids in ice cream (Crowhurst, 1956), and with a potentiometric modification by Hill and Stone (1964). Roeper (1974) has used the formol titration for the estimation of protein nitrogen and casein in skimmed milk. See index for applications.

Colorimetric methods

The biuret reaction giving purple coloration when peptide bonds react with cupric ions at alkaline pH has been employed and it has been reported that low concentrations of reducing sugars do not interfere (Mitsuda and Mitsunaga, 1974). The method has been considerably improved by the use of propan-2-ol and the application of heat (Noll *et al.*, 1974) and a simple method for food proteins comparing favourably with micro-Kjeldahl is described by Ramachudran *et al.* (1984). Folin's reagent (phosphomolybdic-phosphotungstic acid) is reduced by proteins to form a molybdenum blue complex. The reaction was modified and sensitively enhanced by Lowry *et al.* (1951) and has found wide use, especially in biochemical analysis. The method has been automated for milk analysis (Huang *et al.*, 1976) and found suitable for protein in fish meal (Martone *et al.*, 1980).

Sulphonated acid dyes react with proteins at low pH to form an insoluble protein–dye coagulum. If this coagulum is removed by filtration or centrifugation, the amount of dye colour remaining in the supernatant liquid is indirectly proportional to the amount of protein in the sample. The dye reaction with protein is complex and non-uniform, hence the method is highly empirical and requires standardisation and calibration. The method has found use in high volume protein analysis of like samples. Udy (1971) gives a full account and methods are described for milk and milk products (AOAC methods book), meat products (Pearson and Parveneh, 1971), cereals (Hartley, 1975) and frozen dairy desserts (Bruhn *et al.*, 1980). The technique can be made extremely sensitive by extracting the bound dye (McKnight, 1977).

Direct distillation

Because of the presence of the amino acids asparagine and glutamine, which also react as amides, food proteins release ammonia when distilled with excess of strong sodium hydroxide solution. Ronalds (1974) used the technique employing Kjeldahl distillation apparatus for the determination of protein in wheat and barley and Lehmann and Beckmann (1982) for protein in meat products. The Kjeltec DD system is an adaptation of the Tecator Kjeltec automatic distillation unit for this purpose.

Spectroscopic methods

Infra-red absorption in the fundamental region (2·5–16 μm) has been for several years the basis of instruments for the determination of protein (including lactose and fat) in milk, e.g. Grubb Parsons IRMA, Shield Instruments Multispec and Foss Electric Milko-Scan (Kerkhof Mogot *et al.*, 1982). Foss Electric have marketed a similar instrument, the Superscan, to determine protein (with fat and carbohydrate) in meat and meat products after being converted into a milk-like emulsion.

There have been rapid advances in recent years in the development of instruments to measure protein and other principal components in foods, based on near infra-red reflectance (NIR) spectrophotometry in the 750–2500 nm region. The technique has been applied to a very wide range of different foodstuffs (Davies, 1985) and applications have been reviewed by Polesello and Giangiacomo (1983) and Osborne and Fearn (1986).

Low resolution pulsed nuclear magnetic resonance spectroscopy can be applied to the determination of protein content. Copper is used as a relaxation reagent (Tipping, 1982).

Rapid instrumental methods for protein such as those described above require careful calibration using conventional or reference chemical methods. They also generally lack precision and therefore are most useful in process quality control.

In an assessment of the reliability of five methods for the determination of protein in barley and malt (Pomeranz *et al.*, 1977) biuret and dye-binding gave the best agreement with Kjeldahl, infra-red reflectance methods were intermediate and direct alkaline distillation was the poorest.

FAT

The fatty constituents of foods consist of a number of lipid substances (Hannant, 1982). The 'fat' content (sometimes called the ether extract, neutral fat or crude fat), which may be considered as consisting of the 'free' lipid constituents, is that which can be extracted by the less polar solvents such as light petroleum fractions and diethyl ether, whereas the 'bound' lipid constituents require more polar solvents such as alcohols for their extraction. The bound lipids can be broken down by hydrolysis or other chemical treatment to yield free lipid. Hence the amount of extracted lipid found in a foodstuff will depend on the method of analysis used. The extraction of lipid from food has been usefully reviewed by Carter (1977). Kropf (1984) has reviewed the available techniques for rapid fat determinations including those for food processing operations.

Direct solvent extraction methods

The free lipid content, which consists essentially of neutral fats (triglycerides) and free fatty acids, can be conveniently determined in foods by extracting the dried and ground material with a light petroleum fraction or diethyl ether in a continuous extraction apparatus. These are available in numerous designs but are basically of two types. The Bolton or Bailey-Walker type gives continuous extraction whereby condensed solvent drips onto the sample contained in a porous filter container such as a thimble, around which passes the hot solvent vapour. The Soxhlet type gives intermittent extraction with excess of fresh condensed solvent. The efficiency of both methods relies on sample pretreatment and choice of solvent. Employing a modified Bolton apparatus, Manley and Wood (1945) obtained complete extraction of fat from powders within one hour. Tecator market the Soxtec apparatus which is a semi-automatic instrument based on a combination of Bolton and Soxhlet extractions using a choice of solvents for the batchwise rapid determination of fat in foods. Harrison (1939) investigated the use of several solvents on fish meal. He found that extracted material increased with the polarity of the solvent from about 9 per cent using petroleum ether, through hexane, heptane, diethyl ether, carbon disulphide, cyclohexane, benzene, methylene chloride, trichlorethylene, chloroform and acetone to about 16 per cent with dioxan. Complete extraction of neutral fat is impeded by the presence of large amounts of water-soluble substances such as carbohydrates, glycerol and lactic acid. The Foss-Let Fat Analyser is an instrument designed to extract fat from oil seeds by vigorous shaking and pounding with trichlorethylene. The solvent is rapidly filtered into a measuring device containing a float controlled by an adjustable magnetic field, calibrated for fat content. Adjustment of the field until the float rises gives a sensitive indication of the fat concentration. This rapid technique can be used with other foods but for milk products a prior hydrolysis with acid is necessary. Pettinati and Swift (1977) have reported a collaborative study of the determination of fat in meat products by Foss-Let and continuous extraction techniques. They found that the Foss-Let method showed equivalent accuracy and precision to an official AOAC method and was very rapid (7–10 min). The pervasive odour of trichlorethylene is a drawback to this technique.

A useful procedure for extracting fat from semi-solid and wet foods which avoids the initial drying stage is to mix the sample with calcium sulphate, anhydrous sodium sulphate or vermiculite. When the mixture becomes powdery and dry it can be transferred to a Soxhlet thimble in an extraction apparatus.

Extraction in the presence of alcohols causes the release of lipoidal substances bound to proteins and carbohydrates such as phospholipids and glycolipids: hence maximum extraction is obtained by a mixture of polar and non-polar solvents. Southgate (1971) employed hot extraction of foods with 2:1 chloroform:methanol. This procedure co-extracts water and water-soluble substances so the residue after solvent removal and the addition of anhydrous sodium sulphate, needs to be extracted with petroleum ether.

The Bligh and Dyer (1959) technique, as modified by Hanson and Olley (1963), relies on the homogenisation of wet foods with methanol and chloroform in such proportions as to form a single phase miscible with the water in the food. On the addition of further quantities of chloroform and water, two phases separate with the lipid material contained in the chloroform layer. The following procedure may be used or scaled down if a large centrifuge is not available (see also below):

Weigh 2–20 g of prepared sample into a 250 ml centrifuge bottle or a mixing vessel. Add sufficient water to bring total water present to 16 ml, together with 40 ml methanol and 20 ml chloroform. Macerate for 2 min; add a furher 20 ml chloroform and macerate for 30 s; add 20 ml water and macerate again for 30 s. Centrifuge the mixture for 10 min at 2000–2500 r.p.m. Draw off as much as possible of the lower chloroform layer without disturbing the supernatant layers by using a syringe, and filter it through a coarse filter paper.

For the determination of total lipid extract, pipette 25·0 ml of the clear filtrate into a tared dry conical flask or beaker. Evaporate the chloroform to dryness on a steam bath and complete drying in an oven at 105 °C.

This procedure, with a lower final drying temperature (e.g. 60 °C in a vacuum oven), is the preferred method for extracting lipid for gas chromatographic identification, rancidity tests, etc. Karow *et al.* (1984) employed the following procedure to obtain lipid for fatty acid profile analysis (see Chapter 16):

Homogenise 1 g of whole cereal (containing 10 per cent water) with 3·9 ml water and 10 ml methanol in a 30 ml glass centrifuge tube for 20 min. Add 5 ml chloroform and leave to stand 15 min. Add 5 ml chloroform and 5 ml water and mix with the homogenate, then centrifuge for 15 min at 6·78 G. Remove the lower chloroform layer by pipette and filter through glass wool into a 15 ml vial. Place the vial on a heating block at 30 °C and evaporate the solvent under a stream of nitrogen.

A collaborative study of the chloroform–methanol method for fat in various foods has reported good accuracy and precision (Daughtery and Lento, 1983). Khor and Chan (1985) have demonstrated that methylene chloride can be used instead of chloroform in lipid extraction procedures.

Solubilisation extraction methods

Bound lipids can be made free if the food sample is dissolved completely prior to extraction with polar solvents. Dissolution of the food can be achieved by acid or alkaline hydrolysis. In the acid method (Werner–Schmid process, BS 4401: Part 4: 1986: Method B, also known as the Schmid–Boudzynski–Ratzlaff (SBR) method) the material is heated in a boiling water bath with hydrochloric acid to break down the protein and the fat separates as a layer on the top of the acid liquid. The acid concentration during the extraction should be approximately 6 M, e.g. 10 g of milk is treated with 10 ml of concentrated acid, or 1–2 g of a solid food is mixed with 8–9 ml of water and 10 ml of acid. The protein dissolves in the acid and the separated fat can be extracted by shaking at least three times with diethyl ether or diethyl ether and light petroleum mixture. Treatment of the original material with ammonia prior to the addition of acid is advisable with foods such as dried milk and processed cheese. The acid extraction process is less suitable than the alkali method if the material contains a high proportion of sugar. Diethyl ether tends to co-extract some non-lipid matter, hence the lipid in the dried and weighed extract needs to be carefully removed by petroleum ether and the residual non-lipid extract re-dried and re-weighed to give by difference the total fat content in the sample.

The Weibull–Stoldt acid hydrolysis method (Stoldt, 1952), also referred to as the

Weibull–Berntrop (WB) method, used in several European countries as a reference method, overcomes this problem by use of a filtration stage to remove non-lipid or tarry substances. It is prescribed in BS 4401: Part 4: 1986: Method A (ISO 1443-1973) and in Directive 84/4/EEC (OJ No. L15, 18.1.84, p. 28) dealing with the control of feeding stuffs. In this method the weighed sample is heated on a steam bath with dilute HCl then boiled over a flame for 30 min. The sample solution is filtered through a wetted filter paper and washed with hot water. The filter paper is then oven dried and placed directly into a Soxhlet apparatus and extracted with ethyl or petroleum ether or dichloromethane. A modification has been proposed by Ugrinovits (1981) eliminating the lengthy filtration, washing and drying stages.

Acid hydrolysis methods tend to decompose phospholipids and possibly triglycerides to a slight extent so that correlation with chloroform–methanol extraction may be poor with some foods. (For these reasons, fat extracted by acid hydrolysis is not suitable for fatty acid profile by gas chromatography.) However, to all intents and purposes, both acid hydrolysis methods referred to above give a true result for total lipid content by weight, for most foods.

In dissolution using alkali (Rose–Gottlieb method), the material is treated with ammonia and alcohol in the cold and the fat is extracted with ether–light petroleum mixture. The alcohol precipitates the protein which dissolves in the ammonia; the fat can then be extracted with ether. Light petroleum is then added as it reduces the proportion of water and hence also non-fatty soluble substances such as lactose in the extract. Alkali extraction gives very accurate results provided a prescribed technique is adhered to precisely (BS 1741: 1963: 1742: 1951). The method is approved internationally for fat in dairy products (Codex Standard CAC/M 1-1971). It is also suitable for materials which contain much sugar such as sweetened condensed milk.

Separation of solvent containing the extracted fat from samples solubilised by acid or alkali can be facilitated by suitable extraction apparatus such as tubes with siphon or wash bottle type fittings (see Fig. 14.2, p. 539), or by tubes specially designed to assist decanting, e.g. Mojonnier tubes (BS 1743: 1968) see Fig. 14.1, p. 538. A slow speed electrical centrifuge is available which will handle straight-sided fat extraction tubes (Wifug 4000 E).

Hubbard *et al.* (1977) from a range of foodstuffs, Sahasrabudhe and Smallbone (1983) from beef, and de Koning *et al.* (1985) from fish meal, compared the efficiencies of several solvent extraction methods. They all found that chloroform–methanol (or methylene chloride–methanol) procedures were most effective. Maxwell (1984) extracted crude fat using dichloromethane (DCM) or total lipid using DCM–methanol (9 + 1) as elutants in pet food by a simple and rapid dry glass column technique:

2·5 g wet sample is mixed using a pestle and mortar with 10 g anhydrous sodium sulphate, then about 7 g Celite 545 ground in. The dry powder is transferred to a column 300 mm × 25 mm containing a glass wool plug below about 5 g Celite 545–$CaHPO_4$ (9 + 1). 25 ml DCM is poured into the column and when first drop appears, 100 ml more added and the column allowed to run dry into a weighed receiver. For total lipid, the process of elution from the extracted column is repeated using DCM–methanol (9 + 1).

Volumetric methods

These involve dissolving the sample in sulphuric acid and centrifuging out the fat in specially calibrated glass vessels. In the USA the Babcock method is used (see AOAC methods book) and in European countries the Gerber method is commonly employed for the routine determination of fat in milk and dairy products (BS 696), (see Chapter 14). Provided the process is carefully performed, the Gerber method gives results for many different foods which are in good agreement with those obtained using the more lengthy gravimetric procedures (Rosenthal *et al.*, 1985) and is therefore used for example to calibrate infra-red methods. For certain foods, particularly non-dairy products, cleaner separations can be obtained using acetic–perchloric acid mixture instead of sulphuric acid (Salwin *et al.*, 1955). Less corrosive neutral and alkaline reagents have been recommended by Macdonald (1959a, b). The sensitivity of the readings can be increased by using Van Gulik butyrometers (Pearson, 1972).

Physical methods

Apart from the classical methods, techiques have also been devised which measure changes in physical properties due to the presence of fat in solution, e.g. refractive index, specific gravity, impedance and capacitance (Smith, 1969). The automatic infra-red methods described in this chapter under nitrogen and crude protein are also used for the rapid routine determination of fat in certain foods. Low resolution pulsed NMR has been used for the determination of fat in a number of foods and agricultural materials, e.g. meat products (Renon *et al.*, 1985) and moisture and oil content in oilseeds (Gamblin and Agarwala, 1985).

CRUDE FIBRE AND DIETARY FIBRE

CRUDE FIBRE

'Crude fibre' is the insoluble and combustible organic residue which remains after the sample has been treated under prescribed conditions. The most common conditions are consecutive treatments with light petroleum, boiling dilute sulphuric acid, boiling dilute sodium hydroxide, dilute hydrochloric acid, alcohol and ether. This empirical treatment provides a crude fibre consisting largely of the cellulose content together with a proportion of the lignin and hemicelluloses content of the sample. The amounts of these substances in the crude fibre will vary with the conditions employed so for consistent results a standardised procedure must be rigidly followed. The method usually employed is the procedure laid down in the Feeding Stuffs (Sampling and Analysis) Regulations 1982 SI No. 1144, as described below:

REAGENTS

Alcohol—95 per cent v/v.
Diethyl ether.
Hydrochloric acid, 1 per cent v/v—Dilute 10 ml concentrated hydrochloric acid with water to 1 litre.
Light petroleum—boiling range 40–60 °C.
Sodium hydroxide, 0·313 N (0·313 M), 1·25 g NaOH in 100 ml of solution. This solution must be free or nearly free from sodium carbonate.
Sulphuric acid, 0·255 N (0·1275 M), 1·25 g conc. H_2SO_4 in 100 ml of solution.

PROCEDURE

Weigh to the nearest mg about 2·7–3·0 g of the sample (1), transfer to an extraction apparatus and extract with light petroleum (2). Alternatively, extract with light petroleum by stirring, settling and decanting three times. Air dry the extracted sample and transfer to a dry 1000 ml conical flask. Add 200 ml 0·1275 M sulphuric acid measured at ordinary temperature and brought to boiling point, the first 30 or 40 ml being used to disperse the sample, and heat to boiling point within 1 min. An appropriate amount of anti-foaming agent may be added if necessary (3). Boil gently for exactly 30 min, maintaining a constant volume (4) and rotating the flask every few minutes in order to mix the contents and remove particles from the sides.

Meantime, prepare a Buchner funnel (5) fitted with a perforated plate by adjusting a piece of cut cotton cloth (6) or filter paper to cover the holes in the plate so as to serve as a support for a circular piece of suitable filter paper (7). Pour boiling water into the funnel, allow to remain until the funnel is hot and then drain by applying suction. Care should be taken to ensure that the filter paper used is of such quality that it does not release any paper fibre during this and subsequent washings.

At the end of the 30 min boiling period, allow the acid mixture to stand for 1 min and then pour immediately into a shallow layer of hot water under gentle suction in the prepared funnel. Adjust the suction so that the filtration of the bulk of the 200 ml is completed within 10 min. Repeat the determination if this time is exceeded.

Wash the insoluble matter with boiling water until the washings are free from acid; then wash back into the original flask by means of a wash bottle containing 200 ml 0·313 M sodium hydroxide solution measured at ordinary temperature and brought to boiling point. Boil for 30 min with the same precautions as those used in the earlier boiling acid treatment. Allow to stand for 1 min and then filter immediately through a suitable filter paper. Transfer the whole of the insoluble material to the filter paper by means of boiling water, wash first with boiling water then with 1 per cent hydrochloric acid, and finally with boiling water until free from acid. Then wash twice with alcohol and three times with ether. Transfer the insoluble matter to a dried weighed ashless filter paper (8) and dry at 100 °C to a constant weight. Incinerate the paper and contents to an ash at a dull red heat (8). Subtract the weight of the ash from the increase of weight on the paper due to the insoluble material, and report the difference as crude fibre.

NOTES

1. The fineness and heterogeneity of the sample has an important bearing on the result. The regulations require the sample to be ground to pass through a sieve having apertures about 1 mm square.
2. Treatment with light petroleum can be omitted if the amount of fat present is small, e.g. with flour.
3. Frothing can also be reduced by using a 'cold finger'.
4. The flask should be fitted with a Liebig reflux condenser.
5. Special Buchner funnels are obtainable which are divided into two or three parts (Fig. 2.4) so that the residue can be washed off the paper more easily with alkali (Hartley, 1952).

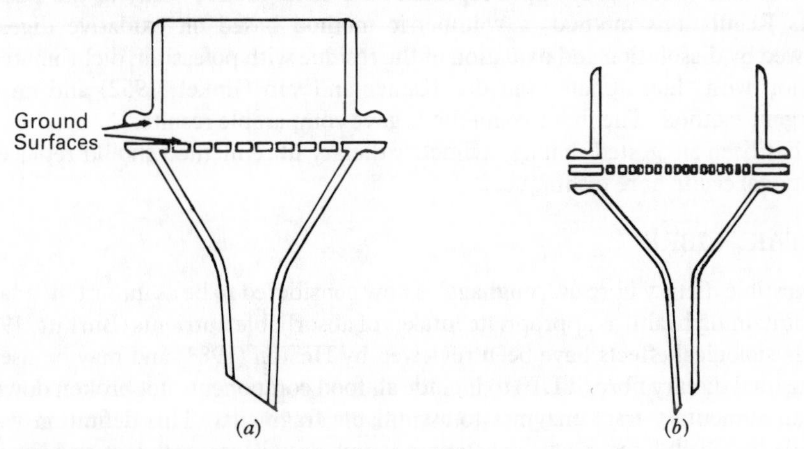

Fig. 2.4 Hartley form of Buchner funnel for filtration of fibre.
(a) 2-section pattern, (b) 3-section pattern.

6. The AOAC procedure prescribes either a stainless steel filter screen or a demountable polypropylene Buchner funnel together with ceramic fibre as the filtering medium (instead of the earlier used asbestos).
7. For the method in general, Whatman No. 54 or 541 filter papers are usually adequate.

8. Instead of preparing a weighted filter paper, the SPA (1943) method for white flour transfers the residue to a dish (previously ignited and weighed) with boiling water. After evaporating off the water on the water bath, the dish is dried in the 100 °C oven, weighed, ignited and re-weighed. Alternatively, a Gooch crucible may be employed.

The above method and the AOAC procedure are based on the original Weende method as is also ISO 5498, a general method for crude fibre determination which permitted different methods of separating the insoluble acid-treated sample. Tecator market the semi-automatic Fibertec equipment for crude fibre determination. When it is necessary to determine the fibre in several samples there are advantages in employing the semi-micro technique of Bredon and Juko (1961). For some foods it may be possible to omit the alkaline digestion (Whitehouse *et al.*, 1945; Jones and Griffith, 1962). For cereals and cereal products containing less than 1 per cent of crude fibre, a modified Scharrer method employing a reagent containing acetic, nitric and trichloroacetic acids, without an alkali treatment, is described in ISO 6541. Zentner (1965) uses a single digestion with a mixture of dimethyl sulphoxide and formic acid for brown bread.

In general, the outer protective coatings of many foods contain considerably more fibre than the softer, more edible inner tissues. The fibre figure can be used therefore for assessing the proportion of shell which is present in foods such as cocoa (p. 379) and pepper (p. 406). Similarly, the fibre figure is of value in calculating the amount of the shells of nuts, the stones of fruits or sawdust which may be present in foods, as all these adulterants have a high fibre content. As the fibre content increases with age in plants, the figure may be of value in assessing the maturity of legumes. In view of the higher amount in the outer tissues of the wheat grain, brown flour gives higher figures than white flour. In consequence a statutory minimum of 0·6 per cent fibre in the dry matter is applied to brown flour and brown bread. The digestibility of animal feeding stuffs varies inversely with the fibre content.

Player and Wood (1980) have reported on a collaborative study of the Feeding Stuffs Regulations method, a volumetric method based on oxidative digestion followed by dissolution and oxidation of the residue with potassium dichromate and titration with thiosulphate (van der Kamer and van Ginkel, 1952) and an acid detergent method. The first two methods gave comparable results.

It has been suggested that a gravimetric dietary fibre method should replace the traditional crude fibre technique.

DIETARY FIBRE

Indigestible dietary fibre or 'roughage' is now considered to be as important a factor in maintaining health as appropriate intakes of absorbable nutrients (Burkitt, 1973). Its physiological effects have been reviewed by Heaton (1983) and may be used to define total dietary fibre (TDF) to include all food components not broken down by human alimentary tract enzymes to assimilable fragments. This definition would include hemicelluloses, pectic substances, gums, mucilages, cellulose and lignin as well as undigested protein and lipid and ingested filth. It should be noted that some of these are soluble materials without a fibrous structure. Alternatively, dietary fibre can be regarded essentially as plant polysaccharides and lignin which are resistant to hydrolysis by the digestive enzymes of man (Trowell *et al.*, 1976), namely non-starch polysaccharides (NSP). Berry (1984) argues that starch made resistant to enzymatic digestion, i.e. resistant starch (RS), by food processing operations should also be included in any physiological definition of dietary fibre.

The lack of a clear chemical definition of dietary fibre has prevented the development of a single method of analysis acceptable to all parties. In effect TDF content has come to mean the quantity present in the foodstuff measured by the analytical technique employed rather than an absolute quantity. Because the empirical methods for crude fibre content are of little use (the results can represent as little as one-seventh of the total dietary fibre of certain foods) a number of methods for dietary fibre have developed. In studies on the nutrition of ruminant animals, Van Soest (1963) reported a procedure involving boiling the sample with 2 per cent cetyl trimethylammonium bromide in 0·5 M sulphuric acid to solubilise the intracellular proteins and yield the insoluble acid detergent fibre (ADF) which is essentially cellulose and lignin.

A more selctive procedure (Van Soest and Wine, 1967) uses 3 per cent sodium lauryl sulphate without acid to determine neutral detergent fibre (NDF).

In combination with an enzymatic digestion, the American Association of Cereal Chemists adopted this procedure as its official method 32-20 for dietary fibre analysis. This method gives a better yield of dietary fibre because of greater recovery of cellulose, hemicelluloses and lignin. None of these detergent methods determine the soluble components of dietary fibre.

However, a combination of enzymatic and gravimetric procedures incorporating a digestion with heat stable α-amylase and a precipitation of soluble dietary fibre fractions by the addition of ethanol has been collaboratively studied (Brosky *et al.*, 1984). It was found to compare favourably with other TDF methods and has been accepted as an AOAC method. An outline of the method is as follows:

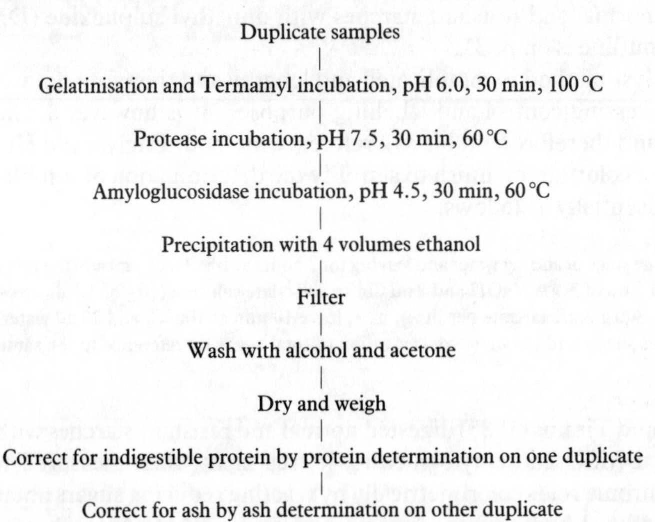

This method will not digest resistant starch which therefore is included in the result for TDF.

An evaluation of the Tecator Fibertec System E for gravimetric TDF methods on duplicate analyses of five samples and a blank per day is reported by Crawford and Williams (1985).

A comparison between detergent and non-detergent methods using HPLC to measure the monosaccharides of the insoluble fractions is described by Nielson and Marlett (1983).

Methods based on selective extractions and chemical and enzymatic hydrolysis are numerous. The dietary fibre data in Paul and Southgate's *The Composition of Foods* (1978) were obtained using a modified Southgate (1969, 1976) technique which in outline is as follows: the dried sample is extracted with aqueous methanol and ether to remove sugars and fat. Starch is hydrolysed enzymatically and the glucose extracted. The residue is separated into cellulose, non-cellulosic polysaccharides and lignin. These compounds are measured colorimetrically with anthrone for hexoses, orcinol for pentoses and carbazol for uronic acids.

The method has proved difficult in operation probably because of incomplete removal of starch. The non-specificity of the colorimetric methods has also been criticised, e.g. orcinol gives a strong reaction with uronic acids. The effectiveness of the neutral detergent fibre and Southgate methods for removing starch have been compared (Marlett and Chesters, 1985).

Englyst *et al.* (1982) described a method in which NSP is measured by gas chromatography of the individual monosaccharides as alditol acetates, following removal of the starch and separation into cellulose and non-cellulosic polysaccharide. Starch is hydrolysed enzymatically using α-amylase and pullulanase after gelatinisation in acetate buffer at 100 °C. This treatment does not disperse and hydrolyse all the starch, particularly that retrograded by food processing (RS). Uronic acids from pectins, hemicelluloses, gums and mucilages are measured colorimetrically. Thus the value for NSP obtained includes RS.

Englyst and Cummings (1984) modified the method to measure NSP directly by dispersing normal and resistant starches with dimethyl sulphoxide (DMSO). This method in outline is on p. 31.

The Englyst method is complicated and lengthy and therefore not well suited for routine processing control and labelling purposes. It is however a comprehensive procedure and therefore suitable as a reference method. Englyst and Hudson (1987) introduced a colorimetric finish to simplify the determination of soluble and insoluble NSP, essentially as follows:

At the end of the stage of adding water and leaving for 2 hours at 100 °C, to 1 ml add 0·5 ml glucose solution (0·5 mg/ml), 0·5 ml of 3·9 M NaOH and 2 ml dinitrosalicylate solution (10 g of 3,5 dinitrosalicylic acid + 16 g NaOH + 300 g NaK tartrate per litre), mix, leave 10 min at 100 °C, add 20 ml water, mix and read absorbance at 530 nm and calculate concentration of total sugar by reference to the same treatment for standards.

Faulkes and Timms (1985) digested normal and resistant starches with heat stable α-amylase, DMSO and amyloglucosidase treatment, then measured total neutral sugars and uronic acids colorimetrically by reacting reducing sugars obtained by acid hydrolysis with 3-hydroxybenzoic acid hydrazide (PAHBAH). Pectins were estimated separately.

Near infra-red reflectance spectrometers have been calibrated using detergent and enzyme methods to measure dietary fibre in cereal products (Baker, 1983; Norvath *et al.*, 1984). An extensive review of analytical developments and problems involved in the determination of dietary fibre is given by Selvendran and DuPont (1984).

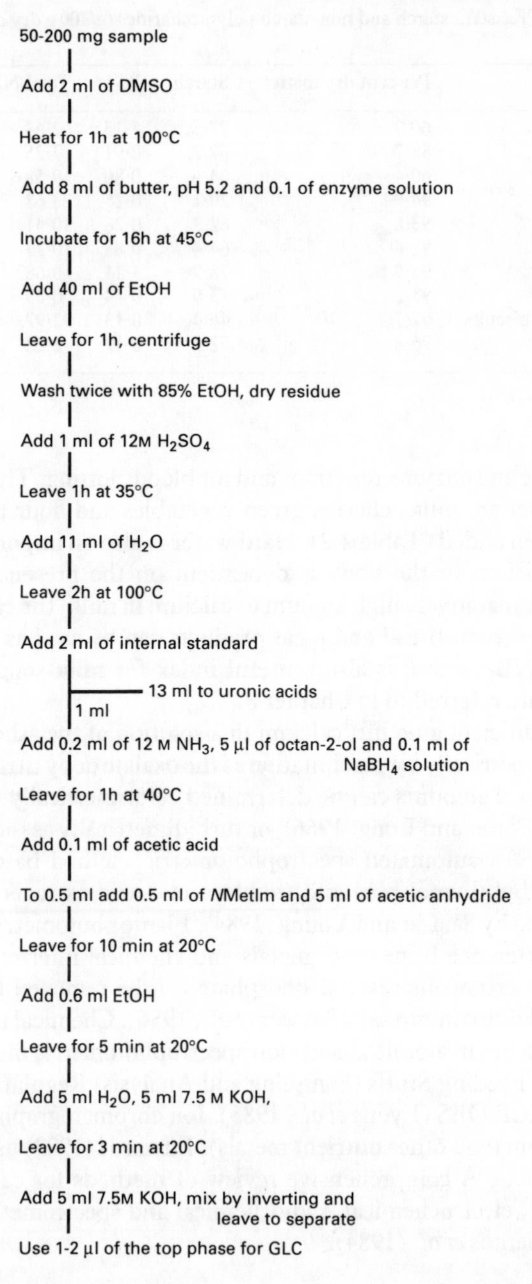

Relative levels of RS and NSP in a range of processed cereal foods are given in Table 2.1.

CALCIUM

Calcium is the most abundant metal element in the body, 99 per cent of it in the bones and teeth. The remainder is essential for the contraction of heart and other

Table 2.1 Content of resistant starch and non-starch polysaccharides (g/100 g dry matter) in a range of cereal products

	Per cent dry matter	Starch	RS	Total NSP	RS:Total NSP
White bread	60·7	77·5	1·15	2·63	1:2·3
Brown bread	58·7	67·7	0·91	7·25	1:8·0
Wholemeal bread	60·4	64·6	0·80	9·58	1:12·0
All Bran (Kelloggs)	94·9	29·1	0·15	23·68	1:158
Weetabix	93·8	62·3	0·26	10·41	1:40·0
Shredded wheat	91·6	64·4	0·88	10·73	1:12·2
Cornflakes	94·9	78·9	3·11	0·68	4·6:1
Rice Krispies	95·2	73·9	0·24	0·89	1:3·7
Digestive wheatmeal biscuits	97·1	46·4	0·13	2·97	1:22·9
Rich tea biscuits	96·8	52·3	0·19	2·22	1:11·7

From Berry (1984).

muscles, for nerve and enzyme functions and for blood clotting. The main sources of calcium in the diet are milk, cheese, green vegetables and flour to which calcium carbonate has been added (Table 2.2). Hard water is also an important source. The absorption of calcium in the body is dependent on the presence of vitamin D. Because of the comparatively high amount of calcium in milk, the calcium content of foods such as ice cream, bread and meat products can be used as an index of milk solids content. (NB, lactose is also a useful index for milk solids.) Methods for calcium in flour are referred to in Chapter 8.

If present in sufficient amount, calcium in a solution of the ashed sample can be determined gravimetrically by precipitation as the oxalate or by titration with EDTA (Moss, 1961). Trace amounts can be determined colorimetrically using glyoxal-bis (2-hydroxyanil) (Potter and Long, 1966), or turbidimetrically as the oxalate (Hunter and Hall, 1953). An automated spectrophotometric method based on the colour formed with methyl thymol blue and trichloracetic acid extracts of calcium from peanuts is reported by Salazar and Young (1984). Flame photometry can also be used but spectral interference from other metals and chemical interference from phosphate can lead to erroneous results. Phosphate can be removed from solution by precipitation as the zirconium salt (Saywer *et al.*, 1956). Chemical interference from phosphate also occurs in atomic absorption spectrophotometric methods (Roos and Price, 1970; The Feeding Stuffs (Sampling and Analysis) Regulations 1982 SI No. 1144) but not in ICP-OES (Lyons *et al.*, 1985). Ion chromatography is a developing method for calcium (and other nutrient metals). Edwards (1983) used it for calcium in oatmeal products. A comprehensive review of methods for calcium (in wines) covering classical, electrochemical, radiochemical and spectrometric procedures is given by Baluja-Santos *et al.* (1984).

Determination of calcium by precipitation as oxalate

Calcium is precipitated as the insoluble oxalate from ammoniacal solutions. Excess of magnesium and phosphate can interfere, therefore strict adherence to the procedure is necessary. Procedures are given in BS 2472: 1966 (ice cream) and the Feeding Stuffs (Sampling and Analysis) Regulations 1982 SI No. 1144. The following is basically the procedure given in the EC Directive on methods of analysis for the

Table 2.2 Calcium content of selected foods, mg per 100 g edible portion

Food	Calcium content	Food	Calcium content
Milk, whole	103	Aubergine	10
Milk, evaporated	260	Cabbage	57
Milk, dried skimmed	1230	Onions	31
Yoghurt, low fat, natural	200	Potatoes	8
Cheese, Cheddar	800	Watercress	220
Eggs	52	Apples	4
Beef, average	7	Figs, dried	280
Fish, white	22	Peanuts, roasted	61
Sardines, canned in oil (fish only)	550	Bread, white	105
		Bread, wholemeal	54
		Rice	4

From *Manual of Nutrition* (MAFF, 1985).

official control of animal feeding stuffs (71/250/EEC) which is identical to that found in SI No. 1144, referred to above.

Ash approximately 5 g of prepared sample at 550 °C. Wash the ash into a 250 ml beaker with 40 ml conc. hydrochloric acid and 60 ml water. Add 3 drops conc. nitric acid and boil for 30 min. Cool and transfer to a 250 ml graduated flask. Make up to the mark, mix and filter. Pipette a volume of the filtrate containing 10–40 mg Ca into a 250 ml beaker. Add 1 ml citric acid solution (30 per cent) and 5 ml ammonium chloride solution (5 per cent). Make up to approximately 100 ml with water and bring to the boil. Add 10 drops bromo-cresol green solution (0·04 per cent) and 30 ml warm saturated ammonium oxalate solution. If a precipitate forms, dissolve it by adding a few drops of conc. hydrochloric acid. Neutralise very slowly with ammonia solution (0·88), stirring continuously until indicator changes colour at pH 4·4–4·6. Place beaker on steam bath for 30 min. Remove the beaker and after 1 hour filter through a fine sintered glass crucible. Thoroughly wash the beaker and crucible then dissolve the precipitate by passing through 50 ml warm sulphuric acid solution (10 per cent v/v). Rinse the crucible and make up the filtrate to about 100 ml with water. Heat the filtrate to 70–80 °C and titrate with 0·02 M potassium permanganate solution (3·161 g/l) until pink colour persists. 1 ml of 0·02 M permanganate ≡ 2·004 mg Ca.

Determination of calcium by atomic absorption spectrophotometry

Reagents

Releasing agent. Moisten 117·3 g lanthanum oxide with water. Slowly add 350 ml conc. hydrochloric acid and shake to dissolve. Cool and dilute to 1 litre with water.

Standard calcium stock solution A. Transfer 2·497 g calcium carbonate to a 1 litre volumetric flask using 100 ml water and slowly add 60 ml of M hydrochloric acid. When the powder has dissolved dilute to the mark with water (1 ml ≡ 1 mg Ca).

Standard calcium working solutions. Dilute 20 ml of A to 200 ml (1 ml ≡ 100 μg Ca) (B). To each of six volumetric flasks add 0, 3, 6, 9, 12 and 15 ml of B and dilute to 100 ml with water (0, 3, 6, 9, 12 and 15 μg Ca per ml).

Procedure

Ash a suitable quantity of sample (see below), moisten with water and carefully add 10 ml diluted hydrochloric acid (1 + 1). Evaporate to dryness on a water bath and continue the heating for a further hour. Cool, add 20 ml water and 10 ml of the diluted hydrochloric acid, boil and filter into a 250 ml volumetric flask. Wash through with hot water. Then cool and make up to volume (C). Set up the atomic absorption spectrophotometer with a hollow cathode lamp using the line at 422·7 nm and a fuel rich flame (air–acetylene or nitrous oxide–acetylene). To a suitable volume of C add releasing agent and water to produce a standard volume of solution to contain 5–10 μg Ca per ml and 10 per cent of releasing agent. Also prepare a similar blank solution, but omitting C. Spray water into the flame and zero the instrument. Spray successively the standard solutions, sample and blank, washing the instrument through with water between each spraying. Plot the mean of 3 readings for each standard solution against the calcium content. Assess the calcium content (allowing for that of the blank) of the sample from the standard curve. Sodium equivalent to that in the sample solution should be added to the calcium standard.

POTASSIUM

Potassium is a major component of animal and vegetable intracellular fluids. In animals it has a complementary action in the functioning of cells with sodium in the extracellular fluids. Apart from its nutritional significance, potassium is a useful index for estimating fruit or fruit juice content of food products (see Chapter 7). The potassium content of some foods is given in Table 2.3, and of fruit juices in Table 7.12.

Table 2.3 Sodium and potassium content of selected foods, mg/100 g edible portion

	Sodium	Potassium
Milk, whole	50	140
Cheese, Cheddar	610	120
Eggs	140	136
Beef, average	70	330
Corned beef	854	134
Pork, average	65	360
Bacon, streaky	1245	183
Chicken	75	290
Sausages, pork	760	160
Haddock, fresh	120	300
Haddock, smoked	1220	290
Butter, salted	780	15
Butter, unsalted	12	15
Margarine	800	5
Potatoes	8	360
Potato crisps	550	1190
Cauliflower	8	350
Peas, frozen	3	190
Peas, canned, processed	380	150
Tomatoes	3	290
Orange juice/oranges	2	180
Canned peaches	1	150
Raisins	52	860
Bread, white	525	110
Bread, wholemeal	560	230
Cornflakes	1160	99
Weetabix	360	420
Coffee, instant	81	3780
Marmite	4500	2600
Chocolate, milk	120	420
Soy sauce	5720	270

From *Manual of Nutrition* (MAFF, 1985).

Potassium in foods can be determined by either atomic absorption spectrophotometry, flame photometry or flame emission spectrophotometry using an atomic absorption spectrophotometer in the emission mode and also by ion chromatography (Edwards, 1983); see Chapter 3. Ion selective potentiometry can be used but not at low pH, e.g. ash dissolved in dilute acid. Chapman and Goldsmith (1982b) used a rapid dry sample addition method for potassiumn in salted peanuts.

Flame photometric procedures are described by Collins and Polkinhorne (1952) and in the EC Directive (71/250/EEC; OJ-L155/13: 12.7.71) covering official methods for the analysis of animal feedingstuffs. In the EC method interference from other elements is prevented by the addition of caesium chloride and aluminium nitrate. Interference from calcium or iron can be removed by boiling first with ammonium oxalate solution. With some liquid foods, analysis by any of the methods can be carried out directly on the food, or after simply centrifuging or filtering, but more often analysis is carried out on a solution of the ash in hydrochloric acid. Ashing in nickel rather than silica quartz or porcelain crucibles is preferred (Rowan *et al.*, 1982). Boland (1965) has reported on a collaborative study of a direct aspiration method for fruits.

Flame photometric method

REAGENTS

Potassium stock solution. Dissolve 5·779 g potassium dihydrogen phosphate (dried for 1 hour at 105 °C) in water and dilute to 1 litre.

Standard dilute potassium solution. Dilute 50 ml stock potassium solution to 1 litre with water (≡ 100 mg/l as K_2O). Alternatively, prepare by dilution from a stock solution containing 0·954 g potassium chloride in 500 ml water (1000 mg/l K).

PROCEDURE

For the direct method take, say, 25 g of jam or 10 g fruit juice or 50 g concentrated soft drink and dilute with water to 500 ml. Mix, filter if necessary, and dilute (e.g. 10 ml/100 ml) to give a solution of suitable concentration (approx. 15 mg/l K_2O in final solution).

Otherwise ash a suitable quantity of sample (e.g. 5 g), transfer the ash to a 400 ml beaker using 100 ml water, add 10 ml conc. hydrochloric acid and boil for several minutes. Cool, dilute with water to 500 ml and filter. Then dilute to a final concentration of solution of approx. 15 mg/l K_2O.

Prepare a series of solutions from the freshly prepared standard dilute potassium solution containing 10, 12, 14, 16, 18 and 20 mg/l K_2O. Using a filter to give a spectral range of 766–770 nm, set the sensitivity of the flame photometer so that the full deflection (100 divisions) is equivalent to 20 mg/l. Spray each standard solution at least three times, checking the sensitivity between each reading against the 20 mg/l solution. From the readings obtained prepare the calibration graph. Reset the instrument at full deflection with 20 mg/l solution and spray the diluted sample solution. Estimate the amount of potassium in the original sample from the calibration graph after taking several readings ($K = 0{\cdot}83 \times K_2O$).

SODIUM

Sodium is present in the extracellular body fluids of animals and is involved in maintaining water balance. It is also essential for muscle and nerve activity. An adult needs about 3 g per day but modern dietary habits can take in 5–20 g per day. The sodium content of unprocessed foods is low but it is used extensively in food processing and in prepared foods as a flavouring. The sodium content of some foods is given in Table. 2.3.

Sodium is most conveniently determined in foods by either atomic absorption spectrophotometry, flame photometry or flame emission spectrophotometry using an atomic absorption spectrophotometer in the emission mode. The comments and references regarding ion chromatography and ion selective potentiometry under potassium, above, apply equally to sodium. Averill (1983) gives a full account of the use of ISE for sodium. Flame photometric procedures are described by Collins and Polkinhorne (1952) and in the EC Directive (71/250/EEC; OJ-L155/13: 12.7.71) on official methods for the analysis of animal feedingstuffs. The latter procedure is prescribed in the Feeding Stuffs (Sampling and Analysis) Regulations 1982 SI No. 1144. Caesium chloride and aluminium nitrate are used in the EC method to prevent interference. With soluble foods and some liquid foods, the filtered or centrifuged solution can be sprayed directly into the flame, but more generally an acid solution of the ash is prepared. Nickel crucibles are better for recovery of sodium compared with quartz porcelain (Rowan *et al.*, 1982).

Flame photometric method

REAGENTS

Sodium stock solution. Dissolve 1·271 g sodium chloride in water and dilute to 1 litre (≡ 500 mg/l Na).

Standard dilute sodium solution. Dilute 10 ml stock sodium solution to 500 ml with water (≡ 10 mg/l Na).

PROCEDURE

For the direct method prepare the solution, macerating if necessary, as described under 'Potassium'. Otherwise ash 5–10 g, carefully add 10 ml water and 5 ml conc. nitric acid, stir and transfer to a 100 ml volumetric flask with water and make up to the mark.

Set up the flame photometer so that the scale reads 0 with distilled water and 100 divisions when the standard dilute sodium is sprayed. Then spray the sample solution (suitably diluted to give a reading within the 0–100 range) and calculate the sodium content by simple proportion.

PHOSPHORUS

Phosphorus is after calcium the most abundant mineral element in the body. It is found in bones and teeth and in cellular fluids. It plays an essential function in the liberation and utilization of energy in animal and vegetable tissue and therefore is widely distributed in all food. The main sources in the diet are milk and milk products, bread and cereal products, meat and meat products and eggs. Phosphate content can be a useful index for estimating fruit and fruit juice content. The phosphorus content of fruit juices is given in Table 7.12. Phosphorus is used as orthophosphate or condensed phosphates as a food additive for various functional purposes in processed foods.

Phosphorus is readily determined in foods as orthophosphate by titrimetry or more usually by colorimetry. Lyons *et al.* (1985) have reported the determination of phosphorus in Kjeldahl digests using ICP-OES, and phosphate species should be readily determinable by ion chromatography. Hamence and Kunwordia (1974) have described a paper chromatographic procedure for the detection of orthophosphate, pyrophosphate and triphosphate. For polyphosphates in meat products, see Chapter 13.

Titrimetric methods

The precipitation of phosphate as a phosphomolybdate has been the basis of titrimetric procedures. In foodstuffs any metaphosphates or pyrophosphates must first be converted to orthophosphate by treatment with nitric acid. The method due to Wilson (1951) based on the precipitation of quinoline phosphomolybdate and described in the seventh edition of this book and in the AOAC methods book, has largely replaced the ammonium phosphomolybdate method (Pearson, 1973). In both these methods the precipitate is collected and dissolved in a small known excess of alkali then back titrated with standard acid.

Colorimetric methods

VANADO-MOLYBDATE METHOD

Hanson (1950) introduced a simple colorimetric method for the determination of phosphate which is based on Mission's reaction. The acid solution containing orthophosphate is treated with an acid reagent containing molybdic acid and vanadic acid and a stable orange–yellow coloured complex of vanadimolybdiphosphoric acid (H_3PO_4, VO_3, $11MoO_3$, nH_2O) is formed. The maximum absorption is at 330 nm, but Donald *et al.* (1956) have reported that satisfactory results can be obtained in the region 420–480 nm, provided the light is monochromatic. The colour development is not markedly affected by the presence of hydrochloric, sulphuric, acetic or citric acid or by fluorides provided they are not present in relatively large quantities. This rapid and trouble free method has been widely introduced and is official in the Feeding Stuffs (Sampling and Analysis) Regulations 1982: SI No. 1144.

REAGENTS

Vanadate–molybdate composite reagents. Dissolve 20 g ammonium molybdate in 400 ml warm water (50 °C) and cool. Dissolve 1·0 g ammonium vanadate in 300 ml boiling distilled water, cool and add 140 ml conc. nitric acid gradually with stirring. Then add the molybdate solution gradually to the acid vanadate solution with stirring and dilute to 1 litre with water.

Standard phosphate solution. Prepare a stock solution contining 3·834 g potassium dihydrogen phosphate (KH_2PO_4) per litre. Dilute 25 ml to 250 ml (1 ml ≡ 0·2 mg P_2O_5).

PREPARATION OF STANDARD GRAPH

To a series of 100 ml volumetric flasks add 0, 2·5, 5, 10, 20, 30, 40 and 50 ml of the standard phosphate solution (≡ 0–10 mg P_2O_5) and dilute each to 50–60 ml with water. Add a few drops of ammonia solution (0·88) and make just acid with nitric acid (1:2). Add 25 ml of the vanadate–molybdate reagent, dilute to the mark and mix. Allow to stand for 10 minutes and measure the optical density in a 2·5 or 10 mm cell at 470 nm.

PROCEDURE

Transfer a suitable volume of solution of the sample (containing 0·5–10 mg P_2O_5) to a 100 ml volumetric flask. If the determination is carried out on the ash, boil the ash with 10 ml of 5 M hydrochloric acid and wash the solution into the 100 ml flask with water, filtering if necessary. Neutralise by the dropwise addition of 0·88 ammonia (the volume of solution at this stage should be 50–60 ml) and then proceed as for the standard graph, i.e. make just acid with dilute nitric acid, add 25 ml of vandate–molybdate reagent, dilute to the mark and measure the optical density after allowing to stand for 10 minutes.

MOLYBDENUM BLUE METHOD

Many methods have been proposed for determining phosphate based on Denigès' reaction. This involves the addition of ammonium molybdate to a solution of an orthophosphate and the phosphomolybdate produced is partially reduced by stanous chloride to a blue compound ('molybdenum blue') which probably has the formula $(MoO_2, 4MoO_3)_2.H_3PO_4$. Alternative reducing agents have been suggested for quantitative purposes such as phenylhydrazine, hydroquinone, sulphur dioxide, hydrazine sulphate and *p*-methylaminophenol. Fogg and Wilkinson (1958) have described a useful technique employing vitamin C.

This colour development has found use in the automatic determination of phosphate using for example Technicon AutoAnalyzer equipment including a filter photometer for absorbance measurements. The method is described in detail by Hanson (1973).

PROCEDURE

Transfer 25 ml of a neutral ash solution containing in all not more than 0·1 mg orthophosphate to a 50 ml graduated flask. Add 5·0 ml molybdate solution (25·0 g/l sodium molybdate in 5 M sulphuric acid) followed by 2·0 ml hydrazine sulphate solution (1·5 g/l). Dilute to mark with distilled water and mix. Immerse flask in boiling water for 10 min, remove and cool in cold water. Measure absorbance at 830 or 680 nm against a reagent blank. Deduce phosphate content from a calibration graph prepared from a solution of potassium dihydrogen phosphate (0·2197 g/l: 1 ml ≡ 0·05 mg P).

EXTRANEOUS MATERIAL (FOREIGN BODIES AND FILTH)

Apart from chemical contamination (Chapter 5) and microbiological contamination, food is frequently contaminated by solid or particulate foreign matter such as sand, soil, insects, glass fragments, plastic, animal and human hairs, textile fibres, metal objects, etc. Such contamination can be an indicator of poor and insanitary manufacturing practices or faulty distribution and storage, and can give rise to health hazards to the consumer.

The FDA of the US have paid considerable attention to the presence of extraneous matter in foods. The 'Filth' Test of the AOAC is mainly concerned with the detection and counting of rodent hairs and fragments of insects (sometimes called 'light filth'). Rodent hairs are invariably present in rodent faeces and it is implied that if hairs are found in flour, say, the original grain was contaminated with rodent pellets which were later ground up with the flour on the rolls.

With the foods which are soluble in water, e.g. sugar products and table jelly, the method merely involves dissolving up the material, filtering and examining any matter remaining in the paper. With cereals (see p. 300), however, the starch and protein should preferably be digested and the extraneous matter separated by shaking with petrol. The petrol preferentially wets the hair and insect fragments, which float up and congregate in the interfacial layer (Kent-Jones *et al.*, 1948). Selected modifications for numerous individual products are described in the AOAC methods book. The journal of the AOAC regularly reports on improved methods and collaborative studies for the determination of filth in various foodstuffs. Methods of identifying the common food-contaminating hairs and insect fragments have been described by Vazquez (1961) and Kurtz and Harris (1962). Smith (1983) has produced a comprehensive scheme for the examination and identification of foreign materials in food, and an illustrated colour atlas for food inspection and quality control is by Sutherland *et al.* (1986).

The examination of spices for heavy filth (dirt, sand, etc.) is given on p. 393, and tomato products for mould on p. 255.

REFERENCES

American Association of Cereal Chemists (1983) *Cereal Laboratory Methods*, 8th ed. St. Paul, Minn: AACC.

Arneth, W. (1984) *Fleischwirtschaft*, **64**, 1086.

Averill, W. F. (1983) *Food Technology*, **37**, 44.

Baker, D. (1983) *Cereal Chemistry*, **60**, 217.

Baluja-Santos, C., Gouzalez-Portal, A. & Bermejo-Martinez, F. (1984) *Analyst*, **109**, 797.

Beljaars, P. R. & Horwitz, W. (1985) *Journal of the Association of Official Analytical Chemists*, **68**, 480.

Berry, C. S. (1984) *FMBRA Bulletin*, (6), 237.

Bligh, E. G. & Dyer, W. J. (1959) *Canadian Journal of Biochemistry and Physiology*, **37**, 91.

Boland, F. E. (1965) *Journal of the Association of Official Analytical Chemists*, **48**, 521.

Bostian, M. Fish, D., Webb, N. & Arey, J. (1985) *Journal of the Association of Official Analytical Chemists*, **68**, 876.

Brammell, W. S. (1974) *Journal of the Association of Official Analytical Chemists*, **57**, 1209.

Bredon, R. M. & Juko, C. D. (1961) *Journal of the Science of Food and Agriculture*, **12**, 196.

Brosky, L. P., Asp, N. G., Furda, I., Devries, J. W., Schweizer, T. F. & Harland, B. F. (1984) *Journal of the Association of Official Analytical Chemists*, **67**, 1044.

Bruhn, J. C., Pecore, S. & Franke, A. A. (1980) *Journal of Food Protection*, **43**, 753.

Burkitt, D. P. (1973) *British Medical Journal*, **1**, 274.

Carter, J. R. (1977) In *Characterisation of oils and fats, Symposium Proceedings* No. **28**. Leatherhead, BFMIRA.

Chapman, B. R. & Goldsmith, I. R. (1982a, b) *Analyst*, **107**, 1014, 1045.

Collins, G. C. & Polkinhorne, H. (1952) *Analyst*, **77**, 430.

Concon, J. M. & Soltness, D. (1973) *Analytical Biochemistry*, **53**, 35.

Crawford, C. & Williams, S. K. (1985) *FMBRA Bulletin* (5), 187.

Croll, B. T., Tomlinson, T. & Whitfield, C. R. W. (1985) *Analyst*, **110**, 861.

Crooke, W. M. & Simpson, W. E. (1971) *Journal of the Science of Food and Agriculture*, **22**, 9.

Crowhurst, B. (1956) *Analyst*, **81**, 123.

Daughtery, C. E. & Lento, H. H. (1983) *Journal of the Association of Official Analytical Chemists*, **66**, 927.

Davies, A. M. C. (1985) *Laboratory Practice*, **34**, 32.

Davis, A. B. & Lai, C. S. (1984) *Cereal Chemistry*, **61**, 1.

De Koning, A., Evans, A. A., Heydenrych, C., dev Purcell, C. J. & Wessels, J. P. H. (1985) *Journal of the Science of Food and Agriculture*, **36**, 177.

De Rham, O. (1982) *Lebensmittel-Wissenschaft und Technologie*, **15**, 226.

Dohan, D. A., Standing, K. G. & Bushuk, W. (1976) *Cereal Chemistry*, **53**, 91.

Donald, R., Schwehr, E. W. & Wilson, H. N. (1956) *Journal of the Science of Food and Agriculture*, **7**, 677.

Doty, W. H., Munsen, A. W., Wood, D. E. & Schneider, E. L. (1970) *Journal of the Association of Official Analytical Chemists*, **53**, 801.

Duckworth, R. B. (ed.) (1975) *Water Relations of Foods*. London: Academic Press.

Edwards, P. (1983) *Food Technology*, **37**, 53.

Englyst, H. N. & Cummings, J. H. (1984) *Analyst*, **109**, 937.

Englyst, H. N. & Hudson, G. J. (1987) *Food Chemistry*, **24**, 63.

Englyst, H. N., Wiggins, H. S. & Cummings, J. H. (1982) *Analyst*, **107**, 307.

Eustace, I. J. & Jones, P. N. (1984) *CSIRO Food Research Quarterly*, **44**, 38.

FAO/WHO (1973) Energy and protein requirements. *Technical Report* No. 522, Ad Hoc Expert Committee. Rome: FAO/WHO.

Faulks, R. M. & Timms, S. B. (1985) *Food Chemistry*, **17**, 273.

Favetto, G., Resnik, S., Chirife, J. & Ferro Fontan, C. (1983) *Journal of Food Science*, **48**, 534.

Florence, E. & Milner, D. F. (1979) *Analyst*, **104**, 378.

Fogg, D. N. & Wilkinson, N. T. (1958) *Analyst*, **88**, 406.

Food Standards Committee (1979) *Report on Water in Food*, FSC/REP/70. London: HMSO.

Fosdick, L. E. & Pike, R. K. (1982) *Journal of the Association of Official Analytical Chemists*, **65**, 178.

Gamblin, P. N. & Agarwala, A. K. (1985) *Journal of the American Oil Chemists' Society*, **62**, 103.

Hadorn, H. (1980) *Mitteilungen aus der Gebiete der Lebensmitteluntersuchung und Hygiene*, **71**, 220.

Hamence, J. H. & Kunwordia, G. H. (1974) *Journal of the Association of Public Analysts*, **12**, 85.

Hannant, G. (1982) *Journal of the Association of Public Analysts*, **20**, 117.

Hanson, N. W. (1973) *Official, Standardised and Recommended Methods of Analysis*. London: The Society for Analytical Chemistry.
Hanson S. W. F. & Olley, J. (1963) *Biochemical Journal*, **89**, 101P.
Hanson, W. C. (1950) *Journal of the Science of Food and Agriculture*, **1**, 172.
Hardman, T. M. (1978) In *Developments in Food Analysis Techniques—1*, King, R. D. (ed.). London: Applied Science.
Harris, C. (1972) *Talanta*, **19**, 1523.
Harrison, R. W. (1939) *Journal of the Association of Official Agricultural Chemists*, **22**, 661.
Hartley, A. W. (1952) *Analyst*, **77**, 53.
Hartley, A. W. (1975) *Journal of the Science of Food and Agriculture*, **26**, 550.
Hatfull, R. S. (1983) *Journal of the Association of Public Analysts***21**, 103.
Havilah, E. J., Wallis, D. M., Morris, R. & Woolnough, J. A. (1977) *Laboratory Practice*, **26**, 545.
Heaton, K. W. (1983) *Human Nutrition: Clinical Nutrition*, **37C**, 151.
Heidelbaugh, N. D., Huber, C. S., Bednarczyk, J. F., Rambout, P. C. & Wheeler, H. O. (1975) *Journal of Agricultural and Food Chemistry*, **23**, 611.
Herschdoerfer, S. M. (ed.) (1967) *Quality Control in the Food Industry*, Vol. 1. London: Academic Press.
Hester, R. E. & Quine, D. E. C. (1976) *Journal of Food Technology*, **11**(4), 331.
Hill, R. L. & Stone, W. K. (1964) *Journal of Dairy Science*, **47**, 1014.
Huang, Y. W., Marshall, R. T., Anderson, M. E. & Charoen, C. (1976) *Journal of Food Science*, **41**, 1219.
Hubbard, W. D., Sheppard, A. J., Newkirk, D. R., Prosser, A. R. & Osgood, T. (1977) *Journal of the American Oil Chemists' Society*, **54**, 81.
Hunter, J. G. & Hall, A. (1953) *Analyst*, **78**, 106.
Isaac, R. A. & Johnson, W. C. (1976) *Journal of the Association of Official Analytical Chemists*, **59**, 98.
Johnson, M. E. & Olson, N. F. (1985) *Journal of Dairy Science*, **48**, 1020.
Jones, D. B. (1931) *US Department of Agriculture Circular* 183.
Jones, D. I. & Griffith, G. (1962) *Nature*, **193**, 882.
Kane, P. E. (1984) *Journal of the Association of Official Analytical Chemists*, **67**, 869.
Karmas, E. (1980) *Food Technology*, **34**, 52.
Karow, R. S., Forsberg, R. A. & Peterson, D. M. (1984) *Cereal Chemistry*, **61**, 196.
Kent-Jones, D. W., Amos, A. J., Elias, P. S., Bradshaw, R. C. A. & Thackray, G. B. (1948) *Analyst*, **73**, 128.
Kerkhof Mogot, M. F., Koops, J., Neeter, R., Slangen, K. J., Van Hemert, H., Kooyman, O. & Wooldrik, H. (1982) *Netherlands Milk and Dairy Journal*, **36**, 115.
Khor, H. T. & Chan, S. L. (1985) *Journal of the American Oil Chemists' Society*, **62**, 98.
Kindstedt, P. S. & Kosikowski, F. V. (1984) *Journal of Dairy Science*, **67**, 879.
Koops, J., Klomp, H. & Elgersma, R. H. C. (1975) *Netherlands Milk and Dairy Journal*, **29**, 169.
Kramer, A. & Twigg, B. A. (1966) *Fundamentals of Quality Control for the Food Industry*, 2nd edn. Westport, Conn: AVI.
Kratochvil, B. G. & Taylor, J. K. (1981) *Analytical Chemistry*, **53**, 925A.
Kratochvil, B. G. & Taylor, J. K. (1982) Technical Note 1153. National Bureau of Standards, US Department of Commerce.
Kropf, D. H. (1984) *Journal of Food Quality*, **6**, 199.
Kurtz, O. L. & Harris, K. (1962) *Micro-analytical Entomology for Food Sanitation Control*. Washington, DC: Association of Official Agricultural Chemists.
Kvaale, O. & Dalhoff, E. (1963) *Food Technology*, **17**, 659.
Kviesitis, B. (1975) *Journal of the Association of Official Analytical Chemists*, **58**, 164.
Lakin, A. L. (1978) In *Developments in Food Analysis Techniques—1*, ed. R. D. King. London: Applied Science.
Landrock, A. H. & Proctor, B. E. (1951) *Food Technology*, **5**, 332.
Lees, R. (1975) *Food Analysis: Analytical and Quality Control Methods for the Food Manufacturer and Buyer*. London: Leonard Hill.
Lehmann, G. & Beckmann, I. (1982) *Fleischwirtschaft*, **62**, 1585.
Lillevik, H. A. (1970) In *Methods in Food Analysis. Physical, Chemical and Instrumental Methods*, 2nd edn, Joslyn, M. A. (ed.). New York: Academic Press.
Lin, R. H. & Randolph, H. E. (1978) *Journal of Dairy Science*, **61**, 870.
Lowry, O. H., Rosebrough, N. J., Far, A. L. & Randall, R. J. (1951) *Journal of Biological Chemistry*, **193**, 265.
Lyons, D. J., Spann, K. P. & Roofayel, R. L. (1985) *Analyst*, **110**, 956.
Macdonald, F. J. (1959a, b) *Analyst*, **84**, 287, 747.
Mackenzie, R. C. (1974) *Analyst*, **99**, 900.
MAFF (1985) *Manual of Nutrition*. London: HMSO.
Manley, C. H. & Wood, E. G. (1945) *Analyst*, **70**, 173.

Markham, R. (1942) *Biochemical Journal*, **36**, 790.
Marlett, J. A. & Chesters, J. G. (1985) *Journal of Food Science*, **50**, 410.
Martone, C. B., Crupkin, M., Barassi, C. A. & Trucco, R. E. (1980) *Journal of the Science of Food and Agriculture*, **31**, 782.
Maxwell, R. J. (1984) *Journal of the Association of Official Analytical Chemists*, **67**, 878.
McGill, D. L. (1981) *Journal of the Association of Official Analytical Chemists*, **64**, 29.
McKnight, G. S. (1977) *Analytical Biochemistry*, **78**, 86.
Ministry of Defence (1965) *Sampling Procedures and Tables for Inspection by Attributes*. Defence Specification DEF–131–A. London: HMSO.
Minor, B. A., Sims, K. A., Bassette, R. & Fung, D. Y. C. (1984) *Journal of Food Protection*, **47**, 611.
Mitchell, J. & Smith, D. M. (1977) *Aquametry*, Part I. New York: Wiley.
Mitsuda, H. & Mitsunaga, T. (1974) *Agricultural and Biological Chemistry*, **38**, 1649.
Morries (1983) *Journal of the Association of Public Analysts*, **21**, 53.
Moss, P. (1961) *Journal of the Science of Food and Agriculture*, **12**, 30.
Nielson, M. J. & Marlett, J. A. (1983) *Journal of Agricultural and Food Chemistry*, **31**, 1342.
Nkonge, C. & Ballance, G. M. (1982) *Journal of Agricultural and Food Chemistry*, **30**, 416.
Noll, J. S., Simmonds, D. H. & Bushuk, W. (1974) *Cereal Chemistry*, **51**, 610.
Norvath, L., Norris, K. H., Horvath-Mosonyi, M., Rigo, J. & Hegedus-Volgyesi, E. (1984) *Acta Alimenta*, **13**, 355.
Okabe, T., Huang, M. T. & Okamura, S. (1973) *Journal of Agricultural Engineering Research*, 18 (1), 59.
Osborne, B. G. (1981) *Journal of Food Technology (London)*, **16**, 13.
Osborne, B. G. & Fearn, T. (1986) *Near Infrared Spectroscopy in Food Analysis*. Harlow: Longman.
Osborne, D. R. & Voogt, P. (1978) *The Analysis of Nutrients in Food*. London: Academic Press.
Pailler, F. M. (1982) *Annales des Falsifications de l'Expertise Chimique et Toxicologique*, **75**, 431.
Pearson, D. (1972) *Food Manufacture*, **47**(4), 45.
Pearson, D. (1973) *Laboratory Techniques in Food Analysis*. London: Butterworth.
Pearson, D. & Parveneh, V. (1971) *Journal of the Association of Public Analysts*, **9**, 39.
Pettinati, J. D. & Swift, C. E. (1977) *Journal of the Association of Official Analytical Chemists*, **60**, 853.
Player, R. B. & Wood, R. (1980) *Journal of the Association of Public Analysts*, **18**, 29.
Polesello, A. & Giangiacomo, R. (1983) *CRC Critical Reviews in Food Science and Nutrition*, **18**, 203.
Pomeranz, Y. & Meloan, C. E. (1978) *Food Analysis: Theory and Practice*. Westport, Conn: AVI.
Pomeranz, Y., Moore, R. B. & Lai, F. S. (1977) *Journal of the American Society of Brewing Chemists*, **35**, 86.
Potter, E. F. & Long, M. C. (1966) *Journal of the Association of Official Analytical Chemists*, **49**, 905.
Prior, B. A. (1979) *Journal of Food Protection*, **42**, 668.
Pyper, J. W. (1985) *Analytica Chimica Acta*, **170**, 159.
Ramachudran, M., Grover, A., Bannerjee, B. D. & Hussain, Q. Z. (1984) *Journal of Food Science and Technology, Mysore*, **21**, 99.
Renon, J. P., Kopp, J. & Valin, C. (1985) *Journal of Food Technology*, **20**, 23.
Roeper, J. (1974) *New Zealand Journal of Dairy Science and Technology*, **9**, 49.
Ronalds, J. A. (1974) *Journal of the Science of Food and Agriculture*, **25**, 179.
Roos, J. T. H. & Price, W. J. (1970) *Journal of the Science of Food and Agriculture*, **21**, 51.
Rosenthal, I., Merin, V., Popel, G. & Bernstein, S. (1985) *Journal of the Association of Official Analytical Chemists*, **68**, 1226.
Rowan, C. A., Zajicek, O. T. & Calabrese, E. J. (1982) *Analytical Chemistry*, **54**, 149.
Sahasrabudhe, M. R. & Smallbone, B. W. (1983) *Journal of the American Oil Chemists' Society*, **60**, 801.
Salazar, A. J. & Young, L. T. (1984) *Journal of Food Science*, **49**, 209.
Salwin, H., Block, I. K. & Mitchell, J. H. (1955) *Journal of Agricultural and Food Chemistry*, **3**(7), 588.
Sandell, D. (1960) *Journal of the Science of Food and Agriculture*, **11**, 671.
Sawyer, R., Tyler, J. F. C. & Weston, R. E. (1956) *Analyst*, **81**, 362.
Scheiner, D. (1976) *Water Research*, **10**, 31.
Scholz, E. (1983) *Deutsche Lebensmittel-Rundshau*, **79**, 302.
Scott, W. J. (1957) *Advances in Food Research*, **7**, 83.
Sekerka, I. & Lechner, J. F. (1978) *Journal of the Association of Official Analytical Chemists*, **61**(6), 1493.
Selvendran, R. R. & DuPont, M. S. (1984) In *Developments in Food Analysis Techniques—3*, ed. R. D. King. London: Elsevier Applied Science.
Singh, V., Sahrawat, K. L. Jambunathan, R. & Burford, J. R. (1984) *Journal of the Science of Food and Agriculture*, **35**, 640.
Smith, P. R. (1969) Review of rapid methods for the estimation of total fat. *Scientific and Technical Survey No.* **56**. Leatherhead: BFMIRA.
Smith, P. R. (1983) *Scheme for the examination of foreign material contaminants in food*. Leatherhead: Food Research Association.

Society of Public Analysts (1943) *Analyst*, **68**, 276.
Southgate, D. A. T. (1969) *Journal of the Science of Food and Agriculture*, **20**, 331,
Southgate, D. A. T. (1971) *Journal of the Science of Food and Agriculture*, **22**, 590,
Southgate, D. A. T. (1976) *Determination of Food Carbohydrates.* London: Applied Science.
Stoldt, W. (1952) *Fette und Seifen*, **54**, 206.
Suhre, F. B., Corrao, P. A., Glover, A. & Anthony, J. M. (1982) *Journal of the Association of Official Analytical Chemists*, **65**, 1339.
Sutherland, J. P., Varnam, A. H. & Evans, M. G. (1986) *A Colour Atlas of Food Quality Control.* London: Wolfe.
Taylor, W. H. (1957) *Analyst*, **82**, 488.
Thung, S. B. (1964) *Journal of the Science of Food and Agriculture*, **15**, 236.
Tipping, L. R. H. (1982) *Meat Science*, **7**, 279.
Troller, J. (1977) *Journal of Food Science*, **43**, 86.
Troller, J. A. & Christian, J. H. B. (1978) *Water Activity and Food.* London: Academic Press.
Trowell, H., Southgate, D. A. T., Wolever, T. M. S., Leeds, A. R., Gassull, M. A. & Jenkins, D. J. A. (1976) *Lancet*, **i**, 967.
Udy, D. C. (1971) *Journal of the American Oil Chemists' Society*, **48**, 29A.
Ugrinovits, M. (1981) *Mitteilungen aus der Gebiete der Lebensmitteluntersuchung und Hygiene*, **72**, 230.
Van der Kamer, J. H. & van Ginkle, L. (1952) *Cereal Chemistry*, **29**, 239.
Van Soest, P. J. (1963) *Journal of the Association of Official Agricultural Chemists*, **46**, 825.
Van Soest, P. J. & Wine, R. H. (1967) *Journal of the Association of Official Analytical Chemists*, **50**, 50.
Vazquez, A. W. (1961) *Journal of the Association of Official Agricultural Chemists*, **44**, 754.
Vincent, K. R. & Shipe, W. F. (1976) *Journal of Food Science*, **41**, 157.
Wall, L. L., Gehrke, C. W., Neuner, T. E., Cathey, R. D. & Rexroad, P. R. (1975) *Journal of the Association of Official Analytical Chemists*, **58**, 811.
Whitehouse, K., Zarow, A. & Shay, H. (1945) *Journal of the Association of Official Agricultural Chemists*, **28**, 147.
Williams, P. C. (1976) *Laboratory Practice*, **25**, 224.
Williams, P. C., Norris, K. H., Johnsen, R. L., Standing, K., Fricioni, R., MacAffrey, D. & Mercier, R. (1978) *Cereal Foods World*, **23**, 544.
Wilson, H. N. (1951) *Analyst*, **76**, 65.
Wood, R. (1985) MAFF Information Bulletin for Public Analysts on EEC Methods of Analysis and Sampling for Foodstuffs, No. 88.
Zentner, H. (1965) *Analyst*, **90**, 698.
Zurcher, K. & Hadorn, H. (1981a) *Deutsche Lebensmittel-Rundschau*, **77**, 343.
Zurcher, K. & Hadorn, H. (1981b) *Mitteilungen aus der Gebiete der Lebensmitteluntersuchung und Hygiene*, **72**, 177.

3

General instrumental methods

The use of instruments and special techniques in the analysis of foods has increased greatly in the last 20 years and instruments are now available which offer faster and cheaper analysis with fewer laboratory staff doing less repetitive work, leading to greater job satisfaction. These newer techniques provide a range of chemical and physical information not previously obtainable. The following sections give introductory outlines to the spectroscopic and chromatographic instrumental methods and techniques now in use in food analysis laboratories. More detailed accounts are given by Willard *et al.* (1974), Bassett *et al.* (1978), Bender (1987), Ingle and Crouch (1988), Ravindranath (1989), Miller (1988) and Strobel (1989), and with particular emphasis on food analysis by Joslyn (1970), MacLeod (1973), King (1978), Stewart and Whitaker (1984), Pomeranz and Meloan (1987), Aurand *et al.* (1987) and Macrae (1988). The comprehensive ACOL (Analytical Chemistry by Open Learning) series on techniques, published by J. Wiley in 1987, is particularly useful.

Descriptions of other instruments and instrumental methods for more specific analytical purposes are to be found throughout the chapters of this book.

SPECTROANALYTICAL METHODS

Electromagnetic radiation may be considered as propagated energy in wave-like form. The range of wavelengths extends from below 1 angstrom to more than 2000 m. The regions of the electromagnetic spectrum are shown in Fig. 3.1 together

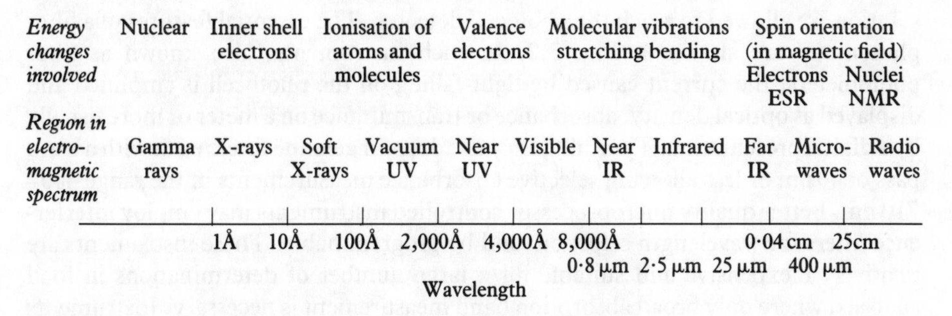

Fig. 3.1 The electromagnetic spectrum and associated energy changes.

with the nature of the atomic and molecular changes which are associated with the radiation. Spectroanalytical methods have been developed to investigate and measure these changes and effects (Ingle and Crouch, 1988). In food analysis, the methods generally used are those which measure emission or absorption of radiation of ultraviolet (UV), visible or infra-red wavelengths.

COLORIMETRY AND SPECTROPHOTOMETRY

Instrumentation

Colorimetry or colorimetric analysis is the measurement of the concentration of a coloured compound in a solution. Since the appearance of colour is due to absorption of certain wavelengths of visible light, the method should more correctly be referred to as absorptiometry or absorptiometric analysis. This avoids the use of 'colorimetry' to describe also the method used for the absolute measurement of colour of solids and liquids. The colour of a solution may be inherent in the dissolved substance, or may be formed by the addition of a reagent which reacts with a colourless compound to form a colour. Within limits, the intensity of the colour formed will be proportional to the concentration of the original coloured or colourless compound in solution and can be measured visually or instrumentally. Solutions in glass tubes or cylinders can be compared by eye. Simple visual colorimeters or colour comparators are viewing devices to hold tubes or cells and use external natural or artificial light as a light source. By comparing the intensity of colour of the transmitted light from test solutions with that from prepared standard solutions or permanent calibrated coloured glass standards, the concentration of the compound can be estimated. The human eye, however, is subject to fatigue and has low sensitivity outside the wavelength range 450–675 nm. Nevertheless, colour comparators being simple to use, rugged and cheap, are very useful in some areas of food and water analysis where approximate determinations are required. The Lovibond Comparator Disc system provides a very wide range of coloured glass standards for many common determinations, including pH, metal ions, organic constituents and quality grading of products by colour or opacity.

Photoelectric colorimeters are simple instruments in which the eye is replaced by a light-sensitive photocell, either a selenium barrier-layer cell or a photo-tube, and the light source is an incandescent lamp. Optical filters are used to isolate spectral regions so that only selected wavelengths of the light transmitted by the coloured solution are allowed to reach the photocell detector. The essential features of a filter photometer are shown in Fig. 3.2. In such instruments, also known as filter photometers, the current caused by light falling on the photocell is amplified and displayed as optical density, absorbance or transmittance on a meter or more usually by a digital presentation. Typically, they use drop-in gelatine filters each with a band pass of 40 nm or less allowing selective absorbance measurements in the range 400–710 nm. Better quality microprocessor controlled instruments may employ interference filters for wavelength selection, and be programmable. These instruments are relatively inexpensive and suitable for a large number of determinations in food analysis, where only broad absorption band measurement is necessary. Instruments of this type, fitted with flow-through cells, are used in flow injection and interrupted

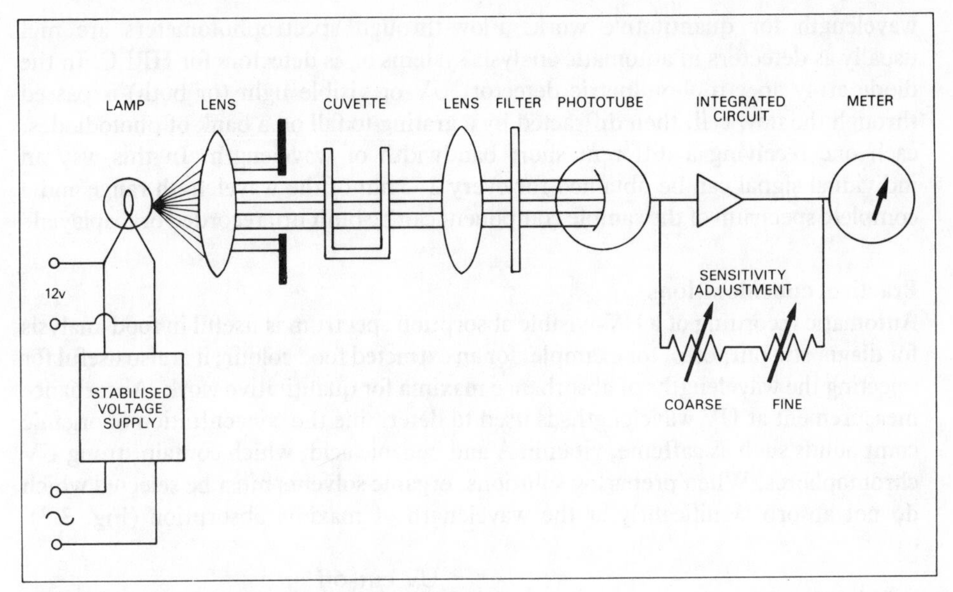

Fig. 3.2 Schematic diagram of a typical filter photometer.

flow automatic analyser systems, including read-out instruments for automated enzyme linked immunosorbent assays (ELISA).

Spectrophotometers are refined absorptiometers, designed to permit the sensitive detection of very narrow wavelength bands of light in the range from UV to near-infra-red, i.e. from 190–1000 nm. Deuterium lamps for the UV and tungsten or tungsten-halide lamps are used as sources, and adjustable slits, interference wedges, prisms and diffraction gratings give narrow (2–14 nm) band pass radiation. Photoemissive tubes, diodes or electron photomultiplier tubes are used to detect the radiation transmitted by the sample solution. Single beam instruments contain only one light path from the source of the detector. Until the 1970s those working on the null-balance principle, such as the Unicam SP500 and SP600 and the Beckmann DU spectrophotometers, were considered to provide more accurate and precise absorptiometric measurements than instruments with direct meter readouts. However, modern advances in instrument design and component technology have led to very accurate and stable direct-reading single and double channel spectrophotometers with meter or digital display and scale expansion facilities to accommodate very weakly or very strongly absorbing solutions.

Recording UV–visible spectrophotometers are designed on a double beam principle in which the samples and reference cells are irradiated by separate but identical beams of varying wavelengths. They contain either one or two photo detectors and beam chopping devices to provide continuous automatic comparison of the transmittance of the sample and reference solutions. As the selected portion of the UV–visible spectrum is automatically scanned, the alternating difference signal from the sample and reference beams is amplified and applied to the pen of the chart recorder which registers the absorbance at each wavelength, thus giving a complete absorption spectrum of the sample solution. In some instruments the signal is also presented as absorbance on a digital display, either continuously or at a selected wavelength for quantitative work. Flow-through spectrophotometers are met usually as detectors in automatic analysis systems or as detectors for HPLC. In the diode array spectrophotometric detector, UV or visible light (or both) is passed through the flow cell, then diffracted by a grating to fall on a bank of photodiodes, each one receiving a different short bandwidth of wavelength. In this way an individual signal can be obtained for every 2–5 nm of the wavelength range and a complete spectrum of the sample component can be built up, recorded or displayed.

Practical considerations

Automatic recording of a UV–visible absorption spectrum is useful in food analysis for diagnostic purposes, for example, for an extracted food colour; it is also useful for selecting the wavelengths of absorbance maxima for quantitative work. Absorbance measurement at UV wavelengths is used to determine the concentration of organic compounds such as caffeine, vitamin A and benzoic acid, which contain strong UV chromophores. When preparing solutions, organic solvents must be selected which do not absorb significantly at the wavelength of maxima absorption (Fig. 3.3).

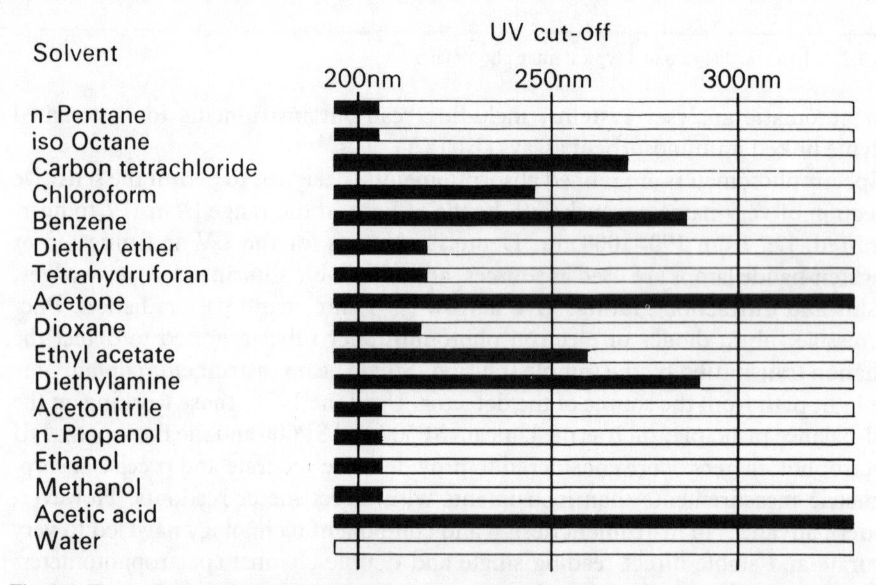

Fig. 3.3 Transmittance of solvents in the ultraviolet region. Cut-off points are at wavelengths where absorbance in a 10 mm cell is approximately unity.

Ultraviolet spectrophotometry is also used in the enzymatic determination of sugars and other food constituents. Methods for the determination of trace metals involving the measurement of coloured complexes in the visible region have been largely replaced by atomic absorption spectrophotometry, although colorimetric procedures are still used in smaller laboratories and in particular for some anions and major cations.

Several factors govern the efficiency of spectrophotometric methods. These include specificity of the colour reaction. This is usually low and clean-up procedures are necessary to remove unwanted substances which would interfere. This can be done in a number of ways, including separation by distillation, adjustment of pH, extraction of the interferent into an organic solvent with or without chemical treatment, complex formation (sequestering the interferent) or isolation of the substance to be determined by complex formation and extraction into an organic solvent. Disposable small plastic column cartridges containing HPLC column phases have revolutionised clean-up (and extraction) techniques in food analysis, and are excellent for sample solution or extract treatment to increase selectivity and sensitivity (see p. 55). The colour development should be intense but too much sensitivity for determinations where it is not needed can lead to a lack of accuracy and precision due to dilution errors. Because many solvents and trace contaminants absorb strongly in the UV, the method should give visible colours with absorbance maxima as free as possible from background absorbance. Readings should be taken at the absorbance maximum.

The coloured entity should obey the Beer–Lambert law as closely as possible over a working range of concentrations so that a reproducible calibration curve of absorbance against concentration can be drawn. The colour development should be relatively rapid. The maximum colour formed should be stable without change for a period long enough to allow for precise measurement. The procedure must give reproducible results by different workers in different laboratories.

Examples of spectrophotometric procedures are given by Bassett *et al.* (1978), and for metals by Sandell (1959) and Crosby (1977). MacLeod (1973) has reviewed the technique. Ingle and Crouch (1988) is a recent text on spectrochemical analysis and Denney and Sinclair (1987) is a self tutor.

ATOMIC EMISSION AND ABSORPTION SPECTROMETRY

Introduction

When a solution of a metallic salt is sprayed as a fine mist into a flame, the solvent in the droplets immediately evaporates, leaving non-ionised salt particles which vaporise. If no side-effects occur, the gaseous salt partially dissociates into gaseous metal atoms. If the gaseous atoms receive sufficient energy from the flame, some will be excited to a higher energy level and on reaching the cooler part of the flame will emit this excitation as radiation at a wavelength characteristic of metal atoms. Measurement of this emission is the basis of flame emission spectrophotometry (FES), which has been made almost obsolete by replacing the flame with a plasma torch of very hot (10 000 K) and energetic, highly ionised argon. This technique is known as plasma emission spectroscopy or inductively coupled plasma-optical emission spectroscopy (ICP). These expensive instruments, by using advanced electronics, optics and banks of photomultiplier detectors, have the capability of determining 20 or more elements simultaneously.

Most metallic elements exhibit atomic emission at UV wavelengths but sodium, potassium, lithium and the alkaline earths emit at visible wavelengths. Measurement of emitted visible radiation is generally known as flame photometry. A much larger number of the gaseous metallic atoms, however, remain energetically in their unexcited ground state. If radiation containing their characteristic resonance (excitation) wavelengths is passed through the flame, the atoms will selectively absorb these wavelengths. Hence the beam of light will be reduced in intensity in proportion to the number of ground state atoms in the flame. Measurement of this light absorption is the basis of atomic absorption spectrometry (AAS).

For most metallic elements AAS is more sensitive than FES but about equal to ICP. The lower flame energies required to excite sodium and potassium atoms have led to the retention of simple flame photometers for their determination. The theory and application of the techniques have been reviewed by Reynolds and Aldous (1970), Kirkbright and Sargent (1974) and Ingle and Crouch (1988). A recent text on ICP is by Moore (1989).

Instrumentation

A schematic diagram of a flame photometer is shown in Fig. 3.4 and a diagram

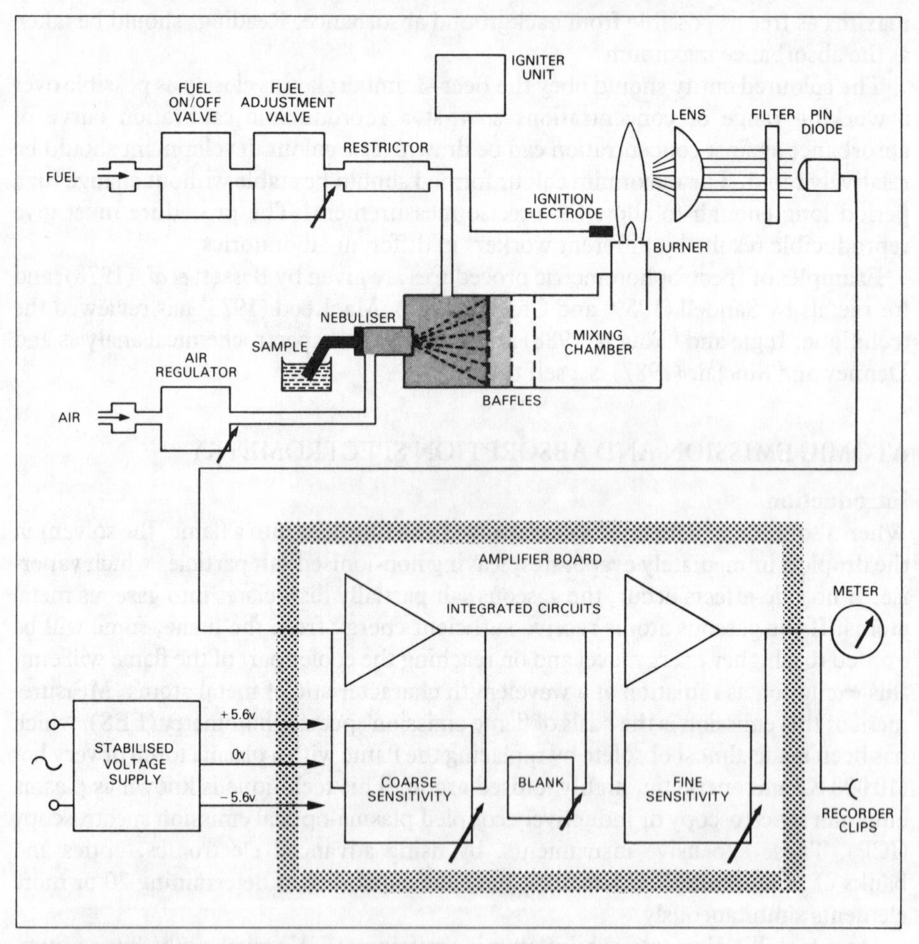

Fig. 3.4 Schematic diagram of a flame photometer.

illustrating atomic absorption spectrometry is shown in Fig. 3.5. Both require regulators, nebuliser-burners, wavelength-selective optics, a sensitive photo-detector and a read-out system. For AAS, a suitable light source for each metal being determined is also required.

The most common resonant line source for AAS is the hollow cathode lamp. This

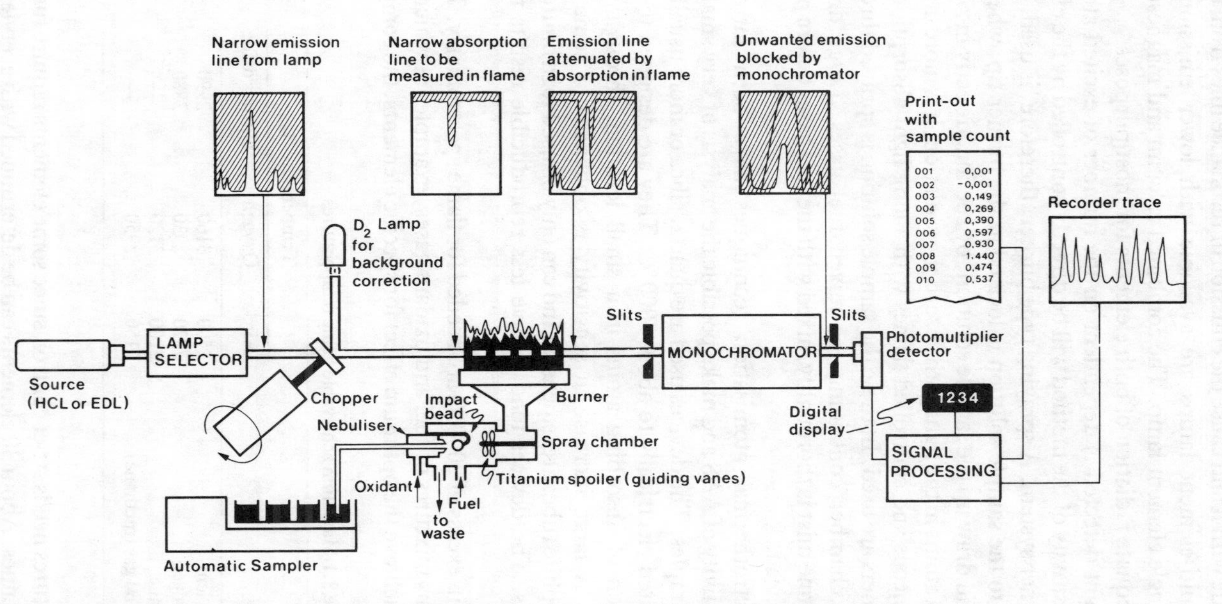

Fig. 3.5 Schematic diagram illustrating the principles of atomic absorption spectrophotometry.

is a borosilicate glass tube containing neon or argon under low pressure with a cathode in the form of a cylinder either composed of, or coated with, the specific element. Under the application of a high voltage, ions of the inert gas accelerate to the cathode, collide with it and cause the emission of resonance lines from the excited metal atoms. Multi-element lamps are available with lower emission for each element than a single element lamp. The combined calcium and magnesium lamps however emit adequate radiation of both elements for most purposes.

In FES, the light detected is dependent on the number of excited atoms in the flame and the sensitivity of the method will be directly controlled by the efficiency of the nebuliser-burner system. A spraying type burner, therefore, is used, in which a capillary leading to the sample solution is located at the burner tip, where the two fuel gases meet and draw up the sample solution, by suction, directly into the flame. In AAS, where all atoms in the light path absorb the radiation, it is more important to produce homogeneous and controlled flames with a long light absorption path and pre-mix type burners are used. In these the sample solution is sprayed through a fine nebuliser into a chamber containing the mixed fuel gases. Large droplets are removed and the fine mist is thoroughly mixed with the fuel gases before passing to a long thin burner head.

Flameless electrothermal atomisation techniques have been introduced to increase the sensitivity of AAS and make possible the analysis of very small samples, including solid samples. The devices used are quartz, silica or more usually graphite tubes or rods heated electrically to about 3000 °C. They are designed to produce a high concentration of absorbing atoms in a small location, through which the radiation passes. A large increase in sensitivity is obtained for some elements. However, the graphite tubes are expensive and can only be used for a limited number of determinations. The devices tend to give less reproducible results than flame techniques.

Temperatures in excess of 2000 °C are needed for flame spectrometry. This is met by burning various mixtures of fuel and oxidant gases, examples of which are given in Table 3.1. To achieve the optimum effect for specific elements, the composition of

Table 3.1 Maximum achievable flame temperatures

Fuel	Temperature °C		
	Air	Oxygen	Nitrous oxide
Acetylene	2500	3160	2990
Hydrogen	2320	2930	2880
Propane	2200	3120	—
Natural gas (methane)	2150	2950	—

the flame gas mixtures needs strict control, since some elements require fuel-rich and others fuel-lean flames. About 30 elements can be determined with acetylene–air but the lower temperature of propane–air is preferred for elements which are more easily converted to the atomic gaseous state. For elements which form refractory oxides, the hotter acetylene–nitrous oxide flame is essential. In all, about 70 metallic elements and metalloids can be determined by AAS.

Flame photometers can operate with natural gas–air mixtures but many of the commercial instruments use propane–air for determination of sodium and potassium in clinical analysis.

In flame photometers an interference filter is used to select a spectral band emission characteristic of the metal being determined. Barrier-layer cells are suitable detectors but in clinical instruments phototubes are also used. In AAS, it is necessary to isolate the resonance line of the radiation that has been absorbed from all other lines emanating from the light source. The wavelength bandwidth for atomic absorption is very narrow and diffraction grating monochromators are used to provide the high resolution needed. Since the requirement is to detect a reduction in signal due to absorption by the sample, increased spectral sensitivity is required and photomultiplier detectors are used. The atoms, which have absorbed the resonance lines in the flame and become excited, will also re-emit the same resonance line radiation in cooler parts of the flame or on collision. Radiation can also arise from excited molecular species containing the element. Such flame radiation can also pass through the monochromator and be detected. This is overcome by modulating the signal from the light source, either electronically or by use of a beam chopping device. The emission from the flame will in effect be continuous, not alternating, and so will be rejected by the electronic detector amplifier and read-out system.

Commercial instruments

An example of a simple flame photometer is shown in Fig. 3.4. This type usually operates on propane, butane or natural gas, has filters for Na, K, Li, Ca and Ba and a photocell detector with a calibrated meter read-out. Many more sophisticated automatic instruments designed for clinical analysis are available, with microprocessor control, digital display and print-out facilities.

There is also a wide range of AAS instruments, many of which have a turret of lamps, each of which can be selected to give multi-elemental analysis. At least 20 different models are currently available, using single or double beam optics. Double beam arrangements compensate for light source fluctuations and detector drift. To eliminate absorbance interference caused by smoke, gaseous molecules and molecular fragments, many instruments incorporate a background correction facility consisting of a deuterium UV continuum light source. Normal double beam optics have been followed by the development of dual double beam instruments, which contain two lamps and two monochromators. Many AAS instruments can also be used as flame emission spectrometers and can thus be used for the determination of sodium and potassium.

Interferences

Interference effects which govern the success of determinations by FES and AAS can be classified as spectral, chemical or physical and have been reviewed by Kirkbright and Sargent (1974). The spectral band width for atomic absorption is very narrow (about 0·005 nm), so spectral overlap from lines other than those selected for analysis is rare. Interference by radiation from the flame is overcome by detection of the modulated source radiation only. In FES, spectral interference from radiation other than that of the test element is more usual because of the simpler filter optics in flame photometers. The use of monochromators, for example, when using AAS instruments in the emission mode, will lessen this. Generally, however, spectral interference is negligible in the flame photometric determination of relatively high concentrations of sodium and potassium, such as is found in food analysis.

Chemical interference is due to factors controlling the production of ground state atoms in the flame. Stable compound formation preventing dissociation can occur in the flame owing to incomplete breakdown of the sample material or the formation of thermally stable substances within the flame. A common example is the interference of phosphate, silicate and sulphate in the determination of calcium. This can be overcome in a number of ways, including the use of a hotter flame such as acetylene–nitrous oxide to provide the required dissociation energy. The AAS determination of tin also requires such a flame to break down the refractory oxide particles formed. If a substance which preferentially combines with the interferent is added, the test element is left free to form gaseous atoms. A common example of such 'release agents' is lanthanum chloride to prevent phosphate interference in the determination of calcium, using an acetylene–air flame. Complexing agents such as EDTA and quinolin-8-ol are also added to sample solutions to prevent refractory phosphate formation in the flame. Chemical interference can also be overcome, and sensitivity increased, by selective extraction of the test solution. Many procedures now use extraction of the test element by chelation into organic solvents (see Chapter 5). The concentration of gaseous atoms in the flame can also be reduced by ionisation. This can be overcome by selecting the coolest flame possible for gaseous atom formation. Ionisation suppressants are metals such as potassium, which are easily ionised in flames and can be added to provide an excess of electrons in the flame, thereby tending to reduce ionisation of the gaseous atoms of the test element.

The viscosity, density, surface tension and volatility of test solutions will influence the efficiency of nebulisation. These so-called 'matrix' effects must be taken into account when making solutions for calibration purposes. The solutions for calibration should be similar in the amount and type of solutes present to the test sample solution. In many cases this is not possible, and calibration by standard addition must be used to overcome matrix effects.

Practical considerations

Instrument manufacturers provide dossiers of analytical methods for various metallic elements and metalloids. Flame atomic absorption spectroscopy data for some of the common elements are shown in Table 3.2. Some elements, such as arsenic, antimony and selenium are difficult to determine by normal AAS, but can be analysed successfully after conversion to their volatile hydrides (Flanjak, 1978; Rees, 1978). The determination of mercury (see Chapter 5) requires special procedures based on either electrothermal atomisation or the cold vapour method of Hatch and Ott (1968). Crosby (1977) has published a comprehensive review of AAS methods for metals and metalloids in food. A number of procedures are available for the AAS determination of several elements in a single food digest (see Chapter 5).

CHROMATOGRAPHIC METHODS

Chromatography is a dynamic physical process in which molecular components in a mixture are separated because of their different affinities for two substances referred to as phases, one being fixed or stationary and the other mobile. The mobile phase can be either liquid, giving rise to a group of techniques known as liquid chromato-

Table 3.2 Atomic absorption spectrometry data for common elements. A–Ac = air–acetylene, A–H = air–hydrogen, A–P = air–propane, N–Ac = nitrous oxide–acetylene

Element	Main resonance line (nm)	Flame		Working range (ng/ml)
		Fuel rich	Fuel lean	
Ag	328·1		A–Ac	1–5
Al	309·3	N–Ac		40–200
As	193·7	A–H		50–200
B	249·8	N–Ac		400–600
Ba	553·6	N–Ac		10–40
Be	234·9	N–Ac		1–5
Bi	223·1		A–Ac	10–40
Ca	422·7	N–Ac		1–4
Cd	228·8		A–Ac	0·5–2
Co	240·7		A–Ac	3–12
Cr	357·9	A–Ac		2–8
Cs	852·1		A–P	5–20
Cu	324·7		A–Ac	2–8
Fe	248·3		A–Ac	2·5–10
Ga	294·4		A–Ac	50–200
Ge	265·2	N–Ac		70–280
Hg	253·7		A–Ac	100–400
In	303·9		A–Ac	15–60
Ir	208·9	A–Ac		40–160
K	766·5		A–P	0·5–2
Li	670·8		A–P	1–4
Mg	285·2		A–Ac	0·1–0·4
Mn	279·5		A–Ac	1–4
Mo	313·3	N–Ac		15–60
Na	589·0		A–P	0·15–0·60
Ni	232·0		A–Ac	3–12
Os	290·9	N–Ac		50–200
Pb	217·0		A–Ac	5–20
Pd	244·8		A–Ac	4–16
Pt	265·9		A–Ac	50–200
Rb	780·0		A–P	1–10
Rh	343·5		A–Ac	5–25
Ru	349·9		A–Ac	30–120
Sb	217·6		A–Ac	10–40
Sc	391·2	N–Ac		15–60
Se	196·0	A–H		20–90
Si	251·6	N–Ac		70–280
Sn	224·6	N–Ac		15–50
Sr	460·7		N–Ac	2–10
Te	214·3		A–Ac	10–40
Ti	364·3	N–Ac		60–240
Tl	276·8		A–Ac	10–50
V	318·5	N–Ac		40–120
W	255·1	N–Ac		250–1000
Y	410·2	N–Ac		200–800
Zn	213·9		A–Ac	0·4–1·6

graphy, or it can be gaseous, in which case the basic technique is known as gas chromatography. Both techniques are of immense value to the food analyst.

LIQUID CHROMATOGRAPHY

Liquid chromatography is a separation technique in which compounds are partitioned between two phases: a stationary phase of large surface area and a mobile liquid phase which flows over the stationary phase. The stationary phase may be an active solid or an immobile liquid. Depending on the selection of mobile phase, particle type used for the stationary phase and the operating conditions, components dissolved in the mobile phase will travel at different rates, resulting in a separation.

Liquid chromatography can be divided into four main types. Adsorption chromatography (liquid/solid or normal phase chromatography) is based primarily on differences in the relative affinity of compounds for the solid adsorbent used as the stationary phase. The separations obtained are determined almost entirely by polar interactions, with the stationary phase being more polar than the mobile phase. Partition chromatography (or liquid/liquid chromatography) is based on the relative solubility characteristics of solutes between the mobile phase and a stationary non-polar liquid phase. The liquid phase is impregnated or, in the case of reversed phase chromatography, chemically bonded to the inert silica support. Ion exchange chromatography is based on the affinity of ions in solution for sites of opposite polarity in the stationary phase. Size exclusion chromatography, which is also known as gel permeation, gel filtration or molecular sieving, is based on the ability of controlled porosity materials to separate the components of a mixture according to the size and shape of the molecules. Snyder and Kirkland (1979) provide an authoritative and comprehensive text on liquid chromatography; see also Miller (1988) and Ravindranath (1989).

The principles of liquid chromatography are used in four analytical techniques, paper chromatography, thin-layer chromatography, column chromatography and high performance liquid chromatography.

Paper and thin-layer chromatography

Paper chromatography is a form of combined partition and adsorption chromatography in which the stationary phase is the absorbed and adsorbed water present in the paper and the support is the paper itself. The mobile phase is a solution consisting of a solvent or a mixture of several liquids, containing water. A spot of the sample extract is dried on the paper. To prevent evaporation, the paper is hung inside a chamber in such a manner that the spot can be irrigated with the mobile phase either downwards by gravity or upwards or horizontally by capillary action. When the mobile phase has saturated the paper over the predetermined distance covering the chromatographic separation, the paper is removed from the chamber and the chromatogram developed by the use of sprayed-on visualising agents. In the case of coloured substances such as food dyes, the separation may be evident without assistance from such agents. Non-polar substances can also be separated by coating the paper with a non-polar compound and using polar solvents as the mobile phase. Paper chromatography is a cheap but generally slow technique, with limitations on the use of spray reagents, some of which may readily attack and destroy paper. The spots tend to be diffuse.

Chromatography on thin sheets of adsorbent is improved by using powdered material adhering to glass. This enables other support materials with different properties to be used, for example silica and Kieselguhr. Apparatus for coating TLC glass plates is available but commercially prepared precoated plates on glass, rigid plastic films and thin aluminium foil are now widely used. More recently, reversed phase coated TLC plates have become available, usually octadecylsilane (C_{18})

bonded to silica. These allow separation and selectivity control on non-polar compounds by modification of the mobile phase composition (as in RP-HPLC). Thin-layer chromatography is far more sensitive, adaptable and rapid than paper chromatography and chromatographic separations are more efficient and can be made semi-quantitative. Plates containing no organic matter can be prepared or purchased allowing the visualisation of spots by charring with sulphuric acid. Visualisation under UV light after spraying the plate with a fluorescing reagent can be very effective when the separated compounds quench or modify the fluorescence. Thin-layer chromatography is a useful method for isolating and identifying minor components of foods. The separated component can be scraped from the plate or cut from the TLC sheet, extracted from the layer material and further analysed, for example by spectrophotometry or GLC. A text on paper and thin-layer chromatography is available (Gasparic and Churacek, 1978).

The technology used in the preparation of microparticulate column packings for HPLC has been applied more recently to TLC with the introduction of so-called high performance thin-layer chromatography (HPTLC). Precoated HPTLC plates are available which show much higher resolution and sensitivity than normal plates. These plates require smaller loadings of sample and more tests per plate can be made with short development runs (30–60 mm), leading to faster analysis time.

A range of densitometers is available for scanning TLC plates to give quantitative information on the separated substances. They are based on four modes of detection: transmission of light, reflection, fluorescence or quenching of fluorescence. Quantitative TLC using densitometry has been reviewed by Shellard (1968). The Iatroscan Analyser is an instrument which combines TLC and flame ionisation detection of the separated compounds. Thin-layer chromatography is carried out on quartz rods covered with a sintered silica gel or alumina layer. The developed rods, after drying, are programmed sequentially through the flame of the FID at preselected speeds. The organic components are pyrolysed and generate an electronic signal which is amplified, recorded and integrated (Sipos and Ackman, 1978).

Column chromatography

Column chromatography was the original form of liquid chromatography first carried out by Tswett in 1906. The stationary phase is usually enclosed in a glass column from 5 to 30 mm in diameter. Many kinds of packaging materials are used, ranging from alumina, silica gel, diatomaceous earths to synthetic resins and polysaccharide substances. All types of liquid chromatography are used. The mobile phase is usually allowed to percolate through the column under gravity. As a technique for chromatographic separation of mixtures of substances, it is little used in food analysis. However, now available is a complete and automatic chromatographic system, designed for biochemical analysis and known as 'fast protein liquid chromatography'. These instruments employ a range of phases in 5 or 10 cm plastic columns operating under moderate pressure.

The use of column chromatography for cleaning up test solutions or for extracting the desired substances from solutions has been revolutionised by the introduction of short, plastic, disposable mini-column cartridges, prepacked with a range of separation phases in relatively large particle size, e.g. Bond-Elut and Sep-Pak cartridges. The test, wash or extraction solutions are drawn rapidly through the cartridges under low vacuum from a water pump. Most clean-up methods, such as liquid–liquid extraction, TLC and glass column chromatography used prior to spectrophotometry, GLC, HPLC, etc., can now be efficiently carried out using these cartridges.

High performance liquid chromatography

By using support materials, small and uniform in size, large surface areas can be obtained leading to highly efficient column chromatographic separations, but columns packed with such materials give high resistance to liquid flow, so that high pressures are required to force mobile phases through them. In the last 15 years, scientific instrument manufacturers have made available suitable columns and column packings, high pressure pumping systems with pulseless delivery and sensitive detectors. These units can be purchased separately or as complete systems. The units are generally mutually compatible, irrespective of manufacture. In consequence, HPLC has become a very widely used technique, complementary to gas chromatography. It is especially of value for the determination of non-volatile compounds, permitting the rapid estimation of additives, contaminants and natural components of foodstuffs. The theory of HPLC is described in the reference books on chromatography given on p. 54 and also by Macrae (1988) and Lindsay (1987). Figure 3.6 shows a schematic diagram of a basic HPLC system.

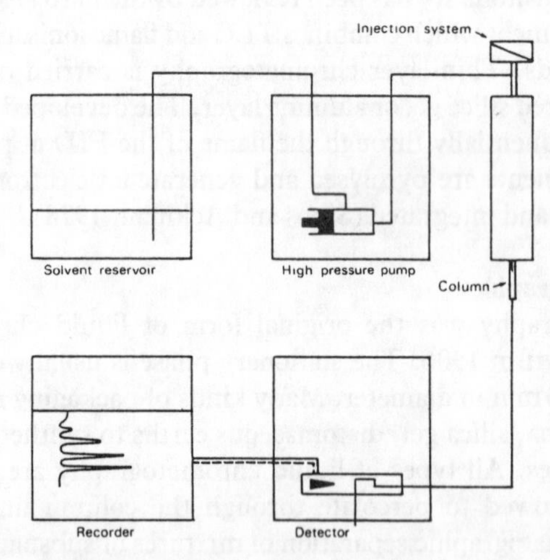

Fig. 3.6 Schematic diagram of a system for high performance liquid chromatography.

Instrumentation

The detectors used in HPLC must be able to detect very small changes in the concentration of organic solutes in the mobile liquid phase, and so must be relatively insensitive to the mobile phase. There are a number of different types which meet this requirement.

Ultraviolet absorption is the most useful detection system and suitable photometers with fixed wavelengths, usually 254 nm or 280 nm, or spectrophotometers of variable wavelength are available, the most sophisticated being multi-diode array detectors, which rapidly sweep the UV or UV–visible absorbance spectrum of eluting substances. Because of the need to detect very small quantities of compounds eluting with low volume separations, these detectors are fitted with flow-through cells of small volume (8–20 μl). Some spectrophotometers can measure absorption at wavelengths less than 200 nm, at which most organic compounds absorb strongly. This may be useful analytically but requires expensive ultra-pure mobile phase solvents. Many organic compounds absorb at wavelengths between 250–300 nm; others with limited or no absorbance in this region can have an appropriate chromophore introduced by chemical derivatisation.

Spectrofluorimeters are also used as HPLC detectors. Fluorescence improves sensitivity and, by selection of excitation and emission wavelengths, the technique can be selective for types of compounds. Non-fluorescing compounds can be detected with high sensitivity by pre-column conversion to fluorescing derivatives (Lawrence and Frei, 1976).

Sensitive refractometers provide a general detection system for fractional percentage composition or greater by measuring, in flow-through cells, the difference in refractive index (RI) between the pure mobile solvent phase and the column eluant containing the chromatographically separated compounds. The difference is amplified electronically and recorded on a chromatogram. For good stability and highest sensitivity precise temperature control is necessary. Refractive index detection is not conducive to varying the composition of the mobile phase (gradient elution, solvent programming) during a chromatographic run.

The mass detector is also a general detector but can be used with solvent programming because the mobile phase is evaporated by hot air, leaving only the separated non-volatile solute particles to be detected by their light scattering. This detector has about the same sensitivity as an RI detector but has a limited range of linearity and suffers somewhat from instability.

Conductivity detection is used in ion chromatography equipment to monitor traces of weak and strong anions and cations. Very sensitive electrochemical detectors are available for the measurement of traces of organic compounds capable of being oxidised or reduced at an electrode. Such substances in food analysis usually contain a UV chromophore and can be determined with greater ease and sufficient sensitivity by UV detection.

Many different types of column and column packing continue to be developed, mainly of three basic types, i.e. for adsorption chromatography, reverse phase and bonded phase chromatography. Most are based on fully porous micro-particulate silica gel, spherical or irregular in shape but finely graded into 10, 5 and sometimes 3 μm diameters. Adsorption chromatography in HPLC is most useful for the separation of non-polar compounds, mixtures of isomers and mixtures of compounds with different functional groups. Reverse phase packings are composed of silica with hydrocarbon chains, usually as octasilane (C_8) or octadecylsilane (C_{18}), chemically bonded to the surface by silylation; they therefore require more polar mobile phases than do adsorbent packings. Various blends of polar solvents containing water and alcohols can be used, giving greater analytical flexibility. Reverse phases tend to remove organic substances from food extracts, to form deposits and blockages at the top of the column, circumvented by employing a short disposable guard column containing the same phase as the analytical column. Adsorbent packings do not retain the contamination, which passes through the column and may interfere with detection. Bonded phases, like reverse phase materials are silylised silica but have selected functional groups chemically attached for specific applications. Nitrile and amino bonded phases are widely used. These seem to exhibit both adsorption and partition effects. Phases with more polar functional groups are available, designed for various ion exchange properties, although ions can also be determined by ion-pair chromatography using reversed phase columns and UV detection with a counter-ion compound added to the mobile phase. For all types of micro-particulate packings, stainless steel tubing, 100–300 mm in length with 3·0, 4·6 or preferably 5·0 mm internal diameter, is widely used. Columns may be packed in the laboratory but the purchase of commercially prepacked columns is now more prevalent, especially with the availability of easy-to-assemble cartridges containing plastic or glass packed columns. More recent developments have been microbore columns and mini-columns, 30–50 mm long, containing 3 μm packings for fast chromatography.

For good separations and symmetrical peaks on the chromatogram, the sample solution must be applied close to the head of the column packing without disturbing it and centrally away from the column walls, in order to minimise wall diffusion effects. Injection can be by septum, valve or sample loop arrangements. A useful feature of HPLC is that the flow of mobile phase can be stopped during injection, and even if necessary during a chromatographic run, without affecting performance. Efficient injection fittings for syringe use can be constructed or purchased but sample loop devices are preferred, since they are convenient, easy to use and highly reproducible.

The pumps used in HPLC are based on a reciprocating piston which displaces a fixed volume of solvent per unit time. Pulsation-free flow is obtained by changing the motor speed in the pumping cycle. Single cylinder pumps are simple in design and easier to maintain. Micro-particulate HPLC columns are commonly operated at flow rates of 16–60 ml/h. Pressures of 3450–34 500 kPa (500–5000 p.s.i.) are usually generated in obtaining such flows. The pressure is dependent on the permeability of the columns, the viscosity of the mobile phase, the particle diameter and the column length.

Practical considerations

The solvent system to be used with a particular column is chosen to provide adequate resolution with convenient retention times. The choice may be aided by thin-layer chromatography data if this is available. High solubility is important, both in the chromatographic process and in ensuring that solute is not precipitated at the column head after injection. Chromatography with a single mobile solvent is unsuitable when compounds of widely different retention times are to be separated. The use of a gradient solvent system in which the composition is gradually changed from one suitable for components with short retention time to one causing rapid elution of components with long retention times, results in a shorter analysis time and improved sensitivity for the slow running components. However, it has the disadvantage that time is required to recondition the column back to the initial mobile phase composition and gradient elution cannot normally be used with some detectors, notably the refractometer. Some changes in baseline will be experienced in all detectors unless they are suitably compensated.

Macrae (1988) has prepared a very complete working manual for HPLC in food analysis, with extensive referencing.

GAS CHROMATOGRAPHY

Introduction

Gas chromatography is a process in which a mixture is separated into its constituent parts by a moving gas phase passing over a fixed solid sorbent. Many of the controlling parameters of liquid chromatography also apply to gas chromatography. Gas–liquid chromatography (GLC) achieves separation by partitioning between a mobile gas phase and a stationary liquid held on a solid support or the wall of the column. Separations by gas–solid chromatography (GSC) are achieved using an immobile solid material with adsorbent, molecular sieving or partitioning properties. The technique of gas chromatography has been described by many authors, for example Trenchant (1969), Littlewood (1970) and Ambrose (1971). The ACOL self-learning text is by Willett (1987).

Instrumentation

A schematic diagram of a gas chromatograph is shown in Fig. 3.7. It consists of five main parts: a supply of carrier gas (the mobile phase) with pressure regulators and flow meters; the injection system; the chromatographic column and oven; the detector and its electronic amplifiers; and the chart recorder and other data presentation equipment.

Carrier gas. The carrier gas, usually provided from high-pressure cylinders, may be nitrogen, helium, argon or hydrogen, depending on the type of separation and detection in operation. Helium is the best for most purposes but oxygen-free nitrogen is probably the most widely used.

Injection system. For optimum chromatographic separations, the sample must be applied to the top of the column as vapour in the smallest possible volume, without decomposition. The normal procedure for liquids or solutions is by injection, using a micro-syringe through a self-sealing septum into a heated metal block, where the sample is rapidly vaporised and passed in heated carrier gas to the head of the column. For good results, injection techniques need to be precisely reproducible, hence gas chromatography demands a large degree of experimental art. Injections of gas samples can also be made by syringe but the most accurate and precise procedure is to use commercially available sample loops and switching valves. Highly reproducible automated headspace injectors can be used to sample vapours, e.g. above samples or solutions in vials. In capillary GLC, the columns have very low sample loading capacities, far less than can be applied by syringe, so that special devices may be necessary to pass only a constant fraction of the sample solution vapour to the column. For example, with a 0·5 μl injection, a 50:1 splitter will pass one part (0·01 μl) to the column with the rest going to waste. For the greatest

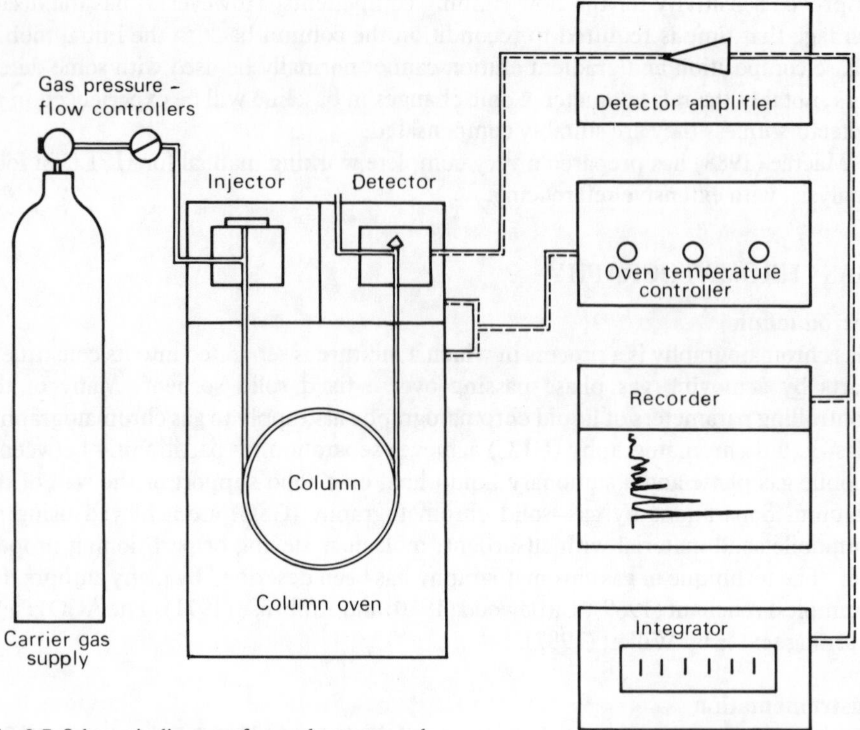

Fig. 3.7 Schematic diagram of a gas chromatograph.

sensitivity, e.g. in trace analysis, there are split-less and on-column injector devices for use with capillary columns. The automatic injection of headspace volatiles and vapours above samples, either solid, liquid or in solution, is a useful and 'clean' technique for certain applications (Charlambous, 1978; Kolb, 1980; Ioffe and Vitenberg, 1984).

Columns. For packed-column gas chromatography in food analysis the column tubing is composed of 3 mm external diameter stainless steel or 2–6 mm internal diameter borosilicate glass. In capillary GLC, very fine bore tubing (0·2–0·5 mm i.d.) in glass or fused quartz (which is flexible yet very durable) is used. The inert support for GLC packed column packings must be a fine granular material of regular particle size and large surface area. Absence of fine powdery materials is essential, otherwise the passage of carrier gas will be impeded, resulting in lengthy analyses. The material must also be strong and not easily fractured into small pieces by mechanical handling. It must be heat stable. The substance most commonly used is diatomaceous earth (e.g. Celite) which can hold 20 per cent loading of stationary liquid phase without becoming sticky or fragile. The most suitable particle size for general use is 80/100 mesh. The choice of stationary liquid phases for GLC is large but for a particular separation the selection of the correct phase may be crucial. The phases can be classified into four main types depending on their ability to form hydrogen bonds:

1. Non-polar (hydrocarbon type) which have little if any hydrogen bond forming

properties, e.g. high melting-point greases, squalane, phenylmethylpolysiloxanes, carboranes.
2. Liquids composed of molecules containing polar or polarisable groups attached to a non-polar skeleton, e.g. dinonyl/phthalate, phenyl silicone, di-(2-ethylhexyl) sebacate.
3. Moderately polar liquid compounds, e.g. polyethyleneglycol adipate, ethylene glycol succinate, polyglycols.
4. Highly polar liquids which easily form hydrogen bonds throughout their structure, e.g. cyanopropyl-phenyl and cyano-ethyl polysiloxanes.

The general rule to be followed when selecting a phase is that like dissolves like; thus polar phases are best for the separation of polar substances. The trade literature of the chromatographic materials suppliers often gives full information on liquid phases, including the loading levels normally used on the inert support. The support is coated by adding to it the correct amount of liquid phase dissolved in a suitable volatile solvent. After gentle mixing, the solvent is removed by evaporation. Many phases for capillary columns (wall coated) are now chemically bonded to virtually eliminate column bleed during use. Liquid phases, coated packings or packed columns for particular instruments and analytical separations are all available commercially.

Packings for GSC depend mainly on surface properties or pore structure for separation and are principally used for the chromatographic analysis of gas mixtures (Cowper and DeRose, 1985) and relatively small molecules. There are three basic types:

1. Inorganic adsorbents based on alumina and silica gel.
2. Molecular sieves based on zeolites which have been dried to leave a network of empty holes the size of which determines the sieving properties. At low temperatures these zeolites strongly retain certain molecules; they therefore find wide use as filters for removing traces of contaminants in gases.
3. Rigid polymer beads. These are microporous aromatic hydrocarbon materials. They behave chromatographically like no other phase and give different compound eluting orders.

Column oven. Except for the analysis of permanent gases, columns are operated at elevated temperatures, and are therefore contained in thermostatically controlled ovens. For the chromatographic separation of a mixture of substances of widely differing boiling points, the temperature of the oven can be programmed to obtain better separations in a shorter time. Increasing the column and carrier gas temperature can have marked effects on the detector so that controls maintaining constant carrier gas flow together with careful selection of column packings are necessary. Because of the nature of capillary columns, they should be operated at a maximum temperature 20–30 °C less than that specified for packed column containing the same stationary phase.

Detectors. The purpose of the detector is to sense the passage of eluting substances in the carrier gas as they pass out of the column. All detectors are designed to produce an electric signal as a result of a physical effect caused by the eluting substance. The signal is then amplified and applied to a recorder to give a trace of the eluting substances (i.e. a chromatogram), or to electronic integrating and data processing equipment. A number of different types of detector are used in food analysis. The most widely used is the flame ionisation detector (FID) which is highly sensitive and has a long linear response range. In this detector the carrier gas is fed into a hydrogen gas stream, which is burned in air. The eluted substances are burned in the flame producing ionic fragments and electrons, which are collected at an electrode. These produce an electric current proportional to the number of carbon atoms in the sample component and for this reason the detector has sometimes been called a carbon counter. The electron-capture detector (ECD) is based on electron absorption by compounds having an affinity for free electrons such as those containing electronegative groups or halogens. The source of the electrons can be a piece of tritium foil or, more usually, the radionuclide ^{63}Ni. The removal of electrons from the baseline standing current by the eluting substances is detected and amplified. The linear response range is short but the detector is extremely sensitive to halogen-containing compounds and is therefore a very selective detector for certain pesticide residues in foodstuffs. The thermal conductivity detector, or katharometer, consists of a heated filament or thermistor to sense changes in the thermal conductivity of the carrier gas caused by the presence of the eluting substances. Helium and hydrogen are the best carrier gases because of their high thermal conductivities. As the thermal conductivity changes, the resistance of the filament changes. Commonly, two or more filaments are used and wired into a Wheatstone bridge circuit, so that as the resistance of some of the filaments change, the bridge circuit produces an electrical output. The katharometer is a simple but only moderately sensitive detector. Unlike FID, it is non-destructive and can therefore be used in preparative and fraction-collecting work. It responds to all inorganic and organic compounds and is probably used nowadays mostly in gas analysis and process control. In pesticide residue work, specific detectors such as the flame photometric detector for phosphorus and sulphur, and the alkali metal flame ionisation detector for nitrogen compounds, are also used. Other detectors for specific purposes are reviewed by Dickes (1979). The mass spectrometer is the most sophisticated detector for gas chromatography and the combination of the two instruments (GC/MS) has found wide use, for example in food flavour research (Land and Nursten, 1978) and the determination of organic contaminants (Gilbert *et al.*, 1985). The ion trap detector is a less expensive special design of mass spectrometer comparable in size to a gas chromatograph. Practical aspects of GC/MS are covered by Message (1984).

Practical considerations

Following injection, a signal from the detector indicates that a substance is passing out of the column after a certain time, dictated by given experimental conditions. Identification of the substance is achieved by reference to the time taken between injections and elution compared to that of known substances (relative retention time). However, gas chromatography alone, even with a mass spectrometer detector, cannot absolutely identify a substance unless the substance is chromatographed on at least two columns of different characteristics.

Gas chromatography is essentially a separation technique but the efficiency of columns and detectors makes possible the accurate and precise measurement of the amounts of constituents separated from a mixture. If the detector response is linear with respect to the concentration of a constituent, the peak area (or height in the case of sharp peaks given by rapidly eluted substances, e.g. in capillary gas chromatography) is proportional to the amount of substances in the injected sample solution. Peak areas are measured by automatic electronic integration. All modern integrators are based on microprocessors and have the adaptability and capability for incorporating complex computing and data processing programmes. Automatic integration can be used for the measurement of the total area of all peaks detected, thereby enabling the calculation of the percentage of one constituent: this is known as normalisation. The use of an internal standard overcomes injection variation and detector response changes due to fluctuation in carrier gas flow and column and detector temperatures. This involves the addition of a reference substance, not already present, to the sample before injection; the substance should have a retention time similar to that of the components of interest but should be clearly separated from it and other peaks. A number of prepared solutions containing concentrations of the component of interest and a fixed amount of the internal standard are chromatographed and a calibration curve is drawn of the concentration of the component against the ratio of the peak heights (or areas) of the component and the internal standard. The same amount of internal standard is added to the unknown test solution and the ratio of peak heights (or areas) again observed. Reference to the calibration graph enables the concentration of the component in the samples to be deduced. The internal standard procedure is also used with complex mixtures in which normalisation of the peaks is not possible.

Modern HPLC has supplanted many analyses previously carried out by gas chromatography but for the analysis of oils and fats, taints, flavours, pesticides and other complex mixtures of volatile components, GLC remains a most important technique in food analysis.

REFERENCES

Ambrose, D. (1971) *Gas Chromatography*. London: Butterworth.

Aurand, L. W., Woods, E. & Wells, M. R. (1987) *Food Composition and Analysis*. New York: Van Nostrand Reinhold.

Bassett, J., Denny, R. C., Jeffrey, G. H. & Mendham, J. (1978) *Vogel's Textbook of Quantitative Inorganic Analysis*. London: Longman.

Bender, G. T. (1987) *Principles of Chemical Instrumentation*. London: W. B. Saunders.

Charlambous, G. (ed.) (1978) *Analysis of Foods and Beverages: Headspace Techniques*. New York: Academic Press.

Cowper, C. J. & DeRose, A. J. (1985) *The Analysis of Gases by Chromatography*. Oxford: Pergamon.

Crosby, N. T. (1977) *Analyst*, **102**, 225.

Denney, R. C. & Sinclair, R. (1987) *Visible and Ultra Violet Spectroscopy*. ACOL series, ed. D. Kearly. Chichester: J. Wiley.

Dickes, G. J. (1979) *Talanta*, **26**, 1065.

Flanjack, J. (1978) *Journal of the Association of Official Analytical Chemists*, **61**, 1299.

Gasparic, J. & Churacek, J. (1978) *Laboratory Handbook of Paper and Thin-Layer Chromatography*. Chichester: Ellis Horwood.

Gilbert, J., Startin, J. R. & Crews, C. (1985) *Journal of the Association of Public Analysts*, **23**, 119.

Hatch, W. R. & Ott, W. L. (1968) *Analytical Chemistry*, **40**, 2085.

Ingle, J. D. & Crouch, S. R. (1988) *Spectrochemical Analysis*. London: Prentice-Hall International.

Ioffe, B. V. & Vitenberg, A. G. (1984) *Head-space Analysis and Related Methods in Gas Chromatography*. New York: Wiley-Interscience.

Joslyn, M. A. (ed.) (1970) *Methods in Food Analysis*. London: Academic Press.

King, R. D. (ed.) (1978) *Developments in Food Analysis Techniques—1*. London: Applied Science.

Kirkbright, G. F. & Sargent, M. (1974) *Atomic Absorption and Fluorescence Spectroscopy*. London: Academic Press.
Kolb, B. (ed.) (1980) *Applied Headspace Gas Chromatography*. London: Heyden.
Land, D. G. & Nursten, H. E. (eds) (1978) *Progress in Flavour Research*. London: Applied Science.
Lawrence, J. F. & Frei, R. W. (1976) *Chemical Derivatisation in Liquid Chromatography*. London: Elsevier.
Lindsay, S. (1987) *High Performance Liquid Chromatography*. ACOL series, ed. D. Kearly. Chichester: J. Wiley.
Littlewood, A. B. (1970) *Gas Chromatography: Principles, Techniques and Applications*. New York: Academic Press.
MacLeod, A. J. (1973) *Instrumental Methods of Food Analysis*. London: Paul Elek.
Macrae, R. (ed.) (1988) *HPLC in Food Analysis*. London: Academic Press.
Message, G. M. (1984) *Practical Aspects of Gas Chromatography—Mass Spectrometry*. New York: Wiley-Interscience.
Miller, J. M. (1988) *Chromatography*. Chichester: J. Wiley.
Moore, G. L. (1989) *Introduction to Inductively Coupled Plasma Atomic Emission Spectroscopy*. London: Elsevier.
Pomeranz, Y. and Meloan, C. E. (1987) *Food Analysis*. New York: Van Nostrand Reinhold.
Ravindranath, B. (1989) *Principles and Practice of Chromatography*. Chichester: Ellis Horwood.
Rees, D. I. (1978) *Journal of the Association of Public Analysts*, **16**, 71.
Reynolds, R. J. & Aldous, K. (1970) *Atomic Absorption Spectroscopy: A Practical Guide*. London: Griffin.
Sandell, E. B. (1959) *Colorimetric Determination of Traces of Metals*. London: Interscience.
Shellard, E. J. (1968) *Quantitative Paper and Thin-Layer Chromatography*. London: Academic Press.
Sipos, J. C. & Ackman, R. G. (1978) *Journal of Chromatographic Science*, **16**, 443.
Snyder, L. R. and Kirkland, J. J. (1979) *Introduction to Modern Liquid Chromatography*, 2nd edn. New York: Wiley-Interscience.
Stewart, K. K. and Whitaker, J. R. (eds) (1984) *Modern Methods of Food Analysis*. Westport, Connecticut: AVI.
Strobel, H. A. (1989) *Chemical Instrumentation*. Chichester: J. Wiley.
Trenchant, J. (ed.) (1969) *Practical Manual of Gas Chromatography*. London: Elsevier.
Willard, A. J., Merritt, L. L. & Dean, J. A. (1974) *Instrumental Methods of Analysis*. London: Van Nostrand.
Willett, J. E. (1987) *Gas Chromatography*. ACOL series, ed. D. Kearly. Chichester: J. Wiley.

4

Food additives

Introduction

Production, preparation, processing and distribution of food is an increasingly complex industry which demands efficient use of resources at all stages. Fertilisers, pesticides, fungicides and growth regulators assist in production and storage of agricultural raw materials. Manufacture, processing, distribution and storage of food products is assisted by the use of chemical additives, processing aids and packaging materials. Thus, the finished product may contain variable amounts of non-nutritive materials which may be present by accident or design. Food contaminants are distinguished from food additives by the existence or absence of an intention that the substance should be present when the food is offered for sale or consumed. Additives are utilised to achieve a specified technological result in the final food. As indicated in Chapter 1, the use of additives in food is controlled by regulations made under the Food Act 1984. The powers of the Minister under Section 4 are framed in the terms of 'requiring, prohibiting or regulating the addition of any specified substance' and 'requiring, prohibiting or regulating the use of any process or treatment'. Thus, specific regulations define a class of additives by function and substances are classified as 'permitted' by inclusion in the schedules appended to the regulations. The term 'additive' is defined under the Food Labelling Regulations (1984) as:

any substance, not commonly regarded or used as food, which is added to, or used as food, which is added to, or used in or on, food at any stage to affect its keeping qualities, texture, consistency, appearance, odour, alkalinity or acidity, or to serve any other technological function in relation to food, and includes processing aids in so far as they are added to, or used in or on, food as aforesaid, but does not include—

(a) vitamins, minerals or other nutrients in so far as they are used solely for the purpose of fortifying or enriching food or of restoring the constituents of food,
(b) herbs or spices when used as seasoning,
(c) hops,
(d) salt,
(e) yeast or yeast extracts,
(f) the total products of any hydrolysis or autolysis of food proteins,
(g) starter cultures,
(h) malt or malt extract,

(i) any substance which is present in food solely as a result of its addition to animal, bird or fish feedingstuffs or its use in a process or treatment carried out in crop husbandry, animal husbandry, veterinary medicine or storage (including any pesticide, fumigant, sprout depressant or veterinary medicine), or

(j) air or water.

The same regulations define 'additive regulations' as:

Colouring Matter in Food Regulations 1973,
Antioxidants in Food Regulations 1978,
Preservatives in Food Regulations 1979,
Emulsifiers and Stabilisers in Food Regulations 1980,
Miscellaneous Additives in Food Regulations 1980.

Whilst the above list is not all embracing with respect to the functionality of additives, the Miscellaneous Additive Regulations include a wide range of materials whose functions are defined out of other specific regulations; for example, see the Sweeteners in Food Regulations 1983 which also refer to bleaching agents, improving agents and to solvents which themselves are the subject of separate regulations.

Assessment of the level of use of permitted additives and the need for new additives is the function of the Food Advisory Committee (formerly the Food Additives and Contaminants Committee had these functions). The committee also considers the restriction, so far as is practicable, of the use of substances of no nutritional value as foods or ingredients of foods. A similar function is carried out on behalf of the European Economic Community (EC) by an independent committee, the Scientific Committee for Food (SCF). One difference of philosophy between the EC and the UK in respect of control of use of additives is that EC additive directives list the approved substances whilst UK regulations list the substances and in some cases the foods in which they may be used, together with the permitted level of usage. It is not correct to say that all additives have no nutritional value, since a number are metabolised and have at least a small calorific value. Attention has been drawn to the fact that some 'natural' foods contain toxicants (see Chapter 7). General reviews on the properties, uses and need for additives in foods have been presented by Anon (1977), Hill (1980), Holland (1980) and Luck (1982).

There is nothing new in the deliberate addition of chemicals to foods since many of the traditional processes of storing and preservation used salt, alcohol, acetic acid, wood smoke or combinations of these. Such processes provided for an extension of availability of otherwise unstable foods and modern manufacturing practice has extended the range and function of additives used in the preparation, processing and storage of these. In all cases a positive need and desired technological function must be demonstrated before the additive is permitted for use in foods.

Safety of additives is assessed from long-term feeding trials on animals, in which the criterion of toxicity employed is often from tests in which the substance is fed to animals in amounts substantially greater than are likely to occur in the human diet. The acceptable daily intake (ADI) is considered in relation to body weight. Before a decision can be made as to which substances are to be recommended for inclusion in a permitted list, additives are placed in one of the following groups by the Committee on Toxicity.

Group A: Additives that the available evidence suggests are acceptable for use in food.

Group B: Additives that on the available evidence may be regarded meanwhile as provisionally acceptable for use in food, but about which further information is necessary and which must be reviewed within a specified time.

Group C: Additives for which the available evidence suggests possible toxicity and which ought not to be allowed in food without evidence establishing their acceptability.

Group D: Additives for which the available evidence suggests probable toxicity and which ought not to be allowed in food.

Group E: Additives for which the available evidence was inadequate to enable an opinion to be expressed as to their suitability for use in food.

Group F: Additives for which no information on toxicity was available.

Substances that are not on permitted lists must not be used for the preparation, manufacture and distribution of foods for sale to the ultimate consumer. In the case of colours, emulsifiers, stabilisers, solvents and most miscellaneous additives, there are no limitations on amounts used in foods although there are prohibitions of their use in certain classes of foods. For preservatives, antioxidants, mineral hydrocarbons and artificial sweeteners, maximum levels of use are prescribed for each named food. Explanations of the provisions of Regulations and the background information on usage of additives may be found in the reports of the advisory committees (Appendices 4 and 5); see also Flowerdew (1979) and Wonnacott (1985) for reviews of UK and EC legislation and the administrative systems relating to the acceptance of food additives. Most of the additive regulations introduced since 1973 have been to implement the provisions of EC Directives (Appendix 3). The international evaluation of the safety of food additives for use on a world-wide basis is undertaken by the Joint FAO/WHO Expert Committee on Food Additives (JECFA); this committee also develops specifications for purity and several reports have been published on both of these aspects. The Codex Alimentarius Commission has also issued lists of additives evaluated for their safety and use in food and a comprehensive text on the technology of food additives has been prepared by Furia (1972, 1980). Apart from the general classes of additives referred to above, those used in conjunction with specific foods are discussed in the relevant chapter, for example bleachers and improvers (flour) and artificial sweeteners (beverages and fruit juices).

The EC has developed a proposal for a Council Directive (86 C 116/02 dated 16.5.86) which sets out criteria for the use of food additives, which are in conformity with the principles of need, toxicological evaluation, purity and limitation of use accepted within the UK. The directive also sets down categories of food additives as:

Colour
Preservative
Antioxidant
Emulsifier
Thickener
Gelling agent
Stabiliser
Flavour enhancer
Acid
Acidity regulator
Anticaking agent

Modified starch
Sweetener
Raising agent
Antifoaming agent
Glazing agent
Flour treatment agent
Firming agent
Humectant
Sequestrant
Yeast nutrient
Foam stabiliser
Enzyme

PRESERVATIVES

Legislation

The addition of preservatives to food is controlled by the Preservatives in Food Regulations 1989 (SI No. 533). These regulations re-enact with amendments the Preservatives in Food Regulations 1979, as amended. They implement Council Directive 64/54/EEC (OJ No. 12, 27.1.64, p. 161/64: OJ/SE 1963–1964, p. 99) on the approximation of the laws of Member States concerning the preservatives authorised for use in foodstuffs intended for human consumption, as last amended by Council Directive 85/585/EEC (OJ No. L372, 31.12.85, p. 43), and Council Directive 65/66/EEC (OJ No. 22, 9.2.65, p. 373/65: OJ/SE 1965–1966, p. 25) laying down specific criteria of purity for preservatives authorised for use in foodstuffs intended for human consumption, as last amended by Council Directive 86/604/EEC (OJ No. L352, 13.12.86, p. 45).

The Regulations—

(a) specify permitted preservatives and prescribe purity criteria for those preservatives (regulation 2(1) and (2) and Schedule 1);
(b) prohibit the sale or importation of food having in it or on it any added preservative except specified foods having in them or on them permitted preservatives within prescribed limits or as otherwise prescribed (regulation 4 and Schedule 2 and Schedule 3 paragraphs 4, 5 and 6);
(c) within prescribed limits, permit the presence in compounded food of permitted preservatives introduced in the preparation of that food by the use of one or more foods specified in Schedule 2 (regulation 5);
(d) prohibit the sale, the importation and the advertisement for sale, for use as an ingredient in the preparation of food, of any preservative other than a permitted preservative (regulation 6(1));
(e) prescribe labelling requirements for permitted preservatives when sold as such (regulation 6(2) and Schedule 3 paragraphs 1, 2, 3 and 6);
(f) prohibit the sale of food specially prepared for babies or young children if it has in it or on it any added sodium nitrate or sodium nitrite (regulation 7);
(g) make provision for the sampling and analysis of citrus fruit for the presence of biphenyl, 2-hydroxybiphenyl and sodium biphenyl-2-yl oxide (regulation 8 and Schedules 4, 5 and 6).

The principal changes effected by the Regulations are—

(a) the inclusion, in implementation of Directives 85/585 and 86/604, of E228 potassium bisulphite as a permitted preservative with specified purity criteria (regulation 2(1) and Schedule 1); and the confirmation that Community controlled wine may contain this and other preservatives to the extent authorised by any Community Regulation (regulations 2(1) and 4(11));
(b) the substitution for low fat spreads, of fat spreads whose fat content does not exceed 70 per cent, as a specified food permitted to contain sorbic acid (regulation 4(2) and Schedule 2);
(c) the extension of the preservatives permitted in fruit or plants, crystallised, glacé or drained (syruped), or candied peel or cut and drained (syruped) peel to include benzoic acid, hydroxybenzoates or sorbic acid and sulphur dioxide (regulation 4(3)(b) and Schedule 2);
(d) subject to prescribed limits, the authorisation for use in prawns and shrimps in brine of the permitted preservatives, sorbic acid, benzoic acid, ethyl 4-hydroxybenzoate, propyl 4-hydroxybenzoate and methyl 4-hydroxybenzoate as well as sulphur dioxide (regulation 4(3)(f) and Schedule 2).

The approach to food analysis will depend on circumstances in which this is to be carried out. When a food containing unknown preservatives is to be examined, a chromatographic screening procedure may be adopted. Tjan and Konter (1972), Adams (1978) and Rajama and Makela (1973) have described screening techniques based on TLC systems. Where the nature of preservatives used is known, specific procedures may be adopted. Selected reviews on individual preservatives have been published by Endean and Harding (1976), Carswell (1977) and Endean (1976a), and Dickes and Nicholas (1976) have described a number of analytical procedures using GLC techniques. Schuller and Veen (1967) and Sawyer and Crosby (1980) reviewed available methods for preservatives. The increasing use of HPLC techniques has been reflected in the development of multicomponent procedures for a range of preservatives, e.g. Lazzarini (1982) has described HPLC conditions for biphenyl, salicylic acid, benzoic acid, hydroxybenzoates and sorbic acid. Methods for sorbic acid and the benzoates in a variety of foods have been described by Aitzetmuller and Arzberger (1984), Gieger (1982), Gertz and Herrmann (1983) and Ali (1985).

INORGANIC PRESERVATIVES

SULPHUR DIOXIDE

Sulphur dioxide may be used in the form of a gas, in solution as sulphurous acid, or as the sulphites of sodium, potassium or calcium; but for the purpose of the regulations the amount present is calculated as sulphur dioxide (SO_2). Sulphurous acid inhibits the growth of mould, yeasts and aerobic bacteria and also prevents the browning of fruits and vegetables due to enzymic reactions. It assists in conserving vitamin C, but inactivates vitamin B by cleavage of the molecule. Sulphur dioxide reacts with the constituents of foods in both reversible and irreversible reactions. When added to food it undergoes association–dissociation reactions in the aqueous phase and an equilibrium is set up between SO_2, H_2SO_3, HSO_3^- and SO_3^{2-}. The equilibrium depends both on the pH of the food and on other reactive species present in the food matrix such as free carbonyl groups in sugars or disulphide groups in proteins (Green, 1976). Wedzicha (1981) and Phillip (1981) have reviewed the uses of SO_2 in the food industry and discussed the mechanisms of antimicrobial and anti-browning reactions. A review of the regulatory position and usage in the USA has been prepared by Applebaum (1985). There are special labelling requirements for foods containing low levels (10–20 mg/kg) in the USA.

FREE AND BOUND SULPHUR DIOXIDE

It is evident from the foregoing that sulphur dioxide when added to food enters into many reactions in which reaction products may be stable or unstable. So far as the Preservative Regulations are concerned, the limit has been held to refer to that amount which is determined in the Monier-Williams method (p. 72). This method will determine the total sulphur dioxide (free and reversibly combined) although, as indicated above, it is claimed that preservative power is related to the concentration of undissociated acid. It is thus apparent that the method required for analysis will depend on the problem to be answered. The enforcement analyst is interested in the total sulphur dioxide but the food technologist will be interested not only in this total but the effective level for preservative action also. For these reasons the literature contains a wide range of analytical methods and manipulation of chemical conditions will permit the determination of free and bound SO_2. Choice of method will also relate to the level to be determined, the physical state of the product and such special factors as the presence of other volatile sulphur compounds or other potential interfering substances in the food. The state of equilibrium in the food will be disturbed during the analysis if the process involves removal of sulphur dioxide, so that analysis designed to determine other than the total amount must be carried out under conditions which minimise the disturbing effects. Losses of sulphur dioxide are accelerated by contact with air and elevation of temperature. It is preferable to submit unopened samples to the laboratory for analysis; where this is not practicable it is necessary to ensure that sub-samples are of sufficient size to fill sample jars. The jars should be closed with an airtight closure and should if possible be preserved in a cool or frozen state (Stadtman *et al.*, 1946; Hearne and Tapsfield, 1956; Jennings *et al.*, 1978).

Qualitative test for sulphur dioxide

Simple tests for the presence of sulphites are based on the ability of sulphur dioxide to decolorise malachite green (AOAC) or *p*-rosaniline (Schmidt, 1960). Alternatively, sulphur dioxide may be reduced by zinc and acid: the hydrogen sulphide liberated is then detected by lead acetate paper. Sulphides naturally present in some vegetables will interfere with the test. A comprehensive general review of methods for estimation of sulphite and sulphur dioxide has been published by Williams (1979); modern methods have been briefly reviewed by Barnett (1985). Parkes (1926) recommended a method based on release of sulphur dioxide with acid and using carbon dioxide generated *in situ* as carrier. The gases are absorbed in a reagent solution which reacts with sulphur dioxide, interference by sulphide being eliminated by the use of copper acetate.

METHOD

Iodine–barium chloride reagent. Dissolve 3 g of iodine in a solution containing 3 g of potassium iodide. Add 2 g barium chloride dissolved in water and make up to 100 ml.

Fit a conical flask with a small bubbler in the form of a small thistle funnel bent twice (similar in shape to the fermentation locks used in wine making). Place a few drops of iodine–barium chloride reagent in the bubbler. Add to the flask 5 g of sample, 0·1 g of copper acetate, a piece of marble and 10 ml of conc. hydrochloric acid and fit on the bubbler. Allow the acid to act on the marble for 10 min and then heat to boiling. The iodine is decolorised and, in the presence of sulphur dioxide, a precipitate of barium sulphate settles in the tube. The formation of a turbidity is inconclusive as it may be due to other substances such as volatile oils (cf. Leather, 1942).

Determination of sulphur dioxide

Quantitative methods for the determination of sulphur dioxide fall into two groups, direct methods and indirect methods. The former include titrimetric, polarographic, electrometric and colorimetric procedures, the latter generally involve separation by distillation in an inert atmosphere followed by absorption of the sulphur dioxide in an oxidising agent, typically iodine or hydrogen peroxide. The estimation is completed by volumetric, gravimetric, colorimetric or electrochemical procedures. Both types of methods may be used for measurement of free and total sulphur dioxide.

Estimation of free sulphur dioxide

The most common method used for direct determination in colourless or lightly coloured solutions is titration with iodine in the presence of excess sulphuric acid. Problems are encountered with drifting end-point due to the dissociation of bound sulphur dioxide (Ingram, 1947) and the measurement of a 'blank' value following combination of the free sulphur dioxide with aldehyde or ketone (Prater *et al.*, 1944). Electrometric detection of the end-point is used for dark or turbid solutions (Ingram, 1947; Tanner and Sandoz, 1972). Eckhoff and Okos (1983) used an aqueous extraction procedure followed by direct titration with iodine in the estimation of free sulphur dioxide in ground corn samples.

Electrochemical techniques have been used for direct measurement of free sulphur dioxide in wines and fruit concentrates. Jennings *et al.* (1978) and Spedding and Stewart (1980) used a gas sensing probe; Bruno *et al.* (1979) describe a polarographic method using lithium chloride in dimethyl sulphoxide as supporting electrolyte. The polarographic method is claimed to be more accurate than the direct iodometric method, with no loss of precision. Both methods have been shown to be applicable for the determination of total sulphur dioxide. Indirect methods for the estimation of free sulphur dioxide are based on the desorption of the gas at room temperature. Nitrogen is bubbled through a fruit juice sample acidified to pH 1 with sulphuric acid. The gas is reabsorbed in alkaline glycerol and measured by titration with iodine. Alternatively, a more sensitive colorimetric assay can be carried out by absorbing in sodium tetrachloromercurate solution followed by reaction with *p*-rosaniline (Lloyd and Cowle, 1963). Precise control of gas flow-rate is necessary to obtain reproducible results. The use of headspace gas chromatography to measure free and molecular sulphur dioxide in liquid foods has been described by Davis *et al.* (1983); the method is based on the direct relationship between gaseous and molecular SO_2 as defined by Henry's Law.

Estimation of total sulphur dioxide

Two basic techniques are used for general application to foodstuffs. The first is based on the adjustment of the sample to pH 13 with alkali to achieve hydrolysis of bound sulphur dioxide followed by acidification to form undissociated sulphurous acid, the second on distillation of sulphur dioxide from an acidified aqueous suspension of the food (pH 1–2) using a stream of inert carrier gas to displace the gaseous sulphur dioxide into an absorbing solution.

DIRECT METHODS

Direct methods for total sulphur dioxide, like those for free sulphur dioxide, are subject to interference from colour, other reactive compounds and recombination reactions. The extent of interference is dependent on the nature of the product and the level of sulphur dioxide present.

The classical direct iodometric titration described for use with light wines has been modified by many workers (see reviews by Schuller and Veen, 1967 and Carswell, 1977). Double titration methods have been described by Potter (1954) and Pearson and Wong (1971). The macerated food is titrated with iodine before and after treatment with hydrogen peroxide. Sulphur dioxide is converted to sulphate by peroxide treatment and the difference in iodine titres represents the sulphur dioxide present. Colorimetric methods have found increasing favour. The method of West and Gaeke (1956) is based on formation of an adduct of sulphur dioxide with sodium tetrachloromercurate, the adduct being used to restore the colour of acid-bleached *p*-rosaniline in the presence of formaldehyde. This method is used for the determination of sulphur dioxide in dried fruits (AOAC) and in raw meat (Karasz *et al.*, 1976).

An ICUMSA study group has compared the rosaniline method (Carruthers *et al.*, 1965) with that using 5,5-dithio-bis-(2-nitrobenzoic acid), DTNB, for determination of sulphur dioxide in white sugar. The group concluded that the rosaniline method was more precise though less rapid.

INDIRECT METHODS

Indirect methods are dependent on the removal of sulphur dioxide by distillation from mineral acid; the choice is usually limited to phosphoric or hydrochloric acid. The gas evolved may be displaced by a stream of inert gas and collected in an absorbing solution. Many variations of detail in procedure have been described. Early methods used iodine as titrant following collection in water. Hydrochloric acid gives a rapid liberation of sulphur dioxide from foods but problems may be encountered with the release of other volatile sulphur compounds which react with iodine and foods containing material such as mustard and onion will give high results. Interference may be prevented by addition of copper acetate prior to the acidification with hydrochloric acid. When phosphoric acid is used, the liberation of sulphur dioxide is slower than with hydrochloric acid, especially in the presence of high amounts of sugars. Monier-Williams (1927c, d) described a method in which the refluxed vapour passes through U-tubes containing hydrogen peroxide which reacts with the sulphur dioxide to form sulphuric acid. The acid produced is titrated with dilute standard alkali. This procedure is preferable to the rapid method using iodine when volatile sulphur compounds are present, as for example, mustard, onion and horseradish.

If gravimetric confirmation is called for, barium sulphate is precipitated by adding barium chloride to the titrated liquid and allowing it to stand overnight in the cold. If heat is used, volatile organic sulphur compounds are also liable to be included (Thrasher, 1966). Interference due to co-distillation of volatile oil can be markedly reduced by previously dispersing the sample in methanol (Tanner, 1963). Meadows (1966) replaces the flask and U-tubes by a double trap absorber combined in a single gas jar. The AOAC initially accepted the Monier-Williams method and Shipton (1954) has made a critical study of this technique and suggested improvements to shorten the procedure considerably. Shipton's main modifications are the use of (1) a vertical apparatus with ground glass joints, (2) a regulated electric heater, (3) a regulated flow of inert gas, (4) the abolition of the necessity of using previously boiled water and (5) the omission of the heating of the condensers at the end of the refluxing. The current edition of the AOAC methods book describes a modified Monier-Williams method incorporating modern glassware and acidometric titration of the absorbing solution (neutralised hydrogen peroxide) to a methyl red end-point. The method is applicable in the presence of other volatile compounds but is not recommended for use with dried onions, leeks and cabbage. The modification due to Tanner (1963) is preferred as a general method by EC, Codex Alimentarius Commission, OIV and IFJU. Figure 4.1 shows the apparatus.

TANNER METHOD

REAGENTS

Phosphoric acid, 88 per cent (d = 1·75).

Hydrogen peroxide solution. 0·2 per cent w/v. Dilute 0·7 ml of 100 vol. hydrogen peroxide to 100 ml. Prepare as required daily.

Sodium hydroxide solution, 0·01 M. Standardise against potassium hydrogen phthalate, dried at 110 °C.

Methanol, AR.

Mixed indicator solution. Mix 50 ml of 0·03 per cent ethanolic solution of methyl red with 50 ml of 0·05 per cent ethanolic solution of methylene blue and filter.

METHOD

Weigh, or pipette, a quantity of sample into the distillation flask as indicated by the table below:

Expected SO_2 content (mg/kg)	Quantity of sample to be taken (g or ml)	Vol. of distilled water to be added (ml)
<10	40–50	20
10–100	20–25	30
>100	5–10	40

Add distilled water to the flask as indicated. Add 50 ml of methanol and mix. Introduce into the distillation receiver 10 ml of hydrogen peroxide, 60 ml of distilled water and a few drops of mixed indicator solution. Add a few drops of 0·01 M sodium hydroxide solution to produce a green colour. Add a similar quantity of neutralised hydrogen peroxide solution to the guard wash bottle. Connect up the apparatus and adjust the nitrogen flow to approximately 60 bubbles per minute. Add 15 ml of phosphoric acid to the funnel and run it into the distillation flask. Heat rapidly to boil the mixture and then simmer

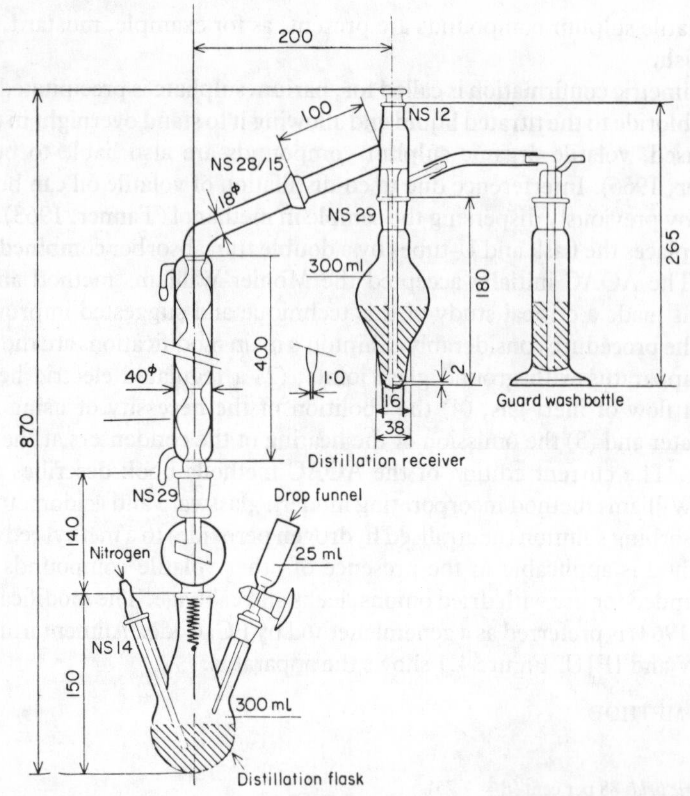

Fig. 4.1 Distillation apparatus for the determination of sulphur dioxide (Tanner). All dimensions in mm.

gently for a total period of 30 min. Detach the receiver from the distillation apparatus and rinse the tube. Titrate the sulphuric acid present with 0·01 M sodium hydroxide solution until the indicator turns green.

CALCULATION

$$\text{Sulphur dioxide content (mg/kg or mg/l)} = \frac{a \times M \times 32 \times 1000}{Q}$$

where a = volume (ml) of sodium hydroxide solution,
M = molarity of sodium hydroxide solution,
Q = weight of sample in g or volume of sample in ml.

The acidity of the distillation medium used has been shown to have a profound effect both on the rate and on the amount of sulphur dioxide released from differing products. Carswell (1977) has summarised the situation and draws attention to a general conclusion that there is no uniformly valid method for all products. Four distillation systems are proposed to meet the requirements of different product types:

1. Reith and Willems (1958) HCl at pH 0·3
2. Zonnevd and Meyer (1960) HCl, methanol at pH 0·3
3. H_3PO_4 at pH 1·5
4. Tanner (1963) H_3PO_4, methanol at pH 2·0

Other techniques which may be employed at the measurement stage include the colorimetric and electrochemical metods referred to earlier. An automated technique using continuous distillation with a rosaniline colorimetric finish has been described by Jennings *et al.* (1978). These workers compared results obtained by use of the Tanner method, the gas probe method and the automated technique. They concluded that the results obtained on a range of foods by the Tanner method and the automated method were comparable. The gas probe method was shown to be applicable to a restricted range of products and problems were found with wine samples due to the effects of alcohol on the electrode membrane. A colorimetric finish to the Monier-Williams method has been described by Wedzicha and Johnson (1979). These workers used a semi-micro distillation unit containing 1–10 ml of liquid sample and 50 ml acid. The distilled sulphur dioxide was collected in a solution of 5,5-dithio-bis-(2-nitrobenzoic acid), DTNB, and the absorbance measured at 412 nm. Recoveries from 91 to 101 per cent were reported for sulphur dioxide contents of ginger ale varying from 8 to 320 mg/l using a 6 minute distillation; little advantage was gained by increasing distillation time. Kearsley and Thanos (1982) used a semi-automatic distillation procedure and claimed a significant improvement in the time needed for analysis. Banks and Board (1982) compared three procedures in application to the analysis of sausage meat. The ion-selective probe was applied directly to aqueous suspensions of sausage meat whilst the DTNB colorimetric procedure was applied to a distillate. These methods were composed with the Monier-Williams procedure. These authors concluded that the ion-selective probe is inappropriate for analysis of the samples studied and that although the other two methods gave satisfactory results, the colorimetric finish proved to be superior in terms of sample preparation, reproducibility, absence of interference and time required for analysis. Beutler (1984) has described an enzymatic procedure using sulphite oxidase, in which the hydrogen peroxide generated is reacted with NADH in the presence of NADH-peroxidase. The decrease in NADH, proportional to concentration of sulphite, is measured by UV absorption; recoveries of 96 per cent added SO_2 were recorded.

NITRATES AND NITRITES

Sodium and potassium nitrate and nitrite have been used for many years in conjunction with sodium chloride as components of dry mixes and brines in the preparation of cured meats (Lechowich *et al.*, 1978; Winstanley, 1980; Przybyla, 1981a). Limited use is also made of nitrates and nitrites in the manufacture of certain cheeses. The Preservatives in Food Regulations 1989 incorporate many of the recommendations of the FACC (1978b) regarding the curtailing of the usage of these salts in the manufacture of cured meats and in cheese. In recognition of the concern regarding the potential to form nitrosamines and to interfere with infant metabolism the regulations also prohibit the presence of added nitrate and nitrite in all foods specially prepared for infants and young children.

The amounts of nitrate and nitrite permitted in cured meats and in cheese are summarised in Table 4.1.

The toxic propensity of nitrates is dependent on their conversion to nitrites by bacterial action, both in the curing operation and in the digestive tract. Nitrites have a more powerful preservative action than nitrates (which equate with common salt), their most important function being the inhibition of growth of *Clostridium botulinum* in cured foods. They also contribute to the colour and flavour characteristics of the products concerned. The pink colour of bacon and ham is due to the conversion

Table 4.1 Limits for sodium nitrate and nitrite: Preservative Regulations 1989

Product	Limit mg/kg, sodium nitrate and sodium nitrite (*x*) of which not more than (*y*) sodium nitrite (Total (*x*) expressed as sodium nitrite) *x*	*y*
Cured meat (including cured meat products) packed in a sterile pack, whether or not it has been removed from the pack	150	50
Acidified and/or fermented cured meat products (including salami and similar products) not packed in a sterile pack	400	50
Uncooked bacon and ham; cooked bacon and ham that is not, and has not been, packed in any hermetically sealed container	500	200
Any cured meat or cured meat product not specified above	250	150
Cheese other than Cheddar, Cheshire, Granapadano or Provolone type or soft cheese	50	5

of myoglobin to form nitrosohaemoglobin. An extensive literature has been developed both on the mode of action and value of nitrites in curing meat products and on the potentially toxic effects due to their *in vivo* reactions to form nitroso compounds. The topic has merited attention at a number of international symposia and summaries of work on all aspects of the problem can be found in the proceedings of two such meetings held in the Netherlands (Krol and Tinbergen, 1974; Tinbergen and Krol, 1977). It should be noted, however, that nitrates are naturally present in edible crops. Different plants accumulate nitrates at varying levels, spinach, beef, radishes, celery, lettuce and turnip tops having significantly higher levels than other vegetables. Estimates have been made of the relative contributions of vegetables, meat and water to the daily diet of individuals (Walker, 1975; Gray and Randall, 1979). A collection of survey data has been compiled in MAFF reports (1978, 1984). Considerable attention has been paid to the generation of survey data on a wide range of foods, including vegetables, fruit juice, cured meats, milk and milk products. Recent publications are summarised in Table 4.2. The levels of nitrate in vegetables

Table 4.2 Recent surveys of nitrate and nitrite content of foods

Foods	Method	Reference
Vegetables (nitrates only)	Nitration, colorimetric	Tanaka *et al.* (1982)
		Rauter and Wolkerstorfer (1982)
Vegetables (nitrates, nitrites)		Anon (1981b)
Fruit, vegetables, cured meat, cheese, milk	Colorimetric	Collet (1984)
Fruit juice, corn, bacon, beer, water	LC	Iskandarani and Pietrzyk (1982)
Vegetables, fresh and cured meats	HPLC	Wootton *et al.* (1985)
Carrot products	Electrochemical, colorimetric	Lange (1981), Rimpel *et al.* (1981)
Casein products	Automated colorimetry	Woollard and Forrest (1980)

vary from 1 to 10 000 mg/kg. The role of nitrate and its influence on the rate of internal corrosion of canned products has been discussed by Hall (1978). Variations of level in the water supply and in the raw produce may make a significant contribution to differing stabilities of packs of similar products.

A critical review of nitrate and nitrite methodology, with special reference to meat products, has been published by Usher and Telling (1975). The performance of alternative clearing agents and the errors arising from the presence of other additives such as ascorbate, phosphate and sulphite are discussed. They conclude that nitrite can be readily and accurately determined in most foods, but that nitrate estimations are subject to greater variations and errors.

Extraction methods

The salts are readily water soluble but, due to the influence of other constituents in the foods, the extracts are relatively unstable. A variety of procedures have been recommended, including the use of both hot and cold water. Losses of nitrate may be minimised by ensuring that the pH of the extract remains above 5. In subsequent deproteinisation stages, the pH must be adjusted to values close to the isoelectric point of the soluble protein, usually between pH 5·5 and 6·5. Carrez solution (zinc acetate and potassium ferrocyanide) is generally regarded as the most effective clearing agent for general use. Fiddler and Fox (1978) carried out a comparative study of 12 methods for use with frankfurters and concluded that the AOAC procedure using hot water digestion and clearing with mercuric chloride was equal to or better than the other techniques. DeSiena *et al.* (1981) and Fox *et al.* (1984) re-examined recommended extraction procedures with a view to optimising the conditions to minimise interference from ascorbate.

Determination of nitrite

Many methods for determination of nitrite rely on variations of the Griess diazotisation procedure, in which an azo dye is produced by coupling a diazonium salt with an aromatic amine or phenol. Care must be taken with the experimental conditions in order to obtain reliable results; factors such as pH, temperature, nature and concentration of reagents all affect the final colour intensity. The diazo compound is usually formed with sulphanilic acid or sulphanilamide; the coupling agent may be l-naphthylamine, l-naphthylamine 7-sulphonic acid (Cleve's acid) or *N*-l-naphthylethylene diamine (NED). The first reagent is now regarded as unsuitable on the grounds of carcinogenicity. Of the other two reagents, Cleve's acid gives colours which allow the continued use of Nessler discs developed for use with l-naphthylamine, whilst NED reagent is preferred by BSI (1976), ISO and EC since it reacts more quickly than Cleve's acid and is less variable in composition. Egginger and Honikel (1979) described a rapid method for application to cured meats based on the use of commercially available reagent kits designed for use in water analysis. Norwitz and Keliher (1985) made a comprehensive study of possible sources of interference in the spectrophotometric determination of nitrite by diazotisation reactions; tolerance limits for 82 variables are tabulated against the use of three commonly used coupling reagent mixes. Amin (1986) has described a new diazo reaction system based on the use of 4-aminobenzotrifluoride and coupling with 1-

naphthol. The reaction compared favourably with other Griess systems in terms of sensitivity and temperature dependence.

Nakamura and Murata (1979) used a benzene solution of 4,5-dihydroxycoumarin to give an unidentified, stable reaction product with absorbance maximum at 410 nm. The range is similar to that for the diazo reaction but the reaction conditions are less complicated.

Chromatographic procedures have been investigated with a view to increasing the sensitivity and limit of detection of the estimation of nitrite and in some cases, of nitrate. Derivation reactions have been used as a preliminary step in the use of electron capture detectors in GLC methods; Funazo *et al.* (1980) and Chikamoto *et al.* (1981) used the diazotisation reaction in the presence of *p*-bromoaniline to form *p*-bromochlorobenzene which is readily partitioned into solvent. Both groups claimed that the limit of detection for the method was 0·01 mg/kg. Electron capture GLC was also used by Wu *et al.* (1984) on the pentafluorobenzyl derivative of nitrite and the results on cured meat compared favourably with the colorimetric method. Tanaka *et al.* (1981a, 1983) used the reaction of nitrite with hydralazine to form tetrazole (5,1-α) phthalazine, which was extracted into toluene and determined by GLC. Applications were reported to meat and fish products and to vegetables. The nitrophenol derivative method has also been used with a GLC finish by Luckas and Lorenzen (1984). Derivatisation reactions have also been combined with an HPLC finish; Noda *et al.* (1982) used the hydralazine reaction whilst Kunugi *et al.* (1983) used xylenol derivatives to estimate nitrite and nitrate simultaneously. Ion chromatography has also been used as a means of estimating more than one anion. Typical procedures have been reported by Tateo *et al.* (1982), who determined chloride and phosphates and the curing salts. Jackson *et al.* (1984) applied ion exchange HPLC to a wide range of foods and claimed good accuracy and precision in comparison with the accepted methods. De Kleijn and Hoven (1984) used ion pair HPLC for the examination of cured meats at levels of 500 mg/kg nitrate and 200 mg/kg nitrite, which levels are no longer in conformity with the requirements of the UK legislation. Fiddler *et al.* (1984) have described the use of a chemiluminescence detector in application to cured meat products, detailed comparison with alternate techniques suggests that chemiluminescence is suitable for consideration as a standard method.

MODIFIED GRIESS–ILLOSVAY REACTION FOR THE DETECTION AND DETERMINATION OF NITRITE

Reagents

1. Dissolve 106 g potassium ferrocyanide trihydrate in water, dilute to 1 litre.
2. Dissolve 220 g zinc acetate dihydrate in water, add 30 ml glacial acetic acid and dilute to 1 litre.
3. Dissolve 50 g disodium tetraborate decahydrate in 1 litre water.
4. Dissolve 2 g sulphanilamide in 800 ml warm water, cool, filter, add 100 ml concentrated hydrochloric acid (stir continuously). Dilute to 1 litre with water.
5. Dissolve 0·25 g of *N*-l-naphthylethylene diamine dihydrochloride in water. Dilute to 250 ml with water. Store in a well stoppered brown bottle and keep in a refrigerator. Renew weekly.
6. Dilute 445 ml concentrated hydrochloric acid to 1 litre water.

   Standard sodium nitrite solution (stock). Dissolve 1 g exactly with sodium nitrite in water, dilute to 100 ml.

   Sodium nitrite working solutions. Prepare daily. Dilute 5 ml stock solution to 1 litre water.

Dilute 5, 10 and 20 ml to 100 ml with water. These solutions contain 2·5, 5·0 and 10·0 μg/ml of sodium nitrite.

METHOD
From the bulk material take a 200 g representative sample of the product. Make the sample homogeneous by passing it through a mechanical mincer fitted with a perforated plate with holes not greater than 4 mm diameter. Analyse uncooked samples immediately, for other samples analyse as soon as possible, but always within 24 hours. The representative sample may be stored before homogenisation for up to 4 days if kept in a closed container at 0 °C. Weigh out to the nearest mg, 10 g of homogenised sample into a 100 ml capacity beaker; macerate with 5 ml borax solution (Reagent 3) and 70 ml of hot distilled water (not lower than 70 °C). Transfer with the aid of 75 ml hot water to a 200 ml wide neck calibrated flask. Heat on a boiling water bath for 30 min and shake repeatedly. Allow to cool, add successively with mixing 2 ml Reagent No. 1, 2 ml Reagent No. 2, make almost to the mark with distilled water. The pH of the supernatant solution should be 8·3; adjust if necessary by dropwise addition of 1 M sodium hydroxide solution or 4 M hydrochloric acid. After the pH check, make to the mark with distilled water. Stand for 30 min. Carefully decant the supernatant liquor through a fluted filter paper (nitrate and nitrite free) to obtain a clear filtrate.

COLOUR MEASUREMENT
Pipette V ml filtrate ($V \ngtr 1$) into a 100 ml one mark flask; add distilled water to approximately 60 ml, add 10 ml Reagent No. 4, 6 ml Reagent No. 6; leave the solution for 5 min in the dark. Add 2 ml Reagent No. 5, mix and leave for 3 min in the dark. Dilute to the mark with water, measure the absorbance of the solution at 538 nm in a 1 cm cell. Prepare a calibration curve using water and 10 ml of each of the diluted standard solutions of sodium nitrite (0, 25, 50 and 100 μg sodium nitrite). Follow the method from the stage marked 'colour measurement' above, omitting the addition of the sample filtrate. From the calibration curve read off the number of μg of sodium nitrite (m_1) equivalent to the absorbance of the test sample. If the sample reading exceeds the absorbance of the top standard solution, a smaller aliquot of test solution must be taken.

The number of mg/kg of sodium nitrite = $200\, m_1/m_0 V$, where m_0 is the mass in g of the test portion, and V is the volume in ml of the aliquot of filtrate taken for test.

Determination of nitrate

Spectrophotometric methods for nitrate are based on three general reaction systems: nitration of aromatic compounds in sulphuric acid, oxidation of alkaloids or aromatic amines in sulphuric acid, and reduction reactions to yield ammonia or nitrite. The two systems based on nitration and oxidation tend to yield erratic results, although the currently recommended AOAC procedure for meat products is based on the nitration of *m*-xylenol in sulphuric acid followed by distillation of the nitro-xylenol into sodium hydroxide to yield a solution for colorimetry.

Electrochemical methods have also been used for determinations of nitrate and nitrite: polarographic methods have been known for many years but little used on a routine basis. Collet (1983) has reported on the use of differential pulse polarography. Fogg *et al.* (1983) have also studied the application of a glassy carbon electrode as a detector in a flow injection system. The use of electrochemical methods has attractions in that electrodes are suitable for rapid, routine screening applications, especially in cases of use with similar samples. Sample preparation for determination of nitrate is generally restricted to extraction with water, removal of chloride with silver sulphate and destruction of nitrite with sulphamic acid. Immersion of the probe then allows for direct determination of nitrate content; a method of standard additions may be used to allow for varying ionic strengths in different sample extracts. Pfeiffer and Smith (1975) and Liedtke and Meloan (1976) have described applications of the electrode to foods. Henshall *et al.* (1977) in a comparison of four methods for determination of nitrate in processed foods concluded that the ion-selective electrode method and the AOAC method based on nitration of *m*-xylenol were unsuitable for general use. They recommended that the Follett and Ratcliffe (1963) method as modified by Evans (1972) to eliminate interference by ascorbate was suitable for general application to a wide range of foods. These authors recommended an automated version of the method of Bloomfield *et al.* (1965), in which nitrate ion interferes quantitatively in the reaction between α-furil-dioxime and rhenium in the presence of tin (II) chloride: when performed manually the method is tedious and requires careful attention to experimental detail but when automated the method was found to have satisfactory precision and to be more convenient for use with large numbers of samples than the cadmium reduction method. However, the latter method (Follett and Ratcliffe, 1963) is recommended by BSI (1976), ISO and by EC for application to meat and cheese products. Nitrate reductase has also been used as a means of producing nitrite from water-soluble nitrate (Hamano *et al.*, 1983).

Reagents

Reagents Nos 1 to 6 detailed in the method for nitrite are used in carrying out the analysis for nitrate. In addition, the following reagents are used:

7. Dilute 20 ml concentrated hydrochloric acid with 500 ml distilled water, add 10 g disodium dihydrogen ethylene-diamine-*NNN'N'*-tetraacetate dihydrate and 55 ml concentrated ammonia solution. Dilute to 1 litre, pH 9·6–9·7.

Standard potassium nitrate solution. Dissolve 1·456 g potassium nitrate in water, dilute to 1 litre.

Method

Preparation of the reduction column (Fig. 4.2). Place 5 zinc rods (15 cm × 5 mm dia.) into 1 litre 30 per cent (w/v) cadmium sulphate solution in a large beaker. Remove the metallic cadmium every 1–2 hours by swirling and after 8 hours decant the solution and wash the deposit twice with 1 litre distilled water. Keep the deposit continuously covered with distilled water, transfer to a laboratory mixer and blend with 400 ml 0·1 M hydrochloric acid. Return the contents of the mixer to the beaker, stir up the cadmium with a glass rod, leave overnight and stir to remove occluded gas. Decant and wash the cadmium twice with 1 litre distilled water. Wash the cadmium into the glass columns of dimensions shown in Fig. 4.2 until a column of metal approximately 170 mm long is obtained. Drain excess water but do not allow the surface of the columns to become exposed to air. Inclusions of gas may be removed with a fine plastic knitting needle. The flow rate for the column should be not greater than 3 ml/min. Wash the column successively with 25 ml 0·1 M hydrochloric acid, 50 ml water and 25 ml of a 1 + 9 dilution of the pH 9·6 buffer (No. 7). Take care not to expose the surface of the column to air. Check the reducing capacity of the column as follows: pipette 20 ml potassium nitrate standard solution and 5 ml of pH 9·6 buffer (No. 7) into the reservoir on the top of the column. Collect the effluent in a 100 ml one mark flask. When the reservoir is nearly empty, wash with about 15 ml of a 1 + 9 dilution of the buffer (No. 7), drain. Repeat with a second portion (15 ml) of buffer, and drain. Fill the reservoir with the diluted buffer and collect effluent until the 100 ml flask is almost to the mark. Pipette 10 ml into a 100 ml flask, make to the mark and determine the nitrite content by the colorimetric method described earlier. If the nitrite content of the effluent is below 0·90 μg/ml sodium nitrite (90 per cent of the equivalence), the column must be discarded and prepared afresh. Prepare the sample extract as described for the determination of nitrite—aliquots of the same deproteinised extract may be used for both assays. Condition the cadmium column by treatment with hydrochloric acid, water and dilute buffer, pipette into the reservoir 20 ml of sample extract and simultaneously add 5 ml of the buffer solution. Collect the effluent and proceed to wash the column with diluted buffer solution as described in the reduction check method. Make up the effluent to 100 ml, pipette an aliquot (v ml) of not more than 25 ml into a 100 ml one mark flask and determine the sodium nitrite content by the colorimetric procedure described.

Calculation

Calculate the nitrate content of the sample (S), expressed in mg/kg sodium nitrate, using the formula:

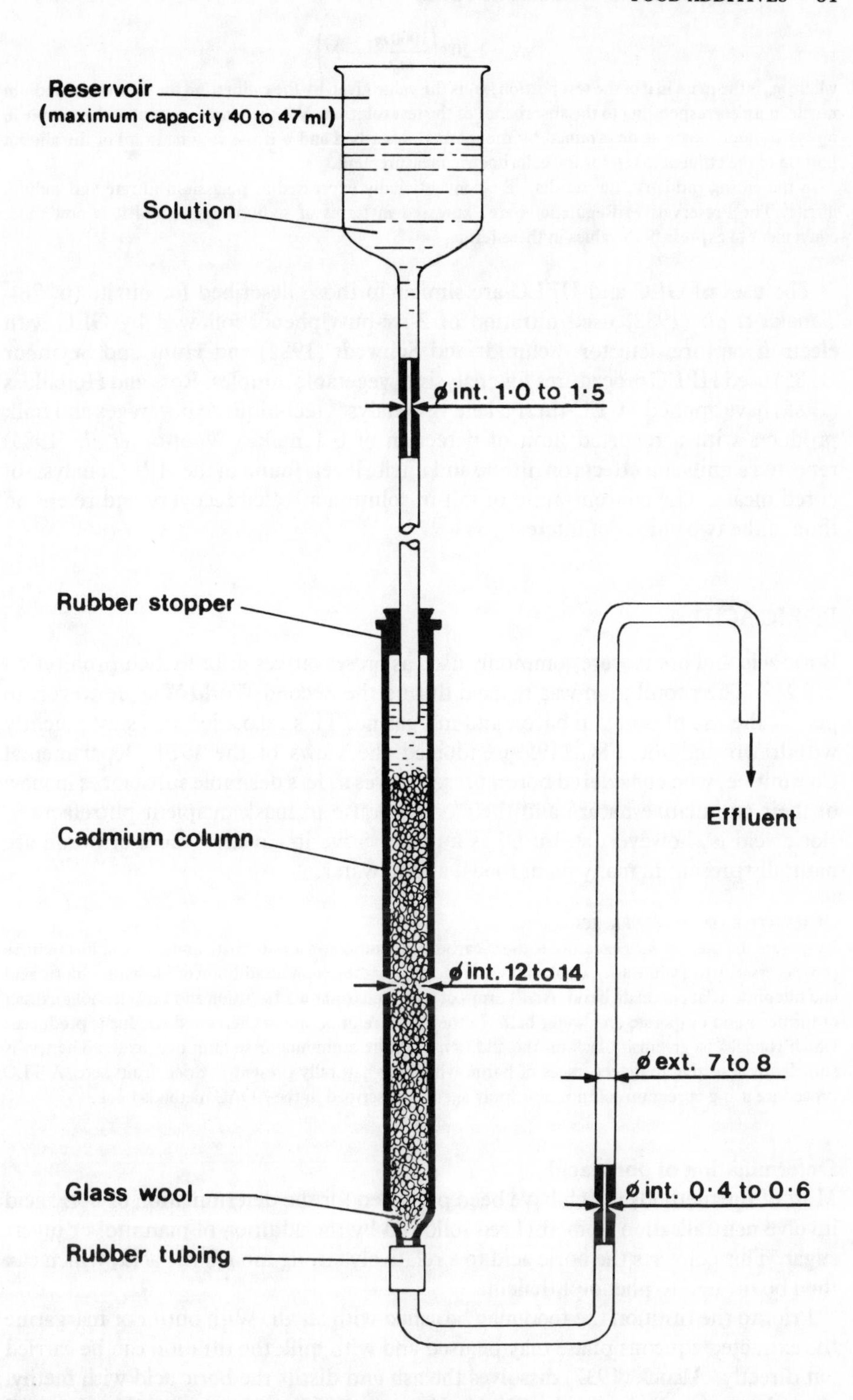

Fig. 4.2 Apparatus for nitrate reduction. All dimensions in mm.

$$S = 1{\cdot}203\left(\frac{1000\,m_1}{m_0 v} - N\right)$$

where m_0 is the mass in g of the test portion, m_1 is the value given by the calibration for the mass of sodium nitrite in μg corresponding to the absorbance of the test solution, N is the nitrite content of the sample in mg/kg sodium nitrite as determined by the method described and v is the volume in ml of the aliquot portion of the effluent taken for the colorimetric measurement.

In the curing industry, the results were conventionally expressed as potassium nitrate and sodium nitrite. The Preservatives Regulations are expressed in terms of sodium nitrite and it is now more convenient to express both values in these terms.

The uses of GLC and HPLC are similar to those described for nitrite (p. 78). Tanaka *et al.* (1982) used nitration of 2-*sec*-butylphenol followed by GLC with electron capture detector. Schmidt and Schwedt (1984) and Hunt and Seymour (1985) used HPLC procedures for analysis of vegetable samples. Ross and Hotchkiss (1985) have applied a GLC-thermal energy analyses technique to beverages and milk products with a reported limit of detection of 0·1 mg/kg. Wootton *et al.* (1985) reported significant effects on nitrate and nitrite levels found in the HPLC analysis of cured meats. The concentration of salt in solution affected recovery and retention time of the two anions of interest.

BORIC ACID

Boric acid and borax were commonly used as preservatives prior to their prohibition in 1925. The prohibition was relaxed during the Second World War, however, to permit the use of borax in bacon and margarine. This relaxation was subsequently withdrawn and the FSC (1959) endorsed the views of the 1924 Departmental Committee, who considered boron preservatives as less desirable substances in view of their cumulative nature and their possible use to mask incipient putrefaction. Boric acid is, however, still used as a preservative in caviar. Traces of boron are naturally present in many plant foods, and in water.

QUALITATIVE TEST FOR BORIC ACID

Evaporate the sample with excess of sodium carbonate solution on a water bath and ignite at low redness (not necessarily to a white ash). Cool, add water, acidify by the dropwise addition of 5 M hydrochloric acid and filter into a flat porcelain basin. Add 4 drops of saturated oxalic acid solution and 1 ml alcoholic extract of turmeric and evaporate on a water bath. In the presence of borates a cherry-red residue is produced, which changes to greenish-black on the addition of dilute ammonia or sodium hydroxide. The test is sufficiently sensitive to detect traces of boron which are naturally present in dried fruits, etc. A TLC procedure using curcumin solution as a spray agent is described in the AOAC methods book.

Determination of boric acid

Most of the methods which have been proposed for the determination of boric acid involve neutralisation to methyl red followed by the addition of mannitol or invert sugar. This converts the boric acid to a relatively strong monobasic acid, which can then be titrated to phenolphthalein.

Prior to the titration the food may be ashed with alkali. With butter or margarine the extracted aqueous phase may be used and with milk the titration can be carried out directly. Alcock (1937) dissolves the ash and distils the boric acid with methyl alcohol vapour. Quinalizarin (Ripley-Duggan, 1953) and carmine (Hatcher and

Wilcox, 1950) may be employed as reagents for the colorimetric determination of boric acid.

Determination by atomic absorption spectrophotometry involves preliminary wet ashing treatment with nitric and sulphuric acids, followed by extraction using 2-ethyl-1,3-hexanediol in methyl isobutyl ketone. A spectroscopic method using acid digestion, chelation and operation of the AA instrument in the emission mode has been described by Franco and Holak (1975) and accepted by the AOAC. Ogawa *et al.* (1979) used a 5 per cent solution of chelate in *n*-hexane/*n*-butyl acetate to extract borate from fish products. A limit of detection of 1 mg/kg was claimed in a colorimetric procedure based on reaction with curcumin. Edwards (1980) has described a method sensitive to 0·6 mg/l, employing azomethine-H in aqueous medium. The method is said to be suitable for automatic analysis.

ORGANIC PRESERVATIVES

BENZOIC ACID

Benzoic acid is usually used as a preservative in the form of the calcium, sodium or potassium salt, but for the purpose of the regulations the amount present is calculated as the acid itself. Weight for weight benzoic acid is only about half as effective as sulphurous acid. Benzoic acid retards the growth of yeasts and moulds, the effective agent being the undissociated acid.

The extent of antimicrobial action is markedly affected by the pH of the medium, the action being optimal between pH 2·5 and 4·0, and diminished at pH values above 5. In addition to the natural occurrence in small berry fruits, benzoic acid levels up to 13 mg/kg have been observed in lactic casein whey (Richardson and Gray, 1981).

Qualitative Test for Benzoic Acid (and Salicylic Acid)

For qualitative purposes it is usually possible to shake out enough of the benzoic or salicylic acid present with ether/light petroleum mixture after acidifying with dilute sulphuric acid. Emulsions are often formed but this can be usually overcome by adopting the following procedure:

Separate the 'watery' ether layer, add more solvent and wash with water. Shake the ether extract with dilute ammonia. Transfer the layer containing the ammonium salt of any organic acid present to another separator, acidify with dilute sulphuric acid and extract with fresh ether/light petroleum mixture. Pour equal volumes of the ether extract into two porcelain dishes and allow the solvent to evaporate off. Examine the residue in one dish by dissolving in 1 ml of water and adding ferric chloride solution. A violet coloration forms in the presence of *salicylic acid*. The amount of benzoic acid normally present in foods is usually insufficient to give a buff precipitate with ferric chloride and it is necessary to use a more sensitive test for its detection, e.g. one of the modifications of *Mohler's test*:

To the residue in the second basin add 5 drops of conc. sulphuric acid and 0·1 g of potassium nitrate and heat on a water bath, stirring occasionally, for at least 30 min. This converts the acid to *m*-dinitrobenzoic acid. Dilute with 1 ml of water and make alkaline with 0·88 ammonia. Transfer to a test-tube and boil in order to break up any ammonium nitrite. Cool and carefully add a few drops of freshly prepared ammonium sulphide solution down the side of the tube so that two distinct layers are formed (the sulphide solution can be prepared by passing H_2S into 0·88 ammonia). In the presence of benzoic acid a reddish brown ring forms at the junction due to the formation of ammonium *m*-diaminobenzoate. The reaction is also given by salicylic acid (tested for previously), cinnamic acid (which may be present as an oxidation product of cinnamic aldehyde from cinnamon oil) and phenolphthalein (which is seldom present). If, however, after the reddish brown ring is formed the layers are mixed and then boiled the aminobenzoate is decomposed to produce a greenish yellow liquid. On the other hand the amido compounds of the interfering acids are stable and no greenish yellow colour is formed.

Although still recognised officially by the AOAC, simple qualitative test procedures are being replaced by multi-detection procedures based on chromatographic separations. Rajama and Makela (1973) used a combined extraction and separation procedure for identification and determination of organic acids. The food is first diluted or homogenised with water; the solution or slurry is applied to a strip of chromatographic paper on a line parallel to one end. Parallel bands of buffer solution (pH 6·5), sodium bicarbonate solution and sodium hydroxide solution are applied to the paper at intervals of about 2 cm from the sample line. The food line is moistened with dilute hydrochloric acid. The end of the paper is dipped into diethyl ether and the solvent is allowed to run as in normal chromatography. Ether soluble acids are concentrated at the lower edge of the bicarbonate line, parahydroxybenzoates at the sodium hydroxide line, and fats are transported to the solvent front. After 30 min development the paper is examined under a UV lamp to locate the preservatives. Extraction with acidified ethanol removes benzoic acid and the hydroxybenzoates. Sorbic acid is removed with water. Quantitative assessment of the preservatives may be made by use of UV absorbance (Almela and Lopez-Roca, 1984), or colorimetric techniques. TLC procedures have been described by a number of workers including Dickes (1965c), Tjan and Konter (1972) and Adams (1978). The latter author has reported on an international collaborative study of a two plate TLC procedure for identification of 11 organic preservatives.

Determination of benzoic acid

The official general methods recommended by the AOAC are based on titrimetry and spectrophotometry after separation either by solvent extraction or by steam distillation.

Direct method. With some foods, e.g. soft drinks and wines, benzoic acid can be shaken out directly with ether/light petroleum or ether/hexane (3:1) after acidifying the sample with dilute sulphuric acid and saturating with salt. The acid is purified by shaking the bulked ether extracts with dilute sodium hydroxide and then treating the alkaline solution at 60 °C with permanganate to remove interfering acids. The excess permanganate is then reduced with sulphur dioxide and the benzoic acid extracted with ether after acidifying the aqueous solution. The solvent must be removed with not more than gentle heat before weighing. It is advisable to confirm the results by dissolving the residue in 2 ml acetone, adding 2 ml water and titrating with 0·05 M sodium hydroxide.

1 ml 0·05 M sodium hydroxide = 0·0061 g benzoic acid.

Cinnamic acid if present is oxidised to benzoic acid by permanganate. Benzoic acid may be selectively extracted at pH 4 in the presence of saccharin.

Very small amounts of benzoic acid can be estimated in ethanol solution by measurement of the UV absorption spectrum, but initial clean-up by extraction from alkaline solution is desirable to remove materials which may interfere with the UV absorption curve of benzoic acid. Endean and Harding (1976) described a silica gel column clean-up applicable to most foods, but they pointed out that vanillin is co-extracted with the acid preservatives and has a spectrum which overlaps that of benzoic acid.

Separation method. With viscous foods, it is advisable to employ steam distillation of volatile acids from phosphoric acid, purification by permanganate treatment and sublimation of the extracted material (Monier-Williams, 1927a, b).

To a 500 ml steam distillation flask add 30–100 g of sample, sufficient water and an excess of salt (40 g/100 ml). Make distinctly acid with phosphoric acid and rapidly steam distil 500 ml into a large porcelain basin containing 10 ml of 1 M sodium hydroxide. Wash down the condenser with 25 ml of approx 0·1 M sodium hydroxide and evaporate down the alkaline distillate to about 20 ml on a water bath. Allow to cool to about 45 °C and add potassium permanganate solution (5 per cent) until a pink colour persists after stirring. Decolorise with sulphur dioxide and add sufficient dilute sulphuric acid to dissolve the precipitated manganese dioxide and make the liquid acid. Transfer to a stoppered 100 ml cylinder, saturate with salt (33 g/100 ml water) and extract the acid with 4 successive portions of 15 ml of ether/light petroleum. Transfer each extract to a 150 mm × 15 mm test-tube held at 30 °C in a water bath, evaporate off the solvent by means of a current of dry air. Wash down any deposit remaining on the sides of the tube into the bottom with a little ether and evaporate off the solvent. Remove the tube from the bath. For most purposes the residue is sufficiently pure to be weighed and/or titrated without further purification. If necessary, the residue may be purified by sublimation, but more usually nowadays the ethereal extract is first concentrated and then made up to a fixed volume with chloroform–ethanol. The solution may then be assayed by TLC procedure (AOAC), UV spectrophotometry, colorimetry, GLC or HPLC. Endean and Harding (1976) described a comprehensive system using solvent extraction followed by TLC for determination of mixed preservatives in a range of foods. Gend (1975) has described an automated distillation procedure with colorimetric finish for application to liquid samples. GLC procedures based on chromatography of methyl esters, or trimethylsilyl derivatives, have been described by Larsson and Fuchs (1974), Fogden *et al.* (1974) and Coelho and Nelson (1983). HPLC techniques have been introduced for the determination of organic preservatives, sweeteners and caffeine in soft drinks. Nelson (1973) used anion exchange columns. Smyly *et al.* (1976) used reversed phase procedures with recoveries better than 99 per cent. Neale and Ridlington (1978) compared four methods for the determination of benzoic acid, sorbic acid and parahydroxybenzoates in soft drinks. The methods included an HPLC procedure, a GLC procedure without derivatisation and two official methods. All methods tested gave statistically equivalent results; the authors claimed that the GLC method had been in routine use for 9 months without the need for a change of columns. The use of HPLC to examine extracts from a wider range of foods has been described by Trifiro *et al.* (1981), Frohlic (1982) and Boehme *et al.* (1984).

4-HYDROXYBENZOATES (PARABENS)

Although allowed to be used in other countries as food preservatives, esters of *p*-hydroxybenzoic acid were not permitted in foodstuffs in the UK until 1962. They are stable and resist hydrolysis and are effective over a wider pH range than benzoic acid. The antimicrobial activity of the 4-hydroxybenzoates increases with the length of the alkyl chain up to the amyl ester. Only the methyl, ethyl and propyl esters (or their respective sodium salts) are allowed to be used in the UK. Factors affecting the choice of paraben preservatives as wide pH range preservatives have been discussed (Anon, 1981a).

Qualitative Test for 4-Hydroxybenzoic Acid

Millon's reagent. Dissolve 3 ml mercury in 27 ml cold fuming nitric acid and dilute with an equal volume of water.

Millon's reagent gives a rose-red colour with 4-hydroxybenzoic acid. The free acid is not steam volatile but the esters are partially so. The test should be applied to the neutral ammonia salt of the acid after extraction with ether from acid solution (esters must be hydrolysed first—see the quantitative test below). As many other aromatic substances with a hydroxyl group attached to the benzene nucleus give colours with Millon's reagent (e.g. salicylic acid gives an orange-red), the test cannot be considered as specified for 4-hydroxybenzoic acid. Salicylic acid can be distinguished, however, by the intense violet colour given with ferric chloride.

Some of the chromatographic methods referred to under 'Benzoic acid' may also be employed for the detection of the *para* acid. Tammilehto and Bucchi (1969) claim that the reaction of *p*-aminobenzoic acid with aminoantipyrine is specific for the free acid. Reversed phase TLC on silanised silica gel has been reported as an efficient system for the separation of individual esters (Rangone and Ambrosio, 1970). Reversed phase HPLC has also been used in application to cured meats (Perfetti *et al.*, 1981). Ehlers and Littman (1984) have reviewed methods for determining parabens esters in salad and mayonnaise products.

Determination of 4-hydroxybenzoic acid

Edwards *et al.* (1937) proposed that the reaction with Millon's reagent could be employed for the colorimetric estimation of 4-hydroxybenzoic acid. The acid is usually present as the ester, which must be hydrolysed before the reagent is added.

Thackray and Hewlett (1964) have pointed out that Millon's reagent lacks reproducibility and stability (apart from possessing other disadvantages) and tentatively recommended a method for determining 4-hydroxybenzoic acid based on the Lemieszek-Chodorowska method. This involves heating with Denigès' reagent followed by addition of sodium nitrite solution. Although the proposed extraction technique does not give complete recoveries, the method can be considered as a useful routine sorting test (cf. Dickes, 1965b).

Reagents

Dilute sulphuric acid. 100 ml conc. sulphuric acid made up to 300 ml with water.

Denigès' reagent. Dissolve 5 g mercuric oxide in 20 ml conc. sulphuric acid and dilute to 100 ml with water.

Method

To 2 g of sample add 60 ml water at 50 °C and adjust to pH 7·5 with sodium hydroxide solution (5 per cent). Heat at 50 °C for 30 min stirring occasionally. Add 2 ml of potassium ferrocyanide solution (15 per cent). Mix carefully, add 2 ml of zinc sulphate solution (30 per cent). Mix, dilute to 100 ml and set aside for 30 min. Filter, and to 50 ml of the filtrate add 1 ml of dilute sulphuric acid. Extract with 3 × 50 ml of diethyl ether. Wash the combined extracts with water (3 × 5 ml/30 s), add 1 drop of phenolphthalein solution (1 per cent) and shake with 3 ml of 0·25 M sodium hydroxide solution. Wash with 3 × 1 ml water, combine the alkaline extracts and washings, heat in a boiling-water bath to remove ether, cool and dilute with water to 10 ml. To 5 ml of this solution add 5 ml of Denigès' reagent. Heat in a boiling-water bath for 5 min. Cool, add 5 drops of aqueous sodium nitrite solution (2 per cent), and allow to stand for 45 min. Measure the absorbance of the pink colour produced at 518 nm. Dissolve 50, 100, 200, 400 and 600 μg of ester in 3-ml quantities of 0·25 M sodium hydroxide, make up to 5 ml and carry out the above method starting from 'add 5 ml of Denigès' reagent. . .' Prepare a calibration graph or calculate a mean extinction coefficient.

It may be necessary to know which ester is present (Dickes, 1965b) but otherwise the results can be reported as the methyl ester.

The same paper also includes a tentative qualitative procedure in which 2 g sample is mixed with 1 ml sulphuric acid (10 per cent). After grinding with anhydrous sodium sulphate, the mixture is stirred with diethyl ether (3 × 25 ml). The combined decanted, filtered extracts are evaporated to dryness and the residue is dissolved by warming with 1 ml 0·1 M sodium hydroxide. After making up to 5 ml the test is completed as in the above method commencing at 'add 5 ml of Denigès' reagent. . .'

Chromatographic methods have been described for determination of the free acid; references to appropriate techniques have been cited in sections dealing with benzoic acid and the *p*-hydroxybenzoate esters.

SALICYLIC ACID

Salicylates were fairly extensively used as food preservatives but their use in the UK was prohibited in 1925.

The use of ferric chloride for estimating salicylic acid in the presence of benzoic acid has been discussed by Nicholls (1928). Salicylates can also be detected by *Jorissen's test:*

To 20 ml of a neutral solution containing about 0·02 g of the extracted acid add 1 ml of copper sulphate solution (1 per cent), 0·01 g of potassium nitrite and acidify with acetic acid (50 per cent). Place in boiling water for some minutes. In the presence of salicylate a red colour is produced. The isomaltol obtained in the steam distillate from coffee and chicory essence gives a violet colour with ferric chloride by Jorissen's test gives a negative results.

Englis *et al.* (1955) have proposed a spectrophotometric method for determining salicylic acid using ether as solvent (absorption peak 236 nm; secondary peak at 306 nm). Salicylic acid is also included in the thin layer and gas chromatographic methods described earlier for benzoic acid, in particular Larsson and Fuchs (1974) used the trimethyl silyl derivative in a gas chromatographic procedure. Official methods described by AOAC and the Nordic Committee on Food Analysis (1968) rely on the ferric chloride colorimetric procedure.

PARACHLOROBENZOIC ACID

Parachlorobenzoic acid appears to be effective for controlling mould and rope in bread and although not permitted in the UK its use is allowed elsewhere. The acid can be separated by steam distillation and extraction and detected by one of the forms of Mohler's test. The material is nitrated with sulphuric acid and potassium nitrate, diluted and made alkaline with ammonia. Then hydroxylamine hydrochloride is layered on top of the liquid in a test-tube so that a green zone appears. The low solubility of *p*-chlorobenzoic acid in water (1 in 5200 at 20 °C) is said to enable a separation to be made from benzoic acid (1 in 350) by treating with water. The acids can be separated by chromatographic methods referred to under 'Benzoic acid'.

DEHYDROACETIC ACID

Dehydroacetic acid (DHA) and its sodium salt (DHA-S) appear to be very effective agents for controlling mould growth. The FSC (1959) stated that the margin between the possible toxic level and the amounts likely to be used commercially is too small to justify the use of DHA or its sodium salt as a food preservative.

For its detection, dehydroacetic acid can be extracted with ether from the food or a chloroform extract of it. After removing the solvent, the residue is dissolved in sodium hydroxide solution and the solution is transferred to a test-tube. On adding salicylaldehyde solution (20 per cent in alcohol), placing in boiling water for 5 min and diluting with water, a red or orange colour is produced in the presence of dehydroacetic acid (AOAC). Sperlich (1960) has reported that the free acid gives an intense blue–violet colour with titanium trichloride. A chromatographic procedure for use with wines has been described by Daniels *et al.* (1983).

PROPIONIC ACID

The Preservative Regulations permit the presence of propionic acid in flour confectionery (max 1000 mg/kg) and Christmas pudding (max. 1000 mg/kg). The addition of propionic acid and its salts to bread is controlled by the Bread and Flour Regulations (1984) (see Chapter 8). Propionic acid or its calcium or sodium salts have been shown to be effective in the control of moulds and rope. It appears to be more effective in acid solution and it is probable therefore that the antimicrobial action is due to the undissociated acid.

Chromatographic methods are generally used to identify propionic acid in extracts prepared by steam distillation. Tjan and Jansen (1971) described a TLC procedure for application to bakery products.

Methods for the determination of propionic acid involve extraction of acids present in the food and separation and identification of the propionic acid from the other lower fatty acids which may be co-extracted. Walker and Green (1965) have recommended a silica gel partition method which is the current official method of the AOAC. Cochrane (1975) reviewed available gas chromatographic methods for the determination of the C_2 to C_6 fatty acids. Laub *et al.* (1983) used butyric acid as internal standard in a simple GLC method. Tonogai *et al.* (1983a) reported a similar procedure using direct solvent extraction with acidified ethylacetate followed by GLC in the analysis of bakery products. Alkali extraction followed by conversion to *p*-nitrobenzyl ester was used by Takatsuki and Sakai (1982). The method was claimed to be equivalent to the AOAC steam distillation procedure. Yabe *et al.* (1983) used methanol extraction, formation of a phenylacyl ester by reaction with *p*-bromophenylacyl bromide. The ester was subjected to HPLC and detected by absorption at 254 nm. Recoveries in excess of 98 per cent from white bread were recorded.

SORBIC ACID

The 1989 Regulations, like the preceding Regulations, take into account the recommendations of the FACC (1977) in extending the range of products to which sorbic acid may be added. It is usually incorporated as a salt and is more effective against yeasts and moulds than bacteria. A review of the properties of sorbates and factors which determine the antimicrobial efficiency in application to typical food products has been prepared by Liewen and Marth (1985a). Sorbic acid has been identified as the cause of the development of off-flavours in Feta cheese due to decarboxylation by moulds (Horwood *et al.*, 1981; Liewen and Marth, 1985b).

For its estimation, sorbic acid can be steam distilled from acid as in the Monier-Williams process for benzoic acid, but omitting treatment of the distillate with permanganate. Carr and Smith (1964), working on prunes, showed that in order to obtain a colourless distillate it was advisable to steam distil from sulphuric acid after saturating the sample with magnesium sulphate. The sorbic acid in the distillate can then be determined either by ultraviolet absorption at pH 2–4 or by oxidation with acid dichromate to malonaldehyde, which reacts with thiobarbituric acid to give a red colour having a maximum absorption at 532 mm.

REAGENTS

Acidified dichromate solution. Mix equal volumes of 0·15 M sulphuric acid and M/600 potassium dichromate solution.

Thiobarbituric acid solution. Dissolve 0·5 g thiobarbituric acid in 25 ml water + 20 ml 0·5 M sodium hydroxide, add 11 ml M hydrochloric acid and dilute to 100 ml with water.

METHOD

Transfer a suitable quantity of sample (e.g. 25–50 g) to a 1 litre double necked flask together with 100 g magnesium sulphate and 100 ml 0·5 M sulphuric acid and steam distill rapidly into 10 ml M sodium hydroxide, collecting approximately 450 ml in 30 min. Do not attempt to apply any heat directly to the flask containing the sample during distillation, otherwise a coloured distillate may result. Transfer the distillate to a 500 ml volumetric flask, add 15 ml 0·5 M sulphuric acid and make up to volume. Determine the sorbic acid by one of the following methods:

Ultraviolet absorption. Make a suitable dilution of the distillate with 0·005 M sulphuric acid. Using water as a reference, measure the extinctions at 262 nm and 280 nm and determine the corrected value for E_{262} from the formula

$$E_{262}^{\text{Sorbic Acid}} = 1{\cdot}418\, E_{262} - 0{\cdot}811\, E_{280}$$

This correction formula is designed to allow for sugar decomposition products derived from fruit in the distillate.

When measurements are taken in the pH range 2 to 4, sorbic acid shows a broad peak from 261 to 264 nm with little variation of the coefficient. By making up the distillate with 0·005 M sulphuric acid a value of $E_{1\,\text{cm}}^{1\%}$ may be taken as 2260 (Melnick and Luckmann, 1954; Alderton and Lewis, 1958).

Colorimetric determination. Pipette 2 ml of distillate into a test-tube and add 2 ml of freshly prepared acidified dichromate solution. Oxidise by heating at 100 °C for 5 min, cool, add 2 ml thiobarbituric acid solution and heat in a boiling water-bath for 10 min. Cool rapidly, dilute with water to 50 ml and measure the absorption of the red colour at 532 nm in a 1 cm cell with reference to water. Calculate the sorbic acid content from the calibration curve.

Preparation of calibration curve. Dissolve 1·0 g sorbic acid in the minimum volume of M sodium hydroxide and dilute to 1 litre with water. Dilute 25 ml to 500 ml with water (50 µg/ml). Then add the following volumes of this solution to a series of 100 ml volumetric flasks: 0, 10, 20, 50, 80 ml, and dilute each to the mark with water. Taken together with the undiluted solution this gives a range from 0 to 50 µg/ml. Pipette 2 ml of each of these solutions into a series of test-tubes and treat as described for the sample solution above (Nury and Bolin, 1962).

Rapid spectrophotometric methods for determining sorbic acid have been described by Lueck and Remmert (1974), Gutfinger *et al.* (1976) and Stafford (1976). Holley and Millard (1980) used direct extraction with iso-octane from an acidified slurry of samples prepared from fermented meat products. The direct ultraviolet absorbance technique was applied with correction for background absorbance after permanganate oxidation. Variants of the above methods are recommended by the AOAC and by the EC (Community Regulation 2984/78, OJ No. L360, 22.12.78).

A number of TLC and HPLC systems have been described for determination of sorbic acid in mixed preservatives (see benzoic acid). Gas chromatographic methods utilising esters or silyl ether derivatives have been described by La Croix and Wong (1972) and Boniforti *et al.* (1961), whilst Graveland (1972) and Noda *et al.* (1973) have described procedures for chromatography of the free acids.

McCalla *et al.* (1977) and Wildanger (1973) have described rapid HPLC techniques using ultraviolet detectors, which offer the advantage of direct on-column injection of liquid samples or simple extracts of foods. The use of HPLC on reversed phase columns has been described by Leuenberger *et al.* (1979) for the determination of mixed preservatives and saccharin in cheese, cakes, ketchups, juices, wines and soft drinks. Liquid foods required no preliminary extraction; complex foods were treated with simple extractions and cleaned up by passing through kieselguhr columns. The preservatives are stripped from the kieselguhr with chloroform–isopropanol and after removal of the solvent the residue is dissolved in methanol and injected on the chromatographic column. Recoveries in excess of 90 per cent were reported for the organic acid preservatives in a range of foodstuffs.

Comparison of HPLC and official spectrophotometric methods has been made by Modi *et al.* (1982). They conclude that the HPLC technique is suitable for official use. Ito *et al.* (1980) and Tame (1980) concluded that the thiobarbituric acid method requires careful attention to detail and that alternate methods are to be preferred.

FORMALDEHYDE

Formalin (36 per cent formaldehyde), a powerful preservative, is not permitted to be added to foods directly. However, under the provisions of the Preservatives Regulations any food may contain up to 5 mg/kg formaldehyde derived from any wet strength wrapping containing formaldehyde-based resin, or of plastic food containers or utensils made from any resin of which formaldehyde is a condensing component. The additive dimethylpolysiloxane may contain up to 1000 mg/kg formaldehyde; food containing dimethylpolysiloxane may thus contain a proportional residue of formaldehyde. Traces of formaldehyde may be found in smoked fish derived from the smoke constituents (Shewan, 1949) and as a result of keeping shellfish in refrigerated storage (Radford and Dalsis, 1982). The definition of 'preservative', however, excludes 'any substance added to food by the process of curing known as smoking'.

Qualitative tests for formaldehyde

In the case of milk, tests for formaldehyde may sometimes be applied directly to the sample. Solid and semi-solid samples should be diluted with water and distilled slowly after acidifying with phosphoric acid. The first few millilitres of distillate should be tested by one of the following tests:

HEHNER'S TEST

Mix 2 ml of distillate with 2 ml of milk in a test-tube. Pour sulphuric acid (90 per cent containing a trace of ferric chloride) down the side of the tube. In the presence of formaldehyde a violet zone forms at the junction of the layers. The violet coloration does not appear usually when relatively large quantities of formaldehyde are present. The milk supplies the tryptophan that must be present for the test to operate.

CHROMOTROPIC ACID TEST

The reagent used is a saturated solution of 1,8-dihydroxynaphthalene-3,6-disulphonic acid in 72 per cent sulphuric acid. For the test, add 1 ml of distillate to 5 ml of the pale straw-coloured reagent in a test-tube, mix and place in boiling water for 15 min. In the presence of formaldehyde a purplish colour is produced.

SCHRYVER'S TEST

Mix 10 ml of distillate with 2 ml of freshly prepared (and filtered) 1 per cent phenylhydrazine hydrochloride solution and add 1 ml of freshly prepared 5 per cent potassium ferricyanide and 4 ml of conc. hydrochloric acid. A red colour is formed in the presence of formaldehyde. A positive reaction is also given by certain methylamines. These latter substances, however, do not give the phloroglucinol reaction which is obtained by adding to an aqueous extract a mixture of 1 per cent of phloroglucinol and an equal volume of 25 per cent sodium hydroxide solution. In the presence of formaldehyde a rose-red colour appears.

SHREWSBURY AND KNAPP'S TEST

The reagent used is prepared by mixing 100 ml of conc. hydrochloric acid with 0·1 ml of conc. nitric acid. With milk, mix 5 ml of sample with 10 ml of reagent and heat in water at 50 °C for 10 min. In the presence of formaldehyde a purple colour is produced.

Determination of formaldehyde

An approximate measure of the amount of preservative present can be obtained by comparing the colour obtained in the above methods with those given by known amounts of formaldehyde. It should be borne in mind that the results obtained tend to be rather lower than the amount originally added due to decomposition after incorporation into the food and to low recoveries in distillates due to polymerisation. It is advisable, therefore, to return the figure as 'Recovered formaldehyde'. The AOAC method for determination of the preservative in maple syrup is based on the reaction between formaldehyde, acetylacetone and ammonia to form a yellow lutidine condensation product. Formaldehyde is normally distilled from the foodstuff before the estimation although in some circumstances the colorimetric reactions may be performed directly.

Hexamethylenetetramine. (Hexamine, $C_6H_{12}N_4$) is permitted in some overseas countries and the 1979 Preservatives Regulations now include a maximum of 25 mg/kg (as formaldehyde) for Provolone cheese and 50 mg/kg for certain marinated fish products. Hexamine is usually estimated by one of the methods described for formaldehyde, to which it is converted on distillation from acid solution. Chromotropic acid reagent is commonly used (Lorenzen and Sieh, 1962; Engst *et al.*, 1969). Brunn and Klostermeyer (1984) reported a chromatographic method for application to cheese. Radford and Dalsis (1982) used HPLC of the 2,4-dinitrophenyl hydrazone derivative to determine formaldehyde residues in shrimp; the limit of detection was estimated as 0·05 mg/kg.

Formic acid. Representations that formic acid should be added to the preservatives permitted in soft drinks were not accepted by the FACC (1972a). For analysis formic acid is usually separated by steam distillation and can be identified on standardised TLC silica-gel plates (Grüne and Nobbe, 1967), and determined by the colour reaction with ferric perchlorate (Guagnini and Vonesch, 1959) or by GLC (Shelley *et al.*, 1963). It can also be determined by the chromotropic acid method after percolating an extract through a column of polyamide powder (Lehmann and Lutz, 1970). The classical gravimetric method following precipitation with mercuric chloride is still a recommended method for determination of formic acid in fish products (AOAC).

DIETHYL PYROCARBONATE (DIETHYL DICARBONATE)

This preservative was developed to replace sulphur dioxide and benzoic acid in soft drinks, wines and beers; in these media most of the additive breaks down into ethanol and carbon dioxide, but in view of the formation of small amounts of urethane in wine it is not permitted in the UK. In the presence of ethanol the pyrocarbonate undergoes solvolysis to yield diethylcarbonate.

Qualitative test for diethyl pyrocarbonate

Diethyl dicarbonate is extracted with diethyl ether and reacted with 4-aminophenazone at 50 °C. Dilution with water, addition of aqueous phenol and ammonia followed by potassium ferricyanide yields a red coloration (Moncelsi, 1970).

Determination of diethyl pyrocarbonate

The colorimetric method above may be used as an assay technique. Gas–liquid chromatography methods are generally used for the estimation of diethyl carbonate residue in wines. Extracts obtained with ether, chloroform or nitric acid are treated with carbon disulphide; Wunderlich (1972) reported on 60–100 mesh Celite 545 loaded with 15 per cent trimethylol-propantripelargonate and 60 mesh Firebrick C-22 loaded with 10–20 per cent Carbowax 20M as column packing materials. Studies on two new spectrophotometric methods for diethyl pyrocarbonate have been carried out by Berger (1975) but, in this case, the methods have been applied to biological fluids. The first method is based on a decolorisation of 5-thio-2-nitrobenzoate, the second on the inactivation of lactate dehydrogenase. Both methods have been used to establish rate constants for the hydrolysis of diethyl pyrocarbonate.

POST-HARVEST FUNGICIDES

Limits of residues of biphenyl, 2-hydroxybiphenyl and thiabendazole are prescribed for specified fruits in the Preservatives Regulations. Biphenyl is effective in preventing mould growth and is normally incorporated into packaging material; 2-hydroxybiphenyl and its sodium salt are effective bactericides and fungicides. Fruits are treated with a solution of the sodium salt in the form of a dip or spray; use is restricted to citrus fruits. Thiabendazole is used on the skin of bananas to control rot. On citrus fruit it is used for control of mould.

Analytical methods for detection and estimation of fungicides have been reviewed by Baker and Hoodless (1974). Schedule 4 of the Preservatives Regulations contains details of sampling rates to be applied to citrus fruits in the enforcement of the provisions in respect of biphenyl and 2-hydroxybiphenyl. The methods of analysis given below are those contained in Schedule 5 of the Regulations; the methods for biphenyl are similar to those recommended by the AOAC.

BIPHENYLS

Qualitative tests for biphenyl and 2-hydroxybiphenyl

The method can be used to detect 5 mg/kg biphenyl and 1 mg/kg 2-hydroxybiphenyl respectively in citrus peel. An extract is prepared from the peel using dichloromethane in an acid medium. The extract is concentrated and separated by TLC using silica gel.

METHOD

Place approximately 80 g of chopped peel into a 250 ml flask. Add 100 ml dichloromethane and 1 ml hydrochloric acid (25 per cent). Heat under reflux for 10 min. Cool and rinse the condenser with 5 ml dichloromethane. Filter the mixture through a fluted paper, collect the filtrate and transfer to an evaporator. Remove solvent at 60 °C to a final volume of 10 ml whilst maintaining the flask in a fixed position. TLC plates (20 × 20 cm) are prepared using a 1:2 slurry of silica gel in water, spread 0·25 mm thick. Dry in warm air. Activate at 110 °C for 30 min. After cooling, divide the plates into lanes 2 cm wide by scribing. Apply 50 μl of the extract from each sample at 1·5 cm from the lower edge of the plate. Develop the plate in a mixture of cyclohexane and dichloromethane (25:95) in a lined tank. Observe the plate under ultraviolet light (254 nm) and note the fluorescent spots. Compare with standards prepared from pure biphenyl and 2-hydroxybiphenyl dissolved in cyclohexane (1 g/100 ml, 1 μl ≡ 10 μg).

CONFIRMATORY TESTS

1. Spray with 0·5 per cent 2,4,7-trinitrofluorenone in acetone, biphenyl gives a yellow spot.
2. Spray with 0·1 per cent 2,6-dibromo-*p*-benzoquinonechlorimine in ethanol, dry in hot air, expose to ammonia vapour, 2-hydroxybiphenyl gives a blue spot.

Determination of biphenyl

The method is accurate to ± 10 per cent for biphenyl levels greater than 10 mg/kg in whole fruit. After distillation in an acid medium, the extract is subjected to TLC on silica gel. The chromatogram is developed and the biphenyl is eluted and determined spectrophotometrically at 247 nm.

METHOD

Shred a 200 g sample of fruit into a mixer. Add 100 ml water, blend at low speed, gradually add 400 ml water and finally blend at high speed for 5 min. Quantitatively transfer the puree to a 2:1 distillation flask with water, rinsing to 1 litre. Add 2 ml conc. sulphuric acid, 1 ml antifoam. Fit a modified Clevenger separator (Fig. 4.3) to the flask, fill the separator with water and add 7 ml cyclohexane to the collecting arm. Distill under reflux for 2 hours. Recover the cyclohexane layer via the stopcock into a 10 ml volumetric flask. Rinse the separator with cyclohexane, collect washings in the volumetric flask and make up to volume. Carry out TLC as described in the qualitative test procedure with the following modifications: use 4·5 cm lanes on the TLC plate and standard amounts of 30, 50 and 70 μg of biphenyl for comparison. Develop the plates to a height of 17 cm in a lined tank, dry in air. Inspect the plates under ultraviolet light (254 nm), mark off rectangles including the areas of silica gel containing the biphenyl: the rectangles should be of equal area. Scrape off the silica gel within each rectangle into individual centrifuge tubes, add 10 ml ethanol to each and mix by shaking for 10 minutes. Centrifuge for 5 minutes at 2500 r.p.m. Decant the supernatant and determine the absorption at 248 nm. Use a control extract prepared from a similar area of the TLC plate free of biphenyl. Plot a calibration curve and calculate in mg/kg (50 μl extract = 1 g sample) (see also Sherma *et al.*, 1983).

Determination of 2-hydroxybiphenyl

The method is suitable for residues of 2-hydroxybiphenyl or its sodium salt in citrus fruits. Recoveries at 12 mg/kg average 80–90 per cent. After distillation in acid medium and extraction by di-isopentyl ether, the extract is purified and treated with 4-aminophenazone to give a red colour. The intensity of the colour is measured at 510 nm.

METHOD

Macerate 250 g of shredded sample with 500 ml water in a high speed blender. Place 150–330 g of the puree into a 1 litre distillation flask with sufficient water to make 500 ml. Add 10 ml orthophosphoric acid (70 per cent w/v) and 0–5 ml antifoam, connect to a modified Clevenger separator. Fill the separator with water and add 10 ml di-isopentyl ether to the collecting tube. Reflux the mixture, maintaining a steady boil, for 6 hours, drain the contents of the separator into a 200 ml separating funnel. Wash the separator and condenser with 60 ml cyclohexane and 60 ml water. Collect the washings in the separating funnel and shake vigorously. Discard the aqueous phase. Add 10 ml of 4 per cent NaOH, shake vigorously for 3 min,

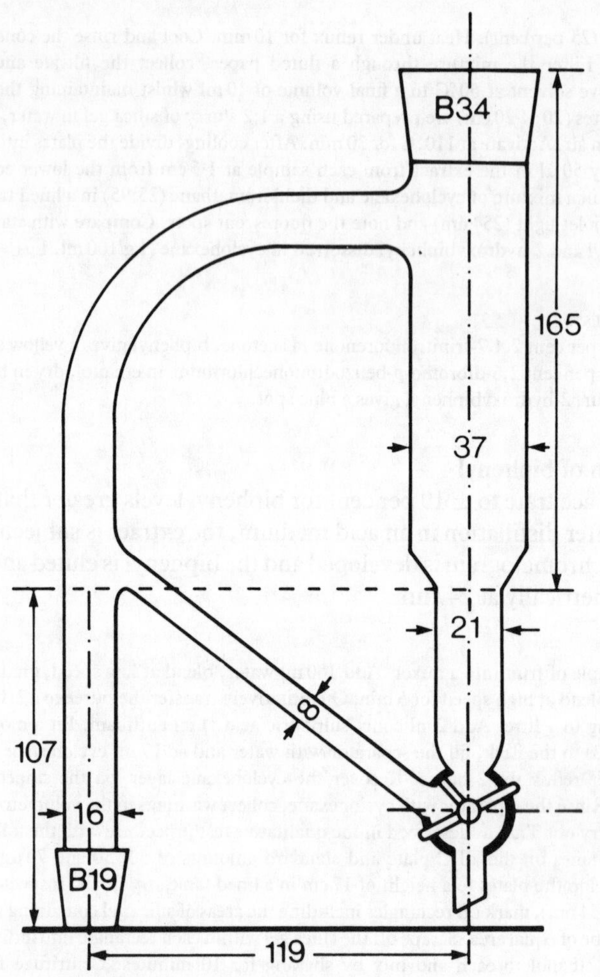

Fig. 4.3 Modified Clevenger type separator. All dimensions in mm.

and collect the aqueous phase in a 100 ml volumetric flask. Repeat the extraction four times, combine the extracts and adjust to pH 9–10 with orthophosphoric acid. Make to volume and add a pinch of silica gel to remove the haze in the solution.

Shake and filter through a dry fluted paper. Place 0·5 and 10 ml aliquots of the filtrate into 25 ml volumetric flasks add 0·5 ml 4-aminophenazone (1 per cent w/v) 10 ml pH 10·4 buffer (0·332 per cent boric acid, 0·4 per cent potassium chloride, 0·186 per cent sodium hydroxide) and 0·5 ml potassium ferricyanide (2 per cent w/v). Make up to the mark and shake vigorously, stand for 5 min and measure absorbance at 510 nm. Prepare a calibration curve for the range 5–50 μg using a solution prepared by dissolving 10 mg 2-hydroxybiphenyl in 1 ml M sodium hydroxide and made up to 1 litre with 0·2 M sodium borate (1 ml = 10 μg 2-hydroxybiphenyl). Untreated citrus fruit has been found to give 'blank' values equivalent to 0·5 mg/kg for oranges and 0·8 mg/kg for lemons.

A critical evaluation of the official methods has been published by Lord *et al.* (1978). The most suitable and commercially available separator conformed to the Type 2 receiver (12·5 ml) specified in BS 756 (1952) for estimation of moisture by the Dean and Stark method. The receiver needed modification by insertion of the three way tap which is now specified in Schedule 6 of the Regulations (Fig. 4.3). These workers also noted that the results obtained in quantitative analysis were critically dependent on procedure, especially in respect of the time of exposure of the TLC plates to ultraviolet light and distillation conditions during the extraction. Player and Wood (1980) reported on collaborative studies of the official methods, that proposed by Lord *et al.* and a modified procedure for 2-hydroxybiphenyl. The conclusions were that the GLC method was equivalent to the official method for biphenyl, but that the official method for hydroxybiphenyl needed modification of the distillation procedure to that now specified.

An alternative GLC procedure for simultaneous determination of both fungicides in a single distillate was also proposed. A glass column packed with 3 per cent OV-17 on Gas Chrom Q coupled to an FID detector was recommended for this determination. Tanaka *et al.* (1978) used a GLC procedure based on separation on a column of 5 per cent PEG-2M on Chromasorb W; these authors also describe a separation system based on standard glassware with ground joints. Improvements in speed of analysis, recoveries and limit of detection were claimed. Farrow *et al.* (1977) have proposed HPLC procedures for the determination of a range of fungicides including all those permitted under the regulations for use on citrus fruits.

Davis and Munroe (1979) proposed the use of ethyl acetete to extract biphenyl from citrus fruit. They claimed that the extracts, when examined by gas chromatography, gave results which were insignificantly different from those obtained by the distillation procedure. The time for analysis was reduced from over 3 hours to less than 15 min. Nakashima *et al.* (1981) have reported the use of HPLC for examination of citrus fruits; recoveries at the statutory limits were reported as 95 per cent for biphenyl and 91 per cent for hydroxybiphenyl. Tanaka *et al.* (1981b) used the reaction between hydroxybiphenyl, *N,N*-diethyl-*p*-phenylene diamine and potassium ferricyanide to form an indophenol dye. The method was claimed to give 95 per cent yields to a detection limit of 0·3 mg/kg.

THIABENDAZOLE

Determination of thiabendazole

Two methods have been proposed for use within the EC for the determination of thiabendazole in citrus fruits and bananas. The first method is described for application to the peel or skins of fruit, the second to the whole fruit. The method applied to whole fruit has been subjected to study and found to be unsatisfactory (Norman *et al.*, 1972a).

METHOD FOR APPLICATION TO PEEL

Thiabendazole is determined by extraction of the peel (or skin) with ethyl acetate in ammoniacal medium. The extract is purified by partition and the thiabendazole determined spectrophotometrically; the results are expressed on the whole fruit basis by making due allowance for the ratio of whole fruit to peel (or skin). The limit of detection is 0·3 mg/kg.

Weigh the fruit (usually four specimens) remove and weigh the peel or skin and determine the ratio of whole fruit to skin. Chop the peel or skin and mix, weigh out 15·00 g ± 0·05 g and place into a homogeniser. Add 2–3 ml ammonia (sp. gr. 0·88) and 100 ml ethyl acetate. Blend at a medium speed for 10 min, allow to stand. Filter through a glass wood plug, collect the filtrate in a 150 ml one mark flask. Repeat the extraction with 50 ml ethyl acetate, filter and combine the filtrates. Make to volume. Place 50 ml of filtrate in a 250 ml separator. Add 20 ml M sodium hydroxide, shake for 5 min and stand until the phases separate completely. Discard the aqueous phase. Repeat the extraction with 20 ml M sodium hydroxide, discard. Wash the organic phase with two 5 ml aliquots distilled water; discard the aqueous phase. Add 10 ml 0·1 M hydrochloric acid to the separator. Shake and stand until the phases separate completely. Collect the acid in a 150 ml separator. Repeat the extraction with 10 ml and 5 ml hydrochloric acid. Combine the extracts and discard the organic phase. Add 5 ml buffer solution (sodium acetate hydrated 3·3 per cent, sodium hydroxide 4·0 per cent, sodium chloride 20·0 per cent in water) and 15 ml ethyl acetate to the acid extract. Shake for 5 min, stand and discard the aqueous phase. Wash with 10 ml distilled water and discard the water. Add to the organic phase 10 ml 0·1 M hydrochloric acid saturated with ethyl acetate. Shake for 5 min, stand until the phases separate completely. Collect the aqueous phase in a 10 mm spectrophotometer cell and plot the absorption spectrum over the range 240–350 nm, retain the organic phase. Use 0·1 M hydrochloric acid saturated with ethyl acetate as the reference solution. If the optical density at the absorption peak exceeds 0·8, transfer the whole of the extract to a 25 ml one mark flask. Extract the organic phase with 10 ml and 5 ml 0·1 M hydrochloric acid saturated with ethyl acetate. Combine the acid extracts, make to volume and repeat the spectrophotometric step.

Calibration

Prepare standard solutions containing 2, 4 and 8 mg/l thiabendazole in 0·1 M hydrochloric acid, from a stock solution containing 160 mg/l thiabendazole in 0·1 M hydrochloric acid.

Plot the absorption spectra of the standard solutions over the range 240–350 nm, using 0·1 M hydrochloric acid as reference solution. Measure the optical density at the maximum (about 303 nm) by reference to a baseline drawn tangentially to the absorption minima at 260 and 334 nm. Prepare the calibration graph and read off the appropriate value for the test solution. After applying the baseline correction, calculate the equivalent content of the original sample.

The identity of thiabendazole in the extract is confirmed by correspondence of the spectrum with that for the standard solutions. The ratios of optical densities 303 nm/316 nm and 303 nm/287 nm are 1·70 ± 0·01 and 1.11 ± 0·01. Deviations from these values are indicative of the probable presence of other fungicides.

A spectrofluorometric method has been proposed but citrus constituents and benomyl interfere. Norman *et al.* (1972b) introduced a TLC method for clean-up of the extracts. Confirmatory tests using enzymatic inhibition have been reported by Tjan and Burgess (1973), whilst Baker and Hoodless (1973) have described a TLC/bioautographic procedure for benzimidazole compounds. HPLC methods using methanol extraction and fluorimetric detection have been reported by Isshiki *et al.* (1980) and Yamada *et al.* (1984). Tafuri *et al.* (1980) used a preliminary derivatisation to form a *p*-nitrobenzyl derivative followed by HPLC; the method was suitable for examination of bananas, pears and citrus fruits. Residues in marmalade at levels to 0·1 mg/kg have been recorded.

THIOUREA

Thiourea ($CS(NH_2)_2$) has been used as a rot and mould suppressant in oranges; MOF Circular MF 2/54, however, stated that its use should be considered as a contravention of the Preservatives Regulations. Thiourea can be detected by its reaction with Grote's reagent (Devani *et al.*, 1973).

Qualitative Test Reagent

Dissolve 10 g sodium nitroprusside in 40 ml ammonia solution. Allow to stand overnight at 0 °C so that all the nitrosoferricyanide decomposes (shown when a few drops no longer produce a red colour when added to a solution of creatinine in M sodium hydroxide). Filter, and add absolute alcohol to the filtrate until no further precipitate forms. Filter, wash the pentacyanoammonioferroate precipitate with absolute alcohol until free from ammonia, and dry in a vacuum desiccator. Store in the dark in a desiccator containing calcium chloride. Prepare a 1 per cent aqueous solution, expose it for 1 day to the air and light and store in a dark bottle in the dark.

METHOD

Shake 10–20 ml orange juice with two thirds its volume of ether, centrifuge and separate the lower layer. Stir in Filter-Cel and transfer to a Hirsch funnel. Keep the vacuum on for a short time and agitate the solution to remove most of the ether. To 5 ml of filtrate add 5 drops of the pentacyanoammonioferroate reagent. If no blue colour appears add a few drops of 0·1 M iodine dropwise, shaking after each addition. About 5 drops are usually necessary for the development of the maximum colour (excess reduces the colour). The presence of thiourea is indicated if a bluish green colour develops.

Determination

The AOAC describes a qualitative test using a modified Grote's reagent which is also employed for the estimation of thiourea in oranges and in frozen peaches. In the quantitative method the blue colour is measured at 610 nm (Winkler, 1961). At low level the colorimetric method is subject to interference from the constituents of the citrus fruits (Toyoda *et al.*, 1979). They propose a method based on an extraction of the citrus peel with ethylacetate/acetone (2:1). The extract is cleaned up on an alumina column, and the thiourea in the eluate is determined as a benzoyl derivative by gas chromatography on a column of Chromosorb G loaded with silicone SE 30. The detection limit is claimed to be 0·08 mg/kg compared with 1 mg/kg for the colorimetric method. Thiourea should be readily detectable by HPLC with ultraviolet detection.

HYDROGEN PEROXIDE

The use of UHT processing, e.g. heating at 140 °C for 5 s, as a means of sterilising liquid foods such as milk, fruit juice and soups has been dependent on the parallel development of techniques of aseptic handling and dispensing into pre-sterilised containers. The containers are usually formed and sterilised at the point of filling; materials used include polypropylene and paper board laminate which may be sterilised by treatment with hydrogen peroxide either singly or in combination with exposure to ultraviolet radiation (Peel and Waites, 1979).

Food which is processed and packaged in this way is expected to be free from residues of hydrogen peroxide. A simple test relies on the addition of a few drops of vanadyl sulphate (1 per cent in sulphuric acid) to 10 ml of milk. A pink or red coloration indicates the presence of peroxide. A number of quantitative methods have been developed in recent years, amongst which the method developed by Iwaida *et al.* (1981) depends on the oxidation of gum guaiac in the presence of peroxidase and is capable of detecting 0·25 μg. Toyoda *et al.* (1982) have described the use of a sensitive oxygen electrode to detect 0·01 mg/l in extracts from a number of foods including milk, yoghurt and fish products. The reaction of peroxide with ammonium thiocyanate in the presence of trichloroacetic acid is capable of detecting 0·05 mg/l (Asai *et al.*, 1982). Perkin (1982) has used the chemiluminescent oxidation of 3-aminophthalhydrazide in the presence of copper as a catalyst to detect levels to 1 μg/l.

ANTIBIOTICS

The FACC (1972a) accepted the view of the Swann Committee that the use of antibiotics in food should be permitted only in exceptional circumstances. The 1962

Regulations permitted the presence of tetracylines in raw fish, nisin in cheese, clotted cream and certain canned food and nystatin for controlling rot in bananas. Of these only nisin is now permitted to be used, although within some EC countries pimaricin has been accepted for application to the rinds of hard cheeses such as Edam and Gouda and to the skins of cured sausages. Within the EC, proposals to permit the use of natamycin in cheese are under discussion. A method of analysis has been proposed, which is based on extraction with methanol followed by cooling to remove fat and spectrophotometric analysis over the range 311–329 nm. Alternatively, the extract may be analysed by HPLC using methanol/water/acetic acid as mobile phase and Lichrosorb RP8 as the analytical column (Tuinstra and Traag, 1982).

Nisin is naturally present in certain cheeses. Its value as a preservative lies in the control of spores of thermophilic strains of *Bacillus* and *Clostridium*, which multiply in cheeses and other products leading to blowing and hard swells. When nisin is added in sufficient amount to control spoilage the residual nisin in processed products is comparatively small, but is sufficient to prevent growth of any remaining spores. It is stable in acid, unstable in alkali, and in the body is destroyed by gut enzymes and appears to be harmless to man. Evidence from the Antibiotics Panel published in the FSC (1959) Report indicated that nisin could be permitted in canned foods where there is no risk of botulism arising. The presence of nisin is therefore permitted in canned food which has either been sufficiently heat processed to destroy *Clostridium botulinum* or has a pH less than 4·5, as well as in cheese and clotted cream. Nisin can be assayed after extraction from processed cheese with hydrochloric acid by a plate diffusion technique using *Micrococcus flavus* as test organism (BS 4020: 1974). For sorting purposes Tramer and Fowler (1964) have described a reverse-phase disc assay technique based on *B. stearothermophilus* spores with overnight incubation at 55 °C. Stumbo *et al.* (1964) have described a sensitive method employing 'heat-damaged' spores of *B. thermophilus* for determining nisin in numerous canned foods.

ANTIOXIDANTS

Legislation

Antioxidants are added to oils and fats to delay the onset of oxidative rancidity, which is the chief cause of spoilage. Care in processing and the selection of storage conditions minimises such deterioration, but the more or less rapid increase in peroxides following an induction period causes considerable trouble. Antioxidants do not render a rancid fat palatable. The mechanisms of lipid oxidation and its inhibition by antioxidants have been discussed by Stuckey (1962) and Coppens (1985), who have noted that antioxidants exhibit synergism, being more potent as mixtures than when used individually. Natural constituents of foods such as tocopherols and ascorbates also have antioxidant properties (Rugraff *et al.*, 1981; Taylor and Barnes, 1981; Bieri, 1984). Bishov *et al.* (1977) noted that phenolic constituents of many foods, and especially spices, contributed to the antioxidant properties of natural food ingredients and participated in the synergism with added antioxidants. Chelating agents such as citric acid may also contribute to antioxidant properties since they inhibit the catalytic effects of traces of copper and iron which may be present in foods. From 1953 antioxidants have been the subject of reports of the FSC and subsequently the FACC. These reports have not, however, substantially altered the statutory limits imposed, although the (1974a) report of the FACC recommended a limited extention of use of specified antioxidants. The Antioxidant in Food Regulations 1978 (SI No. 105) as amended, and which implement provisions of the EC Directives based on 70/357/EEC, permit the presence of various gallates, butylated hydroxyanisole (BHA) and butylated hydroxytoluene (BHT) in dairy products and fats up to certain limits. It should be noted that this directive regards calcium disodium EDTA as an antioxidant for use in mayonnaise and condiment sauces (75 mg/kg) and conserves of various types of beans and lentils (250 mg/kg). Regulations in the UK do not include this usage, for which a derogation exists to the end of 1986. The use of BHA and BHT only are permitted in dehydrated potato products, walnuts, vitamin A preparations and chewing gum base. The permitted list of antioxidants now also includes L-ascorbic acid (and the sodium and calcium salts), ascorbyl palmitate and various tocopherols together with ethoxyquin and diphenylamine as antiscald agents for apples and pears. The regulations prohibit the description or advertisement of any food as being intended mainly for babies or young children if it contains specified antioxidants. The Antioxidant in Foods (Amendment) Regulations 1980 (SI No. 1831) bring into force purity criteria specified in Council Directive 78/664/EEC. These represent amendments of the criteria laid down in the parent regulations.

Detection and determination of antioxidants

A number of reviews of methods for the detection and estimation of antioxidant in fats and food products have been published and many schemes of analysis have been proposed, based on colorimetry, spectroscopy and chromatography (TLC, GLC and HPLC). Significant reviews include those of Cantafora (1976) and Endean (1976b). Methods of analysis for BHT, BHA and the gallates in fat have attracted considerable attention but there are relatively few methods for the other antioxidants. By virtue of their properties antioxidants may be lost during processing and storage of foods. They will also react with peroxides in solvents such as ether and particular attention must be paid to the purification of solvents and to conditions of extraction during the examination of foods.

The extraction procedures used also depend on the nature of the foods and on the antioxidants present. The simple phenolic antioxidants such as BHA and BHT are steam volatile, insoluble in water but soluble in polar solvents; propyl gallate is soluble in water (0·3 per cent) but the higher molecular weight esters tend to be lipophilic. Stuckey and Osborne (1965) have suggested two approaches for the extraction of antioxidant from foods. For high fat foods the fat and antioxidant are removed together by solvent extraction. This is followed by extraction of the antioxidants from fat. For low fat foods, direct solvent extraction processes may be used. In the case of BHA and BHT, steam distillation procedures have been adopted by Dilli and Robards (1977) and by Maruyama *et al.* (1977).

The latter authors used distillation from an acidified salt solution of the food with heptane as a carrier for the antioxidants. Extraction of fat from foods may be carried out by the Soxhlet procedure using petroleum ether or similar solvents. The addition of sodium sulphate may be used to dry moist foods and so aid the extraction of the fat. Chloroform/methanol mixture (1:1) has been preferred by a number of workers as the primary extraction medium; however, when water content of the solvent mixture exceeds 8·5 per cent, phase separation occurs. Thus extracts from moist food will form two layers, the aqueous phase containing water-soluble antioxidants such as propyl gallate and the chloroform layer the fat and other hydrophobic compounds.

Extraction of antioxidant from oils and fats

The most useful solvents for removal of antioxidants from oils and fats are alcohol and acetonitrile. Generally, the fat or oil is dissolved in hydrocarbon solvents and is then subjected to extraction. When acetonitrile is used, the recovery of BHT is generally low, unless many extractions are carried out and dimethyl sulphoxide has been suggested as an alternative solvent for this reason (Phipps, 1973). Many workers have used aqueous ethanol to remove the antioxidants from oils and fats, the extent of removal depending on the alcohol concentration. Use of aqueous methanol has also been advocated by the APA (1963). Peteghem and Dekeyser (1981) used 95 per cent methanol to extract the commonly used antioxidants from lard, shortening and vegetable oil. Endean and Bielby (1975) favoured sequential extraction with acetonitrile and dry methanol; they also claimed, firstly, that BHA, BHT, α-tocopherol, gallates and ascorbyl palmitate could be separated from petroleum ether solutions of oil and fat and, secondly, by using anhydrous solvents these could be removed by evaporation without loss of antioxidant. The AOAC methods book describes a combination of extraction stages to separate propyl gallate, nordihydroguaiaretic acid (NDGA), BHA and BHT from solutions of oil and fats dissolved in petroleum ether. Water removes propyl gallates and acetonitrile the remaining antioxidants from the petroleum ether solution. After dilution of the acetonitrile with water, the antioxidants are back-extracted into petroleum ether and subjected to identity tests. The BHT may be selectively separated by passing the petroleum ether solution through a column of Florisil.

Spectrophotometric procedures for the estimation of individual antioxidants may be applied to the extracts prepared from oils and fats; BHA is conventionally estimated by reaction with diazobenzene sulphonic acid in the presence of sodium nitrite, whilst BHT is estimated by reaction with dianisidine (AOAC). Gallates may be estimated by reaction with ammonium ferrous sulphate or ferrous tartrate (AOAC) to give the characteristic blue colour. However, these specific colorimetric reactions are now being replaced by chromatographic procedures.

GENERAL TLC METHOD

Dissolve 10 g of warm oil or melted fat in 100 ml 40–60° petroleum ether in a beaker, carefully transfer to a 250 ml separator and wash in with petroleum ether. Ensure that the fat or oil is fully dissolved. Add 25 ml acetronitrile saturated with petroleum ether to the separator. Mix carefully by repeated inversion. Run off the acetronitrile into a second separator and repeat the extraction three times. Combine the acetronitrile extracts in a 150 ml round bottom flask. Remove the solvent *in vacuo* in a rotary evaporator at a temperature not exceeding 40 °C. Dissolve the residue in 2 ml methanol, filter if not entirely soluble. Prepare 20 × 20 cm silica gel G plates with a 0·25 mm layer, using 30 g in a slurry with 60 ml 1 per cent citric acid solution. Dry the plates in air, activate at 130 °C for 1 hour, store over activated silica gel in a desiccator. Charge the developing tank with a freshly prepared solvent mixture of petroleum ether/benzene/glacial acetic acid (2:2:1). Line the tank with filter paper, allow to stabilise for 1 to 2 hours in the dark. Place 10 μl and 20 μl spots of extract solution 2 cm apart on a start line on the TLC plate at 2 cm from the bottom edge. Add 5 μl spots of standard antioxidant solution (0·1 per cent m/v in methanol) to the plates at 2 cm intervals along the line. Dry the plate in a stream of hot air and develop in the dark in the chromatographic tank to a distance of 15 cm from the start line. Remove the plate, dry in air. Spray the plate with Gibbs reagent (0·5 per cent m/v 2,6-dichloro-*p*-benzoquinone-4-chlorimine in absolute ethanol). Dry at 103 ± 2 °C for 15 min. Compare the colour and R_F values of the prepared extract with the standards. Cool the plate and place in a tank containing ammonia vapour. Note characteristic colour changes as indicated in the table below:

Antioxidant	R_F	Gibbs reagent	Gibbs reagent followed by ammonia
Propyl gallate	0·12	Brown	Grey–green
Octyl gallate	0·22	Brown	Grey–green
Dodecyl gallate	0·27	Brown	Grey–green
BHA	0·62	Brown–red	Grey
BHT	0·89	Brown–violet	Grey

Other spray reagents have been used, including 0·1 per cent m/v ferric chloride/potassium ferricyanide (1:1) in 80 per cent ethanol (blue spots on yellow–green background) and 0·5 per cent ferric chloride/0·5 per cent α,α′-dipyridyl in methanol, 1:1 (red–brown spots on white background). Endean (1976b) lists 19 further spray reagents, together with the colour responses for the common antioxidants.

Quantitative estimates of the antioxidant concentration may be made by comparison of the spot intensities obtained by varying the amounts of extract (1, 5, 10, 20 μl) run against 5 and 10 μl of solutions containing all the appropriate antioxidants of interest at concentrations of 0·1 per cent in methanol.

Other Methods

Endean and Bielby (1975) have also described similar TLC procedures which include detection and estimation of α-tocopherol and ascorbyl palmitate together with the other antioxidants. An HPLC method for the estimation of gallates, BHA and BHT has been described by Hammond (1978) using methanol extraction followed by deep freeze treatment to remove insoluble fat. The filtered methanol is injected onto a reversed phase column and developed with a linear gradient of methanol varying from 55 to 85 per cent in water; 3,4-xylenol is used as internal standard. Procedures are also described for the confirmation of the estimates of BHT using a developing solvent of methanol/M acetic acid (4:1) on the reversed phase column, which reduces the retention time from 19·5 to 7 min. The extract for the latter method is prepared by treating the oil dissolved in heptane with dimethyl sulphoxide. Ultraviolet detection at 280 nm is used in both methods. Page (1979) has also described an HPLC procedure using solvent extraction, an acidified water/acetonitrile mobile phase with a reversed phase column. Resolution of nine common antioxidants is claimed. The same author (1983) reported on a successful collaborative study of this procedure. Amperometric detection was used in reversed phase HPLC by Kitada *et al.* (1985).

Doeden *et al.* (1979) have described a method using gel permeation of a chloroform solution of oil for estimation of BHA, BHT and *t*-butylquinol (TBHQ) in edible fats and oils. Detection limits of less than 1 mg/kg are claimed for BHT and BHA whilst the limit for TBHQ is approximately 20 mg/kg. Kline *et al.* (1978) have described two GLC methods for determination of 11 antioxidants permitted in the USA. Extracts for estimation of gallates and TBHQ are obtained with 70 per cent ethanol, the residues converted to trimethylsilyl derivates and separated on 3 per cent OV-225 on 80–100 mesh Gas Chrom Q. For the estimation of BHT, BHA, TBHQ and 4-hydroxymethyl-2,6-di-*tert*-butylphenol (Ionox-100), the sample is extracted with ethyl acetate and injected onto 3 per cent OV-17 on 80–100 mesh Gas Chrom Q. Recoveries of antioxidants vary from 90 to 100 per cent at 100 mg/kg for the direct injection procedure and from 70 to 100 per cent for the derivatisation method. Conditions for GC/MS are also described. Carbon disulphide is used as extraction solvent in the AOAC method for the estimation of BHA and BHT in cereal products; GLC on an apiezon column is used in the estimation. Dickes and Nicholas (1976) describe a range of GLC methods for permitted and non-permitted antioxidants. Galensa and Schaefers (1982) used acetonitrile extraction followed by conversion to benzoyl derivatives with benzoyl chloride as a preliminary step to the utilisation of normal and reversed phase HPLC. They report that BHT is not completely benzoylated and that GLC is the preferred technique in this case. Yo *et al.* (1984) used direct injection of vegetable oil into high resolution capillary GC columns for BHA, BHT, TBHQ and Ionox-100 antioxidants; see also Greenberg *et al.* (1984) and Min and Schweizer (1982) for direct GLC methods.

Ethoxyquin is permitted with diphenylamine as an antiscald agent on apples and pears. It is included in the method due to Alicino *et al.* (1963), which is based on solvent extraction and column chromatography; BHA, BHT and ethoxyquin are subsequently determined by ultraviolet spectrophotometry. A fluorimetric method for the determination of ethoxyquin in foods has been discussed by Howard (1972) and Baldi *et al.* (1982a, b), and in fish meal by a spectrophotometric technique (Spark, 1982).

Winell (1976) has described a gas chromatographic method for application to apples. Ethoxyquin is extracted from a homogenate of the sample with hexane, is purified by extraction into acid, and back into hexane from an alkaline phase. The hexane solution is mixed with heptafluorbutyric anhydride. The derivative is chromatographed against tetrahydroquinoline as internal standard using an electron-capture detector. Recoveries of 80–106 per cent are quoted for levels varying from 0·1 to 3·0 mg/kg. Olek *et al.* (1983) described an HPLC technique combined with electrochemical detection. The procedure requires no derivatisation step.

COLOURING MATTER

Legislation

The use of colouring matter in food has been acknowledged to play a part in the consumer acceptability of processed foods. However, the need to monitor the compounds used to colour foods has been recognised since the early days of the century when coloured salts of heavy metals were used to restore the colour of processed foods such as canned peas. Parts of the Preservatives etc. Regulations 1925–1927 provided for the prohibition of use of certain mineral, vegetable and coal tar colours. Subsequently, lists of dyes permitted for use in foods were incorporated into Regulations promulgated in 1957, 1966 and 1973 (SI 1973 No. 1340). The 1973 Colouring Matter in Food Regulations partially implemented the EC Council Directive of 1962 on colouring matters authorised for use in foods. Permitted colours are listed in Part I of Schedule 1 of the Regulations but in order to take into account amendments of the EC Directive this list has been amended several times (SI 1975 No. 1488, SI 1976 No. 2086, SI 1978 No. 1787 and SI 1987 No. 1987).

Part II to Schedule I of the 1973 Regulations specifies the purity criteria for the permitted dyes and Part III prescribes maximum limits for numerous trace metals, free aromatic amines, synthetic intermediates, isomers (4 per cent) and substances extractable by diethyl ether. Schedule 2 prescribes the solvents and diluents that may be used and purity criteria for the diluents. Directive 81/712/EEC is concerned with methods for establishing these qualities.

Various foods are not permitted to contain added colouring matters, namely raw meat, game, poultry, fish, fruit or vegetable; tea, tea essence, coffee, coffee products, condensed or dried milk; milk, cream and certain types of bread. However, the husk of nuts may be coloured. Foods such as bread, cheese (The Cheese (Amendment) Regulations 1974, SI 1974 No. 1122) and butter are only permitted to contain a restricted number of named dyes. The basic dye methyl violet is only to be used for marking raw meat (EC Directive 64/433 on trade in fresh meat). The 1987 Amendment Regulations revoked the use of methyl violet for marking citrus fruit. These Amendments also removed Yellow 2G from the permitted list and prohibited the addition of colouring matter (other than those derived from vitamin sources) to food prepared for babies and young children. The methods of incorporating various types of permitted colouring matters into foods are discussed by Allman (1973).

In an interim report on the review of the 1973 Regulations, the FACC (1979a) made 16 recommendations. The principal proposals were concerned with stricter control on the use of colours, the deletion of certain colours and the need for further toxicological study of a number of the colours currently allowed. Proposals were also made to amplify the purity specifications for permitted colours, which are fully detailed in an appendix to the Report. A survey of commercial dyes against the recommendations was carried out by Hunziker and Zimmerli (1983). A study by the Working Party on the Dietary Intake of Permitted Food Colours is also included as an appendix, indicating that with the exception of one colour, Brown FK, the estimated average intake is well within the limits for published acceptable daily intakes. Methods for the determination of dyestuff intermediates have been developed by a number of authors. A list of recent publications is given in Table 4.3.

Table 4.3 Methods for examination of dyes for intermediates

Name of dyestuff	Technique used	Reference
Erythrosine	HPLC reversed phase	Calvey and Goldberg (1982) Goldberg and Calvey (1982)
Tartrazine	Column chromatography, HPLC	Calvey *et al.* (1981)
Tartrazine	HPLC	Cox and McClure (1982)
Tartrazine, amaranth, Sunset Yellow	Ion-pair HPLC	Lancaster and Lawrence (1982, 1983)
Tartrazine, amaranth, Sunset Yellow Fast Green, Brilliant Blue, Indigo Carmine	HPLC	Tonogai *et al.* (1983)

Newsome (1986) has summarised the legislative position in the USA and provides usage data for the various permitted colours.

GENERAL METHODS FOR IDENTIFICATION

The general scheme for identifying coal-tar dyes present in foods normally involves:

1. Preliminary treatment of the food.
2. Extraction and purification of the dye from the prepared solution or extract of the food.
3. Separation of mixed colouring if more than one is present.
4. Identification of the separated dyes.

Schemes for the extraction and identification of dyes from various foods have been described by the APA (1960), BFMIRA (1963), Stanley and Kirk (1963), Lehmann *et al.* (1970), Gilhooley *et al.* (1972), Graichen (1975) and Boley *et al.* (1979, 1980).

PRELIMINARY TREATMENT OF THE FOOD

Extraction

Dyes can be extracted from foods by adsorption and desorption with polyamide, which retains all acid coal-tar dyes (Lehmann *et al.*, 1970). Unlike the wool dyeing technique which is now rarely used, this procedure does not bring about changes in the dye, so there should be no effect on the subsequent chromatographic and spectrophotometric examination.

Graichen (1975) has described the results of a collaborative study of the use of liquid ion exchangers to elute the dyes from columns prepared by admixture of the food with Celite and one of three aqueous phases. The aqueous phase is selected according to the nature and colour of the food under examination. A solution of amberlite resin in butanol or in hexane is used to elute the dyes from the column. Recently, because of the increasing restrictions on the use of dyes, attention has been drawn to the need to estimate the amounts present in foods. Boley *et al.* (1980) have described a series of extraction procedures based on those of Gilhooley *et al.* (1972) and Graichen (1975), with enzyme pretreatment to release dyes from processed foods with high starch or high protein contents (see Table 4.4). Detailed extraction

Table 4.4 Conditions for enzymic digestion of major food constituents (Boley *et al.*, 1980)

Substrate	Enzyme	Amount (mg)	pH	Temperature (°C)	Examples of use
Protein	Papain	100	7.0	30	Cake, fish and meat products
Fat	Lipase	50	7.7	30	Cake, fish and meat products
Phospholipid	Phospholipase	10	7.3	30	Sponge cake and egg products
Starch	Amyloglucosidase	100	4.5	50	Cereals, luncheon meat, jams, fruit, modified starches
Pectin	Pectinase	50	4.0	50	Jams and fruit
Cellulose	Cellulase	50	5.3	50	Jams and fruit

procedures are described with recoveries approaching 90 per cent for individual food types with the commonly occurring permitted colours. To overcome the poor recovery of indigo carmine, Boley *et al.* (1981) proposed a paired ion separation procedure on a reversed-phase column to isolate the dyestuff from confectionery products. A Sep-Pak C_{18} cartridge was conditioned with methanol–water–cetrimide and the sample solution passed through. The dye was eluted with methanol and 90 per cent recovered. Ion-pair extraction and ion-pair adsorption TLC for food colours is described by Van Peteghem and Bijl (1981).

Separation and identification

Early workers on food colours used paper chromatographic techniques (APA, 1960; Pearson, 1973b). However, these have now been largely superseded by the use of column chromatography, TLC and HPLC; TLC procedures have been described by Dickes (1965a, b), Chapman and Oakland (1968),

Perry and Wooley (1969), Chiang (1969), Hoodless *et al.* (1971a) and Pearson (1973c). Silica gel, cellulose, and polyamide/kieselguhr have been used as stationary phases and many varied solvent mixtures have been described. Pearson (1975) lists 10 solvents, but propanol/ammonia (4:1) was considered to be a satisfactory general system for use with silica gel. Massart and DeClercq (1974) described a procedure based on numerical taxonomy techniques for selection of combinations of TLC stationary phases and solvent systems for use with 26 food dyes. The best systems are trisodium citrate (2 g)/water (85 ml)/0·88 ammonia (15 ml) and propanol/acetic acid/water (6:1:3) with cellulose coated plates. De Clercq and Massart (1974) have described a TLC method based on silica gel G plates developed in a solvent prepared from 1 per cent copper sulphate in dimethyl sulphoxide/isopropanol (30:70). The copper complexation of sulphonated diazo dyes is claimed to reduce tailing of spots encountered in non-complexed systems. Separations using HPLC have been introduced by a number of workers. Chudy *et al.* (1978) used SAS Hypersil and Spherisorb S5W with isopropanol/water/cetrimide/acetic acid mixtures as mobile phase, whilst Boley *et al.* (1980) used methanol/water/cetrimide with the same column material. Young (1984) described an AOAC collaborative study on the use of disposable HPLC phase cartridges for the extraction of synthetic colours from foods and Spears and Marshall (1987) have fully described rapid screening procedures for soluble and insoluble foods, as follows:

PROCEDURE

(I) *Extraction of synthetic colourings from aqueous or readily soluble foods*

Dissolve 10 g of food in 90 ml of water, warming if necessary, or use 10 ml aqueous sample. Determine the pH of the solution and if necessary adjust the pH to 2·5 with 2 M HCl. Affix the longer stem of a Sep-Pak C_{18} sample preparation cartridge (Waters Assoc.), or equivalent, to the syringe barrel and pass 10 ml of methanol through the cartridge from the syringe. Pour about 10 ml of aqueous food sample into the barrel, insert the plunger into the barrel and by holding the syringe in a vertical position squeeze the air out of the syringe and through the cartridge. Continue to pass solution through the cartridge and discard the clear eluent. Remove the syringe plunger and wash the retained colouring with two successive 10 ml aliquots of warm distilled water acidified with a few drops of 2 M HCl, by passing them through the cartridges as described above. The water wash removes sugars, flavourings, fruit acids, and other polar compounds. The colourings can then rapidly be eluted as a concentrated band by passing a solution of methanol containing a few drops of 10 per cent aqueous ammonia through the cartridge. Again, to elute the colourings as a sharp concentrated band, it is necessary to remove air from the syringe by holding the syringe in a vertical position and depressing the plunger before eluting the adsorbed colourings. If the methanol extract appears to be cloudy, add a few drops of 2 M HCl to re-dissolve the colourings. The purified extract can then be identified by TLC. The cartridge can be used again by re-priming with methanol but should be discarded if traces of colourings remain adsorbed.

The Sep-Pak method can be used for extracting colourings from soft drinks, boiled sweets, confectionery containing surface colouring, clear sauces, mouth wash, etc.

(II) *Extraction of synthetic colourings from insoluble and non-homogeneous foods*

Grind 10 g of sample in a pestle and mortar with 20 g of sand and approximately 5 ml of distilled water containing 1 ml of 2 M HCl to form a homogeneous mass. Transfer the mass to a beaker and mix thoroughly with 30 ml of resin-in-butanol solution.

Allow solid material to settle and decant the resin solution through a plug of glass wool into a 250 ml separating funnel. Add 25 ml of 10 per cent ammonia solution to the separating funnel and extract the colourings into the ammonia solution from the resin layer by vigorous inversion of the funnel. Emulsions are dispersed by careful addition of a few drops of absolute ethanol to the interface (pour ethanol down a glass rod). If there is little colour extracted into the aqueous ammonia, add 50 ml of heptane and shake again. Separate the ammonia solution and discard the butanol layer. Rinse the separating funnel with diethyl ether and discard the ether. Return the ammonia solution to the separating funnel and wash the solution with two successive 10 ml portions of diethyl ether to remove residual butanol. Separate the ammonia layer and remove excess ammonia and ether by gently warming under a stream of air or nitrogen. Cool the solution and acidify the extract to pH 3·0 with hydrochloric acid. Concentrate the extracted colours using a Sep-Pak cartridge as described in (I) above.

Confirmation of identity using spectrophotometry

Plots of the absorption spectra in dilute acid (0·1 M HCl), neutral (0·02 per cent ammonium acetate) and dilute alkaline (0·1 M NaOH) solution have been used to confirm identity. Sufficient dye may be removed from the spots separated by paper and thin-layer chromatography. Spectra for a number of currently non-permitted dyes are available. The APA (1960) and Pearson (1967, 1973b) have determined the absorption curves of the water-soluble colours which were listed in the 1957 Regulations (see also Young, 1984). Reference data for the permitted dyes are given in the FACC Report (1979a). Puttemans *et al.* (1983, 1985) and Van Peteghem and Bijl (1981) have described ion-pair formation methods for examination of the commonly used dyes.

Quantitative assessment and stability of food dyes

Boley *et al.* (1979, 1980) used HPLC for separation and quantitation of the dyes. Reynolds *et al.* (1988) reported variable results in a collaborative study on the quantitative determination of synthetic colours in various foods. Knowles *et al.* (1974) have demonstrated that dyes may be unstable in processed food products. They studied the stability of nine red dyes in processed cured meats and noted that all the colours studied were subject to chemical reaction with the foods, that nitrite apparently had a stabilising effect on many colours and that secondary dyes and colourless reaction products were formed during heat processing. It was also found that Red 10B and erythrosin were relatively heat stable in meat systems regardless of the presence of nitrite. Erythrosin is used in certain canned cured meats, especially meat loaf products. Red 2G also breaks down to yield Red 10B in acid foods; this is significant since the reaction product is a non-permitted colour. It should be noted also that dyes extracted from foods may interact in solution (Draper, 1975), and care should be taken to minimise the exposure of extracts to light and to extremes of pH. Fogg and Summan (1984) have used voltametric methods for determination of synthetic colours at levels of 0·1 μg/ml in solution.

Other methods for water-soluble colours

Procedures for estimation of Sunset Yellow and tartrazine in orangeade, tartrazine and Green S in limeade and amaranth and Green S in blackcurrant drink have been described by Fogg and Yoo (1979) using differential-pulse polarography. The advantage claimed is that the estimations can be carried out directly on the soft drinks with minimal sample preparation. The same authors have extended the method to other drinks including dandelion and burdock which contain Chocolate Brown HT, tartrazine and Green S. Fogg and Bhanot (1980) have achieved a shift of the half wave potentials of tartrazine, Yellow 2G and Brilliant Blue FCF to a more negative potential by addition of tetraphenyl-phosphonium chloride to the polarographic cell. Methods for specific dyes have been reported by Hurst *et al.* (1981). These authors determined amaranth in liquorice, whilst Macchiavelli and Andreotti (1984) have described a method for erythrosine (see also Boley *et al.*, 1981).

CARAMEL

Traditionally, caramel is the product formed by subjecting sugar syrup to gentle heating; the resulting brown semi-liquid mass is an ill-defined material with a distinctive flavour and colour. Caramel (E150) is defined in the 1962 EC Colours Directive and thereby in the UK Regulations as 'the products obtained exclusively by heating sucrose or other edible sugars; or water-soluble amorphous brown products, obtained by the controlled action of heat on edible sugars in the presence of one or more of the following chemical compounds: acetic acid, citric acid, phosphoric acid, sulphuric acid, sulphurous acid or sulphur dioxide; ammonium-, sodium-, potassium-hydroxides or gaseous ammonia; ammonium-, sodium-, potassium-carbonates, phosphates, sulphates or sulphites. Thus, by this definition, the starting material may be any of the permitted sugars or sugar products and may include starch and starch hydrolysis products. In the review of the 1973 Colouring Matter in Food Regulations, the FACC (1979a) concluded that the statutory specifications for caramel needed improvement and that the manufacturing methods for products intended for use as colouring agents, needed closer definition. The FACC proposed that four classes should be recognised, namely burnt sugar, caustic, ammonia and ammonium sulphite caramel. Whilst the burnt sugar caramels are regarded primarily as flavourings, the others are colours. In the view of the committee, the principal problem in reviewing the acceptability of the colours lay in the lack of identity of the products and the extent to which the products are chemically homogeneous. From the manufacturer's point of view, the products are defined in terms of the colour intensity, measured in arbitrary units defined by the European Brewing Convention (EBC) and the stability in particular foods.

The broad classes of foods to which caramels are added includes spirits, flour and sugar confectionery and related products, beers and soft drinks, and the manufacturing methods are tailored to confer appropriate product stability in these food types. A summary of manufacturing methods, constitution and properties of the various caramels has been published by Hellwig *et al.* (1981). From the analytical data presented it is evident that the products derived from the various processes will contain significantly different chemical components. Gaunt *et al.* (1977) have considered the problems of specification and relevant toxicological testing of ammonia caramels, and the difficulties identified by these workers are common to the remaining types. In view of the difficulties of providing adequate specifications, attention has been directed to the chemical characterisation of these products and especially to the presence of potentially harmful substances such as 4-methyl imidazole in ammonium sulphite caramel and 2-acetyl-4-tetrahydroxybutylimidazole (THI) in 'ammonia caramels'. Early methods for this determination were based on GLC (Fuchs and Sundell, 1975; Wilks *et al.*, 1977; Cerny and Blumenthal, 1979). Ion-pair chromatography has been applied by Thomsen and Willumsen (1981) using *bis*(2-ethylhexyl)phosphoric acid as an ion-pairing agent and reversed phase HPLC on Nucleosil C8 with aqueous methanol/potassium dihydrogen phosphate as eluant. The limit of detection claimed for the method was 4 μg/kg, with recoveries of 100–104 per cent. When referred to a standard colour intensity of 20 000 EBC units the content of various commercial caramels varied from 6 to 400 mg/kg.

Further detailed analytical studies of ammonia caramels, aimed at establishing constitutional information suitable for use in specifications, has been reported by Patey *et al.* (1985). In these studies the authors used fractionation by ultrafiltration techniques to provide three molecular weight fractions in the ranges < 1000, 1000–10 000 and > 10 000. This work showed that the compounds conferring the main colour intensity lay in the high molecular weight fractions which are of colloidal dimensions. The low molecular weight fractions of caramels of different origin were examined by GLC/MS and showed significant variations in profile. Over 140 compounds were shown to be present in appreciable proportions. Of these only eight compounds were positively identified; these included three hydroxypyridines, three hydroxypyrazines and two tetrahydroxybutylmethyl pyrazines. Mattioni (1964) described a simple technique to detect the presence of caramel in beverages.

To a sample containing not more than 20 per cent ethanol w/v, add 1 g Na_2SO_4 and 2 ml butanol. Agitate a few minutes, then add one drop of a fresh saturated solution of phloroglucinol in conc. HCl. Shake well. A dark–pale red colour in the butanol indicates caramel.

Oil-soluble colouring matters

No oil-soluble colourings are now included in the permitted list. Prior to the 1973 Regulations, however, Oil Yellow GG and Oil Yellow XP were included in the list. Oil Yellow GG can be extracted by sodium hydroxide from its ethereal solution, but XP and many other oil-soluble dyes cannot. For the isolation of the purified colouring, the oil should first be extracted from the food with light petroleum. Then it can be removed with fairly concentrated acid (AOAC) or by solvent partition using *N,N*-dimethylformamide and light petroleum (Mark and McKeown, 1958; Banerjee *et al.*, 1974). Przybylski *et al.* (1960) used chloroform to remove oil-soluble colours from oranges and an alumina column to clean up extracts prior to chromatography by reversed phase procedures.

TLC separations on cellulose layers soaked in liquid paraffin using 2-methoxyethanol/methanol/water (55:15:30) has been used by Hoodless *et al.* (1971b).

Pearson (1973a) has tabulated the principal spectrophotometric peaks for 16 oil-soluble colours which have been used for colouring foods. These were obtained using 0·001–0·002 per cent solutions of the dyes in light petroleum. The main absorption peaks for Oil Yellow GG and Oil Yellow XP were found at 370 and 402 nm respectively.

The following method, in which a solvent extract of the food is purified by adsorption on a column of alumina followed by elution with various solvents will detect artificial oil-soluble colours and some natural pigments.

REAGENTS

Petroleum ether, boiling range 40–60 °C.

Ethanol.

Benzene.

Diethyl ether.

Solvent 1. Petroleum ether 98 ml, acetone 2 ml.

Solvent 2. Petroleum ether 100 ml, acetone 100 ml.
Solvent 3. Acetone 40 ml, ethanol 10 ml.
Solvent 4. Ethanol 40 ml, aqueous ammonia solution (25% m/m) 20 ml.
Anhydrous magnesium sulphate.
Washed sand.
Sulphuric acid 4 M.
Basic aluminium oxide for chromatography. Activate by heating at 400 °C for 1 h.
Silica gel G for TLC plates.

APPARATUS
Aluminium dish 7 cm dia.
Chromatographic column. 200 mm × 10 mm fitted with a tap, plug with glass wool and fill to a height of 100 mm with the activated alumina. Introduce the alumina as a slurry with benzene. Allow the benzene to drain to just above the surface and introduce 50 ml petroleum ether allowing the solvent to drain to just above the surface of the column. Keep the top of the alumina column under a layer of liquid at all times.
TLC plates. Weigh 30 g silica gel G into a 300 ml flask, add 60 ml distilled water, stopper and shake the flask vigorously for 1 min. Spread a 0·25 mm layer on to a 200 × 200 mm plate, dry in air for 30 min, heat at 60 °C overnight and store in a desiccator over silica gel until needed.

Isolation of the colours
Weigh 5–10 g of the food sample in to the aluminium dish, add 25 g of washed sand and grind with a glass rod. Add 5–10 ml ethanol, stir thoroughly and dry overnight at 60 °C. Transfer the contents of the dish to a Soxhlet thimble and extract with petroleum ether for 4 h in a continuous extractor. Remove the flask and drive off the solvent on a steam bath; dissolve 0·5 g of the residue (or the original sample if it is a fat) in 10 ml petroleum ether. Pour the solution on to the top of the alumina column and allow the solvent to drain off at 1 ml/min. As soon as the solution reaches the surface of the alumina rinse the column with 100 ml petroleum ether and allow the solvent to drain off to waste. Stop the flow as the wash solvent reaches the surface of the alumina.

Elution
Elute the column with 50 ml portions of the four solvent mixtures in turn and in the order given above; collect each solvent eluate in a separate 100 ml flask. Remove the solvent in a stream of nitrogen whilst heating the flask on a steam bath, or use a rotary evaporator. Dissolve each residue in 1 ml diethyl ether, stopper the flask and retain for the TLC examination. Carotenes are susceptible to oxidation and care must be taken to minimise the exposure of the extracts to air; it is preferable to flush the storage vessels with nitrogen before inserting the stoppers. If, during the elution stages with ethanolic ammonia, the column becomes red–violet the presence of curcumin is indicated.

Clean-up—treat each flask separately and retain the sequence of the extracts.

To remove residual fat, add 50 ml ethanol and 1·4 g KOH together with a few boiling chips. Boil the flask under reflux for 45 minutes, cool and transfer the contents to a separating funnel containing 100 ml water. Extract the aqueous phase with 50 ml diethyl ether, 25 ml diethyl ether (twice) and combine the ether extracts. Wash the combined extract with 25 ml water and repeat twice, discard the aqueous layer and dry the ether phase with anhydrous magnesium sulphate. Filter, remove the solvent by evaporation and dissolve the residue in 1 ml diethyl ether.

NOTE. The fourth extract is given an additional treatment to remove bixin (from annatto). After the saponification stage, add 100 ml water and acidify with 4 M sulphuric acid, then proceed with the ether extraction stage as detailed above.

Identification
Pipette 4 μl aliquots of the prepared extracts at 2 cm intervals along a line 2·6 cm from one edge of the prepared TLC plate. Develop the plate in a tank containing benzene, allowing the solvent front to travel 17 cm. Dry in air and repeat the separation; compare the sample extracts with standards of the common oil-soluble dyes. For a good separation of Sudan I and Sudan II, a separate developing solution is required. Use 9 parts *n*-hexane and 1 part ethyl acetate. Bixin is converted to norbixin at the saponification

stage; this may be visualised by treatment with Carr Price reagent (25 g antimony trichloride in 75 ml chloroform). Heat the plate for 10 min at 100 °C. A blue stain turning red–brown confirms bixin.

The order of the eluted fractions is as follows:

Solvent 1: Carotenes.
Solvent 2: Aminoanilines.
Solvent 3: Hydroxyanilines.
Solvent 4: Bixin and hydroxyanilines.

Alternative procedures have been described by Baloch *et al.* (1977) and Corradi and Micheli (1981).

NATURAL COLOURS

With increasing restrictions on the use of synthetic food colours, interest in the use of natural colours has grown. Jeffries (1979), Henry (1979) and Counsell and Knewstubb (1983) have described the sources and typical applications of natural colours in processed foods. Methods for stabilisation of natural colours, their susceptibility to processing conditions and enzyme-related losses are discussed by Adams (1981a). The main classes are identified in the Regulations as carotenoids and their ketonic or hydroxylic derivatives, beetroot extracts, anthocyanins, riboflavins, cochineal, chlorophylls and naturally coloured food substances such as paprika, turmeric, saffron and sandalwood together with concentrated extracts of these materials. The carotenoids are oil soluble and are generally determined in assays of vitamin A and pro-vitamin A activity. A review of HPLC methods has been published by Bushway (1985). Riboflavin may be determined by the methods used for the B vitamin complex; HPLC procedures have been described by Mauro and Wetzel (1984) and by Watada and Tran (1985) (see also Chapter 11). The remaining natural colours may be determined by chromatographic procedures or by specific colorimetric methods. The 11th edition of the AOAC methods book includes procedures for the identification of annatto, turmeric and saffron by characteristic colour reactions obtained with extracts from cereal products. Anthocyanins have been extracted from fruit products with acidified methanol; for separation Strack and Mansell (1975) used polyamide chromatography, Torre and Barritt (1977) TLC on microcrystalline cellulose, Manley and Shubiak (1975) HPLC on Pellidon columns with chloroform/methanol, and Wilkinson *et al.* (1977) HPLC with a reversed phase column developed with water/acetic acid/methanol (71:10:19). Bakker and Timberlake (1985) used preparative HPLC on a reversed phase column of Spherisorb-Hexyl and gradient elution with aqueous perchloric acid/methanol mixtures. Survey data on grape skins and port wines is given. Polyamide powder enrichment of food extracts followed by TLC was used by Lehmann *et al.* (1974), whilst reversed phase was used for beetroot red pigments by Vincent and Scholz (1978). An HPLC method has been described by Schwartz and Elbe (1980). Methods for detection of chlorophylls and yellow plant pigments have been described by Shiraki *et al.* (1978), Yoshiura *et al.* (1978) and Csorba *et al.* (1979). The pigments are extracted with acetone and subjected to TLC on silica gel with *t*-butyl alcohol/pentane (1:9) mobile phase. Acetone extracts of foods have been used to identify oleoresin and paprika pigments by Woodbury (1977).

TLC data for 50 natural food dyes on silica gel and cellulose using 18 solvent mixtures are presented by Airaudo *et al.* (1983).

EMULSIFIERS AND STABILISERS

Legislation

In 1956 the Food Standards Committee adopted a Report on Emulsifying and Stabilising Agents submitted by their Preservatives Sub Committee. Most of the substances covered in the Report act as emulsifiers and stabilisers, but some of them are used for other properties they possess, e.g. as crumb-softeners or antistaling agents, or to produce or break foams. The Report added that when emulsifying agents were first used they were few in number and those used were either natural substances, particularly gums or synthetic substances of simple constitution. Progress in chemical synthesis, however, has extended the range of available products with characteristics suitable for specific product application (Wood, 1980; Flack, 1983). A limited list of substances was contained in the Emulsifiers and Stabilisers in Food Regulations 1962 (as amended). These were:

Stearyl tartrate
Complete glycerol esters
Partial glycerol esters
Partial polyglycerol esters
Propylene glycol esters
Monostearin sodium sulphoacetate
Sorbitan esters of fatty acids and their polyoxyethylene derivatives
Cellulose ethers
Sodium carboxymethyl cellulose

Under the 1962 Regulations, foods such as flour and most dairy products were not permitted to contain any emulsifier or stabiliser. Subsequently, the FACC (1970, 1972b) recommended that the permitted list and definitions in the 1962 Regulations should be revised to bring within control an increased number of substances, including agar, alginates, gums, lecithin and pectin, which were previously excluded from the legal definition of emulsifier and stabiliser. The Committee also recommended removal of permission for use of brominated vegetable oils as emulsifiers in soft drinks. Subsequent regulations have resulted in the Emulsifiers and Stabilisers in Food Regulations 1989 (SI No. 876).

These Regulations re-enact with amendments the Emulsifiers and Stabilisers in Food Regulations 1980, as amended.

The amended 1980 Regulations provided in particular for the implementation of Council Directive 74/329/EEC (OJ No. L189, 12.7.74, p. 1), as amended by Council Directive 78/612/EEC (OJ No. L197, 22.7.78, p. 22) and Council Directive 80/597/EEC (OJ No. L155, 23.6.80, p. 23) on the approximation of the laws of Member States relating to emulsifiers, stabilisers, thickeners and gelling agents for use in foodstuffs. They also provided for the implementation of Council Directive 78/663/EEC (OJ No. L223, 14.8.78, p. 7), as amended by Council Directive 82/504/EEC (OJ No. L230, 5.8.82, p. 35) laying down specific criteria of purity for emulsifiers, stabilisers, thickeners and gelling agents for use in foodstuffs, and Council Directive 78/664/EEC (OJ No. L223, 14.8.78, p. 30) laying down specific criteria of purity for antioxidants for use in foodstuffs.

In addition, these Regulations implement Council Directive 86/102/EEC (OJ No.

L88, 3.4.86, p. 40), which further amended Council Directive 74/329/EEC. As well as making a minor change in the serial numbers of pectin and amidated pectin, this amending Directive withdrew authority for the use of polyoxyethylene (8) stearate, polyoxyethylene (40) stearate, lactylated fatty acid esters of glycol and propane-1,2-diol, and dioctyl sodium sulphosuccinate as emulsifiers and stabilisers in food. Pursuant to Council Directive 74/329/EEC, tartaric acid esters of mono and di-glycerides of fatty acids (E472(d)) and mixed acetic and tartaric acid esters of mono and diglycerides of fatty acids (E472(f)) are added to the list of permitted emulsifiers and stabilisers and specific purity criteria are laid down in Schedule 1.

Regulation 4(2) and Schedule 2 Part 1 prescribe a maximum level for guar gum and locust bean gum in food.

Generally, the Regulations:

1. specify permitted emulsifiers and permitted stabilisers and prescribe purity criteria for those emulsifiers and stabilisers (regulation 2 and Schedule 1);
2. prohibit the sale and importation of food which contains any added emulsifier or stabiliser other than those prescribed and regulate the sale of food containing permitted emulsifiers and stabilisers (regulation 4 and Schedule 2);
3. regulate the sale, importation, advertisement and labelling of permitted emulsifiers and permitted stabilisers sold for use as ingredients in the preparation of food (regulations 5 and 6 and Schedule 3);
4. do not apply to any emulsifier and stabiliser, or to any food having in it or on it any emulsifier or stabiliser, which is intended for export (regulation 3).

General reviews of the analysis of emulsifiers and stabilisers have been published by Hibbert (1968), Murphy and Hibbert (1969), Baur (1973), Dieffenbacher and Bracco (1978), Schuster and Adams (1980), Sanderson (1981) and Weiss (1981).

General methods for natural emulsifiers and stabilisers

Schemes for identification of gums and various thickening agents have been published by Ewart and Chapman (1952) and Morley *et al.* (1972). Hunziker and Mizerez (1980) and Hunziker (1984) have described HPLC techniques with gel permeation columns for identification and estimation of water-soluble thickeners. Preuss and Thier (1982, 1983a) have described methods based on GLC derivatives of the constituent sugars and uronic acids present in alginates, pectins and natural gums. Electrophoretic methods have been developed by Pechanek *et al.* (1982) and by Schafer and Scherz (1983). Scherz (1984) used TLC of the polysaccharide reaction products with methanolic HCl as a means of determining alginates in beverages. The AOAC methods book contains colorimetric tests for the common natural gums; these are extracted from foods by suspension in hot water. After any necessary defatting and clearing operations the gum is precipitated with alcohol and subjected to the appropriate test procedure. Hydrolysis of precipitated gums also yields a solution of monosaccharides which responds to the usual copper reducing tests and gives characteristic sugar chromatograms. When starch is present in the food, this is solubilised with calcium chloride and precipitated with iodine before addition of alcohol (AOAC). The scheme described by Morley *et al.* (1972) is based on differential solubility of gums in cetylpyridinium chloride solution and in aqueous alcohol at varying pH and ionic strength. Standard colour tests are applied to the separated gums for confirmation of identity. The scheme is said to be applicable to gums extracted from foods by the AOAC technique. The commonly used colour reagents include:

A. Iodine–potassium iodide in zinc chloride. To 100 ml 60 per cent zinc chloride add a solution containing 10 g potassium iodide and 0·15 g iodine dissolved in 10 ml water.

B. Iodine solution. Dissolve 7 g iodine in a solution containing 5 g potassium iodine dissolved in 5 ml water and dilute to 100 ml with alcohol.

C. Ruthenium red. Dissolve in 10 per cent lead acetate solution to produce a wine-red colour.

Gum tragacanth gives a blue colour with reagent A and a yellow colour on heating on the water bath with 10 per cent sodium hydroxide solution.

Irish moss gives a brown colour with reagent A and brown or lilac with reagent B.

Agar gives a blue–black colour with reagent B. At 40 °C a 0·5 per cent solution of agar gives precipitates with half its volumes of 40 per cent ammonium sulphate solution, and also with one-fifth its volume of basic lead acetate.

Karaya gives a strongly stained pink mass with reagent C.

Carob bean gives a pink or red–brown colour on warming with concentrated sulphuric acid, whereas acacia gives a greenish brown colour.

Properties of other 'natural' compounds are as follows:

Sodium alginate. A 1 per cent solution gives gelantinous precipitates with dilute sulphuric acid, and also with 2·5 per cent calcium chloride solution. With acid ferric sulphate solution a cherry-red colour develops which finally becomes deep purple.

Caseinates of sodium and calcium are usually determined from the phosphoprotein content. The homogenised product is defatted with chloroform/ethanol. Acid-soluble phosphorus containing substances and nucleic acids are removed from the residue by treatment with trichloracetic acid solution (or perchloric acid solution). The insoluble residue is treated with M sodium hydroxide and the phosphorus content of the solution is determined by a standard procedure (Kutscher *et al.*, 1961). Van Baal and Leget (1965) used a similar separation but submitted the final residue to an electrophoretic separation in barbitone buffer. Casein may be extracted from defatted chocolate with sodium oxalate solution in water (3 per cent). After precipitation with tannin the casein may be determined through the nitrogen content of the precipitate (Anon, 1963).

Lecithin is a mixture of phosphatides obtained from soybean and other vegetable oils. Leicithin is insoluble in, but swells with, water. It can be extracted from solid food such as chocolate with cloroform/ethanol. After removal of solvent the P_2O_5 content is determined by wet oxidation and colorimetry.

Lecithin = $P_2O_5 \times 11{\cdot}19$ (AOAC). Pardun (1981) has criticised the values of factors used for various lecithins. He has suggested 11·8 for soya, 13·8 for peanut, 10·7 for egg yolk and 13·1–15·3 for plant lecithins in general.

Veturini (1972) used a TLC procedure for estimation of lecithin in bakery products. Riedl (1981) has discussed the structure and properties of lecithins derived from various sources. Rhee and Shin (1982) used HPLC to estimate the phosphatidylcholine content of soya lecithin as a rapid assay technique.

Carrageenan can be assayed by binding acridine orange to the polyanion and measuring the decrease in green fluorescence from free monomer dye in solution

(Cundall *et al.*, 1973). Fluorescence spectroscopy of the dye complex has been described by Murray and Cundall (1981).

Artificial emulsifiers and stabilisers and their identification and estimation in foods

Stearyl Tartrate

$$\begin{array}{l} HO.CH.COO\,(CH_2)_{17}CH_3 \\ \quad\;\; | \\ HO.CH.COO\,(CH_2)_{17}CH_3 \end{array}$$

Stearyl tartrate is mainly used in bread-making for increasing the dough strength. A method for its determination based on the production of stearyl alcohol following solvent extraction and saponification is given on p. 315.

Complete glycerol esters can be defined as any compound formed by completely esterifying the hydroxyl groups of glycerol with either:

1. a mixture of a single fatty acid and acetic acid; or
2. a mixture of fatty acids and lactic acid; or
3. a mixture of fatty acids and phosphoric acid; or
4. a mixture of fatty acids and diacetyl tartaric acid.

They have the general formula

$$\begin{array}{l} CH_2OR_1 \\ | \\ CHOR_2 \\ | \\ CH_2OR_3 \end{array}$$

where R_1, R_2 and R_3 may be the defined organic acid or phosphoric acid groups.

Hibbert (1968) states that these esters are not true emulsifying agents and are produced with mono- and diglycerides during the interesterification process used for the manufacture of high-ratio shortenings. A peculiarity of most commercial emulsifiers is that they are not always single substances but can be a mixture of substances in amounts which vary according to origin and manufacture (Baur, 1973).

A method for the determination of mono, di and triglycerides in shortenings has been studied by Distler and Baur (1966). Samples are adsorbed on silica gel columns and the glycerides separated by benzene, 10 per cent ether in benzene and ether respectively and the solvent evaporated. GLC of the esters produced by transesterification can be used for the determination of component fatty acids. Hibbert (1968) discussed the use of thin-layer argentation, which depends on the ability of olefinic compounds to form complexes with silver ions, for the separation of various types of glycerides. Complete glycerol esters can be extracted from foods with ether after acid digestion. Dieffenbacher and Bracco (1978) and Martin (1981) prefer chloroform/methanol extraction prior to separation on a silica gel column. They collect three fractions and subject these to TLC on silica gel plates. By combining three developing systems and seven spray reagents they claim that most of the common emulsifiers and stabilisers may be identified. Coustard *et al.* (1982) have described HPLC procedures with mass-spectrometry confirmation of structure of the separated compounds.

Partial glycerol esters can be defined as any compound formed by incompletely esterifying the hydroxyl groups of glycerol with either:

1. any single fatty acid; or
2. any mixture of fatty acids; or
3. any mixture of fatty acids with one of the following organic acids: acetic acid, lactic acid, citric acid, tartaric acid, diacetyl tartaric acid or hydroxystearic acid;

in which the proportion of sodium salt, if any, of any fatty acid present does not exceed 2 per cent (estimated by weight) of the partial glycerol ester.

They have the general formula:

$$\begin{array}{l} CH_2O.R_1 \\ | \\ CHO.R_2 \\ | \\ CH_2O.R_3 \end{array}$$

where at least R_1 and/or R_2 may be hydrogen and R_1, R_2 and R_3 may be organic acid groups. The group includes some of the most commonly used emulsifiers having uses in the dairy, margarine and bakery industries.

Kruty *et al.* (1954) pointed out that monoglycerides can be determined by oxidation with periodic acid. The sample is taken up in 5 per cent dimethylformamide in chloroform and the monoglyceride + glycerol are determined on a 25-ml portion by oxidation with periodic acid in methanol. Sodium bicarbonate and potassium iodide are added and the liberated iodine titrated with sodium arsenite using starch. The glycerol is determined by adding 100 ml water to the 25-ml chloroformic solution and 25-ml aqueous periodic acid. Then the amount of monoglyceride is calculated from the difference between the two titrations (AOCS Fat Analysis Committee, 1957; Baur and Distler, 1966). The IUPAC (1987) method for monoglycerides is based on the oxidation of the 1-monoglycerides by a measured excess of periodic acid in acetic acid solution, followed by addition of potassium iodide and determination of the excess periodic acid with sodium thiosulphate solution.

The lactylated glycerides have been determined by saponifying, acidifying and subsequent determination of the separated lactic acid. Stetzler and Andress (1962) titrated the acid after removal of fatty acids using liquid/liquid extraction. Water-soluble constituents were extracted from a dichloromethane solution of the esters with 5 per cent sodium sulphate solution prior to the determination of water-insoluble lactic acid. Sahasrabudhe *et al.* (1966) employed a silica gel column for the separation. Fett (1961) determined the released lactic acid by conversion to acetaldehyde using sulphuric acid and measuring the purple colour produced with *p*-phenylphenol. Chromatographic and colorimetric techniques have been employed by Kröller (1962, 1964, 1974) for the identification of citric, tartaric and acetyltartaric glycerides in margarine and other foods.

Separations on silica gel (column and thin layer) have been compared with those obtained by the gel exclusion technique with styrene–divinylbenzene gel (Coupek *et al.*, 1976). They concluded that TLC gives good separations of positional isomers but that use of at least two developing systems of differing polarity is necessary to obtain satisfactory characterisation. Gel chromatography has the advantage of speed and simplicity but separation of 1,2- and 1,3-diacylglycerols is not possible. Sahasrabudhe and Legari (1967) and Inoue *et al.* (1981) have described GLC procedures based on use of trimethyl ether derivatives of mono and diglycerides, and Soe (1983) has discussed the use of capillary columns. Sahasrabudhe and Legari (1968) also used silicic acid column chromatography to separate mixtures of emulsifiers into three fractions prior to derivatisation and detailed analysis by GLC. The three fractions comprise triglycerides plus propylene glycol diesters, diglycerides plus propylene glycol monoesters and monoglycerides. Riisom and Hoffmeyer (1978) have described an HPLC procedure for separation of mono, di and triacylglycerols on 10 μm LiChrosorb DIOL with isoctane (95 per cent) and isopropanol (5 per cent). They claim that the method is comparable to or better than the periodic acid oxidation method but has the advantage of speed and ease of operation.

A method for determination of mono, di and triglycerides by column chromatography has been published by IUPAC (1987). Hironez (1984) has used TLC for the determination of emulsifiers in sausages and has presented data on the naturally occurring amounts of mono and diglycerides in sausage products.

Partial polyglycerol esters can be defined as any compound formed by incompletely esterifying the hydroxyl groups of condensed glycerol, which has been formed by causing glycerol to react with itself with the elimination of water, with either:

1. any single fatty acid; or
2. any mixture of fatty acids; or
3. dimerised fatty acids of soya bean oil; or
4. interesterified fatty acids of castor oil.

They have the general formula:

$$R_1(OCH_2.\underset{\displaystyle |\atop\displaystyle OR_3}{CH}.CH_2O)_nR_2$$

where R_1, R_2 or R_3 may be hydrogen and/or fatty acid groups and n does not exceed 3. The wide range of properties possessed by the polyglycerol esters is brought out by Nash and Knight (1967), e.g. emulsifying, stabilising and defoaming. They are recommended for use in oils, margarine and flour confectionery.

As the definition implies, a large number of esters with differing properties may be present so it is advisable to apply a general test rather than attempt to identify a specific substance. Thus Seher (1964) saponified the sample and after acidifying removed fatty matter with ether. The acid solution was then evaporated to dryness and the ethanolic solution of the residue applied to a TLC plate of kieselguhr G. A solvent mixture of ethyl acetate (65 ml), isopropanol (22·7 ml) and water (12·3 ml) was used and the spots detected by spraying with 0·1 per cent sodium iodate followed after 4 min with ethanolic benzidine solution. The glycerols appear as white spots against a blue background. Sahasrabudhe (1967) has shown that polyglycerols and their fatty acid esters can be separated by GLC using the trimethylsilyl ether derivatives and by TLC on silica gel treated with boric acid. Dick and Miserez (1976) have described a GLC method for determination of polyglycerol ricinoleate and of ammonium phosphatides in chocolate. They extracted the sample with toluene, saponified the extract, dissolved the separated fatty acids in pyridine, silylated and analysed by GLC. The polyhydric alcohols were also recovered, silylated and analysed by GLC.

Propylene glycol esters can be defined as any compound formed by completely or incompletely esterifying the hydroxyl groups of propylene glycol with either:

1. any single fatty acid; or
2. any mixture of fatty acids; or
3. any mixture of fatty acids and lactic acid; or
4. alginic acid.

They have the general formula:

$$\begin{array}{l} CH_3.CH.OR \\ \qquad\;\; | \\ \qquad CH_2.OR_1 \end{array}$$

where R and R_1 represent fatty acid groups or hydrogen or lactic or alginic acids. They are used in ice-cream, flour confectionery and shortenings. Propylene glycol esters are often used in conjunction with mono and diglycerides and, as described above, Sahasrabudhe and Legari (1968) separated the various forms on a silicic acid column prior to GLC. Hibbert (1968) states that methods described for the detection of alginates in the AOAC can also be used for propylene glycol alginates. Neissner (1972) has described a general TLC procedure on silica gel using petroleum ether/ethyl ether/acetic acid (60/60/1) for separation of polyhydric alcohol/fatty acid esters. Conacher and Page (1979) reviewed available methods for polyol esters. Direct trimethyl silylation of intact emulsifiers prior to analysis by GLC is considered to be a suitable approach.

Monostearin sodium sulphoacetate appears to be used abroad as an anti-spattering agent for margarine.

Sorbitan esters of fatty acids (spans) and their polyoxyethylene derivatives (tweens). The oil-soluble sorbitan esters (spans) consist of a mixture of the partial esters of sorbitol and its anhydrides with C2–C8 fatty acids. They are used particularly in the baking and ice-cream industries.

A paper chromatographic technique for identifying sorbitan and polyethylene glycol fatty acid esters in bakery additives has been described by Schrepfer and Egle (1958). The esters are extracted with chloroform and saponified. The unsaponifiable matter and fatty acids are removed after acidification with benzene and inorganic ions removed from the neutralised solution by passing down cation and anion exchange resins. The solvent used is *n*-butanol saturated with water and the dried chromatogram is sprayed with lead tetra-acetate.

Various forms of chromatography have been used for the determination of sorbitan monostearate in cakes and cake mixes (Wetterau *et al.*, 1964). The extracted ester is partially purified with silica gel, the polyol is extracted and the isosorbide measured using GLC or other chromatographic methods (Martin *et al.*, 1980; Tsuda *et al.*, 1984).

Murphy and Grisley (1969) pointed out that the isosorbide content of sorbitan monostearate varies between 1·2 and 6·1 per cent according to source. Therefore, unless the source is known, quantitative results by the Wetterau method can be misleading. With that proviso, they extended the method to include GLC analysis of other sorbitan esters. The TMS derivatives of isosorbide are also used in a GLC method applied to desert topping and icecream powders. Kröller (1968) has described a TLC procedure using benzene/methanol (80/20) as developer.

The water-soluble ethylene oxide condensates of the sorbitan esters (tweens) are also used in baking and for ice-cream. Gatewood and Graham (1961) have shown that when polyoxyethylene type compounds are heated with an acidified methanolic solution of 2,4 dinitrophenylhydrazine and alcoholic potash is added, a characteristic purple colour is quantitatively produced with an absorption maximum at 560 nm. Garrison *et al.* (1957) studied two methods for the determination of polyoxyethylene esters in bread and rolls (1) following hydriodic acid cleavage to yield ethyl iodide and ethylene and (2) by precipitation with phosphomolybdic acid and barium chloride; the molybdenum content is determined colorimetrically with ammonium thiocyanate and stannous chloride after dissolving in sulphuric acid. For the determination of polysorbate in bakery products and ice-cream, Hall (1964) used a gravimetric finish following precipitation with phosphomolybdic acid. The method is recommended by the AOAC.

Murphy and Scott (1969) commented on previously published methods and described a procedure in which the polyoxyethylene emulsifier was extracted with chloroform, cleaned up on alumina, separated by TLC and sprayed with a modified Dragendorff's reagent. Kudoh and Yamaguchi (1983) have used cobaltithiocyanate ion exchange resin to absorb non-ionic surfactants as a means of separating the material. Anderson and Girling (1982) used complexation with cobalt salts followed by reaction in chloroform solution with Nitroso-R salt to give an assay procedure sensitive to 5 μg of the surfactant. Slack and Porter (1983) used cupric ammonium sulphate as a complexing agent in a similar procedure to that of Anderson and Girling.

Cellulose ethers can be defined as any compound obtained by replacing the hydrogen atoms in the hydroxyl groups in cellulose by one or more of the following substituent groups: methyl, ethyl, hydroxyethyl, hydroxypropyl. Cellulose ethers are mostly used in the dairy industry.

Mano and Lima (1960) have described procedures for the identification of various ethers and esters of cellulose. Acetone is used to dissolve cellulose esters and ethyl cellulose whilst the remaining ethers can be extracted with water, cellulose remaining insoluble. Two tests are then performed on each extract and residue. First, the material is heated with 85 per cent phosphoric acid and the vapour passed through a filter paper wetted with aniline acetate (aniline in 50 per cent acetic acid). Most of the cellulosic compounds give a pink spot, but cellulose nitrate gives a yellow one. Cellulose acetate and the acetate-butyrate can be detected by paper chromatography of the acids after saponification. Secondly, the material is warmed with benzene and 93 per cent sulphuric acid in a water bath until an intense yellow develops, which rapidly turns reddish. The liquid is cooled rapidly and layered with ethanol. A blue or green ring between the phases indicates a positive reaction for cellulosics except ethylcellulose, which shows a violet ring.

Sodium carboxymethyl cellulose (Na-CMC; CMC) is the sodium salt of a carboxymethyl ether of cellulose $[C_6H_7O_2(OH)_x(OCH_2COONa)_y]_n$

$$\text{where } x = 2{\cdot}00-2.40$$
$$y = 1{\cdot}00-0{\cdot}60$$
$$x + y = 3{\cdot}00.$$

As CMC gives a viscous colloidal solution with water it is used as a thickener, emulsifier and stabiliser in salad dressings, ice-cream, baked goods and many other foods. It is insoluble in ether and ethanol. Its aqueous solution gives a yellow precipitate with uranyl zinc acetate and a blue precipitate with copper sulphate distinguishes it from methyl cellulose, gelatin and various gums. The precipitates formed with copper sulphate and uranyl nitrate have also been used in the estimation of CMC in foods. After isolation from the precipitate with copper sulphate the CMC can be determined with naphthalenediol, with which it gives a red colour.

Hansen and Chang (1968) have shown that CMC may be recovered from a tryptic digest of milk by solvent precipitation of the fraction soluble in 12·5 per cent TCA and can be measured spectrophotometrically on adding sulphuric acid and phenol.

Brominated edible vegetable oils. Brominated vegetable oils, which are no longer included in the permitted list, are bromine addition products of vegetable oils. They are yellow to brown in colour, viscous and oily and tend to have a fruity odour. They are insoluble in water, but soluble in organic solvents and fixed oils. They were mainly used for stabilising the cloud in beverages containing fruit juice.

Brominated oils can be detected and determined from the bromine present. Van Pinxteren (1952) has described a colorimetric method for determining brominated preservatives in soft drinks. The compound is extracted with ether, decomposed and the liberated bromide determined as cyanogen bromide by Koenig's reaction with pyridine, the extinction of the red colour produced being measured at 500 nm (Green and Keen, 1971). Conacher (1973) developed a GLC procedure based on the separation of methyl esters of the brominated oils using methyl tetrabromostearate and methyldibromostearate as reference materials and a Gas Chrom Q column. Lowry and Tinsley (1981) have described an optimised version of this procedure.

GENERAL TLC PROCEDURE FOR EMULSIFIERS

Lipid extraction

To a weighed sample of prepared food (10–20 g) in a macerator, add 50 ml chloroform, 100 ml methanol and 0·5 ml M magnesium chloride solution. Macerate for about two minutes, add 50 ml chloroform and repeat the homogenisation. Transfer the mixture to a large centrifuge tube and centrifuge for 10 minutes at 2000 r.p.m. Decant the upper layer, clarify by filtration if necessary. Repeat the homogenisation step with 50 ml chloroform, centrifuge, decant the liquid phase and combine the two extracts in a separating funnel. Add 90 ml water, shake vigorously. If an emulsion forms add 100 mg sodium chloride to break the emulsion. Separate the lower layer, extract the upper layer with 50 ml chloroform/methanol (2:1 v/v). Shake vigorously, allow to separate, collect the lower layer and combine the two extracts. Discard the upper layer. Evaporate the combined extracts under vacuum at 60 °C for 30 minutes, cool and weigh the extract.

Separation of the extract

Prepare a slurry of 30 g silica gel in 100 ml petroleum ether, wash the gel into a chromatographic column with petroleum ether. A column 2 × 40 cm fitted with a fritted disc and stopcock is suitable. Allow the gel to settle in the column and maintain the liquid level slightly above the solid surface by careful adjustment of the solvent flow rate. Dissolve 1 g of the prepared extract in 10 ml toluene, add the solution to the column and allow the solvent to flow to a level of 1 cm above the silica. Elute successively with: 300 ml of a mixture of ether in petroleum ether (7% v/v); 150 ml chloroform/methanol (2:1 v/v); 100 ml methanol. Add the eluants to the column in 10 ml lots and allow the solvent to run to the surface of the silica before the next addition. Collect three fractions comprising the accumulated eluant for each of the three solvent systems used. Triglycerides are concentrated in the non-polar fraction, whilst the more polar compounds are accumulated into the polar solvents. Remove the solvent from each fraction by evaporation and weigh the residues. Prepare a 5 per cent solution of each of the extracts in 2:1 chloroform/methanol.

Thin-layer chromatography
Equipment and selected reagents
TLC plates, 20 × 20 cm plates precoated with silica gel 60 °F 254.

Solvents
1. Petroleum ether, ethyl ether, acetic acid; 60/40/1.
2. Chloroform, acetone, acetic acid; 94/6/1.
3. Chloroform, methanol, water; 65/25/4.
4. Ethyl acetate, acetic acid, water; 60/14/10.

Spray reagents
1. Eosin: dissolve 0·2 g in 100 ml ethanol.
2. Phosphomolybdic acid: 20% in ethanol.
3. Anisidine–potassium periodate:
   (a) dissolve 0·1 g potassium periodate in 100 ml water.
   (b) dissolve 2·8 g *p*-anisidine in 80 ml ethanol and 70 ml water. Add 30 ml acetone and 1·5 ml M HCl.
   Use these reagents sequentially.
4. Bromocresol green: dissolve 0·04 g in 100 ml ethanol, add 0·1 M sodium hydroxide dropwise to give a blue solution.
5. Silver nitrate:
   (a) dissolve 50 g silver nitrate in 10 ml water, and 200 ml acetone and sufficient water to dissolve the precipitate.
   (b) dissolve 20 g sodium hydroxide in 30 ml water, make up to 1 litre with methanol.
   (c) ammonia solution 10 per cent v/v in water.
   Use these reagents sequentially, heat to 120 °C for 25 minutes.
6. Dragendorff reagent:
   (a) dissolve 1·7 g bismuth(III) nitrate in 20 ml glacial acetic acid and 80 ml hot water. Add 40 g potassium iodide in 100 ml water, 200 ml glacial acetic acid and make up to 1 litre with water.
   (b) dissolve 20 g barium chloride in 80 ml water.
   For use, prepare freshly a mixture comprising: 10 ml of (a), 1 ml orthophosphoric acid, 10 ml ethanol, and 5 ml of (b).
7. Phosphoric acid: prepare a 3 M solution in water. Spray and heat the plate 660 minutes at 120 °C.

Procedure
Before use pretreat the plate by development in chloroform/methanol (2:1 v/v), dry the plate in air and activate by heating at 120 °C for 15 minutes. Select and mark a baseline 25 mm from one edge of the plate, score a finish line at 120 mm from the chosen baseline. Spot 5 μl prepared extracts at 15 mm intervals along the baseline. Evaporate the solvent at room temperature, and place the plate in a chromatographic tank containing the chosen solvent system. Allow the chromatogram to develop to the finish line, remove and dry at 120 °C. Spray the plate with the selected reagent system, heat at 103 °C for 15 minutes unless directed otherwise. Examine the treated plates in daylight and under ultraviolet illumination. Prepare comparison plates using 5 per cent solutions of known emulsifiers to establish R_F values and the colour response for each spray system. Typical performance and reactions are shown in Tables 4.5 and 4.6.

Identification of the acid and alcoholic constituents may be made after saponification, neutralisation and extraction with petroleum ether. The ether contains unsaponifiable materials and the fatty acids, the aqueous phase contains the polyols, hydroxyacids and water-soluble unsaponifiable materials. The TLC

Table 4.5 Emulsifier: group separations

Solvent	Emulsifiers separated
1	Mono and diglycerides
	Esterified monoglycerides
	Ethoxylated monoglycerides
	Polyglycerol esters
2	Glycol esters of fatty acids
	Glycerol esters of fatty acids
	Sorbitan esters of fatty acids
3	Esters of hydroxyacids
	Cholic and desoxycholic acid
	Ethoxylated sorbitan esters
	Sugar esters of fatty acids
	Sulphated alcohols
	Sulphonates
4	Polyethoxylate esters of fatty acids
	Polyethoxylated mono and diglycerides
	Polyglycerol esters
	Sugar esters of fatty acids
	Sulphated alcohols
	Sulphonates

Table 4.6 Spray reagent: emulsifier reactions

Reagent	Emulsifier	Reaction
1	All types	Red, yellow/orange fluorescence
2	All types	Brown on green background
3	Lecithin, sorbitan fatty acid esters	Yellow on pink–brown background
4	Free acidic or basic groups	Acid: yellow on blue background Base: blue on blue–green background
5	All types	Brown
6	Ethoxylated compounds	Orange on yellow background
7	Cholic acid, desoxycholate	Brown; blue–green under UV
	Diacetyl-tartaric acid monoglyceride	Brown on white; yellow under UV
	Ethoxylated glycerides	Brown on white; yellow under UV

procedures described above may be used to identify the constituents; a combination of solvent number 3 and spray reagent 3 is useful for glycols, whilst the organic acids may be detected by reagents 4 and 5.

SOLVENTS

Legislation

Various solvents have been used for (1) the incorporation into food of flavours, essential oils, colours, antioxidants and vitamins, (2) dissolution of raw materials and (3) dilution of concentrates. The possible toxicity of those which tend to remain in the food and are therefore consumed was considered by the FACC (1966). The Pharmacology Panel considered that although there was insufficient information on many of those considered, the available evidence on diethylene glycol monoethylether, hexylene glycol and ethyl citrate suggested that they should not be used in food. Subsequently, The Solvents in Food Regulations 1967 SI No. 1582 and (Amendment) Regulations (SI 1967 No. 1934; SI 1980 No. 1832) permitted the use of a restricted number of solvents, namely ethanol, ethyl acetate, diethyl ether, glycerol, mono-, di- and tri-acetin, isopropyl alcohol and propylene glycol (Anon, 1968; FACC, 1974b). The definition of 'solvent' for legal purposes excludes water, and other permitted additives. Specifications of purity of the acetins are laid down in the regulations, whilst others have to comply with the requirements of EC Directives or those set out by BP or BS.

The FACC (1978a) considered The Solvents in Food Regulations 1967 and classified solvents into three groups: carrier solvents, extraction solvents and processing solvents. Carrier solvents are currently controlled by regulations which it proposed should be extended to include the three groups. The Committee recommended that the permitted list be extended to include:

Light petroleum
Trichloroethylene
Dichloromethane
Dichlorodifluoromethane
1,2-dichloroethane
Ethanol
Methanol
Propan-l-ol
Propan-2-ol
Propane-1,2-diol
Glycerol
Butan-l-ol
Butan-2-ol
Benzyl alcohol
Glycerol acetates
Ethyl acetate
Butyl acetate
Amyl acetate
Diethyl tartrate
Benzyl benzoate
Isopropyl myristate
Ethyl lactate
Mono and diglycerides
Glycerol tributyrate
Glycerol tripropionate
Castor oil
Acetone
Butanone
Diethyl ether
Carbon dioxide

The use of fluid (supercritical) gases as extractants has been developed for application to a wide range of products (Naturex, 1980; Rizvi *et al.*, 1986). A revised definition of 'solvent' has been proposed to take into account the inclusion of a wider range of substances; specific exclusions were proposed for 'natural food substances', other classes of permitted additives, water and flavours. On the advice of the COT, specific limits are proposed for dichlorodifluoromethane (1 mg/kg except where used as a freezant) and methanol (2 mg/kg). Revised specifications and purity criteria are also proposed. The EC Commission has prepared a draft directive on solvents, within which a solvent is defined as 'any substance for dissolving a foodstuff or any component thereof, including any contaminant present in or on that foodstuff'. Specific foodstuffs, water, edible fats and oils, wines and spirits, are excluded. The draft directive for extraction solvents includes two lists of permitted substances for use with foods. The first list contains those which may be used in compliance with good manufacturing practice, namely propane, butane, butyl acetate, ethyl acetate, methanol, propan-2-ol, carbon dioxide, acetone and nitrous oxide. The second list contains substances which are restricted to use with certain foods together with proposed residue limits in the extracted food or food ingredient. A third list of substances for use in the preparation of flavourings from natural materials is also included together with proposals for residue limits. An extraction solvent is defined as 'a solvent which is used in an extraction procedure during the processing of raw materials, of foodstuffs, or of components or ingredients of these products and which is removed but which may result in the non-intentional but technical unavoidable presence of residues or derivatives in the foodstuff or food ingredient.'

For their detection the lower boiling solvents can be distilled from materials such as essences. This effects partial separation and more complete separation can best be achieved by GLC. The determination of light petroleum residues by GLC techniques has been described by a number of authors. Typical methods are due to Downes and Rossell (1984) and Bocca *et al.* (1983). Isopropanol can be detected by qualitative tests as it gives (1) a yellowish white precipitate on boiling with Denigès' mercuric sulphate solution and (2) acetone when gently heated with potassium dichromate solution and conc. sulphuric acid.

On heating with potassium hydrogen sulphate propylene glycol gives a pleasant fruity odour, whereas triacetin and glycerol produce pungent vapours of acrolein. The use of propylene glycol dehydrogenase as a specific method for determination of propylene glycol has been described by Hamano *et al.* (1984).

Lyne *et al.* (1968) have shown that traces of glycerol can be determined from the fluorescence produced on heating with a solution of anthrone in sulphuric acid. Many workers determine glycerol by titration of the formic acid produced on oxidation with periodate (Sporek & Williams, 1954). Methods using HPLC have been described by Nagel *et al.* (1982).

GENERAL METHODS

Residues of solvents remaining on food are generally low and, in general, GLC methods are appropriate for identification and estimation of these additives. Dean *et al.* (1969) have described methods using a specially designed solid sample device. The sample is introduced into a glass capillary, sealed, and placed into the sampler, which is connected to the GC column and the capillary broken by a plunger. Volatile solvents are swept into the GC column. The method is applicable to a wide range of solvents. Headspace gas chromatographic equipment is now commercially available and suitable for this form of analysis (Van Rillaer *et al.*, 1982). Dickes and Nicholas (1976) have described a variety of GC methods for detection of a range of solvents including hydrocarbons, chlorinated hydrocarbons, alcohols and ketones.

MISCELLANEOUS ADDITIVES, INCLUDING LIQUID FREEZANTS

The Miscellaneous Additives in Food Regulations 1980 (SI No. 1834) and (Amendment) Regulations (SI 1982 No. 14) as amended, give a list of additives which, with few exceptions, may be used generally in food without limit. They put into effect recommendations of three FACC reports (1968, 1972c, 1974c) and implement in part a number of EC Council Directives. The list includes additives with a range of functions including acids, anticaking and antifoaming agents, bases, buffers, bulking aids, firming and glazing agents, flavour modifiers, flour bleaching agents and improvers, humectants, liquid freezants, packaging gases, propellants, release agents and sequestrants. Schedule 1 Part I lists the additives (over 130 in all); Part II contains specifications. Schedule 2 lists specified foods to which certain additives may be added with limitations. Schedule 3 is concerned with labelling of permitted additives. Subsequent amendments of the regulations take into account the introduction of the Sweeteners in Food Regulations and transfer of control and specification of the bleaching agents and improvers used in flour. The regulations permit the use of the liquid freezant dichlorodifluoromethane (CFC 12) for preparation of frozen food and lay down a residue limit of 100 mg/kg in the food when thawed to 20 °C. Since many of the additives are used without limit (other than good manufacturing practice) and many are natural constituents of foods, analytical methods are not normally needed except in respect of use in specific commodities. The use of liquid freezants is particularly advantageous to retain free flowing property in the frozen product and to protect items such as soft fruits from excessive damage during the freezing operation (Daly, 1973; Anstrom, 1975). Carter and Kirk (1977) have described a method for the determination of dichlorofluoromethane residues in frozen prawns. The method is based on chromatography of a headspace gas sample generated by heating the foodstuff in a closed vial. The gas is injected onto a column of Porasil B and the freezant is detected by a flame ionisation detector. The authors claim that the method gives satisfactory recoveries (80–100 per cent) at the statutory limit and that the limit of detection is less than 10 mg/kg.

MINERAL HYDROCARBONS

The Mineral Hydrocarbons in Food Regulations 1966 (SI No. 1073) restrict the use of mineral hydrocarbons to specified foods. The prescribed maximum amounts are dried fruit 0·5 per cent, citrus fruits 0·1 per cent, sugar confectionery and bakery products 0·2 per cent, and chewing gum 60 per cent. Also, eggs may be sealed with mineral oil and the rind of whole pressed cheeses may contain paraffin wax; but tolerances are not laid down. The presence of mineral hydrocarbons in confectionery and bakery products is allowed as a result of their use as polishing, glazing or lubricating agents. Any mineral hydrocarbon used must comply with specifications for grade and purity especially in respect of polycyclic hydrocarbons. The FACC (1975) reviewed the use of mineral hydrocarbons and recommends that no new uses in respect of specified foods should be permitted but that the prohibition on their use as solvents in the composition and preparation of food should be discontinued. Control is therefore proposed in respect of a minimal residue remaining after the manufacturing process (see Solvents). The Committee also recommends a reduction in permitted levels to 0·25 per cent (dried fruit), 0·1 per cent (sugar and chocolate confectionery) and 15 per cent (chewing gum). Residues arising from general use as lubricating agents were considered and further limitations proposed by the introduction of new specified foods: table jellies (0·1 per cent), bread and certain flour confectionery products (0·1 per cent). Revised specifications were also proposed. These recommendations were never implemented but more recently (1989) the FACC has recommended that mineral hydrocarbons should be prohibited as food additives.

GENERAL METHODS

The analysis of foods generally relies on digestion in acid followed by extraction of the mineral oil together with natural lipids by use of chloroform. After separation and removal of chloroform by evaporation, the residual oil is dissolved in petroleum ether and passed through an alumina column to separate mineral oil from the natural lipids (Williams, 1966). Mineral oil is readily eluted with petroleum ether; after removal of solvent the identity is confirmed by measurement of refractive index and by infra-red spectroscopy (Coles, 1963). A saponification step to remove excess fat may be necessary if this purification proves to be inefficient; the column separation is then performed on the unsaponified residues. The method is recommended by the AOAC for bakery products and by Codex (CAC/RM 52-1974) for dried fruits. The determination of mineral oil in dried fruit has been described by Radler (1964). Chakravorty (1979) describes an alumina column clean-up followed by TLC using cyclohexane as mobile solvent to detect mineral oil in black peppercorns.

FLAVOURINGS AND FLAVOUR MODIFIERS

The use of flavours, flavourings and flavour modifiers in foods, which in view of the complex nature of the materials concerned, has long been the subject of debate. Many of the products that are used by the flavour industry, in particular the volatile ones, have a long association with perfumery. They can be classified into materials derived from normal foods, materials derived from natural sources that are not normal foods, nature identical but synthetic materials and synthetic substances without natural equivalents. Considerations in selection of flavours have been discussed by a number of authors, notably Przybyla (1981a), Hall and Merwin (1981) and Harper (1982). Collins (1985) has also summarised the shift towards the use of natural and naturally based flavours. Blanchfield (1978) has described the development of particular formulations for specified products, Nursten and Williams (1967) surveyed fruit aromas, and Wasserman (1979) and MacLeod and Seyyadain-Ardebili (1981) have surveyed flavour components of meat. Industry problems in quality control of raw materials, definition of products, nomenclature and maintenance of product stability have been discussed by Bruhn (1982), Heath (1981) and Anon (1982).

The non-volatile materials used as flavours include salt, spices, sugar, sweeteners, protein hydrolysates, glutamate and ribonucleotides. The latter substances are regarded as flavour enhancers as well as flavours in their own right.

Legislation

There are no current regulations on flavouring substances in the UK, although there have been proposals both within the UK and in the EC to develop lists of permitted materials. International legislation has been surveyed by Jenkins (1975). The FACC (1976, 1978d) produced a general report on flavourings and on the glutamate and nucleotide flavour modifiers. The 1976 report recommends that all flavourings, whether natural, synthetic or artificial, should be controlled by a permitted list system. The proposals for control extend to herbs, spices and to natural and nature-identical constituents of foods when used as flavours. Exemptions are proposed for natural food substances and their constituents when used as ingredients, controlled additives and substances present as a result of preparation (e.g. smoking). Definitions of 'flavourings', 'natural food substances' and 'preparation' are proposed. In addition to the use of a similar classification system, the EC proposals include maximum limits for undesirable substances which are associated with flavourings. The list includes 3,4-benzopyrene, beta-asarone, coumarin, HCN, saffrole and thuyone. Methods of analysis have been developed by the International Organisation of the Flavour Industry (IOFI) (see also Liddle and De Smedt, 1976).

Following amendment of the Miscellaneous Additive Regulations the flavour enhancers are controlled substances that are permitted in foods other than those intended for infants and young children.

GENERAL METHODS

The list of flavours is extensive and most are not single substances; many contain tens of identifiable constituents. As a result, analytical control in foods is regarded as a complex problem. Fedeli (1978) and Land and Nursten (1979) have summarised extraction and isolation procedures applicable to foods. Waltking and Goetz (1983) have summarised GLC techniques used for analysis of volatile flavours in oils; extensive reviews of methods and proposals for generally applicable GLC techniques have been described by Bassette (1984), Parliment and Spencer (1981), Pfannhauser *et al.* (1982), Golovnya (1982) and Schrier and Idstein (1985). Specific methods for estimation of flavour enhancers include the use of HPLC (Nyguyen and Sporns, 1984) and enzyme techniques (Ruf and Siepe, 1983; Watanabe *et al.*, 1984; see also Chapter 13).

SWEETENERS

The sweetening of food embraces the use of normal food sugars, starch conversion products (glucose syrups), molasses and honey. In addition to the sensation of sweetness these carbohydrates provide bulk, energy value and, at high concentrations, important preservative action. Alternative non-carbohydrate sweeteners must provide an adequate degree of sweetness to justify their use but they do not necessarily need the other attributes of the permitted sugar products. The non-carbohydrate sweeteners may be divided into two broad classes, namely those with a sweetening effect similar to sucrose—*bulk sweeteners*, and the *intense sweeteners* which have sweetness many times that of sucrose.

The *bulk sweeteners*, which are non-fermentable but are absorbed and metabolised in man by different routes to most food sugars to yield the same number of calories per gram, are of value in the preparation of 'sugar-free' products used by diabetics, in special diets and in the preparation of sweetened foods such as ice-cream that are stored at low temperatures. Examples of these materials include mannitol, sorbitol and other hydrogenated carbohydrates; many of the polyols (polyhydric alcohols) are classified as humectants although they can confer the sensation of sweetness. *Intense sweeteners* include the synthetic products formerly regulated as 'artificial sweeteners', such as saccharin and cyclamate salts, together with a range of materials derived from naturally sweet but rare food materials. In addition to their use in tablet or powder form as 'table top' sweeteners, these products are also utilised in 'diabetic', 'sugar-free' and 'slimming' foods to supplement the sweetening effect of the bulk sweeteners. Although banned from use in the UK, cyclamates are permitted for use in other European countries. There is a considerable difference of opinion on the toxicology of saccharin and cyclamate and local acceptance in various countries depends to some extent on interpretation of toxicological evidence and the presence of reaction by-products, which may differ according to the synthetic route used in the preparation of the material. The toxicological status of the various sweeteners is summarised in the report of the FACC (see below). Crammer and Ikan (1986) have presented a review of the properties of commercially available sweeteners and have compared these with the sweet glycosides derived from the stevia plant, which are permitted for general use in Japan. The comparative sweetness of a number of sweeteners is shown in Table 4.7.

Table 4.7 Sweetness of intense and bulk sweeteners compared with sucrose at 1·0

Sweetener	Sweetness	Other or trade names
Intense sweeteners		
Saccharin sodium	450	
Cyclamate	80	
Aspartame	200	Canderel, Nutra-sweet, Sanecta
Acesulfam-K	100	Sunett
Thaumatin	3000	Talin
Stevioside	150	
Sucralose	600	
Alitame	2000	
Bulk sweeteners		
Hydrogenated glucose syrup	0.75	Lycasin, Malbit, maltitol
Sorbitol	0.5	
Mannitol	0.6	
Xylitol	1.0	
Lactitol	0.4	Lactit, Lactositol, Lactobiosit
Isomalt	0.5	Palatinit

Legislation

In a review of the use of sweeteners in food, the FACC (1982b) recommended that bulk and intense sweeteners be controlled by a permitted list and that the existing Artificial Sweeteners in Food Regulations (SI 1969 No. 1817) be revoked. As a result, new legislation controlling both types was introduced in 1983 as the Sweeteners in Food Regulations (SI 1983 No. 1211). The permitted sweeteners are detailed in Schedule 1 Part I to the regulations and include: acesulfame potassium, aspartame, hydrogenated glucose syrup, isomalt, mannitol, saccharin (Na and Ca salts), sorbitol, sorbitol syrup, thaumatin and xylitol. The Amendment Regulations 1988 SI No. 2112 added lactitol to the permitted list.

Saccharin and saccharin sodium are required to comply with the criteria given in the British Pharmacopoeia monographs; specifications for saccharin calcium are given in Schedule 1 Part II to the regulations. A definition of 'permitted sweetener' is given 'as a substance specified in the schedule to the regulations and which satisfies the purity criteria set out in the schedule'. Amendments are made to existing regulations in respect of the definitions and deletion of the term 'artificial sweetener' wherever it exists. The limit on the use of saccharin in soft drinks is maintained but no other quantitative restrictions on use are made for the remaining sweeteners. Product use restrictions are imposed by the regulations and these are summarised in a guide to the regulations by Jeffries *et al.* (1983). The guide also includes a bibliography of references to description and usage of the various products. Higginbotham (1984) and Pepper (1985) have reviewed the practical uses of high intensity sweeteners, whilst specific product appraisals have been prepared by Smiles (1986) and Torres and Thomas (1981) for polydextrose, von Rymon Lipinski (1985) for acesulfam potassium and Vetsch (1985) for aspartame.

Analytical methods

The bulk sweeteners are in general determined by HPLC as for sugars (q.v.).

General test for intense sweeteners

After extraction with ethyl acetate, qualitative detection of most artificial sweeteners in soft drinks and foods can be carried out by TLC on silica gel plates using *n*-butanol/ethanol/ammonia/water (40:4:1:9) as developing solvent and specific chromogenic reagents (see AOAC methods book). Saccharin can be extracted with diethyl ether from aqueous solution or aqueous extract of food (use clearing agents) after making strongly acid (at least 0·7 M) with hydrochloric acid. Wash combined ether extracts with water.

Qualitative test for saccharin tablets

Mix about half of a powdered tablet with 50 μg resorcinol. Add 10 drops of conc. sulphuric acid and heat gently until it assumes a dark green colour. Cool carefully, add 10 ml water and excess of 5 M sodium hydroxide solution. Saccharin produces a fluorescent green colour.

The AOAC methods book describes methods for the detection of saccharin based on conversion to salicylic acid and the production of a reddish-purple colour by reaction with phenol and sulphuric acid. The AOAC methods book also describes an HPLC method (also determining benzoate and caffeine) for saccharin in soft drinks using a 300 mm RP-18 column (u-Bondapak or equivalent), ultraviolet detection at 254 nm and a mobile phase of 20 per cent acetic acid buffered to pH 3·0 with saturated sodium acetate solution, and modified with 0–2 per cent isopropanol to achieve baseline resolution. The separation of the same three components can be achieved in under 4 min on 10 μm Lichrosorb RP-8 eluting with water/methanol/acetonitrile (49·6:10·8:39.6) (Weyland *et al.*, 1982). Water/methanol/perchloric acid 70 per cent (85:15:0·1) can be used to separate saccharin, benzoate and sorbate. Henning (1983b) used ion-pair HPLC after extraction of saccharin from prepared mustard by homogenisation with 70 per cent methanol and clean-up on a Sep-Pak C_{18} cartridge followed by analysis of an RP column eluted with methanol/water (40:60) containing 5 mM tetraheptyl ammonium bromide.

Many of the artificial sweeteners are now determined in multi-detection methods (Lawrence and Charbonneau, 1988 for all sweeteners) or along with other additives; typically HPLC is used to detect caffeine, aspartame, saccharin and preservatives. General methods of the latter type have been described by Tyler (1984) and Argoudelis (1984) (see below). The former author used a C_{18} Bondapak column and aqueous triethylamine acetonitrile mixtures at pH 4·3 to elute the additives from samples of diet cola.

Aspartame can be determined colorimetrically by reaction with ninhydrin (Oi-Wah *et al.*, 1988) after extraction from the sample into propylene carbonate to remove interference from amino acids. The procedure is as follows: to 1·5 ml of soft drink, sample extract or solution of tablet containing about 300–400 mg/kg of aspartame add 0·5 ml of acetate buffer pH 3·5 (19·35 g sodium acetate trihydrate in 1 litre of 2·5 M acetic acid), mix and add 10 ml propylene carbonate. Shake for 5 min, then centrifuge. Take about 7 ml of lower organic phase and dry with a little anhydrous sodium sulphate. Take 3·0 ml of dried organic phase, add 2 ml of 0·4 per cent ninhydrin in absolute ethanol, mix and stand in boiling-water bath for 20 min, cool and dilute to 10 ml with ethanol. Measure absorbance at 585 nm in a 10 mm cuvette. Prepare a calibration graph using a standard stock solution of aspartame 1000 μg/ml (25 mg of aspartame dissolved in 25 ml propylene carbonate) suitably diluted with propylene carbonate.

Aspartame in tablets and dry powders is readily determined by HPLC after extraction into water, filtration through a 0·45 μm membrane filter and analysis on an RP-8 or RP-18 column with detection at 254 nm.

Aspartame can be determined in soft drinks by HPLC with ultraviolet detection (214 nm) using an amino-bonded RP column (u-Bondapak). Webb and Beckman (1984) used a water/acetic acid/isopropanol mobile phase prepared as follows: add 100 ml glacial acid to about 750–800 ml water. Adjust to pH 3·0 with saturated sodium acetate. Add 30 ml isopropanol and dilute to 1 litre with water. Tsang *et al.* (1985)

used 0·0125 M KH_2PO_4 (pH 3·5)/acetonitrile (90:10). Issaq *et al.* (1986) used a β-cyclodextrin-bonded silica gel column (Cyclobond I) with methanol/1 per cent v/v triethylammonium acetate, pH 4·5 (1:3). This system also determines caffeine and sodium benzoate at 214 nm. If detection is at 254 nm, caramel colouring is detected.

Good separation of aspartame, saccharin, benzoate and caffeine in soft drinks was achieved by Argoudelis (1984) using a strong cation-exchanger column (Partisil-10SCX) eluting with 0·1 M ammonium dihydrogen phosphate (pH 4·5) and detection at 214 nm.

Most of the RP-HPLC techniques for saccharin and aspartame can be tuned to detect and determine acesulfame-K, which elutes more rapidly.

Qualitative and quantitative tests for cyclamates are described in the 8th edition of this book, and also in the AOAC methods book (see also Chapter 7). Hermann *et al.* (1983) have reported an HPLC method for cyclamate in fruit juices after filtration by direct injection on to a column of LiChrosorb RP-8 or RP-18 eluting with water/methanol (88:12) containing 5 mM tetrabutylammonium toluene-*p*-sulphonate (TBTS) buffer with 10 mM glycerine adjusted to pH 5·5 with HCl. Cyclamate does not absorb above 200 nm, but is detected at 267 nm indirectly since it causes a reduction in absorption by replacing the absorbing TBTS. Page and Conacher (1978), Conacher and Page (1979) and Macrae (1988) have reviewed available methods for the determination of the range of commonly used sweeteners. Table 4.8 summarises recent literature on methodology for commonly used sweeteners.

Table 4.8 Methods of analysis for the common sweeteners

Substance	Substrate	Method	Reference
Acesulfam-K	Foods, beverages	HPLC	Grosspietsch and Hachenberg (1980)
Aspartame	Soft drinks, beverages, sweetener mixes	HPLC, UV	Webb and Beckman (1984)
			Tsang *et al.* (1985)
			Issaq *et al.* (1986)
	Product preparations	Ion-pair HPLC	Verzella *et al.* (1985)
	Soft drinks	HPLC	Prudel and Davidkova (1985)
	Foods	Spectrophotometry	Oi-Wah *et al.* (1988)
Cyclamate	Fruit juices	HPLC	Hermann *et al.* (1983)
Dihydrochalcone	General foods	HPLC	Schwartzenbach (1976)
		HPLC	Fisher (1977)
Dulcin	Beverages	Oxidation, spectrophotometry	Veerabhadra Rao *et al.* (1984)
Glycyrrhizin		HPLC	Kitada *et al.* (1980a)
Glycerol	General	Enzyme, HPLC	Goiffon *et al.* (1980)
Polyols	General	HPLC	Brandao *et al.* (1980)
Sorbitol	Foods	GLC	Tsuda and Nakanishi (1981)
	Fruits and fruit products	HPLC	Richmond *et al.* (1981)
	General	Enzyme	Anon (1979)
Saccharin	Foods	HPLC	Kitada *et al.* (1980b)
	Soft drinks	HPLC	Puttemans *et al.* (1984)
	Salads, marinades	Ion-pair HPLC	Henning (1983a, b)
	Soft drinks	Spectrophotometry	Kruger (1980)
	Soft drinks, beans	TLC	Nanase *et al.* (1981)
	Tablets	Spectrophotometric	Guven *et al.* (1984)
Steviosides	Preparations	TLC	Hirokado *et al.* (1980)

In the evaluation of sweeteners it is necessary to include an examination for reaction by-products, breakdown products and, in the case of natural products, the presence of co-extracted materials. In the case of saccharin, attention has been focused on the presence of toluenesulphonamides as reaction by-products. Early methods for the estimation were based on the use of GLC (Roll *et al.*, 1977). Kaeprzak (1978) used GLC methods to survey the quality of commercially available material and concluded that *o*-toluenesulphonamide content was related to the preparative method. Recently, attention has been directed to the use of HPLC for this purpose. Chvalousky and Nondek (1983) used reversed-phase methods, whilst Mooser (1984) reported on the use of an ion-pair HPLC method sensitive to 8 ng *o*-toluenesulphonamide and 16 ng *p*-toluenesulphonamide. The dipeptide sweetener aspartame is unstable in aqueous solution and yields diketopiperazine as the result of hydrolysis and cyclisation. Up to 5 per cent has been reported in aqueous solution. Prudel and Davidkova (1985) have described an HPLC method which is suitable for the detection of aspartame and the hydrolysis products in soft drinks.

OTHER ADDITIVES

Reviews of the uses of a variety of classes of food additives are being carried out; these include a number of currently uncontrolled groups. The list, which is not exhaustive, includes significant additives such as modified starches, polyhydric alcohols, enzymes, clouding agents, and processing aids. A special review of the wide range of processing aids used in the production of beer has been published (FACC, 1978b). This report recommends that only substances classified by the COT in Groups A or B should be used, that adequate specifications are necessary, that residues of gibberellic acid in beer should not exceed 0·05 mg/l and that total usage of silicates should be restricted to 2 g/l. The use of asbestos filters in the clarification of beers is still permitted but further efforts to find alternative materials are encouraged; see also the Report on Asbestos (FACC, 1979b). The FACC (1980a) has recommended classification of 6 modified starches in Group A, 14 in Group B and 3 in Group E. Special proposals were made on limitation of use in foods for infants and young children in respect of starches modified with epichlorohydrin and/or propylene oxide. They also proposed that the degree of substitution be controlled by specification. In a report on bulking aids the FACC (1980b) proposed control of these substances in Regulations under the Food and Drugs Act. Permission for general use in foods other than baby foods was recommended for polydextrose and alpha-cellulose, although the latter is not considered to be suitable for use in bread. These substances are included in the amended Miscellaneous Additive Regulations. The FACC (1982) reported on enzymes and recommended control of their use in foods and recognised the need for development of appropriate specifications.

REFERENCES

Adams, J. B. (1981) *Nutrition and Food Science*, **69**, 12.
Adams, W. S. (1978) *Journal of the Association of Official Analytical Chemists*, **61**, 354.
Airaudo, C. B., Ceri, V., Gayte-Sorbier, A. & Adrianjafiniony, J. (1983) *Journal of Chromatography*, **261**, 273.
Aitzetmuller, K. & Arzberger, E. (1984) *Zeitschrift fur Lebensmitteluntersuchung und-forschung*, **178**, 279.
Alcock, R. S. (1937) *Analyst*, **62**, 522.
Alderton, G. & Lewis, J. C. (1958) *Food Research*, **23**, 338.
Ali, M. S. (1985) *Journal of the Association of Official Analytical Chemists*, **68**, 488.
Alicino, N. J., Klein, H. C., Quattrone, J. J. & Choy, T. K. (1963) *Journal of Agricultural and Food Chemistry*, **11**, 340, 496.
Allman, G. A. (1973) *The Flavour Industry*, **4**, 256.
Almela, L. & Lopez-Roca, J. M. (1984) *Science Alimentation*, **4**, 37.
Amin, D. (1986) *Analyst*, **111**, 103.
Anderson, H. H. & Girling, J. (1982) *Analyst*, **107**, 836.
Anon (1963) *International Chocolate Review*, **18**, 6.
Anon (1968) *British Food Journal*, **70**, 4.
Anon (1977) *Why Additives? The Safety of Foods*. London: The British Nutrition Foundation.
Anon (1979) *Methods of Enzymic Food Analysis*. Lewes, UK: Boehringer Corporation (London) Ltd.

Anon (1981a) *Processed Prepared Food*, **150**, 80.
Anon (1981b) *The Health Effects of Nitrate, Nitrite and N-Nitroso Compounds*. Washington DC: National Academy of Sciences.
Anon (1982) *Food Process Industry*, **51**, 20.
Anstrom, S. (1975) *Food Manufacture*, **50**, 15.
AOCS Fat Analysis Committee (1957) *Journal of the American Oil Chemists' Society*, **34**, 301.
APA (1960) *Separation and Identification of Food Colours Permitted by the Colouring Matters in Food Regulations* 1957. London: APA.
Applebaum, R. S. (1985) *Manufacturing Confectioner*, Nov, 45.
Argoudelis, C. J. (1984) *Journal of Chromatography*, **303**, 256.
Asai, Y., Kuwahira, H., Shimoda, K. & Sato, K. (1982) *Journal of the Food Hygiene Society Japan*, **23**, 438.
van Baal, J. P. W. & Leget, J. N. (1965) *Zeitschrift fur Levensmitteluntersuchung und-forschung*, **127**, 264.
Baker, P. B. & Hoodless, R. A. (1973) *Journal of Chromatography*, **87**, 585.
Baker, P. B. & Hoodless, R. A. (1974) *Pesticide Science*, **5**, 465.
Bakker, J. & Timberlake, C. F. (1985) *Journal of the Science of Food and Agriculture*, **36**, 1315, 1325.
Baldi, M., Maietti, S. & Pietrogrande, M. C. (1982a) *Industrie Alimentari*, **21**, 771.
Baldi, M., Zanoni, L., Pietrogrande, M. C. & Maietti, S. (1982b) *Industrie Alimentari*, **21**, 389.
Baloch, A. K., Buckle, K. A. & Edwards, R. A. (1977) *Journal of Chromatography*, **139**, 147.
Banerjee, T. S., Guha, K. C., Saha, A. & Roy, B. R. (1974) *Journal of Food Science and Technology*, **11**, 230.
Banks, J. G. & Board, R. G. (1982) *Journal of the Science of Food and Agriculture*, **33**, 197.
Barnett, D. (1985) *Food Technology Australia*, **37**, 503.
Bassette, R. (1984) *Journal of Food Protection*, **47**, 410.
Baur, F. J. (1973) *Journal of the American Oil Chemists' Society*, **50**, 85.
Baur, F. S. & Distler, E. (1966) *Journal of the Association of Official Analytical Chemists*, **49**, 816.
Berger, S. L. (1975) *Analytical Biochemistry*, **67**, 428.
Beutler, H. O. (1984) *Food Chemistry*, **15**, 157.
BFMIRA (1963) *Analyst*, **88**, 864.
Bieri, J. G. (1984) *Journal of the American Oil Chemists' Society*, **61**, 1917.
Bishov, S. J., Masuoka, Y. & Kapsalis, J. G. (1977) *Journal of Food Processing and Preservation*, **1**, 153.
Bocca, A., Di Marzio, S. & Milana, M. R. (1983) *Rivista Italiana della Sostance Grasse*, **60**, 441.
Boehme, W., Oehme, U. & Steinwand, M. (1984) *Lebensmittelchemie und gerichtliche Chemie*, **38**, 88.
Boley, N. P., Crosby, N. T. & Roper, P. (1979) *Analyst*, **104**, 472.
Boley, N. P., Bunton, N. G., Crosby, N. T., Johnson, A. E., Roper, P. & Somers, L. (1980) *Analyst*, **105**, 589.
Boley, N. P., Crosby, N. T., Roper, P. & Somers, L. (1981) *Analyst*, **106**, 710.
Boniforti, L., DiStefano, F. & Vercillo, A. (1961) *Bollettini dei Laboratori Chimici Provinciali*, **12**, 505.
Blanchfield, R. (1978) *Food Trade Review*, **48**, 459.
Bloomfield, R. A., Guyon, J. C. & Nurman, R. K. (1965) *Analytical Chemistry*, **37**, 249.
Brandao, S. C. C., Richmond, M. L., Gray, J. I., Morton, I. D. & Stine, C. M. (1980) *Journal of Food Science*, **45**, 1492.
Bruhn, W. (1982) *Dragoco Reports Flavouring Information Service*, **27**, 3.
Brunn, W. & Klostermeyer, H. (1984) *Lebensmittelchemie und gerichtliche Chemie*, **38**, 16.
Bruno, P., Caselli, M., DiFano, A. & Traini, A. (1979) *Analyst*, **104**, 1088.
BSI (1976) British Standard Methods of Test Meat and Meat Products. BS 4401:
Part 7 Determination of Nitrate Content.
Part 8 Determination of Nitrite Content.
Bushway, R. V. (1985) *Journal of Liquid Chromatography*, **8**, 1527.
Calvey, R. J. & Goldberg, A. L. (1982) *Journal of the Association of Official Analytical Chemists*, **65**, 1080.
Calvey, R. J., Goldberg, A. L. & Madigan, E. A. (1981) *Journal of the Association of Official Analytical Chemists*, **64**, 665.
Cantafora, A. (1976) *Rivista della Societa Italiana di Scienza dell'Allimentazione*, 5, 87.
Carr, W. & Smith, G. A. (1964) *Journal of the Association of Public Analysts*, 2, 37.
Carruthers, A., Heaney, R. K. & Oldfield, J. F. T. (1965) *International Sugar Journal*, **67**, 364.
Carswell, D. R. (1977) *Scientific and Technical Surveys No. 103*, Leatherhead Food RA.
Carter, J. R. & Kirk, R. S. (1977) *Journal of Food Technology*, **12**, 49.
Cerny, M. & Blumenthal, A. (1979) *Zeitschrift fur Lebensmitteluntersuchung und-forschung*, **168**, 87.
Chakravorty, K. L. (1979) *Journal of the Association of Public Analysts*, **17**, 125.
Chapman, W. B. & Oakland, D. (1968) *Journal of the Association of Public Analysts*, **6**, 124.
Chiang, H. C. (1969) *Journal of Chromatography*, **44**, 203.
Chikamoto, T., Nagata, S. & Maitani, T. (1981) *Journal of the Food Hygiene Society Japan*, **22**, 113.

Chudy, J., Crosby, N. T. & Patel, I. (1978) *Journal of Chromatography*, **154**, 306.
Chvalousky, V. & Nondek, L. (1983) *Journal of Chromatography*, **268**, 395.
Cochrane, G. C. (1975) *Journal of Chromatographic Science*, **13**, 440.
Coelho, R. G. & Nelson, D. L. (1983) *Journal of the Association of Official Analytical Chemists*, **66**, 209.
Coles, L. E. (1963) *Journal of the Association of Public Analysts*, **1**, 31.
Collet, P. (1983) *Deutsches LebensmittelRundschau*, **79**, 370.
Collet, P. (1984) *Lebensmittelchemie und gerichtliche Chemie*, **38**, 3.
Collins, A. (1985) *Technology Ireland*, **27**, 16.
Conacher, H. B. S. (1973) *Journal of the Association of Official Analytical Chemists*, **56**, 602.
Conacher, H. B. S. & Page, B. D. (1979) *Journal of Chromatographic Science*, **17**, 188.
Coppens, P. (1985) *Food*, May, 49.
Corradi, C. & Micheli, G. (1981) *Industrie Alimentari*, **20**, 372.
Counsell, J. N. & Knewstubb, C. J. (1983) *Food*, August, 18.
Coupek, J., Pokorny, S., Mares, E., Zezulkova, L., Nguyen, T. L. & Pokorny, J. (1976) *Journal of Chromatography*, **120**, 411.
Coustard, J. M., Retho, C., Blanchard, F., Sudraud, G., Caude, M., Rosset, C., Hagemann, R., Gaudin, D. & Virelizier, H. (1982) *Annales Falsifications L'Expertise Chimie*, **75**, 563.
Cox, E. A. & McClure, F. D. (1982) *Journal of the Association of Official Analytical Chemists*, **65**, 933.
Crammer, B. & Ikan, R. (1986) *Chemistry in Britain*, **22**, 915.
Csorba, I., Buzas, Z., Polyak, B. & Boross, L. (1979) *Journal of Chromatography*, **172**, 287.
Cundall, R. B., Phillips, G. O. & Rowlands, D. P. (1973) *Analyst*, **98**, 857.
Daly, J. J. (1973) *Food Process Industry*, **42**, 27.
Daniels, D. H., Warner, C. R., Selims, S. & Joe, F. L. (1983) *Journal of the Association of Official Analytical Chemists*, **66**, 893.
Davis, E. G., Barnet, D. & Hoy, P. M. (1983) *Journal of Food Technology*, **18**, 233.
Davis, P. L. & Munroe, K. A. (1979) *Journal of Agricultural and Food Chemistry*, **27**, 918.
Dean, A. C., Bradford, E., Hubbard, A. W., Pocklington, W. D. & Thomson, J. (1969) *Journal of Chromatography*, **44**, 465.
DeClercq, H. & Massart, D. L. (1974) *Journal of Chromatography*, **93**, 243.
De Kleijn, J. P. & Hoven, K. (1984) *Analyst*, **109**, 527.
DeSiena, A., Jacobs, E. & Romagnoli, R. (1981) *Journal of the Association of Official Analytical Chemists*, **64**, 226.
Devani, M. B., Shishoo, C. J. & Shah, M. G. (1973) *Analyst*, **98**, 759.
Dick, R. & Miserez, A. (1976) *Mitteilungen aus dem Gebiet der Lebensmitteluntersuchung und-Hygiene, Bern*, **67**, 472.
Dickes, G. J. (1965a, b, c) *Journal of the Association of Public Analysts*, **3**, 49, 73, 118.
Dickes, G. J. & Nicholas, P. W. (1976) *Gas Chromatography in Food Analysis*. Sevenoaks: Butterworth.
Dieffenbacher, A. & Bracco, U. (1978) *Journal of the American Oil Chemists' Society*, **55**, 642.
Dilli, S. & Robards, K. (1977) *Analyst*, **102**, 201.
Distler, E. & Baur, F. J. (1966) *Journal of the Association of Official Analytical Chemists*, **49**, 812.
Doeden, W. G., Bowers, R. H. & Ingala, A. C. (1979) *Journal of the American Oil Chemists' Society*, **56**, 12.
Downes, M. J. & Rossell, J. B. (1984) *Journal of the American Oil Chemists' Society*, **61**, 896.
Draper, R. E. (1975) *Journal of the Association of Official Analytical Chemists*, **58**, 614.
Eckhoff, S. R. & Okos, M. R. (1983) *Journal of Agricultural and Food Chemistry*, **31**, 826.
Edwards, F. W., Nanji, H. R. & Hassan, M. K. (1937) *Analyst*, **62**, 178.
Edwards, R. A. (1980) *Analyst*, **105**, 139.
Egginger, R. & Honikel, K. O. (1979) *Flëischwirtschaft*, **59**, 229.
Ehlers, D. & Littman, S. (1984) *Deutsches LebensmittelRundschau*, **80**, 44.
Endean, M. E. (1976a) *Scientific and Technical Surveys*, No. 90. Leatherhead Food RA.
Endean, M. E. (1976b) *Scientific and Technical Survey*, No. 91. Leatherhead Food RA.
Endean, M. E. & Bielby, C. R. (1975) *Technical Circular*, No. 587. Leatherhead Food RA.
Endean, M. E. & Harding, M. S. (1976) *Technical Circular* No. 617. Leatherhead Food RA.
Englis, D. T., Burnett, B. B., Schreiber, R. A. & Miles, J. W. (1955) *Journal of Agricultural and Food Chemistry*, **3**, 964.
Engst, R., Prahl, L. & Jarmatz, E. (1969) *Nahrung*, **13**, 417.
Evans, G. G. (1972) *British Food Manufacturing Industry Research Association, Circular* No. 508.
Ewart, M. H. & Chapman, R. A. (1952) *Analytical Chemistry*, **24**, 1460.
FACC (1966) *Report on Solvents*. FAC/REP/2. London: HMSO.
FACC (1968) *Report on Further Classes of Food Additives*. FAC/REP/7. London: HMSO.
FACC (1970) *Report on the Review of the Emulsifiers and Stabilisers in Food Regulations*, 1962. FAC/REP/9. London: HMSO.

FACC (1972a) *Report on the Review of the Preservatives in Food Regulations*, 1962. FAC/REP/14. London: HMSO.
FACC (1972b) *Supplementary Report on the Review of the Emulsifiers and Stabilisers in Food Regulations*. FAC/REP/16. London: MAFF.
FACC (1972c) *Report on Liquid Freezants of Food*. FAC/REP/15. London: MAFF.
FACC (1974a) *Report on the Antioxidant in Food Regulations* 1966 and 1974. FAC/REP/18. London: HMSO.
FACC (1974b) *Supplementary Report on Solvents in Food*. FAC/REP/17. London: HMSO.
FACC (1974c) *Supplementary Report on the Review of Liquid Freezants of Food*. FAC/REP/19. London: MAFF.
FACC (1975) *Report on the Review of the Mineral Hydrocarbons in Food Regulations*, 1966. FAC/REP/20. London: HMSO.
FACC (1976) *Report on the Review of Flavourings in Food*. FAC/REP/22. London: HMSO.
FACC (1977) *Report on the Review of the use of Sorbic Acid in Food*. FAC/REP/24. London: HMSO.
FACC (1978a) *Report on the Review of Solvents in Food*. FAC/REP/25. London: HMSO.
FACC (1978b) *Report on the Review of Additives and Processing Aids used in the Production of Beer*. FAC/REP/26. London: HMSO.
FACC (1978c) *Report on Nitrites and Nitrates in Cured Meats and Cheese*. FAC/REP/27. London: HMSO.
FACC (1978d) *Report of Remaining Classes of Food Additives used as Ingredients in Food Report on the Review of Flavour Modifiers*. FAC/REP/28. London: HMSO.
FACC (1979a) *Interim Report on the Review of the Colouring Matter in Food Regulations* 1973. FAC/REP/29. London: HMSO.
FACC (1979b) *An Examination of Asbestos in Relation to Food and Drink*. FAC/REP/30. London: HMSO.
FACC (1980a) *Report on Modified Starches*. FAC/REP/31. London: HMSO.
FACC (1980b) *Review of Remaining Classes of Food Additives used as Ingredients in Food Report on Review of Bulking Aids*. FAC/REP/32. London: HMSO.
FACC (1982a) *Report on Use of Enzymes in Foods*. FAC/REP/35. London: HMSO.
FACC (1982b) *Report on Review of Sweeteners in Foods*. FAC/REP/34. London: HMSO.
Farrow, J. E., Hoodless, R. A., Sargent, M. & Sidwell, J. A. (1977) *Analyst*, **102**, 752.
Fedeli, E. (1978) *Rivista Italiana della Sostanze Grasse*, **55**, 253.
Fett, H. M. (1961) *Journal of the American Oil Chemists' Society*, **38**, 447.
Fiddler, R. N. & Fox, J. B. (1978) *Journal of the Association of Official Analytical Chemists*, **61**, 1063.
Fiddler, W., Doerr, R. C., Gates, R. A. & Fox, J. B. (1984) *Journal of the Association of Official Analytical Chemists*, **67**, 525.
Fisher, J. F. (1977) *Journal of Agricultural and Food Chemistry*, **25**, 628.
Flack, E. A. (1983) *Food*, **32**, 32.
Fogden, E., Fryer, M. & Urry, S. (1974) *Journal of the Association of Public Analysts*, **12**, 93.
Fogg, A. G. & Bhanot, D. (1980) *Analyst*, **105**, 234.
Fogg, A. G. & Summan, A. M. (1984) *Analyst*, **109**, 1029.
Fogg, A. G. & Yoo, K. S. (1979) *Analyst*, **104**, 723, 1087.
Fogg, A. G., Chamsi, A. Y. & Abdalla, M. A. (1983) *Analyst*, **108**, 464.
Follett, M. J. & Ratcliffe, P. W. (1963) *Journal of the Science of Food and Agriculture*, **14**, 138.
Fox, J. B., Doerr, R. C. & Gates, R. (1984) *Journal of the Association of Official Analytical Chemists*, **67**, 692.
Franco, V. & Holak, W. (1975) *Journal of the Association of Official Analytical Chemists*, **58**, 293.
Frohlich, D. H. (1982) *Journal of High Resolution Chromatography, Chromatography Communications*, **5**, 158.
FSC (1959) *Report on Preservatives in Food*. London: HMSO.
Fuchs, G. & Sundell, S. (1975) *Journal of Agricultural and Food Chemistry*, **23**, 120.
Funazo, K., Tanaka, M. & Shono, T. (1980) *Analytical Chemistry*, **52**, 1222.
Galensa, R. & Schaefers, F. I. (1982) *Deutsches LebensmittelRundschau*, **78**, 258.
Garrison, R. A., Harwood, V. & Chapman, R. A. (1957) *Journal of the Association of Official Agricultural Chemists*, **30**, 1085.
Gatewood, L. & Graham, H. D. (1961) *Analytical Chemistry*, **33**, 1393.
Gaunt, I. F., Lloyd, A. G., Grasso, P., Gangoli, S. D. & Butterworth, K. R. (1977) *Food Cosmetic Toxicology*, **15**, 509.
Geiger, U. (1982) *Lebensmittelchemie und gerichtliche Chemie*, **36**, 109.
Gend, H. W. (1975) *Zeitung Lebensmitteluntersuchung Forschung*, **158**, 137.
Gertz, C. & Herrmann, K. (1983) *Deutsches LebensmittelRundschau*, **79**, 331.
Gilhooley, R. A., Hoodless, R. A., Pitman, K. G. & Thomson, J. (1972) *Journal of Chromatography*, **72**, 325.
Goiffon, J. P., Blanchere, A. A. & Perez, J. L. (1980) *Annales Falsifications L'Expertise Chimie*, **73**, 17.

Goldberg, A. L. & Calvey, R. J. (1982) *Journal of the Association of Official Analytical Chemists*, **65**, 103.
Golovnya, R. V. (1982) *Journal of Chromatography*, **251**, 249.
Graichen, C. (1975) *Journal of the Association of Official Analytical Chemists*, **58**, 278.
Graveland, A. (1972) *Journal of the Association of Official Analytical Chemists*, 55, 1024.
Gray, J. I. & Randall, C. J. (1979) *Journal of Food Protection*, **42**, 168.
Green, L. F. (1976) *Food Chemistry*, **1**, 103.
Green, M. S. & Keen, G. (1971) *Journal of the Association of Public Analysts*, **9**, 96.
Greenberg, H. J., Hoholick, J., Robinson, R., Kubi, K., Groce, J. & Weber, L. (1984) *Journal of Food Science*, **49**, 1622.
Grosspietsch, H. & Hachenberg, H. (1980) *Zeitschrift fur Lebensmitteluntersuchung und-forschung*, **171**, 41.
Grüne, A. & Nobbe, V. (1967) *Riechstaffe Aromen Korperpflegemittel*, **17**, 501.
Guagnini, O. A. & Vonesch, E. E. (1959) *Anales de la Asociación quimica argentina*, **47**, 41.
Gutfinger, T., Ashkenazy, R. & Letan, A. (1976) *Analyst*, **101**, 49.
Guven, K. C., Izol, T., Ekiz, N. & Guneri, T. (1984) *Analyst*, **109**, 969.
Hall, M. J. (1964) *Journal of the Association of Official Agricultural Chemists*, **47**, 685.
Hall, M. N. (1978) *Campden Food Preservation Research Association, Technical Memorandum, No. 218.*
Hall, R. L. & Merwin, E. J. (1981) *Food Technology*, **35**, 46.
Hamano, T., Mitsuhashi, Y., Tamaka, K., Matsuki, Y., Oji, Y. & Okamoto, S. (1983) *Agricultural and Biological Chemistry*, **47**, 2427.
Hamano, T., Mitsuhashi, Y., Tamaka, K., Matsuki, M., Oji, Y. & Okamoto, S. (1984) *Agricultural and Biological Chemistry*, **48**, 2517.
Hammond, K. J. (1978) *Journal of the Association of Public Analysts*, **16**, 17.
Hansen, P. M. T. & Chang, J. C. (1968) *Journal of Agricultural and Food Chemistry*, **16**, 77.
Harper, R. (1982) *Food Flavours, Ingredients, Processing and Packaging*, **4**, 13.
Hatcher, J. T. & Wilcox, L. V. (1950) *Analytical Chemistry*, **22**, 567.
Hearne, J. F. & Tapsfield, D. (1956) *Journal of the Science of Food and Agriculture*, **7**, 210.
Heath, H. B. (1981) *Royal Society of Health Journal*, **101**, 6.
Hellwig, E., Gombocz, E., Frischenschlager, S. & Petuely, F. (1981) *Deutsches LebensmittelRundschau*, **77**, 165.
Henning, W. (1983a) *Lebensmittelchemie und gerichtliche Chemie*, **37**, 97.
Henning, W. (1983b) *Deutsches LebensmittelRundschau*, **79**, 16.
Henry, B. S. (1979) *Tropical Science*, **21**, 207.
Henshall, J. D., Ongley, M. & Hall, M. (1977) *Campden Food Preservation Research Association, Technical Memorandum No. 179.*
Hermann, A., Damawandi, E. & Ukgmann, M. (1983) *Journal of Chromatography*, **280**, 85.
Hibbert, H. R. (1968) *The Decision and Determination of Emulsifiers and Stabilizers in Foods.* Scientific and Technical Survey No. 53. Leatherhead, Surrey: BFMIRA.
Higginbotham, J. D. (1984) *Food Technology Australia*, **36**, 552.
Hill, F. (1980) *Institute of Food Science and Technology, Proceedings*, **13**, 289.
Hirokado, M., Nakajima, I., Nakajima, K., Mizoiri, S. & Endo, F. (1980) *Journal of the Food Hygiene Society Japan*, **21**, 451.
Holland, G. E. (1980) *National Provisioner*, **183**, 16.
Holley, R. A. & Millard, G. E. (1980) *Journal of the Association of Official Analytical Chemists*, **63**, 1332.
Hironez, K. O. (1984) *Lebensmittelchemie und gerichtliche Chemie*, **38**, 18.
Hoodless, R. A., Pitman, K. G., Stewart, T. E., Thomson, J. & Arnold, J. E. (1971a) *Journal of Chromatography*, **54**, 393.
Hoodless, R. A., Thomson, J. & Arnold, J. E. (1971b) *Journal of Chromatography*, **56**, 332.
Horwood, J. F., Lloyd, G. T., Ramshaw, E. H. & Stark, W. (1981) *Australian Journal of Dairy Technology*, **36**, 38.
Howard, J. W. (1972) *Journal of the Association of Official Analytical Chemists*, 55, 262.
Hunt, J. & Seymour, D. J. (1985) *Analyst*, **110**, 131.
Hunziker, H. R. (1984) *Mitteilungen aus dem Gebeit der Lebensmitteluntersuchung und Hygiene*, **75**, 484.
Hunziker, H. R. & Mizerez, A. (1980) *Mitteilungen aus dem Gebeit der Lebensmitteluntersuchung und Hygiene*, **71**, 87.
Hunziker, H. R. & Zimmerli, B. (1983) *Mitteilungen aus dem Gebeit der Lebensmitteluntersuchung und Hygiene*, **74**, 121.
Hurst, W. J., McKie, J. M. & Martin, R. A. (1981) *Journal of the Association of Official Analytical Chemists*, **64**, 1411.
Ingram, M. (1947) *Journal of the Society of Chemical Industry*, **66**, 50, 105.
Inoue, T., Iwaida, M., Ito, Y. & Tonogai, Y. (1981) *Journal of the Association of Official Analytical Chemists*, **64**, 276.
Iskandarai, Z. & Pietrzyk, D. J. (1982) *Analytical Chemistry*, **54**, 2600.

Issaq, H. J., Weiss, D., Ridlow, C., Fox, S. D. & Muschik, G. M. (1986) *Journal of Liquid Chromatography*, **9**, 1791.

Isshiki, K., Tsumura, S. & Wanatabe, T. (1980) *Journal of the Association of Official Analytical Chemists*, **63**, 747.

Ito, Y., Toyoda, M., Ogawa, S. & Iwaida, M. (1980) *Journal of Food Protection*, **43**, 601.

IUPAC (1987) *Standard Methods for the Analysis of Oils, Fats and Derivatives*, 7th edn. Oxford: Blackwell.

Iwaida, M., Ito, Y. & Toyada, M. (1981) *Journal of Food Hygiene Society Japan*, **22**, 432.

Jackson, P. E., Haddad, P. R. & Dilli, S. (1984) *Journal of Chromatography*, **295**, 471.

Jefferies, D. A., O'Keefe, D. & Flowerdew, D. W. (1983) *A Users Guide to the Newly Permitted Sweeteners (1983)*. Leatherhead, Surrey: Leatherhead Food Research Association.

Jeffries, G. (1979) *Food Processing Industry*, **48**, 33.

Jenkins, J. J. (1975) *Journal of the Association of Public Analysts*, **13**, 1.

Jennings, N., Bunton, N. G., Crosby, N. T. & Alliston, T. G. (1978) *Journal of the Association of Public Analysts*, **16**, 59.

Kaeprzak, J. L. (1978) *Journal of the Association of Official Analytical Chemists*, **61**, 1528.

Karasz, A. B., Maxstadt, J. J., Reher, J. & DeCocco, F. (1976) *Journal of the Association of Official Analytical Chemists*, **59**, 766.

Kearsley, M. W. & Thanos, A. (1982) *Journal of the Association of Public Analysts*, **20**, 145.

Kitada, Y., Tamase, K., Sasaki, M. & Nishikawa, Y. (1980a) *Journal of the Food Hygiene Society Japan*, **21**, 354.

Kitada, Y., Tamase, K., Sasaki, M. & Nishikawa, Y. (1980b) *Journal of the Food Hygiene Society Japan*, **21**, 480.

Kitada, Y., Ueda, Y., Yamamoto, M., Shinomiya, K. & Nakazawa, H. (1985) *Journal of Liquid Chromatography*, **8**, 47.

Kline, D. A., Joe, F. L. & Fazio, T. (1978) *Journal of the Association of Official Analytical Chemists*, **61**, 513.

Knowles, M. E., Gilbert, J. & McWeeney, D. J. (1974) *Journal of the Science of Food and Agriculture*, **25**, 1239.

Krol, B. & Tinbergen, B. J. (1974) *Proceedings of the First International Symposium on Nitrites in Meat Products*. Wageningen: Püdoc.

Kröller, E. (1962, 1964, 1968, 1974) *Fette, Seifen, Anstrichmittel*, **64**, 602; **66**, 456; **70**, 119; **76**, 498.

Kruger, G. (1980) *Lebensmittel-Industrie*, **27**, 264.

Kruty, M., Segur, J. B. & Miner, C. S. (1954) *Journal of the American Oil Chemists' Society*, **31**, 466.

Kudoh, M. & Yamaguchi, S. (1983) *Journal of Chromatography*, **260**, 483.

Kunugi, A., Komoda, Y. & Kurechi, T. (1983) *Journal of the Food Hygiene Society Japan*, **24**, 324.

Kutscher, W., Nagel, W. & Pfaff, W. (1961). *Seitschrift fur Lebensmitteluntersuchung und-forschung*, **115**, 117.

La Croix, D. E. & Wong, N. P. (1972) *Journal of the Association of Official Analytical Chemists*, **54**, 361.

Lancaster, F. E. & Lawrence, J. F. (1982) *Journal of the Association of Official Analytical Chemists*, **65**, 1305.

Lancaster, F. E. & Lawrence, J. F. (1983) *Journal of the Association of Official Analytical Chemists*, **66**, 1424.

Land, D. G. & Nursten, E. H. (1979) *Progress in Flavour Research*. London: Applied Science Publishers Ltd.

Lange, E. (1981) *Lebensmittel Industrie*, **28**, 269.

Larsson, B. & Fuchs, G. (1974) *Swedish Journal of Agricultural Research*, **4**, 109.

Laub, E., Klintrimas, T. & Lichtenthal, H. (1983) *Lebensmittelchemie und gerichtliche Chemie*, **37**, 144.

Lawrence, J. F. & Charbonneau, C. F. (1988) *Journal of the Association of Official Analytical Chemists*, **71**, 934.

Lazzarini, G. (1982) *Bollettini Chimici dei Laboratori Provinciali*, **33**, 27.

Leather, A. N. (1942) *Analyst*, **67**, 52.

Lechowich, R. V., Brown, W. L., Deibel, R. H. & Somers, I. I. (1978) *Food Technology*, **32**, 45.

Lehmann, G. & Lutz, I. (1970) *Zeitschrift für Lebensmitteluntersuchung und-forschung*, **144**, 318.

Lehmann, G., Collet, P., Hahn, H. G. & Ashworth, M. R. F. (1970) *Journal of the Association of Official Analytical Chemists*, **53**, 1182.

Lehmann, G., Moran, M. & Neumann, B. (1974) *Zeitschrift fur Lebensmitteluntersuchung und-forschung*, **155**, 85.

Leuenberger, U., Gauch, R. & Baumgartner, E. (1979) *Journal of Chromatography*, **173**, 343.

Liddle, P. A. P. & De Smedt, P. (1976) *Annales Falsifications L'Expertise Chimie*, **69**, 857.

Liedtke, M. A. & Meloan, G. E. (1976) *Journal of Agricultural and Food Chemistry*, **24**, 410.

Liewen, M. B. & Marth, E. H. (1985a) *Journal of Food Protection*, **48**, 364.

Liewen, M. B. & Marth, E. H. (1985b) *Zeitschrift for Lebensmitteluntersuchung und-forschung*, **180**, 45.

Lloyd, W. J. W. & Cowle, B. C. (1963) *Analyst*, **88**, 394.
Lord, E., Bunton, N. G. & Crosby, N. T. (1978) *Journal of the Association of Public Analysts*, **16**, 25.
Lorenzen, W. & Sieh, R. (1962) *Zeitschrift für Lebensmitteluntersuchung und-forschung*, **118**, 223.
Lowry, R. R. & Tinsley, I. J. (1981) *Journal of the American Oil Chemists' Society*, **58**, 991.
Luck, E. (1982) *Industrie Alimentari*, **21**, 1.
Luckas, B. & Lorenzen, W. (1984) *Lebensmittelchemie und gerichtliche Chemie*, **38**, 76.
Lueck, E. & Remmert, K. H. (1974) *Alimentaria*, **11**, 99.
Lyne, F. A., Radley, J. A. & Taylor, M. B. (1968) *Analyst*, **93**, 186.
Macchiavelli, L. & Andreotti, R. (1984) *Industrie Conservatori*, **59**, 18.
MacLeod, G. & Seyyedain-Ardebili, M. (1981) *Chemical Rubber Company Critical Reviews in Food Science and Nutrition*, **14**, 309.
Macrae, R. (1988) *HPLC in Food Analysis*. London: Academic Press.
MAFF (1978) *The surveillance of food contamination in the United Kingdom*. Food Surveillance Paper No. 1. London: HMSO.
MAFF (1984) *Steering Group on Food Surveillance Progress Report*. London: HMSO.
Manley, C. H. & Shubiak, P. (1975) *Canadian Institute of Food Science and Technology Journal*, **8**, 35.
Mano, E. B. & Lima, L. C. O. C. (1960) *Analytical Chemistry*, **32**, 1772.
Mark, E. & McKeown, G. G. (1958) *Journal of Official Agricultural Chemists*, **41**, 817.
Martin, E. (1981) *Mitteilungen aus dem Gebeit der Lebensmitteluntersuchung und Hygiene*, **72**, 402.
Martin, E., Duret, M. & Vogel, J. (1980) *Mitteilungen aus dem Gebeit dere Lebensmitteluntersuchung und Hygiene*, **71**, 195.
Maruyama, T., Niiya, I. & Imamura, M. (1977) *Shokuhin Eiseigaki Zasshi*, **18**, 283 (Analytical Abstracts (1978), 2F8).
Massart, D. L. & DeClercq, H. (1974) *Analytical Chemistry*, **46**, 1988.
Mattioni, R. (1964) *Bollettini Chimici dei Laboratori Procinciali*, **15**, 539.
Mauro, D. J. & Wetzel, D. L. (1984) *Journal of Chromatography*, **299**, 281.
McCalla, M. A., Mark, F. G. & Kipp, W. H. (1977) *Journal of the Association of Official Analytical Chemists*, **60**, 71.
Meadows, G. S. (1966) *Journal of the Association of Public Analysts*, **4**, 92.
Melnick, D. & Luckmann, F. H. (1954) *Food Research*, **19**, 20.
Min, D. B. & Schweizer, D. (1982) *Journal of Food Science*, **48**, 73.
Modi, G., Chiti, F. & Fiorentino, P. (1982) *Bollettini Chimici dei Laboratori Provinciali*, **33**, 143.
Moncelsi, E. (1970) *Chimicia Industria, Milan*, **52**, 367.
Monier-Williams, G. W. (1927a, b, c, d) *Analyst*, **52**, 153, 229, 343, 415.
Mooser, A. E. (1984) *Journal of Chromatography*, **287**, 113.
Morley, R. G., Phillips, G. O., Power, D. M. & Morgan, R. E. (1972) *Analyst*, **97**, 315.
Murphy, J. M. & Grisley, L. M. (1969) *Journal of the American Oil Chemists' Society*, **46**, 384.
Murphy, J. M. & Hibbert, H. R. (1969) *Journal of Food Technology*, **4**, 227.
Murphy, J. M. & Scott, C. C. (1969) *Analyst*, **94**, 481.
Murray, D. & Cundall, R. B. (1981) *Analyst*, **106**, 335.
Nagel, C. W., Brekke, C. J. & Leung, H. K. (1982) *Journal of Food Science*, **47**, 342.
Nakamura, M. & Murata, A. (1979) *Analyst*, **104**, 985.
Nakashima, K., Nakagawa, T. & Era, S. (1981) *Journal of the Food Hygiene Society Japan*, **22**, 233.
Nanase, Y., Takata, K., Shimaoka, S. & Doki, S. (1981) *Journal of the Food Hygiene Society Japan*, **22**, 526.
Nash, N. H. & Knight, G. S. (1967) *Food Engineering*, **39**, 79.
Naturex, A. G. (1980) *Alimenta*, **19**, 129.
Neale, M. E. & Ridlington, J. (1978) *Journal of the Association of Public Analysts*, **16**, 135.
Neissner, R. (1972) *Fette, Seifen, Anstrichmittel*, **74**, 198.
Nelson, J. J. (1973) *Journal of Chromatographic Science*, **11**, 28.
Newsome, R. L. (1986) *Food Technology*, **40**, 49
Nicholls, J. R. (1928) *Analyst*, **53**, 19.
Noda, H., Minemoto, M., Asahara, T., Noda, A. & Iguchi, S. (1982) *Journal of Chromatography*, **235**, 187.
Noda, K., Kenjo, N. & Takahashi, T. (1973) *Shokuhin Eiseigaku Zasshi*, **14**, 253.
Nordic Committee on Food Analysis (1968) *Standard Methods*, No. 3.
Norman, S. M., Fouse, D. C. & Craft, C. C. (1972a) *Journal of the Association of Official Analytical Chemists*, 55, 1239.
Norman, S. M., Fouse, D. C. & Craft, C. C. (1972b) *Journal of Agricultural and Food Chemistry*, **20**, 1227.
Norwitz, G. & Keliher, P. N. (1985) *Analyst*, **110**, 689.
Nursten, H. E. & Williams, A. A. (1967) *Chemistry and Industry*, **12**, 486.
Nury, F. S. & Bolin, H. R. (1962) *Journal of Food Science*, **27**, 370.

Nyguyen, T. T. & Sporns, P. (1984) *Journal of the Association of Official Analytical Chemists*, **67**, 747.

Ogawa, S., Toyada, M., Tonogai, Y., Ito, Y. & Iwaida, M. (1979) *Journal of the Association of Official Analytical Chemists*, **62**, 610.

Oi-Wah Lau, Shiu-Fai Luk & Wai-Ming Chan (1988) *Analyst*, **113**, 765.

Olek, M., Declercq, B., Caboche, M., Blanchard, F. & Sudraud, G. (1983) *Journal of Chromatography*, **281**, 309.

Page, B. D. (1979) *Journal of the Association of Analytical Chemists*, **62**, 1239.

Page, B. D. (1983) *Journal of the Association of Official Analytical Chemists*, **66**, 727.

Page, B. D. & Conacher, H. B. S. (1978) *Journal of Pure and Applied Chemistry*, **50**, 243.

Pardun, H. (1981) *Fette Seifen Anstrichmittel*, **83**, 240.

Parkes, A. E. (1926) *Analyst*, **51**, 620.

Parliment, T. H. & Spencer, M. D. (1981) *Journal of Chromatographic Science*, **19**, 435.

Patey, A. L., Shearer, G., Knowles, M. E. & Denner, W. H. B. (1985) *Food Additives and Contaminants*, **2**, 107, 237.

Pearson, D. (1967, 1973a, 1973b, 1973c) *Journal of the Association of Public Analysts*, **5**, 37; **11**, 52, 127, 135.

Pearson, D. (1975) *Chemical Analysis of Foods*, 7th edn, p. 53. Edinburgh: Churchill Livingstone.

Pearson, D. & Wong, T. S. A. (1971) *Journal of Food Technology*, **6**, 179.

Pechanek, U., Blaicher, G., Pfannhauser, W. & Woidich, H. (1982) *Journal of the Association of Official Analytical Chemists*, **65**, 745.

Peel, J. L. & Waites, W. M. (1979) UK Patent Application, 7901091.

Pepper, T. (1985) *Food, Flavours, Ingredients, Processing and Packaging*, **7**, 22.

Perfetti, G. A., Warner, C. R. & Fazio, T. (1981) *Journal of the Association of Official Analytical Chemists*, **64**, 844.

Perkin, A. G. (1982) *Journal of Society of Dairy Technology*, **35**, 147.

Perry, A. R. & Woolley, D. G. (1969) *Journal of the Association of Public Analysts*, **7**, 94.

Peteghem, C. H. van & Dekeyser, D. A. (1981) *Journal of the Association of Official Analytical Chemists*, **64**, 1331.

Pfannhauser, W., Eberhardt, R. & Woidich, H. (1982) *Ernahrung*, **6**, 107.

Pfieffer, S. L. & Smith, J. (1975) *Journal of the Association of Official Analytical Chemists*, **58**, 915.

Phillip, G. D. (1981) *Lebensmitteltechnik*, **13**, 460.

Phipps, A. M. (1973) *Journal of the American Oil Chemists' Society*, **50**, 21.

Player, R. B. & Wood, R. (1980) *Journal of the Association of Public Analysts*, **18**, 109.

Potter, E. F. (1954) *Food Technology*, **8**, 269.

Prater, A. N., Johnson, C. M., Pool, M. F. & Mackinney, G. (1944) *Industrial and Engineering Chemistry (Analytical)*, **16**, 153.

Preuss, A. & Thier, H-P. (1982) *Zeitschrift fur Lebensmitteluntersuchung und-forschung*, **175**, 93.

Preuss, A. & Thier, H-P. (1983a) *Zeitschrift fur Lebensmitteluntersuchung und-forschung*, **176**, 5.

Prudel, M. & Davidkova, E. (1985) *Nahrung*, **29**, 381.

Pryzybylski, W., Smyth, R. B. & McKeown, G. G. (1960) *Journal of the Association of Official Analytical Chemists*, **43**, 274.

Przybyla, A. (1981a) *Processed Prepared Foods*, **150**, 95.

Puttemans, M. L., Dryon, C. & Massart, D. (1983) *Journal of the Association of Official Analytical Chemists*, **66**, 1039.

Puttemans, M. L., Dryon, L. & Massart, D. L. (1984) *Journal of the Association of Official Analytical Chemists*, **67**, 880.

Puttemans, M. L., de Voogt, M., Dryon, C. & Massart, D. (1985) *Journal of the Association of Official Analytical Chemists*, **68**, 143.

Radford, T. & Dalsis, D. E. (1982) *Journal of Agricultural and Food Chemistry*, **30**, 600.

Radler, F. (1964) *Food Technology in Australia*, **16**, 475.

Rajama, J. & Makela, P. (1973) *Journal of Chromatography*, **76**, 199.

Rangone, R. & Ambrosio, C. (1970) *Journal of Chromatography*, **50**, 436.

Rauter, W. & Wolkerstorfer, W. (1982) *Zeitschrift fur Lebensmitteluntersuchung und-forschung*, **175**, 122.

Reith, J. F. & Willems, J. J. L. (1958) *Zeitschrift fur Lebensmitteluntersuchung und-forschung*, **108**, 270.

Reynolds, S. L., Scotter, M. J. & Wood, R. (1988) *Journal of the Association of Public Analysts*, **26**, 7.

Rhee, J. S. & Shin, M. G. (1982) *Journal of the American Oil Chemists' Society*, **59**, 98.

Richardson, R. K. & Gray, I. K. (1981) *New Zealand Journal of Dairy Science and Technology*, **16**, 179.

Richmond, M. L., Brandao, S. C. C., Gray, J. I., Markarkis, P. & Stine, C. M. (1981) *Journal of Agricultural and Food Chemistry*, **29**, 4.

Riedl, O. (1981) *Ernahrung*, 5, 340.

Riisom, T. & Hoffmeyer, L. (1978) *Journal of the American Oil Chemists' Society*, 55, 649.

Rimpel, R., Bruchmann, M. & Neismass, R. (1981) *Lebensmittel Industrie*, **28**, 263.

Ripley-Duggan, B. A. (1953) *Analyst*, **78**, 173.
Rizvi, S. S. H., Daniels, J. A., Benado, A. L. & Zollweg, J. A. (1986) *Food Technology*, **40**, 57.
Roll, J. H., Bunton, N. G. & Crosby, N. T. (1977) *Journal of the Association of Public Analysts*, **15**, 27.
Ross, H. D. & Hotchkiss, J. K. (1985) *Journal of the Association of Official Analytical Chemists*, **68**, 41.
Ruf, F. & Siepe, V. (1983) *Alimenta*, **22**, 177.
Rugraff, L., Demanze, C. & Karleskind, A. (1981) *Industrie Alimentari Agriculture*, **98**, 305.
Sahasrabudhe, M. R. (1967) *Journal of the American Oil Chemists' Society*, **44**, 376.
Sahasrabudhe, M. R. & Legari, J. J. (1967) *Journal of the American Oil Chemists' Society*, **44**, 379.
Sahasrabudhe, M. R. & Legari, J. J. (1968) *Journal of the American Oil Chemists' Society*, **45**, 148.
Sahasrabudhe, M. R., Legari, J. J. & McKinley, W. P. (1966) *Journal of the Association of Official Analytical Chemists*, **49**, 337.
Sanderson, G. R. (1981) *Food Technology*, **35**, 50.
Sawyer, R. & Crosby, N. T. (1980). In *Developments in Food Preservatives—1*, ed. R. H. Tilbury. London: Applied Science Publishers Ltd.
Schafer, H. & Scherz, H. (1983) *Zeitschrift fur Lebensmitteluntersuchung und-forschung*, **177**, 193.
Scherz, H. (1984) *Zeitschrift fur Lebensmitteluntersuchung und-forschung*, **179**, 17.
Schmidt, B. & Schwedt, G. (1984) *Deutsches LebensmittelRundschau*, **80**, 137.
Schmidt, H. (1960) *Zeitschrift fur Analytische Chemie*, **178**, 173.
Schrepfer, R. & Egle, H. (1958) *Zeitschrift fur Lebensmitteluntersuchung und-forschung*, **107**, 510.
Schrier, P. & Idstein, H. (1985) *Zeitschrift fur Lebensmitteluntersuchung und-forschung*, **180**, 1.
Schuller, P. L. & Veen, E. (1967) *Journal of the Association of Official Analytical Chemists*, **50**, 1127.
Schuster, G. & Adams, W. (1980) *Zeitschrift fur Lebensmittel Technologie und Verfahrenstechnologie*, **31**, 174.
Schwartz, S. J. & Elbe, J. H. (1980) *Journal of Agricultural and Food Chemistry*, **28**, 540.
Schwartzenbach, R. (1976) *Journal of Chromatography*, **129**, 31.
Seher, A. (1964) *Fete, Seifen, Anstrichmittel*, **66**, 371.
Shelley, R. N., Salwin, H. & Horwitz, W. (1963) *Journal of the Association of Official Agricultural Chemists*, **46**, 486.
Sherma, J., Sielicki, P. J. and Charvat, S. (1983) *Journal of Liquid Chromatography*, **6**, 2679.
Shewan, J. M. (1949) *Chemistry and Industry* No. 28, 501.
Shipton, J. (1954) *Food Preservation Quarterly*, **14**, 54.
Shiraki, M., Yoshiura, M. & Iriyama, K. (1978) *Chemical Letters*, **1**, 103.
Slack, P. T. & Porter, D. C. (1983) *Chemistry and Industry*, **23**, 896.
Smiles, R. (1986) *Food Technology*, **40**, 129.
Smyly, D. S., Woodward, B. B. & Conrad, E. C. (1976) *Journal of the Association of Official Analytical Chemists*, **59**, 14.
Soe, J. B. (1983) *Fette Seifen Anstrichmittel*, **85**, 720.
Spark, A. A. (1982) *Journal of the American Oil Chemists' Society*, **59**, 185.
Spears, K. & Marshall, J. (1987) *Journal of the Association of Public Analysts*, **25**, 47.
Spedding, D. J. & Stewart G. M. (1980) *Analyst*, **105**, 1182.
Sperlich, H. (1960) *Deutsche Lebensmittel Rundschau*, **56**, 70.
Sporek, K. & Williams, A. F. (1954) *Analyst*, **79**, 63.
Stadtman, E. R., Burker, H. A., Hass, V. & Mark, E. M. (1946) *Industrial and Engineering Chemistry*, **38**, 541.
Stafford, A. E. (1976) *Journal of Agricultural and Food Chemistry*, **24**, 894.
Stanley, R. L. & Kirk, P. L. (1963) *Journal of Agricultural and Food Chemistry*, **11**, 492.
Stetzler, R. S. & Andress, T. B. (1962) *Journal of the American Oil Chemists' Society*, **39**, 509.
Strack, D. & Mansell, R. (1975) *Journal of Chromatography*, **109**, 325.
Stuckey, B. N. (1962) *Lipids Oxidation, Symposium Proceedings 1961, Corvallis, Oregon*. Westport, Conn.: AVI Publishing Co.
Stuckey, B. N. & Osborne, C. E. (1965) *Journal of the American Oil Chemists' Society*, **42**, 228.
Stumbo, C. R., Voris, L., Skaggs, B. G. & Heinemann, B. (1964) *Journal of Food Science*, **29**, 859.
Tafuri, F., Marucchini, C., Patumi, M. & Businelli, M. (1980) *Journal of Agricultural and Food Chemistry*, **28**, 1150.
Takatsuki, K. & Sakai, K. (1982) *Journal of the Association of Official Analytical Chemists*, **65**, 817.
Tame, D. A. (1980) *Journal of the Association of Public Analysts*, **18**, 59.
Tammilehto, S. & Bucchi, J. (1969) *Pharmacia Acta Helvetica*, **44**, 138.
Tanaka, A., Nose, N., Suzuki, T., Hirose, A. & Watanabi, A. (1978) *Analyst*, **103**, 851.
Tanaka, A., Nose, N. & Watanabe, A. (1981a) *Journal of the Food Hygiene Society Japan*, **22**, 14.
Tanaka, A., Nose, N., Hirose, A. & Wanatabe, A. (1981b) *Analyst*, **106**, 94.
Tanaka, A., Nose, N. & Iwasaki, H. (1982a) *Analyst*, **107**, 190.
Tanaka, A., Nose, N. & Iwasaki, H. (1982b) *Journal of Chromatography*, **235**, 173.

Tanaka, A., Nose, H., Masaki, H., Kikuchi, Y. & Iwasaki, H. (1983) *Journal of the Association of Official Analytical Chemists*, **66**, 260.
Tanner, H. (1963) *Mitteilungen aus dem Gebiet der Lebensmitteluntersuchung und-Hygiene, Bern*, **54**, 158.
Tanner, H. & Sandoz, M. (1972) *Schweizerische Zeitschrift fuer Obst und Weinbau*, **108**, 331.
Tateo, F., Faleschini, M. L. & Fossati, M. (1982) *Industria Conserve*, **57**, 30.
Taylor, P. & Barnes, P. (1981) *Chemistry and Industry*, **20**, 722.
Thackray, G. B. & Hewlett, A. (1964) *Journal of the Association of Public Analysts*, **2**, 13.
Thomsen, M. & Willumsen, D. (1981) *Journal of Chromatography*, **211**, 213.
Thrasher, J. J. (1966) *Journal of the Association of Official Analytical Chemists*, **49**, 834.
Tinbergen, B. J. & Krol, B. (1977) *Proceedings of the Second International Symposium on Nitrites in Meat Products*. Wageningen: Pudoc.
Tjan, G. H. & Burgess, L. I. (1973) *Journal of the Association of Official Analytical Chemists*, **56**, 223.
Tjan, G. H. & Jansen, J. Th.A. (1971) *Journal of the Association of Official Analytical Chemists*, **54**, 1150.
Tjan, G. H. & Konter, T. (1972) *Journal of the Association of Official Analytical Chemists*, **55**, 1223.
Tonogai, Y., Kingkate, A., Thanissorn, W. & Punthanaprated, U. (1983a) *Journal of Food Protection*, **46**, 284.
Tonogai, Y., Ito, Y. & Iwaida, M. (1983b) *Journal of the Food Hygiene Society Japan*, **24**, 275.
Torre, L. C. & Barritt, B. H. (1977) *Journal of Food Science*, **42**, 488.
Torres, A. & Thomas, R. D. (1981) *Food Technology*, **35**, 44.
Toyoda, M., Ogawa, S., Ho, Y. & Iwaida, M. (1979) *Journal of the Association of Official Analytical Chemists*, **62**, 1146.
Toyoda, M., Ito, Y., Iwaida, M. & Fujii, M. (1982) *Journal of Agricultural and Food Chemistry*, **30**, 346.
Tramer, J. & Fowler, G. G. (1964) *Journal of the Science of Food and Agriculture*, **15**, 522.
Trifiro, A., Bigliardi, D., Bazzarini, R. & Gherardi, S. (1981) *Industria Conserve*, **56**, 22.
Tsang, W. S., Clark, M. A. & Parrish, F. W. (1985) *Journal of Agricultural and Food Chemistry*, **33**, 734.
Tsuda, T. & Nakanishi, H. (1981) *Journal of the Food Hygiene Society Japan*, **22**, 425.
Tsuda, T., Nakanishi, H., Kobayashi, S. & Morita, T. (1984) *Journal of the Association of Official Analytical Chemists*, **67**, 1149.
Tuinstra, L. G. M. Th. & Traag, W. A. (1982) *Journal of the Association of Official Analytical Chemists*, **65**, 820.
Tyler, T. A. (1984) *Journal of the Association of Official Analytical Chemists*, **67**, 745.
Usher, C. D. & Telling, G. M. (1975) *Journal of the Science of Food and Agriculture*, **26**, 1793.
Van Peteghem, C. & Bijl, J. (1981) *Journal of Chromatography*, **210**, 113.
Van Pinxteren, J. A. C. (1952) *Analyst*, **77**, 367.
Van Rillaer, W., Janssens, G. & Beernaert, H. (1982) *Zeitschrift fur Lebensmitteluntersuchung und-forschung*, **175**, 413.
Veerabhadra Rao, M., Kapur, O. P. & Prakasa Sastry, C. S. (1984) *Journal of Food Science and Technology*, **21**, 148.
Venturini, A. (1972) *Industrie Alimentaria*, **11**, 75.
Verzella, G., Bagnasco, G. & Mangia, A. (1985) *Journal of Chromatography*, **349**, 83.
Vetsch, W. (1985) *Food Chemistry*, **16**, 245.
Vincent, K. R. & Scholz, R. G. (1978) *Journal of Agricultural and Food Chemistry*, **26**, 812.
von Rymon Lipinski, G.-W. (1985) *Food Chemistry*, **16**, 259.
Walker, G. H. & Green, M. S. (1965) *Journal of the Association of Public Analysts*, **3**, 87.
Walker, R. (1975) *Journal of the Science of Food and Agriculture*, **26**, 1735.
Waltking, A. E. & Goetz, A. G. (1983) *Chemical Rubber Company Critical Reviews in Food Science and Nutrition*, **19**, 99.
Wasserman, A. E. (1979) *Journal of Food Science*, **44**, 6.
Watada, A. E. & Tran, T. T. (1985) *Journal of Liquid Chromatography*, **8**, 1651.
Watanabe, E., Toyama, K., Karuba, I., Matsuoka, H. & Suzuki, S. (1984) *Journal of Food Science*, **49**, 114.
Webb, N. G. & Beckman, D. D. (1984) *Journal of the Association of Official Analytical Chemists*, **67**, 510.
Wedzicha, B. L. (1981) *Nutrition and Food Science*, **72**, 12.
Wedzicha, B. L. & Johnson, M. K. (1979) *Analyst*, **104**, 694.
Weiss, H. O. (1981) *Lebensmitteltechnik*, **13**, 345.
West, P. W. & Gaeke, G. C. (1956) *Analytical Chemistry*, **28**, 1816.
Wetterau, F. P., Olsanski, V. L. & Smullin, C. F. (1964) *Journal of the American Oil Chemists' Society*, **41**, 791.
Weyland, J. W., Rolink, H. & Doornbos, D. A. (1982) *Journal of Chromatography*, **247**, 221.
Wildanger, W. A. (1973) *Chromatographia*, **6**, 381.
Wilkinson, M., Sweeny, J. G. & Iacobucci, G. A. (1977) *Journal of Chromatography*, **132**, 349.

Wilks, R. A., Johnson, M. W. & Shingler, A. J. (1977) *Journal of Agricultural and Food Chemistry*, **25**, 605.
Williams, J. R. (1979) *Handbook of Anion Determination*. London: Butterworths.
Williams, K. A. (1966) *Oils, Fats and Fatty Foods*, 4th edn. London: Churchill.
Winell, B. (1976) *Analyst*, **101**, 883.
Winkler, W. O. (1961) *Journal of the Association of Official Agricultural Chemists*, **44**, 476.
Winstanley, M. A. (1980) *Food Flavours, Ingredients, Packaging and Processing*, **1**, 26.
Wonnacott, J. (1985) *Nutrition and Food Science*, Nov., 10.
Wood, P. S. (1980) *Process Biochemistry*, **15**, 12.
Woodbury, J. E. (1977) *Journal of the Association of Official Analytical Chemists*, **60**, 1.
Woollard, D. C. & Forrest, L. J. (1980) *New Zealand Journal of Dairy Science and Technology*, **15**, 83.
Wootton, M., Kok, S. H. & Buckle, K. A. (1985) *Journal of the Science of Food and Agriculture*, **36**, 297.
Wu, H.-L., Chen, S. H., Funazo, K., Tanaka, M. & Shono, T. (1984) *Journal of Chromatography*, **291**, 409.
Wunderlich, H. (1972) *Journal of the Association of Official Analytical Chemists*, **55**, 557.
Yabe, Y., Tan, S., Ninomiya, T. & Okada, T. (1983) *Journal of the Food Hygiene Society Japan*, **24**, 329.
Yo, L. Z., Inoko, M. & Matsumo, T. (1984) *Journal of Agricultural and Food Chemistry*, **32**, 681.
Yoshiura, M., Iriyama, K. & Shiraki, M. (1978) *Chemical Letters*, **3**, 281.
Young, M. L. (1984) *Journal of the Association of Official Analytical Chemists*, **67**, 1022.
Zonnevd, D. H. & Meyer, A. (1960) *Zeitschrift fur Lebensmitteluntersuchung und-forschung*, **111**, 198.

5

Contaminants

In contrast to the residues arising from the deliberate use of additives for technological reasons, the presence of 'contaminants' or undesirable substances in foods may be due to a chance incident or a generalised situation deriving from the use of a particular processing method or mode of storage and distribution of the food. The range of substances considered under this heading has increased significantly over the past decade or so as analytical techniques have become more sensitive and more species specific. The knowledge of 'substances present at very low levels' has increased at a faster rate than the ability to assess the level of risk accruing from the 'contaminant' present.

Although it is difficult to legislate against all possible forms of contamination, maximum limits have been prescribed for some trace elements both within the UK and in international legislation. Many of the commodity standards developed by the Codex Alimentarius include limits for heavy metals. Attempts have been made to introduce maximum residue limits for other contaminants such as pesticides and mycotoxins but so far the extent of agreement on appropriate values and on the compounds to be controlled is limited. Most progress has been made by FAO/WHO through the activities of the Codex Committee on Pesticide Residues, which has developed proposals for residue limits on a commodity by commodity basis for most of the agricultural chemicals used on a world-wide scale. Although within the UK the method of control of agricultural and horticultural residues has been effected by control of the substances allowed and by recommendations of treatment conditions including premarket withdrawal of use, the increasing international trade in fresh produce has caused a reappraisal of the situation and attempts are being made to introduce specific limits into legislation. A similar situation exists in respect of the presence of mycotoxins, specifically aflatoxins in foods; the Food Advisory Committee has recommended that legislation be introduced to limit the amount of aflatoxins (B1, B2, G1 and G2) to 10 μg/kg. The following selection of topics is by no means exhaustive of the potential sources and nature of contaminants in foods, the literature on the subject is voluminous and many specialist texts on specific topics exist. Farrer (1985) has summarised the sources of potential chemical hazards in food and has classified these into nine categories consolidated under three headings, namely: raw materials, processing and packaging.

TRACE ELEMENTS

From the viewpoint of the food analyst the term 'trace element' refers to the inorganic elements, mostly metals, which may be present in foods usually in amounts well below 50 mg/kg and which have some toxicological or nutritional significance. Those that are essentially nutritive elements include Co, Cu, Fe, I, Mn and Zn. The non-nutritive elements include Al, B, Cr, Ni and Sn, together with As, Sb, Cd, F, Pb, Hg and Se, which are known to have deleterious effects even when the diet contains less than 10–50 mg/kg. A complicating factor, however, is that elements such as copper and zinc, although essential for life processes when in traces, have an emetic action when ingested in higher amounts. Cumulative poisoning due to ingestion of food containing elements such as lead or arsenic over long periods is probably rare. In addition to the action on the body, however, a few elements tend to destroy vitamin C in fruit products. Moncrieff (1964) has discussed the possible minimum concentrations of such metals as copper, chromium, iron (Fe(II) and Fe(III)) and tin, which are detectable by taste in food products. The routine analysis is mainly concerned with those elements for which limits have been officially prescribed or recommended, i.e. As, Cu, Pb, Sn and Zn.

The presence of the more undesirable trace elements in foods may be due to natural occurrence, e.g. in fish due to the ingestion of water containing industrial effluents or by deposition in the liver of animals. Other sources may include sprays and dusts used as insecticides for crop protection during cultivation, the use of impure chemicals for the manufacture of raw materials or the dissolution of metals from processing equipment or containers, e.g. from tinplate, foils, solders, galvanized iron or cheap enamels or glazes. Reilly (1985) has considered the uptake of adventitious trace elements in the preparation and cooking of foods, whilst Bearfield *et al.* (1983) have discussed the various factors governing the uptake of iron, lead and tin by canned foods. The Food Additives and Contaminants Committee has published a substantial number of reports containing recommendations for statutory and recommended limits (see Appendix 5 for the complete list). Substantial reviews of the trace element content of foods on sale in the UK and on 'total or duplicate diets' have been made by the MAFF Steering Group on Food Surveillance; reference is made to these later in considering the individual elements. In a broadly based review (MAFF, 1985), data on Al, Sb, Cr, Co, In, Ni, Th and Sn is presented. In a wider review of the metal content of canned foods, the FACC (1983) noted that the Fe, Al and Cr content of canned foods does not constitute a hazard to health, the committee considering that there is room for improvement in respect of tin and lead content. Largely as a result of the Committee's recommendations and the survey data statutory limits for the cumulative poisons lead and arsenic now apply to all foods and recommendations are made in respect of other elements such as copper, tin and zinc.

GENERAL CONSIDERATIONS

Sample preparation

The analysis of any material for evidence of contamination from any source requires meticulous attention to detail in a number of respects; for example, in the application of the chosen method of analysis, choice of reagents and especially in the handling and preparation of the sample for analysis. The extent to which a sample can be treated in reduction of bulk and reduction of size to allow for accurate sub-sampling for the selection of a test portion is a trade off between practicality and theoretical necessity. The nature of the foodstuff itself has an influence on the degree of homogenisation that is practicable with readily available equipment. For difficult samples, the choice may be between the preparation of a homogenised slurry in a blender with a weighed amount of a compatible diluent and the use of a pre-drying procedure either by mild heating or freeze drying followed by grinding in appropriate mechanical equipment. When mechanical grinding is used care must be taken to avoid the separation of size fractions and inadvertant distortion of the apparent composition of the sub-sample. It is also possible to introduce contaminant into the sample by attrition of the equipment used. Samples intended for trace element analysis are particularly vulnerable to adventitious contamination from equipment, reagents and the general laboratory environment at all stages of the analytical process and extreme care is essential at all times. Wolf (1982) and Katz (1985) have discussed procedures and precautions for avoidance of contamination during sampling, preparation of test portions and in the analytical stages.

Analytical methods

A review of international standards and methodology recommended by Codex Alimentarius has been presented (Anon, 1980). Crosby (1977), Wolf (1982) and Jones (1984) have published comprehensive reviews of methods of analysis, and several authors have commented that the conventional colorimetric and gravimetric methods are reliable in performance but are slow and tedious in use. The wide availability of instrumental methods, many with multi-element capability, has greatly improved not only the speed at which analyses may be carried out but also in many cases the sensitivity and limit of detection. However, there are many unresolved problems in the universal application of instrumental procedures that are derived from matrix and other interference effects and the results of applying a new technique to a different food matrix should be regarded with caution until interlaboratory studies have verified the application at relevant levels in a range of typical samples. Watson (1986) has commented generally on the availability of methods for trace elements and has indicated the areas needing attention.

The destruction of organic matter

Although it is possible with some readily soluble food materials to apply methods of trace analysis directly to a solution of the food in water or acid, it is usually necessary to destroy the organic matter before proceeding. Basic work on the recovery of trace metals during the destruction processes has been described by Gorsuch (1959, 1970) and ashing methods have been reviewed by Crosby (1977). Both wet and dry methods of combustion have been recommended by the AMC (1976) and various methods of wet and dry ashing have been compared by Isaac and Johnson (1975), Boline and Schrenk (1977), Watson (1984) and Jones (1984).

DRY ASHING

Dry ashing at 430–600 °C (according to the particular element) is a convenient method for the destruction of organic matter. Precautions must be taken, however, against low results which may be due to volatilisation of the element, combination of absorption of the element with ash constituents or the vessel used or incomplete extraction of the ash. Such difficulties can usually be avoided by using an accurately controlled muffle furnace, by adding an ash aid (magnesium oxide or nitrate, sodium carbonate or sulphuric acid) to the food before ashing and by using a suitable acid for the extraction (AOAC). Silica or platinum basins are to be preferred.

WET OXIDATION

Many workers prefer to destroy the organic matter by the wet oxidation method as it is in general more reliable. The necessity for constant vigilance and the possibility of high (and sometimes uncertain) blank values render the method less suitable for routine work, however. The procedure which follows is suitable for most determination.

METHOD

Into a macro-Kjeldahl digestion flask, preferably made of silica, place a suitable quantity of the sample (usually 5 or 10 g), 20 ml conc. nitric acid and up to 20 ml water (depending on the water content of the sample). Boil so that the volume is reduced to about 20 ml, cool and add 10 ml of conc. sulphuric acid. Boil again and add further small quantities of nitric acid *immediately* the liquid begins to blacken. When the addition of nitric acid is no longer necessary (i.e. when the liquid no longer blackens), continue the heating until white fumes are well in evidence. Cool and add 10 ml of saturated ammonium oxalate solution and boil again until copious white fumes are again produced. The oxalate treatment assists in removing yellow colorations due to nitro compounds, fats, etc. so that the final solution is colourless. Every trace of nitric acid must be removed before proceeding. A blank should be prepared at the same time.

For wet oxidation, the AMC Report discusses the use of various other combinations of oxidising agents such as perchloric acid, hydrogen peroxide and potassium permanganate. The procedure can be shortened considerably for some materials by the use of 50 per cent hydrogen peroxide in the presence of sulphuric acid (Taubinger and Wilson, 1965; Down and Gorsuch, 1967; AMC, 1976). The AMC (1977a) method uses hydrogen peroxide and sulphuric acid. Middleton and Stuckey (1954) have described a modified wet oxidation procedure that is especially suitable for animal and fatty materials. The main advantages over the conventional method are that by oxidising the materials at less than 350 °C in covered beakers, the nitric acid is used more economically, there is little likelihood of loss of the element due to volatilisation and the process does not require such constant vigilance. Jones (1984) has summarised the advantages and otherwise of five procedures, using nitric and sulphuric acid mixtures, two using perchloric acid and one combination of sulphuric acid and hydrogen peroxide. Variations of acid techniques using a digestion block are also considered by this author. Holak (1980) reported on a collaborative study of a closed (pressure vessel) digestion system using 0·3 g dry solids and 5 ml nitric acid. After heating in the closed vessel for 2 hours at 150 °C the extract was used for determination of lead, cadmium, copper, zinc, arsenic and selenium. The procedure was found to be satisfactory in all cases except for analysis of copper content. In assessing the results of such studies note should be taken of the levels considered by the authors. For example, the Holak study (loc. cit.) considered two samples with levels of lead at 50 mg/kg and one at 0·3 mg/kg. At a late stage in the study 'typical' value foods were included. More recent developments in wet digestion are in the use of microwave heating of samples and digestion acids in pressure-proof Teflon vessels.

Preparation of solution for AAS

The following applies to the preparation of solutions for examination by atomic absorption spectroscopy trace metals analysis other than for mercury. In general AAS determinations are not normally affected by using different acids but care should be taken to ensure that the same acid and concentration (usually not exceeding 2 M) are used for blank, standard and sample examinations. It is advisable to avoid using porcelain crucibles for dry ashing and recoveries of the metal should be checked. If about 0·2 g of calcium carbonate or 1 g magnesium nitrate is mixed with the sample prior to ashing there is less likelihod of fusion with the vessel and it is easier to dissolve the residue in acid. The ash can usually be dissolved in diluted acids, either HCl–water (1 + 1) or HCl–water–nitric acid (2 + 3 + 1). There are advantages in using a reagent such as ammonium pyrrolidine-1-carbodithioate (ammonium pyrrolidinedithiocarbamate, APDC) for the extraction of metals from aqueous solutions into an organic solvent such as methyl isobutyl ketone that is suitable for direct AAS. The concentration of the metal in the organic phase may be a hundred times that in the aqueous phase, so that smaller amounts can be estimated. Many liquids such as beverages can be aspirated directly after diluting to a soluble solids concentration not exceeding 3 per cent. Some semi-solid foods can be rapidly dealt with by using the procedure described by Simpson and Blay (1966). In the AMC (1973b) modification, 10 g of sample is heated with an acid–water mixture (10 ml HCl + 40 ml water) and then simmered for 5 min. The cooled solution is then transferred to a 100 ml volumetric flask and diluted to the mark with water. The problems associated with handling oily samples are discussed by the AMC (1973b) but it is sometimes possible to aspirate solutions of the fatty matter in an organic solvent directly into the flame.

Sample decomposition methods have been further discussed by Friend *et al.* (1977). Evans *et al.* (1978, 1980) have discussed the possible sources of interference in trace element determination in foods by AAS studies on matrix effects in the analysis for Cd, Pb, Ni, Cu, Fe, Mn and Zu are discussed. Fricke *et al.* (1979) give details of sample preparation and analysis by flame and flameless AAS for 20 common elements. The procedures for sample introduction include solution aerosols and hydride generation. More recently, attempts have been made to carry out analyses directly on pulverised sample material (Langmyhr, 1979; Ottaway *et al.*, 1985).

ARSENIC

Limits for arsenic in food are prescribed by The Arsenic in Food Regulations 1959, as amended. The general limit is 1 mg/kg. Certain foods such as fish are, however, exempted from having to comply with the regulations. Formerly, much of the arsenic contamination found in food resulted from the use of impure sulphuric acid in the manufacture of food ingredients such as sugars (especially glucose), citric, tartaric and phosphoric acids and their salts, and yeast. The levels of arsenic now found in food are normally very low (MAFF, 1982a); exceptions are some foods of marine origin, although the arsenic present in these is organically bound and is relatively non-toxic. Organically combined arsenic will only react with nascent hydrogen (zinc + acid) to form arsine if the material has been previously submitted to wet oxidation. Conversely, any arsine produced by the direct action of zinc and acid on the food can be attributed to inorganic arsenic, which is toxic. The FACC (1984) reviewed the 1959 Regulations and concluded that the general limit of 1 mg/kg in regulations should be maintained but with certain minor modifications in respect of the provisions for special foods. Firstly, the committee recommended that the special provisions for fish and edible seaweed should be extended to all sea food pending a review of the toxicity of marine organo-arsenicals and, secondly, that lower limits should apply in respect of beverages (0·1 mg/kg), concentrated beverages (0·5 mg/kg) and animal and poultry tissues (0·5 mg/kg). The Report made note of the results of surveys conducted by the working party on heavy metals (loc. cit.).

Destruction of organic matter

Destruction of the organic matter preparatory to the determination of arsenic is best carried out by wet oxidation using nitric and sulphuric acids. To avoid loss of the element by volatilisation during the digestion, the material should be heated with nitric acid alone to dissolve as much as possible before the addition of sulphuric acid. For the same reason it is especially important to add the additional amounts of nitric acid *immediately* charring commences. Additional treatment with ammonium oxalate to remove any residual nitrogen compounds is also advisable as their presence may retard the subsequent reduction. Alternatively, the Kjeldahl-type digestion of Williams (1941) is suitable provided the amount of chloride present is low. Methods describing special techniques for fish and fish products have been published by Le Blanc and Jackson (1973) using dry ashing with ashing aids (see also Tam and Lacroix, 1982), and by Uthe *et al.* (1974) using wet oxidation with vanadium pentoxide. With certain chemicals which dissolve in water or acid and liquid foods (e.g. beer, vinegar), the food or its solution can be transferred directly to the Gutzeit bottle and this method is convenient in routine work. See also the general review by Jones (1984) (loc. cit.) and the study by Holak (1980).

Methods available

Reinsch's test in which the material is boiled with hydrochloric acid and copper foil represents a rapid method for the detection of heavy traces of arsenic. Antimony, mercury, bismuth and selenium also give a positive reaction. In the *Marsh–Berzelius method*, any arsenic present reacts with nascent hydrogen to form arsine, which is decomposed by heat.

For the *Gutzeit method*, the arsine forms a yellow to brown stain on a mercuric chloride (or bromide) paper.

The *molybdenum blue colorimetric method*, based on the sensitive reaction usually attributed to Denigès, can also be employed for estimating arsenic.

The *AMC (1975a) method* describes the separation of arsenic, following the destruction of organic matter, from all interference except tin and phosphate by distillation as the bromide. This is completed by a molybdenum blue spectrophotometric method. Any excessive tin must be removed by cupferron extraction and double distillation may be necessary if excessive phosphate is present. Silver diethyldithiocarbamate, which reacts with arsine to produce a red colour that is measured spectrophotometrically at 520–540 nm, has also been used. Hoffmann and Gordon (1963) found that the method is reliable and the results agree well with those obtained by the molybdenum blue procedure. AAS methods are also available and have become generally used. Various techniques have been described in which arsine is evolved by reaction with, for example, sodium tetraborohydrate III followed by measurement in an argon–hydrogen flame (Khan and Shallis, 1968). Rees (1978) has described a simple hydride generation technique that allows determination of arsenic and tin in foods at levels down to 0·1 mg/kg. The AMC method for antimony in biological material may also be used for arsenic (AMC 1980). The hydride generation method has received substantial attention in particular. An extensive collaborative study by Ihnat and Thompson (1980), in which considerable systematic error and imprecision was noted, the preliminary conclusion was that the method needed further study. A related study by Holak (1980) focused attention on the use of closed system digestion procedures using nitric acid as the preferred technique. The combination of closed system digestion and hydride generation with AAS finish is currently (1986 supplement) a 'first action' procedure of the AOAC. Peacock and Singh (1981) reported on a simplified hydride generator system using a boiling tube and syringe injection of sodium tetrahydroborate (III), and these authors recommended the use of thiourea to suppress interference from various transition metals (see also Pierce and Brown, 1977 and Evans *et al.*, 1979).

The hydride generation method has been coupled with colorimetric technique (Maher, 1983) and with inductively coupled plasma emission as a multi-elemental analysis technique. Evans and Dellar (1982) concluded that the method as applied to the anlysis for arsenic lacked sensitivity; a similar study has been reported by Hahn *et al.* (1982).

Procedures for determination of inorganic arsenic in foods have been described by Brooke and Evans (1981) and by Muenz (1984), both groups preferring procedures based on the preliminary distillation of arsenic from hydrochloric acid followed by hydride generation coupled with AAS.

Determination of arsenic (AMC, 1975a)

The molybdenum blue and silver diethyldithiocarbamate methods have been compared (AMC, 1975a), during the course of which the volatility of arsenic during wet combustion was investigated. Although otherwise preferred, the molybdenum blue method was considered to be lengthy and called for a high degree of analytical expertise and an alternative procedure in which the arsenic is first separated by distillation as bromide in a special apparatus was developed. The latter was found to have considerable advantages and was the method of choice. Tin, if present, is removed after wet oxidation by extraction with cupferron (AMC, 1960a). The AOAC recommends both the molybdenum blue and the silver diethyldithiocarbamate procedures as official methods. A combination of dry ashing, hydride generation and silver diethyldithiocarbamate colorimetric method has been studied by Hunter (1986) and is said to be suitable as a diagnostic method for screening tissues from animals suspected to be suffering from arsenic poisoning.

APPARATUS

The special distillation flask assembly (modified from Chaney and Magnuson, 1944) is illustrated in Fig. 5.1 a, b. Before use, traces of tin should be removed by refluxing concentrated HCl for several hours.

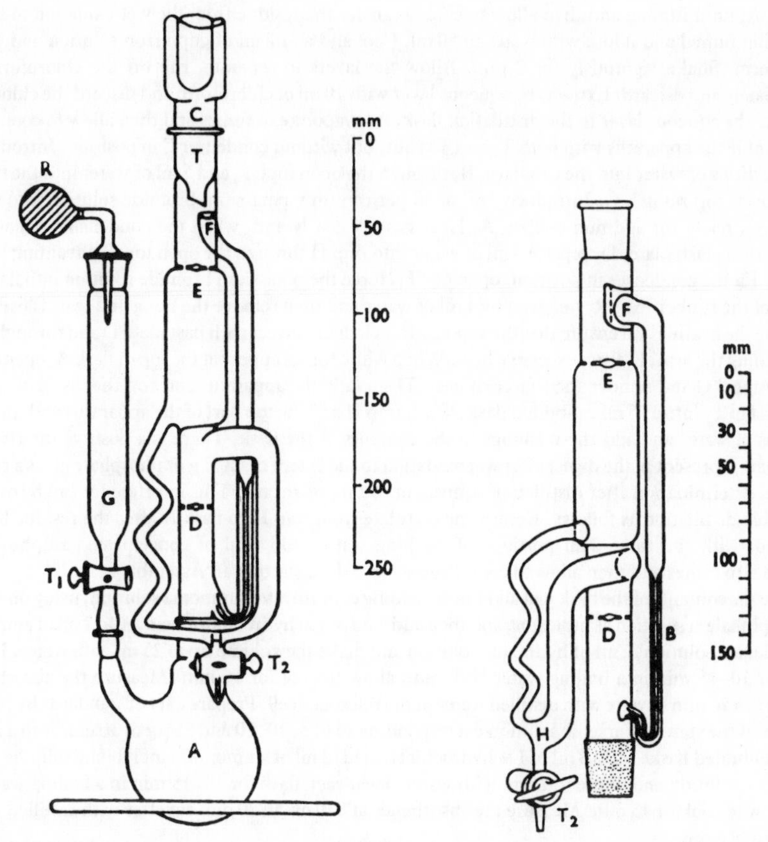

Fig. 5.1 Distillation apparatus for arsenic. a. Complete assembly. b. Details of trap head.

REAGENTS

Sulphuric acid. For foodstuffs analysis.

Nitric acid. For foodstuffs analysis.

Potassium bromide solution. A 30 per cent m/v solution in water.

Ammonium molybdenate solution. A 1 per cent m/v solution in 10 per cent v/v sulphuric acid.

Hydrazine sulphate solution. A 0·05 per cent m/v solution in water.

Standard arsenic solution. Dissolve 4·17 g of analytical-reagent grade sodium arsenate, $Na_2HAsO_4 \cdot 7H_2O$, in water and dilute the solution to 1 litre (solution A). Dilute 10·0 ml of solution A to 1 litre with water (solution B). Prepare solution B freshly as required.

1 ml of solution ≡ 10·0 μg of arsenic.

Ammonia solution, sp. gr. 0·88.

Hydrochloric acid, 1 M.

The following additional reagents are required if tin is present.

Chloroform.

Cupferron solution. A 5 per cent m/v solution in water.

PROCEDURE

Carry out a blank test by the entire procedure, using the exact amounts of reagents used in the test, omitting only the sample. Wet oxidise a known mass, not exceeding 5 g of dry matter, of the sample with 5 ml of concentrated sulphuric acid and a suitable amount of concentrated nitric acid. Adjust the volume to 5 ml with sulphuric acid. Transfer the wet digest to the flask A (Fig. 5.1a), using three 5-ml portions of distilled water to wash out the digestion flask, heat the contents of the flask, without assembling the apparatus, until fuming and then allow to cool. Transfer the residue from the wet oxidation to a 100 ml separating funnel and dilute with water to 50 ml. Cool and add 2 ml of cupferron solution and 10 ml of chloroform. Shake vigorously for 2 min. Allow the layers to separate, run off the chloroform layer immediately and discard. Extract the aqueous layer with 10 ml of chloroform and discard the chloroform. Transfer the aqueous layer to the distillation flask, A, evaporate to fumes and then allow to cool.

Assemble the apparatus with both T_1 and T_2 shut, but without condenser C in position. Introduce two or three drops of water into the capillary, B, through the open top, T, and 5 ml of water into the flask, A, through the top funnel, G. Introduce 3 ml of 30 per cent m/v potassium bromide solution into the tap-funnel, G, ready for addition to flask A. Heat flask A gently and, when the condensing vapour front reaches the splash plate, D, pipette 3 ml of water into trap H through the open top, T, avoiding the side-arm, F. Fit the condenser in position, open tap T_1, force the potassium bromide solution into flask A by means of the rubber teat, R, followed by 1 ml of water and then remove the bung and teat. Close tap T_1, continue the heating and ensure that the vapours follow their correct path past plate D and through trap H by shielding the arm, B, from excessive heat. When white fumes appear at the top of flask A, open the tap, T_1, of funnel G and remove the source of heat. Dismantle the apparatus and run the distillate in tap H through tap T_2 into a 25 ml calibrated flask. Wash trap H and the top part of the apparatus with three 2 ml portions of water and add the washings to the contents of the flask. There is a positive interference if phosphate is present in the digest after wet oxidation to the extent that 0·1 g of phosphate gives a colour in the final determination after distillation equivalent to 1 μg of arsenic. This interference can be overcome by double-distillation as follows. Return the distillate from trap H to flask A after the residue has been removed, with the three 2 ml portions of washing water, add 5 ml of concentrated sulphuric acid, evaporate to fumes and then allow to cool. Proceed as before starting at 'Assemble . . .'.

Make the contents of the flask just alkaline by adding concentrated ammonia solution, using one drop of phenolphthalein solution as indicator, and then add 3 ml of 1 M hydrochloric acid. Add 2 ml of ammonium molybdenate solution, 2 ml of hydrazine solution and make the volume up to 25 ml with water. Heat the flask for 10–15 min in a boiling water bath, and allow to cool for 15 min. Measure the absorbance at 840 nm in a 20 mm cuvette with distilled water in the reference cell. Prepare a set of standards by pipetting aliquots of the standard arsenic solution corresponding to 0, 5, 10, 20 and 25 μg of arsenic into a series of 25 ml calibrated flasks. Add 3 ml of 1 M hydrochloric acid, 2 ml of ammonium molybdate solution, 2 ml of hydrazine solution and dilute to 25 ml with water. Heat each flask for 10–15 min in a boiling water bath and allow to cool for 15 min. Measure the absorbance at 840 nm in 20 mm cuvettes with distilled water in the reference cell.

LEAD

Limits for lead are prescribed in The Lead in Food Regulations 1979, which reduced the former general limit from 2·0 mg/kg to 1·0 mg/kg, with specific limits for scheduled foods varying from 0·2 mg/kg to 10 mg/kg. The limit for foods specially prepared for infants or young children is 0·2 mg/kg. The Lead in Foods (Amendment) Regulations 1985 reduced the prescribed levels for canned food, corned beef and condensed milk. Small amounts of lead occur in many foods naturally but contamination may also occur from the use of lead alloys or compounds for processing materials, including solders, glazes, enamels, wrapping materials and piping or from insecticides, although the latter are now largely replaced by alternative preparations. Alcoholic drinks and soft water are especially liable to pick up lead from piping which conveys them. Levels of lead in food have been surveyed in the reports of MAFF (1972, 1975, 1982b, 1983a) and FAO/WHO (1972); the latter reports indicate that most foods on sale in the UK comply with the Regulations. Significant changes have been achieved in the level of lead in canned baby foods by improvements in can making technology. Similar conclusions have been made by Miles (1982) on products sold in the USA. Further data on food sold in the USA have been reported by Jelinek (1982) and by Schaffner (1981).

Destruction of organic matter

Wet oxidation of the material is suitable for the determination of lead. The difficulties which were formerly experienced due to appreciable amounts of lead in the acids used have been largely overcome by the availability of 'lead-free' reagents from the suppliers of chemicals. Dry ashing with or without an ash-aid at a temperature not exceeding 500 °C is in general also satisfactory. Experiments with lead-212 as radioactive tracer have indicated, however, that (1) recoveries of the metal from dry ashing are influenced by the ash-aid and the temperature used and (2) the method may be more suitable for some foods than others.

Preer *et al.* (1982) compared wet and dry technique and concluded that carefully controlled dry ashing coupled with chelate extraction had the advantage of speed of operation. The choice of preparation method is also linked to sample type and the level anticipated in the food concerned; whatever the method used care must be taken with reagent quality and cleanliness of glassware if the levels sought are those typical of the 'natural' value in commonly occurring foods, i.e. less than 0·1 mg/kg. Sampling problems are also related to the nature of the lead contamination of the food, for example canned foods may contain particulate lead derived from solder splash. Suddendorf *et al.* (1981a) have described a general method for sampling of canned foods which involves the blending of the entire contents of the can with 2 M nitric acid followed by sub-sampling and wet combustion in a mixture of nitric acid, sulphuric acid and hydrogen peroxide.

Methods available

Many methods have been promoted for the separation and determination of lead in food, the measurement technique most commonly used for many years being based on complexometric reaction with reagents such as dithizone—a general method using the reagent is cited by the AOAC and by Codex. Other approaches include polarography and anodic stripping voltammetry; authors favouring the electrochemical techniques use them as multi-element procedures. Borus-Boszormenyi (1980) quotes the procedure for estimation of lead, copper, cadmium and zinc whilst the latest issue of the AOAC manual cites anodic stripping voltammetry as the 'first action' method for lead and cadmium. This recommendation is based on the results of studies such as those reported by Elkins (1982), Capar and Subjoc (1982) and Zink *et al.* (1983a, b). The method has also been used in the analysis of fruit juices and soft drinks with minimal sample preparation (Mannino, 1982, 1983, 1984). An HPLC technique is also being applied in combination with pre-column formation of chelates and use of ultraviolet detection (Ichinoki and Yamazaski, 1985). However, the technique most widely favoured is AAS either in the solution injection format or in combination with graphite furnace atomisation. Whilst solution injection methods have achieved wide acceptance, the use of the graphite furnace mode has been the subject of conflicting claims. Muys (1984) applied the method to samples of meat, eggs and milk powder and claimed satisfactory comparisons with the results obtained in other interlaboratory studies. However, Dabeka (1984) drew attention to the widely varying results obtained in co-operative and collaborative studies on various methods for lead. He showed that in a tightly controlled study on milk powder and infant formulae that a major factor affecting the quality of results obtained with the furnace procedure was connected with background correction methods. These general conclusions were supported by Halls *et al.* (1986) who used closed system pressure digestion in nitric acid of plant materials. The conditions for general application of hydride generation as a preliminary to AAS have been studied by Bonilla *et al.* (1986).

The method given below is a generally accepted procedure developed by the AMC (1975b) and is applied after destruction of the organic matter: the lead is then complexed with ammonium pyrrolidine-1-carbodithioate, the complex extracted with 4-methylpentan-2-one and this solution aspirated into the burner of the spectrometer (see also Evans *et al.*, 1978, loc. cit.).

Determination of lead (AMC, 1975b)

Reagents

4-Methylpentan-2-one. Low in lead grade.

Ammonium pyrrolidine-1-carbodithioate (APDC) solution. Place approximately 1·5 g of APDC in a porosity 4 sintered-glass crucible, wash with 20 ml of acetone and suck dry by means of a water-pump. Weigh 1·0 g of the washed and dried APDC, dissolve it in 100 ml of water containing 0·1 g of mercaptoacetic acid and neutralise the solution to pH 6–7.

Standard lead solution, 10 μg/ml.

Apparatus

Use an atomic absorption spectrometer with an air–acetylene burner and a suitable lamp.

Procedure

Destroy the organic matter in an appropriate amount of sample by wet oxidation (AMC, 1967). When oxidation is complete, dilute the cold digest with 10 ml water, add 2 ml of 10 per cent sodium sulphite solution and boil until white fumes of sulphur trioxide appear. Dilute the digest from 25 g of sample (or other suitable mass depending on the expected lead content) to 50 ml with water, cool and transfer the solution to a separating funnel. Add 2 ml of the APDC solution, shake and allow to stand for 5 min. Accurately add 10 ml of 4-methylpentan-2-one and shake the mixture vigorously for 1 min. Allow the layers to separate, discard all of the aqueous layer and filter the organic layer through a small dry Whatman No. 541 filter-paper into a suitable stoppered flask. Prepare standard solutions so as to cover the required range of lead contents by adding portions of the standard lead solution to 5 ml of sulphuric acid and diluting each standard to 50 ml with water. Add 2 ml of APDC solution and 10 ml of 4-methylpentan-2-one, and proceed as described for the sample solution. Set up the instrument with a lead lamp and the monochromator adjusted to either 217·0 or 283·3 nm, whichever gives the greater signal to noise ratio (see Note 1). Prepare a calibration graph (see Note 2) by aspirating the standard solutions and plotting the mean signal response against lead content. Aspirate the solution obtained from the sample, record the mean signal response and read off the lead content from the calibration graph. If a large number of samples is being examined one or more standard solutions must be re-aspirated at intervals during the course of the analysis.

Notes

1. Aspirate the ketone when setting the zero of the instrument and between all readings. Care must be taken with some burner assemblies when changing solutions as the removal of the ketone renders the flame weak, with a tendency to flash back.

2. The calibration graph will be linear over the range 0–30 μg of lead; above 30 μg the graph will exhibit a slight curvature.

CADMIUM

Cadmium is a relatively toxic metal, which soon after ingestion is liable to cause acute gastritis with vomiting and diarrhoea. Formerly, cadmium was liable to be present in foods due to the use of certain utensils and, in particular, cadmium plated vessels. MAFF (1973b, 1980a, 1983a) reported that cadmium in food arises mainly from natural sources, but it may also be derived from atmospheric discharge into water and subsequent uptake in aquatic animals, enhanced uptake by plants due to the use of superphosphate fertilisers or the disposal of sewage sludge on land. The latest report draws attention to the fact that limits of detection have been improved in recently conducted surveys. The highest levels found in individual foods were in brown crab meat ranging up to 86 mg/kg, the levels in white meat varying up to 2·9 mg/kg. Samples of vegetables grown on land fertilised with sewage sludge showed levels in the region 0·5–1·8 mg/kg. Most foods, however, contain 0·01–0·03 mg/kg according to type.

Determination of cadmium

The AMC (1975a) has described a specific method for cadmium based on extraction as a chelate from an ion-exchange resin and measurement by atomic absorption spectroscopy.

The AOAC method involves extraction with dithizone and back-extraction into acid, followed by atomic absorption spectrophotometry. Anodic stripping voltammetry has also been applied in this determination (Holak, 1980; Capar *et al.*, 1982; Gajan *et al.*, 1982). Adeloju and Brown (1986) in studies on water showed that the organic solvent used in the chelation stage had a significant effect on symmetry of the response peak. Freon was considered to be superior to chloroform as the chelate solvent. Evans *et al.* (1978) concentrated cadmium by extraction of the ammonium pyrrolidine-1-carbodithioate (APDC) chelate into 4-methylpentan-2-one and direct measurement by AAS using the extraction solution; lead and nickel may also be estimated and possible interferences are listed. Dellar (1983) showed that for foods such as offals and shellfish that contain amounts of cadmium in excess of 0·07 mg/kg, the acid digest of the food can sprayed directly into the AAS flame. A historical bibliography of methods for cadmium together with lead and mercury has been prepared by Schuller and Egan (1976).

Determination of cadmium (AMC, 1975c)

The method can be applied to the sulphuric acid extract resulting from any wet oxidation. By suitable adjustment of the amount of sample a wide range of cadmium contents down to sub-microgram amounts can be determined.

REAGENTS (NOTE 1)

Sulphuric acid, density 1·84 g/ml. Analytical-reagent grade.
4-Methylpentan-2-one. For atomic absorption spectrometry.
Potassium iodide solution. 0·1 M.

Amberlite LA-2 solution. A 1 per cent v/v solution of Amberlite LA-2 [*N*-lauryl(trialkylmethyl)amine] in 4-methylpentan-2-one.

Cadmium standard solution. Dissolve 2·282 g of cadmium sulphate ($3CdSO_4.8H_2O$) in 50 ml of water, add 10 ml of sulphuric acid and dilute to 1 litre with water (1 ml of solution ≡ 1000 μg of cadmium). Dilute 10 ml of this solution to 1 litre with water (1 ml of solution ≡ 10 μg of cadmium). Dilute 10 ml of the latter solution to 100 ml with water (1 ml of solution ≡ 1 μg of cadmium).

Destroy the organic matter present in an appropriate amount of sample (see Note 2) by wet oxidation with 10 ml of sulphuric acid and any of the other reagents recommended for this purpose, but preferably with 50 per cent hydrogen peroxide. Observe all the precautions that are necessary to avoid mechanical loss of solution. When oxidation is complete, cool the digest. Add, with care, 20 ml of water and again cool the digest. Transfer the cooled and diluted digest to a 100 ml separating funnel with the minimum amount of water necessary to effect the transfer. Add 5 ml of 0·1 M potassium iodide solution, dilute to about 50 ml with water, and mix well. Then add, by use of a pipette, 10 ml of Amberlite LA-2 solution, and shake the funnel vigorously for 20 s. Allow the phases to separate, run off and reject the aqueous phase, and filter the ketone phase through a dry Whatman No. 541 or equivalent filter-paper in order to remove suspended droplets of aqueous phase. Collect the filtrate in a small, glass-stoppered vessel. Transfer volumes of cadmium standard solution equivalent to 0, 2, 4, 6, 8 and 10 μg of cadmium to a series of separating funnels, each containing a cooled mixture of 10 ml of sulphuric acid, about 20 ml of water and 5 ml of 0·1 M potassium iodide solution. Dilute all of the solutions to about 50 ml with water and extract each with Amberlite LA-2 solution as described under Extraction of Cadmium. Collect the filtrate after each extraction in a separate, small, glass-stoppered vessel (see Note 3).

Set the atomic absorption spectrometer to the optimum conditions for cadmium at a wavelength of 228·8 nm. Aspirate the ketone alone (not the Amberlite LA-2 solution) and set the instrument to zero, adjusting the air flow to give a flame that is just short of luminous (see Note 4). Then aspirate successively the standard, reagent blank and sample solutions and after each reading aspirate the ketone alone, thus carrying out frequent checks on the zero. Prepare a calibration graph by plotting the reading obtained for each standard solution against its known cadmium content. Calculate the cadmium content of the sample by reference to the calibration graph.

NOTES

1. If any of the reagents (or samples) has been in contact with coloured polythene stoppers some cadmium could have been picked up from the pigment. A check should be made on the Amberlite LA-2 solution in ketone in order to ensure that it is free from cadmium.
2. The effective sensitivity required can be controlled simply by adjusting the amount of sample taken.
3. This range of standard will, with most instruments, give a calibration graph with little deviation from linearity. For a 20 g sample the absorbance of the 10 μg standard will be equivalent to 0·5 p.p.m.
4. Care must be taken with some burner assemblies when changing solutions as the removal of the ketone renders the flame weak and there is a tendency to flash back.

MERCURY

Mercury may be present in some foods, e.g. cereals due to the use of organo-mercury fungicides. In the early 1970s there was considerable concern over the presence of toxic methylmercury in fish derived from industrial effluents. Much attention was concentrated on canned tuna and surveys in the USA and by the MAFF (1971, 1973a) and the APA (1971) indicated that 5–25 per cent of the samples contained amounts in excess of the US limit of 0·5 mg/kg (Hg). In the MAFF survey the mercury levels in canned tuna reported ranged from 0·1 to 0·8 mg/kg with a mean value between 0·3–0·4. The Ministry did not consider it necessary to set a limit for canned tuna owing to the low level of consumption in the British diet and the levels found. Although, for convenience, surveys involve determinations of total mercury, in fish most of it is present in the toxic form of methylmercury. In a duplicate diet study on vulnerable fishing communities Haxton *et al.* (1979) concluded that there are unlikely to be any consumers of fish in the UK who are adversely affected by the presence of methylmercury. Ohlin and Andersson (1982) came to similar conclusions in a Swedish study.

Mercury is liable to volatilise during conventional wet oxidation and special precautions must be taken to trap the volatiles during the process. Various methods have been described for the determination of total mercury in biological tissues, including those of the AOAC. In addition to methods using atomic absorption, a method using inductively coupled plasma spectrometry has been reported by Suddendorf *et al.* (1981b). Amalgamation on gold has also been reported as a trapping technique following oxidative liberation of mercury (Konishi and Takahashi, 1983; Koops *et al.*, 1984a). The use of hydrochloric acid in the wet digestion procedure as applied to fish has also been recommended by Louie (1983) and Vibhakar *et al.* (1983). Louie *et al.* (1985) reported the application of the procedure to a range of food samples including fats, oils and high carbohydrate foods. The method described below has been developed by the AMC (1977b) jointly with the MAFF Committee on Analytical Methods (CAM) particularly for fish and includes the separate estimation of mercury present in the form of methylmercury compounds.

Determination of total mercury in fish

The sample is treated with sulphuric acid, nitric acid and hydrogen peroxide; the excess of peroxide is removed by boiling and the addition of potassium permanganate. A portion of the digest is reduced with hydroxylammonium chloride solution and tin(II) chloride solution (see Note 1), then aerated through a cell in an atomic-absorption spectrophotometer fitted with a cold cathode mercury lamp and the absorption measured at 253·7 nm.

APPARATUS

Digestion apparatus.

Heating mantle.

Atomic absorption spectrophotometer. Fitted with a cold cathode mecury lamp.

Ice-bath.

REAGENTS

All materials should be of analytical-reagent grade. However, some (particularly potassium permanganate) can contribute appreciably to the blank and it is therefore recommended that all reagents should be screened for mercury contents before being used in the analysis.

Nitric acid, sp. gr. 1·42.

Sulphuric acid, sp. gr. 1·84. The 'low in lead' grade has been found to be satisfactory.

Hydrogen peroxide, 50 per cent m/v.

Potassium permanganate solution, 6 per cent m/v.

Sodium chloride solution, 15 per cent m/v.

Hydroxylammonium chloride solution, 21 per cent m/v.

Hydroxylammonium chloride–sodium chloride solution. Mix 20 ml of sodium chloride solution and 12 ml of hydroxylammonium chloride solution, and dilute to 100 ml.

Tin(II) chloride solution. Heat 21 g of granulated tin under reflux with 50 ml of water and 50 ml of hydrochloric acid, sp. gr. 1·18, until no more tin will dissolve. Store the solution over a piece of metallic tin.

Standard mercury solutions. Dissolve 0·135 4 g of mercury (II) chloride in 0·1 M nitric acid and dilute to 100 ml with 0·1 M nitric acid.

By pipette, transfer 5 ml of this solution into a 500 ml calibrated flask and dilute to the mark with 0·1 M nitric acid (1 ml ≡ 10 μg of mecury).

By pipette, transfer 5 ml of this solution into a 500 ml calibrated flask. Add 1 ml of 6 per cent potassium permanganate solution and dilute to the mark with 0·1 M nitric acid. Prepare this solution freshly before use (1 ml ≡ 0·1 μg of mercury).

PROCEDURE

Weigh 2·5 g of the homogenised sample into the 200 ml Kjeldahl flask. Add 9 ml of sulphuric acid, sp. gr. 1·84, and attach the upper part of the digestion apparatus to the flask. Heat the flask on the heating mantle and swirl it vigorously until a homogeneous tarry fluid is obtained. Cool the flask in ice and from a pipette add 2 ml of 50 per cent hydrogen peroxide through the top of the condenser. With the flask still in ice and with the apparatus inclined at about 30° from the vertical open tap A so that the peroxide is introduced slowly into the acid mixture. Remove the flask from the ice-bath and swirl it slowly until the reaction begins; it should not be excessively vigorous. As the reaction slows down apply heat to the flask by means of the heating mantle. Close tap A, add 2 ml of nitric acid, sp. gr. 1·42, from a pipette through the top of the condenser and by opening tap A allow the acid to run slowly into the flask while the contents are still hot. After 2 min close tap A and heat the flask until fumes are evolved; run off the condensate that has collected in B into a beaker. Add through the top of the condenser 1 ml of hydrogen peroxide and 1 ml of nitric acid, sp. gr. 1·42, to the flask, close tap A and again heat until fumes are evolved; transfer the condensate that has collected in B to the same beakers as before. Repeat this operation with 0·5 ml portions of both hydrogen peroxide and nitric acid until the digest is a pale straw colour. Cool the condensate in the beaker and return it to the Kjeldahl flask by way of the reflux system. Cool the contents of the flask and add potassium permanganate solution until a permanent pink colour is produced (about 0·5 ml). Tranfer the digest to a 50-ml calibrated flask, rinse the reflux system and Kjeldahl flask with water, add the rinsings to the contents of the calibrated flask, and dilute to the mark with water. Set this solution aside for 24 h before proceeding to the determination by cold-vapour atomic absorption spectrophotometry.

By pipette place a portion, usually 10 ml (see Note 2) of the well mixed digest solution into the aeration test-tube (see Note 3). Add water to bring the volume up to 13 ml, then add 2·0 ml of hydroxylammonium chloride–sodium chloride solution and 0·2 ml of tin(II) chloride solution. Mix well with a flat ended glass rod into a quartz-windowed tube (approximately 10 cm long by 1 cm internal diameter) avoiding undue disturbance of the magnesium perchlorate and silica wool. The tube is placed in the light path of the atomic-absorption spectrophotometer. Set the monochromator at 253·7 nm and adjust the air flow-rate to about 740 ml/min. When mercury evolution ceases, remove the aeration test-tube, add to it a suitable small volume of the standard mercury solution (0·10 μg/ml) chosen to give a peak of similar height to that already obtained for the sample, mix as before, replace the tube in the apparatus, and again aerate through the absorption cell. The sample, reagent and standard solutions should all be at the same temperature. The ratio of responses of the sample and standard gives a measure of the mercury present in the sample aliquot. Make replicate aerations of three aliquots of the sample digest and take the mean result.

Carry out a blank determination using a similar sample material selected for its low mercury content (less than 0·05 mg/kg and preferably less than 0·03 mg/kg) through all stages of the procedure and using the same amounts of sample and reagents as in the test. Also aerate a portion of the standard mercury solution as before and if the apparent mercury content of the blank material is greater than 0·05 mg/kg, carry out an examination of the reagents and replace any that are found to be contaminated with mercury.

NOTES

1. Although the use of two reductants, hydroxylammonium chloride and tin(II) chloride, is not easy to justify on theoretical grounds, in practice some organic matter remains in the final digest, potassium permanganate being added to remove this residue. Some manganese(IV) oxide is therefore precipitated, which is then removed by the addition of hydroxylammonium chloride. Manganese(IV) oxide removal by tin(II) chloride is slow and can lead to interference with mercury evolution, either by adsorption effects or by blocking the aerator. In practice, therefore, it has been found to be important first to add hydoxylammonium chloride solution when precipitated manganese(IV) oxide is present.
2. When small sample sizes (or low levels of mercury) are being handled, the entire sample digest can be aerated through the absorption cell without dilution.
3. A suitable apparatus for this purpose can be made from a Drechsel bottle head, MF 28/3/125, to which

is attached a porosity 4 filter from a filter stick broken off at the filter. This bubbler is connected through a reduction adaptor, B 24/19, to a 35-ml capacity test-tube, MF 24/26.

Determination of methylmercury compounds in fish

In the method developed, an alkaline homogenate of the sample is extracted with toluene, the extract is cleaned up by partition with a cysteine hydrochloride reagent and the final determination of methylmercury chloride is carried out by gas–liquid chromatography. Studies on the method have been reported by James (1983), Hight and Capar (1983) and Awarez *et al.* (1984). A TLC separation method has been described by Margler and Mah (1981), whilst Holak (1982) used the sodium thiosulphate complex of methylmercury chloride in an HPLC separation procedure; see also Schuller and Egan (1976) loc. cit.

APPARATUS

Gas chromatograph. Fitted with an electron capture detector, pulse interval 50 μs. 2 m × 3 mm i.d. glass column containing 2·5 per cent Carbowax 20 M on Chromosorb G, 80–100 mesh, AW, DMCS, operated at 160 °C. Detector temperature 285 °C. Nitrogen carrier gas, 90 ml/min.

Preparation and packing of column. Dissolve 0·5 g of Carbowax 20 M in 200 ml of chloroform and add it to 24·5 g of the Chromosorb G in a round-bottomed boiling flask. Remove the solvent under reduced pressure at a temperature not exceeding 60 °C and remove any residual traces of solvent in a vacuum oven if desired. Pack the dried material into the column applying agitation either mechanically or preferably by means of a vibrator. Condition the column overnight at 185 °C with a nitrogen flow-rate of 100 ml/min.

Centrifuge tubes. Quickfit MF 24/3/8, approximately 80-ml capacity and equipped with well-fitting plastic stoppers.

Centrifuge. To accommodate the specified centrifuge tubes.

Siphons. To fit the specified centrifuge tubes. Werner–Schmid tubes are suitable.

Separating funnels. Capacity 25 and 125 ml.

REAGENTS

All materials should where appropriate be of analytical-reagent grade.

Sodium hydroxide.

Sodium sulphate.

Sodium sulphate solution, 10 per cent m/v.

Sodium hydroxide reagent. A 2 M solution of sodium hydroxide in 10 per cent sodium sulphate solution.

Toluene. Redistil from charcoal to remove substances that will interfere with the gas chromatographic stage.

Hydrochloric acid, sp. gr. 1·18. For 'Foodstuffs Analysis.'

Cysteine hydrochloride. Store in a refrigerator when not in use.

Sodium acetate trihydrate.

Cysteine hydrochloride reagent solution. Dissolve 1·0 g of cysteine hydrochloride and 0·8 g of sodium acetate in 100 ml of 10 per cent sodium sulphate solution.

Methylmercury chloride.

Copper(II) sulphate solution, 1 M.

PROCEDURE

Place 5 g of the homogenised sample into a centrifuge tube and add 25 ml of sodium hydroxide reagent. Mix well, and heat the tube in a bath of boiling water, taking precautions against losses by frothing. Keep the tube in the boiling water until solution is as complete as possible (normally within 30 min). Remove the tube from the boiling water, cool it, add 8 ml of hydrochloric acid, sp. gr. 1·18, cool again, add 25 ml of freshly distilled toluene and 1 ml of 1 M copper(II) sulphate, stopper the tube securely, and shake it for 3 min. Centrifuge the tube at a predetermined safe speed that will give a clear upper toluene layer (10 min is usually sufficient). Siphon the toluene layer into a 125 ml separating funnel, wash the siphon tube with two 5 ml portions of toluene and transfer these without shaking to the separating funnel. Repeat the extraction with two 25 ml portions of toluene, taking the usual precautions to ensure the complete transfer of solvent, wash the tube with toluene as before and collect all the toluene phases and washings in the same separating funnel. Shake the combined toluene extracts successively with 3, 3 and 2 ml portions of cysteine hydrochloride reagent solution and combine these extracts in a 25 ml separating funnel. Add 1 ml of hydrochloric acid, sp. gr. 1·18, and extract with 10 ml of toluene. Run the cysteine solution into a second 25 ml separating funnel, retaining the toluene layer in a separate container. Rinse the first funnel with a further 10 ml of toluene and use this to extract the acidified cysteine solution in the second funnel. Combine the toluene extracts and make them up to 25 ml or other suitable volume that gives a concentration of methylmercury chloride suitable for gas chromatography in the linear range of the detector; dry the toluene solution over a small amount of anhydrous sodium sulphate. Examine 5 and 10 μl volumes of the toluene solution by gas chromatography using the conditions described and adjusting the carrier gas flow-rate to give a suitable retention time and a sharp peak for methylmercury chloride. Run a complete blank on the reagents. Prepare a standard solution of methylmercury chloride in toluene and use it to determine the linear dynamic range of the detector response (linearity) and to determine the minimum detectable amount. Determine the methylmercury chloride content of the blank and sample by comparing their peak heights with those derived from suitable standards.

TIN

In 1952 FSC recommended a limit for tin of 250 mg/kg in canned foods; modern canning techniques including the use of selected lacquers are such that many canned foods now contain less than 100 mg/kg. Following a review on canned foods the FACC (1983) recommended that the guideline limit for such foods on sale to the public should be reduced to 200 mg/kg. In a wider survey, (MAFF, 1985), the Steering Group on Food Surveillance noted that the majority of foods (other than canned foods) contain less than 1 mg/kg of tin (see also Sherlock and Smart, 1984 for a summary of the evidence supplied to the Steering Group). Seow *et al.* (1984) also examined canned fruit juices and compared the results obtained with limits proposed by Codex Alimentarius. Tin will only dissolve rapidly in the food if oxygen is present, so that provided the can is airtight, the major attack takes place during the early period after canning, but tends to slow down when the residual oxygen is used up. Tin plate is most likely to be attacked by acid foods, especially fruits such as rhubarb, prunes and tomato and also asparagus and foods containing sulphur dioxide or salt. Examination of the internal surface of the can is often a useful guide to the extent of the corrosion. 'Sulphur-stains', which are caused by food containing appreciable quantities of sulphur (e.g. meat, fish, swedes), vary in colour from brown to purple, but do not usually imply that much tin is present in the food. The pattern known as 'feathering' should not be ignored, however, and cans showing dull grey patches are usually associated with the presence of tin in the food and the determination of the metal is then advisable. Even amounts of tin below the recommended limit in canned foods may give rise to a metallic taste. Corrosion studies on lacquered and unlacquered tinplate cans have been reported by Bearfield *et al.* (1983) and the protective effect of tin on lead dissolution from solder was confirmed in this study. Studies on lacquer formulations incorporating dispersed tin powder have been reported by Barbieri *et al.* (1982).

Determination of tin

Horwitz (1979) has reviewed commonly used methods for tin. The AMC (1967) considered three methods for tin in foods, covering a range appropriate to the levels of interest in canned foods, one titrimetric and two colorimetric.

The examination of foods for tin content will depend on whether the food has been canned or not, in the former case the method chosen should be suitable for the range 25–500 mg/kg, whilst in the latter case the method should cover the range from 0·01–20 mg/kg. Methods suitable for the lower range include the hydride AAS procedure of Evans *et al.* (1979), or potentially sensitive methods such as the application of anodic stripping voltammetry described by Mannino (1983) and the fluorimetric procedure based on the complexation of tin with diacetylmonoxime nicotinylhydrazone (Rubio *et al.*, 1985).

For canned foods the choice is generally between colorimetric procedures such as that of Kirk and Pocklington (1969), which uses complexation with quercetin in the presence of thiourea to mask the effect of iron(III), or the AAS method described by the AMC (1983). Both methods are given below.

COLORIMETRIC METHOD

Procedure

To a known weight of comminuted sample containing not more than 1000 μg of tin, add 10 ml concentrated nitric acid, mix and after 10 min add 5 ml conc. sulphuric acid. Wet oxidise over a flame adding shots of nitric acid as charring commences. Boil vigorously until white sulphur trioxide fumes appear. Cool, add 20 ml water and transfer to a 50 ml graduated flask. Make up to mark and mix. Pipette 2·0 ml of this solution into a 50 ml graduated flask. Add 0·2 ml of 2, 4-dinitrophenol solution (0·1 per cent in 50 per cent ethanol). Add sodium carbonate solution (10 per cent) dropwise with agitation until the first appearance of a yellow colour. Add 2·5 M HCl dropwise until colourless and add a further 5 ml. Add 3 ml thiourea saturated aqueous solution and 5 ml of quercetin solution (0·2 per cent in ethanol), followed by 25 ml ethanol. Dilute to 50 ml with water and mix. After 30 min, measure the absorbance in a 40 mm cell at 437 nm against a reagent blank. From a calibration graph prepared from a solution of pure tin dissolved in boiling concentrated sulphuric acid, deduce the tin content of the sample.

ROUTINE AAS PROCEDURE (AMC, 1983)

Procedure

Weigh 20·0 g of the sample, to the nearest 0·1 g, after blending if necessary, into a 250-ml conical flask. Add 10 ml of water, distilled or de-ionised, and 10 ml of concentrated hydrochloric acid (see Note). Bring the sample just to the boil if it contains significant amounts of sugar, otherwise boil gently for at least 30 min. Cool, transfer the contents of the flask into a 100-ml calibrated flask, dilute to the mark and mix well. Filter sufficient sample for analysis, usually 20–30 ml.

Aspirate the solutions into a fuel-rich, dinitrogen oxide–acetylene flame, observing the instrument manufacturer's instructions and precautions for use of this flame. Alternatively, an ICP-OES system operated under the manufacturer's instructions may be used. Calibrate the instrument using appropriate dilutions of the stock tin solution in 6 M hydrochloric acid, with the working dilution containing 10 ml of concentrated hydrochloric acid per 100 ml of solution. Usually a 50 mg/l standard is used to calibrate direct reading instruments to read 250 mg/kg accommodating the five-fold mass to volume dilution of the sample.

Note—A run with a reagent blank must be carried out.

A comparative study of the two methods in application to canned milk was reported by Koops *et al.* (1984b), who concluded that the AAS method was less precise than the spectrophotometric method but that the former was preferred on the grounds of convenience and time saving.

Dabeka *et al.* (1985) reported a collaborative study using nitric acid digestion followed by treatment with hydrochloric acid, this latter procedure being considered to be better than the HCl only digestion. The method has been recommended as a 'first action' method for canned foods by the AOAC.

ZINC

The FSC (1953) recommended a general limit for zinc in food of 50 mg/kg, with a maximum of 5 mg/l for ready-to-drink beverages; a statutory limit of 100 mg/kg has been prescribed for gelatin. In certain circumstances the level of zinc derived naturally may exceed 50 mg/kg for herring, shellfish, Crustacea, cereals and animal offals. In a survey report (MAFF, 1981) the Steering Group on Food Surveillance noted that food constitutes the major source of zinc intake, the most significant sources being meat, cereal products and milk.

Determination of zinc

Either wet oxidation or dry ashing may be used prior to the determination of zinc. Although dry ashing at 550 °C is frequently used in routine work, the SAC has recommended overnight ignition at about 450 °C. Colorimetric methods using dithizone are still valid (AOAC), but AAS is normally used, for example in the general method for the determination of trace elements in the same prepared solution referred to below. Capar (1977) reported a collaborative study of a dry ashing, AAS procedure applied to the analysis of shellfish. The method was considered to be satisfactory but background correction was considered to be necessary for accurate analysis. The AOAC has adopted the closed system digestion method as an approved procedure.

COPPER

Statutory limits are prescribed for copper in gelatin (30 mg/kg) and tomato ketchup (20 mg/kg). The FSC recommendations include 2 p.p.m. for most ready-to-drink beverages and the general limit of 20 p.p.m. for most other foods. Continuation of these limits was confirmed in the surveillance report on copper and zinc in foods (MAFF, 1981). The report also contains data on individual foods and 'total diet' samples. The main contributory factors in the diet are cereals, vegetables and meat; offals and shellfish contain higher amounts (10–200 mg/kg), but contribution to dietary intake is minimised by relative consumption levels. Although emetic in large doses, copper is essential for growth. In plants it is necessary for respiration and in vertebrate animals traces of copper are essential for haemoglobin formation. When present in certain foods, however, it tends to act as an oxidation catalyst and as little as 2 mg/kg causes a tallowy flavour to develop in milk and butter and impairs the keeping qualities. Further, the presence of copper promotes the destruction of vitamin C in fruits and vegetables.

Destruction of organic matter

If the material is wet oxidised prior to the determination of copper it is advisable to remove chloride by a preliminary heating with nitric acid before the sulphuric acid is added. High (1947) has shown that dry ashing is also suitable provided nitric acid as well as hydrochloric acid are used for extracting the ash (cf. Middleton, 1965).

Schuller and Coles (1979) recommend the use of nitric acid/sulphuric acid wet combustion and completion of the process with hydrogen peroxide. Alternatively, the closed digestion system has been proposed by Holak (1980).

Determination of copper

Numerous colorimetric reagents have been employed for routine determinations of copper in aqueous solution, e.g. potassium ferrocyanide, dithio-oxamide, sodium diethyldithiocarbamate. Methods involving the extraction of various copper carbamate complexes by organic solvents are now frequently used. The AMC (1963) considered sodium diethyldithiocarbamate together with the possibilities of using other reagents such as zinc dibenzyldithiocarbamate, which is more sensitive, and neocuproine (2, 9-dimethyl-1, 10-phenanthroline), which under selected conditions is virtually specific for copper. The AMC considered, however, that there were certain advantages in using diethylammonium diethyldithiocarbamate. The published method involves the addition of EDTA-citrate to the wet oxidised solution, pH adjustment to 8.5 with ammonia and addition of the carbamate reagent dissolved in carbon tetrachloride. The optical density of the organic layer containing the copper complex is then measured at 436 nm. The interfering metals bismuth and tellurium, if present, can be decomposed by washing with sodium hydroxide (cf. Wyatt, 1953).

By using salts of dibenzyldithiocarbamic acid it is possible to extract the yellow copper complex from acid solution with carbon tetrachloride. Extraction at such a low pH possesses the additional advantage that interference from other metals is negligible. By using the minimum amounts of pure reagents, Abbott and Polhill (1954) were able to reduce the blank to 0·3 μg of copper and determined the metal in oils and fats over the range of 0·02–2 p.p.m. The colorimetric procedure is recognised as an official final action method by the AOAC, but in many laboratories the method has been superseded by AAS.

The AMC (1971b) recommended AAS as the preferred method, whilst Schuller and Coles (1979) used APDC as a chelate for extraction of copper from the acid digest as a preliminary stage in the measurement by AAS. Silva and Valcarcel (1982) recommended 1,2-naphthoquinone thiosemicarbazone for the chelation stage.

Holak (1983), in a further study, used the closed system method followed by AAS or anodic stripping voltammetry. McKinnon and Scollary (1986) have proposed this technique for determination of copper in wine, and concluded that a nitric acid pretreatment is necessary to recover the total copper content of samples.

SODIUM DIETHYLDITHIOCARBAMATE METHOD

Copper reacts with sodium diethyldithiocarbamate in alkaline solution producing a yellow to brown coloration according to the amount of the metal present. If the colour is to be measured in aqueous solution, citrate is added in order to prevent interference due to iron and precipitation of phosphate. The yellow copper carbamate complex can be extracted by carbon tetrachloride; in the presence of citrate the ion is retained in the aqueous solution. Cheng and Bray (1953) eliminated the interference due to 1 mg quantities of iron and most other interfering metals by the addition of EDTA in addition to citrate. EDTA complexes are formed with the bi- and trivalent metals present at pH 7–10. Copper is the only metal of importance, however, which can be sequestered from the EDTA complex by carbamate to form a more stable copper carbamate complex. Bismuth carbamate is also extracted under such conditions, but such interference can be ignored as bismuth is only present in negligible amounts in foods.

ATOMIC ABSORPTION SPECTROSCOPY METHOD

Prior to atomic absorption spectroscopy of copper the methods previously described for preparing the solution are suitable (p. 158). A rapid method suitable for many semi-solid and other foods involves homogenising 5–10 g sample with 40 ml water. Then 10 ml concentrated hydrochloric acid is added, the mixture heated to boiling and then simmered gently for a maximum of 5 min. Levels of copper may be such that AAS examination is possible on the digest directly. Alternatively, after diluting the digest to 100 ml with water and mixing, a suitable volume (e.g. 10 or 20 ml) is filtered for the analysis. Separated oil should be excluded when making up to the mark. Ammonium pyrrolidine-1-carbodithioate (APDC) can be used for the extraction of copper from aqueous solutions (acidity not exceeding 5 M) into an organic solvent, which is suitable for direct atomic absorption spectrophotometry. Before extracting the copper–APDC complex the acidic aqueous solution should be saturated with the solvent (4-methylpentan-2-one or heptan-2-one). The instrument should be set up with a copper hollow-cathode lamp with the monochromator at 324·8 nm. The method described by Evans *et al.* (1980) in which AAS measurements are made in a sulphuric acid digest is applicable to many foods.

SELENIUM

The importance of selenium as an essential nutrient has been recognised in relation to deficiency diseases associated with muscular wastage; the toxicity of the element is well established, however, the balance between necessary dietary intake and levels causing chronic toxic effects being very narrow. In view of the industrial importance of the element and its possible association with industrial exposure and effluents the need to establish adequate methodology has been recognised (Coles, 1974; Hofsommer and Beilig, 1980, 1981; Lloyd *et al.*, 1982). Thorne *et al.* (1978) have published survey data on UK foods and diets.

Determination of selenium (AMC, 1979)

Wet digestion procedures with nitric, perchloric and sulphuric acids are commonly used in the preparation of samples for analysis; care must be taken to ensure that oxidising conditions are maintained throughout the digestion process else selenium may be lost as the volatile hydride. Michie *et al.* (1978) proposed that the losses of selenium could be minimised by carrying out the digestion in a mixture of nitric and perchloric acid whilst ensuring the presence of an excess of nitric acid. On completion of the digestion stage the excess of oxidising acids is removed by heating with sulphuric acid to dense fumes of sulphur trioxide. If a complexometric method is to be used for the analysis the selenium must be present as Se(IV), with treatment of the digest with hydrochloric acid or with hydrogen peroxide. Ting *et al.* (1982) provide further evidence on the digestion methods as applied to tissue samples. The reagent used by most workers is 2,3-diaminonaphthalene (DAN), which forms a fluorescent solvent-soluble complex in acid solution.

METHOD

Destruction of organic matter

Place an accurately weighed amount of sample into a 100 ml Kjeldahl flask containing several acid-washed glass beads. For the purpose of this assay the amount of sample used will depend on two factors, namely the composition of the material and the amount of selenium expected. Samples low in fats and sugar containing 50 per cent or more of water can be digested in quantities up to 10 g, whilst quantities up to 5 g are suitable for samples containing 10–15 per cent of water. Samples with high fat or sugar or containing less than 10 per cent water should be limited to 1–2 g. Add 30 ml nitric–perchloric acid mixture (5 + 1), allow the mixture to stand overnight or heat gently until foaming subsides as the sample is solubilised. Gradually raise the temperature to a gentle boil without foaming, avoid the formation of char by adding small amounts of additional nitric acid. Reduce the volume to approximately one half of the original, cool and add 5 ml each of nitric and sulphuric acids. Boil the contents of the flask gently to white fuming, ensure that no evidence of char appears by cautious addition of 1 ml increments of nitric acid, and continue heating until a clear yellow digest is obtained with dense fumes of sulphur trioxide. Allow to cool.

Reduction of selenium

Dilute the digest with water and add sufficient hydrochloric acid to produce a solution 4 M with respect to HCl. Heat to boiling, cool and dilute to known volume whilst maintaining the acidity constant.

Fluorimetric measurement of selenium

REAGENTS

Ammonia, 17 per cent w/w. Dilute 500 ml ammonia (sp. gr. 0.88) to 1 litre.

Cyclohexane.

Formic acid.

Hydrochloric acid 4 M. Dilute 360 ml HCl (sp. gr. 1.18) to 1 litre.

Hydrochloric acid 0.1 M. Dilute 10 ml HCl (sp. gr. 1.18) to 1 litre.

2,3-Diaminonaphthalene (DAN). Add a few drops of 0.1 M HCl to 1.0 g DAN and stir to a paste. Transfer to a 250 ml separator with 200 ml 0·1 M HCl, shake for 30 seconds. Extract the solution three times with 20 ml portions of cyclohexane, dilute to 1 litre with 0·1 M HCl. Filter to an amber glass bottle, add sufficient cyclohexane to form a protective layer, store in a cool dark place. The reagent is stable for 2–3 weeks.

Hydroxylammonium chloride/EDTA (buffer). Dissolve 0·3 g disodium EDTA and 25 g hydroxylammonium chloride in 1 litre water.

Standard selenium solutions. Add 10 ml nitric acid to 1·0 g selenium, warm to dissolve. Dilute to 1 litre with water (1 ml = 1 mg Se). Prepare dilutions in 0·1 M HCl such that 1 ml = 10.0 μg Se and 1 ml = 0.1 μg Se.

Preparation of calibration graph

Transfer to a series of 150 ml beakers 0, 2·0, 4·0 and 6·0 ml of the dilute standard solution (1 ml = 0·1 μg), add 1 ml perchloric acid, heat on a hot plate to first appearance of fumes. Cool, add 5 ml 4 M HCl, boil gently for 5 min, cool and add to each beaker in turn 20 ml water, 5 ml formic acid and 10 ml EDTA buffer, and using a pH meter adjust the pH of each solution to 1·8 with ammonia solution, add 5 ml DAN solution and transfer to a water bath maintained at 50 °C.

After 30 min remove the beakers from the water bath, cool and transfer to a 250 ml separator fitted with PTFE stopcocks. Adjust volume to 70 ml with water and extract each with 10 ml cyclohexane. Discard the aqueous layers, wash the cyclohexane with 25 ml 0·1 M HCl, allow to separate and discard the acid layer.

Measure the fluorescence of the organic phase with excitation at 369 nm and emission at 525 nm. Correct the readings for 'blank' reagents system.

Measurement of the sample solutions

Take a suitable aliquot of the digest containing not more than 0·6 μg Se, transfer to a 150 ml beaker and treat as for the standards commencing with the addition of 20 ml water. Measure the fluorescent response at 525 nm and compare with the standard graph.

Alternative reagent systems have been discussed by Izquierdo *et al.* (1981) and Koh and Benson (1983), whilst Moreno-Dominguez *et al.* (1983) have used a TLC separation procedure on the DAN–Se chelate system.

Measurement by AAS after hydride generation

The diluted digest is reduced to the hydride with sodium borohydride; the volatile gas is introdued into a carrier gas stream and measured by AAS (the technique is that described for arsenic, p. 146; see also AMC (1980) for a method for Sb). Many authors have studied methods for the determination of the elements forming volatile hydrides as a group and any of the references given in the section on arsenic apply equally to Se, Sb and Te; Godden and Thomerson (1980) and Watson (1981) reviewed the various hydride generation techniques, whilst Rose (1983) and Pahlavanpour *et al.* (1980) have studied the use of hydride generation and inductively coupled plasma techniques.

GENERAL METHODS FOR TRACE ELEMENTS

The methods described above relate mainly to the determination of individual trace elements. It is possible, however, to make up the acid solution from the wet or dry ashed material to, say, 50 ml and use appropriate aliquots for the determination of more than one element. A detailed scheme for the spectrophotometric determination of As, Cu, Pb, Zn, Fe has been described by Strafford *et al.* (1945). Interest has been revived in colorimetric and complexometric techniques with the exploration of the application of HPLC to trace metal analysis (Jones *et al.*, 1985). Smith and Yankey (1982) reported the use of reversed phase HPLC of the dithiocarbamates of Cu, Co, Ni, Pb and Fe.

Evans and co-workers have reported several schemes for the use of AAS in the determination of more than one element in a single digest. These include Cd, Pb and Ni (Evans *et al.*, 1978); Sb, As and Sn as hydrides (Evans *et al.*, 1979); Cu, Fe, Mn, Zn (Evans *et al.*, 1980); In and Tl (Evans *et al.*, 1983); Mo and V (Evans and Caughlin, 1985); and Li, Rb and Sr (Evans and Read, 1985). Jackson *et al.* (1980) have developed procedures for the determination of Cr, Co and Ag. Garcia-Vargas *et al.* (1983) have prepared a general review of AAS procedures whilst Tyson (1984, 1985) has reviewed calibration techniques in AAS and combined flow injection AAS methods. Gillingham (1985) has described the advantages of multi-element standards dispensed into ampoules.

Combination techniques using AAS and anodic stripping voltammetry have also been reported by Jones *et al.* (1977) and Holak (1980) for Cd, Cu, Pb and Zn; Holak (1983) extended the procedure to include Cu, Ni and Cr. Jagner (1982) reviewed the general application of anodic stripping voltammetry methods for trace metal analysis. Kapel and Komaitis (1979) proposed the use of polarographic methods for 12 elements using a common digest and selective extractions with various chelating agents.

The use of inductively coupled plasma optical emission spectroscopy has been considered by Dahlquist and Knoll (1978). Evans and Dellar (1982) and Jones *et al.* (1982) have made critical studies of the matrix effects and limits of detection available for the commonly determined elements and have concluded that the advantages reside with conventional AAS procedures, although the method has not found general favour for food analysis. However, the method is continuing to receive attention (Faires, 1983). Dean *et al.* (1985) have described a combination of graphite rod vaporisation for sample introduction into a plasma system, and concluded that matrix effects were evident in the estimation of Cd, Cu, Mn and Pb, but that the standard additions procedure provided adequate compensation for these effects.

HALIDES

Whilst chlorine, as chloride, is a major constituent of foods, the other halides together with cyanide are minor constituents. Iodine and fluorine are essential trace nutrients but are also harmful if taken in excess, bromine may be present as a residue derived from the use of fumigants and cyanide may arise naturally as a result of the hydrolysis of glycosides particularly in certain fruit and vegetables (see Chapter 7).

FLUORINE

The level of fluorine in foods was formerly controlled by Regulations (1959, 1975), the principal concern being levels in phosphatic materials used as additives in flour and breadmaking. The overall regulations were revoked by the Bread and Flour Regulations (1984) and control of the additives *per se* passed to the Miscellaneous Additives Regulations (1980).

Levels of fluorine in food vary widely; few contain more than 1 mg/kg, but fish may contain 1–10 mg/kg and tea contains 100–150 mg/kg. The major source may be drinking water; soft waters contain little natural fluoride whilst hard waters may contain 10 mg/l. However, fluoride may be added to many water supplies to a controlled level of 1 mg/l.

Available methods

Few methods have been found to be entirely satisfactory for the determination of fluoride, most procedures depending on ashing in the presence of a mineral ash-aid followed by distillation from perchloric acid and measurement by complexation or titration with thorium nitrate in the presence of alizarin indicator (AOAC). Complexometric reagents used include lanthanum–alizarin–fluorine blue (AMC, 1971a) and Solochrome Cyanine R (Dixon, 1970); other reagents for fluorine and the other anions are comprehensively reviewed by Williams (1979). Whilst the colorimetric methods may be insufficiently sensitive to assay the lowest levels of fluorine in foods, electroanalytical methods have been adopted by many workers especially in process control applications.

Read and Collins (1982) have used an ion-sensitive electrode to measure quantities of fluoride ranging from 10 to 50 μg in distillates from various food materials. These authors took special steps to purify the ash-aid material used; in this case specially selected calcium carbonate precipitated by ammonia was treated by heating with perchloric acid and precipitation as hydroxide with sodium hydroxide. After copious washing the hydroxide was filtered and dried at 120 °C.

PROCEDURE

Weigh accurately 10 g (to 0·01 g) prepared food sample into a nickel dish, add 0·5 g prepared calcium hydroxide and 30 ml water, stir to distribute the hydroxide through the sample. Evaporate to dryness, char at 150 °C, place in stainless steel lined muffle furnace and raise the temperature to 600 °C at a rate of 100 °C per hour; ash at this temperature overnight. Cool and transfer the ash to a 200 ml distillation flask with 5 ml portions of water, add 200 mg silver sulphate and 40 ml sulphuric acid (1 + 1 with water), distil in a current of steam by maintaining the distillation flask at 150 °C, collect 200 ml distillate through a column containing 1 g of anion-exchange resin, after completion of the distillation the fluoride may be eluted with 0·2 M sodium acetate into a 25 ml flask. If the amount of fluoride to be distilled is greater than 20 μg the distillate may be collected directly in dilute sodium hydroxide. The fluoride electrode response of the distillate is measured after 1 + 1 dilution with a total ionic strength buffer solution using a reference electrode of Ag/AgCl wire.

Total ionic strength buffer. Dissolve 1 g 1,2-diaminocyclohexane-*N,N,N′,N′*-tetracetic acid in the minimum of dilute sodium hydroxide solution, add to a solution of 82 g sodium acetate and 58 g sodium chloride in 700 ml water, adjust to pH 5·8 with acetic acid and make up to 1 litre. Compare the response with that for a range of solutions containing 0·1–10 μg/ml fluoride.

Nicholson and Duff (1981) have summarised the precautions necessary for successful application of the ion-selective electrode procedure to measurement of fluoride in a variety of substrates, the parameters considered including temperature, illumination of the electrode during the measurement cycle, selection of a constant ionic strength buffer and storage of standard solutions. They showed that temperature dependence effects appeared to be a function of the age and usage of the electrode; some variable effects were also noted according to the illumination conditions of the test cell—ultraviolet light was shown to induce drift in the potential measurements. Beddows and Kirk (1981) have described a procedure for the direct measurement of the fluoride content of milk using the ion-selective electrode, the sample being treated with a solution of sodium citrate to precipitate milk protein prior to the measurement. Dabeka and McKenzie (1981) used a combination of microdiffusion from perchloric acid and selective ion-electrode measurement in the analysis of baby foods over the range 0·2–5 mg/kg. Comparative performance using a number of methods on a range of foods has been reported by Jacobson *et al.* (1982), who concluded that agreement between laboratories was poor and that further method development is needed. Singer and Ophang (1986) reviewed the available methods and reported on the relative performance of fluoride release techniques including ashing and diffusion methods in a market basket survey of foods sold in the USA; the ashing technique was considered to be the most reliable for estimation of total fluoride, with final measurement of fluoride by ion-selective electrode.

BROMINE

The amounts of bromine naturally occurring in foods are generally in the region of 1–5 mg/kg; added bromine may arise from the use of bromate flour improvers and as residue from the use of brominated hydrocarbon fumigants. The compounds most commonly used are methyl bromide and ethylene dibromide. Evidence of the use of fumigants may rest on the use of gas chromatographic techniques. In particular, headspace methods have been used in combination with electron capture detection

(McKay, 1985), confirmation by use of an electrolytic conductivity detector (Heikes, 1985), or use of GC/MS has been reported (Barry and Petzinger, 1985; Gilbert *et al.*, 1985b).

Methods for total bromine involve ashing in the presence of alkali hydroxide (Na or K) and ethanol or MgO. Bromine may be liberated from the ash with potassium dichromate–sulphuric acid and reacted with fluorescein to give the bromo-derivative, with absorbance at 515 nm giving the bromine content (Momokawa *et al.*, 1981). Alternatively, the ash may be treated with sulphuric acid in the presence of acetonitrile and ethylene oxide. The released bromine reacts with the ethylene oxide to give 2-bromoethanol, with the reaction product determined by GC with electron capture detection (Roughan *et al.*, 1983).

IODINE

Iodine is an essential micronutrient, approximately 8 mg iodine being found in the hormone secretions of the thyroid gland, which represents some 20–40 per cent of the total iodine content of the adult body. The level of iodine in plant and animal tissue is generally determined by the environment. Most soils contain low amounts of iodine and cereals, fruits, vegetables and meat products contain 20–50 μg/kg; in contrast, seafoods contain greater amounts by a factor of 10 to 100 and seaweed may contain up to 0·2 per cent. Foods may contain iodine derived from a variety of sources; in particular, the food colour erythrosin is an iodine-containing compound that adds significant amounts to foods such as canned cherries. The use of iodised salt is permitted in some countries, iodates were used as flour improvers and liquid detergents containing iodine are also used as industrial cleaning/sterilising agents. A review of these and other factors influencing the iodine content of foods has been given (Anon, 1981).

Methods for the determination of iodine are generally based on the dry ashing of the food material in the presence of alkali; the measurement may be completed either directly by release of iodine and extraction into carbon tetrachloride or by a catalytic colorimetric method such as that described by Sandell and Kolthoff (1934)—see Williams (1979) and Aumont and Tressol (1986). Moxon and Dixon (1980) reviewed the available methods for the determination of iodine in foods and described an automated procedure based on the catalytic destruction of thiocyanate by nitrite in the presence of iodide; typical data on some UK foods including fish is reported by these authors. Allegrini *et al.* (1981, 1983) reported data obtained in a total diet study carried out in the USA, the latter paper including data obtained by the use of neutron activation.

PESTICIDES

Legislative and regulatory aspects of the control of pesticides are complex issues, and within the UK the approach has evolved from that used in the control of poisons. This approach stems from the need to protect not only the consumer of foods, which may have been treated with biologically active chemicals, but also workers who may be in contact with the materials through the various stages of production, distribution and use. Part of the control procedure is also focused on the correct usage of materials by the agricultural and horticultural industries from the point of view of protection of the general environment. Thus specific formulations are targetted at specified crops at particular times in the production cycle and care is taken to ensure that dosages used are within control limits and that application dates are related to the eventual harvest date. The treatment of animals with agents to combat infestation and animal feed with antibacterial agents also falls within the broad scope of the classification pesticides. The use of pesticides in animal husbandry is controlled by the Veterinary Products Committee, acting under provisions of the Medicines Act 1968. Approved poducts are licensed. The Advisory Committee on Pesticides is responsible for pesticides to be used in agriculture, forestry, food storage, horticulture and timber treatment. The Control of Pesticides Regulations 1986 SI No. 1510, made under provisions of the Food and Environment Act 1985, sets out the control system for the use of pesticides, and The Pesticide (Maximum Residue Levels in Food) Regulations 1988 SI No. 1378 set limits for pesticide residues in foods. These regulations include maximum residue levels for 61 pesticides used on the most important components of the average national diet. The general provisions of the Food Act governing the sale of food that is injurious to health or not of the quality demanded by the consumer, however, still apply.

The philosophy of control by means of limits for specified foods is in line with that developed internationally by the FAO/WHO through the Codex Committee on Pesticide Residues and within the EC through directives on residues in fruit and vegetables (76/895/EEC OJ No. L340/26, as amended 80/428, 81/36 and 82/528), in cereals (86/362/EEC OJ No. L221/37), in animal products (86/363/EEC OJ No. L221/43) and an amendment to Directive 74/63/EEC on animal feeding stuffs (OJ No. L304/38).

From the consumers point of view the interest centres on the potential hazard and toxic effects due to the presence of residues. Materials accepted for use have been subjected to toxicological evaluation and acceptable daily intakes (ADI) are developed by the FAO/WHO Joint Meetings on Pesticide Residues. FAO/WHO (1984a–e, 1985a–d) has published *Guide to the Codex Recommendations Concerning Pesticide Residues* in nine parts, covering residue limits, classification, sampling and analysis. Codex publications are listed in Appendix 2. The Codex maximum limits for pesticide residues are also published as Volume XIII of the Codex Alimentarius (1986, 1988).

By nature the problem of analysis for evidence of pesticide residues is a complex one. The Codex guide lists 149 compounds for which residue limits are proposed or agreed. The amounts present on foods may vary from parts per billion (μg/kg) to several parts per million (mg/kg). In addition to the material used, consideration must also be given to the products of metabolism or chemical breakdown. For example, any crop spray reagents that contain heavy metals will leave residues of varying composition on the surface of the products and an estimate of the heavy metal constituent (As, Cu, Hg and Pb) would provide sufficient evidence of the level of use and acceptability of the product. Whilst toxic organophosphorus insecticides such as malathion will be quickly metabolised to less toxic materials, the organochlorine pesticides remain virtually unchanged and concentrate within the depot fat of treated animals, in wildlife at large and in humans. For this reason and their persistence in the general environment the organochlorine compounds, which were popular in the early years of the use of pesticides, are now regarded as less suitable as general pesticides. Nearly all have been withdrawn.

The combination of residual chemical and food matrix thus provides a multiplicity of analytical problems, the methodology for which is voluminous and far outside the scope of a generally based text. Successful residue analysis consists of a sequence of procedures that depend on the application of principles of good laboratory practice. Telling (1980) has summarised these principles in terms of the skills required of the analyst, essential resources and the principles to be used in selection and application of appropriate methods of sampling and analysis. These basic principles are also described in the publications from FAO/WHO cited above. In addition a number of specialist texts are available, e.g. Zweig (1963–1986), Maybury (1984), Mandava (1985), the *Pesticide Analytical Manual* of the FDA and the IUPAC recommended procedures (Ambrus and Thier, 1986). The Commission Decision (89/610/EEC) describes the reference methods to be used for detecting residues (OJ No. L351, 2.12.89, p. 39).

In general, methods applied are based on the use of solvent separation together with the application of chromatographic procedures and determination using GLC with selective detectors, HPLC and GC/MS. Dickes (1979) and Sherma (1982) reviewed GLC and TLC methods respectively, whilst Ebdon *et al.* (1986) have reviewed the uses of coupled GC/AAS as a multi-purpose and sensitive procedure for application to a wide range of materials and matrices.

The MAFF Committee on Analytical Methods (CAM) has published recommended methods for residues of malathion and dichlorvos in grain (CAM, 1973), volatile fumigants in grain (CAM, 1974), inorganic bromide in grain (CAM, 1976), organophosphorus pesticides in fruit and vegetables (CAM, 1977), organochlorine pesticides in animal fats and eggs (CAM, 1979), a range of organophosphorus pesticides in grain (CAM, 1980), dithiocarbamate pesticides in foodstuffs by a headspace technique (CAM, 1981), a range of organophosphorus pesticides in grain (CAM, 1985) and ethylene thiourea in canned food and vegetables (CAM, 1987).

ARTICLES IN CONTACT WITH FOOD

In the past, the risk of chemical contamination from materials in contact with food during manufacture, processing and packaging was considered to centre on adventitious contamination with lubricants and trace elements such as tin from tin plate, lead from solder and lead and cadmium from glazed enamelware. This latter topic is now covered by EC Directive 84/500. When they were first introduced, plastic packaging materials were thought to be relatively harmless and giving no problems of contamination of foods contained by them. Attention was first focused on polyvinyl chloride (PVC) on account of hazards posed to the health of factory workers in PVC production plants. In the 1970 Report on leaching of substances from packaging materials the FACC pointed to the need for factual information on the nature of materials used in the fabrication of plastics intended for use as food packing material and on the extent to which they might migrate into foods. The problem was initially seen as one of gross migration of materials, especially those that are fat soluble such as plasticisers and antioxidants. An EC General Directive 76/893 first set out a framework for the control of materials and articles intended to come into contact with food during all stages of manufacture, storage and distribution. The general directive proposed control on two fronts, namely the development of a permitted list of materials to be used in the fabrication of plastics and a parallel development of test systems for the assessment of global migration of constituents from the plastic when challenged with a range of test solvents. The provisions of the general directive were translated into UK legislation as the Materials and Articles in Food Regulations 1978 SI No. 1927. These regulations were couched in the language of the general provisions of the Food Act, namely that 'materials and articles in their finished state and intended for contact with food shall not transfer constituents which could endanger human health or alter the characteristics of the food in terms of their nature, substance or quality'. The framework regulations laid the pathway for the introduction of specific requirements as knowledge developed. Squirrell (1981) summarised industry methods for the examination of plastics materials for evidence of additives and process residues.

The problems of testing plastics for global migration and the interpretation of data obtained from food simulant studies have been discussed by many workers including Crosby (1981), Van Battum *et al.* (1982), Rossi (1971, 1981) and McGuinness (1986a, b). EC Directive 85/572 contains a list of food simulant materials for testing plastics and a classification of food to allow choice of appropriate test solvent.

Simulant A:
distilled water or water of equivalent quality.

Simulant B:
3% acetic acid (w/v) in aqueous solution.

Simulant C:
15% ethanol (v/v) in aqueous solution.

Simulant D:
rectified olive oil (Note 1); if for technical reasons connected with the method of analysis it is necessary to use different simulants, olive oil must be replaced by a mixture of synthetic triglycerides (Note 2), or by sunflower oil (Note 3).

NOTES

1. *Characteristics of rectified olive oil*

Iodine value (Wijs)	= 80–88
Refractive index at 25 °C	= 1·4665–1·4679
Acidity (expressed as per cent oleic acid)	= 0·5 per cent maximum
Peroxide number (expressed as oxygen milliequivalents per kg of oil)	= 10 maximum

2. *Composition of the synthetic triglycerides mixture*

Fatty acid distribution

Number of C-atoms in fatty acid residue	6	8	10	12	14	16	18	others
GLC area (per cent)	~1	6–9	8–11	45–52	12–15	8–10	8–12	≤1

Purity

Content of monoglycerides (enzymatically)	≤ 0·2 per cent
Content of diglycerides (enzymatically)	≤ 2·0 per cent
Unsaponifiable matter	≤ 0·2 per cent
Iodine value (Wijs)	≤ 0·1 per cent
Acid value	≤ 0·1 per cent
Water content (K. Fischer)	≤ 0·1 per cent
Melting point	28 ± 2 °C

Typical absorption spectrum (thickness of layer: $d = 1$ cm; reference: water = 35 °C)

Wavelength (nm)	290	310	330	350	370	390	430	470	510
Transmittance (per cent)	~ 2	~ 15	~ 37	~ 64	~ 80	~ 88	~ 95	~ 97	~ 98

At least 10 per cent light transmittance at 310 nm (cell of 1 cm, reference: water 35 °C)

3. *Characteristics of sunflower oil*
Iodine value (Wijs) = 120–145
Refractive index at 20 °C = 1·474–476
Saponification number = 188–193
Relative density at 20 °C = 0·918–0·925
Unsaponifiable matter = 0·5–1·5 per cent

Vinyl chloride

In parallel with the above developments the potential problems arising from the manufacture of PVC were focused on the amounts of residual plastic monomer remaining in the bulk plastic and hence available for transfer to foods when fabricated into containers. A survey of the content of vinyl chloride monomer in foods was carried out and reported by the Steering Group on Food Surveillance (MAFF, 1978), as a result of which and other work EC Directive 78/142 introduced a limit of 1 mg/kg for vinyl chloride in plastic intended to come into contact with food, and EC 80/766 laid down the method of analysis to be used to check plastics materials. A further directive, EC 81/432 contained a provision for testing of food that had been brought into contact with the plastic such that no evidence of transfer of vinyl chloride should be found on application of a method of analysis prescribed by the directive. The prescribed method, which uses headspace GLC, was stated to have a limit of detection of 0·01 mg/kg. The provisions of these directives were implemented in the UK by the Materials and Articles in Food (Amendment) Regulations 1980 and 1982 (SI 1980 No. 1838 and SI 1982 No. 1701).

The development of a method for examination of plastics materials and the results of EC-wide collaborative studies have been described by Rossi *et al.* (1980), the paper also containing full details of the finally agreed method. The procedure for analysis of foods and the results of a collaborative study of the method applied to orange drinks have been described by Biltcliffe and Wood (1982); in most cases the amounts recovered from spiked samples were in the region of 50 per cent of the known amount. The authors concluded that the method could be regarded as a satisfactory procedure for legislative purposes. Jeffs (1980) and Kolb (1984) have summarised various applications of the headspace GLC technique including the determination of vinyl chloride residues in foods. Gilbert *et al.* (1986) have summarised the current legislative position as it affects commercial practice in the assessment of PVC plastics for use as food containment material. A review of the uses of vinylidene chloride was also published by the Steering Group on Food Surveillance (MAFF, 1980a), whilst Gilbert *et al.* (1980) reported on the headspace GLC technique used.

Acrylonitrile

The use of acrylonitrile polymers and copolymers as food containers has been summarised in a review by the Steering Group on Food Surveillance (MAFF, 1982c). Levels of acrylonitrile monomer in resin and in food containers have dropped from 150 mg/kg in 1973 to 8 mg/kg in 1979. Analyses of samples of margarine stored in plastics tubs for acrylonitrile monomer were found to be in the general range 0·01–0·04 mg/kg. The method of analysis generally used for the examination of foods uses headspace GLC with nitrogen-specific detection; procedures used in the UK survey of containers and of samples of fats have been described by Crosby (1982) and by Gilbert and Startin (1982). Page and Charbonneau (1983, 1985) and Page (1985) have described the application of the method to a wider range of foods including honey, cheese, peanut butter and luncheon meats; limits of detection of 4–15 μg/kg were claimed according to sample type.

Styrene

A survey of the uses of polystyrene and related polymers as a packing or construction material and of the residual monomer levels in foods was carried out by the Steering Group on Food Surveillance (MAFF, 1983b). In most polymers and copolymers the minimal residual styrene content is generally 200 mg/kg, although levels of up to 700 mg/kg have been found in materials intended for food contact uses. Although in some cases there is transient contact of foods with thermal insulating containers, e.g. tea and coffee in 'take away' containers, there is evidence of transfer of styrene monomer at levels up to 45 μg/kg. Other foods which showed levels up to 200 μg/kg included yoghurt, dessert products and soft cheeses. The committee concluded that there was no evidence of a likely toxicological hazard from the levels found. Gilbert and Startin (1981) described the procedures used in the survey—headspace GLC coupled with mass spectrometry; Varner *et al.* (1983) have also described an azeotropic distillation procedure and headspace GLC for the analysis of margarines.

POLYNUCLEAR AROMATIC HYDROCARBONS

Contamination of foods with polynuclear aromatic hydrocarbons (PAH) may arise either from environmental pollution or as a result of certain food processing operations such as roasting, grilling or smoke curing. The latter operations may be regarded as being responsible for the generation and introduction of indirect additives to the foods concerned. Polynuclear aromatic hydrocarbons have been the subject of many studies related to possible effects on health due to the carcinogenicity of certain members of this class of compounds. Analytical reviews have been presented by Howard and Fazio (1980) and Fazio (1984); techniques used generally depend on the solubilisation of the food matrix with alcoholic potash followed by solvent extraction and partition between polar and non-polar solvents. Clean-up procedures involve the use of column chromatography and or TLC, and the compounds may be separated by GLC procedures using capillary columns with flame ionisation or mass spectrometric detection. Alternatively, the prepared extract may be examined by HPLC with variable wavelength fluorescence detectors.

Crosby *et al.* (1981) described an HPLC procedure for application to smoked foods, using direct extraction of the foods with acetonitrile, precipitation of protein with lead acetate and saponification of the filtrate with alcoholic potash as the first extraction stage. The solvent extract was subjected to clean-up on a silica gel (Sep-Pak) column followed by TLC on silica gel. Fractions from the TLC plate were then subjected to HPLC with two different bonded phases, octadecylsilane (ODS) and phthalimidopropylsilane (PPS), and methanol–water as mobile phase. Chromatograms were obtained by excitation of the column eluates at 290 nm and 282 nm and measurement of emission at 430 nm and 457 nm respectively. Recovery data for six PAH compounds and values found for bacon, kippers and cheese are given, with detection limits in the region of 0·02 μg/kg being claimed. Frank *et al.* (1984) used HPLC to examine a range of smoked fish and meat products. Lawrence and Weber (1984a) reviewed methods and used an HPLC procedure as a general screening technique, but they considered that capilliary column GLC/FID and GLC/MS were necessary as confirmatory techniques. Dennis *et al.* (1984) obtained data for fish, smoked fish, fish and meat products using the combination of techniques described earlier; HPLC was used to obtain data on processed dairy and vegetable products by Lawrence and Weber (1984b).

NITROSAMINES

The presence of trace amounts of nitrosamines in foods has long been recognised as consequential on the reaction between secondary amines and nitrites. Early work focused attention on the steam volatile nitrosamines, especially *N*-nitrosodimethylamine, which was shown to be the causal agent in the death of poultry fed on a diet based on fish meal. In subsequent animal feeding experiments simple nitrosamines have been shown to be powerful carcinogens; however, there is no clear evidence of a hazard to man (WHO, 1978; Gray, 1980; Havery and Fazio, 1985). The FACC (1978) reviewed the use of nitrates and nitrites as curing agents for meats and cheese, since in the early studies attention had been focused on the use of nitrites as curing agents for meats and meat products. The generation of traces of nitrosamines including *N*-nitrosopyrrolidine and *N*-nitrosodimethylamine in cooked bacon has also been recognised for many years (Sen *et al.*, 1979). However, recognition of the ubiquitous nature of nitrate in foods, particularly vegetables and water, and its ready conversion to nitrite in human saliva has developed a general awareness of human exposure to a range of nitroso compounds generated in the digestion of foods. Walters (1983) and Challis (1985) have summarised the sources of dietary nitrate and nitrite and interrelationships with *N*-nitrosamines derived from these sources.

The development of methods for the determination of the various nitrosamines has generated a voluminous literature. A general review of methods has been prepared by Hotchkiss (1981) and a monograph of methods (Egan, 1978) summarised the early approaches to standardised methods based on separation by distillation from a controlled pH, partition in dichlormethane followed by GLC and detection with a variety of detectors including flame ionisation, selective nitrogen detectors and, latterly, by the use of a chemiluminescent detector. In this detector the nitrosamine is cleaved catalytically to give nitric oxide, which exhibits luminescence in the presence of ozone. This latter procedure, also described by Walker *et al.* (1980), forms the basis of the method recommended by the AOAC for the determination of volatile nitrosamines in fried bacon. Reports of collaborative studies have been presented by Greenfield *et al.* (1982) who claimed the method to be appropriate for the estimation of six nitrosamines at levels varying from 5 to 17 μg/kg. Fiddler *et al.* (1984) proposed an alternate procedure based on dry column separation to that of vacuum distillation from mineral oil adopted by the AOAC. Sen *et al.* (1984)

reported on a collaborative study on non-fat dried milk, a detection limit of 0·3 μg/kg being claimed.

Attention has also been given to the determination of non-volatile nitrosamines. Walters *et al.* (1978) described a general nitroso group method which indicated the level of combined NO present in a sample but gave no evidence of the species present. A method based on the use of prep-column extraction procedures followed by HPLC of the eluate and detection with a thermal energy analyser has been described by Massey *et al.* (1982). Weston (1984) applied the prep-column/HPLC/TEA procedure to the analysis of cereals and cereal products for *N*-nitrosodimethylamine contents and reported values in the region of 1 μg/kg. Tricker *et al.* (1986) have summarised the various methods used for estimation of non-volatile nitrosamines and have described an enzyme release procedure for release of total nitroso compounds from adipose connective tissue of bacon. A general review of methods for volatile and non-volatile nitrosamines has been given by Scanlan and Reyes (1985).

ASBESTOS

Asbestos has been widely used as a material of construction valuable for its insulation and fire-resistant properties and has also been used as the basis of filtration cloth for the clarification of liquids including beverages. Concern has been expressed regarding health risks associated with exposure to asbestos fibres. The subject has been reviewed by FACC (1979), which recommended that all possible action should be taken to reduce contamination in beverages and food to a minimum.

Microscopy provides the most appropriate techniques for the examination for residues in foods, optical microscopy having been used for many years as an environmental screening method. However, it is not generally possible to detect fibres smaller then 2 μm × 0·3 μm, which is normally at the upper end of the range of fibres found in beverages and water. Scanning electron microscopy has a limit in the region of 50 nm, but evidence of fibre morphology may be derived from energy dispersive X-ray spectrometry. Transmission electron microscopy has a resolution of 0·2–0·5 nm and is able to discriminate fibres in the region of 30 nm length with some ease.

Sample preparation for beverages or water is to filter through a Millipore filter followed by ashing to remove organic debris and resuspension in water followed by deposition or a Millipore filter. The sample is coated with carbon using a vacuum evaporator and the support Millipore is removed prior to the microscopic analysis.

MYCOTOXINS

The mycotoxins are toxic compounds of diverse structure produced as metabolic by-products of mould growth. The number of compounds known to have been isolated is in excess of 400, but relatively few are toxic in mammals and fewer have been positively associated with human illness. Historically, the ergot alkaloids derived from the growth of *Claviceps purpurea* on rye and wheat have been recognised as harmful; more recently the by-products of the *Aspergillus* species, aflatoxins, have been implicated as causing liver damage. Heathcote and Hibbert (1978), Bullerman

(1979), Chapman (1983) and Gadsby and Turner (1984) have summarised the conditions leading to mould invasion of foods, the resultant contamination by mycotoxins, the current evidence on problem crops and guideline levels for control of contamination (see also FAO, 1978). In the past decade many surveys of foods for evidence of the presence of mould metabolites have been carried out in response to the incidence of mould spoilage in specific crops. The results obtained in co-ordinated surveys of UK foods were first published by MAFF (1980b), particular attention being paid to the possible contamination of human foods resulting from the consumption of aflatoxin-contaminated feeds by farm animals. Evidence was obtained of the contamination of milk and milk products through the use of mould-contaminated groundnuts as components of cattle feed. Further consideration was also given to the quality of cereals, pulses, dried fruits, fruit juices, edible nuts, meat and meat products and evidence of the presence of ochratoxins, zearalenone and patulin were reported. One of the more important observations in the study was that aflatoxin M, a metabolite of the native contaminant of groundnuts aflatoxins B and G, in milk products was seasonally related and that the incidence of higher levels could be associated with the use of compound feeds. The levels of ochratoxin in pig kidney was also shown to be related to the quality of feed given to the animals during their lifetime. One consequence of the study was the introduction of legal provisions governing the aflatoxin content of animal feeds by the Feedingstuffs Regulations (SI 1982 No. 1143) and the (Amendment) Regulations (SI 1984 No. 51) made under the Agriculture Act. These regulations set limits for aflatoxin B1 varying from 10 μg/kg to 50 μg/kg according to the type of feedingstuff; in addition, there is a specific limit on the aflatoxin B1 content of groundnut and groundnut derivatives of 50 μg/kg.

Further surveys of differing combinations of agricultural crop and mycotoxin contaminant have been published. These include: Gilbert *et al.* (1984) on aflatoxin M1 in milk produced in 1981–83; Kershaw (1985) on aflatoxins in imported edible nuts; Gilbert and Shepherd (1985) on aflatoxins in peanut butters, nuts and nut confectionery; Trenholm *et al.* (1983) on vomitoxin contamination of Ontario wheat in 1980; Gilbert *et al.* (1983) on vomitoxin in UK barley and imported maize in 1983; Osborne and Willis (1984) on tricothecene in UK wheat and imported wheat; and Morgan *et al.* (1986a) on ochratoxin in pigs kidney. As a result of the evidence collected in surveys considerable discussion has ensued on the need for legislatively enforced limits for mycotoxins. The FACC has proposed that a limit for total aflatoxins of 10 μg/kg is appropriate for control purposes in respect of nuts and nut products. The general provisions of the Food Act also govern the safety and quality of foods and specific mention is made of the prohibition of the sale of mouldy foods. This position compares with the provisions covering the trade in animal feeds indicated above.

Sampling and analysis

The problems associated with the enforcement of legislative limits for contaminants in foods such as mycotoxins are concerned as much with sampling as with the method employed in the analysis. Mould contamination is by nature non-homogeneous and hence the amount of mycotoxin is not uniformly distributed throughout the foodstuff. The animal feed regulations contain a sampling provision and a TLC procedure for analytical acceptance. Typical examples of sampling procedures for application to bulk groundnuts out of the shell have been developed by the Tropical Products Institute (TPI) in the UK and by the USDA in the USA. Coker (1984) has reviewed sampling procedures and has given details of the TPI plan together with a review of methods of analysis for the examination of a range of agricultural materials. In summary, the TPI plan demands the collection of a bulk sample of 10·5 kg of material in 105 g sub-samples randomly distributed throughout the batch. The bulk material is mixed and divided into three portions of 3·5 kg and the portions are ground and assayed sequentially for aflatoxin; if any of the 3·5 kg samples contains 10 μg/kg or less the batch is considered to contain less than 30 μg/kg. The USDA plan is more stringent, requiring a sample of 66 kg to be collected in at least 100 aliquots from 20 tonne lots. The sample is divided into three 22 kg samples labelled A, B and C and is analysed sequentially as follows: if sample A is under 16 μg/kg the lot is accepted, if over 75 μg/kg the lot is rejected, for any other value then sample B is examined. If the average value for samples A and B is less than 22 μg/kg the lot is accepted, if over 38 μg/kg the lot is rejected and if the average is in between these values sample C is examined. For acceptance, the average of samples A, B and C must be less than 25 μg/kg. This plan has a tolerance level of 25 μg/kg.

The AOAC devotes a complete chapter to analytical methods for mycotoxins. Screening procedures for aflatoxins are generally based on solvent extraction with chloroform and application of chromatographic methods. As indicated above, the animal feed method uses TLC on silica gel, with development using one of a variety of solvent mixtures such as chloroform(9)/acetone(1), chloroform(94)/methanl(6), diethyl ether(96)/methanol(3)/water(1); the aflatoxins are identified by examination under ultraviolet light and display of characteristic fluorescence.

Methods used in the food surveys indicated above are given in the individual papers, the trend being towards the use of HPLC with fluorescence detection or combined with the use of post-column derivatisation techniques (Hunt *et al.*, 1978a, b, 1979, 1980). Although attractive to many workers, the technique can be matched generally by TLC procedures for screening purposes; it does however offer confirmatory evidence to the screening methods. Schuller *et al.* (1976) reviewed sampling plans and collaboratively studied analytical procedures. General reviews of methods available have been presented by Lee and Goldblatt (1981), Lee and Skau (1981), Scott (1984) and Crosby (1984). A review of minicolumn techniques has been given by Holaday (1981) and generally recommended methods by Egan (1982). Also, attention has been given to the use of immunoassay methods (ELISA) and radioimmunoassay (RIA). Chu (1984) has concluded that sensitivities of 0·1 ng/kg for aflatoxin M1 in milk and 5 ng/kg of aflatoxin B1 in peanuts are attainable; the latter results were confirmed by Morgan *et al.* (1986b). Gilbert *et al.* (1985) have discussed the use of GLC/MS as a confirmatory technique for aflatoxins.

CATTY ODOURS AND TAINTS

Patterson (1968, 1969) has investigated the origin of a catty odour in a meat store, which had been recently painted with a polyurethane paint. Using GLC this was traced to 4-mercapto-4-methylpentan-2-one, which has an offensive catty odour at extreme dilution and is the addition product of the reaction between hydrogen sulphide from the meat and mesityl oxide, an impurity in a paint solvent. This reaction might also arise with other sulphur-containing foods such as eggs and vegetables. Aylward *et al.* (1967) have suggested that under appropriate conditions most of the common vegetables will have a catty taint on heating with a trace of mesityl oxide. The subject of catty odours in a wider range of foods has been the subject of review from BFMIRA (1969a, b).

Taints are essentially off-flavours caused by compounds present in trace amounts. They arise from a wide variety of causes, but more usually arise from contamination by adventitious contact with non-food chemicals during processing or storage. Musty taints in poultry or biscuits may be due to the presence of traces of chlorophenol compounds (Griffiths and Land, 1973). Saxby (1985) has reviewed the methods of detection and isolation that are available for taints in foods. Isolation may be effected by distillation or trapping and identification by gas–liquid chromatography. Kacprzak *et al.* (1984) reported on naphthalene contamination of bran and wheat germ, and Sevenants and Saunders (1984) described a 'medicinal' taint of cake mix due to iodocresol formed by reaction of iodised salt with cresol. A series of case studies on food taints derived from industrial processing has been given by Whitfield (1983), whilst Ramshaw (1984) has presented similar information relating to packaged foods. Whitfield *et al.* (1986) have described procedures for detection of chlorinated anisoles derived from packing using a GC/MS technique.

REFERENCES

Abbott, D. C. & Polhill, R. D. A. (1954) *Analyst*, **79**, 547.
Adeloju, S. B. & Brown, K. A. (1987) *Analyst*, **112**, 221.
Allegrini, M., Boyer, K. W. & Tanner, J. T. (1981) *Journal of the Association of Official Analytical Chemists*, **64**, 1111.
Allegrini, M., Pennington, J. A. T. & Tanner, J. T. (1983) *Journal of the American Dietetic Association*, **83**, 18.
Ambrus, A. & Thier, H. P. (1986) *Pure and Applied Chemistry*, **58**, 1035.
AMC (1960b) *Analyst*, **85**, 629.
AMC (1963) *Analyst*, **88**, 253.
AMC (1967) *Analyst*, **92**, 320.
AMC (1971a) *Analyst*, **96**, 384.
AMC (1971b) *Analyst*, **96**, 741.
AMC (1973a) *Analyst*, **98**, 458.
AMC (1973b) *Analyst*, **98**, 297.
AMC (1975a) *Analyst*, **100**, 54.
AMC (1975b) *Analyst*, **100**, 899.
AMC (1975c) *Analyst*, **100**, 761.
AMC (1976) *Analyst*, **101**, 62.
AMC (1977a) *Analyst*, **102**, 1219.
AMC (1977b) *Analyst*, **102**, 769.
AMC (1979) *Analyst*, **104**, 778.
AMC (1980) *Analyst*, **105**, 66.
AMC (1983) *Analyst*, **108**, 109.
Anon (1980) *Manuals of Food Quality Control.* FAO Food and Nutrition Paper 14/2. Rome: Food and Agriculture Organisation.
Anon (1981) *Bulletin Information Laboratory Coop*, **137**, 4.
Anon (1985) *Manual on Analyical Methods for Pesticide Residues in Foods.* Health Protection Branch, Health and Welfare, Canada, Ottawa.
APA (1971) *Journal of the Association of Public Analysts*, **9**, 65.
Aumont, G. & Tressol, J. C. (1986) *Analyst*, **111**, 841.
Awarez, G. H., Hight, S. C. & Capar, S. G. (1984) *Journal of the Association of Official Analytical Chemists*, **67**, 715.

Aylward, F., Coleman, G. & Haisman, D. M. (1967) *Chemistry and Industry*, 1563.
Barbieri, G., Rosso, S., Milanese, G. & Warwick, M. E. (1982) *Industria Conserve*, **57**, 151.
Barry, T. L. & Petzinger, G. (1985) *Journal of Food Safety*, **7**, 171.
Bearfield, D. W., Greenwood, E. A. & Warwick, M. E. (1983) *British Corrosion Journal*, **18**, 47, 51.
Beddows, C. G. & Kirk, D. (1981) *Analyst*, **106**, 1341.
BFMIRA (1969a) *Food Manufacture*, **44**, 54.
BFMIRA (1969b) *Food Trade Review*, **39**, 47.
Biltcliffe, D. O. & Wood, R. (1982) *Journal of the Association of Public Analysts*, **20**, 55.
Boline, D. R. & Schrenk, W. G. (1977) *Journal of the Association of Official Analytical Chemists*, **60**, 1170.
Bonilla, M., Rodriguez, L. & Camara, C. (1987) *Analytical Atomic Spectrometry*, **2**, 157.
Borus-Boszormenyi, N. (1980) *Nahrung*, **24**, 295.
Brooke, P. J. & Evans, W. H. (1981) *Analyst*, **106**, 514.
Bullerman, L. B. (1979) *Journal of Food Protection*, **42**, 65.
CAM (1973) *Analyst*, **98**, 19.
CAM (1974) *Analyst*, **99**, 570.
CAM (1976) *Analyst*, **101**, 386.
CAM (1977) *Analyst*, **102**, 858.
CAM (1979) *Analyst*, **104**, 425.
CAM (1980) *Analyst*, **105**, 515.
CAM (1981) *Analyst*, **106**, 782.
CAM (1985) *Analyst*, **110**, 765.
CAM (1987) *Analyst*, **112**, 1559.
Capar, S. G. (1977) *Journal of the Association of Official Analytical Chemists*, **60**, 1400.
Capar, S. G. & Subjoc, C. A. (1982) *Journal of the Association of Official Analytical Chemists*, **65**, 1025.
Capar, S. G., Gajan, R. J., Madzsar, E., Albert, R. H., Sanders, M. & Zyren, J. (1982) *Journal of the Association of Official Analytical Chemists*, **65**, 978.
Challis, B. C. (1985) *Proceedings of the Nutrition Society*, **44**, 95.
Chaney, A. L. & Magnuson, H. J. L. (1944) *Industrial and Engineering Chemistry* (Analytical Edition), **12**, 691.
Chapman, W. B. (1983) *Nutrition Bulletin, The British Nutrition Foundation*, **8**, 25.
Cheng, K. L. & Bray, R. H. (1953) *Analytical Chemistry*, **25**, 655.
Chu, F. S. (1984) *Journal of Food Protection*, **47**, 562.
Coker, R. D. (1984) In *Analysis of Food Contaminants*, ed. J. Gilbert. London: Elsevier Applied Science Publishers.
Coles, L. E. (1974) *Journal of the Association of Public Analysts*, **12**, 68.
Crosby, N. T. (1977) *Analyst*, **102**, 225.
Crosby, N. T. (1981) *Food Packaging Materials. Aspects of Analysis and Migration of Contaminants.* London: Applied Science.
Crosby, N. T. (1982) *Analytical Proceedings*, **19**, 428.
Crosby, N. T. (1984) *Food Additives and Contaminants*, **1**, 39.
Crosby, N. T., Hunt, D. C., Philp, L. A. & Patel, I. (1981) *Analyst*, **106**, 135.
Dabeka, R. W. (1984) *Analyst*, **109**, 1259.
Dabeka, R. W. & McKenzie, A. D. (1981) *Journal of the Association of Analytical Chemists*, **64**, 1021.
Dabeka, R. W., McKenzie, A. D. & Albert, R. H. (1985) *Journal of the Association of Analytical Chemists*, **68**, 209.
Dahlquist, R. L. & Knoll, J. W. (1978) *Applied Spectroscopy*, **32**, 1.
Dean, J. R., Ziang, L. & Snook, R. D. (1985) *Analyst*, **110**, 1395.
Dellar, D. (1983) *Analyst*, **108**, 759.
Dennis, M. J., Massey, R. C., McWeeney, D. J., Larsson, B., Eriksson, A. & Sahlberg, G. (1984) *Journal of Chromatography*, **285**, 127.
De Vries, J. W., Broge, J. M., Schroder, J. P., Bowers, R. H., Larson, P. A. & Burns, N. M. (1985) *Journal of the Association of Official Analytical Chemists*, **68**, 1112.
Dickes, G. J. (1979) *Talanta*, **26**, 1065.
Dixon, E. J. (1970) *Analyst*, **95**, 272.
Down, J. L. & Gorsuch, T. T. (1967) *Analyst*, **92**, 398.
Ebdon, L., Hill, S. & Ward, R. W. (1986) *Analyst*, **111**, 1113.
Egan, H. (ed.) (1978) *Environmental Carcinogens—Selected Methods of Analysis.* Vol. 1: Volatile nitrosamines in foods. Lyon: IARC Scientific Publications No. 18.
Egan, H. (1982) *Environmental Carcinogens. Selected Methods of Analysis of Some Mycotoxins.* Lyon: IARC Scientific Publication No. 44, Vol. 5.
Elkins, E. R. (1982) *Journal of the Association of Official Analytical Chemists*, **65**, 965.
Evans, W. H. & Caughlin, D. (1985) *Analyst*, **110**, 681.

Evans, W. H. & Dellar, D. (1982) *Analyst*, **107**, 977.
Evans, W. H. & Read, J. I. (1985) *Analyst*, **110**, 619.
Evans, W. H., Read, J. I. & Lucas, B. E. (1978) *Analyst*, **103**, 580.
Evans, W. H., Jackson, F. J. & Dellar, D. (1979) *Analyst*, **104**, 16.
Evans, W. H., Dellar, D., Lucas, B. E., Jackson, F. J. & Read, J. I. (1980) *Analyst*, **105**, 529.
Evans, W. H., Brooke, P. J. & Lucas, B. E. (1983) *Analytica Chimica Acta*, **148**, 203.
FACC (1970) *Report on the Leaching of Substances from Packaging into Food*. London: HMSO.
FACC (1978) *Report on the Review of Nitrates and Nitrites in Cured Meats and Cheese*, FAC/REP/27. London: HMSO.
FACC (1979) *An Examination of Asbestos in Relation to Food and Drink*. London: HMSO.
FACC (1983) *Report on the Review of Metals in Canned Foods*, FAC/REP/38. London: HMSO.
FACC (1984) *Report on the Review of the Arsenic in Food Regulations*, FAC/REP/39. London: HMSO.
Faires, L. M. (1983) *International Laboratory*, 14.
FAO/WHO (1972) *Evaluation of Certain Food Additives and the Contaminants Mercury, Lead and Cadmium*. 16th Report of the Joint Expert Committee on Food Additives. Geneva: WHO.
FAO (1978) *Mycotoxin Surveillance, A Guideline*. Food Control Series No. 4. Rome: FAO.
FAO (1984a) *Guide to Codex Recommendations Concerning Pesticide Residues*: Part 1—General Notes.
FAO (1984b) *Guide to Codex Recommendations Concerning Pesticide Residues*: Part 3—Guideline Levels for Pesticide Residues.
FAO (1984c) *Guide to Codex Recommendations Concerning Pesticide Residues*: Part 5—Recommended Method of Sampling.
FAO (1984d) *Guide to Codex Recommendations Concerning Pesticide Residues*: Part 6—Portion of Commodities to which Residue Limits Apply.
FAO (1984e) *Guide to Codex Recommendations Concerning Pesticide Residues*: Part 7—Codex Guidelines on Good Practice in Residue Analysis.
FAO (1985a) *Guide to Codex Recommendations Concerning Pesticide Residues*: Part 2—Maximum Limits for Pesticide Residues.
FAO (1985b) *Guide to Codex Recommendations Concerning Pesticide Residues*: Part 4—Codex Classification of Foods and Animal Feedstuffs.
FAO (1985c) *Guide to Codex Recommendations Concerning Pesticide Residues*: Part 8—Recommendations for Methods of Analysis.
FAO (1985d) *Guide to Codex Recommendations Concerning Pesticide Residues*: Part 9—Recommended National Regulatory Practices.
Farrer, K. T. H. (1985) *ASEAN Food Journal*, **1**, 3.
Fazio, T. (1984) *Food Science Technology*, **11**, 395.
Fiddler, W., Pensabene, J. W., Gates, R. A. & Phillips, J. G. (1984) *Journal of the Association of Official Analytical Chemists*, **67**, 521.
Frank, L. J., Salemme, J. & Fazio, T. (1984) *Journal of the Association of Official Analytical Chemists*, **67**, 1076.
Fricke, F. L., Robbins, W. B. & Caruso, J. A. (1979) In *Progress in Analytical Atomic Spectroscopy*, ed. C. L. Chakrabarti. Oxford: Pergamon Press.
Friend, M. T., Smith, C. A. & Wishart, D. (1977) *Atomic Absorption Newsletter*, **16**, 46.
FSC (1953) *Report on Zinc in Foods, REP/8/1953*. London: HMSO.
Gadsby, S. A. & Turner, C. M. (1984) *Journal of Environmental Health*, **92**, 114.
Gajan, R. J., Capar, S. G., Subjoc, C. A. & Sanders, M. (1982) *Journal of the Association of Official Analytical Chemists*, **65**, 970.
Garcia-Vargas, M., Milla, M. & Perez-Bustamante, J. A. (1983) *Analyst*, **108**, 1417.
Gilbert, J. & Shepherd, M. J. (1985) *Food Additives and Contaminants*, **2**, 171.
Gilbert, J. & Startin, J. R. (1981) *Journal of Chromatography*, **205**, 434.
Gilbert, J. & Startin, J. R. (1982) *Food Chemistry*, **9**, 243.
Gilbert, J., Shepherd, M. J., Startin, J. R. & McSweeney, D. J. (1980) *Journal of Chromatography*, **197**, 71.
Gilbert, J., Shepherd, M. J. & Startin, J. R. (1983) *Journal of the Science of Food and Agriculture*, **34**, 86.
Gilbert, J., Shepherd, M. J., Wallwork, M. A. & Knowles, M. E. (1984) *Food Additives and Contaminants*, **1**, 23.
Gilbert, J., Startin, J. R. & Crews, C. (1985a) *Journal of the Association of Public Analysts*, **23**, 119.
Gilbert, J., Startin, J. R. & Crews, C. (1985b) *Food Additives and Contaminants*, **2**, 55.
Gilbert, J., Startin, J. R. & McGuiness, J. D. (1986) *Food Additives and Contaminants*, **3**, 133.
Gillingham, J. T. (1985) *Analytical Proceedings*, **22**, 286.
Godden, R. G. & Thomerson, D. R. (1980) *Analyst*, **105**, 1137.
Gorsuch, T. T. (1959) *Analyst*, **84**, 135.
Gorsuch, T. T. (1970) *The Destruction of Organic Matter*. Oxford: Pergamon Press.

Gray, T. J. B. (1980) In *Developments in Food Preservatives—1*, ed. R. H. Tilbury, p. 53. London: Applied Science Publishers Ltd.
Greenfield, E. L., Smith, W. J. & Malanoski, A. J. (1982) *Journal of the Association of Official Analytical Chemists*, **65**, 1319.
Griffiths, H. M. & Land, D. C. (1973) *Chemistry and Industry*, 904.
Hahn, M. H., Wolnik, K. A., Fricke, F. L. & Caruso, J. A. (1982) *Analytical Chemistry*, **54**, 1048.
Halls, D. J., Mohl, C. & Stoepller, M. (1987) *Analyst*, **111**, 185.
Havery, D. C. & Fazio, T. (1985) *Food Technology*, **39**, 80.
Haxton, J., Lindsay, D. G., Hislop, J. S., Salmon, L., Dixon, E. J., Evans, W. H., Reid, J. R., Hewitt, C. J. & Jeffries, D. F. (1979) *Environmental Research*, **18**, 351.
Heathcote, J. G. & Hibbert, J. R. (1978) *Aflatoxins: Chemical and Biological Aspects.* London: Elsevier.
Heikes, D. L. (1985) *Journal of the Association of Analytical Chemists*, **68**, 431.
High, J. H. (1947) *Analyst*, **72**, 60.
Hight, S. C. & Capar, S. G. (1983) *Journal of the Association of Official Analytical Chemists*, **66**, 1121.
Hoffman, I. & Gordon, A. D. (1963) *Journal of the Association of Official Analytical Chemists*, **46**, 245.
Hofsommer, H. J. & Bielig, H. J. (1980) *Deutsches Lebensmittel Rundschau*, **76**, 419.
Hofsommer, H. J. & Bielig, H. J. (1981) *Zeitschrift fur Lebensmittel und Forschung*, **173**, 32.
Holaday, C. E. H. (1981) *Journal of the American Oil Chemists' Society*, **58**, 931.
Holak, W. (1980) *Journal of the Association of Official Analytical Chemists*, **63**, 485.
Holak, W. (1982) *Analyst*, **107**, 1457.
Holak, W. (1983) *Journal of the Association of Official Analytical Chemists*, **66**, 620.
Horwitz, W. (1979) *Journal of the Association of Official Analytical Chemists*, **63**, 1251.
Hotchkiss, J. H. (1981) *Journal of the Association of Official Analytical Chemists*, **64**, 1037.
Howard, J. W. & Fazio, T. (1980) *Journal of the Association of Official Analytical Chemists*, **63**, 1077.
Hunt, D. C., Bourdon, A. T., Wild, P. J. & Crosby, N. T. (1978a) *Journal of the Science of Food and Agriculture*, **29**, 234.
Hunt, D. C., Bourdon, A. T. & Crosby, N. (1978b) *Journal of the Science of Food and Agriculture*, **29**, 239.
Hunt, D. C., Philp, L. A. & Crosby, N. T. (1979) *Analyst*, **104**, 1171.
Hunt, D. C., McConnie, B. R. & Crosby, N. T. (1980) *Analyst*, **105**, 89.
Hunter, R. T. (1986) *Journal of the Association of Official Analytical Chemists*, **69**, 493.
Ichinoki, S. & Yamazaki, M. (1985) *Analytical Chemistry*, **57**, 2219.
Ihnat, M. & Thompson, B. K. (1980) *Journal of the Association of Official Analytical Chemists*, **63**, 814.
Isaac, R. A. & Johnson, W. C. (1975) *Journal of the Association of Official Analytical Chemists*, **58**, 436.
Izquierdo, A., Prat, M. D. & Aragones, L. (1981) *Analyst*, **106**, 720.
Jackson, F. J., Read, J. I. & Lucas, B. E. (1980) *Analyst*, **105**, 359.
Jacobson, J. S., Troiano, J. J., Cosentini, C. C. & Evans, J. (1982) *Journal of the Association of Official Analytical Chemists*, **65**, 1150.
Jagner, D. (1982) *Analyst*, **107**, 593.
James, T. (1983) *Journal of the Association of Official Analytical Chemists*, **66**, 128.
Jeffs, A. R. (1980) *Applied Headspace Gas Chromatography*, ed. B. Kolb, p. 95. London: Heyden.
Jelinek, C. F. (1982) *Journal of the Association of Official Analytical Chemists*, **65**, 942.
Jones, J. B. (1984) In *Analysis of Food Contaminants*, ed. J. Gilbert. London and New York: Elsevier Applied Science Publishers.
Jones, J. W., Gajan, R. J., Boyer, K. W. & Fiorino, J. A. (1977) *Journal of the Association of Official Analytical Chemists*, **60**, 826.
Jones, J. W. Capar, S. G. & O'Haver, T. C. (1982) *Analyst*, **107**, 353.
Jones, P., Barron, K. & Ebdon, L. (1985) *Analytical Proceedings*, **22**, 373.
Kacprzak, J. L., Geyer, R. & Cooper, C. (1984) *Journal of the Association of Official Analytical Chemists*, **67**, 106.
Kahn, H. L. & Shallis, J. E. (1968) *Atomic Absorption News*, **7**, 5.
Kapel, M. & Komaitis, M. E. (1979) *Analyst*, **104**, 124.
Katz, S. A. (1985) *International Biotechnology Laboratory*, **3**, 10.
Kershaw, S. J. (1985) *Journal of Food Technology*, **20**, 647.
Kirk, R. S. & Pockington, W. D. (1969) *Analyst*, **94**, 71.
Koh, T.-S. & Benson, T. H. (1983) *Journal of the Association of Official Analytical Chemists*, **66**, 918.
Kolb, B. (1984) In *Analysis of Food Contaminants*, ed. J. Gilbert. London and New York: Elsevier Applied Science Publishers.
Konishi, T. & Takahashi, H. (1983) *Analyst*, **108**, 827.
Koops, J., De Graf, C. & Westerbeek, D. (1984a) *Netherlands Milk Dairy Journal*, **38**, 223.
Koops, J., Klomp, H. & Westerbeek, D. (1984b) *Netherlands Milk Dairy Journal*, **38**, 1.
Langmyhr, F. J. (1979) *Analyst*, **104**, 993.
Lawrence, J. F. & Weber, D. F. (1984a) *Journal of Agricultural and Food Chemistry*, **32**, 789.

Lawrence, J. F. & Weber, D. F. (1984b) *Journal of Agricultural and Food Chemistry*, **32**, 794.
Le Blanc, P. J. & Jackson, A. L. (1973) *Journal of the Association of Official Analytical Chemists*, **56**, 383.
Lee, L. S. & Goldblatt, L. A. (1981) *Journal of the American Oil Chemists' Society*, **58**, 928A.
Lee, L. S. & Skau, D. B. (1981) *Journal of Liquid Chromatography*, **4**, 43.
Lloyd, B., Holt, P. & Delves, H. T. (1982) *Analyst*, **107**, 927.
Louie, H. W. (1983) *Analyst*, **108**, 1313.
Louie, H. W., Go, D., Fedczina, M., Judd, K. & Dalins, J. (1985) *Journal of the Association of Official Analytical Chemists*, **68**, 891.
MAFF (1971) *Survey of Mercury in Food.* London: HMSO.
MAFF (1972) *Survey of Lead in Food.* London: HMSO.
MAFF (1973a) *Survey of Mercury in Food. A Supplementary Report.* London: HMSO.
MAFF (1973b) *Survey of Cadmium in Food.* London: HMSO.
MAFF (1975) *Survey of Lead in Food. A Supplementary Report.* London: HMSO.
MAFF (1978) *Survey of Vinyl Chloride Content of PVC for Food Contact and of Foods. Report of the Steering Group of Food Surveillance.* Paper No. 2. London: HMSO.
MAFF (1980a) *Survey of Vinylidene Chloride Levels in Food Contact Materials and in Foods. Report of the Steering Group on Food Surveillance*, Paper No. 3. London: HMSO.
MAAF (1980b) *Survey of Mycotoxins in the United Kingdom. Report of the Steering Group on Food Surveillance*, Paper No. 4. London: HMSO.
MAFF (1981) *Survey of Copper and Zinc in Food. Report of the Steering Group on Food Surveillance*, Paper No. 5. London: HMSO.
MAFF (1982a) *Survey of Arsenic in Food. Report of the Steering Group on Food Surveillance*, Paper No. 8. London: HMSO.
MAFF (1982b) *Survey of Lead in Food. Second Supplementary Report. Report of the Steering Group on Food Surveillance*, Paper No. 10. London: HMSO.
MAFF (1982c) *Survey of Acrylonitrile and Methacrylonitrile Levels in Food Contact Materials and in Foods. Report of the Steering Group on Food Surveillance*, Paper No. 6. London: HMSO.
MAFF (1983a) *Survey of Cadmium in Food. Report of the Steering Group on Food Surveillance*, Paper No. 12. London: HMSO.
MAFF (1983b) *Survey of Styrene Levels in Food Contact Materials and in Foods. Report of the Steering Group on Food Surveillance*, Paper No. 11. London: HMSO.
MAFF (1985) *Survey of Aluminium, Antimony, Chromium, Cobalt, Indium, Nickel, Thallium, and Tin in Food. Report of the Steering Group on Food Surveillance*, Paper No. 15. London: HMSO.
Maher, W. A. (1983) *Analyst*, **108**, 939.
Mandava, N. B. (1985) *Handbook of Natural Pesticides: Methods.* Washington: Chemical Rubber Company.
Mannino, S. (1982) *Analyst*, **107**, 1466.
Mannino, S. (1983) *Analyst*, **108**, 1257.
Mannino, S. (1984) *Analyst*, **109**, 905.
Margler, L. W. & Mah, R. A. (1981) *Journal of the Association of Official Analytical Chemists*, **64**, 1017.
Massey, R. C., Crews, C. & McWeeney, D. J. (1982) *Journal of Chromatography*, **241**, 423.
Maybury, R. B. (1984) *Laboratory Manual for Pesticide Residues Analysis in Animal Products.* Pesticide Laboratory, Food Production and Inspection Branch, Agriculture Canada, Ottawa.
McGuiness, J. D. (1986a) *Food Additives and Contaminants*, **3**, 95.
McGuiness, J. D. (1986b) *Food Additives and Contaminants*, **3**, 103.
McKay, G. (1985) *Journal of the Association of Official Analytical Chemists*, **68**, 203.
McKinnon, A. & Scollary, G. (1986) *Analyst*, **111**, 589.
Michie, N. D., Dixon, E. J. & Bunton, N. G. (1978) *Journal of the Association of Official Analytical Chemists*, **61**, 795.
Middleton, G. & Stuckey, R. E. (1954) *Analyst*, **89**, 138.
Middleton, K. R. (1965) *Analyst*, **90**, 234.
Miles, J. P. (1982) *Journal of the Association of Official Analytical Chemists*, **65**, 1016.
Momokawa, H., Yamada, W., Sato, F. & Sakai, K. (1981) *Journal of Food Hygiene Society Japan*, **22**, 531.
Moncrieff, R. W. (1964) *Perfumery and Essential Oil Record*, 55, 205.
Moreno-Dominguez, T., Garcia-Mareno, C. & Marine-Font, A. (1983) *Analyst*, **108**, 505.
Morgan, M. R. A., McNereney, R., Chan, H. W. S. & Anderson, P. H. (1986a) *Journal of the Science of Food and Agriculture*, **37**, 475.
Morgan, M. R. A., Kang, A. S. & Chan, H. W. S. (1986b) *Journal of the Science of Food and Agriculture*, **37**, 908.
Moxon, R. E. D. & Dixon, E. J. (1980) *Analyst*, **105**, 344.
Muenz, H. (1984) *Lebensmittelchemie Gerichtlische Chemie*, **38**, 96.

Muys, T. (1984) *Analyst*, **109**, 119.
Nicholson, K. & Duff, E. J. (1981) *Analyst*, **106**, 985.
Ohlin, B. & Andersson, A. (1982) *Var Foda*, **34**, 229.
Osborne, B. G. & Willis, K. H. (1984) *Journal of the Science of Food and Agriculture*, **35**, 579.
Ottaway, J. M., Carroll, J., Cook, S., Littlejohn, D., Marshall, J. & Stephen, S. C. (1985) *Analytical Proceedings*, **22**, 192.
Page, B. D. (1985) *Journal of the Association of Official Analytical Chemists*, **68**, 776.
Page, B. D. & Charbonneau, C. F. (1983) *Journal of the Association of Official Analytical Chemists*, **66**, 1096.
Page, B. D. & Charbonneau, C. F. (1985) *Journal of the Association of Official Analytical Chemists*, **68**, 606.
Pahlavanpour, B., Pullen, J. H. & Thompson, M. (1980) *Analyst*, **105**, 274.
Patterson, R. L. S. (1968) *Chemistry and Industry*, 548.
Patterson, R. L. S. (1969) *Chemistry and Industry*, 48.
Peacock, C. J. & Singh, S. C. (1981) *Analyst*, **106**, 931.
Pierce, F. D. & Brown, H. R. (1977) *Analytical Chemistry*, **49**, 1417.
Preer, J. R., Stephens, B. R. & Bland, C. W. (1982) *Journal of the Association of Official Analytical Chemists*, **65**, 1010.
Ramshaw, E. H. (1984) *CSIRO Food Research Quarterly*, **44**, 83.
Read, J. I. & Collins, R. (1982) *Journal of the Association of Public Analysts*, **20**, 109.
Rees, D. I. (1978) *Journal of the Association of Public Analysts*, **16**, 71.
Reilly, C. (1985) *Food Additives and Contaminants*, **2**, 209.
Rose, D. A. (1983) *Analytical Proceedings*, **20**, 436.
Rossi, L. (1977) *Journal of the Association of Official Analytical Chemists*, **60**, 1282.
Rossi, L. (1981) *Journal of the Association of Official Analytical Chemists*, **64**, 697.
Rossi, L., Waibel, J. & Vom Bruck, C. G. (1980) *Food and Cosmetic Toxicology*, **18**, 527.
Roughan, J. A., Roughan, P. A. & Wilkins, J. P. G. (1983) *Analyst*, **108**, 742.
Rubio, S., Gomez-Hens, A. & Valcarcel, M. (1985) *Analyst*, **110**, 43.
Sandell, E. B. & Kolthoff, I. M. (1934) *Journal of the American Chemical Society*, **56**, 1426.
Sawyer, L D. (1985) *Journal of the Association of Official Analytical Chemists*, **68**, 68.
Saxby, M. J. (1985) *Food Manufacture*, **60**, 57.
Scanlan, R. A. & Reyes, F. G. (1985) *Food Technology*, **39**, 95.
Schaffner, R. M. (1981) *Food Technology*, **35**, 60.
Schuller, P. L. & Coles, L. E. (1979) *Pure and Applied Chemistry*, **51**, 385.
Schuller, P. L. & Egan, H. (1976) *Review of Methods of Trace Analysis and Sampling with Special Reference to Food—Cadmium, Lead, Mercury and Methylmercury*. Rome: FAO.
Schuller, P. L., Horwitz, W. & Stoloff, L. (1976) *Journal of the Association of Official Analytical Chemists*, **59**, 1315.
Scott, P. M. (1984) *Journal of the Association of Official Analytical Chemists*, **67**, 366.
Sen, N. P., Seaman, S. & Karpinsky, K. (1984) *Journal of the Association of Official Analytical Chemists*, **67**, 232.
Sen, N. E., Seaman, S. & Miles, W. F. (1979) *Journal of Agricultural and Food Chemistry*, **27**, 1354.
Seow, C. C., Abdul Rahman, Z. & Abdul Aziz, N. A. (1984) *Food Chemistry*, **14**, 125.
Sevenants, M. R. & Sanders, R. A. (1984) *Analytical Chemistry*, **56**, 293A.
Sherlock, J. C. & Smart, G. A. (1984) *Food Additives and Contaminants*, **1**, 277.
Sherma, J. (1982) *Journal of Liquid Chromatography*, **5**, 1013.
Silva, M. & Valcarcel, M. (1982) *Analyst*, **107**, 511.
Simpson, G. R. & Blay, R. A. (1966) *Food Trade Reviews*, **38**, 35.
Singer, L. & Ophang, R. H. (1986) *Journal of Agricultural and Food Chemistry*, **34**, 510.
Smith, R. M. & Yankey, L. E. (1982) *Analyst*, **107**, 744.
Squirrell, D. C. M. (1981) *Analyst*, **106**, 1042.
Strafford, N., Wyatt, P. F. & Kershaw, F. G. (1975) *Analyst*, **70**, 232.
Suddendorf, R. F., Wright, S. K. & Boyer, K. W. (1981a) *Journal of the Association of Official Analytical Chemists*, **64**, 657.
Suddendorf, R. F., Watts, J. O. & Boyer, K. W. (1981b) *Journal of the Association of Official Analytical Chemists*, **64**, 1105.
Tam, G. K. H. & Lacroix, G. (1982) *Journal of the Association of Official Analytical Chemists*, **65**, 647.
Taubinger, R. P. Wilson, J. R. (1965) *Analyst*, **90**, 629.
Telling, G. M. (1980) *Analytical Proceedings*, **17**, 25.
Thorne, J., Robertson, J. & Buss, D. H. (1978) *British Journal of Nutrition*, **39**, 391.
Ting, B. T. G., Nahapetian, A., Young, V. R. & Janghorbian, M. (1982) *Analyst*, **107**, 1495.
Trenholm, H. L., Cochrane, W. P., Cohen, H., Elliot, J. I., Farnworth, E. R., Friend, D. W.,

Hamilton, R. M. G., Standish, J. R. & Thompson, B. K. (1983) *Journal of the Association of Official Analytical Chemists*, **66**, 92.
Tricker, A. R., Perkins, M. J., Massey, R. C. & McWeeney, D. J. (1986) *Food Additives and Contaminants*, **3**, 153.
Tyson, J. F. (1984) *Analyst*, **109**, 313.
Tyson, J. F. (1985) *Analyst*, **110**, 419.
Uthe, J. F., Freeman, H. G., Johnson, J. R. & Michalik, P. (1974) *Journal of the Association of Official Analytical Chemists*, **57**, 1363.
Van Battum, D., Rijk, M. A. H. & Verspoor, R. (1982) *Food and Cosmetic Toxicology*, **20**, 955.
Varner, S. L., Breder, C. V. & Fazio, T. (1983) *Journal of the Association of Official Analytical Chemists*, **66**, 1067.
Vibhakar, S., Nagaraja, K. V. & Kapur, O. (1983) *Journal of the Association of Official Analytical Chemists*, **66**, 317.
Walker, E. A. Griciute, L., Castegnaro, M. & Borzsonyi, M. (1980) *N-Nitroso Compounds: Analysis, Formation and Occurrence.* IARC Scientific Publication No. 31. Lyon: IARC.
Walters, C. L. (1983) *Nutrition Bulletin, The British Nutrition Foundation*, **8**, 164.
Walters, C. L., Downes, M. J., Edwards, M. W. and Smith, P. L. R. (1978) *Analyst*, **103**, 1127.
Watson, C. A. (1981) *Analytical Proceedings*, **18**, 482.
Watson, C. A. (1984) *Trends in Analytical Chemistry*, **3**, 25.
Watson, C. A. (1986) *Analyst*, **111**, 1353.
Weston, R. J. (1984) *Journal of the Science of Food and Agriculture*, **35**, 782.
Whitfield, F. B. (1983) *CSIRO Food Research Quarterly*, **43**, 96.
Whitfield, F. B., Shaw, K. J. & Nguyen, T. H. L. (1986) *Journal of the Science of Food and Agriculture*, **37**, 85.
WHO (1978) *Monograph on the Evaluation of Risk of Chemicals to Humans.* IARC Scientific Publications No. 17. Lyon: IARC.
Williams, H. A. (1941) *Analyst*, **66**, 228.
Williams, W. J. (1979) *Handbook of Anion Determination*. London: Butterworths.
Wolf, W. R. (1982) *Clinical, Biochemical, and Nutritional Aspects of Trace Elements*, p. 427. New York: Alan R. Liss, Inc.
Wyatt, P. F. (1953) *Analyst*, **78**, 656.
Zink, E. W., Moffit, R. A. & Matson, W. R. (1983a) *Journal of the Association of Official Analytical Chemists*, **66**, 1409.
Zink, E. W., Davis, P. H., Griffin, R. M., Matson, W. R., Moffit, R. A. & Sakai, D. T. (1983b) *Journal of the Association of Official Analytical Chemists*, **66**, 1414.
Zweig, G. (1963–1986) (ed.) *Analytical Methods for Pesticides and Plant Growth Regulations*, Vols 1–15. London: Academic Press.

6

Sugars and preserves

This chapter covers 'sugars', i.e. food carbohydrates, and their products and derivatives, together with honey and preserves. Artificial sweeteners and bulk sweeteners are described in Chapter 4. Legislation on sugars and sugar products is described on p. 207 *et seq.*

FOOD CARBOHYDRATES

Properties

Table 6.1 gives some of the properties of the more important food carbohydrates:

Arabinose is a pentose that is widely distributed in plants and occurs to some extent in animal tissues, usually in the form of its anhydrides or pentosans, which yield arabinose (with xyloses) on hydrolysis. When distilled with dilute hydrochloric acid it yields furfural, which gives characteristic colour reactions with resorcinol or phloroglucinol. If other sugars are present arabinose can be estimated in the same manner as pentosans, but if no other reducing sugar is present, copper reduction methods can be employed.

Dextrose. Dextrose or glucose (both are used in this chapter) occurs in a variety of animal and vegetable foodstuffs. *d*-Glucose exists in the anhydrous states (m.p. 146 °C) and in the hydrated form (m.p. 86 °C), which contains one molecule of water. It is produced by the hydrolysis of starch and disaccharides such as sucrose and is, therefore, frequently present in foodstuffs which contain sugar. Dextrose readily reduces Fehling's and Barfoed's solutions and is fermented by yeast. It is converted by glucose oxidase to gluconic acid and hydrogen peroxide.

Fructose. Fructose, laevulose or fruit sugar is commonly present in fruits and is laevorotatory. It exists in considerable quantities in honey and invert sugar and can be produced enzymatically from glucose syrup. The specific rotation of fructose is considerably influenced by temperature. At 20 °C, $[\alpha]_D^{20}$ is −92° but the rotation decreases with rise in temperature and at 87 °C solution of invert sugar exhibits no rotation as the dextrorotation due to dextrose is then neutralised by the laevo-rotation due to the fructose. It also exhibits the phenomenon of mutarotation to a marked extent.

Galactose occurs in a polymerised form in many gums. Commercially, it is a product of the hydrolysis of lactose. It readily crystallises with one molecule of water in the form of a white powder having a m.p. of 120 °C (anhydrous form 167 °C). It exhibits mutarotation and has a high temperature coefficient.

Table 6.1 Properties of food carbohydrates

Name	Formula	*Specific rotation $[\alpha]_D^{20}$	Melting point (°C)	Reaction with Fehling's solution
Monosaccharides				
Arabinose	$C_5H_{10}O_5$	$+190{\cdot}6 \rightarrow +\mathbf{104{\cdot}5}$	160	+
Dextrose	$C_6H_{12}O_6$	$+112{\cdot}2 \rightarrow + \mathbf{52{\cdot}7}$	146	+
Fructose	$C_6H_{12}O_6$	$-132{\cdot}2 \rightarrow - \mathbf{92{\cdot}4}$	103	+
Galactose	$C_6H_{12}O_6$	$+151 \rightarrow + \mathbf{80{\cdot}2}$	167	+
Mannose	$C_6H_{12}O_6$	$-17{\cdot}0 \rightarrow + \mathbf{14{\cdot}2}$	132	+
Disaccharides				
Sucrose	$C_{12}H_{22}O_{11}$	$+ \mathbf{66{\cdot}53}$	160	–
Lactose (anhydrous)	$C_{12}H_{22}O_{11}$	$+34{\cdot}9 \rightarrow + \mathbf{55{\cdot}4}$	252	–
Lactose (hydrated)	$C_{12}H_{24}O_{12}$	$+85{\cdot}0 \rightarrow + \mathbf{52{\cdot}6}$	203	+
Maltose (hydrated)	$C_{12}H_{24}O_{12}$	$+111{\cdot}7 \rightarrow + \mathbf{130{\cdot}4}$	102	+
Trisaccharide				
Raffinose (hydrated)	$C_{18}H_{42}O_{21}$	$+\mathbf{105{\cdot}2}$	118	–
Polysaccharides				
Starch		$+\mathbf{195}$		–
Cellulose				–
Inulin		$-\mathbf{33}$ to $-\mathbf{40}$	178	–
Glycogen		$+\mathbf{196}$		–
Dextrin		$+\mathbf{195}$		–

* Changes due to mutarotation are also shown.

Mannose occurs in complex polysaccharides called mannosans.

Sucrose (saccharose, sugar, cane sugar or beet sugar) is present in a large number of foods. It does not reduce Fehling's solution. In common with other disaccharides it is hydrolysed by acid into monosaccharides. The resulting *invert sugar* contains equivalent amounts of dextrose and fructose. Sucrose is also hydrolysed by β-fructosidase (invertase) to dextrose and fructose.

Lactose, or milk sugar, is present in milk and has only a slightly sweet taste. It melts at 203 °C with decomposition and also occurs as a monohydrate, which is stable at 100 °C but decomposes at 130 °C. This is of interest in connection with the total solids of milk. Thus, although the hydrate is not completely dehydrated until 130 °C is reached, it seems probable the total solids obtained by drying at 100 °C include some anhydrous lactose. Lactose reduces Fehling's solution and is hydrolysed by acids and β-galactosidase yielding equal quantities of dextrose and galactose. The specific rotation of 55·4° relates to the equilibrium of γ form of the α-hydrate and β-anhydride which exists in the freshly prepared solution.

Maltose occurs naturally in plants, leaves and seeds, and in malt and malt extract. It is produced by the hydrolysis of starch and is therefore present in glucose syrup. It exists as a slightly sweet sugar forming a hydrate which is decomposed at or below 100 °C. Maltose is hydrolysed by acid at about 80 °C and by α-glucosidase (maltase) at pH 6·6 to yield two molecules of dextrose.

Raffinose is one of the few trisaccharides of interest to the food analyst. It occurs in beet sugar, molasses and cereals. It does not reduce Fehling's solution. On hydrolysis with strong acids it yields dextrose, fructose and galactose, but with weak acids fructose and the disaccharide melibiose are produced. It is hydrolysed by α-galactosidase (maltase) at pH 4·6 to galactose and sucrose.

Starch is the principal polysaccharide found in food. Its properties and methods for its determination are given in Chapter 9.

The following minor sugars, which may be found in small quantities in certain foods, are important because, for example, the human alimentary tract does not contain specific enzymes for their digestion; they therefore can lead to gastro-intestinal disturbances. Some affect the crystallisation of sucrose.

Monosaccharides	Xylitol, levoglucosan, rhamnose, maltol, ethyl maltol
Disaccharides	Isomaltose, gentobiose, trehalose, cellobiose, neohesperidose
Trisaccharides	Maltotriose, panose, neokestose
Tetrasaccharides	Stachyose

Sweetness of sugars

There are no instruments which measure sweetness. In any attempt to compare the sweetness of various materials it is necessary to rely upon the subjective reactions of individual taste senses. These taste senses depend on whether the body is in need of sugar and become more acute when the body is short of sugar. Reactions may also vary with the state of health. The taste senses are also influenced by the rate at which a material dissolves on the tongue and this is probably the reason that many people consider icing sugar to be sweeter than granulated sugar and the latter to be sweeter than cube sugar. Comparative sweetness is best measured when the sugar is at a low concentration so that the solution is only just sweet.

The apparently greater sweetness of impure sugars has often been attributed to the presence of organic acids or salt. It now seems more probable that the extra sweetness is due partly to fructose, which is over 1½ times as sweet as sucrose, and partly to the illusion created by the 'treacly' flavour of the sugar breakdown products. Mixture of sugars can also behave synergistically, having a greater sweetness than would be expected from the individual constituents.

The *relative sweetness* of some of the more common sugars has been stated as follows: fructose 180, invert sugar 120, sucrose 100, dextrose 70, galactose 32, maltose 32, lactose 16.

Van der Wel *et al.* (1987) have comprehensively reviewed the history of sweeteners, the various receptor theories of sweetness, neurophysiological mechanisms and structure–activity relations.

QUALITATIVE DETECTION OF SUGARS

General tests

Molisch's general test for carbohydrates. Add 2 drops of 20 per cent alcoholic solution of α-naphthol to 2 ml of a 0·1 per cent solution of the substance and pour 2 ml of conc. sulphuric acid down the side of the tube so that a layer forms below the aqueous solution. Carbohydrates, if present, produce a violet colour which is discharged on the addition of excess alkali. The reaction appears to be due to the formation of a condensation product between the α-naphthol and the hydroxymethylfurfural produced by decomposition of the carbohydrate.

Reducing sugars. When a solution of a reducing sugar is heated with an equal volume of mixed Fehling's solution in boiling water the cupric salt is reduced to red cuprous oxide. The reaction is given by lactose and maltose as well as the monosaccharides. Sucrose also gives a precipitate if the solution has been previously hydrolysed to glucose and fructose by heating with acid. Reducing sugars also give a silver mirror when heated with ammoniacal silver solutions.

Monosaccharides. Barfoed's reagent is reduced by monosaccharides, but not appreciably by lactose and maltose. The reagent is prepared by dissolving 6·5 g of crystallised copper acetate in 100 ml of 1 per cent acetic acid solution. For the test, heat 5 ml of the sugar solution with 5 ml of Barfoed's reagent in boiling water for three and a half minutes. Production of a red precipitate of cuprous oxide indicates the presence of a monosaccharide.

Fearon's test for lactose and maltose. To a dilute neutral solution of the sugar add 3 drops of methylamine hydrochloride solution (5 per cent), boil for a few seconds and then add 5 drops of sodium hydroxide solution (20 per cent). A yellow colour changing to carmine indicates the presence of lactose or maltose.

Rubner's test (modified) for lactose and maltose. To 2 ml of an aqueous solution of the sugar add 0·2 g lead acetate, boil and add twice the volume of dilute ammonia. Lactose gives a dense white precipitate, but maltose only a slight precipitate. Then boil gently for 1 minute. Lactose shows a salmon pink colour and maltose an orange yellow.

Paper chromatography

Paper chromatography represents a useful method for the identification of sugars in admixture, particularly where the amount of carbohydrate is small.

If possible, use solutions containing about 0·5 per cent of sugar. Render food extracts free from protein by adding trichloracetic acid solution and filtering. Apply the test and known sugar solutions to Whatman No. 4 paper, dry and employ any of the conventional methods of ascending chromatography. The composition of suitable solvents is given in Table 6.2. The R_F values increase if the proportion of pyridine

Table 6.2 Solvents for paper chromatography of sugars

	Parts by volume							
Solvent constituent	1	2	3	4	5	6	7	8
1 Ethanol	10	—	—	—	—	—	—	15
2 *n*-Propanol	—	65	—	—	—	—	—	—
3 isoPropanol	—	—	75	70	65	—	—	—
4 *n*-Butanol	40	—	—	10	—	45	—	60
5 Ethyl acetate	—	15	—	—	—	—	55	—
6 Pyridine	—	—	—	—	15	35	25	—
7 Ammonia, 0·88	1	—	—	—	—	—	—	—
8 Water	49	20	25	20	20	20	20	25

or water is raised, but butanol and ethyl acetate have the opposite effect. The R_F values of the different sugars are usually in the order pentoses (highest), hexoses, disaccharides, trisaccharides.

$$R_F \text{ value} = \frac{\text{distance moved by solute}}{\text{distance moved by solvent front}}$$

After running the chromatograms, mark the solvent front and dry the paper in a cold air draught. Pass the paper rapidly through a solution prepared by diluting 0·1 ml saturated aqueous silver nitrate to 20 ml with acetone and adding water, dropwise with shaking, until the solid that has separated redissolves. Then spray the dried paper with (or drip through) 0·5 M alcoholic sodium hydroxide (prepared by diluting saturated aqueous NaOH with ethanol). Reducing sugars give dense black spots very rapidly at room temperature and most compounds react within a few minutes. When reduction is complete immerse the paper for a few minutes in 4 M ammonia, wash in running water and dry at 100 °C. If necessary render the spots jet black by exposure to hydrogen sulphide. Silver nitrate is one of the most sensitive reagents for sugars, but it unfortunately reacts with other classes of compounds. Numerous other dipping and spraying reagents that are more selective in action have, however, been employed, such as the following:

1. *Diphenylamine.* Dissolve 1 g diphenylamine in 100 ml acetone and add 1 ml aniline and 6 ml syrupy phosphoric acid.
2. *Aniline phthalate.* Dissolve 0·9 g aniline and 1·6 g *o*-phthalic acid in 100 ml of water-saturated *n*-butanol.
3. *Naphthoresorcinol.* Dissolve 0·1 g naphthoresorcinol in 80 per cent ethanol containing 0·4 M HCl.
4. *Phloroglucin.* Dissolve 0·2 g phloroglucinol in 80 ml ethanol (90 per cent) and mix with 20 ml of 25 per cent trichloracetic acid.

The spray reagents described for TLC separations are also suitable for paper chromatograms.

Thin-layer chromatography

Thin-layer chromatogrphy is usually more rapid and precise than separations on paper. TLC methods for carbohydrates have been reviewed by Ghebregzabher *et al.* (1976). For TLC separations of sugars, the following solvents, spray reagents and plates are suitable (Dickes, 1965, 1966; Hansen, 1975a, b).

Development solvents (proportions by volume)

A. Acetone, *n*-butanol, water: 5, 4, 1.
B. Ethyl acetate, isopropanol, water: 6, 2, 1.
C. *n*-Propanol, water, ethyl acetate, 0·88 ammonia: 24, 12, 4, 1.
D. isoPropanol, acetone, 0·1 M lactic acid (9·0 g/l): 50, 40, 10.

Spray reagents

(a) Add 1 ml of saturated aqueous silver nitrate to 20 ml of acetone, with stirring. Add water dropwise until the solid just dissolves. After development and spraying, place the plates in saturated ammonia vapour for 15 min, then heat at 90 °C for 15 min. Spots show as dark areas on a light ground.

(b) Dissolve 5 g of phosphomolybdic acid in 95 ml of ethanol. After development and spraying, heat the plates at 90 °C for 10 min. Spots show blue–green on a yellow ground.

(c) Add an equal volume of 0·2 per cent ethanolic solution of 1,3-dihydroxynaphthalene to 20 per cent aqueous trichloracetic acid. After development and spraying, heat the plates at 90 °C for 15 min. Spots show blue–violet on a light ground.

(d) Add 1 ml of conc. H_2SO_4 to 5 ml of anisaldehyde dissolved in 50 ml of glacial acetic acid. After development and spraying, heat the plates at 100 °C for 10 min. Spots show blue–violet on a pink ground. This spray is not suitable for plates containing boric acid.

(e) Immediately before use mix 10 ml diphenylamine solution (4 per cent in ethanol) with 10 ml aniline solution (4 per cent in ethanol) and 2 ml orthophosphoric acid. After spraying, place the plate in an oven at 100 °C for 5 min. Remove, respray and return to the oven for a further 10 min.

TLC plates

A wide range of commercial pre-coated ready to use plates (on glass, plastic sheets or aluminium foil) is available, but suitable plates can be prepared as follows:

(a) Apply layers of Kieselgel G (Merck) to glass plates in the usual manner using 0·1 M boric acid solution instead of water.

(b) Apply layers of Kieselguhr G (Hopkin and Williams Ltd) to glass plates using 0·2 M sodium acetate solution instead of water.

(c) Apply Alusil layers of Kieselgel G and Aluminium oxide G in equal proportions.

(d) Silica gel F1500 (or equivalent) commercial plates.

NB Solvent C is recommended for use with Alusil layers (c) for separating maltose (R_F 0·30) from sucrose (R_F 0·39).

METHOD

Apply the test and standard sugar mixture solution about 2 cm from the edge of the thin-layer plate (e.g. 20 × 20 cm) using a micropipette. Add the mobile solvent to the tank lined with filter paper to cause saturation with the vapour of the solvent. After allowing the apparatus to equilibrate, develop the chromatogram to a distance of about 10 cm. The time required for the run is about 40 min. Then dry and spray with one of the reagents above.

The R_F values reported by Dickes for five sugars using solvent A on layer (a) and using Hansen's development solvent D and spray (e) are shown in Table 6.3. With solvent B and layer (b) there is reasonably good separation between different groups of sugars, but differentiation between members of the group is difficult. Inactivation of Kieselgel G is achieved by the incorporation of water in A and lactose and maltose can then be separated on layers impregnated with boric acid.

Table 6.3 R_F values of some sugars using thin-layer chromatography

Sugar	R_F value with solvent A on Kieselgel G impregnated with boric acid	R_F value with solvent D on Silica gel F 1500 plates	Colour with spray (e)
Fructose	0·19	0·41	Red
Lactose	0·34	0·18	Blue–green
Glucose	0·47	0·38	Dark brown
Maltose	0·53	0·25	Blue
Sucrose	0·56	0·31	Light brown
Oligosaccharides in glucose syrups		<0·18	Blue

Gas–liquid chromatography and high performance liquid chromatography

Sugars can be separated and detected qualitatively by both GLC and HPLC but as this is usually carried out as a part of a quantitative determination, procedures used are in the following section.

DETERMINATION OF SUGARS

Introduction

Methods available for the quantitative determination of sugars in foods are based mainly on refractometry, hydrometry, polarimetry, copper reduction, ion-exchange chromatography, HPLC, GLC and spectrophotometry based on enzymatic and colour reactions. The method to be used will depend on several factors, for example the type and number of samples, the accuracy, precision, sensitivity and type of information required, the time, apparatus and quality of staff available.

When pure syrups of known composition are being examined, the refractometer affords a rapid and moderately accurate method of estimating the sugar content. The specific gravity is also of value but is less accurate when measured with a hydrometer. Table 6.4 shows the specific gravities and refractive indices of solutions of sucrose at

Table 6.4 Table of specific gravities* and refractive indices† at 20 °C of solutions of sucrose

Sucrose % m/m	Specific gravity at 20/20 °C	Refractive index n_D^{20}	Sucrose % m/m	Specific gravity at 20/20 °C	Refractive index n_D^{20}
0	1·00000	1·33299	51	1·23727	1·42219
1	1·00389	1·33443	52	1·24284	1·42432
2	1·00779	1·33588	53	1·24844	1·42646
3	1·01172	1·33733	54	1·25408	1·42862
4	1·01567	1·33880	55	1·25976	1·43080
5	1·01965	1·34027	56	1·26548	1·43299
6	1·02366	1·34176	57	1·27123	1·43520
7	1·02770	1·34326	58	1·27703	1·43742
8	1·03176	1·34477	59	1·28286	1·43966
9	1·03586	1·34629	60	1·28873	1·44192
10	1·03998	1·34783	61	1·29464	1·44420
11	1·04413	1·34937	62	1·30059	1·44649
12	1·04831	1·34093	63	1·30657	1·44879
13	1·05252	1·35250	64	1·31260	1·45112
14	1·05677	1·35408	65	1·31866	1·45346
15	1·06104	1·35567	66	1·32476	1·45581
16	1·06534	1·34728	67	1·33090	1·45819
17	1·06968	1·34890	68	1·33708	1·46058
18	1·07404	1·36053	69	1·34330	1·46299
19	1·07844	1·36218	70	1·34956	1·46541
20	1·08287	1·36384	71	1·35585	1·46786
21	1·08733	1·36551	72	1·36218	1·47032
22	1·09183	1·36719	73	1·36856	1·47279
23	1·09636	1·36888	74	1·37496	1·47529
24	1·10092	1·37059	75	1·38141	1·47780
25	1·10551	1·3723	76	1·38790	1·48033
26	1·11014	1·3740	77	1·39442	1·48288
27	1·11480	1·3758	78	1·40098	1·48544
28	1·11949	1·3775	79	1·40758	1·48803
29	1·12422	1·3793	80	1·41421	1·49063
30	1·12898	1·3811	81	1·42088	1·49325
31	1·13378	1·3829	82	1·42759	1·49589
32	1·13861	1·3847	83	1·43434	1·49854
33	1·14347	1·3865	84	1·44112	1·50121
34	1·14837	1·3883	85	1·44794	1·50391
35	1·15331	1·3902	86	1·45480	
36	1·15828	1·3920	87	1·46170	
37	1·16329	1·3939	88	1·46862	
38	1·16833	1·3958	89	1·47559	
39	1·17341	1·3978	90	1·48259	
40	1·17853	1·3997	91	1·48963	
41	1·18368	1·4016	92	1·49671	
42	1·18887	1·4036	93	1·50381	
43	1·19410	1·4056	94	1·51096	
44	1·19936	1·4076	95	1·51814	
45	1·20467	1·4096	96	1·52535	
46	1·21001	1·4117	97	1·53260	
47	1·21538	1·4137	98	1·53988	
48	1·22080	1·4158	99	1·54719	
49	1·22625	1·4179	100	1·55454	
50	1·23174	1·42008			

* National Bureau of Standards; cf. Savage (1972). † International Scale 1936–1959 (Anon, 1937a).

20 °C. More extensive tables can be found in ICUMSA methods (Schneider, 1979) and in the AOAC methods book. Temperature corrections for the refractometer method are given in Table 6.5, and corrections for other dissolved substances are given in Table 6.6. The soluble solids as determined from the refractive index and

Table 6.5 Corrections for determining the percentage of sucrose in sugar solutions by refractometer when readings are made at temperatures other than 20 °C.

Temperature °C	PERCENTAGE SUCROSE 0	5	10	15	20	25	30	40	50	60	70
	Subtract from the percentage sucrose										
10	0·50	0·54	0·58	0·61	0·64	0·66	0·68	0·72	0·74	0·76	0·79
11	0·46	0·49	0·53	0·55	0·58	0·60	0·62	0·65	0·67	0·69	0·71
12	0·42	0·45	0·48	0·50	0·52	0·54	0·56	0·58	0·60	0·61	0·63
13	0·37	0·40	0·42	0·44	0·46	0·48	0·49	0·51	0·53	0·54	0·55
14	0·33	0·35	0·37	0·39	0·40	0·41	0·42	0·44	0·45	0·46	0·48
15	0·27	0·29	0·31	0·33	0·34	0·34	0·35	0·37	0·38	0·39	0·40
16	0·22	0·24	0·25	0·26	0·27	0·28	0·28	0·30	0·30	0·31	0·32
17	0·17	0·18	0·19	0·20	0·21	0·21	0·21	0·22	0·23	0·23	0·24
18	0·12	0·13	0·13	0·14	0·14	0·14	0·14	0·15	0·15	0·16	0·16
19	0·16	0·06	0·06	0·07	0·07	0·07	0·07	0·08	0·08	0·08	0·08
	Add to the percentage sucrose										
21	0·06	0·07	0·07	0·07	0·07	0·08	0·08	0·08	0·08	0·08	0·08
22	0·13	0·13	0·14	0·14	0·15	0·15	0·15	0·15	0·16	0·16	0·16
23	0·19	0·20	0·21	0·22	0·22	0·23	0·23	0·23	0·24	0·24	0·24
24	0·26	0·27	0·28	0·29	0·30	0·30	0·31	0·31	0·31	0·32	0·32
25	0·33	0·35	0·36	0·37	0·38	0·38	0·39	0·40	0·40	0·40	0·40
26	0·40	0·42	0·43	0·44	0·45	0·46	0·47	0·48	0·48	0·48	0·48
27	0·48	0·50	0·52	0·53	0·54	0·55	0·55	0·56	0·56	0·56	0·56
28	0·56	0·57	0·60	0·61	0·63	0·63	0·63	0·64	0·64	0·64	0·64
29	0·64	0·66	0·68	0·69	0·71	0·72	0·72	0·73	0·73	0·73	0·73
30	0·72	0·74	0·77	0·78	0·79	0·80	0·80	0·81	0·81	0·81	0·81

International Temperature Correction Table 1936 (Anon, 1937b)

Table 6.6 Corrections to allow for other dissolved substances

Substance	Percentage in sample	Corrections to be applied to the total soluble solids from the direct refractometric readings %
Invert sugar	20	+0·45
	40	+0·90
	60	+1·35
Glucose solids	20	−0·25
	40	−0·5
Citric acid	1	+0·1
	2	+0·2
	3	+0·3

expressed as sucrose by reference to the tables is most useful industrially for routine control purposes. Data relating to the physical properties of sugar solutions is given by Norrish (1967).

In the routine laboratory analysis of foods, the traditional polarimetric and copper-reduction methods coupled with paper chromatography or TLC can provide most of the information required. However, modern technology and legislative developments often require more specific and sensitive analyses, and these can be supplied by GLC, enzymatic and especially HPLC methods. Laboratories that deal with a large number of samples may also require automatic methods, such as those based on ion-exchange chromatography (Dean, 1978) or enzyme reactions. Southgate (1976) and Lee (1978) have surveyed methods of analysis for sugars and other carbohydrates in foods; Iverson and Bueno (1981) have compared GLC and HPLC methods and Reyes *et al.* (1982) have compared enzymatic, GLC and HPLC methods for sugars in strawberries.

CLEARING AGENTS

For accurate results (particularly in polarimetry) it is often necessary to clarify sample solutions or extracts using one of the following reagents:

Alumina cream is prepared by adding a slight excess of ammonia to a saturated solution of alum. It is either used unwashed or washed by repeated decantation.

Basic lead acetate is efficient, but it should be used in moderation, and it is preferable to add alumina cream afterwards in order to reduce the tendency to absorb sugars (especially fructose). Excess lead should be removed by the addition of sodium sulphate solution.

Neutral lead acetate is less efficient as a clearing agent, but it does not appear to absorb sugars.

Zinc ferrocyanide is a good clearing agent (see p. 297).

Animal charcoal tends to absorb sugars.

Other clearing agents that are used include phosphotungstic acid and sodium tungstate. It should be borne in mind that a correction for the volume of precipitate may be necessary with some samples and when greater accuracy is required. It is advisable to reject the first runnings after filtration of the solution that has been treated with the clearing agent. The colour in food extracts can often be removed or very much reduced by shaking the solution with a little polyamide powder or polyvinylpoly pyrrolidone and filtering (BCL, 1987).

Polarimetric methods

Optical rotation is the angle through which the vibrational plane of linear polarisation of light is rotated when the polarised light passes through a solution containing an optically active compound. Unless otherwise specified, the measurement is carried out with sodium light on a layer 1 dm thick at a temperature of 20 °C.

The *specific rotation* (or specific rotatory power) of the optically active compound measured using the sodium D line (589·44 nm) can be expressed by the formula:

$$\text{Specific rotation at } 20\,^{\circ}\text{C} = [\alpha]_{\text{D}}^{20} = \frac{100\alpha}{lc}$$

where α = observed angular rotation, l = length of tube in decimetres, and c = concentration of the compound in solution in g/100 ml.

Table 6.7 gives the formulae for calculating the specific rotations for various sugars. The specific rotations at 20 °C for various sugars when in solution at various concentrations are given in Table 6.8.

The following specific rotations of other carbohydrates when in solution of about 10–15 per cent concentration at 20 °C using sodium light are sufficiently accurate for most purposes: arabinose +104·5°, galactose +80·2°, raffinose +105·2°, xylose +18·8°, potato starch, dextrin +185·7° (for other starches, see p. 297).

Mutarotation. When a sugar solution is freshly prepared it has a different rotation from that of the same solution after standing some hours. This is due to the gradual

Table 6.7 Formulae for calculating specific rotations

Sugar	Formula
Sucrose	$[\alpha]_D^{20} = 66{\cdot}462+0{\cdot}0087c-0{\cdot}000\,235c^2$
Dextrose	$[\alpha]_D^{20} = 52{\cdot}50+0{\cdot}0188p+0{\cdot}000\,517p^2$
Dextrose	$[\alpha]_{5461A}^{20} = 62{\cdot}032+0{\cdot}042\,57c$
Dextrose	$[\alpha]_{5461A}^{20} = 62{\cdot}032+0{\cdot}0422p+0{\cdot}000\,189\,7p^2$
Fructose	$[\alpha]_D^{20} = -113{\cdot}96+0{\cdot}258q$
Maltose	$[\alpha]_D^{20} = 138{\cdot}475-0{\cdot}018\,37p$
Lactose	$[\alpha]_D^{20} = 52{\cdot}53$
Invert sugar	$[\alpha]_D^{20} = -(19{\cdot}415+0{\cdot}070\,65c-0{\cdot}000\,54c^2)$
Invert sugar temperature correction	$[\alpha]_D^t = [\alpha]_D^{20} + (0{\cdot}283+0{\cdot}001\,4c)(t-20\,°C)$

c = concentration in grams of the sugar per 100 ml.
p = percentage by weight.
q = percentage of water.
t = temperature (°C).
National Bureau of Standards (1942).

Table 6.8 Specific rotations of food sugars at various concentrations

	Specific rotations $[\alpha]_D^{20}$ at the following concentrations			
Sugar	10%	15%	20%	25%
Dextrose	+52·7°	+52·9°	+53·1°	+53·5°
Fructose	−90·7°	−92·0°	−93·3°	−94·6°
Invert sugar*	−19·8°	−20·2°	−20·4°	−20·6°
Lactose (hydrated)	+52·5°	+52·5°	+52·5°	+52·5°
Maltose	+138·3°	+138·2°	+138·1°	+138·0°
Sucrose	+66·5°	+66·5°	+66·5°	+66·5°

*Invert sugar (equimolecular mixture of dextrose and fructose) has a zero rotation at 87 °C

attainment of equilibrium between isomers, which individually have different rotations. Errors due to mutarotation, as this phenomenon is known, can be eliminated by allowing the solution to stand for several hours before making the observations or by adding a few drops of ammonia, which establishes the optical equilibrium at once.

Polarimeters and saccharimeters

The polarimeter is an instrument for measuring the rotation of plane polarised light caused by a solution containing an optically active compound. The instrument consists basically of a light source, a polarising prism to produce plane polarised light, a cell or tube for the sample solution, a rotatable analysing prism to detect the optical rotation caused by the sample solution and a graduated scale to measure the angular rotation. In order to make accurate visual measurement of the degree of rotation, analytical polarimeters incorporate a half-shade device by which matching of two half-fields gives the balance point. The light source is commonly a sodium or mercury vapour lamp to give monochromatic light, and a filter cell containing potassium dichromate is used to eliminate the small amount of general background.

Saccharimeters are similar to polarimeters but have the analysing prism permanently crossed with respect to the polarising prism. Rotation of the polarised light by the sample is measured by compensating wedges of quartz which rotate the light in the opposite direction. These wedges are located between the sample tube and the analysing prism. Moving one of the wedges in or out of the light path until half-fields are matched then corresponds to the rotation caused by the sample. The light source is usually a tungsten lamp. Figure 6.1 shows diagramatically the optical arrangement

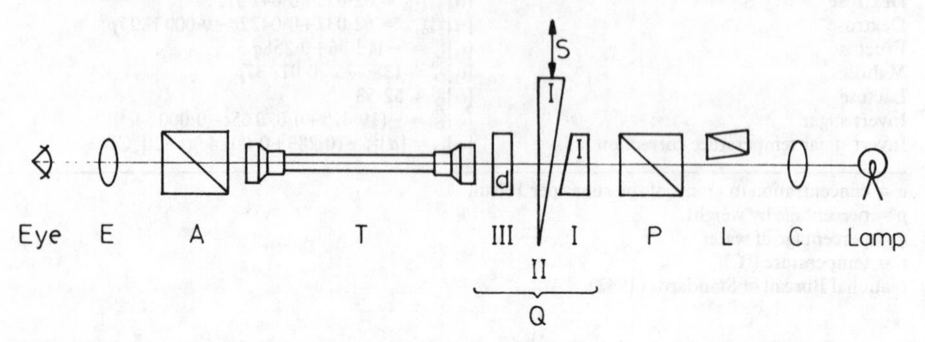

Fig. 6.1 Schematic diagram of a saccharimeter. Key: C, collimating lens; L, Lippich half-shade prism; P, polarising prism; Q, quartz-wedge compensator; T, sample solution tube; A, analyser prism (fixed); E, eyepiece; S, scale and compensator adjustment.

of a saccharimeter. Saccharimeters are also different from polarimeters in that they are calibrated in percentages of sugar, i.e. a stated 'normal weight' of sucrose in 100 ml of solution gives a rotation of 100 divisions on the sugar scale. As various sugar scales had been proposed using different normal weights (Ventzke and Schmidt Haensch scales 26·05 g, i.e. 1° = 0·3466° angular; Laurent and Soleil-Duboscq scales 16·3 g, i.e. 1° = 0·2167° angular), International Commissions in 1900, 1912 and 1932 formulated certain proposals. At a meeting of the Amsterdam Commission in 1932 it was recommended that new saccharimeters be calibrated as degrees sugar (°S) on an 'International Sugar Scale' (ISS):

'It is recommended that the polarisation of the normal solution (26·000 g of pure sucrose dissolved in 100 ml, and polarised at 20 °C in a 200 mm tube, using white light and the dichromate filter as defined by the Commission) be accepted as the basis of calibration of the 100° point on the International Sugar Scale' (1° = 0·34620° angular). The basis of the ISS was the observed angular rotation of 40·765° at 546·227 nm (mercury green line). Newer measurements have shown the rotation of the normal solution to be 40·777° and therefore ICUMSA (1986) have adopted from 1 July 1988 this angular rotation as fixing the 100°Z point on the new ISS, i.e.

$$100°S = 99{\cdot}971°Z$$
$$°Z = 0{\cdot}999\,71°S$$

Using the sodium D line (589·44 nm), 100°Z is 34·626° angular, i.e.

$$°Z = 0{\cdot}346\,26° \text{ angular.}$$

For the purposes of this edition, the °S notation for saccharimeter readings is retained.

In using the relevant formulae, readings must be calculated back so that they are equivalent to those given by a 26 per cent solution using a 200 mm tube.

The accuracy and precision of polarimeters and saccharimeters are very dependent on the eye of the operator. Semi- or fully-automatic instruments using photo-cells have been developed to overcome the visual reading errors. Bellingham and Stanley manufacture a polarimeter with manual setting of an electronic null circuit by zeroing a meter to obtain the readings. A small fully-automatic instrument with electronic digital display and an accuracy of 0·01° angular or 0·02°S, is available from Optical Activity Ltd.

Polarimetric determination of a single sugar. If sucrose is the only optically active substance present, the reading on the saccharimeter of a 26 per cent solution of the sample in a 200 mm tube at 20 °C gives the percentage of sucrose in the sample. For other samples containing another single sugar the formula is:

$$\text{Sugar A(per cent)} = {}^{\circ}\text{S} \times \frac{[\alpha]_D^{20} \text{ for sucrose}}{[\alpha]_D^{20} \text{ for sugar A}}$$

Using a polarimeter reading in angular degrees, prepare a 20 per cent clarified solution of the sample and measure the angular rotation (α) at 20 °C in a 200 mm tube:

$$\text{Sugar B(per cent)} = \frac{\text{'specific rotation' of sample solution}}{\text{Specific rotation of sugar B}} \times 100$$

$$= \frac{\alpha}{[\alpha]_D^{20} \text{for sugar B}} \times \frac{100}{2} \times \frac{100}{20}$$

Polarimetric determination of sucrose in the presence of other sugars. For the estimation of sucrose in the presence of invert sugar in molasses, syrups or similar totally water-soluble sugar products, the Clerget–Hertzfeld double polarisation method is often employed. This depends on the change from dextrorotation to laevorotation when sucrose is hydrolysed by acids or enzymes to a mixture of equal parts of dextrose and fructose.

The AOAC methods book procedure using the saccharimeter is essentially as follows:

Pipette 50 ml clarified 26 per cent m/v solution of the sample into a 100 ml volumetric flask and add 2·3 g sodium chloride and 25 ml water. Dilute to the mark with water at 20 °C and polarise in a 200 mm tube at 20 °C (S_1). Direct reading $D = S_1 \times 2$. Pipette 50 ml of the 26 per cent sample solution into a 100-ml volumetric flask and add 20 ml of water. Then add slowly (while rotating the flask) 10 ml of hydrochloric acid (sp. gr. 1·1029 at 20/4°) and place in a water bath maintained at 60 °C for 10 min, agitating the flask for the first 3 min. Then immerse the flask in water at 20 °C. When the temperature of the contents has fallen to 35 °C dilute nearly to the mark and leave the flask in the water for at least a further 30 min before making it up to the mark. After mixing, polarise in a 200 mm tube at 20 °C (S_2). Inverted reading $I = S_2 \times 2$.

The AOAC formula is based on the change of 132·56°S in the saccharimeter reading at 20 °C for solutions containing the normal weight (26 g of sucrose per cent) from +100° to −32·56° when sucrose is hydrolysed to invert sugar by the above method, i.e.

$$\text{Sucrose (per cent)} = \frac{100\,(D - I)}{132{\cdot}56 - 0{\cdot}0794\,(13 - m) - 0{\cdot}53\,(t - 20)}$$

where t = temperature at which the readings are made and m = g of total solids from original sample in 100 ml inverted solution.

If the original solution contains exactly 26 per cent of the sample and the saccharimeter readings are made at 20 °C, the formula becomes:

$$\text{Sucrose (per cent)} = \frac{100\,(D - I)}{132{\cdot}56}$$

The divisor of 132·56 used in this formula applies to the inversion method described above, but needs to be modified for different conditions of hydrolysis. For example, for overnight or 24 hour acid hydrolysis at room temperature, or for 10 hours at about 30 °C, the divisor becomes 132·66. If inverted using invertase, the divisor becomes 132·1.

A more general or routine procedure for food samples containing sucrose, reducing sugars and water-insoluble matter is to prepare a 5 per cent sample solution by weighing 10 g into a wide necked 200 ml volumetric flask and adding hot water. After mixing and standing, add clearing agents if necessary (Carrez reagents), make up to the mark, shake and filter. Take the saccharimeter reading S_D in a 200 mm tube at 20 °C. Pipette 50 ml of the filtrate into a 100 ml volumetric flask and invert as previously described. Take the saccharimeter reading and multiply it by two to obtain the inverted rotation S_I

$$\text{Sucrose (per cent)} = \frac{26 \times (S_D - S_I) \times 100}{5 \times 132{\cdot}56} = 3{\cdot}92\,(S_D - S_I)$$

NOTES

Because invert sugar has a very high temperature coefficient of rotation it is important to adjust the temperature of the inverted solution to exactly 20 °C before taking the saccharimeter reading. The Clerget method is not suited to the determination of sucrose in foods rich in cereal products containing high levels of oligosaccharides such as raffinose and neokestose, which may also like sucrose give large changes in rotation on being hydrolysed under the acid conditions into their component monosaccharides. The accuracy of a Clerget determination for sucrose can be checked by carrying out a *TI* – *BI* determination (p. 197).

Using angular degrees from readings taken on a polarimeter, the formula converts to

$$\text{Sucrose (per cent)} = \frac{[\alpha_D] - [\alpha_I]}{Q}$$

where $[\alpha_D]$ and $[\alpha_I]$ are the specific rotations of the sample solutions before (D) and after (I) inversion. The value of the inversion division factor Q, which corresponds to the change in angular rotation on hydrolysing sucrose to invert sugar, will depend on the method of inversion, the clearing agent and the light source. The Milk Products Sub-committee of the SPA (1930) gave the following values of Q for inversion at 60 °C and reading at 20 °C.

	Zinc ferrocyanide precipitant Q	Phosphotungstic acid precipitant Q
Sodium light	0·8825	0·8865
Mercury green line	1·0392	1·0439

Polarimetric determination of other sugars in the presence of sucrose. The method may be extended to the estimation of another sugar (X) such as invert sugar or lactose in the presence of sucrose. Using saccharimeter readings, if S is the percentage of sucrose calculated from the Clerget–Hertzfeld formula:

$$\text{Sugar } X(\text{per cent}) = \frac{(D - S)\,66{\cdot}5}{[\alpha]_D^{20} \text{ for sugar } X}$$

Using angular rotations:

$$\text{Sugar } X(\text{per cent}) = \frac{[\alpha_D]_D^{20} - (0{\cdot}665 \times S)\,100}{[\alpha]_D^{20} \text{ for sugar } X}$$

where $[\alpha_D]_D^{20}$ is the specific rotation of the sample from the direct reading (before inversion).

Note that these formulae do not apply when a third optically active substance is present such as in a sample containing glucose syrup as well as sucrose and invert sugar (see p. 214). It should also be borne in mind that other carbohydrates are hydrolysed to a greater or lesser extent according to the experimental conditions. The normal treatment with acid affects, for example, raffinose but does not significantly affect lactose, dextrose, maltose or dextrins, so that the Clerget–Hertzfeld double polarisation method for sucrose is accurate in their presence. Raffinose in, for example, beet sugar can however be approximately estimated by Clerget's acid inversion method (p. 193).

Copper reduction methods

INTRODUCTION

Sugars which possess in their structure free aldehydic or ketonic groups react as weak reducing agents and are termed reducing sugars. These include all the monosaccharides, and the disaccharides maltose, lactose and cellobiose. Disaccharides, such as sucrose, and raffinose and higher oligosaccharides are composed of simple sugars linked through their aldehydic or ketonic groups and are therefore non-reducing carbohydrates (until they are hydrolysed back to their constituent reducing sugars). These properties are used to estimate sugars by the measurement of the reduction of Cu(II) to Cu(I). Fehling's solution consists of alkaline cupric tartrate and is converted to insoluble cuprous oxide when boiled with a solution of a reducing sugar. This forms the basis of a number of procedures. In the Munson and Walker method (see 7th edition of this book, Pearson (1976), or the AOAC methods book), cuprous oxide is filtered off and weighed. In the Lane and Eynon method the reaction is followed titrimetrically using a redox indicator. The Luff–Schoorl method uses an alkaline reagent containing cupric citrate. In this method after boiling with a solution containing reducing sugars, potassium iodide and acid are added after cooling and the liberated iodine, which is equivalent to the unreduced copper, is titrated with sodium thiosulphate. The Somogyi semi-micro method (Pearson, 1976) uses an alkaline copper tartrate reagent and potassium iodate to react on heating with the reducing sugar, the excess iodate being estimated iodometrically with sodium thiosulphate solution after the addition of potassium iodide and dilute sulphuric acid. Alternatively, the non-reduced copper can be determined colorimetrically using an arsenomolybdate reagent.

The copper reducing power of commercial products containing a mixture of reducing sugars, for example glucose syrups prepared by the hydrolysis of starch, is often referred to as 'Dextrose Equivalent' or 'DE' or 'K value'. For example, a syrup containing 42 per cent of reducing sugars expressed as dextrose on the dry basis is termed DE42.

The Directive (79/796/EEC) laying down Community methods of analysis for testing certain sugars (see p. 207) contains several copper reduction methods for the determination of high and low levels of reducing sugars, and these are described in the following sections.

THE LANE AND EYNON TITRATION

The determination of reducing sugars by titration with Fehling's solution is an empirical method and therefore standardised experimental conditions and procedures must be rigidly adhered to in order to obtain satisfactory results. The classical procedure used tables indicating the amounts of invert sugar, dextrose, fructose, maltose or lactose equivalent to volumes of reduced Fehling's solution. In the constant volume technique, which was originally developed at the Laboratory of the Government Chemist (LGC), the conditions for reduction are standardised by the addition of water to give a constant final volume of 75 ml in the titration flask, and the volume of Fehling's solution used is calibrated against a standard reducing sugar solution. By this means the use of tables is eliminated. ICUMSA (1978) have adopted the constant volume technique as official and have dropped the classical procedure.

Fehling's solution, or more correctly the Soxhlet modification of it, consists of equal volumes of the following:

Fehling's 1. Dissolve 69·3 g copper sulphate pentahydrate ($CuSO_4.5H_2O$) in water and make up to 1 litre.

Fehling's 2. Dissolve 100 g sodium hydroxide and 345 g sodium potassium tartrate ($KNaC_4O_6.4H_2O$) in water and make up to 1 litre.

These solutions should be kept separately in amber bottles until required when 50 ml of Fehling's 1 should be pipetted into 50 ml Fehling's 2 and mixed.

Standard invert sugar solution is prepared as follows:

Dissolve 23·750 g pure sucrose in about 120 ml water in a 250 ml volumetric flask, add 9 ml concentrated hydrochloric acid and stand for 8 days at room temperature. Make up to the mark with water. (When inversion is complete, rotation in a 200 mm tube = 11·80° ± 0·05°S.) Transfer 200 ml into a 2 l volumetric flask, add about 200 ml water and with shaking add 71·4 ml of sodium hydroxide solution (40 g/l) containing 4 g benzoic acid. Add about 1 litre water, mix and check with indicator paper that the solution is approximately pH 3. Adjust if necessary and make up to the mark. This produces a stable 1 per cent m/v stock solution of invert sugar which should be diluted 1:4 to give a 0·25 per cent standard solution when required. Alternatively, add 4 g benzoic acid dissolved in 400 ml of hot water to the 200 ml acid solution containing the inverted sucrose and make up to the mark. When 0·25 per cent standard invert sugar solution is required, neutralise 50 ml of the acidic stock solution in a 200 ml flask with M NaOH using litmus paper and make up to the mark. Neutral or alkaline solutions of sugar do not keep and should be prepared when required.

The Lane and Eynon constant volume technique is carried out as follows:

Pipette 20 ml Fehling's solution into a 500 ml conical flask. Add 15 ml water and run in from a burette 39 ml of 0·25 per cent standard invert sugar solution. Add some anti-bumping granules and place over a hot flame. Allow to come to the boil and boil at a moderate rate without moving the flask or altering the flame for exactly 2 min. Add 3–4 drops of methylene blue solution (1 per cent aqueous) 5 seconds before the completion of the boiling period, then to the boiling solution add from the burette small increments (about 0·2 ml) of the standard invert sugar solution until the blue colour lessens, then dropwise until the final disappearance of the blue colour leaving the red coloration due to cuprous oxide. The titration should be completed in 3 min from the commencement of boiling and titration (V_0) should be between 39 and 41 ml with a final volume of 75 ml in the boiling flask.

Prepare an aqueous extract of the sample (C per cent m/v) using clearing agents as necessary, containing 250–400 mg reducing sugars per 100 ml. With the sample solution in the burette, repeat the titration as before using 20 ml Fehling's solution, 15 ml water and 25 ml sample solution. Carry out the titration rapidly in order to find the approximate titre. If on the addition of methylene blue no blue colour is formed, the reducing sugar content of 25 ml sample solution was sufficiently large to reduce all the Fehling's solution and a weaker sample solution should be used. If the final titration exceeds 50 ml, a stronger sample solution (or 10 ml Fehling's solution, restandardised with standard invert sugar solution) should be used. Carry out the final accurate titration (V_1) after adding from the burette, the volume, less 1 ml, of sample solution used in the first titration and sufficient water to give a final volume of 75 ml.

$$\text{Per cent reducing sugars as invert sugar in the sample} = \frac{V_0 \times 25 \times f}{C \times V_1}$$

where f is a correction factor to be applied if large amounts of sucrose are present, e.g. in the anlyses of cane and beet sugar products.

Sucrose content (g in V_1)	0	0·5	1	2	3	4	5
Correction factor f	1·000	0·982	0·971	0·954	0·939	0·926	0·915

Other standard reducing sugar solutions may be used to calibrate the Fehling's solution. For example, if 0·25 per cent pure anhydrous dextrose solution is used, the titration will give the concentration of reducing sugars in the sample expressed as dextrose, and also in the case of glucose syrup, the Dextrose Equivalent, if calculated on the dry basis.

If the mixed Fehling's solution is standardised by adjustment so that 20 ml + 15 ml water is reduced by 40 ml of standard 0·25 per cent invert sugar solution, and the sample solution contains m per cent m/v sample, then

$$\text{per cent invert sugar in the sample} = \frac{1000}{\text{titration} \times m}$$

THE DETERMINATION OF SUCROSE BY THE LANE AND EYNON TITRATION

Sucrose can be determined in the absence of reducing sugars by inverting a portion of the test solution with acid followed by neutralisation with alkali and titration by the Lane and Eynon method, using standard invert sugar solution for calibration.

$$\text{Per cent invert sugar} \times 0{\cdot}95 = \text{per cent sucrose}$$

In the presence of reducing sugars, titrations are carried out on portions of the test solution before and after acid inversion. If BI is the percentage of reducing sugars before inversion expressed as invert sugar and TI is the percentage of total reducing sugars determined after inversion and expressed as invert sugar

$$(TI - BI) \times 0{\cdot}95 = \text{per cent sucrose}$$

The inversion can be carried out in a number of ways.

1. By the procedure used in the Clerget–Hertzfeld polarimetric method (p. 194).
2. Pipette 50 ml of the test solution into a 100 ml graduated flask; add 5 ml conc. hydrochloric acid and let stand overnight. Neutralise with dilute sodium hydroxide using methyl red indicator paper (pH 4·2–6·3), cool and make up to the mark.
3. To a portion of the test solution (containing about 1 per cent total sugars) in a 200 ml graduated flask, add 20 ml inversion acid (50 ml conc. hydrochloric acid

+ 950 ml water). Bring to the boil and boil for exactly 30 s. Cool rapidly and neutralise by the cautious addition of M sodium hydroxide in the presence of both litmus and congo red indicator papers until both are red simultaneously. Make up to the mark.

THE LUFF–SCHOORL METHOD

The Luff–Schoorl method for reducing sugars is an official EC method for the analysis of certain sugar products (Directive 79/796/EEC) and animal feedingstuffs (Directive 71/250/EEC: OJ No. L 155, 12.7.71, p. 21). The reagent is less alkaline than Fehling's solution. In consequence it is a weaker oxidising agent requiring longer boiling with sample solutions compared to the Lane and Eynon technique. It also results in identical calibrations for glucose and fructose, and therefore for invert sugar. It gives equivalent results to those obtained by the Lane and Eynon technique for reducing monosaccharides but glucose syrups, especially those with low dextrose equivalent, tend to yield slightly dissimilar results because of different reaction of the maltose and oligosaccharide with the two reagents. The Luff–Schoorl method does not require titration of a boiling solution as with Lane and Eynon but involves an iodometric determination of the excess Cu(II) with thiosulphate to a sharp end-point with starch.

LUFF–SCHOORL REAGENT

Dissolve 143·8 g anhydrous sodium carbonate in about 300 ml water in a 1 litre volumetric flask. Add with swirling 50 g citric acid monohydrate previously dissolved in 50 ml water. When effervescence ceases add 25 g copper(II) sulphate pentahydrate previously dissolved in 100 ml water. Make up to the mark and mix. After standing overnight, filter if necessary. The strength of the reagent must be accurately 0·1 M copper(II) and 1 M sodium carbonate. This is checked as follows: to 25 ml reagent add 3 g potassium iodide and 25 ml 3 M sulphuric acid solution. Titrate with 0·1 M sodium thiosulphate solution adding soluble starch solution (0·5 per cent) towards the end of the titration which should be 25 ml. Dilute 10 ml of the reagent to 100 ml. Add 10 ml of the diluted reagent to 25 ml of 0·1 M hydrochloric acid solution contained in a conical flask. Place in a boiling water bath for 1 hour. Cool and make up to the original volume with water and titrate with 0·1 M sodium hydroxide solution using phenolphthalein. The titration should be 5·5–6·5 ml. Titrate 10 ml of the diluted reagent with 0·1 M hydrochloric acid solution using phenolphthalein. The titration should be 6·0–7·5 ml. The reagent should be pH 9·3–9·4.

PROCEDURE

Transfer a weighed quantity of prepared samples to a wide-necked 200 ml volumetric flask. Add about 150 ml hot water and keep warm with shaking to extract the water-soluble matter. Clarify by adding 5 ml Carrez I solution (see p. 297) followed by 5 ml Carrez II solution. Make up to the mark, mix and filter. Dilute a portion of the filtrate so that 25 ml contains 15–60 mg reducing sugars.

To 25 ml of reagent in a 300 ml conical flask, add 25 ml of the clarified and diluted sample solution and some anti-bump chips or granules. Fit a reflux condensor or a water-cooled cold finger and bring the contents of the flask to boiling in about 2 min. Simmer gently for exactly 10 min then cool in cold water for 5 min. Add 10 ml fresh potassium iodide solution (30 per cent) followed with care by 25 ml 3 M sulphuric acid solution. A few drops of isopentanol may be added to prevent foaming. When effervescence ceases, titrate with 0·1 M sodium thiosulphate solution until almost colourless. Add 2–3 ml of starch solution and complete the titration (*X* ml). Carry out a reagent blank titration using 25 ml water instead of the sample solution (*Y* ml). Then $(Y - X)$ ml represents the amount of copper(II) reduced by the sugars in the sample aliquot as shown in Table 6.9.

Similarly to the $(TI - BI)$ Lane and Eynon procedure, sucrose can be determined by the Luff–Schoorl method by determination of reducing sugars expressed as invert sugar (or glucose or fructose) before and after acid inversion.

Table 6.9 Thiosulphate equivalents: Luff–Schoorl method

$Na_2S_2O_3$ 0·1 M (ml)	Glucose, fructose, invert sugar (mg)	Lactose (mg)	Maltose (mg)
1	2·4	3·6	3·9
2	4·8	7·3	7·8
3	7·2	11·0	11·7
4	9·7	14·7	15·6
5	12·2	18·4	19·6
6	14·7	22·1	23·5
7	17·2	25·8	27·5
8	19·8	29·5	31·5
9	22·4	33·2	35·5
10	25·0	37·0	39·5
11	27·6	40·8	43·5
12	30·3	44·6	47·5
13	33·0	48·4	51·6
14	35·7	52·2	55·7
15	38·5	56·0	59·8
16	41·3	59·9	63·9
17	44·2	63·8	68·0
18	47·1	67·7	72·7
19	50·0	71·7	76·5
20	53·0	75·7	80·9
21	56·0	79·8	85·4
22	59·1	83·9	90·0
23	62·2	88·0	94·6

Other chemical methods

Methods for determining dextrose and other sugars have been proposed that are based on the oxidation of the sugar to gluconic acid by the action of iodine:

$$C_6H_{12}O_6 + I_2 + 3NaOH \rightarrow HOCH_2(CHOH)_4CO_2Na + 2NaI + 2H_2O.$$

Iodine oxidises aldoses (e.g. dextrose, lactose) but it has little or no effect on ketoses (e.g. sucrose, fructose), so the former sugars can be determined in the presence of the latter (Hinton and Macara, 1924). The same authors (1927) described a modified method for determining aldose sugars employing chloramine-T. Using this reagent the oxidation is slower and more easily controlled than with iodine, and the possible secondary action on sucrose and fructose is more likely to be reduced to negligible proportions.

To a 250 ml bottle, add 25 ml of the neutral clarified solution containing about 0·08 g dextrose (0·15 g lactose or invert sugar), 20 ml of potassium iodide solution (10 per cent) and then 50 ml of 0·025 M (0·05 N) chloramine-T (7·04 g/l). Allow the stoppered bottle to stand in the dark for 1½ hours, then acidify with 10 ml of 2 M hydrochloric acid and immediately titrate with 0·05 M sodium thiosulphate using starch. Carry out a blank on 25 ml water at the same time. The difference between the two titrations represents the amount of 0·025 M chloramine-T required by the sample.

1 ml 0·025 M chloramine-T ≡ 0·0045 g dextrose ≡ 0·009 g lactose hydrate ≡ 0·0085 g inverted sucrose.

Small corrections are necessary when sucrose is present in the solution. Sucrose can be estimated by titrating before and after acid inversion.

Reducing sugars at above pH 10·5 reduce ferricyanide to ferrocyanide, which reacts with ferric ions to form Prussian blue. This has been used as the basis of an automatic procedure (Clegg, 1956). A sodium thiosulphate titration can be used to measure the reduction. Reducing sugars react with 2,3,5-triphenyl tetrazolium to form a cherry-red compound soluble in organic solvents (Crowell and Burnett, 1967). Non-specific colorimetric methods based on condensation reactions in strong acid are useful in automatic sugar analysis systems (Southgate, 1976). Dean (1978) describes a procedure using orcinol in sulphuric acid. A method using anthrone is specific for hexoses (Roe, 1955) both as monosaccharides and from oligosaccharides, after hydrolysis.

High performance liquid chromatography

High performance liquid chromatography (HPLC) has become the most useful and now probably the most used technique for the identification and quantitative determination of sugars in foodstuffs. Sugars do not absorb light at analytical ultraviolet wavelengths and therefore the only generally available and direct method of detection is by refractive index (RI). Unless the more sophisticated and expensive interference type RI detector is used, the normal deflection type is relatively insensitive so in practice the lower limit for the measurement of sugars by HPLC is of the order of about 20 μg per sugar or 0·5–1 per cent in the food sample for each sugar. Refractive index detectors are also very susceptible to changes in column temperature and particularly sensitive to changes in mobile solvent composition. For the latter reason, solvent programming (gradient elution) cannot be used, and since the detectors take many hours to stabilise following setting up or on changing the system, they are at their best when kept continually in use.

The light scattering 'mass detector' is a general bulk property detector for any non-volatile separated solutes including sugars and is about 40 times more sensitive for sugars than RI. However, its non-linearity over the wide working ranges for sugars needed in food analysis limits its use because much standardisation over short ranges of concentration would be needed.

For very low levels of sugars, post-column derivatisation techniques followed by ultraviolet or visible photometric detection can be used, e.g. reaction with tetrazolium blue to form red coloured formazan (Wight and Van Niekerk, 1983). The HPLC detection systems for carbohydrates are reviewed by Honda (1984).

The HPLC separation of food sugars is most usually carried out by strong cation exchange chromatography or by interaction with various forms of surface modified silica gel particles. Plain silica gel can be modified *in situ* by incorporating a small amount of a diamine or a polyamine in the mobile phase. One of the amine groups reacts with the hydroxyl groups on the surface of the silica, whilst the other electron-donating amine group interacts differentially with various sugars to effect the chromatographic separation. Diamines that have been used include 1,4-diaminobutane, ethylenediamine, piperazine and tetraethylenepentamine (Boumahraz *et al.*, 1982). Sugar separations on such *in situ* modified columns is similar to commercial chemically bonded aminopropyl-modified phases now widely used. These phases are usually silica gel reacted with γ-aminopropyltriethyoxysilane. Ethylenediaminopropyltrimethoxysilane or diethylenetriaminopropyltrimethoxysilane may also be used with some advantages, e.g. higher water content in the mobile phase. Aminopropyl-modified silica columns have a limited useful life because of hydrolysis on the column and the formation of Schiff bases between the amine groups and sugars and other carbonyl compounds, from samples. These effects may also on occasions give rise to poor quantitative analysis when using these columns. Hydrolysis can be prevented by using a saturating pre-column. Cross-linking of phases also extends working life. Separations on amine-modified silicas is in order of increasing molecular weight (see Fig. 6.2). The mobile phase is acetonitrile/water usually in the ratio 80/20 or 75/25.

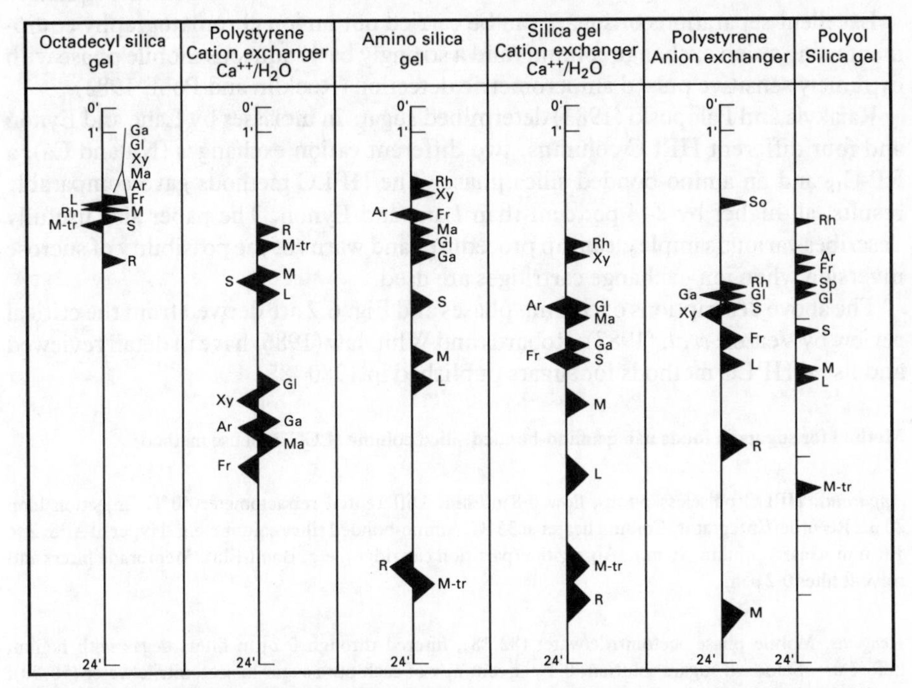

Fig. 6.2 Schematic representation of sugar separations on some HPLC phases, using the most usual mobile phase (from Verzele *et al.*, 1987).

Ar, arabinose	Fr, fructose	Ga, galactose
Gl, glucose	L, lactose	Ma, mannose
M, maltose	M-tr, malto-troise	R, ribose
Rh, rhamnose	S, sucrose	So, sorbitol
Xy, xylose		

Reverse phase HPLC using C_{18} modified silica with water as mobile phase appears to be a useful technique for the determination of oligosaccharides, which separate well. Shorter side chains (C_6 or C_8) decrease the resolution and retention times. Examples include legume seed oligosaccharides (sucrose, raffinose and stachyose) (Wight and Datel, 1986), and starch syrups (Rajakyla, 1986).

Polyol-derivatised silica gel phases have recently been introduced (Verzele and Van Damme, 1986) and are claimed to give very good separations and offer no risk of sugar loss by Schiff base formation, therefore producing trustworthy quantitative analysis.

Cation exchange resins, usually in the calcium form, based on polystyrene or silica are also useful. Using water or 0·01 per cent $CaCl_2.2H_2O$ as the mobile phase and operated at 85 °C, sugars are eluted in descending molecular weight order with fructose after glucose (Fig. 6.2). Columns run at lower temperatures may cause peak broadening due to sugar anomeric form separations. This can be overcome by adding an amine to the mobile phase, e.g. 0·001 M triethylamine. Cation exchange silica columns are robust to high pressure and can also be used with acetonitrile/water mobile phase in which case the elution order is reversed. In long use the silica tends to dissolve in alkaline mobile phase. Polystyrene resins are more pressure sensitive and can have shorter working lives. Exchange columns need not be regenerated regularly.

Excellent separations of sugars can be carried out on ion chromatography equipment using anion exchange columns and a strongly basic aqueous mobile phase with extremely sensitive pulsed amperometric detection (Rocklin and Pohl, 1983).

Rajakyla and Paloposki (1983) determined sugars in molasses by Lane and Eynon and four different HPLC columns, two different cation exchanges (Na and Ca), a RP-C_{18} and an amino-bonded silica phase. The HPLC methods gave comparable results, all higher by 2–4 per cent than Lane and Eynon. The paper also usefully describes various sample clean-up procedures and warns of the possibility of sucrose inversion when ion-exchange cartridges are used.

The above descriptions of column phases and Fig. 6.2 are derived from the critical review by Verzele *et al.* (1987). Robards and Whitelaw (1986) have in detail reviewed and listed HPLC methods for sugars published in 1980–85.

Method for sugars in foods using amino-bonded silica column (LGC in-house method)

Apparatus. HPLC pulseless pump, flow 0·8 ml/min. Differential refractometer 40 °C. Injection loop 20 μl. Recorder/Integrator. Column heater at 35 °C. Amino-bonded silica column, e.g. Hypersil APS, 2 × 100 mm. Guard column, 10 mm. Absorption/partition cartridges, e.g. Bond-Elut. Membrane filters and solvent filter 0·2 μm.

Reagents. Mobile phase acetonitrile/water (82:18), filtered through 0·2 μm filter, degas with helium before use. Standard sugars solution, 0·1 per cent m/v of each pure sugar in acetonitrile/water (50:50), e.g. 100 mg/100 μl, dissolved in water first. (Stable 5 days in refrigerator.) Before use pass few ml through silica and C_{18} cartridges and the 0·2 μm filter.

Resolution. Switch on pump and detector about 3 hours before analysis. Purge reference cell with mobile phase for 15 min after 2 hours. After 3 hours in all, recorder should give stable baseline. Resolution R between glucose and fructose in standard solution must be equal to or greater than 0·8, where

$$R = \frac{2t}{W_f + W_g}$$

where t = peak apex separation between fructose and glucose, mm.
W_f = base width of fructose peak, mm.
W_g = base width of glucose peak, mm.

Procedure. Weigh into 100 ml volumetric flask between 1 and 5 g of homogenised sample to give final sugar extract solution strength 0·01–0·1 per cent m/v for each sugar, where possible. Add 40 ml water, shake mechanically for 45 minutes then add 40 ml acetonitrile and shake 10 min. Bulk to mark with acetonitrile/water (50:50), mix, filter through Whatman 42 or equivalent. Make serial dilutions if necessary to lower a high sugar content. Condition silica and C_{18} clean-up cartridges by passing 2 ml methanol then 2 ml water, then pass about 4 ml of sample filtrate. Pass cleaned-up extract through a 0·2 μm membrane filter and collect in glass vial. Inject and check resolution. Inject standard solution. Collate peak areas for each sugar, and calculate to sample weight. Express to nearest 0·1 per cent m/m for each sugar in sample.

Gas–liquid chromatography
Food sugars can be separated, detected and determined with high sensitivity by gas–liquid chromatography after conversion to volatile and heat-stable derivatives. Trimethylsilyl ether derivatives have been shown to be relatively easy to prepare and are stable in the absence of moisture (Sweeley *et al.*, 1963). Silylation in pyridine is more rapid than the rate of mutorotation (in dimethyl sulphoxide or dimethyl formamide even more so) so only one derivative is formed except for monosaccharides, which occur in food as mutorotated equilibrium mixtures and which therefore give a chromatographic peak for each α and β anomer form. Moreover, gas chromatographic analysis is hindered by overlapping anomer peaks of glucose and fructose. Silylation of easily prepared sugar oximes (Mason and Slover, 1971) removes the difficulties of anomer derivatisation and gives single chromatographic peaks for most sugars. Various procedures using different silylation reagents have been reported (Dickes and Nicholas, 1976; Smith *et al.*, 1978; Dickes, 1979; Prager and Miskiewicz, 1979; Li and Schuhmann, 1980). Robards and Whitelaw (1986) have fully reviewed the development of GLC of carbohydrates and have listed the papers published from 1980–85.

MODIFIED IOCC PROCEDURE
Weigh about 1 g of prepared sample accurately into a beaker. Add 0·3 g mesoinositol (internal standard, IS), 30 ml hot water and mix to extract the sugars. Add 0·5 ml Carrez I solution (p. 297), mix, add 0·5 ml Carrez II solution and mix. Filter into a 100 ml volumetric flask, wash residue with 20 ml of water and bulk filtrate to 100 ml with methanol. Prepare likewise a calibration solution of reference sugars also containing the same amount of internal standard. Evaporate 0·5 ml of the test and calibration sugar solutions in small screw-topped septum vials under a stream of nitrogen. When almost dry add 0·5 ml isopropanol and complete the drying under a stream of nitrogen until a dry solid residue remains. Screw on the septum and inject 0·5 ml oximation reagent (2·5 per cent hydroxylamine hydrochloride in pyridine). Mix and heat for 30 min at 80 °C then allow to cool. Inject 1 ml silylation reagent (trimethylchlorosilane:*N*, *O*-bis-(trimethylsilyl) acetamide, 1:5 by volume) (Trisyl BT, Pierce Chemical Co.). Mix and heat for 30 min at 80 °C and allow to cool. Using a micro-syringe extract 1 µl of the derivatised sugar solution through the septum and inject onto a 1700 mm × 2 mm glass column packed with 5 per cent SE52 on Supelcoport, 80–100 mesh (or equivalent packing) heated isothermally at 150 °C for 2 min then temperature programmed at 2 °C/min to 250 °C; injector/detector (FID) temperature 260 °C with nitrogen carrier gas at 30 ml/min.

For each calibration sugar, the mass factor K is given by:

$$K = \frac{\text{weight of sugar A/100 ml}}{\text{weight of IS/100 ml}} \times \frac{\text{peak area of IS}}{\text{peak area of sugar A}}$$

In the test solution the concentration of each sugar in the sample can be calculated from the above equation using the values of K obtained from the mixed-sugar calibration solution. This TMS–oxime method does not satisfactorily separate glucose and galactose, and some overlapping would occur in solutions containing a complex mixture of pentoses and hexoses. Superior peak resolution of TMS–oximes and a flat base line when temperature programming can be obtained by using glass or fused silica capillary columns (Schaffler and Morel du Boil, 1981). By using cold, on-column injection techniques with these columns, the incidence of sample decomposition in the heated injection port is reduced or eliminated (Traitler *et al.*, 1984). The latter technique also permits the GLC analysis of oligosaccharides containing up to six saccharide units. The GLC of sugars is a part of the Englyst method for dietary fibre (p. 31). The TMS–oxime method is also applicable to sugar alcohols, e.g. sorbitol and mannitol.

Enzymatic methods

The specificity of enzyme reactions has been employed in sugar analysis for many years, for example the polarimetric determination of sucrose before and after hydrolysis catalysed by invertase (β-fructosidase). In the last 15 years a wide range of pure enzyme preparations have become available and has led to the development of several useful specific and sensitive analytical techniques, for example ultraviolet methods based on enzyme reactions on food components (especially on individual sugars, see below), enzyme electrodes (YSI Analyzer: Mason, 1983; Moody *et al.*, 1986), and immobilised enzymes for automatic methods (Bergmeyer, 1970). Enzymatic methods in food analysis have been reviewed by Wiseman (1978).

Glucose (dextrose) is oxidised by β-D-glucose oxidase to gluconic acid and hydrogen peroxide, and can thus be determined by further reacting hydrogen peroxide with reagents to form coloured products or by using selective detectors such as polarographic oxygen electrodes (Okuda and Miwa, 1974). The hexokinase enzymatic procedure for glucose is sensitive and specific and is used as the basis for the ultraviolet determination of other food sugars (BCL, 1987). It involves the conversion of glucose by hexokinase and adenosine triphosphate (ATP) to glucose-6-phosphate (G-6-P). This can be oxidised in the presence of glucose-6-phosphate dehydrogenase by the co-enzyme nicotinamide-adenine-dinucleotide phosphate (NADP) to gluconate-6-phosphate with the simultaneous reduction of NADP to its reduced form, NADPH. The concentration of NADPH, determined by spectrophotometry at 340 or 366 nm, is equivalent to the initial concentration of glucose. Fructose similarly reacts to form fructose-6-phosphate (F-6-P) and this can be converted by the addition of phosphoglucose isomerase (PGI) into G-6-P that reacts with NADP to form a further quantity of NADPH. The increase in optical density is thus proportional to the fructose content.

Sucrose can be hydrolysed with invertase (β-fructosidase) to give glucose and fructose, and the difference between the glucose content determined by the hexokinase ultraviolet methods before and after hydrolysis with invertase corresponds to the sucrose content. Lactose can be hydrolysed in the presence of β-galactosidase to glucose and galactose. Glucose is determined as before, or galactose can be determined by oxidation with nicotinamide-adenine-dinucleotide (NAD) to give galactonic acid in the presence of β-galactose dehydrogenase. NADH is formed in concentration equivalent to the galactose concentration and the lactose content of the sample solution. Maltose can be hydrolysed by maltase (α-glucosidase) to two molecules of glucose, which are then determined via hexokinase. Apart from maltose, α-glucosidase also hydrolyses sucrose, which therefore if present would require a separate determination using invertase. Raffinose is hydrolysed by α-galactosidase to sucrose and galactose. Other α-galactosides, for example galactinol, melibiose and stachyose are also hydrolysed. Starch is hydrolysed by amyloglucosidase to glucose, which can be determined via hexokinase.

Procedures for all these determinations (and for the useful determination of many other food components) that are carried out in 10 mm glass spectrophotometer cells (cuvettes) using test kit combinations are described in a booklet provided free by

Boehringer Mannheim GmbH (BCL, 1987). BCL market test kits for glucose (cat. 716251), glucose/fructose (cat. 139106), glucose/fructose/sucrose (cat. 716260), sucrose/glucose (cat. 139041) and lactose/galactose (cat. 176303), all for food analysis. Enzymatic methods employing the hexokinase ultraviolet technique are microchemical methods using very dilute samples and reagent solutions dosed by micropipettes. Great care and cleanliness is essential therefore if satisfactory results are to be achieved.

GLUCOSE/FRUCTOSE

The following is an adaptation of a BCL method employed at the Laboratory of the Government Chemist using the test kit for blood glucose and an additional reagent for fructose. It is considerably more economical in use than the BCL food analysis test kit for glucose/fructose. The procedure includes the determination of the two sugars in a prepared reference standard solution, which simplifies the calculation and assists in verifying the accuracy of the determination. The reference standard sugars used should be of guaranteed composition, containing minimal amounts of foreign sugars. They should be carefully stored. A larger sample volume than that given in the BCL booklet is also used to improve precision.

REAGENTS

Redistilled water is used throughout. All reagents should be stored at 4 °C. The following applies when using BCL test-kit catalogue No. 124346, and PGI.

I Triethanolamine buffer, 0·3 M. Dissolve the contents of one bottle in 100 ml water.
II NADP. Dissolve the contents in 11 ml water.
III ATP. Dissolve the contents in 11 ml water.
IV HK/G6P-DH. Use the suspension as supplied.
V PGI. Use the suspension as supplied.

Standard glucose and fructose solution. Dissolve about 0·5 g of pure anhydrous dextrose and about 0·5 g of pure anhydrous fructose, both weighed accurately, in water in a 500 ml volumetric flask and make up to the mark. Pipette 10 ml into a 500 ml volumetric flask and make up to the mark to give a solution containing glucose M_g mg/ml and fructose M_f mg/ml.

PROCEDURE

By the use of extraction with warm water and filtering, prepare a diluted aqueous extract of a weighed amount of sample so that the final solution of M_s g/l contains 10–60 mg/l of both glucose and fructose. Clearing agents other than Carrez solutions should not be used. Pipette into clean, dry and sterile 10 mm glass cuvettes quantities (ml) of solutions and reagents according to the following table:

	Blank	Samples	Standard
Reagent I	2·0	2·0	2·0
Reagent II	0·1	0·1	0·1
Reagent III	0·1	0·1	0·1
Test solution	—	1·0	1·0
Water	1·0	—	—

Mix with individual stirring rods for each cuvette, and read absorbance at 340 nm (E_1).

Reagent IV	0·02	0·02	0·02

Mix and read absorbance at 340 nm after 15 min (E_2), and if fructose is required:

Reagent V	0·02	0·02	0·02

Mix and read absorbance at 340 nm after 15 min (E_3).

For glucose

$$E\text{ (sample)} = (E_2 - E_1)\text{ sample} - (E_2 - E_1)\text{ blank}$$
$$E\text{ (standard)} = (E_2 - E_1)\text{ standard} - (E_2 - E_1)\text{ blank}$$

$$\text{per cent glucose in sample} = \frac{M_g}{M_s} \times \frac{E\text{ (sample)}}{E\text{ (standard)}} \times 100$$

For fructose

$$E\text{ (sample)} = (E_3 - E_2)\text{ sample} - (E_3 - E_2)\text{ blank}$$
$$E\text{ (standard)} = (E_3 - E_2)\text{ standard} - (E_3 - E_2)\text{ blank}$$

$$\text{per cent fructose in sample} = \frac{M_f}{M_s} \times \frac{E\text{ (sample)}}{E\text{ (standard)}} \times 100$$

The completeness of the enzyme reactions can be confirmed by calculating the theoretical yield of the sugars in the analysis of the diluted reference standard solution.

For glucose, $0{\cdot}0921 \times E$ (standard) = M_g for 100 per cent recovery
For fructose, $0{\cdot}0927 \times E$ (standard) = M_f for 100 per cent recovery

If the recoveries are outside the range 98–102 and accurate results are required the results should be rejected and the analyses repeated. Substituting values for E (sample) in the last two equations in place of E (standard) gives results without recourse to a reference standard sugars solution.

LACTOSE

The following is a slight adaptation of the BCL enzymatic method based on the hydrolysis of lactose to glucose and β-galactose in the presence of β-galactosidase followed by the oxidation of β-galactose by NAD to galactonic acid catalysed by β-galactose dehydrogenase (Gal-DH). In the latter reaction the amount of NADH formed, measured by absorbance at 340 nm, is equivalent to the original amount of lactose.

Reagents

Redistilled water is used throughout. The following reagents are employed when using BCL test-kit catalogue No. 176303:

I 610 mg lyophilisate, consisting of citrate buffer pH 6·6; NAD, 35 mg; magnesium sulphate and stabilisers. Dissolve contents in 7 ml water.

II 1·7 ml enzyme suspension, β-galactosidase 100 U. Use as supplied.

III 34 ml phosphate buffer consisting of potassium diphosphate, 0·15 mol/l, pH 8·6 with stabilisers. Use as supplied.

IV 1·7 ml enzyme suspension, galactose dehydrogenase 35 U. Use as supplied.

Standard lactose monohydrate solution. Dissolve about 500 mg of pure lactose monohydrate of guaranteed composition, weighed accurately, in water in a 500 ml volumetric flask, make up to the mark and mix. Pipette 10 ml into a 100 ml volumetric flask and bulk to give a solution containing M_1 mg/ml.

Procedure

By the use of extraction with warm water and filtering, prepare a diluted aqueous extract of a weighed amount of prepared sample so that the final solution (M_s g/l) contains 5–10 mg/l lactose. Clearing agents other than Carrez solutions should not be used. Samples containing active bacteria and yeasts, which could provide enzymatic interference, should be sterilised by holding in boiling water for 5 min. Pipette into clean, dry and sterile 10 mm glass cuvettes quantities (ml) of solutions and reagents according to the following table:

	Blank	Samples	Standard
Reagent I	0·2	0·2	0·2
Reagent II	0·05	0·05	0·05
Test solution	—	1·0	1·0

Mix with individual stirring paddles for each cuvette. Let stand 10 min.

Reagent III	1·00	1·00	1·00
Water	2·00	1·00	1·00

Mix and after 2 min read absorbance at 340 nm (E_1).

Reagent IV	0·05	0·05	0·05

Mix, let stand 15 min and read absorbance at 340 nm (E_2).

CALCULATIONS

$$E\ (\text{sample}) = (E_2 - E_1)\ \text{sample} - (E_2 - E_1)\ \text{blank}$$
$$E\ (\text{standard}) = (E_2 - E_1)\ \text{standard} - (E_2 - E_1)\ \text{blank}$$

$$\text{per cent lactose monohydrate in sample} = \frac{M_1}{M_s} \times \frac{E\ (\text{sample})}{E\ (\text{standard})} \times 100$$

Alternatively, the lactose content in the final test solution can be calculated using the extinction coefficient of NADH which equals 6·3 at 340 nm. The relevant equations simplify to:

$$\text{mg/ml in final solution} = 0{\cdot}189 \times E\ (\text{sxample})\ \text{for lactose monohydrate}$$
$$= 0{\cdot}179 \times E\ (\text{sample})\ \text{for anhydrous lactose}$$

The completeness of the enzyme reactions can be confirmed by using the above equation for the diluted reference standard solution:

$$0{\cdot}189 \times E\ (\text{standard}) = M_1\ \text{for 100 per cent recovery.}$$

If the recoveries are lower than 98 per cent, the cause should be sought. If the error is constant, the probable cause is the quality of the reference lactose standard.

SUGARS AND SUGAR PRODUCTS

Legislation

The EC standards for sugar and a number of sugar products used in the manufacture of food are shown in Table 6.10 and have been implemented into UK legislation by the Specified Sugar Products Regulations 1976, SI No. 509 as amended by SI No. 255, 1982. These standards are similar to those recommended by Codex Alimentarius. Most of the official EC methods of analysis are published in Directive 79/796/

Table 6.10 EC standards for sugars—Directive 73/437/EEC (OJ No. L356, 27.12.73, p. 71)

			Semi-white sugar	Sugar or white sugar	Extra white sugar	White soft sugar	Soft sugar	Sugar solution	Invert sugar solution	Invert sugar syrup	Glucose syrup	Dried glucose syrup	Dextrose monohy-drate	Dextrose anhydrous	Lactose
Polarisation	°S	min.	99·5	99·7	99·7	—	—	—	—	—	—	—	—	—	—
Invert sugar	% d.m.	max.	0·10	0·04	0·04	0·3–12	0·3–12	3[a]	3–50[a]	>50[a]	—	—	—	—	—
Dextrose equivalent	% d.m.	min.	—	—	—	—	—	—	—	—	20	20	—	—	—
Dextrose	% d.m.	min.	—	—	—	—	—	—	—	—	—	—	99·5	99·5	—
Loss on drying	%	max.	0·10	0·10	0·10	3	4·5	—	—	—	—	—	—	—	6
Dry matter	%	min.	—	—	—	—	—	62	62	62	70	93	90·0	98.0	—
Conductivity ash[b]	% d.m.	max.	—	—	0·01	0·2	—	0·1	0·4	0·4	—	—	—	—	—
Sulphated ash	% d.m.	max.	—	—	—	—	3·5	—	—	—	1·0	1·0	0·25	0·25	0·8
Sulphur dioxide	mg/kg	max.	15	15	15	40	40	15[c]	15[c]	15[c]	20[d]	20[e]	15	15	—
Colour type[b]	BI units	max.	—	6	2	—	—	—	—	—	—	—	—	—	—
Colour in solution[b]	ICUMSA units	max.	—	—	22·5	—	—	45	—	—	—	—	—	—	—
Colour in solution[f]	Codex units	max.	—	—	—	60	—	—	—	—	—	—	—	—	—
Sucrose + invert	% as sucrose	min.	—	—	—	97	88	—	—	—	—	—	—	—	—
Anhydrous lactose	%	min.	—	—	—	—	—	—	—	—	—	—	—	—	97
pH (10% solution)	20 °C	—	—	—	—	—	—	—	—	—	—	—	—	—	4·5–7·0

[a] Ratio of fructose to dextrose: 1·0 ± 0·1
[b] Method in Regulation EEC/1265/69 (OJ No. L163 4.7.69)
[c] On dry matter (d.m.)
[d] Or 20–400 mg/kg for confectionery and other specified food uses
[e] Or 20–150 mg/kg for confectionery and other specified food uses
[f] Codex Method CAC/RM6—1969

EEC (OJ No. L239, 22.9.79, p. 24). For other official methods specified in the Regulations, see Table 6.10. Internationally recognised methods of analysis have been formulated by ICUMSA (Schneider, 1979) and Codex.

SUGAR

Sugar (sucrose) is obtained from sugar cane (*Saccharum officinarum*), the juices of which contain 8–23 per cent of sugar, and from sugar beet (*Beta alba*), the juices of which contain 13–15 per cent of sugar. The Common Agricultural Policy (CAP) of the EC has given preference to the consumption of sugar from European grown sugar beet which now largely predominates. UK per capita consumption of sugar for 1984 was estimated as 38 kg. A proposed Codex standard for raw cane sugar includes polarisation not less than 96°S, loss on drying not more than 0·6 per cent, and colour not more than 2500 ICUMSA units.

Sugar refining. The conversion of imported raw cane sugar into crystalline sucrose or sucrose solutions is carried out in a number of defined processing steps which include affination, defecation, decolourisation and crystallisation. The production of sucrose from sugar beet uses essentially the same steps but the process can include initially the extraction of sugar juices from the beet. Raw sugar is first mixed with some raw syrup and centrifuged. After washing much of the molasses, the affined sugar is disolved in water and the solution is treated with milk or lime. Carbon dioxide is then bubbled through and impurities (gums, waxes, etc.) are carried down by the precipitate of calcium carbonate that is produced. The cloudy liquid is then filtered and decolorised with animal charcoal. The colourless liquid produced is evaporated under reduced pressure, centrifuged to separate the massecuite from the syrup and the wet crystals are then dried in a current of hot air to produce *granulated sugar*. The EC specifies several grades of white sugar (Table 6.10).

Cube sugar is prepared after casting the residue formed in the vacuum pan into slabs. The moist slabs are dried and cut into cubes.

Icing sugar is prepared by reducing crystalline sugar to a fine powder. Sulphur dioxide up to 20 p.p.m. is permitted. For the retail trade not more than 1·5 per cent tricalcium phosphate is added. Other possible inorganic anti-caking agents are magnesium carbonate or tri-silicate, calcium silicate, sodium-calcium-aluminium silicate and silica gel. For trade purposes 5 per cent of unmodified starch is sometimes used instead. These additions are permitted by the Regulations.

Muscovado sugar is raw, dark coloured unpurified cane sugar.

Demerara sugar is a brown sugar with a characteristic taste. It was originally produced in Demerara, but by a judicial decision in 1913 any sugar resembling it that is a product of the West Indies or Mauritius may be so described. Tin chloride was once used to stabilise the colour of demerara sugar. Barnett (1988) has shown that genuine muscovado sugars can be identified from white sugar coated with molasses by selective dissolution of the crystal coating using 92 per cent ethanol, and also with 80 per cent ethanol which leaves undissolved the sugar core of the crystal. From the potassium and magnesium levels in the 92 per cent extract solution and in the undissolved core, the authenticity of the product can be deduced.

Analysis of sugar

In addition to EC official methods, official and tentative methods for examining sugar products are described by Schneider (1979). Codex methods are given in CAC/RM 1/8–1969. Industrial quality control methods are described by Plews (1970) and Muller (1986).

Solubility. Refined sugar should dissolve in 0·5 part water at 20 °C.

Moisture. Determine the loss after drying 10 g for 3 hours at 103 °C (EC method) or by the Karl Fischer technique (cf. Piper, 1973). The moisture content of white sugar should not exceed 0·1 per cent and preferably be below 0·03 per cent.

Sulphated ash. Add 1 ml conc. sulphuric acid to 10 g sample, ignite gently, then add more acid and re-ignite to constant weight. The sulphated ash of refined white sugar should not exceed 0·02 per cent. In refineries, the ash is determined conductometrically on a standardised instrument by the ICUMSA method. Using a solution containing 5 g/100 ml:

$$\text{per cent Ash} = 0{\cdot}0018 \times \text{Specific conductance (micromhos/cm)}.$$

Details or references are given in Codex methods (CAC/RN 1/8–1969) and EC Regulations Nos 1265/69 and 2103/77.

Sucrose. Determine the polarisation in a saccharimeter using a 26 per cent m/v solution in a 2-dm tube, at exactly 20 °C.

Reducing sugars. Determine by Lane and Eynon's method, preferably using an electrometric end-point detector (Defrates and Castle, 1966) or by the EC official method of back-titration of unreduced copper with EDTA (Knight and Allen, 1960). The official procedure for reducing sugars in semi-white sugar is the Berlin Institute Method whereby under specified conditions the sample solution is used to reduce a complex copper II tartrate reagent. The copper I oxide is then oxidised with iodine solution in excess and the excess iodine titrate with sodium thiosulphate solution. Glucose and fructose may also be determined enzymatically. The presence of invert sugar may affect the keeping properties of white sugar.

Raffinose. The best method of differentiating between cane and beet products is to establish the presence or absence of reffinose. Raffinose is present in all beet products, but absent in products derived from sugar cane. It can be detected and determined by TLC and other methods (Schneider, 1979). The enzymatic determination of raffinose is subject to interference from other minor sugar compounds (p. 204). Ford (1979) estimated raffinose in legume seeds by GLC determination of glucose and melibiose after hydrolysis with invertase. Normal HPLC methods using RI detection are not sufficiently sensitive.

Colour. Dissolve 50 g sample in 50 ml water and pass the solution through a membrane filter. Measure the sugar concentration refractometrically (°Brix), the specific gravity (d), and the absorbance (A) at 420 nm against water preferably in a 10 cm cell (cf. Vane, 1972; Schneider, 1979).

$$\text{Colour (ICUMSA units)} = \frac{A \times 10^5}{l \times {}^\circ\text{Brix} \times d}$$

where l = cell path length (cm).

Refined white sugars manufactured in the UK give colour figures of about 20 or less. EC Regulations 1265/69 and 2103/77 also refer to the Brunswick Institute method, which involves visual comparison of the solid sugar against a standard colour scale.

Colouring matter. After acidification of the sample solution most artificial dyes, if present, will be retained on an iron exchange resin column (p. 104). Ultramarine can be detected by adding dilute hypophosphorous acid to a concentrated solution of the sugar and allowing to stand. If ultramarine is present an unpleasant odour develops within an hour. A further test involves the preparation of a concentrated solution of the sugar in a tall cylinder followed by microscopical examination of any dark specks which fall to the bottom.

Extraneous water-insoluble matter. Dissolve a large sample in water (preferably 1000 g + 1800 ml hot water), pass through a membrane filter and wash with 1000 ml hot water, dry and weigh (Hibbert and Phillipson, 1966). The insoluble matter in good quality white sugar is usually less than 25 mg/kg.

Trace materials. Examine for arsenic, copper and lead.

Sulphur dioxide. Determine by Shipton's method or colorimetrically using rosaniline (Carruthers *et al.*, 1965).

Methods of analysis for other specified sugar products

The following official methods are described in the EC Methods of Analysis Directive (79/796/EEC).

Dry matter. 10 g portions of glucose syrup, dried glucose syrup, dextrose monohydrate and dextrose anhydrous are dried at 70 ± 1 °C in a vacuum oven at a pressure not exceeding 3·3 kPa (34 mbar). The test portions of glucose syrup and dried glucose syrup are first mixed with 10 ml water and a weighed amount of dry kieselguhr. Dry matter in sugar solutions, invert sugar solutions and invert sugar syrups is measured by refractometer at 20 °C and, by reference to tables, expressed as per cent sucrose by weight (°Brix).

Reducing sugars. In sugar solutions, invert sugar solutions and invert sugar syrups, reducing sugars are determined by copper reduction using either the Lane and Eynon constant volume or the Luff–Schoorl procedure. For glucose syrup, dried glucose syrup, dextrose monohydrate and anhydrous dextrose, reducing sugars are determined by the Luff-Schoorl procedure or for expression as Dextrose Equivalent by an elaborate Lane and Eynon constant titre procedure.

Sulphated ash. This is determined in glucose syrup, dried glucose syrup, dextrose anhydrous and dextrose monohydrate by the normal procedure (see under Sugar).

SYRUP, TREACLE, MOLASSES

These materials are viscous sweet liquids containing uncrystallisable sucrose, reducing sugars, dextrin, organic acids, nitrogenous matter and inorganic constituents.

Golden syrup, which contains the highest proportion of sucrose and invert sugar, is prepared from residues obtained during the production of refined sugar. After removing the crystallisable sugar, the residue is boiled with dilute sulphuric acid. After neutralising with chalk, the syrup is filtered through charcoal and concentrated under reduced pressure to at least 83 per cent solids. In order to achieve desirable high viscosity, golden syrup must have a high solids content, much higher than could remain in solution if sucrose was the only sugar present. By controlled partial inversion of sucrose, more than 83 per cent solids is achieved which, although supersaturated, does not crystallise easily.

Molasses are the crude syrups obtained from cane and beet sugar production. It is the mother liquor remaining when the non-sugars (ash and organic non-sugars) has increased to about one-third of the total solids of the syrup. At this point the organic non-sugar content = 1·25 × ash content, approximately.

Table 6.11 shows the percentage composition of some of these products. In recent years, most good-quality samples of golden syrup and treacle have complied with the limits in Table 6.12.

The iron content of golden syrup should preferably not exceed 100 mg/kg.

Syrup, BP contains 66·7 per cent sucrose (m/m) and should comply with the following: optical rotation at 20 °C +56° to +60° and density at 20 °C 1·315–1·333 g/ml.

Analysis of Syrups

Water. Dry on sand after dilution with water and mixing at 70 °C in a vacuum oven for at least five hours. Drying at 100–105 °C can produce high results as fructose tends to decompose at temperatures higher than 75 °C in the presence of water.

Table 6.11 Typical composition of molasses and golden syrup

	Cane molasses		Beet molasses
Sucrose	30·5	48·0	48·5
Reducing sugars	22·0	0·5	2·5
Organic non-sugars	12·0	14·5	14·5
Ash	10·5	11·0	9·5
Water	25·0	26·0	25·0

	Golden syrup			
		2	3	4
Sucrose	31·7	26·1	33·9	34·5
Invert sugar	45·3	47·5	39·8	45·6
Ash	1·6	3·5	1·3	2·2
Water + organic matter (not sugar)	21·4	22·9	25·1	17·7

Table 6.12 Range of values for good quality samples of golden syrup and treacle

	Golden syrup % m/m	Treacle % m/m
Total solids (minimum)	83	83
Total sugars (as invert sugar)	79–83	72–80
Sucrose	31–33	25–33
Reducing sugars (as invert sugar)	47–50	37–50
Sulphated ash (mximum)	1·7	4·0

In refineries it is more usual to measure the refractive index, calculate as pure sucrose and apply corrections for invert sugar and ash:

$$\text{True solids} = \frac{\%\ \text{Refractometric solids}}{1 - (0{\cdot}000\,25 \times \%\ \text{Invert}) + (k \times \%\ \text{Ash})}$$

For cane syrups and molasses, $k = 0{\cdot}0043$.
For beet syrups and molasses, $k = -0{\cdot}0008$.

The formula only applies to syrups containing 20–30 per cent water.

Ash. Determine the sulphated ash.

Sugars. Sucrose and invert sugar can be determined by double polarisation. Glucose syrup, if present, gives a high positive direct rotation and the rotation after inversion is appreciably positive also. If the *K* value is also determined, the equations given in the section on Glucose Syrup can be applied. In the past treacle and golden syrup have been found to contain glucose syrup and several successful prosecutions were reported. There appears, however, to have been no objection taken to similar products, which have been described by other names such as 'crystal syrup' and 'amber syrup'.

The detection of raffinose in syrup indicates the likely presence of products derived from sugar beet.

Trace metals. Examine for arsenic, lead and copper.

Sulphur dioxide. Determine by Shipton's method (Chapter 4).

GLUCOSE SYRUP, ISOGLUCOSE

Glucose syrup (liquid glucose) is defined in EC legislation as a refined aqueous solution of D(+) glucose, maltose and other polymers of D(+) glucose obtained by the controlled partial hydrolysis of edible starch. The Codex definition, which is also used in UK legislation, is a purified concentrated aqueous solution of nutritive saccharides obtained from starch. (For legislative requirements see Table 6.10.) The standards in Codex Stan 9–1981 are the same except that up to 40 p.p.m. of sulphur dioxide is permitted for uses other than in confectionery manufacture. Glucose syrups are often somewhat confusingly referred to as 'glucose' in the food industry, and even worse, syrups prepared by enzyme treatment to contain very high concentrations of glucose are also referred to as 'glucose syrups'. To avoid confusion, 'dextrose' is also used in this chapter where necessary for clarity of expression.

For economic and technical reasons most glucose syrup is at present produced from maize starch but other sources of starch, for example potato and tapioca, and especially wheat, are now also used (Tegge, 1984). Until the 1960s glucose syrup was prepared solely by acid hydrolysis giving syrups with Dextrose Equivalent (DE, see p. 195) values of 30–35. 'Regular' glucose syrup, standardised at 42 DE is still used. Acid hydrolysis is random in nature so that at DE values of less than 30, high molecular weight fractions of starch may persist giving cause to haziness in the syrups. For high DE syrups acid hydrolysis causes sugar degradation with the formation of bitter tasting and coloured compounds.

Since the middle 1960s, selected enzymic conversion has been applied to produce a wide range of starch hydrolysis products (Fullbrook, 1984). Normally, enzymes are not able to react with the intact starch granule so an initial liquefaction process is required in which the starch is broken down into maltodextrins prior to the saccharification process. This liquefaction can be achieved by gentle acid hydrolysis (to DE 5–7) then the addition of α-amylase to obtain DE of about 10–12, or by the use of selected enzymes without acid treatment. Maltodextrins are glucose polymers with an average chain length of 5–10 glucose units per molecule. As powders they are tasteless and odourless at DE of less than 20 and are used as bulking and bodying food ingredients. Their saccharide composition is shown in Table 6.13.

Table 6.13 Typical composition of commercially available starch hydrolysis products

Type of product	Carbohydrate composition %				
	Glucose	Maltose	Maltotriose	Maltotetrose	Higher saccharides
Glucose syrups					
42 DE acid converted	18·5	14·5	12·0	10·0	45·0
42 DE acid–enzyme high maltose	5·9	44·4	12·7	3·3	33·7
42 DE enzyme–enzyme high maltose	2·5	58·0	18·0	2·0	19·5
64 DE acid–enzyme high conversion	37·0	31·5	11·0	5·0	15·5
97 DE dextrose syrup	95	2			3
Maltodextrin powders					
15 DE maltodextrin	0·6	4	7		88·4
20 DE maltodextrin	0·8	5·5	11·0		82·7

The liquefied starch solution is subjected to various enzymic saccharification processes to produce syrups of required characteristics and composition, the principal types being high maltose, high conversion and dextrose syrups. For the production of high maltose syrups, low in glucose content (and therefore low in sweetness), enzyme liquefied starches are treated with α-amylase for 24–48 hours. These syrups show reduced tendency to crystallise and are relatively non-hygroscopic. High conversion syrups of DE 60–70, e.g. Sweetose, are formed by the use of blends of α-amylase and amyloglucosidase for about 40 hours to hydrolyse acid converted DE 42 syrups to predominantly approximately equal mixtures of glucose and maltose. These syrups are used as sweeteners and resist crystallisation at 80–83 per cent solids down to 4 °C. If acid or enzyme liquefied starch is treated with amyloglucosidase for 48–96 hours, almost total conversion to glucose occurs. These dextrose syrups are used for non-yeast fermentations, for the preparation of crystalline dextrose and for the production of fructose syrups. The carbohydrate composition of the generally commercially available glucose syrups are shown in Table 6.13. Glucose syrups usually contain 75–85 (average 80) per cent total solids and 0·3–0·5 per cent of sulphated ash, most of the mineral matter being sodium chloride.

The numerous applications of glucose syrups as food ingredients are related to their DE. The lower DE syrups are more viscous, less likely to form coloured compounds due to browning reactions, less sweet (lower dextrose), less hygroscopic and less fermentable than higher DE syrups. Foods in which glucose syrups (liquid or dried) find extensive use include confectionery, preserves, desserts, ice cream, alcoholic beverages and soft drinks.

Since 1976, immobilised glucose isomerases have been used to convert dextrose syrups into high fructose syrups (isoglucose, isomerised syrup or high fructose corn syrup) containing in the solids about 42 per cent of fructose, 50 per cent of glucose and 8 per cent of oligosaccharides. These syrups have a sweetness comparable to sucrose and are being used to replace sucrose or invert sugar syrup in food formulations, e.g. in fruit drinks. More recently, separation techniques have been applied to high fructose syrup to prepare syrups containing 55 per cent and more, up to 100 per cent of fructose. Blanchard and Geiger (1984) have comprehensively reviewed the technology of high fructose syrups. Council Regulations (EEC) 1785/81 defines isoglucose as the product obtained from glucose or its polymers containing at least 10 per cent fructose in the dry state.

The carbohydrate composition of glucose syrups can be estimated using data from polarimetric and copper-reducing determinations, and specific rotations of assumed sugars present (Pearson, 1976). However, reliable data can only be obtained from specific sugar methods described earlier (TLC, GLC, HPLC and enzymatic methods). By the GLC determination of laevoglucosan, a reversion product formed by the recombination of D-glucose residues under the effect of heat and acid, Kheiri and Birch (1969) have shown that the method of manufacture of glucose syrups can be allocated.

Determination of glucose syrup content of foods containing other sugars

If the type and composition of the glucose syrup present is known, or a sample of it is available, an accurate determination is possible, but if there is no information on the glucose syrup ingredient only an approximate assessment can be made based on assumptions of average composition. A TLC examination will provide useful information on the sugars present in the sample and an approximation of the glucose syrup type and concentration may be assessed by the density of the maltose spot or the dextrin streak compared with those from applied glucose syrup standard solutions. The most common problem is the determination of glucose syrup in the presence of sucrose and inverted sucrose, for example in jams.

If S = percentage sucrose determined by the Clerget method,
D = direct saccharimeter reading,
K = K value determined by the Lane and Eynon method (i.e. BI),
G = percentage of glucose syrup, and
Y = percentage of invert sugar
then on the assumption that the glucose syrup is of average composition (36·5 DE, 179°S for a normal weight in 100 ml, $[\alpha]_D^{20}$ 119)

$$0{\cdot}365G + 1{\cdot}0Y = K$$

$$\text{and } 1{\cdot}79G - 0{\cdot}33Y = D - S$$

Hence an approximate concentration of glucose syrup and invert sugar can be deduced. Some invert sugar will originate from any fruit present, the sucrose of which in cooked products usually is almost entirely inverted. The presence of lactose from milk or milk powder inferferes with the above procedure. Rheinlander (1950) presents a similar scheme for the estimation of a variety of reducing sugars in the presence of sucrose. Analysts wishing to use such equations as those above should if possible insert more accurate DE and °S data for the appropriate glucose syrup being determined.

If the specific methods for sugars are used and in the assumed absence of high fructose syrups, the determination of fructose will on equivalence give the amount of invert sugar (fructose + dextrose) that will have originated from inverted sucrose, either added or in any fruit (also possibly from invert sugar syrup). If an approximate allowance can be made for the sugars provided by the fruit (usually 2–4 per cent of the jam) using data such as that in Paul and Southgate (1978), the approximate original amount of sucrose in the product may be deducted. From the measurement of total dextrose in the sample, and the deduction of dextrose provided by invert sugar, coupled with the compositional data on the possible glucose syrup present (Table 6.13), an estimate of the glucose syrup content of the sample may be approximately deduced. From the evidence of TLC or HPLC, the maltose content may also be used as an index of the glucose syrup content.

At 87 °C, the positive rotation of dextrose equals the negative rotation of fructose. Invert sugar, therefore, shows a zero rotation at 87 °C. The rotation at 87 °C of the solution after inversion is entirely due to glucose syrup. Owing to the expansion of the solution due to the higher temperature employed readings taken at 87 °C should be multiplied by 1·0315.

Methods for the detection and determination of high-fructose glucose syrups and other high fructose products in foods containing other sugars are not yet established (see under honey).

Dextrose and lactose. Dextrose is manufactured from glucose syrup (see above). It is much used in sugar and flour confectionery and in ice-cream. Lactose is extracted by crystallisation from whey. Standards for dextrose (anhydrous and monohydrate) and lactose are summarised in Table 6.10.

HONEY

Honey is the sugary substance produced from the nectar of flowers by the worker bees. It is the oldest sweet food known to man; rock paintings show it being collected from bees nests at least 15 000 years ago.

The world production in 1984 was estimated as 940 000 tonnes, and the total exports on to world markets was about 220 000 tonnes. UK imports in 1984 were 20 900 tonnes coming mainly from Mexico (5930 tonnes; Yucatan, orange blossom), Australia (4900 tonnes; blends, eucalyptus, bluebell, yellow box, clover), and China (2500 tonnes; blends, sunflower, acacia, lime, buckwheat). UK annual production is about 1500 tonnes.

Blended honey is normally graded by colour using the Pfund Honey Grader. The lighter the colour, the higher the quality and value. Table 6.14 gives a suggested international scale for honey colour.

Table 6.14 Draft International colour classification for honey

Colour	Pfund Scale (mm)
White	0–34
Extra Light Amber	35–48
Light Amber	49–65
Pale Amber	66–83
Medium Amber	84–100
Amber	101–114
Dark Amber	115+

The colour of honey varies from almost colourless to nearly black according to its botanical source and to conditions of processing and storage it has undergone.

At normal temperatures honey is frequently supersaturated with respect to glucose and exists as a clear syrup mainly preferred by consumers. On storage, however, coarse granulation or crystallisation can occur, leading in some cases to fermentation. The tendency to granulate depends on a number of factors some of which remain unknown. The ratio of sugar components and moisture have a major effect and White *et al.* (1962) have shown that honeys with a glucose/water ratio of 2:1 or more granulate rapidly, whereas honeys with a ratio of 1·7:1 or less tend to remain liquid.

In times of drought when nectar is scarce, the honeybee collects honeydew, which is a sugary deposit left by certain insects on the green parts of plants. Honeydew honey is an inferior product (see below).

Composition

The main constituents of honey are moisture, glucose (dextrose), fructose, maltose, sucrose, mineral matter and proteins. Proline is the predominant free amino acid. Pollen is invariably present in comb honey, but may be absent in some honeys which have been very finely strained. Enzymes present cause changes in the proportions of the original sugars present and the sucrose may disappear completely on storage. The presence of over about 5 per cent of sucrose may be due to artificial feeding of bees with sugar during the winter or during periods of drought.

Most natural honey has a negative optical rotation, but sucrose and glucose syrup both make the rotation more positive. Technical invert sugar may be present in imitation honeys. High fructose glucose syrup (q.v.) contains fructose and glucose in approximately the same ratio as in honey. Therefore, this bulk sweetener may find use in honey adulteration or the manufacture of imitation honeys.

The ranges of composition figures for various honeys due to Mitchell and co-workers are given in Table 6.15. The composition of honey and methods of analysis

Table 6.15 Ranges of figures obtained on floral honeys from various sources (Mitchell *et al.*, 1954; Kirkwood *et al.*, 1960 and elsewhere)

Water (%)	15·7–26·7
Ash (% d.b.)	0·04–0·93
Nitrogen (% d.b.)	0·05–0·38
Reducing sugars (% d.b.)	85·0–94·9
Glucose (%)	25–40
Fructose (%)	30–45
Sucrose (%)	1–2
Specific rotation $[\alpha_D^{20}]$	−20·4 to +4·8
Dextrin (% d.b.)	1·70–5·22
Free acid (ml 0·1 M NaOH/100 g)	12·9–58
pH	3·6–5·6
Hydroxymethylfurfural (mg/kg)	10–14
Proline (mg/100 g)	51. 17 (range 17–102)

have been reviewed and described by White *et al.* (1962). These authors give the following figures as the averages of 490 American honeys examined: water 17·2 per cent, fructose 38·2 per cent, glucose 31·3 per cent, sucrose 1·3 per cent, maltose 7·3 per cent, higher sugars 1·5 per cent, pH 3·91, free acidity 22 mEq/kg, lactone 7·1 mEq/kg, total acidity 29·1 mEq/kg, lactone/free acid ratio 0·335, ash 0·17 per cent, N 0·041 per cent, diastase value 20·8 (as ml of 1 per cent starch hydrolysed in 1 g in 1 hour.

The physical characteristics and composition of honey are fully reviewed by White (1975) and a detailed literature survey is by Amor (1978).

Legislation

The Honey Regulations 1976 SI No. 1832 as amended, implement Council Directive 74/409/EEC (OJ No. L221, 12.8.74, p. 10). The Regulations prescribe definitions and composition as follows:

Apparent reducing sugar content, calculated as invert sugar	
Honeydew honey or blends containing honeydew honey	not less than 60%
Other honeys	not less than 65%
Moisture content	
Heather honey (*Calluna* sp.) or clover honey (*Trifolium* sp.)	not more than 23%
As above but described as 'bakers' or 'industrial'	not more than 25%
Other honeys	not more than 21%
Other honeys; described as 'bakers' or 'industrial'	not more than 25%
Apparent sucrose content	
Honeydew honey or blends containing honeydew honey, acacia honey, lavender honey and *Banksia menziesii* honey	not more than 10%
Other honeys	not more than 5%
Water-insoluble solids content	
Pressed honey	not more than 0·5%
Other honeys	not more than 0·1%
Ash content	
Honeydew honey or blends containing honeydew honey	not more than 1%
Other honeys	not more than 0·6%

Acidity	not more than 40 mEq/kg
Diastase activity	
Citrus honey or other honey with a naturally low enzyme content	not less than 3 DN units
Other honey	not less than 4 DN units
Hydroxymethylfurfural content	not more than 80 mg/kg

The Codex method for determining diastase activity is included in the Regulations. No official EC methods have yet been finalised for the determination of the other criteria. The Codex standards for honey (CAC/RS 12, 1969), upon which the EC standard are based, include methods of analysis.

Analysis

Sampling. Granulated honeys should be melted by heating in a water bath at 50 °C. Honeys intended for HMF or diastatic activity determination should not be heated.

Total solids and water. The total solids can be estimated by drying in a vacuum oven at 70 °C. Alternatively, the figure can be calculated from the specific gravity of a 20 per cent m/v solution

$$\text{Total solids in solution (\%)} = \frac{\text{sp. gr. of 20\% solution} - 1}{0{\cdot}003\,86}$$

Chataway (1932) and Wedmore (1955) have presented figures relating the refractive index of honey at 20 °C with the water content:

RI	1·5044	1·5018	1·4992	1·4966	1·4940
Water (%)	13·0	14·0	15·0	16·0	17·0
RI	1·4915	1·4890	1·4865	1·4840	1·4815
Water (%)	18·0	19·0	20·0	21·0	22·0
RI	1·4790	1·4765	1·4740	1·4714	
Water (%)	23·0	24·0	25·0	26·0	

Temperature correction = ±0·000 23 per degree above or below 20 °C

The AOAC methods book has a larger table, with refractive indices also at 40 °C.

White (1969) has reviewed chemical and physical methods for moisture in honey. Most genuine honeys contain 13–23 per cent of water.

Ash. Char the sample preferably under an infra-red lamp and complete ashing at 600 °C. The ash of genuine honey seldom exceeds 0·35 per cent. Only a minute trace of sulphate should be present in pure honey.

Total nitrogen. Determine by the macro-Kjeldahl method. The nitrogen in the dry material is usually less than 0·25 per cent.

Diastase activity. Determine by Codex method described in the Regulations.

Proline. Burroughs and Otlet (1986), using a colour reaction with ninhydrin in the presence of formic acid (details in the AOAC methods book), reported a mean value of 51·6 mg/100 g for 102 genuine UK honeys (range 17–102, SD 17·3), which agrees reasonably with earlier American data (White and Rudyj, 1978).

DETECTION OF ADDED SWEETENING MATERIALS

Detection of invert sugar by the determination of HMF. During the manufacture of invert syrup by acid hydrolysis of sucrose, 5-hydroxymethylfurfuraldehyde (HMF) is also formed; HMF and honey adulteration is discussed by White and Siciliano (1980). Lampitt *et al.* (1929) improved the Fiehe's test for invert sugar in honey, which depends on the formation of a red colour when HMF reacts with resorcinol. They showed that small amounts of HMF occur naturally in honey due to acid-catalysed dehydration of the hexose sugars.

The following method due to Winkler (1955) for the determination of HMF in honey is included in the Codex standard.

REAGENTS

Barbituric acid solution. Transfer 500 mg barbituric acid and 70 ml water to a 100 ml volumetric flask, place on a water bath until dissolved, cool and make up to volume with water.

p-*Toluidine solution.* Dissolve 10 g *p*-toluidine (*Beware: carcinogenic reagent*) in 50 ml isopropanol by gentle warming on a water bath in a 100 ml volumetric flask. Cool, add 10 ml glacial acetic acid and make up to 100 ml with isopropanol. Store in the dark.

Oxygen-free distilled water. Pass nitrogen gas through boiling water and cool.

PROCEDURE

Dissolve 10 g unheated honey in 20 ml oxygen-free cold water, transfer to a 50 ml volumetric flask and make up to the mark. Into each of two test tubes add without delay 2·0 ml honey solution and 5·0 ml *p*-toluidine solution. Immediately add 1 ml water (blank) and to the other 1 ml barbituric acid solution. Measure the absorbance of the sample against the blank in a 1 cm cell at 550 nm as soon as the maximum value is reached.

For the calibration use a standard solution of HMF after spectrophotometric assay at 284 nm (molar extinction coefficient = 16 830). Use 0·300 μg for the standard. An approximate figure can be obtained from the equation:

$$\text{HMF (mg/100 g)} = \frac{\text{absorbance}}{\text{cell path length}} \times 19{\cdot}2$$

Winkler (1955) used a geometric correction formula for the spectrophotometric determination and quoted figures for genuine honey of 0–4 mg/100 g and for artificial products 50–150 mg/100 g. The amount of HMF in invert syrup is very variable and depends on the conditions of hydrolysis. Dhar and Roy (1972) claimed improved accuracy in the determination of HMF by employing absorption on charcoal prior to colorimetry. The Winkler method has been criticised for the instability of the developed colour, its temperature dependence and the use of a carcinogenic reagent. An ultraviolet absorption method overcoming these problems and based on the removal of the 284 nm chromophore of HMF has been proposed by White (1979a), and collaboratively studied (White *et al.*, 1979) (see AOAC methods book).

Marini and Righi (1985) reported a simple and rapid HPLC method. Honey (10 g) is diluted to 50 ml with water, then filtered through a 0·45 μm Millipore. The filtrate is injected onto a 250 mm column of 10 μm RP18 and developed isocratically using a mobile phase of methanol/water (10 + 90). The HMF is detected at 285 nm and is compared with prepared standards of about 100 mg/kg (see also Jeuring and Kuppers, 1980).

White *et al.* (1964) have shown that during the storage of honey there is an increase in HMF content concomitantly with losses of diastase and invertase.

Determination of high fructose glucose syrup. Doner *et al.* (1979) used GLC of the trimethylsilyl derivatives of the sugars present to measure the ratio of isomaltose to maltose (R) using cholestane as an internal standard. This ratio is reversed in high fructose glucose syrup (HFGS) compared to genuine honey. The authors give a discriminatory equation, $D = 2{\cdot}73 - 5{\cdot}35\,R$, such that honey is adulterated if D is negative (R more than 0·51), and is genuine if D is positive (R less than 0·51).

A collaborative study of a method based on the difference between the stable carbon isotope ratio $^{13}C/^{12}C$ in honey and HFGS was reported by White and Doner (1978). For method details and interpretation, see AOAC methods book. For data on English honeys, see Burroughs and Otlet (1986).

Kushnir (1979) described a method suitable for a quality control laboratory based on the isolation of oligo- and polysaccharides using carbon column chromatography followed by concentration and examination by TLC (see AOAC methods book). This procedure also identifies adulteration by glucose syrup and invert sugar. It has been collaboratively studied by White *et al.* (1979).

Sugar syrups. Cane sugar syrups and corn syrups can be detected in honey at above 15 per cent adulteration by the carbon isotope ratio mass spectrometry method (see above). Beet sugar is however produced by the more usual photosynthesis pathway and cannot be detected by the method. Stable isotope mass spectrometry in honey analysis is reviewed by Crofts (1987).

Beet sugar adulteration may be identified using galactose oxidase to measure any raffinose present (NB honeydew contains raffinose). Honey with more than 8 and less than 80 mg/100 g of galactose is possibly adulterated (White *et al.*, 1986).

Suspect samples of honey should be analysed for HFGS, HMF, proline and titratable acidity.

FREE ACID AND pH

Natural enzyme activity in honey leads to the production of free acids so that high acidity is indicative of prolonged storage. The pH is related to buffering power of the mineral matter and the lactone content produced by natural hydrolysis of glucose-lactone (Chandler, 1977).

Dilute 10 g with 75 ml water and titrate with 0·1 M sodium hydroxide using phenolphthalein (to pH 8·3 if a meter is used). Express the result as percentage of formic acid or as milliequivalents of acid per kilogram of sample (= titre × 10).

The pH value can be determined on a 10 per cent solution.

DETERMINATION OF CARBOHYDRATES

The principal sugars present in honey are fructose and glucose together with maltose, a little sucrose and dextrin. Doner (1977) in a review on the sugars of honey points out that the 'dextrin' fraction includes a mixture of at least 22 di-, tri- and higher oligosaccharides.

The major individual sugars may be detected and determined by specific methods described earlier in this chapter. White (1979b) has reported on a collaborative study on a selective absorption method and HPLC (see AOAC methods book).

In the Codex standard sucrose and reducing sugars are determined before and after inversion by the Lane and Eynon method.

The polarimetric readings are useful for assessing the genuineness of honey. Formerly it was usual to regard with suspicion any honey which gave a positive optical rotation, as such would be occasioned by the addition either of glucose syrup or sucrose. It is usually assumed that if the specific rotation falls between +5° and −15° and if these figures are only slightly altered after inversion, the sample is probably genuine. If the specific rotation does not fall within these limits (or if a more complete examination is necessary), the copper reducing power should also be determined and calculated as dextrose.

Sucrose (S) can be calculated from the Clerget–Hertzfeld formula (but see p. 194). When the percentage of 'dextrin' is small the percentages of glucose (g) and fructose (f) can be calculated from the polarisation and the K value:

$$g + 0{\cdot}915f = K$$

$$f = \frac{0{\cdot}793K + S - D}{2{\cdot}08}$$

where D = direct polarisation on the International Sugar Scale at 20 °C.

When an appreciable quantity of 'dextrin' is present, the composition may be calculated in terms of sucrose, invert sugar and dextrin, as Clerget's formula gives the cane sugar, and the K value divided by 0·915 gives the invert sugar (or the weight of reduced copper may be calculated directly to invert sugar) (cf. Marshall and Norman, 1938).

Glucose syrup can be estimated from Browne's formula, which is based on the difference in the invert polarisations at 20 and 87 °C. The difference is due to the lowering of the rotation of fructose with increasing temperature and, therefore, depends on the amount of invert sugar present. Glucose syrup if present decreases the difference in the polarisation figures irrespective of the specific rotation. Browne's formula assumes that the average percentage of invert sugar after inversion in honey is 77 per cent:

$$\text{Glucose syrup (\%)} = \frac{100 - 7700(V_{87} - V_{20})}{26{\cdot}7 \times I}$$

where V_{87} and V_{20} are the saccharimeter polarisation figures for a solution containing the normal weight when measured in a 200 mm tube at 87 and 20 °C respectively. I is the percentage of reducing sugars in the sample after inversion, calculated as invert sugar.

Microscopical examination. A microscopical examination of honey should always be made. Genuine honey almost invariably shows some pollen grains and traces of wax beside crystals of sugar. Starch granules should be absent. The honey is diluted with about twice its volume of water and allowed to settle overnight in a conical tube. The sediment, which will include the pollen grains, is removed with a pipette and mounted on microscope slides. Separation with the centrifuge is not so satisfactory (James, 1969). An attempt may be made to identify the pollen grains in order to verify the floral and geographical origin of the honey implied by the label description (Howells, 1969; Sawyer, 1975).

For example, a firm was fined for selling honey with a false description as to weight and for describing imported honey as 'Pure English Honey'. The analyst's evidence was established by the presence of pollen from eucalyptus and other Australian flora (Anon, 1949). Davies (1976) has applied amino acid analysis to the determination of the geographical origin of honey.

Honeydew honey. During periods of prolonged drought bees may have to supplement their normal supplies of nectar from plants by honeydew, which is a sweet, sticky fluid excreted on foliage by leaf-sucking insects (e.g. aphids). Honeydew represents a nuisance as the resultant honey tends to be dark and have an undesirable taste. Compared with floral honey, honeydew honey contains less fructose and glucose but appreciably more reducing disaccharides and higher sugars, including raffinose and melizitose. It has a higher ash, higher pH and is dextrorotatory.

The examination of the occurrence of honeydew in honey has been extensively studied by Kirkwood *et al.* (1960, 1961) and Cowan and Mitchell (1964).

JAM, MARMALADE AND JELLY

The basic principles of jam and marmalade manufacture is the boiling together of fruit (suitably prepared), sugars (sucrose syrup, 63 DE high conversion glucose syrup and to a lesser extent nowadays, invert sugar syrup) and water. For marmalade, the peel is cooked separately prior to mixing in. During the boiling the proportion of solid matter in the mix increases (due to the evaporation of water), a proportion of the sucrose is acid hydrolysed to invert sugar and a gel is produced on cooling. It has been shown that three separate components, namely sugar, pectin and acid, all play a part in forming the gel. Where the fruit is low in pectin or acid, therefore, the deficiency is made up by adding solutions of such ingredients towards the end of the boiling. Jelly manufacture involves boiling the fruit with water and then the extract (after filtration) is boiled with the sugars.

When the process is carried out under atmospheric conditions, whereby the boiling point reached is about 106 °C, the sugar tends to caramelise excessively and the flavour, etc. may be impaired. Such quality deterioration is less likely to occur if the boiling is carried out under vacuum (b.p. about 71 °C). The vacuum process is however less likely to effect removal of sulphur dioxide from sulphited pulps. Sulphited pulps have their natural colour heavily bleached by the SO_2 so added colouring matter is necessary to restore appearance. Artificial colours (Chapter 4) are used but natural colours, particularly anthocyanins from grape skins, and also chlorophyll, crocin, anatto and cochineal are now finding wider use.

Diabetic jams are usually prepared using sorbitol in place of sucrose and glucose syrup. Sorbitol, a polyhydric alcohol, is absorbed from the intestine more slowly than glucose but it is metabolised to give approximately the same energy per unit weight as sucrose. Diabetic jams based on sorbitol, therefore, should not be regarded as, or claimed to be, low in calories. Typical composition of diabetic jams is 2–4 per cent sugars from the fruit and 60–65 per cent sorbitol giving an energy value of 240–260 kcal/100 g, about the same as normal jam.

For further information on the manufacture of jam, marmalade and other preserves, including quality control aspects, see Slater (1986), Broomfield (1988) and Reardon (1988). Analytical data found with various fruits are given in Table 6.17.

Legislation

The Council Directive 79/693/EEC (OJ No. L205, 13.8.79, p. 5) has been implemented by the Jam and Similar Products Regulations 1981, SI No. 1063, as amended (1989), which lay down standards for extra jam, jam, extra jelly, jelly, marmalade and chestnut puree. Additional to the requirements of the Directive, the UK Regulations include standards for UK standard jelly, mincemeat and fruit curds.

For extra jam and extra jelly, no sulphited fruit, pulp, puree, juice or extract may be incorporated, and if more than one fruit is used, the product must not contain apples, pears, clingstone plums, melons, grapes, pumpkins, cucumbers or tomatoes. Minimum standards for fruit content from the Directive (fruit juice or fruit extract content for jelly products) are listed in Table 6.16. Marmalade must not

Table 6.16 Minimum standards for fruit content

	Extra Jam, Extra Jelly (%)	Jam, Jelly (%)
General rule	45	35
Blackcurrants, rosehips, quinces	35	25
Ginger	25	15
Cashew apples[a]	23	16
Passion fruit	8	6

[a] Cashew apples are the yellow or scarlet fleshy pear shaped enlarged peduncles or fruit stalks of the cashew tree.

contain less than 20 per cent of citrus fruit pulp, puree, juice, extract, peel or any combination provided that at least 7·5 per cent is derived from the endocarp. The UK standard jelly is the same as for jelly except that the minimum standards refer to the quantity of fruit from which the juice or extract have been obtained, i.e. a lower standard. Chestnut puree is required to contain not less than 38 per cent of pureed chestnuts.

The minimum specified refractometer solids at 20 °C is 60 per cent except for reduced sugar jam and reduced sugar marmalade, where it is required to be between

30–55 per cent. These two types of product carry the same fruit content standards as for jams and marmalade.

The addition of ascorbic acid is permitted but only as an antioxidant. Colours are permitted only in jam, jelly, UK standard jelly and reduced sugar products. Mono- and diglycerides of fatty acids and a maximum of 10 mg/kg of dimethylpolysiloxane are permitted in all products, but sorbitan monolaurate (max. 25 mg/kg) may only be used in marmalades with little or no peel. Pectin (max. 10 mg/kg) and amidated pectin (max. 5 mg/kg) are permitted in all products except reduced sugar products. Scheduled products labelled for diabetic use may include sorbitol and permitted artificial sweeteners. Reduced sugar products may also contain artificial sweeteners. Sulphur dioxide maximum limits are 10 mg/kg in extra jam, extra jelly or chestnut puree, 50 mg/kg in jelly and 100 mg/kg in other products. (The amending Council Directive 88/593/EEC requires that sulphur dioxide must be declared in the ingredients list if present at more than 30 mg/kg.) Reduced sugar products may contain sorbic acid (max. 750 mg/kg), benzoic acid (max. 500 mg/kg) and various hydroxybenzoates singly or in combination (max. 500 mg/kg).

Analysis

The routine analysis of jam, jelly and marmalade should include the determinations of soluble solids, sulphur dioxide and trace metals and an examination for the type of added colours and fruit present. The fruit content of jam can be assessed from the insoluble solids and potassium contents. The determination of pectin is occasionally required and industrially the proportion of reducing sugars present and the pH value are of considerable importance. Jams and marmalades claiming to contain spirits or wine should be examined by distillation for the alcohol content.

Diabetic preserves should be examined for artificial sweeteners, sorbitol and total carbohydrate.

Soluble solids. The Regulations require the soluble solids to be calculated from the refractive index at 20 °C (uncorrected for insoluble solids). The sucrose conversion table is given on p. 188. The corrections that can be used for reducing sugar and citric acid if the true soluble solids are required are given on p. 189.

Carbohydrates. The sugars present can be identified by TLC or HPLC and determined on the cleared solution, both before and after inversion by a volumetric or polarimetric method. *Reducing sugars* alone are best determined by Lane and Eynon's method by titrating 10 ml of Fehling's solution with a 1·5 per cent solution of the jam. Manufacturers prefer the reducing sugar content to fall within the range 20–40 (calculated as a percentage of the preserve) in order to prevent the separation of crystals during storage. Crystals of sucrose may separate if the amount present is high (reducing sugars below 20 per cent) and invert sugar may form a honey-like mass if the reducing sugar content exceeds 42 per cent. The invert sugar content increases with increased acidity and boiling time. Inversion of sucrose occurs readily in the older open-pan type of jam boiling but in the modern lower temperature vacuum pan techniques, inversion is much reduced leading to the possibility of sugar crystallisation. The modern use of glucose syrups in jams and jellies considerably reduces the tendency to crystallisation. Commonly used is 4 parts of sucrose plus 1 part of glucose syrup solids (from a 55 DE acid converted syrup or a 63 DE acid/

enzyme syrup). The same phenomenon of sugar crystallisation is liable to occur with other preserves, confectionery, table jelly tablets, etc.

The presence of glucose syrup is indicated from the polarimetric readings if the rotation after inversion is positive or only slightly negative. In this case it is necessary to determine the copper reducing power. Then the proportions of sucrose, invert sugar and glucose syrup are given by the formulae on p. 215. Dry glucose solids may be roughly estimated by assuming 20 per cent of water in glucose syrup.

Sorbitol. Sorbitol is difficult and tedious to determine chemically. Hundley and Hughes (1966) have described a titrimetric procedure based on alcoholic extraction of the defatted product followed by hydrolysis in acid and alkali then oxidation with periodate and final titration with thiosulphate. Mannitol is also determined and TLC is needed to identify the sugar alcohols present. Methods for sugar analysis using HPLC can be applied to the determination of sugar alcohols (Schwarzenbach, 1977) but peak overlap from sugar ingredients can be a problem. Commercial 'carbohydrate' columns (polystyrene cation exchange in calcium ion form) and also the new polyol silica gel columns show much improved separations (see p. 201). The sugar alcohols may be separated and determined by GLC of their trimethylsilyl ether (TMS) derivatives (Jones *et al.*, 1966; Moseley *et al.*, 1978). Makinen and Soderling (1980) have determined sorbitol, mannitol and xylitol in fruit using polyacetyl ester derivatives and glass capillary GLC. Sorbitol may be conveniently determined enzymatically by oxidation with nicotinamide-adenine-dinucleotide (NAD) in the presence of sorbitol dehydrogenase to fructose (BCL, 1987). The amount of reduced NAD (NADH), measured at 340 nm, is equivalent to the sorbitol concentration.

FRUIT CONTENT

The minima for fruit in jams, jellies and marmalade prescribed in the Regulations are given above. The fruit content can be assessed from

$$\text{Fruit (\%)} = \frac{\text{Percentage of } A \text{ in the jam}}{\text{Average percentage of } A \text{ in the fruit}} \times 100,$$

where A is any suitable constituent, e.g. insoluble solids, non-sugar solids, seed count, potassium, phosphorus. As acid and pectin are frequently added to make up for deficiencies in the fruit, such values are of doubtful value for calculating the fruit content. Ranges of figures for various determinations on fruits have been the subject of numerous papers, e.g. Lampitt and Hughes, 1928; BFMIRA, 1931; Hughes and Maunsell, 1934; Macara, 1935; Dargie, 1936; Brown, 1938; Hinton and Macara, 1940; Money and Christian, 1950; Money, 1958, 1964; Osborn, 1964; FSC, 1968 (see also Table 6.17). An extensive series of figures on strawberries have been tabulated by Goodall and Scholey (1975). The figures vary widely for each constituent and the consequent calculation of the fruit content using mean values can only be regarded as a reasonable approximation (FSC, 1968). Calculations based on the minimum figure for A usually give absurdly high fruit contents, but Money (1964) states probable ranges for the potassium and phosphorus contents of some fruits in terms of 95 per cent probability (see also Steiner, 1948, 1949).

Insoluble solids. The insoluble solids figure is probably the most commonly used for the calculation of the fruit content of jam. For the determination, stir 20 g of

Table 6.17 Analytical data of the edible portions of some common fruits (Macara, 1931; Money and Christian, 1950)

	Insoluble solids (%)	No. of seeds per 100 g	Refractometer solids (%)	Total solids (%)	Total sugars (%)	Non-sugar solids (%)	Acidity ml 0·1 N per 100 g	Pectin as calcium pectate (%)
Apples								
max.	2·54	—	16·4	18·1	13·0	6·5	270	0·79
min.	1·05	—	6·8	10·8	5·3	2·5	20	0·19
average	1·50	—	12·0	14·0	9·6	4·3	130	0·55
Apricots								
max.	2·49	—	18·5	18·3	11·8	10·4	400	1·32
min.	1·17	—	7·7	7·3	1·6	4·1	81	0·42
average	1·60	—	11·6	12·6	6·0	6·5	242	0·99
Blackberries								
max.	10·0	—	11·4	21·1	10·4	15·8	438	1·19
min.	4·05	—	6·0	12·1	2·4	8·3	75	0·40
average	6·33	3180	8·9	15·4	4·5	10·9	227	0·63
Blackcurrants								
max.	9·72	—	17·6	24·4	10·6	17·7	650	1·79
min.	3·28	—	8·0	13·7	1·6	9·1	121	0·55
average	5·92	4450	13·3	19·7	6·3	13·4	503	1·13
Cherries (Black)								
max.	3·92	—	22·2	25·0	17·3	9·8	169	0·80
min.	1·30	—	12·7	12·9	7·7	3·7	70	0·06
average	1·95	—	17·4	18·7	12·4	6·3	109	0·32
Damsons								
max.	2·45	—	27·3	29·1	13·8	15·6	453	1·35
min.	1·46	—	11·0	13·4	5·6	7·5	226	0·39
average	1·91	—	16·8	18·4	8·5	9·9	320	1·03
Gooseberries								
max.	4·38	—	13·6	15·1	10·2	10·3	415	1·20
min.	1·51	—	4·0	7·9	2·0	3·7	173	0·27
average	2·21	600	7·7	11·0	4·6	6·4	284	0·62
Greengages								
max.	1·99	—	25·0	26·7	14·5	13·9	435	1·42
min.	1·14	—	10·2	10·9	4·0	5·1	80	0·66
average	1·45	—	16·7	16·7	8·5	8·2	192	1·11
Loganberries								
max.	8·00	—	14·9	23·3	7·3	22·2	488	0·68
min.	2·15	—	7·8	13·2	1·1	7·2	151	0·50
average	6·35	—	9·6	16·3	4·3	12·0	375	0·59
Plums (Red)								
max.	1·99	—	18·0	21·5	13·4	10·0	380	1·43
min.	0·66	—	7·4	8·1	2·3	3·6	25	0·34
average	1·17	—	11·7	13·0	7·4	5·6	225	0·79
Raspberries								
max.	7·93	—	12·6	24·8	8·7	17·7	518	0·88
min.	1·40	—	5·0	9·4	1·7	5·8	106	0·10
average	5·12	4190	8·6	13·9	4·5	9·5	283	0·40
Strawberries								
max.	3·40	—	12·0	13·8	9·8	8·1	323	0·90
min.	0·86	—	4·5	6·8	2·8	2·6	81	1·13
average	2·14	—	8·2	10·2	5·6	4·6	136	0·54

sample with 200 ml hot water and boil gently for half an hour. Filter using suction through fine cloth on a Buchner funnel, preferably of the Hartley type. Wash the residue with hot water before scraping it off the cloth and re-boiling with more water. Re-filter through the cloth. Repeat the boiling and filtering and wash the residue thoroughly with hot water. Scrape the insoluble residue into a weighed dish using a little water to assist with the transference if necessary. Evaporate on a boiling-water bath, dry to constant weight and calculate the insoluble solids as a percentage of the sample (Table 6.17).

A useful alternative method involves extracting the preserve with water in a dried weighed Soxhlet thimble in an extraction apparatus. The water is changed after 1 hour. The insoluble residue is treated with acetone before drying in the 100 °C oven to constant weight.

Seed count. Heat 50 g of sample in a porcelain basin with a little water on a water bath and when the insoluble matter has sufficiently disintegrated, count the seeds. Calculate the number of seeds present in 100 g of jam (Table 6.17). The seed count is of assistance for mixed jams, where one fruit contains seeds and the other does not. The individual seeds should also be examined under the lens to confirm that they are genuine. Sets of genuine seeds should be kept for comparison purposes.

Non-sugar solids (Hughes and Maunsell, 1934) represent the difference between the total solids (after drying at 100 °C in a vacuum oven) and the total sugars (as invert sugar).

Lead number (Hinton, 1934), which depends on the acid constituents of the fruit, has also been suggested for determining the fruit content of jams (see also Moir, 1944).

Potassium and phosphorus. Money (1964) determined the potassium content on the jam solution by flame photometry. Phosphorus was determined after dry ashing by the colorimetric molybdenum blue method using hydroquinone-sulphite as reducing agents. Although average figures for these elements in several fruits are tabulated by Money, a warning is given on their application to jams of unknown origin due to the presence of significant amounts in ingredients such as liquid apple pectin, syrup and potassium metabisulphite. Data for the potassium and phosphorus contents of various fruits and fruit juices have also been tabulated by Osborn (1964).

Pectin. Pectins are water-soluble carbohydrate polymers containing long chains of partially methylated polygalacturonic acid. Pectin has the property of forming aqueous gels with sugar and acids. Commercial pectins are extracted mainly from citrus peel and apple pomace and are used to supplement the natural pectin in the fruits used in the manufacture of preserves.

Apples, apricots, plums, blackcurrants and oranges are high in pectin. Blackberries, strawberries, raspberries and pineapple contain moderate amounts whereas peaches and pears are low and cherries are very low in natural pectin.

In extra jam and jams, high methoxyl pectins (more than 50 per cent esterification) are used to supplement fruits. These pectins require more than 60 per cent of soluble solids and about pH 3 to form a set. More than 70 per cent esterified pectins are used for rapid gel formation to aid suspension of fruit particles. Low sugar preserves require low methoxyl pectins (below 50 per cent) and are dependent on available calcium ion levels for gel formation. Amidated low methoxyl pectins require the lowest calcium addition. The gelling rate depends on the ratios of calcium, methoxyl groups and amidation.

King (1925) has described the following method for the gravimetric determination of pectin as calcium pectate:

Weigh 50 g jam into a 500 ml beaker, add hot water, stir and heat on a boiling-water bath to disintegrate the tissues. Add hot 95 per cent alcohol, a little at a time with stirring, until the total volume is 300 ml. Stir the mass at the bottom frequently during 2 hours, keeping at about 50 °C, or until no gelatinous particles are visible. Filter on a coarse paper, preferably using a Hartley funnel. Wash the residue back into the beaker with 300 ml warm alcohol, heat, stir and re-filter. Dissolve the pectin from the residue with boiling water, filter, cool, add a small excess of sodium hydroxide solution (equivalent to 0·02 M) and allow the solution to stand for one hour. Then add acetic acid to give a concentration of free acid of 0·1 M followed by 20 ml of 10 per cent calcium chloride solution. Allow to stand for 1 hour and then boil and filter through a tared filter paper. Wash the gelatinous precipitate with boiling water, then transfer back to the beaker again and boil with 200 ml water. Filter through the same paper, wash with boiling water, dilute acetic acid and water again. Then dry and weigh as calcium pectate.

The separation of pectic substances and the colorimetric estimation of total pectin by reaction with carbazole is described by VdF (1987).

Other gelling agents. The Regulations permit alginic acid, alginates, agar, carrageenan, locust bean gum, guar gum, sodium carboxymethyl cellulose and xanthan gum in reduced sugar preserves only. The presence of diatoms may indicate the presence of agar (see also Parkes, 1921; Coste, 1923; King, 1925).

Acidity. Determine by titrating 10 g of sample with 0·1 M sodium hydroxide using phenolphthalein and calculate the total acidity as a percentage of citric acid (or malic acid if apple predominates). With highly coloured jams such as blackberry and blackcurrant the acidity should be titrated potentiometrically to pH 8·1.

The acidity figure is likely to be of little value for the calculation of the fruit content, as pH adjustment is often practised to improve the set. The substances that are added for this purpose include citric acid, the so-called 'buffer salts' such as sodium bicarbonate and sodium citrate. Industrially, the pH value is usually determined. A high pH often indicates unsatisfactory setting properties and a low pH suggests that the preserve will 'weep' on storage. Most jams have a pH value between 2·9 and 3·4.

Microscopy. The microscopic examination for the type of fruit must be made on the wet insoluble residue obtained after heating a quantity of the sample with water and decanting off the liquid. The microscopic appearance of the common fruit pulps is described by Winton and Winton (1935).

On the addition of iodine, apple pulp usually shows blue patches due to the presence of starch. Partridge (1926) has made the method more sensitive by preparing a 10 per cent extract of the jam with warm water (60 °C) and then straining and centrifuging. The iodine solution is added to the residue, which contains any starch present in a more concentrated form.

Preservatives. Most jam, except extra jam, is manufactured from sulphited pulp containing up to 3000 mg/kg of sulphur dioxide but most of the preservative is lost during boiling (but less so with vacuum cooking). *Sulphur dioxide* can be conveniently determined on 50 g by distillation from acid and titration with iodine.

The Regulations permit sorbic acid, sorbates, benzoic acid and benzoates, and methyl, ethyl and propyl hydroxybenzoates in reduced sugar preserves only (see Chapter 4 for methods).

Trace metals. Jams should be examined for arsenic, copper, lead and, if canned, tin.

Added colour. For artificial colours, add dilute acetic acid to a solution of the jam, extract and identify the colouring matter as described in Chapter 4. Natural colours are increasingly being used.

FRUIT CURDS

Fruit curds are gelled emulsions containing sugar, oil or fat, egg, modified starch, pectin, fruit or fruit juice and colour. In fruit flavour curds the fruit is replaced by a flavouring. Lemon curd is the most popular, followed by orange curd. Traditional lemon cheese contains butter and coagulated egg protein.

Legislation

The Jam and Similar Products Regulations 1981, SI No. 1063, as amended, define fruit curds as an emulsion of edible fat or oil (or both), sugar, whole egg or egg yolk (or both) and fruit, fruit pulp, fruit puree, fruit juice, aqueous extract of fruit or essential oil of fruit, or any combination thereof, with or without other ingredients. Fruit flavour curds are similarly defined except that the fruit ingredients are replaced by flavouring material. Lemon curd and lemon cheese must contain not less than 0·125 per cent of lemon oil and orange curd not less than 0·25 per cent of orange oil. In other products sufficient fruit ingredients for fruit curds or flavouring material for fruit flavour curds are required to be present to characterise the finished product. Mixed fruits and flavours are permitted. Curds and flavour curds are required to contain:

1. not less than 4 per cent of oil/fat;
2. not less than 0·65 per cent of egg yolk solids;
3. not less than 65 per cent of soluble solids by refractometer at 20 °C;
4. not more than 100 mg/kg of sulphur dioxide.

By including 'sugar' in the definitions, the Regulations (which refer to 'sweetening agents' for other products) seem to imply that sucrose is the only permitted sweetening agent, except for sorbitol and permitted artificial sweeteners in curds or flavour curds sold for diabetic use. However, some commercial curds have glucose syrup as an ingredient.

Analysis

The analysis of fruit curds should include the determination of the **soluble solids**, **acidity**, **fat**, **volatile oil** and **egg content** in addition to the examination for **preservatives**, **trace elements** and **added colour**.

Soluble solids. Determine by refractometer (as for jam). Industrially, the proportion of reducing sugars is also of importance as it indicates whether the preserve may crystallise on storage (see Jam). The use of TLC (p. 186) or specific sugar analysis can indicate the presence of glucose syrup or other sweetening matter.

Acidity. Titrate 10 g of sample with 0·1 M sodium hydroxide using phenolphthalein and calculate the total acidity as citric acid.

Fat. Either employ the Rose–Gottlieb method (use 1–2 g plus 9 ml water) or dilute the sample with water and extract the fat several times with ether. The figure obtained will vary with the efficiency of the method used in extracting the egg phospholipid.

Volatile oil. Distil the oil of lemon (or orange) from sodium chloride solution in the BP or other volatile oil apparatus (see p. 395) preferably using 300 g of sample. From the volatile oil figure obtained as v/m (per cent) and the density of the oil (0·85 g/ml), convert the result to m/m (per cent). Alternatively, citrus oils can be determined by titration with brominating solution after co-distillation with isopropanol (Symonds and Dedicoat, 1972):

APPARATUS

Distillation apparatus. 100 ml FB flask with B34 neck; distillation head fitted with 0–200 °C thermometer; condenser; 100 ml conical receiving flask.

Oil bath. At 125–130 °C.

SPECIAL REAGENTS

Brominating solution A (stock solution) (M/60). Dissolve 2·7835 g potassium bromate and 9·9 g potassium bromide in water and dilute to 1 litre.

Brominating solution B (M/240). Dilute 50 ml of A to 200 ml with water.

METHOD

Weigh 10 g sample into the 100 ml FB flask, add 10 ml warm water (40 °C) and swirl. Add anti-bumping granules and 25 ml isopropanol. Connect to the distillation apparatus, and distil from the oil bath. When the temperature in the still head reaches 100 °C, remove the receiving flask. To the distillate add 10 ml dilute hydrochloric acid (1 + 2) and 2 drops methyl orange and titrate slowly with brominating solution B until the indicator colour is destroyed. Carry out a blank determination on the reagents.

1 ml M/240 brominating solution = 0·95 mg lemon oil
= 0·85 mg orange oil.

Recoveries are of the order of 90 per cent. Sulphur dioxide, if present, distils over with the volatile oil and interferes such that 1 ml of B ≡ 0·83 mg SO_2.

Egg content. The method for determining egg in lemon curd is based on the organic phosphorus content (see also under salad cream) determined colorimetrically after wet oxidising the extracted total lipid. The chloroform methanol total lipid extraction procedure described in Chapter 2 is preferred but the following has been used.

Reflux 20 g of sample with 100 ml of alcohol (95 per cent) in a boiling-water bath for 6 hours. Filter on a Buchner funnel. Retain the filtrate and re-extract the residue as before with more alcohol and filter. Evaporate the combined filtrates to dryness in a basin. Wet oxidise the residue in the basin and determine the phosphate in the resultant solution colorimetrically (see Chapter 2).

dried egg (per cent) = 83 × per cent P_2O_5 in sample
egg yolk solids (per cent) = 58 × per cent P_2O_5 in sample

Bagnall and Smith (1945) pointed out that 95 per cent ethanol also extracts 0·08 per cent apparent P_2O_5 from wheat flour and 0·01 per cent from cornflour.

Casson and Griffin (1959, 1961) used choline from enzymatic hydrolysis of the extracted phospholipid as the index for egg content. The choline is separated and cleaned-up on a cation exchange resin column, then determined colorimetrically as the reineckate in acetone. More recently, a draft EC method describes the hydrolysis of the phospholipids by refluxing with barium hydroxide solution, then neutralisation with hydrochloric acid prior to the reineckate precipitation. Work at LGC has shown that using chloroform/methanol extraction, the EC method is satisfactory for the estimation of egg content in lemon curd and other foods including mayonnaise and noodles. Phospholipids from other ingredients will also contribute to the choline content.

Further work at LGC has shown that the egg content of foods can be satisfactorily estimated by gas chromatography of sphingosine, a long chain aminoalcohol constituent of egg yolk sphingolipids, present in whole egg at an average concentration of 25·1 mg/100 g. The separated sphingosine from lemon curd (with added phytosphingosine as internal standard) as the TMS ether derivative is cleaned-up using a Sep-Pak silica cartridge to remove interference from essential oils, prior to GLC.

Starch. Bagnall and Smith (1945) also describe the following procedure for determining starch in fruit curd, which is modified from the method of Illing and Whittle (1944). This method may not be suitable for modern fruit curd products, which contain modified starch.

To the residue on the Buchner funnel obtained in the method above add 75 ml of 0·7 per cent aqueous potassium hydroxide solution and gelatinise the starch by simmering gently for 30 min. Transfer the hot liquid to a 200 ml volumetric flask, cool and make up to volume. Filter. Neutralise 20 ml of filtrate with 5 per cent acetic acid using phenolphthalein. Add 8 ml of 0·1 M iodine solution and 4 ml of *either* 10 per cent potassium acetate solution *or* alcohol and allow to stand until the precipitate settles and then centrifuge. Decant off the supernatant liquid. Rub the residue with a rod and treat with 12 ml of a mixture containing 10 ml of 95 per cent alcohol and 2 ml 0·1 M sodium thiosulphate by addition in several small quantities. When the particles are thoroughly broken up with the rod, add 25 ml of 80 per cent alcohol and filter through a weighed Gooch crucible. Wash the residue with 95 per cent alcohol, dry and weigh the starch. The starch content is usually about 3–4 per cent. Insufficient starch or incompletely gelatinised starch tends to cause fat separation.

Other methods for starch are given in Chapter 8.

Preservatives. Nowadays preservatives are seldom met with in lemon curd. Only sulphur dioxide is permitted. Formerly, salicylic and benzoic acids were occasionally found. Antioxidants may be present from the added edible fat.

Trace elements. Fruit curd should be examined for arsenic, copper and lead.

Added colour. The starch should be precipitated with alcohol and the filtrate examined as for jam.

MINCEMEAT

Mincemeat is manufactured by mixing together apples, dried fruit (currants, sultanas and raisins), mixed peel, sugar, suet, acetic acid and flavourings such as salt, cinnamon, cloves, nutmegs and essence of lemon and caramel colouring. The mixture should contain sufficient soluble solids (about 68 per cent) and acetic acid to prevent growth of micro-organisms.

If the soluble solids content is 70 per cent or more, 0·3 per cent of acetic acid is sufficient. Bakery mincemeats may contain thickeners such as rusk, modified starch or sodium carboxymethyl cellulose. The manufacture of mincemeat is described by Broomfield (1988).

Legislation

The Jam and Similar Products Regulations 1981, SI No. 1063 define mincemeat as a mixture of sweetening agents, vine fruits, citrus peel, suet or equivalent fat and vinegar or acetic acid with or without other ingredients. The compositional requirements are:

1. not less than 30 per cent of vine fruits and citrus peel. Not less than 20 per cent of vine fruits;
2. not less than 2·5 per cent of suet or equivalent fat;
3. not less than 0·5 per cent of acetic acid;
4. not less than 65 per cent of soluble solids by refractometer at 20 °C;
5. not more than 100 mg/kg of sulphur dioxide.

Analysis of mincemeat

The routine analysis of mincemeat should include the determination of **soluble solids** (by refractometer), **fat**, the **fixed and volatile acidity** and an examination for **preservatives, antioxidants** and **trace metals.**

Sampling. The sample should be made homogeneous in a shearing blade type blender.

Fat. The fat can be determined by the Werner–Schmid method using 1–2 g sample plus 9 ml water. Alternatively, the fat can be determined by evaporating an aliquot of an extract obtained by macerating a large sample with chloroform and filtering.

Fixed and volatile acidity. The acetic acid content of mincemeat can be estimated by steam distilling 5 g of sample and titrating the distillate with 0·1 M sodium hydroxide using phenolphthalein as indicator (1 ml 0·1 M NaOH ≡ 0·006 g acetic acid). Alternatively, the volatile acidity can be obtained from the total and fixed acidities. For this, the total acidity is obtained by direct titration of a 5 g sample, and the fixed acidity by titrating a further 5 g sample after evaporating it down several times with water. The difference between the two titrations represents the volatile acidity, which is reported as acetic acid. The fixed acidity is usually calculated as citric acid (1 ml 0·1 M NaOH ≡ 0·007 g).

Preservatives. Sulphur dioxide may be present in mincemeat derived from the fruit used (e.g. apple). Traces of boric acid may be found in mincemeat as it is present naturally in fruits.

Trace metals. Mincemeat should be examined for arsenic, copper and lead. Canned samples should also be examined for tin.

The dried fruit content of mincemeat cannot be determined with any degree of accuracy, but Devey (1959) has claimed some partial success with a method which involves (1) removal of fat, sugar and water by treatment with light petroleum, warm water and acetone respectively, and (2) picking out with forceps the particles of dried fruit and peel, which are then dried, weighed and calculated back in relation to the composition of the fruit.

SUGAR CONFECTIONERY

The term sugar confectionery includes the following more important groups (Williams, 1964):

1. Hard sugar goods, boiled sweets or boilings which contain essentially sugars, acid and flavour. Butterscotch is made from sugars, butter and flavour. Good quality barley sugar contains malt extract.
2. Toffee and fudge, which are made from sugars, milk, fats, flavour (often vanilla) and salt. Cornflour may be used in caramel.
3. Nougat and marshmallow which contains sugars, gelatine or albumin and vanilla flavour. Coconut is often present in nougat.
4. Pastilles, gums and jellies contain sugars, acid, gelatine, gum arabic, agar or glycerine and flavour.

The manufacture of sugar confectionery is described by Lees and Jackson (1973), Lees (1986), Slater (1986), Lees (1987) and Stansell (1988).

Many forms of sugar confectionery contain added colour. The sugars used are often a mixture of glucose and glucose syrups. The types of controls that are applied in the confectionery industry have been discussed by Doolin (1972).

The analysis of sugar confectionery normally includes the determination of moisture and sugars and examination for trace elements, preservatives (sorbic and benzoic acids) and added colours. The preparation of a homogeneous sample for analysis is very difficult for many sugar confectionery products. A shearing blade type blender is most suitable but in many cases it is better to analyse whole sweets in replicate. Boiled sweets are very hygroscopic. If a 20 per cent solution is prepared it can be used for the determinations of refractometric solids and sugar (including glucose syrup) polarimetrically and by Lane and Eynon's method. Moisture is most accurately determined by the Karl Fischer method (Sandell, 1960) after gently heating the sample with absolute methanol under a long reflux condenser. Typical moisture percentages given by various types of confectionery are as follows: hard boilings 2–3·5, lozenges 2, hard gums and toffee 4–6, soft caramel and fudge 10, marshmallow 12–17. The acidity of boilings is usually due to either citric, tartaric, lactic, acetic or malic acids. Fat is determined in toffee and butterscotch (see below). Glycerol in fruit pastilles can be determined by the method of Lloyd (1962) and also enzymatically (BCL, 1987). Products claiming to contain spirits or wine should be examined for alcoholic content by distillation or gas chromatography (Chaveron, 1968).

Butterscotch and other butter confectionery. The Labelling of Food Regulations 1984 states that where the word 'butter' or 'cream' is used for sugar or chocolate, confectionery, e.g. 'butterscotch', 'butta mint', the butterfat content should be not less than 4 per cent by weight. In the analysis the total fat can be determined by the Rose–Gottlieb method commencing by heating with alcohol and ammonia. For the extraction of a larger bulk of fat for the estimation of the butterfat, dissolve the sample in water and extract with ether/light petroleum. With some samples a preliminary treatment with hydrochloric acid may assist with the liberation of the fat but tends to give a darker extract. A purer fat can be obtained by re-extracting the dried fat with ether/light petroleum mixture, filtering and drying again. Alternatively, use the chloroform/methanol extraction procedure described in Chapter 2. Then by using GLC or carrying out the semi-micro Reichert process on the purified fat it is possible to calculate the proportion of butterfat in the extracted fat. The percentage of butterfat in the whole sweet can then be readily calculated from the total fat figure.

REFERENCES

Amor, D. M. (1978) *Composition, properties and use of honey. Scientific and Technical Survey* No. 108. Leatherhead: BFMIRA.

Anon (1937a, b), *International Sugar Journal*, **39**, 22S; **39**, 24S.

Anon (1949) *British Food Journal*, **51**, 39.

Bagnall, D. J. T & Smith, A. (1945) *Analyst*, **70**, 211.

Barnett, M. (1988) *Journal of the Association of Public Analysts*, **26**, 1.

BCL (1987) *Methods of Biochemical Analysis and Food Analysis*. Lewes: Boehringer Corporation (London) Ltd.

Bergmeyer, H. V. (1970) In *Methoden der Enzymatischen Analyse*, ed. H. V. Bergmeyer. Weinhem: Verlag Chemie.

BFMIRA (1931, 1951) *Analyst*, **56**, 35 and **76**, 536.

Blanchard, P. H. & Geiger, E. O (1984) *Sugar Technology Reviews*, **11**, 1–94.

Blanchfield, J. R. (1969) In *Food Industries Manual*, ed. A. H. Woollen. London: Leonard Hill.

Boumahraz, M., Davydov, V. Y. & Kiselev, A. V. (1982) *Chromatographia*, **15**, 751.

Broomfield, R. W. (1988) In *Food Industries Manual*, 22nd edn, ed. M. D. Ranken. Glasgow: Leonard Hill.

Brown, J. F. (1938) *Analyst*, **63**, 262.

Burroughs, L. F. & Otlet, R. L. (1986) *Journal of the Association of Public Analysts*, **24**, 91.

Carruthers, A., Heaney, R. K. & Oldfield, J. F. T. (1965) *International Sugar Journal*, **67**, 364.

Casson, C. B. & Griffin, F. J. (1959, 1961) *Analyst*, **84**, 281; **86**, 544.

Chandler, B. V. (1977) *CSIRO Food Research Quarterly*, **37**, 1.

Chataway, H. D. (1932) *Canadian Journal of Research*, **6**, 352.

Chaveron, H. (1968) *Chocolaterie Confiserie* (246), 23.

Clegg, K. M. (1956) *Journal of the Science of Food and Agriculture*, **7**, 40.

Coste, J. H. (1923) *Analyst*, **48**, 453.

Cowan, A. C. & Mitchell, T. J. (1964) *Analyst*, **89**, 22.

Crofts, L. R. (1987) *Trends in Analytical Chemistry*, **6**, 206.

Crowell, E. P. & Burnett, B. B. (1967) *Analytical Chemistry*, **39**, 121.

Dargie, A. (1936) *Analyst*, **61**, 251.

Davies, A. M. C. (1976) *Journal of Food Technology*, **11**, 515.

Dean, A. C. (1978) *Food Chemistry*, **3**, 241.

Defrates, J. H. & Castle, J. L. (1966) *International Sugar Journal*, **68**, 173.

Devey, J. D. (1959) *Food*, **28**, 205.

Dhar, A. K. & Roy, B. R. (1972) *Analyst*, **97**, 981.

Dickes, G. J. (1965, 1966) *Journal of the Association of Public Analysts*, **3**, 118 and **4**, 45.

Dickes, G. J. (1979) *Talanta*, **26**, 1065.

Dickes, G. J. & Nicholas, P. V. (1976) *Gas Chromatography in Food Analysis*. London: Butterworth.

Doner, L. W. (1977) *Journal of the Science of Food and Agriculture*, **28**, 443.

Doner, L. W., White, J. W. & Phillips, J. G. (1979) *Journal of the Association of Official Analytical Chemists*, **62**, 186.

Doolin, G. S. (1972) *Journal of Milk and Food Technology*, **35**, 424.

Ford, C. W. (1979) *Journal of the Science of Food and Agriculture*, **30**, 853.

FSC (1968) *Food Standards Committee Report on Jams and other Preserves*. London: HMSO.

Fullbrook, P. D. (1984) In *Glucose Syrups: Science and Technology*, ed. S. Z. Dziedzic & M. W. Kearsley. London: Elsevier Applied Science.

Goodall, H. & Scholey, J. (1975) *Journal of Food Technology*, **10**, 39.

Hansen, S. A. (1975a) *Journal of Chromatography*, **105**, 388.

Hansen, S. A. (1975b) *Journal of Chromatography*, **107**, 224.

Hibbert, D. & Phillipson, R. T. (1966) *International Sugar Journal*, **68**, 39.

Hinton, C. L. (1934) *Analyst*, **59**, 248.

Hinton, C. L. & Macara, T. (1924) *Analyst*, **49**, 2.

Hinton, C. L. & Macara, T. (1927) *Analyst*, **52**, 668.

Hinton, C. L. & Macara, T. (1940) *Analyst*, **65**, 540.
Honda, S. (1984) *Analytical Biochemistry*, **140**, 1.
Howells, V. W. (1969) *Journal of the Association of Public Analysts*, **7**, 88.
Hughes, E. B. & Maunsell, A. E. (1934) *Analyst*, **59**, 231.
Hundley, H. H. & Hughes, D. D. (1966) *Journal of the Association of Official Analytical Chemists*, **49**, 1180.
ICUMSA (1978) *Report of the Proceedings of the Seventeenth Session.* Peterborough: ICUMSA.
ICUMSA (1986) *International Sugar Journal*, **88**, 2.
Illing, E. T. & Whittle, E. G. (1944) *Food*, **13**, 32.
Iverson, J. L. and Bueno, M. P. (1981) *Journal of the Association of Official Analytical Chemists*, **64**, 139.
James, G. V. (1969) *Journal of the Association of Public Analysts*, **7**, 128.
Jeuring, H. J. & Kuppers, F. J. E. M. (1980) *Journal of the Association of Official Analytical Chemists*, **63**, 1215.
Jones, J. G., Smith, D. M. & Sasarabudhe, M. (1966) *Journal of the Association of Official Analytical Chemists*, **49**, 1183.
Kheiri, M. S. A. & Birch, G. G. (1969) *Cereal Chemistry*, **46**, 400.
King, J. (1925) *Analyst*, **50**, 371.
Kirkwood, K. C., Mitchell, T. J. & Smith, D. (1960) *Analyst*, **85**, 412.
Kirkwood, K. C., Mitchell, T. J. & Ross, I. C. (1961) *Analyst*, **86**, 164.
Knight, J. & Allen, C. H. (1960) *International Sugar Journal*, **62**, 344.
Kushnir, I. (1979) *Journal of the Association of Official Analytical Chemists*, **62**, 917.
Lampitt, L. H. & Hughes, E. B. (1928) *Analyst*, **53**, 32.
Lampitt, L. H., Hughes, E. B. & Rooke, H. S. (1929) *Analyst*, **54**, 381.
Lee, C. K. (1978) In *Development in Food Analysis Techniques—1*, ed. R. D. King. London: Applied Science.
Lees, R. (1986) *A Basic Course in Confectionery.* Surbiton: Specialised Publications Ltd.
Lees, R. (1987) *Faults, Causes and Remedies in Sweet and Chocolate Manufacture.* Surbiton: Specialised Publications Ltd.
Lees, R. & Jackson, E. B. (1973) *Sugar Confectionery and Chocolate Manufacture.* Aylesbury: Leonard Hill.
Li, B. W. & Schuhmann, P. J. (1980) *Journal of Food Science*, **45**, 138.
Lloyd, W. J. W. (1962) *Analyst*, **87**, 62.
Macara, T. (1931) *Analyst*, **56**, 43.
Macara, T. (1935) *Analyst*, **60**, 392.
Makinen, K. K. & Soderling, E. (1980) *Journal of Food Science*, **45**, 367.
Marini, S. & Righi, G. (1985) *Industrie Alimentari*, **24**, 693.
Marshall, C. R. & Norman, A. G. (1938) *Analyst*, **69**, 315.
Mason, B. S. & Slover, H. T. (1971) *Journal of Agricultural and Food Chemistry*, **19**, 551.
Mason, M. (1983) *Journal of the Association of Official Analytical Chemists*, **66**, 981.
Mitchell, T. J., Donald, E. M. & Kelso, J. R. M. (1954) *Analyst*, **79**, 435.
Money, R. W. (1958) *Journal of the Science of Food and Agriculture*, **9**, 18.
Money, R. W. (1964) *Journal of the Science of Food and Agriculture*, **15**, 594.
Money, R. W. & Christian, W. A. (1950) *Journal of the Science of Food and Agriculture*, **1**, 8.
Moody, G. J., Sanghera, G. S. & Thomas, J. D. R. (1986) *Analyst*, **111**, 605.
Moseley, F. A., Salinsky, J. S. & Woods, R. W. (1978) *Journal of the Association of Official Analytical Chemists*, **61**, 164.
Muller, E. G. (1986) In *Quality Control in the Food Industry*, 2nd edn, Vol. 3, ed. S. M. Herschdoefer. London: Academic Press.
National Bureau of Standards (1942) Polarimetry, saccharimetry and the sugars. *Circular of the National Bureau of Standards* C440. Washington: US Government Printing Office.
Norrish, R. S. (1967) Selected Tables of Physical Properties of Sugar Solutions, *Scientific and Technical Surveys* No. 51, Leatherhead: BFMIRA.
Okuda, J. and Miwa, I. (1974) In *Methods of Biochemical Analysis*, ed. D. Glich. New York: Interscience.
Osborn, R. A. (1964) *Journal of the Association of Official Agricultural Chemists*, **47**, 1068.
Parkes, A. E. (1921) *Analyst*, **46**, 239.
Partridge, W. (1926) *Analyst*, **51**, 346.
Paul, A. A. & Southgate, D. A. T. (1978) *The Composition of Foods.* London: HMSO.
Pearson, D. (1976) *The Chemical Analysis of Foods.* Edinburgh: Churchill Livingstone.
Piper, N. R. (1973) *International Sugar Journal*, **75**, 74.
Plews, R. W. (1970) *Analytical Methods used in Sugar Refining.* Amsterdam: Elsevier.
Prager, M. J. & Miskiewicz, M. A. (1979) *Journal of the Association of Official Analytical Chemists*, **62**, 262.
Rajakyla, E. (1986) *Journal of Chromatography*, **353**, 1.

Rajakyla, E. & Paloposki, M. (1983) *Journal of Chromatography*, **282**, 595.
Reardon, P. (1988) In *Food Technology International Europe*, ed. A. Turner. London: Sterling Publications.
Reyes, F. G. R., Wrolstad, R. E. & Cornwell, C. J. (1982) *Journal of the Association of Official Analytical Chemists*, **65**, 126.
Rheinlander, A. H. (1950) *Analyst*, **75**, 444.
Robards, K. & Whitelaw, M. (1986) *Journal of Chromatography*, **373**, 81.
Rocklin, R. & Pohl, C. (1983) *Journal of Liquid Chromatography*, **6**, 1577.
Roe, J. H. (1955) *Journal of Biological Chemistry*, **212**, 335.
Sandell, D. (1960) *Journal of the Science of Food and Agriculture*, **11**, 671.
Savage, R. I. (1972) *International Sugar Journal*, **74**, 167.
Sawyer, R. W. (1975) *Journal of the Association of Public Analysts*, **13**, 64.
Schaffler, K. J. & Morel du Boil, P. G. (1981) *Journal of Chromatography*, **207**, 221.
Schneider, F. (ed.) (1979) *Sugar Analysis: Official and Tentative Methods*. Peterborough: ICUMSA.
Schwarzenbach, R. (1977) *Journal of Chromatography*, **140**, 304.
Slater, C. A. (1986) In *Quality Control in the Food Industry*, Vol. 3, ed. S. M. Herschdoefer. London: Academic Press.
Smith, P. R., Covency, L. V., Blake, C. J. & English, D. A. (1978) *Determination of sugars by GLC, HPLC and enzymes*. Research Report No. 285. Leatherhead: BFMIRA.
Southgate, D. A. T. (1976) *Determination of Food Carbohydrate*. London: Applied Science.
SPA (1930) *Analyst*, **55**, 115.
Stansell, D. (1988) In *Food Industries Manual*, 22nd edn, ed. M. D. Ranken. Glasgow: Leonard Hill.
Steiner, E. H. (1948) *Analyst*, **73**, 15.
Steiner, E. H. (1949) *Analyst*, **74**, 429.
Sweeley, C. C., Bentley, R., Makita, M. & Wells, W. W. (1963) *Journal of the American Chemical Society*, **85**, 2497.
Symonds, D. C. & Dedicoat, H. (1972) *Journal of the Association of Public Analysts*, **10**, 65.
Tegge, J. (1984) In *Glucose Syrup: Science and Technology*, ed. S. Z. Dziedzic & M. W. Kearsley. London: Elsevier Applied Science.
Traitler, H., Del Vedovo, S. & Schweizer, T. F. (1984) *Journal of High Resolution Chromatography and Chromatographic Communications*, **7**, 558.
Van der Wel, H., Van der Heijden, A. & Peer, H. G. (1987) *Food Reviews International*, **3**, 193–268.
Vane, G. W. (1972) *International Sugar Journal*, **74**, 35.
Verzele, M. & Van Damme, F. (1986) *Journal of Chromatography*, **362**, 23.
Verzele, M., Simoens, G. & Van Damme, F. (1987) *Chromatographia*, **23**, 292.
VdF (1987) In *RSK-Values, The Complete Manual*. Bonn: Association of the German Fruit Juice Industry.
Wedmore, E. B. (1955) *Bee World*, **36**, 197.
White, J. W. (1969) *Journal of the Association of Official Analytical Chemists*, **52**, 729.
White, J. W. (1975) In *Honey—A Comprehensive Survey*, ed. E. Crane. London: Heinemann.
White, J. W. (1979a) *Journal of the Association of Official Analytical Chemists*, **62**, 509.
White, J. W. (1979b) *Journal of the Association of Official Analytical Chemists*, **62**, 515.
White, J. W. & Doner, L. W. (1978) *Journal of the Association of Official Analytical Chemists*, **61**, 746.
White, J. W. & Rudyj, O. N. (1978) *Journal of Apiculture Research*, **17**, 89.
White, J. W. & Siciliano, J. (1980) *Journal of the Association of Official Analytical Chemists*, **63**, 7.
White, J. W., Riethof, M. L., Subers, M. H. & Kushnir, I. (1962) Composition of American honeys. *Technical Bulletin* No. 1261. Washington: US Department of Agriculture.
White, J. W., Kushnir, I. & Subers, M. H. (1964) *Food Technology*, **18**, 153.
White, J. W., Kushnir, I. & Doner, L. W. (1979) *Journal of the Association of Official Analytical Chemists*, **62**, 921.
White, J. W., Meloy, R. W., Probst, J. L. & Huser, W. F. (1986) *Journal of the Association of Official Analytical Chemists*, **69**, 652.
Wight, A. W. & Datel, J. M. (1986) *Food Chemistry*, **21**, 167.
Wight, A. W. & Van Niekerk, P. J. (1983) *Food Chemistry*, **10**, 211.
Williams, C. T. (1964) *Chocolate and Confectionery*, 3rd edn. London: Leonard Hill.
Winkler, O. (1955) *Zeitschrift fur Lebensmitteluntersuchung und-forschung*, **102**, 161.
Winton, A. L. & Winton, K. B. (1935) *The Structure and Composition of Foods*, Vol. 2. New York: Wiley.
Wiseman, A. (1978) In *Developments in Food Analysis Techniques*—1, ed. R. D. King. London: Applied Science.

7

Fruit and vegetable products

GENERAL COMPOSITION OF FRUITS AND VEGETABLES

Fruits. The major part of the edible portion of fresh fruits consists of water (75–95 per cent for most types). Fruits are poor sources of protein (0·2–1·3 per cent as N × 6·25) and oil. The main exceptions to this are the olive, and also the avocado, which may contain as much as 40 per cent oil (Pearson, 1975). Most fruits contain reasonable amounts of carbohydrate. The latter may include varying proportions (according to the fruit, maturity, etc.) of dextrose, fructose and sucrose and possibly starch (e.g. banana, apple). The principal acids present in fruits are citric, tartaric and malic acids. The total acidity falls after picking. The pH of fruits varies from 2·5–4·5 (most types 3·0–3·5). Other constituents of fruits include cellulose and woody fibres, mineral salts, pectin, gums, tannins, colouring matters and volatile oils. Certain fruits, particularly blackcurrants, most citrus fruits and strawberries are good sources of vitmin C. Analytical data of some common fruits are given in Table 7.1; see also Paul and Southgate (1978) for extended data. Wills *et al.* (1983) and Taper *et al.* (1985) have published nutritional data on selected fruits whilst Wrolstad and Shallenberger (1981) have reviewed the literature on glucose, fructose, sucrose and sorbitol content of common fruits; further data on sugars has been published by Richmond *et al.* (1981).

Dried fruits such as currants, raisins, dates and figs contain only about 13–22 per cent of water and in consequence the other main constituents are present in proportionately greater amounts than in the corresponding fresh fruit. Some samples, therefore, contain over 60 per cent of total sugars. Dried fruits are almost devoid of vitamin C but prunes and dried apricots represent useful sources of vitamin A. A restricted range of additives is allowed for use with fresh fruits: the Antioxidant Regulations (SI 1978 No. 105) as amended, specifically permit the presence of 10 mg/kg of diphenylamine as anti-scald agent and 3 mg/kg of ethoxyquin on apples and pears; the Preservative Regulations (SI 1979 No. 752) as amended, permit the presence of specific fungistats on bananas and citrus fruits and the presence of 10 mg/kg sulphur dioxide on grapes as an anti-yeast growth agent.

Fruits, fruit pulp or purée (including tomato pulp, paste and purée) that are not fresh or canned are allowed to contain 350 mg/kg sulphur dioxide or 800 mg/kg benzoic acid (or benzoates). Fruit pieces in stabilised syrup for use in ice-cream or edible ices are allowed to contain 1000 mg/kg sorbic acid. Whilst candied peel is allowed to contain 100 mg/kg sulphur dioxide and 1000 mg/kg sorbic acid, crystallised, glacé or drained fruit are allowed to contain 100 mg/kg sulphur dioxide or 1000 mg/kg benzoic acid (including benzoates) or 1000 mg/kg sorbic acid. Dried fruits, other than prunes and figs, are permitted to contain 2000 mg/kg sulphur dioxide. Dried figs may contain 500 mg/kg sorbic acid as an alternative to 2000 mg/kg sulphur dioxide, whilst in the case of prunes the limits are 1000 mg/kg sorbic acid or 2000 mg/kg sulphur dioxide. The Miscellaneous Additive Regulations (SI 1980 No. 1834) allow the use of calcium EDTA in glacé cherries, whilst dried fruit is permitted to contain up to 0·5 per cent mineral oil. In the USA citrus fruit is permitted to be coated with polyhydric alcohol diesters of oxidatively refined montan wax (SI CFR 2692).

Vegetables. As with fruits, the major part of the edible portion of fresh vegetables consists of water (84–96 per cent for most types—see Table 7.2). Peas, beans and potatoes also contain some protein and carbohydrate (see also Martin-Villa *et al.*, 1982). Potatoes that have been stored at temperatures around 0 °C tend to be over-sweet due to conversion of starch to sugar, but storage at ambient temperatures may produce a 'mealy' potato due to the production of starch from sugar. Some vegetables are comparatively good sources of calcium (especially watercress) and phosphorus. Vitamin C is present to the largest extent in green vegetables and, in view of the comparatively large amounts eaten, other important dietary sources are potatoes, peas and turnip. Carrots are a good source of vitamin A, which is also present in green vegetables and fresh peas. Rhubarb and certain vegetables, e.g. beet and spinach, are rich in oxalate. Data on common foods have been presented by Tabekhia (1980), Kasidas (1980) and Libert and Creed (1985). Dried peas and beans are almost devoid of vitamins A and C but sulphite may be added in the preparation of dehydrated vegetables (moisture content about 5 per cent) in order to retain colour and vitamin C. Although much vitamin C is lost during cooking, it appears to be rather more stable in dehydrated than in fresh vegetables.

Accelerated freeze-dried (AFD) foods are produced by freezing, followed by sublimation of the ice in the food at the highest temperature that will not cause thawing or permit ice-free tissue to be subjected to damage by heat. On reconstitution AFD vegetables and other foods give products that resemble closely the original material in organoleptic and nutritional quality.

GENERAL METHODS OF ANALYSIS

The total energy value of fruits and vegetables is much less important than for other foods. The fruits are generally distinguished from the vegetables in that they are rich in the organic hydroxy-acids including citric, malic, tartaric and ascorbic acid. With few exceptions according to the species of the fruits, one of the first three acids listed predominates and concentrations vary from 1 to 5 per cent. The ascorbic acid content is variable in the region of 0·02 per cent but in the case of currants and some rare fruits (Brand *et al.*, 1982) the content may exceed 0·2 per cent. Fruits and vegetables contain significant amounts of dietary fibre, small amounts of carotenes, B vitamins and minerals. General analysis for quality and compliance with specifications will include determination of: titratable acidity, volatile acidity, water-insoluble solids, mineral impurities, acid-insoluble ash, alkalinity of ash, total solids (or moisture content), pH, soluble solids by refractometer, sugars, ethanol, tin, copper,

Table 7.1 Composition of fruits and fruit products

Food	Description	Edible matter proportion of weight purchased	Water (g)	Sugars (g)	Starch (g)	Dietary fibre (g)	Total nitrogen (g)
Apples, eating	Flesh only, no skin or core	0·77	84·3	11·8	0·1	2·0*	0·04
cooking raw	Flesh only, no skin or core	0·81	85·6	9·2	0·4	2·4	0·05
Apricots fresh raw	Flesh and skin, no stones	0·92	86·6	6·7	0	2·1	0·09
dried raw	Whole fruit	1·00	14·7	43·4	0	24·0	0·76
Avocado pears	10 pears, flesh only; Fuerte variety	0·71	68·7†	1·8	Tr	2·0	0·67
Bananas, raw	Flesh only, no skin	0·59	70·7	1·2	3·0	3·4	0·18
Bilberries, raw	Whole fruit; literature sources	0·98	84·9	14·3	0	—	0·10
Blackberries, raw	Whole fruit	1·00	82·0	6·4	0	7·3	0·20
Cherries, eating raw	Flesh and skin, no stalks or stones	0·87	81·5	11·9	0	1·7	0·09
cooking raw	Flesh and skin, no stalks or stones	0·84	79·8	11·6	0	1·7	0·09
Cranberries, raw	Whole fruit	1·00	87·0	3.5	0	4·2	0·06
Currants, black raw	Whole fruit, no stalks	0·98	77·4	6·6	0	8·7	0·15
red raw	Whole fruit, no stalks	0·97	81·8	4·4	0	8·2	0·18
white raw	Whole fruit, no stalks	0·96	83·3	5·6	0	6·8	0·20
dried	Whole fruit	1·00	22·0	63·1	0	6·5	0·27
Damsons, raw	Flesh and skin, no stalks or stones	0·90	77·5	9·6	0	4·1	0·08
Dates, dried	Flesh and skin, no stones	0·86	14·6	63·9	0	8·7	0·32
Figs, green raw	Whole fruit, no stalks	0·98	84·6	9·5	0	2·5	0·21
dried raw	Whole fruit	1·00	16·8	52·9	0	18·5	0·57
Fruit pie filling, canned	10 cans, blackcurrent, blackberry and apple, gooseberry, apple, cherry	1·00	72·6	23·2	1·9	(1·8)	0·04
Fruit salad, canned	Fruit and syrup	1·00	71·1	25·0	0	1·1	0·04
Gooseberries, green raw	Flesh, skin and pips, no 'tops' or 'tails'	0·99	89·9	3·4	0	3·2	0·18
Grapes, black raw	Flesh only, no skin, pips or stalks	0·81	80·7	15·5	0	0·4	0·09
white raw	Flesh and skin, no pips or stalks	0·95	79·3	16·1	0	0·9	0·10
Grapefruit, raw	Flesh only, no skin, pith or pips	0·48	90·7	5·3	0	0·6	0·10
Greengages, raw	Flesh and skin, no stalks or stones	0·95	78·2	11·8	0	2·6	0·12
Guavas, canned	10 cans, fruit and syrup	1·00	77·6	15·7	Tr	3·6	0·06
Lemons, whole	Whole fruit including skin, no pips	0·99	85·2	3·2	0	5·2	0·12
Loganberries, raw	Whole fruit	1·00	85·0	3·4	0	6·2	0·17
stewed with sugar	Fruit and juice	1·19	77·3	13·4	0	5·2	0·14
Lychees, raw	Flesh only; literature sources	0·60	82·0	16·0	0	(0·5)	0·14
Mandarin Oranges, canned	10 cans; fruit and syrup	1·00	84·3	14·2	0	0·3	0·10
Mangoes, raw	Flesh only; literature sources	0·66	83·0	15·3	Tr	(1·5)	0·08
Medlars, raw	Flesh only, no skin or stones	0·81	74·5	10·6	0	10·2	0·08
Melons, Canteloupe, raw	Flesh only, no skin or seeds	0·59	93·6	5·3	0	1·0	0·16
Mulberries, raw	Whole fruit	1·00	85·0	8·1	0	1·7	0·21
Nectarines, raw	Flesh and skin, no stones	0·92	80·2	12·4	0	2·4	0·15
Olives, in brine	Bottled in brine; flesh and skin, no stones	0·80	76·5	Tr	0	4·4	0·14
Oranges, raw	Flesh only, no peel or pips	0·75	88·1	8·5	0	2·0	0·13

* Apple peel contains 3·7 g per 100 g
† The water content varies from 52 to 79 g per 100 g according to season

Table 7.1 Composition of fruits and fruit products (contd.)

Food	Description	Edible matter proportion of weight purchased	Water (g)	Sugars (g)	Starch (g)	Dietary fibre (g)	Total nitrogen (g)
Passion fruit, raw	Granadilla; flesh and seeds, no skin	0·42	73·3	6·2	0	15·9	0·44
Paw Paw, canned	Papaya; 10 cans, fruit and juices	1·00	80·4	17·0	0	0·5	0·03
Peaches, fresh, raw	Flesh and skin, no stones	0·87	86·2	9·1	0	1·4	0·10
Pears, eating	Flesh only, no skin or core	0·72	83·2	10·6	0	2·3	0·04
cooking	Flesh only, no skin or core	0·77	83·0	9·3	Tr	2·9	0·04
Pineapple, fresh	Flesh only, no skin or core	0·53	84·3	11·6	0	1·2	0·08
Plums, Victoria desserts, raw	Flesh and skin, no stalks or stones	0·94	79·1	9·0	0	2·0	0·08
cooking raw	Flesh and skin, no stalks or stones	0·91	85·1	6·2	0	2·5	0·09
Pomegranate, juice	Juice from fresh fruit	0·56	85·4	11·6	0	0	0·03
Prunes, dried raw	Flesh and skin, no stones	0·83	23·3	40·3	0	16·1	0·39
Quinces, raw	Flesh only, no skin or core	0·69	84·2	6·3	Tr	6·4	0·05
Raisins, dried	Flesh and skin, no stones	0·92	21·5	64·4	0	6·8	0·17
Raspberries, raw	Whole fruit	1·00	83·2	5·6	0	7·4	0·14
Rhubarb, raw	Stems only	0·67	94·2	1·0	0	2·6	0·10
Strawberries, raw	Flesh and pips, no stalks	0·97	88·9	6·2	0	2·2	0·10
Sultanas, dried	Whole fruit	1·00	18·3	64·7	0	7·0	0·28
Tangerines, raw	Flesh only, no peel or pips	0·70	88·7	8·0	0	1·9	0·14

Table 7.2 Composition of vegetables and vegetable products

Food	Description	Edible matter, proportion of weight purchased	Water (g)	Sugars (g)	Starch (g)	Dietary fibre (g)	Total nitrogen (g)
Ackee canned	8 cans, drained contents only	—	76·7	0·8	Tr	2·7	0·46
Artichokes globe boiled	Base of leaves and soft inside parts: boiled 35 min	0·41	84·4	—	0	—	0·18
Artichokes Jerusalem boiled	Flesh only: boiled 20 min	0·85	80·2	—	0	—	0·25
Asparagus boiled	Soft tips only; boiled 25 min	0·20	92·4	1·1	0	1·5	0·54
Aubergine, raw	Eggplant, flesh only	0·77	93·4	2·9	0·2	2·5	0·11
Beans French boiled	Pods and beans; cut up and boiled 30 min	1·00	95·5	0·8	0·3	3·2	0·12
Runner raw	Samples from 6 shops; pod ends and sides trimmed	0·79	89·0	2·8	1·1	2·9	0·36
Broad boiled	Whole beans without pods; boiled 30 min	0·31	83·7	0·6	6·5	4·2	0·66
Butter raw	Whole beans	1·00	11·6	3·6	46·2	21·6	3·06
Haricot raw	Whole beans	1·00	11·3	2·8	42·7	25·4	3·42
Baked canned in tomato sauce	11 cans, 4 brands	1·00	73·6	5·2	5·1	7·3	0·82
Mung, green gram, raw	Literature sources	1·00	12·0	1·2	34·4	(22·0)	3·52
Red kidney, raw	Literature sources	1·00	11·0	(3·0)	(42·0)	(25·0)	3·54
Beansprouts, canned	10 cans, drained contents	0·55	95·4	0·4	0·4	3·0	0·25
Beetroot, raw	Flesh only, no skin	0·82	87·1	6·0	0	3·1	0·21
Broccoli tops, raw	10 samples; predominantly leaves, thick stems removed	0·70	89·0	2·5	Tr	3·6	0·52
Brussels sprouts, raw	10 samples, inner leaves only	0·63	88·1	2·6	0·1	4·2	0·64
Cabbage, red raw	Inner leaves	0·70	89·7	3·5	Tr	3·4	0·27
Savoy, raw	Inner leaves	0·53	89·9	3·3	Tr	3·1	0·53
Spring, boiled	Inner leaves; boiled 30 min	0·59	96·6	0·8	Tr	2·2	0·18
White, raw	7 cabbages; whole cabbage as purchased	1·00	90·3	3·7	0·1	2·7	0·31
Winter, raw	20 cabbages; January King; inner leaves	0·57	88·3	2·7	0·1	3·4	0·45
Carrots, old raw	Flesh only	0·96	89·9	5·4	0	2·9	0·11
young boiled	Purchased with leaves; flesh only, boiled 25 min	0·50	91·1	4·4	0·1	3·0	0·14
Cauliflower, raw	16 cauliflowers; flower and stalk	0·62	92·7	1·5	Tr	2·1	0·30
Celeriac, boiled	Flesh only; boiled 30 min	0·79	90·2	1·5	0·5	4·9	0·26
Celery raw	Stem only	0·73	93·5	1·2	0·1	1·8	0·15
Chicory raw	Stem and young leaves	0·79	96·2	—	0	—	0·12
Cucumber raw	Flesh only	0·77	96·4	1·8	0	0·4	0·10
Endive raw	Leaves only	0·63	93·7	1·0	0	2·2	0·28
Horseradish raw	Flesh of root	0·45	74·7	7·3	3·7	8·3	0·72

Table 7.2 Composition of vegetables and vegetable products (contd.)

Food	Description	Edible matter, proportion of weight purchased	Water (g)	Sugars (g)	Starch (g)	Dietary fibre (g)	Total nitrogen (g)
Laverbread	6 samples; cooked pureed seaweed coated in oatmeal	1·00	87·7	Tr	1·6	3·1	0·51
Leeks raw	Bulb only	0·36	86·0	6·0	0	3·1	0·31
Lentils raw	As purchased	1·00	12·2	2·4	50·8	11·7	3·80
Lettuce raw	28 lettuces; inner leaves of long and headed forms	0·70	95·9	1·2	Tr	1·5	0·16
Marrow raw	10 marrows; flesh only	0·50	93·5	3·0	0·7	(1·8)	0·10
Mushrooms raw	Flesh and stem	0·75	91·5	0	0	2·5	0·64*
Mustard and Cress raw	Leaves and stems	1·00	92·5	0·9	0	3·7	0·26
Okra raw	Literature sources (ladies' fingers)	0·88	90·0	2·3	Tr	(3·2)	0·32
Onions raw	Flesh only	0·97	92·8	5·2	0	1·3	0·15
Parsley raw	Leaves	0·53	78·7	Tr	0	9·1	0·83
Parsnips raw	Flesh only	0·74	82·5	8·8	2·5	4·0	0·27
Peas, fresh raw	Whole peas, no pods	0·37	78·5	4·0	6·6	5·2	0·92
dried raw	Whole peas	1·00	13·3	2·4	47·6	16·7	3·45
split dried, raw	Peas as purchased	1·00	12·1	1·9	54·7	11·9	3·54
Chick, Bengal gram raw	Literature sources	1·00	9·9	(10·0)	(40·0)	(15·0)	3.23
red pigeon, raw	Literature sources	1·0	10·0	(9·0)	(45·0)	(15·0)	3·20
Peppers, green raw	30 peppers; flesh only	0·86	93·5	2·2	Tr	0·9	0·15
Plantain green, raw	Literature sources	0·61	67·0	0·8	27·5	(5·8)	0·16
Potatoes*, old raw	Flesh only	0·86	75·8	0·5	20·3	2·1	0·34
old chips frozen	11 packets as purchased	1·00	73·1	0·4	19·2	1·9	0·35
new boiled	Flesh only; boiled 15 min	0·96	78·8	0·7	17·6	2·0	0·25
instant powder	20 packets	1·00	7·2	2·2	71·0	16·5	1·45
crisps	26 packets, mixed plain and flavoured	1·00	1·7	0·7	48·6	11·9	1·00
Pumpkin, raw	Flesh only	0·81	94·7	2·7	0·7	0·5	0·10
Radishes, raw	Flesh and skin; purchased with leaves	0·50	93·3	2·8	0	1·0	0·16
Salsify, boiled	Flesh only; boiled 45 min	0·63	81·2	—	0	—	0·30
Seakale, boiled	Stem only, boiled 20 min	0·74	95·6	0·6	0	1·2	0·23
Spinach boiled	Leaves; boiled 15 min without added water	0·42	85·1	1·2	0·2	6·3	0·81
Spring greens, boiled	Leaves; boiled 30 min	1·00	93·6	0·9	0	3·8	0·27
Swedes raw	Flesh only	0·86	91·4	4·2	0·1	2·7	0·18
Sweetcorn, on-the-cob raw	16 cobs; kernels only	0·66	65·2	1·7	22·0	3·7	0·66
canned kernels	10 cans; whole contents	1·00	73·4	8·9	7·2	5·7	0·47
Sweet potatoes, raw	Literature sources	0·86	70·0	(9·7)	(11·8)	(2·5)	0·19
Tomatoes, raw	Flesh, skin and seeds	1·00	93·4	2·8	Tr	1·5	0·14
Turnips raw	Flesh only	0·84	93·3	3·8	0	2·8	0·12
Turnip tops, boiled	Leaves; boiled 20 min	0·45	92·8	0	0·1	3·9	0·43
Watercress, raw	Leaves and part of stem	0·77	91·1	0·6	0·1	3·3	0·46
Yam, raw	Literature sources	0·86	73·0	1·0	31·4	(4·1)	0·32

* Detailed nutritional analysis see Finglas and Faulks (1984) and Wills *et al.* (1984).

iron, zinc, benzoic acid and benzoates, sorbic acid and sulphur dioxide. Considerable attention has also been paid to the presence of nitrates in vegetables; typical data has been presented by Lönberg *et al.* (1985) (see also Chapter 4).

From the legal point of view fresh fruits may be required to be examined for mineral oil, added colour, the permitted additives and fumigants. Cranberries and other small berry fruits may contain naturally occurring benzoic acid in amounts up to 0·25 per cent. Fresh fruits and vegetables may be examined for chemical residues arising from agricultural and horticultural practices and from the environment. Dried foods should be examined for preservatives including sulphur dixoide; generally applicable methods are given in Chapters 4 and 5. Park *et al.* (1982) have reported on the use of NIR techniques to measure crude protein, fat, ash and fibre content of vegetables; in general the correlations obtained were regarded as unsatisfactory for wider use. Specific problems of naturally occurring potentially hazardous substances occur with certain food types, e.g. glucosinolates in brassicas, glycoalkaloids in potatoes, tomatoes and related vegetables, the cyanogens in stone fruits and in cassava roots and trypsin inhibitors in species of bean.

Pectin

Many methods have been proposed for the estimation of pectin that depend on extraction and hydrolysis by water, followed by precipitation with alcohol, acetone or as the calcium salts. The method due to Carre and Haynes is as follows:

Extract repeatedly a suitable quantity of pulped fruit or vegetable with cold water, boil the mixed extract and filter. Dilute an aliquot of the filtrate to 300 ml, add 100 ml of 0·1 M sodium hydroxide and allow to stand overnight. Then add 50 ml M acetic acid, followed 5 min later by 50 ml M (2 N) calcium chloride solution. Allow to stand for 1 hour, boil for a few minutes and filter. Wash the residue with boiling water until free from chlorides, again cool with water and filter on a Gooch crucible. Wash, dry and weigh as calcium pectate ($C_{17}H_{22}O_{16}Ca$). Alternatively, the calcium in the precipitate can be titrated directly with EDTA (Holt, 1954).

Meurens (1980) used copper complexation followed by iodometric titration, whilst Robertson (1981) used progressive extraction of alcohol-insoluble solids with water, ammonium oxalate and alkali followed by heating with concentrated sulphuric acid in the presence of sodium tetraborate and reaction with *m*-hydroxydiphenyl to give a chromophore of anhydrogalacturonic acid. Kinter and van Buren (1982) and List *et al.* (1985) employed the same method in studies on fruit juices and commercial preparations. Giangiacomo *et al.* (1982) enzymatically hydrolysed pectin and analysed the galacturonic acid using HPLC; similar methods have been reported by Voragen *et al.* (1982), Walter *et al.* (1983) and De Vries *et al.* (1983). The influence of pectic substances on the determination of the dietary fibre content of fruits and vegetables has been studied by Belo and Lumen (1981), who concluded that the detergent methods consistently underestimate the total pectic substance. Dovell and Harris (1982) optimised the extraction of pectin prior to the determination of residual fibre by enzyme methods; values for oranges, cauliflower, beans and wheat flour were reported.

Pentosans. Besides the pectins, there are other substances present in fruits which yield five-carbon sugars on hydrolysis, the pentosans. These appear to have the general formula $(C_5H_8O_4)_n$ and are formed in increasing quantities as the liquefaction of plant tissue proceeds. A general relationship may exist between the amount of crube fibre and pentosans, which is to be expected as the latter substances are closely allied to cellulose. The pentosans may be estimated by distillation with hydrochloric acid to yield furfuraldehyde, as in the following method due to Tollens.

Distil 2 g of the finely divided material with 100 ml of 12 per cent hydrochloric acid (sp. gr. 1·06) using a distillation flask fitted with a tap funnel. When about 30 ml have passed over, run a similar volume of the same acid into the flask and repeat the distillation. Continue until all the furfural has been distilled off by testing a drop of the distillate with aniline acetate paper: no pink colour should appear. It is usually necessary to distil upwards of 300 ml. To the mixed distillate containing the furfural add a large excess of phloroglucinol solution in dilute hydrochloric acid and allow the mixture to stand for 12 hours. Filter off the precipitate, wash with 150 ml cold water, dry in the oven and weigh. If the weight obtained is a, pentosans = $(a + 0{\cdot}0052) \times F$, where $F = 0{\cdot}8949$ (if $a < 0{\cdot}03$ g), $0{\cdot}8866$ (if $a = 0{\cdot}03$–$0{\cdot}30$ g) or $0{\cdot}8824$ (if $a > 0{\cdot}30$ g).

Douglas (1981) has reviewed available methods. Gas chromatography has been applied to the determination of furfural produced on acid treatment of pentoses (Folkes, 1980). Thomann *et al.* (1982) used chromatography on Sephadex prior to colorimetric analysis. An HPLC method for furfural and HMF with clean-up using C_{18} cartridge is described by Lee *et al.* (1986), namely Radial-Pak C_{18} column, acetonitrile/water (15:85), ultraviolet 280 nm. See also methods described for cereals in Chapter 8.

Wittmann has given the following percentage figures for the pentosan content of various fruits; juniper berries 6·0, raspberries 2·7, elderberries 1·2, grapes (Japanese) 1·6, blackberries 1·2, strawberries 0·9, cranberries 0·75, bilberries 0·8, gooseberries 0·5, currants 0·4.

Vitamin C

The ascorbic acid content of fruits and vegetables can be estimated by macerating the sample, preferably mechanically, with a stabilising agent such a 5 per cent metaphosphoric acid or trichloroacetic acid and titrating the decanted or filtered extract with 2, 6-dichlorophenolindophenol, or fluorimetrically in the presence of *o*-phenylenediamine. The following procedures are based on those described in the AOAC.

TITRATION METHOD

Extraction solution. Dissolve 15 g of phosphoric acid in 40 ml acetic acid and 200 ml water, dilute to 500 ml and filter.

Standard solution. Dissolve 0·05 g ascorbic acid in 45 ml of the extraction solution and make up to 50 ml. Prepare immediately before use.

Indophenol standard solution. Dissolve with shaking 0·05 g 2, 6-dichlorophenolindophenol (sodium salt) in 50 ml water containing 42 mg sodium bicarbonate. Dilute to 200 ml with water. Filter. Standardise by titration against 2 ml of standard ascorbic acid solution added to 5 ml of the extraction solution.

Indicator. Dissolve 0·1 g thymol blue in 10·75 ml 0·02 N sodium hydroxide solution, dilute to 250 ml with water.

Shake juice thoroughly or prepare sample by maceration with a suitable proportion of water. To 50–100 ml of the prepared sample add an equal volume of extraction solution, mix and filter rapidly. Titrate an aliquot containing about 2 mg of ascorbic acid with standard indophenol solution and correct for blank using an equivalent amount of extraction solution.

FLUORIMETRIC METHOD

REAGENTS

o-*Phenylenediamine solution.* 0·05 per cent in water prepared immediately before use.

Extraction solution. 30 g HPO_3 + 80 ml glacial acetic acid made up to 1 litre (keeps 7–10 days at 1–3 °C).

Standard ascorbic acid. 0·1 per cent in extraction solution. 0, 1, 2, 4 ml of this diluted to 50 ml for carbon treatment are equivalent to 0, 10, 20, 40 μg of ascorbic acid in the final solution.

Sodium acetate solution, 50 per cent m/v. Dissolve 500 g sodium acetate trihydrate (or 301 g anhydrous sodium acetate in water) and make volume up to 1 litre.

Boric acid–sodium acetate solution. Dissolve 5 g boric acid in 100 ml 50 per cent sodium acetate solution. Prepare fresh daily.

Acid washed Norit 'A' charcoal. Heat the charcoal with five times its weight of approx. 10 per cent HCl at boiling point. Filtered on Buchner (541 paper), wash free of acid and dry overnight at 110 °C. Keep in a screw-top jar.

METHOD

Follow all steps consecutively without delay. Prepare an extract of the sample in the extraction solution by homogenising or shaking; centrifuge if necessary. Filter through 541 paper. The pH of the extract should be about 1·2. Add 20 ml extract filtrate and 20 ml of each of suitable range of standard solutions to stoppered centrifuge tubes containing 0·4 g charcoal. Shake at intervals for about 5 min, centrifuge and filter. Some samples will give cloudy filtrate and possibly traces of charcoal may pass filter.

Take 2 × 5 ml aliquots from each filtrate. Add one of these to 100 ml flask containing 5 ml boric acid–sodium acetate solution. Stopper, shake well and stand at least 30 min before making up to 100 ml. All samples and standards should have as nearly as possible the same standing time. Add the second 5 ml portion to another 100 ml flask containing 5 ml sodium acetate solution and about 70 ml water, bulk to 100 ml and mix.

Pipette 5 ml from each of 'samples' and 'blanks' into a 15 × 2·5 cm tube. The remaining operations must be conducted in subdued light. Add to each tube 10 ml freshly prepared 0·05 per cent *o*-phenylenediamine. Mix thoroughly and stand in dark at least 35 min then read fluorescence rapidly. (The compound is very light sensitive and readings decrease rapidly if left in light beam.) Transmission maxima 365 and 455 mm.

CALCULATION

If x = μg of ascorbic acid in the final solution
W = weight of sample
V = total volume of extract
Ascorbic acid content = $0{\cdot}4\ Vx/W$ mg/100 g sample

In the above method, which is based on Deutsch and Weekes (1965), vitamin C is first oxidised to dehydroascorbic acid and then reacted with *o*-phenylenediamine to produce a compound which, on activation at 350 nm, fluoresces at 430 nm.

Methods for the estimation of ascorbic acid continue to receive attention; two comprehensive reviews have been produced by Cooke and Moxon (1981) and by Pachla and Reynolds (1985). The latter review covers a broader spectrum of available analytical techniques than the former but a full perspective of the development of methods can be obtained from a study of both reviews.

Titrimetric methods using 2,6-dichlorophenolindophenol remain popular with a number of workers and in a variety of applications. Roehl and Voigt (1984) and Bohrer *et al.* (1984) describe simplified versions of the method for use with highly coloured fruit extracts and with cabbage and sauerkraut respectively. Rymal (1983) used the indophenol reagent on test papers in association with a micro-sampling technique both to provide a field test kit and to improve the performance of the method in analysis of fruit and vegetable extracts. Colorimetric methods using ammonium molybdate and chloranil have been described by Bajaj and Kaur (1981) and by Verma *et al.* (1984). The former authors claim that the molybdate procedure is less subject to interference than the titrimetric method with indophenol and that it gives generally lower results when compared with that procedure. De Vries (1983) has proposed for general use a semi-automated version of the fluorimetric procedure.

The use of HPLC techniques has also received attention from many authors. Rizzolo *et al.* (1984) used a strong anion exchange resin and elution with sodium acetate whilst Ashoor *et al.* (1984) used extraction with EDTA and sulphuric acid prior to chromatography. Reversed phase techniques have been applied to detection of ascorbic acid and dehydroascorbic acid together with their isomers (Haddad and Lan, 1984; Speek *et al.*, 1984; Vanderslice and Higgs, 1988). Wills *et al.* (1983a) compared the performance of HPLC methods with the microfluorimetric and titrimetric methods in application to fresh fruit and vegetables, and concluded that no single method is suitable in all cases.

Naturally occurring constituents that are potentially hazardous

Awareness of the presence of potentialy harmful substances as naturally occurring constituents of certain foods has grown in recent years. Table 7.3 gives a short but incomplete summary of some of the known pharmacologically active substances that have been identified in fruits and vegetables, together with recent references to methods which have been used in surveys. Further general reviews have been published by Liener (1966, 1974), Liener and Kakade (1980) and Rechcigl (1983).

CANNED FRUITS AND VEGETABLES

CAN SIZE AND STANDARDS OF FILL

Standard can sizes are summarised in Table 7.4.

The British Fruit and Vegetable Canners' Association Code of Practice on Canned Fruit and Vegetables came into operation in September 1986 and gives standards of fill, composition and size of canned fruit and vegetables that have been accepted by the trade. The minimum filled weights (mean of 10 cans) of prepared fruits and vegetables are given in Tables 7.5 and 7.8. According to the particular fruit and vegetable and can size, the filled weight of an individual can may be 5 per cent less for the larger cans or 10 per cent less for the smaller cans than the tabulated figures. If there is a deficiency in the average weight of 10 cans (or an indidivual can falls below the appropriate lower minimum), the sample is deemed to satisfy the requirements if a second batch of 10 cans comes within the prescribed standard.

Mannheim and Passy (1981) have discussed the mechanisms of internal corrosion of canned foods and the formulation of protective lacquer for use in can fabrication; lacquer and product must be matched to secure optimum shelf-life and to preserve quality. Barbieri *et al.* (1982) have also reported on the formulation of lacquers suitable for specific products. Fruits and vegetables are amongst the most aggressive products so far as the can is concerned, with combinations of acidity and sulphur-containing amino acids, together with nitrate naturally in the vegetable or in the canning liquor, contributing to the problem. Alternatives to the can include glass bottles and plastics formulations including laminated materials. The use of retortable laminate pouches has been discussed both from the point of view of product quality and materials of construction (Thorpe and Atherton, 1972; Salunkhe *et al.*, 1978; Anon, 1979). Board (1982) has discussed the definition of canned foods in the

Table 7.3 Some fruits and vegetables that are possible sources of naturally occurring hazardous substances

Source	Active agent	Selected references	Method/content
Bananas	5-Hydroxytryptamine	Garcia-Moreno *et al.* (1983)	TLC spectrofluorimetry
Prunus fruits, almonds	Amygdalin	Kajiwars *et al.* (1983)	Chromatography
Cassava	Cyanide	Rao and Hahn (1984)	Automated enzyme
		Nambisan and Jundaresan (1984)	Colorimetry
Beans	Cyanide	Ariga *et al.* (1983)	Enzyme
		Chikamoto and Maitani (1984)	GLC
Legumes	Haemagglutinins		
Beans including soya	Vicine	Marquardt and Frohlich (1981)	HPLC
		Arbid and Marquardt (1985)	HPLC
		Kim *et al.* (1982)	Colorimetry
	Trypsin inhibitor	Kakade *et al.* (1974)	Collaborative study
		Charpentier *et al.* (1984)	Automated colorimetry
Brassica and Cruciferae	Glucosinolates	Sones *et al.* (1984a, b)	Surveys, GLC
		McGregor *et al.* (1983)	Review
		Fenwick *et al.* (1982)	Review
		Spinks *et al* (1984)	HPLC/GLC
	Thiocyanates	Hirosue *et al.* (1983)	Survey GLC
Potatoes	Glycoalkaloids	Coxon (1984)	Review
		Davies and Blinco (1984)	Survey
		Morgan *et al.* (1985)	ELISA and colorimetric
		Clement and Verbist (1980)	Colorimetric assessment
		Van Gelder (1984)	TLC
		Bushway (1982)	HPLC
		Hellenas (1986)	HPLC, ELISA, colorimetry
Tomatoes/solanaceous fruits		Spiess (1981)	Survey
		Jones and Fenwick (1981)	Survey
Rhubarb, Spinach	Oxalates	Tabekhia (1980)	Titrimetry
		Kasidas (1980)	Survey enzyme method
		Yamanaka *et al.* (1983)	Enzyme method
		Treptow (1985)	Survey
Fungi	Mycotoxins	AOAC methods book	Chromatographic
	Hallucinogens	Fischer *et al.* (1984)	HPLC
		Perkal *et al.* (1980)	Chromatographic
		Benoit-Guyod *et al.* (1980)	Pyrolysis GLC

light of modern technology and has proposed that canned food should include heat processed foods in hermetically sealed containers regardless of the material of construction provided that it is safe and is compatible with the food concerned.

CANNED FRUITS

Codex standards have been developed for a wide range of canned fruits, including grapefruit, peaches, pears, pineapple, plums, raspberries, strawberries, apple sauce, mandarin oranges and fruit cocktail; see Appendix 2 for the full list. The various commodity standards relating to these describe methods for the examination of drained weight and calcium content. Canned fruits are also examined for vacuum

Table 7.4 Can sizes

Description of cans	Nominal diameter (inches)	Internal diameter (mm)	Nominal capacity (ml)
5 oz	$2^{11}/_{16}$	65	156
¼DIN	3	73	212
U8	3	73	228
Picnic	$2^{11}/_{16}$	65	236
A1	$2^{11}/_{16}$	65	314
14Z	3	73	403
½DIN	3	73	425
UT	3	73	446
A2	$3^{7}/_{16}$	83	580
A2½	$4^{1}/_{16}$	99	850
A6	$6^{3}/_{16}$	153	2650
A10	$6^{3}/_{16}$	153	3100

(Dilley and Everton, 1966), pH, cut-out syrup strength (by refractometer), added colour and metal contamination. The FACC (1982) conducted a review of the metal content of canned foods. The metals included lead, tin, chromium, iron and aluminium; the report considered the sources of the various metals and the links with can technology. On the basis of evidence provided the committee recommended that the statutory limit for lead should be brought into line with that for corresponding food not in cans, and where appropriate the general limit of 1 mg/kg should apply. The guideline limit for tin in canned food should be reduced to 200 mg/kg. Survey data for metals in foods including canned food have been produced by the MAFF Steering Group on Food Surveillance for copper and zinc (MAFF, 1981), arsenic (MAFF, 1982b), lead (MAFF, 1982a) and cadmium (MAFF, 1983). The problems of pH control in relation to processing parameters for commercial sterility and the balance required to minimise can corrosion and to maximise shelf-life have been discussed in respect of canned tomatoes (Powers, 1976). Sulphur dioxide should not be present in canned fruits as it encourages attack on the tin plate. The drained weights and syrup densities of canned fruits have been discussed by Adam (1965). The drained weight can be taken as the weight remaining 2 min after the sample has been poured onto a sieve having 8 holes/inch. The filled weight can be approximately assessed from the drained weight (Table 7.6). It must be borne in mind, however, that the ratio of the drained to the filled weights varies according to the condition of the fruit, added syrup strength, processing and storage conditions, etc. Other factors being equal, the lighter the syrup density used the higher the drained weight. For the purpose of Codex standards minimum drained weights are expressed on the basis of water at 20 °C that the sealed container will hold when completely filled.

Four designations of syrup strengths are given in Table 7.7, as added at the time of canning and measured by a Brix hydrometer or sugar-scale refractometer at 20 °C. The syrup may contain dextrose and glucose syrup, but the proportion of sucrose must be at least 75 per cent. The fruit contains much less soluble solids than the syrup. During processing therefore the fruit dilutes the syrup. The density of the final syrup (referred to as the 'cut-out' syrup) is in consequence considerably lower than that of the added syrup, e.g. an added syrup of 40° Brix will usually cut-out at

Table 7.5 Minimum filled weight of prepared fruit per can (average of ten cans)

Fruit	5 oz (g)	¼ DIN (g)	U8 or picnic (g)	A1 (g)	14Z (g)	½ DIN (g)	UT (g)	A2 (g)	A2½ (g)	A10 (g)	Other sizes (% of nominal can capacity)
Fruits packed fresh											
Apple purée (unsweetened)	130	190	205	285	360	380	405	525	770	2830	91
Apple purée (sweetened)	140	205	220	305	385	400	430	555	815	2980	96
Apples (solid pack)	115	170	185	255	325	335	360	470	695	2550	81
Apples (sliced in syrup) Apples fresh or re-canned Blackcurrants, Bilberries, Redcurrants, Rhubarb, Gooseberries	80	110	120	170	220	240	250	320	475	1730	55
Apricots, Blackberries, Loganberries, Peaches, Pears, Pineapples, Raspberries, Strawberries	85	115	125	175	225	245	255	335	480	1810	57
Cherries, Greengages, Plums & Damsons	85	120	130	185	235	250	265	345	495	1900	60
Prunes (washed in cold water or equivalent weights by other methods)	45	65	70	95	125	135	140	185	290	1090	32
Re-canned fruits											
Apricots, Peaches, Pears, Pineapple, Fruit Salad, Fruit Cocktail	80	110	120	170	215	225	240	310	460	1700	54

Table 7.6 Drained weight of canned fruit as percentage of filled weight

Canned fruit (packed in 40° Brix syrup)	Drained weight (as pecentage of filled weight)		
	min.	max.	mean
Blackberries	65	90	80
Blackcurrants	75	95	85
Cherries	75	100	90
Damsons	75	100	89
Gooseberries	80	105	94
Greengages	85	100	92
Loganberies	65	90	78
Plums	70	105	86
Raspberries	60	90	81
Rhubarb	75	105	92
Strawberries	55	80	65

Table 7.7 Syrup strengths for canned fruits

	Class A (% m/m)	Class B (% m/m)	Class C (% m/m)
Light syrup	15	15	10
Syrup	30	22	15
Heavy syrup	40	30	20

Class A. Apples (other than purée or solid pack), bilberries, blackberries, blackcurrants, cherries, damsons, gooseberries, greengages, loganberies, plums, raspberries, redcurrants, rhubarb and strawberries.
Class B. Apricots, peaches, pears, pineapple, fruit and fruit cocktail.
Class C. Prunes.

22–25°. The cut-out syrup strength depends on several factors such as the strength of the added syrup, the particular fruit and its stage of ripeness and the ratio of fruit to syrup used. Similarly, the drained weight (the weight of the fruit in the product retained on a sieve after draining) varies considerably according to various factors including the processing conditions. A fall of 5° Brix in the strength of the added syrup causes a drop of about 1 per cent in the drained weight and a fall of about 2·5–3° Brix in the cut-out syrup.

The water activity and pH of canned fruits has been surveyed by Alzamora and Chirife (1983). Other standards of composition contained in the BFVCA Code of Practice are as follows:

Composition of fruit salad. Canned fruit salad shall contain only the following proportions of the total filled weight of fruit: peaches 23–46 per cent, apricots 15–30 per cent, pears 19–38 per cent, pineapple 8–16 per cent, cherries 5–15 per cent. Any composition of mixed fruit which does not conform to the above standard shall not be labelled 'Fruit Salad' or 'Fruits for Salad'.

Composition of fruit cocktail. Canned fruit cocktail shall contain the following fruits in the following proportions of total filled weight of fruit: diced peaches 30–50 per cent, diced pears 25–45 per cent, pieces of or diced pineapple 6–25 per cent, cherries 2–15 per cent and (optional) seedless grapes 6–20 per cent. Any composition of mixed fruit that does not conform to the above standard shall not be labelled 'Fruit Cocktail' or 'Fruits for Cocktail'.

Apple purée (sweetened). Apple purée (sweetened) shall contain not less than 8 per cent of added sugar.

Recommended Codex International Standards for canned tomatoes, peaches, grapefruit, pineapple, plums, raspberries, pears and strawberries include minimum requirements for fill, drained weight and strength of syrup on 'cut-out'. The Codex standard for canned apple sauce includes minimum requirements for refractometer solids, i.e. sweetened 16·5 per cent, unsweetened 7·0 per cent. (See list, Appendix 2, p. 657)

Dickinson and Raven (1962b) studied the stability of erythrosine in cherries packed in plain cans and reported that fluorescein was readily formed by interaction with the tin–iron complex. Complaints occasionally arise alleging that cans of grapes packed in light syrup contain glass. These fragments invariably consists of crystals of argol (cream of tartar derived from the grapes) and Manley and Alcock (1950) consider that their deposition is encouraged by cold storage (see Kagan *et al.*, 1965). Dickinson and Fowler (1955) have also reported on the presence of 'glass-like' crystals of calcium *dl*-tartrate in cans of French cherries. A further cause of complaint is the presence of naringin spots in canned mandarin oranges packed in light syrup. The presence of the mould *Byssochlamys fulva*, which has fairly heat resistant spores, causes disintegration of canned fruit (especially strawberries) due to attack on the pectin (cf. Michener *et al.*, 1966). Carles (1982) has discussed methods for quality assessment of processed fruits and vegetables by means of colour measurements.

Canned fruit pie fillings

Ready prepared fillings for fruit pies have been used by bakers for many years and more recently have become popular as a commodity for retail sale. Blanchfield (1964) has explained that the ingredients used are fruit, sugars and stabiliser (often modified starches) and that their production is more akin to the boiling of jam than the preparation of canned fruit. Most fruit pie fillings contain 30–40 per cent soluble solids (by refractometer), occasionally rising to 50 per cent. James (1968) has given typical formulations used for such fillings for various fruits.

CANNED VEGETABLES

Codex standards have been developed for a range of canned vegetables, including green beans, green and processed peas, asparagus, sweet corn, tomatoes and mushrooms—see Appendix 2.

The commodity standards relating to these describe methods of exmination for drained weight and calcium content together (for peas) with a method for the determination of alcohol-insoluble solids. Most canned vegetables are prepared by blanching, filling into the cans and covering with brine containing 1·25–2·5 per cent salt (with over 2 per cent sugar in addition for peas) before closing and heat processing. The minimum filled weights in the BFVCA Code of Practice (p. 245) are summarised in Table 7.8. Other requirements of the BFVCA code are as follows:

Processed peas. The minimum dry solids content of canned processed peas, expressed on the fluid capacity of the can, shall be 19 per cent (m/v) for any individual can and 19·5 per cent (m/v) for the average of any two or more cans from the same batch. The dry solids content shall be determined by drying 5–10 g of a finely macerated sample of the entire contents of the can in a steam oven for 16–18 h at 98 °C, or for an equivalent drying condition. For a sample of more than one can the dry solids content may be assessed individually for each can and the results averaged.

Butter beans. Canned butter beans are prepared from the seed of white varieties of lima beans, Madagascar beans, or large types of kidney beans (minimum length about 20 mm). The minimum dry solids content of canned butter beans, expressed on the fluid capacity of the can, for any individual can

Table 7.8 Minimum filled weight of prepared vegetables per can (average of 10 cans)

Vegetables	5 oz (g)	¼ DIN (g)	U8 or picnic (g)	A1 (g)	½ DIN (g)	UT (g)	A2 (g)	A2½ (g)	A6 (g)	A10 (g)	Other sizes (% of nominal can capacity)
Beans (green)	65	95	105	140	205	215	270	395	1230	1470	47
Beans (broad) Macedoine, mixed Vegetables, garden peas All diced Vegetables (not greater than 10 mm³)	95	130	140	195	265	280	360	525	1640	1930	62
Beetroot (whole, sliced and diced greater than 10 mm³) Carrots (as for beetroot) Parsnips (as for beetroot) Swedes (as for beetroot) Turnips (as for beetroot) Potatoes (as for beetroot) Celery (as for beetroot)	90	120	130	190	260	275	355	525	1640	1930	61
Spinach (leaf)	100	140	150	205	285	300	385	560	—	—	67
Spinach (purée)	135	200	215	290	400	420	540	795	—	—	94

shall be 18·5 per cent (m/v) and 19 per cent (m/v) for the average of two or more cans from the same batch. The dry solid content shall be determined by drying 5–10 g of a finely macerated sample of the entire content of the can in a steam oven for 16–18 h at 98 °C, or for an equivalent drying condition. For a sample of more than one can the dry solids content may alternatively be sampled individually for each can and the results averaged.

Canellini beans, chick peas and red kidney beans. The minimum dry solids content of canned Canellini beans, chick peas and red kidney beans expressed on the fluid capacity of the can, for an individual can shall be 22·0 per cent (m/v) and 22·5 per cent (m/v) for the average of two or more cans from the same batch. The dry solids content shall be determined in the same way as for butter beans.

Macedoine mixed vegetables. Canned macedoine or mixed vegetables shall consist of at least four vegetables of which no single vegetable should exceed 40 per cent of the total filled weight of vegetables.

STANDARDS OF SIZE

Green peas. Where the following descriptions are used on the label for canned fresh green peas, the sizes of the peas in the cans shall be as stated against each grade:

Smooth varieties

Petits pois extra small	up to 7·5 mm
Petits pois very small	from 7·5 to 8·2 mm
Peas small	from 8·2 to 8·75 mm
Peas medium small	from 8·75 to 9·3 mm
Peas medium	above 9·3 mm

Wrinkled varieties

Petits pois extra small	up to 7·5 mm
Petits pois very small	from 7·5 to 8·2 mm
Garden peas small	from 8·2 to 9·3 mm
Garden peas medium small	from 9·3 to 10·2 mm
Garden peas medium	above 10·2 mm

No size description is necessary for ungraded peas.
All size gradings relate to peas before blanching and canning.

Canned vegetables may be examined for vacuum, pH, salt, sugars, drained weight, added colour, copper, lead and zinc. Drained weights are the weights remaining on a sieve (8 holes/inch) after 2 min and can be related to the approximate filled weight from the data in Table 7.9, which are mainly those due to Adam (1965). Unblanched vegetables have drained weights which are up to 12 per cent less than the corresponding blanched vegetables.

Table 7.9 Drained weight of canned vegetables as percentage of filled weight

Canned vegetable	Drained weight		
	min.	max.	mean
Beans, stringless	95	113	103
Broad beans	98	112	106
Beetroot, sliced	92	104	101
Carrots	91	116	101
Celery	88	101	95
Macedoine	91	123	105
Peas, garden	95	120	105
Peas, processed	116	158	135
Potatoes	100	120	106
Turnips	95	106	102

The Labelling of Food Regulations provide that if any canned peas have been prepared from peas that have been dried or soaked prior to canning they must be described as 'processed peas'. Canned processed peas can be distinguished from canned garden peas by the higher proportion of alcohol-insoluble solids and total solids in the drained product (Houlbrooke, 1963a). In this respect, Morcinek (1981) has proposed the use of ascorbic acid and carotene contents as guideline parameters.

The Preservation in Food Regulations (1979) permit the presence of 100 mg/kg sulphur dioxide in canned peas to which no colour has been added. The sulphur dioxide aids the retention of natural colour in the canning process.

TOMATO PRODUCTS

TOMATO PURÉE, TOMATO PASTE, TOMATO POWDER

Standards for tomato purée were recommended by the Comité International Permanent de la Conserve (CIPC, 1959). Draft recommendations have been discussed by the EC. The Draft EC proposals recommended five gradings according to the concentration of dry solids as determined by refractomer (with a tolerance of 5 per cent):

Type of tomato purée	*% Dry solids (min.)*
Semi-concentrate	12
Concentrate	18
Double concentrate	28
Triple concentrate	36
Sextuple concentrate	55

It is permitted to add salt (max. 10 per cent of dry solids) and spices. Tomato purées governed by these standards may be marketed either without any indication of quality ('standard') or with the indication 'extra'. Various standards for each quality group are proposed, including the following (all figures expressed as percentages on dry solids):

	Standard quality	*'Extra' quality*
Total sugars (as invert sugar) (min.)	45	50
Titratable acidity (as citric acid) (max.)	10	9
Volatile acidity (as acetic acid) (max.)	0·30	0·15
Mineral impurities insoluble in water (max.)	0·10	0·05

Other recommended standards applied to products diluted to 9 per cent refractometer solids relate to colour on the Munsell scale, flavour, odour, consistency, black specks, vegetable impurities and the Howard mould count (80 per cent maximum positive fields for 'standard' and 50 per cent for 'extra').

The Codex standard for processed tomato concentrates includes a specification for tomato purée of 8–24 per cent tomato solids and for tomato paste at least 24 per cent. Seven gradings are 'defined' as to the minimum tomato solids content. The quality criteria differ slightly from those proposed by the EC. Permitted additives include sodium bicarbonate and various acids and the tin content is limited to 250 mg/kg.

The dry solids of tomato purées and pastes contain about 9·5 per cent ash, 13·8 per cent protein, 50–65 per cent total sugar (as invert sugar), 5·8–13·4 per cent total acidity (as citric acid) and 4·8 per cent potassium (as K_2O).

Many aspects of the technology and analysis of purées are discussed by Goose and

Binsted (1973) and by Bigelow *et al.* (1950). Cichowicz *et al.* (1982) have surveyed the quality of raw materials used in the manufacture of tomato products including pastes and purées.

Analysis of tomato purée

Total solids. (a) *By weighing.* Dry a dish containing a rod. Weigh out 2 g of sample, spread out with a little water and dry at 70 °C under partial vacuum (cf. Monier-Williams, 1941). Microwave oven drying has been used as an alternate procedure; Chin *et al.* (1985) report that the method is equivalent to the vacuum oven procedure.

(b) *By refractometer.* Mix 10 g sample with 20 g water and filter through a small paper. After rejecting the first runnings determine the refractive index of a drop of filtrate at 20 °C. Obtain the solid content of the filtrate from the data in Table 7.10 and then obtain the concentration in the original sample by multiplying by 3. Less concentrated samples can be strained directly through muslin.

Table 7.10 Relationship between refractive index and total solids of tomato purée and pulp (Bigelow *et al.*, 1950)

% Total solids*	Refractive index at 20 °C	% Total solids	Refractive index at 20 °C
5·0	1·3398	22·5	1·3651
7·7	1·3433	25·0	1·3690
10·0	1·3468	27·5	1·3731
12·5	1·3502	30·0	1·3772
15·0	1·3538	32·5	1·3816
17·5	1·3575	35·0	1·3860
20·0	1·3611	—	—

* After drying *in vacuo* at 70 °C.

Ash and salt. Ignite 5 g at 550 °C and determine salt on the ash by titration with silver nitrate.

Sugars. Boil 20 g of sample gently with water and make up to 100 ml at room temperature. Filter. Invert 10 ml of filtrate with hydrochloric acid, make up to 200 ml and titrate against Fehling's solution (Lane and Eynon's method). Report the total sugars as invert sugar. Most samples contain very little sucrose so inversion is not necessary in routine work.

Acidity. The total acidity can be determined by direct titration of 2 g with 0·1 M sodium hydroxide using phenolphthalein as indicator. Calculate as citric acid. The volatile acidity can be determined after distillation in steam. Also determine the pH value on a 20 per cent solution of sample.

Insoluble solids. Determine as for jam.

Trace elements. The determination of copper is important as the FSC recommend a maximum of 100 p.p.m. on the dried tomato solids. It may be necessary also to determine arsenic, lead and tin (see also p. 141).

Preservatives. Examine for sulphur dioxide, benzoic acid and hydroxybenzoates.

Colour. This can be assessed by means of the Lovibond Tintometer. Red should predominate over yellow. An excess of blue may indicate scorching. The CIPC state colour standards expressed on the Munsell system. Black specks in a purée indicate inefficient sieving of the raw materials.

Tomato content. The tomato solids can be assessed from the potassium content using flame photometry (p. 35). Darbishire (1965) has suggested that the tomato content can be estimated roughly from the lycopene figure. Fifty ml of a 0·2 per cent aqueous suspension of the purée is shaken vigorously with 25 ml light petroleum (b.p. 80–100 °C) and then for 15 min in a mechanical shaker. The extinction coefficient of the clear extract is then measured in a 1 cm cell at 505 nm using a Unicam SP 600 spectrophotometer ($E^{1\%}_{1\text{cm}} = 2820$). Darbishire found a mean value of 1420 mg/kg for the lycopene in the dry solids of various tomato purées. Studies on the carotenoid content of commercial strains of tomatoes have been reported by Hirota *et al.* (1982). Murphy (1983) has surveyed the lycopene content of commercial product on sale in Victoria, Australia, and recorded values varying from 93·8 to 198·5 mg/kg and a non-significant correlation with organic solids content corrected for sugar. Tomato content can also be estimated from the citric acid content, determined enzymatically (Boehringer, 1979) or by HPLC (Ashoor and Knox, 1984). Heat treatment such as retorting may cause partial decomposition of lycopene, giving low solids results.

Howard mould count. The Howard mould count, in which the percentage of fields containing mould is counted, was first developed in the USA in 1911. This represents an index of the quality of the raw material used for the preparation of tomato purée. Maxima for moulds are applied in various countries (Williams, 1968) and in 1958 the CIPC provisionally approved limits of 60 per cent and 40 per cent positive fields for Standard and Extra Quality purées respectively. The APA (1971) has adopted a maximum limit of 50 per cent (25 per cent for tomato juice). Prosecutions under the Food Act have been successfully undertaken for samples containing 75 per cent positive fields. Of 37 consignments examined by the Port of London in 1967 only 3 showed positive fields in excess of 50 per cent and the mean of all samples was 20 per cent positive fields.

The Howard method of counting is described precisely in the AOAC from which the following is taken. Various aspects of the method have been elaborated by the Continental Can Company (1968). The precision of the Howard mould count technique is dependent on the degree of milling of the tomato pulp in preparing the product (Olson, 1980). The adjustment of products to equalise concentration and particle size has been described by Bandler and Cichowicz (1981) and Aldred (1983) has proposed adjustment for tomato solids content of the product. The recommendations of Bandler and Cichowicz (1981) were adopted by the AOAC (Anon, 1981).

Clean Howard cell, so that Newton's rings are produced between slide and cover glass. Remove cover and with knife blade or scalpel place portion of well-mixed sample upon central disc; with same instrument, spread evenly over disc, and cover with glass so as to give uniform distribution. Use only enough sample to bring material to edge of disc. It is of utmost importance that portions be taken from thoroughly mixed sample and spread evenly over slide disc. Otherwise, when cover slip is put in place, insoluble material, and consequently moulds, may be more abundant at centre of mount. Discount any mount showing uneven distribution or absence of Newton's rings, or liquid that has been drawn across moat and between cover glass and shoulder. Place slide under microscope and examine with such adjustment that each field of view covers 1·5 mm². (This area, which is essential, may frequently be obtained by so adjusting draw-tube that diameter of field becomes 1·382 mm. When such adjustment is not possible, make accessory drop-in ocular diaphragm with aperture accurately cut to necessary size. Diameter of area of field of view can be determined by use of stage micrometer. When instrument is properly adjusted quantity of liquid examined per field is 0·15 mm³.) Use magnification of 90–125×. In those instances where identifying characteristics of mould filaments are not clearly discernible in standard field, use magnification of *c.* 200 × (8 mm objective) to confirm identity of mould filaments previously observed in standard field.

From each of two or more mounts examine at least 25 fields taken in such manner as to be representative of all sections of mount. Observe each field, noting presence or absence of mould filament and recording results as positive when aggregate length of not > 3 filaments present *exceeds* 1/6 of diameter of field. The qualifications for acceptance as a positive field have been significantly modified by the AOAC since 1984 (see AOAC methods book). Calculate proportion of positive field from results of examination of all observed fields and report as per cent fields containing mould filaments.

Although the AOAC method requires a minimum of 25 fields to be examined, if possible it is advisable to count at least 150 unless the sample has an especially low count. Williams (1968) has discussed the interpretation and significance of the Howard mould count and considers that other evidence of spoilage should be considered, e.g. rots caused by bacteria, yeasts, viruses and physiological causes. Williams, Olson and Aldred also comment on the estimation of rot fragments and the relationship with mould count. The method is also given in detail in the AOAC methods book. Russell (1985) has described ruggedness tests on the official method.

Chitin. The estimation of chitin, the insoluble proteinaceous constituent of fungal cell walls, may also be used as a measure of the degree of contamination of tomato products by mould. Chitin content can be estimated by a hydrolysis procedure in which the chitosan formed is then de-aminated to 2,5-anhydromannose, an aldehyde that can be estimated colorimetrically (Jarvis, 1977). Further studies by Bishop *et al.* (1982) and by Cousin *et al.* (1984) confirmed the variability within individual values for Howard mould count and a non-linear relationship between chitin content as measured by glucosamine and the HMC. Lin and Cousin (1985) have reported on the use of HPLC to measure glucosamine after derivatisation with *o*-phthalaldehyde, concluding that further work is needed before such methods can be used for quality control. Recent developments based on the use of ELISA techniques to detect mould antigens have been reported by Notermans and Heuvelman (1985) and by Robertson (1986).

TOMATO POWDER

Tomato powder is also an item of commerce. Parran *et al.* (1972) have described a procedure for its examination for Howard mould count.

TOMATO KETCHUP

Tomato ketchup usually contains tomato purée, sugar, vinegar, salt, onion and garlic or other spices. Thickeners such as carob bean gum and added colour are also often present. In view of the comparatively low acidity, tomato ketchup tends to be attacked by moulds, in particular, and it is therefore filled as hot as possible before capping.

Legal requirements

The Food Standards (Tomato Ketchup) Order, 1949 (SI 1949 No. 1817, as amended by SI 1956 No. 1167) states:

1. The standard for tomato ketchup, catsup, sauce and relish shall be as follows:
(a) Tomato ketchup, catsup, sauce and relish shall contain not less than six per cent by weight of tomato solids derived from clean and wholesome tomatoes or from tomato purée, or its equivalent, made from clean and wholesome tomatoes. (The FSC (1970) recommended that the minimum for tomato solids should be increased to 8 per cent.)
(b) The tomatoes, tomato purée or its equivalent or the tomato ketchup, catsup, sauce or relish shall be so strained, with or without heating, as to exclude seeds or other coarse or hard substances.
(c) Tomato ketchup, sauce and relish shall contain no fruit or vegetables other than tomatoes except onions, garlic and spices added for flavouring purposes.
2. No tomato ketchup, catsup, sauce or relish shall contain any copper in excess of 20 parts of copper per million parts of the tomato ketchup, catsup, sauce or relish, as the case may be. (The FSC (1970) recommended that this statutory limit for copper should be discontinued; in its consideration of the Food Surveillance Working Party Report on Copper and Zinc (MAFF, 1981), the FACC saw no need to change the FSC recommendations on copper.)

The Food Advisory Committee have recommended that the Order be revoked (FdAC/REP/5: 1989).

In the USA the definition of tomato catsup does not permit the addition of artificial colour, preservatives or thickeners.

Analysis of tomato ketchup

In order to ensure compliance with legal requirements, samples of tomato ketchup should be examined for permitted preservatives (sulphur dioxide, benzoic acid and 4-hydroxybenzoate), non-permitted preservatives (e.g. sorbic acid), metallic contamination (especially copper, arsenic and lead), added colour, and for the assessment of the tomato solids.

Tomato solids. The amount of tomato present in tomato ketchup is usually assessed from the potassium content. The potassium content is most conveniently determined by means of a flame photometer (p. 35). Cameron and Bakht (1965) stated that using the latter method the presence of 3 per cent salt does not interfere with the potassium determination. Money (1964) gives an average potassium value of 273 mg/100 g for raw tomatoes using flame photometry. The tomato content can also be assessed from the lycopene or citric acid content (p. 254). The average tomato solids content in tomato ketchup is about 10 per cent (FSC, 1970). The (total ash–salt) percentage figure, which represents about 8·5 per cent of the tomato solids is also useful as a rough check on other results. It is not usually convenient to determine the insoluble solids (average 12·6 per cent) because of the presence of added gums and calculations based on the non-sugar solids (46–54 per cent) must involve an allowance for added sucrose. The percentages quoted are based on tomato solids.

General quality. The general quality of tomato ketchup can be further judged from other determinations such as total sugars (by Lane and Eynon's method), total and volatile acidity, pH, total solids (by drying at 70 °C *in vacuo*) ash, salt, the relative viscosity, colour, ability to withstand storage after incubation for one month at 37 °C, and the Howard mould count. Industrial control tests are discussed by Devey (1958). Quality control methods have been reviewed by Vogez and Weber (1982).

Composition of tomato ketchup

Most samples fall within the following ranges:

	Per cent
Total solids	20–40
Tomato solids	6–17 (min. 6)
Ash	2·9–4·0
Salt	1·6–3·6
Total sugars	10–25
Total acidity (as acetic acid)	0·8–3·0
Copper	Less than 5 p.p.m. (max. 20 p.p.m.)

The growth of moulds and yeasts is usually inhibited if the total acidity (as acetic acid) exceeds 1·2 per cent (see also 'Pickles and Sauces' below).

Canned beans in tomato sauce. LAJAC Code of Practice No. 5 applies to canned bean products described as beans in or with tomato sauce or as baked beans in or with tomato sauce. Canned Bean products so described shall consist only of beans and sauce.

Beans. The beans used shall be white beans of the pea or haricot type and shall be free from mouldy beans and substantially free from badly discoloured or damaged beans.

Sauce. The sauce shall contain tomatoes or tomato purée and sugar in sufficient quantities to satisfy clauses 6 and 7, and may contain onions, garlic, herbs and spices used for flavouring and other ingredients, provided they are declared on the label.

Total solids. The minimum dry solids content of canned beans/baked beans in tomato sauce expressed on the fluid capacity of the can, for any individual can up to but not including, the 8 oz or 8Z, shall be 22·5 per cent (m/v) and 23·55 per cent for the average of two or more cans from the same batch. For can sizes 8 oz and U8 and above the figures shall be respectively 24·5 and 25·5 (m/v).

The total solids content shall be determined by drying 5–10 g of a finely macerated sample of the entire contents of the can in a steam oven for 16–19 h at 98 °C or by any other method which will give equivalent results.

Tomato solids. The tomato solids in the finished pack shall be not less than 1·5 per cent by weight for cans below A2½ and 1·1 per cent for cans A2½ and above that size.

The minimum tomato solids stated above shall be the average of two or more cans taken from the same batch and expressed on the basis of the fluid capacity of the cans.

Sugar. The final products shall contain not less than a total of 5 per cent by weight of sugar (expressed as invert sugar) for all cans up to but not including A2½ size. For all cans including and above A2½ size the figure shall be not less than 4 per cent by weight of sugar. The total minimum sugar content stated above shall be the average of two or more cans taken from the same batch and expressed on the basis of the fluid capacity of the cans.

The finished product shall be free from artifical colouring matter.

The dimensions and capacities of cans are given in Table 7.4.

For the determination of sugars, transfer 10 g minced sample into a 60-ml Soxhlet-type extractor plugged with glass wool using about 90 ml 70 per cent alcohol. Heat the flask on a hotplate and continue the extraction for 30 min after the paste is dull grey and the liquid in the siphon is colourless. Then remove the solvent so that the extract volume is about 20 ml. Filter into a 100-ml volumetric flask, washing the residue and filter paper with water before making up to the mark. Clarification with zinc ferrocyanide may occasionally be necessary. Pipette 50 ml of the solution into a 100-ml volumetric flask, add 10 ml conc. hydrochloric acid and invert at 70 °C for 5 min. Neutralise with sodium hydroxide solution to methyl orange, make up to 100 ml and determine the total sugars (as invert sugar) by the Lane and Eynon method.

The composition of a sample of canned baked beans is given on p. 240.

Beans in tomato sauce should also be examined for freedom from added colour. The asessment of the meat content of canned beans, pork sausage and tomato sauce has been discussed by Thraves (1955) and by Halaby *et al.* (1981).

FROZEN FRUITS AND VEGETABLES

FAO/WHO Codex Alimentarius Commission-recommended international standards have been published for sixteen quick frozen fruit and vegetable products, with a separate publication on methods of analysis to be used (see Appendix 2). The methods describe procedures for net weight, total soluble solids and mineral impurities, all largely based on physical procedures. In addition, the standard for peas and quick frozen corn describe a method for the determination of alcohol-insoluble solids content. Collaborative study of the method proposed has revealed an unacceptable reproducibility when applied to quick frozen corn on the cob and to quick frozen whole kernel corn (Wood, 1982). Biltcliffe *et al.* (1984) reported on a collaborative study of methods proposed by Codex for the determination of loss on drying of quick frozen french fried potatoes. The preferred method was based on 16 h oven drying at 103 ± 2 °C.

The Standards require that the freezing shall be carried out in such a way that the temperature of maximum crystallisation is passed quickly. The quick freezing process shall not be regarded as complete unless and until the product temperature has reached −18 °C (0 °F) at the thermal centre. For peas the alcohol-insoluble solids as determined by a specified method must not exceed 23 per cent m/m (for garden peas 19 per cent m/m). For strawberries the soluble solids of the thawed comminuted sample should be 18–35 per cent m/m (if prepared with dry sugars) or 15–25 per cent m/m (if prepared with syrup). A draft directive which encompasses most of the Codex criteria is being developed by the EC.

OTHER PRESERVED PRODUCTS

CANDIED, DRAINED, CRYSTALLISED AND GLACÉ FRUITS

By incorporating at least about 63 per cent of sugar in the aqueous phase, fruits may be made to keep for a long period. In practice, drying up is the main problem during storage. The method of production varies to some extent with the fruit. Some are lye-peeled in sodium hydroxide solution, whilst in the case of the citrus fruits the peel itself is used. After steeping in sulphurous acid or a salt brine containing bisulphite, the fruit material is washed in water and then successively syruped in increasing concentrations of hot sugar syrups, which may include some glucose syrup. For candied peel, a mixture of lemon and orange peel is used and the sugar is impregnated up to a final soluble solids concentration of about 68 per cent. The APA have expressed the view that 'Foods described as Candied Peel or Cut Peel should contain at least 64 per cent of soluble solids as determined by refractometer at 20 °C'. 'Drained' fruits are produced by draining the candied fruit on wire grids. For 'Glacé' fruit such as glacé cherries, which usually contain added dye, the outside of the drained fruit is washed free of syrup and then covered with a supersaturated sugar solution. 'Crystallised' fruits contain minute sucrose crystals on the surface due to a final immersion in super-saturated syrup. Continuous processes are now sometimes used.

These products should be examined for soluble solids using the refractometer (Forbes, 1963), sugars (see 'Jam', p. 223), preservatives, added dyes and trace metals such as arsenic, copper and lead. The Preservatives Regulations prescribe that crystallised or glacé fruit may contain up to 100 mg/kg sulphur dioxide or up to 1000 mg/kg of benzoic acid or methyl, ethyl or propyl 4-hydroxybenzoate. Stabilised syrup containing fruit pieces may contain 1000 mg/kg sorbic acid, if the product is to be used as an ingredient in edible ices or ice-cream.

Ingleton (1964) has described the manufacture of candied fruit and Grosso (1965) that of candied and glacéd fruit. Preserved and crystallised ginger (Ingleton, 1966) and angelica (from the stalk of an umbelliferous plant, *Angelica archangelica*—not the wild English variety) are prepared by similar methods.

Water activity measurements are of significance in this type of product (see Chapter 2).

FRUIT JUICES

Most fruit juices that appear as such on the retail market are derived from citrus fruits. After expression in a reamer the juice is strained, flash pasteurised, filled into cans or bottles and sealed. Added sugar is sometimes present. Concentrated juice is prepared by distillation under reduced pressure or by freezing (Pollard and Beech, 1966). Sulphur dioxide is invariably present in the concentrated orange juice intended for children and the process is conducted so that there is the minimum loss of vitamin C. Some disparity exists in national legislation as to the varieties of citrus fruit permitted (Mears and Shenton, 1973). International Standards recommended by Codex Alimentarius for fruit juices are summarised in Table 7.11. In addition, the Codex standards include maximum limits for trace elements and give references to appropriate analytical procedures. Standards for concentrated juices and for specific nectars have also been recommended (see Appendix 2). A high proportion of the fruit juice handled in Britain is used in the manufacture of squashes and some juices which have been exported to Britain for this purpose appear to have been sophisticated. For this reason checks on the genuineness of juice should include as many criteria as possible. From the analytical data given in Table 7.12 it is apparent that fruit juices show a wide variation in composition. The figures quoted have been drawn from various sources, including Stern (1943), Morgan (1953, 1954), Money (1966) and Osborn (1964). More recent and comprehensive data, including guide RSK values and ranges for authentic fruit juices and nectars, together with methods of analysis, are published by VdF (1987). Data on apple juices prepared over 3 years has been documented by Mattick and Moyer (1983). Mears and Shenton (1973) have published a comprehensive review of the composition of orange and grapefruit juices. Also, data have been presented on orange juice by Sawyer (1963) and on blackcurrant juice by Barker *et al.* (1961) and Ayres *et al.* (1961, 1962). Further figures have been tabulated by Hulme *et al.* (1965) for citrus fruits. Specific information has also been published on grapefruit juice (Sherratt and Sinar, 1963), for Israeli citrus juices (Ministry of Commerce and Industry of the State of Israel, 1967), and New Zealand citrus juices by Robertson and Kirk (1979). The presumptive specific gravity of concentrated citrus juices has been detailed by Basker (1966). Data on cherries has been presented by Bazzarini *et al.* (1981), and includes amino acid and sugar contents. Sugars in eleven varieties of juice have been surveyed by Li and Schuhman (1983).

Legislation

The Fruit Juices and Fruit Nectars Regulations 1977, SI No. 927, as amended 1982 (SI No. 1311), implement EC Directive 75/726 as amended and prescribe definitions for fruit juice, concentrated fruit juice, dried fruit juice and fruit nectar and the control of labelling and advertisement of these products. They also control the extent to which sulphur dioxide, lactic acid, citric acid and malic acid and the nature of sugars and sugar products that may be added. The use of additional preservatives (SO_2 350 mg/kg, benzoic acid and benzoates 800 mg/kg) is allowed in products not controlled otherwise by the Regulations. King (1980) has reviewed the provisions of the EC Directives and the UK Regulations; a supplementary review of the amendments appeared in Anon (1982).

The analysis of fruit juices

This may include the determination of the following: total solids, acidity, ash, alkalinity of the ash, phosphate, potassium, nitrogen (and the formol titration),

Table 7.11 Some standards recommended by Codex Alimentarius for various fruit juices (preserved exclusively by physical means)

Type of juice	Refractometer solids (% m/m) min.	Added sugar (g/kg) max.	Total titratable acidity (as % m/m anhydrous citric acid) min.	Volatile acids (as g acetic acid/kg) max.	Ethanol (g/kg) max.	Essential oils (ml/kg) max.
Apple	10	—	—	0·4	5	—
Grapefruit	9	50	—	—	3	0·3
Lemon	6	—	4·5	—	3	0·5
Orange	10	50	—	Traces	3	0·4
Grape	15*	—	—	0·4	5	—
Pineapple	10**	25	—	—	3	—
Tomato	4.5***	—	—	—	—	—

* 16 if made from concentrate. ** 13·5 if made from concentrate. *** Exclusive of added salt.

Table 7.12 Analytical data of various fruit juices (from various sources)

Juice	Relative density (20/20)	Total solids (% m/m)	Acidity (as citric acid) (% m/m)	Ash (% m/m)	Alkalinity of ash (as K_2CO_3) (% m/m)	Potassium (as K) (% m/m)	Phosphorus (as P) (% m/m)	Nitrogen (% m/m)	Vitamin C (mg/100 ml)
Orange									
max.	1·045	12·0	3·5	0·63	0·53	0·284	0·023	0·180	80
min.	1·040	10·0	0·4	0·29	0·24	0·089	0·007	0·060	20
mean	1·042	10·8	1·4	0·40	0·35	0·160	0·015	0·103	48
Grapefruit									
max.	1·054	12·1	3·0	0·56	0·45	0·188	0·019	0·094	65
min.	1·040	9·4	0·4	0·34	0·255	0·080	0·007	0·032	35
mean	1·042	10·4	1·6	0·47	0·35	0·126	0·013	0·061	41
Lemon									
max.	1·040	11·2	8·0	0·56	0·415	0·193	0·015	0·084	70
min.	1·030	9·2	5·1	0·29	0·13	0·108	0·005	0·035	30
mean	1·035	10·0	5·3	0·37	0·29	0·146	0·011	0·58	46
Lime									
max.	—	10·7	8·0	0·44	0·37	—	0·013	—	40
min.	—	8·4	4·5	0·33	0·27	—	0·008	—	5
mean	1·035	9·3	7·5	0·38	0·30	—	0·011	—	25
Apple									
max.	1·069	16·9	0·80	0·35	—	0·149	0·029	—	—
min.	1·037	10·1	0·22	0·15	—	0·079	0·007	—	—
mean	1·047	13·0	0·58	0·20	—	0·130	0·015	—	8
Blackcurrant									
max.	1·080	—	5·7	1·5	—	—	0·045	0·120	400
min.	1·033	—	2·8	0·35	—	0·25	0·012	0·021	90
mean	1·055	13·5	3·5	0·75	0·45	0·35	0·026	0·055	200

sugars, vitamin C, trace metals, amino acids and preservatives. Industrially the refractive index, the specific gravity, the pH and volatile oil are usually determined and the juice is examined for enzymic activity. Various standardised procedures for the analysis of fruit juices have been published by the International Federation of Fruit Juice Producers and can be found in VdF (1987). These include procedures for the estimation of chloride, lactic acid, trace metals, L-malic acid and other organic acids, pectin, sodium, potassium, phosphorus, sugars and other sweetening agents and physical features such as colour and viscosity. Methods of analysis are also recommended in the Codex standards relating to fruit juices, concentrates and fruit nectars. Recent reviews of modern methods of analysis including enzyme techniques have been presented by Beutler (1982), Henniger (1984) and Berry (1985). Official methods used for analysis of fruit and vegetable juices in France have been summarised by Merle (1982).

Total solids. Determine preferably by drying at 70 °C under reduced pressure.

Degree of concentration is usually assessed industrially from the refractive index or specific gravity. The following figures are extracted from the Reports of the wartime Citrus Fruit Juices Control:

Degree of concentration	Grapefruit RI (20 °C)	Lemon RI (20 °C)	Orange RI (20 °C)
3×	1·3756–1·3790	1·3681–1·3710	1·3816–1·3860
4×	1·3926–1·3970	1·3796–1·3820	1·3996–1·4040
5×	1·4086–1·4120	1·3911–1·3940	1·4181–1·4230
6×	1·4236–1·4270	1·4041–1·4070	1·4381–1·4430
7×	1·4391–1·4430	1·4149–1·4175	1·4566–1·4610

An accurate method for the determination of the specific gravity of fruit juice concentrate has been described by Basker (1969).

Acidity. For routine purposes it is usually sufficient to determine the titratable acidity (using phenolphthalein). The acidity of juices is usually calculated as the predominant acid, i.e. that of citrus juices as citric acid, apple as malic acid and grape juice as tartaric acid. The flavour of fruit juice is related to the ratio of soluble solids to the total acidity called the 'maturity ratio', which increases as the fruit ripens and is sometimes used for assessing the quality of the juice. The volatile acidity is usually calculated as acetic acid. Methods for determining the individual acids present in fruits are described in the AOAC; typical HPLC techniques have been described by Schwarzenbach (1982), Shaw *et al.* (1983) and Mentasti *et al.* (1985). The use of ion chromatography as a generally applicable method of detecting both inorganic and organic ions has been reviewed by Cox *et al.* (1985).

Ash. The ash is obtained by igniting the sample at 550 °C after a preliminary evaporation. The alkalinity of the ash is then determined by boiling with a measured excess of 0·1 N acid (0·1 M HCl), cooling and back-titrating with 0·1 N alkali (0·1 M NaOH) using methyl orange (1 ml 0·1 N ≡ 0·0069 g K_2CO_3).

Phosphate. For the determination of phosphate in fruit juices, the sample is first made alkaline, evaporated and ashed. The phosphate can then be determined either volumetrically or colorimetrically on the solution obtained by dissolving the alkaline ash in hydrochloric acid (cf. Morgan, 1954).

Potassium is determined on the diluted filtered juice by means of the flame photometer (p. 35).

Nitrogen is determined by the macro-Kjeldahl method. The amino acids are frequently expressed as the formol number (or value), which is based on the formol titration. The latter can be performed potentiometrically to pH 8·15 (Ayres *et al.*, 1961). The formol number can be expressed in terms of millilitres of N(M) sodium hydroxide per litre.

Nitrogenous compounds, which are present in small amounts, have been recommended as indices for the evaluation of fruit juices (p. 272). A rapid HPLC method for determination of amino acid profiles of juices by use of post-column derivatisation with *o*-phthalaldehyde has been proposed by Chaytor (1986).

Sugars. The presence of added sucrose can be detected by determining sugars before and after inversion by copper reduction methods. Glucose syrup can be detected by means of the polarimeter; enzyme or chromatographic methods may also be used (see Chapter 6). Methods have been reviewed by Li and Schuhmann (1983).

Hydroxymethylfurfural (HMF) is formed by dehydration of hexoses, a high concentration being indicative of excessive heat treatment or prolonged storage of concentrates. The IFJU method is based on colorimetric reaction with *p*-toluidine-barbituric acid. Cilliers and Van Niekerk (1948) have used a two column HPLC procedure firstly to extract the HMF from fruit juices and secondly to measure the HMF content. The method is reported to give lower values than that accepted by the International Fruit Juice Union. Similar conclusions using furfural as an indicator of processing and storage abuse were reached by Marcy and Rouseff (1984). These authors used a distillation procedure to separate the furfural and an HPLC technique based on methanol water elution from Zorbax ODS for measurement of the furfural.

Vitamin C

The following titration procedure is similar to that described on p. 11 for general use.

Standard indophenol solution. Dissolve 0·05 g 2,6-dichlorophenolindophenol in water, dilute to 100 ml and filter. To standardise, dissolve 0·05 g pure ascorbic acid in 60 ml of 20 per cent metaphosphoric acid and dilute with water to exactly 250 ml. Pipette 10 ml of this solution into a small flask and titrate with the indophenol solution until a faint pink colour persists for 15 s. Express the concentration as mg ascorbic acid equivalent to 1 ml of the dye solution. The dye solution keeps for a few weeks if stored in the refrigerator, but it should be standardised before use against a freshly prepared solution of ascorbic acid.

METHOD

Pipette 50 ml of unconcentrated juice (or the equivalent of concentrated juice) into a 100 ml volumetric flask, add 25 ml of 20 per cent metaphosphoric acid as stabilising agent and make up to the mark with water. Pipette 10 ml of the solution into a small flask, add 2·5 ml of acetone and titrate with the indophenol solution until a faint pink colour persists for 15 s. Calculate the vitamin C content in the sample as mg per 100 ml (or 100 g). The acetone may be omitted if sulphur dioxide is known to be absent. Its function is to form the acetonebisulphite complex with sulphur dioxide, which otherwise interferes with the titration. Sometimes a small proportion of the ascorbic acid in foods becomes reversibly oxidised during ageing and forms dehydroascorbic acid. If this is suspected first estimate the ascorbic acid as above, then through another portion of the solution pass a stream of hydrogen sulphide for 10 min. Stopper the flask and allow it to stand overnight in a refrigerator. Then remove the hydrogen sulphide by bubbling nitrogen through the mixture and titrate as before. The difference between the two titrations gives a measure of the dehydroascorbic acid. One International Unit of vitamin C = 50 μg ascorbic acid.

HPLC techniques are now widely used (see the discussion on p. 245).

Trace metals and preservatives. The routine examination of fruit juices should include examination for trace elements such as copper, lead and zinc (Snodin, 1973) and for preservatives such as sulphur dioxide, benzoic acid and 4-hydroxybenzoates. Methods are described in Chapter 4.

Enzymic activity. The pectic enzymes present in the raw juice should be destroyed during the pasteurisation process. If they are not destroyed, products into which the juice is incorporated (e.g. squashes) tend to go clear (i.e. the desirable cloudiness is not retained) due to the action of the enzymes on the pectinous substances present. The activity can be assessed by heating a solution containing 1 per cent citrus pectin and 1·2 per cent sodium chloride nearly to boiling point. Then after cooling to 30 °C, transfer 50 ml to a beaker, add 2 ml juice and neutralise to pH 7·0 with 0·02 M sodium hydroxide solution using a pH meter. Keeping the solution at 30 °C, maintain the pH at 7·0 by continuing the titration over a period of 30 min. The acidity produced over the 30-min period is approximately proportional to the pectolytic enzymatic activity.

Ethanol

The Codex Standards allow for the presence of low amounts of alcohol arising from the natural fermentation of juices that are prepared without the use of chemical preservatives. Levels in the region of 3 g/kg may be determined by the traditional distillation and dichromate oxidation procedures such as recommended by the IFJU and AOAC. However, in recent years the enzymic methods and GLC have been widely adopted; GLC using propanol as internal standard has the merit of detecting other volatiles such as methanol and acetaldehyde (Lund *et al.*, 1981; Tanner and Limacher, 1984). Upperton (1985) has assessed both procedures and has concluded that both techniques are appropriate for the estimation of ethanol at naturally occurring levels in fruit juices.

Adulteration

Considerable attention has been paid to methods of analysis aimed at the authentication of juices derived from most varieties of fruit. The literature on the topic has expanded substantially in the past ten years (Petrus and Vandercook (1980) reviewed the available literature and methods used to 1979), and all manner of constituents and techniques have been assessed with a view to a solution of the problem. The general conclusion appears to be that no single parameter has been identified that will give an unequivocal assurance of the authenticity of a juice. The use of multivariate statistical methods has been proposed by a number of workers including Brown *et al.* (1981), Ara and Torok (1980), Richard and Coursin (1982) and Brown and Cohen (1983). The latter authors conclude that there is a need to develop both statistical and analytical techiques if the approach is to become widely used. Isotope ratio methods have been advanced as the most useful procedures to detect the addition of adulterants and diluents to commercial juices. Doner and Bills (1982) and Brause *et al.* (1984) have proposed the application of carbon isotope analysis to detect addition of sugars and oxygen isotope analysis to detect the addition of water. Table 7.13 summarises some literature on the subject published since 1980. Accepted reference data for use in EC has been presented by Bielig *et al.* (1985).

Dimethylpolysiloxane is used as an antifoam agent in the production of juices; limits in the region of 10 mg/kg may be encountered. Kacprzak (1982) has described a method in which the juice is adsorbed on to Florisil, dried and extracted with chloroform. After removal of chloroform the residue is dissolved in methyl isobutyl ketone and the silicon content is determined by atomic absorption spectrophotometry.

Glycols. Propylene glycol and isopropanol may be detected in juices and drinks, and a GLC method has been described by van Rillaer and Beerhaert (1983).

SOFT DRINKS

Soft drinks fall mainly into the following categories:

Ready-to-drink beverages. Most ready-to-drink beverages are carbonated. These are usually prepared from a concentrated syrup containing sugar (and possibly saccharin), fruit juice (or flavouring essence), citric acid and preservative (usually sodium benzoate). Artificial colouring matter is added to some drinks and products such as ginger beer contain a foaming agent of the saponin-type. In the manufacture a definite volume of the syrup is measured into the bottle. The bottle is then filled with carbonated water and closed (cf. Alderson, 1970). The degree of carbonation employed varies according to the product.

Squashes, cordials, crushes. Concentrated soft drinks contain more or less similar ingredients to the ready-to-drink beverages but in greater amounts. They are, however, not carbonated. Lime juice cordial is filtered to a bright clear liquid before bottling, but most other products should remain cloudy on storage provided that pasteurised juice has been used (see above). Squashes containing fruit juice usually contain sulphur dioxide.

Comminuted drinks are prepared from the whole citrus fruit rather than the juice, although some of the insoluble solids are usually removed by sieving. Such products should be described as 'orange drink', 'lemon drink', etc. Benzoic acid is commonly present as preservative. Some technical aspects of comminuted drinks have been discussed by Charley (1963). 'Double' and 'Triple' concentrated drinks are also produced.

'Bitter Orange' and 'Bitter Lemon' have a basis of fruit juice or comminuted fruit and often contain some added bitter principles, e.g. quinine. The FSC Report

Table 7.13 Summary of procedures used to test the authenticity of fruit juices

Substrate juice	Analyses performed	Reference
Apple	Proximate analysis	Koch (1984)
	Sugar, acidity	Millies and Burkin (1984)
	Sugars, phenolics, K, $^{12}C:^{13}C$ ratio	Brause and Raterman (1982)
	Malic acid (presence of fumaric acid in synthetic malic acid)	Junge and Spadinger (1982)
	Proximate analysis	Mattick and Moyer (1983)
	$^{12}C:^{13}C$ ratio	Anon (1980b)
	Sorbitol, asparaginic acid, K and PO_4	Anon (1982)
	Amino acid spectrum	Bielig and Hofsommer (1982)
	Sugars, acids, ash, Na, K, PO_4	Steinhauser and Gherardi (1980)
	pH shift on inoculation with *L. plantarum*	Smolensky and Vandercook (1980)
	Organic acids	Evans *et al.* (1983)
Blackberry	Sugars, acids, anthocyanin pigments	Wrolstad *et al.* (1982)
Blakcurrant	Flavonol glycosides	Siewek *et al.* (1984)
Cherry	Amino acid spectrum	Eski *et al.* (1980)
Cranberry	Sugars, hydroxycinnamic acids	Marwan and Nagel (1982)
Grape	Oxygen isotope	Dunbar (1982)
Strawberry	Flavonol glycosides	Henning (1982)
Citrus (general)	Amino acid profile, proximates	Cohen *et al.* (1984)
		Cohen and Fuchs (1984)
	Acidity, sugars, proximates	Lufschitz (1983)
	Naringin (as indicator of grapefruit juice)	Greiner and Wallrauch (1984)
	Hesperidin, narangin	Siewer *et al.* (1984)
	Sugars, organic acids	Lea and Smith (1985)
	Sugars, organic acids, phenolics	Coppola (1984)
Grapefruit	Naringin, hesperidin	Trifiro *et al.* (1982)
Lemon	Amino acid profiles	Benk (1980)
Orange	Sugars, UV–visible spectra, ^{13}C, ^{18}O, Na and K	Brause *et al.* (1984)
	Sugars by $^{12}C:^{13}C$ ratio	Doner and Bills (1982)
	Proximate analysis, extended analysis, reference values	Bielig *et al.* (1985)
	Proximate analysis	Fuchs and Wretling (1981)
	Arabinose: galactose ratio in water-soluble polysaccharides	Kauschus and Thier (1985)
	Glycosides	Anon (1981)
	pH change after inoculation with *L. plantarum*	Vandercook *et al.* (1980)
	Sugars, citric and isocitric acid, proximate analysis	Anon (1980a)
	Polyphenols	Chandler (1983)
	UV–visible spectra, fluorescence and emission spectra	Petrus and Attaway (1980)
	UV–visible spectra	Saguy and Cohen (1984)
Citrus	Statistical multivariate analysis	Cohen (1985)
		Ara and Torok (1980)
		Richard and Coursin (1982)
		Brown *et al.* (1981)
		Brown and Cohen (1983)

recommended that such products should be regarded as squashes or comminuted drinks (as appropriate) for the purposes of the regulations.

Legal requirements

The Soft Drinks Regulations 1964, as amended, require minimum quantities of fruit juice or potable fruit and added sugars and maximum quantities of unrestricted amounts of saccharin or its calcium or sodium salts, sweeteners permitted by the

Sweeteners in Food Regulations (1983) and specify acids that are permitted to be added, together with detailed labelling requirements. By amendment in 1969 the provisions for use of cyclamates were annulled. The Regulations were further reviewed by the Food Standards Committee in 1976, which recommended the extension of the scope of the regulations to powdered drinks and flavoured syrups together with further labelling requirements. Other recommendations included a caffeine standard for cola drinks of 50–100 mg/l (except decaffeinated drinks).

In general, the regulations prescribe that soft drinks, other than semi-sweet soft drinks, for consumption without dilution shall contain a minimum of 4½ lb of added sugar and a maximum of 56 grains of saccharin per 10 gallons (*equivalent to 45 g and 80 mg/l, respectively*), with a minimum of 3 per cent citrus juice by volume (5 per cent in the case of citrus crushes and soft drinks containing a mixture of citrus and non-citrus fruit juices) or 1½ lb of potable citrus fruit per 10 gallons (*15 g/l*) for comminuted citrus drinks, with certain exceptions. Soda water should contain not less than 5 grains of sodium bicarbonate per pint (*570 mg/l*), dry ginger ale not less than 3 lb of sugar per 10 gallons (*30 g/l*) and brewed ginger beer and similar beverages not less than 2 lb of sugar per 10 gallons (*20 g/l*) but may, in the case of the latter class only, contain up to 80 grains of saccharin per 10 gallons (*114 mg/l*). Soft drinks, other than semi-sweet soft drinks, for consumption after dilution are required to contain a minimum of between 15 per cent and 25 per cent by volume or from 7 lb to 10 lb of comminuted citrus fruit per 10 gallons (*70 to 100 g/l*), according to variety, with a minimum of 22½ lb of sugar and a maximum of 280 grains of saccharin per 10 gallons (*225 g and 400 mg/l, respectively*). Semi-sweet soft drinks for consumption without dilution require between 2¼ lb and 3 lb of sugar and not more than 28 grains of saccharin per 10 gallons (*22·5 g, 30 g and 40 mg/l, respectively*) and those for consumption after dilution 11¼ lb to 15 lb and not more than 140 grains saccharin (*112 g, 150 g and 200 mg/l, respectively*). Special requirements apply for tonic water (minimum for quinine). For the purpose of the regulations the term 'sugar' means any soluble carbohydrate sweetening matter. The regulations make provisions for drinks intended for dilution in respect of composition and amount of saccharin permitted. The requirement is that the drink diluted according to instructions on the label shall comply with regulations in relation to the corresponding soft drink intended for consumption without dilution.

A soft drink (other than soda water) intended for consumption without dilution which is of a description included in Part I of Schedule 2 to the regulations may contain up to the maximum quantity of saccharin if that soft drink also contains not less than 22½ lb of sugar per 10 gallons (*225 g/l*). A soft drink intended for diabetics shall not contain added sugar, but the maximum limits for artificial sweeteners do not apply. Special provisions apply in the case of **'Low calorie' soft drinks**:

1. the standards and limits for sugar and saccharin respectively do not apply;
2. must comply with any relevant minimum standards for fruit or fruit juice;
3. shall comply with a maximum calorie content of 7·5 calories/fl.oz (*26 cal per 100 ml*) if intended for consumption after dilution, or 1·5 calories/fl.oz (*5 cal per 100 ml*) if for consumption without dilution. The conversion factors that are to be used to calculate the calorific (energy) values when expressed as kcal/g are carbohydrate 3·75, protein 4, fat 9, alcohol 7.

Regulation 6 provides that any soft drink may contain the following added acids: ascorbic, citric, lactic, malic, nicotinic, tartaric. Also acetic and phosphoric acids may be added to any soft drink other than a fruit squash, fruit crush or a comminuted citrus drink.

Regulations 7–13 contain many provisions relating to labelling. With a comminuted drink the word 'drink' or 'barley drink' must be immediately preceded by the name of the appropriate citrus fruit. Also if a soft drink does not contain the prescribed minimum amount of fruit or fruit juice, the label must not bear a pictorial device or word suggestive of fruit. Either the name of the suggested fruit followed by the suffix 'ade' may be used or the word 'flavour' can be preceded by the name of the fruit. Any soft drink containing saccharin or other permitted sweetener must bear a specific statement to that effect.

The Preservative Regulations permit soft drinks to contain sulphur dioxide, benzoic acid and 4-hydroxybenzoates up to prescribed limits. These preservatives are also permitted in drinks intended to be frozen before consumption ('freeze drinks').

In a report on the use of additives to produce cloud in soft drinks, the FdAC (1985) considered two classes of materials, namely dispersing agents and clouding agents. The former class have the action of producing cloud by the emulsification of essential oils, whilst the latter class themselves have the effect of cloud formation in otherwise clear solution.

One dispersing agent formerly used was brominated vegetable oil (BVO) but its use was prohibited because of doubts as to its safety (FACC, 1970). The committee reported on representations to permit the use of five substances all of which had inadequate toxicological data; on the matter of clouding agents the committee was not convinced of a need to pemit the use of further additives for the purpose requested. In the event the committee made no recommendations of changes in legislation. Substances considered were:

dispersing agents:
- sucrose di-acetate hexa-iso-butyrate
- glycerol ester of wood rosin
- glycerol ester of partially hydrogenated wood rosin
- colophony/lanolin mixture

clouding agents:
- candelilla wax
- carnuba wax
- wool alcohols

Analysis of soft drinks

In general, the analysis of soft drinks should include the determination of total solids, ash, sugars (before and after inversion by a copper reduction method), or HPLC (Vidal *et al.*, 1985), acidity, pH, preservatives (see above) and trace metals (e.g. Pb, Cu, As, Zn, Fe). In appropriate instances such products should be examined for saccharin and sweeteners, vitamin C, added colouring and the fruit juice content should be assessed. Canned products should be examined for tin content. With shandy-type drinks (p. 264) the alcohol content should be determined by gas chromatography. Tonic water and cola-type drinks require examination for alkaloids and soda water for sodium bicarbonate. Most of the procedures that are required follow the normal pattern of general analyses and only selected special methods are described below. In comparing results against statutory requirements it is useful to bear in mind that 70 grains/10 gallons is equivalent to 100 mg/l.

Sampling

With carbonated beverages the froth tends to make accurate pipetting difficult and it is usually necessary to pour the sample backwards and forwards from one beaker to another before commencing the analysis. A manometric method for the determination of the carbon dioxide is described in the AOAC. Detailed analysis of headspace gases in a variety of beverages has been described by Cook *et al.* (1985). Products containing suspended matter must be inverted several times.

Examination of drinks containing sweeteners and preservatives

The Sweeteners in Food Regulations (1983) permit the use of various sweeteners including sugar alcohols, acesulfame potassium, saccharin and the protein sweeteners aspartame and thaumatin. Although the regulations are silent on the amounts allowed in foods, many of the substances now permitted were developed for use in soft drinks. By virtue of the Soft Drinks Regulations quantitative limits still apply in the case of saccharin. A guide to the products prescribed by the regulations and their uses has been prepared by Jeffries *et al.* (1983). The Preservatives Regulations permit the use of sulphur dioxide, benzoic acid and 4-hydroxybenzoates to prescribed levels.

Sulphur dioxide may be determined by any of the methods described in Chapter 4. Although the sweeteners and benzoic acid preservatives may be determined individually, many of the newer methods developed for analysis of drinks use HPLC and are capable of detecting more than a single additive. Typical of these are the procedures described by Argoudelis (1984) in which aspartame, saccharin, caffeine and benzoic acid are determined; Tyler (1984) used acetonitrile–trimethylammonium phosphate as mobile phase for the separation of the same combination of additives. Herrmann *et al.* (1983) applied HPLC to the detection of cyclamate, saccharin, dulcin and aspartame. The protein derived materials may be detected by amino acid analysers (Jost *et al.*, 1982); in the case of aspartame attention has also been directed towards the detection of the piperazine derivatives including diketopiperazine which occurs as a result of hydrolysis and cyclisation (Scherz *et al.*, 1983; Prudel and Davidkova, 1985; Tsang *et al.*, 1985). Daniels *et al.* (1984) used HPLC and TLC to determine aspartame in dry beverage mixes, and recoveries in the region of 94 to 110 per cent were reported according to the nature of the product. Methods for the detection of dulcin in fruit beverages have been reported by Veerabhadra *et al.* (1984). For further discussion of general methods for sweeteners and preservatives see Chapter 4.

The following simple methods rely on generally available equipment and utilise conventional chemical and chromatographic procedures.

Saccharin

Saccharin can be extracted by making strongly acid with hydrochloric acid and extracting with ether. After removing the solvent it can be detected by the resorcinol test (p. 000), by taste or by conversion to salicylic acid (AOAC).

For the estimation of saccharin in soft drinks it is advisable to remove any fruit oil present. The strongly acidified drink is first shaken several times with diethyl ether. The combined extracts are then shaken with dilute ammonia at least twice. The lower ammoniacal layers are run into a fresh separator, acidifed and extracted several times with ether. The saccharin can be estimated by weighing or titrating the residue, by gravimetric determiation as barium sulphate after fusing with fusion mixture, or by spectrophotometry.

Drinks containing saccharin and benzoic acid

On acidifying a drink containing benzoic acid and saccharin both are extracted with ether. If Monier-Williams' method is employed also, it will give the amount of benzoic acid present and the saccharin content can be deduced by subtraction from the total ether extract.

A simple alternative procedure involves extracting the benzoic acid selectively at pH 4. Then the saccharin is extracted after making the liquid strongly acid. If fruit or fruit juice is present the fruit oil should be removed using ammonia as outlined above. Other acids that might also be extracted can be destroyed by treatment with potassium permanganate.

Benzoic acid. To a suitable volume of sample add 20 ml of pH 4 buffer solution (containing 10 per cent sodium citrate and 6·5 per cent citric acid) and extract the benzoic acid by shaking with three separate portions of 25 ml of ether. Wash each extract with the same 4 ml of water, filter the combined extracts and remove the solvent (remove the last few ml of solvent at room temperature with the assistance of bellows). Dissolve the residue in 2 ml acetone, add 2 ml water and titrate the acid with 0·05 M sodium hydroxide using phenol red as indicator.

$$1 \text{ ml } 0{\cdot}05\ \text{M NaOH} \equiv 0{\cdot}0061 \text{ g benzoic acid}$$

Benzoic acid can also be determined by spectrophotometric determination in ethereal solution (after clean-up with alkali to remove interfering materials). (See Chapter 4 for other methods.)

Saccharin. To the aqueous liquid in the separator add the wash water and 10 ml of conc. hydrochloric acid and extract the saccharin by shaking with three separate portions of 25 ml of ether. Wash the combined extracts with two separate portions of 10 ml of water and then shake the combined wash waters with ether. Combine this ether with the combined ether extracts, filter and remove the solvent. Add 3 ml acetone to the residue, evaporate, add 2 ml water and titrate with 0·05 M sodium hydroxide using bromothymol blue as indicator.

$$1 \text{ ml } 0{\cdot}05\ \text{M NaOH} \equiv 0{\cdot}009\,16 \text{ g saccharin}$$

The presence of saccharin should be confirmed by tasting and qualitative tests (p. 127). Ramappa and Nayak (1983) used a spectrophotometric procedure based on reaction with Azure B.

Cyclamates

In view of the provisions contained in The Soft Drinks (Amendment) Regulations (1969) cyclamates are no longer permitted to be present in soft drinks but their use is allowed in other countries and provision should be made to assess the compliance with UK Regulations. Previously it was permissible to incorporate cyclamates as cyclohexylsulphamic acid or as the calcium or sodium salts.

For their detection, dissolve 0·1 g cyclamate in 10 ml water, add 1 ml conc. hydrochloric acid and 1 ml of 10 per cent barium chloride solution. To the solution, which should be clear, add 1 ml 10 per cent sodium nitrite solution. In the presence of cyclamate a white precipitate of barium sulphate is produced. The reaction is the basis of quantitative methods, in which the barium sulphate is determined gravimetrically. Such methods are less suitable for drinks containing fruit and Rees (1965) used GLC for the analysis of the cyclohexane produced after treatment with nitrite (cf. Dalziel *et al.*, 1972). Shenton and Johnson (1973a) applied semi-automated methods for the determination of cyclohexyl nitrite by diazotisation and coupling with Bratton–Marshall reagent. The same authors (1973b) also suggested adding excess nitrous acid to the sample and then determined the unconsumed nitrous acid colorimetrically with safranine. Some workers precipitate the sulphate with standard barium chloride solution and titrate the excess barium with EDTA. In the following method due to Davies (1966) the end-point detection is improved by using the fluorescent indicator, calcein blue and rhodamine B is used to mask residual fluorescence. Interference due to sulphur dioxide is eliminated by oxidation with hydrogen peroxide in the blank determination. The blank is necessary to eliminate errors due to sulphate or calcium in squashes.

To each of two 250-ml beakers (marked at 150 ml approx.) add 10 ml of sample (or a volume containing 2–35 mg cyclamate), 1 ml of 2 M hydrochloric acid and exactly 10 ml of 0·02 M barium chloride solution. To one beaker add 1 ml hydrogen peroxide solution (2 vol), cover and boil gently for 1 min on a hotplate. Cool, dilute to about 150 ml with water, add 4 ml of 8 M potassium hydroxide solution, place on an inverted ultraviolet lamp (principal wavelength 350 nm) and stir the liquid with an electric stirrer. Add 0·01 per cent rhodamine B solution until there is a slightly red fluorescence and then 4 drops of 0·05 per cent calcein blue solution. Titrate the excess barium in subdued light with 0·01 M EDTA solution (disodium salt) until no blue fluorescence is observed. To the other beaker add 1 ml of 10 per cent sodium nitrite solution *instead of the peroxide*, place on the hotplate and continue as before.

$$\text{Per cent cyclamate} = \frac{\text{Titration difference} \times 100 \times F}{\text{Volume of sample taken}}$$

where $F = 0{\cdot}001\,78$ g for cyclamate anion

= 0·002 012 g for sodium cyclamate
= 0·002 163 g for calcium cyclamate dihydrate.

Saccharin does not interfere in the above method, which can be used for determining cyclamate in a wide range of foods. In the AOAC method cyclamate is hydrolysed by acid treatment under pressure to yield cyclohexylamine, which is extracted with chloroform and reacted with ethanolic *p*-quinone to form a coloured product 2-(cyclohexylamino)-1,4-benzoquinone. Dickes (1965) described a simple TLC procedure to detect saccharin and cyclamate in soft drinks, as follows:

SPECIAL REAGENTS

Mobile solvents

(a) 90 ml acetone + 10 ml 0·88 ammonia
(b) 90 ml ethanol + 10 ml 0·88 ammonia.

Spray reagents (a) 0·005 M silver nitrate (for cyclamates and saccharin). Dissolve 170 mg of silver nitrate in 1 ml of water, add 5 ml of 0·88 ammonia solution and make up to 200 ml with ethanol.

(b) 0·1 per cent α-naphthylamine (for saccharin). Add 5 drops of saturated cupric acetate solution and three drops of glacial acetic acid to 0·1 per cent α-naphthylamine in ethanol. This solution must be prepared and stored with care.

Standard solution. (a) Dissolve 1 g of calcium cyclamate in 50 per cent ethanol and dilute to 100 ml wth the solvent (1 μl = 10 μg of cyclamate).

(b) Dissolve 1 g of the sodium salt of saccharin in 50 per cent ethanol and dilute to 100 ml with the solvent (1 μl = 10 μg of soluble saccharin).

EXTRACTION

Take 20 ml of a soft drink for consumption after dilution plus 90 ml of water (or 100 ml of a soft drink for consumption without dilution), and add 10 ml of 10 per cent sulphuric acid. Carefully extract this solution in a separator with 50 ml of ethyl acetate and filter the latter through a layer of anhydrous sodium sulphate to remove water. Evaporate the extract to 2 ml on a boiling-water bath.

CHROMATOGRAPHY

Transfer 5 μl of the extract and separate suitable aliquots of the standard solutions on to a Kieselgel G thin layer and develop with the acetone–ammonia solvent. If benzoic acid is present, heat the plate to 130 °C for 30 min before applying the silver nitrate spray reagent. Dry and expose to short-wave ultraviolet light for 1 min.

Cyclamates and saccharin show as white spots on a grey background at R_F values 0·20 and 0·50 respectively. Benzoic acid yields a white silver salt with an R_F value slightly less than that of the cyclamates but can be removed by sublimation. To detect saccharin alone, a mauve spot is obtained on spraying with the α-naphthylamine reagent. The R_F value may be affected considerably in the presence of co-extracted impurities.

In the presence of cyclamates, saccharin can be extracted with ethyl acetate after destroying the cyclamate with nitrous acid. The extracted material is precipitated with silver salt and after digesting the precipitate, the saccharin is titrated against standard thiocyanate.

Fruit juice content

Numerous methods have been suggested for the assessment of the fruit juice content of soft drinks, e.g. determination of the ash, alkalinity of the ash, phosphorus, potassium, acidity, formal titration, total nitrogen, albuminoid ammonia, and amino acid profiles. The calculation of the fruit content from such determinations is, however, difficult due to the wide variation in the composition of natural juices (Table 7.12). The estimation of the fruit juice content of drinks may be further complicated if the juice or the product has been sophisticated, e.g. with inorganic substances containing potassium, phosphate, etc. In case of doubt it is advisable to carry out as many of the determinations mentioned above as possible and to compare the results with those obtained with authentic samples of natural juice. Many of the methods described for the analysis of fruit juices can be applied to soft drinks and only certain special procedures are included below. The type of fruit present may be detected by means of specific constituents, e.g. the flavonoids or the natural acids. However, the problem is more complex with modern drinks which tend to be blends prepared from juices from a variety of fruits. A survey of manufacturing procedures in the soft drinks industry has been published by Woollen (1981). Simple quality control techniques include the following methods (see also Benk, 1981; Merle, 1982; Bielig *et al.*, 1985).

Formol titration. The visual formol titration method is useful as a rapid routine check of the fruit content.

Concentrate 200 ml of the beverage in a porcelain dish on a water bath at 80 °C for 2 hours. Cool, neutralise to exactly pH 8 using firstly M and then 0·1 M sodium hydroxide. Add 20 ml of neutral 40 per cent formaldehyde (formalin, previously neutralised to pH 8) and titrate to pH 8 using 0·1 M sodium hydroxide. The titre (ml) divided by 2 represents the 'formaldehyde number' (*F*) and the amount of orange juice in the sample (per cent m/v) is given by the expression 1·05 *F*/1·4). The method can be readily carried out using a pH meter.

The albuminoid ammonia values. Mitra and Roy (1953) have applied the determination of the 'albuminoid ammonia value' in vinegar for assessing fruit juice content. The procedure for vinegar (p. 463) is used with certain modifications. The results are expressed on the m/m basis.

As with vinegar, Mitra and Roy observed a broad correlation between the albuminoid ammonia and total nitrogen figures for various fruit juices and squashes.

Products containing lower amounts of fruit juice gave correspondingly smaller values.

Amino acids

The amino acid composition of juices has received considerable attention; as indices of fruit content specific amino acids have been regarded as having greater significance than the mineral elements. Formol index and albumin ammonia contents are related to amino acid composition, and attempts have been made to raise the indices by addition of nitrogenous compounds that are foreign to the normal fruit. Other methods have been aimed at measurement of individual amino acids and the ratios with formol and other indices have been used as guidelines to authenticity. Those amino acids commonly selected for characterisation of orange juice include measurements of serine and betaine contents; mean values quoted by various authors are in the region of 20 mg/100 ml for serine and 70 mg/100 ml for betaine (see Mears and Shenton, 1973; Petrus and Vandercook, 1980). Chaytor (1986) has demonstrated the absence of serine in commercial orange drinks.

Comminuted Drinks

The FSCs Report on Soft Drinks in 1959 drew attention to the increasing popularity of drinks (both concentrated and ready-to-drink types), that are made by comminuting the whole fruit. Such products are, however, prepared by a number of different processes and there are considerable variations in the proportion of insoluble matter that is rejected and in the ratio of juice to other constituents remaining in the drink. The FSC pointed out that numerical provisions obscure the fact that the standard for comminuted drinks is effectively lower than the standard for citrus fruit squashes. The analysis of drinks containing comminuted fruit follows similar principles to those outlined previously for other types of soft drinks. It should be borne in mind, however, that the preservative present is usually benzoic acid rather than sulphur dioxide.

Born (1957) has described a possible method for distinguishing between a drink made from whole oranges from one made with juice only. The method involves chromatography of a benzene extract (diluted with light petroleum) on an alumina column, followed by spectrophotometry of a well-defined fraction at 325 nm.

CARBONATED DRINKS

Sodium bicarbonate in soda water

Soda water is required to contain at least 5 grains sodium bicarbonate per pint (570 p.p.m.). For this estimation titrate 50 ml sample with 0·05 M hydrochloric acid using methyl orange as indicator (1 ml 0·05 M HCl ≡ 0·0042 g $NaHCO_3$).

Quinine in tonic water and bitter lemon

Indian or quinine tonic water is required to contain at least ½ grain quinine (as quinine sulphate BP) per pint (57 p.p.m.). Quinine sulphate BP is required to contain 99·0–101·5 per cent $(C_{20}H_{24}N_2O_2)_2$, H_2SO_4 calculated on the dried substance. The quinine can be extracted with chloroform after making the drink alkaline with ammonia. The absorbance of the chloroformic extract can then be measured at 335 nm. The method is also applicable to bitter lemon, which contains approx. 30 p.p.m. quinine. The saccharin present provides a synergistic bitter taste. With tonic water, however, the absorbance of the sample can be measured directly at 347.5 nm against a blank containing the other ingredients present, i.e. sugar and acid (Butz and Noebels, 1961). Quinine can also be estimated fluorimetrically (Nuijens and Velden, 1973; Cussonneau and Le Clerc, 1981). Rao *et al.* (1984) described a spectrophotometric method which relies on formation of a complex between quinine and alizarin brilliant violet. The quinine is first extracted into chloroform from an alkaline solution, then back-extracted into sulphuric acid before the colorimetric reaction. Kral and Sontag (1982) have described an HPLC procedure.

Caffeine in cola drinks

Caffeine is present in kola nut, the extract of which is present in cola drinks. The caffeine in drinks of the Coca-Cola and Pepsi-Cola type can be estimated approximately by making the drink alkaline with ammonia, extracting with chloroform and measuring the absorbance at 273 nm ($E^{1\%}_{1cm} = 530$). The method may lead to high results due to interference from other substances extracted and it is preferable to employ simple clean-up on a Sep-Pak cartridge, or HPLC (Argoudelis, 1984; Tyler, 1984). Cola drinks usually contain 60–190 mg or less caffeine per litre.

PICKLES AND SAUCES

Vegetables can be salt-cured in brines containing over 15 per cent salt. Pickled onions are drained, partially desalted in diluted vinegar and finally covered with vinegar to give about 3 per cent salt and 3 per cent acetic acid in the final product. If vegetables are placed in brines containing less than 10 per cent salt, fermentation ensues. Sweet pickles often contain a mixture of chopped vegetables and thick sauce. Thick sauces usually consist mainly of fruits, onions, sugar, vinegar and spices. The commonest thin sauce is Worcester Sauce, which is highly spiced and has a sediment. Original recipes involved a period of maturation. The manufacture of pickles and sauce is described by Molyneux (1971) and Binsted *et al.* (1971).

Pickles and sauces may contain up to 100 mg/kg of sluphur dioxide, which may be present due to the fruit (e.g. apple) or up to 250 mg/kg of benzoic acid or 4-hydroxybenzoates. Pickles and sauces should also be examined for trace metals, e.g. copper, lead, zinc. A standard for pickled cucumbers has been elaborated by Codex (see Appendix 2).

Special attention must be given to the preparation of the sample for analysis. According to the type of material the sample should be minced, mixed in a pestle and mortar or blended mechanically. For the assessment of general quality obtain the total solids (TS) by drying 5 g at 100 °C or preferably by heating *in vacuo* at 70 °C. Also determine the total acidity (TA) as acetic acid by stirring 10 g with water and titrating with 0·2 M sodium hydroxide to phenolphthalein. To another dish add water to a 10 g sample and evaporate to dryness on a water bath. Then stir in more water and evaporate again. Finally, stir up the residue again with water, add phenolphthalein, titrate with 0·2 M sodium hydroxide and calculate the fixed acidity (FA) as acetic acid.

Then, per cent volatile acidity = VA = TA − FA

per cent total volatiles = 100 − TS = TV

per cent volatile acid in the aqueous phase (as acetic acid) = 100 × VA/TV.

This figure is especially important in relation to keeping qualities as it is the acetic acid (the main volatile acid) that is of value for preservation. In general, the volatile acid in the aqueous phase should be at least 3·5 per cent. However, if a sauce has been heat-treated or if sugar and salt are present less acid (say 2·5 per cent) is sufficient for the prevention of spoilage (cf. Bell and Etchells, 1952). The determinations of ash, salt and sugars as with tomato ketchup therefore give further information on the storage characteristics. Techniques using HPLC have been applied to the determination of acids, sugars and sweeteners (see Klein 1982; Andersson and Hedlund, 1983; Ashoor and Welty, 1984; McFeeters *et al.*, 1984). Tests should be made for non-nutritive sweeteners.

NUTS

Composition

Some compositional data of the kernels of various nuts is given in Table 7.14. Green tree nuts contain at least 50 per cent water, but after drying the moisture content may

Table 7.14 Composition of the edible portion of various nuts (after McCance and Widdowson, 1967 and other sources)

Type of nut	Water (%)	N (%)	Oil (%)	Sugar (as invert sugar) (%)	Ash (%)
Almond, sweet	4·7	3·3	53·5	4·3	3·0
Barcelona	5·7	2·1	64·0	3·4	—
Brazil	8·5	2·2	61·5	1·7	3·0
Cashew	4·0	3·2	45·9	—	2·5
Chestnuts	51·7	0·4	2·7	7·0	1·0
Cob	41·1	1·4	36·0	4·7	—
Coconut, fresh	42·0	0·6	36·0	3·7	1·2
Peanuts	4·5	4·5	49·0	3·1	2·4
Walnuts	23·5	2·0	51·5	3.2	1·4

be as low as 4 per cent so that mould formation is inhibited. As the oil content may exceed 60 per cent, rancidity formation is likely unless stored at a low temperature. The protein is often calculated as N × 6·25, but other factors are sometimes used, e.g. for almond N × 5·18, for cashew N × 5·3. The protein content of nuts may vary from about 5 to 30 per cent and the ash is usually less than 3 per cent. Most of the remainder is carbohydrate and fibre. The sweetness is some nuts is due largely to the higher amount of sugars present. The starch content is usually low. Most nuts contain reasonable amounts of calcium, phosphorus and potassium. Certain nuts are good sources of vitamin B_1 and others contain reasonable amounts of vitamin B_6 or niacin. The Codex Alimentarius Commission has published a 'Recommended International Code of Hygienic Practice for Tree Nuts', which relates mainly to sanitary production and handling of almonds and walnuts together with a standard for pistachio nuts (see Appendix 2).

The composition and identification of mixtures of ground almonds, cashew nuts and peanuts have been discussed by Houlbrooke *et al.* (1967). They found that although almonds contain none, the starch in cashew and peanut can be distinguished by microscopical examination. Where the degree of adulteration is considerable it is possible to obtain some idea of its extent from the types of carbohydrate present (Table 7.15). The process of debittering bitter almonds appears to cause a reduction in the water-soluble mineral matter. The essential oil obtained from bitter almonds contains benzaldehyde and hydrocyanic acid.

Salted peanuts are often prepared by blanching and oil roasting at 138 °C (280 °F) before shaking about 2 per cent salt on to them. The moisture content when packed is about 1·3 per cent. Dry roasted types are also manufactured.

PEANUT BUTTER

In the manufacture of peanut butter the peanuts are shelled, dry-roasted at 160 °C (320 °F) and blanched. Salt is added after grinding and other ingredients such as hydrogenated fat (as stabiliser) and dextrose are often incorporated. The results of the analysis of peanut butter samples are given in Table 7.16. Willich *et al.* (1954) using accelerated storage tests showed that peanut butter has good stability even after 2 years storage.

DESICCATED COCONUT

Desiccated coconut is manufactured from the white part of the kernel, which is dried at 60–70 °C (140–170 °F) to a moisture content of less than 2 per cent (cf. Cornelius, 1973). A Codex code (see Appendix 2) recommends that it should be free from pathogens such as salmonellae, and quotes maxima for moisture of 3 per cent and for the acidity of the extracted oil of 0·3 per cent FFA (as lauric acid). Child (1964) considered that the oil content should be between 68 and 72 per cent. Problems associated with the presence of salmonellae in desiccated coconut and methods associated with their destruction have been discussed by Galbraith *et al.* (1960). The Preservatives Regulations permit the presence of up to 50 mg/kg of sulphur dioxide in desiccated coconut. If present, it is more likely to have been introduced by a direct oil-fired drying process rather than being added as a preservative (Williams, 1972).

Mycotoxins in nuts and nut products

Contamination of edible nuts by mould growth is significantly affected by the storage conditions, climate and moisture content of the nuts. In broad terms the critical moisture level for nuts is 9 per cent and for meal 16 per cent although the controlling factor is water activity in the particular situation. Most of the critically affected kernels are discoloured and exhibit fluorescence under ultraviolet illumination. The significance of the consequences of contamination by the mould *Aspergillus flavus* was first recognised in 1960 when over 100 000 turkey poults and ducklings died as a result of feeding with a ration that incorporated mould-damaged Brazilian ground nut meal. Extensive examination by Sargeant *et al.* (1961) showed that, in growth, *Aspergillus flavus* produced aflatoxins, which proved to be acutely toxic and carcinogenic when fed to a wide range of experimental animals. A few incidences of adverse effects in man have been recorded, the evidence for which has been summarised in a food surveillance paper (MAFF, 1980); the report covers a range of food products including peanuts, pistachio, almonds, hazelnuts, walnuts and brazil nuts. Typical survey data (for the years to 1980) are recorded in the report. One of the principal problems in obtaining reliable survey data in situations of this kind is the collection

Table 7.15 The carbohydrate content of almonds, cashew nuts and peanuts (after Houlbrooke *et al.*, 1967)

Nut	Water-soluble sugars (as invert sugar) (%)	Total acid-hydrolysable carbohydrates (as invert sugar) (%)	Total carbohydrates (%)
Almond	4·7	8·5	14·7
Cashew	8·1	26·0	30·2
Peanut	5·8	14·1	20·1

Table 7.16 Composition of various samples of peanut butter

	mean	min.	max.
% Moisture	1·9	1·1	3·2
% Protein (N × 6·25)	29	26	32
% Oil (direct extraction)	48	40	52 (US max. 55)
% Sugars, etc.	7·9	—	—
% Ash	2·5	—	—
% Salt	1·3	0·7	4·2
% Fibre	2·2	—	—

of representative samples, since the contamination is distributed in a heterogeneous manner. Sampling procedures used in the collection of the UK data described in the report were due to Dickens and Satterwhite (1969) and Velasco and Morris (1976). Recent survey data and a summary of methodology used has been published by Gilbert and Shepherd (1985).

Methods used for identification and estimation of mycotoxins in general and aflatoxins in particular have been the subject of considerable activity since the first incident of toxicosis. Early work using TLC was described by Coomes *et al.* (1965) and Agthe *et al.* (1968); these and similar methods relied on visual estimates of the relative fluorescence of the sample extracts with standard aflatoxins by examination under ultraviolet illumination. Pons (1976) described an HPLC procedure using microporous silica and elution with a solvent mixture comprised of water/chloroform/cyclohexane/acetonitrile, in which the aflatoxins were detected by measurement of absorbance at 360 nm. The AOAC method relies on preliminary extraction with water–chloroform, clean-up on a silica column, elution with methanol–chloroform, confirmation and quantitation by TLC. Trucksess *et al.* (1984) used a disposable silica gel column for clean-up of extracts, TLC for estimation followed by confirmation by GC/MS; a detection limit was claimed for aflatoxin B_1 of 2 ng/g. Schuller *et al.* (1982) reviewed methods and legal implications of lowering limits of detection. The Food Advisory Committee recommended in 1986 that a statutory limit of 10 μg/kg be applied to nuts and nut products for human consumption. Total aflatoxin is expressed as the sum of aflatoxins B_1, B_2, G_1 and G_2. The general problem of mycotoxin contamination is discussed further in Chapter 5.

MARZIPAN

Marzipan is made by pounding together 1 part ground almonds and 3 parts sugar (including invert sugar or glucose syrup). Egg yolk is usually included in domestic recipes. A LAJAC Code of Practice issued in 1969 stated that marzipan, almond paste and almond icing shall contain at least 23·5 per cent dry almond substance and no other nut ingredient, and not less than 75 per cent of the remainder shall be solid carbohydrate sweetening matter.

The following ranges of percentage figures for air-dried almonds are useful for calculating the amount present in marzipan (Houlbrooke, 1963b; Monk, 1963).

	min.	max.	average
Oil	50·2	66·7	58·4
Protein	20·0	27·1	23.3
Ash	2·40	3·50	2·92

Monk obtains the almond content of marzipan by multiplying the oil by 1·7 and the protein by 4·35. Firth (1969) gives the following percentage figures. Air-blanched sweet almonds calculated on a dry basis: oil 55·1–64·1 (mean 60·76), protein (N × 6·25) 21·3–28·2 (mean 24·4). Firth, however, considers that the determination of oil in marzipan is difficult and calculates the nut content by subtracting the moisture + sugar from 100 to give the dry sugar-free nut residue. This figure is then adjusted to allow for the sucrose in sweet almonds (3·12–6·05 per cent, mean 4·0 per cent on dry basis). Marzipan should also be examined for other preservatives and added colouring matter.

REFERENCES

Adam, W. B. (1965) *Journal of the Association of Public Analysts*, **3**, 36.

Agthe, C., Lijinsky, W. & Oremus, D. (1968) *Food and Cosmetics Toxicology*, **6**, 627.

Alderson, M. G. (1970) *Food Manufacture*, **45**, 67.

Aldred, J. B. (1983) *Journal of the Association of Public Analysts*, **21**, 73.

Alzamora, S. M. & Chirife, J. (1983) *Journal of Food Science*, **48**, 1385.

Andersson, R. & Hedlund, B. (1983) *Zeitschrift fur Lebensmittel Untersuchung und Forschung*, **176**, 440.

Anon (1979) *Campden Food Preservation Research Association Symposium on Sterilisable Flexible Packaging*.

Anon (1980a) *Food Production and Management*, **103**, 18.

Anon (1980b) *Analytical Chemistry*, **52**, 1269A.

Anon (1981) *Journal of the American Dietetic Association*, **79**, 715.

Anon (1982) *Soft Drinks*, October, 431.

APA (1971) *Journal of the Association of Public Analysts*, **9**, 104.

Ara, V. & Torok, M. (1980) *Industrielle Obst-und Gemuseverwertschrift*, **65**, 297.

Arbid, M. S. S. & Marquardt, R. R. (1985) *Journal of the Science of Food and Agriculture*, **36**, 839.

Argoudelis, C. J. (1984) *Journal of Chromatography*, **303**, 256.

Ariga, T., Kanmura, M., Nakazato, M., Fujinuma, K., Nishijima, M. & Nadi, Y. (1983) *Journal of Food Hygiene Society Japan*, **24**, 289.

Ashoor, S. H. & Knox, M. J. (1984) *Journal of Chromatography*, **299**, 288.

Ashoor, S. H. & Welty, J. (1984) *Journal of Chromatography*, **287**, 452.

Ashoor, S. H., Monte, W. C. & Welty, J. (1984) *Journal of the Association of Official Analytical Chemists*, **67**, 78.

Ayres, A. D., Charley, V. L. S. & Swindells, R. (1961, 1962) *Food Processing and Packaging*, **30**, 413; **31**, 14.

Bajaj, K. L. & Kaur, G. (1981) *Analyst*, **106**, 117.

Bandler, R. & Cichowicz, S. M. (1981) *Journal of the Association of Official Analytical Chemists*, **64**, 570.

Barbieri, G., Rosso, S., Milanese, G. & Warwick, M. E. (1982) *Industria Conserve*, **57**, 151.
Barker, A. G., Billington, A., E. & Charley, V. L. S. (1961) *Food Processing and Packaging*, **30**, 325.
Basker, H. B. (1966) *Journal of the Science of Food and Agriculture*, **17**, 539.
Basker, H. B. (1969) *Analyst*, **94**, 410.
Bazzarini, R., Bigliardi, D., Gherardi, S. & Trifiro, A. (1981) *Industria Conserve*, **56**, 259.
Bell, T. A. & Etchells, J. L. (1952) *Food Technology*, **6**, 468.
Belo, P. S. & de Lumen, B. O. (1981) *Journal of Agricultural and Food Chemistry*, **29**, 370.
Benk, E. (1980) *Flussiges Obst*, **47**, 496.
Benk, E. (1981) *Getraenkeindustrie*, **35**, 923.
Benoit-Guyod, J. L., Benkheder, K., Seigle-Murandi, F. & Duc, D. L. (1980) *Journal of Agricultural and Food Chemistry*, **28**, 1317.
Berry, R. E. (1985) *Food Technology*, **39**, 91.
Beutler, H.-O. (1982) *Flussiges Obst*, **49**, 62.
Bielig, H. J. & Hofsommer, H.-J. (1982) *Flussiges Obst*, **49**, 50.
Bielig, H. J., Faethe, W., Koch, J., Wallrauch, S. & Wucherpfennig, K. (1985) *Confructa*, **29**, 191.
Bigelow, W. D., Smith, H. R. & Greenleaf, C. A. (1950) *Tomato Products*, Research Laboratory Bulletin 27-L. Washington: National Canners Association.
Biltcliffe, D. O., Judd, H. J. & Wood, R. (1984) *Journal of the Association of Official Analytical Chemists*, **67**, 635.
Binsted, R., Devey, J. D. & Dakin, J. C. (1971) *Pickle and Sauce Making*. London: Food Trade Press.
Bishop, R. H., Duncan, C. L., Evancho, G. M. & Young, H. (1982) *Journal of Food Science*, **47**, 437.
Blanchfield, J. R. (1964) *Food Manufacture*, **39**, 40.
Board, P. W. (1982) *Food Production and Management*, November, 12.
Boehringer Corporation (1979) *Methods of Enzymatic Food Analysis*. London: Boehringer Corp.
Bohrer, B., Kury-Herzog, B. & Gierschne, K. (1984) *Industrielle Obst-Gemuseverwert*, **69**, 498.
Born, R. (1957) *Chemistry and Industry* (23), 724.
Brand, J. C., Cherikoff, V., Lee, A. & Truswell, A. S. (1982) *Lancet*, **ii**, 873.
Brause, A. R. & Raterman, J. M. (1982) *Journal of the Association of Official Analytical Chemists*, **65**, 846.
Brause, A. R., Raterman, J. M., Petrus, D. R. & Doner, L. W. (1984) *Journal of the Association of Official Analytical Chemists*, **67**, 535.
Brown, M. B. & Cohen, E. (1983) *Journal of the Association of Official Analytical Chemists*, **66**, 781.
Brown, M. B., Cohen, E. & Volman, L. (1981) *Industria Alimentation Agriculture*, **98**, 181.
Bushway, R. J. (1982) *Journal of Chromatography*, **247**, 180.
Butz, W. H. & Noebels, H. J. (1961) *Instrumental Methods for the Analysis of Food Additives*. New York: Interscience.
Cameron, A. G. & Bakht, A. (1965) *Food Trade Review*, **35**, 46.
Carles, L. (1982) *Alimentation*, **97**, 41, 53.
Chandler, B. V. (1983) *Food Technology Australia*, **35**, 448.
Charley, V. L. S. (1963) *Food Technology*, **17**, 33.
Charpentier, B. A. & Lemmel, D. E. (1984) *Journal of Agricultural and Food Chemistry*, **32**, 908.
Chaytor, J. P. (1986) *Journal of the Science of Food and Agriculture*, **37**, 1019.
Chikamoto, T. & Maitani, T. (1984) *Journal of Food Hygiene Society Japan*, **25**, 530.
Child, R. (1964) *Coconuts*. London: Longman.
Chin, H. B., Kimball, J. R., Hung, J. & Allen, B. (1985) *Journal of the Association of Official Analytical Chemists*, **68**, 1081.
Cichowicz, S. M., Gecan, J. S., Atkinson, J. C. & Kvenberg, J. E. (1982) *Journal of Food Protection*, **45**, 627.
Cilliers, J. J. L & Van Niekerk, P. J. (1984) *Journal of the Association of Official Analytical Chemists*, **67**, 1037.
CIPC (1959) *Food Trade Review*, **29**, 22.
Clement, E. & Verbist, J. F. (1980) *Lebensmittel Wissenschaft und Technologie*, **13**, 202.
Cohen, E. (1985) *Flussiges Obst*, 52, 580, 597.
Cohen, E. & Fuchs, C. (1984) *Flussiges Obst*, **51**, 169, 185.
Cohen, E., Sharon, R., Volman, L., Hoenig, R. & Saguy, I. (1984) *Journal of Food Science*, **49**, 987.
Continental Can Company (1968) *Mold Counting of Tomato Products*. Chicago.
Cook, J. M., Karelitz, R. L. & Dalsis, D. E. (1985) *Journal of Chromatographic Science*, **23**, 57.
Cooke, J. R. & Moxon, R. E. D. (1981) In *Vitamin C*, ed. J. N. Counsell & D. H. Horning, p. 167. London: Applied Science Publishers.
Coomes, T. J., Crowther, P. C., Francis, B. J. & Stevens, L. (1965) *Analyst*, **90**, 492.
Coppola, E. D. (1984) *Food Technology*, **38**, 88.
Cornelius, J. A. (1973) *Tropical Science*, **15**, 15.
Cousin, M. A., Zeidler, C. S. & Nelson, P. E. (1984) *Journal of Food Science*, **49**, 439.

Cox, D., Harrison, G., Jandik, P. & Jones, W. (1985) *Food Technology*, **39**, 41.
Coxon, D. T. (1984) *American Potato Journal*, **61**, 169.
Cussonneau, C. & Le Clerc, A. M. (1981) *Annales Falsificationnes Expertise Chemie et Toxicologie*, **74**, 507.
Dalziel, J. A. W., Johnson, R. M. & Shenton, A. J. (1972) *Analyst*, **97**, 719.
Daniels, D. H., Joe, F. L., Warner, C. R. and Fazio, T. (1984) *Journal of the Association of Official Analytical Chemists*, **67**, 513.
Darbishire, O. B. (1965) *Analyst*, **90**, 439.
Davies, A. M. C. (1966) *Journal of the Association of Public Analysts*, **4**, 11.
Davies, A. M. C. & Blincow, P. J. (1984) *Journal of the Science of Food and Agriculture*, **35**, 553.
Deutsch, M. J. & Weeks, C. E. (1965) *Journal of the Association of Official Agricultural Chemists*, **48**, 1248.
Devey, J. D. (1958) *Food Trade Review*, **28**, 7.
De Vries, J. W. (1983) *Journal of the Association of Official Analytical Chemists*, **66**, 1371.
De Vries, J. W., Rombouts, F. M., Voragen, A. G. J. & Pilnik, W. (1983) *Carbohydrate Polymers*, **3**, 245.
Dickens, J. W. & Satterwhite, J. B. (1969) *Food Technology*, **23**, 90.
Dickes, G. J. (1965) *Journal of the Association of Public Analysts*, **3**, 119.
Dickinson, D. & Fowler, H. D. (1955) *Analyst*, **80**, 706.
Dickinson, D. & Raven, T. W. (1962) *Journal of the Science of Food and Agriculture*, **13**, 650.
Dilley, A. E. & Everton, J. R. (1966) *Laboratory Practice*, **15**, 318.
Doner, L. W. & Bills, D. D. (1982) *Journal of the Association of Official Analytical Chemists*, **65**, 608.
Douglas, S. G. (1981) *Food Chemistry*, **7**, 139.
Dovell, J. C. & Harris, N. D. (1982) *Journal of the Science of Food and Agriculture*, **33**, 185.
Dunbar, J. (1982) *Zeitschrift fur Lebensmittel Untersuchung und Forschung*, **173**, 253.
Eski, A., Reicheneder, E. & Kieninger, H. (1980) *Flussiges Obst*, **47**, 494.
Evans, R. H., Van Soestbergen, A. W. & Ristow, K. A. (1983) *Journal of the Association of Official Analytical Chemists*, **66**, 1517.
FACC (1970) *Report on the Review of the Emulsifiers and Stabilisers in Food Regulations*, 1962, FACC/REP/9, HMSO.
FdAC (1985) *Report on the Review of the use of additives to cloud soft drinks*, FdAC/REP/2, MAFF.
Fenwick, G. R., Heaney, R. K. & Mullin, W. K. (1982) *CRC Critical Reviews in Food Science and Nutrition*, **18**, 123.
Finglas, P. M. & Faulks, R. M. (1984) *Journal of the Science of Food and Agriculture*, **35**, 1347.
Firth, F. O. (1969) *Journal of the Association of Public Analysts*, **7**, 22.
Fischer, B., Luthy, J. & Schlatter, C. (1984) *Zeitschrift fur Lebensmittel Untersuchung und Forschung*, **179**, 218.
Folkes, D. J. (1980) *Journal of the Science of Food and Agriculture*, **31**, 1011.
Forbes, D. G. (1963) *Journal of the Association of Public Analysts*, **1**, 4.
FSC (1970) *Report on the Pre-1955 Compositional Orders*. London: HMSO.
Fuchs, C. & Wretling, S. (1981) *Var Foda*, **33**, 2.
Galbraith, N. S., Hobbs, B. C., Smith, M. E. & Tomlinson, A. J. H. (1960) *Monthly Bulletin of the Ministry of Health*, **19**, 99.
Garcia-Moreno, C., Rivas-Gonzalo, J. C., Pena-Egido, M. J. & Marine-Font, A. (1983) *Journal of the Association of Official Analytical Chemists*, **66**, 115.
Giangiacomo, R., Polesello, A. & Marin, F. (1982) *Industrie Alimentari*, **21**, 386.
Gilbert, J. & Shepherd, M. J. (1985) *Food Additives and Contaminants*, **2**, 171.
Goose, P. G. & Binsted, R. (1973) *Tomato Paste and other Tomato Products*, 2nd edn. London: Food Trade Press.
Greiner, G. & Wallrauch, S. (1984) *Flussiges Obst*, **51**, 626.
Grosso, A. L. (1965) *Confectionery Production*, **31**, 387.
Haddad, P. R., Lan, J. (1984) *Food Technology Australia*, **36**, 46.
Halaby, G. A., Lewis, R. W. & Rey, C. R. (1981) *Food Technology*, **35**, 86.
Hellenas, K.-E. (1986) *Journal of the Science of Food and Agriculture*, **37**, 776.
Henniger, G. (1984) *Flussiges Obst*, **51**, 111, 133.
Henning, W. (1982) *Lebensmittelchemie gerichtlische Chemie*, **36**, 62.
Herrmann, A., Damawandi, E. & Wagmann, M. (1983) *Journal of Chromatography*, **280**, 85.
Hirosue, T., Kawai, H. & Hosogai, Y. (1983) *Nippon Shokuhin Kogyo Gakkaishi*, **30**, 296.
Hirota, S., Sato, H. & Tsuyuki, H. (1982) *Nippon Shokuhin Kogyo Gakkaishi*, **29**, 477.
Holt, R. (1954) *Analyst*, **79**, 623.
Houlbrooke, A. (1963a, b) *Journal of the Association of Public Analysts*, **1**, 16, 33.
Houlbrooke, A., Isles, G. & Russell, R. S. (1967) *Journal of the Association of Public Analysts*, **5**, 126.
Hulme, B., Morries, P. & Stainsby, W. J. (1965) *Journal of the Association of Public Analysts*, **3**, 113.
Ingleton, J. F. (1964, 1966) *Confectionery Production*, **30**, 445; **32**, 527.
James, D. G. (1968) *Food Preservation Quarterly*, **28**, 55.

Jarvis, B. (1977) *Journal of Food Technology*, **12**, 581.
Jeffries, D. A., O'Keeffe, D. & Young, J. N. (1983) *A Users Guide to the Newly Permitted Sweeteners (1983)*, Leatherhead Food Research Association, No. 1.
Jones, P. G. & Fenwick, G. R. (1981) *Journal of the Science of Food and Agriculture*, **32**, 419.
Jost, R., Monti, J. C. & Schaufelberger, U. (1982) *International Journal of Vitamin and Nutrition Research*, **52**, 229.
Junge, C. & Spadinger, C. (1982) *Flussiges Obst*, **49**, 57.
Kacprzak, J. L. (1982) *Journal of the Association of Official Analytical Chemists*, **65**, 148.
Kagan, I. S., Grishina, I. P. & Zyabko, L. P. (1965) *Chemical Abstracts*, **62**, 5810.
Kajiwars, N., Tomiyama, C., Ninomiya, T. & Hosogai, Y. (1983) *Journal of Food Hygiene Society Japan*, **24**, 42.
Kakade, M. L., Rackis, J. J., McGee, J. E. & Puski, G. (1974) *Cereal Chemistry*, **51**, 376.
Kasidas, G. P. (1980) *Journal of Human Nutrition*, **34**, 255.
Kauschus, U. & Thier, H.-P. (1985) *Flussiges Obst*, **52**, 29, 37.
Kim, S. I., Hoehn, E., Eskin, N. A. M. & Ismail, F. (1982) *Journal of Agricultural and Food Chemistry*, **38**, 144.
King, C. (1980) *FFIPP*, March, 15.
Kinter, P. K. & van Buren, J. P. (1982) *Journal of Food Science*, **47**, 756.
Klein, H. (1982) *Industrielle Obst- Gemuseverwert*, **67**, 31.
Koch, J. L. (1984) *Flussiges Obst*, **51**, 510, 545.
Kral, K., Sontag, G. (1982) *Zeitschrift fur Lebensmittel Untersuchung und Forschung*, **175**, 22.
Lea, A. G. H. & Smith, S. J. (1985) *Flussiges Obst*, **52**, 586, 602.
Lee, H. S., Rouseff, R. L. & Nagy, S. (1986) *Journal of Food Science*, **51**, 1075.
Li, B. W. & Schuhmann, P. J. (1983) *Journal of Food Science*, **48**, 633.
Libert, B. & Creed, C. (1985) *Journal of the Horticultural Society*, **60**, 257.
Liener, I. E. (1966) *Toxicants Occurring Naturally in Foods*. Washington, D.C.: National Academy of Sciences.
Liener, I. E. (1974) *Toxic Constituents of Animal Foodstuffs*. New York: Academic Press.
Liener, I. E. & Kakade, M. L. (1980) *Toxic Constituents of Plant Foodstuffs*, 2nd edn. New York: Academic Press.
Lin, H. H. & Cousin, M. A. (1985) *Journal of Food Protection*, **48**, 671.
List, D., Buddruss, S. & Bodtke, M. (1985) *Zeitschrift fur Lebensmittel Untersuchung und Forschung*, **180**, 48.
Lonberg, A. E., Everitt, G. & Mattsson, P. (1985) *Var Foda*, **37**, 316.
Lufschitz, A. (1983) *Food Technology Australia*, **35**, 336.
Lund, E. D., Kirkland, C. L. & Shaw, P. E. (1981) *Journal of Agricultural and Food Chemistry*, **29**, 361.
MAFF (1980) *Survey of Mycotoxins in the United Kingdom*, Food Surveillance Paper No. 4. London: HMSO.
MAFF (1981) Steering Group on Food Surveillance *Survey of Copper and Zinc in Foods*, No. 5.
MAFF (1982a) Steering Group on Food Surveillance *Survey of Lead in Food—Second Supplementary Report*, No. 10.
MAFF (1982b) Steering Group on Food Surveillance *Survey of Arsenic in Food*, No. 8.
MAFF (1983) Steering Group on Food Surveillance *Survey of Cadmium in Food*, No. 12.
Manley, C. H. & Alcock, A. (1950), *Analyst*, **75**, 280.
Mannheim, C. & Passy, N. (1981) *CRC Critical Reviews in Food Science and Nutrition*, **17**, 371.
Marcy, J. E. & Rouseff, R. L. (1984) *Journal of Agricultural and Food Chemistry*, **32**, 979.
Marquardt, R. R. & Frohlich, A. A. (1981) *Journal of Chromatography*, **208**, 373.
Martin-Villa, C., Vidal-Valverde, C. & Rojas-Hildago, E. (1982) *Journal of Food Science*, **47**, 2086.
Marwan, A. G. & Nagel, C. W. (1982) *Journal of Food Science*, **47**, 774.
Mattick, L. R. & Moyer, J. C. (1983) *Journal of the Association of Official Analytical Chemists*, **66**, 1215.
McCance, R. A. & Widdowson, E. M. (1967) *The Composition of Foods*. London: HMSO.
McFeeters, R. F., Thompson, R. L. & Fleming, H. P. (1984) *Journal of the Association of Official Analytical Chemists*, **67**, 710.
McGregor, D. I., Mullin, W. J. & Fenwick, G. R. (1983) *Journal of the Association of Official Analytical Chemists*, **66**, 825.
Mears, R. G. & Shenton, A. J. (1973) *Journal of Food Technology*, **8**, 357.
Mentasti, E., Gennaro, M. C., Sarzanini, C., Baocchi, C. & Savigliano, M. (1985) *Journal of Chromatography*, **322**, 177.
Merle, M. H. (1982) *Annales des Falsifications et Expertise Chimie*, **75**, 549.
Meurens, M. (1980) *Berichte-International Fruchsaft Union Wissenschaft, Technische Komm*, **16**, 217.
Michener, H. D., King, A. D. & Ito, K. A. (1966) *Bacteriological Proceedings*, A15, p. 3.
Millies, K. & Burkin, M. (1984) *Flussiges Obst*, **51**, 629.

Ministry of Commerce and Industry of the State of Israel (1967) *Journal of the Association of Official Analytical Chemists*, 5, 68.
Mitra, S. N. & Roy, S. C. (1953) *Analyst*, **78**, 681.
Molyneux, F. (1971) *Process Biochemistry*, **6**, 17.
Money, R. W. (1964) *Journal of the Science of Food and Agriculture*, **15**, 594.
Money, R. W. (1966) *Journal of the Association of Public Analysts*, **4**, 41.
Monier-Williams, G. W. (1941) *Analyst*, **66**, 319.
Monk, H. E. (1963) *Journal of the Association of Public Analysts*, **1**, 20.
Morcinek, H. (1981) *Industrielle Obst-und Gemuseverwertschrift*, **6**, 616.
Morgan, M. R. A., Coxon, D. T., Bramham, S., Chan, H. W-S., van Gelder, W. M. J. & Allison, M. J. (1985) *Journal of the Science of Food and Agriculture*, **36**, 282.
Morgan, R. H. (1953) *Analyst*, **78**, 323.
Morgan, R. H. (1954) *Food*, **23**, 286.
Murphy, D. (1983) *Journal of the Association of Public Analysts*, **21**, 7.
Nambisan, B. & Jundaresan, S. (1984) *Journal of the Association of Official Analytical Chemists*, **67**, 641.
Notermans, S. & Heuvelman, C. J. (1985) *International Journal of Food Microbiology*, **2**, 247.
Nuijens, J. M. & Velden, H. van der (1973) *Zeitschrift fur Lebensmittel Untersuchung und Forschung*, **153**, 97.
Olson, N. A. (1980) *Food Technology*, **34**, 50.
Osborn, R. A. (1964) *Journal of the Association of Official Agricultural Chemists*, **47**, 1068.
Pachla, L. A. & Reynolds, D. L. (1985) *Journal of the Association of Official Analytical Chemists*, **68**, 1.
Park, Y. W., Anderson, M. J. & Mahoney, A. W. (1982) *Journal of Food Science*, **47**, 1558.
Parran, H. M., Douglas, R. G. & Eisenberg, W. V. (1972) *Journal of the Association of Official Analytical Chemists*, **55**, 73.
Paul, A. A. & Southgate, D. A. T. (1978) *The Composition of Foods*. London: HMSO.
Pearson, D. (1975) *Journal of the Science of Food and Agriculture*, **26**, 207.
Perkal, M., Blackman, G. L., Ottrey, A. L. & Turner, L. K. (1980) *Journal of Chromatography*, **196**, 180.
Petrus, D. R. & Attaway, J. A. (1980) *Journal of the Association of Official Analytical Chemists*, **63**, 1317.
Petrus, D. R. & Vandercook, C. E. (1980) *American Chemical Society Symposium Series, No. 143*, **17**, 395.
Pollard, A. & Beech, F. W. (1966) *Process Biochemistry*, **7**, 229.
Pons, W. A. (1976) *Journal of the Association of Official Analytical Chemists*, **59**, 101.
Powers, J. P. (1976) *CRC Critical Reviews in Food Science and Nutrition*, **12**, 371.
Prudel, M. & Davidkova, E. (1985) *Nahrung*, **29**, 381.
Ramappa, P. G. & Nayak, A. N. (1983) *Analyst*, **108**, 966.
Rao, M. V., Krishnamacharyulu, A. G., Sattigeri, V. D., Manjunath, M. N., Nagaraja, K. V. & Kapuro, P. (1984) *Journal of Food Science and Technology*, Mysore, **21**, 266.
Rao, P. V. & Hahn, S. K. (1984) *Journal of the Science of Food and Agriculture*, **35**, 426.
Rechcigl, M. (1983) *CRC Handbook of Naturally Occurring Food Toxicants*. Florida: CRC Press Inc.
Rees, D. I. (1965) *Analyst*, **90**, 568.
Richard, J. P. & Coursin, D. (1982) *Industria Alimentation Agriculture*, **99**, 411.
Richmond, M. L., Brandao, S. C. C., Gray, J. I., Markakis, P. & Stine, C. M. (1981) *Journal of Agricultural and Food Chemistry*, **29**, 4.
Rizzolo, A., Forni, E. & Polesello, A. (1984) *Food Chemistry*, **14**, 189.
Robertson, A. (1986) in *Proceedings of Symposium, Advances in Immunoassays for Veterinary and Food Analysis*, eds B. A. Morris & M. N. Clifford. London: Elsevier Applied Science Publishers.
Robertson, G. L. J. (1981) *Journal of Food Biochemistry*, 5, 139.
Robertson, J. M. & Kirk, B. L. (1979) *Food Technology in New Zealand*, **14**, 3.
Roehle, J. & Voigt, F. (1984) *Nahrung*, **28**, 579.
Russell, G. E. (1985) *Journal of the Association of Official Analytical Chemists*, **68**, 896.
Rymal, K. S. (1983) *Journal of the Association of Official Analytical Chemists*, **66**, 810.
Saguy, I. & Cohen, E. (1984) *Zeitschrift fur Lebensmittel Untersuchung und Forschung*, **178**, 386.
Salunkhe, D. K., Wu, M. T. & Do, J. Y. (1978) *Journal of Food Quality*, **2**, 75.
Sargeant, K., Sheridan, A., O'Kelly, J. & Carnaghan, R. B. A. (1961) *Nature*, **192**, 1096.
Sawyer, R. (1963) *Journal of the Science of Food and Agriculture*, **14**, 302.
Scherz, J.-C., Monti, J. C. & Jost, R. (1983) *Zeitschrift fur Lebensmittel Untersuchung und Forschung*, **177**, 124.
Schuller, P. L., Stoloff, L. & van Egmond, H. P. (1982) *Environmental Carcinogens—Selected Methods of Analysis*, IARC Scientific Publications No. 44, **5**, Chapter 6.
Schwarzenbach, R. (1982) *Journal of Chromatography*, **251**, 339.
Shaw, P. E., Buslig, B. S. & Wilson, C. W. (1983) *Journal of Agricultural and Food Chemistry*, **31**, 182.
Shenton, A. J. & Johnson, R. M. (1973a, b) *Analyst*, **98**, 745, 749.
Sherratt, J. G. & Sinar, R. (1963) *Journal of the Association of Public Analysts*, **1**, 18.

Siewek, F., Galensa, R. & Herrmann, K. (1984) *Zeitschrift fur Lebensmittel Untersuchung und Forschung*, **179**, 315.

Siewer, F., Galensa, R. & Ara, V. (1984) *Industrelle Obst-und Gemuseverwertschrift*, **69**, 596.

Smolensky, D. C. & Vandercook, C. E. (1980) *Journal of Food Science*, **45**, 1773.

Snodin, D. J. (1973) *Journal of the Association of Public Analysts*, **11**, 47.

Sones, K., Heaney, R. K. & Fenwick, G. R. (1984a) *Food Additives and Contaminants*, **1**, 289.

Sones, K., Heaney, R. K. & Fenwick, G. R. (1984b) *Journal of the Science of Food and Agriculture*, **35**, 712, 762.

Speek, A. J., Schrijver, J. & Schreurs, W. H. P. (1984) *Journal of Agricultural and Food Chemistry*, **32**, 352.

Speiss, A. (1981) *Lebensmittelindustrie*, **28**, 449.

Spinks, E. A., Sones, K. & Fenwick, G. R. (1984) *Fette Seifen Anstrichmittel*, **86**, 228.

Steinhauser, J. & Gherardi, S. (1980) *Flussiges Obst*, **47**, 542.

Stern, I. (1943) *Analyst*, **68**, 44.

Tabekhia, M. M. (1980) *Deutsche Lebensmittel Rundschau*, **76**, 280.

Tanner, H. & Limacher, H. (1984) *Flussiges Obst*, **51**, 168, 182.

Taper, L. J., McNeill, D. A. and Ritchey, S. J. (1985) *Journal of the American Dietetic Association*, **85**, 718.

Thomann, R. & Piechaczek, R. (1983) *Nahrung*, **26**, 915.

Thorpe, R. H. & Atherton, D. (1972) Technical Bulletin No. 21. The Fruit and Vegetable Preservation Research Association.

Thraves, R. (1955) *Food Manufacture*, **30**, 271.

Treptow, H. (1985) *Ernahrung*, **9**, 179.

Trifiro, A., Gherardi, S., Bazzarini, R. & Bigliardi, D. (1982) *Industria Conserve*, **57**, 23.

Trucksess, M. W., Brumley, W. C. and Nesheim, S. (1984) *Journal of the Association of Official Analytical Chemists*, **67**, 973.

Tsang, W.-S., Clarke, M. A. & Parrish, F. W. (1985) *Journal of Agricultural and Food Chemistry*, **33**, 734.

Tyler, T. A. (1984) *Journal of the Association of Official Analytical Chemists*, **67**, 745.

Upperton, A. M. (1985) *Journal of the Institute of Brewing*, **91**, 151.

Vandercook, C. E., Lee, S. D. & Smolensky, D. C. (1980) *Journal of Food Science*, **45**, 1416.

Vanderslice, J. T. & Higgs, D. J. (1988) *Journal of Micronutrient Analysis*, **4**, 109.

Van Gelder, W. M. J. (1984) *Journal of the Science of Food and Agriculture*, **35**, 487.

van Rillaer, W. G. & Beerhaert, H. (1983) *Zeitschrift fur Lebensmittel Untersuchung und Forschung*, **177**, 196.

VdF (1987) *RSK Values, The Complete Manual.* Bonn: Association of the German Fruit Juice Industry.

Veerabhadra Rao, M., Kapur, O. P. & Prakasa Sastry, C. S. (1983) *Food Chemistry*, **12**, 109.

Velasco, J. & Morris, S. L. (1976) *Journal of Agricultural and Food Chemistry*, **24**, 86.

Verma, K. K., Jain, A. & Rawat, R. (1984) *Journal of the Association of Official Analytical Chemists*, **67**, 262.

Vidal, C.-V., Valverde, S., Martin, C.-V., Blanco, I. & Rojas, E.-H. (1985) *Journal of the Science of Food and Agriculture*, **36**, 43.

Vogez, P. & Weber, G. (1982) *International Zeitschrift fur Lebensmittel-Technologie und Verfahrenstechnic*, **33**, 473.

Voragen, A. G. J., Schols, H. A., De Vries, J. A. & Pilnik, W. (1982) *Journal of Chromatography*, **244**, 327.

Walter, R. H., Sherman, R. M. & Lee, C. Y. (1983) *Journal of Food Science*, **48**, 1006.

Williams, H. A. (1968, 1972) *Journal of the Association of Public Analysts*, **6**, 69; **10**, 103.

Willich, R. K., Morris, N. J. & Freeman, A. F. (1954) *Food Technology*, **8**, 101.

Wills, R. B. H., Wilamasiri, P. & Greenfield, H. (1983a) *Journal of the Association of Official Analytical Chemists*, **66**, 1377.

Wills, R. B. H., Scriven, F. M. & Greenfield, H. (1983b) *Journal of the Science of Food and Agriculture*, **34**, 1383.

Wills, R. B. H., Lim, J. S. K. & Greenfield, H. (1984) *Journal of the Science of Food and Agriculture*, **35**, 1012.

Wood, R. (1982) Personal Communication—Report to Codex Committee on Methods of Analysis.

Woollen, A. (1981) *Food Manufacture*, **56**, 38.

Wrolstad, R. E. & Shallenberger, R. S. (1981) *Journal of the Association of Official Analytical Chemists*, **64**, 91.

Wrolstad, R. E., Culbertson, J. D., Cornwell, C. & Mattick, L. R. (1982) *Journal of the Association of Official Analytical Chemists*, **65**, 1417.

Yamanaka, H., Kuno, M., Shiomi, K. & Kikuchi, T. (1983) *Journal of Food Hygiene Society Japan*, **24**, 454.

8

Cereals and flour

CEREALS

All cereals are cultivated grasses of which the seeds yield the grains of commerce; the most important cereals are wheat, oats, barley, rye, maize and rice. The refined products of milling, grits and flours are important sources of carbohydrate but certain of them provide significant amounts of proteins and micronutrients to the diet. Along with the potato, and to a smaller extent other seeds such as the pulses, the cereal grains also form the principal sources of starch for manufacturing purposes. The native starches of the vegetables have characteristic microscopical features which aid recognition of the primary products from milling operations. In spite of the fact that the nature of products that are acceptable as breads is substantially different according to cultural factors, the milled flours from the range of cereals are not all suitable for bread making. In recent years the range of derived cooked products has increased enormously from the common breakfast cereals of the flake and shredded variety to extruded and puffed products commonly seen as snack foods.

Oats

Oats are rich in fats and protein but since the protein lacks elasticity the flour is not suitable for bread making. Oats are prepared by cleaning, drying and storing prior to the removal of husk by grinding. The product can be ground further to produce oat meal or partially cooked and rolled to give rolled oats. The meal contains roughly 12 per cent protein, 73 per cent carbohydrate and 9 per cent fat.

Rye

Rye is commonly grown in climatic regions that are too severe for wheat and in spite of the fact that the proteins do not yield a strong gluten for aerated bread making, rye bread is a staple article of diet in northern Europe. Crisp breads are, however, made from crushed rye grains and are popular in many regions.

Barley

Barley is not eaten extensively as such, although malted grain is a source of malt flour and extracts. The principal use of the grain is in animal feeding and in the production of fermentation products.

Maize

The grain forms an important part of the diet in the central and southern regions of Africa and of North America. It is unsuitable for bread making, but the ground grain is used to make 'grits' and cooked products such as tortilla. The nutritional value of the cereal is substantially affected by the culinary methods used in its preparation. Maize is a significant source of manufactured products including breakfast cereals, snack foods, cornflour and starch syrups.

Rice

The main cereal of the orient and sub-tropical regions, it is nevertheless a significant contributor to the diet of many other regions. As a grain, rice is a poor source of protein and fat and because of the method of preparation the polished grain loses valuable micronutrients such as vitamin B_1. The grain is used in further processing to yield a range of puffed products and a ground flour.

Wheat

Wheat is the most important cereal source of flour for bread making, first grown in the Middle East but now widely grown throughout the world.

The wheat grain has the following average percentage composition: endosperm 85, bran 12·5, germ 2·5. The composition of wheat flour, however, varies considerably according to the class of wheat, its country of origin, and the proportion of outer parts removed by the particular milling process (Elias, 1972; Nelson, 1985). As the outer portions contain more protein, fat, fibre and ash than the starchy endosperm, the proportion of each of these constituents decreases as the extraction percentage gets less. Also, wholemeal flour is brownish in colour, whereas the finest grades (containing the least fibre, etc.) are white. Wheat flour differs from other flours in containing a considerable proportion of gluten and to this its special suitability for bread making is due. The composition of the gluten present has a bearing on the 'strength' and water-holding properties of the flour. The two proteins forming the greater part of the gluten are glutenin and gliadin, and while the latter appears to be identical in strong and weak wheats, the former exists in different varieties. The figures given in Table 8.1 have been extracted from various sources. Comparative data for other common cereals are given in Table 8.2.

Legislation

The main statutory standards for flour derived from wheat and no other cereal are prescribed in The Bread and Flour Regulations 1984 (SI 1984 No. 1304). All flour (other than wholemeal, wheat malt flour and self-raising flour which has a calcium content of not less than 0·2 per cent) is required to contain in 100 g not less than 235 mg and not more than 390 mg of added calcium carbonate. All flour, regardless of type, is required to contain not less than the following amounts of micronutrients in 100 g: iron 1·65 mg, vitamin B_1 0·24 mg and nicotinic acid (or nicotinamide) 1·60 mg. Wholemeal flour, which is defined as the whole of the product derived from the milling of clean wheat, will contain these amounts naturally and no addition will be required. In cases where nutrient additions are necessary the materials used must comply with standards of purity laid down in the regulations. Brown flour must contain at least 0·6 per cent crude fibre in the dry matter; crude fibre is defined as the

Table 8.1 Composition of unfortified wheat flours or various extraction rates and wheat germ and wheat bran

	Flour (72%)	Flour (80%)	Wholemeal (95–100%)	Wheat germ	Wheat bran
Moisture (%)	13–15	13–14·5	13–14	9–12	14
Protein (%)(N × 5·7)	8–13	8–14	10–15	25–30	12–16
Fat (%)	0·9–1·4	1·0–1·6	1·5–2·5	8·5–11·0	3 0–4·0
Carbohydrates (%)	65–70	64–70	60–68	39–45	—
Fibre (%)	0·1–0·3	0·2–0·4	1·8–2·5	2·0–2·5	9–12
Ash (%)*	0·3–0·5	0·6–0·9	1·2–2·0	4·0–4·5	4·0–6·0
Calcium (mg Ca/100 g)*	15	20	30	—	—
Iron (mg Fe/100 g)	1·2	1·7	2·5	—	—
Vitamin B_1 (mg/100 g)	0·10	0·25	0·40	2·1	0·70
Nicotinic acid (mg/100 g)	0·8	1·3	6	6·8	23·2
Riboflavin (mg/100 g)	0·03	0·05	0·12	0·45	0·24
Limiting amino acid	Lysine	Lysine	Lysine		

* Most flour (except wholemeal) is fortified with chalk (see below), which increases the ash and introduces the equivalent of approx. 125 mg Ca per 100 g.

Table 8.2 Typical composition of the main whole cereal grains

	Wheat	Rice	Maize	Millet	Oats	Rye	Barley
Moisture (%)	13·4	11·2	13·0	12·0	13·3	13·4	14·9
Protein (%)	12·1	8·0	9·5	10·1	13·0	11·0	10·0
Fat (%)	2·0	1·8	4·3	3·3	7·5	1·9	2·5
Crude fibre (%)	1·9	8·9	2·2	—	10·3	1·9	4·4
Ash (%)	1·8	5·0	1·2	1·5	3·1	2·0	2·6
Calcium (mg/100 g)	30	15	12	30	60	50	35
Iron (mg/100 g)	3·5	2·8	5·0	6·2	3·8	3·5	4·0
Vitamin B_1 (mg/100 g)	0·40	0·25	0·33	0·40	0·50	0·27	0·27
Nicotinic acid (mg/100 g)	0·0	1·0	1·5	3·5	1·3	1·2	7·0
Riboflavin (mg/100 g)	0·12	0·10	0·13	0·12	0·14	0·10	0·20
Energy (MJ/100 g)	1·40	1·49	1·19	1·44	1·61	1·30	1·40
Limiting amino acid	Lysine	Lysine Threonine	Tryptophan Lysine	Lysine	Lysine	Lysine	

'organic matter contained in the dried defatted residue obtained by digesting bread or flour successively with boiling acid and boiling alkali'. No detailed method of analysis is laid down in the regulations. Where flour contains the product of milling other cereals then the name of the food must be such that it indicates the presence of the other cereal or cereals. There is no requirement to indicate in the name of the food the presence of the prescribed nutrients or technological additives such as prepared wheat gluten or malt flours derived from barley or wheat. The word 'self-raising' must appear in the name of any flour which contains more than 0·4 per cent available carbon dioxide as determined by the method in the regulations. The regulations contain a schedule of permitted additives for bread and flour and restrictions as to their use. The provisions for flour and self-raising flour are summarised in Table 8.3. Useful information on the background to legislation is given in the FSC (1960, 1974) Reports on Bread and Flour and by Spencer (1961).

An EC Draft Regulation, which provided for five types of soft wheat flour graded

Table 8.3 Additives permitted in flour and self-raising flour

Additive	Type of flour	Amount permitted UK regulations (mg/kg)	Codex standard
E150 Caramel	Wholemeal, brown		
E170 Calcium carbonate	All flour except wholemeal	2000	
E220 Sulphur dioxide E223 Sodium metabisulphite	All flour intended for use in the manufacture of biscuits or pastry, except wholemeal	200 (as SO_2)	200
E300 Ascorbic acid	All flour except wholemeal	200	300
E335 mono-Potassium L-(+)-tartrate E341 Calcium tetrahydrogen diorthophosphate	Self-raising flour) flour intended for use in the manufacture of buns or scones		
E341 triCalcium diorthophosphate	All four except wholemeal	600	
E450(a) diSodium dihydrogen diphosphate	Self-raising flour; flour intended for use in the manufacture of buns or scones		
500 Sodium hydrogen carbonate	Self-raising flour; flour intended for use in the manufacture of buns or scones		
516 Calcium sulphate	All flour except wholemeal	4000	
541 Sodium aluminium phosphate, acidic	Self-raising flour; flour intended for use in the manufacture of buns or scones		
575 D-Glucono-1,5-lactone	Self-raising flour; flour intended for use in the manufacture of buns or scones		
920 L-Cysteine hydrochloride	(a) All flour used in the manufacture of biscuits, except wholemeal or flour to which E220 or E223 has been added (b) Other flour, except wholemeal	300 75	90
924 Potassium bromate	All flour except wholemeal	50	50
925 Chlorine	All flour intended for use in the manufacture of cakes except wholemeal	2500	
926 Chlorine dioxide	All flour except wholemeal	30	
927 Azodicarbonamide	All flour except wholemeal	45	45
alpha-Amylases, Proteinases	All flour		GMP
Benzoyl peroxide	All flour except wholemeal	50	
Mono calcium phosphate			2500
Lecithin			2500

according to the ash content, is summarised in Table 8.4. Work on the Regulation has been suspended.

The Codex Alimentarius Commission has a committee on cereals, pulses and legumes. Standards developed so far include those for wheat flour, maize grains, whole maize meal and degermed maize meal. A summary of the principal requirements of these standards is given in Table 8.5. Notes on the permitted additives in wheat flour are given in Table 8.3.

Analysis of flour

The routine analysis of flour may include the determination of **moisture, ash, added chalk, sulphur dioxide, oil, protein, acidity, iron, thiamine** and **nicotinic acid**, an examination for **improvers** and **bleaching agents** and a **microscopical examination**.

Industrially, certain other types of analysis are of some importance, namely examination of the gluten, physical tests on the dough produced from the flour, determination of the particle size, maltose, colour and grade figures and the 'filth' test. The methods that are described below are essentially those which are of the greatest general interest. For further in-depth information on these and the other methods of cereal analysis the reader should consult Kent-Jones and Amos (1967), the reference work of the American Association of Cereal Chemists (1972) and the International Association for Cereal Chemistry (ICC). Some standard methods have been described by the Wheat Commission (1939). Aspects of dough rheology are discussed by Bloksma (1972).

Sampling. Thorough mixing is most important, especially with fortified flour. The Ad Hoc Flour Sampling Group (FSC, 1974, Appendix 6) reported that manual sampling, if properly carried out, will give adequately representative results for added nutrients provided that the feeding and mixing of nutrients is satisfactory. Data are presented that show there is difficulty in homogeneous mixing of the prescribed additives in commercial flour milling.

The Bread and Flour Regulations 1984 contain sampling provisions for application at the flour mill and at a dockside. In the former case the sample is removed from the flour stream either mechanically or manually at a point near to the discharge from the milling process. Six consecutive sub-samples are removed at intervals of 10 to 15 minutes. The complete admixture of sub-samples is deemed the sample for the purposes of the Act. At the dockside a sub-sample is taken from each of not less than six containers in a consignment. The combination of sub-samples forms the official sample. These requirements are in accordance with the recommendations of the Flour Sampling Group.

Moisture. The Wheat Commission method for determining moisture involved drying 5 g of sample for 5 h at 100 °C in a dish having diameter of 40–70 mm and a height not more than 60 mm (and provided with a lid). The AOAC recommends heating for 1 h at 130 °C or 5 h in a vacuum oven at 98–100 °C. Industrially, more rapid methods are employed using higher drying temperatures, e.g. the Carter–Simon method (which involves drying for 15 min at 155 °C). Such methods may give results over 1 per cent higher than those obtained using the Wheat Commission method, but they are often sufficiently accurate when considered on a comparative basis and can usually be correlated with the standard method. The results obtained

Table 8.4 Proposed EC grading of soft wheat flour according to the ash content

Description of soft wheat flour	Ash content as percentage of dry matter
Type 1	not more than 0·50
Type 2	0·51–0·65
Type 3	0·66–0·90
Type 4	0·91–1·20
Type 5	1·21–2·10

from drying methods represent the amount of 'free' moisture produced at the drying temperature. For example, some of the water present remains 'bound' and is probably associated with the proteins. Distillation methods using 25 g of sample and toluene or xylene as the immiscible solvent are said to give a figure that is equivalent to the total water present. In view of the different results obtained according to the technique employed, it is especially important to consider the method of determination when interpreting figures for the moisture content of flour. The ICC reference method No. 109 for cereals and flour requires the sample to be heated at 50 °C at reduced pressure in the presence of phosphorus pentoxide. The drying time is a minimum of 100 h with a repeat interval of 48 h.

The development of rapid methods of analysis for application to cereals and other foods has followed two distinct paths: firstly, the use of rapid drying techniques based on microwave heating, and secondly, the use of electromagnetic effects (Kropf, 1984), including a variety of spectroscopic methods especially those based on near infra-red transmittance or reflectance techniques. Davis and Lai (1984) describe the optimisation of microwave-oven procedures in application to wheat flours. Variables examined included power input and sample size. Correlation with the air oven method was obtained in 8 minute drying times with 3 g samples of hard wheat flours and 2 g samples of soft wheat flours; in both instances a power input of 415 W gave the best comparison.

The literature on multiple analytical methods based on near infra-red spectroscopy (NIR) is voluminous. Osborne and Fearn (1986) have produced an extensive theoretical and practical review of the available techniques and instrumentation and their application to the analysis of a variety of foods. Considerable attention has been paid to the application of NIR to the analysis of cereals and cereal products in both industrial and laboratory environments. In all applications of this type it is necessary to calibrate the instrument used by comparison of analyses of the same sample using standardised procedures, and considerable effort has to be expended in obtaining an accurate calibration for use within each laboratory. In a five year study Osborne *et al.* (1982) showed that the calibration sets for flours from hard and soft wheat could be combined with some loss of accuracy but that for each year's crop a bias factor had to be applied. Some of the bias recorded was attributed to instrumental drift but this could not account for all the correction needed between the individual years in the survey.

Moist flour with a moisture content in excess of about 13 per cent (Wheat Commission method) is liable to attack by micro-organisms, mites and insects.

Ash. The outer portions of the wheat grain contain considerably more mineral

Table 8.5 Summary of main provisions of Codex standards for cereals (1986)

	Wheat flour	Maize*	Maize meal	Degermed maize meal
Sources	*Triticum aestivum* L. *Triticum compactus* Host.	*Zea mays indentata* L. *Zea mays indurata* L.	*Zea mays* L.	*Zea mays* L.
Moisture (%)	≯15·5	≯15·5	≯15·0	≯15·0
Ash (%)	—	—	≯3·9	≯1·0
Crude Fat (%)	—	—	≮3·1	≮2·25
Acidity of Fat (mg KOH/100 g)	≯30	—	—	—
Protein (%)	≮ 7·0 (N × 5·7)		≮8·0 (N × 6·25)	≮7·0 (N × 6·25)
Methods				
Sampling	ISO 2170–1980 ICC 130 ISO 6644–1981 AACC 64–60	ISO 950–1981	ISO 2170–1980 ICC 130 AOAC (1980) 10·092 AACC 64–60	ISO 2170–1980 ICC 130 AOAC (1980) 10·092 AACC 64–60
Moisture	ISO 712–1979 ICC 110/1 AOAC (1980) 14·004 AACC 44–15A	ISO 6540–1980 ICC 110/1 AOAC (1980) 14·004	ISO 712–1979 ICC 110/1 AOAC (1980) 14·004 AACC 44–15A	ISO 712–1979 ICC 110/1 AOAC (1980) 14·004 AACC 14–15A
Ash	AOAC (1980) 14·006 ISO 217–1980 AACC 08–01		AOAC (1980) 14·006 ISO 217–1980 AACC 08–01	AOAC (1980) 14·006 ISO 217–1980 AACC 08–01
Protein	ICC 105/1 AOAC (1980) 2·055 AACC 46–11		ICC 105/1 AOAC (1980) 14·026 AACC 46–11	ICC 105/1 AOAC (1980) 14·026 AACC46–11
Crude fat			AOAC (1980) 14·067 ISO 5986–1983	AOAC (1980) 14·067 ISO 5986–1983
Fat acidity	AOAC (1980) 14·070 AACC 02–01A			

* The Maize Standard contains limits for physical defects and provisions for homogeneity of colour between grains.

matter (about 5 per cent) than the inner parts, so the ash figure of white flour is considerably less than that of wholemeal flour. In the absence of added mineral matter, therefore, the ash figure gives an indication of the grade—'Patent' gives an ash of about 0·3–0·4 per cent, 'Straight-run' (72 per cent extraction) about 0·45 per cent, lower grades 0·5–0·65 per cent or more and 'Wholemeal' 1·2–1·8 per cent (cf. Tables 8.1 and 8.2). If added chalk is present as required by the Bread and Flour Regulations for most flour, it is more difficult to assess the grade from the ash figure. In view of its direct relationship with the ash, the colour of untreated flours can be assessed and controlled from the ash (Staudt, 1965). Correlations between ash content and reflectance at 546 nm has been studied by Gillis (1963). He concluded that the ash content of a flour may be obtained within a few minutes by use of the reflectance method. Use of near infra-red reflectance measurements to estimate ash and protein content of flour mill streams has been reported by Watson *et al.* (1976); correlation coefficients of 0·968 for ash and 0·998 for protein were claimed.

The Wheat Commission method for determining the ash involves igniting 3–5 g of the flour in a silica dish at 600 °C to constant weight. The ISO, AOAC and Codex techniques prefer heating at 55 °C in a silica dish. In view of its simplicity this method is preferred in commercial circles. The ICC method No. 104 requires the sample to be ignited at 900 °C. Whilst this latter temperature is favoured by many EC countries, the method requires use of platinum dishes and there are difficulties in maintaining a uniform temperature of 900 °C in a muffle furnace. The ash of wheat flour consists principally of the phosphates of potassium and magnesium. On analysis the ash of flour gives mean percentages of 49 (as P_2O_5), 37 (as K_2O), 6 (as MgO) and 5·5 (as CaO). Such figures obviously do not apply if the flour contains added chalk. Further, it should be borne in mind that, as the bran contains more magnesium and less potassium and calcium than the centre of the grain, such proportions vary with the extraction rate.

The Wheat Commission also lays down the procedure for determining the **acid-soluble ash**.

PROCEDURE

Add 3–4 ml of 10 M hydrochloric acid to the ash, place on a steam bath and heat until dry. Continue heating for an hour. Add 25 ml of 2 M hydrochloric acid, heat for 5 min on a steam bath, decant through an ashes filter paper. Re-extract the residue in the dish with a second aliquot of the dilute acid. Transfer the residue to the filter paper, wash thoroughly with hot water. Place the paper and the residue in the original dish and ignite, cool, weigh, calculate the insoluble ash content.

Dvorak *et al.* (1962) have described a method for determining sand in the original flour (as distinct from silica derived from the ash), which is based on its isolation with carbon tetrachloride in a modified separating funnel. The method is also used to assess the ash content of the flour used in self-raising flour (AOAC).

Oil or fat. The outer portions of the wheat grain contain considerably more oil than the rest, so that the whitest flours contain the least (about 1 per cent). Wheat germ oil consists mainly of linoleic, oleic, palmitic and linolenic acids (see Paul and Southgate, 1978). The oil has an iodine value of 115–125 and a saponification value of about 185. Treatment, particularly with chlorine dioxide, causes destruction of vitamin E (tocopherols), for which methods are discussed in Chapter 16.

The fat in flour is usually determined either by direct extraction with light petroleum or by extraction with mixed ethers after acid hydrolysis. The varying amounts of fat extracted by different methods have been compared by Herd and Amos (1930), who concluded that direct extraction measures the free fat only, that acid hydrolysis methods give the total fat content (free + combined) and the usual alcohol method for estimating lipids gives the fat + unhydrolysed lipids. (See also Chapter 2 and Kropf (1984) for instrumental methods.)

For the determination of oil by the direct method, it is preferable to use a Soxhlet or Bailey–Walker extractor with light petroleum (b.p. 40–50 °C) as solvent. The material should be mixed with an equal weight of finely ground sand before placing in the extraction thimble.

For the acid hydrolysis method stir 2 g of flour with 2 ml of alcohol, add 7 ml of conc. hydrochloric acid and 3 ml of water. Heat at about 80 °C for 30 min. Cool, add 10 ml of alcohol to the hydrolysed mixture, extract three times with 25 ml of ether, followed by 25 ml light petroleum, combine the solvent extracts, remove the solvent by evaporation and weigh the residue.

Protein. Wholemeal flours contain more protein than white flour, prepared from the same grist. Also, in general, 'strong' flours contain more protein than 'weak' flours.

The total nitrogen can be determined on approximately 1·5 g accurately weighed by the Kjeldahl method or on 0·2–0·5 g if a semi-micro modification is employed. The crude protein is usually calculated using the factor N × 5·7. Following a comprehensive study of the amino acid composition of various Canadian wheats, however, Tkachuk (1966a, b) considered that 5·62 was a more accurate factor to employ for flour. Improvements to the classical Kjeldahl method have been devised and commercial instruments, which claim varying degrees of automation of the procedure, have been developed (see also Chapter 2). A commercial semi-automated instrument that uses a temperature controlled digestion block and a microprocessor controlled distillation and titration unit has been described by Hjalmarsson and Akesson (1983). When applied to samples of ground wheat and barley the results of analysis by the traditional procedure and the instrumental method were indistinguishable in terms of accuracy and precision. Techniques using NIR have also been widely applied to the determination of protein in the routine quality control of input cereals and the product streams in flour mills (see Osborne and Fearn, 1981, 1986; Bolling and Zwingelberg, 1984). The latter authors have reported on the continuous in-line monitoring of flour in a mill with a continuous print out of protein content of the flour produced. Fearn (1986) has discussed the statistical errors in the calibration of an NIR instrument in application to the determination of protein in flours; in particular he points out that the calibration is an empirical relationship and that there are significant interactions between moisture content, surface texture and particle size and the reflectance due to protein in any sample.

Craney (1972) reported a correlation coefficient of 0·995 (with standard error ±0·24 per cent protein) between the Kjeldahl method and the biuret procedure when applied to ground wheat. The biuret method depends on the formation of a blue–violet complex between cupric ions and protein. Absorbance is measured at 550 nm. Udy (1956) described a dye-binding method in which the wheat proteins react with Orange G at pH 2·2 to form an insoluble complex. The protein content is then assessed from the colorimetric measurement of unbound dye at 470 nm.

Greenaway (1972) compared the Kjeldahl method with biuret and dye-binding procedures for the testing of wheat. He concluded that agreement between Kjeldahl and biuret methods was good but that the correlation with dye binding was non-linear and not satisfactory for control purposes at low protein contents. A modified correlation curve was proposed to improve the agreement between all methods. Williams and Butler (1982) concluded that satisfactory correlations were confined to within season comparisons; significant variations were ascribed to climatic conditions during maturation. Ronalds *et al.* (1984) have established correlations between Kjeldahl protein and the nitrogen content by distillation from alkali. Flour quality assessments have been made based on estimation of acetic acid soluble protein using the alkali distillation procedure.

Gluten. The method usually employed for the estimation of gluten is not altogether satisfactory, and consists of adding to 20 g of flour about 15 ml of water so as to make a stiff dough. This is allowed to stand in a beaker of water and is then squeezed in the fingers and gently kneaded under a stream of running water until all the starch is washed out and the wash water is clear. The residue of moist gluten is squeezed dry in the fingers and weighed, or alternatively the gluten is rolled on a hardened filter paper, spread into a thin layer and dried in a moisture oven. When the weight is almost constant, the result is returned as dry gluten. The 'strength' of the flour depends upon both the amount and the physical qualities of the gluten. Crude gluten obtained in this way contains albumin, globulin, glutenin, gliadin and proteose. More consistent results can be obtained if a commercial Gluten Washer is employed. In the ICC method No. 106 the dough is prepared with a sodium chloride solution and the dough ball is washed with the same solution. A mechanical press is used to remove excess moisture. The weighed quantity of gluten multiplied by 10 is the wet content. Redman and Burbridge (1979) have compared the performance of two gluten washers and have shown that significantly different results were obtained with the two machines. The differences were explained in terms of the efficiency of starch removal.

Attempts have been made to assess the baking qualities of flour by detailed studies on the gluten proteins. Bottomley *et al.* (1982) used sodium dodecyl sulphate (SDS) extracts and polyacrylamide gel electrophoresis to separate glutenins, gliadins, albumins and globulins. They suggested that a correlation could be found between molecular size of the glutenin fraction and the bread making quality. Further studies using reversed phase HPLC and size exclusion chromatography have been reported by a number of authors, including Preston (1982), Huebner and Bietz (1984a, b, 1985) and Bietz *et al.* (1984). Kaiser and Krause (1985) have also reviewed the available techniques for protein fractionation and have assessed their value in the identification of the source and quality of proteins. Wheat variety identification by gel electrophoresis has also been studied by Salmon and Burbridge (1985). Osborne (1984) reported that NIR spectroscopy cannot provide a measurement of the bread making quality of wheat beyond that which may be predicted from the protein content.

The regulations allow for the addition of dried gluten in small quantities to flour for bread making to improve the bread making quality; Stevens (1983) has described a simple physical separation method to detect its presence. The method relies on the specific gravity of gluten (1·35) being sufficiently different from endosperm and starch granules (1·45 and 1·5 respectively) to allow separation by flotation on a 30/70 mixture of toluene and carbon tetrachloride. The particles of dried gluten are distinguishable from the natural protein fraction by size and shape and by response to staining with Congo Red solution. The method is claimed to be applicable to levels of addition down to 0·1 per cent.

Skerritt *et al.* (1985) have studied methods for the detection of gluten proteins in foods in general, especially in relation to the need to control 'gluten-free' foods prepared for use by coeliac disease patients. A monoclonal antibody technique has been developed that is capable of detecting one part of gluten in 10 000 parts of starch. A gel electrophoresis method that relies on the detection of gliadin has been reported by McCausland and Wrigley (1976). The method is suitable for the enforcement of a limit of 0·3 per cent gluten in uncooked foods, but it is less reliable in application to baked goods.

Fibre. The percentage of crude fibre in wheat flour varies from 0·1 (white flour, 72 per cent extraction) to about 2 (wholemeal). The Bread and Flour Regulations therefore require wheatmeal or brown flour to contain at least 0·6 per cent m/m crude fibre calculated on the dry matter.

The general method for determining crude fibre as laid down in the Feeding Stuffs (Sampling and Analysis) Regulations 1982 is described on p. 26. The SPA (1943) suggested for the determination of fibre in flour that the procedure described in the 1932 Regulations could be modified by omission of the initial extraction with petroleum spirit, initially 'creaming' 3–5 g of flour by slow addition to 20 ml of cold sulphuric acid 0·1275 M followed by addition of 180 ml of boiling acid using a reflux condenser for the maintenance of constant volume during the digestion, transfer of the fibre, after the washing with ether, from the filter paper to a dish (previously ignited and weighed) with boiling water. Removal of the water by evaporation on a water bath, drying at 100 °C, weighing the fibre followed by ignition and correction for the 'ash' content, completes the procedure. Baker (1977) has described a modification of the acid-detergent method, using a buffered hydrochloric acid–potassium chloride solution mixed with detergent. In all cases the method gives higher values for fibre than currently accepted methods. The author claims that his method removes starch and protein without breaking down cellulose (see also van Soest, 1963a, b, 1973a, b). Baker and Holden (1981) have reported typical data on ready-to-eat breakfast cereals using the neutral detergent fibre method.

Player and Wood (1980) have reported on a series of collaborative studies aimed at assessing the comparative performance of two crude fibre methods and an acid-detergent method. In this work the methods reported on were the Feeding Stuffs Regulations method, a volumetric procedure based on the method of van de Kamer and van Ginkel (1952) and the acid-detergent method of van Soest (1963a, b). The study confirmed that the acid-detergent method yields a higher value for 'fibre', but that the method is less satisfactory with regard to repeatability and reproducibility than the other two methods. There is little evidence of significant differences between the two crude fibre methods in respect of the performance characteristics. The authors noted that the Feeding Stuffs Regulations method has a relatively low sensitivity at low levels of crude fibre but improves as the level increases. The opposite effect was noted with the volumetric method.

During the past few years emphasis has turned from an interest in estimation of crude fibre to the nutritionally significant measure of dietary fibre (Heaton , 1983). Development of suitable analytical methodology has been hindered by confusion surrounding the definition of the nature and constitution of dietary fibre (see Chapter 2 for further information on these problems). Under the definition proposed by Trowell *et al.* (1976, 1978) the feature which distinguishes the polysaccharides of fibre from those of starch is the resistance to amylase and disaccharidases. The resistant polysaccharides include cellulose, hemicellulose and pectin, which constitute the major components of plant cell wall material. Lignin is a complex of branched polyphenols, which constitute some 20 per cent of the cell wall structure of plant materials; tannins are also present in some vegetable materials and they are phenolic in nature. The various starch and cellulose derivatives used as additives or bulking aids will also have some degree of resistance to enzyme digestion as will the gums, mucilages, algal polysaccharides, agar and carageenans that are widely used as thickening and emulsifying agents. Schweizer *et al.* (1984) have examined the composition of various fibre fractions from wheat bran. Numerous studies have reported on the various approaches to the estimation of dietary fibre for application to cereal products (Heckman and Lane, 1981; Asp, 1983; Asp *et al.*, 1985). Cummings *et al.* (1985) have summarised the 'state of the art' in methodology and reported on a collaborative study of five procedures for dietary fibre, one for crude fibre and one for lignin; the methods varied from analysis of the alcohol-insoluble residue, a neutral detergent method, and three variations on enzyme digestion procedures proposed by Englyst *et al.* (1983), and Asp *et al.* (1985). These authors concluded that none of the methods tested has proved to be suitable for determination of dietary fibre for enforcement purposes; the procedure of Englyst *et al.* appears to be the most accurate and informative but the method needs simplification and further study. The Englyst procedure (see p. 31) uses GLC to measure the sugars produced by hydrolysis of the non-starch polysaccharide fraction. Neilson and Marlett (1983) have proposed the use of HPLC for the measurement of the neutral sugar fractions; NIR procedures have also been proposed by Baker (1983) and Horvath *et al.* (1984). The problems of calibration and correlation are discussed by both authors. Bell (1985) has proposed that the dietary fibre contents of raw and cooked wheat products can be correlated with the total pentose contents. The pentose content of the sample is determined by reaction of the acid hydrolysate with phloroglucinol. The response is compared with standard solutions of xylose. When compared with the Southgate procedure over a period of months, the author claimed that good correlations were obtained for wheat flour, bread and rye flour and that the method satisfied ruggedness criteria for routine use. Methods for dietary fibre are described in some detail in Chapter 2.

Data on dietary fibre content of cereals and cereal products have been produced by Chen and Anderson (1981), Englyst *et al.* (1983) and Wenlock *et al.* (1985). In all cases the data are based on enzyme digestion procedures and values are given for cellulose, lignin, non-cellulose hexoses, pentoses and uronic acids. Miller and Judd (1984) have discussed the problems associated with assessment of metabolisable energy value of foods and the effect on digestibility of foods with a high fibre content.

Acidity. Three different procedures have been proposed for the determination of the acidity of flour: by titration of a water extract or an alcoholic extract, or alternatively the FFA of the extracted fat is determined.

Water extract method. Shake 18 g of flour with 200 ml of CO_2-free water in a conical flask and place in a water bath at 40 °C for one hour with the flask loosely stoppered. Filter and titrate 100 ml of clear filtrate with 0·05 M sodium hydroxide solution to phenolphthalein. The acidity of the water extract increases during storage and is calculated as lactic acid or potassium dihydrogen phosphate (1 ml 0·05 M NaOH ≡ 0·0068 g KH_2PO_4). The acidity of brown flour is higher than that of white flour.

Alcohol extract method. Add 10 g of flour to 100 ml of neutral 90 per cent alcohol and allow to stand for 24 h with occasional shaking. Then titrate 50 ml of the supernatant liquid with 0·05 M alcoholic sodium hydroxide solution to phenolphthalein. The acidity of the alcohol extract is often calculated as sulphuric acid. Recently milled flours give figures between 0·03 and 0·04 per cent (as H_2SO_4).

The FFA or acid value as determined on the light petroleum extract of the flour tends to rise more rapidly as the flour deteriorates during storage than the acidity of the aqueous or alcoholic extracts. Morrison (1963a, b) studied changes in FFA composition of wheat flour during 476 weeks' storage and found 8–15 per cent FFA in normal flour lipids rising to 25–70 per cent in old flour. Flours containing over 33 per cent FFA are considered unsuitable for use.

pH. Shake 10 or 12 g of flour with 100 ml of water and allow to stand for at least 30 min. Filter and determine with pH of the filtrate either visually or with a pH meter. The pH of flour usually falls within the range of 6·0–6·8. Bleaching of flour with chlorine gas causes a fall in the pH value.

Starch. A number of methods have been proposed for the determination of starch. In many cases it is sufficient to estimate starch content as the difference between 100 and the sum of moisture, sugar, protein, fat, fibre and ash. The estimate, which includes contributions from non-starch material such as hemicellulose is subject to the cumulative errors of the other determinations, but in many cases is of equivalent precision to the direct determination.

Direct methods may be classified as hydrolytic or non-hydrolytic. In the hydrolytic methods starch is hydrolysed to glucose, which is then determined. The hydrolysis may be by means of acid, enzymes or a combination of these. Glucose may be determined by titration (Kent Jones and Amos, 1967), by anthrone reaction (Clegg, 1956), by oricinol reaction (Dean, 1978), by polarimetry or by glucose oxidase. The use of enzyme methods for determination of sugars is described in Chapter 6.

HYDROLYTIC METHOD FOR STARCH

Extract 3 g of finely powdered flour several times with ether in a Soxhlet extractor. Wash on a filter with 10 per cent alcohol, followed by 95 per cent alcohol. Drain the residue, wash into a flask with about 50 ml water and heat for 15 min in a boiling-water bath stirring constantly so that all the starch is gelatinised and a homogeneous mixture is obtained. Cool the solution to 55 °C, add about 0·03 g of diastase dissolved in a little water and keep the mixture at 55–60 °C for at least an hour. A test of a drop of the solution gives no blue colour with iodine. Raise the temperature to 100 °C, wash and make up the filtrate and washings to 250 ml. Heat 200 ml filtrate with 20 ml hydrochloric acid in a boiling-water bath for 2½ h. Cool, neutralise with sodium carbonate, dilute to 500 ml and estimate the dextrose by the Lane and Eynon process. Starch = 0·90 × dextrose.

There are many variations of the above technique that attempt to take into account the presence of other cellulosic materials, which may yield glucose with acid. Different enzyme treatments may be used to take account of non-starch fractions. Trop and Grossman (1972) described a method claimed to be suitable for measurement of as little as 2 per cent starch in coloured foods. The method is based on acid hydrolysis and oxidation of glucose with glucose oxidase. The amount of oxygen consumed is measured polarographically. Alternatively, the hydrogen peroxide produced in the reaction may be estimated by reaction with *o*-dianisidine or *o*-phenylenediamine (Osborn and Voogt, 1978). An automated method for determination of sugars, amylose and starch in potato has been described by McCready *et al.* (1974). Soluble sugars are removed with 80 per cent ethanol. Reducing sugars are determined by reaction with dinitrosalicylic acid, glucose is oxidised with glucose oxidase and sucrose with invertase. Starch is determined in the alcohol-insoluble residue by hydrolysis in trifluoroacetic acid and reaction with dinitrosalicylate. HPLC techniques are now being used to determine individual sugars (Hunt *et al.*, 1977). Blake and Coveney (1978) have evaluated acidic and enzymic hydrolytic methods for determination of starch and modified starch. They concluded that both acidic hydrolysis and enzymic hydrolysis gave good results for native starches but that low recoveries were obtained with modified starch. Dispersion of the starch in dilute alkali before enzymic hydrolysis was found to give better yields in the latter case.

DISPERSION METHODS

These methods depend on the dispersion of the starch in an appropriate solution followed by determination by polarimetry or reaction with colorimetric reagents. Alternatively, the starch may be precipitated after dispersion and clarification and determined gravimetrically.

Many starch dispersants have been used. The ones commonly accepted are hydrochloric acid (Ewers, 1905; Rask, 1927a, b; ICC method No. 123) and acidified calcium chloride (AOAC; ICC method No. 122; Fraser and Hoodless, 1963). In the polarimetric procedures it is necessary to correct for the sugars present in the product by pre-extraction of the sample with cold water or dilute alcohol. In the polarimetric procedures it is necessary to adopt a figure for the specific rotation of starch; literature values vary from +180° to +220°. Specific rotations in calcium chloride for maize, wheat, tapioca and potato are in the range +201° to +204°. The AOAC methods book recommends +203° for wheat starch with acid calcium chloride dispersant. Ewers (1908, 1910) treated starch with hydrochloric acid and hot water. Ferrocyanide, molybdate or tungstate solutions were used as clearing agents and the specific rotation of cereal starches were found to be in the range +181° to +186°. The Ewers method has been widely studied and many authors have stressed the need to standardise the variables. Dudas (1971, 1972, 1973) has investigated such variables as analysis time, degree of stirring and alternative clearing agents. He concluded that Carrez solution was the best of the available clearing agents and that other variables need careful standardisation. The principles upon which the Ewers method is based have been criticised by Ulmann and Richter (1961) but nevertheless the method is favoured in the EC as a general procedure for determination of starch in cereals, cereal-based foods and mixed feeds.

GENERAL POLARIMETRIC METHOD FOR DETERMINATION OF STARCH

The method is applicable to cereal-based foods and feeds, and comprises two determinations. In the first, the sample is treated with warm, diluted hydrochloric acid. After clarification and filtering, the optical rotation of the solution is made by polarimetry. In the second, the sample is extracted by 40 per cent ethanol. After acidifying the filtrate with hydrochloric acid, clarifying and filtering, the optical rotation is measured under the same conditions as in the first determination.

Reagents

Hydrochloric acid 7·6 M: (27·84 per cent w/v sp. gr. 1·125).

Hydrochloric acid 0·3094 M: (1·128 per cent w/v).

Carrez solution I: Dissolve 21·9 g zinc acetate dihydrate in water containing 3 g acetic acid, make up to 100 ml with water.

Carrez solution II: dissolve 10·6 g potassium ferrocyanide trihydrate in water, make up to 100 ml with water.

Ethanol 40 per cent v/v.

Procedure

Determination of the total optical rotation (P or S) (note 1). Weigh to the nearest mg 2·5 g of the finely divided sample and place in a 100 ml volumetric flask. Add 25 ml hydrochloric acid (0·3094 M), shake to obtain a proper distribution of the sample for analysis, and add a further 25 ml hydrochloric acid. Immerse the flask in a boiling-water bath, and during the first 3 min shake vigorously and steadily to avoid agglomeration. The quantity of water in the water bath must be sufficient to allow the bath to be maintained at boiling-point when the flask is immersed in it. The latter should not be removed from the bath whilst being shaken. After 15 min precisely, remove the flask from the bath, add 30 ml cold water and cool immediately to a temperature of 20 °C.

Add 5 ml of Carrez solution I and shake for 1 min. Then add 5 ml of Carrez solution II and shake again for 1 min. Make up to volume with water, mix and filter. If the filtrate is not perfectly clear (which is rare), begin the analysis again using a larger quantity of Carrez solutions I and II, for example, 10 ml.

Then measure the optical rotation of the solution in a 200 mm tube with the polarimeter or the saccharimeter.

Determination of the optical rotation (P or S) of substances soluble in 40 per cent ethanol. Weigh to the nearest mg 5 g of the sample, place it in a 100 ml volumetric flask and add about 80 ml ethanol (see note 2). Let the flask stand for 1 hour at room temperature; during this time, shake vigorously six times so that the sample for analysis mixes thoroughly with the ethanol. Then make up to volume with ethanol, mix and filter.

Using a pipette, place 50 ml of the filtrate (=2·5 g of the sample) in a 250 ml Erlenmeyer, add 2·1 ml hydrochloric acid (7·6 M) and shake vigorously. Fit a reflux condenser to the Erlenmeyer and immerse the latter in a boiling-water bath. After 15 min precisely, remove the Erlenmeyer from the bath, decant its contents into a 100 ml volumetric flask, rinsing with a little cold water, and cool to a temperature of 20 °C. Clarify using Carrez solutions I and II, make up to volume with water mix, filter, and measure the optical rotation as indicated in a 200 mm tube.

Calculation of Results

The level of starch as a percentage of the sample is calculated as follows:

By the polarimeter

$$\text{Percentage of starch} = \frac{2000(P-P')}{[\alpha]_D^{20}}$$

P = total rotation in degrees

P' = rotation in degrees given by substances soluble in 40 per cent ethanol

$[\alpha]_D^{20}$ = specific rotation of pure starch. The generally accepted values for this factor are as follows:

+ 185·9°: rice starch
+ 185·7°: potato starch
+ 184·6°: maize starch
+ 182·7°: wheat starch
+ 181·5°: barley starch
+ 181·3°: oat starch
+ 184·0°: other types of starch, and also starch mixtures in compound feedingstuffs.

By the saccharimeter

$$\text{Percentage of starch} = \frac{2000}{[\alpha]_D^{20}} \times \frac{(2N \times 0{\cdot}665)(S-S')}{100} = \frac{26{\cdot}6N(S-S')}{[\alpha]_D^{20}}$$

S = total rotation in saccharimetric degrees
S' = rotation in saccharimetric degrees given by substances soluble in 40 per cent ethanol
N = the weight in g of sucrose in 100 ml water giving a rotation of 100 saccharimetric degrees in a 200 mm tube.

REPEATABILITY

The difference between the results of two parallel determinations carried out on the same sample must not exceed 0·4 per cent in absolute value for starch levels lower than 40 per cent, and 1·1 per cent in relative value for starch levels equal to, or higher than, 40 per cent.

NOTES

1. When the sample contains more than 6 per cent carbonates, calculated as calcium carbonate, these carbonates must be destroyed before the determination of the total rotation, by treatment with an exactly equivalent quantity of diluted sulphuric acid.
2. In the case of products with a high level of lactose, such as powdered lactoserum or skim milk powder, proceed as follows after adding 80 ml ethanol. Fit a reflux condenser to the flask and immerse the flask in a water bath at 50 °C for 30 min. Then allow to cool and continue the analysis as indicated.

The formation of adducts with iodine has been used as a means of purifying starch and in the correction of polarimetric readings. The starch–iodine complex may be destroyed by addition of alcohol and thiosulphate. Nedeltscheva *et al.* (1975) have described a spectrophotometric method based on the reaction of dextrinised starch with iodine. The method is claimed to be of suitable accuracy and speed for use in industrial quality control and for general laboratory use. A general literature review of the available analytical methods has been prepared by Blake (1978).

Maltose figure. The 'maltose figure' is a measure of the sugar produced by enzymic action on the starch present when the flour is incubated with water (the *diastatic activity*). Industrially, the maltose figure is of considerable interest as it gives an indication of the gas producing powers of the corresponding dough. For its determination, Blish and Sandstedt (1933) incubated the flour in the presence of an acetate buffer (pH 4·7) at 30 °C for one hour and the sugar produced is then determined volumetrically using alkaline ferricyanide. The result is expressed as mg maltose produced by 10 g of flour. Figures in excess of 350 mg maltose denote the presence of flour from malted or sprouted grain. Unlike the original procedure of Rumsey, in which the flour was incubated with water at 27 °C and the sugar determined using Fehling's solution, the Blish–Sandstedt method seems to be little affected by any chalk that may be present. A modification of the Blish–Sandstedt method has been proposed by Hildebrand and McClellan (1938) and the method is also recommended by the AOAC. Kent Jones and Amos (1967) have shown that the Blish–Sandstedt value is approximately 1·5 times that given by the Rumsey method. Starch damage of sound flour is reflected by the maltose value. However, if the maltose value is high the main cause will be a high α-amylase activity. Methods of estimating starch damage have been reviewed by Seidemann (1975).

Mineral additions can be separated by shaking the flour with carbon tetrachloride (see 'Self-Raising Flour', p. 311).

Chalk. The justification for addition of chalk to flour has its origin in the theory that phytic acid inhibited absorption of calcium. To counteract this apparent need most types of flour are required to contain 235–390 mg of chalk per 100 g. Wholemeal flour is exempted from this requirement. This is somewhat anomalous as a higher extraction rate results in an increase in the phytic acid content (FSC, 1960, 1974). The addition is also not necessary in the case of wheat malt flour nor with self-raising flour with a calcium content (from ACP in practice) of at least 0·2 per cent m/m. The necessity and advisability of adding chalk to flour has been reviewed (FSC, 1974), but retained in the current regulations. Two difficulties arise in connection with the estimation of the added chalk: the uneven distribution in the flour, and the uncertainty as to the accuracy of the blank. Surveys have shown that a considerable proportion of samples examined contained amounts of chalk outside the statutory range (FSC, 1974, Report of the Ad Hoc Flour Sampling Group in Appendix 6). The chalk content of flour may be determined either as total calcium or by release of carbon dioxide by the addition of acid. When total calcium methods are used allowance must be made for the amount naturally present in the flour; this may vary from 15 (white) to 24 (wheatmeal) and 30 (wholemeal) expressed as mg Ca/100 g. Several methods are available for measurement of total calcium in the ash, including precipitation as oxalate and titration with potassium permanganate (AOAC, Greer *et al.*, 1942), and atomic absorption spectroscopy (Chapter 3). A colorimetric method for determination of calcium without ashing of the flour has been described by Sawyer *et al.* (1956). This method is based on reaction of calcium with chloranilic acid. Although the reaction is not specific for calcium, the interfering ions present in flour are relatively small and likely to be constant. The method is suitable for routine analysis especially as a quality control procedure in milling. Gas evolution techniques are convenient for measurement of added chalk in flour. Fraser and Weston (1950) have described a volumetric procedure that is carried out in a Chittick apparatus: 20 g of flour is shaken mechanically with 70 ml water and 20 ml of 5 M hydrochloric acid; the volume of carbon dioxide evolved is measured in a constant pressure system using water filled manometers. The determination must be carried out in a room at a controlled temperature.

Tyler and Weston (1956) have pointed out that as creta (chalk) contains about 230 mg/kg of fluorine its addition at a rate of 14 oz per sack contributed about 0·7 mg/kg to the flour. Methods for determination of phytate using HPLC have been developed by Oberleas (1983), Lee and Abendroth (1983) and Camire and Clydesdale (1982). Typical data for a variety of cereal goods have been produced by Torelm and Bruce (1982) and for flours and bread by Daniels and Fisher (1980).

Iron. The Bread and Flour Regulations require all flour to contain at least 1·65 mg of iron per 100 g, such nutrient to be added (not with wholemeal flour) in the form of ferric ammonium citrate or green ferric ammonium citrate (conforming to the purity criteria in the Miscellaneous Additives in Food Regulations 1980), ferrous sulphate or dried ferrous sulphate (conforming in each case with the BP 1980 specification), or iron powder (as defined in Schedule 5 to the regulations, which specifies a minimum of 90 per cent iron, at least 95 per cent of which must be soluble in warm (37 °C) 0·2 per cent m/m HCl). Ferric ammonium citrate is more readily absorbed in the body than powdered or reduced iron (MOH, 1968).

Added iron can be detected qualitatively by wetting the flour with potassium thiocyanate (5 per cent in 1 M HCl) and comparing the red colour after 15 min with that produced by a non-enriched flour similarly treated.

For the determination of iron in flour, the FSC (1960) recommended the use of the method of Pringle (1946) using *o*-phenanthroline:

METHOD

Weigh out a suitable quantity of the sample (containing 0·05–0·4 mg Fe) into a silica basin, add 10 ml of glycerol–alcohol mixture (1:1) and, after a preliminary heating at the mouth of the muffle furnace, or under an infra-red lamp, ash overnight at 600 °C. Cool, add 1·0 ml conc. nitric acid, evaporate, replace the basin in the furnace for an hour. Cool, add 5 ml of 5 M hydrochloric acid to the ash, place on a steam bath for 15 min and filter through a hardened filter paper into a 100 ml volumetric flask. Add 3 ml of diluted hydrochloric acid (1 in 100) to the basin, bring to the boil on a hot plate, filter. Repeat this process four times, wash the basin and filter with hot water. Make the filtrate up to 100 ml, mix and pipette 10 ml into a 25 ml volumetric flask. Add 1 ml of 2 per cent sulphur dioxide solution and titrate with 2 M sodium acetate using a small piece of congo red paper (colour change from blue to pink). Add 2 ml of 0·25 per cent aqueous *o*-phenanthroline solution and make up to the mark. After allowing the red colour to develop overnight, read the optical density in a 4-cm cell at 520 nm against a blank prepared in the same manner as the test solution.

CALIBRATION

Dissolve 0·7024 g ferrous ammonium sulphate, $Fe(NH_4)_2(SO_4)_2.6H_2O$, in water, add two drops of hydrochloric acid and dilute to 1000 ml. Dilute 50 ml to 1000 ml (1 ml = 0·005 mg Fe). Pipette aliquot amounts of this solution (over the range 0–0·05 mg Fe) into a series of 25 ml flasks and add the appropriate reagents as described above. The AOAC methods book recommends the use of α–α′ dipyridyl as complexing agent. In many laboratories the determination of iron is now carried out by atomic absorption spectroscopy. Crosby (1977) has reviewed appropriate methods as applied to a solution prepared from the ash or by digestion of the original material in acid.

Filth tests. Flour should be examined for evidence of insect and rodent infestation. Smith (1983) has noted that cereals and cereal products generally contain a wide range of commonly occurring contaminants, industrial detritus and evidence of infestation. He has described a systematic scheme for examination of foods for foreign materials. Emphasis is placed on the use of microscopic techniques, physical separation and examination under differing conditions of illumination. Earlier texts that are still of value for confirmation of sources of infestation include Busvine (1965), Hinton and Corbet (1963), Hughes (1961), Doring and Beilig (1978) and Doering (1983). Detailed filth test procedures are given in the AOAC methods book.

IMPROVERS AND BLEACHING AGENTS

The additives permitted in flours are summarised in Table 8.3. It should be noted that there are restrictions on the type of flour to which many of these additions may be made; the product most commonly excluded is wholemeal flour, although there are some restrictions to speciality products such as biscuit- and cake-making flour. Many of the additives are used both in the mill and in the bakery; in the latter case commercial formulations containing mixtures of bread improvers are widely used. In such circumstances there is the possibility that the finished product could contain more than the statutory limit of a particular compound. In the case of bromates a Code of Practice has been agreed by the National Association of British and Irish Millers and the Bakery and Allied Trades Association. The Code sets a maximum level of bromate to be added at the mill at 15 mg/kg and at 30 mg/kg in the case of additions through bread improvers at the bakery. The safety of bromate has more recently been questioned and its use is likely soon to be banned. During storage flour increases in 'strength' and there is a slow improvement in baking quality through oxidation effects on the gluten. The addition of improvers renders similar effects in a short time (a few hours). The action of permitted oxidising agents has been discussed by Collins (1985), who illustrates the synergistic effects of some of the commonly used mixtures and relates these to the bread-making processes. Numerous substances have been added to flour and bread in the past but the list has now been curtailed by the advent of the current regulations. Improvers included acid phosphates, alkali persulphates, bromate, ascorbic acid (oxidising effect through dehydroascorbic acid), L-cysteine hydrochloride and azodicarbonamide.

Bleaching agents that have been used include oxides of nitrogen, chlorine, nitrogen trichloride, chlorine dioxide and benzoyl peroxide. The bleaching effect is to reduce the xanthophyll pigments but not the bran particles.

EXAMINATION OF FLOUR FOR IMPROVERS AND BLEACHING AGENTS

For the detection of improvers it is sometimes possible to effect separation of the mineral additives from the flour by shaking with carbon tetrachloride (p. 311).

Monocalcium phosphate (ACP) may be added to flour and would be found in the sediment, which can then be examined for calcium and phosphate (p. 31). The FSC (1974) recommended that ACP should be deleted from the permitted list of improvers in flour, but in view of its value as an anti-rope agent and as a yeast nutrient that its continued use in bread would be permitted. The current regulations reflect this recommendation.

Persulphates of ammonium or potassium were used as improvers for flour at a concentration of about 110 mg/kg (calcium sulphate being used as diluent). These additives are no longer permitted in flour under the Bread and Flour Regulations. Bread made from flour containing ammonium persulphate is likely to contain about 90 mg/kg of ammonium sulphate.

Persulphates can be detected by making 20 g of flour into a paste with 20 ml water and pouring an alcoholic solution of benzidine (1 per cent) over it. Alternatively, the reagent is poured over a wetted slab of compressed flour on a Pekar board (Kent-Jones and Amos, 1967). Particles of persulphate in contact with the benzidine reagent are revealed as deep blue specks. Other reagents such as methylene blue and potassium iodide have also been proposed for the detection of persulphates. A quantitative method for determining ammonium persulphate in flour has been described by Auerbach *et al.* (1949; AACC, 1972). The method relies on the oxidation of reduced fluorescein.

Potassium bromate is usually used at a concentration of 10–15 mg/kg in flour produced at the mill. The improving action does not take place until the flour is made into dough and its use results in the bread containing 10 mg/kg of potassium bromide.

Bromates can be detected by pouring over a slab of compressed and wetted flour a solution containing 0·5 per cent potassium iodide in 2 M hydrochloric acid. The appearance of blackish spots suggest the presence of bromate. An iodometric titration method for the determination of bromate has been described by Armstrong (1952) and is recommended by the AOAC (see also Barrett *et al.*, 1971).

Vitamin C acts similarly and is said to be as effective as bromate when used at a concentration of 20–80 mg/kg (in 80 per cent extraction flour). The regulations allow up to 200 mg/kg in all flours except wholemeal. Vitamin C can be detected by using a 0·1 per cent solution of 2,6-dichlorophenolindophenol in the Pekar board test. White spots appear in a few minutes in the presence of the vitamin. Lookhart *et al.* (1982) have described an HPLC method applicable to flours that is claimed to be sensitive to the levels normally found in bread-making flours.

L-Cysteine. For effective use the material is used at levels within the range 10–80 mg/kg. In combination with bromate and ascorbate it forms the system for Activated Dough Development (ADD), which avoids the use of high speed mixing as is used in the Chorleywood process (Axford and Elton, 1960). The estimation may be carried out by extraction with methanol followed by titration with *o*-hydroxymercuribenzoic acid (Barrett *et al.*, 1971).

Chlorine acts as a bleaching agent and an improver, but is no longer used for bread flour. Flour intended for the manufacture of cakes may be treated with up to 2500 mg/kg. Chlorine combines with the natural pigments, the starch and the protein in flour with a reduction in nutritive value of the gluten. Treatment also causes a lowering of the pH of the flour.

Chlorine dioxide (Dyox) is applied as a gas only at the mill. The product is applied in admixture with air at 4 per cent v/v. The gaseous mixture may contain not more than 1 per cent free chlorine. It acts both as an improver and bleaching agent; the bleaching effect produces a decrease in yellow pigmentation and some destruction of vitamin E, but a contrary increase in the flour grade colour (Collins, 1985). The improving action is modified according to the bread-making process and is affected synergistically by the other oxidising improvers. The level of use is generally 20 mg/kg, well within the statutory limit. Evidence of the use of chlorine may be obtained by solvent extraction of the fat from 30 g of flour. After removal of solvent the residue gives a characteristic flame test with copper wire. Total chlorine in the fat may be estimated by extracting 500 g of flour with 700 ml petroleum ether. After removal of solvent the fat is heated with fusion mixture (sodium and potassium carbonate plus potassium nitrate) at 525 °C for 1 hour. The chloride is determined by titration with silver nitrate.

Benzoyl peroxide may be used in flours up to a maximum of 50 mg/kg. In practice it is used at levels in the region of 30 mg/kg and is commonly incorporated in a commercial mixture that contains 18 per cent benzoyl peroxide, 78 per cent calcium sulphate and 4 per cent magnesium carbonate. It acts in a similar way to chlorine dioxide with similar effects. Sharratt *et al.* (1964) found no significant pathological changes in rodents that had been fed on breadcrumbs prepared from treated flour.

Flour treated with benzoyl peroxide may contain benzoic acid in addition to the original material. Tests for total benzoic acid are normally applied using conversion to salicylic acid after extraction with acetone in the presence of sodium hydroxide (Knight and Kent-Jones, 1953 and AOAC). A TLC procedure sensitive to 0·1 mg/kg has been reported by Nagayama *et al.* (1982). A GLC method has been reported by Ogawa *et al.* (1984).

Azodicarbonamide may be present at levels up to 45 mg/kg. It is normally used in a blend with bromate, for which it is a satisfactory replacement since it is three to five times more effective on a weight/weight basis. Azodicarbonamide is difficult to detect and control analytically in admixture with flour.

Sulphur dioxide. Flour intended for the manufacture of biscuits may contain up to 200 mg/kg sulphur dioxide. It is incorporated as the gas or as sodium bisulphite at a level of about 20 mg/kg (as SO_2).

VITAMINS IN FLOUR

Vitamin B_1, Thiamine, Aneurine. The Bread and Flour Regulations prescribe that all flour shall contain at least 0·24 mg of vitamin B_1 per 100 g. Many of the methods that have been proposed for the determination of thiamine are based on Jansen's thiochrome method. Under standard conditions, the net fluorescence of the thiochrome produced by the oxidation of the vitamin (using alkaline ferricyanide) is directly proportional to its concentration over a given range. The fluorescence can be measured either by reference to a standard quinine solution in a null-point instrument or directly in a spectrofluorimeter.

The following method is that recommended by the AMC of the SPA (1951). A similar procedure is recommended by the AOAC as a general method for foods and by the ICC as standard method No. 117 for application to cereal products.

APPARATUS

Base-exchange tubes. These are to be made of glass, the upper part being not less than 15 cm long and 0·8–1·0 cm in internal diameter and the lower part being narrow-bore tubing of suitable length. The lower end may be fitted with a tap or other method of controlling the rate of flow. A reservoir to contain at least 30 ml may be attached to the upper part.

Oxidation vessels. These may be stoppered measuring cylinders, stoppered bottles, separating funnels or large boiling tubes of about 100 ml capacity. Fluorescent grease must not be used to lubricate taps; glycerin or silicone may be used.

Fluorimeter. For filter instruments the exciting radiation must be within the range 300–400 nm; it is most conveniently obtained at suitable intensity by the use of a high-pressure mercury-vapour lamp, in conjunction with a primary filter with maximum transmission at 365 nm. A secondary filter transmitting mainly light between 400 and 450 nm (maximum transmission 435 nm) is placed between the fluorescent solution and the photo-cell. Chance OX1, 1·5–2·0 mm thick, has been found suitable for the primary filter, and Chance OB2 (blue) 1·5–2·0 mm thick, for the secondary filter. In many laboratories a modern spectrofluorimeter with double monochromators is used.

REAGENTS

All chemicals should be of analytical reagent quality or the purest otherwise obtainable.

Ethanol. Redistilled in all-glass apparatus.

Sodium hydroxide, 15 per cent solution. Dissolve 15 g of sodium hydroxide in water and dilute to 100 ml.

Potassium ferricyanide, 1 per cent solution. Dissolve 1 g of potassium ferricyanide in water and dilute to 100 ml. This reagent is stable for at least a week if kept cool and in the dark, preferably in a brown bottle.

Alkaline potassium ferricyanide solution. Dilute 3 ml of 1 per cent potassium ferricyanide to 100 ml with cool 15 per cent sodium hydroxide solution. Prepare just before use.

Hydrochloric acid, approximately 0·2 M solution. Dilute 17 ml of concentrated hydrochloric acid to 1 litre with water.

Sodium acetate, 2·5 M solution. Dissolve 205 g of anhydrous sodium acetate (CH_3COONa) or 340 g of trihydrate ($CH_3COONa.3H_2O$) in water and dilute to 1 litre.

Isobutyl alcohol (water saturated). Steam distil commercial isobutyl alcohol in an all-glass apparatus. The fluorescence of the distillate should be not more than that of a solution containing 0·1 μg of quinine sulphate per ml in 0·05 M sulphuric acid.

Enzyme solution. Prepare a fresh solution daily from a suitable source of phosphatase [Taka-diastase (diluted with lactose), Mylase P and Clarase have been found suitable]. Suspend, with thorough shaking, 6 g of the enzyme preparation in 2·5 M sodium acetate solution and dilute to 100 ml with additional sodium acetate solution. Each batch of material used as a source of phosphatase should be tested for thiamine by this fluorimetric method and the necessary correction applied.

Potassium chloride, 25 per cent solution. Dissolve 250 g of potassium chloride in distilled water and dilute to 1 litre. The reagent is stable indefinitely.

Acid potassium chloride solution. Dilute 8·5 ml of concentrated hydrochloric acid to 1 litre with 25 per cent potassium chloride solution. The reagent is stable indefinitely.

Base-exchange silicate. This consists of an artificial zeolite in the form of a granular powder of 60–90 mesh size, tested for its suitability for adsorbing and eluting thiamine under the given conditions (Decalso F, or Permutit T, have been found suitable).

Place a convenient quantity (100–500 g) of the base-exchange silicate in a suitable beaker, add sufficient hot 3 per cent acetic acid solution to cover the material and maintain the temperature at about 100 °C for 10–15 min stirring frequently. Allow the mixture to settle and decant the supernatant liquid. Repeat the washing three times with hot 25 per cent potassium chloride solution, wash with boiling water until the last washing gives no reaction for chloride. Dry the material at approximately 100 °C and store in a well-closed container.

Stock thiamine solution (100 μg per ml). Dissolve 50 mg thiamine in sufficient 0·2 M hydrochloric acid to make 500 ml. This solution is stable for several months if stored in a refrigerator.

Standard thiamine solution. Dilute 5 ml of stock thiamine solution, warmed to room temperature, to 100 ml with water. Transfer 10 ml of this dilution to a flask containing 200 ml of approximately 0·05 M sulphuric acid and 12·5 ml of sodium acetate solution and dilute to 250 ml with distilled water. The final concentration of thiamine is 0·2 μg per ml. This solution is stable for at least a week if stored in a refrigerator.

Stock quinine sulphate solution (100 μg per ml). Dissolve 0·025 g of quinine sulphate BP in sufficient 0·05 M sulphuric acid to make 250 ml. This solution is stable for 12 months if stored in a dark brown bottle in a refrigerator.

Quinine standard (1 μg per ml). Dilute 10 ml of stock quinine sulphate solution to 1 litre with 0·05 M sulphuric acid. This solution is stable for 3 months if stored in a brown bottle in a refrigerator. Any solution that has been exposed to ultraviolet light in the fluorimeter should be discarded.

Acetic acid, 3 per cent solution. Dilute 30 ml of acetic acid to 1 litre with distilled water.

Bromocresol green indicator. Triturate 100 mg of bromocresol green with 7·2 ml of 0·05 M sodium hydroxide and dilute with sufficient water, free from carbon dioxide, to make 200 ml.

Nitrogen gas in cylinders. If desired, an air current can be used instead.

EXTRACTION

1. *Sampling.* The flour to be assayed should pass a No. 39 BS sieve or equivalent, and should be well mixed just before withdrawal of the sample, to ensure homogeneity.
2. Accurately weigh or pipette into a large boiling tube a sample (not more than 5 g) estimated to contain not more than 50 μg of thiamine. Add 65 ml of approximately 0·1 M hydrochloric acid or sulphuric acid. Digest the sample for 30 min in a bath of boiling water, with frequent mixing. The liquid must remain at a pH below 4·5 during the digestion. If at the end of the digestion it is not distinctly acid to the bromocresol green indicator, the extract should be discarded and a further quantity of the sample extracted with more concentrated acid.
3. Cool the extract to below 50 °C and adjust the pH to between 4 and 4·5 by addition of 2·5 M sodium acetate solution, using bromocresol green as external indicator. Add 5 ml of freshly prepared enzyme suspension, mix, and incubate at 45–50 °C for 3 hours, or at 37 °C overnight with addition of a drop of sulphur-free toluene.
4. Cool to room temperature, centrifuge the mixture until the supernatant liquid is clear and quantitatively transfer the supernatant liquid to a 100 ml volumetric flask. Wash the residue by centrifuging successively with 10 ml, 10 ml and 5 ml of 0·1 M hydrochloric acid or 0·05 M sulphuric acid. Add the washings to the supernatant liquid and dilute the whole to 100 ml with water. This is the 'original extract'.

PURIFICATION

1. Plug the bottom of an adsorption column with glass wool, which should be lightly packed, and fill the column with 5 g of activated base-exchange silicate suspended in water. Allow the water to drain almost entirely, but leave enough to cover the base-exchange silicate, and pour in 5 ml of 3 per cent acetic acid. Allow to drain as before.
2. Transfer 25 ml of the 'original extract' to the column by means of a pipette. Discard the filtrate that has

percolated through the column. Wash the column with three successive portions, about 10 ml each, of boiling water and discard the washings.
3. After washing the column, pour through 10 ml of almost boiling acid potassium chloride solution from a supply kept hot in boiling water. Collect the eluate in a stoppered 25 ml graduated cylinder. Add a second 10 ml portion when all of the first portion has entered the base-exchange silicate and collect the eluate in the same cylinder. When this second portion has drained through, cool the eluate to room temperature, dilute to 25 ml with acid potassium chloride solution and mix well. This is the 'sample eluate'. For assay in null-point instruments the solution should contain approximately 0·2 μg/ml. If in excess of this concentration adjust the volume to *V* ml with potassium chloride solution. If a spectrofluorimeter is used a calibration curve may be constructed over the range 0·2–1·0 μg/ml. The eluate volume is then less critical.

OXIDATION TO THIOCHROME

In this and all subsequent stages undue exposure of the solution to direct daylight or other source of ultraviolet light must be avoided.

1. Pipette 5 ml of sample eluate into each of two oxidation vessels.
2. Start a stream of nitrogen or air bubbling through the solution in vessel number 1, add 5 ml of alkaline potassium ferricyanide solution and then add 25 ml of water-saturated isobutyl alcohol, the current of nitrogen or air still being continued. Shake (or continue vigorous bubbling) for 90 s.
3. Start a stream of nitrogen or air bubbling through the solution in vessel number 2, then add 5 ml of 15 per cent sodium hydroxide solution, following by 25 ml of water-saturated isobutyl alcohol, then continue as in step (2). This is the 'extract blank'.
4. Repeat steps (1), (2) and (3) with 5 ml of standard thiamine solution in place of the sample eluate. The solution from step (3) is the 'standard blank'.

To avoid changes in experimental conditions the oxidation of all solution used in a given assay should be carried out in immediate succession. Similar precautions must be taken in the reading of their fluorescence.

MEASUREMENT OF FLUORESCENCE

1. After the solutions have stood for a few minutes to allow complete separation, add 1 ml of ethanol to the upper layer in each vessel and stir the upper layer gently until it is clear, taking care to avoid disturbing the aqueous layer.
2. Take off each upper layer into a cuvette and measure its fluorescence against that of the quinine standard, if a null-point fluorimeter is being used, or directly if a spectrofluorimeter is being used. The blank should exhibit only faint fluorescence.

CALCULATION

Calculate the thiamine content of the sample after correction of the readings for the non-thiochrome blank fluorescence by the following formula:

$$T = \frac{I - I_b}{S - S_b} \times \frac{1}{5} \times \frac{25}{V} \times \frac{100}{w}$$

where I = intensity of fluorescence for sample
I_b = intensity of fluorescence for sample blank
S = intensity of fluorescence for standard
S_b = intensity of fluorescence for standard blank
V = eluate volume
w = weight of sample.

'RECOVERY' EXPERIMENT

Repeat the above procedure, including the steps of extraction, purification, conversion to thiochrome, separation of thiochrome solution and measurement of fluorescence, with a 'recovery' experiment, made by adding, to another portion of the sample that is the same weight as that previously taken, a volume of the stock thiamine solution containing an amount of thiamine similar to the amount expected in that weight of sample. The recovery should be not less than 80 per cent for a valid assay.

In the routine analysis of fortified flour the simplified procedure described by the MRC (1943) may be used. In this method thiamine is extracted by treatment of the flour with dilute hydrochloric acid. Enzyme digestion and purification on the silicate column is not generally necessary in such cases. Pippen and Potter (1975) have described modified conditions to minimise losses of thiamine in the purification procedure. Davidek *et al.* (1977) have shown that thiamine and its phosphate derivatives can be separated on a Sephadex gel column. Kirk (1977) has described a semi-automated method based on the thiochrome method and using continuous flow procedures for application to foods. Defibaugh *et al.* (1977) and Gregory and Kirk (1978) have reviewed and evaluated chemical, microbiological and automated methods for estimation of thiamine in foods. Kennedy and McLeary (1981) have proposed the use of thiaminase to hydrolyse the vitamin as a means of correcting for non-thiochrome fluorescence without recourse to column clean-up methods. The use of HPLC techniques has figured prominently in modern method development. Osborne and Voogt (1978) have described a general procedure for application to most foods. In this method the vitamin is extracted by acid hydrolysis and the extract is treated with papain and diastase to hydrolyse protein and to liberate the vitamin. The solution is filtered and the vitamin is separated on a silica gel column using disodium hydrogen phosphate as eluant. After post-column derivatisation to thiochrome the vitamin is detected fluorimetrically.

Fellman *et al.* (1982) used an HPLC method to determine both thiamine and riboflavin in a number of foods, including wheat flour. In this case a conventional extraction procedure is followed by conversion of thiamine to thiochrome. The extract is then purified on a C_{18} Sep-Pak column by washing with phosphate buffer at pH 7 followed by a 5 per cent methanol phosphate buffer and after elution with aqueous methanol the extract is separated on a C_8-Radial-Pak column with 37 per cent methanol phosphate buffer at pH 7. Both vitamins are detected with a fluorescence detector. The detection limits claimed for the vitamins are 0·5 ng and 1·0 ng respectively. When compared with the AOAC procedures for a limited number of samples the method is claimed to give satisfactory performance. Further HPLC methods have been proposed by Mauro and Wetzel (1984) and Wehling and Wetzel (1984). Following a survey of available methods Nicholson *et al.* (1984) concluded that further studies are required to validate all aspects of the application of HPLC technique.

Nicotinic acid (niacin), nicotinamide and riboflavin. The Bread and Flour Regulations prescribe that all flour shall contain at least 1·60 mg of nicotinic acid or nicotinamide per 100 g; the FSC (1960) (Analytical Panel) stated that the determination of nicotinic acid is satisfactory if the recommended method of the SPA (1946b) using *Lactobacillus arabinosus* is employed and 'any chemical method or alternative microbiological method selected should be shown to give results strictly comparable with this method'. The water-soluble B vitamins may be assayed microbiologically. The general principle of the methods depends on the selection of a micro-organism for which the vitamin being assayed is an essential nutrient.

A basal medium is prepared that is adequate for the growth of the selected organism in all respects except for the vitamin (Snell and Strong, 1939). The procedure is that an extract is prepared from the food either by digestion with dilute acid or with enzymes. Varying amounts of extract are then added to a fixed volume of basal medium in a test tube. The contents of the tubes are made up to a fixed volume with water. After capping and sterilising in an autoclave, the tubes are cooled and inoculated with a suspension of the assay organism in sterile saline solution. After incubation the extent of growth of the organism is assessed either by titration of the acid produced or by measurement of the turbidity of the solution after agitation of the tubes and suspension of the colony in the medium. The amount of vitamin in the original sample is assessed by comparison of the growth in the extract tubes with that in a series of standards. Table 8.6 summarises the organisms used and assay ranges for the common B vitamins.

Table 8.6 Microbiological assay of vitamins, organisms and assay range

Vitamin	Organism		Assay range (ng/tube)
Nicotinic acid (niacin)	*Lactobacillus plantarum*	ATCC 8014	0–125 ng
Thiamine	*Lactobacillus viridescens*	ATCC 12706	0–25 ng
B_6	*Saccharomyces carlsbergensis*	ATCC 9080	0–10 ng
B_{12}	*Lactobacillus leichmannii*	ATCC 7830	0–0·2 ng
Pantothenic acid	*Lactobacillus plantarum*	ATCC 8014	0–100 ng
Biotin	*Lactobacillus plantarum*	ATCC 8014	0–0·25 ng
	Streptococcus zymogenes	ATCC 10100	0–10 ng
Riboflavin	*Lactobacillus casei*	ATCC 7469	0–200 ng
	Streptococcus faecalis	ATCC 8043	0–5 ng
Folic acid	*Lactobacillus casei*	ATCC 7469	0–1 ng

Bell (1974) has summarised the available microbiological methods for assay of nicotinic acid and the other water-soluble vitamins. He describes a shortened procedure for routine use. Extreme care must be taken with preparation of sample extracts to maintain the pH of the growth medium and in the addition of the measured volumes of sample extracts to the assay tubes. Since a 16 hour incubation period is used, accurate temperature control is necessary. Preferably a water bath at 37 °C should be used instead of an air oven incubator. Sawyer (1965) has described procedures to be adopted in routine laboratory operations using the conventional titrimetric methods for measurement of growth of the organisms.

A chemical method for determining nicotinic acid in foods has been proposed by Dennis and Rees (1949) which is based on the yellow colour developed when pyridine is treated with cyanogen bromide in the presence of an aromatic amine.

CHEMICAL METHOD FOR NIACIN

The following procedure is applicable to a wide range of products varying from pharmaceuticals to foods and fortified feedingstuffs. Methods vary according to the nature of the material but in general acidic extraction methods are commonly used. The AOAC recommend treatment with calcium hydroxide for the release of bound forms of the vitamin from cereal products.

REAGENTS

Standard niacin: dissolve 50 mg niacin in 25 per cent ethanol, dilute to 500 ml with the ethanol solution and store in a refrigerator in low actinic glass container. For the working standard, dilute 2 ml to 50 ml with water (1 ml = 4 μg niacin).

Sulphanilic acid: mix 10 g sulphanilic acid with 85 ml water, dissolve by addition of 1 ml increments of concentrated ammonia solution. Adjust the pH of the solution to pH 4·5 with 1:1 hydrochloric acid (bromocresol green indicator), dilute to 100 ml and filter.

Cyanogen bromide: weigh 50 g cyanogen bromide by difference into a large flask containing 450 ml warm water (40 °C), transfer to a 500 ml volumetric flask and make up to volume, store in a dark bottle in a refrigerator. Note that all operations with cyanogen bromide must be carried out under a hood, and protective gloves must be worn.

Dilute ammonia solution: dilute 5 ml ammonia to 250 ml with water.

PROCEDURE

Weigh 30 g sample into a large conical flask, add 200 ml 0·5 M sulphuric acid, mix, and autoclave for 30 min in an autoclave set at a 1 bar (15 p.s.i.) pressure (121 °C). Cool, add 10 ml 50 per cent sodium hydroxide solution and adjust to pH 4·5 with further dropwise additions of the sodium hydroxide solution. Bromocresol green may be used as an outside indicator. Make the solution up to 250 ml and filter. If the sample contains cereal the bound nicotinic acid may be released by holding the pH at 13 for 15 to 30 min after the hydrolysis step in the autoclave. Pipette 40 ml filtrate into a 50 ml volumetric flask containing 17 g ammonium sulphate, mix, dilute to the mark and filter. Pipette 40 ml of the 4 μg/ml solution of niacin into a 50 ml volumetric flask containing 17 g ammonium sulphate, dilute to the mark.

COLOUR DEVELOPMENT

Prepare two tubes for each sample and standard, one of each pair is used as a reference blank for the test solution. Pipette 1 ml of the standard into a pair of tubes, into the other pair pipette 1 ml of sample extract. To each blank tube add 5 ml water. Proceed with each of the following steps for each tube to completion of the reaction and measurement of the corresponding absorbance before starting the next reaction. Swirl the tube and add 0·5 ml dilute ammonia solution, swirl, add 2 ml sulphanilic acid, swirl and add 0·5 ml dilute hydrochloric acid (1 plus 5 water), read the absorbance at 450 nm within 2–3 min. For the test extract and the standard, add 0·5 ml ammonia solution, swirl, 5·0 ml cyanogen bromide, swirl, after 30 s add 2·0 ml sulphanilic acid, swirl and immediately add 0·5 ml water. Mix and read the absorbance within 2–3 min. Correct the readings for the appropriate blank and calculate the content of the sample extract solution by comparison with the standard.

Varied reagent strengths can be used to carry out the colour reaction. The automatic version of the above method is also described by the AOAC.

An automated version of this method for application to cereal products has been described by Gross (1975). Egberg (1979) has described collaborative studies in which the microbiological and chemical methods were used. Several laboratories used the automated technique in the studies and the author reported that the results by the three methods were comparable but that the automated chemical procedure was more repeatable within laboratories and more reproducible between laboratories than the other two methods.

Dimethylbarbituric acid has been used as a colorimetric agent in an automated procedure for the estimation of niacin in grain and cereal products by Holtz (1984). The method is claimed to give good accuracy and precision when compared to the established methods.

Procedures using HPLC have also been applied to the determination of niacin; a typical procedure using a reversed phase separation on a C_{18} column with aqueous methanol/phosphate extractant has been described by Tyler and Shrago (1980).

Vitamin B_2, riboflavin. The vitamin occurs widely in foods in free and bound forms as flavin mononucleotide (FMN) or riboflavin-5′-phosphate and as flavin adenine dinucleotide (FAD). Both forms are associated with proteins as co-enzymes. Whilst the vitamin may be determined by fluorimetric assay in cases where the potency is high, the more commonly used procedures, for foods containing natural amounts in the range 0·1–0·5 mg/100 g, include microbiological assay or HPLC methods. Extraction of the vitamin from the sample is by acid hydrolysis, to yield a mixture of free riboflavin and FMN. In the case of microbiological assays this stage is all that is required to prepare the assay solution; in the case of HPLC it is necessary to hydrolyse the extract further with enzyme to convert the FMN to free riboflavin (Lumley and Wiggins, 1981). These authors obtained equivalent results on a number of foods with microbiological assay and an HPLC method using reversed phase column packing and a citrate aqueous methanol eluant. Fellman *et al.* (1982) used a similar procedure to determine riboflavin and thiamine on a single extract prepared from flour and other foods.

Other water-soluble vitamins. Similar considerations apply to the estimation of the remaining water-soluble vitamins as described above for thiamine, niacin and riboflavin. Microbiological assays are still widely used for the estimation of folacin, pyridoxine (B_6), biotin, pantothenic acid, choline and cyanocobalamin (B_{12}). Appraisals of the microbiological methods have been prepared by Christie and Wiggins (1978) and Voigt *et al.* (1979). Walsh *et al.* (1979) compared the microbiological assay of pantothenic acid and a radioimmunoassay method in application to cereals, breads, fruits and vegetables. These authors concluded that the RIA method gives acceptable assays but that cost comparisons favoured the microbiological method. Voigt *et al.* (1979) described computer programs for the evaluation of microbiological assay data. The software described includes test routines for the identification of the presence of inhibitors or stimulants in the sample extracts.

Methods using HPLC have also been developed for these vitamins. A series of routine methods have been described by Osborne and Voogt (1978) and by Wills *et al.* (1977). Vitamin B_6 has received attention from a number of authors and comparative methods have been described by Vanderslice *et al.* (1981) and Gregory (1980). Both these authors reported the use of reversed phase methods and compared their results with microbiological assays. The former authors gave typical results obtained on a range of breakfast cereals. Gregory *et al.* (1982) compared HPLC and microbiological assays of folacin in cereal products. They concluded that both methods gave equivalent results on cereal products. A similar study on infant formula has been reported by Hoppner and Lampi (1982). In this case reverse phase chromatography on octadecyl-silica and elution with aqueous acetonitrile was used.

The methods for vitamins described above have generally wider application than cereal products. In particular they are appropriate in the assay of infant foods based on cereal and milk products. The application to meat-based products may require modifications to the methods, in particular to the extraction methods. The original references should be consulted for appropriate details.

Vitamin E, which is present in wheat germ oil, can be determined by the methods described in Chapter 16.

SELF-RAISING FLOUR

Composition

Most self-raising flours consist essentially of a mixture of flour, sodium bicarbonate (about 1·15 per cent) and an acid phosphate (about 1·5 per cent ACP or ASP). The flour used is usually derived from a soft wheat. The FSC (1960) Report stated, however, that there is no reason why any type of flour should not be used, provided that the appropriate description is used. The Bread and Flour Regulations prescribe a minimum of 0·40 per cent of available carbon dioxide as determined by the method defined (see below). No maximum is prescribed for residual carbon dioxide. Self-raising flour has to comply with the general requirements for flour. The permitted raising agents are detailed in Table 8.3.

Analysis of self-raising flour

The analysis of self-raising flour should include a **microscopical examination,** the determination of **the total and residual carbon dioxide, ash,** the possible contaminants and the qualitative detection of the acid ingredient. In addition, the quantitative estimation of the ingredients may be required (on the same lines as described for baking powder).

Available carbon dioxide. Although the actual method for the final determination of the carbon dioxide is left to the individual analyst, the Bread and Flour Regulations define available carbon dioxide and prescribe the methods of analysis for estimation of total carbon dioxide and residual carbon dioxide.

The available carbon dioxide is the difference between the total weight of carbon dioxide in the flour and the weight of residual carbon dioxide in the flour.

Total carbon dioxide is determined by ascertaining the weight evolved when the self-raising flour is treated with excess of dilute sulphuric acid, the evolution being completed either by boiling for 5 min or by means of reduced pressure.

Residual carbon dioxide is determined by taking not less than 5 g of the self-raising flour, which is mixed to a smooth paste with distilled water, and mixed with a further quantity of distilled water of not less than 20 times the weight of the flour taken. The mixture is heated in a boiling-water bath for 30 min and vigorously stirred for the first 5 min and for approximately half a minute at intervals of approximately 5 min. The mixture is then boiled for 3 min, and vigorously stirred during the period, and then transferred to the apparatus for determination of carbon dioxide through which air that contains no carbon dioxide is passed for not less than 10 min. The residual carbon dioxide is the weight evolved when the self-raising flour is further treated with an excess of dilute sulphuric acid, the evolution being completed either by boiling for 5 min or by means of reduced pressure.

The carbon dioxide can be determined gravimetrically, volumetrically (by back-titration after absorption in barium hydroxide) or gasometrically (Chittick method), i.e. by similar methods to those employed for baking powder (p. 339), but approximately 20 g of sample should be taken for the test. In view of the relatively large bulk of flour used, the addition of 15–20 glass beads (4–6 mm diameter) is useful for shaking the flour with the acid and thereby ensuring completion of the reaction.

Sodium bicarbonate. In view of the relatively low bicarbonate content of self-raising flour, the presence of chalk makes an appreciable contribution to the total carbon dioxide as determined in the usual way. Stephenson and Hartley (1955) have shown, however, that if a dilute solution of acid sodium pyrophosphate (ASP) is added to a self-raising flour containing chalk the carbon dioxide is released from the bicarbonate but not from the calcium carbonate. In the method, 17 g of the self-raising flour is placed in the flask of a Chittick gasometric apparatus (p. 341) and the solution in the gas burette is brought to the zero mark. Then 45 ml of fresh ASP solution (2·5 per cent w/v) is added to the flour and the mixture is shaken for 2 min and allowed to stand for 5 min before equalising the pressure in the gas burette using the levelling bulb. Then the volume of gas produced is read off and corrected for temperature and pressure. After the reagent blank has been determined for the method, the carbon dioxide produced by the bicarbonate can be calculated. The 'total' carbon dioxide due to both the bicarbonate and the chalk can be determined in the apparatus in the usual way using 17 g of flour and 45 ml of dilute sulphuric acid (1 + 5). As the increase in the volume of gas produced by the chalk is relatively small, both determinations should be performed a number of times in order to obtain a reasonably accurate difference figure. For the modified calculations that are necessary for these methods the original paper should be consulted.

The addition of chalk is not compulsory for self-raising flour with a calcium content of at least 0·2 per cent such as would be present if ACP is used as the acid ingredient.

Mineral additions. A useful method for detecting or estimating the added mineral matter is to shake 50 g of the flour with 200 ml of *dry* carbon tetrachloride in a separator having a short delivery tube and fitted with a tap containing a hole of wide bore. On allowing the mixture to stand, the flour floats on the surface and the added bicarbonate and acid ingredients fall to the bottom and may be tapped off into a basin. After evaporating off the liquid, the residue can be examined qualitatively for added ingredients by the usual tests, e.g. bicarbonate, phosphate, tartrate and calcium. This method enables such ingredients to be examined in their original state, i.e. without any interaction of the ingredients (unlike the ash). For the estimation of the additions, the residue in the separator should be given two further shakings with carbon tetrachloride (see Chapter 9).

BREAD

Composition

Bread is composed of dough (from wheat or rye flour), yeast, water and other ingredients, which has been fermented and subsequently baked. Two methods are commonly employed for the mass production of bread. The first is the continuous Chorleywood process (Axford *et al.*, 1963). This involves mechanical working of the dough, fast chemical oxidation with ascorbic acid, the addition of fat and extra water to the dough, omitting the pre-ferment, but raising the level of yeast. The addition of L-cysteine hydrochloride to the list of permitted improvers resulted in the introduction of the second manufacturing method, namely the activated dough development process, which confers some of the advantages of the Chorleywood process without the necessity of employing high speed dough mixers (see Collins, 1985). Table 8.7 shows the typical composition of a range of breads (Knight *et al.*, 1973; Paul and Southgate, 1978). Acidity changes from about 0·1–0·5 per cent (as lactic acid) as the bread changes from fresh to stale. Surface active agents such as GMS and polyoxyethylene stearate are added as anti-staling agents. The need for these and other

Table 8.7 Composition of various breads

Type	Water (%)	Protein (N × 5·7) (%)	Fat (%)	Sugars (%)	Starch and dextrins (%)	Dietary fibre (%)	Energy (kJ/100 g)
Wholemeal	40·0	8·6	2·7	2·1	39·7	8·5	918
Brown	39·0	8·9	2·2	1·8	42·9	5·1	948
Hovis	40·0	9·7	2·2	2·4	42·7	4·6	968
White	39·0	8·0	1·7	1·8	47·9	2·7	991
toasted	24·0	9·6	1·7	2·1	62·8		1265
dry crumb	9·7	11·6	1·9	2·6	74·9	—	1508
Currant	37·7	6·4	3·4	13·0	38·8	—	1063
Malt	39·0	8·3	3·3	18·6	30·8	—	1054
Soda	34·2	8·0	2·3	3·0	53·3	2–3	1122
Rolls, brown crusty	28·6	11·5	3·2	2·1	55·1	—	1229
brown, soft	31·0	11·7	6·4	1·9	46·0	—	1194
white crusty	28·8	11·6	3·2	2·1	55·1	—	1231
white soft	28·8	9·8	7·3	1·9	51·7	—	1291
starch reduced	8·5	44·0	4·1	1·6	44·1	—	1631

additives in modern bread manufacture is discussed in relation to the changes in quality of wheat available for flour milling (Anon, 1978).

Many emulsifiers or dough conditioners are added in the form of general bread improvers. These may also contain yeast, oxidising agents, soya flour, sugars and fats. Glyceryl monostearate is long established in bread as a crumb softener and anti-staling agent but has been largely replaced by DATEM (diacetyl tartaric esters of monoglycerides) esters. Stearyl lactylates are also used, giving slightly better loaf volume and crumb softness effects than DATEM esters.

Although micro-organisms seldom cause trouble with flour, certain bacteria and moulds are liable to grow on bread. *Bacillus subtilis* may survive the baking process and, particularly if the bread is cooled slowly or wrapped before cooling, is liable to cause **rope** (cf. Amos and Kent-Jones, 1931). Sticky yellow–brown spots appear in the crumb and an odour resembling decaying fruit is apparent. In extreme instances the middle of the loaf becomes brown and, on breaking, the crumb can be pulled out into threads (hence the name). **Bleeding bread** is the name given to a condition due to *Serratia marcescens* (*B. prodigiosus*), which causes a bright red colour to appear in the crumb. Another form of infection is due to the 'red bread mould', *Monilia sitophila*, which gives rise to a salmon pink colour. From time to time outbreaks of this occur in bakeries, but most often the fungus attacks wrapped, particularly sliced, loaves after the bread has been despatched. In general mould growth is encouraged by storage in warm humid conditions.

Legislation

The main statutory standards for bread are prescribed in The Bread and Flour Regulations 1984 SI No. 1304. The ingredients that may be used are specified for each type of bread. White bread must be prepared from dough made from wheat flour, yeast and water that has been fermented and baked. Additional ingredients from a restricted list may be present and in some instances the amounts permitted for use without affecting the name of the food are limited in proportion to the weight of flour used to make the dough. Permitted ingredients are: salt, edible oils and fats, milk and milk products, sugar, prepared wheat gluten, wheat germ, poppy seeds, sesame seeds, caraway seeds, cracked wheat, cracked or kibbled malted wheat, flaked malted wheat, kibbled malted rye, cracked or kibbled malted barley, acetic acid, vinegar, malt extract, malt flour and liquid or dried egg. Non-wheat flours from rice and soya bean may be present to a maximum combined level of 2 per cent of the weight of flour. The same maximum level is also allied to any or all of oatmeal, oatflakes and cracked oat grain. Starch other than modified starch may be added to the extent of 225 mg/kg of flour.

Brown bread must contain a minimum of 0·6 per cent fibre (on the dry weight of the bread). All the ingredients allowed for white bread may be used but additionally the limit for soya flour is 5 per cent. Wheat germ bread must contain a minimum of 10 per cent processed wheat germ and the additives allowed in brown bread. Wholemeal bread must be prepared from wholemeal flour without addition of other cereal flours. In addition, the ingredients used in white bread are permitted except that the name of the food must indicate the presence of milk and milk products, liquid or dried egg, wheat germ, rice flour or oat products where these are used.

The ingredients of soda bread are limited to those for white bread with the addition of sodium bicarbonate. Provision is made for the sale of partially baked bread provided that the name indicates the need for further cooking. The regulations also contain a schedule of additives that are specifically permitted in bread in addition to those allowed in bread and flour. Claims on labels are subject to the general provisions of the Food Labelling Regulations.

BUN LOAVES, CHOLLAS AND BUTTERCREAM

Following negotiations with the Baking Industry Committee on Flour Confectionery, LAJAC Code of Practice No. 7 on bun loaves, chollas and buttercream was published in 1971. Although under the reorganisation of local government LAJAC was disbanded the Code is still observed by industry and the enforcement authorities. It states that bun loaf should contain (figures expressed as percentages of the flour on the w/w basis)

1. at least 5 per cent of liquid whole egg (or equivalent of dried egg) *or* added fat *or* both of these ingredients, *and*
2. at least 10 per cent in all of either or both of these ingredients, together with sugar.

Chollas possessing the same degree of enrichment should be regarded as Bun Loaves.

Buttercream is a filling or decoration prepared from butter and sugar, with or without other ingredients. It must contain at least 22·5 per cent butterfat and no other fat, added as such. It must be presented as one word, i.e. not as 'butter cream' nor hyphenated as 'butter-cream'.

ANALYSIS OF BREAD

In general, the analysis of bread should include the determination of moisture, fat (by acid hydrolysis, p. 291), protein, fibre, ash and salt and a microscopical examination. Carbohydrate is generally estimated by difference. Examination for permitted additives and antioxidants (carry over from fats) may also be necessary. It is preferable to carry out most of the analyses on an air dried sample. Then by taking into account the weight loss, the results can be calculated back to the original material.

Moisture. The moisture content of bread should preferably be determined by the AOAC method (FSC, 1974):

Weigh the loaf on receipt (A) on a semi-accurate balance to the nearest 0·1 g. Cut the loaf into 2–4 mm slices, spread on paper and allow to dry in a warm room overnight so that the bread is crisp and brittle. The sample should be in equilibrium with the atmosphere so that the moisture content remains constant during grinding. Then return *quantitatively* the air-dried bread to the balance and re-weigh (B). Grind the sample to pass through a No. 20 sieve, mix and transfer to an airtight container. Determine the total solids by drying 2 g at 130 °C for one hour (C per cent) or at 98–100 °C for 5 hours in a vacuum oven at a pressure less than 25 mmHg.

$$\text{Per cent moisture in original bread} = 100 - \frac{BC}{A}$$

Knight *et al.* (1973) found a mean moisture content of whole Chorleywood Baking Process (CBP) loaves including crust of 39·0 per cent compared with 38·5 per cent for those produced by the conventional process. The FSC (1974) noted that the moisture content of individual loaves varies widely and figures in excess of 40 per cent have been reported. Many countries apply a maximum limit of 38 per cent water to the whole loaf. The Australian limit of 45 per cent is based on 5 g samples taken from any part of the loaf (in practice from the centre). Partly baked loaves (as sold) contain about 45 per cent moisture.

The use of near infra-red reflectance spectroscopy for estimation of moisture, protein and fat has been described by Osborne *et al.* (1984b) and by Duvenage (1986). The former authors carried out analyses on slices of bread cut directly from the loaf and on air dried samples. Coefficients of variation for moisture contents were indistinguishable for both methods of sample preparation. The latter author preferred the use of air dried samples for routine measurements.

Ash. The ash of white bread should not normally exceed 2 per cent (brown bread slightly higher) and the acid-insoluble ash should be no more than 0·2 per cent. The added salt makes up approximately half of the total ash. The method described for flour may be used.

Protein. Ordinary bread contains about 13–13·5 per cent protein (N × 5·7) in the dry matter. Bread described as 'high protein' or 'protein' must contain not less than 22 per cent protein on the dry matter. The Kjeldahl procedure described for flour may be used for the analysis.

Crude fibre. The general method described for flour may be used. Similar considerations apply in respect of dietary fibre as are discussed earlier.

Improvers and bleachers undergo changes during the baking process but in some instances the derived product can be detected. For instance, the presence of traces of benzoic acid in the bread usually indicates that the flour used had been treated with benzoyl peroxide (Knight and Kent-Jones, 1953). During baking, persulphates are converted to the sulphate and bromates to the corresponding bromide. The presence of these anions does not necessarily prove that improvers have been used, however, as they may be derived from the wheat or other ingredients. The losses and changes which occur during baking make the accurate estimation of improvers and bleachers a matter of some difficulty.

An ion chromatography method for measurement of residual bromate has been described by Watanabe *et al.* (1982). The method is claimed to have a detection limit of 1 mg/kg. A GLC procedure for residual bromide sensitive to 0·01 mg/kg has also been described by Oyamada *et al.* (1983).

Emulsifiers and stabilisers. Only certain emulsifiers and stabilisers are permitted to be present in bread (see Schedule 3 of the Regulations). Thin-layer chromatography may be used for identifying the emulsifiers present. Extract 10 g dry crumb with chloroform/methanol (2:1), evaporate and dissolve the residue in diethyl ether. Use silica gel plates with two solvent systems to identify the emulsifiers in the ether solution [solvent I (benzene/ether/ethanol/acetic acid, 50:40:2:0·2 v/v), solvent II (chloroform/methanol/water, 65:25:4 v/v)]. Develop the first chromatogram in solvent I; develop the second chromatogram in solvent II to 10 cm, dry then develop in solvent I to 15 cm. Spray both plates with 25 per cent v/v sulphuric acid and heat to 200 °C. Compare with standard emulsifiers. Martin (1981) used TLC to separate 19 emulsifiers, whilst Slack and Porter (1983) reported a colorimetric procedure for the estimation of the permitted emulsifiers based on complexation with copper and measurement at 260 nm. A gravimetric method for estimation of diglycerides, based on precipitation with phosphomolybdic acid and barium, has been proposed by Lawrence *et al.* (1981).

Williams (1964) has described a method for the detection and approximate determination of stearyl tartrate in bread, which is based on the production of stearyl alcohol on saponification. In this method, 50 g of the air-dried ground sample is extracted with solvent after acid hydrolysis. The lipid material is then saponified, the isolated unsaponifiable fraction is dissolved in 1 ml chloroform and 20 μl is applied to a TLC silicic acid plate, together with standards containing 1–20 μl of a 0·7 per cent chloroformic solution of stearyl alcohol. Separation is effected with a 9 + 1 mixture of ethyl ether and light petroleum (b.p. 40–60 °C) and the plate is sprayed with 0·2 per cent dichlorofluoroscein in alcohol and examined in ultraviolet light. Cetyl alcohol, however, behaves in a similar manner. The use of GLC methods has been reviewed by Conacher and Page (1979), Haken (1979) and Tsuda *et al.* (1984). Martin *et al.* (1980) proposed a systematic procedure for separation and estimation of emulsifiers based on column chromatography, TLC and GLC.

Propionic acid. The methods described in Chapter 4 are appropriate. Tonogai *et al.* (1983) described a GLC procedure applied to an acidified ethyl acetate extract of bread. Yabe *et al.* (1983) extracted with methanol and formed a phenylacyl ester of propionic acid, which is then separated by HPLC; 99 per cent recoveries are reported at the level of 100 mg/kg.

Phosphate can be determined on an acid solution prepared from the ash by the molybdenum blue method (Chapter 2).

Iron can be determined as described for flour (p. 299).

B vitamins. The thiochrome method for flour is also suitable for the determination of vitamin B_1 in bread. The treatment with phosphatase during the preparation of the extract is necessary with bread in order to free combined thiamine—purification of the extract by the silicate column procedure is also necessary. The loss of vitamin B_1 during bread making is about 15 per cent. Microbiological methods and HPLC are appropriate for the determination of the other B vitamins (p. 306).

Milk. To justify the description 'milk bread', it has been regarded in the past that at least 6 per cent whole milk solids should be present in the dry matter. The data in Table 8.8 were presented by the FSC (1974) and show that milk bread contains slightly greater amounts of protein, fat and calcium than ordinary bread.

Table 8.8 Composition of white bread with and without the addition of whole milk or milk powder at different levels

	Bread containing:					
	No addition	Whole milk solids			Skim milk solids	
Nutrient	of milk	5·5%	4·2%	3·0%	4·2%	2·1%
Protein (g)	7·8	8·8	8·5	8.5	9·2	8·5
Fat* (g)	0·7	2·1	1·8	1·4	0·7	0·7
Carbohydrate (g)	54·7	53·6	54·0	54·0	54·3	54·3
Calcium (mg)	102·0	145·0	134·0	127·0	148·0	127·0
Iron (mg)	1·45	1·42	1·43	1·44	1·44	1·45
Thiamine (mg)	0·162	0·169	0·166	0·162	0·173	0·166
Riboflavin (mg)	0·025	0·085	0·071	0·056	0·088	0·056
Nicotinic acid (mg)	1·49	1·45	1·46	1·47	1·47	1·48

* Each figure shows the amount present per 100 g of bread. No allowance is made for the contribution of any other fats that might be added to the bread.

The fact that milk contains lactose, but the other constituents do not, can be used for the tentative detection and determination of milk in bread. Chromatographic methods have been described by Dickes (1966) and by Hunt *et al.* (1977) (see Chapter 6). Enzyme methods suitable for determination of lactose in the presence of other sugars are also described in Chapter 6.

Orotic acid. A colorimetric method for estimating the non-fatty milk solids in milk bread based on the orotic acid (2, 6-dihydroxypyrimidine-4-carboxylic acid) content has been described by Archer (1973). The mean orotic acid content of non-fatty milk solids is 62·5 mg/100 g (range 48·0–74·5 mg/100 g).

REAGENTS

Clearing agent I: zinc sulphate, 23 per cent m/v solution.

Clearing agent II: potassium hexacyanoferrate (II), 15 per cent m/v solution.

p-*Dimethylaminobenzaldehyde solution (DAB):* 3 per cent m/v in propan-l-ol.

Standard orotic acid solutions. Dissolve 50 mg orotic acid in a mixture of 1 ml of 0·88 ammonia and 10 ml water and dilute to 500 ml with water. Dilute to 100 ml with water and dilute 2·5, 5, 10 and 15 ml of this solution to 50 ml to produce solutions containing 2·5, 5, 10 and 15 μg orotic acid per 5 ml.

PROCEDURE

Determine the total solids of the milk bread sample after air-drying as previously described. Weigh 5 g of dried sample into the beaker of the homogeniser, add 100 ml water and mix at the maximum speed for 1 min. Filter the supernatant liquor through a 15 cm Whatman No. 541 paper, rejecting the first 10 ml; 5 ml is required for the determination.

Into a series of glass stoppered test-tubes add by pipette 5 ml of test solution (containing 2–15 μg orotic acid), 5 ml of each of the standard orotic acid solutions, and 5 ml of water to act as the blank. Add to each tube 1·5 ml of saturated bromine water and allow the mixture to stand at room temperature for not more than 5 min. As the addition of bromine water is made to the series of tubes the time will vary slightly between each, the time of reaction is not critical provided it is between 1 and 5 min. Add 2 ml 10 per cent ascorbic acid solution to each tube and place the tubes in a water bath at 40 °C for 5 min. Add to each 3 ml DAB and return to the bath for 10 min. Cool to room temperature, add to each tube 4·0 ml *n*-butyl acetate and shake vigorously for 15 s. Transfer the upper separated layers to dry test-tubes containing 1 g anhydrous sodium sulphate, mix gently, add another gram of anhydrous sodium sulphate, mix gently and allow to separate. Transfer the clear butyl acetate layer to 1 cm cell and measure the optical density at 461–462 nm against the blank. By calculation or calibration graph calculate the weight of orotic acid per 5 ml of sample extract, and hence the amount in the dry sample. For converting to milk assume that skim milk powder contains 62·5 mg orotic acid per 100 g.

The proportion of butterfat in bread can be assessed from the semi-micro Reichert process or preferably by gas chromatography (Phillips and Sanders, 1968; Dickes, 1979) when applied to the extracted fat (see p. 587).

BISCUITS, CAKE AND FLOUR CONFECTIONERY

Composition

According to the FSC (1964) Report on Food Labelling the term flour confectionery means a product that is suitable for consumption without further preparation, the characteristic ingredient of which is ground cereal whether or not containing carbohydrate sweetening matter, etc., but does not include biscuits or bread.

Many authorities consider that the presence of butter in cake should not be claimed unless at least 5 per cent butterfat is present in the crumb. Also if the unqualified designation 'Butter Cake' is used or claims are made such as 'made with butter' all the added fat should consist of butterfat (see FSC, 1974). In the Second Report on Labelling the FSC (1979) drew attention to the exemption from the requirement to list on the label the ingredients of bread, flour, flour confectionery and biscuits. The committee recommended that the exemption for these products be withdrawn. The Food Labelling Regulations (1984) took account of this recommendation.

Analysis

The analysis of flour confectionery should include examination for preservatives, antioxidants, emulsifiers and added colour. The general quality can be assessed by determining the moisture (dry at 100 °C), fat and sugar (see below), ash, salt, protein and egg (see p. 352). Chocolate products should be examined for cocoa. An examination for mould and extraneous matter should always be performed.

The following are methods especially adapted for the analysis of flour confectionery:

Fat. Weigh 2 g minced sample into a Mojonnier or other suitable extraction tube, add 2 ml alcohol and 10 ml diluted hydrochloric acid (25 conc.: 11 water by vol.) and maintain at 65 °C for 30 min. Cool, add 10 ml alcohol and shake with 25 ml diethyl ether. Then add 25 ml light petroleum, shake, allow to separate, decant the upper layer into a weighed flask. Repeat the extractions at least twice, remove the solvent and dry at 100 °C, cool and weigh. If the presence of butter is claimed in cake it should contain at least 5 per cent butterfat in the crumb (see above). To check for this, extract a quantity of fat directly from the cake and examine by the semi-micro Reichert and Polenske methods or by GLC.

Sugars. Blend 5 g minced sample mechanically with 150 ml water, transfer to a 250 ml volumetric flask with water, add dialysed iron or zinc ferrocyanide clearing agents, make to the mark. Shake, filter and invert 50 ml filtrate in a 100 ml volumetric flask at 70 °C. Determine the sugars before and after inversion as invert sugar and sucrose respectively using Lane and Eynon's method, or by a suitable chromatographic or enzyme procedure (Chapter 6) and allow for insoluble matter by multiplying by 0·99. Calculate the soluble solids from the following formula:

$$\text{Per cent soluble solids (in aqueous phase)} = \frac{\text{per cent total sugars}}{\text{per cent total sugars} + \text{per cent moisture}} \times 100.$$

Mould is liable to grow on cakes which have less than about 63 per cent soluble solids in the aqueous phase.

Cocoa contents. According to LAJAC Code of Practice No. 1, where the word 'chocolate' is used in the description of flour confectionery, the product shall contain at least 3 per cent of dry non-fat cocoa solids in the moist crumb. This requirement does not apply if the word 'chocolate' is immediatley succeeded by 'flavoured' or to chocolate covered or decorated baked products, the crumb of which contains no chocolate or chocolate colouring, e.g. eclairs. The cocoa content can be assessed by the method of Moir and Hinks (1935) in which the 'total alkaloids' are extracted with chloroform and determined by the Kjeldahl procedure. Factors of 3·36 and 100/3·15 are used for conversion of nitrogen into alkaloid, and alkaloid into fat-free cocoa-matter respectively.

Chapman *et al.* (1963) have described a rapid method, in which the total alkaloids are determined spectrophotometrically after adsorption on Fuller's earth. The method is applicable to cocoa and most chocolate products.

REAGENTS

Zinc acetate solution, 21·9 g zinc acetate + 3·0 ml acetic acid in 100 ml.
Potassium ferrocyanide solution, 10·6 per cent w/v in water.
Fuller's earth. Special grade for adsorption (BDH Ltd).
0·1 M *sodium hydroxide solution.*
Molar hydrochloric acid.

METHOD

Weigh out 1 g of cake-mix or air-dried powdered cake (or 0·1 g of cocoa) into a 100 ml beaker. Add 25 ml distilled water in small portions, mix to a smooth paste. Bring to the boil with stirring, taking precautions against frothing, place on a water bath for 15 min. Transfer to a 50 ml volumetric flask with water. Cool. Pipette 1 ml of zinc acetate solution into the flask, mix by rotation, add 1 ml of potassium ferrocyanide solution by rotation mix, avoid inclusion of air bubbles. Dilute to 50 ml, mix. Transfer to a centrifuge tube, centrifuge at 2500 r.p.m. for 5 min and decant through a No. 4 filter into a clean dry beaker. Pipette 30 ml of filtrate into a dry centrifuge tube containing 0·5 g of Fuller's earth, mix thoroughly with a glass rod, finally rinse the rod with a few millilitres of water. Centrifuge at 2500 r.p.m. for 5 min or until a clear supernatant liquid is obtained, decant the liquid to waste without disturbing the residue. Add approximately 15 ml of water to the residue, mix with a glass rod. Rinse and centrifuge as before. Discard the clear liquid, avoid loss of solid. Transfer the residue of Fuller's earth to a dry 50 ml volumetric flask with 0·1 M sodium hydroxide, dilute to the mark with 0·1 M sodium hydroxide. Shake vigorously for 2 min, transfer to a centrifuge tube and centrifuge at 2500 r.p.m. for 10 min or until clear. Decant the clear liquid into a dry 100 ml beaker, avoid any inclusion of Fuller's earth. Pipette a 40 ml aliquot into a 100 ml volumetric flask, pipette 14 ml of M hydrochloric acid, mix, dilute to 100 ml with water and mix well. Measure the ultraviolet absorption between 270 and 274 nm at 1 nm intervals against distilled water and also that of a blank containing 40 ml 0·1 M sodium hydroxide + 14 ml M hydrochloric acid made up to 100 ml. Theobromine has a maximum absorption $E^{1\%}_{1\text{cm}}$ of 548 at 272 nm and the sample should have a maximum at 272 ± 1 nm.

$$\text{Total alkaloids w/w} = \frac{E\,(272\text{ nm, 1 cm})\,(\text{sample} - \text{blank}) \times 50 \times 50 \times 100}{548 \times 30 \times \text{aliquot taken} \times \text{wt taken}}$$

$$= \frac{E\,(272\text{ nm, 1 cm})\,(\text{sample} - \text{blank}) \times 0{\cdot}38 \text{ for 40 ml aliquot}}{\text{Weight taken}}$$

Dry fat-free cocoa = total alkaloids × 33·3.

The factor 33·3 is derived from an overall range 30·8–38·7. Other instrumental techniques, including HPLC procedures for estimation of theobromine in cocoa products are described in Chapter 10.

Biscuits

The manufacture of biscuits has been discussed by Wade (1965), Wade and Watkin (1966) and Stafford (1968). The common basic ingredients of biscuits are flour, fat, sugar and syrups, aerating chemicals and milk or water.

Most types are prepared from one of the following classes of dough: soft sweet (e.g. 'Digestive'), hard sweet (e.g. 'Rich Tea') and laminated (e.g. 'Cream Crackers'). Biscuits were classified in former MOF Code of Practice No. 20, which is given in full in the FSC (1974) Report. Proposals have also been the subject of discussion by LAJAC and the EC.

Biscuits can be examined for moisture, fat, nitrogen, sugars, ash and salt as described for flour confectionery. The analytical data given in Table 8.9 have been reported by Paul and Southgate (1978) and by Pearson and Goussous (1968). Methods using NIR have been proposed for quality control of biscuit mixes and doughs (Osborne *et al.*, 1984a).

Table 8.9 Composition of various biscuits

Type	Moisture (%)	Fat* (%)	Protein (N × 5·7) (%)	Sugars (%)	Ash (%)	Salt (%)	Energy (kJ/100 g)
Cream Crackers	4·3	16·3	9·6	Trace	—	1·5	1857
'Digestive'	4·5	20·5	9·6	16·4	—	1·1	1981
Plain	5·2	13·2	7·4	15·8	—	—	1877
Sweet	0·7	30·7	5·5	25·0	—	—	2282
Water	4·5	12·5	10·7	2·3	—	—	1859
Shortcake	2·2	23·5	5·1	20·0	1·0	0·4	2124
Crispbread rye	6·4	2·1	9·4	3·2	—	0·5	1367
wheat starch reduced	4·9	7·6	45·3	7·4	—	1·5	1642
Ginger nuts	3·4	15·2	5·6	35·8	—	0·8	1923
Matzo	6·7	1·9	10·5	4·2	—	0·05	1634
Oatcake	5·5	18·3	10·0	3·1	—	3·1	1855
Sandwich	2·6	25·9	5·0	30·2	—	0·6	2151
Semi Sweet	2·5	16·6	6·7	22·3	—	1·0	1925
Short sweet	2·6	23·4	6·2	24·1	—	0·9	1966
Shortbread	5·0	26·0†	6·2	17·2	—	0·7	2115
Wafers–filled	2·3	29·9	4·7	44·7	—	0·2	2242

NB The remainder consists mainly of starch.
* Methods of determination used vary between different workers.
† Mostly butterfat.

PASTA PRODUCTS

Macaroni and similar products are produced essentially by kneading a mixture of semolina derived from durum wheat with water into a dough. The mass is then extruded through dies under pressure or rolled, and cut into various shapes and finally dried to a moisture content of about 13 per cent. Also in 1949 the former MOF published the following descriptions in a Code of Practice:

1.	Macaroni	Tubular shape.
2.	Spaghetti	Solid rods not less than 1/16″ diameter
3.	Vermicelli	Flat or rod shape not more than 1/16″ diameter or width.
4.	Plain Noodles	Flat ribbon shape, containing no egg.
5.	Egg Noodles	Flat ribbon shape, containing egg.
6.	Farfals	Ground, granulated or shredded alimentary paste.
7.	Macaroni elbows	Tubular elbow shape approximately between ¾″ and 1½″ overall length.
8.	Spaghetti Pearls	Spaghetti, short-cut, to not more than ⅛″
9.	Macaroni Rings	Macaroni, short-cut to not more than ⅛″
10.	Macaroni Pearls	Macaroni, short-cut to not more than ⅛″ and not more than ⅛″ diameter.
11.	Macaroni Products, rice shaped	Extruded products in the shape of rice.
12.	'Fancy Shapes'	Should be described under their normal classification, e.g. 'Shells', 'Letters', etc.

Note. Under the descriptions 1, 2, 3 and 4, short lengths (of one inch or so) should be qualified by the word 'cut', e.g. 'Cut Macaroni'.

In a USFDA Standard of Identity, egg macaroni and noodles must contain at least 5·5 per cent solids of egg or egg yolk (as percentage of TS). In the UK it has been suggested that egg noodles should contain at least 3 per cent of egg solids.

A method for determination of egg content of pastas based on estimation of cholesterol content using GLC has been proposed by Fostel (1984) and using HPLC by Hurst *et al.* (1985). Electrophoretic methods to examine the protein fractions have been proposed for the authentication of durum wheat pasta; Burgdon *et al.* (1985) studied the adulteration with soft wheat flour, whilst Joudrier *et al.* (1981) used the technique to demonstrate addition of barley flour at levels down to 2 per cent. Salmon (1985) has made a general study of the applications of electrophoresis techniques to wheat and wheat products. She has demonstrated that durum wheat flour can be distinguished more readily by examination of the polyphenol oxidase enzymes. Pasta samples are extracted with 6 per cent urea solution and the clarified extract is loaded on to an alkaline tris/hydrochloric acid buffered gel. The enzyme fractions are located with buffered catechol solution. The method is claimed to detect additions of non-durum wheat at a level of approximately 5 per cent.

Canned spaghetti in tomato sauce, according to Paul and Southgate (1978) contained (in 6 cans of one brand examined) 83·1 per cent water, 1·7 per cent protein, 3·4 per cent sugar, 8·8 per cent starch and dextrins (as glucose) and 0·7 per cent fat.

OTHER CEREAL PRODUCTS

RYE FLOUR

An EC draft regulation provides for four types of rye flour that are graded according to ash content. Rye can be detected by microscopical examination, which reveals the large round starch grains in addition to smaller sizes which resemble those of wheat (also some rye grains show a stellate hilum). Rye flour is occasionally contaminated with **ergot**, a parasite that is the sclerotium of *Claviceps purpurea* and originates in the ovary of *Secale cereale.* Ergot is slightly curved, hard and dark purple in colour. Its microscopical characters are somewhat ill-defined, but chemical tests have been devised that are based on the fact that ergot tends to give a coloured extract. In employing such tests it is essential that comparisons are made with the uncontaminated material. The following method is that recommended by the AACC:

Shake 10 g flour with 20–30 ml ether and 1·2 ml of sulphuric acid (5 per cent) and allow to stand for 6 hours in a closed flask. Filter and wash the residue with ether until 40 ml of filtrate is obtained. Add 1·8 ml saturated sodium bicarbonate. If ergot is present, the lower layer becomes violet. Ergot can also be detected by extracting a large flour sample with warm alcohol and examining the extract spectrophotometrically after concentrating to about 5 ml. In the presence of ergot the solution is reddish brown and characteristic absorption bands appear at 538 and 499 nm (and faintly at 467 nm).

PEARL BARLEY

Pearl barley consists of the whole barley grain with the husk removed. It is sometimes faced with talc, which can be estimated by Kržižan's process (p. 323). Any dusty portions should be examined microscopically for meal mites, which are particularly liable to attack pearl barley that has undergone several months' storage.

SOYA MEAL

Soya is a pulse, not a cereal. The soya bean is a rich source of oil and protein. Both components are used as food ingredients. Dehulled soya bean is dried and milled to yield flour that is used as a source of lipoxygenase enzyme in a compound dough conditioner containing fats, emulsifiers, antioxidants and α-amylase. Although initially developed as a source of vegetable oil, the defatted flour (meal) has been recognised as a valuable high protein feedingstuff. Further purification of the high protein fraction reduces the carbohydrate level and produces soya concentrate and isolate. Toasted defatted soy flour is also used in bakery products and as a milk replacer. The protein concentrates may be mixed with wheat flour and potato starch to yield a binder for use with meat products such as sausages. Soya flour is also used as emulsifier in products such as salad dressings since it contributes emulsion stabilising properties together with protein oil and lecithin as thickening constituents which act as replacement for egg products. Tables 8.10 and 8.11 show typical data for the various fractions. The material is also a good source of B vitamins and in this respect is equivalent to wholemeal flour. Enzyme-rich, highly dispersible, defatted flour is used as a bread improver.

Table 8.10 Typical composition of soya products

	Soya beans (%)	Soya meal (%)	Soya flour (full fat) (%)	Soya concentrate (%)	Soya isolate (%)
Water	10·5	8·1	6·5	4	4
Oil	18·0	7·5	20·5	0·5	Trace
Protein (N × 6·25)	35·0	42·1	42·5	69	92
Fibre	4·6	5·0	3·5	5·0	—
Ash	4·8	5·5	5·0	7·7	3
Carbohydrate	27·1	31·8	22·0	14	1
Lecithin (as P_2O_5)	0·26	0·30	0·28	—	—

Table 8.11 Typical composition of soya flours (per cent, on dry basis)

	Defatted flour	Low fat flour	Full flat flour
Oil	2	7	22
Protein (N × 6·25)	53	48	43
Carbohydrate + fibre	39	39	29
Ash	6	6	6

The use of protein products derived from soya is discussed in Chapter 13.

Soya is almost devoid of starch, but it can be identified microscopically by the presence of 'hour-glass' cells. These are best seen on the fibrous residue separated by boiling the sample with dilute acid and alkali as in the determination of fibre. Provided the product being examined has not been heated during manufacture, soya can be detected by the **urease test.** The test depends on the fact that soya contains the enzyme urease which liberates ammonia from urea:

Shake 3–5 g of the material with 5 ml of 2 per cent urea solution and wipe the top part of the tube dry with a cloth. Put a piece of moistened red litmus paper in the top of the tube, cork lightly and incubate at 37 °C for 2 hours. If soya is present the litmus paper turns blue, due to the ammonia liberated by the urease. See also Chapter 13.

In the absence of lecithin from egg or other products, soya can be approximately determined from the phospholipid content. It can also be assessed from the manganese content. Mohr *et al.* (1984) have described an immunochemical method for detection of soya protein in bakery products and bread. Correction factors to account for baking losses are also proposed.

RICE

Definitions of various types of rice are given in the EC Regulation 359/67/EEC as amended. For example, husked rice is defined as rice grains from which the husks have been removed, but which are still enclosed in the pericarp. In a further Regulation a maximum moisture content of 15 per cent is specified. Also paddy rice is the rice of which the grains are still enclosed in the floral husk and for this a maximum moisture content of 14·5 per cent is specified (Regulation No. 362/67/EEC as amended by 867/67 and 1555/71).

The 'facing' and polishing of rice during the preparation for sale was first described in 1909. Small quantities of blue pigment and of oil were employed to improve the colour and translucency. Talc, when used for facing rice, was used in conjunction with glycerine. In the early part of the century the practice had become general in the trade and a maximum of 0·5 per cent was recommended for extraneous mineral matter. The use of such mineral additions has, however, now largely disappeared. In the case of ground rice the ash is usually between 0·2 and 0·4 per cent and samples that have a figure in excess of this range should be regarded with suspicion and the acid-insoluble ash should be determined also. An unfaced rice usually gives a readily fusible ash, i.e. it gives a glassy residue when heated over an Argand burner, whereas the ash of faced rice, containing more silica, is not readily fusible and is flaky. Ground rice should also be examined microscopically.

Talc in whole rice can be determined by the following method due to Kŕiźan:

Shake 20 g of sample with dilute ammonia and dilute hydrogen peroxide. Heat to about 60 °C so that the gas formed causes the particles of talc to come away from the surface. Decant off the liquid containing the talc, wash the grains several times with water and add these washings to the decanted liquid. Heat the liquor with hydrochloric and chromic acids to oxidise suspended meal, filter off the talc, wash, ignite and weigh. In unpolished rice the residue does not normally exceed 0·025 per cent.

Compared with milled rice, hulled rice is rich in phosphoprotein and lecithin, which is valuable nutritionally. The differences in the phosphorus and potassium contents of whole and polished rice can be seen from the following figures:

	Phosphorus (per cent P_2O_5)	Potassium (per cent K_2O)
Whole rice	0·27	0·25
Polished rice	0·09	0·07

Particularly in countries where rice is the staple diet, the presence of vitamin B_1 in the outer coating of the grain is of considerable nutritional importance. It has been established that a diet consisting of polished rice is liable to cause beri-beri, due to the consequent low intake of the antineuritic vitamin.

CANNED CEREAL MILK PUDDINGS

Canned milk puddings are marketed containing about 87 per cent milk, 6 per cent sugar and 7 per cent of one of the following cereals: macaroni, semolina, sago, tapioca or rice. Sodium citrate, bicarbonate or phosphate is often added as a stabiliser. Objection may be taken to (1) products containing over 10 per cent added water and (2) products labelled with a prominent declaration of 'milk', if prepared from dried milk.

Analysis

Sampling. Render homogeneous in a mechanical blender.

Water. Dry 5 g on sand at 100 °C, preferably in a vacuum oven.

Fat. Determine by a modification of the Werner–Schmid process (Chapter 2). Commence by mixing 4 g with 10 ml conc. hydrochloric acid and 5 ml water and heat in a water bath at 60 °C for 30 min. It has been suggested that rice pudding prepared from full cream milk should contain at least 2·8 per cent milk fat (or 2 per cent if made from a mixture of full cream and separated milk).

Protein. Determine on 4 g by the macro-Kjeldahl method. Calculate as N × 6·38. With rice puddings 80–85 per cent of the total protein present is derived from milk.

Ash. Ignite 5 g at 550 °C.

Sugars. Determine the lactose and sucrose on a clarified extract before and after inversion by Lane and Eynon's method. Pim (1964) removed the starch with 70 per cent alcohol before centrifuging and clarifying with lead acetate and alumina cream followed by potassium oxalate. Dalley and Wood (1963) determine the lactose by the chloramine-T method. See also general methods for sugars in Chapter 6.

Starch + Fibre. Estimate by difference (see calculation methods below).

Ranges of typical composition figures for canned rice puddings are given in Table 8.12. The composition of puddings containing other cereals have been discussed by Houlbrooke (1964) and Pim (1964, 1967).

Calculation of composition

Following useful work by Dalley and Wood (1963), Markland (1963) and Houlbrooke (1964), Pim (1964, 1967) has made a comprehensive study of the composition of ingredients used in the preparation of the

Table 8.12 Ranges of composition data obtained from canned rice puddings (per cent)

Total solids	21·9–23·9
Fat	2·8–3·3
Nitrogen	0·49–0·54
Sucrose	4·5–6·9
Lactose (anhydrous)	3·6–3·9
Ash	0·60–0·72
Starch + Fibre	5·5–7·5

various canned cereal milk puddings (Table 8.13). The milk content can be calculated from the milk N, obtained by subtracting the mean cereal N from the total N and applying Vieth's ratio. The milk content can also be calculated directly from the lactose using Veith's ratio, except in the case of macaroni and semolina for which an allowance for reducing power of the cereal should be made. The reducing power of the cereal was determined after sterilisation in the can as used in the manufacturing process. Pim also eliminated errors due to the formation of lactose–protein complexes during processing by using a control can of milk sterilised at the same time as ones containing other ingredients. The final calculations of the milk content in canned cereal milk puddings depends on using reliable mean figures for milk and the correctness of Vieth's ratio. Whereas Pim considers that the presumptive minimum of 8·5 per cent should be used for milk solids-not-fat content Houlbrooke used a figure of 8·8 per cent and consequently reported that several of the puddings examined contained added water. Seasonal variations in the ratio of lactose to protein in the milk are minimised, however, by taking the mean of the two figures for milk, calculated from milk protein and milk lactose respectively, as the actual milk content of the pudding.

Cereal content of pudding

Per cent Starch + per cent Fibre = S

S = per cent TS − per cent (Fat + Protein + Sucrose + Anhydrous lactose + Ash)

Per cent Cereal in pudding

$= \frac{100S}{A}$, where A is the mean per cent (Starch + Fibre) in Table 8.13.

Milk content of pudding

1. *From Nitrogen*

$$\text{Per cent Milk N} = \text{per cent Total N} - \frac{\text{per cent Cereal} \times B}{100}$$

where B = mean per cent N in cereal (see Table 8.13)

per cent *Milk* = per cent *Milk* N × 200.

2. *From Lactose*

$$\text{Per cent Milk lactose hydrate} = \text{per cent Total `lactose hydrate'} - \frac{\text{per cent Cereal} \times C}{100}$$

where C = mean per cent reducing sugars in cereal calculated as lactose hydrate (see Table 8.13).

Per cent *Milk* = per cent *Milk lactose hydrate* × 21·72.

Added water

Per cent added water = 100 − (per cent Milk + per cent Cereal + per cent Sugar), where per cent Milk = mean of (1) and (2) = M.

Fat content of milk used

Allowance has to be made for the fat in the cereal used:

Table 8.13 Mean values of constituents of cereals used in canned milk puddings (After Pim)

	A Starch + Fibre (%)	B Nitrogen (%)	C Reducing sugars (calculated as lactose hydrate) (%)	D Fat or oil (%)
Macaroni	70·0	2·3	3·3	1·9
Rice	77·0	1·1	0·0	0·9
Sago	83·5	0·08	0·0	0·2
Semolina	71·0	2·0	1·6	1·6
Tapioca	84·5	0·07	0·0	0·1

$$\text{Per cent Fat in original milk} = \frac{100F_s - \text{per cent Cereal} \times D}{M} \times 100$$

where F_s = per cent fat in sample.

D = mean per cent fat in cereal (see Table 8.13).

M can be calculated as above or be taken as 100 − (per cent Cereal + per cent Sucrose).

Pim (1967) gives modified formulae for calculating the composition of cereal puddings containing added cream.

BABY FOODS

Prepared infant foods are based either on prepared cereal or milk powder. There are no prescribed formulae for the prepared cereal products in UK legislation although there are restrictions on the nature and extent of use of additives, e.g. colours, antioxidants, curing agents and preservatives, levels of contaminants, e.g. lead, allowed in foods specifically marketed for infants and young children. There are, however, specific formulae prescribed in the Skimmed Milk with Non-Milk Fat Regulations (SI 1960 No. 2331 as amended) (see Chapter 14). These regulations and the Condensed Milk and Dried Milk Regulations (SI 1977 No. 928 as amended) require that products based on skimmed milk are to be labelled 'Unfit for Babies' unless they have been approved for infant feeding by a panel set up by COMA. The overall legislative position in the UK is summarised in the Food Standards Committee Report on Infant Formulae (FSC REP 73 1981). This report includes a summary of conclusions of a COMA working party on infant foods together with nutritional guidelines for the composition of artificial feeds compared with that of mature human milk and the recommendations of the FACC regarding the use of additives in infant formulae.

The Codex Alimentarius Commission has developed international standards for composition and labelling of a number of foods for infants and young children:

Infant Formula	CODEX STAN 72–1981
Canned Baby Foods	CODEX STAN 73–1981
Processed Cereal-Based Foods	CODEX STAN 74–1981
Gluten-Free Foods	CODEX STAN 118–1981

Methods of analysis proposed for use in the enforcement of the provisions of the Codex Standards are essentially those of the AOAC. These have been reviewed by

Tanner (1982), Yeung (1982) and Granade (1982) in relation to the nutritional and compositional standards prescribed.

REFERENCES

American Association of Cereal Chemists (1972) *Approved Methods of the AACC*. Minnesota: AACC.
Amos, A. J. & Kent-Jones, D. W. (1931) *Analyst*, **56**, 572.
Anon (1978) *Food Manufacture* (March), 37.
Archer, A. W. (1973), *Analyst*, **98**, 755.
Armstrong, A. W. (1952) *Analyst*, **77**, 460.
Asp, N. (1983) *In Focus*, **6**, 5.
Asp, N., Furola, I., Dekries, J., Schweizer, T. F. & Harland, B. F. (1985) *Journal of the Association of Official Analytical Chemists*, **68**, 677.
Auerbach, M. E., Eckert, H. W. & Angell, E. (1949) *Cereal Chemistry*, **26**, 490.
Axford, D. W. E. & Elton, G. A. H. (1960) *Chemistry and Industry*, 1257.
Axford, D. W. E., Chamberlain, N., Collins, T. H. & Elton, G. A. H. (1963) *Cereal Science Today*, **8**, 265.
Baker, D. (1977) *Cereal Chemistry*, **54**, 360.
Baker, D. (1983) *Cereal Chemistry*, **60**, 217.
Baker, D. & Holden, J. M. (1981) *Journal of Food Science*, **46**, 396.
Barrett, S. Croft, A. G. & Hartley, A. W. (1971) *Journal of the Science of Food and Agriculture*, **22**, 173.
Bell, B. M. (1985) *Journal of the Science of Food and Agriculture*, **36**, 815.
Bell, J. G. (1974) *Laboratory Practice*, **23**, 235.
Bietz, J. A., Burnouf, T., Cobb, L. A. & Wall, J. S. (1984) *Cereal Chemistry*, **61**, 124, 129.
Blake, C. J. (1978) *Scientific and Technical Survey* No. 109. Leatherhead Food R. A.
Blake, C. J. & Coveney, L. V. (1978) *Research Reports*, No. 293. Leatherhead Food, R. A.
Blish, M. J. & Sandstedt, R. M. (1933) *Cereal Chemistry*, **10**, 189.
Bloskma, A. H. (1972) *Cereal Science Today*, **17**, 380.
Bolling, H. & Zwingelberg, H. (1984) *Getreide Mehl und Brot*, **38**, 3.
Bottomley, R. C., Kearns, H. F. & Schofield, J. D. (1982) *Journal of the Science of Food and Agriculture*, **33**, 481.
Burgdon, A. C., Ikeda, H. S. & Tanner, S. N. (1985) *Cereal Chemistry*, **62**, 72.
Busvine, J. R. (1965) *Insects and Hygiene*, 2nd edn. London: Methuen.
Camire, A. L. & Clydesdale, F. M. (1982) *Journal of Food Science*, **47**, 575.
Chapman, W. B., Fogden, E. & Urry, S. (1963) *Journal of the Association of Public Analysts*, **1**, 59.
Chen, W. J. L. & Anderson, J. W. (1981) *American Journal of Clinical Nutrition*, **34**, 1077.
Christie, A. A. & Wiggins, R. A. (1978) *Developments in Food Analysis Techniques*, ed. R. D. King. London: Applied Science Publishers Ltd.
Clegg, K. M. (1956) *Journal of the Science of Food and Agriculture*, **7**, 40.
Collins, T. H. (1985) *Bulletin Flour Milling and Breadmaking Research Association*, No. 6, 219.
Conacher, H. B. S. & Page, B. D. (1979) *Journal of Chromatographic Science*, **17**, 188.
Craney, C. E. (1972), *Cereal Chemistry*, **49**, 496.
Crosby, N. T. (1977) *Analyst*, **102**. 225.
Cummings, J. H., Englyst, H. M. & Wood, R. (1985) *Journal of the Association of Public Analysts*, **23**, 1.
Dalley, R. A. & Wood, E. C. (1963) *Journal of the Association of Public Analysts*, **1**, 38.
Daniels, D. G. H. & Fisher, W. (1980) *Bulletin Flour Milling and Breadmaking Research Association*, No. 3, 131.
Davidek, J., Pudil, F. & Seifert, J. (1977) *Journal of Chromatography*, **140**, 316.
Davis, A. B. & Lai, C. S. (1984) *Cereal Chemistry*, **61**, 1.
Dean, A. C. (1978), *Food Chemistry*, **3**, 241.
Defibaugh, P. W., Smith, J. S. & Weeks, C. E. (1977) *Journal of the Association of Official Analytical Chemists*, **60**, 522.
Dennis, P. O. & Rees, H. G. (1949), *Analyst*, **74**, 481.
Dickes, G. J. (1966) *Journal of the Association of Public Analysts*, **4**, 45.
Dickes, G. J. (1979) *Talanta*, **26**, 1065.
Doering, V. (1983) *Gordian*, **83**, 62.
Doring, U. & Bielig, H. J. (1978) *Gordian*, **78**, 72, 244.
Dudas, F. (1971) *Die Stärke*, **23**, 390.
Dudas, F. (1972) *Die Stärke*, **24**, 367.
Dudas, F. (1973) *Die Stärke*, **25**, 263.
Duvenage, E. (1986) *Journal of the Science of Food and Agriculture*, **37**, 384.

Dvorak, J., Bryndac, J. & Hronikova, M. (1962) *Prumsyl potravin*, **13**, 428.
Egberg, D. C. (1979) *Journal of the Association of Official Analytical Chemists*, **62**, 1027.
Elias, D. G. (1972) *Process Biochemistry*, **7**, 38.
Englyst, H. N., Anderson, V. & Cummings, J. H. (1983) *Journal of the Science of Food and Agriculture*, **34**, 1434.
Ewers, E. (1905) *Zeitschrift Offentl Chemie*, **11**, 407.
Ewers, E. (1908) *Zeitschrift Offentl Chemie*, **14**, 8, 50.
Ewers, E. (1910) *Zeitschrift Offentl Chemie*, **15**, 8.
Fearn, T. (1986) *Analytical Proceedings*, **23**, 123.
Fellman, J. K., Artz, W. E., Tassinari, P. D., Cole, C. L. & Augustin, J. (1982) *Journal of Food Science*, **47**, 2048.
Fostel, H. (1984) *Ernahrung*, **8**, 723.
Fraser, J. R. & Hoodless, R. A. (1963) *Analyst*, **88**, 558.
Fraser, J. R. & Weston, R. E. (1950) *Analyst*, **75**, 402.
FSC (1960) *Report on Bread and Flour*, FSC/REP/43, Food Standards Committee. London: HMSO.
FSC (1964) *Report on Food Labelling*, FSC/REP/48, Food Standards Committee. London: HMSO.
FSC (1974) *Second Report on Bread and Flour*, FSC/REP/61, Food Standards Committee. London: HMSO.
FSC (1979) *Second Report on Food Labelling*, FSC/REP/69, Food Standards Committee. London: HMSO.
Gillis, J. A. (1963) *Cereal Science Today*, **8**, 40.
Granade, A. W. (1982) *Journal of the Association of Official Analytical Chemists*, **65**, 1491.
Greenaway, W. T. (1972) *Cereal Chemistry*, **49**, 609.
Greer, E. N., Mounfield, J. D. & Pringle, W. J. S. (1942) *Analyst*, **67**, 352.
Gregory, J. F. (1980) *Journal of Agricultural and Food Chemistry*, **28**, 486.
Gregory, J. F. & Kirk, J. R. (1978) *Journal of Agricultural and Food Chemistry*, **26**, 338.
Gregory, J. F., Day, B. P. F. & Ristow, K. A. (1982) *Journal of Food Science*, **47**, 1568.
Gross, A. F. (1975) *Journal of the Association of Official Analytical Chemists*, **58**, 799.
Haken, J. K. (1979) *Advances in Chromatography*, **17**, 163.
Heaton, K. W. (1983) *Human Nutrition: Clinical Nutrition*, **37C**, 151.
Heckman, M. M. & Lane, S. A. (1981) *Journal of the Association of Official Analytical Chemists*, **61**, 1339.
Herd, C. W. & Amos, A. J. (1930) *Cereal Chemistry*, **7**, 257.
Hildebrand, F. C. & McClellan, B. A. (1938) *Cereal Chemistry*, **15**, 107.
Hinton, H. E. & Corbet, L. A. (1963) *Common Insect Pests of Stored Food Products*, 4th edn, British Museum Economic Series No. 15. London: British Museum.
Hjalmarsson, S. & Akesson, R. (1983) *International Laboratory*, **13**, 70.
Holtz, F. (1984) *Zeitschrift Lebensmittel Untersuchung und Forschung*, **179**, 29.
Hoppner, K. & Lampi, B. (1982) *Journal of Liquid Chromatography*, **5**, 953.
Horvath, L., Norris, K. H., Horvath-Mosonyi, M., Rigo, J. & Hegedus-Volgyesi, E. (1984) *Acta Alimentaria*, **13**, 355.
Houlbrooke, A. (1964) *Journal of the Association of Public Analysts*, **2**, 54.
Huebner, F. R. & Bietz, J. A. (1984a, b) *Cereal Chemistry*, **61**, 449, 544.
Huebner, F. R. & Bietz, J. A. (1985) *Journal of Chromatography*, **327**, 333.
Hughes, A. M. (1961) *The Mites of Stored Food*, MAFF Technical Bulletin. London: HMSO.
Hunt, D. C., Jackson, P. A., Mortlock, R. E. & Kirk, R. S. (1977) *Analyst*, **102**, 917.
Hurst, W. J., Aleo, M. D. & Martin, R. A. (1985) *Journal of Agricultural and Food Chemistry*, **33**, 820.
Joudrier, P., Woaye-Hune, C. & Gobin, C. (1981) *Annales de Falsification des Expertises Chimie et Toxicologie*, **74**, 523.
Kaiser, K. P. & Krause, I. (1985) *Zeitschrift Lebensmittel Untersuchung und Forschung*, **180**, 181.
Kennedy, C. A. & McCleary, B. V. (1981) *Analyst*, **106**, 344.
Kent-Jones, D. W. & Amos, A. J. (1967) *Modern Cereal Chemistry*, 6th edn. London: Food Trade Press.
Kirk, J. R. (1977) *Journal of the Association of Official Analytical Chemists*, **60**, 1234.
Knight, R. A. & Kent-Jones, D. W. (1953) *Analyst*, **78**, 467.
Knight, R. A., Christie, A. A., Orton, C. R. & Robertson, J. (1973) *British Journal of Nutrition*, **30**, 181.
Kropf, D. H. (1984) *Journal of Food Quality*, **6**, 199.
Lawrence, J. F., Iyengar, J. R. & Conacher, H. B. S. (1981) *Journal of the Association of Official Analytical Chemists*, **64**, 1462.
Lee, K. & Abendroth, J. A. (1983) *Journal of Food Science*, **48**, 1344.
Lookhart, G. L., Hall, S. B. & Finney, K. F. (1982) *Cereal Chemistry*, **59**, 69.
Lumley, I. D. & Wiggins, R. A. (1981) *Analyst*, **106**, 1103.
Markland, J. (1963) *Journal of the Association of Public Analysts*, **1**, 54.
Martin, E. (1981) *Mitteilungen aus dem Gebeit der Lebensmitteluntersuchung und Hygiene*, **72**, 402.

Martin, E., Duret, M. & Vogel, I. (1980) *Mitteilungen aus dem Gebeit der Lebensmitteluntersuchung und Hygiene*, **71**, 195.
Mauro, D. J. & Wetzel, D. L. (1984) *Journal of Chromatography*, **299**, 281.
McCausland, J. & Wrigley, C. W. (1976) *Journal of the Science of Food and Agriculture*, **27**, 1203.
McCready, R. M., Ducay, E. D. & Gauger, M. A. (1974) *Journal of the Association of Official Analytical Chemists*, **57**, 336.
Miller, D. S. & Judd, P. A. (1984) *Journal of the Science of Food and Agriculture*, **35**, 111.
MOH (1968) *Iron in Flour*, Reports on Public Health and Medical Subjects No. 117. London: HMSO.
Mohr, E., Scheve, H. & Amboss, M. (1984) *Gordian*, **84**, 66.
Moir, D. D. & Hinks, E. (1935) *Analyst*, **60**, 439.
Morrison, W. R. (1963a, b) *Journal of the Science of Food and Agriculture*, **14**, 245, 870.
MRC (1943) *Biochemical Journal*, **37**, 433.
Nagayama, T., Nishijima, M., Kamimura, H., Yasuda, K., Saito, K., Ibe, A., Ushiyama, H. & Naoi, Y. (1982) *Shokuhin Eiseigaki Zasshi*, **23**, 253.
Nedeltscheva, M., Stoikov, G. & Popova, S. (1975) *Stärke*, **27**, 298.
Neilson, M. J. & Marlett, J. A. (1983) *Journal of Agricultural and Food Chemistry*, **31**, 1342.
Nelson, J. H. (1985) *American Journal of Clinical Nutrition*, **41**, 1070.
Nicholson, I. A., Macrae, R. & Richardson, D. P. (1984) *Analyst*, **109**, 267.
Oberleas, D. (1983) *Cereals Foods World*, **28**, 352.
Ogawa, S. Susuki, H., Toyoda, M., Ito, Y. & Iwaida, M. (1984) *Nippon Shokuhin Kogyo Gakkaishi*, **31**, 356.
Osborne, B. G. (1984) *Journal of the Science of Food and Agriculture*, **35**, 106.
Osborne, B. G. & Fearn, T. (1981) *Bulletin Flour Milling and Breadmaking Research Association*, No. 4, 185.
Osborne, B. G. & Fearn, T. (1986) *Near Infrared Spectroscopy in Food Analysis*. London: Longman.
Osborne, D. R. & Voogt, P. (1978) *The Analysis of Nutrients in Foods*. London and New York: Academic Press.
Osborne, B. G., Douglas, S., Fearn, T. & Willis, K. H. (1982) *Journal of the Science of Food and Agriculture*, **33**, 736.
Osborne, B. G., Fearn, T., Miller, A. R. & Douglas, S. (1984a) *Journal of the Science of Food and Agriculture*, **35**, 99.
Osborne, B. G., Barrett, G. M., Cauvain, S. P. & Fearn, T. (1984b) *Journal of the Science of Food and Agriculture*, **35**, 940.
Oyamada, N., Veno, S., Kubota, K. & Ishizaki, M. (1985) *Journal of the Food Hygienic Society of Japan*, **26**, 13.
Paul, A. A. & Southgate, D. A. T. (1978) *The Composition of Foods*. London.
Pearson, D. & Goussous, F. (1968) *Journal of the Association of Public Analysts*, **6**, 95.
Phillips, A. R. & Sanders, B. J. (1968) *Journal of the Association of Public Analysts*, **6**, 89.
Pim, F. B. (1964, 1967) *Journal of the Association of Public Analysts*, **2**, 62; **5**, 61.
Pippen, E. L. & Potter, A. L. (1975) *Journal of Agricultural and Food Chemistry*, **23**, 523.
Player, R. B. & Wood, R. (1980) *Journal of the Association of Public Analysts*, **18**, 29.
Preston, K. R. (1982) *Cereal Chemistry*, **59**, 73.
Pringle, W. J. S. (1946) *Analyst*, **71**, 490.
Rask, O. S. (1927a, b) *Journal of the Association of Official Analytical Chemists*, **10**, 108, 473.
Redman, D. G. & Burbridge, K. (1979) *Bulletin Flour Milling and Baking Research Association*, No. 5 (October), 191.
Ronalds, J. A., O'Brien, L. & Allen, S. J. (1984) *Journal of Cereal Science*, **2**, 25.
Salmon, S. E. (1985) *Flour Milling and Baking Research Association, Bulletin*, **6**, 247.
Salmon, S. E. & Burbridge, K. M. (1985) *Flour Milling and Baking Research Association, Bulletin*, **2** and **6**, 78 and 218.
Sawyer, R. (1965) *Proceedings of the First International Congress on Food Science and Technology*, Vol. 3. London: Gordon Breach Science Publishers.
Sawyer, R., Tyler, J. F. C. & Weston, R. E. (1956) *Analyst*, **81**, 362.
Schweizer, T. F., Frolich, W., Del Vedovo, S. & Besson, R. (1984) *Cereal Chemistry*, **61**, 116.
Seidemann, J. (1975) *Nahrung*, **19**, 497.
Sharratt, M., Frazer, A. C. & Forbes, O. C. (1964) *Food and Cosmetics Toxicology*, **2**, 527.
Skerritt, J. H. & Smith, R. A. (1985) *Journal of the Science of Food and Agriculture*, **36**, 980.
Slack, P. T. & Porter, D. C. (1983) *Chemistry and Industry*, **23**, 896.
Smith, P. R. (1983) *Scheme for the Examination of Foreign Material Contaminants in Foods*, Leatherhead Food Research Association.
Snell, E. E. & Strong, F. M. (1939) *Industrial and Engineering Chemistry, Analytical Edition*, **11**, 346.
SPA (1943, 1946a, 1946b, 1951) *Analyst*, **68**, 276; **71**, 398, 401; **76**, 127.

Spencer, M. (1961) *Food Processing and Packaging*, **13**, 53.
Stafford, H. R. (1968) *Process Biochemistry*, **3**, 58.
Staudt, E. (1965) *Milling*, **144**, 86.
Stephenson, W. H. & Hartley, A. W. (1955) *Analyst*, **80**, 461.
Stevens, D. J. (1983) *Flour Milling and Baking Research Association, Bulletin*, **5**, 225.
Tanner, J. T. (1982) *Journal of the Association of Official Analytical Chemists*, **65**, 1488.
Tkachuk, R. (1966a, b), *Cereal Chemistry*, **43**, 207, 223.
Tonogai, Y., Kingkate, A., Thanissorn, W. & Punthanaprated, U. (1983) *Journal of Food Protection*, **46**, 284.
Torelm, I. & Bruce, A. (1982) *Var Foda*, **34**, 79.
Trop, M. & Grossman, S. (1972) *Journal of the Association of Official Analytical Chemists*, **55**, 1191.
Trowell, H., Southgate, D. A. T., Wolever, T. M. S., Leeds, A. R., Gassull, M. A. & Jenkins, D. A. (1976) *Lancet*, **i**, 967.
Trowell, H., Godding, E., Spiller, G. & Briggs, G. (1978) *American Journal of Clinical Nutrition*, **31**, 1489.
Tsuda, T., Nakanishi, H., Kobayashi, S. & Morita, T. (1984) *Journal of the Association of Official Analytical Chemists*, **67**, 1149.
Tyler, J. F. C. & Weston, R. E. (1956) *Analyst*, **81**, 375.
Tyler, T. A. & Shrago, R. R. (1980) *Journal of Liquid Chromatography*, **3**, 269.
Udy, D. C. (1956) *Cereal Chemistry*, **33**, 190.
Ulmann, M. & Richter, M. (1961). *Stärke*, **13**, 67.
Van der Kamer, J. H. & van Ginkel, L. (1952) *Cereal Chemistry*, **29**, 239.
Vanderslice, J. T., Maire, C. E. & Yakupkovic, J. E. (1981) *Journal of Food Science*, **46**, 943.
Van Soest, P. J. (1963a, b) *Journal of the Association of Official Analytical Chemists*, **46**, 825, 829.
Van Soest, P. J. (1973a, b) *Journal of the Association of Official Analytical Chemists*, **56**, 781, 825.
Voigt, M. N., Ware, G. O. & Eitenmiller, R. R. (1979) *Journal of Agricultural and Food Chemistry*, **27**, 1305.
Wade, P. (1965) *Food Trade Review*, **35**, 47.
Wade, P. & Watkin, D. (1966) *Food Trade Review*, **36**, 44.
Walsh, J. H., Wyse, B. W. & Hansen, R. G. (1979) *Journal of Food Biochemistry*, **3**, 175.
Wanatabe, I., Tanaka, R. & Kashimoto, T. (1982) *Shokuhin Eiseigaku Zasshi*, **23**, 135.
Watson, C. A., Etchevers, G. & Shuey, W. C. (1976) *Cereal Chemistry*, **53**, 803.
Wehling, R. L. & Wetzel, D. L. (1984) *Journal of Agricultural and Food Chemistry*, **32**, 1326.
Wenlock, R. W., Sivell, L. M. & Agater, I. B. (1985) *Journal of the Science of Food and Agriculture*, **36**, 113.
Wheat Commission (1939) *Examination and Analyses Byelaws*, S. R. & O. 1939 No. 850. London: HMSO.
Williams, E. I. (1964) *Analyst*, **89**, 289.
Williams, P. C. & Butler, C. D. (1982) *Journal of the Science of Food and Agriculture*, **33**, 614.
Wills, R. B. H., Shaw, C. G. & Day, W. R. (1977) *Journal of Chromatographic Science*, **15**, 262.
Yabe, Y., Tan, S., Ninomiya, T. & Okada, T. (1983) *Journal of the Food Hygiene Society Japan*, **24**, 329.
Yeung, D. L. (1982) *Journal of the Association of Official Analytical Chemists*, **65**, 1500.

9

Starch products; baking powders; eggs; salad cream

STARCH

Composition and properties

Starches are major constituents of many foods. They are the natural carbohydrates, which form the nutrient reservoirs of plants and have the general formula $(C_6H_{10}O_5)n$, where *n* is probably not less than 1000. Starches are readily convertible to reducing sugars by hydrolysis, either by heating with acid or by reaction with enzymes such as diastase. When heated with water, starch grains swell and at about 70 °C burst to form a viscous paste. When heated, starch has no melting point; on gentle heating it swells, chars and forms caramelised products. It is insoluble in organic solvents and has limited solubility in cold water, and does not reduce Fehling's solution until hydrolysed. Its optical rotation in dispersed solution is variable and depends on the source (Chapter 8). The density of starch varies between 1·50 and 1·53 according to source. Some properties of commercial starches are shown in Table 9.1.

Table 9.1 Properties of some commercial starches, typical values

	Potato starch	Maize starch	Wheat starch	Waxy maize starch
Diameter range (μm)	5–100	3–26	1–40	3–26
Granule (10^{-6} m) mean	30	15	10	15
Amylose: Amylopectin ratio	21:79	28:72	28:72	0:100
Amylose degree of polymerisation (DP)	3000	800	800	–
Amylopectin degree of polymerisation (DP)	2 000 000	2 000 000	2 000 000	2 000 000
Gelatinisation temp. (°C)	60–65	75–80	80–85	65–70
Peak viscosity (Brabender units) 5% starch concentration	3000	600	300	800
Swelling power at 95 °C	1153	24	21	64
Solubility (%) at 95 °C	82	25	41	23

Source: O'Mahony (1985).

Most starches contain amylose and amylpectin; amylose consists of straight unbranched chains of glucose units linked by α-1,4-glucosidic bonds. The amylose fraction is slightly soluble in water and gives the characteristic blue colour with iodine. Amylopectin is a highly branched glucose polymer that gives a red–purple reaction with iodine; the branches are linked by α-1,3- or α-1,6-glucosidic bonds. Starch components react differently to enzyme systems. β-Amylase will hydrolyse the straight chains to maltose. Thus the products of starch hydrolysis are maltose and a complex dextrin, which contains the branch roots of the amylopectin. This branched dextrin reacts with iodine to give a purple product. Hydrolysis with α-amylase results in attack at the branch points of the chains to yield low molecular weight dextrins that do not react with iodine.

The physical and chemical properties of starch may be varied by use of a wide range of techniques to suit particular food products and processing conditions. These include pregelatinisation, oxidation and chemical modification of the hydroxyl groups by the formation of ethers or esters (Table 9.2). Reviews of the chemistry, production, properties and analysis of starch products have been published by Sandstedt (1965), Ingleton (1970), Radley (1976a, b), O'Dell (1979), Fawcett (1985) and O'Mahony (1985).

Uses of starch

In view of their wide range of properties starches and starch derivatives (modified starches) may be found in many foods. They act as inert fillers in preparations such as baking powders and as thickeners in processed foods such as sauces, soups, gravies, creams (including ice cream), puddings and confectionery. Starch may be used as a substitute for more expensive ingredients such as egg and vegetable oil in the preparation of salad dressings. Many modern snack foods are prepared from starch by heating, extrusion and deep frying of partially dried product. It is unlikely that a single starch possesses all the properties required by the food manufacturer in preparation of a range of foods. However, it is possible to choose not only from the natural starches (wheat, maize, potato, rice and sorghum) but also from the variations of properties afforded by heat treatment and chemical modification (O'Dell, 1979). Since the amounts of starch and starch derivatives used in food varies widely it has been argued that the materials should be regarded as normal ingredients of foods. However, in view of the fact that many modified starches are subject to significant chemical action the view has been expressed that special controls are necessary. The EC Commission has proposed a directive to control the nature of modified starches for use in foods (OJ No. C31/6, 1.2.85). The term 'modified starch' is defined as 'the products obtained by one or more chemical treatments of edible starches, which may have previously undergone a physical or enzymatic treatment'. Table 9.2 summarises the nature of treatments used in preparation of starch derivatives (see also FACC, 1980).

Microscopical examination

Starches and cereals are usually identified by their microscopic characteristics. To a minute quantity on the slide is added a drop of alcohol and then a large drop of glycerine and water (50:50) before covering with the coverslip. After removing any excess liquid with filter paper the sample should be examined first with a low power objective and then with a magnification of 400. The striations and hilum can be observed more clearly by modifying the illumination. When there is any doubt as to whether or not particular grains are starch, a little very dilute iodine solution run under the coverslip, or examination with crossed polaroids enables the point to be

Table 9.2 Modified starches—summary of types and treatment

Denomination	Treatment	Specification
White or yellow dextrin, roasted or dextrinated starch	Dry heat with 0·15% HCl or 0·17% H_3PO_4	pH 2·5 to 7·0
Acid treated	HCl not more than 7% H_2SO_4 not more than 2% H_3PO_4 not more than 7%	pH 4·8 to 7·0
Alkali treated	NaOH or KOH not more than 1%	pH 5·0 to 7·5
Oxidised	NaOCl not more than 5·5% (as Cl_2)	Not more than 1% Carboxyl Not more than 0·5% NaCl
Bleached	(a) Peracetic acid and/or H_2O_2 containing not more than 0·45% active oxygen, or (b) NaOCl not more than 0·82% available chlorine, or (c) $NaOCl_2$ not more than 0·5%, or (d) SO_2 or sulphites, or (e) $KMnO_4$ not more than 0·2%, or (f) $NH_4S_2O_7$ not more than 0·075%	Not more than 0·1% added Carboxyl No reagent residues Not more than 50 mg/kg residual Mn
	Esterification with:	
Monostarch phosphate	Permitted phosphates	Not more than 0·5% residual phosphate (as P) in potato or wheat starch. Not more than 0·4% for other sources
Acetylated	Acetic anhydride not more than 10% Vinyl acetate not more than 7·5%	Not more than 2·5% acetyl groups
Acetylated distarch adipate	Acetic anhydride not more than 0·10% and not more than 0·12% adipic anhydride	Not more than 2·5% acetyl groups
Distarch phosphate	Sodium trimetaphosphate, or not more than 0·1% phosphorus oxychloride	Not more than 0·14% residual phosphate (as P) in potato or wheat starch Not more than 0·04% for other sources
Distarch phosphate	Not more than 0·1% sodium trimetaphosphate or phosphorus oxychloride combined with not more than 10% acetic anhydride or 7·5% vinyl acetate	Not more than 2·5% acetyl groups and not more than 0·14% residual phosphate (as P) in potato or wheat starch. Not more than 0·04% for other sources
Hydroxypropyl	Propylene oxide not more than 10%	Not more than 0·1% mg/kg propylene chlorhydrins
Phosphated distarch phosphate	Combined treatment as for mono- and distarch phosphate	Not more than 0·5% residual phosphate (as P) in potato or wheat starch. Not more than 0·4% for other sources
Hydroxypropyl distarch phosphate	Sodium trimethyl phosphate or phosphorus oxychloride, combined with etherification with not more than 10% propylene oxide	Not more than 0·1% mg/kg propylene chlorhydrins and not more than 0·14% residual phosphate in potato and wheat starch (as P), or not more than 0·04% for other sources.

easily decided. The microscopical appearance of the starches are illustrated in Figs 9.1–9.9 (Greenish and Collin, 1904; Williams, 1966). Differentiation between most of the starch grains is relatively easy after a little experience, but distinguishing between the large grains of wheat (max. 45 μm) and barley (max. 40 μm) is difficult. Size comparisons can be made, however, using lycopodium (Wallis, 1965). The ultrastructure of starches has been further explored by use of electron microscopy. Detailed studies of individual starches and of the effects of processing have been reported by Vaughan (1979).

Heating causes starch gelatinisation (loss of birefringence), so that the appearance of the grains is different in many processed products. Gelatinisation is occasioned by a decrease in molecular organisation of the granule and this occurs together with swelling in starches heated in water to 58–78 °C. When heated above about 85 °C the outline of the grains becomes faint, but observation is sometimes assisted by the addition of dilute iodine solution. Illustrations of the microscopical appearance of gelatinised cereal grains have been shown by MacMasters *et al.* (1957) and of damaged wheat starch by Sandstedt and Schroeder (1960). Also, Sandstedt and Mattern (1960) have shown that the proportion of damaged starch in flour can be assessed from its more rapid digestion by amylases compared with undamaged starch. Flint (1982) has reviewed the preparation techniques used for identification of starches in starch- and lipid-containing snack foods. Iodine and osmium tetroxide vapour staining allows the extent of starch gelatinisation to be studied without the introduction of artefacts commonly encountered with other preparation and staining methods. Chiang and Johnson (1977) have described a method for the estimation of gelatinised starch in starch products by use of glucoamylase digestion and reaction with *o*-toluidine. Total starch content is also measured by gelatinising the whole sample in sodium hydroxide. The test procedure for gelatinised starch is then followed. Many physical properties of doughs and baked bread are affected by the quantity of damaged starch present (see Chapter 8).

CEREAL BASED INSTANT FOODS

Composition

Many of these foods consist of a cornflour (or modified starch) base with added flavour and colour. Other starches such as farina (from potato) for these products are also widely used. The products also contain flavouring and colouring ingredients, additives such as emulsifiers and stabilisers, modified proteins and modified fats. Examples include instant whips, instant cereal products, cake mixes and the traditional products such as blancmange and custard powders.

Custard powders are usually composed of cornflour, flavouring (salt, vanilla extract) and a yellow dye. On occasion the cornflour has been partially replaced by other starches such as potato or arrowroot. Wheat flour has also been used but tends to produce a gel with a less desirable consistency in the final product. As a result of a case heard in 1925 it would appear that an offence is committed if a starch preparation containing no egg is sold as 'custard' not qualified by the word 'powder'. Egg is not normally present in commercial custard powder. If egg is present, the protein cells and oil will be revealed in the microscopical examination and it can be estimated from the solvent soluble phosphorus and cholesterol content. Sweetened

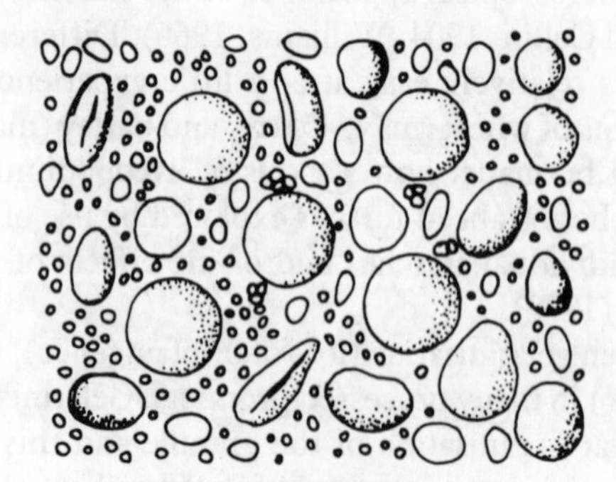

Fig. 9.1 Wheat starch.

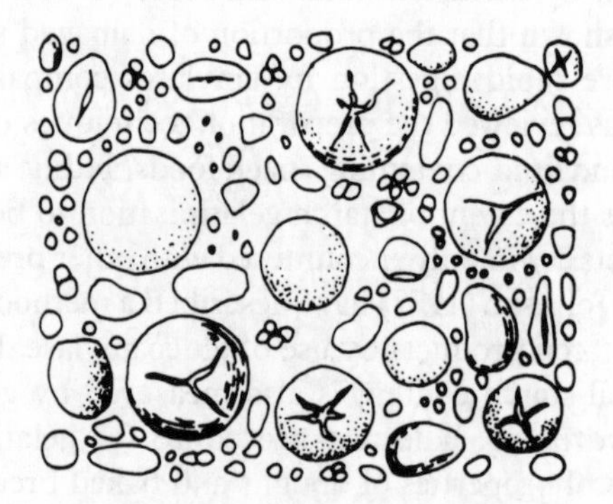

Fig. 9.2 Rye starch.

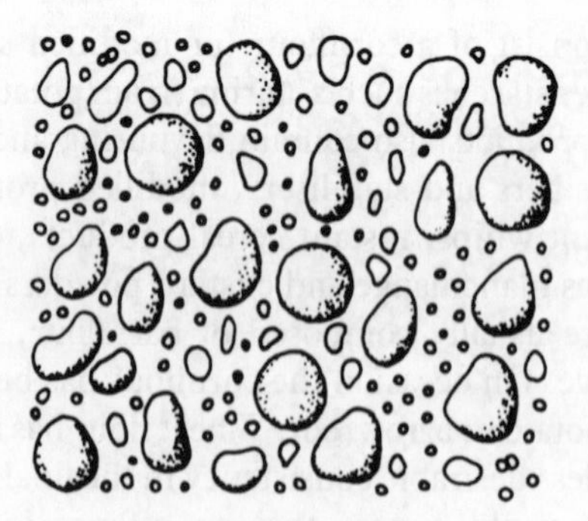

Fig. 9.3 Barley starch.

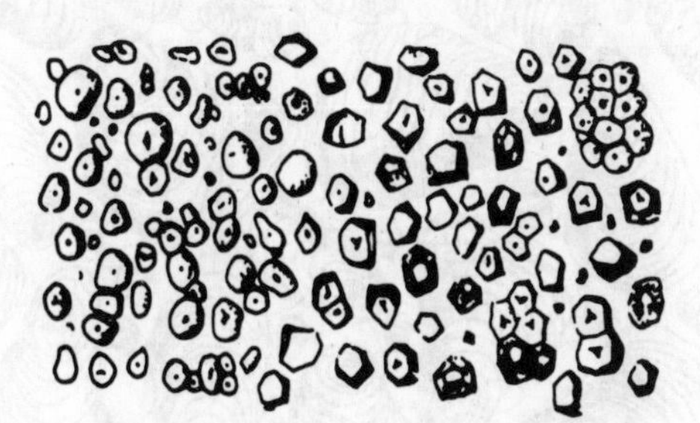

Fig. 9.4 Maize starch (cornflour).

Fig. 9.5 Oat starch.

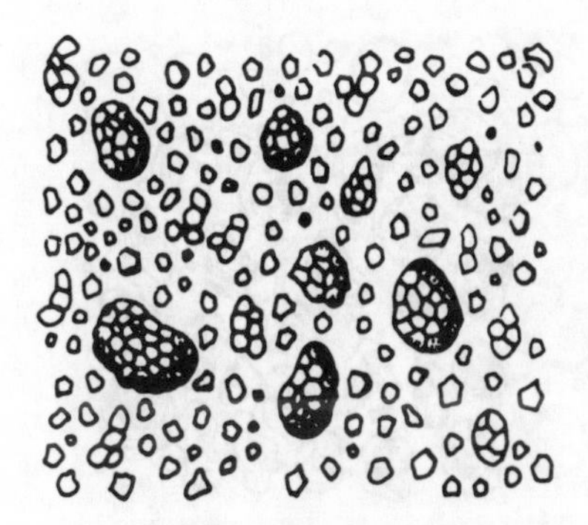

Fig. 9.6 Rice starch.

Fig. 9.7 Potato starch.

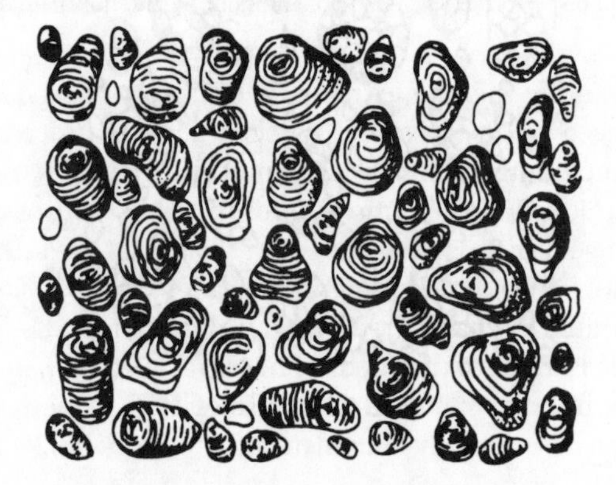

Fig. 9.8 Arrowroot starch.

Fig. 9.9 Tapioca.

custard powders may contain dextrose, sucrose or saccharin. Samples should be examined for ash, sulphur dioxide, added colour, moisture and vanillin content.

Canned custard is prepared essentially from milk, sugar and custard powder. Consequently, the product contains some starch, vanillin and possibly added colour and an emulsifier. Egg or albumen are further possible ingredients. The methods of analysis described for canned creamed rice can be employed for such products, together with the detection of colours and emulsifiers and the determination of vanillin and trace elements. Egg content can be assessed from the solvent soluble phosphorus or cholesterol content.

Analysis

Moisture is conveniently determined by a drying method. If the Carter–Simon rapid method of drying for 15 min at 155 °C is employed the moisture content should not exceed 15 per cent.

Vanillin. Mix 20 g of sample with 198 ml of 50 per cent isopropyl alcohol to produce a 10 per cent extract (approx.). Filter, evaporate 100 ml of filtrate to about 25 ml, transfer to a 100 ml volumetric flask, add 4 ml of lead acetate solution (5 per cent of neutral and basic) and make up to the mark. Filter and pipette a suitable aliquot of the filtrate (say 5 ml) into a 50 ml volumetric flask, add 5 ml Folin–Denis reagent (10 per cent sodium tungstate, 2 per cent phosphomolybdic acid and 5 per cent orthophosphoric acid/water), make up to the mark with saturated sodium carbonate solution and allow to stand for exactly 10 min. Filter. Measure the blue colour of the filtrate against standard vanillin solutions (similarly prepared) at 610 nm. A blank should be performed using 20 g of cornflour or the equivalent of the farinaceous matter present.

Alternatively, vanillin may be extracted with solvent and determined by GLC. Schlach and Dicecco (1974) have described a general procedure for use with a number of foods including ice cream, cake mix and vanilla extracts. These authors describe three extraction procedures according to the nature of the sample. For essences methylene chloride is used whilst for wet foods the sample is acidified and shaken with ethanol. After centrifuging, the ethanol extract is poured into water and the vanillin is separated by shaking with methylene chloride. Soxhlet extraction with methylene chloride is used for dry foods. The extract is chromatographed against *m*-methoxyphenol as internal standard. The use of HPLC methods with ultraviolet detection have been described by Herrman and Stoeckli (1982), Dalang *et al.* (1982) and Wallace (1983).

The presence of ethyl vanillin in vanillin is indicative of the use of synthetic essence (Martin *et al.*, 1964). Martin *et al.* (1977) have published data for assessment of the authenticity of vanilla extracts from the common geographical sources in Africa, Asia and America. The AOAC methods book described a chromatographic method for determination of both compounds and coumarin.

Most custard powders contain 0·1–0·6 per cent vanillin and coumarin derived from vanilla extracts and substitutes have occasionally been found. It is not considered to be a desirable food ingredient.

BAKING POWDER

Composition

Baking powder consists essentially of an acid ingredient mixed with an approximately equivalent amount of sodium bicarbonate and diluted with starch. Occasionally salt is present. The acidic ingredients that are most commonly employed are acid sodium pyrophosphate (ASP) and acid calcium phosphate (ACP). For many years acid calcium phosphate used for food purposes was manufactured from rock phosphate and liable to contain relatively high levels of contaminants, including fluorine. Modern ACP is no longer directly derived form rock phosphate and is subject to the detailed specifications in the Miscellaneous Additives in Food Regulations 1980 (SI 1980 No. 1889). Other ingredients that have been suggested for baking powder are adipic acid, sodium aluminium sulphate, citrates, calcium lactate and glucono-δ-lactone (see p. 502). The advantage of the latter is that it produces no acid or chemical taste (FSC, 1974). Specific purity criteria for the permitted materials are prescribed in Part II of the Miscellaneous Additives in Food Regulations.

The factors that affect the rate of reaction between various acid ingredients and sodium bicarbonate have been discussed by Conn (1965) and Morris and Read (1945). In general, cream of tartar and ASP react slowly but tartaric acid and ACP act rapidly. Some powders contain mixtures of ASP and ACP and others of cream of tartar and tartaric acid. The presence of ingredients that react more quickly also means a more rapid loss of available carbon dioxide during storage. In food manufacture, however, the separate ingredients are often weighed out just prior to the process. The acid phosphates which are then used sometimes have the acidity reduced by the addition of starch and such mixtures are often referred to as **'Cream Powders'**. A common mixture of this type contains 64 per cent of ASP, so that the acidity is approximately equivalent to that of cream of tartar (2 parts acid to 1 part bicarbonate). The 'strength' of an acid ingredient is expressed as the 'neutralising value' (see below).

Golden raising powder is essentially a mixture of baking powder and a yellow food colour.

Legislation

Formerly, the Food Standards (Baking Powder and Golden Raising Powder) Order, 1944 (SR and O, 1944 No. 46; 1946 No. 157) prescribed standards for available carbon dioxide and residual carbon dioxide shown below:

	* Available carbon dioxide (minimum) (per cent)	* Residual carbon dioxide (maximum) (per cent)
Baking powder	8	1·5
Golden raising powder	6	1·5

* As determined by the prescribed method.

The FSC (1970) stated that the Order was redundant. However, it remained in force until it was revoked by the Bread and Flour Regulations 1984. The FSC also noted that the trade and enforcement authorities were in favour of the retention of a minimum standard.

Analysis

Starch and mineral content may be separated by flotation (see p. 311).

The analysis of mineral ingredients should include the determination of **acidity**, which can be determined by boiling 0·84 g of the acid phosphate or cream powder with 20 g of salt and 50 ml water and titrating with 0·5 M sodium hydroxide to phenolphthalein. The titration multiplied by 5 gives the number of parts of sodium bicarbonate equivalent to 100 parts of acid phosphate (the neutralising value).

The analysis of baking powder and golden raising powder should include a **microscopical examination**, the determination of **the total and residual carbon dioxide**, the **ash**, possible contaminants and the qualitative detection of the acid ingredient. In the determination of lead it is preferable to extract from acid solution with diethyldithiocarbamate. In alkaline solution the lead is co-precipitated with calcium phosphate. In addition, the original composition may be required and it is then necessary to determine the alkalinity of the ash, phosphate, tartrate, etc. The aluminium in alum baking powders can be determined by atomic absorption spectrophotometry (Holak, 1970). Golden raising powder should also be examined for the added colour present.

Available carbon dioxide. Although the actual method for the final determination of the carbon dioxide was left to the individual analyst, the Food Standards (Baking Powder and Golden Raising Powder) Order defined available carbon dioxide as the difference between total carbon dioxide and residual carbon dioxide. The order prescribed the methods of analysis. In principle the method and definitions are similar to those prescribed for self-raising flour in the Bread and Flour Regulations 1984.

Residual carbon dioxide

Treat a sample of 2 g of baking powder or golden raising powder, with 25 ml of water and evaporate to dryness on a boiling-water bath. Repeat the treatment to dryness with a further 25 ml of water and evaporate. The residual carbon dioxide is the weight evolved when the reacted sample is treated with excess of dilute sulphuric acid at room temperature, followed either by boiling or by vacuum.

Total carbon dioxide is determined by measurement of the weight evolved when the baking powder or golden raising powder is treated with excess of dilute sulphuric acid at room temperature, the evolution being completed either by boiling for 5 min or by means of reduced pressure.

Determination of carbon dioxide

The carbon dioxide can be determined gravimetrically, volumetrically or gasometrically.

GRAVIMETRIC METHOD

The gravimetric train method is regarded as a reference procedure for the determination of the carbon dioxide. The gas produced from the sample is aspirated through U-tubes fitted with taps and packed with Sofnolite or Carbosorb to absorb the carbon dioxide, which is weighed as such. The air passing through the apparatus is first freed from carbon dioxide by passage through a tower containing sticks of potassium hydroxide. The reaction of the sulphuric acid with the sample is normally completed by boiling under reflux and the gas is dried by passage through tubes containing pumice and conc. sulphuric acid before reaching the absorption tubes.

VOLUMETRIC METHOD

The volumetric method has the advantage that the apparatus needed is composed of readily available pieces of simple equipment. The carbon dioxide produced by the addition of acid to the sample is absorbed in barium hydroxide solution and the excess alkali is back-titrated with oxalic acid solution.

The apparatus used consists of a large Buchner-type filter flask containing a test-tube of convenient size (Fig. 9.10).

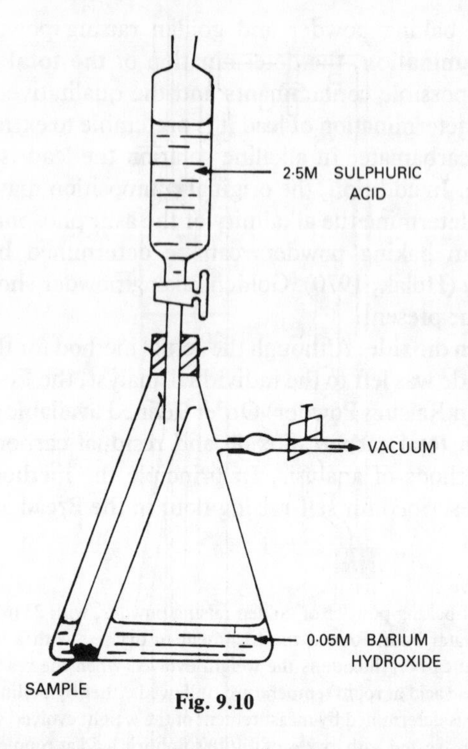

Fig. 9.10

Weigh out 0·25 g of baking powder into a dry test-tube. Pipette 50 ml of 0·05 M barium hydroxide solution into a filter flask (500–700 ml) and insert the tube containing the sample. Attach a tap funnel to the neck of the flask by means of a rubber bung so that the delivery end of the funnel fits inside the neck of the test-tube. Evacuate the flask via a short piece of rubber pressure tubing attached to the side tube of the flask. Close the evacuated flask by means of a screw clip fitted to the piece of tubing. Then run excess of 2·5 M sulphuric acid drop by drop from the tap funnel into the test-tube containing the sample avoiding inclusion of air. Close the tap and allow to stand overnight so that the carbon dioxide produced reacts completely with the barium hydroxide. In the morning remove the funnel, wash the liquid on the outside of the test-tube into the flask with a little distilled water and back-titrate the unused alkali with 0·05 M oxalic acid using phenolphthalein as indicator (*a* ml). Carry out a blank titration (*b* ml).

$$\text{Total } CO_2 \text{ (per cent)} = \frac{(b - a) \times 0{\cdot}0022 \times 100}{\text{weight taken}}$$

For the residual carbon dioxide treat 2 g of the sample in the manner prescribed in the Order. Transfer the residue to a suitable test-tube with the assistance of a little water, insert the tube in a filter flask and determine the residual carbon dioxide in the same way.

$$\text{Total } CO_2 - \text{Residual } CO_2 = \text{Available } CO_2.$$

The volumetric method is less suitable for the analysis of self-raising flour.

GASOMETRIC METHOD

The volume of carbon dioxide produced on addition of acid is measured in gas burettes. After correcting the volume to standard temperature and pressure, the weight of gas evolved from the baking powder is calculated. The Chittick method gives rapid results with both baking powder and self-raising flour:

The form of Chittick apparatus shown in Fig. 9.11 has a burette D graduated in millilitres at 20 °C with marks at 25 ml intervals from 25 ml to zero and from zero to 200 ml. Prepare the displacement solution,

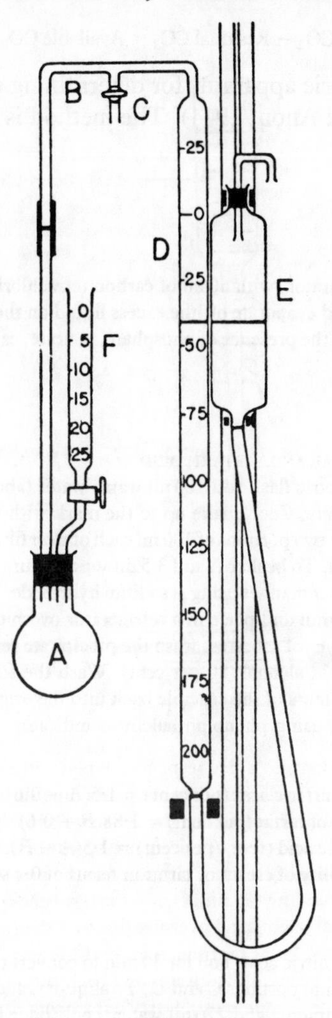

Fig. 9.11

dissolve 100 g sodium chloride in 350 ml water and add 1 g sodium bicarbonate, 2 ml of 0·5 per cent aqueous methyl orange solution and enough diluted sulphuric acid (1 + 10) to make just acid (pink). Check all joints on the Chittick apparatus for tightness and grease the taps. Introduce into the gas burette by means of the levelling bulb E a suitable volume of displacement solution. For the determination of the total carbon dioxide, weigh out 0·5–1·0 g baking powder or golden raising powder (or 10 g self-raising flour) into a dry flask A, add several large glass balls and connect to the apparatus. Close the tap on burette F and open tap C. By moving E, bring the displacement solution to the upper 25 ml mark on burette D, after allowing some minutes for the apparatus to equilibrate with the room temperature. Close C and lower E so that the level in D is about 10 ml above the zero mark. By means of F slowly add exactly 25 ml diluted sulphuric acid (1 + 10) to flask A. At all times during the decomposition keep the solution in E at a lower level than in D. Allow to stand for 2 min, shake A vigorously. Stand for 15 min, level the displacement solution in E and D and read off the volume of carbon dioxide produced. Correct the volume to STP, calculate the weight of gas produced (density of CO_2 = 1·976 g/l at 0 °C and 760 mmHg) and express as a percentage of the sample. Check the apparatus periodically using a known weight of sodium bicarbonate.

For the residual carbon dioxide determination treat the sample as prescribed in the statutory regulation, transfer the residue obtained to flask A and continue as for total carbon dioxide.

$$\text{Total } CO_2 - \text{Residual } CO_2 = \text{Available } CO_2.$$

A simple form of gasometric apparatus for determining carbon dioxide has been illustrated by Hughes (1942; Anon, 1944). The method is described in the AOAC methods book.

Ingredients

Qualitative tests

Shake about 4 g of sample in a separator, with 60 ml of carbon tetrachloride and transfer the deposit of baking chemicals to a small dish and evaporate off the excess liquid on the water bath. Examine the dry residue qualitatively for evidence of the presence of phosphate, tartrate, calcium, sodium and potassium.

Quantitative examination

DETERMINATION OF CREAM OF TARTAR AND TARTARIC ACID

To 2·5 g sample in a 250 ml volumetric flask add 100 ml warm water (about 50 °C), stand for 30 min at room temperature, shake occasionally. Cool, make up to the mark with water, shake vigorously, filter through a large fluted paper. Pipette two portions of 100 ml each of *clear* filtrate into two 250 ml beakers (A and B) and evaporate to about 20 ml. To beaker A add 3·5 ml M potassium hydroxide, mix well, add 2 ml acetic acid. Treat beaker B in a similar manner using M sodium hydroxide. Treat each mixture separately. Cool to 15 °C, stir vigorously for 1 min and place in a refrigerator overnight. Filter the precipitate on a Gooch crucible containing a thin layer of asbestos, wash the precipitate remaining on the sides of beaker through the filter with 75 ml ice-cold alcohol (80 per cent). Wash the sides of the crucible with 25 ml alcohol, suck dry. Transfer the contents of the crucible back into the original beaker with 100 ml of hot water and titrate with 0·1 M alkali using phenolphthalein as indicator. If the titrations are A and B respectively:

$$\text{Total tartaric acid (per cent)} = 1{\cdot}5(A + 0{\cdot}6)$$
$$\text{Cream of tartar (per cent)} = 1{\cdot}88(B + 0{\cdot}6)$$
$$\text{Tartaric acid (free) (per cent)} = 1{\cdot}5(A - B).$$

The figure 0·6 represents the solubility of cream of tartar in terms of 0·1 M alkali.

PHOSPHATE AND CALCIUM

Dissolve 1 g of sample in hot dilute nitric acid, boil for 30 min to convert pyrophosphate into orthophosphate and filter. Divide the filtrate into portions C and D. To aliquot C add molybdic acid solution (100 g molybdenum trioxide in 144 ml ammonia and 270 ml water, cool, pour into 1148 ml water and 489 ml nitric acid), warm to 68 °C, dissolve the washed phosphomolybdate precipitate in slight excess of 0·5 M sodium hydroxide and titrate the excess with 0·5 M HCl (p. 36). Calculate the phosphate as P_2O_5 (and to the equivalent of acid phosphate—see p. 343). Alternatively, pyrophosphates can be determined by the procedure of Czech and Fuchs (1962).

To determine the *calcium* in aliquot D, make the cold solution just neutral with ammonia, add ammonium acetate (considerable excess) and acetic acid, boil and add ammonium oxalate. Filter off the precipitate of calcium oxalate (which may contain traces of iron and aluminium), wash, dissolve in acid and titrate hot with 0·02 M (0·1 N) permanganate (see p. 32). 1 ml 0·02 M permanganate ≡ 0·002 004 g Ca ≡ 0·002 804 g CaO. Also calculate the calcium as ACP (see p. 343).

SULPHATE

Boil 3 g of sample with dilute hydrochloric acid to hydrolyse the starch and precipitate the sulphate as barium sulphate in the usual way. Calculate the calcium (as CaO) equivalent to the calcium sulphate present and deduct it from the total calcium (as CaO). Calculate the phosphate equivalent to the remaining calcium.

STARCH

Starch in baking powder is commonly estimated by difference but may be determined by dissolving the raising ingredients in the minimum quantity of cold dilute hydrochloric acid, filtering off the crude starchy matter and weighing it directly. Alternatively, the starch can be determined polarimetrically or by the diastase method.

CALCULATION OF THE ORIGINAL COMPOSITION OF BAKING POWDERS

Although the amount of available carbon dioxide falls during storage, the original composition of a baking powder (when manufactured) can be assessed reasonably accurately. The main figures required for the calculation are the residual carbon dioxide and the amount of the acid ingredient(s) present, e.g. phosphate as P_2O_5, calcium, cream of tartar, tartatic acid. The figures for ash and the alkalinity of the ash are often useful also for checking the calculation. Having determined the amounts of the acid constituents present, the bicarbonate is given by:

Total $NaHCO_3$ = $NaHCO_3$ equivalent to the acid present (available) + $NaHCO_3$ equivalent to the residual CO_2.
The amount of filler is usually obtained by difference:

Filler (per cent) = 100 − Total $NaHCO_3$ (per cent) − Total acid ingredients (per cent).

Salt is occasionally present and this must be allowed for in the calculation of filler and also when using the ash figure for configuration of the general results. The moisture content of the filler can also be calculated by assuming that all the moisture in the baking powder is associated with the starch present. The direct determination of moisture in baking powder presents certain difficulties. Drying over concentrated sulphuric acid in a vacuum desiccator to constant weight is probably the most satisfactory for general use. Heating the sample (as in conventional drying and distillation methods) causes decomposition of the bicarbonate. It may be possible to allow for such losses by performing blanks using pure bicarbonate. Lim has pointed out, however, that if ASP is present a further correction is necessary in the distillation method, which may be caused by pyrophosphate hydrolysis. Interference due to the same cause also appears to produce errors if the moisture is determined by the Karl Fischer technique.

The amount of bicarbonate equivalent to the acid constituents can be calculated from the following equivalents for 1 part of bicarbonate:

Cream of tartar	2·24
Tartaric acid	0·893
Pure ASP	1·32
Commercial ASP	1·36
Pure ACP	1·045
Commercial ACP	1·25

Commercial ACP contains approximately 57·0 per cent P_2O_5 and 18·4 per cent Ca.

With baking powders containing cream of tartar and tartaric acid the calculations follow similar lines. The theoretical chemical equations can be employed and the final result can often be readily checked from the ash and the alkalinity of the ash on the assumption that the bicarbonate is converted on ignition to Na_2CO_3, that the cream of tartar goes to K_2CO_3 and the tartaric acid gives no ash.

EGGS

Composition

Hens eggs on the average weigh 57 g and consist of 57 per cent white, 32 per cent yolk and 11 per cent shell. This compares with an average weight for duck eggs of 78 g, consisting of 52 per cent white, 37 per cent yolk and 11 per cent shell. Table 9·3 gives the average composition figures for the edible contents.

Table 9.3 Average composition of egg products

	Duck			Hen		
	Whole egg (%)	Yolk (%)	Albumen (%)	Whole egg (%)	Yolk (%)	Albumen (%)
Solids	24·5–26·0	47·0–50·0	11·0–12·5	30·3	55·0	12·8
Protein	11·8–12·2	15·5–17·5	9·0–11·0	13·9	17·0	11·0
Lipid	11·0–11·6	30·0–32·0	0 –0·04	14·4	35·7	0·05
Ash	0·8–1·0	1·0–1·6	0·5–0·7	1·0	1·2	0·7
Carbohydrate	0·3–0·6	0·2–0·5	0·4–0·9	1·2	1·1	1·0
Chloride (NaCl)	0·3	0·3	0·3	—	—	—
Phosphorus (P_2O_5)	0·5	1·38	0·05	—	—	—
pH	7·7	6·3	9·1	—	—	—
Sp. gr.	1·048	1·026	1·055			

The shell of hen and duck eggs contains about 3·5 per cent protein, 1·5 per cent water and 95 per cent of inorganic matter (mainly calcium carbonate with traces of magnesium carbonate and calcium phosphate).

The chemical composition of eggs has been reviewed by Parkinson (1966) and by Petersen (1978); data are also presented in food tables by Paul and Southgate (1978). The main proteins that have been identified in egg white are ovalbumin (over 50 per cent), conalbumin, ovomucoid, globulins and ovomucin.

Galyean and Cotterill (1979) fractionated egg white using DEAE cellulose chromatography followed by polyacrylamide gel electrophoresis. Minor changes in protein structure following spray drying were noted.

The yolk contains simple proteins (livetins) and phosphoproteins (vitellin and vitellenin). Most of the phosphoproteins are present in loose combination with phospholipids as lipoproteins. The lipids of egg yolk include about 62 per cent triglycerides, 21 per cent lecithin, 7 per cent cephalin, 4 per cent cholesterol and 0·5 per cent FFA. Typical fatty acid percentage composition for egg total lipid is:

C14:0	0·6	C16·0	25·8	C16:1ω9	0·8
C16:1ω7	3·4	C17·0	0·3	C17:1ω8	0·3
C18:0	8·6	C18·1ω9, 11	41·1	C18:1ω7	2·4
C18:2ω6	11·2	C18·3ω3	0·5	C18:1t	0·7
C20:1ω9	0·4	C20·4ω6	2·2	others, <0·3	1·7

Polyunsaturated fatty acid composition varies with hen's diet.

Legislation

The EC has made three regulations on marketing standards for eggs. Regulation 2772/75 sets three quality classes, Class A (fresh eggs), Class B (second quality or preserved eggs) and Class C (non-graded manufacturing eggs). Class A may be described as 'EXTRA' if they meet certain fairly stringent conditions. The seven weight grades for Class A and B eggs are as follows (Class C eggs need not be weight graded:

Grade	*Weight per egg (g)*
1	70 or over
2	65–70
3	60–65
4	55–60
5	50–55
6	45–50
7	under 45

The Eggs (Marketing Standards) Regulation 1985, SI No. 1271, as amended by SI No. 1184, 1987, provide the enforcement of Regulation (EC) No. 3494/86 (which amends Regulation 2772/75) by redefining eggs as follows:

1. 'Eggs' means hen eggs in shell suitable for direct human consumption or for use in the food industries, except for incubated eggs.
2. 'Industrial eggs' means hen eggs in shell other than those defined as 'eggs', including incubated eggs.
3. 'Grade C eggs' shall be eggs which do not meet the requirements applicable to eggs in grades A or B. They may only be passed to egg processing plants or to industry.

Regulation (EC) No. 1943, 1985 defines minimum criteria to be met by producers for eggs labelled as free range, semi-intensive, deep litter and brown.

Examination

The shell is not impervious, but contains numerous minute pores. These are essential for the interchange of oxygen, etc. between the growing embryo and the atmosphere, but unfortunately also permit the entry of bacteria and moulds. The yolk of the egg should be firm and surrounded by a strong vitelline membrane. The white consists of three distinct layers, namely inner thin white (30 per cent), thick white (50 per cent) and outer thin white (20 per cent). In a newly-laid egg the thick white is attached to the shell membranes at the ends and the yolk is anchored to the thick white by the chalazae. The vitelline membrane is semi-permeable and during storage water diffuses slowly from the white to the yolk. When a good-quality egg is cracked on to a plate the yolk is tall and compact and stands centrally in a thick layer of white. A thin white surrounding a flattened yolk indicates a lower quality egg. The height of the albumen can be expressed in Haugh units (Brant *et al.*, 1951).

Eggs are examined by holding against the aperture of a candling lamp, broad end uppermost, before twirling to and fro round the long axis. The internal quality is best judged from the visibility, ease of movement and shape of the yolk. As the internal structure weakens, the closer can the yolk approach the shell and the more visible it becomes on twirling. During storage air replaces the evaporated water and in general the air space should not exceed 7 mm in depth. Candling should also reveal blemishes such as blood spots, blood eggs and meat spots (MAFF, 1958). Candling will also detect incubated eggs that will not hatch (incubator reject eggs). At present the addition of such eggs (less than 6 days old) to egg products is allowed in the EC but is forbidden in the USA and Canada (but see below, under egg products). The sealing of eggs with waterglass and its detection are discussed by Nicholls (1931). Preservation of eggs by treatment with epoxide vapour and storage in sterile mineral oil has been described by Wilcox (1971).

The pH of the white of a newly-laid egg is about 7·8, but due to carbon dioxide diffusing out rises to 9·3 after 3 days and remains constant. Changes in the ammoniacal nitrogen and fat acidity during storage of course affect the odour and taste. Since duck eggs are often laid near ponds and stagnant pools they are more likely to be contaminated by *Salmonella* species than are hens eggs. Romanoff and Romanoff (1949) have published data on eggs from a range of avian species.

EGG PRODUCTS

Legislation and standards

A UN/ECE (United Nations Economic and Social Council and Economic Commission for Europe) standard for hens' egg products is set out in Table 9.4.

Table 9.4 Standards for egg products

	Egg solids (% min.)	Fat (% min.)	FFA of fat as oleic acid (% max.)
Liquid and frozen whole egg	24·0	10·0	—
Liquid and frozen egg yolk	43·0	27·0	—
Liquid and frozen egg albumen	10·5	—	—
Dried whole egg	95·0	30·0	3·5
Dried egg yolk	95·0	56·0	3.5
Pan dried egg albumen	84·0	—	—
Spray dried egg albumen	92·0	—	—

Council Directive 89/437/EEC (OJ No. L212, 22.7.89, p. 87) on the hygiene and health problems affecting egg products production and marketing requires (effectively from December 1991) *inter alia*, the monitoring for residues of products with pharmacological or hormonal action, antibiotics, pesticides, detergents or other harmful or deleterious substances, and the meeting of certain microbiological and chemical criteria. The chemical requirements are: 3-hydroxybutyric acid, not more than 10 mg/kg d.b. (see p. 350); lactic acid, not more than 1000 mg/kg d.b.; succinic acid, not more than 25 mg/kg d.b. (d.b.; on the dry solids basis).

Liquid egg. The manufacture and treatment of liquid egg has been described by Heller (1964) and by Wilcox (1971). During the preparation of batches of liquid egg for manufacture there is some possibility of the presence of *Salmonella* due to the presence of a small proportion of infected eggs. Such organisms can, however, be destroyed by pasteurisation at 64·4–65·5 °C (148–150 °F) without coagulating the egg or impairing the baking qualities (Knight, 1963). Monsey and Jones (1979) showed that egg white can be pasteurised by heating at 57·2 °C for 2½ minutes. Shrimpton *et al.* (1962) showed that the reduction of activity of α-amylase by heat could be made the basis of a simple routine test for assessing the adequacy of heat treatment. For the test the sample is incubated at 44 °C with starch. Any active amylase (present in unheated or insufficiently pasteurised egg) is able to break down the starch, so that the addition of iodine will not give a blue colour. When the enzyme has been destroyed due to adequate pasteurisation, the starch is unaffected and the normal blue colour develops. Subsequently, the Liquid Egg (Pasteurisation) Regulations, 1963 (SI No. 1503) required that liquid egg should be heated to at least 64·4 °C for at least 2½ min and the cooled product should comply with the α-amylase test.

The regulations contain detailed conditions for the pasteurisation process, sampling, transport and the analytical method for checking α-amylase activity (see below). They also apply to the separated components of egg (yolk and white) whether frozen, chilled or otherwise preserved.

Liquid egg and the separated or blended products may be stabilised by the addition of salt and preservatives such as benzoates or sorbates. Other additives may include vegetable oils as anticoagulants, acids and bases as pH adjusters, and phosphates as colour stabilisers (generally for yolk), emulsifiers and thickeners. Liquid egg contains about 0·4 per cent carbohydrates. If present in the spray-dried product, this can cause browning reaction on storage. Liquid egg and liquid albumen is subjected to a desugaring process using glucose oxidase/catalase before spray drying to reduce the glucose level to less than 0·01 per cent in the final product. Conversely, the aerating properties of egg may be stabilised by sugaring at the rate of 1 part sugar to 2 parts of egg followed by pasteurisation and vacuum evaporation to yield a concentrated product.

Frozen whole egg. The product is prepared using freezing temperatures between −24 °C and −40 °C. Such temperatures, combined with air circulation, cause rapid freezing, which is important as it reduces bacterial multiplication and produces a more uniform product. After freezing the product is stored below −9·5 °C. The product should contain less than 75 per cent water, at least 10 per cent of oil (extract) and the FFA of the fatty extract should not exceed 3 per cent as oleic acid. Boric acid has been used as a preservative in frozen egg products.

Dried egg. Dried eggs may be produced by spray-drying or the AFD process (Anon, 1968a, b). The moisture content of fresh samples should not exceed 5 per cent but it may rise during storage to as much as 9 per cent. The other main constituents present usually fall within the following percentage ranges: protein (N × 6·68) 40·4–50·5, chloroform extract (fat, lecithin, etc.) 40–46, ash 3·2–4·2, chloroform soluble phospholipid as P_2O_5 1·05–1·40.

The separated components are also available as dried products: dried yolk generally contains 3–4 per cent moisture, 31·0–32·5 per cent protein and 60–64 per cent lipids; dried albumen contains 5–7 per cent moisture, 82–84 per cent protein and 0·5 per cent lipid. Anti-caking agents may be incorporated into dried egg mixes.

Analysis of egg products

Sampling. No simple rules can be adopted to cover the range of egg products since conditions differ widely. Eggs in shell are sampled as individual units and a composite sample made by breaking out. Liquid egg is packed in different sized containers; in general a 300–500 g sample is drawn and combined in a single sample jar. The sample is kept in a freezer or packed in solid CO_2 until ready for analysis. Frozen egg is withdrawn from bulk with a corer or auger; composite samples are prepared from approximately 300–500 g and the temperature of the sample is raised to approximately 40 °C to allow mixing before analysis. A 300–500 g sample of dried egg is prepared by mixing the contents of a suitable number of small packets; for larger containers, small amounts of sample are withdrawn and combined to give a composite sample. The powder is mixed by passing through a domestic flour sifter, lumps are broken and the powder mixed. Flake albumen is ground to pass through a 60 mesh sieve. The combined sample is kept in a hermetically sealed jar in a cool place.

Total solids. Four methods are available for determination of total solids.

The AOAC method (5 hours in a vacuum oven at 98–100 °C) is generally regarded as the most accurate procedure. Liquid products are pre-dried on a steam-bath before the vacuum oven treatment. The weight taken for analysis varies (2 g for dried egg, 2·5 g for yolk, 5 g for liquid and 10 g for albumen). In the less accurate 'quick oven' method, the sample is dried for 1 hour. Infra-red drying instruments are used in quality control laboratories, and are suitable for routine use provided they are carefully calibrated; analysis takes 10–40 minutes per sample. In plant quality control of total solids of liquid egg is carried out by refractometric reading; a pocket instrument calibrated over the range 0–50 per cent sugar solids or an Abbe refractometer is used. Pour a well mixed 20 ml aliquot of the liquid egg into a 25 ml stoppered measuring cylinder. Add by pipette 1 ml 0·88 ammonia to the cylinder. Mix by inversion without aeration. Allow to stand for 5 min. Mix by inversion two or three times. Transfer two or three drops of the clarified sample to the refractometer prism and read, note the temperature of the instrument. (Preferably the pocket instrument and sample should be maintained at the same temperature by immersion in a water bath.) Liquid yolk is diluted 1:1 with water before clarification; liquid albumen may not need clarification before reading. Preferably the temperature of the sample should be adjusted to 20 ± 0·2 °C; alternatively, a temperature correction must be applied according to the formula $R_{20} = R_T - 0{\cdot}08\,\Delta T$

where R_{20} = refractometer solids at 20 °C

R_T = refractometer solids at T °C.

$\Delta T = 20° - T°$.

For pasteurised whole egg, ammonia clarified (15 drops of 0·88 NH_4OH to 20 ml of egg), the linear relationship (± 0·5 per cent at 20 °C) using a sugar refractometer is as follows:

Refractometer reading	Whole egg solids (per cent)
24·0	22·2
25·0	23·3
26·0	24·4
27·0	25·5
28·0	26·6

Lipids. Chloroform/ethanol (1:1) is used in the AOAC direct extraction procedure. Place 12 ml liquid whole egg (5 ml liquid yolk) into a 100 ml volumetric flask. Add 25 ml mixed solvent very slowly, shaking constantly until all the protein is coagulated and all lumps dispersed. Then add 60 ml solvent. Shake at 5 min intervals for 1 hour, make to mark. Stand until clear. For liquid albumen, pre-dry a 60 ml sample on a steam bath. Dry at 98–100 °C for 90 min in an oven. Cool in a desiccator. Weigh 5 g of the solids into a 100 ml volumetric flask and treat as above. In the case of dried egg products take 3 g whole egg, 2 g whole yolk, 10 g whole albumen. Treat with chloroform/ethanol as above. Transfer a 50 ml aliquot of the clear solvent into a beaker. Evaporate to dryness on a steam bath. Oven dry for 15 min at 98–100 °C. Dissolve the residue in 10 ml chloroform. Filter through a plug of cotton wool in the stem of a small funnel into a dry tared 100 ml beaker. Wash with 10 ml for 90 min. Weigh and calculate the lipid content.

The Soxhlet procedure may be applied to dried products. Chloroform/ethanol is used for extraction. Alternatively, the AOAC acid hydrolysis procedure (preferred by UN/ECE) may be used, the value obtained being approximately 90 per cent of that obtained by the direct chloroform/ethanol method. The acid hydrolysis method, three chloroform/methanol extraction procedures, a hexane/isopropanol procedure and a fats-by-difference calculation have been compared by Fletcher *et al.* (1984).

The isolation of egg phospholipids by TLC and silicic acid column chromatography is described by Moschidis *et al.* (1984), and HPLC by Hanson *et al.* (1981). The free lipids are essentially triglyceride, whilst 'bound lipid' contains the phospholipid.

Egg oil has the following characteristic figures: refractive index at 20 °C, 1·4655–1·4670; saponification value, 188–198; iodine value, 60–70; unsaponifiable matter, up to 4 per cent; phosphorus (as P_2O_5) 1·4 per cent. The fatty acid composition is given on p. 344.

Lipid phosphorus. The amount of organic phosphorus extracted varies with the solvent used (Manley and Lobley, 1948). Extract 3–10 g dried sample for 2–3 hours with warm chloroform in a continuous extractor. The extract is then dried and weighed to ascertain the fat content. The extract is wet oxidised using 2–5 ml of sulphuric acid and nitric acid, and the phosphorus determined as phosphate (p. 36). For a further method employing methanol see the section on Salad Cream.

Protein. The Kjeldahl method is used; the AOAC recommends mercury/potassium sulphate mixture as catalyst but alternatively a mixture of 96 per cent sodium sulphate (anhydrous), 3·5 per cent copper sulphate and 0·5 per cent selenium dioxide may be used. The following conversion factors have been recommended: white, N × 6·70; yolk, N × 6·62; whole egg, N × 6·68.

A correction for non-protein nitrogen may be applied. This takes into account the nitrogen that occurs mainly in phospholipid. The nitrogen in lipid is considered to be 0·6 per cent. Lipid nitrogen as protein = $0{\cdot}006 \times N_F \times$ Fat per cent where N_F = appropriate nitrogen factor.

Glucose. An enzymic procedure or Somogyi titrimetric method may be used (Chapter 6). Deproteinisation is necessary but traces of sugars tend to be lost in the process. Desugared eggs normally contain less than 0·1 per cent glucose but this level cannot be measured.

Free fatty acids in dried eggs. The sample (15 g dried whole egg or 11 g dried yolk) is mixed with 5 g sodium sulphate (anhydrous) and extracted with 100 ml chloroform/ethanol (1:1). An aliquot of the solvent is titrated with 0·1 M sodium hydroxide to a phenolphthalein end-point. The result is calculated in terms of oleic acid and for good quality dried egg the value (calculated as oleic acid) should not exceed 3 per cent of the weight of fat.

Ash. Ashing is normally carried out in stages; liquid products are pre-dried on a steam bath followed by vacuum oven and 3 g of dried solids are transferred to a tared dish. The dish is placed in a muffle furnace and the temperature is raised to 250 °C for 1 hour and then by gradual stages by 300 °C (20 °C/hour). When fuming ceases the char is broken with a fine probe and the ashing is completed at 500 °C.

α-Amylase. The sample is subjected to the test that is detailed in the Schedule to the liquid egg (Pasteurisation) Regulations 1963, to determine the efficacy of the pasteurisation process. The regulations require the sample to be tested as soon as possible after receipt in the laboratory. Any sample that shows evidence of deterioration should not be subjected to the test.

SAMPLE PREPARATION

The sample to be examined is prepared for the test as follows:

1. Pasteurised whole egg—the original sample.
2. Pasteurised liquid yolk—dilute 5 ml yolk with 10 ml water.
3. Dried whole egg—mix 20 g dried egg with 60 ml water, take 15 ml for the test.
4. Dried yolk—mix 10 g dried yolk with 50 ml water, take 15 ml for the test.

REAGENTS

Starch solution. Weigh an amount of soluble starch, of analytical reagent quality and of known moisture content, equivalent to 0·7 g of dry starch. Mix the starch to a thin cream with cold water. Transfer the cream to about 50 ml of boiling water, boil for 1 min and cool by immersion in cold water. Add three drops of toluene and dilute with water to 100 ml in a volumetric flask. The solution is stable for 14 days.

Solution of iodine. Dissolve 0·1269 g iodine and 3·6 g potassium iodide in 1 litre of distilled water. Prepare freshly before use. The solution may be made by dilution from a stronger solution with appropriate adjustment of potassium iodide concentration.

Solution of trichloroacetic acid: Prepare a 15 per cent m/v solution of trichloroacetic acid in water.

METHOD

Weigh out 15·0 g of the prepared sample into a small flask. Add 2·0 ml of the starch solution and mix thoroughly. Place this mixture for 30 min in a water bath maintained at 44 ± 0·5 °C. Remove the mixture and cool. Add 5·0 ml of the mixture to 5·0 ml of trichloroacetic acid solution and shake thoroughly. Add 15 ml of water, shake again, filter, reject the first runnings. Add 10 ml of the clear filtrate to 2 ml of iodine solution contained in a test-tube.

INTERPRETATION

The sample passes the α-amylase test if the filtrate in the solution of iodine immediately turns a blue–violet colour. For this purpose colours more blue–violet than 3 of a standard Lovibond Comparator Disc 4/26, or of a comparable spectrophotometric standard are taken as satisfactory.

The colour should be compared in an all-purpose Lovibond Comparator using a 25 mm cell. The reagents and procedure can be checked by preparing two control tubes at the same time. In one of these (1) the egg is replaced with an equivalent amount of water and in the other (2) the starch is replaced with an equivalent amount of water. Tube (1) should be a deeper blue and (2) a lighter blue than any shade on the disc. The test is inapplicable to samples that contain sugar, citric acid or citrates as these substances interfere with the reaction.

pH. The pH is measured using a standard pH meter with glass/calomel electrode combination. Sample preparation is as follows:
Liquid egg products—measure pH directly.
Dried whole egg—mix 20 g with 60 ml water.
Dried albumen—mix 10 g with 80 ml water.
Dried yolk—mix 20 g with 100 ml water.
Frozen egg product—bring to room temperature and measure pH directly.

Incubator reject eggs. The presence of 3- or β-hydroxybutyric acid is an indicator of the presence of incubator reject egg in egg products. Proposed methods are GLC (Robinson *et al.* 1975) and enzymatic assay (Parry *et al.*, 1980; BCL, 1987). Elenbaas *et al.* (1986) have shown that the two techniques give comparable results, with detection limit of 0·5 mg/kg of 3-HBA. Uijttenboogaart *et al.* (1986) found that incubator reject eggs held in incubators for longer than six days contained about 30 mg/kg 3-HBA, whereas table eggs contained about 0·7 mg/kg.

NIR. Osborne and Barrett (1984) have described the use of infra-red transmission spectroscopy for the analysis of protein, total lipid and total solids content of liquid egg products.

Solubility of dried whole egg. The solubility of dried whole egg can be assessed by the Haenni method. In the modification described by Hawthorne (1944), 1 g of the powder is placed with 5 ml of 5 per cent sodium chloride solution (m/v) in a stoppered tube, which is then shaken thoroughly in a standardised manner. The refractive index of the dispersed sample and of the salt solution are both measured.

$$\text{Haenni value} = y = (N_D^{25} \text{ of sample soln} - N_D^{25} \text{ of solvent}) \times 1000.$$

From Hawthorne's graph:

$$\text{Solubility} = \frac{\log_{10} y - 0{\cdot}445}{0{\cdot}01}$$

Salt is used as it gives clearer solutions than water (cf. Fryd and Hanson, 1944). Alternatively, the following direct method may be used:

METHOD

Weigh 1 g of sample into a dry beaker, add 10 ml of water, allow the mixture to soak for 3 h and then heat on a water bath at 50 °C for 30 min. Transfer the mixture to a centrifuge tube without washing the beaker and centrifuge for at least 1 min. Pour off the upper layer without disturbing the sediment and then wash the beaker with 10 ml of water at 50 °C. Pour the water into the centrifuge tube and, after mixing, re-centrifuge. Pour off the upper water and wash the sediment with a further 10 ml of water. After this transfer the sediment to a tared filter or basin, dry and weigh. A good-quality fresh sample usually shows a solubility of over 90 per cent, but there is a marked reduction during storage and with increasing moisture content.

SALAD CREAM AND MAYONNAISE

Composition

The average composition of salad cream sold in the UK has been reported by McCance and Widdowson (1960) and by Paul and Southgate (1978). The values are compared in Table 9.5.

Table 9.5 Composition of salad cream

	Water (%)	Sugars (%)	Total nitrogen (%)	Lipid (%)	Phosphorus (mg/%)	Salt (%)
1960	47·4	10·3	0·52	36·0	90	2·1
1978	52·7	13·4	0·30	27·4	90	2·1

Legislation

The Salad Cream Regulations, 1966 (SI 1966 No. 1051) define salad cream as any smooth, thick, stable emulsion of vegetable oil, water, whole egg or egg yolk and an acidifying agent, with or without the addition of one or more of the following substances, namely vinegar, lemon juice, salt, spices, sugar, milk, milk products, mustard, edible starch whether modified or not, edible gums and other minor ingredients and permitted additives, and the expression includes mayonnaise, but does not include horse-radish sauce, horse-radish cream, ice-cream, imitation cream or mustard sauce sold as such.

The regulations require salad cream to contain not less than 25 per cent edible vegetable oil, and not less than 1·35 per cent egg yolk solids (both calculated on a m/m basis).

Although the UK standards for mayonnaise are identical with those for salad cream, the former is required to conform to higher standards in some other countries (Table 9.6).

Table 9.6 Standards for salad cream and mayonnaise in various countries

Country	Minimum oil content (%) Salad cream	Mayonnaise
UK	25	25
USA	30	65
Switzerland	60	75
Belgium	50*	80

* Also lower minimum for egg yolk

The Food Advisory Committee (FdAC/REP/5: 1989) have observed that the Regulations do not satisfactorily standardise the product and the nature and oil contents vary widely from product to product. As a consequence the FdAC recommend revocation of the Regulations.

Analysis. The analysis of salad cream should include the determination of the oil, egg-yolk solids and trace elements. The oil should be identified by the fatty acid profile in conjunction with the usual oil values and examined for evidence of rancidity. Imported samples may contain added preservative.

In quality control the following determinations may be required: total solids (dry on sand), acidity (as acetic acid), sugar, ash, salt, efficiency of emulsification (under the microscope the bubbles of oil should be small and regular in size), viscosity, colour and stability.

The manufacture and control of salad cream has been discussed by Devey (1959a, b). A review discussing the influence of egg yolk, volume of phases, mustard, mixing, water hardness and viscosity on mayonnaise emulsions is given by Harrison and Cunningham (1985).

Oil. Use a modified Werner–Schmid method.

Shake 1 g of sample with 10 ml of 5 M hydrochloric acid and place in a beaker of water at 70 °C. Boil the water and remove the tube after half an hour at boiling point. Add water, cool and shake with 25 ml of ether followed by 25 ml of light petroleum. Transfer the organic solvent layer to a weighed flask. Extract at least three times with solvent, combine the extracts, remove the solvent and dry the oil for 40 min in the 100 °C oven. Weigh and calculate the oil content.

Davies and Brocklehurst (1985) have proposed the use of NIR methods for estimating the oil content of mayonnaise and mayonnaise salads. The precision of the method is claimed to be better than ±1 per cent at an oil content of 40 per cent (m/m).

Oil values. The acid extracted oil from the determination of oil (above) should not be used for assessing the edibility of the oil. For this assessment mix a separate portion of the sample with excess of anhydrous sodium sulphate and extract the oil by stirring in fresh ether. Filter the ether extract into a flask, and make up to a convenient volume in a volumetric flask. Determine the concentration of the extract by evaporating a known volume. Then determine the iodine and rancidity values on aliquot parts of the solution (air-dried where necessary). Alternatively, the general method described by Pearson (1965, 1974) using chloroform may be employed. The preferred oil extraction method is the chloroform/methanol procedure described in Chapter 2. The oil may be characterised by the determination of traditional values or by determining the fatty acid profile by gas chromatography (Chapter 16).

Egg-yolk solids. The accurate estimation of egg-yolk solids in salad cream is somewhat difficult. It is probably most frequently determined from the phospholipid content. The following method, employing methanol, appears to give reasonably satisfactory results (Manley and Lobley, 1948; Wojtowicz and Stauffer, 1968).

Reflux 20 g of salad cream with 100 ml of absolute methanol for 6 hours, stand overnight. Filter, re-extract the residue with methanol and wash through the filter with more solvent. Evaporate the combined filtrates containing the extracted organic phosphorus to dryness on the water bath. Transfer the residue to a digestion flask, wet oxidise with sulphuric–nitric acid and wash into a 100 ml volumetric flask. Make up to 100 ml at room temperature, mix and filter. Pipette 50 ml of filtrate into a further 100 ml volumetric flask, neutralise with ammonia (0·88), make just acid with dilute nitric acid, add 25 ml of vanadate–molybdate reagent and dilute to the mark. Mix and measure the absorbance at 270 nm (Chapter 2).

$$\text{Egg-yolk solids} = P_2O_5 \times 56$$
$$\text{Dried egg} = \text{Egg-yolk solids} \times 1{\cdot}48$$

The methanol extracts 1·20 per cent P_2O_5 from dried egg compared with 0·15 per cent from soya flour.

Alternatively, the phospholipid may be extracted by the total lipid method described in Chapter 2. Following addition of alcoholic potash and ignition to yield an alkaline ash the P_2O_5 content is determined colorimetrically.

A method for determining egg in salad cream has been proposed by Daubney and Sexton (1950). This is based on the colorimetric determination of the choline derived from egg lecithin by means of acid hydrolysis and conversion of the choline into the pink reineckate, which is soluble in acetone. For calculation purposes it is assumed that dried egg yolk contains 22 mg choline/g (Schwarz, 1961a, b). Casson and Griffin (1959, 1961) employ the enzyme lecithinase D for the hydrolysis of the phospholipids. The choline liberated is separated by adsorption on a cation-exchange resin and, after elution, is determined colorimetrically as the reineckate at 520 nm.

Punwar (1975, 1976) and Fostel (1984) have studied GLC methods for estimation of cholesterol in a number of multicomponent foods and ingredients. The lipid is extracted with chloroform/methanol and saponified. The unsaponifiable residue is extracted with benzene and after removing solvent the sterols are dissolved in dimethyl formamide. The trimethylsilyl ethers are separated and estimated by gas chromatography against 5,α-cholestane as internal standard. The coefficient of variation ranges from 5·6 to 23 per cent according to the nature of the product. The highest value recorded was for analysis of samples of mayonnaise. Chemin Douaud and Karleskind (1979) have described the use of a method based on the use of cholesterol oxidase and catalase. The lutidine formed in the reaction is measured colorimetrically. The determination of egg content is subject to the product containing no other ingredients with significant cholesterol content. Paul and Southgate (1978) report 2·57 per cent cholesterol in egg-yolk solids. Sweeney and Weihrauch (1977) have reviewed a number of methods for determination of cholesterol and have recorded the cholesterol content of a range of dairy products.

Armandola (1963) has described a method for the assessment of the egg content of noodles. This is based on the ratio of oleic acid to linoleic acid found by gas chromatography of the methyl esters of the fatty acids. For further discussion on the determination of egg content of foods, see p. 229.

Trace elements. A full analysis of salad cream should include the examination for arsenic, copper, lead and zinc. In view of the large amount of oil present it is advisable to employ a cold wet oxidation process, or where possible dry ashing for the destruction of organic matter.

REFERENCES

Anon (1944) *Analyst*, **69**, 19.
Anon (1968a) *Food Processing and Marketing*, **37**, 173.
Anon (1968b) *Food Trade Review*, **38**, 43.
Armandola, P. (1963) *Technica molitoria*, **14**, 94.
BCL (1987) *Methods of Biochemical Analysis and Food Analysis*. Lewes: Boehringer Corporation (London) Ltd.
Brant, A. W., Otte, A. W. & Norris, K. H. (1951) *Food Technology*, **5**, 356.
Casson, C. B. & Griffin, F. J. (1959, 1961) *Analyst*, **84**, 281; **86**, 544.
Chemin Douaud, S. & Karleskind, A. (1979) *Revue francaise des Corps Gras*, **26**, 313.
Chiang, B. Y. & Johnson, J. A. (1977) *Cereal Chemistry*, **54**, 429.
Conn, J. F. (1965) *Bakers' Digest*, **39**, 66.
Czech, F. W. & Fuchs, R. J. (1962) *Cereal Science Today*, **7**, 34.
Dalang, F., Martin, E. and Vogel, J. (1982) *Mitteilungen aus dem Gebeit der Lebensmitteluntersuchung und Hygiene*, **73**, 371.
Daubney, C. G. & Sexton, G. E. W. (1950) *Analyst*, **75**, 305.

Davies, A. M. C. & Brocklehurst, T. F. (1985) *Journal of the Science of Food and Agriculture*, **37**, 310.
Devey, J. D. (1959a, b) *Food Trade Review*, **29**, 4, 16.
Elenbaas, H. L., Muuse, B. G., Haasnoot, W., Rutjes, B., Stouten, P., Uijttenboogaart, T. G. & Steverink, T. G. (1986) *Journal of Agricultural and Food Chemistry*, **34**, 663.
FACC (1980) *Report on Modified Starches*, FAC/REP/31. London: HMSO.
Fawcett, P. (1985) *Food Flavourings, Ingredients, Processing, Packaging*, 7, 20.
Fletcher, D. L., Britton, W. M. & Cason, J. A. (1984) *Poultry Science*, **63**, 1759.
Flint, F. O. (1982) *Food Microstructure*, **1**, 145.
Fostel, H. (1984) *Ernahrung*, **8**, 723.
Fryd, C. F. M. & Hanson, S. W. F. (1944) *Journal of the Society of Chemical Industry*, **63**, 3.
FSC (1970) *Food Standards Committee Report on the Pre-1955 Compositional Orders*. London: HMSO.
FSC (1974) *Food Standards Committee Second Report on Bread and Flour*. London: HMSO.
Galyean, R. D. & Cotterill, O. J. (1979) *Journal of Food Science*, **44**, 1345.
Greenish, H. G. & Collin, E. (1904) *Anatomical Atlas of Vegetable Powders*. London: Churchill.
Hanson, V. L., Park, J. Y., Osborn, T. W. & Kiral, R. M. (1981) *Journal of Chromatography*, **205**, 393.
Harrison, L. J. & Cunningham, F. E. (1985) *Journal of Food Quality*, **8**, 1.
Hawthorne, J. R. (1944) *Journal of the Society of Chemical Industry*, **63**, 6.
Heller, C. L. (1964) *Food Trade Review*, **34**, 43.
Herrman, A. & Stoeckli, M. (1982) *Journal of Chromatography*, **246**, 313.
Holak, W. (1970) *Journal of the Association of Official Analytical Chemists*, **53**, 877.
Hughes, E. B. (1942) *Chemistry and Industry*, **61**, 103.
Ingleton, J. F. (1970) *Confectionery Production*, **36**, 757.
Knight, R. A. (1963) *Food Trade Review*, **33**, 49.
MacMasters, M. M., Wolf, M. J. & Seckinger, H. L. (1957) *Journal of Agricultural and Food Chemistry*, **5**, 455.
MAFF (1958) *The Testing of Eggs for Quality*. London: HMSO.
Manley, C. H. & Lobley, H. (1948) *Analyst*, **73**, 30.
Martin, G. E., Feeney, F. J. & Scaringelli, F. P. (1964) *Journal of the Association of Official Analytical Chemists*, **47**, 561.
Martin, G. E., Ethridge, M. W. & Kaiser, F. E. (1977) *Journal of Food Science*, **42**, 1580.
McCance, R. A. & Widdowson, E. M. (1960) *The Chemical Composition of Foods*. London: HMSO.
Monsey, J. B. & Jones, J. B. (1979) *Journal of Food Technology*, **14**, 381.
Morris, R. W. & Read, M. G. (1945) *Journal of the Society of Chemical Industry*, **64**, 291.
Moschidis, M. C., Demopoulos, C. A. & Kritikou, L. G. (1984) *Journal of Chromatography*, **292**, 473.
Nicholls, J. R. (1931) *Analyst*, **56**, 383.
O'Dell, J. (1979) *International Flavours and Food Additives*, **10**, 26.
O'Mahony, M. P. (1985) *Food Flavourings, Ingredients, Processing, Packaging*, 7, 23.
Osborne, B. G. & Barrett, G. M. (1984) *Journal of Food Technology*, **19**, 349.
Parkinson, T. L. (1966) *Journal of the Science of Food and Agriculture*, **17**, 101.
Parry, A. E. J., Robinson, D. S. & Wedzicha, B. L. (1980) *Journal of the Science of Food and Agriculture*, **31**, 905.
Paul, A. A. & Southgate, D. A. T. (1978) *The Composition of Foods*, 4th edn. London: HMSO.
Pearson, D. (1965, 1974) *Journal of the Association of Public Analysts*, **3**, 76; **12**, 73.
Petersen, D. W. (1978) *Journal of the American Dietetic Association*, **72**, 45.
Punwar, J. K. (1975) *Journal of the Association of Official Analytical Chemists*, **58**, 804.
Punwar, J. K. (1976) *Journal of the Association of Official Analytical Chemists*, **59**, 46.
Radley, J. A. (1976a) *Examination of Starch and Starch Products*. London: Applied Science Ltd.
Radley, J. A. (1976b) *Industrial use of Starch and its Derivatives*. London: Applied Science Ltd.
Robinson, D. S., Barnes, E. M. & Taylor, J. (1975) *Journal of the Science of Food and Agriculture*, **26**, 91.
Romanoff, A. L. & Romanoff, A. J. (1949) *The Avian Egg*. New York: Wiley and London: Chapman & Hall.
Sandstedt, R. M. (1965) *Cereal Science Today*, **10**, 305.
Sandstedt, R. M. & Mattern, P. J. (1960) *Cereal Chemistry*, **37**, 379.
Sandstedt, R. M. & Schroeder, H. (1960) *Food Technology*, **14**, 257.
Schlack, J. E. & Dicecco, J. J. (1974) *Journal of the Association of Official Analytical Chemists*, **57**, 329.
Schwarz, K. (1961a, b) *Deutsche Lebensmittel-Rundschau*, **57**, 201, 232.
Shrimpton, D. H., Monsey, J. G., Hobbs, B. C. & Smith, M. E. (1962) *Journal of Hygiene, Cambridge*, **60**, 153.
Sweeney, J. P. & Weihrauch, J. L. (1977) *Chemical Rubber Company Critical Reviews of Food Science and Nutrition*, 8, 131.
Uijttenboogaart, T. G., Steverink, T. G., Elenbaas, H. L., Haasnoot, W., Muuse, B. G. & Stouten, P. (1986) *Journal of Agricultural and Food Chemistry*, **34**, 667.

Vaughan, J. G. (1979) *Food Microscopy*. London: Academic Press.
Wallace, E. M. (1983) *Journal of Chromatographic Science*, **21**, 139.
Wallis, T. E. (1965) *Analytical Microscopy*, 3rd edn. London: Churchill.
Wilcox, G. (1971) *Eggs, Cheese and Yoghurt Processing*. New Jersey: Noyes Data Corporation.
Williams, K. A. (1966) *Oils, Fats and Fatty Foods*, 4th edn. London: Churchill.
Wojtowicz, P. F. & Stauffer, L. J. (1968) *Journal of the Association of Official Analytical Chemists*, **51**, 590.

10

Beverages and chocolate

TEA

Manufacture. Tea consists of the prepared leaves of various species of *Thea*, the commonest being *T. sinensis*, *T. bohea*, *T. viridis* and *T. assamica.* There are four stages in the manufacture of black tea, namely withering (exposure on racks to soften the leaf), rolling (rupturing the cells to release the juice for fermentation), fermentation (at about 27 °C to produce changes in the tannin) and firing (drying using hot air). Green teas are withered at high temperature, so that the enzymes responsible for fermentation are destroyed and the fermentation process is omitted. Oolong tea is only partially fermented. Apart from the country of origin and the individual garden marks, teas are usually graded as Orange Pekoe, Broken Orange Pekoe, Broken Orange Pekoe Fannings, Pekoe, Broken Pekoe, Broken Pekoe Souchong, Fannings and Dust. The terms 'Pekoe' and 'Orange' refer to the young leaf. Good teas are usually 'tippy', i.e. they contain a good proportion of 'Tips', which are the small golden pieces of leaf bud from the tip of the shoot. 'Fannings' have a smaller particle size and are often used in tea bags. Teas as sold in the retail trade are blends of up to 20 different types of tea. Blending is carried out after evaluation by experienced tasters taking into account origin, flavour, colour, body, pungency, price of the leaf and target price for the blend. The hardness of the water in the area in which the tea is likely to be infused may have an effect on appearance and flavour of the brew. The taster may develop special blends for sale in specific localities.

Tea bags have become the most popular form of retail packaging. Most of the popular brands sold contain about 3 g (just over 1/10 oz) of Fannings per bag.

Details of the manufacture, technology and biochemistry of tea have been published by Harler (1963), Eden (1976) and Millin (1987, 1988). Jain and Takeo (1984) have reviewed the nature of known enzyme systems in tea and have categorised their functions into four groups associated with flavour characteristics and plant metabolism.

Legislation

Apart from maximum limits prescribed for contaminants, there are no compositional standards laid down in the UK for tea. Requirements in overseas regulations are quoted in Table 10.1.

Imported tea comes from India, Pakistan, Sri Lanka and certain regions of Africa. Other sources are China, Japan, Formosa and Indonesia.

Table 10.1 Standards for tea in countries other than in the UK

	Canada	Australia	Switzerland	USA	Iraq
Total ash (%)	8 (max.)	—	—	4–7	5–7·5
Water-soluble ash (%)	2·75 (min.)	—	—	—	3 (min.)
Extractives (%)	30 (min.)	30 (min.)	—	—	30 (min.)
Stalks (%)	—	—	22 (max.)	—	10 (max.)
Caffeine (%)	—	—	—	—	1·0 (min.)

Composition

The chief constituents of tea are moisture, tannin, nitrogenous matter (including caffeine), oil, wax, inorganic matter (especially potassium salts) and fibre. Table 10.2 gives the ranges of analytical figures obtained with genuine black and green teas and spent tea.

Table 10.2 Analytical figures for black tea, spent tea and green tea (as percentages)

	*Genuine Black Tea	†Spent Black Tea	Green Tea
Moisture	3·9–9·5	—	6·1–9·2
Total ash	4·9–6·5	—	5·2–7·2
Water-soluble ash	3·0–4·2	Below 1	2·6–4·1
Acid-insoluble ash	0·1–0·4	—	0·05–0·9
Alkalinity of soluble ash (as K_2O)	1·2–2·0	Below 0·3	1·2–1·6
Extractives	30–50	10	33–45
Caffeine	1·9–3·6	Below 1	1·5–4·3
Tannin	7·3–15·1	—	—
Total nitrogen	5·0–6·2	—	—
Crude fibre	14–18	—	9–15
Ether extract	10–11	—	

* Figures quoted cover ranges for teas from India, Sri Lanka (Ceylon), China and Japan. In general, teas from China and Japan contain less extractives than other types (Manley, 1965).
† Spent tea (or exhausted tea) consists of the dried leaves after they have been infused.

The constituents that are of importance in relation to the overall flavour of tea are the caffeine, tannin and the volatile oil. In good teas the ratio of caffeine to tannin is about 1:3. Wood *et al.* (1964a, b), Wood and Roberts (1964) have correlated opinions of tea tasters with the results of chemical analysis and considered that colour and strength were associated with theaflavin and thearubigin and briskness with theaflavin plus caffeine. According to Stahl (1962) the flavour of tea is dependent on a number of volatile substances including terpenoid and sulphur containing compounds, fatty acids, alcohols, esters and carbonyl compounds (Wickremasinghe and Swain, 1964). Although over 80 compounds have been identified using gas chromatography (Bondarovich *et al.*, 1967; Pierce *et al*, 1969; Renold *et al.*, 1974) there is little evidence of which ones have the greatest organoleptic significance. Mahanta *et al.* (1985) associated volatile flavour components with the degradation of lipids during processing. Brewed tea is a significant source of fluoride (Monier-Williams, 1950). Michie and Dixon (1977) have studied the distribution of trace element content of the leaf between the infusion and spent leaf. They showed that the infusion is a significant source of manganese but that the leaf acted as a scavenger for toxic elements such as soluble lead salts.

Lead. Although aluminium foil replaced lead foil for the lining of tea chests some years previously, Williams (1964) reported the presence of 10–350 p.p.m. of lead in certain teas which came into the Port of London between 1959–61, notably in those from China, Formosa and Indonesia. Williams showed that the average amount of lead in teas imported from 10 countries in 1962 varied from 1·9 to 4·2 p.p.m. (see also Dixon and Michie, 1977).

Adulteration

In the past, tea was admixed with foreign leaves, spent tea, iron filings, etc. and green tea was often 'faced' with prussian blue, etc. but apart from some samples that contain excessive stalk, and others that are musty, adulteration is virtually a relic of the past (Williams, 1964).

Analysis of tea

A fairly complete picture of the quality of a tea can be obtained by carrying out an organoleptic and microscopical examination and also the determination of moisture, ash, water-soluble ash, alkalinity of soluble ash, acid-insoluble ash, total nitrogen, extractives, tannin, caffeine, crude fibre, ether extract, stalks, arsenic, copper and lead. For the routine analysis of tea and determination of the **ash**, **alkalinity of soluble ash**, **extractives**, **stalks**, **copper** and **lead** together with a **microscopical examination** should reveal most forms of adulteration. ISO has developed a number of standard methods for sampling, analysis and sensory evaluation of teas. Decaffeinated tea should also be examined for extraction solvent residues; **methylene chloride** is normally used for the extraction.

Organoleptic examination. Place 3 g of tea in a mug covered with a lid. Pour in boiling water, replace the lid and allow the infusion to stand for 5–6 min, corresponding to the extraction of the maximum caffeine/tannin ratio. Pour the liquor into a cup without handle to assess the taste. Shake the infused leaf from the mug onto the inverted lid. Procedures for sensory testing are given in BS 6008:1985 (ISO 3103–1980). The desirable characteristics vary according to the district of origin but in general the liquor should be bright and clear, have a slight pungency ('briskness') and a reddish colour. With milk no greyish yellow should be obtained. The infused leaf colour should be bright red or coppery (not brownish).

Microscopical examination. Soak the infused leaves in chloral hydrate solution (50 per cent) for several hours and then bleach them with hypochlorite and examine under the microscope. With green tea, however, the leaves are ready for microscopical examination after heating them in chloral hydrate solution for a short time. Teas after clearing should be treated with phloroglucin and hydrochloric acid in order to stain the sclerenchymatous matter. The main diagnostic characters are the long hair and the somewhat peculiarly shaped idioblasts (both of which stain with phloroglucin and HCl), the stomata surrounded by tangentially elongated cells and the cluster crystals of calcium oxalate (Fig. 10.1). An examination for filth is also appropriate in assessment of the quality of teas (AOAC); see also Lim (1981) for a method for use in identification of light filth material.

Sample preparation. ISO 1572 (BS 6049:Part 1:1985). For determination of moisture content, thoroughly mix the sample of leaf. For other determinations, grind the leaf in a high speed mill to pass a 0·56 mm sieve.

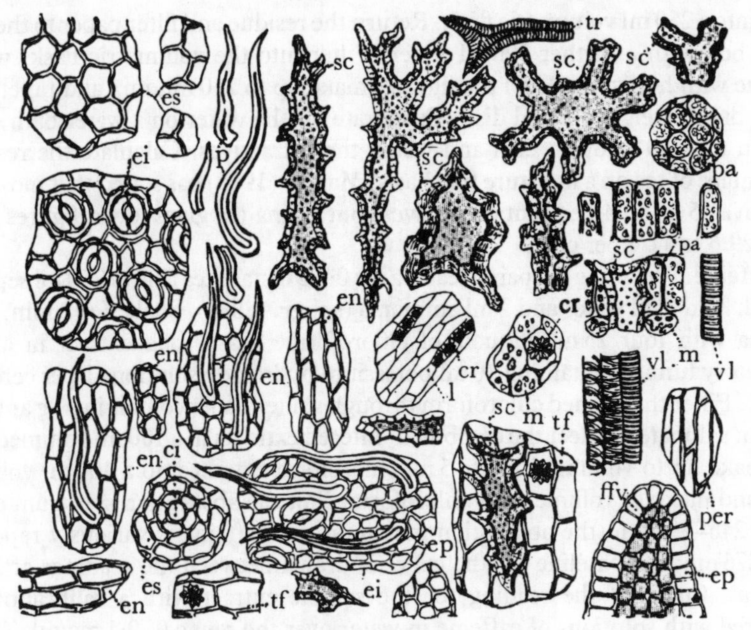

Fig. 10.1 Microscopical features of tea

cr, Crystals.
er, Lower epidermis.
en, Neural epidermis.
ep, Apex of marginal tooth.
es, Upper epidermis.
ffv, Debris of fibrovascular bundles.
l, Bast with cluster crystals.
m, Spongy parenchyma.
p, Simple hairs.
pa, p′a′, Palisade cells.
per, Pericycle, slightly lignified.
sc, Idioblasts from the mesophyll, cortical tissue, and the pitch of the stem.
tf, Cortical tissue.
tr, Tracheids.
vl, Vessel.

Moisture. Dry 5 g at 100 °C for 5 h. ISO 1573–1975 (BS 6049:Part 2:1985) prescribes 103 ± 2 °C for 6 hours. A good quality tea seldom has a moisture content in excess of 7 per cent. Tea with a moisture content in excess of 11 per cent is liable to go mouldy and consequently produce a musty infusion.

Ash. ISO 1575 (BS 6049:Part 4:1985). Carbonise 5–10 g prepared sample in a platinum dish at 525 °C, moisten the charred mass with hot water and re-ignite at 525 °C so that a greyish to brownish ash (not green) is produced. The total ash should not exceed 7 per cent. The main constituents of the ash are potassium (27–36 per cent as K_2O) and phosphorus (14–18 per cent as P_2O_5). The AOAC methods book prescribes use of olive oil to aid carbonisation.

Water-soluble ash, alkalinity of the soluble ash, acid-insoluble ash. Determine by the usual methods (Chapter 2). ISO 1576, 1577, 1578 (BS 6049:Parts 5, 6 and 7:1985). In view of the high content of manganese an aqueous solution of the ash may be pink due to the formation of permanganate. This colour will interfere in the titration for alkalinity. It may be avoided by the hot water and re-ignition treatment indicated above. The alkalinity of the soluble ash, which is usually calculated as K_2O, is useful for the detection of the presence of exhausted leaves (Table 10.2).

Water extractives. ISO 1574 (BS 6049:Part 3:1985). Boil 2 g tea (ground to pass through a 0·56 mm sieve) with 100 ml water for an hour under a reflux condenser, filter into a 250 ml volumetric flask. Return the residue and filter paper to the boiling flask, boil with a further 100 ml water. Filter into the volumetric flask, wash the residue with hot water. Cool the filtrate, make up to 250 ml, mix and pipette 50 ml into a dried, weighed metal dish. Evaporate off the water on a water bath, dry the dish in the 100 °C oven, cool and weigh the extractives. Calculate the result as a percentage of tea on a moisture free basis. Manley (1965) has stated that most Indian teas give 35·5–46·4 per cent extractives, but China tea gives lower figures ranging from 29·6 to 41·4 per cent.

Caffeine. Weigh 1 g prepared leaf tea (or 0·3 g instant tea) into a 250 ml separatory funnel. Add 3 ml water and 2 ml ammonia (sp. gr. 0·880). Shake for 2 min. Extract the tea with four 25 ml aliquots of chloroform. Wash each extract in a second separatory funnel containing 10 ml potassium hydroxide solution (1 per cent w/v in water). Filter the washed chloroform through a filter paper containing 1 g anhydrous sodium sulphate. Collect the combined, filtered extracts in a 100 ml volumetric flask and make up to volume. Pipette 5 ml chloroform extract into a 100 ml volumetric flask and make to volume with chloroform. Plot the absorption spectrum over the range 246–298 nm (the absorption maximum for caffeine is variously reported as 272–276 nm; the baseline for the curve is drawn between the readings at 246 and 298 nm). Compare the reading for the sample extract with a calibration curve prepared with solutions of caffeine in water over the range 0–0·1 mg/ml. A simple titrimetric procedure using chloramine-T has been described by Mayanna and Jayaram (1981).

For instant teas, coffee and coffee products the presence of free amino acids resulting from the breakdown of protein may necessitate the use of clean-up procedures before spectrophotometry. Conacher and Rees (1964) described the use of magnesium oxide and celite. Two columns of Celite, one basic one acidic, are used in the AOAC method, whilst Lehmann and Neumann (1973) described the use of a polyamide column. The use of GLC methods has been described by Newton (1969) and Vitzthum *et al.* (1974). Caffeine is extracted by the ammonia/chloroform method and GC is carried out using a nitrogen selective detector, which is capable of detecting 1 ng of caffeine. Strahl *et al.* (1977) have compared the spectrophotometric procedure with a gas chromatographic method. They reported that the spectrophotometric method gave higher results than the GLC method when applied to coffee but not for tea. Since the spectrophotometric method will include other purines it is likely that chromatographic procedures will give lower values. Procedures using HPLC have been applied to the determination. Madison *et al.* (1976) used a sulphonated cation exchange resin with 0·01 M nitric acid as mobile phase after extracting the caffeine with methylene chloride. In the authors' laboratory the following HPLC method has been developed using Spherisorb S50DS as column packing with a methanol mobile phase.

METHOD

Weigh 0·3 g prepared sample into a 100 ml volumetric flask. Add 50 ml methanol. Place the flask in a hot water bath for 30 min. Cool, make up to volume with methanol, and mix. Filter. Pipette 25 ml filtrate into a 100 ml volumetric flask. Add 50 mg sodium benzoate (internal standard). Make up to volume with methanol. Prepare a 250 × 4 mm stainless steel chromatographic column with Spherisorb S50DS packing. Equilibriate with methanol at a flow rate of 1·5 ml/min. Inject 10 μl of the prepared extract. Develop the chromatogram for 4 min by observation of the optical density of the effluent at 280 nm.

Calibrate by reference to standard solutions containing 0–100 mg caffeine and 500 mg sodium benzoate/litre of methanol.

Reid and Good (1982) evaluated a number of bonded phase silica materials and solvent systems for elution of caffeine and theobromine. C_{18} bonded silica is the preferred column packing material with chloroform as the eluting solvent. A similar procedure has been described by Terada and Sakabe (1984). Ashoor *et al.* (1983) have described an HPLC method similar to that given above but applied to decaffeinated beverages including tea and coffee; see also Blauch and Tarka (1983) and Ferrara *et al.* (1982).

Crude fibre. See Chapter 2.

Light petroleum extract. Dry 2 g of sample at 100 °C and extract with light petroleum for 5 hours in a continuous extractor.

Stalk. Boil 5 g for 15 min with 500 ml water. Transfer to a large basin and pick out the stalks with forceps. Dry the leaves and stalks separately in the 100 °C oven and weigh. The proportion of stalk should preferably be below 25 per cent. It is useful to examine the wet leaves and stalks for adulteration under a hand lens when the parts are being separated.

Tannin is usually determined by Lowenthal's permanganate oxidation process. The following is the modification recommended by the Tea Research Station, Assam, India (AOAC, 11th edition).

REAGENTS

Indigo carmine. Dissolve 1·5 g indigo carmine in 1 litre water containing 50 ml conc. sulphuric acid; 25 ml of this solution should require 4·0 ml 0·008 M potassium permanganate when titrated to a faint pink. The permanganate should be standardised against oxalic acid.

Gelatine. Soak 25 g gelatine for 1 h in saturated sodium chloride solution. Warm until dissolved, cool and make up to 1 litre with saturated sodium chloride solution.

Acid sodium chloride solution. Add 25 ml conc. sulphuric acid to 975 ml saturated sodium chloride solution.

PROCEDURE

Boil 5 g prepared sample with 400 ml water for 1 h, filter through cotton wool into a 500 ml volumetric flask and make up to the mark with water after cooling (A). Pipette 100 ml extract (A) into a 250 ml volumetric flask, add 50 ml gelatine solution and make up to the mark with acid sodium chloride solution. Pour into a conical flask containing 20 g kaolin, shake for 15 min and filter (filtrate B).

$$25 \text{ ml B} \equiv 10 \text{ ml A} \equiv 0{\cdot}1 \text{ g sample}$$

Pipette 10 ml of extract A into a litre flask, add 25 ml indigo carmine solution and diltute to 750 ml. Run in 0·008 M potassium permanganate solution from a burette, 1 ml at a time with shaking, so that the colour changes from blue through yellow to a faint pink (A ml). Titrate similarly 25 ml of filtrate B (B ml). Subtracting the appropriate blanks from A and B (usually 4·0 and 4·5 ml, see reagent preparation) the tannin titre is $(A - 4{\cdot}0) - (B - 4{\cdot}5)$.

1 ml 0·2 M potassium permanganate ≡ 0·0416 g tannin (gallotannic acid), so that 1 ml 0·008 M ≡ 0·001 664 g tannin.

The results are expressed on the dry basis.

Smith (1913) has described an alternative method in which the tannin is precipitated completely by saturated cinchonine sulphate solution. Cinchonine tannate contains 55 per cent tannin.

Methylene chloride. Page and Charbonneau (1984) have described a headspace GLC procedure for the determination of solvent residues in decaffeinated products.

The sample is first frozen before preparation and grinding, an analytical sample is introduced into saturated sodium sulphate solution and the gaseous phase is equilibrated and analysed on 80–100 mesh Porasil A. The effluent gas is passed through an electrolytic conductivity detector and the response is compared with methylene bromide as internal standard. Levels in the region of 0·05–20 mg/kg were reported in commercial samples.

INSTANT TEA

Tea is also sold as a dried extract called 'Instant Tea'. The analysis is normally carried out by methods described for instant coffee. Typical analyses are 2·4 per cent moisture, light petroleum extract 0·3 per cent, ash 7·8 per cent, water-soluble ash 5·5 per cent, alkalinity of soluble ash 4·5 ml N acid per g, caffeine 3·2 per cent, total reducing sugars (as invert sugar) 16·8 per cent and insoluble matter 0·16 per cent (Chang *et al.*, 1973). Cold water-soluble instant tea has been found to have an ash content of 20 per cent. There appears to be a significant loss in flavour and theaflavins during the manufacture (infusion and drying) of instant tea (Anon, 1968). The loss on drying should not exceed 5 per cent. Some imported products contain added artificial colouring matter.

COFFEE

Manufacture

Most commercial coffee is derived from the prepared seeds of *Coffea arabica* and *C. canephora*, of which *robusta* is the commonest variety. The fruits are often called 'cherries', which they resemble in appearance and each contains two hemispherical seeds. The external layers (pulp, skin, mucilage, parchment, etc.) are removed by a variety of processes including washing, drying and fermentation. After removing worthless beans the coffee is graded. It arrives as 'green' coffee, but many types have a greenish, bluish or brownish tinge. On the flattened side there is a longitudinal cleft. The raw green beans are roasted by heating at 180–230 °C for 15–20 min and the seed increases in size, due to the production within the bean of carbon dioxide, which acts as a preservative until it is released by grinding. Ground coffee is often packed under vacuum or in the presence of an inert gas. Unlike other canned products, a bulging can does not indicate spoilage, but rather that there is a release of carbon dioxide. For detailed information on all aspects of coffee science and technology see Clark and Macrae (1985) or Clifford and Wilson (1985). Useful information is also included in reviews by Clarke (1987) and Millin (1988). Stoltz and Masters (1979) have described technological developments in the manufacture of instant coffee.

Legislation

The Coffee and Coffee Products Regulations 1978 (SI 1978 No. 1420) extended the range of products controlled by the 1967 Regulations and implemented the EC Directive 77/436/EEC (OJ No. L172, 12.7.77) relating to coffee extracts and chicory extracts. The regulations prescribe definitions and reserved descriptions for the products, labelling requirements and the use of certain food additives.

The Coffee and Coffee Products (Amendment) Regulations 1982 (SI No. 254), implemented the EC Directive (OJ No. L327, 24.12.79), which lays down methods of analysis for determination of the caffeine content of decaffeinated coffee extracts and the dry matter contents of various products governed by the directive. Directive 85/573/EEC (OJ No. L372, 31.12.85) as implemented by the Coffee and Coffee Products (Amendment) Regulations 1987 (SI No. 1986) amended 77/436/EEC in respect of the labelling and composition of coffee and chicory extracts. The main provisions relate to: standardisation of weights of packages sold in retail; the use of the term 'concentrated' for certain liquid extracts (see Table 10.3); limits the use of the term 'decaffeinated' to coffee extracts containing less than 0·3 per cent by weight of the coffee based dry matter; use of the term 'roasted with sugar' only for liquid extracts prepared from raw materials that have been roasted in sugar; 'with sugar', 'preserved with sugar', 'with added sugar' to be used with liquid extracts to which sugar has been added to the raw material after roasting; amended the composition limits for chicory based extracts (see footnotes to Table 10.3); placed limits on the presence of 'other substances' in coffee and chicory extracts. The amendments also deleted the unenforceable stipulation for extraction ratio of coffee extract products.

Coffee is defined as the dried seed of the coffee plant whether or not the seed has been roasted or ground or roasted and ground. Chicory is also defined as the root of *Cichorium intybus*. The schedules to the regulations prescribe compositional standards for designated products in terms of proportions of coffee and chicory in the case of the ground materials and in terms of coffee and chicory based dry matter for the prepared products. In addition, certain products are allowed to contain regulated sugar products (liquid coffee extracts 12 per cent maximum, liquid extracts with chicory 35 per cent maximum and fig 25 per cent maximum). The compositional requirements of the regulations are summarised in Table 10.3. Decaffeinated coffee or coffee mixture must not contain more than 0·1 per cent of its coffee based dry matter as anhydrous caffeine whilst decaffeinated extracts must not contain more than 0·3 per cent of the coffee based dry matter as anhydrous caffeine. Anti-caking agents are permitted in dried (instant) products intended for use in vending machines. Solvent residues are permitted in decaffeinated products. Chicory and coffee essence is permitted to contain benzoic acid and the parabens esters at a level of 450 mg/kg. Clarke (1979) and Martin (1977) have discussed international standards and legislation for green coffee, instant (soluble) coffee and decaffeinated coffee. Green coffee is important in international trade; Brazil, Columbia, Mexico, Guatemala, El Salvador, East and Central Africa are important sources. Processing of the coffee bean generally reflects national tastes; hence there is little interest in international standards for roasted coffee.

Composition. Raw green coffee contains water, proteins, caffeine, oil, various carbohydrates and acids (mainly soluble and non-volatile), trigonelline and mineral matter. Roast coffee contains reducing sugars, caramelised sugars, hemicellulose, fibre, proteins, non-volatile acids (caffeic, chlorogenic, citric, malic, oxalic, quinic, tartaric), caffeine, oil, trigonelline and the ash, in which the main elements present are potassium, phosphorus and magnesium. Chlorogenic acid content has been suggested as a quality indicating factor (Trugo and Macrae, 1983). The pyrolysis volatiles from proteins are of some importance in relation to coffee flavours. The sulphur-containing amino acids methionine and cysteine have been identified

Table 10.3 Summary of legal requirements as amended for various coffee products

	Roasted coffee (%)	Dry matter from named ingredients (%)
Coffee and chicory (French coffee)	51 (min.)	
Coffee with fig flavouring (Viennese coffee)	85 (min.)	
Dried extracts (Instant X)		95 (min.)
Coffee		
Chicory		
Coffee and chicory[a]		
Coffee and fig		
Paste extracts		70–85
Coffee		
Chicory		
Coffee and chicory[a]		
Coffee and fig		
Liquid extracts		
Coffee		15–55[b]
Chicory		25–55[c]
Coffee and chicory[a]		15–55[d]
Coffee and fig		15–55[d]
Essence		
Chicory and coffee		20 and 5[e] (minima)

[a] Or 'chicory and coffee' where chicory ingredient exceeds coffee.
[b] Product may contain 12 per cent (max.) added sugar products. May be termed 'concentrated' if coffee based dry matter exceeds 25 per cent.
[c] Product may contain 35 per cent (max.) added sugar products. May be termed 'concentrated' if chicory based dry matter exceeds 45 per cent.
[d] Product may contain 25 per cent (max.) added sugar products.
[e] Product may contain added sugar products.

together with indole and tryptophan. The volatiles include a large number of compounds such as acids, alcohols, aldehydes (which predominate), diacetyl, furfural, hydrogen sulphide, ketones, mercaptans, and phenols (Rhoades, 1960; Nurok *et al.*, 1978; Tressl *et al.*, 1978; Scholze and Maier, 1982; Engelhardt *et al.*, 1984). Staling of ground coffee is now considered to be related to loss of volatile aroma constituents, which is accelerated by uptake of mositure, rather than to be due to rancidity of the coffee oil. Quality indices for stored coffee beans from various sources have been reported by Walkowski (1981), Cros *et al.* (1980) and Shimoda *et al.* (1985). Changes in volatile flavour components on storage have been correlated with organoleptic qualities by Radtke-Granzer and Piringer (1981); they proposed a lower limit of 89 mg/kg for the sum of 2-methyl propanol, 3-methyl butanal, diacetyl and 2-methyl furan as indicative of a satisfactory product. Analytical data for coffee, chicory and dandelion from various sources are given in Table 10.4 (Clifford, 1975).

The fixed oil of coffee has the following characteristics: iodine value 95–100, saponification value 170–200, unsaponifiable matter 5–11 per cent, sp. gr. 0·935–0·945, refractive index 1·465–1·470.

Adulteration. Formerly, coffee was sometimes adulterated with roasted chicory, dandelion and other roots, acorns, figs, date stones and cereals. Most of the adulterants have considerably higher water extracts than coffee. Also the roots tend to yield a higher acid-insoluble ash, contain more fibre and pentosans, but contain

Table 10.4 Analytical figures (as percentages) for raw and roasted coffee and roasted chicory and retail samples of dandelion root and dandelion and chicory.

	Raw coffee (green)			Roasted coffee			Roasted chicory			Roasted Dande-lion	Roasted Dande-lion & chicory
	min.	max.	aver.	min.	max.	aver.	min.	max.	aver.		
Moisture	8·2	13·8	10·3	0·3	5·6	2·2	2·5	12·0	5·5	6·9	7·3
Total ash	3·0	4·5	4·0	3·4	4·9	4·3	4·0	6·7	5·0	15·7	9·6
Water-soluble ash	—	—	2·9	3·0	3·6	3·2	1·6	3·3	2·8	1·5	0·9
Ditto (as % of total ash)	—	—	—	65	85	75	—	—	55	8·8	9·4
Alky of sol. ash (as K_2CO_3)	—	—	—	1·9	3·2	2·4	—	—	—	0·4	0·1
Acid-insoluble ash (as % of total ash)	—	—	—	—	—	Ngl.	10	35	20	—	—
Crude fibre	—	—	—	10·5	15·3	13·0	—	—	6·9	12·5	15·1
Tannin	—	—	9·0	—	—	4·6	—	—	—	1·5	2·7
Total nitrogen	—	—	2·7	2·1	3·3	2·6	—	—	1·4	0·87	1·17
Caffeine	1·1	1·8	1·3	0·9	1·8	1·2	—	—	Nil	Nil	Nil
Starch	—	—	—	0·9	3·5	2·3	—	—	2·1	—	—
Ether extract	11·4	13·7	12·2	8·0	14·2	13·5	0·9	3·9	2·1	1·3	2·3
Extractives of dry sample*	25	34	—	23	33	*	70	78	*	50·5	61·6

* Depends to some extent on extraction method (see discussion below).

less oil. Some data for roasted chicory and dandelion roots are given in Table 10.4. Lutman (1982) has described a method based on estimation of total monosaccharides of coffee with barley. Typical data for different varieties of coffee beans are given.

Analysis of coffee or coffee and chicory

BS 5752 in various parts describes methods for coffee and coffee prducts. For **green coffee**, Part 1:1987 (ISO 1446) is moisture content, basic reference vacuum oven method; Part 2:1984 (ISO 1447) is the routine moisture method, loss at 130 °C; Part 4:1985 (ISO 4149) is sensory examination, foreign matter and defects; Part 5:1985 (ISO 4150) is particle sizing by sieve; Part 7:1988 (ISO 6673) is loss in mass at 105 °C; and Part 8:1986 (ISO 6667) covers insect damaged beans.

The determinations that are necessary for the examination of samples of roasted coffee and of coffee and chicory are similar to those for tea and can be readily deduced from Table 10.4. For these and for the determinations of trace metals, the usual procedures can be followed and only those that are especially applicable are included below. For routine purposes, all samples should be examined microscopically and the amount of soluble extractives should be determined.

Microscopical examination. Boil the sample with water, clear (if necessary) with chloral hydrate solution and then stain with phloroglucinol and hydrochloric acid. The main microscopical features of coffee are the longitudinal and transverse sclerenchymatous fibres of the pericarp (Fig. 10.2; cf. Vaughan, 1970). Chicory has exceptionally large vessels, up to 115 μm with rather short pits. Dandelion shows concentric rings due to groups of sieve tubes and the vessels have long pits and are rather more numerous than in chicory.

Moisture. Dry 5 g at 100 °C to constant weight. The moisture content of roasted coffee should preferably not exceed 5 per cent. ISO have defined procedures for

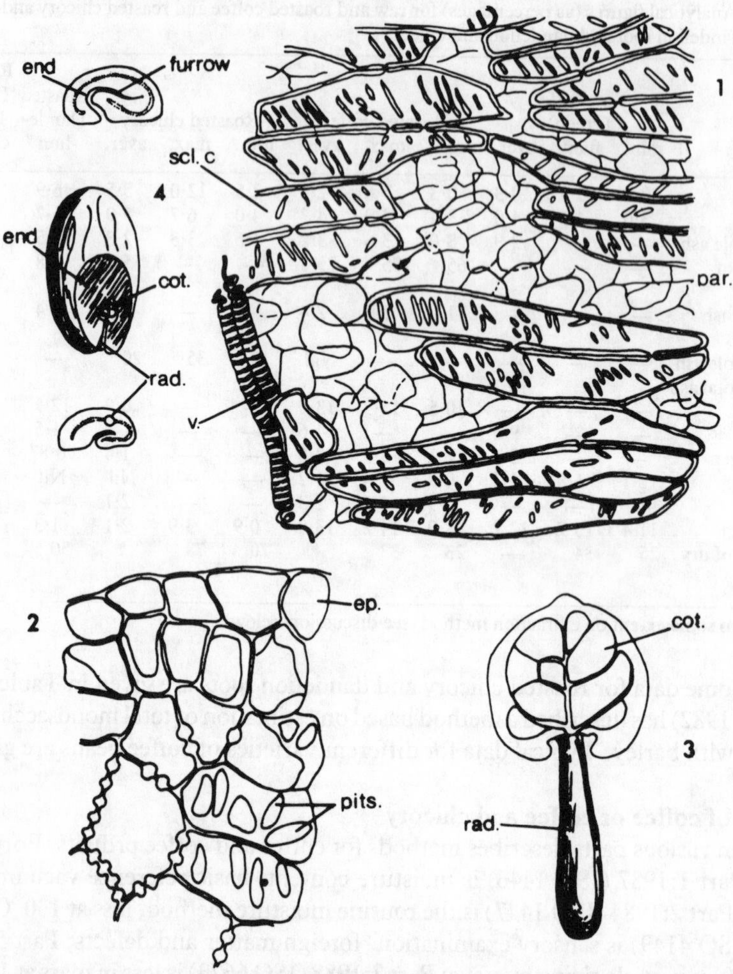

Fig. 10.2 Microscopical features of coffee. 1, Portion of the seed coats, surface view; **par.**, collapsed parenchymatous tissue; **scl. c.**, sclerenchymatous cells; **v.**, vessel, ×220. 2, Transverse section of outer part of endosperm; **cp.**, epidermis, ×220. 3, Embryo of seed; **cot.**, cotyledon; **rad.**, radicle. 4, Sections of seed; **cot.**, cotyledon; **end.**, endosperm; **rad.**, radicle.

determination of the moisture content of green coffee (ISO 1447) and of instant coffee (ISO 3726). The method uses vacuum oven drying.

Chicory and fig. The most reliable method for detecting chicory in ground coffee is by means of the microscope. A simple but satisfactory test may be made by gently sprinkling the powdered sample on the surface of water in a tall cylinder. The coffee floats on the surface, but the chicory begins to sink down within a few seconds. Most of the falling chicory particles leave behind them a trail of colour due to the large amount of caramel they contain.

Most of the methods that have been proposed for the determination of chicory in a mixture with ground coffee are based on the fact that the hot water extract of chicory is nearly three times that of coffee. The following method gives reasonably good results:

Boil 5 g of the dried sample with 200 ml water for 15 min and pour the liquor into a 400 ml beaker. Reboil the residue with 50 ml water, mix the extracts, cool and filter into a 250 ml volumetric flask. Make the filtrate up to 250 ml, pipette 50 ml into a weighed dish, evaporate to dryness on the water bath, dry in the 100 °C oven and weigh.

From the weight of the residue, calculate the percentage of soluble matter (E) in the original dried sample. Then if E_1 and E_2 are the average water extracts of coffee and chicory respectively:

$$\text{Chicory (per cent)} = \frac{(E - E_1)}{(E_2 - E_1)} \times 100.$$

For many years, figures of 24 and 70 were recommended as the average values of E_1 and E_2 respectively. However, higher average figures are considered to be more appropriate (see below). It should be borne in mind also that the proportion of soluble matter extracted depends on the particle size of the sample used and the extraction technique employed. Hughes and Wise (1934) therefore recommended the use of a standardised procedure.

Using the prescribed extraction method, the authors found that the average specific gravities ($d^{1}_{15\cdot5}$) of 5 per cent extracts of dry roasted coffee and chicory were 1·0058 and 1·0143 respectively. The sp. gr. can be calculated from the total solids of the extract by means of the expression: sp. gr. $= 1 + 0{\cdot}004S$, where S is the total solids of the extract (as per cent of m/v).

In a later paper Hughes and Wise (1955) pointed out that, on the average, 5 per cent extracts of the types of coffee and chicory then being used had a higher sp. gr., i.e. they yield a higher proportion of soluble matter than the types used formerly. The average figures given in the two papers for the extracts of dry coffee and chicory using the above extraction method are as follows:

Sample	sp. gr. ($d^{15\cdot5}_{15\cdot5}$) of 5% dry matter	Equivalent to: soluble matter (as % m/m of sample)	Reference date
Coffee	1·0058	29	1934
Coffee	1·0064	32	1955
Chicory	1·0143	71·5	1934
Chicory	1·0154	77	1955

Vree and Yeransian (1973) have reported on a collaborative study of methods for determining soluble solids in roasted coffee.

Most of the carbohydrate in chicory root is present in the form of inulin (a polyfuranose that yields fructose on acid hydrolysis) and fig that is rich in sugars. A method based on the colorimetric determination of ketose sugars has been proposed for the assessment of the relative proportions of chicory and fig in coffee preparations (Kazi, 1979). His method is basically as follows:

REAGENTS

Solution A: Resorcinol in 95 per cent ethanol, 0·05 g/100 ml. Stable for one week in amber glass at 4 °C.
Solution B: Ferric ammonium sulphate in conc. HCl, 0·0216 g/l.

CALIBRATION

Using the colour development procedure, prepare graph of absorbance against fructose with standard fructose solutions in water, in the range 0–0·005 per cent.

PROCEDURE

Weigh 4·00 g roasted and ground coffee product into a 500 ml conical flask. Add 80 ml water and boil under reflux for 10 min. Filter hot extract through a fast filter paper into a 200 ml volumetric flask. Add 10 ml alumina cream clearing agent, mix, make to mark with water, mix and filter.

To a glass stoppered tube, add 3 ml Solution A then 3 ml Solution B, add 1·00 ml of cleared and filtered extract, and to another tube add Solutions A and B with 1 ml of water as blank. Close tubes, heat for 40 minutes in a water bath at 80 °C, cool quickly to room temperature in running water, measure absorbance at 480 nm in a 10 mm cell, and deduct reagent blank. Deduce ketose sugar content from calibration graph, and calculate coffee, chicory or fig content assuming on average the following ketose sugar contents: roasted coffee 0·53 per cent, roasted chicory 48·0 per cent, and roasted fig 23.0 per cent.

Kazi showed that his method is ten times less variable than the water extraction methods. For more information on the detection of chicory, see p. 373.

Caffeine. Extract 10 g of the sample with boiling water under reflux for 5 h, evaporate the extract down to fairly low bulk and transfer to a 100 ml volumetric flask (total volume 85 ml). Add 5 ml of each of the zinc ferrocyanide clearing agents (see p. 000), make up to the mark, mix and filter into a 100 ml measuring cylinder. Measure the volume of filtrate, transfer to a separator, add ammonia and extract the caffeine with chloroform. Column clean-up of the extract with Celite is recommended by the AOAC and in BS 5752:Part 3:1988 (ISO 4052). The caffeine content of the extract is measured spectrophotometrically (see under instant coffee). Newton (1969) has reviewed the methods for determination of caffeine in coffee and has described a collaborative study for the spectrophotometric procedure. He concluded that the spectrophotometric procedure was acceptable for routine use. (See the comments and references given for caffeine in tea; similar considerations apply also for decaffeinated coffee and coffee products.) The use of HPLC methods is also described under instant coffee (p. 373).

Contaminants. Samples of coffee may require to be examined for arsenic, copper and lead. Atkinson *et al.* (1950) have studied the possible sources of the relatively high amounts of arsenic that have been found in chicory root.

LIQUID EXTRACTS

Liquid coffee extracts consist of a mixture of a concentrated decoction of coffee with added sugar. Specified products, sometimes referred to as essences, also include mixtures with chicory or fig extracts.

Legislation

The legal requirements for coffee extracts are summarised in Table 10.3. Specified sugar poducts are also permitted in these products. Methods of analysis for determining the composition of coffee and chicory extract were laid down in the directive 79/1066/EEC and these were incorporated in the Coffee and Coffee Products (Amendment) Regulations 1982 (see p. 363).

Composition

Attempts have been made to assess the amounts of coffee and chicory that have been used in the preparation of liquid extracts from the determinations of caffeine—and other purines, oligo and monosaccharides, total solids, ash and ash constituents. Tatlock and Thomson (1910) observed that whatever the degree of extraction of coffee, the ratio of caffeine and ash to total extract is constant. The coffee equivalent to the caffeine found is first calculated on the average basis of 1·25 per cent. Then the ash due to the coffee is taken as 4·2 per cent of the coffee and the excess of ash calculated as due to chicory, which is assumed to have an average ash of 5·2 per cent. Clarke and Walker (1974) have reported on the interrelationship of potassium and other mineral constituents of green, roasted and instant coffee. Pazola and Cieslak (1979) have discussed changes in carbohydrate content of roasted rye, barley and chicory with a view to assessing their value as coffee replacers.

From the analysis of decoctions of various samples of coffee, Edwards and Nanji (1937, 1938) found that the following ratios applied

	Ratios for genuine coffee decoctions	Suggested limiting ratio (max.)
Extractives/caffeine	12·9:1 to 20·6:1	24:1
Extractives/ash	about 6:1	7:1
Ash/caffeine	2·57:1 to 3·04:1	4:1

The authors state that the above limiting values are generous and therefore, if they are exceeded, adulteration is probable. It should be borne in mind, however, that the presence of added sugar causes an increase in the extractives figure and in the first two ratios, but that the ash/caffeine ratio remains virtually unaltered. For the application of the use of the above ratios for the detection of adulteration and the calculation of the proportion of chicory that may have been used for preparing decoctions, the original papers should be consulted.

Chicory roots contain a relatively large proportion of inulin, which yields fructose on hydrolysis. It has been suggested, therefore, that this constituent may be useful for assessing the amount of chicory present; this approach is only suitable if the mixture contains no added sucrose (Newman, 1981). Smith (1981) has proposed an HPLC method for determination of 5-hydroxymethyl furfural, which may be used as an index of chicory content (see also under instant coffees, below).

Analysis of liquid coffee extracts

The analysis of liquid coffee extracts may include the following: the determination of **total solids, ash, sugars, caffeine,** and **trace elements, As Cu, Pb**. Industrially, the solids content is usually assessed from the sp. gr. or the refractive index. The method prescribed by the regulations is the vacuum oven drying method, 16 hours at 70 °C. The pH value is also determined.

Caffeine. Mix 20 ml of coffee extract with 50 ml water in a 100 ml volumetric flask. Add 5 ml of each of the zinc ferrocyanide clearing agents (p. 000), dilute to the mark and shake. Allow to stand for 5 min and filter through a rapid filter paper. Measure the volume of filtrate, transfer to a separator, add 10 ml of 0·88 ammonia and extract five times with 25 ml portions of chloroform. Complete the method as detailed under coffee and instant coffee. However, HPLC procedures are preferable (see below).

Sugars. The sugars may be determined after clarifying with Carrez solution by the Luff-Schoorl or by Lane and Eynon procedure (see Chapter 6). Chromatographic methods may also be used. It should be borne in mind that roasted coffee and chicory contain on the average about 0·7 and 20 per cent of reducing sugars, calculated as dextrose.

SOLID COFFEE EXTRACTS ('INSTANT COFFEES')

Solid coffee extracts are prepared essentially by drying coffee decoctions obtained from a series of percolating columns. The system is operated at pressures up to 300 p.s.i. and at 180 °C. Under these conditions hydrolysis of component polysaccharides occurs giving rise to greater solid extraction yields than in domestic coffee liquor preparation. The liquid extracts are spray dried but the best quality instant coffees are prepared by freeze drying. A few products have been made by mixing glucose syrup with the decoction before drying and solid extracts of coffee and chicory are also prepared. The product tends to be hygroscopic and is packed into airtight tins in rooms that are maintained under conditions of low humidity. For the technology and manufacture of instant coffee, see under the references on p. 362.

Legislation

The legal requirements for instant coffees are summarised in Table 10.3. Instant coffees are required to contain at least 95 per cent of dried water-soluble solids of coffee or coffee and chicory or coffee and fig.

Composition

Most instant coffees give analytical figures which fall within the following ranges: moisture 1·3–4·1 per cent (should be less than 5 per cent); ash 7–14 per cent; alkalinity of soluble ash 8–14 ml 0·1 N acid per g; caffeine 2·0–8·0 per cent and total sugars (as reducing sugars) 12–32 per cent. Extracts prepared from coffee and chicory give lower figures for ash, alkalinity of soluble ash and caffeine. It has been suggested that dried coffee and chicory extract should contain at least 1·0 per cent caffeine and dried coffee and fig extracts at least 2·7 per cent. A consumer survey (Anon, 1973) found a variation in caffeine content for various brands of instant coffee of 2·0–4·3 per cent. The mean percentage caffeine content for the three types were as follows: powders 3·1, granules 3·0 and freeze-dried brands 3·8 (overall average 3·2 per cent). The four 'mild' brands examined gave a higher mean figure of 3·7 per cent.

Quality assessments of both instant coffee and the raw materials for manufacture have been reported by Wang *et al.* (1983). They have used a headspace sampling technique, collection of volatiles on porous polymer and GLC with FID and a sulphur sensitive detector to identify aroma linked volatiles. Differences between off-flavour coffee and acceptable raw material were explained in terms of loss of a range of volatile compounds, which included diacetyl, pentanones and pyrazine.

Analysis of instant coffee

Instant coffee may be examined for solubility, moisture, ash, alkalinity of the soluble ash, sugars, caffeine, arsenic, copper and lead; also, if suspected, for the presence of undeclared ingredients such as chicory extract.

BS 5752 describes, in various parts, methods for instant coffees: Part 3:1988 (ISO 4052) for caffeine content; Part 6:1988 (ISO 3726) for loss in mass at 70 °C under vacuum for 16 hours; Part 9:1986 (ISO 7534) for insoluble matter; Part 10:1986 (ISO 7532) for particle size; and Part 11:1987 (ISO 8460) for free-flow and compacted bulk densities.

The **moisture** content is determined by the dry matter method described in the Regulations and the BS/ISO standards. The same sources describe the method for **insoluble matter**, which is based on filtration of 500 ml solution containing 5 g of sample in specified apparatus that includes a metallic 100 μm microfilter disc, which is then dried and weighed. Reynolds *et al.* (1983) have reported on a collaborative study of the method.

Caffeine. The following method, basically that of the AOAC/BS/ISO, is prescribed in the Amendment Regulations 1982 (see p. 363) for the control of decaffeinated coffee extracts (dried, liquid or paste).

REAGENTS

Sulphuric acid, 2 M solution.
Sodium hydroxide, 2 M solution.
Celite 545 or equivalent.
Ammonia solution, approximately 4 M (1 volume of concentrated ammonia solution to 2 volumes of water).
Diethyl ether, pure or repurified by chromatography on a column of basic aluminium oxide of activity grade 1.
Pass 800 ml of diethyl ether through a column filled with 100 g of aluminium oxide. Keep in dark bottles until used.
(Diethyl ether, recently distilled and free of peroxides, can be used instead of diethyl ether purified by chromatography.) Saturate the diethyl ether with water.
Caffeine (1,3,7-trimethyl-2,6-dihydroxypurine), pure, anhydrous.
Chloroform, pure or repurified by chromatography over aluminium oxide and saturated with water.

SPECIAL APPARATUS

Chromatographic columns approximately 250 mm long, 21 mm internal diameter (column I) and 17 mm internal diameter (column II), fitted with stopcocks.
Ultraviolet spectrophotometer with 10 mm silica cells.

PROCEDURE

Preparation of the test portion.
Weigh, to the nearest 0·1 mg, about 0·5 g of dried coffee extract sample, between 0·5 and 0·7 g of coffee extract paste sample or between 0·8 and 3·2 g of liquid coffee extract sample. (These last two weights should be chosen to give test portions containing approximately 0·5 g of dried coffee extract.) Transfer sample to a 100 ml beaker add 5 ml of ammonia solution and warm for two minutes on a boiling-water bath. Add 6 g of Celite and mix carefully.

Filling for the columns
Column I (alkaline column, 21 mm i.d.)
Mix carefully, by kneading with a flexible spatula blade, 3 g of Celite and 2 ml of the sodium hydroxide solution, until homogeneous (see note below). A slightly wet powder will be obtained. Transfer this powder, in small portions (of approximately 2 g), into a chromatographic column, the lower part of which is packed with a wad of cotton wool or glass wool. Tamp down the mixture after each addition, without excessive force, using a glass rod, one end of which is flattened to the internal diameter of the column, until a perfectly homogeneous and compact layer is obtained. A small wad of cotton wool or glass wool may be placed on the top of the Celite.

Note: Column packing material may be prepared in bulk quantities in advance and stored in closed containers.

Transfer the Celite–sample mixture into the column. Dry the sample beaker twice with portions of about

1 g of Celite transferring this Celite into the column. Tamp down to obtain a homogeneous layer and place a wad of cotton wool or glass wool on the top of the mixture.

Column II (acid column, 17 mm i.d.)
Mix 3 g Celite with 3 ml of 2 M sulphuric acid as described for column I. Transfer the powder to the 17 mm column on to a wad of cotton wool or glass wool. Tamp the column, place a wad of cotton wool or glass wool on the top of the column.

Chromatography
Mount the columns one above the other so that the effluent from column I can drip directly into column II. Pass 150 ml of diethyl ether through the two columns. Keep open the stopcock in column I. Adjust the stopcock of column II so that a quantity of supernatant liquid remains above the column packing. Remove column I. Pass 50 ml of diethyl ether through column II, using the initial portion to wash the tip of column I and passing this portion also into column II. Discard the effluent from column II.

Note: Used diethyl ether may be recovered by shaking it with iron(II) sulphate.

Pass a stream of air through column II with the stopcock open (for example by using an inflated rubber blower), until no more diethyl ether drips from the column and the air flow from the stopcock carries only a faint smell of diethyl ether (see note below). Elute column II with 45 to 50 ml of chloroform. Collect the eluate in a 50 ml volumetric flask, make up to volume with chloroform and mix carefully.

The flow rate of the diethyl ether and the chloroform under conditions of natural flow should be between 1·5 and 3 ml/min. If this rate is exceeded, channelling through the packing should be suspected.

Note: This step should be carried out in a well-ventilated fume-cupboard to prevent both the possibility of inhalation of solvent vapours and the possibility of an explosion.

Spectrophotometric measurement
Measurement of the test solution
Avoiding error from chloroform evaporation, measure the absorbance of the solution of caffeine in chloroform using silica cells, against chloroform at 276 nm (absorption maximum). Measure also the absorbance at 246 nm (absorption minimum) and at 306 nm in order to verify the purity of the caffeine obtained.

If the absorbance at 276 nm exceeds 1·3, repeat the measurement on a diluted portion of the test solution. In this case, take into account the dilution factor; the appropriate factors of the formulae will have to be adjusted accordingly. If the absorbance measured at 276 nm is lower than 0·2, repeat the determination using a larger test portion.

Preparation and measurement of a reference solution
Prepare a reference solution of caffeine in the following manner:

Weigh, to the nearest 0·1 mg, 100 ± 20 mg of pure anhydrous caffeine. Place in a 1000 ml volumetric flask, dissolve in chloroform, and make up to volume. With a volumetric pipette take 5 ml of this solution and make up the volume to 50 ml with chloroform.

Measure the absorbance of this solution as described for the test extract. The corrected absorbance of the reference solution should be in the region of 0·4.

Number of determinations
Carry out at least two determinations on the same test sample.

Blank test
Carry out a blank test on the reagents following the procedure described above but omitting the test portion. Before using recovered reagents, repeat the blank test to verify their purity.

EXPRESSION OF RESULTS
Formulae and method of calculation

The caffeine content, in per cent by mass of dry matter on the sample, is equal to:

$$\frac{5 \times 10^5 \times C \times A_1}{A_2 \times m \times p}$$

where:

C is the concentration of caffeine in the reference solution in g/ml;
A_1 is the corrected absorbance of the purified extract [i.e. absorbance at 276 nm − 0·5 × (absorbance at 246 nm + aborbance at 306 nm)];
A_2 is the corrected absorbance of the caffeine reference solution [i.e. absorbance at 276 nm − 0·5 × (absorbance at 246 nm + absorbance at 306 nm)];
m is the mass in g of the test portion;
p is the dry matter content, expressed as a percentage by mass, of the sample, as determined by methods 2 or 3, Annex II of the Amendment Regulations 1982.

Repeatability
The difference between the results of the two determinations carried out simultaneously or in rapid succession on the same sample, by the same analyst, under the same conditions, shall not exceed 0·01 g caffeine per 100 g sample on a dry matter basis.

In the authors' laboratory the following HPLC method for **caffeine** has proved to be satisfactory.

PROCEDURE
Dissolve 1·00 g of sample in 100 ml of hot water. Filter 20·0 ml through a Millipore filter (0·45 μm) under vacuum and apply to a Bond Elut C_{18} cartridge (or equivalent) under vacuum. Elute the caffeine with 5·0 ml of mobile phase (0·005 M sodium acetate: tetrahydrofuran, 95:5 at pH 5), collect in a 10 ml flask and make to volume. Inject 20 μl onto a Spherisorb ODS, C_{18}, 5 μm packed column 25 cm long × 4 mm i.d. Elute with the mobile phase at 1 ml/min, observe absorbance at 280 nm, calibrate with standard caffeine solutions, 0–1 mg caffeine in 10 ml mobile phase.

For routine purposes, the HPLC step can be eliminated and the absorbance of the eluant from the cartridge measured at 280 nm on a spectrophotometer.

Blanch and Tarka (1983) and Trugo *et al.* (1983) described the HPLC determination of caffeine and theobromine in coffee, tea and instant hot cocoa mixes. Terada and Sakabe (1984) described HPLC for theobromine, theophylline and caffeine in food products. Ashoor *et al.* (1983) used HPLC for caffeine in decaffeinated coffee, tea and beverage products.

Chicory. The detection or determination of chicory (or fig) in instant coffees may be carried out by the Kazi (1979) method, described on p. 367, but without the initial boiling extraction step. Spray dried chicory extract powders contain close to 71 per cent and spray dried coffee extracts about 1–2 per cent of ketose sugars as determined by this method. The method is very susceptible to the presence of sugars containing fructose (e.g. in essences), and in samples of unknown composition the possible presence of added sugar should be ascertained, e.g. by TLC or HPLC. These techniques, if indicating the presence of high levels of fructose compared with other sugars, will indicate the possible presence of chicory but not, of course, fig.

Chicory can also be detected (by HPLC) by the presence of 5-hydroxymethyl furfural (Smith, 1981). Its content is however variable, depending on roasting and processing conditions.

Solvent residues. Decaffeinated instant coffee may need to be examined for the presence of residues of non-permitted extraction solvents (see comments and references on p. 120).

COCOA

Cocoa is derived from the prepared seeds of *Theobroma cacao*, which is a native of tropical America. Most commercial cocoa comes from Ghana, Nigeria, Brazil, the West Indies and Sri Lanka. The mature orange to red fruit is 15–20 cm long, has a shape resembling a pointed vegetable marrow, and contains 30 to 60 seeds embedded in a mucilaginous pulp. After fermentation in boxes, the separated seeds ('beans') are dried to a moisture content of 6–8 per cent. The colour and taste of the final beans is influenced considerably by the method of fermentation. Cocoa beans are examined both visually and chemically to assess the manufacturing quality. Some examinations are carried out on the individual beans. In the 'cut test' (ISO R 1114) the beans are cut longitudinally and both halves are examined. The number of defective beans in a sample is counted. Defects include mould, insect damage, germination, poor colour and poor fermentation. Beans that have not been fermented for one reason or another will not produce the characteristic flavour; unfermented beans are grey (slaty) whilst underfermented beans have shades of violet. Attempts have been made to produce an objective test for fermentation by extraction of the pigment and measurement of the absorption spectrum; however, the presence of slaty beans cannot be detected by this means. The IOCCC have studied the use of ammoniacal nitrogen content as an index of fermentation. Pettipher (1986) has used a Sep-Pak C_{18} cartridge to concentrate anthocyanins from an acidic extract of cocoa beans; after removal from the column with acid methanol the anthocyanins were measured spectrophotometrically. The author stated that the method can be used to detect grossly unfermented cocoa but that further evaluation is necessary to establish its usefulness in quality control.

Manufacture

The beans are 20–30 mm long, flattened ovoid in shape and chocolate brown in colour. The kernel consists of two folded chocolate coloured cotyledons, which can be easily broken into angular fragments (nibs).

In the traditional manufacture of cocoa, the beans are roasted at 93–140 °C for ¼–2 hours and during roasting the moisture content falls to 2·5–4 per cent.

The seedcoat or shell, which represents about 12 per cent of the weight of the roasted bean, becomes loosened preparatory to its removal by winnowing. (In the modern manufacture of cocoa, the beans may be subjected to a short infra-red heating which loosens the shell for easy removal to leave the cocoa nib.) The nib is roasted to develop the characteristic flavour derived from condensation of aldehydes formed from amino acids and the formation of pyrazine derivatives via Maillard reactions. Roasted nib is ground to a semi-plastic substance, which is referred to as chocolate liquor or cacao mass, containing 50–58 per cent of fat (cocoa butter). Cacao mass (heated below 93 °C), is mixed with concentrated solution of potassium or sodium carbonate and most of the moisture is removed at about 110 °C. The resultant product is pressed to remove part of the fat (cocoa butter) to produce cocoa press cake, which is then pulverised to form cocoa powder. If the cocoa butter is not partially removed, the resultant beverage is too rich and there is excessive separation of globues of fat. The colour of the final product varies with the fat content, i.e. the higher the fat content the richer and darker is the colour of the cocoa powder.

'Alkalisation' is the name given to the treatment of nib, mass or powder with alkali. It can be carried out before or after roasting. Bicarbonate, carbonate, or hydroxide of potassium, sodium, ammonium or magnesium, or a combination, are used. Alkalisation causes, *inter alia*, the hydrolysis of starches and natural glycosides, and the modification of tannins and polyphenolic compounds to less stringent substances. By careful control a wide range of colours (orange to almost black) and flavour levels can be developed. Alkalisation also improves dispersability.

The manufacture of cocoa and chocolate is discussed by Chatt (1953), Wieland (1972), Minifie (1980), Wood and Lass (1985) and Beckett (1988).

Legislation

The Cocoa and Chocolate Products Regulations 1976 (SI 1976 No. 541) and the Cocoa and Chocolate Products (Amendment) Regulations 1982 (SI 1982 No. 17) implemented the EC Directive No. 73/241/EEC as amended (74/411, 74/644, 75/155, 76/628, 78/609, 80/608). They prescribe in Schedule I, definitions, reserved descriptions and compositional requirements for cocoa, cocoa products, drinking chocolate and specified chocolate products. Schedule II lists permitted acids, bases, emulsifiers and additional ingredients together with prescribed levels for their use.

Cocoa is defined as the fermented and dried seed of the cocoa tree (*Theobroma cacao* L.), which is a native of tropical America. An International Standard for cocoa beans has been developed by Codex Alimentarius (Codex Stan 141–1983). This prescribes limits for mouldy beans (4 per cent), slaty beans (8 per cent), insect damaged beans (6 per cent), moisture (7·5 per cent). The standard also provides for cocoa nib, cocoa mass, cocoa press cake and prescribes limits for additives and contaminants. The provisions of the standard are near to but not identical with those in the UK regulations in respect of the derived cocoa products. The UK regulations give the following definitions. Cocoa nib is the cotyledon of roasted or unroasted cocoa beans with a residue of germ or shell not exceeding 5 per cent and an ash content not exceeding 10 per cent, both expressed on a dry defatted matter basis. **Cocoa dust** is a mixture of nib and shell containing not less than 20 per cent fat on the dry matter. **Cocoa press cake** and **cocoa powder** are the products derived from nib or mass containing not more than 9 per cent moisture and not less than 20 per cent cocoa butter on a dry basis. **Sweetened cocoa** is a mixture of sucrose and cocoa powder containing not less than 32 per cent cocoa. **Drinking chocolate** is a mixture of sucrose and cocoa powder containing not less than 25 per cent cocoa. In addition, a series of fat reduced products based on fat reduced cocoa butter calculated on the dry matter are defined. The amount of alkali in cocoa nib, cocoa mass and press cake is limited to 5 per cent calculated as potassium carbonate. Prescribed amounts of lecithins and ammonium phosphatides are allowed in the cocoa products.

Codex Stan 105–1981 covers cocoa powders (not less than 20 per cent cocoa butter on the dry basis; not more than 7 per cent of moisture), and dry cocoa–sugar mixtures (sweetened cocoa, not less than 25 per cent of cocoa butter on the dry basis; sweetened cocoa mix, not less than 20 per cent of cocoa butter on the dry basis).

Composition

The main constituents of cocoa are moisture, tannin (including the anthocyanin pigment 'cacao red'), nitrogenous matter (proteins and purines including theobromine and caffeine), starch and other carbohydrates, cocoa butter and inorganic matter (especially potassium and phosphorus). Compositional data is given in the reference texts given under 'Manufacture'.

Table 10.5 gives analytical figures for roasted cocoa nib and shell and also commercial cocoa.

Table 10.5 Analytical data for cocoa nib, cocoa shell and commercial cocoa (all figures as percentage of the whole sample)

	Cocoa nib			Cocoa shell			Commercial cocoa		
	min.	max.	aver.	min.	max.	aver.	min.	max.	aver.
Moisture	2·3	3·2	2·7	3·7	6·6	4·9	2·7	7·8	5·0
Fat	48	57	52	1·7	5·9	2.8	19	28	25
Total ash	2·6	4·2	3·3	7·1	20·7	10·5	3·8	8·9	5·4
Water-soluble ash	0·7	1·9	1·2	2	5·7	3·7	3·2	5·7	4
Alkalinity of sol. ash (as K_2CO_3)	1	2·3	1·7	3·5	4·1	3·8	1·3	4·8	2·5
Acid-insoluble ash	Nil	0·07	0·02	0·05	11·2	2·5	0·01	1·4	0·24
Crude fibre	2·2	3·2	2·6	12·8	19·2	16·6	2·3	5·2	3·9
Total nitrogen	2·2	2·5	2·4	1·7	3·2	2·3	2·1	3·7	2·7
Theobromine	0·8	1·3	1·0	0·2	0·9	0·5	0·8	1·6	1·2
Caffeine	0·1	0·7	0·4	0·04	0·3	0·16	0·04	0·3	0·2
Starch	6·5	9	8·1	3·4	5·2	4·1	8·7	13·5	11·2
Cold water extract	9	14	11	20	25	23	15	22	19
Amount passing through 100 mesh sieve	—	—	—	—	—	—	85	97	92

Analysis of cocoa

The routine analysis of cocoa will normally include the microscopical examination and the determination of **moisture, ash, alkalinity of the ash, fat, crude fibre, arsenic, copper** and **lead**. Other determinations that may be necessary are those of **soluble ash, acid-insoluble ash, starch, nitrogen, theobromine, caffeine** and the **cold water extract**.

Microscopical examination. The most characteristic microscopical features of cocoa are the brown pigment, the almost colourless small starch grains (about 7 μm), the 'club-shaped' hairs and the fat crystals (Fig. 10.3).

The fact that the husk contains considerably more mucilage, spiral vessels and sclerenchymatous cells than the nib can be used to assess approximately the proportion of shell present. The pieces of mucilage are stained pink by ruthenium red (0·02 per cent in 10 per cent lead acetate).

Jackson (1965) has described microscopical methods for quantitative assessments of shell, filth, etc. in cocoa. White and Shenton (1980) have published an annotated bibliography of microscopic techniques as applied to cocoa, chocolate and sugar confectionery.

Kaffka *et al.* (1982) have discussed the application of NIR methods in quality control. Quality control of raw materials for confectionery manufacture has been discussed by Kleinert (1982); the importance of sensory evaluations, appearance, aroma, flavour and texture is stressed. Headspace chromatography has been applied to assess quality at the various processing stages including fermentation and roasting

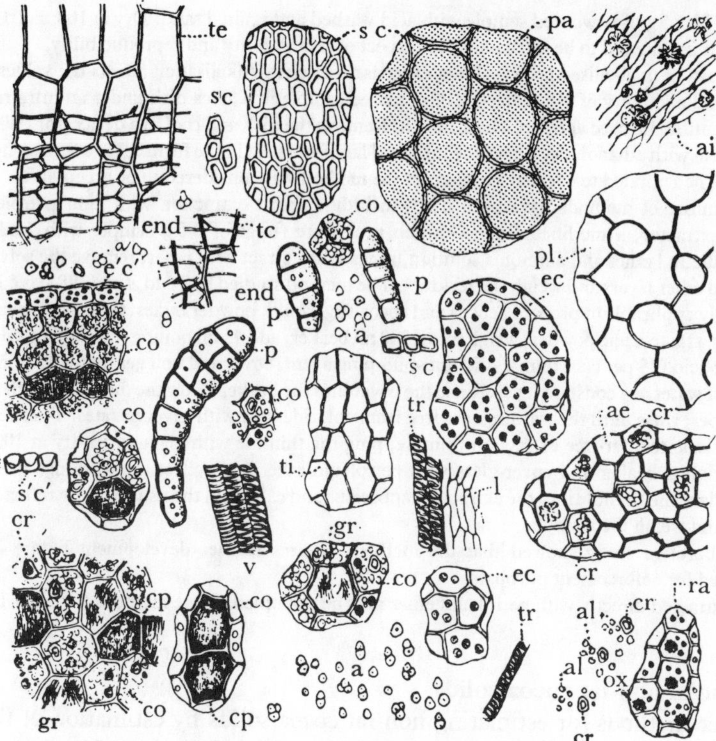

Fig. 10.3 Microscopical features of cocoa

a, Starch grains.
ae, Outer layer of endosperm.
ai, Inner layer of endosperm.
al, Aleurone grains.
co, Cotyledon.
cp, Pigment cells containing cocoa-red.
cr, Crystals of fat.
ec, Epidermis of cotyledon, surface view.
e′c′, Epidermis of cotyledon, profile.
end, Inner epidermis of pericarp.
gr, Crystals of fat.
l, Bast from fibro-vascular bundles.
ox, Calcium oxalate crystals.
p, Pluricellular hairs.
pa, pl, Parenchyma of seed-coat.
ra, Cells of radicle.
sc, Sclerenchymatous layer of seed-coat, surface view.
s′c′, Sclerenchymatous layer of seed-coat, in profile.
te, Outer epidermis of seed-coat to which the inner epidermis of the pericarp (end) is adhering.
ti, Inner epidermis of seed-coat.
tr, v, Vessels, etc., from fibro-vascular bundle.

of beans (Zeigleder, 1981, 1982). The same author (1984) has also proposed an HPLC method for estimation of pyrazines and has correlated the results with aroma and degree of roasting. The degree of fermentation has also been assessed by treating beans with methanol and HCl and measurement of the optical density of the extract (Gur'Eva and Tserevitinov, 1983).

Methods of analysis for enforcement of the Community Directive on Cocoa and Chocolate Products are being developed (a draft directive that is concerned with analysis for evidence of foreign fats appeared in 1986). A selection of the simple procedures have been studied collaboratively and reported by Judd *et al.* (1984). The methods, which are briefly summarised below, were originally developed and published by IOCCC, and include those for moisture, fat and ash on cocoa powder and cocoa mass.

Moisture. Mix 5 g of powdered sample with acid washed and ignited sand. Dry at 103 ± 2 °C for 4 h. The method is considered to be satisfactory in respect of repeatability and reproducibility.

Ash. The ash of unalkalised cocoa seldom exceeds 5 per cent. Alkalisation affects the values for ash, soluble ash and alkalinity of ash. Char 5 g of prepared sample in a silica dish under an infra-red lamp, transfer to a muffle furnace at 600 °C, heat for 2 h. Remove the dish and cool in a desiccator. When cool, moisten the ash with ethanol, dry under the infra-red lamp and heat in the furnace for a further hour, cool and weigh. When applied to cocoa mass samples the method was considered to be acceptable.

Fat. A number of methods have been used, including direct extraction with light petroleum in a continuous extractor, a modified Rose–Gottlieb procedure (treat of 0·5 g sample with 5 ml ethanol followed by diethyl ether and ammonia solution, repeatedly extract with ether, remove the solvent from the combined ether layers and weigh the residue). The method studied by Judd *et al.* (1984) is considered to be generally applicable to prepared cocoa and cocoa/sugar/milk powder mixes. Grind sufficient sample to yield 1 g of a fat to a fine powder, weigh into a 300 ml beaker, add 45 ml boiling water followed by 55 ml hydrochloric acid (25 per cent m/m). Add anti-bumping agent, cover and boil gently for 15 min keeping the volume of water at a constant level. Rinse the cover into the beaker, filter the digest through a wet fat-free filter paper, rinse until the washings are free from chloride (test with silver nitrate). Transfer the wet filter and residue to a fat-free extraction thimble, plug the thimble with glass-wool, dry at 100 ± 1 °C. Rinse the original beaker and cover-glass with petroleum ether and collect the washings in a Soxhlet apparatus. Place the thimble into the extraction apparatus and extract in the usual way by reflux. Remove the solvent and weigh the residue.

The collaborative study showed that the method requires further development before it can be recommended for enforcement purposes.

The fat extracted directly with petroleum ether may be examined by the methods for cocoa butter (p. 387).

Estimation of non-fat cocoa solids

The earlier methods for estimating non-fat cocoa solids by estimation of theobromine and caffeine content were critically reviewed by Holmes (1950). A procedure commonly employed is that described by Moir and Hinks (1935), whose modification for chocolate cake is given on p. 318. A factor of 3·26 is used for the conversion of nitrogen into 'total alkaloid'. This is based on the assumption that the alkaloids in cocoa nib consist of theobromine and caffeine in the ratio of 9:1. The authors found that roasted fat-free dry blended cocoa nibs contained on the average 3·15 per cent of 'total alkaloids'. The amount of fat-free cocoa solids in a product such as chocolate was calculated by dividing the percentage of 'total alkaloids' by 0·0315. Harral (1940) converted the theobromine to caffeine by shaking with methyl sulphate and alkali, both alkaloids being extracted in the form of caffeine, which is more readily extracted with chloroform than theobromine.

Chapman *et al.* (1963) described a method for determining total alkaloids involving adsorption on Fuller's earth, removal with alkali and measurement of the ultraviolet absorption after acidifying (p. 318). For cocoa, 0·1 g sample is employed and dry fat-free cocoa is expressed as total alkaloids × 33·3.

A method based on extraction of purines with boiling water, cooling, clarification with lead acetate, removal of excess lead as carbonate followed by spectrophotometric measurement of the filtered extract at 245, 272 and 306 nm (Hadorn and Zuercher, 1965) has been used for many years.

Gerritsma and Koers (1953) described a relatively simple method for cocoa residues. The theobromine is extracted by shaking with chloroform and determined titrimetrically.

To a 500 ml bottle add 2 g of sample, 270 ml of chloroform and then 10 ml of ammonia solution (10 per cent m/m NH_3). Stopper and shake vigorously for 5 min. Add 12 g of anhydrous sodium sulphate, shake again and allow to stand overnight. Then filter and wash through with 100 ml of chloroform. Distil off the solvent and remove the final traces in the 100 °C oven. Add 50 ml of water to the extracted theobromine and boil. Cool, exactly neutralise with 0·1 M sodium hydroxide (1 or 2 drops), using phenol red (0·5 ml of 0·1 per cent solution in alcohol) as indicator and add 20 ml of 0·1 M silver nitrate (theobromine + $AgNO_3$ → theobromine – Ag + HNO_3). Titrate the acid which is liberated with 0·1 M sodium hydroxide to the red colour of the indicator (pH 7·4). 1 ml 0·1 M NaOH ≡ 18·0 mg theobromine.

Pusey (1977) has studied the above method for estimation of cocoa solids. He showed that an unknown sample can be assayed to within plus 52 per cent and minus 14 per cent. However, when the theobromine content of the raw material is known the estimate is within ±6·7 per cent.

Chromatographic procedures for the measurement of cocoa purines have received considerable attention in recent years. Paper chromatography has been applied by Sjoberg and Rajama (1984) who claimed that the results obtained compared favourably with those by HPLC. The use of TLC has been studied by Stangelberger *et al.* (1983). Kreiser and Martin (1978) using HPLC concluded that the measurement of purines is of little value except to a manufacturer checking his own product; nevertheless further studies using a variety of HPLC methods have been supported (Anon, 1981) and reported. Typical methods are those of Jurgens *et al.* (1980), and Trugo *et al.* (1983). Various studies on cocoa and cocoa products showing the range of values for raw material and commercial articles have been recorded by DeVries *et al.* (1981), Woollard (1982), Herrmann *et al.* (1982) and Craig and Nguyen (1984). Values for theobromine and caffeine reported by Zoumas *et al.* (1980) are shown in Table 10.6. Reversed phase chromatography using C_{18} columns with water/metha-

Table 10.6 The methylxanthine (theobromine and caffeine) contents of commercial cocoa and chocolate products, per cent (Zoumas *et al.*, 1980)

	Cocoa mass	Cocoa powder	Plain chocolate	Milk chocolate
Theobromine				
max.	1·73	2·66	0·63	0·19
min.	0·82	1·46	0·36	0·13
average	1·22	1·89	0·46	0·15
Caffeine				
max.	0·42	0·35	0·12	0·05
min.	0·06	0·08	0·02	0·01
average	0·21	0·21	0·07	0·02

nol and water/methanol/acetic acid as eluants are favoured by several authors; the extraction procedure used in many cases is treatment of the sample with hot water and precipitation of protein with trichloracetic acid. Microbore HPLC has also been employed by Hurst *et al.* (1985), who report good precision and economy of reagents by comparison with conventional methods.

Shell. Hand separated cocoa beans contain approximately 12 per cent shell and nib from roasted beans contains 1–2 per cent shell (legal limit 5 per cent on fat-free dry solids). Methods for determination of shell content of nib depend either on chemical determinations of substances that occur at differential amounts in shell and nib or microscopic assessment of characteristic structures in the shell. Amongst the chemical methods Doppler (1926) determined pectin and Mosiman (1945) determined methanol, methoxygalacturonic acid and pectin. Winkler (1956) reviewed chemical methods and recommended a procedure based on galacturonic acid estimation. Reeskamp (1959) rejected the method because of variability. Lockwood (1939) based his assessment of shell content on the determination of fibre. He reported 5·7 per cent fibre in nib and 18·5 per cent in shell (on dry fat-free basis). More recently interest has been shown in the determination of behenic acid tryptamide (BTA). Wurziger and Harms (1970) identified two components behenic (C_{22}) acid and lignocerin (C_{24}) acid-5 hydroxytryptamide as the characterising components at levels of 0·5 mg per cent in nib and 1·3–5·8 mg per cent in shell. Methods for fibre and pectic acid are described in the AOAC. Methods based on microscopical examination and counts of characteristic cells are by no means new. Three kinds of cells have been identified; slime cells, stone cells and spiral cells. The methods for spiral cells have been studied by Jackson (1968). The coefficient of variation for the method is estimated as 38 per cent when differences in type of bean are taken into account. Stone cell count methods have received similar study by van Brederode and Reeskamp (1957) and Jackson (1970). Van Brederode and Reeskamp noted the effect of grinding on the number of groups of stone cells and the number of cells in each group. They showed that a relationship exists between the cell groups, group size and shell content for a specified powder size. Jackson (1970) noted that the cell count and group count procedures gave approximately equivalent results. A limited study carried out in the author's laboratory using stone cell methods and fibre methods showed that there was a significant difference between the results obtained on common samples by counting and by estimation of fibre. The results by fibre analysis tended to be lower than by the count procedure. Schafer (1983) used α-amylase degradation to assess non-starch polysaccharides.

Starch. Evidence of addition of starch is obtained by microscopical examination. Starch can be estimated using diastase:

De-fat 8 g of the sample and weigh out 5 g of the dry residue into a flask. Add 100 ml of 12 per cent alcohol, shake, filter and wash with 20 ml of industrial alcohol. Wash the extracted residue into a flask with 50 ml of ammonia-free water, heat in boiling water for 20 min and cool. Add 0·1 g of diastase (+ water) and maintain at 50–55 °C for 2 h.* Then cool, make up to 250 ml and filter. Mix 200 ml of filtrate with 20 ml hydrochloric acid (sp. gr. 1·125) and heat in boiling water for 3 h. Cool, neutralise (just) with sodium hydroxide solution and make up to 250 ml. Then estimate the reducing sugars using Fehling's solution and calculate as dextrose (dextrose × 0·90 = starch). Using this method, cocoa containing no added farinaceous matter shows a starch percentage of about 7 per cent.

* It is useful to transfer one drop to a test-tube and add iodine to ensure that the hydrolysis has been completed.

Sugars or other carbohydrate sweeteners are present in drinking chocolates. For their estimation, prepare a 5 per cent solution clarified with zinc ferrocyanide and determine the sugars present in the filtrate by any appropriate method described in Chapter 6. Lactose, if present from milk constituents, may be determined most conveniently by HPLC or enzymatically.

Adulteration. Fry (1984) has used evidence afforded by the presence of *d*-inositol and (+)-piritol as indicative of the presence of carob in cocoa powder.

CHOCOLATE

Manufacture

Chocolate is manufactured from the cocoa products (nib, mass, cocoa and butter) together with sugar and other ingredients according to type. Plain (dark) chocolate is prepared by rolling together a mixture of cocoa nib, sugar and cocoa butter between steel rollers. The warm chocolate and extra cocoa butter and any added flavours are processed (conching) at about 70 °C in a longitudinal or circular conche. Conching improves viscosity and texture and improves flavour by driving off unwanted volatiles such as acetic, propionic and butyric acids. Moisture is driven off and oxidation of astringent flavour components takes place. After tempering, the liquid chocolate is run into warmed and polished moulds and set by cooling. The special characteristics of chocolate are due to the unique properties of the cocoa butter which melts just below body temperature and the fineness of the sugar and cocoa particles. For milk chocolate, the milk is pre-condensed with sugar and mixed with cocoa liquor and dried to form milk crumb after drying. The milk crumb is then mixed with cocoa butter and manufactured in the same way as plain types, except that conching is carried out at lower temperatures and for a shorter time. The butterfat present makes milk chocolate softer and more easily melted. On storage chocolate that has been incorrectly manufactured, packed or stored is liable to get a white bloom of fat on the surface or a grey film of stickiness (sugar bloom) due to absorption of moisture by the sugar. Methods of chocolate manufacture are discussed by Howat (1968), Cook (1972) and Wieland (1972). Rossell (1983) and Anon (1984) have given accounts of the nature and use of cocoa fat and the effects on the properties of a range of cocoa based confectionery products by the incorporation of fats from other sources. Rossell discussed the desired attributes of block and moulded chocolate and contrasted these with the attributes of the material used for coatings and couvertures, which may contain low amounts of cocoa butter and significant quantities of modified vegetable fats. He gives typical product recipes and discusses the requirements of the legislation in relation to the various product types.

Legislation

The composition of chocolate and chocolate products is defined in Part II to Schedule 1 of the Cocoa and Chocolate Products Regulations (SI 1976 No. 541) (the provisions for gianduja nut chocolate were amended by the 1982 (Amendment) Regulations), the main requirements of which are summarised in Table 10.7. Edible substances are permitted as additional ingredients within prescribed limits (Table 10.8). They are defined as 'any substance suitable for use as a food, which is wholly a natural product, whether or not that substance has been subjected to any process or treatment, or any vitamin or mineral, or any flavouring substance which does not impart the flavour of chocolate or milk fat'. It does not include cocoa or chocolate products (as defined) or the acids, bases or emulsifiers that are separately permitted in chocolate products. The 1982 (Amendment) Regulations deleted polyoxyethylene (20) sorbitan monostearate and sorbitan monostearate from the list of permitted emulsifiers.

The preamble to the EC Directive draws attention to the use of vegetable fats other than cocoa butter in chocolate products and states that the situation will have to be re-examined in the light of future developments. The use of such fats (cocoa butter substitutes or equivalents) has been the subject of controversy in the UK (APA, 1966). A proposal for a Council Directive (COM (83) 787) has been published (OJ No. C32, 7.2.84, p. 3) updating earlier legislation and allowing for the use of vegetable fats and other national differences in chocolate formulation. Table 10.9

Table 10.7 Compositional requirements for chocolate; cocoa and chocolate products regulations (as amended)

	Total dry cocoa solids (min. %)	Dry non-fat cocoa solids (min. %)	Cocoa butter (min. %)	Fat (min. %)	Milk solids (%)	Milk fat (min. %)	Sucrose[b] (max. %)
Chocolate	35	14	18	—	—	—	—
Plain chocolate[a]	30	12	18	—	—	—	—
Milk chocolate A[a]	25	2·5	—	25	14	3·5	55
Milk chocolate B[a]	20	2·5	—	25	20	5	55
Couverture chocolate[a]	35	2·5	31	—	—	—	—
Couverture Milk chocolate	25	2·5	—	31	14	3·5	55
Gianduja nut[a] chocolate	32[c]	8[c]	—	—	≯ 5[g]	—	—
Gianduja nut[a] milk chocolate	25[d]	—	—	—	10[d]	—	—
Cream chocolate[a]	25	2·5	—	25	3, < 14[e]	7	55
Skimmed milk[a] chocolate	25	2·5	—	25	14	0, < 3.5[f]	55
White chocolate[a]	—	—	20	—	14	3·5	55
Chocolate vermicelli	32	—	12	—	—	—	—
Milk chocolate vermicelli	20	—	—	12	12	3	66

[a] Filled chocolates; consist of a filling (except flour confectionery or biscuits) wholly or partly enclosed in 25 per cent (min.) of one or more of the chocolate product indicated.
[b] Sucrose may be replaced by an equal amount of dextrose, fructose, lactose or maltose or by any combination to a maximum of 5 per cent of the total weight of the product (except for dextrose alone when the maximum level of replacement is 20 per cent of the total weight of the product).
[c] Calculated on the weight of product excluding nuts. The product must contain finely ground hazel nut (20 per cent min., 40 per cent max.); may contain almonds, hazel nuts and other nuts either whole or broken to a maximum of 60 per cent total nut content in the whole product.
[d] As for [c] above except that the minimum ground hazel nut is 15 per cent.
[e] Minimum and maximum levels for milk solids not fat.
[f] Minimum and maximum levels for milk fat.
[g] The dry milk solids shall contain not more than 1·25 per cent butterfat.

demonstrates the typical differences between UK and continental chocolate composition. Compositional standards defined by Codex (Codex Stan 87–1981) are similar to those of the UK legislation. Codex also adopted a standard for white chocolate (called cocoa butter confectionery in 1985). This standard has provisions that are similar to those for white chocolate detailed above except that the emulsifiers now deleted by UK regulations are included.

Composition

Analytical data for a number of products are detailed in Table 10.10, and in view of the wide range of products on the market the figures are for guidance only. Paul and Southgate (1978) also list additional information for chocolate and confectionery bars; typical amino acid and fatty acid data are also presented.

Analysis

The comments in respect of EC legislation regarding methods of analysis for cocoa (p. 377) apply to the analysis of chocolate. The quality of chocolate can be assessed

Table 10.8 Permitted edible ingredients in prescribed chocolate products. The whole provision applies to chocolate, plain chocolate, milk chocolate, cream chocolate, skimmed milk chocolate, white chocolate, couverture chocolate and couverture milk chocolate, except that paragraph (2) does not apply to couverture chocolate and couverture milk chocolate

Permitted edible ingredient	Min.Z (%)	Max. (%)
1. Any edible substance in clearly visible and discrete pieces, except flour or starch added as such or any fat or fat preparation not derived exclusively from milk.	5	40
2. Any edible substances not in clearly visible and discrete pieces, except for flour or starch added as such, sucrose, dextrose, fructose, lactose, maltose or any fat or fat preparation not derived exclusively from milk; or	—	30
3. Any vegetable fat or preparation of vegetable fat not derived from cocoa beans or parts of cocoa beans; or	—	5
4. Any partial coating or decoration; or	—	10
5. Any combination of one or more of the substances specified in (1) above with one or more of the substances specified in (2) to (4) provided the specified maximum limits are not exceeded; or	—	40
6. Any combination of two or more substances specified in paragraphs (2) to (4) provided the specified limits are not exceeded.	—	30

Table 10.9 The composition of UK and continental European chocolate (Nickless and Sidaway, 1980)

	UK		Europe	
	Milk	Plain	Milk	Plain
Sucrose	45	55	45	45
Skim milk powder			5	
FCMS[a]	28		14	
Cocoa liquor[b]	12	32	16	45
Milk fat		3		
Cocoa butter equivalent	5	5		
Cocoa butter	9·5	4·5	19·5	9·5
Lecithin	0·5	0·5	0·5	0·5

[a] Full cream milk solids contain 28 per cent fat.
[b] Cocoa liquor contains 55 per cent fat.

by determining the following: **moisture, fat, sugars, nitrogen, 'total purines', ash, water-soluble ash, crude fibre** and **trace metals** (e.g. **As, Pb** and **Cu**).

In addition, the nature of the fat is assessed by chromatographic procedures (see Chapter 16). The presence of 'foreign' proteins is also detected by the use of electrophoretic and immunological methods. The factory control of chocolate processing is discussed by Krauss (1972).

Prior to the analysis the sample should be grated and immediately transferred to an airtight jar. In the case of chocolate and milk chocolate it may be necessary to melt and temper the sample before shredding. The draft EC directive contains a detailed procedure for this tempering process (see p. 377). The **moisture** can be determined by drying at 100 °C and the **fat** by continuous extraction using light petroleum. Extraction of fat from milk chocolate using boiling hydrochloric acid gives higher values for milk-fat content than does direct extraction with light petroleum (Vinken-

Table 10.10 Analytical data for plain, milk and nut chocolate (as percentages)

	Plain chocolate	Milk chocolate	Nut chocolate
Moisture	0·8–2·3	0·8–1·8	1·1
Total ash	1·0–2·5	1·8–2·4	1·5
Water-soluble ash	0·7–1·1	—	—
Fat	30–40	30–40	36·6
Sucrose	37–55	35–47	45·2
Lactose	—	7–11	6·9
Starch	2–7	—	—
Total nitrogen	0·6–1·3	0·8–1·5	1·12
Theobromine	0·4–0·7	0·2–0·5	—
Crude fibre	0·8–1·5	0·4–1·0	—
*Values on the fat**			
Reichert value	—	4·3,5·6	4·8
Polenske value	—	1·2,1·1	1·3
Kirschner value	—	4·0,4·3	4·2

* Results from individual samples.

bourg *et al.*, 1970). In the collaborative study by Judd *et al.* (1984) the method performed satisfactorily. Pulsed NMR has been proposed as a rapid method for determination of fat content without prior solvent extraction (Leung *et al.*, 1985). The method correlated well with the estimation of solid fat content by dilatometry. **Lactose** and **sucrose** can be estimated on a clarified solution by titrimetric or other procedures (see Chapter 6). The amount of **non-fatty milk solids** in milk chocolate is conventionally assessed from the **lactose** and **calcium** content and from the **orotic acid** figure, which remains unchanged by heating during manufacture (Motz, 1972). Orotic acid is brominated to produce dibromobarbituric acid. Ascorbic acid is added to remove excess bromine. When *p*-dimethyl-aminobenzaldehyde in pentanol is added a yellow colour is produced and its intensity is measured at 461 nm after extraction with butyl acetate. The AOAC method for determination of milk protein in chocolate is based on its solubility in sodium oxalate solution. In many cases the milk protein is not completely soluble and results may be low especially when the milk has been subjected to heat treatment during the preparation of the crumb.

Methods for assessment of the presence of foreign proteins derived from milk and milk products and from nuts and nut products have received attention from a number of authors. In common with the developments in the examination of meat and meat products, immunological and electrophoretic methods have been the subject of collaborative study. Klein *et al.* (1983, 1984) and Guenther *et al.* (1983) have reported studies on the double gel diffusion method and immunoelectrophoresis in relation to products containing milk powder, whey powder, soya flour, hazelnut and peanut paste. Garrone *et al.* (1984) reported the detection of hazelnut in chocolate products using an electrophoretic method; the method demonstrates the usual problem of calibration in the case of application to commercial products in which the source of the raw material is unknown.

The identification and estimation of butterfat in cocoa products has depended for many years on the Reichert–Polenske method. Current practice, however, is to use the estimation of butyric acid by gas chromatographic methods. Withington (1967), Harrison and Withington (1970) and Hadorn and Zuercher (1968) have described suitable procedures (see also Dickes and Nicholas, 1976; Dickes, 1979). The method has been studied twice in a collaborative trial. Biltcliffe and Wood (1982) reported that the procedure of Phillips and Saunders (1968) is suitable for control purposes (for full details of the method, see p. 587). The use of triglyceride analysis is discussed under Cocoa Butter (p. 387). Cholesterol content of milk fat has also been used to indicate the presence and amount of milk fat present in vegetable fat. The cholesterol index is recommended by the IOCCC. Cleverley and Dawes (1963) have described a method based on differential infra-red spectroscopy for detecting the presence of palm kernel, hydrogenated and pure coconut oils, butterfat and tallow in fat extracted from chocolate. The amount of dry fat-free cocoa matter is assessed by the methods for cocoa. Schmid (1976) has described a method for the determination of phosphatides in cocoa products. This method is based on extraction and subsequent precipitation of phosphatides with cobalt thiocyanate. The phosphorus content of the complex is then determined and the phosphatide content is calculated. Hurst and Martin (1980) have used a solvent extraction method, clean-up on a Sep-Pak cartridge and HPLC to measure lecithin in chocolate. Ugrinouits (1983) has reviewed the available methods. Ingles *et al.* (1985) have examined chocolate and other proteinaceous products for the presence of the biogenic amines. Histamine and *p*-tyramine were detected in drinking chocolate but no evidence of their presence was found in milk chocolate, whilst histamine was detected in dark chocolate. These findings are not consistent with those of Hurst and Toomey (1981).

COCOA BUTTER AND SUBSTITUTES

Manufacture

Cocoa butter is the fat obtained by prescribed processes from cocoa products. Press cocoa butter is the fat obtained by pressure from cocoa nib or cocoa mass; expeller cocoa butter is obtained by a torsion process from cocoa beans or combination of cocoa beans with nib or press cake; solvent extracted cocoa butter is obtained by extraction of beans or other cocoa material with permitted solvents. Cocoa butter from any of these processes may also be refined by neutralisation and deodorised by steam or treatment with adsorbents. The physical properties of the chocolate are controlled by the melting properties of the cocoa butter, by the tempering process and by addition of minor amounts of other fats. The manufacturing process also requires strict control of the raw materials. Currently, control is achieved by examining melting properties, fatty acid profile (by GLC), the triglyceride structure (also by GLC) and the ultraviolet absorbance of the fats used.

Legislation

Four types of cocoa fat are defined in Schedule I Part 1 of the Cocoa and the Chocolate Product Regulations. Control values for acidity and unsaponifiable matter are prescribed in the regulations together with defined processes for production and treatment of the fats. These are summarised in Table 10.11. The provisions of UK regulations are similar to those of the Codex Standard for Cocoa Butters (Codex Stan 86–1981) with the exception that the Codex standard prescribes the traditional constants (refractive index, melting point, slip point, acidity, iodine value, saponification value and unsaponifiable matter). Typical data are quoted in Table 10.12.

Table 10.11 Cocoa butters—UK legislation

Product description	Raw material	Process	Permitted treatments	Unsaponifiable matter (% max.)	Acidity (% max.) as oleic acid
Press cocoa butter	Cocoa nib and press cake	Pressure	Filtration, centrifugation, steam deodorisation	0·35	1·75
Expeller cocoa butter	Cocoa beans, nib and press cake	Torsion	Filtration, centrifugation steam deodorisation	0·5	1·75
Refined cocoa butter	Cocoa beans, nib press cake	Pressure, torsion, solvent	Neutralisation, decolorisation, and the other approved methods	0·5	1·75
Cocoa fat	Cocoa beans or parts of beans	Not defined	Not defined	Not defined	Not defined

Table 10.12 Data for cocoa butter (including Codex limits)

Colour of fat	Yellowish-white
Specific gravity (15·5 °C/15·5 °C)[a]	0·950–0·975
Refractive index at 40 °C	1·456–1·459
Melting point (°C)[a]	31–35°
Iodine value	33–42
Saponification value	188–198
Unsaponifiable matter (%)	0·5 (max.)
Acid value (usual range)	0·5–1·75
Reichert value	0·2–1·0
Polenske value	Up to 0·5
Kirschner value	0·2
Titre (°C)	48–51

[a] When cocoa butter is melted it does not, on cooling, regain its normal sp.gr. and m.p. for some time and 72 h should elapse between the time of melting and observing these constants.

Composition

Cocoa butters of different origin vary, within narrow limits, in the type of fatty acids which they contain. Paul and Southgate (1978) record the three major constituents as 16:0, 18:0 and 18:1. Minor constituents include 20:0 and 18:2. Small variations in the chemical composition have considerable influence on their melting properties. For economic reasons the industry has developed the use of other vegetable fats (substitutes (CBS) and equivalents (CBE)) tailored when necessary to simulate as far as possible the properties of genuine cocoa butter. These include illipe butter, coconut oil, palm kernel oil, and hydrogenated rearranged fats, all of which are used to extend or replace the more expensive cocoa product. Wolf (1978), Finke (1979), Anon (1977), Tonnesman (1977), Laning (1981), Anon (1984) and Pease (1985) have discussed the development and properties of the hydrogenated rearranged vegetable fats used in manufacture of a variety of chocolate products. See also Nickless and Sidaway (1980) and Rossell (1983) for an account of the composition and properties of cocoa butter and the effects of the physical properties of the product by inclusion of fats from other sources. The technological properties of cocoa butter have also been evaluated by means of differential scanning calorimetry (Merken *et al.*, 1982). Timms and Parekh (1980) have studied the technological properties of milk fat derivatives in admixture with cocoa butter, characterising the products using NMR (see also Petersson *et al.*, 1985).

Analysis

The analysis of fats is described in Chapter 16. As indicated earlier the traditional tests are still widely used for characterisation and control but the industry now places reliance on the fatty acid profile and the triglyceride stucture. Padley and Timms (1978) and Finke (1979) first described the techniques for determination of carbon number of the C50–C54 triglycerides by GLC and the interpretation of results of analysis. The topic has received considerable attention since these first proposals; variations on the technique and a variety of applications have been reported by a number of authors including Padley and Timms (1980), Nickless and Sidaway (1980), Finke (1980, 1982), Young (1984), Gegiou and Staphylakis (1985), Chaveron and Verdoia (1984) and Verdoia-Dermont and Young (1985). The method has also been proposed for inclusion in a draft EC Directive for control of chocolate and chocolate products. A successful collaborative study of the procedure proposed in the draft directive has been carried out in the UK. The use of HPLC analysis of the triglyceride fractions has also been reported by Schulte (1981), Shukla *et al.* (1983, 1985) and Phillips *et al.* (1984), these authors claiming that the HPLC method has the advantage of simplicity over that using GLC and equivalent results are claimed. A new HPLC procedure using light scattering detection for the separation of triglycerides offers a possible method for the estimation of CBE in chocolate (Palmer and Palmer, 1989).

The sterol content of the unsaponifiable fraction of fats aids the distinction between natural and hardened fats. Dick and Miscrez (1980) and Homberg and Bielefeld (1982) used GLC whilst Staphylakis and Gegiou (1985) used TLC in combination with GC/MS to examine free and esterified sterols in cocoa butter. In combination with examination of the fatty acid fractions of the esters the method is claimed to give a means of characterising cocoa butters.

REFERENCES

Anon (1968) *Which* (April), 100.

Anon (1973) *Which* (November), 337.

Anon (1977) *Food Engineering*, **49**, 30.

Anon (1981) AIFC/OICC *Newsletter*, **6**, 12.

Anon (1984) *Journal of the American Oil Chemists' Society*, **61**, 468.

APA (1966) *Journal of the Association of Public Analysts*, **4**, 65.

Ashoor, S. H., Seperich, G. J., Monte, W. C. & Welty, J. (1983) *Journal of the Association of Official Analytical Chemists*, **66**, 606.

Atkinson R. M., Dickinson, D. & Harris, F. J. T. (1950) *Journal of the Science of Food and Agriculture*, **1**, 264.

Beckett, S. T. (1988) (ed.) *Industrial Chocolate Manufacture and Use.* Glasgow: Blackie.

Biltcliffe, D. O. & Wood, R. (1982) *Journal of the Association of Public Analysts*, **20**, 69.

Blauch, J. L. & Tarka, S. M. (1983) *Journal of Food Science*, **48**, 745.

Bondavovich, H. A., Giammarino, A. S., Renner, J. A., Shephard, F. W., Shingler, A. J. & Giantureo, M. A. (1967) *Journal of Agricultural and Food Chemistry*, **15**, 36.

Chang, R. L., Hsu, H. T. & Li, C. F. (1973) *Journal of Food Technology*, **8**, 217.

Chapman, W. B., Fogden, E., Urry, S. (1963) *Journal of the Association of Public Analysts*, **1**, 59.

Chatt, E. M. (1953) *Cocoa: Cultivation, Processing, Analysis.* London: Interscience.
Chaveron, H. & Verdoia, C. (1984) *Annales des Falsifications et Experimentale Chimie*, **77**, 401.
Clarke, R. J. (1979) *Food Chemistry*, **4**, 81.
Clarke, R. J. (1987) In *Quality Control in the Food Industry*, Vol. 4, ed. S. M. Herschdoerfer. London: Academic Press.
Clarke, R. J. & Macrae, R. (1985) (eds) *Coffee.* Vol. 1, *Chemistry*; Vol. 2, *Technology*; Vol. 3, *Physiology*; Vol. 4, *Agronomy*; Vol. 5, *Related Beverages*; Vol. 6, *Commercial and Technico-Legal Aspects.* London: Elsevier Applied Science.
Clarke, R. L. & Walker, L. J. (1974) *Journal of the Science of Food and Agriculture*, **25**, 1389.
Cleverley, B. & Dawes, S. N. (1963) *New Zealand Journal of Science*, **6**, 555.
Clifford, M. N. (1975) *Process Biochemistry*, **10**, 20.
Clifford, M. N. & Willson, K. C. (1985) (eds) *Coffee—Botany, Biochemistry and Production of Beans and Beverage.* London: Croom Helm.
Conacher, H. B. S. & Rees, D. I. (1964) *Analyst*, **89**, 806.
Cook, L. R. (1972) *Chocolate Production and Use.* New York: Books for Industry.
Craig, W. J. & Nguyen, T. T. (1984) *Journal of Food Science*, **49**, 302.
Cros, E., Fourny, G., Guyot, B., Rouly, M. & Vincent, J-C. (1980) *Cafe-cacao-The*, **24**, 203.
Dick, R. & Miscrez, A. (1980) *Mitteilungen aus dem Gebeit der Lebensmitteluntersuchung und Hygiene*, **71**, 499.
Dickes, G. J. (1979) *Talauta*, **26**, 1065.
Dickes, G. J. & Nicholas, P. V. (1976) *Gas Chromatography in Food Analysis.* London: Butterworth.
Doppler, C. L. (1926) *Zeitschrift fur Lebensmittel Untersuchung und Forschung*, **51**, 126, 462.
Eden, T. (1976) *Tea*, 3rd edn. London: Longmans.
Edwards, F. W. & Nanji, H. R. (1937, 1938) *Analyst*, **62**, 841; **63**, 323.
Engelhardt, U., Peters, K. & Maier, H. G. (1984) *Zeitschrift fur Lebensmittel Untersuchung und Forschung*, **178**, 288.
Ferrara, A., Reale, P., Calaminici, M. G. & Iaccarino, T. (1982) *Bollettino dei Laboratori Chimici Provinciali*, **33**, 55.
Finke, A. (1979) *Gordian*, **79**, 112.
Finke, A. (1980) *Deutsche Lebensmittel Rundschau*, **76**, 162, 187, 384.
Finke, A. (1982) *Deutsche Lebensmittel Rundschau*, 78, 389.
Fry, J. C. (1983) *Leatherhead Food Research Association, Technical Circular*, 422.
Garrone, W., Antocucci, M. & Bonu, U. (1984) *Milchwissenschaft*, **39**, 464.
Gegiou, D. & Staphylakis, K. (1985) *Journal of the American Oil Chemists' Society*, **62**, 1047.
Gerritsma, K. W. & Koers, J. (1953) *Analyst*, **78**, 201.
Guenther, H. O., Kliene, E., Finke, A. & Baudner, S. (1983) *Lebensmittelchemie gerichtle Chemie*, **37**, 105.
Gur'Eva, K. B. & Tserevitinov, O. B. (1983) *PMCA Research Notes*, **7**, 3.
Hadorn, H. & Zuercher, K. (1965) *Mitteilungen aus dem Gebiete der Lebensmitteluntersuchung und Hygiene*, **56**, 491.
Hadorn, H. & Zuercher, K. (1968) *Mitteilungen aus dem Gebiete der Lebensmitteluntersuchung und Hygiene*, **59**, 369.
Harler, C. R. (1963) *Tea Manufacture.* London: Oxford Univ. Press.
Harral, J. C. (1940) *Analyst*, **65**, 408.
Harrison, D. & Withington, D. F. (1970) *Journal of the Association of Public Analysts*, **8**, 68.
Herrmann, A., Lotscher, E. & Wagmann, M. (1982) *Mitteilungen aus dem Gebeit der Lebensmitteluntersuchung und Hygiene*, **73**, 121.
Holmes, K. E. (1950) *Analyst*, **75**, 457.
Homberg, E. & Bielefeld, B. (1982) *Deutsche Lebensmittel Rundschau*, **78**, 73.
Howat, G. R. (1968) *Process Biochemistry*, **3**, 31.
Hughes, E. B. & Wise, W. (1934) *Journal of the Society of Chemical Industry*, **53** 189T.
Hughes, E. B. & Wise, W. (1955) *Chemistry and Industry*, 549.
Hurst, W. J. & Martin, R. A. (1980) *Journal of the American Oil Chemists' Society*, **57**, 307.
Hurst, W. J. & Toomey, P. B. (1981) *Analyst*, **106**, 394.
Hurst, W. J., Snyder, K. P. & Martin, R. A. (1985) *Journal of Chromatography*, **318**, 408.
Ingles, D. L., Back, J. F., Gallimore, D., Tindale, R. & Shaw, K. J. (1985) *Journal of Agricultural and Food Chemistry*, **36**, 402.
Jackson, M. (1965) *Manufacturing Confectioner*, **45**, 45.
Jackson, M. M. (1968) *Journal of the Association of Official Analytical Chemists*, **51**, 725.
Jackson, M. M. (1970) *Journal of the Association of Official Analytical Chemists*, **53**, 476.
Jain, J. C. & Takeo, T. (1984) *Journal of Food Biochemistry*, **8**, 243.
Judd, H. J., Percival, M. D. & Wood, R. (1984) *Journal of the Association of Public Analysts*, **22**, 81.

Jurgens, U., Grunherr, K. von & Blosczyk, G. (1980) *Lebensmittelchemie gerichtle Chemie*, **34**, 109.
Kaffka, K. J., Norris, K. H., Kulcsar, F. & Draskovits, I. (1982) *Acta Alimentaria*, **11**, 271.
Kazi, T. (1979) *Food Chemistry*, **4**, 73.
Klein, E., Finke, A., Baudner, S. & Gunther, H. O. (1983) *Zucker und Susswar Wirtschaft*, **36**, 65.
Klein, E., Finke, A., Baudner, S. & Gunther, H. O. (1984) *Review Chocolate Confectionery and Bakery*, **9**, 22.
Kleinert, J. (1982) *Zucker und SusswarWurtschaft*, **35**, 333.
Krauss, P. (1972) *Manufacturing Confectioner*, **52**, 38.
Kreiser, W. R. & Martin, R. A. (1978) *Journal of the Association of Official Analytical Chemists*, **61**, 1424.
Laning, S. J. (1981) *The Manufacturing Confectioner*, June, 52.
Lehmann, G. & Neumann, B. (1973) *Zeitschrift fur Lebensmittel Untersuchung und Forschung*, **153**, 363.
Leung, H. K., Anderson, G. R. & Norr, P. J. (1985) *Journal of Food Science*, **50**, 942.
Lim, F. F. (1981) *Journal of the Association of Official Analytical Chemists*, **64**, 287.
Lockwood, H. C. (1939) *Analyst*, **64**, 92.
Lutman, F. (1982) *Bollettino dei Laboratori Chimici Provinciali*, **33**, 573.
Madison, B. L., Kozarek, W. J. & Damco, C. P. (1976) *Journal of the Association of Official Analytical Chemists*, **59**, 1258.
Mahanta, P. K., Hazarika, M. & Takeo, T. (1985) *Journal of the Science of Food and Agriculture*, **36**, 1130.
Manley, C. H. (1965) *Journal of the Association of Public Analysts*, **3**, 101.
Martin, C. R. A. (1977) *British Food Journal*, **79**, 166.
Mayanna, S. M. & Jayaram, B. (1981) *Analyst*, **106**, 729.
Merken, G. V., Vaeck, S. V. & Dewulf, D. (1982) *Lebensmittel Wissenschaft und Technologie*, **15**, 195.
Michie, N. D. & Dixon, E. J. (1977) *Journal of the Science of Food and Agriculture*, **28**, 215.
Millin, D. J. (1987) In *Quality Control in the Food Industry*, Vol. 4, ed. S. M. Herschdoerfer. London: Academic Press.
Millin, D. J. (1988) In *Food Industries Manual*, 22nd edn, ed. M. D. Ranken. Glasgow: Blackie.
Minifie, B. W. (1980) *Chocolate, Cocoa and Confectionery: Science and Technology*, 2nd edn. Westport: AVI.
Moir, D. D. & Hinks, E. (1935) *Analyst*, **60**, 439.
Monier-Williams, G. W. (1950) *Trace Elements in Food*. London: Chapman and Hall.
Mosiman, G. (1945) *Mitteilungen aus dem Gebiete der Lebensmitteluntersuchung und Hygiene*, **36**, 284.
Motz, R. J. (1972) *Analyst*, **97**, 866.
Newman, J. R. (1981) *Journal of the Association of Public Analysts*, **19**, 59.
Newton, J. M. (1969) *Journal of the Association of Official Analytical Chemists*, 52, 653.
Nickless, H. & Sidaway, A. F. (1980) In *Fats and Oils: Chemistry and Technology*, eds R. J. Hamilton & A. Bhati. London: Applied Science.
Nurok, D., Anderson, J. W. & Zlatkis, A. (1978) *Chromatographia*, **11**, 188.
Padley, F. B. & Timms, R. E. (1978) *Chemistry and Industry*, 918.
Padley, F. B. & Timms, R. E. (1980) *Journal of the American Oil Chemists' Society*, **57**, 286.
Page, B. D. & Charbonneau, C. F. (1984) *Journal of the Association of Official Analytical Chemists*, **67**, 757.
Palmer, A. J. & Palmer, F. J. (1989) *Journal of Chromatography*, **465**, 369.
Paul, A. A. & Southgate, D. A. T. (1978) *The Composition of Foods*. London: HMSO.
Pazola, Z. & Cieslak, J. (1979) *Food Chemistry*, **4**, 41.
Pease, J. J. (1985) *Journal of the American Oil Chemists' Society*, **62**, 426.
Petersson, B., Anjou, K. & Sandstrom, L. (1985) *Fette Seifen Anstrichmittel*, **87**, 225.
Pettipher, G. L. (1986) *Journal of the Science of Food and Agriculture*, **37**, 289.
Phillips, A. R. & Saunders, B. J. (1968) *Journal of the Association of Public Analysts*, **6**, 89.
Phillips, F. C., Erdahl, W. L., Schmit, J. A. & Privett, O. S. (1984) *Lipids*, **19**, 880.
Pierce, A. R., Graham, H. N., Glassner, S., Madlin, H. & Gonzalez, J. G. (1969) *Analytical Chemistry*, **41**, 298.
Pusey, M. S. (1977) *Analyst*, **102**, 69.
Radtke-Granzer, R. & Piringer, O-G. (1981) *Deutsche Lebensmittel Rundschau*, **77**, 203.
Reeskamp, C. J. (1959) *Journal of the Association of Official Analytical Chemists*, **42**, 408.
Reid, S. J. and Good, T. J. (1982) *Journal of Agricultural and Food Chemistry*, **30**, 775.
Renold, W., Naef-Mueller, R., Keller, U., Willhalm, B. & Ohloff, G. (1974) *Helvetia Chimica Acta*, **57**, 1301.
Reynolds, S. L., Thorpe, S. A. & Wood, R. (1983) *Journal of the Association of Public Analysts*, **21**, 47.
Rhoades, J. W. (1960) *Journal of Agricultural and Food Chemistry*, **8**, 136.
Rossell, J. B. (1983) *Food*, October, 28.
Schafer, H. (1983) *Journal of Agricultural and Food Chemistry*, **31**, 1375.
Schmid, M. J. (1976) *Fette Seifen Anstrichmittel*, **78**, 35.
Scholze, A. & Maier, H. G. (1982) *Lebensmittelchemie gerichtle Chemie*, **36**, 111.

Schulte, E. (1981) *Lebensmittelchemie gerichtle Chemie*, **35**, 108.
Shimoda, M., Wada, K., Shibata, K. & Osajima, Y. (1985) *Nippon Shokuhin Kogyo Gakkaishi*, **32**, 377.
Shukla, V. K. S., Nielsen, W. S. & Batsberg, W. (1983) *Fette Seifen Anstrichmittel*, **85**, 274.
Shukla, V. K. S., Nielson, W. S. & Batsberg, W. (1985) *PMCA Research Notes*, **92**.
Sjoberg, A.-M. & Rajama, J. (1984) *Journal of Chromatography*, **295**, 291.
Smith, H. L. (1913) *Analyst*, **38**, 312.
Smith, R. M. (1981) *Food Chemistry*, **7**, 41.
Stahl, W. H. (1962) *Advances in Food Research*, **11**, 202.
Stangelberger, J., Steiner, I., Kroyer, G. & Washuttl, J. (1983) *Lebensmittelchemie gerichtle Chemie*, **37**, 70.
Staphylakus, K. & Gegiou, D. (1985) *Lipids*, **20**, 723.
Stoltz, A. & Masters, K. (1979) *Food Chemistry*, **4**, 31.
Tatlock, A. & Thomson, B. (1910) *Journal of the Society of Chemical Industry*, **29**, 138.
Terada, H. & Sakabe, Y. (1984) *Journal of Chromatography*, **291**, 453.
Timms, R. E. & Parekh, J. V. (1980) *Lebensmittel Wissenschaft und Technologie*, **13**, 177.
Tonnesmann, B. O. M. (1977) *Manufacturing Confectioner*, **57**, 45.
Tressl, R., Bahri, D., Koeppler, H. & Jensen, A. (1978) *Zeitschrift fur Lebensmittel Untersuchung und Forschung*, **167**, 108, 111.
Trugo, L. C. & Macrae, R. (1983) *Analyst*, **108**.
Trugo, L. C., Macrae, R. & Dick, J. (1983) *Journal of the Science of Food and Agriculture*, **34**, 300.
Ugrinouits, M. (1983) *Alimenta*, **22**, 7.
Van Brederode, H. & Reeskamp, C. J. (1967) *Zeitschrift fur Lebensmittel Untersuchung und Forschung*, **105**, 461.
Vaughan, J. G. (1970) *The Structure and Utilization of Oil Seeds.* London: Chapman and Hall.
Verdoia-Dermont, C. & Young, C. C. (1985) *Annales des Falsifications et Experimentale Chimie*, **78**, 119.
Vinkenbourg, C., Vos, H. J. & Mantzouranis, A. (1970). *Revue Internationale de la Choclaterie*, **25**, 446.
Vitzthum, O. G., Barthels, M. & Kwasny, H. (1974) *Zeitschrift fur Lebensmittel Untersuchung und Forschung*, **154**, 135.
Vree, P. H. Yeransian, J. A. (1973) *Journal of the Association of Official Analytical Chemists*, **56**, 1126.
Vries, J. W. De, Johnson, K. D. & Heroff, J. C. (1981) *Journal of Food Science*, **46**, 1968.
Walkowski, A. (1981) *Lebensmittelindustrie*, **28**, 75.
Wang, T. H., Shanfield, H. & Zlatkis, A. (1983) *Chromatographia*, **17**, 411.
White, G. W. & Shenton, A. J. (1980) *Journal of the Association of Public Analysts*, **18**, 129.
Wickremasinghe, R. L. & Swain, T. (1964) *Chemistry and Industry*, 1574.
Wieland, H. (1972) Cocoa and chocolate processing. *Food Processing Review* No. 27. New Jersey: Noyes Data Corporation.
Williams, H. A. (1964) *Journal of the Association of Public Analysts*, **2**, 8.
Winkler, W. O. (1956) *Journal of the Association of Official Analytical Chemists*, **39**, 688.
Withington, D. F. (1967) *Analyst*, **92**, 705.
Wolf, J. A. (1978) *Candy Snack Industry*, **143**, 50.
Wood, D. J. & Roberts, E. A. H. (1964) *Journal of the Science of Food and Agriculture*, **15**, 19.
Wood, D. J., Bhatia, I. S., Chakraborty, S., Choudhury, M. N. D., Deb, S. B., Roberts, E. A. H. & Ullah, M. R. (1964a, b), *Journal of the Science of Food and Agriculture*, **15**, 8, 14.
Wood, G. A. R. & Lass, R. A. (1985) *Cocoa.* 4th edn. Harlow: Longman.
Woollard, D. C. (1982) *New Zealand Journal of Dairy Science and Technology*, **17**, 63.
Wurziger, J. & Harms, U. (1970) *Gordian*, **70**, 376, 538, 470, 473.
Young, C. C. (1984) *Journal of the American Oil Chemists' Society*, **61**, 576.
Zeigleder, G. (1981) *Zucker und SusswarWirtschaft*, **34**, 105.
Zeigleder, G. (1982) *Review Chocolate Confectionery and Bakery*, **7**, 17.
Zeigleder, G. (1984) *Susswaren Technische Wirtschaft*, **28**, 422.
Zoumas, B. L. Kreiser, W. R. & Martin, R. A. (1980) *Journal of Food Science*, **45**, 314.

11

Herbs and spices. Salt

There is no clear cut distinction between herbs and spices. Herbs are fragrant non-woody plants of which the leaves, stems, flowers and seeds are used, usually fresh, for the mild delicate flavouring of food dishes, whereas spices are the dried parts of aromatic plants, usually tropical, including seeds, flowers, leaves, bark or roots, used for distinctively seasoning or spicing dishes. There are many well illustrated books available on herbs and spices. Noteworthy are those by Dowell and Bailey (1982) covering all food ingredients, and Heal and Allsop (1983) on spices containing much useful information. Standard reference texts are by Pruthi (1980) and Purseglove *et al.* (1981). Herbs and spices are incorporated in foods only in small amounts but they make important contributions towards the odour and flavour due to the presence of the volatile oil (essential oil) and fixed oil. Many instances of adulteration have been reported. Adulteration or substitution were often practised in the past. Foreign leaf has been found in herbs due to either deliberate addition or careless gathering (grass, weeds, etc.). Spices have been adulterated with foreign spices, the exhausted spice, cereals, ground stones and shells from fruits and nuts, sand, sawdust, etc. Most adulterants can be detected microscopically, but care must be exercised in distinguishing between a deliberate sophistication and relatively small amounts derived from matter that remained on the grinding stones from the previous grinding. Spices frequently give a high microbiological count, but the organisms usually present are predominantly spores (Dyett, 1963).

There are statutory standards for mustard and curry powder and the BSI (following ISO standards and recommendations) has published several specifications for spices. US standards and those prescribed in the BP and BPC for spices used for pharmaceutical purposes also serve as a useful guide to quality and have been retained in the tables. The USDA Agriculture Handbook No. 8–2 lists the nutritional composition of the herbs and spices described in this chapter. The volatile oils contain principally terpene hydrocarbons (Table 11.1) and oxygenated terpenoids (Table 11.2).

Oleoresins, the concentrated solvent extracts of herbs and spices, are becoming widely used as flavouring materials in composite foods instead of the natural and variable botanical material (Heath, 1986). These oleoresins are fluid or viscous semi-fluids and contain the volatile oils and also non-volatile fixed oil and resinous components, their quantity being dependent on the solvent used. Dichloroethane and dichloromethane are the solvents most widely used at present but more recently

Table 11.1 The terpene hydrocarbons in the essential oils of some herbs and spices

Essential oil	Total terpenes (%)	Principal terpenes
Allspice (Pimento)	25–30	α-phellandrene caryophyllene
Anise	10–20	α-pinene, α- & β-phellandrene, *p*-cymene
Caraway	40–50	limonene, α-pinene
Cardamom	50–55	limonene, α-pinene
Cinnamon/Cassia	20–30	caryophyllene, α-phellandrene, *p*-cymene
Celery	80–50	α-limonene, selinene
Clove	10–20	β-caryophyllene
Coriander	30–35	α- and β-pinene γ-terpinene
Dill	35–45	limonene
Fennel	20–30	α-pinene, α-phellandrene, dipentene
Ginger	about 90	zingiberene, β-pinene
Marjoram	about 40	α-terpinene
Nutmeg/Mace	about 60	α-pinene, camphene dipentene, *p*-cymene
Parsley	about 50	α-pinene
Pepper, Black	70–80 (mono)	α-thujene, α-pinene, camphene
	20–30 (sesqui)	α-phellandrene D-3-carene, α-terpinene, terpinolene
Rosemary	85–90	α- and β-pinene camphene
Sage	about 65	α-pinene
Thyme	60–80	β-pinene, *p*-cymene γ-terpinene, camphene

From Heath (1986).

Table 11.2 Oxygenated terpenoids in the essential oils of some herbs and spices

Compound	Present in	Sensory characters
Monoterpene alcohols		
Borneol	Nutmeg, rosemary, ginger	Pungent, camphoraceous, mint, burning
Carveol	Caraway, spearmint	Typical caraway-like
Carvacrol	Thyme, oreganum	Warmly pungent, smoky
Geraniol	Coriander	Rose-like
Linalool	Nutmeg, coriander	Floral, sweet
Menthol	Peppermint	Mint-like, fresh cooling
Terpineol	Cardamom, anise, cinnamon leaf	Sweet, woody
Sesquiterpene alcohols		
Zingiberol	Ginger	Strongly pungent, ginger-like
Monoterpene aldehydes		
Citral	Ginger	Strongly lemon-like
Citronellal	*Oreganum* spp.	Lemony, rose-like
Monoterpene ketones		
Carvone	Caraway, dill (*d*-form)	Strongly caraway-like
	Spearmint (*l*-form)	Spearmint-like
Fenchone	Fennel	Sweet, camphoraceous, burning taste
Thujone	Sage	Characteristic

From Heath (1986).

supercritical carbon dioxide (at temperatures above 70 °C and pressures 90–300 bar), behaving as an efficient non-polar solvent leaving no solvent residues, has been introduced.

General methods of analysis

Methods of sampling and analysis of spices and condiments have been published in various British Standards (BS 4585) and in the BP and BPC. Van den Dool (1980) has given a useful update on the quality control and standard of spices.

Before the main analysis, a cursory examination should be made for the presence of waste material (extraneous matter) such as excess stalks, foreign organic matter, insects, rodent excrement, etc. (BS 4540). In addition, whole spices must be ground thoroughly before the analysis is commenced. The routine analysis of most herbs and spices may include the determination of **total ash, acid-insoluble ash, fixed ether extract, volatile oil, arsenic** and **lead, filth, alcohol extract** and a **microscopical examination**.

Moisture should preferably be determined by distillation from toluene (BS 4585: Part 2), using a Dean and Stark type apparatus (see BS 756). Drying (preferably at 70 °C in a vacuum oven) can also be employed. Although volatile oil is also lost during the drying, the result obtained is often sufficiently accurate for comparison purposes. Sanna (1986) has reported the results of a collaborative study of a water entrainment distillation method. The method was satisfactory using toluene for most spices containing 7–12 per cent moisture, but vacuum oven drying was suggested as the best method for spices containing large amounts of sugar.

Total ash, water-soluble ash and **acid-insoluble ash** are all determined by the conventional methods (see BS 4584: Part 3). A high ash or acid-insoluble ash usually suggests inferior quality.

Crude fibre assesses the amount of outer protective tissue. It is particularly used for assessing the quality of white pepper.

Filth. BS 4585: Part 14:1983 (ISO 1208–1982) describes filth as mineral matter (sand, soil) and matter of animal origin (insect fragments, rodent hairs and excreta). Mineral matter is separated by stirring the sample in a beaker with excess of carbon tetrachloride. The supernatant liquid is poured through a Buchner funnel leaving the sand, soil, etc. in the beaker. The process is repeated and finally the residue is transferred from the beaker to an ashless filter paper and examined for filth. If there is an appreciable residue, it is ignited in a platinum dish and the sand and soil weighed.

The light filth is separated from the mineral matter–free spice sample, with or without pancreatin treatment, by flotation with petroleum ether in a Wilman trap flask (see AOAC methods book) and examined microscopically.

Extraneous matter. BS 4585: Part 1:1970 (ISO 927) describes extraneous matter as the foreign substances designated in the specification for a named spice or condiment and can be separated by hand from 100–200 g of sample.

Non-volatile ether extract is determined in a continuous extraction apparatus on 2–3 g of the powdered sample using diethyl ether as solvent. After 1 hour's extraction the material should be removed, re-ground and returned to the apparatus for a further 2 hours' extraction (cf. BS 4585: Part 6). After removal of the solvent, the extract is heated at 100 °C for three-quarters of an hour and then cooled and weighed.

Volatile oil. If in the determination of fixed ether extract (above) the solvent is first allowed to evaporate off at room temperature and then the extract is dried over sulphuric acid for one day, the weight obtained represents the total ether (fixed + volatile oil). If the total ether extract is then dried at 100 °C the weight lost represents the **volatile ether extract** (per cent m/m). Although this method may have to be used if only a small quantity of sample is available, the European Pharmacopoeia method, the same as BS 4585: Part 15:1985 and ISO 6571–1984, gives more satisfactory results. It involves distilling the oil over with boiling water, condensing and collecting the oil in a measured volume of xylene in a graduated tube. The apparatus is shown in Fig. 11.1.

Clean the apparatus with a proprietary detergent or by successive washings with acetone, water and chromic acid (equal volumes of concentrated sulphuric acid and saturated potassium dichromate solution), rinse thoroughly and drain. Add the specified amount of sample (25–50 g) and water (300–600 ml) plus anti-bump granules to a 500 or 1000 ml round bottom flask. Through the side arm beneath the condensor introduce water until it flows through the return arm to the flask. Half fill the steam trap on the return arm with water and pipette 1·0 ml of xylene through the side arm and insert the stopper. Heat the flask and regulate the distillation to 2–3 ml/min. This can be ascertained by removing water during the distillation by means of the 3-way tap until the water level is at the bottom of the 5 ml bulb, then closing the tap. The time taken to fill the bulb is noted, then the tap re-opened to continue the return of the distillate to the flask. After the prescribed time (2–5 h) the volume of organic phase is read and gives the ml of volatile oil per sample weight taken inclusive of the 1 ml xylene added.

Sage and Fleck (1934) have described an apparatus for the determination that can be constructed from ordinary laboratory glassware. The results obtained from the distillation methods, although comparative, are probably lower than the amount actually present in the spice due to some extent to the solubility of the oil in the water. The result is also affected by the particle size.

Various components of the volatile oils of spices have been separated using gas chromatography by Datta *et al.* (1961). Comparison of peak heights indicated that the method may be employed to identify the geographical region from which the spice comes. The technique was also applied to black pepper oil by Jennings and Wrolstad (1961) (see also under pepper). The SAC (1971, 1973) and AMC (1984) have also published reports relating to GLC separations of essential oils. The screening of the quality of aromatic herbs and spices by capillary GLC (50 m; OV–1) using automatic headspace analysis has been described by Chialva *et al.* (1983).

Alcohol extract

Shake 5 g of sample frequently with 100 ml of 95 per cent alcohol in a closed flask for 6 h. Then allow to stand for 18 h and filter. Evaporate 20 ml of filtrate to dryness in a tared basin, dry at 105 °C, cool and weigh. Calculate the result as a percentage of the original sample (cf. BS 4585: Part 4).

Microscopical examination should be carried out on all samples of ground herbs and spices. The prepared slides should be examined first with the low-power objective (10 ×) and then under the high power (40 ×). A set of authentic samples should be available for comparison purposes.

A '*water-slide*' should first be prepared by dispersing the sample with a drop of alcohol and then adding one to two drops of glycerol solution (30 per cent in water) before sliding on the coverslip. The water-slide is particularly useful for detecting starch. The presence of starch can be confirmed by adding a drop of very dilute iodine solution, which produces the usual dark blue colour. Unless light coloured particles are seen in the cursory examination it is usually unnecessary to prepare a water-slide of leafy material (unless powdered).

A '*cleared-slide*' is prepared by gently boiling the material with chloral hydrate solution (prepared by warming 80 g of crystals with 50 ml of water) either on the slide or in a tube until the particles look fairly transparent. Chloral hydrate has a twofold action: (1) it removes material such as starch, thereby concentrating the other tissues and (2) it removes colouring matter, etc. from the tissues so that their outlines can be seen much more clearly. Sclerenchymatous tissue can be stained red by warming the cleared material with excess of phloroglucin solution (1 per cent in 90 per cent alcohol) followed by a drop of conc. hydrochloric acid (cf. Charlett, 1956a).

The following have produced useful works in the study of the microscopical

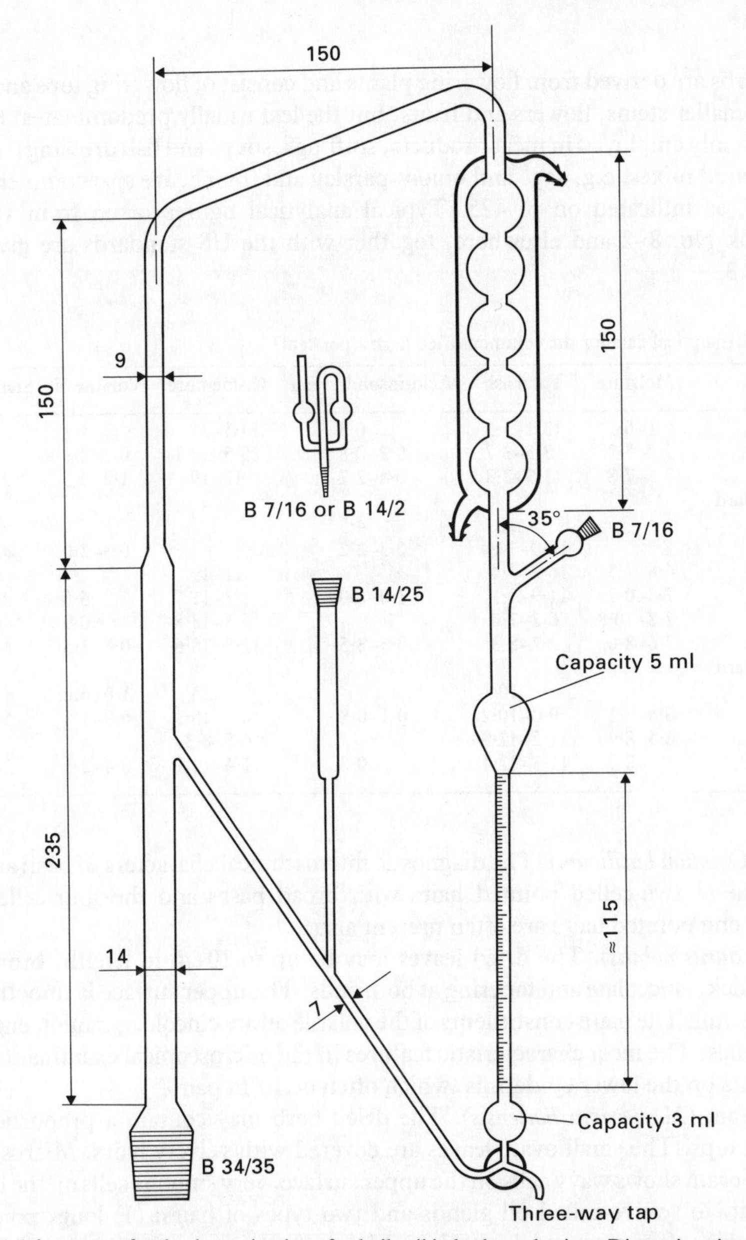

Fig. 11.1 Apparatus for the determination of volatile oil in herbs and spices. Dimensions in millimetres.

examination of foods: Greenish and Collin (1904); Jackson and Snowdon (1968); Parry (1962); Winton and Winton (1939); and Wallis (1967).

Contaminants. Herbs and spices may be examined for arsenic and lead. Fumigant residues may be of interest. Reeves *et al.* (1985) determined residues of methyl bromide and inorganic bromide in several spices before and after fumigation using GLC and an ash/filtration method respectively.

HERBS

Dried herbs are derived from flowering plants and consist of flowering tops and may include smaller stems, flowers and fruits, but the leaf usually predominates. Herbs are commonly employed in meat products, stuffings, soups and fish dressings. Many dry prepared mixes, e.g. sage and onion, parsley and thyme, are also commercially available, as indicated on p. 425. Typical analytical figures taken from USDA Handbook No. 8–2 and elsewhere, together with the US standards are given in Table 11.3.

Table 11.3 Analytical data for the common dried herbs (per cent)

Herb	Moisture	Total ash	Acid-insoluble ash	Crude fibre	Volatile oil	Stalks etc.
Basil	6·1–6·7	13·1–5·5	0·3	14·5–21	1·8	
Bay	5·2–5·7	3·5–3·7	0·2–0·8	25·5–27·1	0·7–2·3	10
Marjoram	7·5–7·8	11·9–2·3	0·4–2·7	17–19	1·2–2·5	2–12
US standard max.		16	4·5			10
Mint		10·0–14·4	0·3–2·2		0·3–3·0	4–15
Oregano	6·8–7·5	6·9–7·4		11–19		
Parsley	7·3–0·7	11·9–3·1	1·0–6·0	9·7–11	6·6	2–3
Rosemary	7·8–10·8	6·2–7·0		16·5–18·8	0·9	
Sage	7·6–8·0	7·7–8·3	0·5–3·5	17·5–18·6	0·6–1·5	4–14
US standard max.		10	1	25	1·0 (min.)	12
Savory	8·8–9·1	9·0–10·2	0·1–0·8	15·3	0·9–1·5	5–15
Tarragon	6·5–8·9	11·2–12·9		6·5–8·3		
Thyme	7·4–8·2	11·3–12·1	0·8	11·4–12·1	0·4–2·5	2–8

Basil (*Ocimum basilicum*). The diagnostic microscopical characters of basil are the warty, one-or two-celled pointed hairs with broad bases and the four-celled oil glands. Long pointed hairs are often present also.

Bay (*Laurus nobilis*). The dried leaves may be up to 10 cm in length, brown in colour, thick, lanceolate and tapering at both ends. The upper surface is smooth, the lower one dull. The main constituents of the volatile oil are cineole, geraniol, eugenol and terpenes. The most characteristic features in the microscopical examination are the stomata on the lower epidermis, which often occur in pairs.

Marjoram (*Marjorama hortensis*). The dried herb may contain a proportion of flowering tops. The small ovate leaves are covered with velvety hairs. Microscopically marjoram shows wavy cells on the upper surface, very sinuous cells on the lower side, eight- to twelve-celled oil glands and two types of hairs: (1) long, pointed, multicellular, often warty and with broad bases, and (2) glandular hairs on short stalks. The US standards for marjoram are quoted in Table 11.3.

Mint, spearmint, garden mint. This is the dried leaves or broken or rubbed dried leaves of any single cultivar of *Mentha spicata*, *viridis* or related hybrids. It is greyish green or dull green in colour and ISO 2256–1984 specifies moisture 13 per cent max., and on the dry basis, total ash 12 per cent max., acid-insoluble ash 2·5 per cent max. and volatile oil 0·5 per cent min. The volatile oil of mint contains about 50 per cent of menthol. Instances of the adulteration of mint with ailanthus leaves (tree of heaven) have been reported. The pieces of ailanthus are a rather lighter green than the pieces of mint leaf and may be further distinguished by microscopical examination (Anon, 1934).

Concentrated mint sauce contains vinegar, some sugar and possibly saccharin, salt and colouring and further vinegar (and sometimes sugar also) is added before use. Sugar is included in the unconcentrated types. Mint sauce prepared with mint adulterated with hazel leaf was the subject of a court case heard in 1960.

Oregano, Mexican sage, wild marjoram. Oregano and similar names are colloquially used throughout the world for numerous plants in various families. The commercial European product consists of the dried leaves of *Oreganum vulgare* spp. *vulgare*, whole cut or ground, coloured light greyish green to olive green. ISO 7925 1985 requires not more than 3 per cent broken stalk and other parts of the plant, 12 per cent moisture, 12 per cent total ash on dry matter, 2 per cent acid-insoluble ash on dry matter, and not less than 1·5 per cent volatile oil on the dry matter. The volatile oil does not contain carvacrol or thymol but consists mainly of numerous hydrocarbons including caryophyllene, germacrene and *cis*-ocimene (Lawrence, 1984a).

Parsley (*Carum petroselinum*). Dried parsley usually has a bright green colour. Its volatile oil is readily lost, particularly during the drying. Microscopically, parsley does not show either hairs or crystals. The upper epidermal cells are polygonal but those on the lower epidermis are wavy. The epidermal cells tend to be striated.

Rosemary (*Rosmarinus officinalis*). The diagnostic microscopical characters of rosemary are the thin-walled, branching hairs sometimes with globular heads and the eight-celled oil glands.

Sage (*Salvia officinalis*). Sage differs considerably in appearance according to the country of origin. Dried English sage is green, whereas that from Cyprus is a pale bluish-grey colour, hard and leathery and the highly esteemed type from Dalmatia is grey and very fluffy. All types are pubescent (covered with hairs). Sage from the Mediterranean area usually contains rather more volatile oil than the English varieties included in Table 11.1. Dalmatian sage may contain up to 4·5 per cent oil, which differs in character from the oil from English sage. The characteristic flavour substance in the oil appears to be the ketone thujone. Microscopically, the main features are the numerous long, slender, whip-like, non-warty, pointed hairs, the short hairs with globular glands and the oil glands (mostly twelve-celled) (Fig. 11.2). Button (1953) described the adulteration of sage with cistus leaf, which can be detected microscopically by the star-shaped hairs. Many samples of English sage tend to give figures for the ash and acid-insoluble ash in excess of the US maximum (see Table 11.3). The BPC (1934) prescribed maxima for foreign organic matter (3 per cent) and ash (8 per cent).

Savory (*Satureia*). The diagnostic microscopical characters of savory are the warty, pointed hairs (many jointed) and the very large 12-celled oil glands. The chief odorous constituent appears to be carvacrol.

Tarragon (*Artemesia dracunculus*). The slender shoots used for making tarragon vinegar and as an ingredient of fines herbes have small narrow leaves of a thin delicate texture and dull olive in colour. The leaves have a peculiar and characteristic pungent taste. A similar taste is possessed by *Tagetes lucida*, a species of marigold, which is sometimes used as a substitute for true tarragon. The characteristic essential oil disappears on drying the herb.

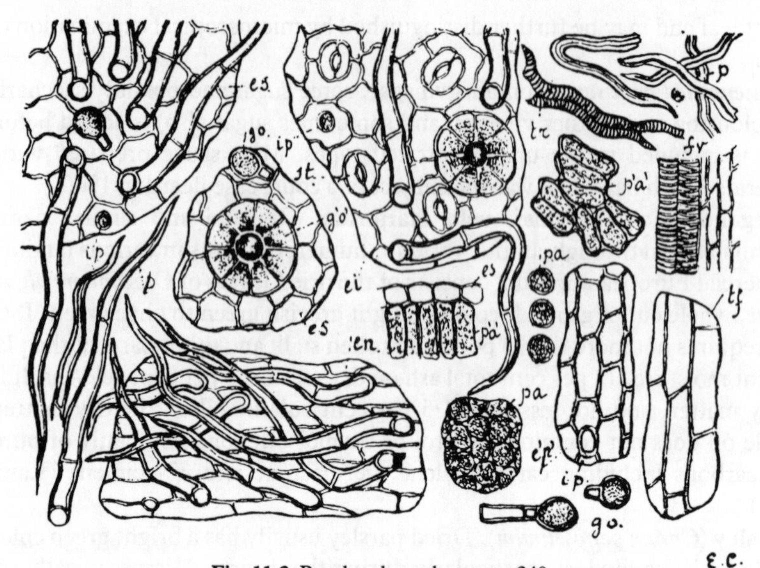

Fig. 11.2 Powdered sage leaves. × 240

ei, Lower epidermis.
en, Neural epidermis.
e′p′, Epidermis of petiole.
es, Upper epidermis.
fv, Vessels from midrib.
go, Unicellular gland.
g′o′, Octocellular gland.
ip, Scar of hair.
p, Simple hairs.
pa, p′a′, Palisade cells.
st, Stomata with the ostiole at right angles to the walls separating two cells from one another.
tf, Cortical tissue of midrib.
tr, Tracheids.

Thyme (*Thymus vulgaris*). Whole thyme is the small (1–5 mm long) dried leaves, ash green to brownish grey in colour, together with the purplish pink to brownish dried flowers. The diagnostic microscopical characters are the one- to two-celled warty hairs, the small stomata and the numerous twelve-celled oil glands. ISO 6754–1985 specifies filth 2 per cent max.; stems etc. 5 per cent max.; moisture 12 per cent max.; total ash 14 per cent max. d.b. (dry matter basis); acid-insoluble ash 5 per cent max. d.b.; crude fibre 30 per cent max. d.b.; volatile oil 1 per cent min. The most important constituents of the volatile oil are the phenols thymol and carvacrol. **Lemon thyme** (*Thymus citriodurus*) is greener than thyme proper when dried. It has a lemon odour in addition to that of thyme proper. The volatile oil content of lemon thyme is about 1 per cent.

BARK

Cinnamon (*Cinnamomum zeylanicum*). Cinnamon bark, also referred to as Ceylon cinnamon, although it is also produced in Seychelles and Madagascar, is imported as long, brown quills 6–9 mm in diameter. Other forms in trade are quillings (broken quills), featherings, chips and powder (ground cinnamon). Table 11.4 gives the typical ranges for the analytical figures together with the BS 6409:1983 (ISO 6539–1983) standards. Cinnamon oil (distilled from the bark) contains 60–75 per cent (m/m) of cinnamaldehyde. An iodimetric method using chloramine-T for total alde-

Table 11.4 Analytical data and standards for cinnamon (per cent)

	Typical range	BS Specification	
		Sri Lanka	Seychelles Madagascar
Moisture	7·8–10·5	12 max.	12 max.
Total ash	4·1–5·7	5 max. d.b.	7 max. d.b.
Water-soluble ash	1·4–2·6		
Acid-insoluble ash	0·1–0·7	1·0 max. d.b.	2·0 max. d.b.
Volatile oil (v/m)	0·7–1·4	min. d.b. whole 1·0 powder 0·7	min. d.b. whole 0·4 powder 0·3
Fixed oil	1·3–1·7		
Starch	16–22		
Extraneous matter		1 max.	1 max.

d.b. = dry basis.

hydes in cinnamon bark and leaf oils (and in cassia bark oil) is described by Shankaranarayana *et al.* (1981). The BP Addendum 1982 describes a GLC method for eugenol using vanillin as internal standard. The oil also contains 4–10 per cent of phenols (mainly eugenol). Microscopically, cinnamon can be identified by the long bast fibres, with thick walls, the sclerenchymatous cells (often thickened more on one side than the other), the small starch grains and the minute calcium oxalate crystals (Fig. 11.3).

Cassia comes from the bark of *Cinnamomum cassia*, *C. loureirii* and *C. burmanii* as quills, whole or in pieces and as ground powder. It is produced in China, Indonesia and Vietnam. The powder is yellowish to reddish brown and the BS 6326:1982 (ISO 6538–1982) standards are shown in Table 11.5. Cassia closely resembles cinnamon (it is so called in the United States) and has been used as an adulterant or substitute for the latter. Chemically, however, the volatile oil of cassia contains no eugenol and more cinnamaldehyde (at least 75 per cent). Also Datta (1961) has shown that cinnamon contains less mucilage but gives a higher proportion of ash of the mucilage than cassia. For genuine cinnamon the former should not exceed 4 per cent (cf. cassia 8–11 per cent) and the latter should be at least 17 per cent (cf. cassia 6–8 per cent). The sample is defatted with ether and macerated successively with ethanol and then 5 per cent acetic acid. The filtered extract is reduced in volume and the mucilage precipitated with ethanol and dried at 100 °C. Microscopically, cassia has wider fibres (diameter over 40 μm, whereas those in cinnamon are usually under 30 μm), contains cork and larger starch grains. Care must be taken in interpreting the results as some poorer quality samples of cinnamon may resemble cassia.

Betts (1965), however, has shown that the two barks can be differentiated by TLC separations based mainly on the detection of eugenol in the oils present. The technique used differs slightly from that described on p. 414 for umbelliferous fruits. The plates are spread with magnesium silicate (TLC, Woelm) after mixing with absolute ethanol (12 g to 50 ml) and air-dried. Alternatively, use a suitable commercially pre-prepared plate. The plates are spotted with an extract prepared by macerating 500 mg bark with 2·5 ml acetone. After running the solvent, eugenol (present in cinnamon, but not in cassia) is detected from the slate-blue spot (R_F 0·45) by spraying with Folin and Ciocalteu's reagent for detecting phenols. A spot due to

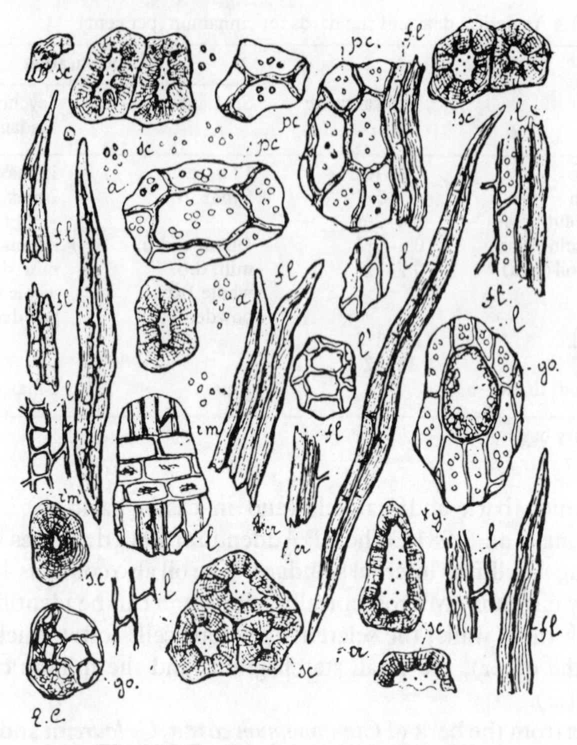

Fig. 11.3 Powdered cinnamon bark. × 240

a, Starch grains.
cr, Minute crystals of calcium oxalate.
fl, Blast fibres with very thick walls, entire or broken.
go, Secretion cells.
l,l′, Bast in longitudinal and transverse section.
pc, Cortical parenchyma.
rm, r′m′, Medullary rays in radial and tangential section.
sc, Sclerenchymatous cells.
vg, Sieve-tubes, collapsed.

Table 11.5 Standards for cassia (per cent)

	BS Specification		
	China	Indonesia	Vietnam
Moisture max.	12·0	12·0	12·0
Total ash max. d.b.	4·0	6·0	4·5
Acid-insoluble ash max. d.b.	0·8	2·0	2·0
Volatile oil min. d.b.			
whole	1·7	1·0	3·0
ground	1·3	0·8	—

d.b. = dry basis.

cinnamaldehyde (present in both barks), R_F 0·55, is revealed by spraying further with diphenylhydrazine solution. The method also assists in identifying another possible substitute, namely *C. burmannii*, which microscopically (like cinnamon) shows thin fibres, but the calcium oxalate is present as small prisms instead of acicular crystals.

BULBS

Onion, garlic. Dried onion is available in the powdered or kibbled form. The composition of the fresh and dried forms is given by Sherratt (1943). BS 6205:1981 for dehydrated onion and BS 6194:1981 for dehydrated garlic both require extraneous matter max. 0·5 per cent. For slices, flakes, rings, pieces, kibbled powder and grits, both specifications require total ash max. d.b. 5·5 per cent; acid in soluble ash max. d.b. 0·5 per cent; moisture max. 8·5 per cent (except for powder and grits max. 6 per cent). The kibbled form has occasionally been attacked by the flour moth (*Ephesia*). Microscopically, both onion and garlic powders show abundant parenchymatous cells. Garlic powder shows sclerenchymatous cells and rhombic crystals of calcium oxalate, both of which are usually absent in onion.

FLOWERS

Capers (*Capparis spinosa*) are cultivated in the Mediterranean area and the wilted flower buds are pickled in salt and vinegar and packed in bottles.

Cassia buds are the dried, unripe fruits of *Cinnamomum cassia* and *C. loureirii*. They vary from 6 to 14 mm in length, the width of the cup varying from 4 to 6 mm. The odour resembles cinnamon and the taste is similar to cassia bark (Parry, 1962).

Cloves (*Eugenia caryophyllus*) are dark reddish-brown flower buds, 13–21 mm long. Headless cloves are those consisting of the receptacle and sepals but without the dome-shaped head. Khoker cloves are pale brown, mealy or wrinkled due to fermentation caused by inadequate drying. Mother cloves are the ovoid brown ripe berries (fruits) and clove stems are the fragments of stalks and tendrils. BS 6097:1981 (ISO 2254–1980) describes three grades, the standards of which are shown in Table 11.6.

Table 11.6 Standards for cloves (per cent)

		Grades		
		1	2	3
Headless cloves	max.	2	5	—
Tendrils, stalks and mother cloves	max.	0·5	4	6
Khoker cloves	max.	0·5	3	5
Extraneous matter	max.	0·5	1	1
Whole cloves				
Moisture	max.	12	12	12
Volatile oil ml/100 g	min. d.b.	17	17	15
Powdered cloves				
Moisture	max.	10	10	10
Total ash	max. d.b.	7	7	7
Acid-insoluble ash	max. d.b.	0·5	0·5	0·5
Crude fibre	max.	13	13	13
Volatile oil ml/100 g	min. d.b.	16	16	14

d.b. = dry basis.

The principal constituents of cloves include the volatile (essential) oil (containing typically 90 per cent of total phenols, principally eugenol), tannin and a crystalline substance, caryophyllin. A monograph on the essential oil of clove bud (Malagasy) including identified fingerprint capillary GLC profiles on non-polar and polar phases is given by the AMC (1984); g/ml 1·044–1·057; refractive index 1·528–1·538; optical rotation 100 mm tube 0 to −1·5°; miscibility with 70 per cent ethanol 1 in 2 (all at 20 °C). The most likely adulterant is exhausted clove, which is shrunken and may be detected by the low volatile oil and low ether and alcohol extracts. Genuine whole cloves sink in water, but exhausted cloves float. Microscopically, cloves show rosette crystals of calcium oxalate, epidermal cells with stomata from calyx teeth, oil glands (and oil drops) and triangular pollen grains (Fig. 11.4). *Clove stalks* seldom yield more than 7 per cent of volatile oil but the crude fibre may exceed 20 per cent. Microscopically, the stalks show prisms of calcium oxalate and sclerenchymatous cells. Mother cloves contain starch.

Saffron consists of the whole or powdered three stigmas joined to the upper section of the style of a small autumn flowering crocus (*Crocus sativus*). Its colour is bright orange red and it is a permitted food colouring matter in the UK. Saffron has a strongly perfumed bouquet with a honey-like aroma, and a pungent bitter-honey taste. It is cultivated around the Mediterranean and in the near and far East; the best quality now comes from Spain.

Saffron yields 0·5–1 per cent essential oil characterised by picrocrocin. The intense colouring capacity is due to crocin. BS 6145:1981 (ISO 3632–1980) lays down standards and test methods for saffron in filaments and powder form and shows anatomical microscopical structures. The supernatant liquid from 4 mg of saffron extracted into 100 ml water should have minimum $E^{1\%}_{1\text{cm}}$ at 440 nm = 110 (filament) or 150 (powder) for category 1 type (Mancha), and 80 or 120 respectively for category 2 type (Rio).

Saffron, being extremely expensive, is often adulterated with synthetic colours such as tartrazine, amaranth, sunset-yellow and ponceau 4R, or with turmeric or safflower. These can be detected as follows:

Take 0·1–0·2 g saffron or a few ml of an extract, add 50 ml boiling water, and keep hot for 1 h. Adjust pH to 4–5 with acetic acid or ammonia, add 1 g polyamide powder and shake for 1 min. Transfer mixture to a glass column containing a filter frit, porosity 2. Wash contents of drained column with 10 × 10 ml hot water, then elute most of the colour from saffron and turmeric with aliquots of methanol. The remaining colours on the column may be eluted with portions of methanol:ammonia (95:5). Evaporate the eluates to dryness and examine by TLC (p. 104).

FRUITS

PEPPER

Black pepper consists of the dried whole unripe berries of *Piper nigrum*. The whole peppercorns are almost spherical, about 5 mm in diameter and the black surface is deeply reticulated.

White pepper consists of the dried whole almost ripe berries of *Piper nigrum* from which the outer part of the pericarp has been removed by friction after previously fermenting or soaking in water. The whole peppercorns are spheroids and on average are slightly smaller than black peppercorns. The colour is greyish-white and there are 10 to 16 white lines running from the apex to the base like meridians.

Decorticated white pepper consists of white pepper with all the coats down to the

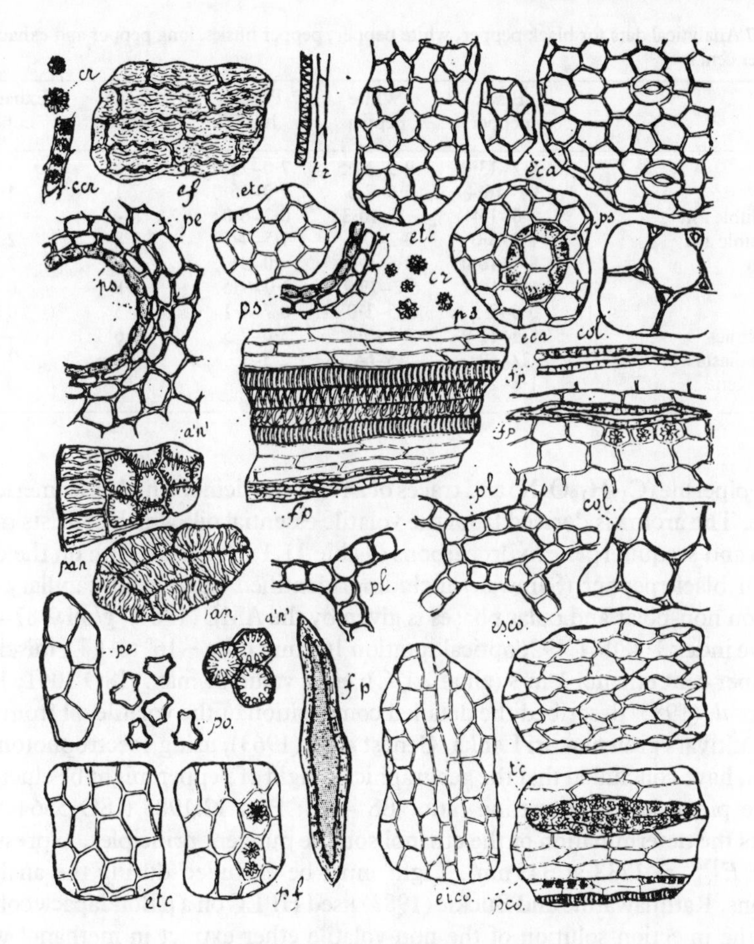

Fig. 11.4 Powdered cloves. × 240

an, a′n′, Cells from the anthers, surface view and in profile.
col, Collenchymatous layer.
ccr, Crystal cells.
cr, Rosette crystals of calcium oxalate.
eca, Epidermis of calyx teeth.
ef, Epidermis of filament.
eico, Lower epidermis of corolla.
esco, Upper epidermis of corolla.
etc, Epidermis of lower portion of clove.
fp, Fibres from the pericycle.
pco, Parenchyma of the corolla with numerous calcium oxalate crystals.
pe, Cells of outer layers of parenchyma.
pf, Parenchymatous cells from filament.
pl, Cells of lacunous portion of parenchyma.
po, Pollen grains.
ps, Oil-glands.
tc, Crystal cells.
tr, Vessels, etc., from calyx-teeth and corolla.

perisperm removed. This treatment has the effect of reducing the amount of crude fibre and ash (both usually below 1 per cent, cf. Table 11.7).

Migonette pepper or **shot pepper** is a rough ground mixture of black and white corns.

Green pepper is the unripe fresh corns, bottled or canned in brine or vinegar, or freeze dried.

The pungency of pepper appears to be due to 5–9 per cent of the non-volatile

Table 11.7 Analytical data for black pepper, white pepper, pepper husks, long pepper and exhausted cubebs (per cent)

	Black pepper	White pepper	Pepper husks	Long pepper	Exhausted cubebs
Moisture	8·7–13·0	9·5–14·5	7·0–11·0	9·5	7·7
Total ash	3·1–16·4	0·6–2·5	7·0–15·0	7·1	4·5
Acid-insoluble ash	0–1·4	0–10·3	0·2–6·0	—	—
Water-soluble ash	1·5–3·0	0·2–0·8	1·5–4·5	3·8	2·6
Crude fibre	8·5–16·0	0·5–5·0	20–31	5·5	2·3
Fixed oil	6·1–10·7	6·2–19·7	3·0–915	6·8	6·1
Volatile	0·6–2·5	0·5–1·0	0·7–1·1	1·5	0·4
Alcohol extract	8·5–11·9	7·2–9·4	5·0–7·5	8·6	—
Starch (by diastase)	22–40	45–64	5–13	41	7
Total nitrogen	1·8–2·5	1·6–2·3	2–3	2	1

alkaloid piperine ($C_{17}H_{19}O_3N$) and traces of a resin chavicine, which is isomeric with piperine. The aroma is derived from the volatile essential oil, which consists mainly of mono and sesquiterpene hydrocarbons (Table 11.1). A monograph on the essential oil of black pepper (Sarawak) including identified fingerprint capillary GLC profiles on non-polar and polar phases is given by the AMC (1984); g/ml 0·87–0·89; refractive index 1·480–1·492; optical rotation 100 mm tube −16° to +4°; miscibility with 95 per cent ethanol 1 in 3 (all at 20 °C); ester value 11 max. (ISO 3061; 1979). Buckle *et al.* (1985) described the detailed composition of the volatile oil from three pepper cultivars grown in Sri Lanka. Genest *et al.* (1963), using spectrophotometric methods, have concluded that the geographical origin of pepper might be elucidated from the piperine to piperettine ratio. BS 4585: Part 12:1983 (ISO 5564–1982) describes the determination of the ethanol soluble pungent principles, expressed as piperine $E^{1\%}_{1cm}$ = 1238 at 343 nm). Light must be excluded during the analytical operations. Rathnawathie and Buckle (1983) used HPLC on a μBondapack column, eluting the injection solution of the non-volatile ether extract in methanol with a methanol:water, 50:50 mobile phase.

Microscopically, black pepper shows numerous layers (Fig. 11.5).

PERICARP

1. Epidermis of small polygonal cells with dark brown contents, occasional stomata and small prismatic crystals of calcium oxalate.
2. An interrupted sclerenchymatous layer of cells with dark contents.
3. Several layers corresponding to the mesocarp, the outer layers of parenchymatous tissue and the inner layers lignified. Oil cells and a little starch are also present among these layers.
4. An inner sclerenchymatous layer of lignified beaker-shaped cells.

SEED COAT

5. Three layers of collapsed cells, one brown, one yellow and one colourless.

PERISPERM

6. The outer layers consist of polygonal cells with aleurone grains. The inner layers consist of elongated cells that contain numerous aggregates of small starch grains (4–6 μm). Oil cells are also present in the perisperm.

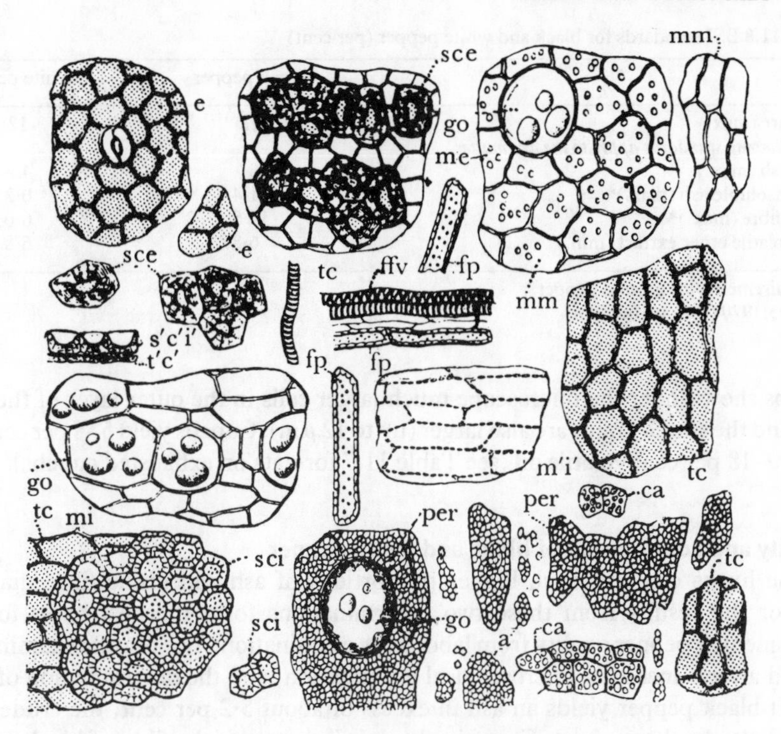

Fig. 11.5 Microscopical features of black pepper

a, Starch grains.
ca, Outer layers of perisperm containing aleurone.
e, Outer epidermis of pericarp.
ffv, Debris of fibro-vascular bundle.
fp, Pitted fibrous cells from bundle.
go, Oil-cells.
me, Outer portion of mesocarp.
mi, Inner portion of mesocarp.
m'i', The same in longitudinal section.
mm, Cells from central part of mesocarp.
per, Cells of perisperm containing starch.
sce, Cells of the outer sclerenchymatous layer.
s'c'i', Cells of the inner sclerenchymatous layer, in surface view.
tc, The brown layer of seed-coat, in surface view.
t'c', The same in profile.

White pepper shows the same microscopic appearance as black, except that the outer parts of the pericarp are missing (epidermis and sclerenchymatous layer). As coloured cells are present in the seed coat, most samples of genuine white pepper contain dark particles. It is necessary to consult the analytical figures (particularly the crude fibre), however, in order to obtain a more quantitative measure of excessive husk (see Table 11.7). The previous BSI standards for black and white pepper are shown in Table 11.8.

Long pepper (*Piper officinarum*, *P. longum*) takes the form of spikes up to 6·5 cm long but the individual fruits on the spike are not unlike black pepper. Compared with black pepper, long pepper shows (under the microscope) relatively few stone cells and the larger starch grains of the perisperm (up to 10 μm) show a characteristic white appearance under polarised light.

Cubebs or **Tailed pepper** (*Piper cubeba*) consists of globular fruits up to 6 mm diameter, dark brown and with a reticulated pericarp. Compared with pepper,

Table 11.8 BSI standards for black and white pepper (per cent)

	Black pepper	White pepper
Moisture (max.)	12	12
The following standards apply to the dry matter		
Total ash (max.)	8	4
Acid-insoluble ash (max.)*	1·4	0·2
Crude fibre (max.)*	17·5	6·0
Non-volatile ether extract (min.)	6·8	6·8

* Requirements for ground pepper.
BS 4595:1970 (now withdrawn).

cubebs show under the microscope much larger cells in the outer layer of the seed coat and the starch grains are also larger (up to 12 μm). Cubebs yield 5–8 per cent ash and 10–18 per cent volatile oil (see Table 11.7 for data on exhausted cubebs).

Quality and adulteration of black and white pepper

As the husks contain much higher proportions of ash and crude fibre than the interior the results from these two determinations form the best bases for the assessment of pepper quality from laboratory examination. The routine examination should always include a microscopical examination, and the determination of total ash. If black pepper yields an ash in excess of about 5·2 per cent, the crude fibre should also be determined. Excessive husk is indicated if the fibre of black pepper exceeds about 15 per cent.

With white pepper, the colour of the powder itself (particularly if looked at after mounting on a slide preparatory to the microscopic examination) often gives a good indication of the proportion of husk present. In view of the heavy decortication that is commonly carried out, the ash of white pepper now seldom exceeds about 1·2 per cent and the standards quoted in Table 11.7 seem especially liberal. The crude fibre figure, which is correspondingly low in most samples, is a better indication of the quality of white pepper. Excess husk is indicated if the crude fibre exceeds 5 per cent:

$$\text{Excess husk (per cent)} = (\text{per cent fibre} - 5) \times 4.$$

A high crude fibre figure may also be due to the presence of other adulterants such as fruit stones and nut-shells. The typical fibre figures (as percentages) for such materials are: olive stones ('pepperette' or 'poivrette') 60, date stones 65, walnut shells 52, coconut shells 56 and almond shells 49. Microscopically, such adulterants contain (like pepper husks) a high proportion of stone cells. Other possible adulterants of pepper that can also be detected microscopically are: cereals (especially rice starch in view of its similarity to pepper starch), ginger, capsicum, long pepper, cubebs and turmeric. The colouring of pepper with, say, 0·2 per cent of turmeric should not, however, be looked upon as adulteration. Turmeric may be estimated by comparison of the colour of the alcohol extract with that of dilutions of an alcoholic tincture of freshly ground turmeric. A GLC method for benzyl glucosinate can be used to detect the adulteration of black pepper with papaya seed (Curl and Fenwick,

1983). Pepper has in the past been adulterated with inorganic materials, e.g. magnesium carbonate (Dunn and Bloxam, 1937).

CAPSICUM, CHILLIES, RED PEPPERS—CAYENNE PEPPER, PAPRIKA

Red peppers are derived from various species of *Capsicum* and are grown in many parts of the world. They are fully described by Purseglove *et al.* (1982). The whole pods are often referred to as **chillies**. The pods vary considerably in size, pungency and colour. Thus, in general, the strong cayenne types seldom exceed 3·5 cm in length and the colour is orange red, whereas the less pungent paprika forms (mainly grown in Europe and eaten raw in salads or cooked as a vegetable) may be as much as 12 cm in length, green or yellow in colour and when fully ripe dark red. They are sometimes treated with mineral oil to improve the lustre. Although the names used for the various types are not standardised the following represents the main types which are met.

Cayenne pepper. The capsicums that are usually used for the preparation of cayenne pepper are grown in Africa, India, the East Indies, China and America. The US regulations specify that cayenne pepper is the dried ripe fruit of *C. frutescens* L., *C. baccatum* L., or some other small-fruited species of *Capsicum*.

Capsicum, BPC is derived from *C. annuum* v. *minimum* and is cultivated mainly in Africa and South America.

Paprika is usually derived from *C. annuum* and grown in central Europe and more recently in USA. The fruits are dried and ground to the brilliant red powder, paprika. Hungarian paprika are long, pointed and more pungent fruits and are used in goulash dishes.

Pimiento is another form of paprika that is grown in USA and Spain (Spanish paprika).

Ranges for analytical data and standards are summarised in Table 11.9.

Capsicum fruits contain a fatty oil, red carotenoid pigments (mainly capsanthin) and traces of a liquid alkaloid (all non-pungent). Capsicum seeds are richer in oil than the rest of the fruit.

The characteristic pungent taste of capsicums and chillies is due to 0·05–1·0 per cent capsaicin ($CH_3OC_{17}H_{24}.NO_2H$). Traditionally, pungency assessment has been measured by the Scoville Index (BS 4585: Part 7:1977, ISO 3513–1977) in which ethanol extracts of ground chillies are diluted with a sucrose solution until pungency is just perceived by taste. A GLC method for assessing pungency is described by Todd *et al.* (1977). Pungency assessment is critically reviewed by Maga (1975).

A Joint Committee of the Pharmaceutical Society and the SAC (1964) recommended a method in which capsaicin is extracted by an ether–alkali partition extraction procedure and is determined by a spectrophotometric difference method. Alternatively, the extracted capsaicin can be determined by direct spectrophotometry or by a colorimetric method based on the coupling reaction between diazobenzenesulphonic acid and capsaicin.

Microscopically, cayenne pepper shows a large number of structures derived from the pericarp, seed, calyx and stalk. Of these the most striking features are the numerous drops of orange-coloured oil, the polygonal beaded cells of the outer epidermis of the pericarp, the somewhat sinuous cells with thickened and pitted

Table 11.9 Typical composition and standards for *Capsicum* species

	Cayenne/red pepper			Chillies/capsicum			Paprika		
	Range	US standards	Canadian standards	BPC	BSI	ISO	Range	US standards	Canadian standards
Moisture (%)	3·7–9	10·0 max.			11 max.			12·0 max.	
Total ash (%)	5·1–6·4	8·0 max.	8 max.	8 max.	10 max. d.b.	8 max. d.b.	5·6–7·6	8·5 max.	8·5 max.
Water-soluble ash (%)	2·8–4·9						4·7–5.5		
Acid-insoluble ash (%)	0·05–0·25	1·0 max.	1·25 max.		1·6 max. d.b.	1·25 max. d.b.	0–0·35	1·0 max.	1 max.
Crude fibre (%)			28 max.			28 max. d.b.			23 max.
Alcohol extractive (%)	20–30								
Non-volatile ether extract (%)	15·5–22		15 min.		15 min. d.b.	15 min. d.b.			18 min.
Capsaicin (%)				0·5 min.					
Nitrogen (%)	1·8–2·3					2 min. d.b.	2·2–2·8		
Starch (%)	0·5–1·5		1·5 max.						
Foreign organic matter (%)				1 max.					
Calyces/pedicels (%)				3 max.					
Extractable colour (ASTA units)								110·0 min.	
Scoville pungency		30 000 –55 000							

US Federal Specification No EE–S–63H, 1975; Canadian Government Standards B.07.014 and B.07.015; BS 5048 (now withdrawn); ISO R972; BPC 1973; d.b. = dry basis.

walls of the inner epidermis of the pericarp and the very large sclerenchymatous cells of the epidermis of the seed coat with very thick sinous walls (Fig. 11.6). All the capsicum-type red peppers show similar characters, although the paprikas tend to have larger epidermal cells.

The most likely adulterant or substitute of cayenne pepper is exhausted capsicum, which can be detected by the low non-volatile ether extract, which should be at least 15 per cent. Microscopical examination should reveal the presence of the other possible adulterants, e.g. farinaceous materials, ginger, mustard and turmeric.

ALLSPICE, PIMENTO, JAMAICA PEPPER

The fruits of *Pimenta officinalis* are dark reddish brown, brittle and rough almost globular berries resembling large peppercorns, about 6–9·5 mm in diameter. The range for the analytical figures of allspice are shown in Table 11.10.

BS 4594:1970 (now withdrawn) recommended for the whole spice a maximum of 12·0 per cent moisture; and on the dry material, maxima of 4·5 per cent ash and 0·4 per cent acid-insoluble ash and a minimum of 3·5 per cent v/m of volatile oil. Additional requirements for ground allspice are maxima of 8·5 per cent non-volatile ether extract and 27·5 per cent fibre and a minimum of 2·8 per cent volatile oil (all on dry basis).

The yellowish volatile oil contains about 70 per cent of eugenol, which accounts for the close similarity of the odour to that of cloves. Eugenol and methyl eugenol in methanol extracts of the berries can be determined by reverse phase HPLC with ultraviolet or electrochemical detection (Smith and Beck, 1984). Microscopical examination shows variously sized and shaped sclerenchymatous cells (the most characteristic are pointed), one-celled conical thick-walled hairs, starch (many compound grains), oil glands and prismatic and cluster crystals (Fig. 11.7).

A method for light filth (insect fragments, rodent hairs) in ground allspice has been collaboratively studied (Dent, 1980) and is now an AOAC method.

Table 11.10 Typical range of analytical values for whole allspice (per cent)

	Moisture	Ash	Water-soluble ash	Fixed oil	Volatile oil	Alcohol extract	Crude fibre	Total nitrogen	Starch (by diastase
Max.	10·5	4·8	2·7	7·7	5·2	14·3	24·0	1·0	3·8
Min.	9·4	4·1	2·3	4·3	3·0	7·4	20·0	0·8	1·8

UMBELLIFEROUS FRUITS

The umbelliferous fruits (often called 'seeds') are all rather similar in structure and are called cremocarps. They split into two one-seeded portions (mericarps). The arrangement of the ridges on the rounded dorsal surface of the mericarps is often useful for identification purposes. Table 11.11 gives ranges for some of the analytical figures obtained from the seven umbelliferous fruits that are used as spices, together with the US standards and those that have been prescribed in Britain for pharmaceutical purposes (BP and BPC).

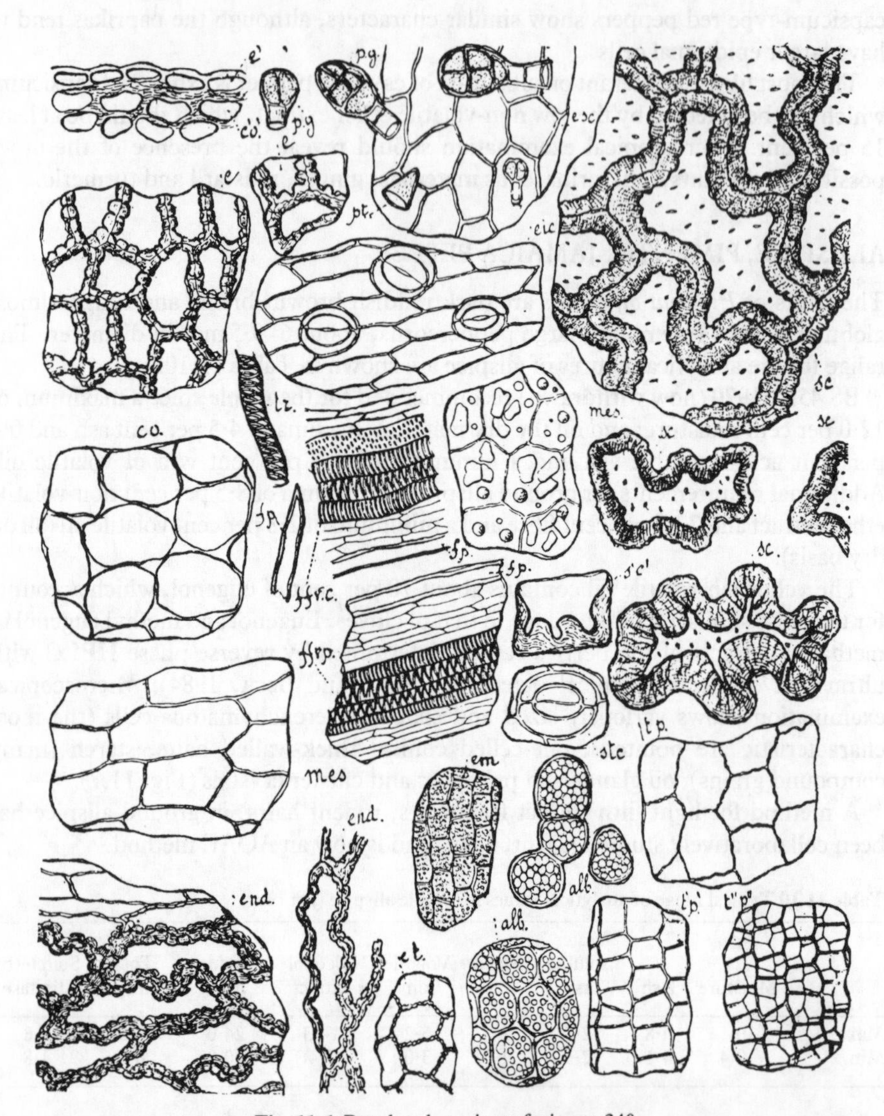

Fig. 11.6 Powdered capsicum fruits. × 240

alb, Cells of endosperm of seed.
co, Collenchymatous hypoderma of pericarp, surface view.
c′o′, The same, in profile.
e, Outer epidermis of pericarp, surface view.
e′, The same, in profile.
eic, Lower epidermis of calyx.
esc, Upper epidermis of calyx.
em, Tissue of embryo.
end, Inner epidermis of pericarp.
et, Epidermis of placenta.
ffvc, Fibro-vascular bundle from calyx.
ffvp, Fibro-vascular bundle from periparp.
mes, Parenchymatous tissue of pericarp.
pg, Glandular hairs from calyx.
pt, Simple hairs from calyx.
sc, Sclerenchymatous epidermal cells of seed-coat, surface view.
s′c′, The same, in profile.
stc, Stoma from calyx.
tp, t′p′, t″p″, Parenchymatous tissue of seed-coat.
tr, Vessel.

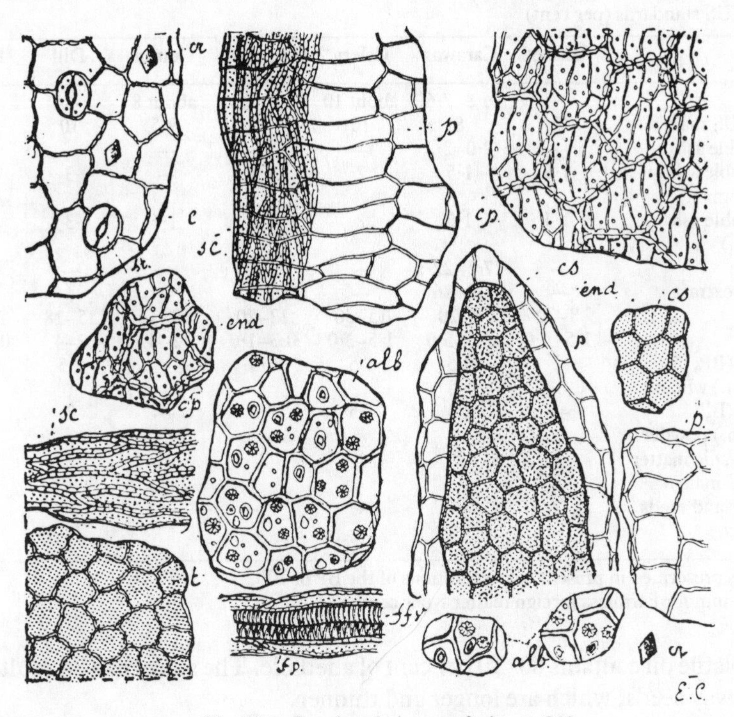

Fig. 11.7 Powdered pimento fruits. × 240

a, Starch grains from seed.
alb, Endosperm.
cl, Dissepiment.
cr, Crystals.
e, Epidermis of pericarp, surface view.
e′, The same, in profile.
ffv, Fibro-vascular bundle.
fp, Fibrous cells from bundle.
go, Glasnds.
m′e′, Outer layer of parenchyma of pericarp in profile.
mi, Inner layer of parenchyma.
p, Hair from epidermis of pericarp.
sc, Sclerenchymatous cells of pericarp.
te, Outer layer of seed-coat.
, Middle-layer of seed-coat.

Aniseed (*Pimpinella anisum*). The spice consists of two seeds each 2–4 mm long, ovoid with five fine light ribs against a grey–green background when fresh but brown when stale. The fine stalk, which passes through the centre of the fruit, may be present. Microscopically, the main diagnostic characteristics are the short, thick walled conical hairs, the polygonal striated epidermal cells, the aleurone grains and calcium oxalate rosettes.

ISO 7386–1984 describes three grades of purity containing, respectively, extraneous matter max. 1, 3 and 4 per cent and shrivelled, immature, damaged and broken fruits max. 3, 4 and 10 per cent. Chemically, the standards depend on the area of origin: (1) west Mediterranean countries and (2) east Mediterranean countries and India. For both areas moisture max. 12 per cent; total ash max. d.b. 10 per cent; acid-insoluble ash max. d.b. 2·5 per cent. For area 1, referring to grades respectively, volatile oil min. 2·5, 2·0 and 1·0 per cent; for area 2, 2·0, 1·5 and 1·0 per cent.

Table 11.11 Ranges for the analytical figures of the umbelliferous fruits together with the British (BP, BPC*) and US standards (per cent)

	Aniseed	Caraway	Celery	Coriander	Cumin	Dill	Fennel
Total ash	—	4·8–7·6	about 10	—	about 8	—	—
Total ash (US max.)	9	8	10	7	9·5	10	9
Water-soluble ash	—	2·0–2·2	—	—	—	—	—
Acid-insoluble ash (BP, etc. max.)	1·5	1·5	2	1·5	—	3	1·5
Acid-insoluble ash (US max.)	1·5	1·5	2	1·5	1·5	3	2
Crude fibre	—	17·5–22·3	—	—	—	—	—
Cold water extract	—	20–26	—	—	—	—	22–27
Fixed oil	8–20	8–20	15–30	12–20	10–14	15–18	12–20
Volatile oil	1·5–4·0	2·5–5·9	1·5–3·0	0·3–1·0	2–4	2–4	0·8–4·0
Volatile oil (BP, BPC*min.) whole	2	3·5	1·5	0·3	—	2·5	1·2
Volatile oil (BP, BPC*min.) powder	—	2·5	1·5	0·2	—	2	1
Foreign organic matter (BP, BPC max.)	1	2	1	2	2†	2	1·5
Other fruits and seeds (BPC max.)	2	—	4	—	—	—	—

* Standards prescribed in present or past editions of the BP or BPC.
† US maximum for harmless foreign matter 5 per cent.

The volatile oil contains 80–90 per cent of anethole. The spice may be adulterated with caraway seeds, which are longer and thinner.

Caraway (*Carum carvi*). Whole caraway consists of the mericarps, which have been split from the nearly mature fruits by threshing after drying. They are 4–6 mm long and bear five longitudinal pale to dark yellow ribs on a brown background. ISO 5561–1981 requires extraneous matter max. 1 per cent; moisture max. 13 per cent; total ash max. d.b. 8 per cent; acid-insoluble ash max. d.b. 1·5 per cent; volatile oil min. 2·5 ml/100 g. The volatile oil contains about 50 per cent of carvone (see Tables 11.1 and 11.2). Exhausted seeds can be detected from the much lower figures that are obtained for the cold water extract (below 15 per cent), the fixed oil and the volatile oil. Microscopically, the main diagnostic characters are the different shaped sclerenchymatous cells, the striated epidermis and the aleurone grains. Hairs are absent. The light brown Levant (or Mogador) caraway and also Indian dill have been substituted for genuine caraway; these contain higher amounts of apiole.

Celery (*Apium graveolens*). The 'seeds' are brown, sub-spherical and very small. Although the volatile oil consists largely of terpenes, the characteristic odour of celery appears to be due to a combination of a large number of compounds. The most outstanding feature of the microscopic appearance are the elongated 'crossing' cells, which are not unlike those of fennel.

Celery salt usually contains a mixture of salt and ground celery seed. A minimum of 20 per cent celery seed has been prescribed in Canada. The salt used usually contains a proportion of magnesium carbonate. Calcium stearate is also present on occasion.

Coriander (*Coriandrum sativum*). The fruits are almost spherical, up to 6 mm in diameter, brown and glabrous. Each cremocarp contains ten straight secondary ridges, which alternate with ten, primary, wavy, less conspicuous ridges. The volatile oil contains up to 70 per cent of the alcohol *d*-linalool. The oil distilled from the unripe fruit has a disagreeable odour. Under the microscope, coriander shows layers of lignified sclerenchyma, which cross one another and this results in a characteristic appearance that can be likened to a parquet floor (Fig. 11.8). Hairs are absent.

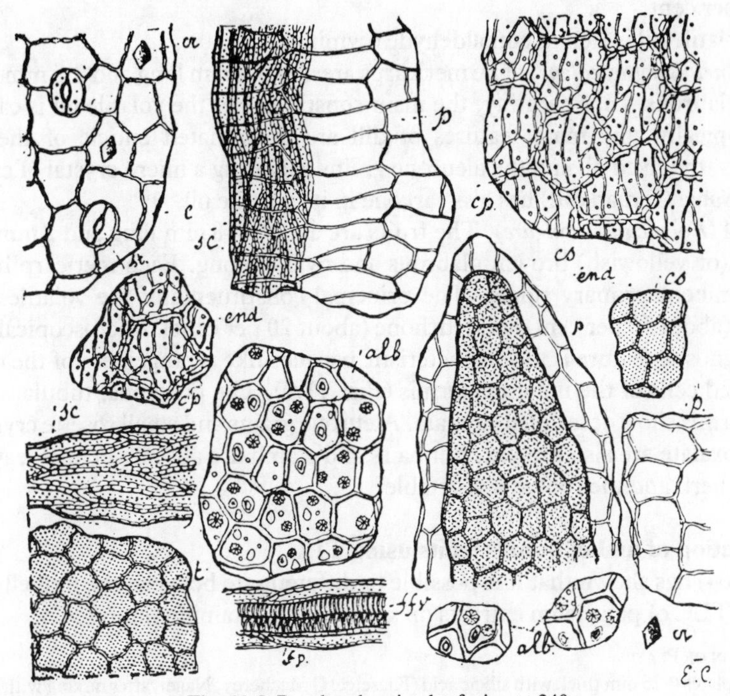

Fig. 11.8 Powdered coriander fruits. × 240

alb, Endosperm.
cp, Large sclerenchymatous cells of inner part of pericarp.
cr, Crystals of calcium oxalate from epidermal cells.
cs, Fragment of secretory duct.
e, Outer epidermis of pericarp.
end, Inner epidermis of pericarp.
ffv, Fragments of fibro-vascular bundle.
fp, Sclerenchymatous fibres.
p, Parenchymatous tissue from pericarp.
sc, Fibrous sclerenchymatous layer of pericarp.
st, Stomata.
t, Seed-coat.

ISO 2255–1980 describes two grades, 1 and 2, containing respectively, extraneous matter max. 1·5 and 3·5 per cent and sound whole fruit min. 93 and 88 per cent. Both grades for whole seed, moisture max. 10 per cent; volatile oil min. d.b. 0·4 ml/100 g, and for ground spice, moisture max. 10 per cent, total ash max. d.b. 7 per cent; acid-insoluble ash max. d.b. 1·5 per cent; volatile oil min. d.b. 0·2 ml/100 g.

Cumin (*Cuminum cyminum*). The seeds occur as paired or separate carpels (mericarps) and are 3–6 mm long. Each mericarp, ochre grey to light brown in colour, has five thin pale primary ribs and four wider darker secondary ribs. Cumin resembles caraway seeds but are lighter in colour and unlike caraway have minute bristles. Microscopically, cumin shows branches of pluriseriate hairs, variously shaped sclerenchymatous cells, aleurone grains and calcium oxalate rosettes. The spice is available as dried seeds or ground to a brownish green powder.

ISO 6465–1984 describes three commercial grades, each limited to a maximum of 5 per cent of broken fruit, and respectively 1, 3 and 5 per cent max. of extraneous matter. The corresponding chemical standards are moisture max. 9, 10, 13 per cent; total ash max. d.b. 9·5, 12, 13 per cent; acid-insoluble ash max. d.b. 1·5, 3, 5 per cent; non-volatile ether extract min. d.b. 15, 15, 12 per cent; volatile oil min. 2·5, 1·5, 1·5 per cent.

The volatile oil is rich in cumaldehyde (cyminol).

Dill (*Anethum graveolens*). The mericarps are about 4 mm long and 2·5 mm broad, flat and glabrous. Like caraway, the main constituent of the volatile oil is carvone. Microscopically, the main features of dill are the striated cuticle of the outer epidermis, the fixed oil and the aleurone grains enclosing a micro crystal of calcium oxalate. *Sowa* (Indian dill) has less carvone in its volatile oil.

Fennel (*Foeniculum vulgare*). The fruits are about 4–8 mm long and 2 mm wide, greenish (or yellowish) brown, glabrous and oval–oblong. Each mericarp has five pale prominent primary ridges. The principal constituents of the volatile oil are anethole (about 50 per cent) and fenchone (about 20 per cent). Microscopically, the most diagnostic feature is the characteristic parquet-like arrangement of the narrow thin-walled cells of the inner epidermis (Fig. 11.9). The polygonal tubular cells of the epidermis show occasional stomata. Aleurone grains and small rosette crystals of calcium oxalate are also present, but hairs and starch are absent. Fennel leaves are used as a herb and the root as a vegetable.

Identification of umbelliferous fruits using TLC

Betts (1964) has shown that it is possible to differentiate between the umbelliferous fruits by TLC of petroleum extracts on silicic acid containing fluorescein.

Preparation of Plates

Spread the plates 0·25 mm thick with silicic acid (Kieselgel G Macherey, Nagel) after mixing with 0·05 per cent aqueous fluorescein sodium (25 g to 50 ml). Heat the plates at 105 °C for 30 min and store over desiccant. Alternatively, use pre-prepared plates, e.g. Merk Kieselgel $60F_{254}$ coated aluminium sheets.

Method

Shake 0·5 g powdered sample with 5 ml light petroleum (b.p. 60–80 °C). Allow the suspension to settle and apply 0·03 ml of extract to the TLC plate by 10 successive applications with a glass rod so that the spot is approx 1 cm in diameter. Similarly, apply extracts of authentic specimens of Umbelliferae for comparison purposes. Add the mobile solvent (1:1 chloroform–benzene) to the tank and line with paper wetted with it. Allow the solvent to rise about 15 cm from the baseline.

Examine the plate in ultraviolet light and note any dark fluorescence-quenching spots against the bright yellow background of fluorescein. Then treat briefly with bromine vapour to convert fluorescein to eosin and examine again under ultraviolet light for any persistent fluorescein fluorescence of unsaturated substances. Then spray with a saturated solution of 2,4 dinitrophenylhydrazine in M hydrochloric acid so that ketones and aldehydes appear as orange spots. After air-drying, spray with sulphuric acid containing 1 per cent vanillin after which compounds such as khellin and thymol rapidly give coloured spots, but others (e.g. fenchone) may take several hours to appear. Betts quotes the following average R_F values for the spots from various umbelliferous fruit extracts:

Anise	0·17, 0·38
Caraway	0·43
Coriander	0·26
Cumin	0·58, 0·55
Dill	0·42
Fennel	0·72, 0·57, 0·46, 0·41, 0·38

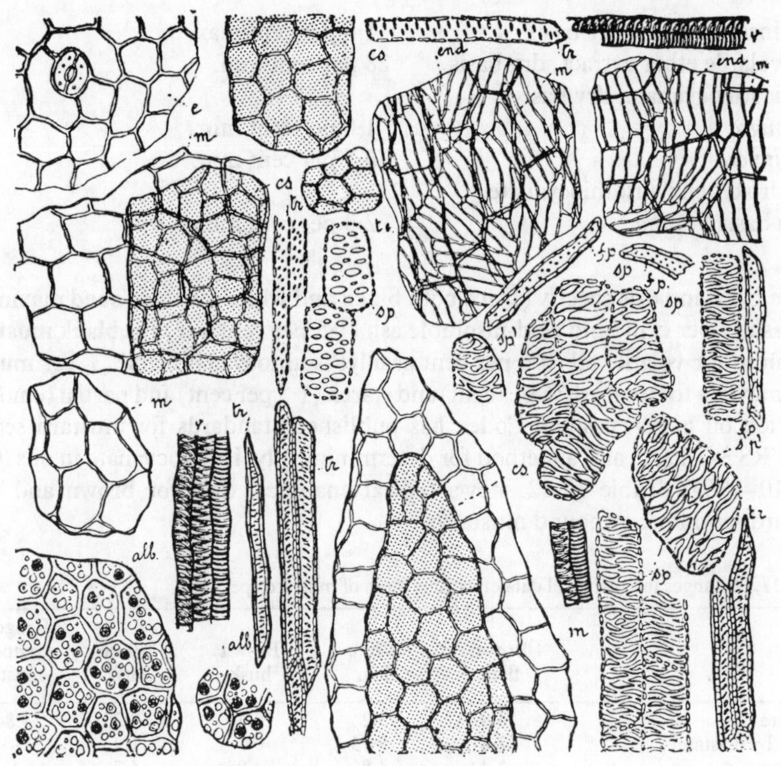

Fig. 11.9 Powdered fennel fruit. × 240

alb, Endosperm.
cs, Vitta.
end, Inner epidermis of pericarp.
fp, Fibres from the fibro-vascular bundles.
m, Cells of mesocarp.
sp, Spiral and reticulate cells.
s′p′, The same isolated, some intact, others broken.
te, Seed-coat.
tr, Tracheids from the bundles in the mesocarp.
v, Vessels, etc. from the carpophore and pedicel of the fruit.

SEEDS

Mustard products, used both as a spice and a condiment, are derived from the ground seeds of white mustard (yellow mustard) *Sinapsis alba*, brown mustard (Indian mustard) *Brassica juncea*, and black mustard *B. nigra*. Mustard seeds contain 20–40 per cent of fixed oil (see p. 416). The more pungent or hot mustards are those with the fixed oil removed. The glycoside sinigrin in black and brown mustard is hydrolysed in the presence of water by the enzyme myrosinase to the pungent, sharply irritating and lachrymatory volatile oil, which is essentially allyl isothiocyanate. White mustard, which contains the glycoside sinalbin, similarly produces the non-volatile but very pungent *p*-hydroxybenzyl isothiocyanate.

BS 6302:1982 (ISO 1237–1981) specifies for mustard seed:

Damaged or shrivelled seeds	2 per cent max.
Extraneous matter	0·7 per cent max.
Loss on drying (103 °C, 2 g, 3 hours)	10 per cent max.
Total ash, dry basis	6·5 per cent max.

Acid-insoluble ash, dry basis	1·0 per cent max.
Non-volatile ether extract, dry basis	28 per cent min.
Allyl isothiocyanate, dry basis	
B. nigra	1·0 per cent min.
B. juncea	0·7 per cent min.
p-Hydroxybenzyl isothiocyanate, dry basis *S. alba*	2·3 per cent min.

The US standards specify (1) for both **black and white mustard seed** maxima for total ash (5 per cent) and acid-insoluble ash (1·5 per cent), (2) for **black mustard** a minimum for volatile oil (0·6 per cent as allyl isothiocyanate) and (3) for **mustard flour** maxima for total ash (6 per cent) and starch (1·5 per cent) and partial removal of the fixed oil is permissible. Codex has published standards for mustard seed oil (CAC/RS 34–1970) and a method for determining allyl isothiocyanate in oils (CAC/RM 10–1969). Table 11.12 shows typical analytical data for brown and white mustards and for compound mustards.

Table 11.12 Ranges for analytical data of various types of mustard (per cent)

	Brown flour	White flour	Brown husk	White husk	Ranges for compound mustards
Moisture	4·9	5·2	10·3	7·25	4·8–5·3
Starch (by diastase	2·8	5·15	0·7	1·2	3·3–14·3
Nitrogen	4·4	4·8	2·95	4·7	4·1–5·3
Fixed oil	39·8	38·1	14·5	22·8	24·0–38·8
Allyl isothiocyanate	1·5	*	—	—	0·5–0·9
Total sulphur	1·5	1·5	0·85	1·2	1·14–1·57
Ash	4·1	4·1	4·6	4·25	3·7–4·5
Phosphorus (as P_2O_5)	1·8	1·47	0·75	1·3	1·51–2·02
Pentosans	3·0	3·3	9·5	7·4	1·1–3·6
Crude fibre	3·7	3·4	15·7	13·2	1·4–4·2

* Volatile oil of white mustard flour is less than 0·1 per cent.

The Food Standards (Mustard) Order (S R & O 1944, No. 275) requires that '**mustard, compound mustard** or **mustard condiment** shall be of such composition as to yield not less than 0·35 per cent of allyl isothiocyanate after maceration with water for two hours at 37 °C and shall consist of a blend of brown and white mustard flours with or without amylaceous flours and/or spices; provided that the proportions of amylaceous flours and spices (if any) shall not together exceed 20 per cent by weight.'

Prepared mustards in paste form in a wide range of pungency, for which there is no specific legislation, are now widely used and take about 80 per cent of the mustard market in the UK. The finest mustards use black seeds whereas prepared mustard of the traditional English type is made from brown and white seeds. The pastes can have smooth or whole grain texture. Added ingredients may include vinegar, sugar, wheat flour, salt, spices, citric acid, wine and added colouring matter, natural or artificial. Dijon mustard is made from dehusked black seeds, is pale yellow and variable in pungency. Bordeaux mustard is prepared without dehusking black seeds, is dark coloured and often flavoured with tarragon. Meaux mustard is made from partly crushed and partly ground black seeds. German mustard is a smooth blend of vinegar and black seeds. American hamburger mustard is made from white seeds only and is pale yellow, very mild and thin in consistency. Mustards prepared with vinegar lose pungency due to hydrolysis of the isothiocyanates. To retain pungency pH control in manufacture and formulation is necessary. The preservation and stability of prepared mustards of the hot English type is due to a balance between acidity, sugar and moisture content, although mustard itself has some antibacterial properties.

The routine examination of powdered mustard should include a microscopical examination and the determination of ash, fixed oil and allyl isothiocyanate. If the microscopical examination indicates the presence of wheat flour, the starch determination should also be considered. Other determinations that may be necessary are moisture, acid-insoluble ash, crude fibre (measure the amount of husk), nitrogen, phosphorus and sulphur.

In the following method, which closely resembles that in BS 6302, the allyl isothiocyanate is produced by the enzyme in the presence of water and the volatile compound is then distilled over and allowed to react with silver nitrate in ammoniacal solution.

Weigh out 5 g of mustard into a 250 ml distillation flask (with side tube) and add 100 ml of water. Stopper the flask (neck and side tube) and incubate at 37 °C for 2 h or 70 °C for 15 min. Remove the stoppers and add 20 ml of alcohol (95 per cent), mix and stand for 45 min. Connect the flask to a condenser, replace the cork in the neck and distil about 65 ml into a 100 ml volumetric flask (wide neck) containing about 10 ml of ammonia solution (1 vol. 0·88 ammonia + 2 vol. water) taking care that the end of the condenser delivery tube dips below the surface of the ammonia. Add exactly 20 ml of 0·1 M silver nitrate solution to the distillate, mix and allow to stand overnight. Then heat in a boiling-water bath in order to coagulate the precipitate (about 30 min), cool and make up to 100 ml with water. Filter. To 50 ml of filtrate add 5 ml of conc. nitric acid and titrate with 0·1 M potassium thiocyanate using 5 ml of 10 per cent ferric ammonium sulphate as indicator.

1 ml 0·1 M silver nitrate ≡ 0·004 95 g allyl isothiocyanate.

CALCULATION

5 g taken; 20 ml 0·1 M $AgNO_3$ added to distillate and make up to 100 ml.
50 ml filtrate (≡ 2·5 g) require 7·25 ml 0·1 M KCNS.

$$\text{Allyl isothiocyanate} = \frac{(\frac{20}{2} - 7{\cdot}25) \times 0{\cdot}004\,95 \times 100}{2{\cdot}5} = 0{\cdot}54 \text{ per cent}$$

$$\text{Crude volatile oil} = 0{\cdot}54 \times \tfrac{100}{93} = 0{\cdot}58 \text{ per cent.}$$

Following collaborative analyses, Andersen (1970) recommended the use of GLC for the determination. It is now an AOAC method:

Add 6·0 g of freshly ground seed or powder to 150 ml ethanol in water (5 per cent). Stopper, stir or mix 90 min continuously in 37 °C water bath. Distil about 70 ml into 100 ml vol. flask containing 20 ml ethanol in water (5 per cent). Dilute to volume with water. Inject 4–10 µl on to 4 m × 4 mm i.d. 5 per cent Carbowax 4000 on Fluoropak 80, 20–40 mesh (or equivalent) heated at 145 °C. Compare with standard, i.e. 30 µl allyl isothiocyanate in 50 ml ethanol in water (10 per cent) in 100 ml vol. flask. Shake until dissolved and dilute to volume with water.

Sahasrabudhe and Mullin (1980) describe a GLC method for allyl- and phenylethyl-isothiocyanate in horse radish.

BS 6302 describes a colorimetric method for *p*-hydroxybenzyl isothiocyanate based on conversion by alkaline hydrolysis of the sinalbin glycoside into glucose, the hydrogen sulphate of sinapin and *p*-hydroxybenzyl isothiocyanate.

Weigh 5 g of ground sample into a beaker, add 100 ml water at 70 °C and 100 mg calcium carbonate. Cover, keep at 70 °C for 15 min, cool, add 20 ml NaOH (about 1 M), mix and stand 15 min. Adjust to pH 6–6·5 with HNO_3 1 M and transfer to a 250 ml volumetric flask. Add 2 ml potassium ferrocyanide (106 g/l) and 2 ml zinc acetate (219 g/l containing 3 g acetic acid) with shaking. Make up to the mark and pipette in a further 2·0 ml water to take into account the insoluble matter. Mix, filter away from bright sunlight. In a 50 ml volumetric flask add 5·0 ml filtrate and 5 ml ferric alum (200 g/l in 0·5 M sulphuric acid). Dilute to the mark, mix and measure absorbance at 450 nm. The test should be repeated with the addition of 2 drops of mercuric chloride solution (50 g/l) to correct for any absorbance due to phenols present.

Dilute 5 ml of 0·1 M potassium or ammonium thiocyanate to 1 litre and prepare calibration graph by taking 5–25 ml aliquots into 50 ml volumetric flasks and develop colour as above.

$$p\text{-hydroxybenzyl isothiocyanate per cent dry basis} = 2{\cdot}84 \times \frac{m}{10^6} \times \frac{250}{5} \times \frac{100}{M} \times \frac{100}{100 - H}$$

where m = µg thiocyanate from calibration graph
M = weight of sample (g)
H = moisture content (per cent)

During storage mustard becomes moist—conditions which encourage production of allyl isothiocyanate, which tends to be lost by volatilisation.

Microscopical examination should preferably be carried out on the defatted material obtained after treatment with ether and then alcohol. Black and white mustard resemble one another closely in microscopical appearance (Figs 11.10 and 11.11). The sclerenchymatous layer in black mustard, however, is dark yellowish-brown and the upper surface shows a darker coloured polygonal network, whereas such reticulations are usually not discernible in white mustard. In addition, white mustard has an epidermis containing large polygonal cells with a colourless, striated mucilage, but that of black mustard is not striated. White mustard shows a hypodermal layer of large collenchymatous cells. Both types of mustard contain aleurone grains and in the unextracted material numerous round oil globules. A few minute rounded starch grains are seen occasionally due to the presence of unripe seeds. The microscopical examination should also show the presence of added starch and adulterants such as pepper, capsicum and turmeric. The last can also be detected by TLC on a prepared extract (see p. 425).

Starch is best determined by an enzymatic method such as that of Holm *et al.* (1986). The Ewers polarimetric method (p. 296) may suffice. Wheat flour can be assumed to contain 72 per cent of starch.

The other determinations that may be necessary such as ash, fixed oil, etc. are carried out by the usual procedures. The crude fibre is useful for assessing the proportion of husk removed (Table 11.12). The proportion of black and white mustard present can be assessed approximately from the allyl isothiocyanate figure and from the amount of sulphur associated with nitrogenous constituents other than allyl isothiocyanate.

Prepared mustard is described in US standards as a paste composed of a mixture of ground mustard seed and/or mustard flour and/or mustard cake, with salt, a vinegar, and with or without sugar (sucrose), spices, or other condiments. The fat- and salt-

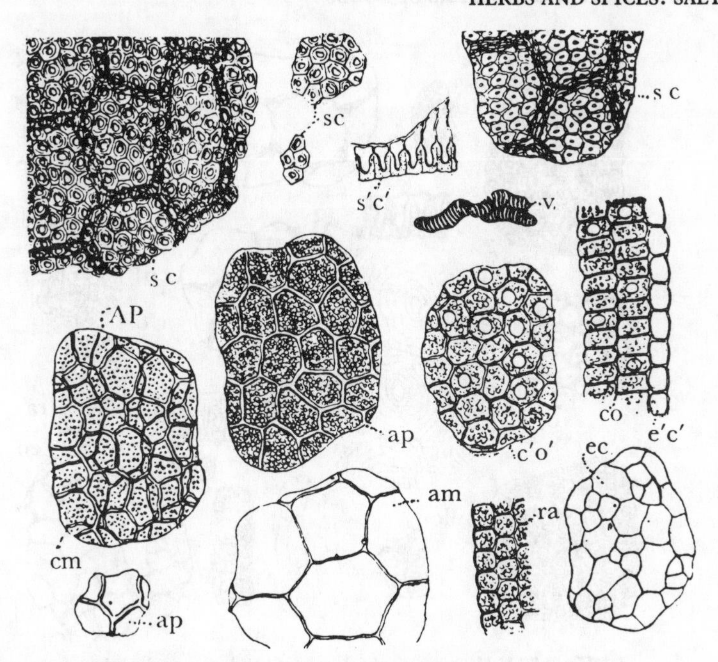

Fig. 11.10 Microscopical features of black mustard seeds

am, Mucilaginous epidermis without distinct concentric striations.
ap, Aleurone layer, isolated.
AP, The same together with the membranous layer, **cm.**
co, Margin of cotyledon in transverse section, with epidermis, **e′c′**.
ec, Epidermis of cotyledon, surface view.
ra, Radicle
sc, Sclerenchymatous layer surface view; the cells are brown and exhibit the characteristic polygonal network.
s′c′, The same in profile, showing the characteristic thickenings.
v, Debris of raphe.

free solids should contain a minimum of 5·6 per cent of nitrogen and not more than 24 per cent of carbohydrates (as starch) nor more than 12 per cent of crude fibre.

Prepared mustards should be examined microscopically and for water, ash, chloride, ether extract, allyl isothiocyanate, nitrogen, starch, acidity and preservatives (e.g. sulphur dioxide, benzoic acid). Products of this type examined gave the following percentage figures: water 55–74, ash 4·7–8·4, acidity (as acetic acid) 2·0–3·2, allyl isothiocyanate 0·15–0·28.

Nutmegs and **mace** are both derived from *Myristica fragrans*, the former being the kernel of the seed and the latter being the dried arillus (an extra seed coat). True Banda nutmegs are about 2·5 × 2 cm, ovoid and the surface is light brown, finely pitted and with reticulate markings. In comparison, Bombay nutmegs (*M. malabarica*) and Macassar (or Papua) nutmegs (*M. argentea*) are longer and narrower and tend to lack aroma. Nutmegs are now cultivated in both East and West Indies.

Table 11.13 gives the ranges for the analytical figures of nutmeg and mace together with the BP and US standards.

The US standards permit a thin coating of lime as a protection from insect attack. On occasion, damaged nuts have been repaired by moulding the pieces with clay. Such factitious nutmegs have a high ash and yield little volatile oil. Ground nutmegs should be examined for insect fragments.

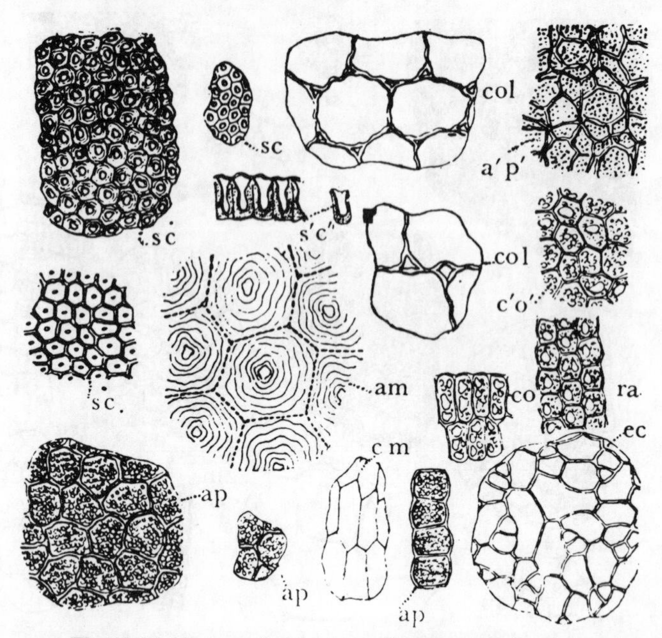

Fig. 11.11 Microscopical features of black mustard seeds

am, Large mucilaginous cells of epidermis in surface view, showing concentric striation.
ap, Isodiametric cells of proteid layer, containing aleurone grains.
a′p′, The same with membranous layer of seed-coat adhering.
cm, Membranous layer.
co, Cells from outer part of cotyledons.
c′o′, Cells from inner part of cotyledons.
col, Collenchymatous cells.
ec, Epidermis of cotyledons.
ra, Tissue of radicle.
sc, Sclerenchymatous layer, surface view.
s′c′, The same in profile.

Table 11.13 Analytical data and standards for nutmet and mace (per cent)

	Nutmeg			Mace	
	Range	BPC	US	Range	US
Moisture	4–8	—	—	3·5–7	—
Ash	1·8–4·5	3 (max.)	5 (max.)	1·6–2·5	3 (max.)
Water-soluble ash	1–2	—	—	0·9–1·7	—
Acid-insoluble ash	0–0·3	—	0·5 (max.)	—	0·5 (max.)
Fixed oil	30–40	—	25 (min.)	24–33	20–30
Volatile oil (v/w)	5–15	5 (min.) whole 4 (min.) powder	—	4–15	—
Alcohol	10–16·5	—	—	21·5	—
Crude fibre	2–3·7	—	10 (max.)	4·7–7·3	10 (max.)
Nitrogen	1·1–1·4	—	—	0·85–1·15	—
Starch	7·5–12	—	—	—	—

The volatile oil (g/ml 0·883–0·917, optical rotation 100 mm tube + 8° to +25°; refractive index 1·475–1·488, miscibility with 90 per cent ethanol 1 in 3, all at 20 °C), residue on evaporation 3 per cent max., consists mainly of camphene and pinene (see Tables 11.1 and 11.2). A monograph on the volatile essential oil of East Indian nutmeg, including identified fingerprint capillary GLC profiles on non-polar and polar phases is given by the AMC (1984). West Indian nutmeg oil has lower monoterpene content and higher level of oxygenated compounds, notably safrole and the toxic substance myristicin. Expressed oil of nutmeg is a yellow, aromatic fat (m.p. 25–43 °C) consisting mainly of trimyristin. Microscopically, powdered nutmeg shows solid fat, simple and compound starch grains and the characteristic brown cells of the perisperm.

'True' Banda mace, when dried, consists of flattened, irregular golden-yellow pieces, about 2·5–5 cm in length and brittle. The typical analytical ranges are given in Table 11.13. The fixed and volatile oils of mace tend to resemble those of nutmeg in composition. Microscopically, however, mace shows irregular grains of amylodextrin (which stain red with iodine), large, pointed thick-walled epidermal cells and there is no starch.

Bombay mace and Macassar (or Papua) mace, which are considered inferior, can often be detected from the higher fixed ether extract (over 50 per cent) and the higher alcohol extract (over 35 per cent) (cf. Table 11.13). Further, if the spice is exhausted with light petroleum and then the extracted material is treated with ether, Bombay mace yields a much higher ether extract (30 per cent) than genuine mace (about 2–4 per cent).

Cardamon (*Ellettaria cardamomum*). The seeds of true cardamom are about 4 mm long, reddish-brown, irregularly angular and wrinkled. The fruit is up to 2 cm long, greenish, cream or pale buff in colour, ovoid or oblong and may be longitudinally striated or nearly smooth. The BP prescribes the following standards for cardamom fruits: maximum ash 6 per cent (US 8 per cent), maximum acid-insoluble ash 3·5 per cent (US 3 per cent) and minimum volatile oil 4 per cent. The seeds contain much starch and they yield about 7 per cent extractive to 45 per cent alcohol. BS 4596:1981 (ISO 882–1980) recommends for cardamoms a maximum of 13·0 per cent moisture and, on the dry material, a maximum of 9·2 per cent ash and a minimum of 4·0 per cent v/m volatile oil. Microscopically, the diagnostic features are the sclerenchymatous cells containing a nodule of silica and the small starch grains in the cells of the perisperm, which also contain prismatic crystals of calcium oxalate.

Low grade substitutes (false cardamom) come from the related species *Amomum subulatum* (Nepal or large cardamom) and *Amomum korarima* (Ethiopian cardamom). Their volatile oils, unlike true cardamom, contain very little α terpinyl acetate.

Fenugreek (*Trigonella foenum-graecum*). The seeds are up to 6·5 mm long, 3 mm wide and 2·5 mm thick, hard, yellowish-brown, oblong to rhomboidal in outline, flattened with a deep oblique furrow on one side. They are produced in the Mediterranean basin countries, India, Pakistan, Ethiopia and Afghanistan. Freshly ground fenugreek should have an odour with sweet maple-like character, reminiscent of celery. The flavour is farinaceous and very bitter. BS 6377:1983 (ISO 6575–1982) specifies extraneous matter 4 per cent max., moisture 11·0 per cent max., and on the dry basis, total ash 5·0 per cent max., acid-insoluble ash 1·5 per cent max., cold water soluble extract 30·0 per cent min. Fenugreek seeds contain about 6 per cent of fixed oil, 28 per cent of mucilage and also trigonelline and choline. Microscopically, fenugreek shows epidermal palisade cells, which are about five times as long as they are wide, hour-glass cells with bar-like thickenings, mucilage, aleurone grains and starch. Fenugreek is mainly of interest as the principal odorous constituent of curry powder.

ROOTS AND RHIZOMES

Ginger is derived from the rhizome of *Zingiber officinale*. A number of different physical forms are available commercially, e.g. unpeeled, peeled, scraped, uncoated or coated whole rhizomes, hands and fingers being pieces of rhizomes, splits, slices and ratoons (second year crop). The attractive whitish appearance of some gingers is due to treatment with calcium carbonate or sulphate, which also assists in warding off attack by insects and fungi. Some gingers are treated with lime to soften the skins before peeling and have a calcium content in excess of 1 per cent (as CaO) whereas for natural ginger the figure is usually below 0·5 per cent. Bleaching by sulphurous acid or bleaching powder is sometimes practised. The UK regulations allow up to 150 mg/kg of sulphur dioxide in dried root ginger.

The trade in dried ginger is dominated by Jamaica, India, Nigeria and Sierra Leone, with China and Australia emerging as large producers. India produces 50 per cent of the world's supply. Jamaican clean peeled whole rhizomes are regarded as the highest quality available, light buff in colour, 6–9 cm long with a volatile oil content of 1–1·3 per cent and non-volatile ether extract of about 4·4 per cent. Nigerian is of lesser quality and is exported mainly as splits (un-peeled whole rhizomes split longitudinally). It is very pungent, with a camphoraceous note, volatile oil 2–2·5 per cent, non-volatile ether extract about 6·5 per cent. Sierra Leone ginger is either coated or rough scraped whole rhizomes, and a large proportion of the crop is ratoon ginger; volatile oil 1·6 per cent, non-volatile ether extract about 7 per cent. Indian ginger is mainly Cochin or Calicut, rough scraped whole rhizomes, sometimes coated, bleached or limed. Cochin is about 12 cm long, light brown to yellow–grey whereas Calicut is orange to red–brown; volatile oil 1·9–2·2 per cent, with a lemon-like aroma, and non-volatile ether extract about 4·3 per cent. Australian ginger comes from Queensland and has a lemon-like aroma and flavour. Its quality is intermediate between that of Jamaica and Sierra Leone and has a volatile oil content of up to 4·4 per cent. Chinese ginger is marketed in whole peeled form and as coated slices, often bleached with sulphur dioxide. A comprehensive account of ginger is given by Lawrence (1984) and Purseglove *et al.* (1981).

Standards and typical analytical data for ginger are given in Table 11.14. The crude fibre is useful for assessing the degree to which the rhizomes have been scraped. The presence of exhausted (spent) ginger is indicated if the cold water and alcohol extracts are low. In addition, the water-soluble ash represents only about 25 per cent of the total ash instead of the usual 40–50 per cent for genuine ginger.

In view of the wide variations found in natural gingers, the standards should not be applied too meticulously. In general the cold water extract is the most useful criterion of quality.

The aroma of ginger is mainly derived from the volatile oil which contains cineole, terpenes (*d*-camphene, β-phellandrene and zingiberene), citral and borneol (see Tables 11.1 and 11.2). The pungency is due to an oleoresin containing gingerol, shogaol and other pungent compounds. These may be determined by isocratic

Table 11.14 Analytical data and standards for ginger (per cent)

	BP	BS Unbleached	BS Bleached	US	Canada	Usual ranges for genuine ginger
Moisture	—	12 (max.)	12 (max.)	12·0 (max.)	10 (max.)	8·4–13·9
Total ash	6 (max.)	*8 (max.)	*12 (max.)	7·0 (max.)	7·5 (max.)	3·2–7·6
Water-soluble ash	1·7 (min.)	—	—	—	2 (min.)	1·0–3·7
Acid-insoluble ash	—	—	—	1·0 (max.)	2 (max.)	—
Calcium (CaO)	—	*1·1 (max.)	*2·5 (max.)	—	—	—
Volatile oil (v/m)	—	*1·5 (min.)	*1·5 (min.)	1·5 (min.)	—	1·0–3·1
Fixed oil and resin	—	—	—	—	—	2·8–7·5
90% alcohol extract	4·5 (min.)	—	—	—	—	4·5–8·1
Crude fibre	—	—	—	8·0 (max.)	9 (max.)	1·7–6·5
Nitrogen	—	—	—	—	—	1·0–1·5
Cold water extract	10 (min.)	—	—	—	*13·3 (min.)	7·0–14·0
Starch	—	—	—	42·0 (min.)	*45 (min.)	48·5–53·0
Gingerol	—	—	—	—	—	0·9–2·5
Sieve test, passing US30	—	—	—	95 (min.)	—	—

* On dry mass
Standards: BP 1973, BS 4593:1970 (now withdrawn); US Fed Spec No EE–S–631H, 1975; Canada: 1964.

HPLC and ultraviolet detection at 270 nm (Baranowski, 1985). After extracting with acetone, Chen *et al.* (1986) found 0·65–0·85 per cent of gingerol compounds in green ginger and 1·10–1·56 per cent in dry ginger by reverse phase HPLC.

Microscopically, ginger shows (1) abundant starch grains, which are either sack-shaped, ovoid or simple, are faintly striated, have an eccentric hilum (length 12–50 μm), and exhibit an asymmetric cross when examined under polarised light, (2) thin-walled parenchymatous cells, (3) the cells containing oleoresin and (4) the sclerenchymatous fibres and vessels, neither of which give the characteristic pink colour with phloroglucin and hydrochloric acid (Fig. 11.12). Possible adulterants, which should normally be revealed in the microscopical examination, are capsicum, turmeric and foreign starches.

Preserved ginger is produced in China, Hong Kong and Australia by boiling tender, freshly peeled rhizomes after which they are packed in sugar syrup. Crystallized ginger is produced similarly but pieces are dried and dusted in powdered sugar.

Horseradish (*Cochlearia armoracia*). After peeling, horseradish root is placed in vinegar in order to retain the pungency, and usually packed in jars. It is pale yellow in colour and has a pungent taste and a mustard-like odour, which is due to allyl and phenylethyl isothiocyanate production from the hydrolysis of the glucoside sinigrin by the enzyme myrosinase (cf. black mustard). Sahasrabudhe and Mullin (1980) determined the isothiocyanates in air and freeze dried root by GLC.

Horseradish shows a somewhat persistent peroxidase activity (as indicated by the usual test—addition of 1 per cent guaiacol in 50 per cent alcohol and 1 per cent hydrogen peroxide solution produces a red coloration).

Horseradish is used for the manufacture of horseradish cream, which is rather like salad cream and contains shredded horseradish, egg yolk, milk, oil, vinegar, salt and mustard.

Turmeric (*Curcuma domestica*). The prepared primary rhizome is pear-shaped ('bulb' turmeric), whilst the secondary rhizome tends to be cylindrical ('finger' turmeric). The latter type has the greater tinctorial value. On analysis, turmeric

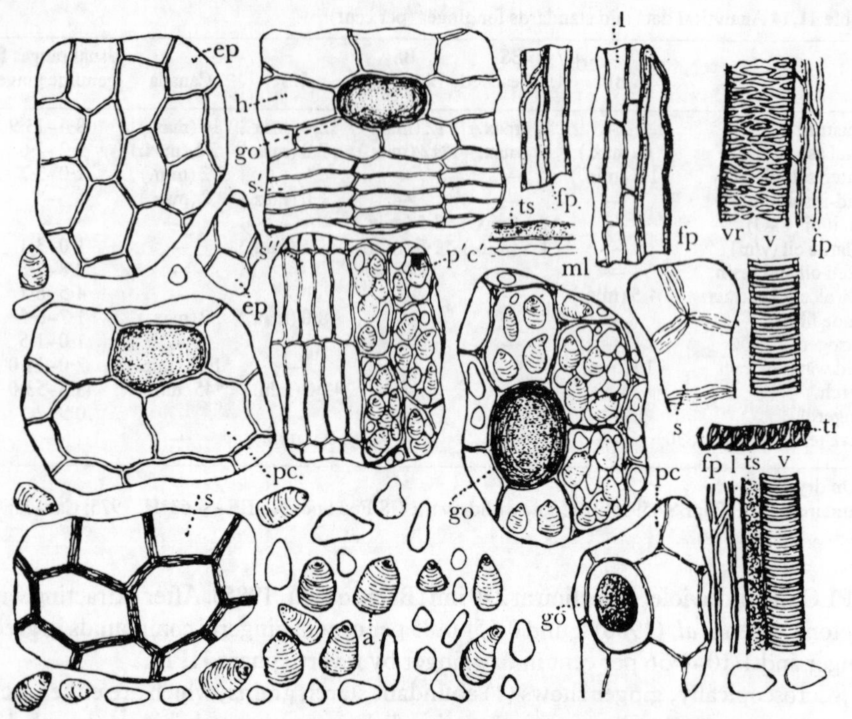

Fig. 11.12 Microscopical features of ginger rhizome

a, Starch grains.
ep, Epidermis.
fp, Sclerenchymatous fibres.
go, Oleoresin cxells.
h, Hypoderma.
l, Bast.
ml, Parenchyma of stele.
pc, Cortical parenchyma, the walls somewhat swollen by treatment with potash.
p′c′, The same in longitudinal section.
s, s′, s″, Cork, in surface view, transverse and longitudinal section respectively.
tr, Vessels.
ts, Elongated secretion cells.
v, vr, Vessels.

yields 6·5–7·5 per cent ash, 7–9 per cent fixed oil, 3–5 per cent volatile oil and much starch. As the rhizomes are steamed or boiled during their preparation, the yellow coloured starch appears somewhat gelatinised under the microscope. Turmeric has a bitter taste, but is used perhaps rather more for imparting colour, due to the presence of curcumin, a yellow crystalline substance. BS 6417:1983 (ISO 5562–1983) specifies extraneous matter (chaff, leaves, stones, vegetable matter) 2 per cent max., and for ground turmeric, moisture 10 per cent max., and on the dry basis, total ash 9 per cent max., acid-insoluble ash 1·5 per cent max., colouring matter 2 per cent minimum expressed as curcumin. The spectrophotometric method for curcumin is described in BS 4585:1983 (ISO 5566–1983) and is based on ethanolic extraction and $E^{1\%}_{1cm}$ for curcumin = 1607 at 425 nm. Turmeric is considered to be inferior as a spice, but is invariably present in curry powder and sometimes in ground mixed spice. It has been used as an adulterant of ginger, mustard and pepper and can be detected chemically by preparing an alcoholic tincture and using it for the usual test for borates. Alternatively, curcumin, indicating the presence of turmeric, can be detected by TLC in methanolic extracts of the spice compound using silica gel plates and chloroform–acetic acid (80:20). Curcumin is detected by ultraviolet fluorescence after spraying with a mixture of boric and oxalic acids.

A very high lead content has been reported in the past in turmeric due to the use of lead chromate as colouring. Samples should also be examined for artificial colours and microscopically for the presence of foreign starches.

PRODUCTS OF HERBS

Mixed herbs usually contain three or four herbs selected from thyme, savory, sage, marjoram, parsley and rosemary.

Stuffings contain the appropriate herbs together with breadcrumbs, suet, salt and sometimes pepper and colouring. Sage and onion stuffing was required by an official Order (SR & O 1942 No. 1863, now revoked) to contain at least 2 per cent of sage and 3 per cent of dried onion. The same revoked Order prescribed a minimum of 2 per cent of thyme for parsley and thyme stuffing. The analysis of stuffings has been discussed by Sherratt (1943).

PRODUCTS OF SPICES

Curry powder usually contains four to six spices selected from fenugreek, turmeric (about 24 per cent), cinnamon, cassia, coriander, fennel, ginger, capsicum, pepper, mustard and pimento. Other possible spices are cardamoms, mace, cummin, dill or a piece of bay leaf. Salt is also usually present. ISO 2253–1974 permits the declared presence of edible starchy material and not more than 5 per cent of salt; spices and condiments not less than 85 per cent (as does the Food Standards (Curry Powder) Order), moisture 10 per cent max. and on the dry basis, volatile oil 0·4 per cent min., non-volatile ether extract 7·5 per cent min., acid-insoluble ash 1 per cent max. and crude fibre 15 per cent max. No artificial colouring matter. The Lead in Foods Regulations 1979 limit for curry powder is 10 p.p.m.

Ground mixed spice usually contains about five ground spices selected from cloves, ginger, coriander, fennel, cinnamon, cassia, capsicum and turmeric. Sugar is also occasionally present. Mixed spice usually yields a total ash of 4·1–4·8 per cent and an acid-insoluble ash of 0·3–1·2 per cent. The arsenic content is usually negligible but lead may be present (up to 5 p.p.m.).

Pickling spice consists of a mixture of whole spices (not ground). Most samples contain coriander, pimento, chillies, ginger (broken pieces), white mustard, cloves, black pepper and mace. Occasionally cinnamon, black mustard and white pepper are present.

Liquid and powdered seasoning extracts are found to be convenient for the incorporation of flavour into foods such as soup mixes and meat products. The liquid seasonings are virtually essences of the spice oils. The solid extracts usually contain salt or dextrose as diluent or adsorbent for the oleoresin. Emulsifying agents may also be present. The technology of such products has been discussed by Forrow (1962) and Heath (1963).

SALT

Apart from its flavouring properties common salt (sodium chloride) has been a valued food chemical and traded commodity for thousands of years as a preservative, particularly for meat and fish. More recently, the levels of salt in the human diet has come under scrutiny, since the excessive amounts have been linked with physical disorders such as high blood pressure.

The primary sources of salt are either from brines, e.g. from the oceans, salt lakes and brine streams, or from rock salt deposits. In the UK practically all salt is obtained by solution mining, i.e. pumping water into underground rock salt beds and collecting the brine so formed, which is then concentrated by vacuum evaporation. Pure dried vacuum salt is a fine even-grained salt of high purity in the form of cubic crystals, mostly 0·2–0·5 mm size, and consists typically on the dry basis as 99·96 per cent NaCl, 0·03 per cent Na_2SO_4, 0·008 per cent soluble alkalinity as Na_2CO_3, 0·002 per cent insoluble matter mainly $CaCO_3$, and 5 p.p.m. of anti-caking agent (sodium ferrocyanide), with a moisture content of 0·005–0·01 per cent. Undried vacuum salt is slightly less pure and contains 2·5–4 per cent moisture. Dendritic salt is a vacuum salt in the form of fine branched or star-like crystals, which have rapid solubility and a high capacity for holding moisture, and consists on the dry basis of 99·7 per cent NaCl, 0·25 per cent Na_2SO_4, 0·05 per cent total alkalinity and up to 20 p.p.m. of the crystal modifier sodium ferrocyanide, with a moisture content of about 1·5–3·5 per cent. These forms of salt are those chiefly used in the food industry. Sea salt is less pure than vacuum salt and retains many of the trace elements present in sea water. The production of iodised salt, used as an aid in preventing goitre, has been described by Holman (1956). In the old Code of Practice CP11 (1949), iodised salt was also to be described as prepared or free-running, and to contain magnesium carbonate and potassium iodide with the iodine content 433–725 μg/oz. Table salt is pure dried vacuum salt with added magnesium carbonate (0·3–1 per cent) to assist free running. Household, culinary or cooking salts are types of vacuum salt.

Standards

The Codex Standard for good grade salt, not at present finalised, lays down general standards for salt from all sources. The sodium chloride content is to be not less than 97 per cent on the dry matter, but no moisture limit is specified. Anti-caking agents in the form of coating agents (carbonates, phosphates, silicates and fatty acid salts) to be not more than 20 g/kg altogether, or in the form of crystal modifiers (ferrocyanides) not more than 10 mg/kg as ferrocyanide except for dendritic salt (20 mg/kg max.). Emulsifier (polysorbate) and processing aid (polydimethylsiloxane) are both limited to 10 mg/kg.

BS998:1969, vacuum salt for butter and cheese making and other food uses, is currently under review but is not expected to change significantly. It will become the standard for pure vacuum salt. Current provisions are, on the dry basis, sodium chloride 99·6 per cent min., alkalinity as Na_2CO_3 300 p.p.m. max., matter insoluble in water 300 p.p.m. max., sulphate as Na_2SO_4 3000 p.p.m. max., Fe 10 p.p.m. max., Cu 2 p.p.m. max., As 1 p.p.m. max., Pb 2 p.p.m. max. (will reduce to 1 p.p.m.), Ca 100 p.p.m. max., Mg 100 p.p.m. max., anti-caking additives

15 p.p.m. max., and moisture 0·2 per cent max. for dried salt and 4·0 per cent max. for undried salt.

Analysis

The draft Codex method refers to other ISO and ECSS standards for methods of analysis. The NaCl content is calculated from the remaining chloride content after the calculation of the $CaSO_4$, $CaCl_2$, $MgSO_4$, Na_2SO_4, $MgCl_2$ and KCl content, expressed on the dry matter basis.

BS998:1969 describes methods, several of them in need of updating, e.g. to the use of atomic absorption spectrophotometry for metals.

Sodium chloride. Dissolve 0·200 g in 35 ml water, add 15 ml dilute nitric acid, 5 ml of triacetin (as coating agent instead of toxic nitrobenzene) and 50 ml 0·1 M silver nitrate and shake thoroughly. Then add ferric ammonium sulphate solution and back titrate with 0·1 M ammonium thiocyanate until the reddish-brown colour formed does not fade after 5 min.

$$1 \text{ ml } 0{\cdot}1 \text{ M silver nitrate} = 0{\cdot}005\,844 \text{ g NaCl.}$$

BS998:1969 prescribes direct titration of the chloride with silver nitrate using 0·1 per cent dichlorofluorescein in 95 per cent ethanol as indicator, which changes from fluorescent green to pink.

Calcium and magnesium. For traces in vacuum salt, BS998 and the Codex standard describe EDTA titrations. For table salt, the calcium can be first precipitated with diammonium phosphate, filtered, ignited and weighed as the pyrophosphate ($Mg_2P_2O_7 \times 0{\cdot}21847 = Mg$). For routine testing, table salt can be dissolved in excess standard acid and back-titrated with standard alkali. Light magnesium carbonate used in table salt is hydrated basic magnesium carbonate and gives a residue of 42·0–45·0 per cent on ignition at 900 °C or higher.

Potassium ferrocyanide. BS998 describes a 15 p.p.m. limit test based on the development of a ferric/ferrous ferrocyanide colour compared with a standard.

Alkalinity. BS998 describes the titration of 50·0 g of salt, dissolved in 250 ml freshly boiled water, followed by filtration on a Buchner funnel to remove insoluble matter, with 0·1 M HCl using a screened methyl orange indicator (0·25 g methyl orange and 0·15 g xylene cyanol FF in 50 ml of ethanol 95 per cent, dilute to 250 ml with water).

$$\text{Alkalinity p.p.m. as } Na_2CO_3 = 5{\cdot}3 \times 10^5 \times \frac{\text{ml } 0{\cdot}1 \text{ M HCl}}{50{\cdot}0\,(100 - \text{per cent } H_2O)}$$

Insoluble matter. This is determined by dissolving 100·0 g of salt in 500 ml freshly boiled water and passing the solution through a dried and tared No. 2 porosity sintered glass filter. After washing the filter free of salt it is dried and reweighed.

Iodide. Most of the older methods for determining iodide in iodised salt involve oxidation of the iodide to iodate, acidifying, the addition of KI and the titration of the free iodine with thiosulphate. Bromine is the oxidising agent in the following method due to Andrew and Mendeno (1935):

Dissolve 50 g of iodised salt in water in a 250 ml volumetric flask, make up to the mark, mix and filter. Collect exactly 200 ml filtrate and transfer to a conical flask, make slightly acid to methyl orange and add 1 ml of saturated bromine water. Add anti-bump granules and boil until salt begins to separate, then add sufficient water to dissolve the solids after cooling. Add 2 ml of M HCl and 0·2 g pure KI and titrate the liberated iodine with 0·005 M sodium thiosulphate using starch.

$$1\ \text{ml}\ 0{\cdot}005\ \text{M}\ Na_2S_2O_3 = 0{\cdot}000\,105\,8\ \text{g iodine} = 0{\cdot}000\,138\,4\ \text{g KI}$$
$$\text{p.p.m.} \times 28{\cdot}35 = \mu\text{g/oz}$$

Gonzalez Diaz *et al.* (1980) describe an indirect spectrophotometric method based on the reaction between palladium and 3-(2-thiazolylazo)-2,6-diaminotoluene (2,6-TADAT), which forms a blue complex. Palladium preferentially couples with iodide so the colour is reduced. A simple, one-step HPLC method for iodide in iodised salt using ultraviolet detection is described by De Kleijn (1983).

REFERENCES

AMC (1984) *Analyst*, **109**, 1343.
Anderson, D. L. (1970) *Journal of the Association of Official Analytical Chemists*, **53**, 1.
Andrew, R. L. & Mandeno, J. L. (1935) *Analyst*, **60**, 801.
Anon (1934) *Analyst*, **59**, 33.
Asp, N. G. (1986) *Starke*, **38**, 224.
Baranowski, J. D. (1985) *Journal of Chromatography*, **319**, 471.
Betts, T. J. (1964) *Quarterly Journal of Pharmacy and Pharmacology* (Supplement) **16**, 131T.
Betts, T. J. (1965) *Quarterly Journal of Pharmacy and Pharmacology*, **17**, 520.
Buckle, K. A., Rathnawathie, M. R. & Brophy, J. J. (1985) *Journal of Food Technology*, **20**, 599.
Button, D. F. H. (1953) *Analyst* (1953), **78**, 679.
Charlett, S. M. (1956a, 1956b) *Food Manufacture*, **31**, 114, 198.
Chen, C. C., Kuo, M. C. & Ho, C. T. (1986) *Journal of Food Science*, **51**, 1364.
Chialva, F., Dogha, G., Gabri, G. & Ulian, F. (1983) *Journal of Chromatography*, **279**, 333.
Curl, C. L. & Fenwick, G. R. (1983) *Food Chemistry*, **12**, 241.
Datta, A. B. (1961) *Journal of the Association of Official Agricultural Chemists*, **44**, 639.
Kleijn, J. P. de (1983) *Deutsche Lebensmittel-Rundschau*, **79**, 184.
Dent, R. G. (1980) *Journal of the Association of Official Analytical Chemists*, **63**, 1266.
van den Dool, H. (1980) *Perfumer & Flavourist*, **5**, 3.
Dowell, P. & Bailey, A. (1982) *The Book of Ingredients*. London: Michael Joseph.
Dunn, J. T. & Bloxam, H. C. L. (1937) *Analyst*, **62**, 121.
Dyett, E. J. (1963) *Chemistry and Industry*, 234.
FACC (1972) *Report on the Review of the Preservatives in Food Regulations* 1962. London: HMSO.
Food Standards Committee (1970) *Report on the pre-1955 Compositional Orders*. London: HMSO.
Forrow, K. G. (1962) *Food Manufacture*, **37** (1), 45.
Genest, C., Smith, D. M. & Chapman, D. G. (1963) *Journal of Agricultural and Food Chemistry*, **11** (6), 508.
Gonzalez Diaz, V., Tallo Gonzalez, C. R. & Garcia Montelongo, F. (1981) *Analyst*, **106**, 1224.
Greenish, H. G. & Collin, E. (1904) *Anatomical Atlas of Vegetable Powders*. London: Churchill.
Heath, H. B. (1963) *Reports on the Progress of Applied Chemistry*, **48**, 512. London: Society of Chemical Industry.
Heath, H. B. (1986) *Food FIPP*, **8**, 55.
Heal, C. & Allsop, M. (1983) *Cooking with Spices*. Newton Abbott: David & Charles.
Holm, J., Bjorck, I., Drews, A. & Asp, N. G. (1986) *Starch/Stärke*, **38**, 224.
Holman, J. C. M. (1956) *Chemistry & Industry*, 1514.
Jackson, B. P. & Snowdon, D. W. (1968) *Powdered Vegetable Drugs*. London: Churchill.
Lawrence, B. M. (1984a) *Perfumer and Flavourist*, **9**, 1.
Lawrence, B. M. (1984b) *Perfumer and Flavourist*, **9**, 41.
Maga, J. A. (1975) *Critical Reviews in Food Science and Nutrition*, **6**, 177.
Parry, J. W. (1962) *Spices—their Morphology, Histology and Chemistry*. London: Food Trade Press.
Pruthi, J. S. (1980) *Spices and Condiments: Chemistry, Microbiology, Technology*. New York: Academic Press.
Purseglove, J. W., Brown, E. G., Green, C. L. & Robbins, C. L. (1981) *Spices: Tropical Agriculture Series*, Vols 1 & 2. Harlow, Essex: Longman.
Rathnawathie, M. & Buckle, K. A. (1983) *Journal of Chromatography*, **264**, 316.
Reeves, R. G., McDaniel, C. A. & Ford, J. H. (1985) *Journal of Agricultural and Food Chemistry*, **33**, 780.
Sage, C. E. & Fleck, H. R. (1934) *Analyst*, **59**, 614.

Sahasrabudhe, M. R. & Mullin, W. J. (1980) *Journal of Food Science*, **45**, 1440.
Sanna, L. A. (1986) *Journal of the Association of Official Analytical Chemists*, **69**, 834.
Shankaranarayana, M. L., Raghavan, B., Abraham, K. O. & Natarajan, C. P. (1981) *Journal of Food Quality*, **4**, 35.
Sherratt, J. G. (1943) *Analyst*, **68**, 200.
Smith, R. M. & Beck, S. (1984) *Journal of Chromatography*, **291**, 424.
Society for Analytical Chemistry (1964, 1971, 1973) *Analyst*, **89**, 377; **96**, 887; **98**, 616.
Todd, P. H., Bensinger, M. G. & Biftu, T. (1977) *Journal of Food Science*, **42**, 660.
Wallis, T. E. (1967) *Textbook of Pharmacognosy*, 5th edn. London: Churchill.
Winton, A. L. & Winton, K. B. (1939) *Structure and Composition of Foods*, Vol. 4. New York: Wiley.

12

Fermentation products

Introduction

Alcoholic liquors differ from many other food products since they are subjected to detailed legislation on two counts: the first relates to the descriptive, compositional and marketing requirements of the Food Act (1984), The European Communities Act (1972) and specific provisions of subordinate Regulations. The second relates to the importance of the products as sources of internal revenue: the manufacture, storage, sale and importation of alcoholic beverages is controlled by the Alcoholic Liquor Duties Act (1979) as amended (SI 1979 No. 241) and subordinate regulations made by the Commissioners of Customs and Excise. The controlled products are: spirits, beer, wine, made-wine and cider as defined in the Act.

WINES

Manufacture

Wines are essentially the products of fermentation of hexose sugars of the expressed juice (must) of fruit (usually the grape) by intact yeast cells to form ethyl alcohol and carbon dioxide. There are three main classes:

1. Light wines or table wines produced by spontaneous fermentation due to organisms on the fruit surface, e.g. Burgundy, Claret, Hock, Moselle.
2. Sparkling wines, which are given a second fermentaion in the bottle so that a visible excess of carbon dioxide is produced, e.g. Champagne, Asti Spumante. With inferior types carbonation replaces the secondary fermentation.
3. Fortified wines containing about 20 per cent v/v alcohol, made by the addition of spirits to wines, e.g. Port, Sherry, Madeira, Marsala.

Wines can be further sub-classified according to country, area, variety of grape used, colour and degree of sweetness (Marrison, 1957). Red table wines are made from black grapes. White wines are made from either black or white grapes, but if the former are used the skins which otherise impart colour are removed prior to fermentation. Rosé (pink) wines are often prepared from lightly pressed red grapes and rapid fermentation at a slightly elevated temperature. Quality aspects of red wines are discussed by Somers (1975). Must is treated with sulphite to inhibit bacterial fermentation and is added to white wine to prevent oxidative browning.

Port is a sweetened purplish red wine fortified with brandy (or possibly silent spirit) to give about 20 per cent alcohol. Before the must is completely fermented it is fortified, the yeasts die and the sweet jeropiga is added. White port is also prepared from the juice of white grapes. Madeira is an excellent keeping wine which owes its characteristic flavour to the estufado system of maturation in hot rooms. Marsala is made in Italy.

Sherry is a fortified wine produced in an area around Jerez de la Frontera in Spain. It is prepared from the small sweet Palomino grape, which contains suitable yeasts for the fermentation on the skin. Dry sherry with or without flor yeast is matured by the Solera (fractional blending) system. Neutral grape alcohol is used for fortification. Fino sherry is a pale dry sherry with flor containing about 17·5 per cent alcohol. Amontillado sherry is a medium dry darker type containing a little more sugar and alcohol. Oloroso sherry is a dry sherry without flor, sweetened and fortified with alcohol to 18–20 per cent. Amoroso sherry is a little paler and sweeter. The production of sherry is described by Goswell (1968), who also discusses the prevention of deposition in the product of potassium and calcium tartrates.

Legislation

Within the EC, control of the regional designation, marketing, quality, description, composition and process operations on wines is established by numerous Council and Commission Regulations covering detailed aspects of these factors. The principal Council Regulations (822/87/EEC on the common organisation of the market in wines; 823/87/EEC laying down special provisions relating to quality of wines grown in specific regions and 997/81/EEC, which prescribes specific rules for the description and presentation of wines and grape musts) are being progressively amended to take account of the expansion of wine imports from outside the Community. Commission Regulations 1108/82/EEC lay down methods of analysis to be applied to wine.

In the UK the Common Agricultural Policy (Wine) Regulations 1987, SI No. 1843 provide for the enforcement of the numerous EC Regulations (as specified in Schedules 1 and 2) concerned with the production and marketing of wine and related products, and the arrangements for enforcement of the Regulations.

The EC Regulations limit the use of the word 'wine' alone to products made from fresh grapes or grape must. In the Second Report on Food Labelling (FSC, 1979) (FSC/REP/69) the Food Standards Committee said that clarification of labelling provisions was necessary in respect of the use of the word 'wine', especially in relation to those products obtained by fermentation of fruit other than grapes. This proposal was confirmed in the Food Labelling Regulations 1984 (SI No. 1305). Specific provisions are also made for the declaration of alcoholic strength and for the exclusion from requirement to indicate 'minimum durability'. The committee also noted that it is impossible to discover any criterion—other than usage—which protects any particular geographical name for a wine. With the exception of 'Port' and 'Madeira', which are already protected by the Anglo-Portuguese Commercial Treaty Acts of 1914 and 1916, any wine producer in any country may use a distinctive name provided it is preceded by his own country's name in adjectival form. However, the word 'sherry' can only be used for sherry produced in Spain with the single exception of the Cypriot products, which can be described as Cypress sherry. The word 'champagne' can now only be used for the product from the

Champagne region of France. Council Decision 89/146/EEC established reciprocal agreements between the EC and Austria regarding the use of the designations of quality and the geographical regions to be associated with specific names of the major European producers of wine.

The question of international agreement on nomenclature of intoxicating liquors is under consideration by a Committee of Experts on the Production and Marketing of Vine Products and Spirits set up by the Council of Europe. The compositional requirements of the Regulations are summarised in Table 12.1. Terms used to express alcoholic strength of wines are defined in Annexe II to 822/87/EEC (OJ No. L 84, 27.3.87, p. 1), and are shown below.

ACTUAL ALCOHOL STRENGH BY VOLUME

The number of volumes of pure alcohol contained at a temperature of 20 °C in 100 volumes of the product at that temperature.

POTENTIAL ALCOHOL STRENGTH BY VOLUME

The number of volumes of pure alcohol at a temperature of 20 °C capable of being produced by total fermentation of the sugars contained in 100 volumes of the product at that temperature.

TOTAL ALCOHOL STRENGTH BY VOLUME

The sum of the actual and potential alcohol strengths.

NATURAL ALCOHOL STRENGTH BY VOLUME

The total alcohol strength by volume of a product considered before any enrichment. The first three strengths may also be expressed in terms of mass.

Alcoholometry

Consequent on the implementation of Council Directive 76/766/EEC (OJ No. L262, 27.9.76, p. 149), the system of alcoholometry to be used in the UK is based on the procedures of OIML. The Alcohol Tables Regulations 1979 (SI No. 132) define the terms alcoholic strength by volume and mass that are to be used in the labelling of alcoholic liquors under the provisions of the Food Labelling Regulations 1984, SI No. 1305, and in addition, authorise the preparation and use of 20 °C alcoholometric tables. A laboratory alcohol table has been drawn up and published by HM Customs and Excise (1979); the table has been constructed in accordance with the general formula recommended by OIML. The use of the table is authorised for Revenue purposes by The Alcoholic Liquors (Amendment of Units and Method of Measurement) Regulations 1979 (SI No. 241). An extract of the alcohol table is given in Table 12.2.

Composition

Dry wines contain less than 0·2 per cent sugar, sweet wines up to about 6 per cent and champagne may contain as much as 16 per cent. The pH is important during fermentation and storage. Most wines have a total acidity (as tartaric acid) from 0·3–0·55 per cent and a volatile acidity (as acetic acid) from 0·03–0·35 per cent. A 'vinegary' taste is synonymous with poor quality and limits of 0·12 per cent for white

Table 12.1 Summarised compositional requirements of EC Wine Regulations, principally 822/87, except 823/87 where indicated

Characteristic	Type of Wine	Standard	Regulation
Actual alcoholic strength	EC Table wine (Zones A & B)	∢ 8·5%	Annexe I, 13
	EC Table wine Zones CI(a)(b), CII, CIII	∢ 9·0%	Annexe I, 13
	Semi-sparkling wine*	∢ 7·0%	Annexe I, 17
	EC Quality sparkling wine psr	∢ 10%	823/87. Art. 8.4
	EC Quality sparkling wine psr of the aromatic type	∢ 6%	823/87. Art. 8.4
	Imported wine (general)	∢ 9%	Art. 70.1(b)
	Imported wine (specified)	∢ 8·5%	Art. 70.2(a)
	EC Liqueur wine	∢ 15%, ∢ 22%	Annexe I, 14
Total alcoholic strength	Semi-sparkling wine*	∢ 9%	Annexe I, 17
	EC Table wine	∢ 15%	Annexe I, 13
	EC Specified wine	∢ 17%	Annexe I, 13
	Imported wine	∢ 15%	Art. 70.1(b)
	EC Liqueur wine	∢ 17·5%	Annexe I, 14
	EC Quality wine psr	∢ 9%	823/87 Art. 8.4
	EC Specified white quality wines psr	∢ 8·5%	823/87 Art. 8.4
Natural alcoholic strength	EC Liqueur wine	∢ 12%	Annexe I, 14
	EC Quality liqueur wine	Not specified	Annexe I, 14
Total acidity as tartaric acid	EC Table wine	∢ 4.5 g/l (60 mmol/l)	Annexe I, 13
	Imported wine	∢ 4.5 g/l (60 mmol/l)	Art. 70.1(b)
Volatile acidity	EC White and rosé wine	∢ 18 mmol/l	Art. 66.1
	EC Red wine	∢ 20 mmol/l	Art. 66.1
Sulphur dioxide general limit	Red	∢ 160 mg/l	Art. 65.1
	White and Rosé	∢ 210 mg/l	
Wines with residual sugar content ∢ 5 g/l	Red	∢ 210 mg/l	Art. 65.2
	White and Rosé	∢ 260 mg/l	
	Spätlese and quality white psr	∢ 300 mg/l	
	Auslese and superior white	∢ 350 mg/l	
	Beerenauslese and specified quality	∢ 400 mg/l	
Excess pressure due to carbon dioxide	Sparkling wine	∢ 3 bar at 20 °C	Annexe I, 15
	Aerated sparkling wine	∢ 3 bar at 20 °C	Annexe I, 16
	Semi-Sparkling wine	∢ 1 bar ∢ 2·5 bar at 20 °C	Annexe I, 17
	Aerated semi-sparkling wine	∢ 1 bar ∢ 2·5 bar at 20 °C	Annexe I, 18

psr = produced in specified regions: Regulations 823/87.
Zones A, B and C are defined in Annexe IV of Regulation 822/87.
* and aerated semi-sparkling wine.

and 0·14 per cent for red table wines (expressed as acetic acid) are prescribed in the USA. In addition to acetic and other acids, ethyl acetate is probably an important contributor to spoilage (Amerine, 1954).

Examples of analytical data given by some of the more important wines are shown in Table 12.3. Minor volatile components (mainly acids, alcohols and esters) in wine have been identified by GLC after extraction with trichlorfluormethane (Hardy and

Table 12.2 Relationship between density in air of alcohol solutions at 20 °C and the proportions of ethanol

Density in air kg/m^3 at 20 °C	Percentage ethanol	
	by volume at 20 °C	by mass
788·16	100·00	100·00
790·0	99·64	99·41
800·0	97·51	96·07
810·0	95·08	92·52
820·0	92·37	88·79
830·0	89·42	84·92
840·0	86·25	80·93
850·0	82·89	76·86
860·0	79·35	72·73
870·0	75·65	68·55
880·0	71·79	64·31
890·0	67·76	60·01
900·0	63·54	55·65
910·0	59·12	51·22
920·0	54·48	46·68
930·0	49·53	41·99
940·0	44·15	37·02
950·0	38·08	31·60
960·0	30·92	25·39
970·0	22·32	18·14
980·0	13·14	10·57
990·0	5·00	3·98
997·0	0·00	0·00

Table 12.3 Composition of some wines

	Port	Sherry	Claret	Burgundy	Hock	Champagne
Specific gravity	0·995–1·050	0·992–1·015	0·990–1·020	0·995–1·001	0·990–1·008	1·040–1·055
Alcohol (g/100 ml)	13·5–20·0	13·5–20·5	7·5–12·5	7·5–12·5	7·5–12·5	10·0–14·0
% Total solids	3·3–13·0	2·0–9·6	2·0–3·5	2·0–3·5	1·5–2·5	9·5–18·0
% Free volatile acid (as acetic acid)	0·05–0·10	0·15–0·23	0·09–0·15	0·2–0·35	0·05–0·15	0·03–0·20
% Fixed acid (as tartaric acid)	0·35–0·55	0·25–0·50	0·30–0·50	0·3–0·60	0·25–0·45	0·30–0·45
% Ash	0·25–0·35	0·35–0·55	0·2–0·3	0·2–0·40	0·10–0·25	0·25–0·45
% Sugars	2·5–12.0	2·0–7·0	0·0–0·7	0·03–0·55	0·0–0·35	8·5–16·0
% Phosphoric acid (P_2O_5)	0·03–0·05	0·03–0·05	0·03–0·04	0·02–0·03	0·02–0·03	0·03–0·05
% Glycerol	0·3–1·3	0·4–1·0	0·3–1·0	0·3–1·0	0·25–1·3	0·3–1·0

Ramshaw, 1970). The analytical protocols used and aroma standards required for descriptive analysis of wines have been reported by Noble *et al.* (1984). The technique was applied to wines from the Bordeaux region in attempts to make quality assessments. The authors concluded that although significant flavour differences could be detected the assessment of quality remains a subjective judgement.

Analysis

EC Council Regulation on Analysis of Wine 2390/89/EEC prescribes the analytical information required from an authorised (official) laboratory on wine imported into the EC. The analysis includes determination of total alcoholic strength by volume, the actual alcoholic strength by volume, total dry extract, total acidity, volatile acidity, citric acid, sulphur dioxide and evidence of the authenticity of the species of grape used in production of the wine or must. An examination for the presence of non-permitted preservatives, sweeteners, food dyes and other alcohols is normally carried out. There are seldom sampling problems with ordinary types but sparkling wines and others showing froth should be de-gassed as described under 'Beer'. Methods for the analysis of wines are given in detail by Amerine and Ough (1988) and also by Linskens and Jackson (1988). Methods for the application in enforcement of EC Regulations are set out in the Annex to Regulations 2984/78. Both 'standard' and 'simplified' methods are described (see also Annex to the Compendium of International Methods for the Analysis of Wines, prepared and supplemented under the 1954 International Convention for the Unification of Methods for the Analysis and Appraisal of Wines).

Alcoholic strength. The alcoholic strength is ascertained by use of gravimetric methods.

METHODS

Relative density. Determine by pyknometer, accurate hydrometer or gravity bottle at 20 °C. Most wines have a specific gravity of at least 0·990.

Alcohol. Measure out 100 ml sample in a volumetric flask at 20 °C and wash into the distillation flask with 50 ml water, neutralise any acidity with M sodium hydroxide and distil slowly into the same 100 ml volumetric flask. Collect 90–95 ml, make up to 100 ml with water at 20 °C and determine the sp. gr. at 20 °C/20 °C preferably using a specific gravity bottle or pyknometer. Estimate the alcohol content as percentage by volume from the Laboratory Tables (see above Alcoholometry). The 'standard' method in the EC Regulations involves double distillation and measurement of the gravity of the distillate by pyknometer.

Cooke (1974) has shown that it is possible to obtain the alcoholic strength of wines from values of their refractive indices and relative densities at normal ambient temperatures. Dixon *et al.* (1963) have described a field test for obtaining the spirit strength of wines. The kit consists of a hand refractometer and special hydrometer ('saccharometer') to give the 'R' and 'D' readings respectively, from which the percentages of alcohol and sugar can be obtained by applying simultaneous equations. In these methods alcohol content is expressed in terms of 'Proof Spirit'; this system of measurement has now been officially discarded as a result of the implementation of EC Council Directive, 76/766/EEC (see Alcoholometry). A chemical method has been described by Caputi and Wright (1969). Automated methods for determination of ethanol have been described by a number of workers. The principles upon which methods are based vary. Automated distillation procedures have been described by Lidzey *et al.* (1971). Stockwell and Sawyer (1970) used a method in which a propan-2-ol and propan-1-ol mixture was used as an internal standard, blended automatically with the sample and the resulting solution injected automatically on to a gas chromatographic column packed with porapak. Dupont (1978) has described an enthalpimetric procedure based on automated dilution of the sample with aqueous sodium chlorate solution. Tep *et al.* (1978) have also studied the same method and have described procedures for minimising interference. Results on red, white and rosé wines were claimed to be equivalent to those obtained by the pyknometric method with a precision better than 0·1 per cent. An automated 'R' ands 'D' procedure has also been described and compared with other available methods by Geiss *et al.* (1979). Pilone (1985) and Junge (1985) have reported collaborative studies of official methods for analysis of wines, including titrimetric and distillation procedures. The precision of the distillation/gravimetric procedure for determination of ethanol was described as excellent. Gibson and Woodward (1986) have described an automated system using an enzyme alcohol oxidase to determine the alcohol content of beers and wines; the method is claimed to give suitable results for control analysis.

Methanol, isopropanol and the higher alcohols. Traditionally, the other alcohols have been determined in distillates prepared from wines by the application of chemical tests. Methanol is determined by the Denigès reaction after oxidation to formaldehyde with acid permanganate, the formaldehyde being determined by Schiffs reagent or by chromotropic acid. Isopropanol is oxidised with potassium dichromate to yield acetone, which is precipitated with acid mercuric sulphate. Higher alcohols are determined by condensation reaction with *p*-dimethylaminobenzaldehyde.

Gas chromatographic procedures are now considered to be more appropriate for the identification and estimation of the alcohols and other volatile components. Typical methods and data on commercial wines have been described by Sponholtz (1982), Shinohara and Watanabe (1981), Schreier (1980), and Postel (1981). The use of HPLC and specific enzyme methods is also appropriate for determination of acetaldehyde and the fruit acids (Joyeux Lafon-Lafourcade, 1979; McCloskey and Mahaney, 1981; Okamoto *et al.*, 1981).

Glycerol. Glycerol is a by-product of alcoholic fermentation. The ratio of the proportion of glycerol to alcohol produced varies according to the yeast, the freshness of the grapes, the sugar content and sulphite dosage. Amounts of glycerol between 0·3 and 1·5 per cent have been reported in various wines, but the results obtained vary according to the procedure used. The methods proposed for the determination up to about 1952 have been reviewed by Amerine (1954). The AOAC described two procedures (1) by direct weighing and (2) by oxidation with dichromate. Amerine (1954) has tabulated figures found for amounts of glycerol (see also Table 12.3) in various wines as well as the alcohol/glycerol ratios.

Although excessive glycerol is indicative of adulteration its presence in normal amounts in wines appears to impart smoothness to the taste. Avellini *et al.* (1977) have reported on the collaborative study of four procedures for the determination of glycerol. The methods studied included fluorimetric, volumetric, colorimetric and gas chromatographic methods. These authors concluded that the fluorimetric method is especially useful for routine analysis since it is specific for glycerol and is unaffected by preliminary treatment of the wine. The fluorimetric method used depends on the oxidation of glycerol with periodate to formaldehyde followed by reaction with 2,4 pentanedione and ammonia to yield the fluorescent chromophore of 1,2 diacetyl-1,4-dihydrolutidine (see also Ubigli, 1979).

Sugars. Exactly neutralise 100 ml wine with sodium hydroxide, evaporate to half that volume to remove alcohol, cool, add zinc ferrocyanide (p. 296), make up to 100 ml and filter. Red wines are first clarified with lead subacetate. Determine the sugars present before and after inversion by titrimetric or other accepted methods. The Luff–Schoorl method (p. 198) is preferred by OIV.

Extract. Pipette 50 ml of a dry wine or 25 ml of a sweet wine into a metal dish, evaporate on a water bath and dry in the 100 °C oven for 2 h. Alternatively, the extract can be assessed from the gravity of the de-alcoholised sample as described under 'Beer' (see also Wagstaffe, 1974). The indirect method, based on the gravity of the solution of de-alcoholised wine, is preferred for sweet wines because of the inaccurate results obtained by drying fructose in the presence of organic acids.

Ash. If a platinum dish is used for determining the extract the ash can be obtained, after weighing, by ignition at as low a temperature as possible. The EC Regulation directs that the incineration is carried out at 500–550 °C. Special directions are also given for the measurement of total and partial alkalinity of the ash.

Acidity. The acidity is expressed either in terms of milliequivalents per litre or as specified acids (see Junge, 1985 for a report of collaborative studies).

METHODS

Total acidity. Boil 25 ml sample under reflux for 20 min to expel carbon dioxide. Then wash down the condenser with water and titrate with 0·1 M sodium hydroxide either potentiometrically or using bromothymol blue. Calculate as milliequivalents per litre (EC) or as tartaric acid.

Volatile acidity. Boil 50 ml sample under reflux for 20 min. Cool, wash down the condenser with water and steam distil. Adjust the heat so that the volume in the distilling flask is approximately 25 ml and collect at least 200 ml distillate. Titrate the distillate with 0·1 M sodium hydroxide using phenolphthalein and calculate the volatile acidity as milliequivalents per litre (EC) or as acetic acid. High volatile acidity is not always due to *Acetobacter*.

Fixed acidity. Calculate the fixed acidity by difference as milliequivalents per litre (EC) or as tartaric acid.

McCloskey (1980a) has described a method for determination of acetic acid based on the use of acetic kinase, the recovery of acetic acid being claimed at 100 per cent.

Directions for determining pH are also given in the EC Regulation.

Tartaric acid. With most fruit tartaric acid is the predominant fixed acid. The amount present changes during ripening. In wine it is present partly as the free acid, partly as cream of tartar with a little calcium tartrate. Tartrates separate in deposits of old wine. The total amount can be determined by precipitation as potassium hydrogen tartrate after addition of acetic acid, potassium acetate and potassium chloride to the sample (AOAC). The tartaric acid may be determined either by titration or colorimetrically after separation by anion exchange resin.

Brunelle *et al.* (1967) have determined fixed acids in wines after precipitation as insoluble lead salts. Trimethylsilyl derivatives of the acids are then subjected to gas chromatography. Succinic, malic, tartaric and citric acids can be determined in this way. Grape wines contain tartaric and malic acids, but other fruit wines do not contain tartaric acid. Ion chromatography methods are also appropriate for the determination of the fruit acids (Woo and Benson, 1983). Lazaro *et al.* (1986) have described a flow injection analysis method using a colorimetric reaction of tartrate and vanadate to determine tartaric acid in Jerez wines. They claimed that no separation of the wine acids is necessary before the analysis is carried out. Steiner *et al.* (1984) used an HPLC method on the esterified acids but reported low recoveries of tartaric acid. McCloskey (1980b) has proposed an enzyme assay with malate dehydrogenase and glutamate oxaloacetate transaminase as an accelerator. The method gives rapid analysis for levels up to 2000 mg/l as malate.

Sulphites are commonly used in the sulphuring of casks and may sometimes be added to wine. For the estimation the following direct process gives reasonably accurate results with white wines:

Add 50 ml of wine to 25 ml M sodium hydroxide solution and, after allowing the mixture to stand for 15 min, add 10 ml of dilute sulphuric acid (1 + 3) and a few drops of starch solution. Then titrate with 0·05 M iodine solution (1 ml 0·05 M iodine ≡ 0·0016 g SO_2). For coloured liquids the distillation method described in Chapter 4 should be employed.

Only a portion of the sulphur dioxide is active against micro-organisms. If sulphuric acid is added first to the sample the free sulphur dioxide can be directly titrated with iodine (Chapter 4).

Sorbic acid is permitted in wines up to a maximum of 200 mg/l. Methods for its estimation are discussed in Chapter 4.

Diethyl pyrocarbonate (DEPC) has been developed for use as a processing aid and antimicrobial agent in wine, beer, cider, perry, etc., but its use is not permitted in the UK. When used it is rapidly hydrolysed with formation of carbon dioxide and ethanol. Current evidence indicates that when DEPC is used in wine urethane is produced. Methods for determining DEPC involving GLC are also discussed in Chapter 4.

Diethylene glycol has been used to adulterate wines to simulate more expensive wines with a higher natural sugar content, e.g. Auslese and similar wines. A suitable GLC method of analysis used at LGC and MAFF is as follows:

APPARATUS AND REAGENTS

Bis(TMS)trifluoroacetamide (BSTFA).

Acetonitrile and hexane, AR grades.

Microlitre syringes, 1 μl, 5 μl and 100 μl.

2 ml glass vials with screw-top cap and PTFE-faced septum.

Capillary gas chromatograph fitted with a 25 m × 0·25 mm CP Sil 5 CB (ex Chrompack), or equivalent, fused silica column.

METHOD

Set up chromatograph with split injection (5:1 ratio) and FID. Set carrier gas (helium) flow rate to 1·8 ml/min. Operate isothermally at 100 °C with injector and detector at 250 °C and 300 °C respectively. Diethylene glycol retention time is about 5 min.

Place 5 μl of wine in a 2 ml vial and add 100 μl BSTFA. Heat at 60 °C for 15 min and then dilute reaction mixture with 100 μl acetonitrile. Inject 1 μl into capillary GC. Eliminate possible cross-contamination of injection syringe by rinsing in sequence with BSTFA, hexane and acetonitrile. Construct calibration curve by adding diethylene glycol to blank wine and monitor system by analysing spiked wines at regular intervals. Detection limit is about 0.01 g/l. (Analysis can be confirmed by selected ion monitoring GC/MS, using ions at m/z 191 and 235.)

Cider. Steuart (1955) based a test for the detection of cider in grape wine on the relative proportions of sorbitol present (2 mg/100 ml in Bordeaux wines and 120–510 mg/100 ml in cider).

SPIRITS

Manufacture

Spirits are produced by the distillation of fermented liquors such that most of the alcohol and various other volatile substances pass over into the distillate. Spirits are unlikely to undergo microbiological spoilage due to the high alcohol content. Although almost any fermented beverage can be made into a potable spirit the only ones of major commercial importance are brandy, gin, rum, whisky and vodka.

As spirits are produced by distillation they contain less extractives than other alcoholic liquors; with the exception of British vodka (to which it is added) they contain no glycerine, but a high percentage of alcohol together with certain esters which, though present in small amount, are of much importance as they confer the characteristic flavour of the particular spirit. The principal congeners in potable spirits have been detected by GLC; Dickes (1979) has summarised the accumulated evidence. Simpkins (1985) has further examined the use of congener profiles to characterise distilled spirits, showing that the profiles must be used with caution and that supplementary evidence is often necessary to confirm authenticity.

Legislation

The Food Act, 1984 states that in proceedings taken under S.2 it shall be a defence, in respect of diluted whisky, brandy, rum or gin, for the defendant to prove that the spirit in question had been diluted with water only and that its strength was still not lower than 37·2 per cent v/v. There are, however, provisions in The Labelling of Food Regulations, that pre-packed brandy, gin, rum, and vodka that has an alcoholic content below 37·2 per cent v/v may be sold provided the label bears a declaration of the alcohol content together with a statement that the produce is 'under strength'. A similar provision exists in the case of whisky except that the appropriate strength is 40 per cent v/v. The requirement does not apply to brandy where strength has fallen below 37·2 per cent v/v only through maturing in cask. The declaration of alcoholic strength is subject to definitions in The Alcohol Tables Regulations 1979 (p. 432).

EC Regulation 1576/89 lays down the general rules on the definition, description and presentation of spirit drinks (see Table 12.4).

Table 12.4 General rules on the definition, description and presentation of spirit drinks; Regulation 1576/89/EEC

Type	Minimum actual alcoholic strength (% v/v)	Maximum methanol content (g per litre of 100% alcohol)
Brandy (wine)	36·0	2·0
Fruit brandy (e.g. Calvados, Kirsch)	37·5	10·0
Grape marc	37·5	10·0
Gin	37·5	—
Rum	37·5	—
Vodka	37.5	—
Whisky/whiskey	40·0	—

Composition

Brandy is the spirit distilled from wine made off the skins of selected grape varieties, principally the Folle blanche, a non-sweet type grown in the Charente area of France. The resultant wine is slowly distilled in pot-stills to produce *brouillis*, several of which are combined and re-distilled. The main fraction is about 70 per cent v/v and the blended Cognac is matured in oak casks, which themselves assist in the formation of the characteristic bouquet. Changes during maturation in the cask are discussed by Simpson (1971). Good Cognac brandies are sold as 'Three Star', 'Extra Old', VOP and VSOP (in order of increasing price). Armagnac brandy is considered to be of lower quality than Cognac. Brandies are also produced in other areas of France and in Spain, Greece, Germany, S. Africa, Australia and the USA.

LAJAC (1963) Code of Practice No. 2 states that the names Cognac and Armagnac should only be used in the description of brandies produced from those regions. Fruit brandies derived from one fruit (other than grapes) shall be described as 'X Brandy, a spirit distilled from fermented Xs', where 'X' is the name of a fruit such as plum. The Code also states that no exception will be taken to established names such as Cherry Brandy and Apricot Brandy provided they contain at least 20 per cent genuine brandy (from grape). The description 'Brandy' or 'Fruit Brandy' should only be applied to potable spirits derived from the distillation of the fermented juice of fresh grapes or the selected juice of other fruits. Machinery for granting certificates of age for brandy does not extend beyond 5 years (in the case of the UK, 3 years). Any brandy declared to be older than this must depend on the good faith of the shipper.

The composition of brandy varies widely, but the specific gravity is usually about 0·94 and the total solids between 0·1 and 0·7 per cent. Data compiled by Cox are given in Table 12.5. More recent figures for brandy compiled from various sources with special reference to the congenerics are quoted in Table 12.6. Singer (1966)

Table 12.5 Data obtained from samples of brandy

	1	2	3	4	5	6
Specific gravity	0·9566	0·9481	0·9477	0·9427	0·9414	0·9471
% Alcohol v/v at 20 °C	41·8	44·0	45·2	47·2	47·8	43·3
% Total solids	1·904	0·646	1·024	0·757	0·608	0·132
% Ash	0·012	0·012	0·006	0·005	0·004	0·004
% Total acidity (as acetic acid)	0·046	0·039	0·072	0·054	0·041	0·004
= g per 100 l alcohol	110	89	160	115	86	9
*Total esters (ethyl acetate)	580	463	564	502	335	112
*Volatile esters	122	57	88	101	123	44
*Furfural	6	2	3	4	5	0
*Aldehyde	69	41	71	76	29	24
*Higher alcohols	146	315	782	675	611	374

* g per 100 l alcohol.

graded brandies according to the ratio of the contents of propanol to 2-methyl-1-propanol and of 2-methyl-1-butanol to 2-methyl-1-propanol. Askew and Lisle (1971) determined congeners in 86 samples of brandy from 15 countries, but found no consistent value or ratio which could be used to characterise brandy as a class of spirituous beverages. Hogben and Mular (1976) made similar studies of Australian brandy and brandies imported into Australia. These authors concluded that limiting values for the ratios of the congeners could be applied to Australian brandy which is produced under Government Supervision. Simpkins (1985) extended this study in respect of the control of illicit distillation of spirits. Barnett and Einsmann (1977) made similar studies on US spirits. They concluded that 1-hexanol content was a useful parameter for distinguishing between brands. Components of brandy respon-

Table 12.6 Ranges of figures for other alcohols in brandies and whiskies as g per 100 l ethanol

Type of spirit	3-Methyl-1-butanol	2-Methyl-1-propanol	Propanol	Methanol
Branded Cognacs	102–394	36–129	14–48	14–187
Unbranded Cognacs	119–310	42–101	15–42	—
Inferior Cognacs	134–177	56–66	31–38	500
Branded whisky blends	74–83	61–100	21–36	
Superior blended whiskies	91, 94	90, 89	37, 34	
Immature all-malt whiskies	162–265	76–170	27–40	
6–10-year-old all-malt whiskies	164–220	88–138	21–50	
Immature grain whiskies	0·8–12·5	41–72	22–49	
6-year-old grain whiskies	4·3–19	49–126	21–49	
Bourbon 2–4 years	220–446	42–150	—	
Bourbon 6–10 years	245–506	41–198	—	

sible for the characteristic flavour have been identified as ethyl esters of octanoic, decanoic and dodecanoic acids (Guymon and Crowell, 1969; Beaud and Ramuz, 1978).

Gin is a diluted highly rectified spirit, flavoured with juniper berries and extracts from other plants. The basic spirit is made from maize and rye grains together with malted barley or rye. The fermented mash is distilled in a patent still. The juniper berries together with materials such as coriander, cassia and orange peel are added to the still prior to the second distillation. Unlike most other spirits gin is virtually devoid of congenerics as the secondary constituents that contribute to the taste and odour are not derived from the same source as the alcohol. Compounded gin is by comparison an inferior product, being made by adding essential oils to spirit. Old Tom gin contains added sugar and sloe gin is prepared by steeping sloes in gin and often adding sugar also. Unlike most wines and other spiritis, gin deteriorates during storage due to oxidation of the oils present. Various aspects of gin manufacture have been discussed by Simpson (1966).

Gin has a specific gravity of about 0·955 and the total solids of unsweetened gin seldom exceeds 0·1 per cent. Normal figures for the congenerics expressed as g/100 l ethanol are as follows: aldehydes 2, esters 5, acids 1, fusel oil 7. Data compiled by Cox are included in Table 12.7.

Table 12.7 Data obtained from gin and rum

	Rum 1	Rum 2	Rum 3	Gin 4	Gin 5
Specific gravity	0·9436	0·9375	0·9403	0·9531	0·9504
% Alcohol v/v at 20 °C	45·5	49·1	47·2	40·0	41·2
% Total solids	0·385	0·420	0·295	0·095	0·013
% Ash	0·041	0·039	0·025	0·008	0·009
% Total acidity (as acetic acid)	0·102	0·095	0·082	0·015	0·020
= g per 100 l alcohol	224	193	174	—	—
*Esters	241·3	275·0	190·3	—	—
*Furfural	10·5	9·6	6·8	—	—
*Aldehydes	27·8	35.2	24·5	—	—
*Higher alcohols	174·5	190·3	160.1	—	—

* g per 100 l alcohol.

The principal alcoholic constituents of the congeners are 3-methyl-1-butanol, 2-methyl-1-propanol and propanol; no evidence of the presence of methanol has been reported. Clutton and Evans (1978) identified the principal constituents of gins as being consistent with the use of juniper, coriander, cinnamon and cassia as flavouring ingrediets.

Rum is a spirit distilled from fermented sugar cane juice or molasses. It is principally made in Jamaica and other areas of the Caribbean. Molasses is diluted to contain about 12 per cent sugar, mixed with 'dunder' (residue in the still from a previous distillation) and allowed to ferment for a few days either spontaneously or with the addition of selected yeasts. Pot stills are being replaced by continuous types. The colourless distillate is aged in casks for at least 3 years during which time colour is taken up from the wood. Rum is diluted before bottling and differs from whisky or brandy in the preponderance of esters, particularly ethyl butyrate and acetate, over higher alcohols. Colourless rums have become increasingly popular in recent years.

Rum usually has a specific gravity of 0·87–0·94 and the total solids usually varies from 0·4 to 3·0 per cent. Some analytical figures reported on three samples of rum by Cox are included in Table 12.7. More recent figures for Jamaica and Demerara rums are included in Table 12.8.

Table 12.8 Rum, scotch whisky and brandy (all results expressed as g per 100 l ethanol)

	Jamaica rum	Demerara rum	Scotch whisky	Cognac
Total acidity (as acetic acid)	80–180	90–240	30–70	5–100
Volatile acidity (as acetic acid)	50–100	50–140	10–25	—
Fixed acidity (as acetic acid)	—	—	20–50	—
Esters	120–200	40–70	10–20	70–100
Aldehydes	Up to 40	0–20	100–270	2–35
Furfural	Up to 10	Trace	10–20	2–3
Fusel oil	140–200	100–160	—	—
Total solids	530–1600	700–4000	—	—

The major fusel oil component of rum has been reported as 3-methyl-1-butanol, with lesser amounts of propanol and 2-methyl-1-propanol. The quality of rum is largely dependent on the amounts of ethyl esters of fatty acids of even carbon number from C_8 to C_{16} (Stevens and Martin, 1965).

Vodka is a distilled liquor originally produced in Russia and neighbouring countries from wheat, potato, rye and possibly barley or maize. It is now produced in the UK, Spain and the USA in large quantities. Its manufacture resembles that used for the production of other spirits such as grain whisky, but the rectification tends to be higher. The commoner brands of vodka on sale in the UK are labelled as containing between 37·2 and 79·7 per cent v/v. The product does not differ markedly from diluted ethanol and its identifying characteristics are sometimes further reduced by filtration through charcoal. In order to comply with the Immature Spirits Regulations, British vodka is 'flavoured' by addition of a small amount of glycerol.

Whisky. Although unblended malt whiskies are commonly drunk in Scotland, most Scotch whisky is a blend of malt and grain whiskies (Schedule 7(1) of the Finance Act, 1969). Malt whisky is made entirely from malted barley. A high diastatic power in the malt is necessary in order to obtain maximum conversion of starch to sugar. A distinctive flavour is produced in the malt during heating over peat. The wort is not boiled and is cooled to about 21 °C (70 °F) before adding distiller's yeast and fermenting for 2–3 days. The resultant 'wash' containing about 10 per cent alcohol is then doubly distilled in copper pot stills. The distillate containing 63–70 per cent alcohol is then matured in sherry casks for at least 3 years. Scotch grain whisky is made from unmalted maize (and sometimes rye and oats), which is gelatinised by cooking before mashing with a proportion of barley malt. After distillation in a Coffey (patent) still the whisky is matured in oak casks. Grain whisky tends to contain less congenerics than malt whisky. Various aspects of the manufacture of 'Scotch' whisky have been discussed by Simpson (1968), Pyke (1965) and Taylor (1968). Irish whiskeys are produced by three distillations (Court and Bowers, 1970). Bourbon, which is made from rye or maize, can be distinguished from whisky by its comparatively high phenyl ethyl alcohol content (Kahn and Conner, 1972).

Whisky usually has a specific gravity of 0·94–0·95 and the total solids varies from 0·05 to 0·3 per cent. Figures compiled from various sources are quoted in Tables 12.6, 12.8, 12.9 and 12.10. Changes in the congeners of whisky samples examined since 1908 are tabulated in the Report of the Government Chemist for 1965.

Table 12.9 Data for various samples of whisky. Mean figures given in parentheses

	Mature grain whisky	Proprietary whisky blends	Mature malt whisky
Obscuration (% v/v ethanol)	Up to 0·7	0·2–1·0	0·6–2·0
Total aldehyde (as g acetaldehyde/100 l ethanol)	2–5(3)	8–12(10)	6–25
Total esters (as g ethyl acetate/100 l ethanol)	10–16(14)	17–35(28)	37–85
Furfural (g/100 l ethanol)	Up to 0·1	1·2–1·9(1·5)	1·7–8·5
Total higher alcohols (g/100 l ethanol)	70–150(110)	150–220(190)	300–450
Total acidity (g acetic acid/100 l ethanol)	3·5–8·0(5)	15–23(18)	11–28

Gas chromatographic procedures have been used in attempts to distinguish between the various types of whisky. Hall (1976) used 3-methyl-1-butanol, 2-methyl-1-propanol ratios to distinguish grain whisky and bourbon. Duncan and Philp (1966) used three differing sets of GC conditions in the analysis of malt and grain whiskies. Two separate columns were used for analysis of lower esters and diethyl acetal, and the alcohols and higher esters, by direct injection. A third column was used for the estimation of carbonyls separated from spirits by formation of bisulphite compounds. Significant differences were noted between malt and grain whiskys especially in the content of 3-methyl-1-pentanol (see Table 12.6). Saxberg *et al.* (1978) have used pattern recognition techniques to distinguish various qualities of whisky from analysis of the volatile components. Lehtonen and Soumalainen (1979) have obtained data on congeners from a range of whiskies by GLC procedures. Samples from America, Canada, Spain and Japan were compared with Irish

Table 12.10 Data for various brands of Scotch whisky

	Sample number					
	1	2	3	4	5	6
Marked	70°PS	70°PS	70°PS	—	—	5 year old
Alcohol (° Proof UK)	71·3	71·5	72·4	77·1	76·4	75·7
Alcohol (% v/v)	40·8	40·9	41·5	44·1	43·7	43·3
Total solids (%)		—		—		0·090
Volatile acidity (g acetic acid/ 100 l ethanol)	30·6	25·5	28·7	9·0	10·4	38·8
Fixed acidity (g acetic acid/ 100 l ethanol)	7·1	14·1	9·0	1·0	0·6	5·5
Total esters (g ethyl acetate/ 100 l ethanol)	28·0	35·3	34·8	17·6	16·9	32·7
Total aldehyde (g acetaldehyde/ 100 l ethanol	11·4	10·5	9·7	4·9	3·1	9·3
Higher alcohols by GLC (g/100 l ethanol)	129	130	186	41	44	193
Furfural (g/100 l ethanol)	—	—	—	—	—	1·0
Total secondary products (g/100 l ethanol)	—	—	—	—	—	290
Total volatile impurities (g/100 l ethanol)	—	—	—	—	—	284
Methyl alcohol (g/100 l ethanol)	—	—	—	—	—	9·5

and Scotch whisky. MacNamara (1984) used coated fused silica capillary columns to examine malt whisky for the major congeners; typical data for 26 constituents is given.

Examination of complaint samples of whisky has been discussed by Cochrane (1986); a procedure for concentrating the flavours using Freon II as a solvent prior to use of GC-odour assessment and GC/MS is described. Lehtonen *et al.* (1984) used HPLC to estimate unsaturated aldehydes and correlated the content with off flavours in spirits.

Analysis of spirits

As in the manufacture of many other products, quality control of spirits is generally achieved by close control of ingredients and process conditions. Organoleptic appraisal is of especial value where blending of a number of production batches is used to obtain a uniform finished product (Brandt, 1964). Detailed analysis is of value in quality control procedures and in assessment of the maturation processes utilised for particular spirits such as whisky and brandy. However, in view of the high level of revenue duty placed on a relatively cheap product, analysis assumes a particularly important function in the control and monitoring of the distribution of spirits and in detection of illegal practices. These practices may include illicit distillation and, in view of sales from open bottles, can include operations such as dilution, adulteration, and substitution. Since duty on spirits is assessed by alcohol content, the accurate determination is of prime importance, the generally accepted procedures depending either on determination of the specific gravity of the undistilled sample and the alcohol free extract (obscuration) or on the determination of the specific gravity of a distillate. Other determinations that may assist in assessment of a spirit are those of acidity, ash, tannins and the congeners (esters, furfural, aldehydes and other alcohols).

Although classical methods of analysis give useful information on the composite components of potable spirits, they have now been largely superseded by GLC methods. Several reports of collaborative study of classical methods and alternative GLC procedures have been published by the Research Committee on Analysis of Potable Spirits (RCAPS, 1970, 1972, 1974a, b, 1977). The reports have been collected and expanded into a single publication which includes details of test procedures and results of collaborative studies (Anon, 1979). Dickes and Nicholas (1976) have surveyed gas chromatographic methods for the analysis of spirits, and recommended preferred procedures for determination of congeners. Dickes (1979) has updated this review. Work on illicit spirits has been reviewed in the Report of the Government Chemist (Anon, 1972) (see also Simpkins, 1985).

Alcohol. The alcohol content of spirits is best determined by direct distillation of 50 ml of sample, washing into the distilling flask with not more than 5 ml water. By determining the relative density in air of the distillate at 20 °C after making up to the original volume the percentage of alcohol in the sample can be found by reference to the Laboratory Alcohol Table (see alcoholometry). The method is described in detail by the Research Committee (RCAPS, 1970). In addition the Fourth Report (RCAPS, 1974b) discusses the sources of error in measuring the strength of alcohol in distillates using various forms of pyknometer. Commercial instruments for measurement of the specific gravity of distillates are now available. The instrument widely used in industry measures the change in period of oscillation of a filled tube when subjected to electromagnetic excitation (Leopold *et al.*, 1977). If the total solid matter is small such as in gin and vodka the alcohol content can be assessed approximately by determining the specific gravity of the sample without distillation; for an accurate determination it is necessary to make a suitable correction (obscuration). The soluble matter increases the gravity and the alcoholic content as indicated by the relative density of the sample itself will be lower than the true figure. This difference between the apparent and real alcohol contents is referred to as the 'obscuration'. Each 1 per cent m/v of extract increases the relative density by 0·0041, and the calculated correction for the relative density is then deducted from the observed value on the undistilled sample. The alcohol content of the original sample is then derived by interpolation in the alcohol table.

Alternatively, the obscuration can be obtained by the Balling method. The method depends on a relationship which states that, in a sample of obscured spirits the relative density of the distillate is equal to the relative density of the sample divided by the relative density of the extract. Tables have been constructed on the basis of this empirical relationship and these are used to relate the relative densities of the sample and its extract directly to the obscuration. The relative density of the extract is obtained by evaporating down a known volume to low bulk so that all the alcohol is removed. Then the liquid is made up to the original volume of the sample and the relative density measured. The obscuration of the spirit is then obtained by entry into the Balling table at the appropriate points. Ranges of obscuration—corrections expressed in terms of alcohol content for different spirits are given in Table 12.11.

The obscuration correction determined on a single sample by the Balling method

Table 12.11 Obscurations of spirits

Spirit	Obscuration (per cent v/v)
Brandy	1–2·2
Gin	Usually negligible
Rum	1–3·5
Whisky	0–0·8
Vodka	Less than 0·5

may be applied to correct hydrometer readings for further samples taken from the same shipment or production batch. Wagstaffe (1976) published a critical evaluation of this obscuration procedure, also describing verification of alcohol hydrometers (Wagstaffe, 1980) for use in assessment of spirit strength.

Methanol. Minute traces of methanol are naturally present in spirits; it has been used as an adulterant, but significant quantities are found especially in the lower grades of brandy. Methods for its estimation were the subject of the Second Report of the Research Committee (RCAPS, 1972). A colorimetric procedure described is based on the AOAC method (Mathers, 1958). The methanol is separated from most of the ethanol by fractional distillation and oxidised by acid permanganate to formaldehyde, which reacts with chromotropic acid. The red colour formed is measured at 575 nm. The GLC methods described in the First Report (RCAPS, 1970) for higher alcohols are also suitable for the estimation of methanol in spirits. A separate modified GLC procedure for estimating methanol alone is included in the Second Report of the Committee. Water, methanol, acetaldehyde and methyl formate are eluted in that order from a column with porous polymer beads (Poropak Q) as stationary phase. Other components of spirits have longer retention time and do not interfere. Methyl formate is used as internal standard. Methanol can be detected in distillates from spirits by Denigès' test.

Extract (total solids). Wash the residue in the flask (after the distillation of the alcohol above) into a weighed platinum dish, evaporate on the water bath and then in the 100 °C oven, cool and weigh.

Ash. Ignite the residue from the determination of extract at a dull red heat.

Acidity. The methods for determining acidity in spirits included in the Third Report of the Research Committee (RCAPS, 1974) are essentially those described in the AOAC. The end-point of the titration should preferably be detected with the aid of a pH meter. A suitable visual indicator is phenolphthalein.

Total acidity

To 300 ml water adjusted to pH 7·8 add 50 ml of sample and titrate to pH 7·8 with 0·1 M sodium hydroxide (titre V_1 ml). Calculate the total acidity as both acetic and tartaric acid.

Fixed acidity

Evaporate 50 ml of sample to dryness in a flat shallow dish on a boiling-water bath and heat the residue for a further 30 min. Then stir in 50 ml water (pH 7·8) and titrate to pH 7·8 with 0·1 M sodium hydroxide (titre V_2 ml). Calculate the fixed acidity as tartaric acid.

Volatile acidity

From the above titres:

Volatile acidity (g acetic acid/100 l of alcohol in sample) = $1200\,(V_1 - V_2)/S$, where S = the percentage of alcohol (v/v) in the sample. Using this method the Committee reported the following results: whisky (4 samples) 25–28; rum (4 samples) 9–63; brandy (3 samples) 21–38. Using GLC, Nykanen *et al.* (1968) showed that rum contained more volatile acids than whisky or brandy. Also the proportion of acetic acid in the total volatile acids tended to be higher in whisky and rum than in brandy.

Esters. Methods for the estimation of esters in spirits are described in the Third Report of the Research Committee (RCAPS, 1974). The esters are hydrolysed with alkali and the excess is titrated with acid.

METHOD

Fill a 250 ml volumetric flask to the mark with the sample at 20 °C and transfer to a distillation flask with the aid of 20 ml water. Distil and collect nearly 250 ml of distillate in the same flask and make up to volume at 20 °C with water. Procure two flasks (300–500 ml) fitted with reflux condensers. To one flask add 100 ml of distillate and to the other 100 ml water. Then treat both sample and blank as follows: Add 3 drops phenolphthalein and titrate with 0·1 M sodium hydroxide. Then add exactly 10·0 ml of the alkali. Add glass beads and heat under reflux for one hour. Cool rapidly and titrate with 0·05 M sulphuric acid. Let the difference in titre be V ml. If the alcoholic strength of the sample is S per cent v/v, the total esters = 880 V/S as g ethyl acetate/100 l of alcohol in sample. Using this method the Committee reported the following results: whisky (5 samples) 26–36; rum (6 samples) 8–76; brandy (5 samples) 73–84.

The Committee have also described an alternative procedure, which is essentially the method of Blessinger *et al.* (1965) adopted by the AOAC. The esters are converted to hydroxamic acids, which are measured colorimetrically as the ferric salts.

Modifications of the method proposed include the use of a neutralised distillate from a direct titration of volatile acidity and carrying out the saponification under pressure in a screw-stoppered silver flask. For whisky Duncan and Philp carry out the reactions in 1-oz McCartney bottles. Various esters in potable spirits have been identified by GLC, which tends however to give lower results than chemical methods for esters, as not all the components are measured (Martin *et al.*, 1964, 1974; Duncan and Philp, 1966; Guymon and Crowell, 1969).

Total aldehyde. The method for total aldehyde in spirits recommended in the Third Report of the Research Committee (RCAPS, 1974) is the direct bisulphite procedure of Guymon and Crowell (1963) as adopted by the AOAC (Burroughs and Sparks, 1961).

REAGENTS

A. Mix 15 g potassium metabisulphite ($K_2S_2O_5$) with 70 ml conc. HCl and dilute to a litre with water.
B. Dissolve 200 g sodium phosphate (Na_3PO_4) and 4·5 g EDTA in water and dilute to a litre.
C. Dilute 250 ml conc. HCl to 1 litre with water.
D. Mix 100 g boric acid with 170 g sodium hydroxide, add water to dissolve and dilute to 1 litre.

PROCEDURE

Place 300 ml boiled and cooled water in a large flask. Add 10 ml of A and, by pipette, 50 ml of a distillate of the sample containing about 50 per cent alcohol. Stopper, swirl to mix and allow to stand for 15 min. Then add 10 ml of B, swirl to mix and allow to stand a further 15 min in the stoppered flask. Add 10 ml of C and 10 ml fresh starch solution (0·2 per cent). Swirl to mix, add enough iodine (approx. 0·1 M) so that the excess bisulphite is *just* destroyed and bring the solution to a faint blue end-point. Add 10 ml of D and titrate the liberated bisulphite with 0·05 M iodine to the same faint blue end-point.

Total aldehyde (including acetal) as g acetaldehyde/100 l of ethanol

= Titre × 2·2/S (where S is the alcoholic strength of the sample in per cent v/v).

Using a similar procedure the Committee reported the following results (as g/100 l of alcohol): whisky (4 samples) 10·2–15·2; rum (4 samples) 22·6–31·2; brandy (3 samples) 22·1–29·6. Schiff's regent was used in earlier colorimetric procedures for the determination of aldehydes, which restore the colour of magenta previously decolorised with sulphurous acid (ISO R1388: 1970). A modified procedure suitable for the examination of several samples simultaneously has been described by Duncan and Philp (1966).

Furfural. Hewitt's method has been most commonly employed for the estimation of furfural in alcoholic distillates. This is based on the pink colour produced with aniline acetate.

REAGENTS

Aniline acetate solution. Mix equal volumes of redistilled aniline, glacial acetic acid and water. Prepare as required.

Standard furfural solution. From a 1 per cent v/v solution of furfural in 50 per cent alcohol prepare 0·1 per cent, 0·01 per cent, and 0·001 per cent solutions in 50 per cent alcohol.

PROCEDURE

Place 10 ml of sample distillate containing 50 per cent alcohol in a stoppered 15 ml tube. Prepare similar tubes containing 0·5 ml upwards of the standard 0·001 per cent furfural solution, the volume in each case being made up to 10 ml with 50 per cent alcohol. Add 0·5 ml aniline acetate solution to each tube and mix. Measure the optical density of the solutions at 540 nm after allowing the colour to develop for 15 min.

The procedure recommended by the AOAC is based on the measurement of the ultraviolet absorption spectrum of furfural after steam distillation of the spirit. This method is considered to give a substantial systematic error by the Research Committee on Potable Spirits (RCAPS, 1977). The error is considered to be due to co-distillation of other ultraviolet absorbing species. The committee also regard the aniline acetate procedure to be unsuitable for general use in view of the toxic hazard presented by the use of aniline. A gas chromatographic procedure, based on the use of 15 per cent Carbowax 1540 on 80–100 mesh Chromosorb W-AW DCMS with heptanol as internal standard, was recommended by the committee as a more specific and accurate method for estimation of furfural (see also Duncan and Philp, 1966). Jeuring and Kuppers (1980) used reversed phase HPLC for the estimation of furfural and hydroxymethyl furfural in brandy.

Higher alcohols. Higher alcohols were originally estimated by the Allen–Marquardt titrimetric method as modified by Schidrowitz after oxidation with dichromate. Later workers have described modifications of the Komarowsky colorimetric reaction, in which the higher alcohols and an aldehyde react in hot sulphuric acid to give coloured products (Osborn and Mott, 1952; Boruff, 1959; Singer and Stiles, 1965; AOAC). The following procedure using 4-hydroxybenzaldehyde-3-sulphonic acid is Method B recommended by the Research Committee (RCAPS, 1970) in their First Report.

REAGENTS

Sodium 4-hydroxybenzaldehyde-3-sulphonate. An aqueous 4 per cent m/v solution.

Mixed pentanols. Isoamyl alcohol, AR grade (mainly 3-methylbutan-1-ol) is suitable.

Standard higher alcohol solutions.

1. Prepare the following mixtures:

   7·5 ml of 2-methylpropan-1-ol and 2·5 ml of mixed pentanols.
   5·0 ml of 2-methylpropan-1-ol and 5·0 ml of mixed pentanols.
   2·5 ml of 2-methylpropan-1-ol and 7·5 ml of mixed pentanols.

2. From each of these foregoing mixtures, and also from 2-methylpropan-1-ol and from mixed pentanols (reagents 4 and 5), prepare solutions of 5 g of (total) higher alcohols in 100 ml of 40 per cent v/v aqueous ethanol.
3. Further dilute each of the five solutions so obtained with 40 per cent v/v aqueous ethanol to give five sets of dilute standard solutions of (total) higher alcohols, each set containing 25, 50, 75, 125 and 250 g of (total) higher alcohols per 100 l of solution.

PROCEDURE

Preparation of sample. If necessary, dilute the sample with water or ethanol to an ethanol concentration of 40 ± 1 per cent v/v. To 50 ml of the diluted sample add about 20 ml of water and distil. Collect at least 47 ml in a 50 ml flask, mix, dilute to volume with water and mix well.

Blank test and calibration curves. Transfer 0·20 ml of each dilute standard of a set, and also of a 40 per cent v/v aqueous ethanol solution as a blank, to a series of six 10 ml volumetric flasks (see Note 4) and then add in order:

0·20 ml of sodium 4-hydroxybenzaldehyde-3-sulphonate solution.

2·0 ml of concentrated sulphuric acid. Mix by swirling.

Place the flasks in a cold water bath and heat the water to boiling. Boil for 30 min. Remove the flasks and allow them to cool. Make the volume to 10 ml with concentrated sulphuric acid. Mix well. Measure the optical densities (OD) at 445 nm and at 560 nm, using the blank as a reference (see Note 3). Repeat for each of the other four sets of dilute standards (see Note 4). Plot the OD at 445 nm against the higher alcohol concentration for each set, to give a group of five straight calibration lines. Measure the slopes of the lines as grams of higher alcohols per 100 l of solution per unit of OD. Determine the average ratio of OD at 560 nm to OD at 445 nm for each set of standards. Plot the slopes against the OD ratios.

Estimation of total high alcohols in the sample. Proceed as in the 'blank test' using duplicate 0·10 ml aliquots of prepared sample and of any one of the 25 dilute standard higher alcohol mixtures. Use 0·10 ml of 40 per cent v/v aqueous ethanol solutions as the blank. Calculate the average of the results of each pair of measurements.

CALCULATION

Let A = OD at 445 nm of the prepared sample.

B = OD at 560 nm of the prepared sample.

P = OD at 445 nm of the standard when measured at the same time as the prepared sample.

R = OD at 445 nm of the standard when measured at the time the standard lines were established.

1. Calculate the ratio B/A and from the plot of the slopes in the 'blank test' find the slope corresponding. Let this slope be S.
2. The higher alcohol content of the sample, H, is given by $H = 2{\cdot}5\ (SAR/P)$ g/100 l of ethanol in the sample.

If the original sample contains V per cent v/v of ethanol, where V is less than 40, and ethanol has been added to raise the strength, then $H = 100\ (SAR/VP)$ g/100 l of ethanol in the sample.

Special cases. For some products the proportion of 2-methylpropan-1-ol in the higher alcohols is fairly constant and can be matched by a single set of standards, i.e. the slope, S, can be taken always to be the same. Typical ratios of pentanols to 2-methylpropan-1-ol are: malt whisky, 1·7:1; blended whisky, 1:1; rum, 3:1; cognac brandy, 2·6:1.

NOTES

1. Curved calibration lines may be obtained with filter instruments having a large band width.
2. A satisfactory grade of ethanol is one which gives no perceptible colour when tested according to the procedure for Estimation of High Alcohols in Potable Spirits (Colorimetric Method—for highly rectified spirits) (Method D—also described in the same Report).
3. Cells with a 10 mm light path are satisfactory for measurements at 445 nm. The optical density at 560 nm may be much higher and cells with a light path of 2 mm may be necessary.
4. Any other convenient number of standards may be handled at one time.

An abridged version of the above method for occasional use is described by the Research Committee, who also include in their report four GLC methods for the higher alcohols in spirits. Separate instructions are given in each GLC method for the support, stationary phase, packing, conditioning and the operating conditions (Duncan and Philp, 1966; RCAPS, 1970; Martin *et al.*, 1974; AOAC). Singer (1966) showed that the ratios of the different alcohols were more characteristic than the total higher alcohol content; see also Askew and Lisle (1971), Hogben and Mular (1971), and Barnett and Einsmann (1977).

Tannins. The AOAC method of determining tannins measures spectrophotometrically at 760 nm the reduction of molybdotungstate reagent (Folin–Denis). The result is expressed as tannic acid. A modification suitable for British whiskies has been recommended by Duncan and Philp (1966).

Stability. Warwicker (1960, 1963a, b) has investigated the factors which affect the stability of spirits. The main reasons for instability were found to be partly inherent, partly due to contamination during maturation and partly to faulty filtration. Gin and vodka are not matured, but the necessity of using clean spirit and avoiding prolonged contact with metals is stressed.

Permanganate time (Alcohol for Compound Spirits). Permanganate time is a measure of the oxidisable impurities in alcohol of high quality such as that used in the manufacture of gin and vodka. The results of the test may be expresed as the time taken to reduce the colour of permanganate added, either to a standard level or to remove it completely. Alternatively, it is used as a limit test in which colour must remain for a specified time. The ISO procedure (ISO R 1388; 1970) is based on the reduction of permanganate colour to match that of a colour standard. In their Fifth Report, the Research Committee (RCAPS, 1977) listed the various reaction conditions used by 13 different authorities. They concluded that the method is suitable only as an internal quality control test method.

BEER

Manufacture

Beer is prepared by fermentation at about 15 °C with yeast of a solution of carbohydrates (wort) mainly derived from cereals. The main stages in the brewing process are the preparation of a fermentable extract (sweet wort) from malted barley and other cereal grains, boiling the wort with hops to introduce flavour, fermentation of the hopped wort by yeast, separation of yeast cells from the beer and, finally, maturation and conditioning. The finished beer is distributed in barrels or filled after filtration and pasteurisation into bottles, cans or casks.

Originally, the principal source of carbohydrate was malted (germinated and dried) barley. However, in view of the high enzyme activity and the high protein content of malt it has become the practice to use unmalted cereals and cereal preparations (adjuncts) with low protein content. These practices have allowed brewers to reduce the cost of materials and to reduce the net protein content of the fermentation liquor with consistent improvement in the stability of the finished beer.

A selected yeast, normally *Saccharomyces cerevisiae* or *Saccharomyces carlsbergensis* is used for the fermentation stage. The yeast cells propagate and by utilising sugars, amino acids and other nutrients they produce alcohol and carbon dioxide. In the traditional UK brewing process the yeast aggregates around carbon dioxide gas bubbles and accumulates on the surface of the fermenting wort. Bottom fermentation at a lower fermentation temperature is more commonly practised in Europe and the USA. After fermentation beer is clarified by adding finings containing isinglass which remove the sludge of tannin, protein and yeast.

The main types of malt liquors can be classified as follows:

1. Bitter or pale ale, a high quality product containing the most hops.
2. Mild ale, which is sweeter and contains less bitter.
3. Stout, brewed mainly in Dublin and London, which is dark in colour due to the use of caramelised and black malt.
4. Lager, which is brewed more especially on the Continent by bottom fermentation and tends to contain more extractive and less alcohol than other types.

The history and development of the brewing process has been summarised by the FSC (1977). Current industrial costs and developments have been reviewed (Anon, 1977). Methods used in continuous brewing have been reviewed by Royston (1968). The use of unmalted barley and enzyme preparations have been discussed by Wieg *et al.* (1969). Yates (1979) has described brewing methods and styles of beer produced in various countries throughout the world.

Legislation

The development of legislation relating to beer has been mainly concerned with aspects of taxation. Currently, beer is controlled for Customs purposes by the Alcoholic Liquor Duties Act 1979 as amended SI 1979 No. 241. Beer is defined in section 1(3): 'Beer includes ale, porter, stout and any other description of beer, and any liquor which is made or sold as a description of beer or as a substitute for beer and which on analysis of a sample there at any time is found to be of a strength exceeding 1·2 per cent (v/v at 20 °C), but does not include:

1. Black beer the worts whereof before fermentation were of a specific gravity of 1200° or more; or
2. Liquor made elsewhere than on licensed premises of a brewer for sale which on analysis of a sample at any time is found to be of an original gravity not exceeding 1016° and to be of strength not exceeding 1·2 per cent'.

The materials that yield fermentable carbohydrate which are to be used in the brewing of beer and upon which duty is to be charged, are defined in Section 37. These are: malt or corn sugar, any saccharine substance, extract or syrup; rice; flaked maize (and any equivalent substance); any other substance capable of use in brewing.

Section 3(1) of the Act defines 'gravity' as any liquid with the specific gravity at 20 °C; 'degrees of gravity' as specific gravity multiplied by 1000 and 'original gravity' as the gravity before fermentation. The Commissioners of Customs and Excise retain the right to specify the methods of analysis to be used in the determination. The methods relating to determination of the original gravity of beer are set out in Regulation 11 of the Beer Regulations 1978 (SI 1978 No. 893, amended by SI 1978 No. 1186). This Regulation required that the analysis be carried out at 60 °F but this has been amended by the Alcoholic Liquors (Amendment of Enactments Relating to Strengths and to Units of Measurement) Regulations 1979 (SI 1979 No. 241), which require computation of alcoholic content of any liquor by volume at 20 °C. A new 'spirit indication' table for use in determination of original gravity of worts is included in this regulation.

In its 1977 Report the Food Standards Committee summarise the general legislative provisions for control of beer under the then Food and Drugs Act 1955, and discussed proposals made in drafts of an EC directive; the latter proposals having been withdrawn. The Committee made recommendations that the composition, labelling and advertising of beer should be controlled by regulations made under the Act. A definition of beer was proposed as 'the liquid product of the alcoholic fermentation of a wort prepared wholly or mainly from cereal-derived raw material and sugary raw materials, yeast and drinking water, with or without the addition of hops or bitter materials derived from hops and which contain not less than 2 per cent of alcohol by volume at 20 °C'. (Compare with the Customs definition above.) Proposals were also made in respect of the proportions of malted barley and other cereals present in specified beers, the declaration of strength, conditioning, carbonation, use of descriptive terms and labelling. So far, no legislation has been proposed for implementation of the recommendations.

A review of additives and processing aids used in the production of beer has been carried out by the FACC, Report No. 26, 1978. The committee recommended that adequate specifications should be prepared for all additives used in beer; substances permitted should be those classified by the COT as Group A or B; residues of gibberellic acid should be limited to 0·05 mg/l; total usage of silicates should not exceed 2000 mg/l; and that attempts should be made to find a suitable alternative for asbestos filters. The committee also lists additives used in the brewing process that may leave residues in the final beer.

Composition

Some examples of analytical data given by various types of beers are shown in Table 12.12. The overall ranges of figures for lager quoted in a report of the Consumers' Association (1973) were original gravity 1029°–1081° and 2·9–9·2 per cent alcohol.

Table 12.12 Data obtained from various types of beer

	India pale ale	Mild ale	Bock	Lager beer	Irish stout
Specific gravity	1012·5	1010·1	1020·5	1014·5	1016·0
% Alcohol (m/m)	4·30	3·15	4·50	3·20	4·30
% Total solids	4·95	3·55	6·80	5·38	5·70
% Acidity (as acetic acid)	0·15	0·08	0·12	0·17	0·17
% Ash	0·26	0·20	0·29	0·20	0·23
Original gravity	1053·0	1042·1	1063·5	1046·5	1057·5

Analysis of beer

The estimations commonly required to be carried out on beer are those for specific gravity, alcohol, extract, acidity, preservatives, arsenic and lead. A historical survey of methods has been published by Hudson (1960). The Institute of Brewing Analysis

Committee and the European Brewery Convention recommend appropriate methods for use by the industry.

Sample preparation. For all determinations other than measurement of carbon dioxide the sample is de-gassed by repeated pouring from one large beaker to another. The de-gassed beer is finally filtered through a dry filter paper. An individual, unopened container must be used for determination of carbon dioxide alone; a further sample is used in the other analyses.

Carbon dioxide. The measurement is normally carried out by use of a piercing device attached to a pressure gauge and absorption burette. Different designs are used for examination of bottled and canned beer (AOAC) (see also Gray and Stone, 1938; Barker and Parkinson, 1955; Seidemann and Woop, 1982).

Acidity. Difficulty may arise in the titration of dark beers; in such circumstances use of a pH meter is to be preferred.

Fixed acidity. Evaporate 20 ml prepared sample to dryness on a boiling-water bath and then successively evaporate after adding water. Finally, stir with water and titrate with 0·1 M sodium hydroxide using phenolphthalein as indicator (*F* ml). Calculate the fixed acidity as lactic and acetic acid.

Volatile acidity. Dilute 20 ml prepared sample with water and titrate with 0·1 M sodium hydroxide using phenolphthalein as indicator (*T* ml). Then ($T - F$) ml calculated as acetic acid represents the volatile acidity.

Alcohol. For accurate analysis the distillation procedure is required.

Alcohol. Transfer 75 or 100 ml prepared sample measured at 20 °C to a distillation flask, rinsing in with 30–40 ml water. Distil into the same flask as used for measuring out the original sample and make up the distillate to 75 or 100 ml with water at 20 °C. Determine the relative density at 20 °C (G) preferably using a bottle of pyknometer and estimate the alcohol in the sample (Table 12.2). The **Spirit indication** equivalent to the alcohol content is given by the expression $1000\,(1 - G)$.

Extract gravity. Cool the residue in the distillation flask, dilute to the original volume (again in the same measuring flask) with water at 20 °C and measure the relative density (G_1). Express as degrees $1000\,G_1$.

Original gravity of the wort. The original gravity (OG) of a beer wort is the gravity prior to fermentation. Excise duty on the beer is levied according to the OG of the wort and the method to be used for its determination is prescibed in the Beer Regulations (as amended). In order to calculate the OG it is necessary to determine the extract gravity, the alcohol content (as spirit indication) and the acidity of the beer. If the total acidity exceeds 0·1 per cent as acetic acid the spirit indication derived from the sp. gr. of the distillate is modified according to Table 12.13. The 'gravity lost' equivalent to the corrected spirit indication is then determined by reference to Table 12.14. (See the amended Beer Regulations, 1978.) The original gravity OG is the sum of 'gravity lost' and the extract gravity ($1000\,G$); a worked example is illustrated below:

Example

Relative density of distillate (G)	0·9935
Relative density of extract (G_1)	1·0204
Acidity as acetic acid (%)	0·31
Relative density of water × 1000	1000·0
Relative density of distillate (G) × 1000	993·5
Spirit indication = 1000 − 993·5 =	6·5
Allowance for 0·21 excess acid (Table 12.12)	0·28
Corrected spirit indication	6·78

Table 12.13 Corrections to be made to spirit indication of beer for excess acid

Acetic acid	Corresponding degrees of spirit indication									
	0·00	0·01	0·02	0·03	0·04	0·05	0·06	0·07	0·08	0·09
0·0	—	0·02	0·04	0·06	0·07	0·08	0·09	0·11	0·12	0·13
0·1	0·14	0·15	0·17	0·18	0·19	0·21	0·22	0·23	0·24	0·26
0·2	0·27	0·28	0·29	0·31	0·32	0·33	0·34	0·35	0·36	0·37
0·3	0·39	0·40	0·42	0·43	0·44	0·46	0·47	0·48	0·49	0·51
0·4	0·52	0·53	0·55	0·56	0·57	0·59	0·60	0·61	0·62	0·64
0·5	0·65	0·66	0·67	0·69	0·70	0·71	0·72	0·73	0·75	0·76
0·6	0·77	0·78	0·80	0·81	0·82	0·84	0·85	0·86	0·87	0·89
0·7	0·90	0·91	0·93	0·94	0·95	0·97	0·98	0·99	1·00	1·02
0·8	1·03	1·04	1·06	1·07	1·08	1·09	1·10	1·11	1·13	1·14
0·9	1·15	1·16	1·18	1·19	1·21	1·22	1·23	1·25	1·26	1·28
1·0	1·29	1·31	1·33	1·35	1·36	1·37	1·38	1·40	1·41	1·42

Gravity lost from Table 12·13	30.0
Extract gravity (G_1) × 1000	1020·4
Original gravity of wort (OG)	1050·4

Hudson (1975) has described the method in detail and has reported on a collaborative study.

Essery and Hall (1956) have shown that the OG of a normal beer can be rapidly assessed by measurement of refractive index and the relative density of the prepared sample.

Calculations of this type based on measurements of refractive index and gravity give a less precise measure of the OG than the distillation procedure, but the method is sufficiently accurate for routine control purposes (Olshausen, 1952). Sawyer and Dixon (1968a, b) have devised an automatic sorting method for assessing the OG of beer. It should be noted that all these methods relate to the procedure in use before 1980 when all operations were carried out at 15·5 °C (60 °F). The Spirit Indication Table is that determined by Thorpe and Brown (1914) and included in earlier regulations. Halsey (1985, 1986) has described the use of NIR analysis for the quality control measurements of ethanol and the original gravity of beers. Cutaia (1984) has proposed that analysis by GLC using *n*-propanol as internal standard is an appropriate technique for determination of ethanol. Mason (1983) used an amperometric enzyme electrode system to measure ethanol in beer and other food products. He claimed that the method gives a satisfactory analysis in 2 min. The use of HPLC procedures has also been proposed by Morawski *et al.* (1983) as monitoring methods during fermentation processes.

The following ranges of OG for beers were quoted by the Consumers' Association (1967): light ales 1030·8–1038·7, best pale ales 1040·3–1051·3, strong ales (nip bottles) 1068·2–1077·2, brown ales 1031·0–1037·5, stouts 1043·3–1044·5, lagers 1029·7–1034·4.

It is normally only possible to ascertain if water has been added to beer if the OG as delivered from the brewery is known.

Total solids. Either determine by drying 25 ml at 100 °C in a silica or platinum dish or calculate from the extract gravity (see above):

$$\text{TS (per cent m/v)} = \frac{1000\,(G_1 - 1)}{3{\cdot}86} - 1{\cdot}1 \times \text{per cent Ash.}$$

Table 12.14 Degrees of gravity lost against spirit indications. Beer Regulations (1978) Amended by The Alcoholic Liquors (Amendment of Units and Methods of Measurements) Regulations 1979

Spirit indication	Degrees of gravity	Spirit indication	Degrees of gravity	Spirit indication	Degrees of gravity	Spirit indication	Degrees of gravity
0·0	0·00	4·0	17·22	8·0	35·31	12·0	54·06
0·1	0·42	4·1	17·67	8·1	35·77	12·1	54·55
0·2	0·85	4·2	18·11	8·2	36·22	12·2	55·04
0·3	1·27	4·3	18·56	8·3	36·68	12·3	55·53
0·4	1·70	4·4	19·00	8·4	37·14	12·4	56·03
0·5	2·12	4·5	19·45	8·5	37·59	12·5	56·52
0·6	2·54	4·6	19·90	8·6	38·05	12·6	57·01
0·7	2·97	4·7	20·35	8·7	38·51	12·7	57·51
0·8	3·39	4·8	20·80	8·8	38·97	12·8	58·00
0·9	3·81	4·9	21·25	8·9	39·43	12·9	58·50
1·0	4·24	5·0	21·70	9·0	39·88	13·0	58·99
1·1	4·66	5·1	22·16	9·1	40·34	13·1	59·48
1·2	5·08	5·2	22·61	9·2	40·80	13·2	59·98
1·3	5·51	5·3	23·06	9·3	41·26	13·3	60·47
1·4	5·33	5·4	23·51	9·4	41·72	13·4	60·97
1·5	6·35	5·5	23·96	9·5	42·18	13·5	61·46
1·6	6·78	5·6	24·42	9·6	42·64	13·6	61·96
1·7	7·20	5·7	24·87	9·7	43·10	13·7	62·45
1·8	7·62	5·8	25·31	9·8	43·56	13·8	62·94
1·9	8·05	5·9	25·76	9·9	44·02	13·9	63·44
2·0	8·47	6·0	26·21	10·0	44·48	14·0	63·97
2·1	8·91	6·1	26·66	10·1	44·94	14·1	64·47
2·2	9·35	6·2	27·11	10·2	45·41	14·2	64·96
2·3	9·79	6·3	27·57	10·3	45·88	14·3	65·46
2·4	10·23	6·4	28·02	10·4	46·35	14·4	65·96
2·5	10·66	6·5	28·48	10·5	46·82	14·5	66·46
2·6	11·10	6·6	28·93	10·6	47·29	14·6	66·96
2·7	11·53	6·7	29·39	10·7	47·77	14·7	67·47
2·8	11·97	6·8	29·84	10·8	48·24	14·8	67·97
2·9	12·40	6·9	30·29	10·9	48·72	14·9	68·47
3·0	12·84	7·0	30·75	11·0	49·20	15·0	68·97
3·1	13·27	7·1	31·20	11·1	49·68	15.1	69·48
3·2	13·71	7·2	31·66	11·2	50·16	15·2	69·98
3·3	14·15	7·3	32·12	11·3	50·65	15·3	70·49
3·4	14·58	7·4	32·57	11·4	51·13	15·4	70·99
3·5	15·02	7·5	33·03	11·5	51·62	15·5	71·50
3·6	15·46	7·6	33·48	11·6	52·10	15·6	72·01
3·7	15·90	7·7	33·94	11·7	52·59	15·7	72·51
3·8	16·34	7·8	34·40	11·8	53·08	15·8	73·02
3·9	16·78	7·9	34·85	11·9	53·57	15·9	73·53
						16·00	74·04

Ash. Determine by igniting the total solids at as low a temperature as possible.

Chloride. Evaporate 25 ml beer with 1 g chloride-free lime, ignite at a low temperature and titrate the residue with 0·1 M silver nitrate.

Bitter substances. Earlier methods involved acidifying the beer and extracting the bitter substances with organic solvents such as chloroform, ether and benzene. Later procedures have been based on the more specific determination of the hop substances responsible for the bitterness such as isohumulones. These have employed, for example, spectrophotometry (Rigby and Bethune, 1955), TLC (Aitken *et al.*, 1967), paper, column and gas chromatography (Stocker, 1961), and

HPLC (Anderson 1983). An International Bitterness unit has been approved by the European Brewery Convention (Bishop, 1967, 1968a, b).

Flavour components. The volatile constituents of wort and beer have been the subject of considerable research. Fusel oils are major contributors to flavour (Morgan, 1965). Tressl *et al.* (1978) have studied hop aroma constituents in lager beers by capilliary GC coupled with mass spectrometry. Diacetyl is recognised as a major contributor to the flavour of beers and the determination by colorimetric and GLC methods has been the subject of study by many workers. West (1978) has reported on collaborative studies of three methods recommended by the American Society of Brewing Chemists. The colorimetric method relies on reaction with *o*-phenanthroline to yield 2,3-dimethyl-chinoxaline, which is measured at 340 nm. Buijten and Holm (1979) have reported on an automated technique for colorimetric determination of diacetyl following continuous flow distillation. General headspace sampling techniques for examination of flavour components by GLC have been described by Williams and Strauss (1977) and by Takeoka and Jennings (1984).

Tannins. Certain tannins appear to have some association with beer stability. Owades *et al.* (1958) devised a procedure involving the spectrophotometric examination of a methanolic solution of the tannins at 270 nm, following successive extractions with iso-octane and ethyl acetate (Trommsdorff, 1962). Roston and Kissinger (1981) used HPLC.

Colour. The colour of beer can be assessed by measuring the extinction at 465 nm (Bishop, 1966). The AOAC recommends measurement at 430 nm, with measurement at 700 nm to check for clarity of the filtered sample. In brewing laboratories beers are also examined for nitrogen, foam characteristics, esters (Hudson, 1960) and diacetyl (Owades and Jakovac, 1963).

Preservatives. Beer may contain up to 70 mg/kg of sulphur dioxide, which can be determined by the *p*-rosaniline method (AOAC). The Preservatives Regulations also permit beer to contain up to 70 mg/kg of either benzoic acid or methyl, ethyl or propyl 4-hydroxybenzoate. The methods described in Chapter 4 may be used.

Trace metals

Arsenic. In the early part of the century numerous instances were reported of beer contaminated with arsenic due to the use of arsenical fuel for drying the malt or hops, and arsenical pyrites in the sulphuric acid used in the manufacture of the glucose. The Arsenic in Food Regulations prescribe a maximum for arsenic in alcoholic beverages of 0·2 p.p.m. For routine purposes arsenic can be determined by placing the sample directly into the bottle prior to the production of arsine (Chapter 5).

Lead. Traces of lead are liable to be present in beer if it has been in contact with the metal and the FSC Report on Lead recommended in 1954 that the use of lead pipes and lead containers should be discontinued in all breweries, bottling plants and licensed premises. The determination of lead is described in Chapter 5.

Copper is commonly used in brewery plant and methods for its determination in beer have been compared by Brenner *et al.* (1963) (see also Chapter 5).

Aluminium is used for beer cans and its significance and determination is discussed by Stone *et al.* (1963).

Tin. Beer is also packed in tinned cans and amounts of tin in beer of 0·02 p.p.m. may give rise to haze and a concentration above 0·1 p.p.m. may be objected to on aesthetic grounds. A rapid method for its determination employing cathode ray polarography has been described by Rooney (1963).

Zinc in beer can be determined by the spectrophotometric method described by Stone (1962), or by atomic absorption spectrophotometry (see Chapter 5).

Iron. The presence of iron in beer may give rise to haze or undesirable flavours. It can be determined by the *o*-phenanthroline colorimetric method (p. 299).

Cobalt has been used as a froth stabiliser at levels up to 1 mg/kg. It can be determined spectrophotometrically or by atomic absorption spectrophotometry.

Spoilt beer. Any beer spoilt during manufacture or distribution may be eligible for refund of excise duty paid. Examination includes a determination of acidity due to secondary 'acetic' fermentation; a procedure for this analysis has been described by Sawyer *et al.* (1970) and modified by Williams *et al.* (1982). Changes in flavour have been discussed by Laws and Peppard (1982).

SHANDY

Canned and bottled shandy usually consists of a mixture of two-thirds lemonade and one-third beer and as it contains less than 1·2 per cent v/v alcohol is therefore outside the provisions of the licensing law. The Labelling of Food Regulations prescribe that products described as 'shandy', 'cider shandy', 'shandygaff' (with ginger beer), 'cider and ginger beer', etc., should contain at least 0·9 per cent v/v alcohol. Section 73 of the Alcoholic Liquor Duties Act 1979 also controls the mis-labelling of goods of this type as beer. In the Second Report on Food Labelling the FSC (1979) recommended that products such as lager and lime should be brought within the minimum alcohol requirements that applies to prepacked shandies. Specific provisions also apply in the case of drinks described as 'alcohol-free' (namely ≯ 0·5 per cent by volume) and 'dealcoholised' (namely ≯ 0·5 per cent by volume) unless the strength is indicated as a maximum level in conjunction with the words 'not more than'.

CIDER AND PERRY

Manufacture

Cider is the product obtained from the fermentation of apple juice or possibly from that of apples and pears. In the UK suitable apples with a relatively high tannin content (classed as bittersweet, bittersharps, sharps and sweets) are grown in several south-western counties. The expressed juice undergoes controlled yeast fermentation at 15 °C for 3 weeks. A dominant yeast species is sometimes used rather than less desirable ones indigenous to the fruit (Bowen and Beech, 1964). The cider is blended (and possibly diluted) particularly in relation to acidity (≡ 0·5 per cent malic acid) preferably prior to storage. The carbonated cider is finally sterilised by filtration or flash pasteurisation at 88 °C. Various aspects of cider manufacture have been discussed by Buckle (1967), including compositional data of both apple juice and cider. Perry is derived mainly from pear juice.

Legislation

The manufacture of cider and perry is controlled by the Cider and Perry Regulations 1976 (SI 1976 No. 1207). Cider is also defined (with perry) in the Alcoholic Liquor

Duties Act 1979 as a liquor of a strength less than 8·7 per cent alcohol obtained by fermentation of apple or pear juice without the addition at any time of alcoholic liquor or any other colouring or flavouring substance.

The Labelling of Food Regulations permit the use of the word 'vintage', used in or in conjunction with the name of the drink, provided the apples or pears were harvested in a designated 'vintage year'. In the second report on Food Labelling the FSC (1979) recognised that the control on the word 'champagne' in connection with cider and perry be deleted since EC regulations would control use of the name. In the first Report on Labelling the committee recommended that it should be permissible to describe products with a strength in excess of 8·7 per cent alcohol as 'apple wine'. No explicit provision appears to have been made to meet such a case.

Composition

Most cider samples give figures that fall within the following ranges: sp. gr. 1·004–1·002, acidity 0·32–0·50 per cent (as malic acid), tannin 0·18–0·27 per cent, alcohol 4–8 per cent v/v (higher figures given by champagne ciders), total solids 2·4–8·0 per cent, ash 0·20–0·35 per cent, alkalinity of ash 0·026–0·127 per cent (as K_2CO_3), phosphate 0·013–0·023 per cent (as P_2O_5).

Analysis

The methods of analysis used for cider are similar to those described for beer and fruit juices; examination normally includes estimation of preservatives. Tannin may be determined colorimetrically at 760 nm by the reaction with Folin–Denis reagent (p. 450, see AOAC). In recent years much work has been carried out on the fusel oils and flavour constituents of ciders. Pollard *et al.* (1965) examined the distillates of cider and perry finding 2- and 3-methyl-1-butanol within the range 74–142 mg/l in cider, 86–104 mg/l in perry and 2-methyl-1-propanol within the range of 14–82 mg/l in cider and 36–43 mg/l in perry. Williams *et al.* (1978) and Williams and Tucknot (1978) used Porapak Q for pre-concentration of flavour volatiles in cider followed by GC/MS identification of a wide range of substances present in the volatile fraction. Whiting and Coggins (1975) have described a method for separation and determination of phenolic constituents in cider. They reported widely varying amounts of 5-chlorogenic acid, 4-*p*-coumarylquinic acid and catechins according to the cultivar used in production of the cider.

VINEGAR

Manufacture of malt vingar

The brewing operations fall into three main stages involving mashing, fermentation and acetification. The preliminary stages of manufacture are similar to those for brewing of beer. During mashing, malted barley and other cereals are milled and mixed with hot water in a 'mash tun'. The mixture is allowed to stand for a period of time during which the starches are converted to sugars by the diastase enzymes of the malted barley. This process is allowed to proceed until the starch has been substantially converted to sugars. The sweet liquor is drained from the spent grain through a perforated false bottom in the mash tun. In the fermentation process, the sweet liquor from the mash tun is collected in a fermentation vessel and a culture of an appropriate type of yeast is added. Fermentation results in the conversion of the fermentable sugars to ethyl alcohol and carbon dioxide and continues until it is substantially complete. This is the first fermentation referred to in the Kat v. Diment judgment of 1950. The yeast mass is then separated from the alcoholic liquor which, after a storage period, is passed to the next process stage, acetification. This stage involves the oxidation of the alcohol in the liquor to acetic acid by atmospheric oxygen under the agency of micro-organisms of the genus *Acetobacter*. The process is carried out in specially constructed vessels called acetifiers, designed to allow the high degree of aeration necessary for maximum efficiency of acetification. This is the 'second fermentation' referred to in the Kat v. Diment judgment. A final filtration process is carried out before pasteurisation to produce a clear vinegar, which is then allowed to mature.

As produced for the retail market the acidity is about 5 per cent as acetic acid. Patented processes have applied the technique of batch deep culture on a continuous basis. By controlling the feeding of substrate into the acetifier so that it matches the oxidation rate an almost complete conversion to acid can be achieved (White, 1966). The effect of such processes on the composition has been discussed by Morgan and Voelcker (1963). Apart from such advances, enzyme systems derived from fungi such as *Aspergillus* and *Rhizopus* have been proposed for saccharifying the starch. Such systems were accepted by the FSC (1971) together with methods for converting starch using mineral acid as used in glucose syrup manufacture.

Distilled malt vinegar is a colourless liquid prepared by distilling malt vinegar. The product only contains the volatile constituents of the vinegar from which it is derived. It is commonly used in the manufacture of pickled onions.

Spiced vinegars such as those of tarragon, garlic and elder are prepared by steeping the leaves or spices in an ordinary vinegar. The presence of the flavouring material does not appear to affect the analytical figures.

Spirit vinegar, also referred to as white or alcohol vinegar, is produced by the oxidation of an alcoholic fluid. The latter is derived by yeast fermentation of a substrate such as molasses, which is distilled prior to the oxidation. It is diluted to an acidity of 4–5 per cent for use in pickles.

Wine vinegar is mainly produced on the Continent. The minimum acetic acid content applied in the EC is 6 per cent. It is prepared by the acetous fermentation of grape juice or of white or red wines and has a characteristic aroma. Wine vinegars usually contain 0·1–0·4 per cent potassium acid tartrate and small amounts of alcohol and reducing sugars.

Cider vinegar is not as common in Britain as in the USA. It is laevorotatory, yields a strongly alkaline ash rich in potassium and often gives a very high oxidation value. Malic acid is not necessarily present. Cider vinegar often contains added caramel.

Non-brewed condiment. Artificial non-brewed vinegars usually contain about 4·5 per cent m/v of synthetic acetic acid and are coloured with caramel. In 1950 the High Court of Justice ruled in the Kat v. Diment appeal case that the description non-brewed as applied to a solution of synthetic acetic acid was a false trade description and an offence against the Merchandise Marks Act. The trade subsequently decided to describe such artificial products as 'Non-Brewed Condiment' (see also Circular MF 15/50). The minimum of 4 per cent acetic acid applies to non-brewed condiment as well as to brewed vinegar.

Concentrated solutions of acetic acid. Non-brewed condiments are often prepared from 80 per cent acetic acid or vinegar essence or a concentrated non-brewed condiment containing added caramel.

Compositional standards

The agreement between the SPA and the Malt Vinegar Brewers' Federation in the 1930s that all vinegars including artificial products should contain at least 4 per cent w/v has invariably been accepted by the courts. The standard was later confirmed by the FSC (1971), who recommended the following definitions (cf. FSC, 1966).

Vinegar is the liquid produced from a suitable raw material containing starch or sugar or starch and sugar by the process of double fermentation, alcoholic and acetous, and which contains at least 4 per cent w/v acetic acid.

Malt vinegar is the vinegar produced without intermediate distillation by the process of double fermentation, alcoholic and acetous, from malted barley, with or without the addition of cereal grain, the starch of which has been converted to sugars solely by the diastase of the malted barley.

Grain vinegar is the vinegar produced without intermediate distillation by the process of double fermentation, alcoholic and acetous, from any cereal grain the starch of which has been converted to sugars by a process other than solely by the diastase of malted barley.

Spirit vinegar is the vinegar made by the acetous fermentation of the alcoholic distillate from the product of alcoholic fermentation of suitable raw material solutions containing sugars. Thus the term 'spirit vinegar' must not be applied to any product obtained by the acetous fermentation of synthetic alcohol.

The Preservatives in Food Regulations permit cider or wine vinegar to contain up to 200 mg/kg of sulphur dioxide. Other types of vinegar may only contain up to 70 mg/kg of sulphur dioxide. Both legal and commercial standards for vinegar are discussed by White (1971).

Analysis of vinegars and non-brewed condiments

The main types of examination that may be required to be carried out on samples of vinegar and artificial products are given below. In order to check compliance with legal and other requirements it is advisable to examine samples for the following: **total acidity, ash, preservatives, arsenic, copper, lead, zinc, alkaline oxidation value, iodine value** and a qualitative test for **mineral acid**. Analytical figures that were accepted for many years as typical for various types of vinegar are shown in Table 12.15. In general, modern methods of production tend to produce rather lower values and these are reflected in the results for malt vinegars purchased between 1960 and 1968 and quoted in Table 12.16. Morgan and Voelcker (1963) quoted the following figures for malt vinegars produced by 'modern' high yield proceses: oxidation value 505–1770, iodine value 180–1600, ester value 72–152, TS 0·8–2·0 per cent, ash 0·12–0·37 per cent, N 434–791 mg/l (see also Paul and Southgate, 1978 for data on mineral constituents).

Organoleptic examination. The odour and taste are useful for assessing the type of vinegar. The colour can be assessed spectrophotometrically. Any deposit should be examined for vinegar eels (*Anguillula aceti*), which closely resemble potato eelworms.

Table 12.15 Analytical figures of vinegars and non-brewed condiment (various sources)

	Malt vinegar	Non-brewed condiment	Cider vinegar	Wine vinegar	Spirit vinegar (concentrated)
Relative density	1·013–1·022	1·007–1·022	1·013–1·024	1·013–1·021*	1·015–1·020
Total acidity as acetic acid (%)	4·3–5·9	4·1–5·3	3·9–9·0	4·4–7·4*	11·5–12·2
Fixed acidity as acetic acid (%)	0·2–0·4	Negligible	0·1–0·2	—	—
Malic acid (%)	—	—	0·07–0·16	—	—
Total solids (%)	1·4–3·5	0·10–0·45	1·9–3·5	1·4–3·2*	0·15–0·60
Total ash (%)	0·18–0·45	0·02–0·05	0·20–0·45	0·15–0·69*	0·02–0·05
Alkalinity of ash (ml 0·01 M/ml)	—	—	2·2–4·0	—	—
Total nitrogen (%)	0·04–0·14	0–0·2	—	—	0·003–0·03
Phosphoric acid as P_2O_5 (%)	0·05–0·12	0–0·02	0·04–0·30	—	0·005
SO_3 (%)	0·05–0·12	0–0·01	—	—	—
Sodium chloride (%)	0·15–0·25	0·01–0·12	—	—	—
Total sugars (%)	—	—	0·15–0·7	0·22–0·56*	—
Alcohol (% v/v)	—	—	0·03	—	0·15
Oxidation value	500–1800	0–20	Up to 3500	600–2000	90–650†
Alkaline oxidation value	70–180	0–10	—	60–180	3–20†
Iodine value	380–1500	0–25	—	380–1000	5–30†
Ester value	30–140	0–15	—	50–220	0–20†

* Analysis by the Paris Municipal Laboratory.
† Calculated to 5% acetic acid.

Table 12.16 Malt vinegar analytical figures obtained from 69 samples purchased from 1960–1968

	Mean	Min.	Max.
Total solids (% m/v)	1·36	0·30	2·84
Ash	0·22	0·06	0·76
Total acidity (% m/v) (as acetic acid)	4·93	4·05	6·91
Formol titration (0·1 M alkali/10 ml)	0·85	0·6	2·7
Oxidation value	789	382	1460
Alkaline oxidation value	112	46	255
Iodine value	595	164	1328
Ester value	91	22	203

Relative density. A vinegar containing 5 per cent acetic acid should have a relative density of about 1·019.

Total acidity. Dilute 10 ml of sample with water in a porcelain basin, add phenolphthalein and titrate with 0·5 M sodium hydroxide, stirring constantly.

$$\text{Total acidity} = \text{Titration} \times 0{\cdot}3 = \text{per cent m/v as acetic acid.}$$

The total acidity should not be less than 4 per cent. Good quality products usually contain at least 5 per cent. The activities of micro-organisms and vinegar eels cause a fall in the acidity.

Volatile acidity. Steam distil 10 ml sample in a semi-micro Kjeldahl distillation apparatus and titrate the distillate with 0·5 M sodium hydroxide using phenolphthalein. Alternatively, evaporate 10 ml sample at least five times with water, titrate the fixed acidity and calculate the volatile acidity as acetic acid by difference. Ashoor and Welty (1984) have used HPLC to determine acetic acid.

Total solids. Evaporate 25 ml sample in a platinum dish and dry to constant weight in an oven at 100 °C. This volatile acid, which tends to remain partially in the total solids, can be removed by three evaporations with water.

Ash. Ignite the total solids at as low a temperature as possible. If a high ash figure is obtained, salt should also be determined.

Nitrogen. Determine by the Kjeldahl method on 25 ml of sample. The formol titration can also be applied.

Phosphate. Determine on the ash or preferably after wet oxidation.

Metals. Examine the sample for arsenic, copper, lead and zinc.

Mineral acid. Traces of sulphate are usually present in vinegar but if the amount exceeds 0·03 per cent (as H_2SO_4) it gives rise to some suspicion. Added mineral acid can be detected by the following simple test:

Mix 2 ml sample with 2 ml alcohol and add 2 drops of methyl orange. A red colour indicates a low pH due to added acid. Other indicators that may be employed are methyl violet and metanil yellow. The pH of products containing 4 per cent acetic acid seldom falls below 2·9 for a malt vinegar and 2·5 for an artificial product.

The presence of traces of mercury is said to be strongly indicative of the presence of commercial 80 per cent acetic acid. Added sulphuric acid can be estimated by Hehner's method, which is based on the fact that the ash of genuine vinegar is alkaline in reaction. The presence of added mineral acid is improbable if the ash is alkaline to litmus.

Preservatives. Brewed products should be examined for sulphur dioxide and sorbic acid. Borate has on occasion been detected in vinegar.

Caramel. Shake 100 ml sample with 50 ml ether and apply Fiehe's resorcinol test to the residue obtained after allowing the ether to evaporate. Care must be taken in interpreting the results as furfural is sometimes naturally present as in cider vinegar (see Chapter 6).

Methods for differentiating between different types of vinegar

Apart from the taste and odour the figures for nitrogen, phosphoric acid and ash give some indication of the type of vinegar. A reasonably good quality malt vinegar derived from malted or unmalted barley gives figures for nitrogen and phosphoric acid (as P_2O_5) of not less than 0·05 per cent each. Where rice has been used in conjunction with malt the nitrogen figure tends to be considerably higher than the phosphoric acid. The total ash figure also gives some indication as to the starting material, e.g. products from molasses have a low ash and grain converted by acid hydrolysis leads to a high ash figure. The formol titration represents a useful rapid method for deciding whether a vinegar is brewed or not.

The most useful method for routine purposes for differentiating between the various types of vinegar are based on the values obtained on the volatile substances present, i.e. the oxidation, alkaline oxidation, iodine and ester values (Tables 12.15 and 12.16). Brewed vinegars give comparatively high values, but those for artificial products are low as the latter are almost devoid of volatile reducing substances. Morgan and Voelcker (1963) indicated, however, that the newer methods of production of malt vinegar involving more efficient acetification have tended to cause a change in previously accepted values.

It was formerly considered that the presence of malic acid was a useful criterion for cider vinegar. However, malic acid is likely to disappear by way of malo-lactic fermentation, either concurrently with yeast fermentation or during storage. This gives rise to lactic acid, which would be readily oxidised by the various *Acetobacter* species during acetification. Normally, the malic acid has disappeared after yeast fermentation and is most unlikely to remain after acetification. In general, other substances peculiar to apple juice also tend to break down and there are no known tests available for differentiating cider vinegars from such criteria. From the normal routine anlyses it may be noted that cider vinegars often tend to give comparatively high figures for total acidity, the formol titration, oxidation value and sorbitol (Table 12.15).

Vinegars can be more specifically identified by use of gas chromatographic evidence.

In addition to the substances mentioned above, brewed vinegars contain significant proportions of vitamin B_1, riboflavine, nicotinic acid, pantothenic acid and pyridoxine, all of which are absent in artificial products.

A bibliography on the literature covering the analysis and adulteration of vinegar has been compiled by Nagarathnamma and Siddappa (1964). Kearsley and Gibson (1981) and Kearsley (1981) reviewed recent proposals for methods of differentiation of brewed vinegars from non-brewed condiments. They concluded that measurements of osmotic pressure and freezing point depression (FPD) on the distillate from neutralised samples give a significant differentiation of non-brewed condiment from conventional vinegars.

The range of values of FPD found for non-brewed condiment was 0–0·007 °C, whilst the corresponding values for the different types of brewed vinegar varied from 0·020 to 0·175 °C. Characterisation of the type of brewed vinegar was not possible using FPD. Examination of the ultraviolet absorption spectra of distillates of malt, cider and wine vinegars also showed differences between these and spirit vinegar and non-brewed condiment.

Chemical characteristics

Formol titration. In general, brewed vinegars give a definite formol titration but artificial products do not. The method is therefore useful as a rapid sorting test, but it is not specific for differentiating between brewed and non-brewed products as distilled vinegars do not give a formol titration.

Add phenolphthalein to 10 ml sample, *nearly* neutralise the acid with 0·5 M sodium hydroxide and then make *exactly* neutral with 0·05 M sodium hydroxide. Add 5 ml formalin (previously made *exactly* neutral to phenolphthalein), mix well and after standing for 5 min, titrate the acidity produced with 0·05 M or 0·1 M sodium hydoxide. Most brewed vinegars give a formol titration equivalent to 0·5–3·0 ml of 0·1 M NaOH on a 10 ml sample. Artificial products and distilled vinegars give a negligible formol titre.

Albuminoid ammonia. Mitra (1953) showed that the albuminoid ammonia value (commonly determined in water analysis) can be used for differentiating between malt vinegars and artificial products. There is a broad correlation between the total nitrogen and the albuminoid ammonia and the latter figure is very high when compared with those obtained with artificial products:

	Albuminoid ammonia p.p.m. (as N m/v)	Total nitrogen p.p.m. (m/v)
Malt vinegars	208–400	440–840
Artificial condiments	0–4·0	11–34

Examination of distillate. Methods for differentiating between various types of vinegar based on the oxidation, iodine and ester values have been usefully developed by Edwards and Nanji (1938). The modified distillation technique described below is that due to Illing and Whittle (1939). The alkaline oxidation value method of Lyne and McLachlan (1946), which may have certain advantages over the other values, can also be determined on the same distillate.

Whitmarsh (1942) studied the individual volatile reducing substances that affect the oxidation and iodine values and found that alcohol and acetyl methyl carbinol appear to be mainly responsible for the distinctive values obtained with malt vinegar.

DISTILLATION

Distil 60 ml of sample from a 350 ml flask fitted with a small tap funnel. When 45 ml of distillate has come over add 15 ml water to the flask down the tap funnel and distil a further 15 ml to give a total volume of distillate of 60 ml.

Oxidation value is the number of millilitres of 0·002 M potassium permanganate used by 100 ml of sample in 30 min under standard conditions.

To a 250 ml glass-stoppered bottle add 5 ml of distillate from a malt or wine vinegar (or 10 ml from a dilute spirit vinegar or artificial product), 10 ml dilute sulphuric acid (1 + 3) and exactly 15 ml of 0·02 M potassium permanganate. Allow to stand at about 18 °C for 30 min and then add 5 ml of 10 per cent potassium iodide solution. Titrate the liberated iodine with 0·02 M thiosulphate (a ml) using starch near the end-point. Carry out a blank at the same time (titration b ml).

$$\text{If 10 ml distillate used, oxidation value} = 20\,(b - a)$$
$$\text{If 5 ml distillate used, oxidation value} = 40\,(b - a)$$

Alcohol and acetyl methyl carbinol are the principal contributors to the oxidation value.

Alkaline oxidation value is the number of parts by weight of oxygen required to oxidise 100 000 parts of sample under standard conditions.

To a 250 ml glass-stoppered bottle add 2 ml of distillate, 100 ml water, 10 ml of 10 per cent sodium hydroxide solution and exacly 10 ml of 0·02 M potassium permanganate. Allow to stand for 30 min and then acidify with 10 ml dilute sulphuric acid (1 + 3). Add 0·5 g of potassium iodide and titrate the liberated iodine with 0·02 M thiosuphate (a ml) using starch near the end-point. Carry out a blank at the same time (titration b ml).

$$\text{Alkaline oxidation value} = 8\,(b - a)$$

About a minute after adding the permanganate a deep green colour develops with malt and wine vinegars, but with spirit vinegars and artificial products the liquid remains a pink or purple colour. Under the conditions of the determination Morgan and Voelcker (1963) have pointed out that acetyl methyl carbinol and diacetyl give identical values and are more powerful reducing agents than alcohol or ethyl acetate. Alcohol can however make a significant contribution to this value.

Iodine value is the number of millilitres of 0·01 M iodine absorbed by 100 ml of sample under standard conditions.

To a 250 ml glass-stoppered bottle add 5 ml of distillate from malt vinegar (or 10 ml from a spirit vinegar or artificial product) and make just neutral to litmus with 10 M potassium hydroxide. Add 10 ml M potassium hydroxide and exactly 10 ml 0·1 M iodine. Allow to stand in the dark for 15 min and then add 10 ml dilute sulphuric acid (1 + 3) and titrate the liberated iodine with 0·02 M thiosulphate (*a* ml) using starch near the end-point. Carry out a blank at the same time (titration *b* ml).

$$\text{If 10 ml distillate used, iodine value} = 20\,(b - a)$$
$$\text{If 5 ml distillate used, iodine value} = 40\,(b - a)$$

Iodine values are mostly influenced by the presence of acetyl methyl carbinol and diacetyl as the concentrations of alcohol and ethyl acetate in vinegar are insufficient to affect the results.

Ester value is the number of millilitres of 0·01 M potassium hydroxide required to saponify the esters contained in 100 ml of sample under standard conditions.

To a flask (suitable for fitment to a reflux condenser) add 25 ml of distillate and make just alkaline with M potassium hydroxide to phenolphthalein. Just remove the pink colour by the dropwise addition of 0·02 M hydrochloric acid. Then add exactly 10 ml 0·1 M potassium hydroxide and heat under reflux for 2 h on a boiling-water bath to saponify the esters. Cool and titrate with 0·02 M hydrochloric acid (*a* ml) after adding additional phenolphthalein and, if necessary, 100 ml water (e.g. with malt vinegar, which tends to produce a reddish-brown colour that obscures the end-point unless the liquid is diluted). Carry out a blank at the same time commencing with 25 ml water instead of the distillate (blank titration = *b* ml).

$$\text{Ester value} = 8\,(b - a)$$

Although alcohol gives no ester value, acetyl methyl carbinol and diacetyl give figures as well as ethyl acetate itself.

Pearson and Ntiforo (1972) examined the possibilities of omitting the distillation step in determining the above values. This was achieved by taking the differences between the figures obtained (1) directly on the sample and (2) after evaporating several times and making up to the original volume. It was shown that using this technique the alkaline oxidation value was the most satisfactory for distinguishing between malt vinegar and non-brewed condiment. The contribution to the values of any caramel present is minimal. For wine vinegars iodine values appear to give a better differentiation than other figures using the non-distillation procedure (cf. Mecca, 1964).

Gas chromatography. Kahn *et al.* (1966) found that using GLC it was possible to resolve alcohols, esters, acids and 3-hydroxy-2-butanone. They compared data obtained by analysis of cider vinegars and a range of commercial brands. The technique has also been used by Morgantini (1962) with particular reference to the determination of ethanol, the presence of which is especially significant in spirit vinegar. Soumalainen and Kangasperko (1963) using GLC have confirmed that, apart from ethyl acetate, constituents such as diacetyl, various isomeric butanols and pentanols and acetoin occur in wine vinegars but not spirit vinegar. Also using GLC, Aurand *et al.* (1966) identified a total of 25 volatile components from cider, wine and tarragon vinegars. Also 11 volatile components were identified from five samples of distilled vinegar. Acetaldehyde, acetone, ethyl acetate and ethyl alcohol were found in all the vinegars examined. Hundley (1968) examined wine and cider vinegars for sorbitol. None was found in authentic wine vinegar but cider vinegar was found to contain amounts varying from 1120 to 6380 mg/l.

REFERENCES

Aitken, R. A., Bruce, A., Harris, J. O., Seaton, J. C. (1967) *Journal of the Institute of Brewing*, **73**, 528.
Amerine, M. A. (1954) *Advances in Food Research*, **5**, 430.
Amerine, M. A. & Ough, C. S. (1988) *Methods for the Analysis of Musts and Wines*, 2nd edn. New York: Wiley Interscience.
Anderson, B. M. (1983) *Journal of Chromatography*, **262**, 448.
Anon (1972) *Report of the Government Chemist*, p. 104. London: HMSO.
Anon (1977) *British Food Journal*, **79**, 197.
Anon (1979) *Methods of Analysis of Potable Spirits*. London: Laboratory of the Government Chemist.
Ashoor, S. H. & Welty, J. (1984) *Journal of the Association of Official Analytical Chemists*, **67**, 8855.
Askew, B. & Lisle, D. B. (1971) *Journal of the Science of Food and Agriculture*, **22**, 102.
Aurand, L. W., Singleton, J. A., Bell, T. A. & Etchells, J. L. (1966) *Journal of Food Science*, **31**, 172.
Avellini, P., Damiani, P. & Burini, G. (1977) *Journal of the Association of Official Analytical Chemists*, **60**, 536.
Barker, C. J. & Parkinson, T. L. (1955) *Journal of the Institute of Brewing*, **61**, 512.
Barnett, J. H. & Einsmann, J. R. (1977) *Journal of the Association of Official Analytical Chemists*, **60**, 297.
Beaud, P. & Ramuz, A. (1978) *Mittilungen aus dem Gebiete der Lebensmittel-untersuchung und Hygiene*, **69**, 423.
Bishop, L. R. (1966, 1967, 1968a, b) *Journal of the Institute of Brewing*, **72**, 443; **73**, 525; **74**, 252, 254.
Blessinger, D. T., La Roe, E. G. & Conner, H. A. (1965) *Journal of the Association of Official Analytical Chemists*, **48**, 1126.
Boruff, C. S. (1959) *Journal of the Association of Official Analytical Chemists*, **42**, 331.
Bowen, J. F. & Beech, F. W. (1964) *Journal of Applied Bacteriology*, **27**, 333.
Brandt, B. S. (1964) *Laboratory Practice*, **13**, 717.
Brenner, M. W., Mayer, M. J. & Blick, S. R. (1963) *Proceedings of the American Society of Brewing Chemists*, 165.
Brunelle, R. L., Schoeneman, R. L. & Martin, G. E. (1967) *Journal of the Association of Official Analytical Chemists*, **50**, 329.
Buckle, F. J. (1967) *Process Biochemistry*, **2**, 37.
Buijten, J. C. & Holm, B. (1979) *Journal of Automatic Chemistry*, **1**, 91.
Burroughs, L. F. & Sparks, A. H. (1961) *Analyst*, **86**, 381.
Caputi, A. & Wright, D. (1969) *Journal of the Association of Official Analytical Chemists*, **52**, 85.
Clutton, D. W. & Evans, M. B. (1978) *Journal of Chromatography*, **167**, 409.
Cochrane, G. C. (1986) *Analytical Proceedings*, **23**, 357.
Consumers' Association (1967, 1973) *Which?* p. 365; 85.
Cooke, J. R. (1974) *Analyst*, **99**, 306.
Court, R. E. & Bowers, V. H. (1970) *Process Biochemistry*, **5**, 17.
Cutaia, A. J. (1984) *Journal of the Association of Official Analytical Chemists*, **67**, 192.
Dickes, G. J. (1979) *Talanta*, **26**, 1065.
Dickes, G. J. & Nicholas, P. V. (1976) *Gas Chromatography in Food Analysis*. London: Butterworth.
Dixon, B. E., Sexton, G. E. W. & Mayne, T. G. (1963) *Chemistry and Industry* (47), 1873.
Duncan, R. E. B. & Philp, J. M. (1966) *Journal of the Science of Food and Agriculture*, **17**, 208.
Dupont, P. (1978) *Annales Nutrition Aliment*, **32**, 905.
Edwards, F. W. & Nanji, H. R. (1938) *Analyst*, **63**, 410.
Essery, R. E. & Hall, R. D. (1956) *Journal of the Institute of Brewing*, **62**, 153.
FSC (1966) *Food Standards Committee Report on Claims and Misleading Descriptions*, London: HMSO.
FSC (1971) *Food Standards Committee Report on Vinegars*. London: HMSO.
FSC (1977) *Report on Beer*, FSC/REP/68, London: HMSO.
FSC (1979) *Second Report on Food Labelling* FSC/REP/69. London: HMSO.
Geiss, E. Kupka, H. & Nestler, V. (1979) *Zeitschrift fur Analytische Chemie*, **295**, 148.
Gibson, T. D. & Woodward, J. R. (1986) *Analytical Proceedings*, **23**, 360.
Goswell, R. W. (1968) *Process Biochemistry*, **3**, 47.
Gray, P. P. & Stone I. (1938) *Industrial and Engineering Chemistry, Analytical Edition*, **10**, 15.
Guymon, J. F. & Crowell, E. A. (1963) *Journal of the Association of Official Analytical Chemists*, **46**, 276.
Guymon, J. F. & Crowell, E. A. (1969) *American Journal of Enology and Viticulture*, **20**, 76.
HM Customs and Excise (1979) *Laboratory Alcohol Table*. Density/Strength at 20 °C for laboratory use. RDC 80/267/04 Kings Beam House, Mark Lane, London.

Hall, P. S. (1976) *Journal of the Association of Public Analysts*, **14**, 41.
Halsey, S. A. (1985) *Journal of the Institute of Brewing*, **91**, 306.
Halsey, S. A. (1986) *Analytical Proceedings*, **23**, 126.
Hardy, P. J. & Ramshaw, E. H. (1970) *Journal of the Science of Food and Agriculture*, **21**, 39.
Hogben, R. & Mular, M. (1976) *Journal of the Science of Food and Agriculture*, **27**, 1108.
Hudson, J. R. (1960) *Development of Brewing Analysis—A Historical Review*. London: Institute of Brewing.
Hudson, J. R. (1975) *Journal of the Institute of Brewing*, **81**, 318.
Illing, E. T. & Whittle, E. G. (1939) *Analyst*, **64**, 329.
Jeuring, H. J. & Kuppers, F. J. E. M. (1980) *Journal of the Association of Official Analytical Chemists*, **63**, 1215.
Joyeux, A. & Lafon-Lafourcade, S. (1979) *Annales des Falsifications Experimental Chimie*, **72**, 317, 321.
Junge, C. (1985) *Journal of the Association of Official Analytical Chemists*, **68**, 141.
Kahn, J. H. & Conner, H. A. (1972) *Journal of the Association of Official Analytical Chemists*, **55**, 1155.
Kahn, J. H., Nickol, G. B. & Conner, H. A. (1966) *Journal of Agricultural and Food Chemistry*, **14**, 460.
Kearsley, M. W. (1981) *Journal of the Association of Public Analysts*, **19**, 121.
Kearsley, M. W. & Gibson, W. J. (1981) *Journal of the Association of Public Analysts*, **19**, 83.
LAJAC (1963) *Journal of the Association of Public Analysts*, **1**, 101.
Laws, D. R. J. & Peppard, T. L. (1982) *Food Chemistry*, **9**, 131.
Lazaro, F., Luque de Castro, M. D. and Valcarcel, M. (1986) *Analyst*, **11**.
Lehtonen M. & Soumalainen, H. (1979) *Process Biochemistry*, **14**, 5.
Lehtonen, P., Laakso, R. & Puputti, E. (1984) *Zeitschrift fur Lebensmittel Untersuchung und Forschung*, **178**, 487.
Leopold, H., Jelinek, R. & Tilz, G. P. (1977) *Biomedizinische Technik*, **22**, 231.
Lidzey, R. G., Sawyer, R., Stockwell, P. B. (1971) *Laboratory Practice*, **20**, 213.
Linskens, H. F. & Jackson, J. F. (1988) (eds) *Wine Analysis*. Berlin: Springer-Verlag.
Lyne, F. A. & McLachlan, T. (1946) *Analyst*, **71**, 203.
MacNamara, K. (1984) *Journal of High Resolution Chromatography*, **7**, 641.
Marrison, L. W. (1957) *Wines and Spirits*. London: Penguin.
Martin, G. E., Schoeneman, R. L. & Schlesinger, H. L. (1964) *Journal of the Association of Official Agricultural Chemists*, **47**, 712.
Martin, G. E., Dyer, R. H. & Buscemi, P. C. (1974) *Journal of the Association of Official Analytical Chemists*, **57**, 610.
Mason, M. (1983) *Journal of the Association of Official Analytical Chemists*, **66**, 981.
Mathers, A. P. (1958) *Journal of the Association of Official Analytical Chemists*, **41**, 121.
McCloskey, L. P. (1980a) *American Journal of Enology and Viticulture*, **31**, 170.
McCloskey, L. P. (1980b) *American Journal of Enology and Viticulture*, **31**, 212.
McCloskey, L. P. & Mahaney, P. (1981) *American Journal of Enology and Viticulture*, **32**, 159.
Mecca, F. (1964) *Bollettino dei Laboratori Provinciali*, **15**, 578.
Mitra, S. N. (1953) *Analyst*, **78**, 499.
Morawski, J., Dincer, A. K. & Ivie, K. (1983) *Food Technology*, **37**, 57.
Morgan, K. (1965) *Journal of the Institute of Brewing*, **71**, 166.
Morgan, R. H. & Voelcker, E. (1963) *Food Processing and Packaging*, **32**, 207.
Morgantini, M. (1962) *Bollettino dei Laboratori Chimici Provinciali*, **13**, 117.
Nagarathnamma, M. & Siddappa, G. S. (1964) *Indian Food Packer*, **18**, 8.
Noble, A. C., Williams, A. A. & Langron, S. P. (1984) *Journal of the Science of Food and Agriculture*, **35**, 88.
Okamoto, M., Ohtsuka, K., Imai, J. & Yamada, F. (1981) *Journal of Chromatography*, **219**, 175.
Olshausen, J. J. (1952) *Brewers Digest*, **27**, 45.
Osborn, G. H. & Mott, E. O. (1952) *Analyst*, **77**, 260.
Owades, J. L. & Jakovac, J. L. (1963) *Proceedings of the American Society of Brewing Chemists*, 22.
Owades, J. L., Rubin, G. & Brenner, M. W. (1958) *Proceedings of the American Society of Brewing Chemists*, 66.
Pearson, D. & Ntiforo, A. O. (1972) *Journal of the Association of Public Analysts*, **10**, 114.
Pilone, G. J. (1985) *Journal of the Association of Official Analytical Chemists*, **68**, 188.
Pollard, A., Kieser, M. E., Stevens, P. M. & Tucknott, O. G. (1965) *Journal of the Science of Food and Agriculture*, **16**, 384.
Postel, W. & Meier, B. (1981) *Zeitschrift fur Lebensmittel Untersuchung und Forschung*, **173**, 85.
Pyke, M. (1965) *Journal of the Institute of Brewing*, **71**, 209.
Research Committee on the Analysis of Potable Spirits (RCAPS) (1970, 1972, 1974a, b, 1977) *Journal of the Association of Public Analysts*, **8**, 81; **10**, 49; **12**, 40, 45; **15**, 47.
Rigby, F. L. & Bethune, J. L. (1955) *Journal of the Institute of Brewing*, **61**, 325.

Rooney, R. C. (1963) *Analyst*, **88**, 859.
Roston, D. A. & Kissinger, P. T. (1981) *Analytical Chemistry*, **53**, 1695.
Royston, M. G. (1968) *Process Biochemistry*, **3**, 58.
Sawyer, R. & Dixon, E. J. (1968a, b) *Analyst*, **93**, 669, 680.
Sawyer, R., Dixon, E. J., Lidzey, R. G. & Stockwell, P. B. (1970) *Analyst*, **95**, 957.
Saxberg, B. E. H., Duewer, D. L., Booker, J. L. & Kowalski, B. L. (1978) *Analytica Chimica Acta*, **103**, 201.
Schreier, P. (1980) *Journal of Agricultural and Food Chemistry*, **28**, 926.
Seidemann, J. & Woop, K. (1982) *Lebensmittelindustrie*, **29**, 66.
Shinohara, T. & Wanatabe, M. (1981) *Agricultural and Biological Chemistry*, **45**, 2903.
Simpkins, W. A. (1985) *Journal of the Science of Food and Agriculture*, **36**, 367.
Simpson, A. C. (1966, 1968, 1971) *Process Biochemistry*, **1**, 355; **3**, 9; **6**, 25.
Singer, D. D. (1966) *Analyst*, **91**, 127, 790.
Singer, D. D. & Stiles, J. W. (1965) *Analyst*, **90**, 290.
Somers, T. C. (1975) *Food Technology in Australia*, **27**, 49.
Soumalainen, H. & Kangasperko, J. (1963) *Zeitschrift fur Lebensmittel Untersuchung und Forschung*, **120**, 353.
Sponholtz, W. R. (1982) *Zeitschrift fur Lebensmittel Untersuchung und Forschung*, **174**, 458.
Steiner, W., Muller, E., Frohlich, D. and Battaglia, R. (1984) *Mitteilung aus dem Gebeite der Lebensmitteluntersuchung und Hygiene*, **75**, 37.
Steuart, D. W. (1955) *Journal of the Science of Food and Agriculture*, **6**, 387.
Stevens, R. K. & Martin, G. E. (1965) *Journal of the Association of Official Analytical Chemists*, **48**, 802.
Stocker, H. R. (1961) *Schweizerische Brauerei-Rundschau*, **72**, 267.
Stockwell, P. B. & Sawyer, R. (1970) *Analytical Chemistry*, **42**, 1136.
Stone, I. (1962) *Wallerstein Laboratories Communication*, **25**, 329.
Stone, I., Gantz, C. S. & Saletan, L. T. (1963) *Proceedings of the American Society of Brewing Chemists*, 149.
Takeoka, G. & Jennings, W. (1984) *Journal of Chromatographic Science*, **22**, 177.
Taylor, I. C. (1968) *Highland Whisky*. Inverness: An Comunn Gaidhealach.
Tep, Y., Brun, S., Lebeouf, J. P. & Bard, M. (1978) *Annales, Nutrition Aliment*, **32**, 899.
Thorpe, E. & Brown, H. T. (1914) *Journal of the Institute of Brewing*, **20**, 569.
Tressl, R., Friese, L., Fendesack, F. & Koeppler, H. (1978) *Journal of Agricultural and Food Chemistry*, **26**, 1422.
Trommsdorff, H. (1962) *Monatsschrift fur Brauerei*, **15**, 210.
Ubigli, M. (1979) *Vini Italia*, **21**, 250.
Wagstaffe, P. J. (1974) *Analyst*, **99**, 537.
Wagstaffe, P. J. (1976) *Analyst*, **101**, 292.
Wagstaffe. P. J. (1980) *Bulletin OIML*, June, 1.
Warwicker, L. A. (1960, 1963a, b) *Journal of the Science of Food and Agriculture*, **11**, 709; **14**, 365, 371.
West, D. B. (1978) *Journal of the Association of Official Analytical Chemists*, **61**, 100.
White, J. (1966, 1971) *Process Biochemistry*, **1**, 139; **6**, 21.
Whiting, G. C. & Coggins, R. A. (1975) *Journal of the Science of Food and Agriculture*, **26**, 1833.
Whitmarsh, J. M. (1942) *Analyst*, **67**, 188.
Wieg, A. J., Hollo, J. & Varga, P. (1969) *Process Biochemistry*, **4**, 33.
Williams, A. A. & Tucknott, O. G. (1978) *Journal of the Science of Food and Agriculture*, **29**, 381.
Williams, A. A., May, H. V. & Tucknott, O. G. (1978) *Journal of the Institute of Brewing*, **84**, 97.
Williams, J. G., Holmes, M. & Porter, D. G. (1982) *Journal of Automatic Chemistry*, **4**, 176.
Williams, P. J. & Strauss, C. R. (1977) *Journal of the Institute of Brewing*, **83**, 213.
Woo, D. J. and Benson, J. R. (1983) *International Laboratory*, **13**, 22.
Yates, S. (1979) *Chemistry and Industry*, 887.

13

Flesh foods

MEAT

Introduction

The major conventional meat yielding species of the world include cattle, buffalo, sheep, pig, goat, deer, horse, and the various game and poultry species.

Meat is traditionally considered to be one of the prime protein sources and in the eyes of most western consumers is considered to be essential to health and well being. In the UK, the consumption of meat has been linked historically with wealth. Reports of the National Food Survey Committee have shown changes in the consumption pattern. Recent surveys have indicated trends from fresh meat to convenience foods containing meat. Developments in the use of meat as a primary food and as an ingredient in processed products have been reviewed by Bruce (1979) and in relation to regulations by the FSC (1980). Trends in the production of processed pork meat have been discussed by Marks (1985). Increased use has been made of new packaging techniques, especially those based on mechanical treatment and the use of 'film' packaging. Modern poultry meat processing is reviewed by Jones (1986). Cole and Lawrie (1974) and Lawrie (1977) have reported on factors determining and influencing the organoleptic quality, composition, nutritive qualities and methods of preservation of fresh meat. The importance of proper selection of breeding stock on yields and quality is stressed by Lawrie (1977). Those quality characteristics that are considered important in fresh meat include freshness, tenderness, colour, flavour, juiciness and water holding capacity. Muscle structure and the biochemical changes in the tissues, which are related to the relative quality and value of the meat, are discussed by Tarrant (1975). The yield of protein from carcasses is increased by application of mechanical deboning techniques. Methods used have been summarised by Noble (1976) whilst the composition, properties and methods of utilisation of deboned poultry and red meat have been discussed by Froning (1976) and Field (1976). Modification by the use of emulsion extenders has a significant effect on the water holding capacity of meat proteins. Several types of meat derived proteins and non-meat ingredients have been utilised or suggested for incorporation into meat. Studies on the technological properties of mechanically recovered meat (MRM) by Rejt and Pisula (1982) have shown in particular that both the water binding capacity and pH are affected (see also below). Mittal and Usborne (1985) reviewed both the materials used and the process techniques. Quality standards, including chemical and microbiological criteria for retail meats have been reviewed by Holland (1979). The advantages and disadvantages of microbiological standards for perishable foods are detailed; alternative approaches, including attention to good hygienic practice and process control, are discussed. The need for attention to improved hygiene standards has assumed greater significance with the introduction of mechanical recovery of meat tissue from carcass bones (Anon, 1982a).

Composition

The basic composition of meat varies between different types and cuts. Meat is structurally muscle fibre held in bundles by connective tissue through which blood vessels, nerves and fat cells are abundantly distributed. The fibre sheath also encloses water-soluble protein and other nitrogenous compounds together with mineral salts.

Proteins are without doubt the most important constituents of the edible parts of meat animals. The edible complex usually consists of the proteins actin and myosin, together with smaller quantities of collagen, reticulin and elastin. There are also quantities of respiratory pigments—myoglobin, nucleoproteins, enzymes and other miscellaneous components, including the B vitamins.

The inedible portion of the animal contains a considerable amount of protein such as collagens (skin, tendon, connective tissue), the keratins (hair, horn, hoof), elastin (ligaments) and the blood proteins. Lusbough and Schweigert (1960) have tabulated data for individual fresh meat cuts of beef, pork and lamb. The figures of primary interest for analytical control are 'lean meat content' and 'meat equivalent' of a product. Typical modern data for the major and minor constituents of different meat cuts and offal are given by Paul and Southgate (1978) and Moss *et al.* (1984). A selection of data is given in Table 13.1. The composition of lean, subcutaneous fat and rind in commercial hybrid gilt pigs (young sows) of one level of fatness and weight, undertaken by AMC (1986) in a preliminary study on the nitrogen factor for pork (see p. 489), is shown in Table 13.2. The composition of fresh pork cuts used for the production of Wiltshire cured bacon has been reviewed by Harrison (1983). Fresh muscle meat contains essentially no carbohydrate, no fibre and approximately 1 per cent ash. Organ meats contain glycogen and glucose up to approximately 4 per cent; liver contains three to four times the amount of iron in lean meat. Dransfield *et al.* (1983) prepared sixteen samples of minced lean beef from two breeds, Galloway and Charolais. The prepared samples were examined at eight laboratories by recommended procedures. These authors found that the differences between replicate determination within laboratories were often larger than is suggested for reference methods. The differences between laboratories were within expectation. Sawyer (1984) examined the data and demonstrated that the water/protein ratios obtained for the two breeds were significantly different.

Data for MRM (Mechanically Recovered Meat) has been reported by a number of authors. Bijker *et al.* (1983) gave data for pork whilst Meech and Kirk (1986) have examined and reported on MRM prepared in a number of machines and from various species. These latter authors concluded that the machine type and operating practices had an influence on the composition of product obtained within a single species. Significantly different compositional characteristics were obtained within a species according to the location and type of bones used. They concluded that the direct

Table 13.1 Composition of raw carcass meat (after Paul and Southgate)

	Water (%)	Nitrogen (%)	Protein (N × 6·25) (%)	Fat (%)	Water protein ratio	Protein on fat-free (PFF) (%)	Nitrogen on fat-free (NFF) (%)
Beef							
Lean	74·0	3·25	20·3	4·6	3·64	21·3	3·41
Fat	24·0	1·40	8·8	66·9	2·76	26·6	4·2
Brisket	62·2	2·68	16·8	20·5	3·70	21·1	3·37
Forerib	57·4	2·56	16·0	25·1	3·59	21·4	3·42
Mince	64·5	3·01	18·8	16·2	3·43	22·4	3·59
Rump	66·7	3·02	18·9	13·5	3·53	21·8	3·49
Sirloin	59·4	2·65	16·6	22·8	3·58	21·5	3·43
Stewing steak	68·7	3·23	20·2	10·6	3·40	22·6	3·61
Topside	68·4	3·13	19·6	11·2	3·49	22·1	3·52
Veal	74·9	3·37	21·1	2·7	3·55	21·7	3·46
Pork							
Lean	71·5	3·29	20·7	7·1	3·45	22·3	3·54
Fat	21·1	1·08	6·8	71·4	3·10	23·8	3·78
Belly	48·7	2·44	15·3	35·5	3·18	23·7	3·78
Loin	54·3	2·55	15·9	29·5	3·42	22·6	3·62
Leg	59·5	2·66	16·6	22·5	3·58	21·4	3·43
Lamb							
Lean	70·1	3·33	20·8	8·8	3·37	22·8	3·65
Fat	21·2	0·99	6·2	71·8	3·42	22·0	3·51
Breast	48·3	2·67	16·7	34·6	2·89	25·5	4·08
Chops	49·5	2·34	14·6	35·4	3·39	22·6	3·62
Cutlet	48·7	2·35	14·7	36·3	3·31	23·1	3·69
Leg	63·1	2·86	17·9	18·7	3·53	22·0	3·52
Shoulder	56·1	2·50	15·6	28·0	3·60	21·7	3·47
Scrag	55·1	2·50	15·6	28·2	3·57	21·7	3·48
Chicken							
Breast	75·4	3·74	23·4	1·0	3·22	23·6	3·78
Drumstick	77·0	3·26	20·4	3·1	3·77	21·1	3·37
Thigh	77·8	3·18	19·9	3·5	3·99	20·6	3·30
Wing	75·4	3·49	21·8	3·3	3·46	22·5	3·61
Back	76·1	3·14	19·6	5·2	3·88	20·7	3·31
Skin	57·1	2·24	14·0	31·2	4·08	20·3	3·26
Fat	33·8	0·86	5·4	63·3	6·26	14·7	2·35
Turkey							
Skin	49·6	2·03	12·7	36·9	3·91	20·1	3·22
Meat and skin	72·0	3·29	20·6	6·9	3·50	22·1	3·53
Light meat	75·2	3·71	23·2	1·1	3·24	23·5	3·75
Dark meat	75·9	3·24	20·3	3·6	3·74	21·1	3·36
Duck							
Meat only	75·0	3·15	19·7	4·8	3·81	20·7	3·31
Meat, fat and skin	43·9	1·80	11·3	42·7	3·88	19·7	3.14
Rabbit							
Loin and leg meat	74·6	3·50	21·6	4·0	3·45	22·5	3·64

determination of MRM content by the use of chemical indices seemed unlikely with the present state of knowledge. Histochemical studies indicated that the microstructure of MRM is distinctively different compared with meat and the presence of ground bone in red meat is indicative of the incorporation of MRM. However, the presence of

Table 13.2 The composition of lean tissue, subcutaneous fat and rind in pork joints (AMC, 1986)

	Moisture (%)			Fat (%)			Nitrogen (%)		
	Lean	Subfat	Rind	Lean	Subfat	Rind	Lean	Subfat	Rind
Leg	74·6	1964	50·5	2·5	7470	11·0	3·36	0·94	6·35
Rump loin	72·0	1366	54·6	4·6	8174	6·5	3·45	0·61	6·39
Rump belly	71·6	2064	56·6	6·2	7374	7·0	3·22	0·82	5·99
Rib loin	68·4	1266	50·0	9·8	8274	11·6	3·27	0·63	6·31
Rib belly	66·7	1863	53·9	13·1	7476	7·5	3·02	0·92	6·28
Collar	69·0	1462	44·6	10·8	7974	17·9	2·97	0·75	6·14
Hand	71·7	2069	51·5	7·3	7175	15·4	3·12	1·09	5·39

bone in the product derived from poultry meat is not obviously demonstrated. The nature of MRM has been investigated by Kirk and Pickering (in press). Compositional data on chicken meat has been reviewed by Demby and Cunningham (1980), Hamm *et al.* (1980), Meiners *et al.* (1982), Grey *et al.* (1983), Thomas *et al.* (1984) and Grey *et al.* (1985). The reviews include information on the effects of processing on the nutritive value of poultry flesh and on the distribution of nutrients within the carcase.

Legislation

The UK and EC legislation on meat and meat products is listed by Kirk and Lumley (1988). With the exception of EC regulations relating to water in poultry and for meat in intra-Community trade, there are no specific compositional regulations for fresh and frozen meat; the basic requirements of the Food Act relating to description, labelling and wholesomeness apply. Sinar *et al.* (1976) have suggested limits for fat and moisture content of minced beef. These authors advanced an opinion that the fat content of minced beef should not exceed 25 per cent. Cassidy and Fryer (1989), from an analytical survey of 64 samples in Somerset, also found that minced meat should not contain more than 25 per cent of fat and 20 per cent of connective tissue but recommended that products designated as 'lean' or 'extra lean' should contain proportionately less fat. Council Directive 88/657/EEC lays down requirements for the production of and intra-Community trade in minced meat, meat in pieces of less than 100 g and meat preparations (fresh, chilled or frozen, uncooked, containing other ingredients, seasonings or additives). The Directive comes into force in January 1992, and lays down microbiological standards and compositional standards, as follows:

	Fat (%)	Collagen: meat protein ratio
Lean mince	≯ 7	≯ 12
Beef mince	≯ 20	≯ 15
Mince containing pig meat	≯ 30	≯ 18
Mince of other species	≯ 25	≯ 15

The minced meat may not be obtained from scrap cuttings or trimmings, or be prepared from head meat, shin, injection areas, diaphragm, central part of belly muscles and bone scrapings. Council Regulation 2967/76/EEC lays down standards and methods for the water content of frozen and deep frozen poultry. Enforcement of the Regulations within the UK is provided for by the Poultry Meat (Water

Content) Regulations 1984 SI No. 1145, which specify that the method of analysis shall be the Annex III procedure.

The EC Regulation contains methods for determining the water absorption during processing, by direct weight gain, drip loss, and chemical analysis. The drip loss method is not appropriate for the testing of poultry treated with polyphosphates. Poultry treated in this way is temporarily excluded from the extraneous water limit of 7·4 per cent prescribed in the Regulations. Dutschke (1979) has commented critically on the water in poultry regulation from the point of view of efficacy of achieving the stated objectives and from the point of view of enforcement. He pointed out that the stated objective at the outset of discussion was to improve hygienic practices in processing, but that the regulation itself is concerned with protection of consumers from fraudulent incorporation of excess water derived from the use of immersion chilling techniques. This author also criticised the inclusion within the regulation of alternative methods of analysis that are based on diverse concepts. Jonas (1981) reported on collaborative trials to examine the commercial application of the methods described in the Council Regulation; he concluded that the Annex I test procedure to measure weight gain during washing and chilling was a reliable process control method. Neither the Annex II drip test nor a modified room temperature test gave good estimates of the extraneous water content. A favourable opinion was expressed on the Annex III procedure based on determination of water/protein ratio; this was with the proviso that the amount of offal packed with the carcass is accounted for in the analysis. The limit for extraneous water is dependent on an agreed calculation of physiological water content. This has been shown to have a variation of approximately 5 per cent. Similar conclusions to those of Jonas were reached by Hall and Eames (1981). Commission Regulation (EEC) 2632/80 amended the Annexes to the Council Regulation and describes a new Annex I (to replace Annex III) and Annex II (to replace Annex IV).

The general problem of introduction of water into foods (including meats) by use of polyphosphates has been discussed by the FSC (1978). The problem of distinguishing between fresh whole meat cuts and those treated with water was also considered in the Report on Meat Products (FSC, 1980). The committee concluded that, in view of the allowed use of tendering agents to improve the quality of cheap cuts of meat, the prohibition of the use of water in the treatment of meat was impracticable. They proposed that such treatment would convert fresh meat into a meat product and that appropriate labelling of the resulting products would be necessary. Thorne (1978), Weiland (1972) and Gerrard (1963) have discussed the use of tenderisers based on proteolytic enzymes in solid or liquid forms. Solid preparations of enzymes such as papain and trypsin contain 4 per cent of active ingredient, 50 per cent sodium chloride and 46 per cent dextrose; additional flavouring materials such as monosodium glutamate may be used. Meat treated in this way must be described as 'tenderised'.

The Meat (Treatment) Regulations 1964 (SI 1964 No. 19) prohibit the use of ascorbic acid, erythorbic acid, nicotinic acid and nicotinamide to fresh meat. These additives help maintain the fresh colour of meat, but adverse physiological reactions have been noted in consumers of meat treated with nicotinic acid (Kendrick and Watts, 1969). Certain classes of food additives are specifically prohibited for use with fresh meat. These are: antioxidants (other than tocopherols); preservatives; colours and the colouring agents detailed above. Other classes of additives are not specifically prohibited, but their misuse could lead to prosecution under Section 2 of the Food Act. The FSC (1980) also recommended that no provision should be made for the replacement of part of the lean meat content of whole meat cuts with vegetable protein products.

The use of growth promoting feed additives (and thereby the resultant levels in meat) are controlled by various regulations made under the Medicines Act 1968. Directive 81/602/EEC requires member states to ensure that meat and meat products are subject to controls in respect of residues of pharmacologically active substances. A proposal for a directive laying down sampling schedules was laid on 8 May 1985. Residues of veterinary drugs and their metabolites may arise in animal products including meat, offal, eggs and milk. A wide variety of drugs is used in animal husbandry; anthelmintic agents, antibiotics and other antimicrobial compounds are used to treat and prevent diseases caused by micro-organisms. The range of controlled substances includes agents which act as growth promoters by improving energy conversion of animal feeds. Such agents may act hormonally (anabolic substances) or by reduction of pathogenic activity in the gut. Proposals for methods of analysis have received considerable attention in recent years both in the UK and in the EC. Fresh poultry may not be treated with antibiotics, preservatives or tenderisers.

The hygienic production of meat is controlled by the provision of the Slaughterhouse (Hygiene) Regulations 1977 (SI 1977 No. 1805) and in the case of poultry the Poultry Meat (Hygiene) Regulations 1976 (SI 1976 No. 1209) as amended. Handling of meat throughout the distribution chain is covered by the provisions of the Food Hygiene (General) Regulations 1970 SI No. 1172, as amended. The hygiene of meat in intra-Community trade is covered by EC Directives.

Analysis of meat

PREPARATION OF SAMPLES FROM RAW MEAT

In view of the structure of meat it is somewhat difficult to obtain by mincing a representative and homogeneous sample that is suitable for analytical work. However, the powerful food processor/homogeniser equipment now in general culinary use has solved this problem and readily produces a fine representative paste, provided there is sufficient sample (Beebe *et al.*, 1989). For smaller samples a shearing type bottom or top drive macerator can be used (see also Chapter 2).

ASSESSMENT OF SPOILAGE OF RAW MEAT

The changes that take place during the spoilage of meat have been the subject of a review by Dainty (1971). In spite of improvements in methods of microbiological examination a detailed investigation into meat quality occupies several days. The acceptability of beef can be assessed from the following methods:

Total volatile nitrogen (TVN). The TVN is related to protein breakdown and can be determined by the modified Lucke and Geidel macro-distillation method described under fish (p. 510); see also Kearsley *et al.* (1983) for a modified distillation procedure with improved performance and shortened distillation time. Most beef can be considered as acceptable if the TVN and TVN/FF figures do not exceed 16·5 and 19·7 mg N/100 g respectively (Pearson and Muslemuddin, 1968, 1971). The hypoxanthine index (see p. 512) has been proposed as a method for assessment of the freshness of raw meat by Pizzocaro (1978). He showed a linear correlation with organoleptic score and proposal limiting values of 220–230 mg/kg for beef muscle tissue.

Extract-release volume. The water holding capacity of freshly slaughtered meat is high, but it drops markedly within a few hours and then increases slowly during further storage. The reverse trends occur in the volume of filtrate obtained from a macerate known as the extract-release volume (ERV), which can be measured by the following method (Pearson, 1969b).

Weigh 15 g minced sample into the 100 ml vortex beaker of a laboratory homogeniser. Add 60 ml extraction reagent (50 ml 0·2 M KH_2PO_4 and 3·72 ml 0·2 M NaOH diluted to 200 ml with water, check pH is 5·8), blend for 2 min. Pour the homogenate into a filter paper (Whatman No. 1, 18·5 cm diameter) and measure the volume (ERV) collected in 15 min. For acceptability the ERV of beef should be at least 17 ml.

Free fatty acids of the extractable fat

Macerate the sample mechanically with chloroform, filter. Re-filter through a paper containing anhydrous sodium sulphate. Determine the fat content of the filtrate by evaporating a known volume. Mix 25 ml filtrate with 25 ml neutral alcohol and titrate the FFA with 0·1 M sodium hydroxide using phenolphthalein. Most samples of beef can be considered as acceptable if the FFA does not exceed about 1·2 per cent (as oleic acid on the extracted fat).

Vasundhara *et al.* (1983) have described a GLC procedure to measure changes in the free fatty acid profiles in canned mutton products. The procedure was claimed to be sufficiently sensitive to allow assessments to be made in respect of the quality of meat at the canning stage.

Peroxide value of the extractable fat

The chloroform filtrate can also be used for the assessment of peroxide values (Pearson 1965, 1974). Most fresh beef samples give peroxide values of 0·0–1 mmol/kg of extracted fat and 5 might be taken as a critical acceptability limit.

Thiobarbituric acid value (TBA)

The extent of oxidative rancidity of meat lipids may be measured by steam distilling from an acid slurry of minced meat. The amount of malonaldehyde in the distillate is determined by reaction with acidified thiobarbituric acid. The reaction product gives an absorbance maximum at 530–535 nm (for method, see p. 642). The separation of malonaldehyde by distillation has been criticised, since the application of heat may induce the development of TBA sensitive substances. Tomas and Funes (1987) avoided distillation by using meat exudates obtained by centrifugation and the modified Matsushita reaction (Asakawa and Matsushita, 1980).

Procedure. 1·5 ml TBA reagent (0·5 g TBA + 0·3 g sodium dodecylsulphate in 100 ml water) and 1·5 ml of 0·2 M glycine-HCl buffer pH 3·6, are added to 0·3 ml exudate (deproteinated with aqueous trichloracetic acid). Heat mixture in boiling water for 15 min, cool in ice and measure absorbance at 532 nm in a 10 mm cell.

An HPLC procedure has been developed by Kakuda *et al.* (1981). The effect of albumen on the TBA value of comminuted poultry meat has been discussed by Negbenebor *et al.* (1985).

Growth promoting agents. Considerable attention has been paid to the methods of analysis for estimating low levels of residues of the various classes of growth promoting agents and their metabolites. In many cases residue concentrations are less than 1 mg/kg and in some the levels are in the region of 1 μg/kg. Adequate test methods for the estimation of residues of the hormones trenbolone, hexoestrol and the related stilbenes are available. Harwood *et al.* (1980) used radioimmunoassay techniques for hormones in tissue samples. This approach has been used successfully since 1981 in a monitoring scheme for the detection of residues of stilbene and thyrostatic anabolic agents in meat intended for sale in the UK. The use of screening methods for the detection of antimicrobial agents has received attention from a number of workers. Corry *et al.* (1983) have studied and improved the efficiency of a widely used method. The use of TLC and HPTLC procedures for sulphonamides and chloramphenicol have been studied both in the EC and by the Analytical Methods Committee. Johannes *et al.* (1983) used glucuronidase to free chloramphenicol residues from meat, the extract being purified on a C_{18} column and examined by TLC using acid stannous chloride and ultraviolet to visualise. Several workers have used GC or GC/MS of TMS derivatives (see Nelson *et al.*, 1983 and Peleran *et al.*, 1984). Horwitz (1981) reviewed the available methods for sulphonamides. He showed that substantial effort was needed to improve the chromatographic procedures. Petz (1982) published an HPLC procedure applicable to eggs, milk and meat for the estimation of chloramphenicol and sulpha drugs using acetonitrile extraction and HPLC separation on a C_{18} column. The AMC (1986) has produced reports on the determination of dimetridazole, halofuginone and monensin in chicken tissues. The former study was based on the HPLC method developed by Hobson-Frohock and Reader (1983). Smyth *et al.* (1986) have reviewed recent developments in HPLC techniques for the analysis of veterinary antimicrobials.

The following methods have been used for routine survey and control purposes.

Trenbolone acetate. Holder *et al.* (1979) described a procedure in which the tissue samples are first homogenised with benzene and after drying and removal of solvent the residue is extracted with acetonitrile and hexane. The acetonitrile fraction is dried with sodium sulphate and the solvent removed at 60 °C in a rotary evaporator. The residue is transferred to a column of silica gel with 5 × 1 ml portions of benzene. After washing with 20 ml benzene the TBA is eluted with 60 ml of 10 per cent acetone in benzene. The HPLC analysis of extracts is carried out on reversed phase ODS column loaded with C_{18} hydrocarbon using 75:25 methanol:water as eluant. Detection is by ultraviolet absorbance at 345 nm.

Zeranol. Using the extraction method described by Hunt *et al.* (1978) samples are homogenised in 200 ml of acetonitrile and 1 per cent aqueous potassium chloride. The extract is filtered under vacuum and the filtrate is washed with 2 × 50 ml of 2,2,4-trimethyl pentone. The acetonitrile layer is recovered at each stage, 25 ml of water is added and the anabolic residues are extracted into chloroform 50, 10 and 10 ml aliquots. The combined chloroform extracts are dried over anhydrous sodium sulphate and the solvent is removed at 50 °C in a rotary evaporator. The extract is further purified by solution in chloroform and partition on silica gel. The zeranol is removed with 150 ml diethyl ether and dried at 40 °C. The residue is dissolved in chloroform and analysed by HPLC using 10 μm silica particles and 50:1 chloroform:acetic acid. Zeranol is detected at 280 nm.

Thiouracil and derivatives. Ground tissue is extracted with methanol, the fat is removed by extraction with petroleum ether. Amines and amino acids are then removed with an acidic ion-exchange resin (Dowex 50 WX 8). Polar lipids are removed from the neutralised extract with diethyl ether. The antithyroids are reacted with 7-chloro-4-nitro-benzo-2-oxa-1,3-diazole. The product is acidified and extracted with diethyl ether. After evaporation of the ether the residue is subjected to two-dimensional

TLC analysis on silica gel using chloroform:propionic acid (95:5). The plate is examined under ultraviolet illumination at 366 nm and after spraying with a solution of cysteine in ammoniacal ethanol. (EC paper 1123/VI/78.)

Sulphonamide residues. Ground tissue is extracted in the presence of an internal standard, sulphapyridine, with ethyl acetate. The sulphonamides are partitioned into a glycine buffer (pH 12). The pH of the aqueous phase is adjusted to 5·25 with HCl/phosphate buffer, after defatting with hexane and extraction with methylene chloride the solvent is removed and the residue is recovered in methanol. Separation of the residues from the co-extractives is carried out on silica gel plates containing a preadsorbent spotting layer. Visualisation is achieved after development in chloroform:*tert*-butanol (80:20) by the use of fluorescamine solution and observation under ultraviolet (see also AOAC methods book).

Chloramphenicol. Monochloramphenicol is used as internal standard in the method prescribed by the USDA for examination of meats. The tissue is homogenised and extracted with ethyl acetate. After removal of solvent the residue is dissolved in 4 per cent aqueous sodium chloride. The aqueous phase is partitioned three times against hexane and the solvent is discarded. Chloramphenicol is extracted with ethyl acetate and after removal of solvent the residue is converted into trimethyl silyl derivatives and the mixture is resolved by GLC on 100/120 mesh Gas Chrom Q loaded with 3 per cent OV–101. Detection is by electron capture.

PRESERVATION AND PRESERVATIVES

Meat is normally preserved by placing in a cold store (1–5 °C), by chilling (−1·5 °C) or freezing (lower than −10 °C), by canning or by dehydration (maximum moisture 7·5 per cent). Apart from the use of certain substances for curing, the Preservatives Regulations do not allow meat to be preserved with artificial preservatives. For the estimation of sulphur dioxide in meat the distillation method or colorimetric procedure is suitable (Chapter 4). Formalin, salicylates, benzoates and borates are the principal non-permitted preservatives that have been found in meat.

MEAT PRODUCTS

The range of ingredients used in processed meat products is wide and includes the basic meat materials, cereals, other nitrogenous ingredients and the technical adjuvants necessary to provide the properties which characterise the individual products. Meat products may be divided into two groups: those which are formulated from raw meats and that retain the perishable nature of raw meats, e.g. sausages; and those that have been subjected to procedures, such as drying, smoking, salting, curing or other methods of preservation, including canning. The concept of processed meats as opposed to preserved meats has been discussed by Hannan (1974). He drew attention to the definition of process in the Meat (Treatment) Regulations 1964, and outlined the manufacturing methods used in the UK. Bruce (1979) has reviewed the nature of modern meat products and has discussed the uses of carcass by-products, mechanically deboned meats and the range of non-meat ingredients used in their formulation. Useful texts on American practices are by Pearson and Dutson (1988) on meat by-products and by Pearson and Tauber (1984) on processed meats.

Composition

Typical analytical data for processed meats and other recipe products have been published by Paul and Southgate (1978); specific references to data on selected classes of products are quoted in the following sections.

One hundred and eighty volatile compounds have been isolated and identified as contributors to the flavour of cooked meats by Rhodes (1974). No single substance could be identified as having a characterising odour. The conclusion appears to be that the typical cooked meat odour is a summation of many compounds and the effect of feedstuffs, carcass conditioning and cooking methods are all considered in the studies.

Legislation

Five sets of regulations relating to the composition and labelling of meat and spreadable fish products were revoked in 1984 and replaced with a single statutory instrument, namely the Meat Products and Spreadable Fish Products Regulations 1984 SI No. 1566, as later amended by Regulations SI No. 987, 1986. For interpretation, they rely heavily on the 1984 Labelling of Food Regulations, and were evolved following a lengthy process of review and consultation on the recommendations in several reports by the FSC (see Appendix 4). These reports were concerned not only with the provisions in the statutes dating from 1967 but also with the related technological developments in the production of meat products that had taken place in the intervening years. Such developments included the modernisation of practices in curing and preserving of meats by injection, tumbling and reforming prior to sterilisation or pasteurisation. The use of process aids, non-meat protein rich ingredients and various techniques of recovery of waste meat by-products.

Scope of the Regulations

As a result, the Meat Products and Spreadable Fish Products Regulations contain a mixture of compositional standards and labelling provisions that do not stand alone but when a particular product is being considered due regard must be paid not only to the Labelling Regulations but also to the general provisions of the Food Act 1984 (nature, substance and quality demanded; labels or advertising must not be misleading) and to Section 1 of the Trade Descriptions Act 1968 (false trade description). Provisions of The Labelling Regulations have been discussed in Chapter 1. Principal attention in the case of a specific product is directed to the name of the food, list of ingredients, information on storage, use and durability.

The Amendment Regulations 1986 were introduced to allow for fat associated with meat over and above the minimum meat content of any product to be present without the need for a declaration of added fat; to exempt certain whole and cured meats from the need to declare a minimum meat content; to clarify compositional standards in relation to the calculation of lean meat content and to the related labelling provisions; and to provide for a minimum meat content requirement for meat and vegetable pies (or puddings) of 10 per cent when cooked and 8 per cent when uncooked.

Definitions

Schedule 3 defines a meat product as any food that consists of meat or of which meat is an ingredient but does not include:

1. Raw meat to which no ingredient, or no ingredient other than proteolytic enzymes has been added.
2. Uncooked chickens (and other poultry) and cuts and offals thereof, to which no ingredient has been added other than additives, water, self-basting preparations or seasonings.
3. Haggis, black pudding, white pudding.
4. Brawn, collard head.
5. Sandwiches, filled rolls and similar bread products, which are ready for consumption without further processing, other than products containing meat which are sold under the name 'burger', 'economy burger', or 'hamburger' whether or not qualified by other words.
6. A food for which the name of the food is 'broth', 'gravy' or 'soup', whether or not qualified by other words.
7. Stock cubes and similar flavouring agents.
8. Products commonly known in Scotland as 'potted head', 'potted meat' and 'potted hough'.
9. Any product containing the fat, but no other meat, of any animal or bird.

Meat is defined as the flesh, including fat, and the skin, rind, gristle and sinew in amounts naturally associated with flesh used, or any animal or bird which is normally used for human consumption, and includes any part of the carcass listed below that is obtained from such an animal or bird, but does not include any other part of the carcass:

Mammalian species
Diaphragm, liver, pancreas, tail meat, heart, thymus, kidney, tongue head meat (muscle meat and associated fatty tissues only).

Avian species
Gizzard, liver, heart, neck.
It should be noted that skin, rind, gristle and sinew may be used in amounts naturally associated with the flesh used but that the parts of carcass listed above, which in the former regulations were controlled 'offals', may be used in any quantity. Labelling provisions are required to take account of this fact.

A spreadable fish product is one which is described as a 'paste', 'pate' or a 'spread' or any other spreadable product which has fish but not just fish oil as an ingredient.

Name of the food
Schedule 4 lays down a number of reserved descriptions for meat and spreadable products. This means that the names listed in the schedule can only be used where the compositional requirements of the regulations are met. There are fewer reserved descriptions than in the previous legislation. A summary is given in Table 13.3. It should be noted that there are no minimum meat content standards for canned meat as such but that there is a requirement to declare minimum meat or fish content.

It is a general tenet of the regulations that the name of a food may be qualified by other words provided that the use of other words does not make the name misleading nor is capable of misleading the consumer as to the true nature of the product. Conditions are also laid down in respect of presentation, appearance and use of ingredients in the case of portions which are cuts, joints, slices of raw, cooked or cured meats. In the case of cooked meats the use of seasonings, herbs and spices need not be declared in the name of the food whilst in the case of cured meats, curing salts and prescribed flavouring ingredients the same condition applies. This condition does not relieve the manufacturer of the need to declare a full list of ingredients as required by the labelling regulations. Indications of any treatment that can materially affect the product and its further storage in the home must be declared, for example, meats which have previously been frozen and thawed must bear a notice indicating the fact and a warning not to refreeze. Similarly, the use of enzymes must be designated by the use of the word 'tenderised'.

Declaration of meat content
As a general rule, meat products (other than those which have the appearance of cuts, joints, portions and slices) and spreadable fish products must be labelled with the minimum percentage of meat or fish that the product contains. There are rules that relate to the declaration in the case of meat in a liquid medium such as sausages in brine. Then, as well as the minimum meat content, there must be an indication that the meat content of the product is calculated excluding the weight of the brine. Products with meat content of less than 10 per cent may be labelled with the actual content or with a 'less than 10 per cent' statement. Any product that has lost water during cooking or processing must be labelled with 'not less than 100 per cent' meat or fish. In the case of corned meat the phrase '100 per cent corned X' is to be used. Where corned meat is used as an ingredient in another food then that food must be marked with the statement 'minimum X per cent corned Y', where Y is the name of the meat.

Raw or cooked meat or cooked cured meat with more than 10 per cent added water must bear an indication of meat content and a declaration 'with added water' where water has been added.

Added water
The presence of water need not necessarily be declared in the name of the food, but where it has been added to products which have the appearance of cuts, slices, joints, etc. of raw, cooked or cured meat then there are additional marking and labelling requirements:

1. In cases where there is no declaration of the presence of ingredients in the name but water has been added.

Table 13.3 Regulated meat products, minimum statutory requirements

Product description	Minimum meat content (%)	Qualifications
Burger (used alone or as part of another word)	80	At least 65% of meat content to be lean meat At least 80% of product to be named meat If cured meat is named at least 80% of product to be raw meat as used for curing Meat content of burger in a bread roll relates to meat fill
Economy burger (whether or not 'burger' is part of another word)	60	At least 65% of meat content to be lean meat At least 60% of product to be named meat If cured meat is named at least 60% of product to be raw meat as used for curing Meat content of burger in a bread roll relates to meat fill
'Hamburger' (whether or not hamburger forms part of another word)		All meat to be beef, pork or beef and pork Type of meat to be indicated Other requirements as for 'burgers'
'Chopped meat' or 'Chopped cured meat'	90	At least 65% of meat content to be lean meat Word meat may be replaced by the type of meat
'Corned beef' or other named corned meat	120	Food must consist wholly of corned meat At least 96% of meat content to be lean meat All meat used must be of named type
'Luncheon meat'	80	At least 65% of meat content to be lean meat The name of the type of meat may be used
'Meat pies' (cooked) over 200 g 100–200 g under 100 g 'Scottish pie' (cooked)	25 21 19 20	At least 50% of required meat content to be lean Type of meat may replace the word 'meat' Includes 'meat pudding', 'Melton Mowbray pie', 'game pie' unless qualified by type of meat and another food
'Meat pies' (uncooked) over 200 g 100–200 g under 100 g 'Scottish pie' (uncooked)	21 18 16 17	At least 50% of required meat content to be lean Type of meat may replace the word 'meat' Includes 'meat pudding', 'Melton Mowbray pie', 'game pie' unless qualified by type of meat and another food
'Meat pies' name qualified by another food e.g. 'beef and onion'	12·5 10·5	Cooked Uncooked At least 50% of meat content to be lean meat
'Pasty', 'Pastie', Bridie' and 'sausage roll'	12·5 10·5	Cooked Uncooked At least 50% of meat content to be lean meat
'Pork' sausage, link, chipolata or sausage meat	65	80% of meat must be pork meat At least 50% of meat content must be lean meat
'Beef' sausage etc.	50	At least 50% of meat to be beef At least 50% of meat content to be lean meat
'A' and 'B' sausage etc.	50	At least 80% of meat to be first named meat At least 50% of meat content to be lean meat If name is qualified by liver and/or tongue at least 30% of meat to be liver and/or tongue
Paste containing meat or meat and fish (combined content)	70	At least 65% of meat content to be lean meat If preceded by name of type of meat or cured meat must be characterised by the named meat

Table 13.3 Regulated meat products, minimum statutory requirements (contd.)

Product description	Minimum meat content (%)	Qualifications
Pate containing meat or meat and fish (combined content)	70	At least 50% of meat content to be lean meat If preceded by name of type of meat or cured meat must be characterised by the name meat Inclusion of the name of another food renders the as a non-reserved name
Spread containing meat or meat and fish (combined content)	70	The name must be preceded by the name of a type of meat or cured meat or fish or mixture of meat/cured meat or fish At least 65% of the required meat content must be lean meat

a. Raw or cooked meat, the maximum water content must be declared—'with not more than X per cent added water'.
b. Uncooked cured meat, where more than 10 per cent of water has been added, the amount in excess of 10 per cent must be indicated in 5 per cent steps.
c. Cooked cured meat must be labelled in respect of all added water, the declaration to be in 5 per cent bands.

2. Where the food bears an indication of ingredients in its name and also contains added water.
   a. Raw or cooked meat, or
   b. Cooked cured meat or uncooked cured meat of which more than 10 per cent is added water.
   Must be marked with the words 'with added water'.

The labelling regulations also require that added water of more than 5 per cent must be declared in its correct position in the list of ingredients.

Ingredients and additives
The ingredients list must contain ingredients or additives in the correct descending order of use. Additives other than flavourings must be declared by category and by specific name or serial number. The dispensation on flavourings is subject to further developments on the EC Directive on Flavouring Agents. Where fat cannot be taken into account in the declaration of minimum meat content its presence must be taken into account in other ingredients.

Offals
The following parts of a carcass may not be used in an uncooked product and are not regarded as meat: brains, feet, intestine (large and small), lungs, oesophagus, rectum, spinal cord, spleen, stomach, testicles and udder.

If any of the above parts are used in a cooked product they may be declared specifically or generically. These requirements have been modified by prohibition into the use of certain bovine offals arising from the incidence of bovine spongiform encephalopathy (BSE) (see SI No. 2061/1989).

Meat content
Where the declaration of minimum meat content is required, in a number of products a fixed proportion of the meat content must be lean meat; the regulations contain no definition of 'lean meat' but state that 'lean meat content' means the 'total weight of lean meat free, when raw, of visible fat'. (NB The FSC Report on Meat Products (1980) recommended that lean meat should mean meat free from trimmable fat and connective tissue, containing no more fat or connective tissue than is naturally associated with the particular cut or trimmed meat used, provided that neither the fat nor the connective tissue content exceeds 10 per cent. Best quality meat cuts contain rather less than this stipulation.)

The meat content is calculated on the basis that the total weight of raw meat and the total weight of any solid bone naturally associated with the meat (only if the fact that bone is declared in the name of the food, either expressly or implied). In the case of a sausage, the weight of any detachable edible skin is excluded.

In the case of a partly or wholly dehydrated product, which has to be reconstituted before consumption, the meat content is worked out on the basis of the product when reconstituted.

There are no EC directives that prescribe compositional standards for meat products (for minced meat, meat in pieces and meat preparations, see p. 472). However, Council Directive 88/658/EEC controls conditions of manufacture and quality of certain preserved meat products. The directive defines controlled products as those prepared wholly or partly from meat that has undergone treatment to ensure a certain degree of preservation by heat, salting, or drying (either singly or in combination). A meat product is defined as a product prepared from or with meat that has undergone treatment such that the cut surface shows that the product no longer has the characteristics of meat.

Novel proteins

The FSC (1974) reviewed the use as food of all unconventional sources of protein, including textured vegetable protein and those produced from micro-organisms. The committee made tentative recommendations for control of the level of use of non-meat proteins in meat products. These proposals were subsequently modified by the FSC (1980) in the Report on Meat Products. The committee recognised the use of non-meat protein products for functional purposes at levels up to a limit of 2 per cent calculated as dry weight on the total weight of the meat product. In circumstances where the limit of 2 per cent is exceeded the Committee proposed that the name of the product should contain some reference to the use of non-meat protein and that the non-meat protein product should be declared in terms of lean meat equivalent. The maximum level of use was recommended to be fixed at 33·3 per cent in products where non-meat protein replaces a proportion of the statutory amount of lean meat. The 1980 report includes tabulated data of calculated lean meat equivalents of novel proteins. When used in controlled products these are based on a factor of 4·5 times the protein content.

The current Meat Product Regulations (see p. 478) contain no specific reference to novel proteins or proposals for control and rely on labelling provisions to take account of the inclusion of 'foreign proteins' in manufactured meat products.

CURED MEAT

Manufacture

Historically, curing meant the addition of sodium chloride to meat for the purpose of preservation. The characteristic pink colour is due to the reaction of nitrite with myoglobin. Nitrite is present with nitrate as an impurity in commercial curing salt and is responsible for the characteristic flavour of cured meats, such as bacon and ham. Together with salt, nitrite has an inhibiting effect on the growth of *Clostridium botulinum*.

Other additives that are used in modern curing methods include sugars, ascorbate, phosphates, organic acids, ribonucleotides, glutamate, spices, liquid smokes and glucono-δ-lactone. The various additives have effects on pH, colour, flavour and water retention in the products. Water is also stabilised in the final product by the addition of agar, alginates or carrageenan in appropriate cases (Schweigert, 1974).

The traditional curing treatment was by rubbing dry salt mixture into the meat or by immersion in a brine containing 15–25 per cent salt. Both methods required a lengthy maturation period. Modern methods of wet curing involve the injection of a weaker brine directly into the flesh (pumping) so that the curing salt is distributed through the arterial system. A shorter period of curing is required in this case and maturation is completed in the distribution chain in the case of products such as bacon and ham. Langen (1982) has discussed curing techniques and injection technology in relation to quality of the finished product. The manufacturing process for moulded, film wrapped ham products has been described in detail (Anon, 1985). The advantages of improved shelf-life and process control to yield a uniform product are considered and compared with the 'classical' canning technologies.

The smoking process is also combined with curing to yield specific products such as smoked bacon, ham and cured sausages; smokes are now added in a liquid form as sprays but both procedures result in the deposition of mild antiseptics such as phenols and formaldehyde on the surface of the meat. Corned beef is a form of compressed cured meat prepared by coarse chopping, cooking, curing with salt and nitrite and canning with a full botulinum cook. Products prepared by in-can processing of raw meat, curing ingredients and agar with or without cereal filler are also available.

The technology of curing has been reviewed by Kroĺ and Tinbergen (1974) and the manufacture of bacon and ham is thoroughly covered by Hogan (1978). The combination of modern curing methods with those of restructuring by tumbling in the presence of phosphates followed by pressing and cooking have been discussed by Krause *et al.* (1978), Korhonen *et al.* (1978), Winter (1978) and Bertling (1979).

Legislation and standards

As indicated in the earlier discussion on meat products legislation, the position on cured products is that the ingredient listing and water content declaration have an impact. Ham, bacon and other cured products, under conditions prescribed, are required to carry a declaration of the amount of water added together with the list of ingredients. Where the name of the product is changed so as to indicate the presence of water and other ingredients the amount added is not prescribed. The product is, however, subject to a minimum meat content declaration. Cooper (1985) has discussed the complexity of the requirements of the regulations in so far as the manufacturer is concerned.

The composition of modern hams has been the subject of considerable research effort. Gardner (1978) studied 57 samples of modern hams and 24 samples of traditionally cured hams. He showed highly significant differences between the two types on the basis of water/protein ratio, protein content on fat-free products and calculated meat content using a nitrogen conversion factor of 3·45 (p. 488). Harrison (1978) surveyed a total of 271 samples of ham of three types, namely 'off the bone', 'prepacked sliced' and 'non-prepacked sliced'. The added water content in the three categories varied up to 25 per cent, 30 per cent and 35 per cent respectively. Hall (1978) presented similar data on bacon, and on canned and prepackaged ham. Bertling (1977) analysed 469 samples of ham manufactured in Germany. Distribution data of water, protein and fat content are presented and the author proposes limiting values for protein, fat, water/protein ratio and fat/protein ratio for genuine hams. The values suggested are: meat protein, not less than 17·0 per cent; fat, not more than 14·0 per cent; fat/meat protein ratio, not more than 0·7 per cent; water/protein ratio, not more than 4·0 per cent and connective tissue protein in meat protein, not more than 7·0 per cent. In general 80 per cent of the samples examined agreed with the proposed limits.

Weber and Hopke (1981) confirmed the use of water/protein ratio as a quality control procedure for raw hams, although the USDA Food Inspection Service proposed the use of 'protein on fat free' basis for assessment of the compliance of cured pork products with US Federal Regulations. Sawyer (1984) pointed out that the US proposals lacked sensitivity at higher levels of water addition (Anon, 1983), especially in respect of the 'high water content' hams. Hannan (1983) has presented information on factors that may influence the assessment of cured meat products, especially in respect of the water content. Brinker (1979) has discussed the impact of legislative proposals on the labelling of meat products, especially in the context of the use of added protein materials.

Data for a selection of typical products on sale in the UK are shown in Table 13.4.

Table 13.4 Composition of cured meat products

	Water (%)	Fat (%)	Protein (%) (N × 6·25)	Ash (%)
Bacon	13–68	8–80	5–20	1·5–4·0
Ham cured	30–64	5–49	15–27	2·0–4·0
Corned beef (Argentine)	54–57	9–14	23–31	2·0–3·0
Luncheon meat	52–56	23–29	11–15	2·0–2·3
Ham canned	67–77	2–6	14–19	3·0–5·5
Pork shoulder canned	70–77	2–5	14–17	3·5–5·0

Codex standards (see Appendix 2) have been developed for canned corned beef (CAC/RS 88–1976), cooked cured ham (CAC/RS 96–1978), cooked cured pork shoulder (CAC/RS 97–1978) and for cooked cured chopped meat (CAC/RS 98–1978) (Table 13.5). The FSC (1980) commented on the Codex Standard for corned

Table 13.5 Summary of compositional criteria in Codex standards

Corned beef	≮ 21% total protein
Cooked cured ham	≮ 16·5% meat protein (fat free)
	≮ 18·0% meat protein (fat free) on average
Cooked cured pork shoulder	≮ 16·0% meat protein (fat free)
	≮ 17·5% meat protein (fat free) on average
Cooked cured chopped meat (with binder)	≮ 85% ingoing meat content
	≯ 30% fat
Cooked cured chopped meat (without binder)	≮ 90% ingoing meat content
	≯ 25% fat

beef and on the modern processing techniques that yield a corned beef type of product with low raw meat equivalents. Hauser (1975) has published compositional data on South American and European corned beef. Examples of products containing dried skim-milk powder, polyphosphate and soya protein are included in the survey.

The Preservatives in Food Regulations 1979 were amended in 1982 to take account of recommendations of the FACC (1978a) that four classes of controlled meat products be established for the purpose of regulating the use of nitrates and nitrites. These requirements, outlined in Table 13.6, differ from those limits

Table 13.6 Preservatives in cured meats. Requirements of the Preservatives in Food Regulations

Product	Sum of nitrate and nitrite as sodium nitrite (mg/kg)	Sodium nitrite (mg/kg)
Uncooked bacon and ham; cooked bacon and ham not in hermetically sealed container	500	200
Cured meat and cured products, packed in sterile pack	150	50
Acidified and/or fermented cured products, not in sterile pack	400	50
Any other cured meat or cured meat product	250	150

prescribed in the Codex Standards enumerated above. The Codex Standards also include prescribed levels of use for specified additives in each of the defined products. These are tabulated in Table 13.7. Hamence and Kunwardia (1974) have

Table 13.7 Additives permitted in Codex meat product standards

	Canned corned beef	Cooked cured ham and pork shoulder	Cooked cured chopped meat
Sodium nitrite[b] (mg/kg)	50	125	125
Sodium nitrate[b] (mg/kg)	—	500	—
Ascorbic acid[a] (mg/kg)	500	500	500
Guanylic acid[a,c] (mg/kg)	—	500	500
Inosinic acid[a,c] (mg/kg)	—	500	500
Glutamic acid[a,c] (mg/kg)	—	2000	5000
Phosphates as [a,b](P_2O_5) (mg/kg)	—	3000	3000
Citric acid[a] (mg/kg)	—	GMP	GMP
Glucono-δ-lactone (mg/kg)	—	—	3000
Agar, gelatine, alginates, carrageenan	—	GMP	—
Natural flavours	—	GMP	GMP
Natural smoke solutions	—	GMP	—

[a] As sodium salts.
[b] Also includes potassium salts.
[c] The flavour modifiers recommended for control by the FACC (1978b) are monosodium glutamate, guanosine-5′-disodium phosphate, inosine-5′-disodium phosphate and sodium-5′-ribonucleotide.

reported on the use of polyphosphates in cured meats and other meat products. The FACC (1978b) considered the use of flavour modifiers and concluded that a permitted list consisting of monosodium glutamate, guanosine 5′-disodium phosphate, inosine 5′-disodium phosphate and sodium 5′-ribonucleotide should apply to foods other than those specifically prepared for young children. These compounds also appear in the list of additives permitted in the Codex canned cured pig meat products. The FACC report also indicated levels of these controlled additives that may be present naturally in meats, meat extracts and yeast extract. The use of flavour modifiers was permitted by schedule amendments to the Miscellaneous Additives in Food Regulations 1980 (see SI 1982 No. 14).

SAUSAGES

Meat utilised in sausage manufacture consists generally of trimmings from various carcase dressing operations, either in butchery or manufacture of processed meats. The balance of lean, fat and connective tissue (skin, etc.) in the mixture is achieved by appropriate sorting and blending of the trimmings used. Fresh sausage meat is prepared by blending the meat components, usually pork and/or beef, with wheat rusk, seasoning (salt, spices or spice extracts), ice and sulphite preservative.

Some products contain flavour enhancers, vitamin C, colouring matter, polyphosphates and other emulsifiers. The emulsifying ingredients may be derived from solubilised proteins derived from soya, milk, cereals and ground nuts; emulsified rind is also used by some manufacturers (Marshall, 1960; Hornsey, 1969). The flavour and quality of the sausage depends, firstly, on the emulsification of the fat by soluble protein, and secondly, on the stability after mixing, during storage and cooking. Modern sausage manufacturing processes have been summarised by Martin (1980) and Hannan (1974). Both authors contrasted the methods of manufacture of the British sausage with that of the continental or semi-dry product, which is cured and fermented. Karmas (1972) has surveyed various recipes, methods of manufacture and uses of additives in the production of frankfurters. Most natural casings are prepared from the intestines of cattle, sheep and pigs whilst the so called artificial casings are manufactured by solubilisation and regeneration of protein film. Skinless sausages are moulded in artificial casings, which are peeled off before packing.

Composition

The results of analysis of fresh beef and pork sausages are quoted in Table 13.8.

Table 13.8 Range of figures found from the analysis of samples of beef and pork sausage purchased from retail establishments (1968) and (1978)

	Beef sausage				Pork sausage			
	1968 min.	1968 max.	1968 mean	1978 mean	1968 min.	1968 max.	1968 mean	1978 mean
% Water	40·6	52·5	48·2	50·3	32·8	54·0	44·4	45·4
% Fat	16·6	33·3	24·9	24·1	22·7	37·2	31·2	32·1
% Protein (N × 6·25)	7·5	12·7	9·4	9·6	8·6	13·8	10·2	10·6
% Ash	2·2	2·9	2·5	2·5	1·1	3·2	2·1	2·4
% Carbohydrate (diff.)	7·8	18·1	13·2	13·5	3·6	18·2	10·6	9·5
% Defatted meat[a]	25·0	57·9	37·3	35·7	29·0	56·8	40·7	43·7
Total meat[a]	51·8	75·9	61·5	59·8	62·0	91·1	71·8	75·8
% Salt	1·4	2·0	1·7	1·9	0·5	2·0	1·6	1·8
Sulphur dioxide (mg/kg)	45	450	233	—	60	435	198	—
No. of samples			22	20			47	18

[a] Calculated on assumption that protein supplements such as dried milk are absent.

Composition varies widely, since the manufacturer is required to comply with a minimum legal standard that leaves a wide margin for the formulation of 'quality' product. A consumer survey in 1974 of a number of meat products including pork sausage and beef sausage gave average percentages of 41 and 37 for defatted lean, 31 and 27 for fat and 28 and 36 for filler, respectively. Compositional data on the wide range of sausages available in the late 1980s, including sausages sold as 'low fat', are not available. In 1985 an LGC study of nine different brands of pork sausages gave the following range of results: moisture 46·5–56·0 per cent, fat 15·5–30·7 per cent, ash 1·72–3·28 per cent, nitrogen 1·53–2·30 per cent, protein 9·6–14·4 per cent, apparent total meat 55–80 per cent, collagen/protein 6·6–22·7 per cent.

According to FSC (1980), the average fat content of pork sausages was 25·8 per cent. In Leicestershire from 1985 to 1988 the fat content of pork sausages, other than those sold as 'low fat', was on average 19–22 per cent, ranging from 5–35 per cent in 1985 to 14–32 per cent in 1988.

Legislation

As with other meat products, sausages are controlled by the current Meat Products Regulations (see Table 13.3). It is apparent that 'beef sausages' and 'pork sausages' are required to contain at least 50 and 65 per cent total meat (as defined) respectively. Where the type of meat is included in the description, the meat content must include the following minimum percentages of the named meat: beef sausage 50; liver sausage 30; tongue sausage 30 and other types (including pork sausage) 80. The regulations also require any sausage to include a lean meat content of at least 50 per cent of the minimum meat content. In practice lean meat usually contains about 4–12 per cent intramuscular fat (see calculation of meat content later). Under the 1979 Preservative Regulations fresh sausages may contain up to 450 mg/kg sulphur dioxide.

MEAT CONTENT

As indicated on p. 479, meat products, whether subjected to compositional standards or not, the labelling requirements are defined in terms of meat content with subsidiary qualification regarding lean meat, fat and other protein-containing ingredients, including offals. The regulations concentrate on the main ingredient 'meat' and not on any subsidiary constituent. From the manufacturing point of view, the prescribed minimum content assists in the quality control of products by recipe check. From the analytical standpoint, meat consists essentially of water, protein and fat and the practical approaches to analytical control are based on these components. Difficulties in the interpretation of results arise from the ability (or lack of it) to distinguish with sufficient accuracy between nitrogenous components derived from different sources and the variable composition of the raw materials.

Commercial meat products frequently contain as ingredients other raw materials, which contribute either directly or indirectly as nitrogen-containing additives to the gross nitrogen content of the product. Ingredients such as milk powder, wheat and other cereal flours, soya flour and high protein fractions of these products are all used in varying amounts. Yeast extracts and other hydrolysed proteins legitimately contribute to flavour. Nitrogen from these materials cannot readily be distinguished directly from that derived from meat protein, although correction factors based on indirect estimates may be applied in some cases.

'Lean' meat is sometimes regarded mistakenly as chemically fat-free meat, but its use should be applied only to meat free of visible or trimmable fat, as inferred in the Meat Products Regulations.

Analytical meat content

The Stubbs and More (1919) method, with minor modifications, is used to estimate the meat content of meat products. The method is based on an estimation of fat-free meat from total nitrogen content, corrections being applied for the contribution from any cereal filler or other nitrogenous material present. If no correction or incomplete correction is made for other non-meat nitrogenous ingredients, the Stubbs and More calculation may be said to yield 'apparent' fat-free or total meat content. The Analytical Methods committee of the Society for Analytical Chemistry has recommended factors (N_F) for conversion of the nitrogen content of various species into the equivalent fat-free raw meat content (Table 13.9). The values of N_F are the mean nitrogen content of the lean tissue expressed on the fat-free basis. There are no agreed factors for conversion of nitrogen into cooked, cured or processed meat. The meat content of such products may be expressed in terms of 'fat-free raw meat equivalent'. Alternatively, allowances may be made for cooking loss or for the inclusion of nominal amounts of curing solution (FSC, 1980).

Table 13.9 Mean 'nitrogen factor' (N_F) used to convert 'meat nitrogen' into fat-free raw meat

Variety of meat	N_F	Reference to AMC study in *Analyst*
Kidney	2·7	1966, **91**, 538
Tongue, ox or pig	3·0	1967, **92**, 326
Blood	3·2	1968, **93**, 478
Veal	3·35	1965, **90**, 256
Ox liver, pork	3·45	1964, **89**, 630; 1961, **86**, 557; 1986, **111**, 969
Turkey, dark meat	3·5	1965, **90**, 581
Beef	3·55	1968, **93**, 422
Chicken, dark meat	3·6	1963, **88**, 583
Pig liver, turkey, whole	3·65	1964, **89**, 620; 1965, **90**, 581
Chicken, whole	3·7	1963, **88**, 583
Chicken breast, turkey breast	3·9	1963, **88**, 503; 1965, **90**, 581
Lamb	3·6	(see text)

The use of nitrogen factors for calculation of meat content has been the subject of review, especially in relation to the widened labelling provisions of the Meat Product Regulations. In particular, attention has been directed to the factors for individual anatomical portions of chicken, turkey and fresh pork. Many of the factors published by the SAC are recognised as compromise values derived from a large number of analyses and with significant variability due to the nature of the raw material. Utilisation of much of the earlier data has been directed towards application of the derived factors to comminuted products such as sausages rather than to individual cuts of meat.

Work on chicken and turkey carried out at the Food Research Institute, Norwich, has concentrated attention on the development of data on anatomical portions and attempts have been made to discern the effects of age and sex on the factors. Thomas *et al.* (1984) reported data on chicken carcass meat and for the edible offals and blood. Age of the birds examined varied from 21 to 364 days. The general conclusions were that the currently recommended nitrogen factor for raw breast meat is satisfactory but that in the case of products made from dark meat there is a need to distinguish between the generality of 'dark meat' and thigh and drumstick. The authors propose 3·35 and 3·38 respectively as compared with the factor of 3·6 currently accepted for dark meat. Factors were reported for liver (3·53), heart (2·59), gizzard (3·08) and blood (2·47). Grey *et al.* (1983) reported similar conclusions for turkeys and again concluded that further work is necessary to develop factors acceptable for wider use (see also Anon, 1984).

The composition of fresh pork has received attention from a number of workers. A detailed study on pork intended for bacon production has been reported on by Harrison (1983). He has concluded that the use of a single factor of 3·45 for the calculation of fat-free raw pork significantly overestimates the lean in back, belly and middlecut pork and underestimates the lean in fore meat. However, he demonstrated that the factor 3·45 is appropriate for comminuted products derived from various cuts. Revised N_F factors proposed by Harrison for combined lean and fatty tissue are: hind (3·48), back (3·68), belly (3·53), fore end (3·25), whole (3·47). The factors for tissue including rind are significantly higher than those defined above. In a larger study, the Meat Factors Sub-committee of the AMC (1986) has shown that similar conclusions can be drawn and that further investigations are required to refine the data in respect of the effects of breed, sex, weight/age and level of fatness of the carcass. Interim N_F proposals are:

	Lean and subcutaneous fat	Lean, fat and rind
Collar	3·35	3·50
Hand	3·35	3·60
Leg	3·45	3·60
Rib belly/rump belly	3·45	3·70
Rib loin/rump loin	3·60	3·80

The combined lean meat, subcutaneous fat and rind of the whole pig carcass yields a factor of 3·65 whilst the middle cuts in natural proportions yield a factor of 3·75. For the entire carcass, lean plus subcutaneous fat, the N_F remains at 3·45.

As yet there is no agreed nitrogen factor for lamb but calculations from published data on the composition of lambs' meat suggest that a factor of 3·6 is a near approximation.

The typical general calculation using the Stubbs and More procedure depends on the determination of water, protein, fat and ash content of the sample. Carbohydrate content (C) may be determined directly or by difference.

$$C = 100 - (\text{per cent water} + \text{per cent fat} + \text{per cent protein} + \text{per cent ash})$$

The nature of the cereal filler can be detected by identifying the starch and other features by microscopical examination. Appropriate values for nitrogen content of cereal fillers have been described by the SAC. The values recommended are: 2·0 per cent for wheat rusk (SAC, 1965) and 1·8 per cent for pearl barley (SAC, 1968); potato starch and cornflour have negligible nitrogen content. Table 13.10 gives the nitrogen content of some other materials commonly used in meat products.

Assuming the filler is wheat rusk:

$$\text{Per cent filler nitrogen} = N_c = 0{\cdot}02C.$$

If N_T = per cent total nitrogen in sample:

$$\text{Per cent fat-free meat} = DM = \frac{N_T - 0{\cdot}02C}{N_F} \times 100$$

where N_F is the appropriate nitrogen factor for the meat present (Table 13.9).

$$\text{Per cent total meat} = \text{Fat-free meat} + \text{fat}.$$

Table 13.10 Nitrogen contents of various non-meat nitrogen-containing products commonly found in meat products

Non-meat products	Nitrogen (%)
Rusk	2·0
Casein	15·8
Sodium caseinate	14·8
Soya isolate	14·5
Texturised soya	8·0
Soya flour	8·0
Monosodium glutamate	8·3

The calculation of meat content of meat products has been reviewed by Pearson (1975), who drew attention to a number of refinements of the calculation that he devised to allow for intramuscular fat and connective tissue in 'lean'. All the refinements of the basic calculation make assumptions on the level of fat in lean tissue and protein in fatty tissue and according to Cottam (1975) the average level of defatted meat in sausages is enhanced by approximately 1·0 per cent if the formulae elaborated by Pearson are used.

Procedures that have been proposed for the estimation of muscle meat content are based either on estimation of non-protein nitrogenous constituents or on differential solubility of proteins. Total creatinine as an index for meat content has been studied by Dahl (1963) and Gunther *et al.* (1969). The use of the creatinine index has been questioned by Baltes (1968), who pointed out that synthetic creatinine has been used to boost the apparent meat content of comminuted products. Further proposals to counter the use of synthetic materials by use of the carbon isotope ratios have been discussed. Improvements to the method have been proposed by Chicot *et al.* (1983) and by Gabor *et al.* (1984) but the technique seems to be most appropriate for quality control purposes on known formulations. Anderson (1981) used the measurement of actin by an isotope dilution procedure as a method of estimating meat content of extended meat products. He concluded that the method required calibration against products of known formulation if it is to be used successfully.

Rangeley and Lawrie (1976) have suggested that 3-methylhistidine and methylysine contained within the polypeptide structure of actin and myosin could be used as an index of lean meat content. However, in a second paper the same authors (1977) reported difficulties in the determination of the 3-methylhistidine index for samples containing large amounts of fillers and binders; a rapid HPLC method has been described by Skurray and Lysaght (1978). Jones *et al.* (1982) described a fluorimetric procedure for determining 3-methylhistidine; acid hydrolysates obtained by heating the product under reflux in 6 M HCl are neutralised and treated with fluorescamine. The reaction products are examined by HPLC and compared with standard 3-methylhistidine solutions taken through the procedure. The technique is considered to be suitable for quality control of soup formulations, but the authors state that more data are necessary before the method can be used as a procedure to measure meat content.

A method based on selective solubility of non-protein and connective tissue is favoured by German workers. A recent example is that of Herrmann *et al.* (1976). It is claimed that after 1 hour of autoclaving at 130 °C and at pH 9 that all extraneous nitrogen compounds are solubilised and that the residue correlates with lean meat.

Hoffmann (1979) has discussed the problem of direct determination of meat protein content of continental sausage products. The ratio water/protein (Feder Number) is used for assessment of the analytical quality of whole meat products. Typical values are shown in Table 13.1 and see also data obtained for pork by Perrin and Ferguson (1968). For cured meat products an upper limiting value of the water/protein ratio of 4·0 is generally accepted in Europe (Herschdoerfer and John, 1960; Bertling, 1977). In the case of canned meat products to which gelatine has been added, e.g. canned cured meats, the meat content is influenced by the level of curing ingredients used and by the amount of gelatine used to solidify exuded juices. The amounts of gelatine used in small packs (0·5–1 kg) generally correspond to 0·7 per cent as protein. Where this is a declared additive a deduction of 0·7 per cent from the found protein content is regarded as a suitable quality control procedure.

Quality of meat products is also assessed by the quality of the meat used, one of the most important criteria of which, toughness, is influenced by the amount of connective tissue in the cuts used. Hydroxyproline is present in large amounts in collagen, the major protein of skin and connective tissue, but only to a limited extent in muscle protein and plant protein (Table 13.11). Stegeman and Stalder (1967) have described

Table 13.11 Hydroxyproline content of some proteins

Source	Hydroxyproline content per cent, dry fat-free basis
Purified collagen	13·4–14·5
Purified tendon	11·2–13·2
Tendon with muscle	12·3–13·3
Cooked skin	11·0–12·0
Skeletal muscle	0·002–0·07
Plant materials	Negligible

a colorimetric procedure that has been widely used to determine collagen in meat products. A semi-automated version of the technique has been described by Arneth (1974). Studies on the colorimetric methods, including the ISO (1978): No. 3496 procedure, have been reported by Jonas and Wood (1983) and by Csiba (1984). The latter author proposed a modification of the BS/ISO procedure (see p. 500) using acetate–citrate buffer to adjust the pH of the hydrolysate of the sample. He claimed that there was no significant difference in performance between the ISO method and the modified technique; hence in view of the simplification the modified procedure is preferred. This has been confirmed in the authors' laboratory. Coomarswamy (1972) has given examples of the calculations necessary to correct the estimation of the meat content of sausage in cases where excessive use of rind is suspected. Histological procedures for estimation of collagen have been reported by Hildebrandt *et al.* (1977), who stress the need for computer aided systems with projection microscope observation if the method is to be used routinely. Flint and Firth (1983) and Flint and Pickering (1984) used a picric acid–sirius red stain to distinguish collagen, muscle tissue and processed rind in both raw and processed meat products. Collagen data on specific products have been reported by a number of workers, including Board *et al.* (1978), Pailler (1979) and Fey (1977). German legal standards for canned sausage products have been discussed by Prandl (1979).

Lord and Swan (1984) have reported on the examination of cheaper cuts of meat, namely shin beef and diaphragm, and on 24 commercial samples of beefburgers. Using standardised methods of analysis these authors reported that 22 of the samples contained connective tissue/lean meat ratios of less than 20 per cent, the average being 17 per cent and the maximum 28 per cent. These values compare with the literature values of 10–20 per cent connective tissue in well trimmed lean meat.

Meat species identification

After meat has been removed from the carcass, deboned and further processed, the anatomical characteristics of the meat become disguised and it is increasingly difficult for the consumer to recognise the species by normal organoleptic assessment. Meats of a similar colour and texture, such as chicken and rabbit, horsemeat and beef or other bovine species, become almost indistinguishable. The adulteration and partial substitution of beef by other species including condemned meat and kangaroo meat has received considerable publicity in recent times. The incorporation of undeclared species has a religious significance in certain communities; the concern about 'unclean' meat extends not only to the flesh and by-products but also to the methods of slaughter.

The problem of species recognition (speciation) is more difficult as the product is increasingly processed or mixed with other ingredients, especially when the added ingredients are themselves rich in protein.

Whilst horsemeat has long been a common substitute for beef, early attempts at identification were based on chemical titre using the high glycogen content of horseflesh. The protein precipitin test of Castledine and Davies (1968) is an early example of the use of serological reaction methods. Hayden (1979) and Atasi *et al.* (1970) demonstrated the occurrence of cross-reactions of anti-horse serum with other related species. Neither the glycogen titre nor the precipitin test are suitable for processed products. The high ratios of unsaturated fatty acids have been used to characterise the fat of meat species such as horse meat, although it has been shown that the detailed composition of fat is influenced by the diet of the animal. Both GLC and TLC methods have been used in the examination of fats (see Chapter 16). Sayed and Dashlauty (1979) reported the use of TLC on silver nitrate treated silica gel as a method to demonstrate the absence of pork fat in canned meat and sausages. At LGC, pork fat in beef fat has been determined by HPLC of the triglycerides (Palmer and Palmer, 1989). Casas *et al.* (1984) have described an immunoelectrophoretic method for examination of raw meat products and claim that the method can detect the soluble protein from pig meat in admixture with beef, horse and chicken. Kim and Shelef (1986) characterised and identified raw beef, pork, chicken and turkey meats and mixtures thereof semi-quantitatively by gel electrophoresis of sacroplasmic proteins. At LGC, down to 1 per cent of lamb in beef mince has been estimated by isoelectric focusing electrophoresis of extracted soluble proteins coupled with laser densitometry. Patterson *et al.* (1984) described an improved enzyme-linked immunosorbent assay (ELISA) for the detection of beef, sheep, horse, kangaroo, pig, camel, buffalo and goat meat. A limit of detection of 1 per cent is claimed for a positive identification of closely related species in mixtures of raw flesh.

King and Shaw (1984) reviewed the techniques generally favoured for the identification of species in formulated meat products, especially those that have been cooked and pasteurised. Commercially available kits utilising the ELISA method have been described (Anon, 1982b). The then available antisera were of limited value for cooked meats although Hayden (1981) had demonstrated the properties of some antisera to heat stable antigens. King and Shaw concluded that no one technique is suitable for unequivocal identification of species but that a combination of ELISA and isoelectric focusing methods could give evidence of the presence of foreign species at approximately 10 per cent levels in a range of processed products. Kang'ethe *et al.* (1986) applied the immunodiffusion method to a range of products from 14 different bovine species available on the market in Kenya.

Griffiths and Billington (1984) attempted the ELISA assessment of antibodies to beef blood serum to determine indirectly the apparent beef content of joints and model mixutes. They concluded that other meats, rusk and soya flour displayed negligible interference. The apparent meat content was found to depend critically on the cut of meat used, variations from 54 per cent to 176 per cent according to the anatomical source were observed and the authors suggested that the method was more suitable as a qualitative test for beef. Their work demonstrated the limitations of immunoassays of samples with variable residual blood serum content using antisera raised against blood serum proteins. More recently, much improved assays for one meat in another have been developed by using purpose-made, purified antibodies raised against more specific proteins or protein types. An ELISA kit for beef, pork and poultry in cooked products is now available (Cortecs Diagnostics, Deeside, Clwyd, Wales) and at LGC an ELISA procedure has been developed using an antiserum raised against lamb muscle fibre insoluble proteins (Beeches Biotech, Taunton, Somerset) to detect and determine lamb meat muscle in beef mince. Martin *et al.* (1988) have reported an ELISA method for chicken in raw meat mixtures, down to 1 per cent, using a purified antiserum raised against chicken sarcoplasmic proteins. This paper has a useful full list of references on the subject of meat speciation.

Useful background and instructive texts on electrophoresis are by Melvin (1987) and Andrews (1986) and on immunoassays by Voller *et al.* (1979), Johnstone and Thorpe (1987) and Morris *et al.* (1988). Histological examination has been used to detect offals in meat products (Kelch, 1961; Hole *et al.*, 1964) and the presence of mechanically deboned meat in meat products (Bijker *et al.*, 1985). The use of the anserine: carnosine ratio has been proposed for the detection of chicken in meat products. Tinbergen and Slump (1976) reported that the ratio for chicken (2·2–5·5) is sufficiently different from that for beef (0·06–0·2) and for pork (0·02–0·1) to allow the use of the ratio as an indicator of the presence of chicken meat in beef and pork products.

Added protein ingredients

Apart from the adventitious presence of proteins in the normal ingredients of recipe products, the use of protein from blood, milk, egg, cereal and soya as binding ingredients in comminuted meats has been a long-established practice. When used as binders the added proteins improve flavour and texture and also improve the retention of water and fat. Milk powder and milk proteins have been used for many years in sausage and comminuted products, and typically 6 per cent of dried skimmed milk is used in commercial sausage formulations. The presence of skimmed milk powder at this level can be detected by the increased salt-free ash content and by the calcium and the lactose content of the product. In making the allowance for skimmed milk powder it is assumed that the powder contains 1·84 per cent calcium (as CaO), 5·81 per cent of nitrogen and 50 per cent of lactose (McVey and McMillin, 1940). Conventionally, lactose was determined as reducing sugar following fermentation of the other sugars present with yeast. However, direct methods using HPLC or enzyme procedures are now considered more appropriate (Bahl, 1971; Chapter 6).

Milk derived proteins (whey and caseinates) are now widely used. Detection and determination depends on direct methods of analysis, the most common being based on electrophoresis. Olsman *et al.* (1969) described a routine method for the electrophoresis on starch gel slabs of protein solubilised in 2-mercaptoethanol/urea. Since 1970, polyacrylamide gels have found favour. Hofmann (1979) described a system using dodecyl sulphate polyacrylamide gels, on which were obtained patterns characterising meat, egg, soya and milk protein.

Manz (1981) has used serological methods in attempts to differentiate egg and meat protein in known mixtures. He reported little success with mixtures of egg yolk protein and muscle protein. In addition to mechanically recovered meat, another form of animal protein is finding increasing use in meat products. Collagens derived from bone and rind are extracted and sold as 'extenders' for manufactured products. Copeman (1981) has described the manufacturing method, composition and use of commercially available products. He noted that the products have 10–15 per cent protein and illustrated the effect on hydroxyproline content of beef sausage mix formulated with 1–4 per cent bone collagen (4–16 per cent meat equivalent). He discussed the use of the material in relation to the provisions of the legislation applying in 1981.

Protein products isolated from vegetable sources have been used as meat extenders and meat replacers. The FSC (1974) Report on Novel Proteins summarised early developments, whilst Duda (1974) described the processing technology of soya beans and other protein-rich vegetable materials including the characteristics, composition and uses of flours, isolates and texturised products as meat extenders and meat analogues. Seigel *et al.* (1979) have described the preparation of hams by injection of pork muscle with curing brine containing solubilised soya protein. Protein hydrolysates are also used as flavouring ingredients in many meat products (Manley and Swaine, 1979; Murray, 1979). One commercial product derived from fungal protein has been launched as a wholly meat replacement product. The fungal material is harvested from fermentation on a carbohydrate medium and is characterised by a high content of chitin.

In view of the many constituent proteins in meat analogues and extenders, the analytical determination of foreign protein in meat products is a complex one. More than 20 years ago it was possible to correct for soya protein in the calculation of meat content, since it could reasonably be assumed that the soya product would be present as a flour with 10 per cent non-fermentable carbohydrate. Lythgo *et al.* (1941) used the procedure for control purposes. However, as the products have become increasingly sophisticated reliance on this and other indirect indices has become less reliable. This situation parallels that described above for estimation of milk powder content. In view of the effects of the high protein content of these on the apparent meat content as determined through the nitrogen content, there has been a high level of activity in analytical research in the past 15 years. Whilst large particles of textured protein in meat products may be detected organoleptically, their quantification is less easy and when used in comminuted products organoleptic identification is not so easy. Analytical methods used may be divided into five main categories: microscopy; serology; electrophoresis; amino acid/peptide analysis and indirect methods.

Soya products such as grits, flours and concentrates may be recognised under the microscope by the typically shaped palisade and hour glass cells. Reusse (1971) used polarised light to aid detection. Stereological point counting techniques have been used in conjunction with differential staining of the proteins with toluidine blue to estimate soy protein in meat products (Flint and Meech, 1978). Coomaraswamy and Flint (1973) used periodic acid/Schiff reagent to stain the carbohydrate components of soya and Procion Brilliant Blue to counterstain the protein. Flint and Johnson (1979) have used similar procedures to recognise commercial wheat gluten in meat products. Griffiths *et al.* (1981) reviewed the microscopical technique of Flint and Meech and concluded that the method gives low accuracy in determination of certain types of soya material, is time-consuming and has a high cost per determination. However, they point out that microscopic inspection yields information on the types of material present and when used in conjunction with other methods can give useful results.

Gel electrophoresis has received much attention. The method is applicable to solubilised protein, but increasing difficulty is observed with its application to severely heat denatured product. Proteins are extracted from the meat matrix with solutions containing urea (5–10 M) in combination with mercaptoethanol and, in some cases, dodecyl sulphate. The extracting solution is usually buffered to a pH in the range 7–9, although 5 M urea in 30 per cent acetic acid has been used. Most workers now use polyacrylamide gels with an appropriate transport medium containing buffer and extractant at an alkaline pH. Identification and quantitative estimation is made by staining the protein bands with dye. Measurements are made by scanning microdensitometric techniques. Quantitative estimates of soya protein content are subject to normal experimental errors and also depend on the use of standard materials which approximate in form and treatment to those in the samples. At high process temperatures the proteins become progressively denatured, more difficult to extract and the band patterns become faint and loose, characterising fine structure. Armstrong *et al.* (1982) used haemocyanin as an internal standard and claimed to be able to detect down to 0·5 per cent soya protein in uncooked and pasteurised meat products. The method has been examined by collaborative study and inter-laboratory variance was found to be unacceptably high for the method to be used generally. As a control procedure where ingredient material is available the method gives satisfactory performance (Olsman *et al.*, 1985). This study also included a comparison with the ELISA method. The general conclusions are discussed below. The technique of isoelectric focusing, whereby proteins are separated on polyacrylamide gel by difference in isoelectric point by the use of pH gradient, has also been investigated by Llewellyn (1977).

Serological methods rely on the highly specific interaction between an antigen, such as a specific protein, and antibodies present in an antiserum raised against the antigen in an experimental animal. All the methods depend on the specific reaction between antigen and antibody to form an insoluble complex. The principle of application of the method is relatively old; Glyn (1939) used a precipitation test to detect soya flour in sausages, Ferguson *et al.*(1942) refined the technique to make it semi-quantitative, and Degenkolb and Hingerle (1969) developed a routine screening procedure based on the precipitation reactions of anti-soya antisera with saline extracts of meat products. Developments of the method include the use of diffusion techniques with agar gel as a support medium. The procedure has been coupled with the electrophoresis technique to minimise non-specific antigen–antibody reactions, which can arise from the proteins from other sources such as legumes (Koh, 1978). There are now many variations on these techniques available, all dependent on the production of antisera from protein that matches that to be used in products. As in the electrophoresis techniques the effects of process denaturation must be taken into account in the evaluation of results.

Hitchcock *et al.* (1981) have applied the established immuno-diagnostic procedure using enzyme-linked immunosorbent assay to the estimation of soya protein. This method is economical in the use of antibody preparation, which is obtained by injection of representative sample of the protein into a rabbit. Anti-antibodies are raised in sheep against the rabbit antisera and then chemically linked to an enzyme. Measurement of antigen concentration using the enzyme-linked system may then be made in a number of ways. The authors used an indirect procedure in which PVC microtitration plates are first sensitised with antigen. Solutions of antigen and excess rabbit antibody are incubated and transferred to the sensitised wells.

After a further incubation and washout of unbound antibody–antigen an excess of enzyme-linked anti-antibody is added to each well, and again incubated. After washout the retained enzyme complex (inversely proportional to the sample concentration) is assayed colorimetrically with an enzyme reactive species. The method is relatively simple to operate provided supplies of antibody and anti-antibody can be obtained and it appears to be useful as a specific semi-quantitative method. In common with many other techniques of the kind it relies on close matching of the reference protein with that actually used in preparation of the product. Griffiths *et al.* (1984) assessed the commercially available reagents for the ELISA procedure. Experiments were carried out on model meat products containing commercial soy protein ingredients. A method suitable for collaborative study was developed and formed the subject of a paper by Crimes *et al.* (1984) who reported on the collaborative study of the procedure developed. The conclusions of the study were that there are differences in response of the method to different soya products and in common with other methods this is one of the inherent weaknesses of the technique. (These authors, however, seem not to have appreciated the fact that if the immunoassay is calibrated against a soya additive of known nitrogen content, the contribution of nitrogen from the actual unknown soya additive in the meat product can be readily calculated and allowed for in the estimation of meat content.) The performance characteristics are, however, superior to the other techniques based on microscopy and gel electrophoresis. When compared with the electrophoresis procedure (Olsman *et al.*, 1985) the conclusions were that both methods are suitable for qualitative analysis but that more work is necessary to make the methods suitable for surveillance purposes.

Methods based on chromatographic analysis of amino acids and peptides have been used by a number of workers. Computer profiling technique has been applied to the amino acid chromatogram of fully hydrolysed meat products (Lindquist *et al.*, 1975; Olsman, 1979; Lindberg *et al.*, 1985). The procedure is subject to ambiguity of interpretation, especially with complex mixtures or proteins of similar amino acid pattern are involved. Biological variation and processing damage also causes some complication.

Bailey (1976) has described a method based on the analysis of the characteristic soluble peptides formed by enzyme cleavage of the proteins. The sample is first defatted, dried and heat denatured to a constant level by autoclaving, and the proteolytic digest is analysed on an Aminex-A-5 column. Elution pattern characteristics of the protein are observed and quantitative assessments of the relative amounts of protein present can be made. A further modification of the method has been made by Llewellyn *et al.* (1978); in this procedure the proteolytic digest is fractionated by an ultrafiltration membrane before column analysis. These workers reported characteristic peaks that could be correlated with soya protein and muscle meat protein contents. The method is cumbersome for routine use and has a relative limit of detection of 5 per cent soya protein in the total protein. The estimated accuracy is within the range ± 25 per cent of the amount present. Agater *et al.* (1986) have published a critical evaluation of the method and concluded that the peptide procedure for estimation of lean meat content appeared to have little advantage over the 3-methylhistidine method, and that the ELISA method is to be preferred for estimation of soya protein.

There have been many attempts at indirect analysis of soya protein content; the search for an index has included polysaccharides, oligosaccharides, protein bound sugars, free amino acids, free peptides, phytates, saponins, sterols and metals. No one index has been found to be entirely satisfactory for general use. In practice the presence of different ingredients in a product helps to confuse the issue.

A substantial review of the technology of novel proteins including methods of analysis has been issued as a study group report by the EC (1978). Critical analytical reviews have been published by Eldridge (1981), Griffiths *et al.* (1981), Olsman and Hitchcock (1980), Olsman (1975, 1977), Llewellyn (1979, 1982) and Smith (1975).

General methods for the analysis of meat and meat products

Preparation of the sample. To prevent loss of moisture the sample must be a reasonable size in relation to the capacity of the chopper to be used. After preparation, the chopped material must be kept in glass or similar air- and water-tight container. Remove any bone from the sample and chop thoroughly in an electrically driven food chopper with rotating blades. Mix thoroughly, carry out analysis as soon as possible but no later than 24 hours after preparation and storage in a filled, airtight container. Where delay is unavoidable, the moisture content must be determined. The sample may be stabilised by drying *in vacuo* <60 °C or by freeze drying. Samples of fat may be prepared by extraction of the dried material with petroleum ether or hexane and removal of the solvent by spontaneous evaporation. The extracted fat may be examined for characterising values provided the sample is not heated and it is kept in a refrigerator or in a deep freeze.

For the analysis of composite products (meat in jelly, meat pies and sausage rolls). Weigh the whole product, separate the component parts (meat and jelly or meat and pastry), weigh the two constituents and prepare as separate samples. The analysis of the whole product is obtained by simple proportion of the results of analysis of the constituent parts. Note that for pastry the protein content is obtained by use of the multiplier 5·7 for the nitrogen content. Pearson (1973) has presented a set of equations for calculation of the meat content of filled pastry products.

Moisture. Dry at 103 °C 15 g acid-washed sand in a flat metal dish (60 mm diameter × 25 mm) together with a thin glass rod with flat end. Cool and weigh the dish and contents. Weigh in 5 g prepared sample. Add 5–10 ml ethanol and mix the mass thoroughly with the glass rod. Place the dish on a water bath at 60–80 °C, remove the alcohol by evaporation and stir occasionally. Heat the dried mass at 103 ± 2 °C for 2 hours in a drying oven. Cool in a desiccator and weigh. Repeat drying and weighing until successive weighings differ by less than 0·1 per cent of the mass of the test portion (ISO 1442; BS 4401 pt 3).

Ash and salt. Mix 5 g test portion with 1 ml magnesium acetate (150 g/l) in a preheated and preweighed dish (60 mm diameter × 25 mm). Heat the dish on a steam bath for 30 min and incinerate from ambient temperature in a muffle furnace controlled at 550–660 °C. Incinerate for 1 hour at the controlled temperature. Check for white ash, cool and weigh the dish. Correct the weight of ash for the residue (MgO) from 1 ml of magnesium acetate (ISO 936; BS 4401 part 1). Salt in the ash is determined by the Volhard method (see Chapter 2). Alternatively, digest 3 g of sample with 5 ml 0·5 M silver nitrate and 15 ml nitric acid. Decolorise by adding small aliquots potassium permanganate solution and boiling. Add 25 ml water, boil, cool, dilute to 150 ml, and add 25 ml ether. Titrate the excess silver nitrate with ammonium thiocyanate using ferric ammonium sulphate indicator (AOAC) (see also BS 4401 pt 6).

Total phosphorus and polyphosphate. For total phosphorus incinerate 3 g sample to a white ash at 550–600 °C. Cool. Dissolve the ash in dilute hydrochloric acid. Heat to dryness on steam bath then dissolve the residue in dilute hydrochloric acid (1 M). Boil and make up to 100 ml. Determine the phosphorus contents as P_2O_5 by the colorimetric molybdenum blue method (Chapter 2).

Alternatively, digest 3 g test sample for 5 min in nitric acid (20 ml). Cool, then add 5 ml sulphuric acid. Heat and complete digestion by dropwise addition of nitric acid until colourless. Heat until white fumes appear. Cool, then add 15 ml water and boil for 10 min. Complete the determination by the colorimetric procedure (AOAC) or by precipitation as quinoline phosphomolybdate (ISO 2294; BS 4401 pt 10).

Polyphosphates: detect by maceration of 50 g meat product with 15 ml water and 10 g trichloracetic acid. Refrigerate for 1 hour. Filter and shake the filtrate with ether and decant. To the aqueous phase add an equal volume of ethanol. Develop the polyphosphates by TLC on cellulose plates with isopropanol/trichloracetic acid/ammonia. Detect using ammonium molybdate followed by 1-amino-2-naphthol-4-sulphonic acid in metabisulphite (BS 4401 pt 15).

The use of phosphates in meat products has been reviewed by Iles (1973). Addition of phosphates can be suspected if the salt-free ash content is high. Since combinations of sodium and potassium salts are used the use of sodium or potassium contents for diagnosis must be used with caution (Grey *et al.*, 1979). Truman and Dickes (1976) concluded that the ratio of P_2O_5 to protein content is a useful indicator of the use of polyphosphates in the treatment of chickens. Confirmation of polyphosphate by a chromatographic method has been described by Hamence and Kunwardia (1974). O'Neill and Richards (1978) reported an NMR method for the specific identification of polyphosphates in meat and fish products.

Extractable (free) fat. Place the dried portion from the determination of moisture into an extraction thimble. Extract for 4–5 hours in a Soxhlet extraction apparatus (see Chapter 2) with *n*-hexane or light petroleum (40–60 °C). Remove the solvent by evaporation and dry at 103 ± 2 °C for 1 hour. Cool and weigh the fat. Repeat the extraction process with a second flask and fresh solvents. The additional fat extract should not exceed 0·1 per cent (ISO 1444; BS 4401 pt 5).

Total fat. Boil 3–5 g of sample with 50 ml 4 M hydrochloric acid. Dilute and extract the fat with light petroleum or *n*-hexane using the 'wash bottle' technique. Collect the extract in a weighed flask and remove the solvent by evaporation. Dry and weigh the fat (SAC, 1974). Alternatively, pour the hydrolysed mass into a filter paper, wash with hot water, dry the filter paper containing the residue of the sample, roll and insert into an extraction thimble. Complete the method by the Soxhlet technique (ISO 1443). Other methods have been reviewed by Smith (1969) and Pearson (1972).

Nitrogen. Determine by the macro-Kjeldahl method (ISO 937–1978; BS 4401 pt 2, 1980) on 1·5 g prepared sample using anhydrous potassium sulphate (15 g) and copper sulphate (0·5 g) as catalyst and 25 ml conc. sulphuric acid for the digestion. Digest for a minimum of 2 hours (90 min after clarification). Cool. Add excess sodium hydroxide solution and distil into boric acid. Titrate the ammonia with 0·1 M hydrochloric acid. The protein content of the sample is calculated as N × 6·25. Rapid automatic or semi-automatic Kjeldahl type equipment using block digestors are now widely used, see Chapter 2.

Starch. Weigh 25 g prepared sample into a 500 ml beaker. Add 300 ml hot ethanolic potassium hydroxide (5 per cent w/v). Cover with a watch glass. Heat on a boiling-water bath for 1 hour with stirring. Decant through a filter paper and wash the starch residue with hot ethanol. Transfer the residue to a 250 ml beaker with 100 ml hot M hydrochloric acid. Cover the beaker and immerse in boiling water for 2½ hours. Cool, neutralise, transfer to a 200 ml volumetric flask and make to the mark. Clear with potassium hexacyanoferrate and zinc acetate. Complete the determination by volumetric method (Chapter 8). The methods described by the SAC (1974) and in ISO 5554 (BS 4401 pt 12) are similar and use a volumetric procedure involving liberation of iodine and titration with thiosulphate. A polarimetric method has been described by Fraser and Holmes (1958); an enzymatic method using an automatic analyser to determine glucose has been developed by Arneth (1977). This method is applied to a hydrolysate prepared by reaction with perchloric acid.

Eustace and Johnson (1984) and Eustace and Jones (1984) have discussed the analytical procedures generally applied to boneless meat and to meat products to test compliance with the requirements of the Australian export regulations. These regulations require that bulked packed meat is labelled with the chemical lean meat content. Chemical and instrumental methods for the estimation of fat, protein and moisture are reviewed in respect of performance characteristics and convenience of use.

In the case of the estimation of fat content, the direct determination of specific gravity of a 700–800 g sample of lean meat is considered to yield a good value for quality control purposes. The accuracy and precision of the Foss-Let analyser, which measures the specific gravity of a solvent extract, are also regarded as acceptable for the method to be used with meat and meat products.

Methods for the determination of water that are not regarded as suitable for routine control purposes include: capacitance measurement, azeotropic distillation and the Karl Fischer titration. Variations on the conventional oven-drying procedures include infra-red heating and microwave heating. These methods are considered to be equivalent in performance to oven-drying. In particular, the microwave method is considered to be an effective and economical procedure for control since it is also possible to measure the fat content of the sample on the same test sample. Multi-component methods based on infra-red transmission or on near infra-red reflectance have also received considerable attention. The transmission method yields values for fat, water, protein, carbohydrate and ash; all values are acceptable and in particular the method is considered to be well suited for estimation of the fat and protein content. The near infra-red method is considered to give satisfactory values for fat, but the standard error of the protein determination is considered to be unsatisfactory. The indirect methods for fat content have been evaluated by Casey and Crosland (1982). These authors chose the 'Infra-Tester' procedure as a rapid and inexpensive method of estimating the fat content of beef mince. Further assessments of the multi-component procedures have been published by Bostian *et al.* (1984), Davies *et al.* (1984), Lanza (1983), Kruggel *et al.*

(1981) and Bjarno (1982). All the studies reported confirmed that the main problem with either of the available infra-red spectroscopic techniques is with the standard error of the protein measurement.

Hydroxyproline. The following method (ISO 3496; BS 4401 pt 11) is widely used for the determination of hydroxyproline and estimation of collagen content.

REAGENTS

Buffer pH 6·0. Dissolve in water 50 g citric acid monohydrate, 26·3 g sodium hydroxide, 146·1 g sodium acetate trihydrate. Dilute to 1000 ml, mix with 200 ml water and 300 ml propan-1-ol.
Chloramine-T reagent. Dissolve 1·41 g *N*-chloro-*p*-toluene-sulphanamide, sodium salt in 10 ml water. Add 10 ml propan-1-ol and 80 ml buffer solution.
Colour reagent. Dissolve 10 g *p*-dimethylamine benzaldehyde in 35 ml perchloric acid (60 per cent m/m), add slowly 65 ml propan-2-ol.
Hydroxyproline standard. Dissolve 50 mg hydroxyproline in water. Add 1 drop 6 M hydrochloric acid. Dilute to 100 ml. For use, dilute 5 ml to 500 ml then 10, 20, 30 and 40 ml to 100 ml.

Weigh 4 g prepared sample into a 200 ml flask, add 100 ml of a solution of tin(II) chloride in 6 M hydrochloric acid (7·5 g/l). Reflux for 16 hours, filter the hot hydrolysate into a 200 ml volumetric flask. Wash the reflux flask and filter paper three times with hot 6 M hydrochloric acid (10 ml aliquots). Make to the mark with water and mix.

Transfer 25 ml of the digest (or an equivalent to 100–500 μg of hydroxyproline) to a 250 ml beaker, adjust the pH to 8 ± 0·2 with 10 M sodium hydroxide solution. Filter the contents of the beaker into a 250 ml flask, wash the filter with 3 × 30 ml portions of water, make to the mark.

Pipette 4 ml of the solution to a test tube, add 2 ml chloramine-T solution, stand for 20 min. Add 2 ml colour reagent, mix thoroughly, cap the tube and heat at 60 ± 0·5 °C for 20 min. Cool for 3 min, measure the absorbance at 558 ± 2 nm. Correct the absorbance for a reagent blank using water in place of the hydrolysate. Prepare a calibration graph with 4 ml aliquots of hydroxyproline solutions containing 0·5, 1·0, 1·5 and 2·0 μg/ml respectively. (See also Coomaraswamy, 1972 and Josefowicz *et al.*, 1977.

Nitrate and nitrite. Evans (1972) has reviewed methods for determination of nitrate and nitrite in meat products. Removal of ascorbic acid from the extracts is necessary if this is present in amounts exceeding 20 mg/kg. Sen and McPherson (1978) have described a study of the effects of interference by ascorbic acid in the determination of nitrate and nitrite by the colorimetric procedure using diazotisation with sulphanilamide followed by coupling with *N*-(l-naphthyl)ethylene diamine. They show that the interference by high levels of ascorbic acid in extracts from meat can be avoided if a combined colorimetric reagent consisting of equal volumes of 1 per cent sulphanilic acid in 30 per cent acetic acid and 0·1 per cent *N*-(l-naphthyl)ethylene diamine in 60 per cent acetic acid is used (Sen and Donaldson, 1978). Fox *et al.* (1984) demonstrated that nitrite determination using the procedure could be improved by using an alkaline extract to diminish the effect of ascorbic acid. They proposed an alternative colorimetric reagent based on sulphanilamide and l-naphthylamine, but pointed out that l-naphthylamine has the disadvantage of being a potential carcinogen (see Chapter 4, also BS 4401 pts 7 and 8 for the currently accepted methods.)

Ascorbic acid. Extract 50 g prepared sample with 100 ml metaphosphoric acid–acetic acid solution prepared by dissolving 30 g metaphosphoric acid in 1000 ml water containing 80 ml glacial acetic acid. Mix thoroughly in a laboratory homogeniser. Centrifuge and decant the supernatant through an acid washed filter paper (Whatman 541). Titrate 2 ml of extract with a solution of 2,6-dichloro-*N*-*p*-hydroxyphenyl-*p*-benzoquinone monoimine (30 g in 200 ml water) until a permanent pink colour persists for at least 1 min. 0·1 ml ≡ 7 μg total ascorbate.

If erythorbic acid is present the titre will include this isomer. Check for the presence of erythorbic acid by TLC on cellulose treated with metaphosphoric acid solution. Develop in water saturated butanone and detect by use of the titrant as spray reagent. A method for detection and estimation of the two isomers has been reported by Bunton *et al.* (1979). The method used TLC separation and the microfluorimetric procedure of Deutsch and Weeks (1965) for the estimation.

Sulphur dioxide. Suitable methods have been described in Chapter 4. The decolorisation of malachite green is used as a qualitative test (AOAC). The distillation procedure is preferred for estimation.

Urea and ammonium salts. A clinical test strip procedure using urease to release ammonia has been described by Honikel and Sauer (1975). The sample is extracted with physiological saline. After homogenisation and centrifugation the clear extract is adjusted to pH 8·2 with sodium hydroxide. The extract is pipetted on to a reaction stick, which is placed into a closed cell and incubated. The amount of urea present is proportional to the stain length on the test stick.

Glutamic acid

An enzyme colorimetric procedure has been recommended by ISO and BSI (ISO 4134; BS 4401 pt 14), suitable for cured meats (Table 13.7), meat extracts, bouillon cubes, etc. It is available as a test combination kit from Boehringer Mannheim.

REAGENTS

Perchloric acid 1·0 M. Dilute 8·6 ml perchloric acid 70 per cent (m/m) to 100 ml with water.
Potassium hydroxide 2 M. Dissolve 56·1 g of potassium hydroxide in water, dilute to 500 ml.
Buffer solution pH 8·5. (1) Dissolve 1·86 g octylphenol-decaethyleneglycolether (Triton X–100) dilute to 100 ml. (2) Dissolve 0·86 g dipotassium hydrogen phosphate and 0·007 g potassium dihydrogen orthophosphate in water. Dilute to 100 ml. Mix 20 ml (1) with 5 ml (2).
Nicotinamide adenine dinucleotide (NAD). Dissolve 0·025 g NAD in 5 ml water.
Iodonitrotetrazolium chloride (INT). Dissolve 0·030 g INT in 50 ml water and store in a brown flask.
Diaphorase. Dissolve 0·003 g lyophilised diaphorase in 1 ml water.
Glutamate dehydrogenase (GIDH). 10 mg/ml solution.
L(+) glutamic acid; dissolve 0·050 g of L(+) glutamic acid in 25 ml water. Adjust to pH 7 with potassium hydroxide solution, dilute to 50 ml. Dilute 1 + 49 for use.

METHOD

Weigh 50 g of the prepared sample into a laboratory homogeniser. Add 100 ml cold (0 °C) M perchloric acid and homogenise. Transfer the homogenate to a 100 ml centrifuge tube and centrifuge at 3000 r.p.m. for 10 min. Move the fat layer. Decant the supernatant through a fluted paper into a 200 ml flask. Discard the first 10 ml of the filtrate. Transfer 50 ml of filtrate to a 100 ml beaker. Adjust to pH 10 with potassium hydroxide and make to 100 ml in a one mark flask. Cool in ice water. Filter through a fluted paper, discarding the first 10 ml of the filtrate.

Pipette 25 ml (*V* ml) into a 250 ml volumetric flask, dilute to the mark with water (maximum concentration of L(+) glutamic acid 30 mg/l). Adjust the temperature of test solution and the pH 8·5 buffer to 20–25 °C. Pipette in turn into each of two photometric cells: 2·5 ml buffer, 0·20 ml NAD, 0·20 ml INT, and 0·05 ml diaphorase. Into one cell pipette 0·50 ml water and into the other 0·50 ml of test extract. Mix the contents of each cell with a plastic spatula and read the absorbance against the air at 492 nm. Retain the cells for the enzyme reaction.

$$A_1 = \text{absorbance of test solution.}$$
$$A_{1B} = \text{absorbance of blank.}$$

Pipette 0·05 ml GIDH solution on the plastic spatula, mix with the contents of one of the cells. Repeat the operation with the second cell. Read the absorbance of each cell at 492 nm after 10 min and every 2 min until a constant rate of increase of absorbance per unit time is obtained. Plot the absorbance against time for each cell. Extrapolate the linear portion of the curve to zero time. Read the extrapolated absorbance values.

$$A_2 = \text{absorbance } (T \equiv 0) \text{ of the test solution.}$$
$$A_{2B} = \text{absorbance } (T \equiv 0) \text{ of the blank solution.}$$

Repeat the operations above using 0·5 ml standard L(+) glutamic acid in place of the 0·5 ml test extract. The appropriate values are:

$$A'_1 = \text{absorbance of standard}$$
$$A'_{1B} = \text{absorbance of blank}$$
$$A'_2 = \text{absorbance } (T \equiv 0) \text{ of standard}$$
$$A'_{2B} = \text{absorbance } (T \equiv 0) \text{ of blank.}$$

CALCULATION

$$\Delta A = (A_2 - A_1) - (A_{2B} - A_{1B})$$
$$\Delta A' = (A'_2 - A'_1) - (A'_{2B} - A'_{1B})$$

$$\text{L(+) glutamic (per cent m/m)} = \frac{\Delta A}{\Delta A'} \times \frac{1}{V \times m}\left(100 + \frac{M \times m}{100}\right)$$

where V = volume in millilitres of filtrate taken to give test extract.
M = moisture content of sample as percentage m/m.
m = mass in grams of test portion.

Glucono-δ-lactone. (e.g. in cured meats (p. 482) and Table 13.7).
An enzyme ultraviolet procedure has been recommended by ISO and BSI (ISO 4133; BS 4401 pt 13) and is available as a test combination kit from Boehringer Mannheim.

REAGENTS

Perchloric acid 0·4 M. Dilute 17·3 ml perchloric acid (70 per cent m/m) to 500 ml with water.
Potassium hydroxide 2 M. Dissolve 56·1 potassium hydroxide in water. Dilute to 500 ml.
Buffer solution pH 8·0. Dissolve 2·64 g glycylglycine and 0·284 g magnesium chloride hexahydrate in 150 ml water. Adjust to pH 8 with potassium hydroxide. Dilute to 200 ml with water.
Nicotinamide adenine dinucleotide phosphate (NADP). Dissolve 50 mg NADP disodium salt in 5·0 ml water.
Adenosine-5-triphosphate (ATP). Dissolve 250 mg ATP disodium salt and 250 mg sodium hydrogen carbonate in 5 ml water.
6-Phosphogluconate dehydrogenase (6-PGDH). Commercial suspension containing 2 mg 6-PGDH/ml from yeast.
Gluconate kinase (GK). Suspension containing 1 mg/ml from *E. coli*.

METHOD

Weigh 50 g of the prepared sample into a laboratory homogeniser. Add 100 ml cold (0 °C) 0·4 M perchloric acid and homogenise. Transfer the slurry to a 100 ml centrifuge tube. Centrifuge at 3000 r.p.m. for 10 min. Move the fat layer. Decant the supernatant through a fluted filter paper into a 200 ml conical flask and discard the first 10 ml. Transfer 50 ml of the filtrate to a 100 ml beaker. Adjust to pH 10 with potassium hydroxide and make up to 100 ml in a volumetric flask with water. Cool in ice for 20 min. Filter through a fluted paper. Discard the first 10 ml. Pipette 25 ml of the filtrate (V ml) into a 250 ml volumetric flask. Dilute to the mark with water (maximum concentration of D(+) gluconate is 400 mg/l). This is the prepared extract.

Pipette into each of two photometric cells: 2·5 ml of pH 8 buffer, 0·10 ml NADP, 0·10 ml ATP. Into one of the cells pipette 0·20 ml of prepared extract; into the other 0·20 ml water. Pipette 0·05 ml of 6-PGDH suspension on to a plastic spatula, mix with the contents of one of the cells. Repeat the operation with the second cell. Read the absorbance of each cell against air at 365 nm after 5 min. Retain the cells for the reaction.

A_1 = absorbance of test solution.
A_{1B} = absorbance of blank.

Pipette 0·01 ml of GK suspension on to the plastic spatula. Mix with the contents of one of the cells. Repeat the operation with the other cell. Read the absorbance of each cell at 365 nm after 10 min and every 2 min until a constant rate of increase of absorbance is obtained. Plot the absorbance against time and extrapolate the linear part of the curve back to zero time.

A_2 = absorbance (T ≡ 0) of the test solution.
A_{2B} = absorbance (T ≡ 0) of the blank solution.

CALCULATION

$$\Delta A = (A_2 - A_1) - (A_{2B} - A_{1B})$$

Glucono-δ-lactone per cent by mass in prepared sample

$$= \frac{15\,058 \times \Delta A}{V \times m}\left(100 + \frac{M \times m}{100}\right)$$

Where V = volume in millilitres of filtrate to make prepared extract.
M = moisture content of prepared sample per cent m/m.
m = mass in grams of test portion.

Total creatinine

REAGENTS

Picric acid solution. Prepare a 1·0 per cent m/v solution. Standardise against 0·1 M sodium hydroxide using phenol red and filter. Store for at least 4 days in the dark.

Standard creatinine solution. Dissolve 0·4008 g creatinine zinc chloride (dried at 100 °C) in 0·1 M hydrochloric acid. Transfer to a 250 ml volumetric flask and make up to the mark with the 0·1 M acid. Pipette 10 ml into a 100 ml volumetric flask and dilute to the mark with water (1 ml = 0·1 mg creatinine). Use the solution soon after preparation.

PROCEDURE

Weigh 10–25 g prepared sample into a 150 ml beaker. Add 5–10 ml cold ammonia free water. Triturate with a further 50 ml cold water. Stir at 3 min intervals over 15 min. Stand for 2–3 min and decant through a filter paper into a 500 ml volumetric flask. Repeat the extraction three times with 50 ml portions cold ammonia free water and four times with 25 ml portions cold water. Transfer the insoluble material to the filter. Wash with three portions (10 ml) of cold water. Drain thoroughly, mix and dilute the extract to volume.

Measure 150 ml extract onto a 250 ml beaker. Evaporate to 40 ml on a steam bath stirring occasionally. Neutralise to phenolphthalein (spotting plate). Add 1 ml 0·1 N acetic acid and boil gently 5 min. Filter. Wash beaker four times with hot water. Wash the residue on filter thoroughly and drain.

Evaporate filtrate and washings to 5–10 ml. Transfer to a 50 ml volumetric flask. Add 10 ml 2 N hydrochloric acid and mix. Hydrolyse for 2 hours in a boiling-water bath or alternatively for 20 min in an autoclave at 117–121 °C.

Cool, then add 10 ml of 2 M sodium hydroxide and dilute to 1:1 with water. Pipette 5 ml into a dry 100 ml volumetric flask and add 15 ml water. Also, commence the blank by adding 20 ml water to a similar flask. To each add 20 ml of picric acid solution and 2·5 ml of 2 M sodium hydroxide and place on a water bath at 21 ± 1 °C for 15 min, swirling frequently. Then make up to the mark. Mix and measure the test solution against the blank at 500 nm in a 1 cm cell within 15 min. Change the blank after every three readings as the optical density alters with temperature. Prepare the calibration graph over the range 0–1·0 mg creatinine by following the same procedure as for the test. For this add to a series of 100 ml volumetric flasks 0, 1, 2, 3, 4, 5, 6, 7, 8, 9 and 10 ml of standard creatinine solution, water to give a total volume of 20 ml and then 20 ml picric acid solution. Construct the graph relating optical density to the creatinine concentration in mg/100 ml.

The analysis of meat extracts is carried out on a 10 ml aliquot of a 10 per cent m/v solution commencing with the hydrolysis step. The free creatinine content of meat extracts is determined by omitting the hydrolysis step and carrying out the colorimetric reaction with picric acid (AOAC). Creatinine × 1·16 = free creatinine. Other methods have been described by Hadorn and Kok (1962) and Dahl (1963). The latter method includes a separation of creatinine and arginine on cation exchange resin. The author showed that significant quantities of arginine are present in certain offals and that other interfering substances are present in milk powder. Typical values for a range of tissues are given. A method using creatinine kinase has been described by Ettel and Tuor (1977). The results obtained on a variety of samples including meat, soups and sauces compared favourably with those obtained by three alternative procedures. An HPLC method for application to meat extracts, bouillons and stock cubes has been described by Battaglia (1978); see also Csiba (1984) for a modification of the neutralization step (p. 491).

pH. adjust the temperature of the test portion of prepared sample to 20 ± 2 °C. With suitable electrodes the measurement may be made by direct insertion. If the prepared sample is dry, homogenise with an equal quantity of CO_2 free water. Immerse the electrodes and take the reading at 20 ± 2 °C. (Chapter 2).

Water activity. A discussion of the principles of measurement of a_w is given in Chapter 2; methods have been reviewed by Prior (1979). A method of calculation of a_w values of dry sausage from salt and moisture content has been described by Demeyer (1979).

FISH

Introduction

Commercial fish mostly comes within the following classes: (1) round white fish, e.g. cod, haddock, whiting; (2) flat white fish, e.g. plaice, halibut; (3) fatty fish, e.g. herring, mackerel and (4) freshwater or anadromous fish, e.g. salmon and trout. All of these four classes are teleosts, i.e. have bony skeletons. The elasmobranchs, e.g. skate, dogfish, on the other hand have cartilaginous skeletons. Shellfish include the crustacea and molluscs. The crustacea have jointed outer skeletons, e.g. crab, shrimp and prawn. The molluscs include the snail-like univalves such as the periwinkle and whelk and the bivalves with double-hinged shells, e.g. oyster, mussel, cockle.

Fishing practice has changed radically in the past few years. Much of the deep sea fish harvested is processed, filleted and frozen at sea rather than transported as 'fresh' fish in ice. Fillets may be skinned and frozen into block. During processing white fish is dipped in polyphosphate solution to reduce drip loss. It may be treated with ascorbic acid as an antioxidant. Mechanically deboned fish flesh (mince) may be added to the fillet blocks as filler when the material is to be used in production of fish fingers. Prepared minced fish is also produced from cheaper fish varieties. Polyphosphate and sodium chloride are added to minimise protein denaturation (Keay, 1976). Mince blocks are used to prepare cheaper quality fish fingers and other homogenised products such as fish cakes and fish sticks (Ghadi and Lewis, 1977). Baker (1978) and Mann (1977) have discussed the utilisation of underused fish species by way of the preparation of minced fish blocks as a raw material for the preparation of moulded products. An extension of storage life and quality, over that obtained by chilling in ice, is obtained by storage at −18 °C or below. Blocks of minced fish are more susceptible to oxidation than the same species in the form of fillets in block. The condensed phosphates are widely used to assist in the preservation of minced fish blocks although the oxidative rancidity is inhibited more successfully by use of BHA, BHT and the other phenolic antioxidants. Excessive use of phosphates and salt can lead to a rubbery texture on prolonged storage. Alginates have been used to bind moulded products prepared from minced fish. Denaturation is minimised by 'fast freezing' through the temperature range 0–5 °C. The material is also protected from 'freezer burn' by spraying with water to form an ice film (glaze) over the pieces, fillets or blocks. Assessment of denaturation by cell fragility testing has been described by Love and Muslemuddin (1972a, b). The muscle is homogenised with formaldehyde solution and the optical density of the homogenate is determined. Salfi *et al.* (1985) have reported a method for the differentiation of fresh and frozen-thawed fish muscle using an electrophoretic method to detect mitochondrial aspartate aminotransferase in muscle extracts. Fresh muscle extracts demonstrated a single isozyme band whilst the frozen-thawed extracts developed additional bands which migrated to the anode.

Composition

The main constituents of fish are water, protein and fat. Some examples of the compositional data for various species are shown in Table 13.12. From this it is apparent that the teleostean types can be roughly classified in relation to the fat content. With the fatty fishes such as the herring the fat content varies considerably according to the age and size of the fish and the season, being at a maximum in the autumn and lowest at the end of the winter. A study of the composition tables in the literature shows that the sum of the percentages of water, fat, protein (N × 6·25) and ash in fish amounts usually to between 100·2 and 100·6. This discrepancy is caused by the use of the conventional factor for the conversion of nitrogen to protein of 6·25, whereas the factor for the major protein in fish, myosin, is closer to 6. The SAC (1966, 1971) has recommended nitrogen factors (on the whole fish including fat) for cod (2·85) and coal fish or saithe (2·90), and in 1973 published factors or ranges for the N content, without making any specific recommendations, for many other species of fish (Table 13.13) (see also Table 13.17 and page 516 for FAC, 1987 data). In general fish is a better source of vitamins and minerals than meat. White fish flesh contains little of vitamins A and D, but fatty fish are rich in D. The amounts of B-group vitamins present in fish is about the same as in meat. Fish is a good source of phosphorus and magnesium. Compared with lean meat, it contains 100 times as much iodine, less iron and about the same amount of copper (Paul and Southgate, 1978).

Legislation and standards

The main interaction of legislation with production and sale of fish relates primarily to Public Health aspects of harvesting, handling and processing. The principal enactments are Public Health (Shellfish) Regulations 1934 (SR and O 1934, No. 1342), printed as amended by SI 1948 No. 1120) and the Therapeutic Substances (Preservation of Raw Fish) Regulations 1964 (SI 1964 No. 883). The provisions of the Food Hygiene Regulations are supplemented by two codes of practice (Nos 3 and 4, 1960) on the handling of fish during transportation and retail. Due to the high incidence of *C. botulinum* in commercial trout farms, the DHSS have issued an advisory memorandum on handling, processing, distribution and cooking of fresh, frozen and smoked trout. The appropriate designations of commercial fish are listed in Schedule 1 of the Food Labelling Regulations 1984.

Restrictions on the use of additives with raw fish are essentially similar to those for raw meat. Colours and certain antioxidants (the gallates, BHA, BHT and ethoxyquin) are not permitted; specified preservatives may be used only in products scheduled in the Preservatives Regulations. These include marinated herring and mackerel, to which the benzoates may be added, and canned crab meat, prawns, shrimps and scampi, to which SO_2 may be added. Other additives are not specifically prohibited, but their misuse could lead to an offence under Section 2 of the Food Act. Fish, canned fish, dried fish and shellfish are scheduled products in the Lead in Food Regulations. The general limit in the Arsenic in Food Regulations does not apply to fish, which naturally contain more than 1 mg/kg. The philosophy of the exemption for 'naturally occurring' arsenic in sea food was discussed by the FACC in its review of the relevant regulations (Report No. 39: 1984). A number of Codex standards have been developed for fish and fish products. These are listed in Table

Table 13.12 Percentage composition of the edible portion of fish

Class	Species	Water (%)	Fat (%)	Protein N × 6·25 (%)
White round	Cod, haddock, whiting	79–84	0·1–0·9	15–20
White flat	Plaice, sole	77–81	0·5–4·0	16–19
White flat	Halibut	75–80	0·5–9·5	15–19
Fatty	Herring	60–75	7–30	14–20
Fatty	Mackerel	60–75	2–20	17–23
Fatty fish	Eel	57–82	2–28	17
Freshwater	Salmon	67	0·3–15	16–25
Shellfish	Crab, fresh	73	5	20
Shellfish	Lobster, boiled	72	3	22
Shellfish	Prawns, cooked	70	2	23
Shellfish	Shrimps, cooked	62	2	24
Shellfish	Mussels boiled	79	2	17
Shellfish	Oysters	86	1	11
Elasmobranch	Skate	77–82	0·2–2	18–24
Elasmobranch	Dogfish	75	4–6	20
White fish roe	Cod, haddock roe	73	1–2	21·5
Fatty fish roe	Herring roe	73	3·3	22

Extensive tables relating to the composition of fish have been prepared by Vinogradov (1953) and Love *et al.* (1959). Dyer *et al.* (1977) have reported on the proximate analysis of frozen fish fillets and steaks. Paul and Southgate (1978) have also reported data on raw and cooked fish.

13.14. Although there are a few compositional requirements, the standards list additives that are allowed in production of frozen products: polyphosphates (0·5 per cent m/m as P_2O_5) for prevention of drip loss, ascorbic acid (0·1 per cent) as antioxidant and in the case of shellfish, sulphur dioxide (0·01 per cent m/m in the raw product) as preservative.

Inspection of fish

If fish is presented in frozen form it is usual to attempt a measurement of thaw loss (glaze). Since there is difficulty in interpreting the test result due to water migration during storage there are no universally accepted methods for this measurement; the AOAC recommends a cold water spray procedure followed by a 2 min drain period and weighing to obtain the thawed weight of the product. The glaze is equivalent to gross weight minus the thawed drained weight as a percentage of the gross weight. Other procedures include controlled thawing followed by mopping with absorbent towels to remove adhering moisture, or overnight thaw at a controlled temperature and humidity. The problem of measurement of water pick up in processing and of glaze added has been discussed by the FSC (1978) in its Report on Water in Food; see also p. 516 regarding glazed and coated fish products. The committee recommended that the industry be encouraged to develop appropriate codes of practice and that they conform to agreed labelling practices. The thawed separated fish or degutted whole fish as appropriate, is inspected by candling for the presence of parasites and evidence of disease. Parasites such as worms and lice have occasionally been reported in fish, but as they tend to reside in the stomach and intestine they are normally removed during gutting. Diseases causing tumours, ulcers and boils have also been reported occasionally in fish, but poisoning in humans due to the ingestion of fresh (unprocessed) fish is rare. Shellfish have been the cause of outbreaks of poisoning,

Table 13.13 Nitrogen contents of raw fish species (SAC, 1973). See also Table 13.17

Common name	Range of nitrogen contents (%)[a]
Anchovy	2·45
Brill	3·17
Catfish (rockfish)	2·58–3·14
Cod	2·21–3·20[b]
Crab	2·4
King	1·10–2·36
Norwegian	2·14–3·43
Crayfish	3·85
Dab	2·05–3·04
Dogfish	2·02–3·24
Eel	1·86–3·41
Flounder (fluke)	2·69–2·77
Haddock	2·33–3·25
Hake	2·64–2·98
Halibut	2·78–3·27
(Greenland, black, mock)	1·98
Herring	2·56–3·41
Lemon sole	2·63–2·94
Ling	2·63–3·56
Lobster	2·70
Mackerel	2·57–3·66
Pilchard	2·45–3·20
Plaice	2·51–3·02
Pollack (lythe)	3·06–3·46
Porbeagle	3·68–4·32
Prawn (deep water)	2·38–2·86
Redfish	2·69–3·17
Saithe (coal fish, coley)	2·51–3·46[c]
Salmon:	
Atlantic	3·60
Sockeye	2·83–3·50
Chinook	3·06
Pink	3·20
Sardine	3·07
Shark (basking)	2·43
Skate (ray, roker)	2·92–3·87
Shrimp	1·68–3·78
Trout	2·80–3·67
Tuna	3·84
Whiting	3·04–3·18
Whiting (blue)	2·22–2·55
Witch	2·33–2·52
Wrasse	3·08–3·15

[a] Where a range is not given the figure is the average result of a number of determinations.
[b] SAC recommended mean for cod 2·85.
[c] SAC recommended mean for coley 2·90.
Note: unlike nitrogen factors for meat, which are expressed on the fat-free basis, the nitrogen content (nitrogen factors) for fish are not on the fat-free basis.

however, and Reynolds (1968) proposed an action limit of 15 *E. coli*/g of flesh. Blackwood (1978) has discussed the monitoring of the microbiological quality of fish at all stages from harvesting to domestic use. Beckers *et al.* (1981) have carried out a comparative study of the microbiological quality of frozen pre-cooked and peeled

Table 13.14 Codex Standards for fish and fishery products

Canned Pacific salmon	3–1981
Canned shrimps and prawns	37:1981
Canned tuna and bonito	70:1981
Canned sardines	94:1981
Canned crab meat	90:1981
Quick frozen gutted Pacific salmon	36:1981
Quick frozen fillets of cod and haddock	50:1981
Quick frozen fillets of flat fish	91:1981
Quick frozen fillets of ocean perch	51:1981
Quick frozen shrimps or prawns	92:1981
Quick frozen fillets of hake	93:1981
Quick frozen lobsters	95:1981
Canned mackerel and Jack mackerel	119:1981
Codes of hygienic practice	
Fresh fish	CAC/RCP 9:1976
Canned fish	CAC/RCP 10:1976
Frozen fish	CAC/RCP 16:1978
Shrimps and prawns	CAC/RCP 17:1978
Molluscan shellfish	CAC/RCP 18:1978
Lobsters	CAC/RCP 24:1979
Smoked fish	CAC/RCP 25:1979
Salted fish	CAC/RCP 26:1979
Minced fish	CAC/RCP 27:1983
Crabs	CAC/RCP 28:1983

shrimp and prawns from South East Asia and from the North Sea. The authors concluded that the product from both sources would not pass the draft end-product standards proposed by FAO/WHO. They also noted that ammonia content is a valuable objective quality index for shrimp from the North Sea, but that many of the samples taken from South East Asia showed evidence of spoilage before cooking and in these circumstances the cooking procedure will considerably reduce ammonia content. The problem of occurrence of *C. botulinum* in trout reared on trout farms has been noted on p. 505. Consumption of scombroid species (mackerel) has been associated with allergic reactions said to be due to the levels of histamine in the flesh. Kietzmann (1978) has discussed the formation of histamine by the action of bacterial histidine decarboxylase. The bacteria responsible are *Proteus morganii*, *E. coli* and *Clostridium welchii*.

As fish spoils, various physical changes become apparent, namely the odour deteriorates, the surface becomes dull and the slime on it becomes opqaue, the flesh softens, the eye pupil becomes turbid and a reddish coloration develops along the backbone. Baines and Shewan (1965) have presented organoleptic scales for assessing the raw odour and cooked flavour of white fish. In spite of the extensive search by the world's fish research stations for a rapid scientific method that would be suitable for the routine assessment of the degree of spoilage in fish, most inspectors still rely largely on appearance, odour, etc. In borderline cases, however, chemical determinations may be of considerable value.

Quality assessment of filleted fish includes examination for evidence of the presence of bones. The thawed fillets may be examined by feeling with the finger tips; alternatively bones may be separated from tissue by digestion of the tissue, treatment with dilute acid or alkali or by enzymes.

Young and Whittle (1985) have demonstrated tristimulus colour measurement methods for application to minced fish as an aid to quality assessment and grading of product. They point out that developments in instrument miniaturisation could lead to wider application of the technique as an industrial control procedure.

Analysis of raw fish

SAMPLE PREPARATION

The method of preparation of the sample depends on the form in which the fish is received and the analysis to be carried out. For proximate analysis of frozen fish first place it in a plastic bag and thaw by immersion in an agitated water bath held at approximately 20 °C. Judge the completion of thawing by gently squeezing the bag until no hard core or ice crystals can be felt. Render the thawed or wet fish into a uniform mass either by double mincing or the use of an electrically driven rotary chopper. Store the comminuted sample in a completely filled airtight container. Analyse as quickly as possible. When storage is necessary keep at a temperature not exceeding 4 °C.

For estimation of amines and total volatile bases, for frozen fish, thaw at a temperature not exceeding 4 °C; mince or chop in a rotary homogeniser. Ensure that the product remains at or about 4 °C by pre-cooling the equipment. Mix the sample and analyse immediately.

ASSESSMENT OF SPOILAGE OF RAW FISH

Fish stored in ice spoils as a result of bacterial and enzyme action, which results in the formation of volatile bases, in particular trimethylamine (TMA), dimethylamine (DMA) and ammonia (Wong *et al.*, 1967). Trimethylamine oxide is reduced during spoilage to TMA whilst ammonia is the end-product of protein breakdown. The amounts of TMA and total volatile nitrogen (TVN) present in the fish are commonly used as indices of fish spoilage. As a result of the analysis of 262 samples Klein and Stoya (1979) have concluded that no correlation was apparent between sensory characteristics and TMA content but that there was a significant correlation between TMA and TVN. Hiltz *et al.* (1976) have noted on storage of hake at −10 °C that DMA production was associated with deterioration in organoleptic quality, including texture of the fish. Lipid hydrolysis and oxidation was also noted.

Using the AMC method detailed below white fish is generally considered to be fresh if the TVN is less than 20 mg N/100 g. If the TVN figure reaches 30 mg N/100 g most authorities consider the fish to be stale, whilst at a level of 40 mg N/100 g the fish is regarded as unfit for consumption (Lang, 1979). With fatty fish the rancidity of the fat is considered to be a more useful indicator of fitness for consumption. Fish oil may be extracted by maceration with chloroform. Peroxide values on the extracted oil in the range 10–20 ml of 0·002 M thiosulphate/g correspond to incipient spoilage. An increase in the acid value due to hydrolysis of the fat is also used as an indicator of spoilage. Charnley and Davies (1944) used a maximum value of 2·75 in the acceptance grading of canned herring. Fresh water fish quality has been assessed generally from fat rancidity values. The flesh contains little trimethylamine oxide so that volatile nitrogen consists essentially of ammonia (Pearson and Muslemuddin, 1968).

Burt (1977) has discussed the limited value of the fish spoilage indices based on determination of volatile amines. He points out that the test measures the effects of the action of spoilage micro-organisms rather than the effects of ageing, since they ignore autolytic aspects of deterioration. He has studied the degradative changes in fish muscle tissues associated with the hydrolysis of adenosine phosphate and has shown that the hypoxanthine content rises throughout the commercially important periods of storage at chill temperatures. Burt also points out that hypoxanthine itself has a bitter flavour and that the degradative pathway involves inosine 5′-monophosphate at an intermediate stage. The latter compound is a flavour enhancer, hence the hypoxanthine index has a special significance in relation to the organoleptic assessment of fish staleness.

In view of the allergic response of certain individuals to histamine derived from histidine, these two measurements are considered to be of importance in the screening of fatty fish in the assessment of fitness for consumption.

DETERMINATION OF TOTAL VOLATILE BASES AND TRIMETHYLAMINE

The method recommended by the AMC (1979) for determination of TVN and TMA is based on a semi-microdistillation procedure. Extracts or solutions are made alkaline with sodium hydroxide. The bases are steam distilled into standard acid and back-titrated with standard alkali. Formaldehyde is added to the neutralised mixture and the acid released is equivalent to the volatile bases other than trimethylamine.

Procedure

Weigh 100 ± 0·5 g of prepared sample into a homogeniser with 300 ml of 5 per cent m/v trichloroacetic acid. Run the homogeniser to obtain a uniform slurry, filter or centrifuge to obtain a clear extract.

By pipette, transfer 5 ml of the extract to a semi-microdistillation apparatus. Add 5 ml 2 M sodium hydroxide solution. Steam distil. Collect in 15 ml 0·01 M standard hydrochloric acid. Add indicator solution (1 per cent rosolic acid in 10 per cent v/v ethanol). Titrate to a pale pink end-point with 0·01 M sodium hydroxide. Add 1 ml 16 per cent m/v neutralised formaldehyde for every 10 ml liquid in the titration flask. Titrate the liberated acid with 0·01 M sodium hydroxide.

Calculation

$$\text{Total base nitrogen} = \frac{14(300 + W) \times V_1}{500} \text{ mg/100 g}$$

$$\text{Trimethylamine nitrogen} = \frac{14(300 + W) \times V_2}{500} \text{ mg/100 g}$$

where V_1 ml = volume standard acid consumed in the first titration.
V_2 ml = volume standard acid released for the second titration.
W = water content of the sample g/100 g.

A colorimetric method for determination of trimethylamine has been described by the AMC (1979). The method uses the extraction procedure with trichloracetic acid described for the determination of TVN. Other bases are complexed with formaldehyde. Toluene is added followed by a solution of potassium hydroxide. Trimethylamine is extracted into the toluene layer and reacted with picric acid to yield a coloured complex. Botta *et al.* (1984) examined six different methods for determination of TVN in relation to assessment of the quality of fresh cod. They concluded that extraction with magnesium sulphate combined with steam distillation gave the most reliable results. Extraction with trichloracetic acid and steam distillation was the next preferred option in terms of reliability but on grounds of cost the method could be the one chosen for routine use. A GLC method for the determination of methylamines has been described by the AMC (1979). Ritskes (1975) described a GLC procedure for determination of methylamines and trimethyloxide after reduction with titanous chloride. Lundstrum and Racicot (1983) reviewed the GLC methods for determination of diethylamine and triethlyamine in seafoods. The method preferred by these authors includes perchloric acid extraction, neutralisation with potassium hydroxide and extraction with benzene. The separation is carried out on a porous polymer column and the compounds are detected using a selective flame ionisation detector sensitive to nitrogen and phosphorus compounds. Recoveries varying from 93 per cent to 110 per cent were reported over the range 0–30 mg N/100 g of fish.

A gas phase sensor was used by Storey *et al.* (1984) to measure volatile amines obtained by acid extraction and subsequent addition of alkali. They showed that the procedure lacked the precision required for a standard method but that the speed of operation made the method suitable for quality control purposes. Automated techniques using distillation and colorimetry have been described by Ruiter and Weseman (1976).

Ward *et al.* (1979) have demonstrated the use of a selective ion electrode for the determination of ammonia in shrimp species; ammonia is also a significant degradation product of the elasmobranch species, which contain high natural levels of urea (Pearson and Muslemuddin, 1969). Parris (1984) described an improved fluorimetric HPLC method for ammonia and volatile amines in meat and fish tissues.

NON-VOLATILE BASES AND OTHER DECOMPOSITION PRODUCTS

The presence of histamine is also an indicator of decomposition and has been linked to scombroid poisoning. The natural level in fresh fish is less than 5 mg/100 g and the higher values in decomposing fish are due to the decarboxylation of histidine. Most countries do not have regulatory limits on the allowable levels of histamine in foods. The USA has recently prescribed a two-stage limit for tuna; the defect level is 10 mg/100 g if there are odours associated with decomposition and 20 mg/100 g if based on histamine evidence alone. Sweden has imposed a limit of 20 mg/100 g for fish offered for sale. The amount that is considered to constitute a hazard to health is of the order of 50 mg/100 g. Liquid chromatographic methods have been proposed by Mietz and Karmas (1977), Schmidtlein (1979), Hui and Taylor (1983), Walters (1984), Lonberg (1984) and Ingles *et al.* (1985). Post-column derivatisation is used to give fluorescent species and to attain limits of detection in the nanogram range. It is likely that the use of the ELISA immunoassay method will prove feasible.

Ingles *et al.* (1985) give levels of the biogenic amines 2-phenylethylamine, *p*-tyramine, histamine, putrescine and cadaverine, proposed as initiators of migraine, in fish and other suspect foods such as cheese, chocolate, yeast products and beverages. Foo (1977) has described a simple paper chromatographic procedure for determination of both histamine and histidine. The fish is first extracted with 5 per cent trichloracetic acid. The prepared extract is defatted, concentrated and suspended in aqueous methanol. After chromatography on paper with 2-methylpropan-

2-ol/methylethylketone/ammonia/water as solvent, histidine and histamine are located by use of a ninhydrine spray reagent. The method is sensitive to 0·5 μg.

Hypoxanthine is a degradation product of adenosine triphosphate and the level present reflects evidence of autolytic deterioration. The AMC recommends the use of xanthine oxidase to convert hypoxanthine into uric acid, which may be measured by the absorbance at 290 nm. Burt *et al.* (1968) have described a version of this method for use with the Autoanalyser. Lawrence and Gould (1980) have described a test paper adaptation of the same method; test strips were standardised to correlate with the estimate of the number of days of storage on ice or, more informatively, a 'freshness index'. Burns and Ke (1985) reviewed available methods and used reverse phase liquid chromatography on perchloric acid extracts of cod and crab meat. Xanthine oxidase treatment was used to confirm hypoxanthine by removal of the appropriate peak in a second test sample. The procedure was claimed to give an equivalent performance to the AMC method and to have a lower limit of detection. Ke *et al.* (1984) studied the measurement of thiobarbituric acid reactive species (TBARS) used as an indicator of oxidative deterioration of fatty foods in application to samples of tuna, herring, cod and mackerel. Limits are proposed for the acceptability and grading of various species. Storey and Mills (1976) have reviewed the methods for measurement of fish quality.

Identification of fish species

Most fish can be readily distinguished by conformation, colour, and by the scale characteristics (Newton, 1979). However, in the case of fish fillets and fish products recourse must be made to other procedures. Electrophoretic methods have been widely used for species identification. Mackie (1968, 1969, 1972) has shown that uncooked and cooked fish may be identified by adaptations of the method of extraction prior to gel electrophoresis (Coduri *et al.*, 1979). The limitations of the procedure in relation to identification of highly denatured fish protein, especially pressure cooked fish, has been discussed by Mackie and Taylor (1972). The isoelectric focusing method has been applied to fish species by Lundstrom and Roderick (1979), Lundstrom (1979), Sattaneo *et al.* (1981), Laird *et al.* (1982) and Boe (1983). Laird *et al.* (1982) considered that isoelectric focusing on polyamide gel gave greater resolution of proteins than electrophoresis on agarose gels. Although isoelectric focusing is preferable for application to cooked fish, they considered that disc electrophoresis is adequate for most routine purposes. Lundstrom (1983) confirmed the opinions of Laird *et al.* in equivalent work aimed at the characterisation of shellfish. Terio *et al.* (1985) applied similar techniques to the identification of sardine species.

FISH PRODUCTS

Introduction

The variety of fish products available is widening with the development of convenience and frozen foods. Typical compositional data for some traditional products is given in Table 13.15. Fish are normally canned in brine, vegetable oil or sauce. The most significant species are salmon, sardines, pilchards, sild, herrings, brisling, tuna, crab and shrimps. Methods of manufacture of canned salmon have been

Table 13.15 Composition of canned fish and fish products

	Water (%)	Fat (%)	Protein (%)	Ash (%)	Salt (%)
Canned					
Caviar	40–60	7–17	23–28	5–12	—
Pilchards	64	15	19	2	1·4
Salmon	57–70	5–10	19–24	2–4	1·4
Sardine (drained ex-oil)	58	13	23	5	1·7
Sardine in oil	49	28	20	4	1·3
Sardine in sauce	65	12	19	4–5	1·8
Tuna in oil	55	22	20	2	1·1
Crab	79	1	18	2	1·4
Shrimp	75	1	21	3	2·5
Fish cake	70	1	11	2	1·3
Fish fingers	57–65	6–10	10–14	1–2	0·7
Fish paste	67	10	15	3	1·5
Smoked salmon	65	5	25	6	4·7
Smoked cod	78	1	18	5	3·0

described by Dewberry (1961). The traditional method of production of smoke cured fish was by dry salting or immersion in brine followed by smoking over wood (preferably oak shavings). Most types are smoked below 38 °C but temperatures over 90 °C may be used for speciality products. Smoke curing can be effected by the use of electrostatic deposition or by dipping in smoke solutions followed by air drying. Wood smoke contains formaldehyde, acetone, acids, phenols, tar, alcohols and diacetyl. Salting causes loss in weight of up to 30 per cent and the smoking process 5–25 per cent. Salt content of finished fish varies from 2 to 15 per cent according to type. Modern techniques of processing include the use of dyes in association with light smoking treatment. Dyes commonly used include Brown FK, tartrazine and amaranth.

Breaded and battered fish products have a substantial commercial significance. They include whole fish fillets, fish pieces, 'bite size' portions, fish fingers (or sticks) and rissoles (including fish burgers and fish cakes). Most of the products are sold in frozen form and the manufacturing technology has advanced significantly in recent years. The fish cores may be raw or pre-cooked according to the type of product being manufactured. Fish fingers and fish sticks are the most sophisticated products since they may be prepared from a single species or a mixture of species. The raw material is usually a frozen block of fillets or pieces with or without minced flesh. Some products are made mainly from minced flesh. The frozen block contains polyphosphates as water binding agents together with other additives as antioxidant and acidifying agents. Minced fish blocks contain addition additives to bind and condition the material to make it suitable for the end use (Keay, 1976). In the manufacture of fish fingers the frozen block is sawn into appropriate shapes by band saw to form the 'core'. The core is sprayed, breaded and lightly cooked; alternatively, the core may be pre-fried and then treated with coating material. It is claimed that addition of minced fish to the core aids the binding of the coating material. Generally, the ratio of core to total product weight will vary from 40 to 70 per cent according to the type of product. Fish cakes are made from pre-cooked fish with a filler. Normally, potato is used together with seasoning, flavour additives and spices.

The moulded cakes are fried and coated in batter and or bread crumbs. Fish cakes are now generally sold quick frozen.

The most novel fish product is surimi, which is the insoluble protein from minced fish (usually Alaska pollack) and is used to manufacture marine food analogues such as imitation crab meat.

Legislation and standards

The Meat Products and Spreadable Fish Products Regulations 1984 (SI 1984 No. 1566) as amended lay down compositional standards expressed in terms of raw meat or fish. The main requirement for pastes and spreads is a minimum of 70 per cent raw fish and in the case of a mixed meat and fish spreads and pastes the sum of the meat and fish content must be 70 per cent (see Table 13.3, p. 480). Fish is defined as 'the edible portion of any fish including edible molluscs and crustacea'. A code of practice has been drawn up by the industry, which specifies the minimum named meat content for certain fish pastes or patés statutorily required to contain at least 70 per cent fish. The agreement is summarised in Table 13.16.

Table 13.16 Code of Practice, agreement on minimum content of named fish in fish pastes

Description of fish paste	Minimum named fish content expressed as % of finished product
Anchovy	25% real anchovies
Bloater	20% smoked cured herrings
Brisling and Tomato	40% brisling
Buckling and Tomato	25% buckling
Crab	10% crab or crab tomali
Herring and Shrimp	40% herring, 2·5% shrimp or 1% shrimp concentrate or 0·5% shrimp extract[a]
Herring and Tomato	40% herring
Kipper	20% smoked cured herring
Lobster	10% lobster or lobster tomali
Lobster and Tomato	10% lobster or lobster tomali
Pilchard and Tomato	40% pilchard
Prawn	12·5% prawn meat
Salmon	25% salmon
Salmon and Crab	25% salmon, 5% crab
Salmon and Anchovy	25% salmon, 2·5% anchovy
Salmon and Shrimp	25% salmon, 2·5% shrimp or 1% shrimp concentrate or 0·5% shrimp extract[a]
Salmon and Tomato	25% salmon
Salmon and Lobster	25% salmon, 5% lobster
Sardine	25% sardine
Shrimp[b]	12·5% shrimp
Smoked Herring	20% smoked cured herrings
Smoked Salmon	10% smoked salmon
Smoked Roe	40% roe
Smoked Trout	15% smoked trout
Salmon and Cucumber	25% salmon
Crab and Lobster	5% crab, 5% lobster
Scampi	12·5% scampi

[a] Shrimp concentrate or shrimp extract contributes its own analytical content to the total fish content, e.g. in using 1% shrimp extract the paste must contain in addition 69% fish, making the required minimum of 70% fish content by analysis.

[b] This paste must contain at least 7·5% shrimp meat and the remainder may be either shrimp meat, shrimp concentrate or shrimp extract in the undermentioned proportions: shrimp meat 5%, shrimp concentrate 2% and shrimp extract 1%.

A minimum raw fish content of 35 per cent is prescribed for fish cakes by the Food Standards (Fish Cakes) Order 1950 (SI 1950 No. 589); the FSC have twice recommended that the minimum fish content be raised to 40 per cent. The Order does not contain a definition of fish. Burgess *et al.* (1970) have pointed out that fish cakes are frequently manufactured from belly flaps, which may contain 25 per cent skin and bone. The bone can be softened by pressure cooking until it can be crumbled and included in the mixture.

There are no prescribed compositional standards for any other fish products. The Codex standards enumerated in Table 13.14 contain no compositional criteria. However, a standard for fish sticks, fish fingers and like products is under development. The FSC (1966, 1970) have commented on the labelling of these products. The FAC (1987) recommended that some coated and ice glazed products should be controlled by regulations, e.g. fish fingers should be a reserved description and they should contain not less than 55 per cent of fish for battered products and not less than 60 per cent for breadcrumbed. Products named using the term 'fillet' should be made from blocks of fillets, with or without added polyphosphate, and products containing more than 10 per cent of minced fish should be so described. All coated (battered, breadcrumbed) products should bear a minimum fish content declaration and all glazed fish products should bear an indication of net weight of fish core prior to glazing.

Proposals for fish product regulations following some of the FAC recommendations were issued by MAFF in July 1988. Declaration of minimum fish content will be required for all products covered by the proposed regulations but no specific standards are set for fish fingers. The proposals include revocation of the Food Standards (Fish Cakes) Order 1950. As yet no regulations have appeared.

Assessment of fish content

The examination of fish pastes and fish cakes to determine fish content uses procedures similar to those for meat products except that it must be remembered that the nitrogen factors are not expressed on the fat-free basis, so yield actual fish content (including fish oil/fat). The Stubbs and More factors for fish and shellfish species are listed in Table 13.13; where potato is used as filler, as in fish cakes, the nitrogen correction is equivalent to 1·5 per cent of the carbohydrate content. Analysis of the separated core of a fish cake or other coated product may give low results for the fish content of the whole product since in some circumstances (e.g. poor manufacture or storage) soluble fish protein can migrate to the coating.

A LAJAC Code of Practice (No. 3) defines a procedure for calculation of the meat content of Norwegian canned crab products. In this procedure 15 per cent protein ($N \times 6{\cdot}25$) is taken as the equivalent of crab meat (including the natural fat content). Correction may be applied for any cereal filler which may be present by assuming 2 per cent nitrogen in the carbohydrate content. Wood (1977) has drawn attention to the difficulties of analysing shrimp pastes for fish content since the regulations are defined in terms of raw fish, whereas shrimp is usually cooked before it reaches the manufacturer. Aitken (1979) has also commented on the problems arising from the use of the Stubbs and More factors in the analysis of products manufactured from ill-defined fish species.

The AMC (1978) has studied the available methods for assessment of the fish content of fish fingers. Two basic methods are applicable, one (AOAC) depending on stripping the coating and weighing the core, the other on the use of the Stubbs and More procedure on the whole product after homogenisation. There are a number of varieties of the stripping procedure. Antonacopoulos (1977) has described a method in which the sample is conditioned at −18 °C, the coating is removed by scraping with a scalpel 15 min after the fish finger is removed from the refrigerator. The AOAC method requires the fish finger to be dipped in a water bath at 30 °C for 10–30 seconds and dried with a paper towel before the coating is removed with a blunt knife. A variety of methods of calculating the result of the analysis is available when the stripping procedure is used. They depend on mass ratio, per cent solids ratio, per cent nitrogen ratio and on nitrogen/protein ratio between the core and the whole product. The AMC have concluded that the traditional Stubbs and More procedure when applied to the whole product gives the most consistent results. However, the method is subject to bias if other nitrogenous material is used in preparation of the coating. Results obtained by the stripping method depend on the efficiency of removal of the coating and are subject to operator variation. The mass ratio calculation also assumes that the core weight is equivalent to fish content. This may not necessarily be the case with all products, especially those manufactured from treated fish or minced fish blocks. The examination of seafood cocktails and prepared salads has been considered by King and Regan (1977) and by Priebe (1979a).

The Working Group to the FAC on analytical methods (see FAC, 1987) recommended use of the Stubbs and More method for the estimation of fish content and that suitable nitrogen factors should be agreed between industry and public analysts. The Group observed that a single nitrogen factor appears to be possible for unspecified white fish (see Table 13.17). They also recommended the use of a modified

Table 13.17 Nitrogen factors for white fish (from FAC, 1987)

Species	Source	No. of samples	Mean N (%)	Range	Standard deviation	Coefficient of variation (%)
Cod	AMC	295	2·871	2·21–3·20	0·137	4·8
	Torry	182	2·906	2·64–3·29	0·101	3·5
Coley/	AMC	95	2·923	2·52–3·27	0·165	5·6
Saithe	Torry	256	2·926	2·54–3·32	0·137	4·7
European Hake	Torry	183	2·871	2·52–3·28	0·143	5·0
Haddock	Torry	361	2·962	2·52–3·31	0·167	5·6
Ling	Torry	271	3·020	2·70–3·36	0·118	3·5
Plaice	Torry	182	2·665	1·91–3·19	0·207	7·8
Whiting	Torry	365	2·912	2·35–3·35	0·184	6·3

Note: It is assumed that fish catches will conform to a normal distribution and therefore the nitrogen content in the great majority of samples analysed will fall close to the mean.

Codex method for the estimation of glaze on shrimps and prawns, and that a standard method for the determination of glaze on fish should be agreed. The results of a subsequent collaborative study of various methods of determining ice glaze (not yet published) indicated that no single method is suitable for different levels of ice glaze in the range of products examined.

General methods for the analysis of fish products

General analysis of fish products relies on the normal determination of proximate composition. Products may also require analysis for salt, additives (colours, preservatives, phosphates, thickeners) and trace elements, in particular lead, mercury and arsenic. Arsenic levels in fish vary widely. In general the values in white fish vary from 0·1 to 5 mg/kg, in crab up to 20 mg/kg, in shellfish up to 120 mg/kg, whilst prawns and shrimps have been found to contain up to 200 mg/kg. Coulson *et al.* (1935) have published evidence which is claimed to demonstrate that arsenic is present in a form that is naturally excreted.

Cans of fish, particularly crustacea, may be found to contain small crystals of magnesium ammonium phosphate (struvite), which may be mistaken for glass fragments. Characterisation of product may depend on the microscopic examination of scales, bones or pieces of shell and extraction and analysis of oil or fat by traditional methods or by gas chromatography. Fish oils are highly unsaturated, having high iodine values and saponification values. The electrophoretic methods of species identification have been used with processed fish products (Mackie, 1972; Mackie and Taylor, 1972).

The AMC (1979) has published recommended methods for sample preparation and determination of the proximate composition of fish products.

Moisture. Weigh 5 g prepared sample into a pre-dried and weighed metal dish containing 15 g acid washed sand. Add 5–10 ml ethanol, mix with the sample mass using a small glass rod. Heat on a water bath at 60–80 °C, stirring occasionally. Evaporate to apparent dryness. Place the dish in an oven at 103 ± 2 °C for 2 hours. Cool in a desiccator. Weigh, repeat the heating and cooling steps until the successive weighings agree to within 0·002 g and calculate moisture loss.

Total fat. Digest 3–5 g prepared sample with 50 ml 4 M hydrochloric acid in a 250 ml conical flask. Boil gently for 1 hour, maintaining the liquid volume at 50 ml. After 1 hour add 150 ml hot water. Moisten a fluted hardened filter paper (Whatman No. 54 or equivalent) in a glass funnel. Filter the contents of the conical flask. Wash the residue into the filter paper with hot water. Continue washing until the runnings do not give an acid reaction to test paper. Dry the filter paper on a watch glass in an oven. Roll the paper and place in an extraction thimble. Complete the determination by extraction with hexane or light petroleum 40–60 °C using the Soxhlet method.

Extractable fat. According to the anticipated fat content, take a 10–50 g portion of the prepared sample, mix with anhydrous sodium sulphate, transfer to an extraction thimble. Extract for 6 hours with light petroleum 40–60 °C in a continuous extractor. Remove the solvent by evaporation and with the extracted, dry fat.

Note. For fatty fish, the sample is oven-dried and extracted without mixing with sodium sulphate.

Ash. Heat a 5 g prepared sample in a pre-weighed dish with 1 ml 25 per cent m/v magnesium acetate solution on a water bath. When dry, incinerate the mass at 550–600 °C for at least 3 hours (to white ash). Cool in a desiccator and weigh the ash. Carry out a blank determination on 1 ml magnesium acetate solution (using the same pipette). Correct the weight of ash observed on the sample for the blank. Calculate the ash content.

Chloride. Weigh an amount of sample expected to contain 30–50 mg chloride (as sodium chloride) into a 250 ml conical flask. Add 30 ml water and two boiling beads. Warm the contents. Add 20 ml 0·05 M silver nitrate and 10 ml concentrated nitric acid. Boil gently for 5 min. Cool to about 80 °C. Cautiously add a slight excess of saturated permanganate solution. Boil gently, repeat the additions of potassium permanganate solution until the solution remains coloured after 5 min boiling. Add sucrose to destroy the excess permanganate. Cool for 30 s. Add 0·1 g urea. Cool for 10 min in running water. Add 5 ml acetone, 2 ml ammonium iron(III) sulphate (5 per cent m/v in 10 per cent nitric acid). Titrate with 0·05 M potassium thiocyanate. Carry out a blank omitting the sample. Calculate the chloride content, expressed as sodium chloride if appropriate to the sample.

Nitrogen. Apply the Kjeldahl procedure to 2 g prepared sample using 15 g potassium sulphate and 0·5 g copper(II) sulphate as catalyst.

Total phosphorus. Dry in a dish a 2·5 g sample with 5 ml magnesium acetate solution (25 per cent m/v). Incinerate at 550–600 °C (white ash). Cool and moisten the ash with 5 ml water. Add 25 ml 0·5 M sulphuric acid and 50 ml water. Transfer to a 250 ml volumetric flask, and make up to the mark with water. Carry out the determination of phosphate using the molybdovanadate procedure (Chapter 2). Calculate as P_2O_5.

Polyphosphate cannot readily be estimated in fish since it rapidly hydrolyses on storage to orthophosphate.

GELATINE

Manufacture

Mainly because of its jelly-forming properties, gelatine is present as an ingredient in a number of foods. In the manufacture the collagen present in waste material such as hides, skin and bones is converted to gelatine by simmering with water. Two processes are employed for the manufacture, namely acid extraction and lime extraction (Ames, 1949; Plowright, 1963).

Lime gelatine usually has a higher ash, higher calcium content and its solution has a higher pH than an acid gelatine. The higher calcium content of lime gelatine tends to form insoluble salts in some foods in which it is used, e.g. calcium tartrate crystals may be formed if tartaric acid is present and calcium phosphate is sometimes deposited in jars of lambs' tongues in jelly. The two types are best distinguished from the isoelectric point (IEP), which is about pH 8 for an acid gelatine and about pH 5 for a lime gelatine (Ward, 1951).

Composition

The ultimate composition of gelatine is approximately: carbon 50·5, hydrogen 6·7, nitrogen 17·9, sulphur 0·6 and oxygen 24·3 (all as percentages).

In determining gelatine (dry protein) the appropriate Kjeldahl factor is Total N × 5·55. Manufactured gelatine contains 12–17 per cent moisture and up to 2·5 per cent ash. Tannin solution and platinum salts precipitate gelatine, but iron, aluminium, lead, copper and gold salts do not. It is soluble in acetic acid but insoluble in alcohol or ether.

Gelatine that is used for food purposes usually comes within the following ranges of composition: moisture 12–17 per cent; ash 0·5–2·5 per cent; calcium (as CaO 0·1–0·6 per cent; iron (as Fe) 10–150 p.p.m. and pH 4·0–6·3.

Legislation

The Food Standards (Edible Gelatine) Order 1951 (SI 1951 Nos 1196, 2240), now revoked, required a 3 per cent solution of gelatine to set to a jelly and prescribed the following maxima: ash 3·25 per cent, arsenic 2 p.p.m. (mg/kg), copper 30 p.p.m. (mg/kg), lead 5 p.p.m. (mg/kg) and zinc 100 p.p.m. (mg/kg). The Preservatives Regulations prescribe a maximum of 1000 mg/kg of sulphur dioxide for gelatine and 3000 mg/kg sorbic acid for gelatine capsules.

Analysis

The routine examination of gelatine normally includes the simple **setting test,** together with the determination of **ash, trace elements (arsenic, copper, lead, zinc), sulphur dioxide** and the **acidity**. Industrially, other common determinations are those for **calcium, iron** and **moisture** and the assessment of various physical properties are also of considerable importance, e.g. **jelly strength, water absorption, viscosity, melting point, setting point, colour, clarity** and the **pH value**. Methods for chemical analysis of gelatine are described in BS 757: 1975. A **bacteriological examination** is also of special importance industrially, particularly if the gelatine is to be incorporated in preserved fish products.

Moisture. Weigh out 1 g into a metal dish, add 10 ml water and allow to soak. Place on the water bath to dissolve the gelatine. Evaporate off the water. Complete the drying in an oven at 100 °C for 18 ± 1 h. The moisture content is not as useful a measure for assessing quality as with other materials. A low moisture content may imply poor quality due to excessive heating during the drying process, causing protein degradation and consequent loss of jelly strength.

Ash. Incinerate carefully 5 g at 495 ± 5 °C. Treatment of the residue may be necessary to obtain a white ash.

Trace elements. Determine arsenic, copper, lead and zinc. The BS recommends colorimetric procedures; AAS methods may be used (Chapter 5).

Calcium and iron. Ash 10 g. Boil the ash with 10 ml of 5 M hydrochloric acid and filter the extract into a 100 ml volumetric flask. Make the filtrate up to the mark and use 50 ml for the estimation of the calcium by dissolving the oxalate precipitate in sulphuric acid and titrating with permanganate (p. 32). A calcium figure in excess of about 0·3 per cent (as CaO) usually indicates the properties of a lime gelatine.

Ferric iron imparts a reddish-brown tinge to the gelatine solution. Use 5–25 ml of the solution for the estimation of iron by a suitable colorimetric method.

Sulphur dioxide. The British Standard recommends two variations of the Monier-Williams method. Method 2, which uses a compact distillation apparatus with a vertical condenser, is subject to error if the gelatine contains acetic acid or any other volatile acid. The error may be avoided if an iodometric technique or colorimetric procedure is used for the end measurement (Chapter 4).

Acidity. Dissolve 2 g in 50 ml water on the water bath and titrate with 0·1 M sodium hydroxide using phenolphthalein. Formerly the BP required that gelatine for pharmaceutical purposes should not require more than 5 ml 0·1 M alkali.

pH value is determined on a 1–2 per cent solution. The pH value may vary from about 4·0 to 6·3. Higher values in this range suggest the presence of lime gelatine. For the determination of the *isoelectric point*, prepare a series of tubes containing 2 per cent solutions of the gelatine and including varying proportions of 0·02 N acid and alkali so that successive tubes differ by 0·2 per cent acidity (as H_2SO_4). The IEP is then obtained by measuring the pH of the tube with the maximum turbidity (Ames, 1944).

Jelly strength can be measured in absolute units on the Bloom Gelometer (BS 757: 1975) or more easily using the Boucher Jelly Tester (Koprowski, 1951). Industrially, a comparative measure against an agreed standard is often sufficient (BS 757: 1975). *Colour* (degree of 'brownness') and *clarity* (turbidity) are also important industrially and can be approximately assessed in the solution prepared for the jelly strength (Saunders and Ward, 1953). Solutions for jelly strength must be prepared by soaking in cold water for 1–3 h and then dissolving at 60 °C (higher temperatures tend to cause loss of strength).

PROTEIN EXTRACTS AND HYDROLYSATES

MEAT EXTRACT

Meat extract has been used for many years to impart flavour to foods. Traditionally, it is a by-product of corned beef manufacture and a significant proportion of UK imported extract originated in South America. During corned beef processing, batches of the boned meat are cooked in the same tank of water. The yield of soluble extractives varies from 2 to 4 per cent depending on the meat used and the process conditions. When the meat solids content reaches about 5–6 per cent in the liquor, the fat is skimmed and the liquid is concentrated to a solids content of approximately 80 per cent. During the concentration stage some creatine is converted to the anhydride creatinine. During storage, however, hydrolysis occurs and the less soluble creatine may crystallise out. For commercial purposes the quality of meat extract is generally assessed by the total creatine content, which should preferably exceed 10 per cent on dry matter basis.

A definition used in the USA is as follows:
'Meat extract is the product obtained by extracting fresh meat with boiling water and concentrating the liquid by evaporation after removal of the fat. It contains at least 75 per cent of total solid matter, of which not more than 27 per cent is ash and not over 12 per cent sodium chloride. The fat should not exceed 0·6 and the nitrogen be not less than 8 per cent. The nitrogenous compounds contain not less than 40 per cent of meat bases and not less than 10 per cent of creatine and creatinine'

In view of the widespread use of meat extract and the relatively low yield from meat processing, more advanced direct extraction methods are now in use. Noyes (1969) and Karmas (1970) have described processes for the hydrolysis of meat from a variety of species. The product may be prepared from whole carcasses or from trimmings, according to the quality required. Meat is finely divided and mixed with two volumes of hot water. The mass is boiled and filtered. The washed water-insoluble residue is further hydrolysed, either with enzyme followed by mineral acid or with mineral acid alone and under pressure. After neutralisation the product is filtered, concentrated to remove salt by crystallisation, centrifuged and further concentrated to yield a hydrolysed meat extract. Various blends of water-soluble and acid-soluble fractions are then prepared to give distinctive meat flavours. Meat may be liquefied by high pressure treatment with steam. Similarly, commercial extracts are prepared by high pressure steam extraction of ground bones.

Composition

A semi-fluid meat-based product sold by retail consists of meat extract, stock, hydrolysed protein, yeast extract, dried beef, caramel, starch, salt and other flavourings. This product contains about 39 per cent water, 6·2 per cent nitrogen, 14 per cent ash, 11·2 per cent chloride, 0·7 per cent fat and 1 per cent creatine plus creatinine. A brand of compressed cubes on the market contains hydrolysed protein, beef stock, flour, yeast extract, caramel, salt, beef extract, beef fat, desiccated beef, spices and onion powder. A series of stock cubes gave the following analysis (as percentages): water 6·5–14·7; organic matter 34·4–66·2; mineral matter 25·7–59·1; salt 18·8–50·3; fat 0·9–2·4; carbohydrates 3·4–40·4, total nitrogen 1·7–8·2 (= protein 10·9–51·4), creatine plus creatinine 0·7–2·0. The cubes appear to contain up to 15 per cent of added gelatine. A wider range of flavoured products is now sold and typical composition ranges are 8·5 per cent moisture, 4·5–5·0 per cent nitrogen, 29–32 per cent ash, 24–27 per cent salt and 3–6 per cent fat.

YEAST EXTRACTS

Yeast from breweries, mainly *Saccharomyces cerevisiae*, is used in the preparation of yeast extract. A slurry of yeast cells undergoes autolysis spontaneously. Proteases cause the breakdown of proteins into polypeptides and amino acids. When extraction is complete the liquor is further processed to remove insoluble cellular material before concentration into a paste containing about 70 per cent solids.

Composition
General analysis of autolysed yeast extract gives the following composition: water 12·4 per cent; organic soluble solids 34·7 per cent; organic insoluble matter 12·2 per cent; ash 40·7 per cent; salt 38·5 per cent and total nitrogen 7·3 per cent. The main contribution to flavour enhancement of foods by these products is due to the ribonucleotide content. There is no evidence of creatine and creatinine. The manufacture and composition of commercial yeast extracts has been reviewed (Anon, 1962; Holder, 1977). Comparative analyses indicating differences between meat and yeast extracts have been recorded by Crush (1964a, b) and by Wood (1956). Yeast extracts contain high levels of biogenic amines (histamine, etc.), thought to give rise to migraine in some individuals (Ingles *et al.*, 1985).

FISH PROTEINS

Deodorised fish protein is now available as a bland protein concentrate for use in food materials. The product may be used as a stock for hydrolysis to yield hydrolysed protein. When mixed with yeast to form a slurry the protein is hydrolysed by the proteolytic enzymes in the yeast and the product is used as a food supplement (Noyes, 1969; Sikka *et al.*, 1979). Production of fish sauces and typical analyses have been reported by Gildberg *et al.* (1984).

Surimi is isolated and deodorised fish muscle fibres (mainly from pollack), and is used to manufacture reformed fish products such as crab sticks.

HYDROLYSED VEGETABLE PROTEIN

Hydrolysed vegetable proteins may be prepared by methods similar to those used for meat and yeast proteins. The flavour of the product is controlled by the degree of hydrolysis of the protein and by the blend. In general a 60–85 per cent hydrolysis (35–50 per cent of the total nitrogen as α-amino acid) yields a product with optimal flavour with monosodium glutamate content varying from 9 to 14 per cent (Hansen, 1974). When spray-dried, a partial hydrolysate of wheat gluten is a valuable bread improver. Other products are used as flavours and flavour-enhancing ingredients in soups, bouillon and other processed foods. Hydrolysed protein pastes have the typical composition of moisture 12–24 per cent, ash 40–50 per cent, salt 30–40 per cent and total nitrogen 5–7 per cent. Powder products have moisture 2–6 per cent, total nitrogen 9 per cent and salt 31 per cent.

Analysis of extracts and protein hydrolysates
Suitable methods for the partial analysis of meat extract and similar products have been published by the SAC (1974) and SPA (1951). The determination of water, ash, chloride, total nitrogen and total creatine and creatinine are sufficient for general quality assessment. If a more detailed analysis is required, particularly for the detection of adulterants and non-meat ingredients, further determinations may be necessary for phosphate, fat, soluble and insoluble nitrogen, amino nitrogen, purines, gelatine, nicotinic acid, starch and trace elements.

Sampling. Blend pasty samples prior to weighing out by warming so that sediment is incorporated. Prepare a 10 per cent (m/v) stock solution using hot water to ensure solution of all soluble materials.

Water. Dry 10 ml of stock solution in a nickel dish fitted with a lid first on the water bath and then to constant weight at 100–103 °C.

Ash. Evaporate 10 ml of stock solution in a platinum dish on the steam bath. Char on a low flame and complete the ashing at 550 °C.

Chloride. Evaporate 20 ml of stock solution with 10 ml of 5 per cent sodium carbonate solution and ignite at a temperature not greater than dull redness. Extract with hot water, filter and wash. Retain the filtrate. Return the filter paper and residue to the dish. Moisture with a little carbonate solution and ignite to a white ash. Dissolve the ash in dilute nitric acid. Filter and wash the paper thoroughly with water. Combine the filtrates and determine the chloride by the Volhard method.

Total nitrogen. Determine on 5 ml of stock solution by the macro-Kjeldahl method. The digestion time should proceed for 3 h after clearing.

Gelatine. Weigh out into a 250 ml beaker 10 g of meat extract. Add 125 ml water. Bring to the boil. Stir constantly. Add 0·5 ml glacial acetic acid. Digest the mixture on the steam bath for 15–30 min. Filter through a No. 4 Whatman paper into a 250 ml volumetric flask. Wash with hot water. Cool the filtrate. Make up to 250 ml. Pipette 25 ml into a porcelain basin. Add 0·25 ml formalin. Mix thoroughly with a glass rod. Concentrate to a thick consistency. Add 0·25 ml formalin, mix thoroughly and spread the mixture evenly over the basin to within 2·5 cm of the rim. Bake hard on a boiling steam bath for 2 hours. Extract the contents of the dish twice with 100 ml of diluted formalin (2·5 ml formalin diluted to 100 ml with water) at 40 °C. Allow 1 hour for each extraction, maintain at 40 °C. Filter each washing through a No. 54 Whatman paper. Break up the complex during the final extraction. Loosen the complex. Transfer to the filter paper. Wash with a further 100 ml of the dilute formalin solution at 40 °C. Determine the nitrogen in the gelatin–formaldehyde complex by the Kjeldahl method (gelatine = N × 5·55).

Creatine and creatinine. The method described under meat products is suitable.

Purines. Chromatographic methods have been described using cation exchange resin separation with ultraviolet absorption measurement of the separated purines. The use of HPLC techniques using similar principles has been described by Matsumoto *et al.* (1977) and by Clifford (1976). The former authors extract sample with 5 per cent perchloric acid and digest at 100 °C with concentrated acid. The neutralised hydrolysate is chromatographed on Aminex A-4 using 0·05 M ammonium phosphate and 5 per cent ethanol. The effluent is monitored for ultraviolet absorption at 254 nm.

FLAVOUR ENHANCERS

Monosodium glutamate (MSG) and the ribonucleotides disodium inosinate and disodium guanylate are used as flavour enhancers, particularly with salt in meat products and soups and sauces. Up to 1·5 per cent MSG may be present and larger amounts may result in certain physiological effects, known as 'Kwok's Quease' (Anon, 1969). When used with the ribonucleotides in the ratio 19:1, 30 per cent less flavour enhancer produces the same flavouring effect (FACC, 1978b). The Food Chemicals Codex specifications apply to these products.

Fernandez-Flores *et al.* (1969) have described a procedure for the estimation of MSG in many food products and which is currently the AOAC method. All food substances are extracted with water if insignificant amount of starch is present, or with acetone–water mixture if starch is present. An aliquot of the concentrated extract is chromatographed on a column of Dowex 50W–X8 (H^+ form). Glutamic acid is retained on the column and eluted with M hydrochloric acid. The glutamic acid content of the eluate is estimated by the formol titration, from which the MSG content can be calculated. A paper chromatographic method has been described by Bailey and Swift (1970). A similar system for the nucleotides is described in the Food Chemical Codex.

Conacher *et al.* (1979) have reviewed the methods for estimation of monosodium glutamate. They have proposed a procedure based on the GLC separation of trimethylsilyl derivatives of the amino acids. These authors claim that their method is quicker and more accurate than those currently recommended by AOAC.

As a result of work which showed that mice injected with large doses of MSG developed brain damage, three large US producers of baby foods stated in 1969 that they would voluntarily withdraw it as an ingredient in baby foods. Hydrolysed proteins are now used as alternative ingredients in these products. The FACC (1978b) reported on flavour-enhancing ingredients and recommended that the three flavour-enhancing agents should not be permitted in foods specially prepared for babies or young children. These recommendations were enacted in the Miscellaneous Additives in Food (Amendment) Regulations 1982. No restrictions were placed on their use in other foods.

REFERENCES

Agater, I. B., Bryant, K. J., Llewellyn, J. W., Sawyer, R., Bailey, F. J. & Hitchcock, C. H. S. (1986) *Journal of the Science of Food and Agriculture*, **37**, 317.

Aitken, A. (1979) *Journal of the Association of Public Analysts*, **17**, 135.

AMC (1978) *Analyst*, **103**, 973.

AMC (1979) *Analyst*, **104**, 434.

AMC (1986) *Analyst*, **111**, 969.

Ames, W. M. (1944) *Journal of the Society of Chemical Industry*, **63**, 200.

Ames, W. M. (1949) *Journal of the Society of Leather Trades Chemists*, **3**, 407.

Anderson, P. J. (1981) *Journal of Food Science*, **46**, 1111.

Andrews, A. J. (1986) *Electrophoresis*. Oxford: Oxford University Press.

Anon (1962) *Food Trade Review*, **32**, 31.

Anon (1969) B.N.F. *Information Bulletin* (3), 6.

Anon (1982a) *Meat*, June, 27.

Anon (1982b) *Meat*, July, 17.

Anon (1983) *The National Provisioner*, **188**, 22, 72.

Anon (1984) *Food Trade Review*, March, 119.

Anon (1985) *Meat Industry*, March, 25.

Antonacopoulos, N. (1977) *Deutsches Lebensmittel Rundschau*, **73**, 315.

Armstrong, D. J., Reichert, S. H. & Riemann, S. M. (1982) *Journal of Food Technology*, **17**, 327.

Arneth, W. (1974) *Fleischwirtschaft*, **54**, 86.

Arneth, W. (1977) *Fleischwirtschaft*, **57**, 1023.

Asakawa, T. & Matsushita, S. (1980) *Lipids*, **15**, 773.

Atasi, M. Z., Tarbowski, D. P. & Pauli, J. H. (1970) *Biochimica et Biophysica Acta*, **221**, 253.

Bahl, R. K. (1971) *Analyst*, **96**, 88.

Bailey, B. W. & Swift, H. L. (1970) *Journal of the Association of Official Analytical Chemists*, **53**, 1268.

Bailey, F. J. (1976) *Journal of the Science of Food and Agriculture*, **27**, 827.

Baines, C. R. & Shewan, J. M. (1965) *Laboratory Practice*, **14**, 160.

Baker, R. C. (1978) *New York Life Science Quarterly*, **11**, 12.

Baltes, W. (1968) *Zeitschrift fur Lebensmittel Untersuchung und Forschung*, **138**, 203.

Battaglia, R. (1978) *Deutsche Lebensmittel Rundschau*, **74**, 337.

Beckers, H. J., van Schothorst, M., van Spreekens, K. J. A. & Oosterhuis, J. J. (1981) *Zentralblatt fur Bakteriologie und Hygiene, I. Abt. Orig. B.*, **172**, 401.

Beebe, R. M., Lay, E. & Eisenberg, S. (1989) *Journal of the Association of Official Analytical Chemists*, **72**, 777.

Bertling, L. (1977) *Fleischwirtschaft*, **57**, 715.

Bertling, L. (1979) *Fleischerei*, **30**, 895.

Bijker, P. G. H., Koolmees, P. A. & van Logtestijn, J. G. (1983) *Meat Science*, **9**, 257.

Bijker, P. G. H., Koolmees, P. A. & Tuinstra-Melgers, J. (1985) *Archiv Lebensmittel Hygiene*, **36**, 49.

Bjarno, O-C. (1982) *Journal of the Association of Official Analytical Chemists*, **65**, 696.
Blackwood, C. M. (1978) *Journal of the Canadian Institute of Food Science and Technology*, **11**, A42.
Board, P. W., Montgomery, W. A. & Rutledge, P. J. (1978) *Journal of the Science of Food and Agriculture*, **29**, 569.
Boe, B. (1983) *Food Chemistry*, **11**, 127.
Bostian, M. L., Webb, N. B. & Hadden, J. P. (1984) *Journal of Food Science*, **49**, 1347.
Botta, J. R., Lauder, J. T. & Jewer, M. A. (1984) *Journal of Food Science*, **49**, 734.
Brinker, A. (1979) *Journal of the American Oil Chemists' Society*, **56**, 211.
Bruce, B. M. (1979) *International Flavours and Food Additives*, **10**, 106.
Bunton, N. G., Jennings, N. & Crosby, N. T. (1979) *Journal of the Association of Public Analysts*, **17**, 105.
Burgess, G. H. O., McLachlan, T., Talterson, I. N. & Windsor, M. L. (1970) *Analyst*, **95**, 471.
Burns, B. G. & Ke, P. J. (1985) *Journal of the Association of Official Analytical Chemists*, **68**, 444.
Burt, J. R. (1977) *Process Biochemistry*, **12**, 32.
Burt, J. R., Murray, J. & Stroud, G. D. (1968) *Journal of Food Technology*, **3**, 165.
Casas, C., Tormo, J., Hernandez, P. E. & Sanz, B. (1984) *Journal of Food Technology*, **19**, 283.
Casey, J. C. & Crosland, A. R. (1982) *Journal of Food Technology*, **17**, 567.
Cassidy, W. & Fryer, J. (1989) *British Food Journal*, **91**, 22.
Castledine, S. A. & Davies, D. R. A. (1968) *Journal of the Association of Public Analysts*, **6**, 39.
Cattaneo, P., Soncini, G., Brenna, O. & Cantoni, C. (1981) *Industrie Alimentari*, **203**.
Charnley, F. & Davies, F. R. E. (1944) *Analyst*, **69**, 302.
Chicot, J. P., Lang, P., Lanteaume, M. T. & Chevrier, J. P. (1983) *Annales Falsifications L'Expertise Chimique de Toxicologique*, **76**, 73.
Clifford, A. J. (1976) *Advances in Chromatography*, **14**, 1.
Coduri, R. J., Bonatti, K. & Simpson, K. L. (1979) *Journal of the Association of Official Analytical Chemists*, **62**, 269.
Cole, D. J. A. & Lawrie, R. A. (1974) *Meat, Proceedings of the Twentyfirst Easter School in Agricultural Science*. University of Nottingham: Butterworth.
Conacher, H. B. S., Iyengar, J. R., Miles, W. F. & Botting, H. G. (1979) *Journal of the Association of Official Analytical Chemists*, **62**, 604.
Consumers' Association (1974) *Which?* (November), 330.
Coomaraswamy, M. (1972) *Journal of the Association of Public Analysts*, **10**, 33.
Coomaraswamy, M. & Flint, F. O. (1973) *Analyst*, **98**, 542.
Cooper, L. (1985) *Meat Industry*, March, 18.
Copeman, D. M. (1981) *The Monthly Review*, **90**, 34.
Corry, J. E. L., Sharma, M. R. & Bates, M. L. (1983) *Society Applied Bacteriology Series No. 18*. London: Academic Press.
Cottam, J. (1975) *Journal of the Association of Public Analysts*, **13**, 133.
Coulson, E. J., Remington, R. E. & Lynch, K. H. (1935) *Journal of Nutrition*, **10**, 255.
Crimes, A. A., Hitchcock, C. H. S. & Wood, R. (1984) *Journal of the Association of Public Analysts*, **22**, 59.
Crush, K. G. (1964a) *Journal of the Science of Food and Agriculture*, **15**, 550.
Crush, K. G. (1964b) *Journal of the Science of Food and Agriculture*, **15**, 555.
Csiba, A. (1984) *Acta Alimentaria*, **13**, 189.
Dahl, O. (1963) *Journal of the Agricultural and Food Chemistry*, **11**, 350.
Dainty, R. H. (1971) *Institute of Food Science and Technology Proceedings*, **4**, 178.
Davies, A. M. C., Gee, M. G. & Grey, T. C. (1984) *Journal of Food Technology*, **19**, 175.
Degenkolb, E. & Hingerle, M. (1969) *Archiv Lebensmittel Hygiene*, **20**, 73.
Demby, J. H. & Cunningham, F. E. (1980) *Worlds Poultry Science Journal*, **36**, 25.
Demeyer, D. (1979) *Fleischwirtschaft*, **59**, 973.
Deutsch, M. J. & Weeks, C. E. (1965) *Journal of the Association of Official Analytical Chemists*, **48**, 1248.
Dewberry, E. B. (1961) *Food Manufacture*, **36**, 185.
Dransfield, E., Casey, J. C., Boccard, R., Touraille, C., Buchter, L., Hood, D. E., Joseph, R. L., Schon, I., Casteels, M., Cosentino, E. & Tinbergen, B. J. (1983) *Meat Science*, **8**, 79.
Duda, Z. (1974) *Vegetable Protein Meat Extenders and Analogues*. Rome: Food and Agriculture Organisation of the United Nations.
Dutschke, G. (1979) *Fleischwirtschaft*, **59**, 1687, 1864.
Dyer, W. J., Hiltz, D. F., Hayes, E. R. & Munro, V. G. (1977) *Journal of the Canadian Institute of Food Science and Technology*, **10**, 185.
EC (1978) *Report of the Study Group on Vegetable Proteins in Foodstuffs for Human Consumption, in Particular in Meat Products*. Luxembourg: EUR 6026 Commission of the European Communities.
Eldridge, A. C. (1981) *Journal American Oil Chemists' Society*, **58**, 483.
Ettel, W. & Tuor, A. (1977) *Deutsche Lebensmittel Rundschau*, **73**, 357.
Eustace, I. J. & Johnson, B. Y. (1984) *Food Technology in Australia*, **36**, 556.

Eustace, I. J. & Jones, P. N. (1984) *CSIRO Food Research Quarterly*, **44**, 38.
Evans, G. G. (1972) *Technical Circular* No. 508. Leatherhead Food RA.
FAC (1987) *Report on Coated and Ice-Glazed Fish Products.* FDAC/REP/3. London: HMSO.
FACC (1978a) *Report on the Review of Nitrate and Nitrite in Cured Meats and Cheese.* FAC/REP/27. London: HMSO.
FACC (1978b) *Report on the Review of Flavour Modifiers*, FAC/REP/28. London: HMSO.
Ferguson, C. S., Racicot, P. A. & Rane, L. (1942) *Journal of the Association of Official Analytical Chemists*, **25**, 533.
Fernandez-Flores, E., Johnson, A. R. & Blomquist, V. H. (1969) *Journal of the Association of Official Analytical Chemists*, **52**, 744.
Fey, R. (1977) *Zeitschrift fur Lebensmittel Untersuchung und Forschung*, **164**, 233.
Field, R. A. (1976) *Food Technology*, **30**, 38.
Flint, F. O. & Firth, B. M. (1983) *Analyst*, **108**, 756.
Flint, F. O. & Johnson, R. F. P. (1979) *Analyst*, **104**, 1135.
Flint, F. O. & Meech, M. V. (1978) *Analyst*, **103**, 542.
Flint, F. O. & Pickering, K. (1984) *Analyst*, **109**, 1505.
Foo, L. Y. (1977) *Journal of the Association of Official Analytical Chemists*, **60**, 183.
Fox, J. B., Doerr, R. C. & Gates, R. (1984) *Journal of the Association of Official Analytical Chemists*, **67**, 692.
Fraser, J. R. & Holmes, D. C. (1958) *Analyst*, **83**, 371.
Froning, G. W. (1976) *Food Technology*, **30**, 50.
FSC (1966) *Report on Claims and Misleading Descriptions*, FSC/REP/50. London: HMSO.
FSC (1970) *Report on pre 1955 Compositional Orders*. FSC/REP/55. London: HMSO.
FSC (1974) *Report on Novel Protein Foods.* FSC/REP/62. London: HMSO.
FSC (1978) *Report on Water in Food.* FSC/REP/70. London: HMSO.
FSC (1980) *Report on Meat Products.* FSC/REP/72. London: HMSO.
Gabor, E., Gaspar, O. & Vamos, E. (1984) *Acta Alimentaria*, **13**, 13.
Gardener, G. A. (1978) *Journal of Food Technology*, **13**, 359.
Gerrard, F. (1963) *Meat Trades Journal*, **189**, 89.
Ghadi, S. V. & Lewis, N. F. (1977) *Fleischwirtschaft*, **57**, 2155.
Gildberg, A., Espejo-Hermes, J. & Magno-Orejana, F. (1984) *Journal of the Science of Food and Agriculture*, **35**, 1363.
Glyn, J. H. (1939) *Science*, **89**, 444.
Grey, T. C., Robinson, D. & Jones, J. M. (1979) *Journal of Food Technology*, **14**, 587.
Grey, T. C., Robinson, D., Jones, J. M. & Pritchard, R. (1983) *Journal of the Association of Public Analysts*, **21**, 1.
Grey, T. C., Robinson, D. & Jones, J. M. (1985) *Journal of the Association of Public Analysts*, **23**, 77.
Griffiths, N. M. & Billington, M. J. (1984) *Journal of the Science of Food and Agriculture*, **35**, 909.
Griffiths, N. M., Billington, M. J. & Griffiths, W. (1981) *Journal of the Association of Public Analysts*, **19**, 113.
Griffiths, N. M., Billington, M. J., Crimes, A. A. & Hitchcock, C. H. S. (1984) *Journal of the Science of Food and Agriculture*, **35**, 1255.
Gunther, F., Burckhardt, O. & Oostinga, I. (1969) *Fleischwirtschaft*, **49**, 474.
Hall, P. S. (1978) *Journal of the Association of Public Analysts*, **49**, 474.
Hall, P. S. & Eames, A. E. (1981) *Journal of the Association of Public Analysts*, **19**, 65.
Hamence, J. H. & Kunwardia, G. H. (1974) *Journal of the Association of Public Analysts*, **12**, 85.
Hamm, D., Searcy, G. K. & Klose, A. A. (1980) *Journal of Food Science*, **45**, 1478.
Hannan, R. S. (1974) In *Meat*, eds D. J. A. Cole and R. A. Lawrie, p. 205. London: Butterworths.
Hannan, R. S. (1983) *Meat Industry*, February, 7.
Hansen, L. P. (1974) *Vegetable Protein Processing.* New Jersey: Noyes Data Corporation.
Hardon, H. J. & Kok, H. A. (1962) *Mitteilungen aus dem Gebiet der Lebensmitteluntersuchung und Hygiene*, **53**, 1.
Harrison, A. J. (1978) *Journal of the Association of Public Analysts*, **16**, 123.
Harrison, N. (1983) *Journal of the Association of Public Analysts*, **21**, 59.
Harwood, D. J., Heitzmann, R. J. & Jouquey, A. (1980) *Journal of Veterinary Pharmacology and Therapeutics*, **3**, 245.
Hauser, E. (1975) *Mitteilungen aus dem Gebiet der Lebensmitteluntersuchung und Hygiene*, **66**, 74.
Hayden, A. R. (1979) *Journal of Food Science*, **44**, 494.
Hayden, A. R. (1981) *Journal of Food Science*, **46**, 1810.
Herrmann, C., Thoma, H. & Kotter, L. (1976) *Fleischwirtschaft*, **56**, 87.
Herschdoerfer, S. M. & John, S. M. (1960) *Proceedings 6th European Meeting of Meat Research Workers.*
Hildebrandt, G., Konigsman, R. & Kretschmer, F. J. (1977) *Fleischwirtschaft*, **57**, 689.

Hiltz, D. F., Lall, B. S., Lemon, D. W. & Dyer, W. J. (1976) *Journal of the Fisheries Board of Canada*, **33**, 2560.
Hitchcock, C. H. S., Bailey, F. J., Crimes, A. A., Dean, D. A. G. & Davis, P. J. (1981) *Journal of the Science of Food and Agriculture*, **32**, 157.
Hobson-Frohock, A. & Reader, J. A. (1983) *Analyst*, **108**, 1091.
Hofmann, K. (1979) *Fleischwirtschaft*, **59**, 1413.
Hogan, W. J. (1978) *The Complete Book of Bacon*. London: Northwood Publications.
Holder, C. L., Blakemore, W. M. & Bowman, M. C. (1979) *Journal of Chromatographic Science*, **17**, 91.
Holder, M. G. (1977) *Food Processing Industry*, **46**, 38.
Hole, N. H., Roberts, M. P. & Bryce-Jones, K. (1964) *Analyst*, **89**, 332.
Holland, G. C. (1979) *Journal of Food Protection*, **42**, 675.
Honikel, K. O. & Sauer, H. (1975) *Fleischwirtschaft*, 55, 1724.
Hornsey, H. C. (1969) *Food Manufacture*, **44**, 48.
Horwitz, W. (1981) *Journal of the Association of Official Analytical Chemists*, **64**, 104, 814.
Hui, J. Y. & Taylor, S. L. (1983) *Journal of the Association of Official Analytical Chemists*, **66**, 853.
Hunt, D. C., Bourdon, A. T. & Crosby, N. T. (1978) *Journal of the Science of Food and Agriculture*, **29**, 239.
Iles, N. A. (1973) *Phosphates in Meat and Meat Products*, Scientific and Technical Surveys No. 81. Leatherhead, Food RA.
Ingles, D. L., Back, J. F., Gallimore, D., Tindale, R. & Shaw, K. J. (1985) *Journal of the Science of Food and Agriculture*, **36**, 402.
Johannes, B., Koerfer, K., Schad, J. & Ulbricht, I. (1983) *Archiv Lebensmittel Hygiene*, **34**, 1.
Johnstone, A. & Thorpe, R. (1987) *Immunochemistry in Practice*, 2nd edn. Oxford: Blackwell Scientific Publications.
Jonas, D. A. (1981) *Journal of Food Technology*, **16**, 683.
Jonas, D. A. & Wood, R. (1983) *Journal of the Association of Public Analysts*, **21**, 113.
Jones, A. D., Shorley, D. & Hitchcock, C. H. S. (1982) *Journal of the Association of Public Analysts*, **20**, 89.
Jones, J. M. (1986) *Journal of Food Technology*, **21**, 663.
Jozefowicz, M. L., O'Neill, I. K. & Prosser, H. J. (1977) *Analytical Chemistry*, **49**, 1140.
Kakuda, Y., Stanley, D. W. & van de Voort, F. R. (1981) *Journal of the American Oil Chemists' Society*, **58**, 773.
Kang'ethe, E. K., Gathuma, J. M. & Lindqvist, K. J. (1986) *Journal of the Science of Food and Agriculture*, **37**, 157.
Karmas, E. (1970) *Meat Product Manufacture*. New Jersey: Noyes Data Corporation.
Karmas, E. (1972) *Sausage Processing*. New Jersey: Noyes Data Corporation.
Ke, P. J., Cervantes, E. & Robles-Martinez, C. (1984) *Journal of the Science of Food and Agriculture*, **35**, 1284.
Kearsley, M. W., El-Khatib, L. & Gunu, C. O. K. A. (1983) *Journal of the Association of Public Analysts*, **21**, 123.
Keay, J. N. (1976) *Proceedings of Conference on the Production and Utilisation of Mechanically Recovered Fish Flesh*. Torry Research Station, Aberdeen.
Kelch, F. (1961) *Fleischwirtschaft*, **41**, 13, 301, 473.
Kendrick, J. L. & Watts, B. M. (1969) *Journal of Food Science*, **34**, 292.
Kietzmann, U. (1978) *Lebensmittelchemie gerichchtle Chemie*, **32**, 85.
Kim, H. & Shelef, L. A. (1986) *Journal of Food Science*, **51**, 731.
King, F. J. & Regan, P. J. (1977) *Journal of the Association of Official Analytical Chemists*, **60**, 963.
King, N. L. & Shaw, F. D. (1984) *CSIRO Food Research Quarterly*, **44**, 1.
Kirk, R. S. & Lumley, I. D. (1988). In *Developments in Meat Science—4*, ed. R. Lawrie. London: Elsevier Applied Science.
Klein, H. & Stoya, W. (1979) *Lebensmittelchemie und gerichtle Chemie*, **33**, 92.
Koh, T. Y. (1978) *Journal of the Canadian Institute of Food Science and Technology*, **11**, 124.
Koprowski, W. S. (1951) *Analyst*, **76**, 732.
Korhonen, R. N., Reagan, J. O., Carpenter, J. A., Campion, D. R. & Stribling, K. V. (1978) *Journal of Food Science*, **43**, 856.
Krause, R. J., Ockerman, H. W., Krol, B., Moermann, P. C. & Plimpton, R. (1978) *Journal of Food Science*, **43**, 853.
Krol, B. & Tinbergen, B. J. (1974) *Proceedings of the International Symposium on Nitrite in Meat Products* held at the Central Institute for Nutrition and Food Research TNO, Zeist, The Netherlands, September 10–14, 1973. Wagengingen: Centre for Agricultural Publishing and Documentation.
Kruggel, W. G., Field, R. A., Riley, M. L., Radloff, H. D. & Horton, K. M. (1981) *Journal of the Association of Official Analytical Chemists*, **64**, 692.

Laird, W. M., Mackie, I. M. & Ritchie, A. H. (1982) *Journal of the Association of Public Analysts*, **20**, 125.
Lang, K. (1979) *Archives Lebensmittel Hygiene*, **30**, 215.
Langen, B. (1982) *Food Processing Industry*, August, 35.
Lanza, E. (1983) *Journal of Food Science*, **48**, 471.
Lawrence, R. & Gould, M. F. (1980) *Proceedings Institute of Food Science and Technology*, **13**, 23.
Lawrie, R. A. (1977) *Meat Science*, **1**, 1.
Lindberg, W., Ohman, J. & Wold, S. (1985) *Analytica Chimica Acta*, **171**, 1.
Lindquist, B., Ostgren, J. & Lindberg, I. (1975) *Zeitschrift fur Lebensmittel Untersuchung und Forschung*, **159**, 15.
Llewellyn, J. W. (1977) *Proceedings of the Analytical Division Royal Society of Chemistry*, **14**, 75.
Llewellyn, J. W. (1979) *International Flavours and Food Additives*, **10**, 115.
Llewellyn, J. W. (1982) *In Developments in Food Proteins—1*, ed. B. J. F. Hudson. London: Applied Science Publishers Ltd.
Llewellyn, J. W., Dean, A. C., Sawyer, R., Bailey, F. J. & Hitchcock, C. H. (1978) *Journal of Food Technology*, **13**, 249.
Lonberg, A. E. (1984) *Var Foda*, 218.
Lord, D. W. & Swan, K. J. (1984) *Journal of the Association of Public Analysts*, **22**, 131.
Love, R. M. & Muslemuddin, M. (1972a, b) *Journal of the Science of Food and Agriculture*, **23**, 1229, 1239.
Love, R. M., Lovern, J. A. & Jones, N. R. (1959) *The Chemical Composition of Fish Tissues*, DSIR Food Investigation Special Report No. 69. London: HMSO.
Lundstrom, R. C. (1979) *Journal of the Association of Official Analytical Chemists*, **62**, 624.
Lundstrom, R. C. (1983) *Journal of the Association of Official Analytical Chemists*, **66**, 123.
Lundstrom, R. C. & Racicot, L. D. (1983) *Journal of the Association of Official Analytical Chemists*, **66**, 1158.
Lundstrom, R. C. & Roderick, S. A. (1979) *Science Tools*, **26**, 38.
Lushbough, C. H. & Schweigert, B. S. (1960) *The Science of Meat and Meat Products, American Meat Institute Foundation*, p. 185. San Francisco: W. H. Freeman and Company.
Lythgo, H. C., Ferguson, C. S. & Racicot, P. A. (1941) *Journal of the Association of Official Analytical Chemists*, **24**, 799.
Mackie, I. M. (1968) *Analyst*, **93**, 458.
Mackie, I. M. (1969, 1972) *Journal of the Association of Public Analysts*, **7**, 83; **10**, 18.
Mackie, I. M. & Taylor, T. (1972) *Analyst*, **97**, 609.
Manley, C. H. & Swaine, R. L. (1979) *Food Product Development*, **13**, 26.
Mann, P. J. (1977) *Food Engineering*, **49**, 106.
Manz, J. (1981) *Fleischwirtschaft*, **61**, 1580.
Marks, H. F. (1985) *Meat Industry*, March, 15.
Marshall, R. (1960) *Food Trade Review*, **30**, 60.
Martin, C. R. A. (1980) *British Food Journal*, **82**, 38.
Martin, R., Azcona, J. I., Tormo, J., Hernandez, P. E. & Sanz, B. (1988) *International Journal of Food Science & Technology*, **23**, 303.
Matsumoto, M., Aoyagi, Y. & Sugahara, T. (1977) *Journal of the Japanese Society of Food and Nutrition*, **30**, 155.
McVey, W. C. & McMillin, H. R. (1940) *Journal of the Association of Official Analytical Chemists*, **23**, 811.
Meech, M. V. & Kirk, R. S. (1986) *Journal of the Association of Public Analysts*, **24**, 13.
Meiners, C., Crews, M. G. & Ritchey, S. J. (1982) *Journal of the American Dietetic Association*, **81**, 435.
Melvin, M. (1987) *Electrophoresis*. ACOL series, ed. D. Kearley. Chichester: J. Wiley.
Mietz, J. L. & Karmas, E. (1977) *Journal of Food Science*, **42**, 155.
Mittal, G. S. & Usborne, W. R. (1985) *Food Technology*, **4**, 121.
Morris, B. A., Clifford, M. N. & Jackman, R. (1988) (eds) *Immunoassays for Veterinary and Food Analysis—1*. London: Elsevier Applied Science.
Moss, M., Holden, J. M., Ono, K., Cross, R., Slover, H., Berry, B., Lanza, E., Thompson, R., Wolf, W., Vanderslice, J., Johnson, H. & Stewart, K. (1984) *Journal of Food Science*, **48**, 1767.
Murray, J. C. F. (1979) *Proceedings Institute of Food Science and Technology*, **12**, 121.
Negbenebor, C. A. & Chen, T. C. (1985) *Journal of Food Science*, **50**, 270.
Nelson, J. R., Copeland, K. F. T., Forster, R. J., Campbell, D. J. & Black, W. D. (1983) *Journal of Chromatography*, **276**, 438.
Newton, R. T. (1979) *Journal of the Association of Official Analytical Chemists*, **62**, 722.
Noble, J. (1976) *Quick Frozen Foods*, **34**, 185.
Noyes, R. (1969) *Protein Food Supplements*. New Jersey: Noyes Data Corporation.
Olsman, W. J. (1975) *Methods for Detection and Quantitative Assay of Plant Proteins in Meat Products*. R4867. Zeist: Central Instituut Voor Voedingsonderzoek.

Olsman, W. J. (1977) *Methods for Detection and Quantitative Assay of Plant Proteins in Meat Products*. RS5463. Zeist: Central Instituut Voor Voedingsonderzoek.
Olsman, W. J. (1979) *Journal of the American Oil Chemists' Society*, **56**, 285.
Olsman, W. J. & Hitchcock, G. H. S. (1980). In *Developments in Food Analysis Techniques*, Vol. 2, ed. R. D. King. London: Applied Science Publishers.
Olsman, W. J., Houtepen, W. H. C. & Van Leeuwen, C.M (1969) *Zeitschrift fur Lebensmittel Untersuchung und Forschung*, **141**, 253.
Olsman, W. J., Dobbelaere, S. & Hitchcock, C. H. S. (1985) *Journal of the Science of Food and Agriculture*, **36**, 499.
O'Neill, I. K. & Richards, C. P. (1978) *Chemistry and Industry*, No. 2, 65.
Pailler, F. M. (1979) *Annales Falsifications L'Expertise Chimique det Toxicologique*, **72**, 335.
Palmer, A. J. & Palmer, F. J. (1989) *Journal of Chromatography*, **465**, 369.
Parris, N. (1984) *Journal of Agricultural and Food Chemistry*, **32**, 829.
Patterson, R. M., Whittaker, R. G. & Spencer, T. L. (1984) *Journal of the Science of Food and Agriculture*, **35**, 1018.
Paul, A. A. & Southgate, D. A. T. (1978) *The Composition of Foods*. London: HMSO.
Pearson, D. (1965, 1969b, 1974) *Journal of the Association of Public Analysts*, **3**, 76; **7**, 120; **12**, 73.
Pearson, D. (1969a, 1972) *Food Manufacture*, **44**, 42; **47**, 45.
Pearson, D. (1973) *Laboratory Techniques in Food Analysis*. London: Butterworth.
Pearson, D. (1975) *Analyst*, **100**, 73.
Pearson, A. M. & Dutson, T. R. (1988) (eds) *Edible Meat By-products. Advances in Meat Research*, Vol. 5. London: Elsevier Applied Science.
Pearson, D. & Muslemuddin, M. (1968, 1969, 1971) *Journal of the Association of Public Analysts*, **6**, 117; **7**, 50; **9**, 28.
Pearson, A. M. & Tauber, F. W. (1984) *Processed Meats*, 2nd edn. Westport: AVI.
Peleran, J. C. & Bories, G. F. (1984) *Chemical Abstracts*, **100**, 2160a.
Perrin, C. H. & Ferguson, P. A. (1968) *Journal of the Association of Official Analytical Chemists*, **51**, 971.
Petz, M. (1982) *Deutsche Lebensmittel Rundschau*, **78**, 396.
Pizzacaro, F. (1978) *Rivista di Zootecnia e Veterinaria*, No. **4**, 263.
Plowright, D. G. (1963) *Food Manufacture*, **38**, 182.
Prandle, O. (1979) *Fleischwirtschaft*, **59**, 1232.
Priebe, K. (1979a, b) *Fleischwirtschaft*, **59**, 1239, 1658.
Prior, B. A. (1979) *Journal of Food Protection*, **42**, 668.
Rangeley, W. R. D. & Lawrie, R. A. (1976) *Journal of Food Technology*, **11**, 143.
Rangeley, W. R. D. & Lawrie, R. A. (1977) *Journal of Food Technology*, **12**, 9.
Rejt, J. & Pisula, A. (1982) *Meat Science*, **6**, 185.
Reusse, U. (1971) *Archiv fur Lebensmittel Hygiene*, **22**, 136.
Reynolds, N. (1968) *Public Health Inspector*, **76**, 524.
Rhodes, D. N. (1974) *Proceedings 10th Anniversary Symposium of the Institute of Food Science and Technology* 5.
Ritskes, T. M. (1975) *Journal of Food Technology*, **10**, 221.
Ruiter, A. & Weseman, J. M. (1976) *Journal of Food Technology*, **11**, 59.
SAC (1961, 1965, 1966, 1968, 1971, 1973) *Analyst*, **86**, 560; **90**, 579; **91**, 540; **93**, 476; **96**, 744; **98**, 456.
SAC (1974) *Official, Standardised and Recommended Methods of Analysis*, 2nd edn. London: HMSO.
Salfi, V., Fucetola, F. & Pannunzio, G. (1985) *Journal of the Science of Food and Agriculture*, **36**, 811.
Sattaneo, P., Soncini, G., Brenna, O. & Cantoni, C. (1981) *Industrie Alimentarie*, **20**, 3.
Saunders, P. R. & Ward, A. G. (1953) *Journal of the Science of Food and Agriculture*, **4**, 523.
Sawyer, R. (1984) *In Control of Food Quality and Food Analysis*, eds G. G. Birch and K. J. Parker. London and New York: Elsevier.
Sayed, L. & Dashlauty, A. (1979) *La Rivista Italiana delle Sostanze Grasse*, **56**, 52.
Schmidtlein, H. (1979) *Lebensmittelchemie und gerichtle Chemie*, **33**, 81.
Schweigert, R. G. (1974) *Meat and Meat Products*. Mini-Symposium Proceedings, IFST, March 1974, 10.
Sen, N. P. & Donaldson, B. (1978) *Journal of the Association of Official Analytical Chemists*, **61**, 1389.
Sen, N. P. & McPherson, M. (1978) *Journal of Food Safety*, **1**, 247.
Siegel, D. G., Tuley, W. B. & Schmidt, G. R. (1979) *Journal of Food Science*, **44**, 1272.
Sikka, K. C., Ranjeet Singh, Gupta, D. P. & Duggal, S. K. (1979) *Journal of Agricultural and Food Chemistry*, **27**, 946.
Sinar, R., Hooke, G. F. & Conchie, E. C. (1976) *Journal of the Association of Public Analysts*, **14**, 127.
Skurray, G. R. & Lysaght, V. A. (1978) *Food Chemistry*, **3**, 111.
Smith, P. R. (1969) *A Review of Rapid Methods for the Estimation of Total Fat*. Scientific and Technical Surveys No. 56. Leatherhead: BFMIRA.
Smith, P. R. (1975) *Proceedings Institute of Food Science and Technology*, **8**, 154.

Smyth, W. F., Ayling, C. & Smyth, J. G. (1986) *Analytical Proceedings*, **23**, 84.
SPA (1951) *Analyst*, **76**, 329.
Stegemann, H. & Stalder, K. (1967) *Clinica Chimica Acta*, **18**, 267.
Storey, R. M. & Mills, A. (1976) *Process Biochemistry*, **11**, 25.
Storey, R. M., Davis, H. K., Owen, D. & Moore, L. (1984) *Journal of Food Technology*, **19**, 1.
Stubbs, G. & More, A. (1919) *Analyst*, **44**, 125.
Tarrant, P. V. (1975) *Nutrition and Food Science*, **41**, 11.
Terio, E., Tiecco, G. & Tantillo, G. (1985) *Industrie Alimentarie*, **24**, 8.
Thomas, M. L., Grey, T. C., Jones, J. M., Robinson, D. & Stock, S. W. (1984) *Journal of Food Technology*, **19**, 11.
Thorne, S. (1978) *Nutrition and Food Science*, No. **52**, 13.
Tinbergen, B. J. & Slump, P. (1976) *Zeitschrift fur Lebensmittel Untersuchung und Forschung*, **161**, 7.
Tomas, M. C. & Funes, J. (1987) *Journal of Food Science*, **52**, 575.
Truman, R. W. & Dickes, G. J. (1976) *Journal of the Association of Public Analysts*, **14**, 5.
Vasundhara, T. S., Kumudavally, K. V. & Sharma, T. R. (1983) *Journal of Food Protection*, **46**, 1050.
Vinogradov, A. P. (1953) *The Elementary Composition of Marine Organisms* (Translation). New Haven: Yale Univ.
Voller, A., Bidwell, D. E. & Bartlett, A. (1979) *The Enzyme-linked Immunosorbent Assay (ELISA). A Guide with Abstracts of Microplate Applications*. Guernsey: Dynatech Europe.
Walters, M. J. (1984) *Journal of the Association of Official Analytical Chemists*, **67**, 1040.
Ward, A. G. (1951) *Food*, **20**, 255.
Ward, D. R., Finne, G. & Nickelson, R. (1979) *Journal of Food Science*, **44**, 1052.
Watson, D. H. (1983) *Food Chemistry*, **12**, 167.
Weber, H. & Hopke, H-U. (1981) *Fleischwirtschaft*, **61**, 200.
Weiland, H. (1972) *Enzymes in Food Processing and Products*. New Jersey: Noyes Data Corporation.
Winter, F. F. (1978) *Fleischerei*, **29**, 54.
Wong, N. P., Damico, J. N. & Salwin, H. (1967) *Journal of the Association of Official Analytical Chemists*, **50**, 8.
Wood, E. C. (1977) *Journal of the Association of Public Analysts*, **15**, 99.
Wood, T. (1956) *Journal of the Science of Food and Agriculture*, **7**, 196.
Young, K. W. & Whittle, K. J. (1985) *Journal of the Science of Food and Agriculture*, **36**, 383.

14

Dairy products I

COMPOSITION AND ANALYSIS OF MILK

Composition of milk

'Cows milk' has been defined as the secretion, excluding colostrum, that can be gained by normal milking methods from the lactating mammary gland of the healthy, normally fed cow. Milk can be considered as containing three basic components, namely water, fat and non-fatty solids (NFS) or solids-not-fat (SNF). The organic matter in the non-fatty portion consists mainly of casein and whey proteins together with lactose and lactic and citric acids. The average composition of milk from the domestic cow, goat, ewe and other mammals is given in Table 14.1.

Table 14.1 Average percentage composition of the milk of the cow and other mammals (from Lampert, 1970 and elsewhere)

	Water	Protein	Fat	Lactose	Ash
Cow	87·4	3·3	3·9	4·6	0·72
Human	86·5	2·0	4·1	7·2	0·21
Ass	89·8	1·9	1·4	6·2	0·45
Buffalo	82·4	4·7	7·4	4·6	0·78
Camel	87·6	3·4	3·0	5·1	0·71
Cat	83·0	7·0	4·5	4·8	0·60
Dog	74·5	3·1	10·2	11·3	0·80
Elephant	85·6	3·2	8·1	7·4	0·63
Ewe	81·6	5·6	7·5	4·4	0·90
Goat	86·9	3·3	4·5	4·6	0·79
Llama	86·5	3·9	3·1	5·6	0·80
Mare	89·8	2·0	1·5	6·1	0·41
Porpoise	41·2	11·2	45·8	1·1	0·57
Rabbit	68·5	12·9	13·6	2·4	2·55
Reindeer	66·1	10·1	19·8	2·5	1·45
Seal	34·0	12·0	54·0	none	0·53
Sow	80·6	6·1	7·6	4·7	0·92
Vixen	81·8	6·3	6·2	4·2	1·31
Whale	69·8	9·4	19·4		0·99
Zebu	86·2	3·0	4·8	5·3	0·70

The composition of standard UK dairy products including milk, cream, butter and yogurts are given in a booklet obtainable from the National Dairy Council. A detailed survey of the content of B complex vitamins and vitamin C in retail milks (and condensed milk products and yogurts) is reported by Scott and Bishop (1986).

Scott *et al.* (1984a) report the contents of vitamins A, D_3, C and B complex in pasteurised bulk milk showing regional, seasonal and breed effects. Scott *et al.* (1984b) report the content of vitamin C, riboflavin, folic acid, thiamine, B_{12} and B_6 in home delivered and refrigerator stored pasteurised milk.

Apart from pathological conditions in cows that give rise to the production of abnormal milk (Rook, 1961) there are a number of factors which affect the composition. Average figures do not of course reflect the day-to-day variations that occur in milk taken from individual animals. Also many of the factors discussed below are affected indirectly by the level and type of feed.

Factors affecting the composition of milk

Large variations, especially in the fat content, are given by the different breeds of cows (Table 14.2). In addition to the figures quoted, within each breed there is a wide range of composition.

Table 14.2 Lactation trends by breed, per cent m/m

	Fat			Protein		
	1976	1985	1986	1976	1985	1986
Ayrshire	3·85	3·96	3·97	3·32	3·34	3·33
British Friesian	3·68	3·90	3·93	3·22	3·23	3·23
British Holstein	3·68	3·84	3·87	*	3·17	3·16
Dairy Shorthorn	3·57	3·69	3·74	3·28	3·27	3·28
Guernsey	4·46	4·67	4·69	3·55	3·55	3·56
Jersey	4·90	5·27	5·33	3·79	3·80	3·81

* Insufficient data.
(From Report of the Breeding and Production Organisation 1986/87; Milk Marketing Board).

The general seasonal trends in composition of bulk milk are well defined and published annually by the Milk Marketing Board. In England and Wales (1986–87) the fat content was highest in October and November (4·08 per cent) and lowest in June (3·83 per cent). The SNF was high in September, 8·99 per cent, falling to a minimum of 8·67 per cent in February to April, then rising in May and June to 8·88–8·96 per cent, but fell slightly by 0·1 per cent in July–August. For UK retail milk Florence *et al.* (1985) showed similar seasonal fluctuations. These changes are a reflection of the effects of other factors, particularly the seasonal variations in feeding and the stage of lactation. Seasonal changes vary also according to the locality.

Colostrum, the thick liquid secreted immediately after parturition, differs markedly in composition from normal milk. The total solids content of colostrum may be as high as 25 per cent and consists mainly of protein.

The fat, SNF and protein content of milk are at a maximum in early lactation, then fall to a minimum for SNF and protein after 6 weeks and for fat at about 10 weeks and finally increase until the end of lactation. Converse changes occur in the lactose, the amount of which is low in colostrum, rises markedly in the first week, remains steady until the sixth month and then diminishes again until the end of lactation. The fat and SNF contents decrease with successive lactations, there being a drop of about 0·1 per cent in both in each of the second, third and fourth lactations after which the fall is less pronounced. Milk from individual cows may show a day-to-day variation.

Such fluctuations may be influenced by the mental and physical condition. Excitement, worry or discomfort is liable to have an adverse effect on both the quality and quantity produced. There is a progressive increase in fat content during the milking. The first drawn milk contains about 1 per cent fat whereas the strippings may contain over 10 per cent. Also, morning milk has a lower fat content than that of evening milk due to the interval between milkings being longer at night than during the day (Table 14.3).

Table 14.3 The effect of milking interval on fat content and yield of milk (after Johansson and Claesson, 1957)

Milking interval (h)	Fat (%)	Yield (lb)
2	6·0	4·6
4	4·6	9·3
6	4·5	13·3
8	4·1	15·7
10	3·6	18·7
12	3·6	18·7

The phospholipid content of milk also varies with season and region over the range 0·23–0.38 per cent of the total milk solids.

Milk proteins

Two main forms of protein occur in milk, micellular proteins (75–85 per cent), which are caseins associated with calcium, phosphate and citrate, having an open, random structure and which precipitate at pH 4·6; and serum or whey proteins (15–22 per cent) differing in their molecular, physical and functional properties but having globular structure and that are soluble at pH 4·6. There is also 2–4 per cent of a proteose peptone fraction.

The percentage composition of the casein fraction is α_{S1}-casein, 39–46; α_{S2}-casein, 8–11; β-casein, 25–35; κ-casein, 8–15; and γ-casein, 3–7. The percentage composition of the whey protein fraction is β-lactoglobulin, 7–12; α-lactoglobulin, 2–5; serum albumin, 0·7–1·3; and immunoglobulins, 1·9–3·3 (Brunner, 1981).

Milk protein fractions are important food ingredients, functional additives and industrial chemicals (Morr, 1982; Swartz and Wong, 1985).

Acid casein is obtained by adjusting skim milk to pH 4·6 with mineral acid or by lactic fermentation. The precipitated curd is separated, washed and dried, and is low in calcium and phosphate, which remains in solution.

Rennet casein is obtained as a coagulate caused by the specific action of chymosin, a proteolytic enzyme in rennet. It retains the calcium–phosphate complex and can be solubilized by sequestering the calcium by the addition of sodium phosphate.

Caseinates are manufactured by solubilizing casein with alkalis or sequestering agents and drying the resultant product. Sodium caseinates form translucent solutions of high viscosity and water binding properties, whereas calcium caseinates form opaque solutions with low viscosity. They are ingredients of coffee whitener products.

The labelling and composition of the above products are controlled by the Caseins and Caseinates Regulations (SI 1985 No. 2026) arising from Council Directive 83/417/EEC (OJ No. L237, 26.8.83, p. 25), and ISO/IDF/AOAC methods of test are published as various parts of BS 6248.

Whey powder is prepared by drying the serum left after the separation of casein from skim milk and contains 70–80 per cent of lactose and 9 per cent of protein. *Whey protein concentrates* are prepared by removing lactose and soluble ions from whey by ultrafiltration, to yield powders containing 35–80 per cent of protein. *Lactalbumin* is precipitated out when whey is heated to 90 °C and consists of insoluble denatured whey proteins. The industrial isolation of whey proteins is described by Marshall (1982).

Co-precipitates are formed by co-reacting casein and denatured whey proteins. The precipitated material will have varied functional properties depending on the heat applied, pH and calcium content.

Analytical methods for whey powder and similar products are described in various parts of BS 1743 (see under dried milk).

Acidity and pH. Cows milk is acid to phenolphthalein, alkaline to methyl orange, but amphoteric to litmus due to the presence of phosphates. The pH is usually between 6·4 and 6·6. The total acidity of freshly drawn milk is usually about 0·14 per cent (as lactic acid). On storage, the acidity increases due to the action of micro-organisms, and a sour taste is perceptible when this reaches about 0·3 per cent. Milk is decidedly sour at 0·4 per cent and when the acidity reaches 0·6 per cent it curdles at ordinary temperatures. A different spoilage pattern is set up in heat-treated milk due to the change in the microbial flora.

Mineral matter

The total ash of genuine milk varies from about 0·71 to 0·75 per cent and for normal milk constitutes about 8·3 per cent of the non-fat solids. The principal inorganic constituents in milk are listed in Table 14.4.

Table 14.4 The inorganic constituents of retail whole milk (bulked, pasteurised) (Florence *et al.*, 1985)

		Ash (%)	Ca	Mg	Na (mg/100 g)	K	P
Non-Channel Islands milk	Mean	0·72	114	12	55	144	92
	S.D.	0·01	1	1	2	3	1
Channel Islands Milk	Mean	0.75	131	13	54	143	100
	S.D.	0·02	5	1	5	8	2

Vieth's ratio

For milk containing a normal amount of SNF (8·4–8·7 per cent) the ratio of lactose (hydrated):protein:ash is almost exactly 13:9:2. This is known as Vieth's ratio and as it is unaffected by the removal or addition of water, it is useful for detecting abnormalities in liquid, condensed and dried milk and cream (p. 556). From the data of Harding and Royal (1974) for bulked milk in England and Wales for the period

1947–1970, Vieth's ratio was 12·7:8·9:2. For retail milk produced in 1983–84, Vieth's ratio for non-Channel Island milk was 13:8·7:2; and for Channel Island milk was 13:9·8:2 (from data of Florence *et al.* 1985).

Legislation and standards

The 'Codex code' referred to in this and the following chapter is the Code of Principles concerning Milk and Milk Products, International Standards and Standard Methods of Sampling and Analysis for Milk Products of the Codex Alimentarius Commission, CAC/MI–1973, 7th Edn (FAO/WHO, Rome).

The EC standards prescribed in Regulation 1411/71/EEC as amended by Regulations 1556/74/EEC and 566/76/EEC were implemented by the Drinking Milk Regulations 1976, SI No. 1883. These regulations restrict the types of milk for human consumption in the UK to raw milk, non-standardised whole milk produced in the UK and containing not less than 3·0 per cent milk fat, standardised whole milk imported from other Member States with a fat content not less than a guideline percentage fixed for each year, semi-skimmed milk with a fat content between 1·5 and 1·8 per cent and skimmed milk with a fat content of not more than 0·3 per cent.

Heat treatment and labelling of the last two types of milk is controlled and methods of test prescribed by the Milk and Dairies (Semi-Skimmed and Skimmed Milk) (Heat Treatment and Labelling) Regulations SI 1986, No. 722. The EC legislation has not affected the standards and sale of milk defined by the Milk and Dairies (Channel Island and South Devon Milk) Regulations 1956, SI No. 919, to contain at least 4 per cent milk fat. Section 36 of the Food Act 1984 prohibits the addition of water, colouring matter, preservatives, dried milk and condensed milk to milk. It also prohibits the addition of separated milk, or a mixture of separated milk and cream to unseparated milk and the sale of any product designated as milk whose preparation has included the use of separated, dried or condensed milk.

The Milk (Special Designation) Regulations 1986, SI No. 723, prescribe the conditions under which producers' and dealers' licences may be granted in relation to milk that is designated as 'Untreated', 'Pasteurised', 'Sterilised' and 'Ultra Heat Treated'. Provisions for sampling and the official methods of test are prescribed. These regulations require that 'Untreated Milk' shall come from a herd accredited as free from brucellosis and shall not at any stage have been treated by heat or in any manner likely to affect its nature or qualities. Untreated milk is required to satisfy the methylene blue test (p. 551). The regulations require that 'Pasteurised Milk' shall be either retained at 62·8–65·6 °C for at least 30 min or be retained at not less than 71·7 °C for at least 15 s. The milk is then to be cooled immediately to not more than 10 °C following either of these two methods of pasteurisation. Pasteurised milk is required to satisfy the methylene blue and phosphatase tests (pp. 551, 553). The regulations require that 'Sterilised Milk' shall be filtered or clarified, homogenised and thereafter heated to and maintained at not less than 100 °C for such a period that it will comply with the turbidity test (see p. 554). Milk that has been sterilised by a continuous-flow method has to satisfy the prescribed colony count test. The regulations require that 'Ultra Heat Treated Milk' shall be retained at a temperature of not less than 132·2 °C for not less than 1 s and shall then be filled and sealed aseptically into the sterile containers in which it is to be supplied to the consumer. Ultra heat treated milk treated by the injection of steam is required to contain the same amount of milk fat and MSNF before and after treatment; UHT milk is required to satisfy the prescribed colony count test.

The Importation of Milk Regulations 1988, SI No. 1803, re-enact with amendments the 1983 regulations and make provisions for the implementation of Directive 85/397/EEC (OJ No. L226, 24.8.85, p. 13) on intra-Community trade in heat-treated milk. The regulations allow the importation of bulk and heat-treated milks under strictly defined conditions and accompanied by an authorised certificate. A directive is in preparation for methods of analysis to enforce the regulations. It will include methods for sampling, total solids, fat, nitrogen, specific mass, freezing point, phosphatase activity, peroxidase activity, turbidity test, aerobic colony count at 30 °C and 21 °C, coliforms, somatic cells, detection of antibiotics and sulphonamides, bacterial lipopolysaccharide test, pyruvate content and pathogens.

The Milk (Special Designation) (Amendment) Regulations 1988 SI No. 1805 and the Milk and Dairies (Semi-skimmed and Skimmed Milk) (Heat Treatment and Labelling) (Amendment) Regulations 1988 SI No. 1804) state that any milk heat-treated in another member state of the EC cannot be further heat-treated unless accompanied by a certificate prescribed by the Importation of Milk Regulations 1988, and if the temperature of the milk exceeds 6 °C.

Sheep and goat milk (and their products) are now widely marketed in the UK, mainly in health food shops. They are not subject to any specific legislation but the Department of Agriculture and Fisheries in Scotland have issued Codes of Practice on their hygienic control.

The quality of flavoured milk-based drinks (at least 85 per cent of cows milk) is controlled and the methods of test prescribed by the Milk-based Drinks (Hygiene and Heat Treatment) Regulations 1983, SI No. 1508.

Analysis of milk

Methods and techniques for sampling milk and milk products for microbiological, chemical, physical and sensory analysis are described in BS 809: 1985 (ISO 707–1985). BS 1741: 1963, which describes methods of chemical analysis of liquid milk, is currently under revision and will be published in separate parts. BS 1741: Part 1: 1987 is the general introduction and includes the preparation of laboratory samples. The standard methods published by the International Dairy Federation also comprehensively cover milk and dairy product analysis (see *Journal of the Association of Official Analytical Chemists*, 1984, **67**, 1152–1171).

Prior to taking portions for each analytical determination, the milk sample should be thoroughly mixed by continuous slow inversions of the sample bottle or by slowly pouring it into a beaker and back into another beaker and repeating the process many times.

In the routine examination of milk it is convenient to take the lactometer (density) reading and, after determining the **fat** by the Gerber method, the **total solids** and hence the **non-fatty solids** content (both as per cent m/m) can be calculated. The results obtained are then confirmed by gravimetric methods on any doubtful samples and the determination of the **freezing point** is then also advisable while the milk remains reasonably fresh. The development of sourness can be assessed by determining the **acidity** of the milk. The assessment of the **degree of pasteurisation** from the results of the **phosphatase test** is also a common routine procedure for milks that are sold as 'pasteurised'. Other determinations that may be required are those of **protein, lactose, ash, chloride, copper** and **citric acid** and examination for **dirt, added dyes** and **preservatives**, and residues of **detergents** and **antibiotics**. The **methylene blue test** and the **bacteriological examination** must be carried out on samples taken under aseptic precautions.

For commercial purposes, the rapid and automatic determination of fat, protein, lactose and total milk solids is carried out on specially designed infra-red absorption instruments (e.g. Milko-Scan and Multispec), using absorption bands 3·5 μm or 5·7 μm for fat, 6·5 μm for protein, 9·6 μm for lactose and 4·3 μm for water. The technique is fully reviewed by Briggs *et al.* (1987). An evaluation of a Milko-Scan instrument is given by Van de Voort (1980), and its use with ewes, goats and buffalo milk is described by Harris (1986) and El-Salam *et al.* (1986). The wide versatility of the newer near infra-red reflectance instruments in dairy product analysis is described by Kennedy *et al.* (1985).

Specific gravity. The specific gravity of milk varies according to the proportions of fat (sp.gr. 0·93), milk solids not-fat (sp.gr. 1·614) and water. The density of milk can be conveniently measured by means of the lactometer, which is a special hydrometer calibrated over the range 1·025–1·035 (25·0–35·0° as lactometer degrees). For the determination, the milk should be reasonably fresh and be thoroughly but gently mixed, avoiding incorporation of air. The temperature of the sample being measured must always be taken.

If the percentage of fat is determined, e.g. by the Gerber method, the total solids figure can be calculated from the modified **Richmond's formula** (BS 734: 1985).

$$T = 0{\cdot}25\,D + 1{\cdot}22\,F + 0{\cdot}72 \qquad (1)$$

where D = the BS density hydrometer reading at 20 °C, T = total solids (per cent) and F = fat (per cent). SNF = $T-F$. For routine purposes the figures are usually reported to the nearest 0·05 per cent. The correction for densities measured at temperatures other than 20 °C is approximately +0·24 for each 1 °C above 20 °C. O'Keefe (1967) criticised formula (1) as it overestimates the total solids by 0·096 and suggested that the one previously used ($T = 0{\cdot}25\,D + 1{\cdot}21\,F + 0{\cdot}66$) is correct.

The specific gravity of milk taken more than an hour after milking (when all air bubbles have disappeared) is lower than that observed subsequently (**Recknagel's phenomenon**). During the first 12 h after milking the specific gravity (observed) of the milk may rise as much as 0·0013. The change has been attributed to alterations in the casein and the physical condition of the fat. Richmond's formula applies to aged and cooled milks with the Recknagel effect virtually completed. The BS method, however, requires that the density is measured after the milk has been warmed to 40 °C, maintained at that temperature for 5 min (with gentle mixing) and cooled to 20 °C. This ensures that all milks are examined under the same conditions. If, however, the milk contains fat in the solid state, the BS requires that the density is measured after first cooling to 4·5 °C (or lower) for at least 6 h (or to 7·2 °C for at least 16 h) and then raising the temperature of the sample to 20 °C. After such treatment the following further modified formula is applicable:

$$T = 0{\cdot}25\,D + 1{\cdot}22\,F + 0{\cdot}55 \qquad (2)$$

Slide rules and other devices based on formula (1) above have been devised. Also

BS 734 includes extensive tables that relate the fat content and the SNF. Corrections for temperature to be applied to the hydrometer reading are included.

Total solids. The following is an outline of the gravimetric method for determining the total solids of milk (BS 1741: Part 2; ISO 6731; IDF 22: 1963):

Weigh out about 5 g of mixed milk into a weighed round flat-bottomed metal dish (about 7 cm diameter) and provided with a closely fitting lid, which is weighed with the dish. Place the uncovered dish on a boiling-water bath for 30 min or until most of the moisture is driven off. Then wipe the bottom of the dish and transfer it to a well-ventilated oven at 102 ± 2 °C. Place the lid next to the dish in the oven. Dry for 2 h in the oven and then cover the dish with the lid, cool for 30 min in a desiccator and weigh. Then heat the dish and lid as before for 1 h periods in the oven, cool, weigh, etc., until the loss in weight between successive weighings does not exceed 1 mg. If the acidity of milk exceeds 0·20 per cent (as lactic acid) some of the acid present may volatilise on drying and neutralising of the milk with 0·05 M strontium hydroxide is advisable. If this is done a deduction must be made from the weight of the residue obtained equivalent to 0·004 28 g for each ml of 0·05 M strontia used.

Fat. Rapid volumetric methods are often used for routine purposes for determining fat in milk. In the UK, the **Gerber method** is commonly used:

BS 696: Part 1: 1988 describes the necessary apparatus and Part 2 covers the method of determination and gives specifications for the reagents. The test is performed in butyrometer tubes that are closed with special metal and rubber stoppers. For the test, pipette the following liquids into a milk butyrometer, ensuring that they do not mix with one another: 10 ml of sulphuric acid (density 1·812–1·818 at 20 °C ≡ 90 per cent m/m approx.), 10·94 ml of mixed milk (special pipette) and 1 ml of fat-free amyl alcohol (density 0·809–0·813 at 20 °C). Close the tube with a stopper, mix the contents thoroughly and immediately centrifuge at 1100 r.p.m. for 4 min. Transfer the tube (stopper downwards) to a water bath at 65 °C for at least 3 min and read off the percentage fat directly from the scale (extending from the bottom of the upper meniscus to the flat bottom of the fat column). Readings to an accuracy of 0·05 per cent are usually adequate for routine purposes. As it is difficult to separate the small fat globules in homogenised (e.g. 'sterilised') milk, it is advisable to re-centrifuge after warming in the 65 °C bath until the reading reaches a maximum. Other modifications for skim milk, semi-skimmed milk, separated milk, buttermilk and whey are given in BS 696: Part 2.

By adding a trace of ferric chloride to the sulphuric acid used, a violet colour is produced if the milk contains formaldehyde (cf. p. 547). Numerous reagents have been proposed for replacing the sulphuric acid used in the Gerber method by less corrosive liquids. Those which incorporate synthetic detergents are preferred (Houston, 1955). Such a method due to Macdonald (1959a) is as follows:

Macdonald's neutral reagent can be prepared using 5 per cent m/v trisodium citrate, 5 per cent m/v sodium salicylate, 1 per cent m/v EDTA (di-sodium salt), 1·1 per cent m/v Tween 85, 3 per cent v/v *n*-butyl alcohol and 25 per cent v/v methylated spirit (industrial ethanol) in water. Mix gently by inversion before use. For the method, pipette 10 ml Macdonald's reagent into a milk butyrometer (as used in the Gerber method) and 10·94 ml milk. Stopper, invert twice and place in water at 65 °C for 5 min. Then thoroughly mix by shaking and invert 2 or 3 times while shaking. Immediately centrifuge at 1100 r.p.m. for 5 min, return the butyrometer to the water bath at 65 °C for at least 3 min and read off the percentage fat from the scale. Return the tube to the 65 °C water bath, re-centrifuge and read again. All determinations should be carried out in duplicate. The method can be applied to other dairy products (Macdonald, 1959b).

In the USA the Babcock milk test bottle is used (see the AOAC methods book). It is much larger than the Gerber tube, taking 18 g of milk and 17·5 ml sulphuric acid (sp. gr. 1·82) and after dissolution and centrifuging, soft water is added to bring the fat column into the graduated neck. The Gercock test bottle is a Babcock test bottle with a Gerber calibrated neck, and its use is described in the 2nd Supplement (1986) to the AOAC methods book. Bradley (1986) has reported on a collaborative study, which found a precision for fat content equivalent to that of the Rose–Gottlieb method.

The gravimetric **Rose–Gottlieb** method is prescribed as the reference method for determining fat in milk (BS 1741: Part 3: 1987; ISO 1211–1984; IDF 1B: 1983). The procedure is given in full detail in Directive 79/1067/EEC laying down Community methods of analysis for testing preserved milk products (OJ No. L327, 24.12.79, p. 29). The BS/ISO/IDF method gives preference to the use of Majonnier fat extraction flasks (BS 5522: 1977; ISO 3889–1977) (see Fig. 14.1), but suitable centrifuges are

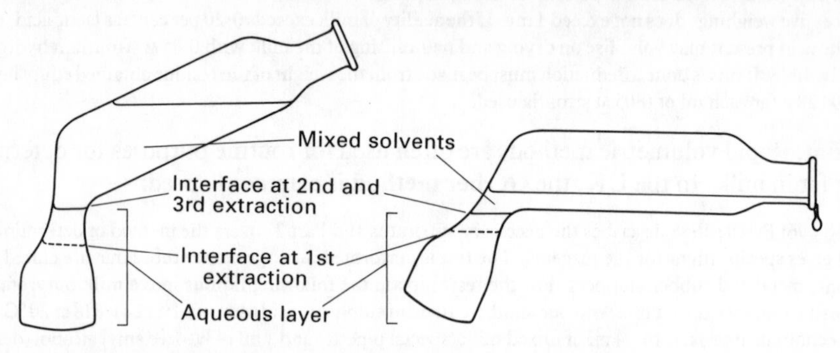

Fig. 14.1 Majonnier tube showing decantation of fat solvent.

not readily available. Alternatively, fat extraction can be carried out with tubes fitted with siphon or wash-bottle fittings. A suitable slow speed electrical centrifuge is available (Wifug Lab Centrifuges, UK). Suitable extraction tubes with ground-glass joints and adjustable blow-off fittings can be easily constructed from standard laboratory glassware (Fig. 14.2). Although the procedures differ slightly according to the type of tube used, the following can be adapted for each.

Weigh out 10 g of milk into the tube, add 1 ml of 0·88 ammonia and mix. Then add 10 ml of alcohol (95 per cent) and again mix well. Add 25 ml peroxide-free diethyl ether, stopper the tube and shake vigorously for 1 min. Add 25 ml of light petroleum (b.p. 40–60 °C) and shake vigorously for 30 s. After separation is complete (after centrifuging or standing for at least 30 min) transfer the fat solution into a suitable flask (previously dried at 100 °C, cooled and weighed). To the tube add two successive lots of 5 ml of mixed 'ethers' and transfer (without shaking) to the flask. Add 5 ml ethanol, mix, then repeat the extraction (with 15 ml of ether and 15 ml of light petroleum) and the subsequent operations. Repeat the extraction with the ethers once more without adding more ethanol. Distil off the solvents from the flask, dry the fat for 1 h at 100 °C, cool and weigh. If any non-fatty matter appears to be present wash out the fat from the flask with light petroleum, dry, reweigh and correct the result accordingly.

The gravimetric **Werner–Schmid method** can be carried out in the extraction tubes used for the Rose–Gottlieb process. This method, and also the Weibull–Berntrop method (IDF 126A–1988), is particularly applicable for sour milk and other milk-based foods not suitable for fat determination by the Rose–Gottlieb method.

Weigh out 10 g of the milk into the tube, add 10 ml of conc. hydrochloric acid and immerse in boiling water until all the casein has dissolved. At this stage the mixture should be brown (or violet) in colour and the fat will be seen to collect on the surface. Cool the tube in running water. Extract the fat by shaking with 30 ml of diethyl ether and blow the extract, after allowing the layers to separate, into a weighed flask. Separation is often assisted by the addition of a little alcohol. Repeat the extraction three times more and

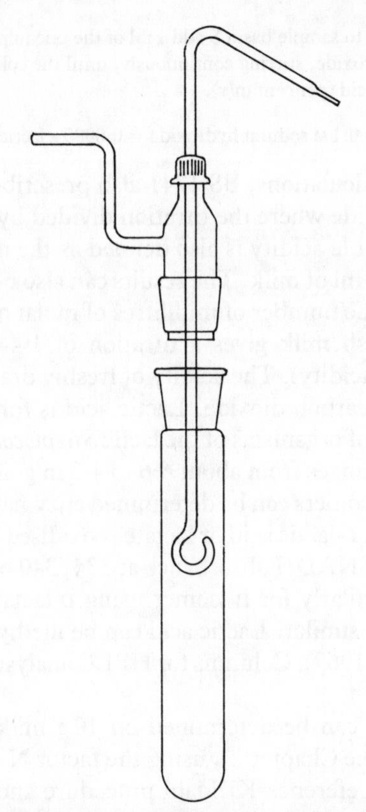

Fig. 14.2 Quickfit test tube and adjustable Dreschel bottle head adapted for use in the estimation of fat. (Courtesy of Corning Medical and Scientific (Europe).)

distil off the solvent. Then dry the fat at 100 °C, cool and weigh. Wash out the fat from the flask with light petroleum, carefully leaving any insoluble residues in the flask. Dry the flask in the oven, cool and reweigh and correct the result accordingly.

A method for total lipids or neutral and polar lipids in milk by solvent elution from a column is described by Maxwell *et al.* (1986). To a column of 5 g of Celite 545/ $CaHPO_4.2H_2O$ (9:1) is added a mixture 5 g of milk (or 2 g of cream), 20 g of anhydrous Na_2SO_4 and 10 g of Celite 545. Total lipid is extracted with about 150 ml dichloromethane/methanol (9:1) or neutral lipids with 100 ml DCM followed by 150 ml DCM/methanol (9:1) for the polar lipids.

Titratable acidity. The acidity of milk is usually determined by direct titration (BS 1741: Part 10: Section 10.1: 1989).

REAGENTS

Phenolphthalein indicator solution. Dissolve 1 g of phenolphthalein in 110 ml of ethyl alcohol (95 per cent v/v), add 80 ml water and 0·1 M sodium hydroxide solution until one drop gives a faint pink coloration. Make up to 200 ml with distilled water.

Reference colour solution. Dissolve 1·5 g of cobalt(II) sulphate hydrate in water and dilute to 100 ml.

METHOD

Pipette 10 ml of milk into each of two basins C and S. To the colour control basin C add 1 ml of reference colour solution and stir. Then to sample basin S add 1 ml of the phenolphthalein indicator solution and titrate with 0·1 M sodium hydroxide, stirring continuously, until the colour matches the pink tint of C. Calculate the acidity as lactic acid (per cent m/v).

$$1 \text{ ml } 0{\cdot}1 \text{ M sodium hydroxide} = 0{\cdot}0090 \text{ g lactic acid}$$

For convenience in calculations, BS 1741 also prescribes titration of the acidity with M/9 sodium hydroxide where the titration divided by 10 equals lactic acid per cent. In BS 1741, titratable acidity is also defined as the millilitres of 0·1 M NaOH required to neutralise 10 ml of milk. The results can also be expressed as 'degrees of acidity' which equal to the number of millilitres of molar alkali per litre of milk.

Usually, 10 ml of fresh milk gives a titration of 1·4–2·0 ml of 0·1 M sodium hydroxide (≡14–20° of acidity). The acidity of freshly drawn milk is mainly due to phosphates, casein and carbon dioxide. Lactic acid is formed during souring due principally to the action of organisms of the lactic *Streptococcus* group on the lactose. The pH of fresh milk changes from about 6·6 to 4·3 in going sour.

Lactic acid as D and L isomers can be determined enzymatically. In the presence of L-lactate dehydrogenase, L-lactic acid or lactate is oxidised by NAD to pyruvate and NADH. The increase of NADH absorbance at 334, 340 or 365 nm is equivalent to the L-lactic content. Similarly for D isomer using D-lactate dehydrogenase (BCL, 1984). IDF 69A: 1984 is similar. Lactic acid can be methylated and determined by GLC (Salwin and Bond, 1969). Columns for HPLC analysis of organic acids are also available.

Protein. The protein can be determined on 10 g milk by the macro-Kjeldahl method or its variants (see Chapter 2), using the factor N × 6·38. BS 1741: Part 5: Section 5.1 describes a reference Kjeldahl procedure and Section 5.2 describes a routine method based on block digestor/semi-automatic equipment. Karmar and Van Boekel (1986) have calculated the protein N factor for milk as 6·34 (and casein 6·34, paracasein 6·29, rennet whey protein 6·45 and acid whey protein 6·30) from the evaluation of protein structures.

The **formol titration** can be used for the rapid determination of protein in fresh milk. The method depends on the fact that when formaldehyde is added to neutralised milk, free acid (which can be titrated with alkali) is produced in proportion to the amount of proteins present. The protein content is then obtained by multiplying the titration by an empirical factor, which depends on the ratio of casein to albumin and also the particular technique employed. The method proposed by Pyne (1932) in which interference due to calcium is prevented by the addition of oxalate, gives reasonably accurate results:

To 10 ml of milk add 0·5 ml of 0·5 per cent phenolphthalein indicator and 0·4 ml of neutral saturated potassium oxalate. Mix. Allow to stand for a few minutes and neutralise with 0·1 M sodium hydroxide to the standard pink colour described above. Add exactly 2 ml of formalin. Mix. Allow to stand for a few minutes and titrate the new acidity produced with 0·1 M sodium hydroxide to the same pink colour (titration *a* ml). Then titrate separately 2 ml of the formalin + 10 ml of water with the 0·1 M alkali (*b* ml) as blank. Then the protein content of the milk (equivalent to N × 6·38 from the Kjeldahl method) is 1·7 $(a-b)$ per cent.

If the oxalate is omitted, the first titration gives the acidity and a higher formol factor is usually appropriate, i.e. protein = 1·95 $(a-b)$ per cent. For further information on formol titration methods and factors employed, refer to p. 000 and Taylor (1957).

Dye-binding methods rely on the ability of dyes to combine with polar groups of proteins of opposite ionic charge. The insoluble complex is then removed by centrifugation or filtration and the concentration of unbound dye is assessed from a spectrophotometric curve relating optical density and protein content as determined by the reference Kjeldahl procedure. For manual methods see IDF 98A: 1985, ISO–5542 and the AOAC methods book. The commonest dyes employed are amido black and orange G and both are included in a simple procedure described in the eighth edition of this book. A collaborative study of results obtained on the Pro-Milk Automatic Instrument is described by Sherbon (1975). Dye-binding methods for quality control purposes have now been largely replaced by infra-red methods (p. 536).

Casein nitrogen. Casein accounts for 80 per cent of the total N and can be precipitated from milk by dilute acids and enzymes such as pepsin and rennin. The following method for the determination of casein N is based on the direct method described in BS 1741: 1963.

Weigh 5 g milk into a 100 ml beaker. Add 50 ml warm water (40 °C) and 0·5 ml acetic acid solution (10 per cent m/v). After 10 min add 0·5 ml M sodium acetate. Mix. Cool and decant into a 9 cm pleated filter paper. Wash the precipitate 3 times by decantation into the paper and wash it twice whilst on the paper. Drop the washed paper with precipitate into a digestion flask and determine the casein N by the Kjeldahl method (p. 17).

Ash is determined by incinerating the dried residue of 10 g of milk at a temperature of 525 ± 25 °C (BS 1741: Part 9: 1988). Any black particles that remain can usually be removed by wetting the ash with distilled water, then evaporating off the water on the boiling-water bath and re-igniting. An ash figure below 0·73 per cent may be due to the presence of added water. The inorganic constituents of milk are shown in Table 14.4. The chloride content of normal milk seldom exceeds 0·13 per cent when calculated on the whole milk (national average 0·105 per cent). Milk derived from cows suffering from mastitis (mastitis milk), however, when compared with normal milk, shows increased chloride and soluble nitrogen figures, together with a reduction in the amounts of lactose and casein (and consequently the NFS). The *casein number* [(casein N per cent/total N per cent) × 100] of mastitis milk falls below the normal value of 80. Also whereas the *chlorine–sugar number* [(chlorine per cent/lactose per cent) × 100] of normal milk is about 2·3, with mastitis milk that ratio rises above 3. When mastitis is suspected, the milk from individual quarters of the udder should be examined separately. Cows suffering from mastitis are usually treated with antibiotics, which may be found as a residue in the milk produced (p. 549) if the treatment is not carried out correctly.

Chloride can be determined on the ash or by direct titration methods as proposed by Davies (1932, 1938) and Sanders (1939). The following procedure is essentially similar to that in BS 1741: Part 8: 1988.

REAGENTS

Silver nitrate solution. 0·05 M.

Indicator solution. Dissolve 50 g ammonium ferric sulphate in 100 ml water and add 1 ml conc. nitric acid.

Potassium thiocyanate solution, approx. 0·05 M, standardised as follows:

Pipette 10 ml of the 0·05 M silver nitrate solution into a flask and add 30 ml water, 1 ml conc. nitric acid and 1 ml indicator solution. Titrate with the potassium thiocyanate solution and calculate the molarity.

PROCEDURE
Pipette and weigh 10 ml milk into a 250 ml flask and then add by pipette 10 ml 0·05 M silver nitrate. Add 10 ml conc. nitric acid and a few anti-bump granules and boil gently for a few minutes. The liquid should then be pale yellow. Cool, add 60 ml water and 1 ml indicator solution and titrate the excess silver nitrate with the potassium thiocyanate solution. Perform a blank determination by repeating the procedure with 10 ml water instead of the milk

1 ml 0·050 M potassium thiocyanate ≡ 0·001 773 g chloride.

An ion-selective electrode method for the determination of chloride in milk has been described by Sweetsur (1974).

Lactose. In BS 1741: Part 7: Section 7.2: 1988, for routine purposes lactose is determined volumetrically with chloramine-T. If Lane and Eynon's method (p. 196) is employed prepare the solution by weighing 20–30 g milk into a 250 ml volumetric flask. Add 50 ml water, 5 ml of each of the Carrez reagents (p. 296), mix and make up to the mark.

BS 1741: Part 1: Section 7.1 describes a reference polarimetric method after clarifying with Carrez reagent. Methods for the polarimetric determination have been discussed by Biggs and Szuarto (1963). Lactose may also be determined by GLC and HPLC and enzymatically, based on the ultraviolet detection of NADH formed by enzymic oxidation (see Chapter 6). Kleyn and Trout (1984) have reported on a collaborative study of the enzymatic (BCL, 1984) procedure. A simple and rapid colorimetric method for lactose in milk (and whey) is given by Abu-Lehia (1987). It is important to make clear when reporting results whether they are expressed as lactose anhydrous or monohydrate (lactose × 1·0526 = lactose monohydrate). The determination of lactose in milk by the Lane and Eynon method appears to give higher results than methods based on chloramine-T, polarimetry or enzymes.

Citric acid is best determined enzymatically (BCL, 1984). It can also be determined by HPLC by reverse phase ion exchange and on proprietary columns for organic acids (see HPLC suppliers catalogues).

Lactulose (see p. 550). Lactulose yields galactose and fructose on acid or enzymic hydrolysis. It can also be determined enzymatically, by GLC with other sugars as the TMS derivative (see p. 203) or by HPLC using a sulphonic acid ion exchange column and RI detection. The IDF and ISO are preparing standard methods.

ADULTERATION AND CONTAMINATION OF MILK

The quality of milk is occasionally reduced by the accidental or deliberate abstraction of fat or addition of water. Extraneous water in milk is quite often caused by faulty cleansing of dairy apparatus and bottling plant. Other possible adulteration can be from improper use of additives, e.g. preservatives, colours, thickeners and from contaminants such as detergents, disinfectants, antibiotics and dirt.

Although the quality of milk produced from an individual cow may fall below the legal minimum for fat content, that from a herd seldom does and modern methods of bulk tanker transport and storage reduces the possibility even further. Apart from milk from diseased cows the ratios of the various constituents remain fairly constant. Vieth, for instance, in about 1884, showed that the ratio of lactose:protein:ash was remarkably close to 13:9:2 (cf. Markland, 1963; Hatfull, 1964; Pim, 1964, 1967; see p. 533 for more recent values). Vieth also showed that the ash represents 8 per cent of the SNF. A higher proportion of albumin than 0·6 per cent suggests the presence of colostrum in the milk or disease in the animal.

The amount of added or extraneous water in milk can only be assessed from the freezing point of the milk. It is feasible but unlikely that milk might be adulterated with a solution of a substance that is isotonic with milk and therefore the mixture would still give a normal freezing point. The addition of an inorganic salt would usually be detected from the proportionately high ash. Volatile ammonium compounds, on analysis, would increase the apparent protein content. It is not impossible that milks with a normal freezing point depression (FPD) may have been sophisticated with condensed milk or skimmed milk powder.

Freezing point of milk. Cows may on occasion for a variety of reasons produce milk with an unnaturally low SNF: therefore the determination of SNF itself will not provide positive evidence for the presence of extraneous water. It is well established however that even though the SNF may vary widely, for biophysical reasons the osmotic pressure and hence the freezing point of milk is nearly a constant. Kessler (1984) has calculated the contributions of the various constituents of milk to its freezing point (see Table 14.5). Henningson (1963) stated that the lactose and

Table 14.5 Milk constituents and their effects on freezing point (Kessler, 1984)

Percentage of milk ingredients	Freezing point decrease in Hortvet °C
87·5 water	
(3·33 proteins)	
2·66 casein	−0·000 001
0·67 whey protein	−0·000 407
3·78 fat	−0·000 000
4·65 lactose	−0·289
[0·80 salts)	
0·085 NaCl	−0·062
0·019 $NaHCO_3$	−0·0095
0·072 KCl	−0·041
0·013 K_2SO_4	−0·0045
0·083 KH_2PO_4	−0·026
0·075 K_2HPO_4	−0·0275
0·052 $CaHPO_4$	−0·016
0·062 $Ca_3(PO_4)_2$	−0·021
0·052 $K_3C_4H_5O_7$	−0·014
0·059 $Mg_3(C_4H_5O_7)_2$	0·0075
0·137 $Ca_3(C_4H_5O_7)_2$	−0·0145
0·091 salts incl. proteins	−0·0000
Sum of salts	−0·2435
Lactose	−0·289
Total freezing point decrease	−0·533

chloride account for 75 per cent of the FPD and discussed the factors that cause variation, e.g. season, feed, breed, time of milking, access to water and weather. Macdonald (1948) stated that the average freezing point of bulk milk is −0·544 °C (i.e. FPD of 0·544 °C or 544 m°C) when determined by the Hortvet technique. Wood (1972) considered that the minimum FPD should be taken as 0·540 °C rather than

0·530 °C as had been previously used for legal purposes and for the estimation of the amount of extraneous water present.

NB Freezing point depressions measured or calibrated by Hortvet apparatus (FPD(H)) give marginally larger values than true equilibrium freezing point depressions (FPD(T).

$$\text{FPD(T)} = 0{\cdot}9656\ \text{FPD(H)}$$

In 1979–80, a technical committee of the British Standards Institution revised BS 3095 describing methods of the determination of the FPD of milk and as a concomitant part of the work a survey of herd milk in England and Wales was carried out in order to study the effects on FPD of the natural variations in milk quality including any influence of season, breed and herd size. A programme was organised in which the milks from 400 herds were tested, the samples having been taken at both the evening and the morning milking. Samples were also taken from the farm bulk tank. A separate survey of 48 herds in Scotland was also carried out.

Statistical analysis of the FPD results obtained by Public Analysts using the Hortvet method indicated that the average FPD was 543·3 m°C with a standard deviation (SD) of ±4·6 m°C, showing that 95 per cent of samples of authenticated herd milk can be expected to have FPD of between 534 and 552 m°C. The results also showed that in two herds out of three, the samples from evening milkings had a higher FPD than those from morning milkings. The average FPD varied from season to season, being largest in October and smallest in February. The difference is at most 3·1 m°C for authenticated milk and 2·8 m°C for bulk milk samples. The daily volume of milk produced from a herd had no bearing on the measured FPD. The value of titratable acidity varied from 0·12 to 0·18 per cent expressed as lactic acid and there was no significant relationship between the titratable acidity and FPD.

These facts are included in the latest 1988 revision of BS 3095. In Part 2, simplified conclusions on the interpretation of the above data are recommended, as follows:

(a) If the FPD(H) of the sample of herd milk is 535 m°C or greater it can be accepted as being free from extraneous water.
(b) Although an individual sample of herd milk whose FPD(H) is 534 to 530 m°C may contain extraneous water, the FPD alone is insufficient justification to presume it contains extraneous water.
(c) It is likely that herd milks having an FPD(H) of 529 m°C or less contain extraneous water. The lower the FPD value the greater the likelihood that the herd milk contains extraneous water.
(d) Herd milk with an FPD(H) of less than 525 m°C will almost certainly contain extraneous water apart from very exceptional cases.
(e) Whilst repeat sampling ('appeal to herd' procedure) is applicable in all cases for which it is practicable, it is strongly recommended for samples with a FPD(H) of 529 to 525 m°C.

    If an authenticated repeat sample has been taken from a herd milk within 48 h, the suspect milk should not be considered to have contained extraneous water unless FPD of the repeat sample is at least 7 m°C greater than that of the suspect sample.
(f) If the titratable acidity of a sample of herd milk (determined at approximately the same time as its FPD) does not exceed 0·18 g of lactic acid per 100 ml of milk, the FPD of the milk should be accepted as valid even though some development of acidity may have occurred.

If the titratable acidity exceeds 0·18 g, but does not exceed 0·30 g, an approximate estimate of the FPD may be obtained from the corrective formula (p. 546).

The Standard states that the mixing of milk from different herds would be expected to cause the FPD of the mixed milk progressively to approach the national average, but when milk from a small number of herds all from within a restricted zone is bulked it would be unwise to carry the assumption too far. The survey did not provide evidence to indicate the degree of mixing that would justify use of a narrower range of FPDs in the calculation of extraneous water in such bulked milk. The data did, however, show that unadulterated bulked herd milk, like herd milk, is unlikely to have an FPD(H) of less than 530 m°C.

The above data and interpretations are in good agreement with an earlier similar survey carried out in the USA and Canada where FPD determinations were made by thermistor cryoscope (Henningson, 1969). Recommendations following this survey stated that milk with a freezing point of −0.525 °C or lower should be regarded as free of added water have been adopted as official by the AOAC (Nielsen, 1976).

Milk freezes in cold weather. If during thawing the liquid portion in a partially frozen milk is drawn off it yields milk of better quality in terms of SNF and fat content. Kessler (1984) has shown the effect of several different heat and other treatments on the freezing point of milk (see Table 14.6). Bruce and McIntosh (1982) showed that pasteurisation in a plate heat exchanger (72 °C for 15 s) has no effect on FPD.

Table 14.6 The effect of treatment of milk on its freezing point (Kessler, 1984)

Product and technological treatment	pH	Freezing point °C (Hortvet)	Deviation in m°C
Raw milk (test material)	6·67	−0·526	—
Direct UHT—145 °C (C* = 0·7)		−0·517	+ 9
Indirect UHT—145 °C (C* = 0·7)		−0·523	+ 3
Raw milk (test material)	6·68	−0·524	—
Heated for 30 s at 74 °C		−0·524	0
Heated for 2·8 s at 85 °C		−0·522	+ 2
Heated for 303 s at 95 °C		−0·523	+ 1
Pre-heating indirect UHT			
78 °C/33·6 s heated		−0·524	0
95 °C/50·3 s at 139 °C		−0·521	+ 3
Raw milk (de-gassed)			
(test material)	6·62	−0·525	—
O_2-saturated†	6·62	−0·529	− 4
N_2-saturated†	6·64	−0·529	− 4
CO_2-saturated†	5·90	−0·596	−71
$N_2/O_2/CO_2$-saturated†	5·90	−0·590	−65
Stored at 4 °C for 24 h	6·58	−0·534	− 9
Stored at 4 °C for 48 h	6·33	−0·539	−14

† The milk was gased for 15 min at 10 °C.
C* = Sterilisation index.

Determination of the freezing point depression of milk

The apparatus and techniques for determining the freezing point depression of milk are described in detail in BS 3095: Part 1: 1988. The Standard describes two methods based on the work of Hortvet (1921, 1922); one is a modernised version of the original Hortvet mercury-in-glass thermometer measurement of temperature difference and the other is a method using commercial thermistor cryoscope instruments. Work at the National Institute for Research in Dairying has shown that the two techniques give the same results although the cryoscope method shows rather better precision. Most laboratories that routinely carry out freezing point determinations on milk now use thermistor cryoscopes, and therefore the Hortvet method is not described in this edition.

THERMISTOR CRYOSCOPE METHOD

Determine the titratable acidity of the sample. If the acidity exceeds 0·3 per cent m/v as lactic acid, the determination of freezing point should be abandoned. This method is fundamentally the same as the Hortvet method but uses a different volume of sample (2·5 ml), a different temperature measuring device and a different procedure for initiating freezing. Commercial thermistor cryoscopes for milk analysis are manufactured by Fiske Associates and Advanced Instruments Inc., and for successful use the manufacturers instructions must be followed implicitly. BS 3095 states that the cooling bath shall be maintained at −6·5 to −8·5 °C, not fluctuating more than ±0·2 °C and that the cryoscope be calibrated so that it gives FPD values that are the same as would be obtained by the Hortvet method. The instrument must be plateau-seeking and not time-set. Standard sodium chloride calibration solutions are used (Table 14.7).

Table 14.7 Freezing point depression of salt solutions

Nominal FPD m°C Hortvet	Sodium chloride g/l 20 °C	FPD m°C true
0	0·000	
27	0·421	26·1
422	6·859	407·5
480	7·818	463·5
500	8·149	482·8
520	8·480	502·1
525	8·563	506·9
530	8·646	511·8
535	8·728	516·6
540	8·811	521·4
545	8·894	526·3
550	8·977	531·1
560	9·143	540·7
621	10·155	599·6

Two calibration solutions are used; one with FPD close to that of the sample and another with FPD at least 100 m°C smaller. The instrument is adjusted until it gives the correct indication for FPD for both calibration solutions; thereafter the reading given by the milk sample is the FPD.

CALCULATION OF PROPORTION OF ADDED WATER

If the acidity is between 0·18 and 0·30 per cent m/v as lactic acid, correct the FPD in m°C (H) by means of the formula:

$$\text{FPD (corrected)} = \text{FPD (determined)} - 340\,(\text{per cent lactic acid} - 0{\cdot}18)$$

For the purpose of calculating the proportion of added water present in milk the modified calculation due to Elsdon and Stubbs (1936) is used:

$$\text{Extraneous water (per cent m/m)} = \frac{\text{FPD}_m - \text{FPD}_s}{\text{FPD}_m} \times (100 - T)$$

where T = total solids (per cent m/m), FPD_m is the freezing point depression of genuine milk and FPD_s is the freezing point depression of the sample.

FROZEN STORAGE OF MILK SAMPLES

Because liquid milk cannot be stored long at room or refrigeration temperatures before the acidity rises to above 0·30 per cent as lactic acid, thus preventing the determination of FPD, and because the use of chemical preservatives is not acceptable for several reasons, the Technical Committee of the BSI investigated the effect of freezing milk samples prior to the determination of FPD. The experiments showed that the FPD of pasteurised and raw milk stored in plastic bottles fitted with airtight caps in a domestic deep freezer (−18 to −24 °C) were changed only slightly over long periods. BS 3095: Part 3: 1988 gives guidance on frozen storage of samples. Samples are placed in a refrigerator at about 5 °C overnight to thaw, then conditioned at room temperature for 3 hours prior to the determination of FPD. The Standard gives the following corrections to be applied:

Weeks of frozen storage	Correction m°C (H)
3 or less	0
More than 3 but less than 9	+1
More than 9 but not more than 12	+2

DIRT IN MILK

The reference method for the determination of dirt in milk is described in BS 4938: Part 1: 1982 and is based on centrifugation after standing for 72 hours in a sedimentation vessel. Milk with a moist dirt content of 30 μl/l or more is regarded as unacceptable. Because this method takes four days to complete, it is probably never carried out. BS 4938: Part 2: 1982 describes a rapid method for visible dirt content based on filtering 200 ml of sample within 3 min through a filter disc held in a specified piece of apparatus. The amount of visible dirt in mg/l is obtained by visual comparison with photographic standard discs obtainable as BS 4938: Part 2P: 1982.

The BS states that on limited evidence, the reference method gives results for moist dirt in μl/l which are ten times larger than results for visible dirt in mg/l by the rapid method. Therefore, milk containing 3 mg/l or more of visible dirt is unacceptable.

PRESERVATIVES IN MILK

Preservatives have seldom been reported as being present in milk in recent years. In view of their effectiveness as preserving agents there is always the possibility that boric acid or formaldehyde may be found on occasion. Other preservatives that were formerly used are hydrogen peroxide, benzoic acid, fluoride, salicylate, β-naphthol, sodium carbonate and glycerine.

Boric acid. For the detection of boric acid or borax, make about 5 ml of the milk just alkaline with lime water, evaporate and ignite the residue, acidify the ash with hydrochloric acid and drop into the liquid a piece of turmeric paper. After a few minutes transfer the paper to a clean dish and dry it on the water bath. In the presence of boric acid the paper turns pink or red according to the amount present (see also p. 82).

Formaldehyde can be detected and determined by the methods described in

Chapter 4. It can be detected when the routine Gerber method for the determination of fat is carried out using sulphuric acid that contains a trace of iron. A violet colour is produced as in Hehner's test when formalin is present. Formaldehyde can be steam distilled from milk (50 g) in the presence of 5 ml H_3PO_4 (85 per cent), converted to its 2,4-dinitrophenyl hydrazone and extracted into chloroform. Van Schalm (1953) applied HPLC to the hydrazone taken up in methanol, by reverse phase chromatography using methanol/water (35:65) as mobile phase, whereas Buckley *et al.* (1986) applied GLC with FC detection. Both methods were capable of determining less than 0·1 p.p.m. of formaldehyde.

Benzoic acid, propyl-hydroxybenzoate and sorbic acid in milk have been determined spectrophotometrically (Wahbi *et al.*, 1977), and can readily be determined by HPLC.

CONTAMINANTS IN MILK

Hypochlorite may be present in milk due to its being used for disinfecting utensils. Hypochlorites may be detected by the following test, which was originally proposed by Wright and Anderson (1938) and is included in BS 1741: Part 11.

To 3 ml milk (cooled to 0·5 °C) in a test-tube add 3 ml of 73·5 per cent sulphuric acid containing 0·025 per cent stannous chloride (also cooled to 0·5 °C). Shake in a freezing mixture of ice and salt, allow to stand in the mixture for 3 min and then centrifuge at 2500 r.p.m. for 3 min. When examined under ultraviolet light, the presence of a yellow fluorescence is indicative of the presence of hypochlorite. The test detects the presence of chlorate not hypochlorite (stock solutions of hypochlorite used in the dairy industry must contain at least 0·7 per cent of sodium chlorate, although the level slowly decreases with storage).

Quaternary ammonium compounds (QAC) may be present in milk due to some residual detergent solution remaining after bottle washing. The following test due to Miller and Elliker (1951) detects about 5 mg/kg in milk and is included in BS 1741: Part 11.

Indicator solution. Prepare a stock solution by dissolving 0·05 g eosin in 100 ml acetone. Shake 10 ml of stock solution with 90 ml of tetrachloroethane and 1 g of citric acid and filter before use.

Buffer. Dissolve 25 g citric acid in 100 ml water and adjust to pH 3·5 with 50 per cent sodium hydroxide solution (approx. 12 ml required).

PROCEDURE

To a centrifuge tube add 1 ml milk, 5 ml water, 1 ml indicator solution and 0·2 ml buffer and shake hard for 10 s. Centrifuge for 5 min at 3200 r.p.m. If QAC is present the bottom layer assumes a red or pink colour. Samples containing about 1 mg/kg of QAC show a faint pink colour. If the colour is deep pink or red, the amount of QAC present can be approximately determined by titration with a standard anionic detergent solution (see BS 1741).

A qualitative test based on the formation of a blue bromophenol blue–QAC complex soluble in ethylene chloride is described in the AOAC methods book. A useful table is given of the optical properties of 12 different QAC reineckate salts.

Iodophors, which are complexes of iodine with non-ionic detergents, can be detected by the presence of free iodine down to about 5 p.p.m. (BS 1741: Part 11).

REAGENTS

Hydrochloric acid conc.

Colour reagent: 2-phenylbenzo(h)chromen-4-one (7,8-benzoflavone) 5 g/l in equal volumes of 95 per cent ethanol and diethyl ether.
Hydrogen peroxide 60 g/l.

PROCEDURE

To 1 ml of milk cooled in ice water, add 1 ml conc. HCl, mix and cool. Add 5 drops of colour reagent then 5 drops of peroxide and mix by shaking. Lilac or blue colour within a few seconds indicates the presence of more than 5 p.p.m. iodine. Compare with milk free from iodophors.

Artificial colours. Unless described as 'Flavoured', milk must not contain added colours. Formerly annatto, caramel and coal-tar dyes were added. *Annatto* can be detected by pouring a few millilitres of milk into a flat dish and adding sodium bicarbonate solution. Then a strip of filter paper is inserted. After some hours the paper is stained brown in the presence of annatto and turns pink on the addition of a drop of stannous chloride solution. *Coal-tar dyes* can be detected by methods described in Chapter 4. The presence of an azo dye usually gives a pink colour on the addition of hydrochloric acid to the milk or, better, to the coloured curd.

Sucrose can be detected by paper chromatography, TLC, HPLC or by formation of HMF after acid hydrolysis.

Antibiotics in milk

Antibiotics are frequently used for the treatment of cows suffering from mastitis and traces may be present in the milk for three days after the last treatment. A Report of the Milk and Milk Products Technical Advisory Committee (1963) drew attention to the possible health hazard due to the presence of such residues in milk. Apart from the possible ill effects on the health of individuals due to sensitisation, allergy, etc., the presence of antibiotics in milk can also cause difficulties in the production of cheese and yogurt. The Committee made various recommendations including intensification of work on the control of mastitis; the formulation of preparations with an excretion time of less than 48 hours; labelling of preparations to show the minimum time that milk from treated animals should be withheld; continuation of the search for a suitable marker—not necessarily a dye—capable of rapid detection; checks on milk by buyers and food and drugs authorities; and a publicity campaign directed at producers.

DETECTION OF PENICILLIN

Bishop and White (1984) have given the history of antibiotic testing and have reviewed current methods including cylinder plate methods, Charm test, Delvotest-P and disc assay methods. A suitable reference method employing the measurement of zones of inhibition of *Bacillus stearothermophilus* (disc assay) is described in BS 4285: Section 5.3: 1987, or FIL IDF 57: 1970.

In general, antibiotic residues in milk are now detected by means of their inhibitory effect on sensitive bacterial cultures. Inhibition is observed by the failure of a suitable redox or pH indicator to change colour when the culture is incubated in the milk sample.

The TTC test of BS 4285: 1968 (Wright and Tramer, 1961) employs *Streptococcus thermophilus* and triphenyl tetrazolium chloride (colourless), which is reduced in antibiotic-free milk to formazan (red). For practical purposes the TTC test has been replaced by more convenient commercially available test kits such as 'Interest Accuspheres' (Intervet Laboratories Ltd, Science Park, Milton Road, Cambridge CB4 4BH). The milk sample is heated to 95 °C to inactivate naturally occurring inhibitory substances, then cooled. The *S. thermophilus* culture, bromocresol purple indicator and appropriate nutrients are added as a 'sphere'. Upon incubation at 45 °C for 4 hours acid is produced in antibiotic-free milk that turns the indicator yellow. In the presence of an antibiotic the indicator remains blue. The failure point of the test is equivalent to a penicillin level of 0·02 i.u./ml. A rather similar test, possibly the most used at present abroad, is the 'Delvotest' (Gist-Brocades, N.V., Postbus 1, Delft, The Netherlands); it employs *B. stearothermophilus* var. *calidolactis*, bromocresol purple and an incubation of 2½ hours at 63–66 °C. The enzyme penicillinase, which specifically inactivates penicillin, may be added to one of a pair of duplicate tests in any of the above methods in order to differentiate penicillin, the most commonly encountered antibiotic, from other inhibitory substances.

The TTC test also detects very small amounts of tetracyclines. Some authorities have expressed the view that the most potentially dangerous antibiotic likely to be present in milk is chloramphenicol. A method for the determination of the nitrofurans (nitrofurazone with nitrofurazolidine) has been described by Ratcliffe (1963). A microbiological assay for penicillin residues using *Sarcina lutea* is stated to measure 0·01 i.u./ml (Katz and Fassbender, 1978). Penicillin G, V and ampicillin can be determined in milk by HPLC using LiChrosorb RP-18 column and a mobile phase of methanol/water/0·2 M phosphate buffer (5:13:2) containing 11 mM sodium 1-heptane sulphonate. The milk is pretreated using a C_{18} Sep-Pak cartridge and detection is at 200 nm (Terada and Sakabe, 1985). A comprehensive account of methods for the detection of inhibitors in milk is in the IDF Bulletin, No. 220: 1987. A rapid test for penicillin G is in the form of an ELISA immunoassay test kit (Sure-Test, Cambridge Veterinary Sciences), sensitive to less than 0·01 i.u./ml (Anon, 1989).

Milk samples must be kept deep frozen before examination if the test for antibiotic residues cannot be applied immediately. Penicillin rapidly becomes inactivated in milk at room temperature and even at refrigeration temperatures there is appreciable loss within a few days. Dickes (1973) reported that there is no decrease in penicillin concentration in milk when stored at −10 °C over 18 months.

HEAT TREATMENT OF MILK

According to Andrews (1984), pasteurised, UHT and sterilised milk can be differentiated by their lactulose content. Lactulose yields galactose and fructose on acid or enzymic hydrolysis. For general guidance the maximum levels of lactulose should be: pasteurised milk 1–2 mg/100 ml, UHT not more than 40 mg/100 ml and sterilised about 70 mg/100 ml. Levels of more than 40 mg/100 ml in UHT milk may suggest the incorporation of recycled UHT milk (see also p. 542).

The Milk (Special Designation) Regulations 1986 SI No. 723 define pasteurised milk as milk retained at a temperature of not less than 62.8 °C and not more than 65·6 °C for at least 30 min then immediately cooled to not more than 10 °C, or milk retained at a temperature of not less than 71·7 °C for at least 15 s then immediately cooled to not more than 10 °C. Sterilised milk is defined as milk that is filtered or clarified, homogenised and thereafter heated to and maintained at not less than 100 °C until it complies with the prescribed turbidity test. Ultra Heat Treated milk is milk that is retained at not less than 132·2 °C for not less than 1 s.

The Regulations prescribe that the methylene blue test be applied to untreated and pasteurised milk. Industrially, the unofficial resazurin test is frequently employed in routine control work. The phosphatase test is prescribed for assessing the efficiency of pasteurisation and the turbidity test is applied to sterilised milk. Ultra Heat Treated milk and sterilised milk which is heated by a continuous flow method are required to comply with a colony count test; UHT milk treated by steam injection must contain the same percentages of milk fat and MSNF before and after treatment.

Very similar regulations, 1986 SI No. 722, cover the heat treatment and labelling of semi-skimmed and skimmed milk.

Sampling. 'Provisions as to Sampling' are stated in Part I of the Third Schedule to the Regulations. Any sample that does not arrive at the testing laboratory on the day on which it was taken shall be discarded.

The methylene blue reduction test for untreated milk and pasteurised milk assesses the bacterial quality of the milk and hence the keeping qualities. The test depends on the fact that the reducing activities of the micro-organisms and any reducing substances in the milk cause a fall in the redox potential and the change is assessed visually by means of methylene blue. Most of the reducing activity of the bacteria is used for removing the dissolved oxygen in the milk (for respiration) and the dye is not reduced until all the oxygen is removed. The methylene blue test is more rapid than plate count methods and the results are more reproducible.

The Methylene Blue Test for Untreated Milk and Pasteurised Milk is described in Part II of the Third Schedule to the Regulations:

TREATMENT OF SAMPLE

1. (1) On arrival at the testing laboratory the sample of milk shall at once be removed from the insulated container. Thereafter it shall be stored as follows:

(a) a sample taken at any time during the period from 1st May to 31st October, inclusive, in any year shall be kept at atmospheric shade temperature until 9.30 a.m. on the following day.

(b) a sample taken at any time during the period from 1st November to 30th April, inclusive, in any year shall be kept in its original container or in a sterile sample bottle of at least 80 ml capacity at atmospheric shade temperature until 5·0 p.m. on the day of sampling and thereafter at a constant temperature at 18·3 ± 1 °C until 9.30 a.m. on the following day.

(2) If during the period of storage at atmospheric shade temperature to which a sample is subjected this temperature at any time exceeds 21 °C the test shall not be applied.

(3) The test shall be begun between 9.30 and 10.0 a.m. on the day after the sample is taken.

REAGENT—METHYLENE BLUE

2. (1) Tablets manufactured under arrangements made by the Minister shall be used for the test (see Circular FSH 3/72, MAFF). A solution shall be prepared aseptically by adding one tablet to 200 ml of cold, sterile, glass-distilled water in a sterile flask, shaking until the tablet is completely dissolved, and making up the solution to 800 ml with cold, sterile, glass-distilled water. The resultant solution shall be stored in a stoppered sterile flask in a cool, dark place, and shall not be used if—

(*b*) it has been exposed to sunlight, or

(*b*) a period of two months has elapsed since the date of preparation.

(2) The amount of methylene blue required for a day's work shall be poured off from the stock bottle into a suitable glass container. The pipette used for transferring the methylene blue solution to the tubes of milk shall not be introduced into the stock bottle.

APPARATUS

3. (1) Test tubes shall conform to BS 3218: 1982, nominal size 150/16, and shall be accurately marked

at 10 ml. They shall be plugged with cotton wool or covered with closely fitting aluminium caps or stored in such a way as to prevent contamination.

(2) Pipettes shall be 1·0 ml straight-sided blow out delivery pipettes, and shall be plugged with cotton wool at the upper end.

(3) Glassware and rubber stoppers shall be sterile immediately before use.

(4) The water bath shall be fitted with a reliable automatic thermoregulator capable of maintaining the water at a temperature of 37.5 ± 5 °C.

METHOD OF CARRYING OUT THE TEST

4. (1) The sample shall be mixed thoroughly by inverting and shaking and the milk shall be transferred to a test-tube up to the 10 ml mark in such a manner that one side of the interior of the test-tube is not wetted with milk. 1 ml of methylene blue solution shall be added without letting the pipette come into contact with the milk in the tube or with the wetted side of the interior of the tube. After a lapse of 3 s, the solution remaining in the tip of the pipette shall be blown out. The test-tube shall be closed with a rubber stopper, aseptic precautions being taken, and shall then be inverted twice slowly, so that the whole column of contained air rises above the level of the milk. Within a period of 5 min the test-tube shall be placed in a water bath. The water in the bath shall be kept above the level of the milk in the test-tube, and its temperature, which shall be between 37 °C and 38 °C, shall be maintained as nearly uniform as possible by means of a reliable automatic thermoregulator. The interior of the bath shall be kept completely dark.

(2) A control tube shall be used for comparison with each batch of experimental tubes to indicate when decolorisation is complete. The control tube shall be prepared by immersing in boiling water for 3 min a stoppered test-tube containing 1 ml of tap water and 10 ml of mixed milk having a fat content and colour similar to that of the milk being tested.

(3) The milk shall be regarded as decolorised when the whole column of milk is completely decolorised or is decolorised up to within 5 mm of the surface. A trace of colour at the bottom of the tube may be ignored provided that it does not extend upwards for more than 5 mm.

INTERPRETATION

5. The test shall be deemed to be satisfied by milk which fails to decolorise methylene blue in 30 min.

The resazurin test also assesses the reducing activities of the mirco-organisms present in the milk. The resazurin dye used changes from blue in the oxidised state through various shades of pink to colourless. As changes in the colour (or tint) are seen in a shorter time than with methylene blue, the unofficial resazurin test is frequently employed industrially as a rapid platform test and for routine grading. The resazurin test appears to be more sensitive than methylene blue for revealing 'faults' in certain milks, e.g. for milks rich in weakly reducing organisms, for aged milks with a large number of cells. The following procedure is suitable for routine purposes (cf. BS 4285: 1968, to reappear in BS 4285: Part 5.1).

Resazurin reagent. Dissolve one standard resazurin tablet (manufactured by Bengers Ltd and Boots Pure Drug Co.) in 50 ml cold sterile water (approx. 0·005 per cent).

PROCEDURE

Mix and aerate the sample by shaking 25 times, each shake being an up-and-down movement with an excursion of about 30 cm. Pour the sample into a sterile test-tube up to the 10 ml mark and add 1 ml fresh solution from a sterile pipette. Close aseptically with a rubber stopper, invert the tube twice and place in a covered water bath at 37·5 °C for 10 min. If 1, 2 or 3 h incubation periods are used (see below), invert the tube every ½ h.

If there is complete reduction (white colour) record as 'O'. If the colour is very pale pink, pink and white mottling or a deeper pink band at the top above a paler pink below record as '½'. Otherwise invert the tube and match against mixed milk as blank on resazurin disk 4/9 in a Lovibond comparator. Milks are 'rejected' if the reading obtained is 3½ or less.

If a 1 h incubation is employed the reading is close numerically to the time required in hours to reduce methylene blue. A further modification is the examination of mixed evening and morning milk for complete reduction of the dye after 2 h in summer and 3 h in winter (Thomas *et al.*, 1963; Thomas and Makinson, 1964; Fischer, 1965).

The phosphatase test for pasteurised milk. The principle on which the test is based is that although phosphatase enzymes are invariably present in raw milk, they are inactivated during pasteurisation. It has also been shown that this enzyme is more difficult to destroy than the most heat-resistant pathogenic organisms that were once found in milk, namely tubercle bacilli.

The prescribed phosphatase test was devised by Aschaffenburg and Mullen (1949) and involves incubation of the milk with disodium *p*-nitrophenol phosphate under alkaline conditions. If the milk contains phosphatase a yellow colour is produced due to the formation of *p*-nitrophenol. The degree of destruction of the phosphatase in the milk during pasteurisation is then assessed by comparing the colour produced with the standard colours on an APTW comparator disc (Tramer and Wight, 1950). A positive reaction may be caused by contamination of heat-treated milk with raw milk.

The Phosphatase Test for Pasteurised Milk is described in Part III of the Third Schedule to the Regulations.

EXAMINATION OF SAMPLE

1. The sample of milk shall be examined as soon as possible after arrival at the testing laboratory. If it is not examined immediately on arrival at the testing laboratory, it shall be kept at a temperature of between 3 °C and 5 °C until examined. The sample shall be raised to room temperature immediately before being tested.

PRECAUTIONS

2. The following precautions shall be taken:
(*a*) A sample which shows evidence of taint or souring shall not be tested.
(*b*) All glassware shall be clean immediately before use.
(*c*) A fresh pipette shall be used for each sample of milk. Pipettes shall not be contaminated with saliva.
(*d*) The test shall not be carried out in direct sunlight.
(*e*) Distilled water shall be used throughout.

REAGENTS

3. (1) Whenever possible, reagents of analytical quality shall be used.
(2) The buffer–substrate solution shall be prepared as follows:
(*a*) Buffer solution: 3·5 g of anhydrous sodium carbonate and 1·5 g of sodium bicarbonate shall be dissolved in distilled water, and made up to one litre.
(*b*) Substrate: Disodium *p*-nitrophenyl phosphate. The solid substrate shall be kept in a refrigerator.
(*c*) Buffer–substrate solution: 0·15 g of the substrate shall be placed in a 100 ml measuring cylinder, and made up to 100 ml with the buffer solution. The solution shall be stored in a refrigerator and protected from light. It shall give a reading of less than the standard marked 10 on the comparator disc APTW or APTW7 when viewed in transmitted light through a 25 mm cell in the 'all purposes' comparator, distilled water being used for comparison. The solution shall not be used for more than one week.

APPARATUS

4. The following apparatus shall be used:
(*a*) A Lovibond 'all purposes' comparator complete with stand for work in reflected light.
(*b*) A Lovibond comparator disc APTW or APTW7.

(*c*) Two fused glass cells, 25 mm depth.
(*d*) A water bath or incubator capable of being maintained at 37·5 ± 0·5 °C.
(*e*) A pipette to deliver 5·0 ml.
(*f*) A supply of 1·0 ml straight-sided pipettes of an accuracy equal to that of NPL grade B.
(*g*) A 1000 ml graduated flask.
(*h*) A 100 ml measuring cylinder.
(*i*) A supply of test-tubes conforming to British Standard 625: 1959, nominal size 150/16, with rubber stoppers to fit.

CARE OF APPARATUS

5. (1) New glassware shall be cleaned and free from contamination from substances which may interfere with the test.

(2) After use, each test tube shall be emptied, rinsed in water, well washed in hot water containing soda, rinsed in warm water, rinsed in distilled or de-ionised water and finally dried.

(3) If after treatment in accordance with sub-paragraph (1) hereof a test tube does not appear to be clean, the treatment shall be repeated with the addition that after being rinsed in warm water it shall be soaked in 50 per cent commercial hydrochloric acid and then rinsed again in warm water before being rinsed in distilled or de-ionised water and finally dried.

(4) Glassware used for the test shall not be used for any other purpose and shall be kept apart from all other apparatus in the laboratory.

METHOD OF CARRYING OUT THE TEST

6. 5 ml of the buffer–substrate solution shall be transferred to a test-tube using a pipette and the test-tube shall be stoppered and brought to a temperature of 37 °C. 1 ml of the milk to be tested shall be added, the test-tube stopper replaced and the contents well mixed by shaking. The test-tube shall then be incubated for exactly 2 h at 37 °C. One blank prepared from boiled milk of the same type as those undergoing the test shall be incubated with each series of samples. (Where the sample consists of highly coloured milk, such as homogenised milk or milk from Channel Island cows, a separate blank of such milk shall be prepared.) After incubation the test-tube shall be removed from the water bath and its contents shall be well mixed. The blank shall be placed on the left-hand ramp of the stand and the test sample on the right. Readings shall be taken in reflected light by looking down on to the two apertures with the comparator facing a good source of daylight (preferably north light). If artificial light is needed for matching, a 'daylight' type of illumination must be used. The disc shall be revolved until the test sample is matched. Readings falling between two standards shall be recorded by affixing a plus or minus sign to the figure for the nearest standard.

INTERPRETATION

7. The test shall be deemed to be satisfied by milk which gives a reading of 10 μg or less of *p*-nitrophenol/ml of milk.

Kleyn and Yen (1970) have recommended the use of phenolphthalein monophosphate as substrate. Phosphatase activity can reappear in stored UHT milk (and cream).

Raw goats milk contains appreciably less alkaline phosphatase than cows milk, and therefore the above method does not give a correct monitoring of the degree of pasteurisation of goats milk. Williams (1986) modified the method for this purpose by using 3 ml of sample and a 4-hour incubation period at 37·5 °C.

The turbidity test for sterilised milk. When milk is heated at 80 °C or above, all the albumin becomes denatured and if solutions of inorganic salts or acids are added the albumin separates with the casein. In the test proposed by Aschaffenburg for sterilised milk the sample is shaken with ammonium sulphate, filtered and the filtrate is then heated. Any albumin that is still present in solution due to insufficient heat treatment during manufacture will reveal itself as a turbidity in the heated filtrate. If the milk has been sterilised adequately, however, all the albumin will have been precipitated and no turbidity will be produced. The test is not prescribed for UHT milk, which gives a slight turbidity. The officially prescribed 'Turbidity Test for Sterilised Milk' is described in Part IV of the Third Schedule to the Regulations:

EXAMINATION OF SAMPLE

1. The sample of milk may be examined at any time after delivery to the testing laboratory but shall be at room temperature when the test is begun.

REAGENT

2. Ammonium sulphate A.R. shall be used.

APPARATUS

3. The following apparatus shall be used:
(*a*) Conical flasks of 50 ml capacity.
(*b*) Graduated cylinders of 25 ml capacity.
(*c*) Test-tubes conforming to British Standard 625: 1959, nominal size 150/16.
(*d*) Filter funnels of 6 cm diam.
(*e*) Beakers of 400 ml capacity.
(*f*) 12·5 cm No. 12 Whatman folded filter papers.

METHOD OF CARRYING OUT THE TEST

4. 4 ± 1 g of ammonium sulphate shall be weighed into a 50 ml conical flask. 20 ± 0·5 ml of the milk sample shall be measured out and poured into the conical flask, the flask being shaken for 1 min to ensure that the ammonium sulphate dissolves. The mixture shall be left for not less than 5 min and then filtered through a folded filter paper into a test-tube. When not less than 5 ml of a clear filtrate have collected, the tube shall be placed in a beaker of water, which has been kept boiling, and kept therein for 5 min. The tube shall be transferred to a beaker of cold water, and when the tube is cool, the contents shall be examined for turbidity by moving the tube in front of an electric light shaded from the eyes of the observer.

INTERPRETATION

5. The test shall be deemed to be satisfied when a sample of milk treated as in paragraph 4 hereof gives a filtrate showing no sign of turbidity.

A test has been proposed by Fellows (1953) that is perhaps more suited to industrial purposes as it assesses the extent to which sterilised milk has been heated. This measures the 'reducing capacity' of sterilised milk. Lactulose is formed by the heating of milk (see p. 550).

Bacteriological examination. Milk that has been sterilised by a continuous flow method or treated by the ultra heat temperature method is required to satisfy the colony count test prescribed in Part V of the third schedule to the regulations. Similar requirements apply to semi-skimmed milk and skimmed milk in their regulations.

Various bacteriological methods for the examination of milk are described in various parts of BS 4285: 1984 *et seq.* and in a number of IDF standards. In the past the primary object of pasteurisation was the destruction of *Mycobacterium tuberculosis*. Following its eradication from the dairy herd, the object of pasteurisation now is to eliminate other contaminants that may be present, such as *Brucella*, *Salmonella dublin*, *Campylobacter* and pathogens causing mastitis such as streptococci and staphylococci. Pasteurisation is also commercially important because it increases shelf-life. Lumps of fat on the milk surface ('Bitty' or 'Broken' cream) is caused by aerobic spore formers (*Bacillus cereus, B. mycoides*), which produce lecithinase and disrupt the fat globule membrane.

CREAM

Legislation

For legal purposes, 'cream' means that part of milk rich in fat that has been separated by skimming or otherwise; and 'clotted cream' (usually described as Devonshire Cream or Cornish Cream) means cream that has been produced and spearated by the scalding, cooling and skimming of milk or cream. Mechanical separation gives products containing up to about 65 per cent fat. The composition of creams in the UK have been surveyed by Florence *et al.* (1985) and are given in Table 14.8. The manufacture of cream is described by Lampert (1970) and Rothwell (1968). Legal standards for cream are prescribed in The Cream Regulations 1970, SI No. 752, which follow closely the recommendations in the 1967 FSC Report on Cream and prescribe minimum fat contents for various types (Table 14.9).

The regulations also permit various additives, including certain stabilisers and thickeners, and whipped cream may contain up to 13 per cent sugar. With certain exceptions cream has to be described as 'pasteurised', 'ultra heat treated' or 'UHT', or 'untreated'. The 1982 FSC Report on cream has recommended removal of the specific milk fat content standards for sterilised creams and whipped cream. The Report proposed modified definitions for various creams and that untreated cream should be labelled 'raw, unpasteurised'.

The FSC Report was followed in 1986 by proposals for new regulations. In these cream is redefined as that part of milk rich in fat that has been separated by skimming or otherwise, the final composition of which may have been adjusted by the addition of milk or skimmed milk. Clotted cream is also redefined as cream that has been produced by the scalding, cooling and skimming of milk, or by the scalding of cream. Whey cream is introduced as that part of cheese whey rich in fat that has been separated by skimming or otherwise. The proposed standards for milk fat content for the permitted names of creams are listed in Table 14.9. The proposals list various additives allowed within specified limits. In whipped cream, aerosol cream and non-retail sale whipping cream, gelatine together with other thickeners up to 0·3 per cent may be present. These creams and also non-retail sale cream for flour confectionery manufacture may contain up to 13 per cent of sucrose. UHT single and half cream may contain up to 1·5 per cent of sodium caseinate.

Although the proportion of fat present in cream varies greatly, the relative proportions of the other ingredients remain the same as in the milk from which the cream was derived. This applies to the ratio of SNF to water unless water has been lost by heating as in the preparation of clotted cream. It is useful to compare the SNF (as determined analytically) with that calculated from: $SNF = 0{\cdot}102 \times water$. If the SNF is substantially lower than that calculated, watering is indicated. As Vieth's ratio still applies it is also useful to check that the lactose, protein and ash are 0·54, 0·37 and 0·38 of the SNF respectively.

Richmond has shown how the constituents of the SNF decrease as the amount of fat rises and suggested the use of the following formulae.

Table 14.8 Composition of creams (Florence *et al.*, 1985)

Cream		g/100 g					mg/100 g					
		Moisture	Fat	Protein	Lactose	Ash	Cl	Ca	Mg	P	Na	K
Single	Min.	70·0	17·0	2·4	3·5	0·55	72	80	8·1	69	40	102
	Max.	75·5	25·0	3·0	4·1	0·63	86	100	11·1	81	61	128
	Mean	73·7	19·1	2·6	3·9	0·59	80	91	9·2	75	50	115
Double	Min.	43·4	45·8	1·4	2·4	0·33	46	46	5·1	45	31	40
	Max.	49·5	52·2	1·9	2·8	0·39	57	56	7·6	53	49	80
	Mean	47·5	48·0	1·7	2·6	0·36	51	50	5·8	50	39	64
Whipping	Min.	50·8	36·6	1·8	2·5	0·38	54	52	6·0	51	34	57
	Max.	57·6	44·5	2·2	3·2	0·48	68	69	7·5	63	61	97
	Mean	55·4	39·3	2·0	3·0	0·43	59	62	6·8	58	42	79
UHT	Min.	72·7	17·8	2·5	3·6	0·65	71	77	8·5	72	44	162
	Max.	75·2	19·9	2·9	4·0	0·74	79	93	9·8	79	64	208
	Mean	73·7	19·0	2·6	3·8	0·70	76	85	9·0	74	53	181
UHT half	Min.	79·6	12·0	2·6	4·1	0·63	87	93	9·2	76	53	123
	Max.	80·2	13·0	3·0	4·2	0·66	92	103	10·6	84	57	130
	Mean	79·9	12·3	2·8	4·1	0·65	89	96	9·8	79	55	127

$$S = \frac{10 - 0{\cdot}1F}{1{\cdot}08}; S = 10{\cdot}1 - 0{\cdot}1\,T; F = 1{\cdot}102\,T - 10{\cdot}2$$

where S = SNF (per cent) F = fat (per cent) and T = total solids (per cent).

The content of the B complex vitamins and vitamin C in retail creams are reported by Scott and Bishop (1988).

Table 14.9 Standards for cream

	1967 regulations % milk fat minimum	1986 proposals % milk fat	
		Not less than	Less than
Clotted cream	55	55	—
Double cream	48	48	—
Whipping, whipped cream	35	35	48
Sterilised cream	23	—	—
Cream	18	—	—
Single cream	18	18	35
Half cream	12	12	18
Sterilised cream	12	—	—
Whey cream	—	18	—

Analysis of cream

In view of the statutory requirements it is important to determine the **fat content** of cream. Other possible determinations are those for the **total solids, protein, lactose** and the **acidity**. Whipped cream, especially, should be examined for **sucrose** and **stabilisers**. It may be necessary to examine samples for **preservatives, colouring matters** and **foreign fats**. Methods for the determination of total solids, fat and titratable acidity in cream are given in BS 1741.

Fat can be determined accurately by the Rose–Gottlieb process (BS 1741: Part 4: 1987; IDF 16A: 1971) using 1–2 g cream + 9 ml warm water and then adding ammonia and alcohol as for milk (p. 538).

For routine purposes the Gerber method as follows is more convenient (BS 696: Part 2):

Using special cream butyrometers weigh out 5 g of cream into a stoppered weighing funnel (specifications for the tubes, funnel, etc., are given in BS 696: Part 1). Pipette 10 ml Gerber sulphuric acid into a cream butyrometer. Then wash all the cream from the funnel into the butyrometer with 6 ml of hot water (at least 70 °C) and add 1 ml of amyl alcohol. Add more hot water to bring the level of the contents to 5 mm below the shoulder of the tube. Stopper, mix, place in a water bath at 65 °C for 3 to 10 min, centrifuge at 1100 r.p.m. for 5 min and return to the 65 °C bath for 3 min. Then read off the percentage of fat directly on the scale. The method is given in detail in Part 2 of BS 696.

To characterise the fat, extract the total lipid with chloroform/methanol as described in Chapter 2. The extracted lipid can be examined by the methods given under 'Butterfat'.

Other determinations on cream that may be necessary are summarised below:

Total solids. Dry 2 g as for milk, p. 537.

Protein. Determine by the macro-Kjeldahl method.

Lactose and sucrose. Determine by the Lane and Eynon process or polarimetrically (before and after inversion) on a solution obtained by clearing with the Carrez zinc ferrocyanide reagents. HPLC or enzymatic analysis is particularly useful for the determination of lactose, especially if other reducing sugars are present. A qualitative test for sucrose is given below.

Ash. Dry 5–10 g on a steam bath, carefully char the dry residue (flame or infra-red lamp) and ash at 525 °C (BS 1741: Part 9: 1988).

Chloride. BS 1741: Part 8: 1988 is a Volhard method. Sample is boiled with nitric acid, as for milk (p. 541).

Acidity. (BS 1741: Section 10.2: 1989). Weigh 10 g cream into each of two porcelain basins. To the control basin C add 10 ml water and 2 ml dilute cobalt sulphate solution. To the sample basin S add 2 ml phenolphthalein solution and titrate with 0·1 M alkali. Details of the various solution strengths are given on p. 539. The ratio of acid to SNF should be of the same order as for fresh milk.

Thickeners. *Starch* may be detected by the addition of iodine.

Gelatine may be detected by Stokes test. Mix together 10 ml of cream, 20 ml of water and 20 ml of Stokes reagent (dissolve mercury in twice its weight of conc. nitric acid and dilute to 25 times the volume with water). Shake, allow to stand for 5 min and filter. To the filtrate add an equal volume of saturated picric acid solution. If gelatine is present, a yellow precipitate is produced.

Sucrose can be detected by Cayaux's test. Mix together 15 ml of cream, 0·1 g of resorcinol and 1 ml of conc. hydrochloric acid and heat the mixture on the water bath. The presence of sucrose is indicated if a red colour is produced. For the quantitative determination of sucrose see p. 562.

Sodium bicarbonate produces a higher than usual ash and alkalinity of the ash. The pH is also raised.

Sodium phosphate should be looked for in sterilised cream. If the sodium content exceeds about 41 mg per 100 g, the phosphate content should also be determined. Figures for the normal composition of the ash are given under 'Milk' (see also Table 14.8).

Methods for various emulsifiers and stabilisers are given in Chapter 4. Whipped cream, sterilised or UHT treated cream may contain calcium chloride, sodium or potassium orthophosphate, carbonate or citrate to help form and maintain the emulsion (maximum 0·2 per cent).

Preservatives and colouring matters were formerly found in cream. Boric acid and formalin were the commonest preservatives and annatto the main colour used. Nisin may be added to clotted cream.

Bacteriology and heat treatment of cream

Apart from non-heated cream derived from raw milk, i.e. cream from the farm, all cream must conform to the Milk and Dairies (Heat Treatment of Cream) Regulations 1983 SI No. 1509 as amended. The Regulations require that pasteurised cream has been heated at 63 °C for 30 min or at 72 °C for 15 s or at equivalent heat treatments and must satisfy the prescribed coliform test or phosphatase test or both tests. The phosphatase test is similar to that described for milk (p. 553) except that 15 ml of buffer–substrate solution and 2 g of cream is used, and Carrez clearing reagents are added after incubation. The prescribed method includes instructions on how to deal with apparent reactivation of phosphatase.

Sterilised cream must have been heated at 108 °C for 45 min or equivalent treatment to kill all micro-organisms and spores and must satisfy the prescribed colony count test. Ultra heat treated cream must have been heated at 140 °C for at least 2 s or an equivalent treatment and must also satisfy the colony count test. Also, UHT cream treated by the injection of steam must have the same percentage of total solids before and after treatment.

Clotted cream is heat treated sufficiently during preparation to destroy food poisoning organisms and those liable to cause off-flavours and souring, but handling and transport subsequent to production tend to encourage re-contamination.

The methylene blue test provides a satisfactory screening test for assessing the hygienic quality of fresh cream. The test is performed in a similar way to that described for ice-cream (p. 605), commencing by adding 1 ml methylene blue solution to 7 ml ¼-strength Ringer's solution together with cream added to the 10 ml mark. Cream samples not decolorising the dye in 4 h at 37 °C are considered as satisfactory and those in ½–4 h as fairly satisfactory.

Reconstituted and imitation cream

Section 48 of the Food Act 1984 places restrictions on the use of the word 'cream' and defines 'reconstituted cream' as a substance that resembles cream in appearance and contains no ingredient not derived from milk except water. This definition is virtually the same as that used previously for 'artificial cream', which is considered to be a rather misleading title (FSC *Report on Artificial Cream*, May 1951). Section 48 also defines 'imitation cream' as a substance that resembles cream in appearance and is produced by emulsifying edible oils or fats with water. Also in May 1951 the FSC *Report on Synthetic Cream* recommended that imitation cream should contain at least 25 per cent of fat and/or oil in the form of triglycerides.

No simple method exists whereby a well-emulsified reconstituted cream can be definitely differentiated from genuine cream. Determination of the lactose, protein, ash and fat may disclose an abnormality in the relative amounts present if the reconstitution has been made in incorrect proportions.

The type of fat present may be characterised by GLC on the extracted lipid. The most common emulsifying agents used are methyl cellulose, sodium alginate and glyceryl monostearate.

CONDENSED MILK

Legislation

The Condensed Milk and Dried Milk Regulations 1977 SI No. 928 implemented Council Directive 76/118/EEC (OJ No. L24, 30.1.76, p. 49) on the approximation of the laws of Member States relating to certain partly or wholly dehydrated preserved milk for human consumption. The EC legislation has introduced a wider range of products and the Regulations have in consequence prescribed definitions and reserved descriptions as shown in Table 14.10.

Table 14.10 Statutory standards for condensed milk products

Reserved description	Milk fat (%)	Total milk solids (%)
Sterilised condensed milk		
Unsweetened condensed high-fat milk	more than 15·0	26·5 min.
Evaporated milk	9·0 min., 15·0 max.	31·0 min.
Unsweetened condensed milk	7·5 min., 15·0 max.	25·0 min.
Unsweetened condensed partly skimmed milk—retail sale	4·0 min., 4·5 max.	24·0 min.
—other sale	more than 1·0, 7·5 max.	20·0 min.
Unsweetened condensed skimmed milk	1·0 max.	20·0 min.
Condensed milk with added sucrose		
Sweetened condensed milk		
—retail sale	9·0 min.	31·0 min.
—other sale	more than 8·0	28·0 min.
Sweetened condensed partly skimmed milk—retail sale	4·0 min., 4·5 max.	28·0 min.
—other sale	more than 1·0, 8·0 max.	24·0 min.
Sweetened condensed skimmed milk	1·0 max.	24·0 min.

No more than 25 per cent of the total milk solids content (TMS) of any product is allowed to be derived from dried milk and not more than 0·3 per cent of the SNF of any product is to consist of lactic acid or lactate calculated as lactic acid. Various permitted additives are listed. In general sodium or potassium carbonates, citrates, ortho- and diphosphates, and calcium chloride are permitted at 0·2 per cent max. for products containing not more than 28·0 per cent TMS, and at 0·3 per cent max. for products containing more than 28·0 per cent TMS. The total added phosphate content calculated as P_2O_5 is limited to 0·1 per cent for products with TMS not exceeding 28·0 per cent, and 0·15 per cent for products with TMS exceeding 28·0 per cent. There are special additive provisions for products subjected to UHT treatment. Sweetened, condensed products are required to contain not more than 0·02 per cent added lactose, and any condensed milk products may contain added vitamins.

The Regulations were amended by SI No. 1066: 1982 to implement the Community methods of analysis (see below). They were further amended by SI No. 2299: 1986, implementing labelling changes prescribed in Directive 83/635/EEC (OJ No. L357, 21.12.83, p. 37), and also by SI No. 1959: 1989 implementing methods of sampling for analysis prescribed in Directive 87/524/EEC (OJ No. L306, 28.10.87, p. 24).

Manufacture

In the traditional production of evaporated milk the following processes are involved: pre-warming of the standardised milk, vacuum evaporation, homogenising, stabilising (heat stability is often increased by addition of phosphates or citrates), filling, sealing, sterilising at 116 °C for 15–20 min and cooling. For the manufacture of sweetened condensed whole milk, the milk used is standardised by the addition of cream so that the ratio of fat to NFS is 1:2·44. The subsequent operations in the process are pre-heating at 77–82 °C, addition of sucrose (to give a final concentration of 42–45 per cent), vacuum evaporation at 54–60 °C, rapid cooling to about 30 °C, seeding with lactose (to produce rapid crystallisation of small crystals of lactose so that a smooth texture is obtained), filling and sealing. Modern methods vary and milk powder can be used to replace some of the milk. It will be noted that sweetened condensed milk (unlike unsweetened evaporated milk) is not heat sterilised after filling into the can but keeps well, even after opening the can, because of the comparatively high sugar concentration. Blown cans due to contamination with yeasts should not occur since the 1977 Regulations require that any milk, skimmed milk, cream or dried milk used in the preparation of sweetened condensed milk products must have been subjected to heat treatment at least equivalent to pasteurisation. Table 14.11 shows the results of the analysis of typical samples of condensed milk products.

Analysis of condensed milk

The analysis of canned condensed milk should include the determination of **total solids, fat, protein** and **sugars**. Other determinations that may be required are those of **ash, acidity, lactic acid, citric acid** and **tin**. Methods for the analysis of condensed milk have been described by the BSI (BS 1742: 1951), the SPA (1927, 1930, 1932) and the IDF, and have been adopted by the EC in Directive 79/1067/EEC

Table 14.11 Analytical data for condensed milk products

	Sweetened				Unsweetened		
	Full cream		Skimmed		Full cream		Skimmed
	1	2	3	4	5	6	7
Moisture (%)	25·0	25·1	26·6	27·0	66·1	67·4	76·6
Milk solids (%)	33·3	32·9	27·2	26·3	33·8	32·5	23·3
Fat (%)	10·6	9·6	0·2	0·2	9·2	9·1	0·7
Lactose monohydrate (%)	12·2	13·0	14·9	14·2	13·3	12·7	12·5
Protein (%)	8·5	8·4	9·6	9·5	9·1	8·7	8·3
Sucrose (%)	41·5	41·9	46·1	46·5	—	—	—
Ash (%)	1·93	1·86	2·45	2·31	2·05	1·94	1·78
Acidity (as lactic acid) (%)	0·32	0·28	0·35	0·30	0·35	0·40	0·41
Specific gravity	1·30	—	—	—	1·08	1·08	—

laying down Community methods of analysis for testing preserved milk products (OJ No. L327, 24.12.79, p. 29).

Preparation of sample

For unsweetened products shake the can well. Open the can and transfer the contents to the sample jar making sure that the product is homogeneous. If not, warm the sample jar in a water bath at 40 °C and shake vigorously. For sweetened products warm the unopened can in a water bath at 30–40 °C for 30 min. Open the can and thoroughly mix the contents with a spatula, and when homogeneous pour into the sample jar.

Total solids. Place 25 g of prepared sand (passing 500 μm but retained by 180 μm test sieve (BS 410: 1986), acid washed and ignited) and a glass stirring rod in a metal dish alongside its lid in an oven at 98–100 °C and dry to constant weight. Replace the lid before removing the dish from the oven. Allow the dish to remain for 45 min in a desiccator before weighing. Tilt the sand to one side of the dish, place on the clear space about 1·5 g of the prepared sweetened condensed milk sample, or 3 g of unsweetened condensed/evaporated milk and weigh rapidly. Add 5 ml of water in the case of sweetened condensed milk, or 3 ml with unsweetened condensed/evaporated milk and mix with the rod, heating if necessary to dissolve lactose, mix the diluted milk thoroughly with the sand and leave the rod in the dish. Place the dish on a boiling-water bath for 20 min, carefully stirring during the early period. Lay the rod flat in the dish, which should be protected from direct contact with the metal of the bath. Transfer the dish with its lid alongside to a well-ventilated oven at 98–100 °C. After 1½ hr, cover the dish and place in the desiccator for 45 min and weigh. Return the dish to the oven, heat for one hour with the lid off, then remove and weigh as before. Repeat this process until the loss of weight between successive weighing does not exceed 1 mg.

The reference method ISO 6731 for milk, cream and evaporated milk (cf. BS 1741: Part II) does not specify the use of sand. The IDF reference method is 15B: 1988.

Note. Because of the presence of additives, the total milk solids content is not precisely the same as total solids in evaporated milk, or total solids less sucrose in the case of sweetened condensed milk.

Fat. Determine the milk fat content by the Rose–Gottlieb method (p. 538) using 2–2·5 g for sweetened and 4–5 g for unsweetened products.

Fat may also be estimated by the rapid Gerber process using ordinary milk butyrometers:

Use 10·94 ml of a 20 per cent solution (m/v) of the sample and carry out the method as described for milk (p. 537):

$$\text{Fat (per cent)} = \frac{\text{Milk butyrometer reading} \times 100}{20}$$

The tube should be heated to 65 °C before centrifuging and re-centrifuging is always advisable.

Sucrose. Sucrose in sweetened condensed milk products can be determined by polarisation before and after acid inversion.

Transfer to a 100 ml beaker approx. 40 g of sample, accurately weighed, add 50 ml of hot distilled water (80–90 °C), mix, transfer to a 200 ml volumetric flask, washing in with successive quantities of distilled water at 60 °C until the total volume is 120–150 ml. Mix, cool and then add 5 ml of the dilute ammonia (100 ml 0·88 per litre). Again mix and allow to stand for 15 min. Add a sufficient quantity of the dilute acetic acid to neutralise exactly the ammonia added and mix again. Add, with gentle mixing, 12·5 ml Carrez 1 clearing agent (p. 296) and mix, followed in the same manner by 12·5 ml Carrez 2 reagent. Bring the contents of the flask to 20 °C and add distilled water (at 20 °C) up to the 200 ml mark. Up to this stage all additions of water or reagents should be made in such a manner so as to avoid formation of air bubbles and all mixing should be made by rotation of the flask rather than by shaking. If bubbles appear before completion of the dilution to 200 ml, any traces of occluded air should be removed by careful attachment of the flask to a vacuum pump and by rotation of the flask. Close the flask with a dry stopper and mix thoroughly by shaking. Allow to stand for a few minutes and then filter through a dry filter paper, rejecting the first 25 ml or so of filtrate. For the direct polarisation, determine the rotation of the filtrate at 20 °C.

For the invert polarisation, pipette 40 ml filtrate into a 50 ml volumetric flask and add 6 ml 6·34 M hydrochloric acid. Immerse the entire bulb of the flask for 12 min in a water bath maintained at 60 °C, mixing by rotary movement during the first 3 min, in which time the contents of the flask should have attained the temperature of the bath. Cool, dilute to 50 ml at 20 °C with water, mix and allow to stand for one hour before determining the invert rotation at 20 °C.

If W = weight of sample taken (g).
F = percentage of fat in the sample.
P = percentage of protein (N × 6·38) in the sample.
V = volume in ml to which the sample is diluted before filtration, in the preparation of filtrate.
v = correction in ml of precipitate produced during clarification.
D = observed direct polarimeter reading.
I = observed invert polarimeter reading.
l = length in decimetres of polarimeter tube.
Q = inversion divisor factor

$$\text{Then } v = \frac{W}{100}[(F \times 1{\cdot}08) + (P \times 1{\cdot}55)]$$

and the percentage of sucrose in the sample

$$= \frac{D - (^5\!/_4 \times I)}{Q} \times \frac{V - v}{V} \times \frac{V}{l \times W}$$

Values of Q, the change in specific rotation of sucrose on inversion, divided by 100, are given on p. 194 for various sources of light and the use of zinc ferrocyanide clarification. Variation of over 0·2 °C from 20 °C in the invert reading necessitates a temperature correction (Pearson, 1976).

Total Milk Solids (in Sweetened Condensed Milk) = Total Solids − Sucrose.

Simplified procedures for determining sucrose in condensed milk have been described by Wood (1954) and Murphy (1960). Sucrose may also be determined by the specific methods described in Chapter 6.

Lactose. If sucrose is determined polarimetrically, the lactose content can be calculated from the direct reading (see p. 193). Lactose can also be determined by the specific methods described in Chapter 6 or by copper reduction as follows. Weigh 10–12 g of sample into a 250 ml volumetric flask, dilute with 200 ml hot water and allow to stand for at least 30 min. Cool, add 4 ml of Carrez 1 solution, mix and add 4 ml of Carrez 2 solution. Dilute to the mark, filter and determine the lactose by Lane and Eynon's method using 25 ml of mixed Fehling's solution. The result may be calculated as hydrated lactose. The correction for the volume of precipitate that should be deducted from the total 250 ml is similar to that given for condensed milk above. Vieth's ratio applies to condensed milk as for liquid milk.

Ash. Determine by igniting 5 g of sample.

Protein. Determine on about 2 g of sample by the macro-Kjeldahl method and calculate the result as N × 6·38.

Acidity. The difference between 100 and the sum of the constituents is usually about 0·5–0·6 per cent and this discrepancy is largely due to citric and lactic acid. Citric acid can be determined by the methods referred to on page 542. Lactic acid and lactates can be determined specifically by enzymatic analysis (BCL, 1984, cf. BS 1743: Part 8: 1987, ISO 8069–1986, IDF 69A: 1984) and also by HPLC.

The total acidity obtained by direct titration and using phenolphthalein as indicator is usually between 0·25 and 0·45 per cent (as lactic acid).

Other determinations. If the container shows internal corrosion, the condensed milk should be examined for tin. Unsweetened condensed milk is rather more liable to attack the tin plate than sweetened products. The presence of thickeners in condensed milk would now appear to be a relic of the past. If suspected, samples should be examined for gelatine and starch by the methods described under 'Cream'. Very small amounts of carrageenan have been used to improve stability.

DRIED MILK PRODUCTS

Introduction

Formerly, skimmed milk and buttermilk, the by-products from cream and butter production, and whey from cheese-making were largely either run to waste or used for animal feeding. The introduction of efficient mechanical drying of these by-products and of whole milk coupled with legislation to encourage it, has led to wider utilisation of these milk products. Milk and its liquid by-products can now be converted to powders that can be easily transported anywhere and stored for long periods to provide nutritious foodstuffs in themselves or as ingredients for the manufacture of other foods such as baby food powders and filled milks. Over-production of milk can also be dealt with by manufacturing butter and skimmed milk powder, as is carried out under the Common Agricultural Policy of the EC.

Manufacture

Dried milk is usually prepared by spray-drying or roller-drying. Spray-dried milk is produced by atomising a spray of hot pre-condensed milk (TS 40–50 per cent) inside a large chamber where it meets hot air entering at up to 198 °C for whole milk and up to 260 °C for skimmed milk. The milk dries instantaneously and the falling powder is quickly removed. Roller-dried milk is produced by running or spraying pre-condensed milk (up to 24 per cent TS) on to revolving metal drums or rollers, which are heated internally to about 150 °C. The milk dries almost instantaneously on the roller and the powder is removed by scrapers and rapidly cooled. Over-heating produces too dark a product and is avoided in both processes. When manufactured the moisture content of dried milk is about 3 per cent but water is absorbed during storage. Much useful information on the technology and control of the manufacture of milk powders is included in the reviews of Davis (1968) and King (1966).

Because of its superior bacteriological quality and longer shelf-life, roller-dried powder has been preferred in the past for infant feeding purposes, but nowadays spray-drying has almost entirely superseded roller-drying because of the overall superior quality of the product especially as regards solubility. Nevertheless, spray-dried skimmed milk powder is difficult to dissolve satisfactorily in water, especially in the domestic kitchen. This problem has been successfully overcome by the introduction of 'instant' skimmed milk powder, which consists of agglomerations of powder particles into large and porous clusters or granules. These are formed by controlled drying of a slurry of the powder so that specific lactose crystallisation occurs. These granules when added to water rapidly disintegrate and thereby assist dissolution of the powder particles. All skimmed milk powder sold by retail is now of the 'instant' type. It is possible to differentiate between roller and spray-dried powders by microscopical examination. When mounted in water, spray powder appears as regular approximately spherical particles varying in size (about 10–130 μm). The particles contain pockets of entrapped air. Roller powder, however, appears as irregularly shaped particles that vary in size.

Powders for infant feeding are prepared from various milk powders carefully formulated to replace breast milk and provide all the nutrients required by babies (see p. 571).

Legislation

As described in the previous section dealing with condensed milk products, the Condensed Milk and Dried Milk Regulations 1977 SI No. 928 implemented the Council Directive 76/118/EEC, which also dealt with the standards for dried milk products. The Regulations as amended prescribe reserved descriptions and composition as shown in Table 14.12 and also prescribe the methods of test.

Table 14.12 Statutory standards for dried milk products

Reserved description	Dried milk product: Milk fat (%)	Dried milk product: Moisture (%)
Dried high-fat milk or high-fat milk powder	more than 42·0, 65·0 max.	5·0 max.
Dried whole milk or whole milk powder	more than 26·0, 42·0 max.	5·0 max.
Dried partly skimmed milk or partly skimmed milk powder	more than 1·5, 26·0 max.	5·0 max.
Dried skimmed milk or skimmed milk powder	1·5 max.	5·0 max.

Not more than 0·3 per cent of the SNF of any of the products is to consist of lactic acid or lactate expressed as lactic acid. The same miscellaneous additives permitted in condensed milk are allowed. Ascorbic acid, sodium ascorbate or ascorbul palmitate at a maximum level of 0·05 per cent expressed as ascorbic acid are allowed as antioxidants. Products containing at least 1·5 per cent milk fat may contain up to 0·5 per cent lecithin as an emulsifier. Powders for vending machines may also contain up to 1·5 per cent of any permitted anti-caking agent. Vitamin fortification is permitted. Labelling instructions must include a declaration of the drying process used, whether the product is an 'instant' type, and milk fat content of the product after dilution for those containing at least 1·5 per cent milk fat in the powder.

The quality standards for skimmed milk powder brought in by an EC Intervention Agency (in UK the Intervention Board for Agricultural Produce) (Regulation 2248/86/EEC, OJ No. L196, 18.7.86, p. 26) include the following: fat, 1·0 per cent max.; moisture, 3·5 per cent max.; titratable acidity, 19·5 ml of 0·1 M sodium hydroxide per 10 g of SNF max.; lactate, 150 mg/100 g max., additives, nil; phosphatase test, negative; insolubility index, 0·5 ml max.; scorched particles, disc B minimum (15·0 mg); total colony count, 40 000/g max.; detection of buttermilk, negative and detection of whey, negative.

Analysis of dried milk and dried milk products

Typical composition figures obtained with full-cream and skimmed milk powders are given in Table 14.13. Vieth's ratio applies as for liquid milk (cf. Halliday *et al.*, 1960).

Table 14.13 Composition of full cream and skimmed-milk powders

	Full cream milk powder	Skimmed-milk powder
Moisture (%)	3–7	3–7
Milk fat (%)	24–28	0·5–1·5
Protein (%)	23–27	33–37
Ash (%)	5–7	7–9·5
Lactose (%)	35–39	48–53
Citric acid (%)	1–1·5	1·5–2
Lactic acid (%)	—	1·5–2
Calcium (%)	0·8–1	1·1–1·3
Phosphorus (%)	0·6–0·8	0·9–1·0
Sodium (%)	0·3–0·4	0·4–0·5
Potassium (%)	1–1·3	1·5–1·7
Chloride (%)	0·8–0·9	1·1–1·2
Vitamin B_1 (mg/kg)	2–6	3–5
B_2 (mg/kg)	about 10	about 16
B_6 (mg/kg)	2–3	2–3
B_{12} (mg/kg)	20	30
Nicotinic acid (mg/kg)	about 6	about 12
Ascorbic acid (mg/100 g)	about 10	about 6

Official methods of analysis for moisture, fat, lactic acid and lactates and phosphatase activity for the enforcement of the Regulations are laid down in Directive 79/1067/EEC (OJ No. L327, 24.12.79, p. 29). The specified control methods for Intervention skimmed milk powder (Regulation 625/78/EEC, OJ No. L84, 31.3.78, p. 19) are mainly IDF, ISO or ADMI (1971) derived. The main determinations, both chemical and physical, to be performed on dried milk and dried milk products, including dried whey, dried buttermilk and infant formulae, are now being produced in various parts of the revised BS 1743. These methods are identical to relevant jointly produced IDF/ISO/AOAC methods.

Moisture. Weigh 2 g of sample into a metal dish fitted with a lid and dry the uncovered dish at 102 ± 1 °C for 2 h and then to constant weight (EC). The IDF method (FIL-IDF 26: 1964) includes a drying temperature of 102 ± 2 °C for 1 g of sample.

Sherbon *et al.* (1978) showed in several studies that drying at 100 °C in a fan convected oven gave results not significantly different from those obtained in a vacuum oven.

Fat. Direct solvent extraction of the fat gives low results. For reference purposes the total lipid **Rose–Gottlieb method** is recommended.

Weigh 1–1·5 g sample into a Rose–Gottlieb or Mojonnier tube, add 10 ml water and disperse the sample by shaking and warming. Add 2 ml 0·88 ammonia, shake and warm to dissolve the protein. Cool, add 10 ml ethanol, mix and extract with mixed ethers as for liquid milk (p. 538).

A collaborative study of the Rose–Gottlieb method prescribed in Directive 79/1067 found the stated precision data to be unachievable (Crudgington, 1985).

A more rapid result can be obtained by employing the **Gerber process** (BS 696; Part 2).

To a butyrometer (8 per cent milk butyrometer for full cream or powders containing more than 10 per cent of fat; 4 per cent butyrometer for SMP) add 10 ml of Gerber sulphuric acid and cold water so that a layer is formed on top of the acid about 6 mm deep. Add 1·69 ± 0·01 g of sample by means of the dry steamless funnel, followed by 1 ml of amyl alcohol. Then add sufficient hot water (about 70 °C) to bring the level of the liquid up to about 5 mm below the shoulder, allowing all air entrained in the neck to escape. Stopper the tube, mix, place in a water bath at 65 °C for 3 to 10 min and centrifuge at 1100 r.p.m. for 5 min. It is advisable to return the butyrometer to the 65 °C bath and centrifuge again.

$$\text{Percentage of fat} = 6{\cdot}66 \times \text{butyrometer reading.}$$

Double the weight of SMP can be taken if a skimmed milk, double quantity, butyrometer is used.

The reagent proposed by Macdonald (p. 537) can also be employed, commencing by dispersing 1·69 g of sample in 5 ml hot water.

Rancidity of the fat should be checked by determining the peroxide value on an extract obtained with chloroform/methanol (p. 640).

Protein. Determine on 1 g sample using the macro-Kjeldahl method (protein = N × 6·38). Alternatively, the protein can be estimated using a dye-binding procedure (O'Connell and McGann, 1972, see p. 540).

Ash. Ignite 2 g sample at 525–550 °C. If the alkalinity of the ash is determined (p. 13) express the result as Na_2CO_3 or as ml of 0·2 M HCl required per 100 g.

Lactose. Determine volumetrically, polarimetrically or specifically on a solution obtained after clearing with the zinc ferrocyanide reagents as described for condensed milk and milk. IDF 79A: 1989 describes two enzymatic methods.

Acidity can be determined as for liquid milk after stirring 10 ml hot water (80 °C) with 1 g sample (S). Use a similar mixture for the control (C) together with 1 ml reference colour solution and rapidly titrate S to phenolphthalein as described on p. 539. The acidity should not exceed about 1·5 per cent m/m expressed as lactic acid in the powder.

The BS 1743: Part 7.2: 1982 (ISO 6092–1980) routine method is similar but expresses the result as ml 0·1 M NaOH per 10 g of SNF, taking a weight of sample containing exactly 5 g SNF dissolved in 50 ml water at 20 °C and left to stand 20 min before titration. The colour match is made in 100 ml conical flasks.

BS 1743: Part 7.1: 1982 (ISO 6091–1980) is a reference method very similar to the above routine method except that the end-point is taken at pH 8·40 using a pH meter and glass electrode. No reference colour solution is used.

Lecithin. Added lecithin in instant powders can be extracted with more than 90 per cent recovery by Soxhlet extraction with petroleum ether, without extracting milk phospholipid (Wewala and Baldwin, 1982).

Vitamins. Added vitamins A and D can be estimated by HPLC, e.g. AOAC methods book, Wickrowski and McLean (1984), Sertl and Molitor (1985), and Grace and Bernhard (1984). Reynolds and Judd (1984) used the same HPLC system for vitamin D after a Sep-Pak cartridge clean-up as for vitamin A. Indyk (1982) described a rapid and simple spectrophotometric method. Mills (1985) found that the Carr Price method (p. 645) gave comparable results for vitamin A with HPLC methods. Barnett *et al.* (1980) have analysed infant feeds and dairy products for vitamins A, D_2 or D_3, E and K_1 by reversed phase HPLC (see also Chapter 16).

Solubility. The assessment of the solubility figure of milk powder is essentially empirical depending as it does on factors such as the method of drying, the drying temperature, the acidity and the method of carrying out the test. The results obtained therefore tend to be comparative. Most spray-dried powders are almost 100 per cent soluble, whilst the solubility of roller-dried powders is usually 80–95 per cent. The various earlier published methods have been reviewed by King (1966). Waite and White (1949) considered that the method of Parsons (1949) although originally proposed for roller powders was the most suitable for most types of dried milk:

Shake 4 g powder with 32 ml hot (50 °C) water for 10 s and place in a water bath (50 °C) for 5 min. Shake for 1 min making 5 double excursions of 30 cm/s. With skimmed powders use this liquid cooled to 20 °C for the determination of total solids. (In the case of full-cream and half-cream powders, centrifuge the hot reconstituted milk in a 25 ml tube for 10 min at 2000 r.p.m. Cool in a refrigerator and remove the fat layer after running a needle around the cake of fat. Warm to 20 °C, break up the deposit with a rod and shake the corked tube hard to produce apparent homogeneity.) In either case weigh about 2 ml (weight L_1) of liquid into a weighed metal dish with lid. Also centrifuge the reconstituted milk for 10 min and weigh about 2 ml (weight L_2) into another dish. Dry both dishes on a water bath and then in a 100 °C oven for 1½ h, cool and weigh. If the weights of the dried solids corresponding to L_1 and L_2 are S_1 and S_2 respectively, then

$$\text{Solubility (per cent)} = \frac{100\ L_1 S_2}{L_2 S_1}$$

It is now considered more appropriate to assess the insolubility rather than the solubility of milk powders.

The original ADMI method to determine Solubility Index in a more standardised form as BS 1743: Part 3: 1988 (ISO 8156–1987, IDF 129A: 1988) and called Insolubility Index, involves adding to 10 g of skimmed milk powder (or dried buttermilk) or 13 g of dried whole milk, dried partly skimmed milk or infant formulae or 7 g of dried whey, 100 ml of distilled water at 24 °C in a specially designed homogeniser jar. After adding 3 drops of silicone anti-foaming agent the contents are homogenised for exactly 90 s. After 5–15 min standing 3 more drops of silicone are added and the contents mixed with a spoon and poured into specially graduated conical centrifuge tubes to the 50 ml mark. The tubes are centrifuged for 5 min at specified revolutions per min depending on the diameter of the centrifuge head (e.g. 1075 r.p.m. for 10 inch and 800 r.p.m. for 18 inch head). Using a suction device, for example a pasteur pipette fitted with a rubber bulb, the supernatant liquid in each tube is removed to the 10 ml mark of the tube. 20 ml of water at 24 °C is added to each tube and a stirring rod used to disperse the sediment. Each tube is filled to the 50 ml mark with water at 24 °C, stoppered and inverted five times and again centrifuged for 5 min. The volume of sediment in a tube in ml to the nearest scale division gives the Insolubility Index.

In his review, King (1966) discusses the effects of the various methods of manufacture (including film drying under vacuum and foam-spray drying) on the dispersibility and reconstitutability. Powders that are very wettable and dispersable and dissolve in water in a few seconds without agitation may have inferior storage properties due to higher moisture content.

Scorched or burnt particles (BS 1743: Part 19). The ADMI method involves mixing 25 g of spray dried skimmed milk (or buttermilk) or 32·5 g of spray dried whole milk for 60 s with 250 ml of water and 0·5 ml of a defoaming agent using a Waring blender or similar homogeniser followed by filtration through a specially designed filter assembly containing 1¼ inch diameter cotton discs or pads. The residue on the disc is washed by passing 50 ml of water through the filter. The disc or pad is then dried somewhat in a low temperature oven and the amount of scorched particles on it compared with the ADMI photographic standards. For roller dried products, hot water containing sodium citrate or EDTA is used.

Bulk density. Place 20 g of dried milk in a 100 ml measuring cylinder and drop the cylinder 150 mm on to a soft pad ten times. Level off the powder and measure the bulk density in g/ml. Spray-dried powders give figures in the region of 0·5 g/ml.

Presence of whey. The detection of lactose crystals is suggestive of the presence of whey. Regulation 2248/86 prescribes a size exclusion HPLC method for glycomacropeptides based on Olieman and Van den Beden (1983), for rennet whey. Greiner *et al.* (1985) described a rapid immunoturbidimetric method using an antibody to whole bovine whey.

Other methods to be covered by BS 1743 include dispersibility and wetting time, heat class, lactic acid and lactates, neutralisers, nitrate and nitrite, phosphatase activity and Na, K and Ca.

CULTURED MILK PRODUCTS

A full review of the production methods for the three main cultured milk products, i.e. yogurt, cream and buttermilk, and the propagation of their cultures, is given by Ashton (1963).

SOUR CREAM

Sour (cultured) cream now finds wide use in catering, e.g. in 'dips'. Methods of preparation vary but essentially a pasteurised cream mix containing 18–20 per cent milk fat and 7–9 per cent of MSNF (corrected with SMP if necessary) is heated to about 80 °C for 10–30 minutes, cooled to 70 °C, thoroughly homogenised then cooled to 20–30 °C and 1–3 per cent of culture added, followed by a small addition of rennet solution to give a smooth and firm final texture. Gelatine, alginate or starch may also be added as stabilisers. Since disturbance of the culture affects the final texture, incubation at about 22 °C for 12–15 hours to about 0·6–0·7 per cent acidity as lactic acid, is carried out in the retail packaging.

YOGURT

Yogurt is the Turkish name for milk from cows, goats, sheep or buffalo fermented by a lactic culture. It is known in the Caucasus as matzoon and katyk, in Greece as tiaouriti, in Egypt as laban, in Bulgaria as naja, in Italy as gioddu and in India as dahi. In the UK plain unflavoured yogurt is much less preferred to flavoured, especially fruit, cows milk yogurts. Liquid yogurts for drinking are also available. The Food Standards Committee Report in 1975 led in 1983 to the Code of Practice for the composition and labelling of yogurt produced by the UK Dairy Trade Federation and LACOTS. In this, yogurt is defined as the acidified coagulated cows milk product made from, or any combination of, milk, partly skimmed milk, skimmed milk, concentrated milk, concentrated partly skimmed milk, concentrated skimmed milk with or without cream, dried whole milk, dried partly skimmed milk, dried skimmed milk, reconstituted dried milks, whey, unfermented buttermilk, any milk proteins, butter and butter oil, in which after pasteurisation, lactic acid has been produced within the product by bacterial cultures consisting of *Lactobacillus bulgaricus* with or without *Streptococcus thermophilus* with which may be used other suitable lactic bacteria. Yogurt should contain the appropriate live organisms. Yogurt may, however, be subject to heat treatment after fermentation.

The Code of Practice requires that all yogurt should have a minimum MSNF of 8·5 per cent and a minimum milk protein content of 3 per cent. Low fat yogurt should contain 0·5–2 per cent of fat. Very low fat or skimmed milk yogurt should contain less than 0·5 per cent. Fruit yogurt unqualified by the word 'flavour' or 'flavoured' should contain at least 5 per cent of the named fruit, whole, puréed or in pieces. Natural yogurt should contain no added colour, preservative, stabiliser or thickener and should not be heat treated after fermentation, but may contain added and declared vitamins, sugar or flavouring foodstuff. Yogurt subjected to final heat treatment should be described as pasteurised, sterilised or UHT as appropriate. Up to 0·5 per cent of stabilisers and thickeners may be present, and up to 1·0 per cent only by the use of gelatine, pectins, starch or modified starch.

Most flavoured yogurt in the UK is of the stirred type, produced in bulk before the addition of fruit, etc. before packaging. Set yogurt, in which the fermentation is carried out in the retail packaging, is more prevalent in continental Europe.

Most of the yogurt in the UK contains 0·5–1·5 per cent of milk fat. The SNF of the prepared milk is increased by up to 3 per cent by the addition of skimmed milk powder. Up to 0·5 per cent of stabiliser is added to prevent syneresis, and the mixture homogenised. It is heated to 90–95 °C for about 5 min then cooled to about 43 °C and about 2 per cent of starter added. This is incubated at about 43 °C for 3–4 hours, cooled to 4–5 °C then blended with flavours/fruit ingredients and packaged. There is a trend nowadays towards final pasteurisation to produce ambient or long-life yogurts. The heat treatment for set yogurt is similar, but the inoculated milk is packed into containers before incubation. Pasteurised products cannot be described as yogurt in many continental European countries, and are referred to as fermented desserts, etc. A working party of interested organisations in the UK are reviewing this problem and other matters covered by the Code of Practice.

The composition of natural and flavoured yogurts (Paul and Southgate, 1978) is shown in Table 14.14. Titratable acidity for fresh yogurt should be 0·85–0·95 per

Table 14.14 The composition of cows milk yogurt

	Natural	Flavoured/fruit/nut
Moisture (%)	86	73–79
Fat (%)	1–0	0·9–2·6
Protein (%)	5·0	4·8–5·2
Lactose (%)	4·6	3·2–4·8
Total carbohydrate (%)	6·2	14–18
Calcium (%)	0·18	0·16–0·18
Vitamin B_1 (mg/100 g)	0·05	0·05
Vitamin B_2 (mg/100 g)	0·26	0·23–0·27
Vitamin C (mg/100 g)	0·4	0·4–1·8

cent as lactic acid (pH 4·4–4·5); IDF 47: 1969 specifies not less than 0·7 per cent. Drinking yogurts prepared from separated milk, fruit juice and added sugars contain 0·4–0·8 per cent fat and 7·0–7·4 per cent MSNF. Sheep and goats milk yogurt, for which there is no specific legislation or Codes of Practice, is now marketed in the UK.

For the examination of cultured milks such as yogurt the methods described for fresh milk can be employed with certain modifications (Davis, 1970). Fat can be determined by the Werner–Schmid method, or more rapidly by the Gerber or Macdonald (p. 537) methods using 11·3 g of sample in a milk or skim-milk butyrometer. Other determinations are those for total solids (after neutralisation with alkali, which also measures acidity and from which the SNF can be obtained), and for sugars. The assessment of the fruit content is difficult as milk not only gives an appreciable formol titration but contains potassium and phosphorus. Treatment with alcohol, ammonia and then hot water may give some idea of the insoluble solids present, however. The determination of citric and malic acids, e.g. enzymatically or by HPLC, can provide an estimate of the fruit content. Flavoured yogurt may contain added colours, which should be identified. With real-fruit types, examination for sorbic acid and benzoic acid is advisable. The Preservatives in Food Regulations permit fruit yogurt to contain up to 60 mg/kg of benzoic acid or methyl, ethyl or propyl 4-hydroxybenzoate (cf. Williams, 1972). Apart from the chemical analyses some attention should be paid to examination for syneresis and the presence of moulds and yeasts.

KEFIR, KEPHIR

Kefir is prepared in south west Asia from cows, mares or goats milk using kefir grains, which are the gelatinous particles formed in the fermentation. The acid and alcoholic fermentation produced by *Streptococcus lactis*, *Lactobacillus bulgaricus* and lactose-fermenting yeasts produce a gassy and foamy product containing about 0·7 per cent alcohol (cf. Lampert, 1970; Lang and Lang, 1973).

KOUMISS, KUMISS

Koumiss is rather similar to kefir and is prepared in Russia from mares or ass' milk, which is high in lactose but can also be made from cows milk. The acidity varies from 0·7 to 1·9 per cent lactic acid corresponding to about 0·9–2·8 per cent alcohol (cf. Lang and Lang, 1973). In certain areas a brandy-like beverage called araka is prepared by distilling koumiss.

BUTTERMILK

Originally, the term buttermilk was applied to the acid by-product of the churning of sour cream in buttermaking. Later some buttermilk resulted from butter production using sweet cream as the raw material. The composition of sweet-cream buttermilk is similar to skimmed milk, 91·0 per cent water, 0·4 per cent fat, 4·5 per cent lactose, 3·4 per cent nitrogenous matter, 0·04 per cent lactic acid and 0·7 per cent ash. That from sour cream contains 0·6 per cent lactic acid and 3·4 per cent lactose. A certain proportion of water is often added in buttermaking, but an addition of over 25 per cent in buttermilk can be regarded as fraudulent. Spray dried buttermilk powder consists of 6–10 per cent moisture, 23–33 per cent protein, 43–46 per cent lactose, 4–6 per cent butterfat, 7–10 per cent ash and 5–6 per cent lactic acid. According to Lampert (1970), practically all buttermilk now sold to consumers in the USA is cultured buttermilk.

Cultured buttermilk is usually prepared from pasteurised skim milk inoculated with a starter culture with the possible additions of cream and citric acid. It is incubated at 22 °C until an acidity of about 0·8 per cent as lactic acid is reached. Traces of salt are sometimes added to reduce the sharpness of the acidity. In the USA a minimum standard of 8·5 per cent SNF has been suggested for cultured buttermilk. IDF 47: 1969 specifies 0·6 per cent lactic acid minimum and 1 per cent fat maximum.

FILLED MILK AND INFANT FORMULAE

The Skimmed Milk with Non-Milk Fat Regulations (SI 1960 No. 2331) and Amendment Regulations (SI 1976 No. 103 and SI 1981 No. 1174) impose requirements as to the labelling and advertising of certain specified foods that have the appearance of milk, condensed milk or dried milk, and which contain skimmed milk and non-milk fat. Such products are often referred to as 'filled milks', but the Codex code states the term is misleading within the meaning of Article 4. The regulations require labels and advertisements for these foods to bear the words 'Skimmed milk with non-milk fat'. The regulations also prohibit (subject to certain savings) the labelling or advertising of the specified foods and beverages containing skimmed milk, in a manner suggestive of milk or anything connected with a dairy interest.

The labels for such products are required to bear the words 'Unfit for babies' or 'Not to be used for babies' except that where the kind and amount of fat used is specified, the words 'Should not be used for babies except under medical advice' may be substituted.

None of these declarations about baby feeding is required in the case of branded infant formulae products named in the second schedule to the regulations as amended. Such products are, however, required to conform to various standards for protein, fat and vitamins A, C and D and also the following:

1. shall contain polyunsaturated fatty acids of the *cis–cis* form to the extent of not less than 12 per cent of the total fatty acids present in such food;
2. shall not contain any protein other than protein derived from milk; and
3. shall not contain any ingredient of no nutritional value.

The 1981 FSC Report on Infant Formulae recommended specific legislation to prohibit the marketing of products providing the sole source of nourishment for the young infant (i.e. substitute breast milk) unless approved. The 1984 FAC Report on the 1960 Regulations recommended simplification and relaxation of labelling controls but continuation of the 'unfit for babies' warning statements. The regulations should continue to list approved infant formulae products until specific regulatory controls for them are introduced. A proposal for a Council Directive, COM (86) 564 (OJ No. C285, 12.11.86, p. 5) on laws relating to infant formulae and follow-up milks is currently under consideration.

The general analysis of these products follows the same lines as those described for liquid, condensed and dried milks, in particular, various parts of BS 1743.

For tests on the fat the lipid should be extracted with chloroform/methanol as described in Chapter 2 and examined for rancidity, physical constants, etc. and for the absence of butterfat by GLC. The presence of casein represents strong evidence for the presence of milk protein.

REFERENCES

Abu-Lehia, I. H. (1987) *Food Chemistry*, **24**, 233.
ADMI (1971) *Standards for Grades of Dry Milks including Methods of Analysis*. Bulletin 916. Chicago: American Dry Milk Institute Inc.
Andrews, G. R. (1984) *Journal of the Society of Dairy Technology*, **37**, 92.
Anon (1989) *Dairy Industries International*, **54**, 23.
Aschaffenburg, R. & Muller, J. E. C. (1949) *Journal of Dairy Research*, **16**, 58.
Ashton, T. R. (1963) *Journal of the Society of Dairy Technology*, **16**, 68.
Barnett, S. A., Frick, L. W. & Baine, H. M. (1980) *Analytical Chemistry*, **52**, 610.
BCL (1984) *Methods of Enzymatic Food Analysis*. Lewes: Boehringer Corporation (London) Ltd.
Biggs, D. A. & Szuarto, L. (1963) *Journal of Dairy Science*, **46**, 1196.
Bishop, J. R. & White, C. H. (1984) *Journal of Food Protection*, **47**, 647.
Bradley, R. L. (1986) *Journal of the Association of Official Analytical Chemists*, **69**, 831.
Briggs, D. A., Johnson, G. & Sjaunja, L. O. (1987). In *Rapid Indirect Methods for Measurement of the Major Components of Milk*. Bulletin of the International Dairy Federation, No. 208.
Bruce, A. M. & McIntosh, L. J. (1982) *Journal of the Association of Public Analysts*, **20**, 95.
Brunner, J. R. (1981) *Journal of Dairy Science*, **64**, 1038.

Buckley, K. E., Fisher, L. J. & Mackay, V. G. (1986) *Journal of the Association of Official Analytical Chemists*, **69**, 655.
Crudgington, D. R. (1985) *Journal of the Association of Public Analysts*, **23**, 103.
Davies, W. L. (1932) *Analyst*, **57**, 79.
Davies, W. L. (1938) *Journal of Dairy Research*, **9**, 327.
Davis, J. G. (1968) Dairy products. In *Quality Control in the Food Industry*, vol. 2, ed. S. M. Herschdoerfer. London: Academic Press.
Davis, J. G. (19970) *Dairy Industries*, **35**, 139.
Dickes, G. J. (1973) *Journal of the Association of Public Analysts*, **11**, 36.
El-Salam, M. H. A., Al-Khamy, A. F. & El-Etriby, H. (1986) *Food Chemistry*, **19**, 213.
Elsdon, G. D. & Stubbs, J. R. (1936) *Analyst*, **61**, 383.
Fellows, P. V. (1953) *Dairy Industries*, **18**, 309.
Fischer, B. M. (1965) *Journals of the Society of Dairy Technology*, **18**, 230.
Florence, E., Knight, D. J., Owen, J. A., Milner, D. F. & Harris, W. M. (1985) *Journal of the Society of Dairy Technology*, **38**, 121.
Grace, M. L. & Bernhard, R. A. (1984) *Journal of Dairy Science*, **67**, 1646.
Greiner, S. P., Kellen, G. J. & Carpenter, D. E. (1985) *Journal of Food Science*, **50**, 1106.
Halliday, J. H., Burden, E. H. W. J. & Lamont, J. J. (1960) *Analyst*, **85**, 839.
Harding, F. & Royal, L. (1974) *Dairy Industries*, **39**, 372.
Harris, W. M. (1986) *Analyst*, **111**, 37.
Hatfull, R. S. (1964) *Journal of the Association of Public Analysts*, **2**, 109.
Henningson, R. W. (1963) *Journal of the Association of Official Agricultural Chemists*, **46**, 1036.
Henningson, R. W. (1969) *Journal of the Association of Official Analytical Chemists*, **52**, 142.
Hortvet, J. J. (1921) *Journal of Industrial and Engineering Chemistry*, **13**, 198.
Hortvet, J. J. (1922) *Journal of the Association of Official Agricultural Chemists*, **5**, 172.
Houston, J. (1955) *Journal of the Society of Dairy Technology*, **8**, 47.
Indyk, H. (1982) *New Zealand Journal of Dairy Science and Technology*, **17**, 257.
Johansson, I. and Claesson, O. (1957) In *Progress in the Physiology of Farm Animals*, ed. J. Hammond. London: Butterworth.
Karmar, A. H. & Van Boekel, M. A. J. S. (1986) *Netherlands Milk and Dairy Journal*, **40**, 315.
Katz, S. E. & Fassbender, C. A. (1978) *Journal of the Association of Official Analytical Chemists*, **61**, 918.
Kennedy, J. F., White, C. A. & Browne, A. J. (1985) *Food Chemistry*, **16**, 115.
Kessler, H. G. (1984) *Milchwissenschaft*, **39**, 339.
King, N. (1966) *Dairy Science Abstracts*, **28**, 105.
Kleyn, D. H. & Trout, J. R. (1984) *Journal of the Association of Official Analytical Chemists*, **67**, 637.
Kleyn, D. H. & Yen, W. (1970) *Journal of the Association of Official Analytical Chemists*, **53**, 869.
Lampert, L. M. (1970) *Modern Dairy Products*. London: Food Trade Press.
Lang, F. & Lang, A. (1973) *Food Manufacture*, **48**, 23.
Macdonald, F. J. (1943, 1948, 1959a, b) *Analyst*, **68**, 171; **73**, 423; **84**, 287, 747.
Markland, J. (1963) *Journal of the Association of Public Analysts*, **1**, 54.
Marshall, K. R. (1982) In *Developments in Dairy Chemistry—1*, ed. P. F. Fox. London: Applied Science Publishers.
Maxwell, R. J., Mondimore, D. & Tobias, J. (1986) *Journal of Dairy Science*, **69**, 321.
Milk and Milk Products Technical Advisory Committee (1963) *Antibiotics in Milk in Great Britain* (Report of the Milk Hygiene Sub-Committee). London: HMSO.
Miller, D. D. & Elliker, P. R. (1951) *Journal of Dairy Science*, **34**, 273.
Mills, S. (1985) *Journal of the Association of Official Analytical Chemists*, **68**, 56.
Morr, C. V. (1982) In *Development in Dairy Chemistry—1*, ed. P. F. Fox. London: Applied Science Publishers.
Murphy, F. E. (1960) *Analyst*, **85**, 720.
Nielsen, V. H. (1976) *American Dairy Review*, **38**, 20.
O'Connell, J. A. & McGann, T. C. A. (1972) *Laboratory Practice*, **21**, 552.
O'Keefe, M. G. (1967) *Journal of Dairy Research*, **34**, 211.
Olieman, C. & Van den Beden, J. W. (1983) *Netherlands Milk and Dairy Journal*, **37**, 27.
Parsons, A. T. (1949) *Journal of Dairy Research*, **16**, 377.
Paul, A. A. & Southgate, D. A. T. (1978) *The Composition of Foods*. London: HMSO.
Pearson, D. (1976) *The Chemical Analysis of Foods*. Edinburgh: Churchill Livingstone.
Pim, P. B. (1964, 1967) *Journal of the Association of Public Analysts*, **2**, 62; **5**, 61.
Pyne, G. T. (1932) *Biochemical Journal*, **26**, 1006.
Ratcliffe, R. J. M. (1963) *Journal of the Association of Public Analysts*, **1**, 78.
Reynolds, S. L. & Judd, H. J. (1984) *Analyst*, **109**, 489.
Rook, J. A. F. (1961) *Dairy Science Abstracts*, **23**, 251, 303.

Rothwell, J. (1968) *Process Biochemistry*, **3**, 19.
Salwin, H. & Bond, J. F. (1969) *Journal of the Association of Official Analytical Chemists*, **52**, 41.
Sanders, G. P. (1939) *Journal of Dairy Science*, **22**, 841.
Scott, K. J. & Bishop, D. R. (1986) *Journal of the Society of Dairy Technology*, **39**, 32.
Scott, K. J. & Bishop, D. R. (1988) *Journal of the Science of Food and Agriculture*, **43**, 193.
Scott, K. J., Bishop, D. R., Zechalko, A., Edwards-Webb, J. D., Jackson, P. A. & Scuffam, D. (1984a) *Journal of Dairy Research*, **51**, 37.
Scott, K. J., Bishop, D. R., Zechalko, A. & Edwards-Webb, J. D. (1984b) *Journal of Dairy Research*, **51**, 51.
Sertl, D. C. & Molitor, B. E. (1985) *Journal of the Association of Official Analytical Chemists*, **68**, 177.
Sherbon, J. W. (1975) *Journal of the Association of Official Analytical Chemists*, **58**, 770.
Sherbon, J. W., Mickle, J. R. & Ward, W. D. (1978) *Journal of the Association of Official Analytical Chemists*, **61**, 550.
SPA (1927, 1930) *Analyst*, **52**, 402; **55**, 111.
Swartz, M. L. & Wong, C. J. (1985) *Cereal Foods World*, **30**, 173.
Sweetsur, A. W. M. (1974) *Analyst*, **99**, 960.
Taylor, W. H. (1957) *Analyst*, **82**, 488.
Terada, H. & Sakabe, Y. (1985) *Journal of Chromatography*, **348**, 379.
Thomas, S. B. & Makinson, P. E. (1964) *Dairy Industries*, **29**, 432.
Thomas, S. B., Jones, T. I. & Griffiths, D. G. (1963) *Dairy Industries*, **28**, 812.
Tramer, J. & Wight, J. (1950) *Journal of Dairy Research*, **17**, 194.
Van de Voort, F. R. (1980) *Journal of the Association of Official Analytical Chemists*, **63**, 973.
Van Schalm, K. J. (1983) *Netherlands Milk and Dairy Journal*, **37**, 59.
Wahbi, A. A. M., Abdine, H. & Blaih, S. M. (1977) *Journal of the Association of Official Analytical Chemists*, **60**, 1175.
Waite, R. & White, J. C. D. (1949) *Journal of Dairy Research*, **16**, 379.
Wewala, A. R. & Baldwin, A. J. (1982) *New Zealand Journal of Dairy Science and Technology*, **17**, 251.
Wickrowski, A. F. & McLean, L. A. (1984) *Journal of the Association of Official Analytical Chemists*, **67**, 62.
Williams, D. J. (1986) *Australian Journal of Dairy Technology*, **41**, 28.
Williams, H. A. (1972) *Journal of the Association of Public Analysts*, **10**, 99.
Wood, E. C. (1954) *Analyst*, **79**, 779.
Wood, E. C. (1972) *Journal of the Association of Public Analysts*, **10**, 68.
Wright, R. C. & Anderson, E. B. (1938) *Analyst*, **63**, 252.
Wright, R. C. & Tramer, J. (1961) *Journal of the Society of Dairy Technology*, **14**, 85.

15

Dairy products II

BUTTER

Introduction

Butter is the product obtained by churning the cream that has been separated from warm cows milk. It is then 'worked', i.e. excess water is removed and the mass is rendered homogeneous usually with the addition of salt. Flavoured butter is prepared from sour cream, but sweet cream butter is prepared without the use of a starter. In the UK butter tends to be prepared from comparatively fresh milk, but in some other countries the higher acidity of the cream used is often partially neutralised with bicarbonate. Methods used for the manufacture of butter by both batch and continuous processes have been reviewed by Rothwell (1966) and Lampert (1970). Williams (1984) has compared the performance of on-line meters for moisture (dielectric) and salt (gamma radiation back-scatter) in continuous butter makers, with standard chemical methods.

Butter consists of unaltered fat globules and moisture droplets embedded in a continuous phase of butterfat. The textural properties are largely dependent on the proportion of liquid to solid fat in the continuous phase. The flavour of butter is due to diacetyl and to a lesser extent to butyric, acetic, propionic and formic acids, acetaldehyde and acetoin. Butter produced from acid cream contains much more diacetyl than when the raw material is fresh cream, but the amount can be increased by the addition of culture.

Basically, butter contains butterfat, water and curd, the last consisting of casein, lactose and mineral matter. Most samples now sold contain 80–84 per cent of fat, close to 16·0 per cent water, 0·03–1·8 cent salt and about 1 per cent MSNF. The vitamin A content of butter is very variable being dependent on various factors such as season, climate and feeding of the animal. Paul and Southgate (1978) indicate a range for vitamin A (retinol) of 520–970 μg/100 g and for carotene 350–650 μg/100 g.

Legislation

Statutory requirements for butter are prescribed in the Butter Regulations, 1966, SI No. 1074 as amended. These state that butter may contain one or more of the following substances, namely annatto, α-, β and γ-carotene, synthetic β-carotene, or turmeric, or salt or lactic acid cultures.

Regulation 4(1) provides that butter shall contain not less than 80 per cent milk fat, not more than 2 per cent milk solids other than fat, and not more than 16 per cent water, with the provision that 'salted butter' may contain less than 80 per cent but not less than 78 per cent milk fat if the amount by which the milk fat content percentage falls below 80 per cent, does not exceed the amount by which the percentage of salt in such butter exceeds 3 per cent. In addition to most of the above standards the Codex code (see p. 534) permits the addition of annatto, β-carotene and curcumin for colouring, and up to 2000 mg/kg of neutralising salts.

For butter oil, anhydrous butterfat and EC CAP concentrated butter containing identification markers, see p. 614.

Analysis of butter

The routine analysis of butter should include the determination of water and salt. Also the filtered anhydrous fat may be examined by GLC or the Reichert and Polenske pocesses to ascertain that the fat consists of genuine butterfat. The fat may also be required to be analysed for rancidity and other values (BS 684). Methods of sampling are included in BS 809: 1985 (ISO 707) and reference methods of analysis and rapid procedures in various parts of BS 5086. Sampling is covered in detail in the Codex code (p. 534), which also includes methods for the determination of salt, and for the acid value and refractive index of the fat.

Sampling. Place the sample in a screw-capped jar, warm in an oven at 32–35 °C and shake vigorously from time to time in order to produce a homogeneous fluid emulsion.

Water (BS 5086, Part 2). Weigh out 2–6 g of sample into a metal dish and dry in a 102 °C oven to constant weight (approx. 2–3 h). Marschke *et al.* (1979) have compared infra-red moisture balance and microwave oven methods with a standard oven drying method for the rapid determination of moisture in butter.

Curd + salt (Solids-not-fat, BS 5086, Part 2). Melt the residue from the moisture determination by gentle heat, add light petroleum, stir and filter through a tared filter paper or a weighed Gooch or sintered glass crucible. The filtrate can be used for the fat determination below. Wash the dish and filter with more light petroleum until the filtrate is free from fat. Dry the paper or crucible in an oven at 100 °C and weigh the curd + salt.

Salt. Transer the curd + salt together with filter paper or crucible to a beaker containing hot water, cool and titrate with 0·05 M silver nitrate solution using potassium chromate as indicator (see also BS 5086, Part 5). IDF 12B: 1988 describes a Mohr titration method in which 5 g of butter + 100 ml boiling water + 2 ml potassium chromate 5 per cent solution is titrated hot or after cooling with 0·1 M silver nitrate until an orange tint persisting for 30 s is obtained.

$$1\ \text{ml}\ 0{\cdot}1\ \text{M}\ AgNO_3 = 0{\cdot}005\,84\ \text{g NaCl}.$$

Alternatively, the Volhard method may be applied to the curd + salt or to the original sample (BS 5086, Part 4). For the latter add 10 ml water to 2 g, add 25 ml 0·05 M silver nitrate solution and gently boil with 10 ml conc. nitric acid. Add 0·3 g urea, mix, cool, add 1 ml nitrobenzene and titrate the unused silver nitrate with 0·05 M potassium thiocyanate using ferric alum (cf. BS 5086).

$$1\ \text{ml}\ 0{\cdot}05\ \text{M KCNS} = 0{\cdot}002\,922\ \text{g NaCl}.$$

NB Unlike nitrobenzene, triacetin is not toxic, but is inexpensive, less odorous and equally effective in coating the precipitate.

Salt in butter acts as a preservative by retarding the growth of most bacteria and moulds. There is evidence that some butters that contain over 2 per cent salt do not necessarily keep as well as ones that contain much smaller amounts. Unsalted butter contains less than 0·1 per cent salt.

Curd. The curd, which almost corresponds in genuine butter to the MSNF, is usually obtained by subtracting the amount of salt from the curd plus salt figure. It consists essentially of protein (casein), lactose and lactic acid. Protein can be determined on the curd plus salt residue by the Kjeldahl method using the factor N × 6·38.

Lactose can be determined on the curd by the Lane and Eynon process or any of the other methods described in Chapter 6. Enzymatic analysis is particularly suitable for low levels of lactose. Butter usually contains between 0·18 and 0·40 per cent lactose.

Fat is usually obtained by difference (BS 5086, Part 2):

$$\text{Fat (per cent)} = 100 - (\text{per cent water} + \text{per cent salt} + \text{per cent curd}).$$

Alternatively, the fat can be extracted with light petroleum (as in the curd + salt estimation above) and weighed after removing the solvent, or the Gerber method may be employed to give an approximate figure using 2·5 g in a cream butyrometer.

Ash. The cream used in butter manufacture is sometimes partially neutralised with sodium carbonate or bicarbonate before use. Such treatment produces higher than average figures for the ash, ash less salt and the alkalinity of the ash.

Titratable acidity. Shake 18 g of butter with 90 ml of hot previously boiled water and titrate hot with 0·02 M sodium hydroxide solution using 1 ml of 1 per cent phenolphthalein as indicator.

$$\begin{array}{c}\text{Titratable acidity}\\ \text{(as per cent lactic acid)}\end{array} = \frac{0{\cdot}02\ \text{M NaOH required (ml)}}{100}$$

pH of serum. BS 5086: Part 7: 1985 (ISO 7238–1983) describes the separation of serum (aqueous phase) by melting 100 g of butter in a beaker, then pouring the molten sample into two centrifuge tubes which are then stoppered and centrifuged gently with the stoppers downward. With the stoppers still downward the tubes are then cooled in an ice bath. Then by removing the stoppers over a beaker, the serum is collected, and pH measured with a combination electrode after adjusting to 20 °C.

Boric acid can be detected by the tumeric test. Melt 5–10 g of butter and pour away the upper fatty layer. Then pour the aqueous liquid remaining into a flat porcelain dish, add 3 drops of saturated oxalic acid solution, 2 drops of 5 M hydrochloric acid and conc. alcoholic extract of turmeric and evaporate on the water bath. In the presence of boric acid a reddish residue is produced.

Other preservatives such as benzoates, sulphites, nitrates and fluorides have on occasion been reported as being present in butter.

Added colours. Divide an ethereal solution of the isolated butterfat into two tubes. To one tube (A) is added 1–2 ml hydrochloric acid (1 + 1) and to (B) 1–2 ml 10 per cent sodium hydoxide solution. If azo dyes are present, tube A will show a red colour, but B will show no colour. If annatto or other vegetable colour is present there is no colour in A, but a yellow colour appears in B. For further information on vegetable colours reference should be made to the AOAC methods book.

Trace metals. It has been shown that the presence of 1–2 mg/kg of copper in butter encourages fat rancidity with the production of a fishy or tallowy taint in a few days. Rancidity may also be encouraged by the presence of other transition metals such as chromium, iron, manganese and nickel, but the action is less marked. The rancidity formation appears to increase with increasing acidity. For milk and milk products BS 6394: Part 1: 1983 (ISO 5738) describes a colorimetric method for copper using sodium diethyldithiocarbamate (see p. 159), and Part 2 (ISO 6732) describes a colorimetric method for iron using *o*-phenanthroline (see p. 299). Atomic absorption spectrophotometry has largely replaced colorimetric methods (see AOAC methods book and Chapter 5).

Diacetyl is one of the substances produced during cream ripening. It is produced by oxidation of its precursor, acetylmethylcarbinol and has a pronounced butter odour and taste. It has been shown that in general the greater the acidity of the cream used the greater is the amount of diacetyl in the butter and the stronger is its flavour. For its estimation the diacetyl can be isolated by cream distillation. In the Pien method the distillate is acidified after the addition of 3,4,3′,4′-tetra-aminodiphenyl and the absorption of the yellow complex formed is measured in the region of 350 nm (cf. Parodi, 1965). The method of Prill and Hammer (1938) involves formation of dimethylglyoxime from diacetyl in the presence of hydroxylamine (Owades and Jakovac, 1963; Walsh and Cogan, 1974). An improved method for diacetyl in dairy products containing α-acetolactic acid, such as starter and buttermilk, is described by Veringe *et al.* (1984).

EXAMINATION OF BUTTERFAT

The usual range of constants for genuine butterfat is given in Table 15.1 (see also p. 586).

Table 15.1 Data for butterfat

	Max.	Min.	Mean
Specific gravity at 37·8 °C/37·8 °C	0·913	0·910	0·912
Refractive index at 40 °C	1·4561	1·4524	1·4548
RI on Zeiss-butyro scale	45·5	40·0	43·5
Saponification value	232	222	228·5
Iodine value	40	26	36
Reichert value	32·8	24·5	30
Polenske value	3·5	1·5	2·8
Kirscher value	27	21	24
Baryta value	−0·7	−23·8	−9·6
Mean molecular weight of fatty acids	267	258	269

The rather wide variation in the analytical figures obtained with genuine butterfat makes it somewhat difficult to assess accurately the concentration of butterfat in mixed fats and fatty foods and conversely the concentration of other fats present. Butterfat in its triglycerides contains varying percentages of butyric, caproic, caprylic, capric, lauric, myristic, stearic, palmitic and oleic acids, together with small quantities of other acids (Dickes and Nicholas, 1972). Table 15.2 shows the fatty acid distribution in the fat of cow, goat and human milk (cf. Paul and Southgate, 1978). Until recently the most common methods for assessing the proportion of butterfat in mixtures were based on the presence of a comparatively high proportion of glycerides containing steam volatile water-soluble acids, particularly butyric acid.

More recent and far superior methods using chromatography are also mostly based on the relatively high proportion of butyric acid in butterfat. The iodine and saponification values are of no help in detecting adulteration of butterfat as other fats do not alter the figures to any extent. The measurement of physical constants are used much less now than formerly.

PREPARATION OF SAMPLE

The sample of fat used must be clear and free from water and non-fatty matter. It can be obtained by filtration of the melted fat as described for the Reichert process (below).

Physical constants

Density can be determined by weighing the melted fat and then water in a pyknometer at 37·8 °C. Due to the high proportion of glycerides of low molecular

Table 15.2 Fatty acid composition of cow, goat and human milk fat (per cent m/m in the total fatty acids)

	Cow		Goat	Human	
	Range	Mean		Range	Mean
C 4:0	2·6–3·9	3·2	2·1		0
C 6:0	1·5–2·3	2·0	2·4		0
C 8:0	0·9–1·4	1·2	3·2		Trace
C10:0	2·5–3·2	2·8	9·1	0·5–2·0	1·4
C12:0	3·1–4·0	3·5	4·5	3·3–8·2	5·4
C14:0	10·4–12·4	11·2	11·3	5·6–8·5	7·3
C14:1	1·1–1·6	1·4	Trace		Trace
C15:0		1·1			
C15:1		0·7			
C16:0	24·1–32·0	26·0	27·0	20·2–26·8	26·5
C16:1	2·1–3·1	2·7	2·4	2·4–5·7	4·0
C17:0		1·0			
C17:1		1·1			
C18:0	9·2–13·2	11·2	9·6	4·7–10·3	9·5
C18:1	22·0–30·7	27·8	26·0	35·4–46·4	35·4
C18:2	0·8–1·9	1·4	2·3	4·1–13·0	7·2
C18:3	0·6–2·5	1·5		0·4–3·4	0·8
C20:1					0·5

weight, butterfat has a higher density than most other fats. The difference between the density of butterfat and other fats is almost at a maximum at 37·8 °C. If the measurement is made at other temperatures the correction to be applied is 0·0007 for each °C.

Formerly, the **refractive index** of butterfat at 40 °C gave a valuable indication of its purity. Fats used nowadays in margarine manufacture however tend to give somewhat similar figures. Nevertheless, work in Holland has reported values obtained at 40 °C, together with measurements of the dependence of temperature and wavelength of the refractive index (Walstra, 1965). Corrections should be applied if the acid value of the fat equals or exceeds 2 (see Codex code, p. 534). The readings for butterfat are often expressed on the Zeiss-butyro scale (Z). The relationship between the two scales is given by the following formula:

$$\mathrm{RI} = 1{\cdot}4220 + 0{\cdot}000\,815Z - 0{\cdot}000\,001\,4Z^2$$

Chemical examination

Chemical tests are most commonly used for assessing the rancidity of butterfat and the proportion of other fats in butterfat. The latter tests are the outcome of the classic work of Hehner and Angell, who, in 1872, first devised a method that aimed at the estimation of butyric acid. Later work showed that better results could be obtained by determining the soluble and insoluble volatile acids (corresponding to the Reichert (or Reichert–Meissl) and Polenske values respectively). The historical development of these processes has been surveyed by Davis and Macdonald (1953). The method is empirical and therefore for consistent results it is essential to adhere strictly to the standardised procedure adopted by the SPA (1936) and by the BSI (BS 684, Part 2.11). A semi-micro modification is given on p. 582.

Isolation of butterfat for analysis

Heat a portion of butter in a beaker to a temperature of 50–60 °C until the fat separates from the water and curd. Filter the fat layer through a paper into a dry vessel in an oven at a temperature above the solidifying point of the fat. If necessary, re-filter the filtrate under the same conditions, until it is clear and free from water. Liquefy the fat completely and mix before taking samples for analysis.

Determination of volatile acids in butterfat (Reichert, Polenske and Kirschner values)

The following method is modified from BS 684, Part 2.11 and is reproduced by permission of the British Standards Institution, 2 Park Street, London, W1Y 4AA, from whom copies of the complete standard may be obtained.

A. DEFINITIONS

The Reichert value is the number of millilitres of 0·1 M aqueous alkali solution required to neutralise the water-soluble volatile fatty acids distilled from 5 g of the fat under the precise conditions specified in the method.

The Polenske value is the number of millilitres of 0·1 M aqueous alkali solution required to neutralise the water-insoluble volatile fatty acids distilled from 5 g of the fat under the precise conditions specified in the method.

The Kirschner value is the number of millilitres of 0·1 M aqueous alkali solution required to neutralise the water-soluble volatile fatty acids which form water-soluble silver salts distilled from 5 g of the fat under the precise conditions specified in the method.

B. APPARATUS

(i) Graduated cylinder 100 ml capacity, complying with BS 604 (ISO 4788).
(ii) Pipette of 50 ml capcity, complying with BS 1583, Class B.
(iii) Distillation apparatus as shown in Fig. 15.1.

C. REAGENTS

Glycerol.

Sodium hydroxide, 50 per cent (m/m) solution. Dissolve sodium hydroxide in an equal weight of water and store the solution in a bottle protected from carbon dioxide. Use the clear portion free from deposit.

Sulphuric acid, dilute. Dilute approximately 25 ml concentrated sulphuric acid to 1 litre and adjust until 40 ml neutralise 2 ml of the 50 per cent sodium hydroxide solution.

Pumice powder, ground, passing through a BS 50 mesh test sieve and remaining on a BS 90 mesh test sieve (BS 410: 1986).

Phenolphthalein indicator, 0·5 per cent solution in 95 per cent (v/v) ethanol (industrial methylated spirit is suitable).

Ethanol, 95 per cent (v/v), neutralised to phenolphthalein immediately before use.

Sodium hydroxide, approximately 0·1 M solution, accurately standardised.

Barium hydroxide, approximately 0·05 M solution, accurately standardised.

Silver sulphate, powdered.

D. PROCEDURE

Weigh 5 ± 0·01 g of the fat prepared as described above into a Polenske flask. Add 20 g of glycerol and 2 ml of the 50 per cent sodium hydroxide solution. Protect the burette containing the latter from carbon dioxide, and wipe its nozzle clean from carbonate deposit before withdrawing solution for test; reject the first few drops withdrawn from the burette. Heat the flask over a naked flame, with continuous mixing, until the fat is saponified, including any drops adhering to the upper parts of the flask and the liqud becomes perfectly clear; avoid overheating during this saponification. Cover the flask with a watch glass.

Make a blank test without fat, but using the same quantities of reagents and following the same procedure, again avoiding overheating during the heating with sodium hydroxide; such overheating would be indicated by darkening of the solution.

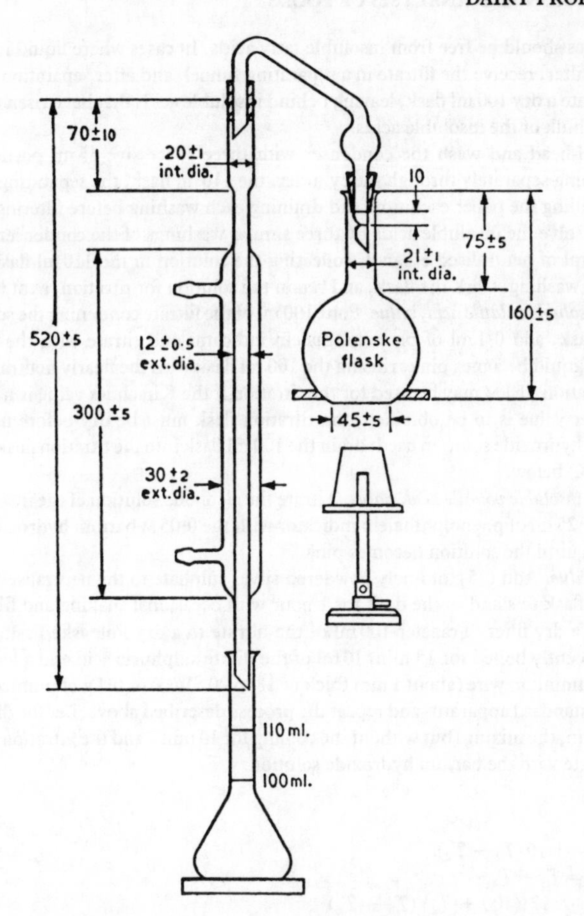

Fig. 15.1

Measure 93 ml of boiling water*, which has been vigorously boiled for 15 min, into a 100 ml graduated cylinder. When the soap is sufficiently cool to permit addition of the water, without loss, but before the soap has solidified, add the water, draining the cylinder for 5 s, and dissolve the soap. If the solution is not clear (indicating incomplete saponification) or is darker than light yellow (indicating overheating), repeat the saponification with a fresh sample of the fat.

Add 0·1 g of powdered pumice, followed by 50 ml of the dilute sulphuric acid, and connect the flask at once with the distillation apparatus. Heat the flask without boiling its contents, until the insoluble acids are completely melted, then increase the flame and distil 110 ml in between 19 and 21 min. Keep the water flowing in the condenser at a sufficient speed to maintain the temperature of the issuing distillate between 18 and 21 °C.

When the distillate reaches the 110 ml mark, remove the flame and replace the 110 ml flask by a cylinder of about 25 ml capacity, to catch drainings. Close the 110 ml flask with a stopper, and without mixing the contents place it in water at 15 °C for 10 min so as to immerse the 110 ml mark. Remove the flask from the water, dry the outside and invert the flask carefully, avoiding wetting the stopper with the insoluble acid. Mix the distillate by four or five double inversions, without violent shaking, filter through a dry 9 cm open-texture filter paper (a Whatman No. 4 or 41 paper is suitable) which fits snugly into a funnel. Reject the first runnings and collect 100 ml in a dry flask; cork the flask and retain the filtrate for titration as at (R)

* The temperature of the water in a cylinder rinsed out and refilled with boiling water is generally between 85 °C and 90 °C; 90 ml of cold water retained on draining, 93 ml altogether. Addition of hot water avoids loss of time in dissolving the soap.

below. The filtrates should be free from insoluble fatty acids. In cases where liquid insoluble fatty acids pass through the filter, receive the filtrate in a separating funnel, and after separation draw off the lower (aqueous) layer, into a dry 100 ml flask, leaving behind insoluble acids that have risen to the surface. Add these to the main bulk of the insoluble acids.

Detach the still-head and wash the condenser with three successive 15 ml portions of cold water, passing each washing separately through the cylinder, the 110 ml flask, the separating funnel if used and the filter, nearly filling the paper each time and draining each washing before filtering the next. Discard the washings. Dissolve the insoluble acids by three similar washings of the condenser, the cylinder, and the filter with 10 ml of neutralised ethanol, collecting the solution in the 110 ml flask and draining the ethanol after each washing. Cork the flask, and retain the solution for titration, as at (P) below.

(R) *Reichert, or soluble volatile acid, value.* Pour 100 ml of the filtrate containing the soluble volatile acids into a titration flask, add 0·1 ml of phenolphthalein indicator and titrate with the barium hydroxide solution until the liquid becomes pink, rinsing the 100 ml flask with the nearly neutralised liquid toward the end of the titration (0·1 M may be used for the titration if the Kirschner value is not required).

If the Kirschner value is to be obtained, the titration flask must be dry before use; note the actual volume of barium hydroxide solution used; drain the 100 ml flask into the titration flask, close with a cork, and continue at (K) below.

(P) *Polenske, or insoluble volatile acid, value.* Titrate the alcoholic solution of the insoluble volatile acids after addition of 0·25 ml of phenolphthalein indicator with the 0·05 M barium hydroxide or 0·1 M sodium hydoxide solution until the solution becomes pink.

(K) *Kirschner value.* Add 0·5 g of finely powdered silver sulphate to the neutralised solution from (R) above. Allow the flask to stand in the dark for 1 hour with occasional shaking and filter the contents in the dark through a dry filter. Transfer 100 ml of the filtrate to a dry Polenske flask, add 35 ml of cold distilled water, recently boiled for 15 min, 10 ml of the dilute sulphuric acid and a loosely-wound 5 mm coil of 30 cm of aluminium wire (about 1 mm thick or 18 to 20 SWG) or 0·1 g of pumice powder. Connect the flask with the standard apparatus and repeat the process described above, i.e. the distillation of 110 ml in from 19 to 21 min, the mixing (but without the cooling for 10 min), and the filtration and the titration of 100 ml of the filtrate with the barium hydroxide solution.

E. CALCULATIONS

Reichert value = $1{\cdot}10\,(T_1 - T_2)$.

Polenske value = $T_3 - T_4$.

$$\text{Kirschner value} = \frac{121(100 + T_1)\,(T_5 - T_6)}{10\,000}.$$

where T_1 = ml of 0·05 M barium hydroxide solution used for sample under (R).
T_2 = ml of 0·05 M barium hydroxide solution used for blank under (R).
T_3 = ml of 0·05 M barium or 0·1 M sodium hydroxide solution used for sample under (P).
T_4 = ml of 0·05 M barium or 0·1 M sodium hydroxide solution used for blank under (P).
T_5 = ml of 0·05 M barium hydroxide solution used for sample under (K).
T_6 = ml of 0·05 M barium hydroxide solution used for blank under (K).

Semi-micro method for determining Reichert, Polenske and Kirschner values

A semi-micro modification of the standard Reichert process, which is carried out on a 1 g sample, can be used. The IOCCC procedure is described in Regulation 924/74/EEC, which deals with trade arrangements applicable to certain processed agricultural products (OJ No. L111, 24.4.74, p. 1). Apart from the saving in materials the semi-micro method can be completed in a shorter time than the standard process. The method is particularly suitable where the amount of fat available for analysis is small, e.g. the fat extracted from butter confectionery.

APPARATUS

The shape of the apparatus is similar to that used for the standard Reichert process, but the still-head and condenser are scaled down to about one-third the size. The distilling flask has a capacity of 60–100 ml and the receiving flask is graduated at 22 ml.

REAGENTS

Sodium hydroxide solution. 50 per cent by weight.

Dilute sulphuric acid. Dilute 25 ml of conc. sulphuric acid to 1 litre and adjust so that 40 ml neutralises 2 ml of the 50 per cent sodium hydroxide solution.

PROCEDURE

For the *Reichert value*, weigh 1 g of filtered fat into the distilling flask, add 0·4 ml of the sodium hydroxide solution and 4 ml glycerol and saponify the fat by gently heating over a small flame, with continuous mixing, until the liquid clears (as in the standard process). Partially cool and dissolve the soap by the careful addition of 20 ml of hot water. Add 10 ml of the dilute sulphuric acid, swirl round and confirm that the liquid is acid by the addition of litmus paper. Add pumice powder and connect the flask to the distillation apparatus. Distil so that 22 ml of distillate are collected in 10–12 min. Remove the 22 ml flask at the end of the distillation and place a small beaker under the condenser. Immerse the 22 ml flask in cold water, filter through a small filter paper and, if only the Reichert and Polenske values are to be determined, titrate 10 ml filtrate with 0·02 M sodium hydroxide solution using phenolphthalein. If the Kirschner value is also to be determined, titrate 20 ml filtrate with 0·01 M barium hydroxide solution. If 10 ml filtrate is titrated:

$$\text{Reichert value} = 2{\cdot}2\ (\text{titre} - \text{blank}).$$

If 20 ml filtrate is titrated:

$$\text{Reichert value} = 1{\cdot}1\ (\text{titre} - \text{blank}).$$

For the *Polenske value* wash the condenser twice with 3-ml portions of cold water and pass each wash separately through the beaker, the 22 ml flask and the filter paper that was used for filtering the Reichert portion. Then dissolve the washed water-insoluble acids by washing three times with 3-ml portions of neutral alcohol, passing each wash through the condenser, the beaker, the 22 ml flask and finally through the filter paper into a small flask. Titrate the combined alcoholic washes with 0·02 M sodium hydroxide (or equivalent) using phenolphthalein.

$$\text{Polenske value} = (\text{titre} - \text{blank}).$$

For the *Kirschner value*, add 0·1 g of silver sulphate to the titrated solution from the Reichert determination and place in the dark for 1 h with occasional shaking. Filter, pipette 20 ml of the filtrate into the distilling flask and add 7 ml of water, 2 ml of the dilute sulphuric acid and two small pieces of aluminium wire. Distil and collect 22 ml of distillate. Stopper the 22 ml flask, mix the contents by inversion, filter and titrate 20 ml of filtrate with 0·01 M barium hydroxide using phenolphthalein (titration = a ml). If the blank titration is b ml and r was the number of ml of 0·01 M barium hydroxide added for the Reichert titration:

$$\text{Kirschner value} = \frac{(a - b) \times 1{\cdot}21 \times (20 + r)}{20}.$$

A semi-micro process has also been described in detail by Dyer *et al.* (1941).

Interpretation and use of R, P, K values

In the routine examination of butter the estimation of the Reichert and Polenske values generally suffices. With margarine, when the question of the presence or amount of butterfat arises, the Kirschner value is of next importance.

In the Reichert process genuine butterfat tends to flash clear suddenly during the saponifications, but margarine fat seldom does. Also, during the distillation the appearance of the insoluble acid is of interest. Thus in the presence of much coconut oil these are generally in the form of oily drops, whereas if there is much palm-kernel oil they take the form of white flakes. A genuine butter usually gives not less than 24·5 for the Reichert value. Siberian, south Russian butters and those from badly fed cows have been known to fall below this figure, occasionally even down to 20, but a sample giving a Reichert value of 24 or less should always be the subject of further investigation. The majority of genuine butters give a value of 26 or more, so that if the value found is, say, 23, there may be an addition of other fat or margarine. Consideration of the Polenske value is often of help at this stage. Coconut and palm-kernel oils are almost the only common fats that give appreciable Reichert or Polenske values but these are now rarely used in the UK as ingredients of a margarine mixture. There exists in genuine butter an approximate proportionality between the Reichert and Polenske and Kirschner values (Table 15.3).

Table 15.3 Relationship between the Reichert, Polenske and Kirschner values for butterfat

Reichert	Polenske	Kirschner
32	3·5	26·4
31	3·2	25·8
30	3·0	25·0
29	2·9	24·8
28	2·6	24·0
27	2·4	23·3
26	2·0	21·8
25	1·8	20·9
24	1·7	20·5
23	1·6	20·0

These values do not often fluctuate more than, say, 0·5 in the Polenske, and 0·6 in the Kirschner processes; hence a low Reichert value with a high Polenske value is suspicious and suggests the presence of another fat, whereas a low Reichert value with a correspondingly low Polenske value may be simply due to the origin of the butter. Since the Kirschner value is approximately a measure of butyric acid, it is often a great help in deciding whether a low Reichert value is due to abnormality or to adulteration. As an example, an unknown butter gives a Reichert value of 23, and Polenske value of 3·6. There is the possibility that it is abnormal, or that it is a mixture of 80 per cent butter having Reichert 26 and Polenske 2 with 20 per cent of palm-kernel oil, which has a Reichert value of 6, and Polenske value of 10. The Kirschner value was found to be 17·6, which is too low for a genuine butter. The normal Kirschner value corresponding to Reichert 23 is 20. Thus the Reichert and Polenske figures cast suspicion on the sample, and the Kirschner value showed that it is adulterated. If the Reichert–Polenske–Kirschner values are in the correct ratio, it is not safe to assume adulteration just because they are somewhat low. Positive proof of the adulterant should be sought by the phytosterol acetate test, the determination of iso-oleic acid, the various colour reactions of particular oils (Pearson, 1976) or more preferably by GLC of the fatty acids and sterols or by triglyceride profiles (e.g. Timms, 1980).

The Kirschner value is also a most useful indicator of the proportion of butter in margarine. Bolton *et al.* (1912) have shown that the proportion of butterfat in a margarine composed of fats other than coconut or palm-kernel is in agreement with the expression

$$K = 0{\cdot}244\,B + 0{\cdot}28, \text{ or } B = \frac{K - 0{\cdot}28}{0{\cdot}244}.$$

When palm-kernel or coconut oil is present, the proportion of butterfat in margarine may be deducted from the Reichert, Polenske and Kirschner values using the equation

$$\text{Butterfat (per cent)} = \frac{K - 0{\cdot}1P - 0{\cdot}24}{0{\cdot}244}.$$

Similarly, Williams (1949) has shown that the proportion of butterfat in admixture with coconut oil and palm-kernel oil can be deduced from graphs based on formulae of this type.

The above expressions are not strictly applicable when the amount of butterfat is much over 10 per cent, but as this is the maximum permissible in margarine it is unusual to find samples containing more than this quantity. Elsdon and Smith (1925) prefer to use the equation

$$\text{Butterfat (per cent)} = \frac{K - 0{\cdot}2 - 0{\cdot}1P}{0{\cdot}235}$$

and give a series of corrections according to the exact value of P. The most satisfactory results are obtained if the analyst works out his own data, using known mixtures.

Laruelle *et al.* (1976) have investigated the fatty acid composition of Reichert and Polenske fractions in butterfat by GLC.

The determination of RPK values is empirical and time consuming, and is approaching obsolescence because chromatographic methods are more specific and sensitive, and most food laboratories are now equipped with gas chromatographs.

Colorimetric methods

Bassette and Keeney (1956) have described a colorimetric procedure for the determination of butterfat based on the reaction of organic esters with hydroxylamine under alkaline conditions to form hydroxamic acid derivatives of the originally esterified organic acids. A red complex is formed in acid solution between the hydroxamic acids and ferric ions and its intensity is measured absorptiometrically. The method is described fully by Pearson (1976).

A method based on the colorimetric determination of tocopherol, which is present in comparatively small amounts in butterfat when compared with vegetable oils, has been described by Mahon and Chapman (1954).

Chromatographic methods for butterfat

Many workers have studied the fatty acid composition of butterfat by gas chromatographic separation of the methyl (Smith, 1961; Luddy *et al.*, 1968), ethyl (Anselmi *et al.*, 1960; Withington, 1967), butyl (Iyer *et al.*, 1967) or propyl (Grosjean and Fouassin, 1968) esters. Smith *et al.* (1978) investigated the unsaturated fatty acid isomers in the lipids of butter (and margarine) by separation of the methyl esters using argentation TLC into saturated, monoeic, dienoic and polyenoic fractions followed by GLC of each fraction.

Iverson and Sheppard (1977, 1986) have described the preparation of butyl esters for the GLC determination of fatty acids in butter. They found that conversion to the butyl esters was quantitative using either the boron trifluoride, sulphuric acid or sodium butoxide technique. Their 1986 paper claims complete recovery of the volatile short chain acids and gives fatty acid data on 50 samples. From fatty acid methyl ester data obtained by temperature programmed GLC, Muuse *et al.* (1986) list the composition of 1156 samples of Dutch butterfat. The distribution of fatty acids in butterfat, and in goat and human milk fat are given in Table 15.2. Adulteration can be deduced from the ratios of the proportions of certain fatty acids. The ratio of C12 to C10 acid is about 1·1 for butterfat, but over 8 for margarine fats (Dickes and Nicholas, 1972). The C14/C12 ratio is approximately 3–4 for butterfat. It should be borne in mind that such ratios may be affected by the animal's feed or oxidation of the fat during storage of the butter. The short chain fatty acids of milk fat from C4 to C10 are derived from synthesis in the mammary gland. The C12–C16 fatty acids are derived from both mammary gland synthesis and the animal diet, whereas the longer chain C18 fatty acids originate almost entirely from the diet. Kuksis and McCarthy (1964) have pointed out that the controlled addition of lard and palm or coconut oil can result in a chromatographic pattern resembling the glyceride distribution in butterfat. The seasonal and regional variations in the fatty acid composition of milk fat have been described by Hall (1970).

As with the Reichert value it is also possible to identify butter from other fats by the relatively high proportion of butyric acid (3·2–4·0 per cent; average 3·6 per cent) in butterfat. The determination of butyric acid can thus be used to estimate the butterfat and therefore milk, milk solids or cheese content of foodstuffs. Butyric acid in butterfat and oils and fat can be determined by temperature programmed gas chromatography of its methyl ester, methyl butyrate. IUPAC (1987) has issued a number of procedures for the preparation of fatty acid methyl esters and general guidance for the application of GLC to determine the qualitative and quantitative composition of a mixture of fatty acid methyl esters. Further details are given in Chapter 16. The following methylation procedure for lipids containing butterfat involves the reaction of the extracted lipid with methanolic potassium hydroxide as an intermediate stage of saponification. It is essentially that of Christopherson and Glass (1969).

PROCEDURE

In a test tube weigh to the nearest mg 1 g of the extracted lipid. Add 10 ml heptane and if required for quantitative purposes 1 ml of internal standard solution (0·5 per cent or 0·1 per cent methyl pentanoate in heptane, depending on whether butyric acid content is more or less than 1 per cent). Add 0·5 ml of methanolic potassium hydroxide 2 M, stopper and mix the content by inversions until the solution becomes clear (about 20 s). Allow the solution to become turbid as glycerol separates and decant the upper layer of heptane containig the methyl esters into another test tube, and store this injection solution in a refrigerator. Where methyl butyrate is determined the following conversion factors apply:

Per cent m/m methyl butyrate in fat

$$= \text{per cent m/m butyric acid in fat} \times \frac{102}{88}$$

$$= \text{per cent m/m butyric acid in the total fatty acids} \times \frac{102 \times 0{\cdot}95}{88}$$

(assuming fat contains 95 per cent m/m fatty acids)

$$= \text{per cent m/m methyl butyrate in the total methyl esters} \times 1{\cdot}005$$

(where 1·005 is the ratio of three times the average molecular weight of the component methyl esters to the average molecular weight of butterfat triglycerides)

Per cent butterfat = 24 × per cent methyl butyrate in the fat.

Phillips and Sanders (1968) have described a relatively simple procedure whereby the fat is saponified and after acidification in aqueous solution and filtering, the filtrate contains the water-soluble fatty acids in which butyric and caproic acids predominate. Kuzdzal-Savoie and Kuzdzal (1968) proposed a similar procedure; for details and collaborative study results of the two methods, see Pocklington and Hautfenne (1986).

The following, Method 2.310 (IUPAC, 1987) is a slight adaptation of the Phillips and Sanders procedure:

COLUMN

2 m × 6 mm o.d. glass containing 10 per cent FFAP on Chromosorb W (AW-HP, 80–100 mesh). Plug the ends of the column with glass wool previously washed with phosphoric acid. Condition the column for at least 48 hours at 225 °C or until peaks for *n*-valeric and *n*-butyric acids show little tailing. The peaks should be fully resolved and on a baseline flat enough to allow reliable peak height measurement to within 0·5 mm. The performance of the column can be improved by injections of 1 μl of phosphoric acid solution (2·5 per cent m/v) at the working conditions. Set the column oven at a temperature not greater than 150 °C and a suitable nitrogen carrier gas flow to give a retention time for *n*-butyric acid of about 5 min. Adjust the hydrogen and air flow to the flame ionisation detector to give maximum sensitivity at a detector temperature 20 °C or more above that of the column oven.

PROCEDURE

Weigh accurately into a 50 ml beaker 100–110 mg of the fat or extracted lipid from the foodstuff (chloroform/methanol). Add 3 ml of ethanolic potassium hydroxide solution 0·5 M, cover the beaker with a watch glass and place it on a steam bath. After 10 min or when no fat globules can be seen on the liquid surface, remove the watch glass and continue heating until the ethanol has completely evaporated. Allow the beaker to cool and add 5 ml of water. Cover with a watch glass and agitate carefully to completely dissolve the soap, with warming if necessary. Add 5 ml phosphoric acid solution (5 per cent m/v) and swirl gently to coagulate the precipitated fatty acids. Filter through a fluted filter paper and pipette 5 ml of the filtrate into a stoppered test-tube containing 2 ml of internal standard solution (0·25 mg/ml aqueous *n*-valeric acid). Inject 1 μl of the final solution onto the GLC column in its operating condition. From the chromatogram obtained, measure the peak heights of *n*-butyric and *n*-valeric acids.

CALIBRATION AND CALCULATION

Prepare a series of calibration solutions containing 0–5 ml water, 2 ml of internal standard solution and 0–5 ml of *n*-butyric acid standard solution (0·4 mg/ml aqueous) so that the final volume in each case is 7 ml. Inject 1 μl of each calibration solution and plot a graph of the weight of butyric acid in mg against the peak height ratio of *n*-butyric acid:*n*-valeric acid. From sample solution injections calculate the peak height ratio (R) and from the calibration graph deduce the equivalent weight of butyric acid (M mg). Then where W mg is the weight of the fat or extracted lipid taken initially,

$$\text{Per cent } n\text{-butyric acid in the fat} = \frac{2M \times 100}{W}$$

and assuming that butterfat contains on average 3·6 per cent butyric acid,

$$\text{per cent butterfat in the fat or extracted lipid} = \frac{2M \times 100 \times 100}{W \times 3{\cdot}60}$$

Butyric acid may also be determined by HPLC as its phenylethyl ester (Christie *et al.*, 1987) or as its *p*-bromo phenylacyl ester (Boley and Colwell, 1987). The latter method gives results very similar to the Phillips and Saunders GLC method.

Adulteration of butterfat

The adulteration of butter can be detected by identifying the presence of sitosterol, which occurs in vegetable oil but is absent in butterfat (Eisner *et al.*, 1962; La Croix, 1970; Thorpe, 1970). Pyle *et al.* (1976) have shown by GC/MS that the major sterols in margarine are β-sitosterol, stigmasterol, Δ^7-stigmasterol and campesterol, whilst cholesterol was the major sterol in butter (see Chapter 16 for the GLC of sterols).

Hydrogenated oils can be detected from the determination of the solid unsaturated fatty acids ('iso-oleic acid') (cf. Pearson, 1976). Elaidic acid, which is the *trans*-isomer corresponding to oleic acid, C18:1, and is termed 'iso-oleic acid' does not occur in butter, but constitues 25–50 per cent of the acids in hardened oils. A collaborative study using packed column GLC for *trans* unsaturation in margarine is reported by Gildenberg and Firestone (1985). Hydrogenated oils do not necessarily contain traces of nickel even though the metal has been used as catalyst.

The phytosterol acetate test, which involves the preparation of the sterol acetates and the determination of their mixed melting point (Pearson, 1976), gives an indication of the presence of vegetable oil or hydrogenated vegetable fats when admixed with animal fats. The latter contain cholesterol but no phytosterols and even though a butter may give a slight reaction in the Halphen test due to the cow having been fed on cotton-cake, such butter does not show the presence of any phytosterols. The presence of hydrogenated animal or vegetable fats can also be detected by the infra-red absorption of isolated *trans* unsaturated fatty acids (IUPAC, 1987; see also p. 633).

Microscopic examination of butter and margarine sometimes yields valuable information. Butter that has not been melted is devoid of crystalline structure, whereas margarine fat or renovated butter often shows crystalline forms. A fresh piece of the fat is pressed out under the cover slip on a microscope slide and examined with a low power objective, a Nicol prism and analyser being used. If extraneous light is carefully excluded, with pure butter all is dark, but with margarine or renovated butter bright portions and indefinite crystalline forms may usually be seen. Care should be taken to ascertain that the crystals are not those of salt by examining the slide without the polarising apparatus.

Effect of storage

Although butter cannot be classed as a highly perishable food the fat does deteriorate during chilled storage. Butter that has to be kept for some considerable time, as in the case of shipment over long distances, is stored at temperatures about −15 to −11 °C in order to reduce the activity of lipolytic enzymes. The free fatty acids (FFA) of the fat of fresh butter do not usually exceed 0·3 per cent as oleic acid and the peroxide value is less than 10 mEq/kg. Initially, the main spoilage reaction is usually hydrolysis, but this may be superseded by other reactions. Thus although an FFA figure in excess of 0·7 per cent would almost invariably be undesirable, some quite rancid butters give normal FFA figures. Complaint samples should therefore be examined for as many rancidity values as possible. The grading of butters according to various rancidity figures and pH are discussed by Piracox (1967). The separation of free fatty acids from fat in milk and milk products by mini-column chromatography, followed by determination by GLC, has been described by Deeth *et al.* (1983). The determination of FFA as *p*-bromo phenacyl esters by HPLC was given by Reed *et al.* (1984). The reactions involved in the deterioration of butter have been reviewed by Pont (1960). Using accelerated storage tests, Pont (1964) has shown that Swift test values show poor correlation with the keeping quality of butterfat and also with that of biscuits containing the butterfat.

MARGARINE AND OTHER FAT SPREADS

Composition

Margarine is a water-in-oil emulsion, consisting of a continuous oil phase and a finely dispersed discontinuous aqueous phase. The continuous phase contains flavour components, colouring matter and vitamins and also fat crystals. The physical properties of the margarine are primarily influenced by the type and quantity of the fat crystals. For bacteriological reasons the size of the water droplets must be small, and this is achieved by the addition of emulsifiers, usually soya lecithin and monoglycerides. This is particularly important in the low salt margarines preferred in continental European countries. To achieve the required palatability and mouthfeel, the emulsification should produce about 95 per cent of the water globules having diameter of 1–5 μm. A gram of margarine will contain 10 to 20×10^9 water globules. The control of the droplets size by emulsifiers also assists in preventing spattering when margarine is heated above 100 °C.

Annato and carotene are the commonly used colouring matters. Numerous blends of refined edible oils and fats, natural, modified, interesterified, fractionated or hydrogenated are used. The best quality margarines are manufactured usually from vegetable oils but in the UK in recent years approximately half of all the oils used have been partially hydrogenated marine oils. Animal fats are still used in some speciality margarines.

Some margarines contain 10 per cent butter and the required buttery flavour can be enhanced with diacetyl and acetyl methyl carbinol. Of more recent introduction because of the increase in refrigerator ownership are tub margarines, which contain relatively soft fat blends (low solid fat content) and are spreadable at 0–5 °C. The correlation between saturated fats and coronary heart disease and the desirability for more essential (polyunsaturated) fatty foods in the diet of health conscious consumers has led to the development and increasing consumption of soft tub margarines made from vegetable oils rich in linoleic acid (C18:2) such as sunflower seed oil. Numerous butter and margarine alternatives (low fat spreads) are now marketed to meet the increasing demand for low calorie foods. Most of these spreads contain half the fat of butter and margarine, and the fat may be blends of vegetable oils, partially hydrogenated vegetable oils and butter, butter oil, buttermilk or cream.

The manufacture of margarine is described by Opfer (1978). Microbiological aspects have been discussed by Muys (1969).

The fatty acid isomer, tocopherols and sterols composition of US margarines and spreads have been compiled by Slover *et al.* (1985). Using packed and capillary GLC, Druckrey *et al.* (1985) have reported on the fatty acids in Danish margarines. The composition of various Australian margarines have been described by Wills *et al.* (1982). Typical fatty acid composition of the various types of margarine and a low fat spread, all manufactured in the UK, are shown in Table 15.4. For comparison, Table 16.3 on page 620 gives the fatty acid composition of natural oils and fats used in food manufacture. Typical sterol composition of margarines is shown in Table 15.5.

Most samples of margarine contain over 80 per cent fat, nearly 16 per cent water, 0·6–1·5 per cent salt, about 1 per cent SNF (excluding salt) often described as 'curd', and the proportions of vitamins A and D required by regulation.

Legislation and standards

The Margarine Regulations, 1967 require all margarine to contain at least 80 per cent fat of which not more than one-tenth by weight may be milk fat, and not more than 16 per cent water and any margarine sold by retail to contain in each ounce 760–940 i.u. vitamin A and 80–100 i.u. vitamin D.

Labelling and advertising requirements include declaration of butter content (if claimed to be present), restrictions on the prominence given to the word 'margarine' in relation to the brand name or the words 'butter', 'cream' or 'milk', and that there must be no suggestion indicating anything connected with a dairy interest.

The Erucic Acid in Food Regulations 1977, SI No. 691 (see p. 000), apply particularly to margarine because of the wider use of rapeseed oil in margarine manufacture.

The Codex standard for margarine (CAC/RS 32–1969) also recommends a minimum fat content of 80 per cent and a maximum water content of 16 per cent. Control of the levels of vitamins (A, D and E) is left to the national legislation in each individual country. The other permitted additives include salt, sugars, various added colours and flavours, emulsifiers, preservatives (max. 1000 mg/kg of sorbic + benzoic acid), antioxidants (maxima prescribed) and various acids and alkalis. Maxima for trace metals are also prescribed (0·1 mg/kg for As, Cu and Pb and 1·5 mg/kg for Fe). The standard is supported by prescribed methods of analysis, e.g. Reichert, Polenske and Kirschner values, determination of fat, water, salt and vitamins.

Following recommendations in the Food Standards Committee *Report on Margarine and other Table Spreads*, FSC/REP/74, 1981, proposals for new regulations were issued in 1988. These cover controls on composition and labelling, removal of the restrictions on dairy interest connotations (since this is covered by EC Regulation 1898/87 on the protection of dairy designations), and vitamin fortification.

Analysis of margarine and spreads

The analysis of fat spreads follows closely on the lines described for butter. In view of the statutory requirements it is important to determine the water, fat and vitamin A contents. The rendered, filtered fat should give a low Reichert value although it will be raised slightly if butter is present as in some brands. The proportion of butterfat (max. 10 per cent) in the fat can be assessed from the Reichert–Polenske–Kirschner values or by the determination of butyric acid by GLC (p. 587). Rancidity values should be determined on the fat, which from freshly manufactured margarine should give an FFA figure of about 0·16 per cent calculated as oleic acid.

Examination for colours, preservatives and antioxidants may sometimes be advisable (see Codex code). The presence of carotene should be revealed in the vitamin A determination. A method for the detection of annatto and coal-tar dyes is given on p. 549. The commonest emulsifiers used are monoglycerides and lecithin. The examin-

Table 15.4 Typical fatty acid composition of margarine types

Type of margarine	% m/m in the total fatty acids											
	12:0	14:0	16:0	16:1	17:0	18:0	18:1	18:2	18:3	20:1	22:1	Trans
Packet												
Mixed oils (animal/fish/veg)	1·2–1·8	3–7	17–21	3–9	1·0	5–15	20–44	5–6	0·9–1·4	0·6–10	0–10	9–40
Vegetable oils	11	5	25	0·7	0·2	7	37	11	1·7	0·6	0·4	9
Vegetable oils high in PUFA	—	0·3	17	0·2	0·1	6	26	42	7	0·3	0·2	6
Tub												
Mixed oils (animal/fish/veg)	—	4–6	16	5–8	0·9–1·6	5–6	20–30	10–24	3	5–8	2–8	35
Vegetable oils	1·8	1·0	10–13	0·2–0·7	0·1	7–9	20–40	30–55	2–4	0·2–0·9	0–1	9–13
Vegetable oils high in PUFA	0–0·7	0·3–0·8	8–9	0·2	—	8–19	20–26	54–56	0·4–1·8	0·2–0·4	0·1	9–13
Low fat spread												
Vegetable oil	2·2	1·0	12	0·6	0·2	6	42	28	5	0·5	0·8	16

PUFA = Polyunsaturated fatty acids. Trans = Total isolated *trans*-isomer determined by infra-red absorption.

Table 15.5 Typical sterol content of packet and tub margarines (mg/100 g of lipid)

Type of margarine	Cholesterol	Campesterol	Stigmasterol	β-Sitosterol	Brassica-sterol	Other
Mixed oils (animal/fish/veg)	130–260	20–60	0–15	25–80	0–20	0–30
Vegetable oils	6–26	30–70	0–35	80–200	0–10	0–25
Vegetable oils high in PUFA	0–5	30–50	20–50	130–200	0	20–35

PUFA = polyunsaturated fatty acids.

ation of margarine for regulatory purposes is discussed by Fox and Minchinton (1972). Industrially, the examination of margarine for copper and physical properties is of special interest. The solid or crystalline fat content of the hydrogenated oils and the prepared oil blends is determined by low resolution NMR (Bosin and Marmor, 1968, see also Chapter 16).

LIPIDS

Information may be required on the nature of the fats present, whether they are of animal, vegetable or marine origin. Usually GLC of fatty acid methyl esters and of sterols gives this information. No single packed GLC column is capable of providing the full major fatty acid profile, but chromatography on a non-polar column such as Apiezon L and on a polar column such as butanediol succinate (BDS) or polyethylene glycol adipate (PEGA) will between them provide the fatty acid profiles shown in Table 15.4. Hydrogenation results in the loss of double bonds, some migration along the carbon chain and some stereochemical transformation from *cis*- to *trans*-isomers. Hence each unsaturated fatty acid in margarine lipids may incorporate a number of positional and geometric isomers. Fuller isomer information is obtainable by capillary GLC using wall-coated open tubular columns (Slover and Lanza, 1979). A collaborative study on GLC methods for *trans* unsaturation in margarine lipids is reported by Gildenberg and Firestone (1985). *Cis*- and *trans*-isomers in butter and margarine have been determined by a combination of argentation TLC and packed column GLC (Smith *et al.*, 1978). Similar procedures are used for the determination of erucic acid, *cis*-22:1ω9 (p. 636) and by IUPAC (1987) for *trans*-18:1 fatty acids. GLC methods for oil and fat analysis are given in Chapter 16.

It will be observed from Tables 15.4, 15.5 and 16.3 that a relatively high concentration of 16:1, 20:1 and 22:1 and high cholesterol but low brassicasterol is indicative of the presence of fish oils. The presence of brassicasterol and significant quantities of 20:1 and 22:1 indicates the use of rapeseed oil. A blend of animal and vegetable oils will show high cholesterol, moderate levels of 16:1 and in the absence of rapeseed oil, low 20:1 and 22:1. Vegetable oil margarines, particularly of the tub type, contain high concentrations of 18:2 and little 14:0 and 16:1. They also contain significant levels of β-sitosterol but little cholesterol.

The ratio of polyunsaturated fatty acids (PUFA) to saturated fatty acids is of nutritional interest. The predominant PUFA in dietary fats that is physiologically important is γ-linoleic acid (*cis-cis* 18:2 ω6) The total level of PUFA may be determined by GLC fatty acid analysis or by the determination of the *cis*, *cis*-methylene interrupted polyunsaturated fatty acids after saponification of the lipid by measuring the absorbance at 234 nm of the conjugated diene hydroperoxides produced by oxidation of the acids catalysed by the enzyme lipoxidase (Sheppard *et al.*, 1978).

The data provided by packed column GLC as shown in Table 15.4 gives no indication of the levels of *cis*- and *trans*-isomers. The *cis*-isomers in natural polyunsaturated vegetable oil fatty acids contain only non-conjugated (isolated) double bonds. On hydrogenation or other modification, these may change to isolated double bonds in *trans*-configuration, which absorb infra-red radiation at about 970 cm^{-1} (10·3 nm). This forms the basis of a method (IUPAC, 1987) for the determination of total isolated *trans*-isomers expressed as methyl elaidate (*trans*-18:1) in the methyl esters.

VITAMINS

A statutory method for determining the vitamin A content of margarine was formerly prescribed in the now revoked Food Standards (Margarine) Order, 1954. This prescribed that the vitamin A content should be calculated as the sum of the vitamin A present as such or as its esters plus 0·8 times the β-carotene equivalent of any carotenes present; any α-carotene being considered as equivalent in potency to half its weight of β-carotene; and when red palm oil is used as a source of carotenes, the β-carotene equivalent shall be taken as 53·5 per cent of the total carotenes.

The former prescribed method in the 1954 Order involves saponification of the margarine followed by extraction of the unsaponifiable matter with diethyl ether. The unsaponifiable matter is passed through a process of double column chromatography, which separates carotene from vitamin A. The optical density of the carotene solution is measured over the spectral range 440–450 nm and the vitamin A fractions, after being identified by the Carr–Price reaction, are combined in light petroleum. The optical density of the solution is measured at the wavelength of maximum absorption, normally at 324 nm.

The method recommended by Codex is that described in the AOAC methods book. In this procedure the unsaponifiable matter is passed through an adsorption chromatography column consisting of two segments of activated alumina separated by a central segment of alkaline alumina that removes interferences not capable of removal by normal alumina. The top section prevents caking of the column packing and initiates the separation of vitamin A from carotene and other interfering substances. The bottom segment of the column is non-fluorescent and permits the observation of the chromatographic separation and elution of vitamin A when irradiated with ultraviolet light.

In brief the method is as follows:

Alumina is activated by heating it in a muffle furnace at 600 °C for 3 hours. After cooling, water is added dropwise with shaking until the alumina contains 3 per cent m/m of water. Alkaline alumina is prepared by mixing alumina with equal weight of 10 per cent m/m aqueous potassium hydroxide. After decanting the excess liquid, the moist alumina is dried overnight at 100 °C. After cooling, water is added dropwise with shaking until the alkaline alumina contains 3 per cent m/m of water.

The column is a glass tube, 10 mm i.d. × 90 mm long, with top funnel end and a sealed-in medium porosity sintered glass disc above the lower drawn-out stem. Activated alumina is added with tapping to form the lower segment of height 10 mm, followed by 20 mm of alkaline alumina, and 40 mm of activated alumina for the top segment.

10 ± 0·1 g of margarine is saponified with 75 ml of 95 per cent alcohol and 25 ml of potassium hydroxide solution (50 per cent m/m, 780 g/l) by boiling for 5 min and allowing to stand for 20 min. The unsaponifiable matter is extracted with diethyl ether in a separating funnel. The ether extracts are washed with water and finally dried with anhydrous sodium sulphate. The dried ether extracts are evaporated

over a steam bath to a volume of about 25 ml, then transferred to a small beaker and evaporated until an oily solvent-free residue is left using nitrogen gas to remove the last vestige of solvent. When cool, 5 mm of petroleum ether 40/60 are added and the solution transferred quantitatively to a 10 ml volumetric flask and diluted to volume. Applying a slight vacuum to the column, 5 mm of petroleum ether 40/60, followed by 5 ml of sample solution, then a further 5 mm of petroleum ether are added to the column. As the last portion of solvent disappears, 5 ml portions of 16 per cent ether in petroleum ether are added until carotene is totally eluted, observed by the lack of yellowishness in 1 ml portions of eluate. The carotene fractions are later bulked and evaporated over steam to about 2 ml and finally at 30–40 °C with a stream of nitrogen. The carotene is dissolved in 5 ml petroleum ether and quantitatively transferred to a 10 ml volumetric flask and diluted to volume. Elution of the column is continued with 5 ml portions of 16 per cent ether in petroleum ether whilst observing the chromatographic separation of vitamin A by noting its fluorescence when irradiated with ultraviolet light. The vitamin A band should elute completely within 20 min of the start. The eluate following carotene and prior to vitamin A is discarded. Vitamin A is collected in a beaker, the start and finish being monitored closely by ultraviolet fluorescence. The solvent is removed as for carotene and if the residue is colourless it is dissolved in 5 ml absolute alcohol and transferred quantitatively to a 10 ml volumetric flask and diluted to volume with alcohol. If the residue is coloured, the AOAC method describes further clean-up on a magnesium oxide column.

The absorbance (A) of the carotene solution is measured in a 10 mm cell at 450 nm. If W = g sample/ml of solution,

$$\text{carotene content in } \mu g/g = \frac{A \times 4{\cdot}17}{W}$$

$$\text{or carotene content in i.u. of vitamin/g} = \frac{A \times 6{\cdot}95}{W}$$

The absorbance of the vitamin A solution is measured in a 10 mm cell at 325 nm, then when A_{325} is the absorbance at 325 nm

$$\text{vitamin A content in } \mu g/g = \frac{A_{325} \times 5{\cdot}5}{W}$$

$$\text{or vitamin A content in i.u. of vitamin A/g} = \frac{A_{325} \times 18{\cdot}3}{W}$$

A rapid result can be obtained by measuring the absorption at 620 nm of the colour produced on adding the antimony chloride reagent to a solution of the unsaponifiable matter, namely the Carr–Price method, p. 645.

Budowski and Bondi (1957) showed that by converting vitamin A to anhydrovitamin A in benzene solution and in the presence of toluene-*p*-sulphonic acid as catalyst, the increase in extinction at 399 nm is proportional to the vitamin A present. The method is applicable, without further purification, to unsaponifiable extracts of a variety of food including butter and margarine.

REAGENTS

Dissolve 15 mg toluene-*p*-sulphonic acid monohydrate in 100 ml redistilled benzene by heating under reflux. Distil off 10 ml solvent to remove water, cool whilst protecting from moisture, and adjust the volume to 100 ml with dry benzene. Activate before use by repeating the distillation.

PROCEDURE

Saponify the prepared fat and extract the unsaponifiable matter in the usual way. Make up the final extract in dry benzene so that 1 ml contains 5–50 i.u. Pipette 2 ml of the extract into each of two small stoppered cylinders. To one add 8 ml benzene (B) and to the other 8 ml of catalyst solution (C). Mix, allow to stand for 1 min only and deactivate the catalyst in C by shaking with 1 g sodium carbonate for 1 min. Then centrifuge and measure the optical density at 399 nm in a 1 cm cell against the blank B. If E_D is the increase in optical density produced by dehydration,

$$\text{Vitamin A in i.u. per ml of the solution of the unsaponifiable matter} = \frac{E_D}{0{\cdot}0122}$$

(0·0122 is equivalent to the increase in optical density caused by one i.u of vitamin A.)

Vitamins A and D are usually incorporated into margarine as a 'mastermix' concentrate consisting of synthetic vitamin A acetate or palmitate dissolved in vegetable oil together with ergocalciferol in the correct ratio. The determination of oil-soluble vitamins is also referred to in Chapter 16 where advances using HPLC techniques for the estimation of vitamins A and D are described.

CHEESE

Manufacture

About 250 000 tonnes of cheese are produced annually in the UK, approximately 75 per cent being Cheddar. In 1987, imports of cheese into the UK amounted to 172 000 tonnes consisting of 61 000 tonnes of Irish Cheddar, and (in '000 tonnes) Dutch 25, German 25, French 14, Danish 11 and Canadian 3.

Most types of cheese are made from curd produced from the coagulation of souring milk by *rennin*, an enzyme obtained from the inner lining of the fourth stomach of the calf. The difference in texture, properties and composition arise due to differing methods of production particularly during the ripening. Most of the recognised stages are included in the manufacture of Cheddar cheese (Scott, 1967a, b, 1972). Starter is added to pasteurised milk so that an acidity of 0·2 per cent lactic acid (pH 5·9–6·3) is produced. Rennet, the commercial extract containing rennin, is added and the curd formed at 29–31 °C. The curd or junket is cut and some of the whey (sweet whey) is removed by means of heat and acid development. After milling the curds are salted and pressed into moulds releasing more whey. Mild treatment produces a moister, softer cheese, whilst drier harder varieties with better keeping qualities are produced by applying more heat after cutting. Table 15.6

Table 15.6 The distribution of milk components in curd and whey, per cent m/m (after Davis and Macdonald, 1953)

	Milk	Curd	Whey
Water	87·3	6·5	80·8
Fat	3·7	3·5	0·2
Lactose	4·7	0·3	4·4
Casein	3·0	2·6	0·4
Albumin	0·4	Trace	0·4
Ash	0·7	0·1	0·6

shows the distribution of the various constituents of milk when converted into whey and curd. In making Cheddar, 10–15 per cent of the milk is converted into cheese, the remainder is whey, i.e. 1 kg of Cheddar cheese produces 9 litres of whey. In the UK, some 2 billion litres of whey are produced annually. Much of this is converted by drying and evaporation techniques, membrane processing such as ultrafiltration and reverse osmosis, and also demineralisation by ion exchange and electrodialysis, into a wide range of useful food ingredients (see p. 532). Because of the scarcity and expense of rennet, several alternative enzyme preparations have been tried. Wong *et al.* (1977) have compared the nutrient composition of Cheddar cheeses made from six commercial milk-clotting preparations of animal and fungal origin.

Varieties of cheese have been classified into types by Davis (1965).

Cream cheese from cream is normally unripened and has a buttery texture and Cottage cheese is coagulated by acid (pH 4·3–4·6) and produces acid whey. The composition of sweet and acid whey is given in Table 15.7. Processed cheese is

Table 15.7 The composition of sweet and acid whey, per cent m/m

Constituent	Sweet whey	Acid whey
Total solids	6·35	6·9
Moisture	93·65	93·5
Fat	0·3	0·04
Protein	0·8	0·75
Lactose	4·65	4·9
Minerals	0·5	0·8
Lactic acid	0·05	0·4

prepared by heating a cheese (usually a mild, hard variety) and emulsifying with water and salts such as sodium phoshate or citrate or sodium potassium tartrate. The heating improves the keeping qualities and wrapping in aluminium foil or plastic laminates reduces losses of moisture and deterioration of texture. The softer forms are called cheese spread. The addition of proteolytic enzymes can halve the normal maturation time of Cheddar and other cheeses (Law and Wigmore, 1983; Kirby *et al.*, 1987).

Legislation and standards

Following Food Standards Committee Reports in 1949, 1956 and 1962, statutory requirements for cheese (mainly applicable to retail sale) were prescribed in The Cheese Regulations 1970 (SI No. 94) together with The Cheese (Amendment) Regulations 1974 (SI No. 1122) and 1984 (SI No. 649), both dealing with additives and ingredients.

General standards for certain types of cheeses are included in the Codex code (CAC/MI-1973) for milk and milk products, which also includes recommended methods of analysis. The degree of acceptance of the recommended Codex standards for cheese by individual governments are included in a further code of principles (CAC/CI-C25, 1972).

Composition

The composition of various types of cheese is shown in Tables 15.8, 15.9 and 15.10. Other data can be found in Lampert (1970), IDF (1971) and Paul and Southgate (1978). The proportion of fat in the dry matter is important when considering the quality of cheese. If this is much below 50 per cent it suggests that the product has been prepared from partly skimmed milk. Another criterion of genuineness is that the fat/protein ratio should not be less than 1:1 (it is often as high as 1·25:1).

Table 15.8 Proposed compositional requirements for named varieties of cheese

Variety of cheese	Min. % milk fat in dry matter	Max. % water calculated on total weight of cheese	Min. % milk fat calculated on total weight of cheese
Blue Stilton	48	42	28
Brie	45	56	20
Caerphilly	48	46	26
Camembert	40	56	18
Cheddar	48	39	29
Cheshire	48	44	27
Coulommiers	40	56	18
Cream Cheese	90	50	45
Cream Cheese, Double			65
Danablu	50	47	27
Derby	48	42	28
Dunlop	48	44	27
Edam	40	46	22
Baby Edam	40	47	21
Loaf Edam	40	46	22
Baby Loaf Edam	40	47	21
Emmental (*Emmentaler*)	45	40	27
Gloucester	48	44	27
Double Gloucester	48	44	27
Gouda	48	43	27
Baby Gouda	48	45	26
Gruyere (*Greyerzer/Gruviere*)	45	38	28
Lancashire	48	48	25
Leicester	48	42	28
Parmesan	32	32	22
Roquefort	52	38	32
Saint Paulin	40	56	18
Wensleydale	48	46	26
White Stilton	48	46	26

Richmond has devised the following formula for calculating from the composition of the cheese the probable composition of the milk from which it was made:

$$\text{Fat in milk} = \frac{100\,F}{35{\cdot}4\,P + F} + 0{\cdot}25$$

where F and P are the percentages of fat and protein respectively in the cheese.

Although much depends on the variety, samples containing the higher levels of lactic acid, ash and salt may give rise to complaints of a taste of 'burning'. More generally however the distinctive flavour of cheese is due to a subtle balance of many constituents including the lactic acid, free fatty acids, products of protein degradation, alcohols, ketones and esters.

Histamine is a biologically active amine and ingestion of large quantities can result in food poisoning symptoms. Cheese and other fermented foods may contain significant quantities of histamine. Using a fluorimetric method, Chambers and Staruszkiewicz (1978) have reported on the analysis of 31 varieties of cheese.

The formation of mould growth in cheese is well known. The use of sorbic acid, nisin, effective packaging and control of temperature and humidity during storage assists in preventing or delaying the onset of such troubles.

Table 15.9 General composition of British cheeses (g/100 g)

Constituent		Lancashire (22)*	Leicester (27)	Caerphilly (20)	Cheshire (33)	Wensleydale (23)	Double Gloucester (22)	Blue Stilton (13)	White Stilton (3)	Derby (14)	Cottage (19)	Cheddar (15)
Moisture	Min.	38·1	34·3	39·3	37·1	38·6	34·8	32·5	44·4	35·2	75·6	33·9
	Max.	44·5	42·3	43·9	44·1	45·5	40·3	42·8	48·3	40·4	81·0	41·3
	Mean	41·7	37·5	41·8	40·6	41·5	37·3	38·8	45·8	38·0	79·1	36·0
Fat	Min.	29·0	30·2	29·4	29·8	28·8	31·5	31·0	30·2	30·5	2·4	30·0
	Max.	34·5	37·0	33·5	34·4	33·5	36·6	42·0	32·2	36·1	5·2	37·4
	Mean	31·0	33·7	31·3	31·4	31·5	34·0	35·0	31·3	33·9	3·9	34·4
Dry matter	Min.	55·5	57·7	56·1	55·9	54·5	59·7	57·2	51·7	59·6	19·0	58·7
	Max.	61·9	65·7	60·7	62·9	61·4	65·2	67·5	55·6	64·8	24·4	56·1
	Mean	58·3	62·3	58·2	59·4	58·5	62·7	61·2	54·2	62·0	20·9	64·0
Crude protein	Min.	22·0	22·3	20·9	21·8	21·7	23·1	18·5	19·3	23·2	11·6	24·0
	Max.	25·3	26·9	25·4	25·6	24·8	27·5	25·1	20·9	26·5	16·2	27·3
	Mean	23·3	24·3	23·2	24·0	23·3	24·6	22·6	19·9	24·2	13·8	25·5
Ash	Min.	2·5	2·8	2·3	2·3	2·4	2·8	3·1	2·4	2·7	1·1	2·5
	Max.	3·7	4·4	3·5	3·7	3·7	3·9	4·2	2·8	3·6	1·6	4·3
	Mean	3·0	3·4	2·8	2·9	2·8	3·3	3·6	2·6	3·2	1·3	3·6
Lactic acid	Min.	1·1	0·9	1·3	1·0	1·1	1·1	0·7	0·7	1·1	0·15	0·9
	Max.	1·7	1·5	1·6	1·6	1·6	1·6	1·1	1·1	1·3	0·40	1·3
	Mean	1·4	1·2	1·4	1·4	1·4	1·2	1·0	0·9	1·2	0·23	1·1
Salt	Min.	1·1	1·0	0·9	0·9	0·9	1·0	2·4	1·8	0·9	0·8	1·1
	Max.	2·2	2·3	1·9	2·0	2·1	2·1	2·9	2·0	1·7	1·4	2·4
	Mean	1·5	1·6	1·3	1·4	1·2	1·5	2·7	1·9	1·4	1·0	1·8

* Number of samples in survey. From Florence *et al.* (1984).

Table 15.10 Mineral composition of British cheeses (mg/100 g)

Mineral		Lancashire (22)*	Leicester (27)	Caerphilly (20)	Cheshire (33)	Wensleydale (23)	Double Gloucester (22)	Blue Stilton (13)	White Stiton (3)	Derby (14)	Cottage (19)	Cheddar (15)
Sodium	Min.	445	375	370	379	335	450	865	710	370	320	385
	Max.	860	895	655	780	925	905	1295	810	745	485	973
	Mean	589	653	511	551	503	588	1064	750	536	393	673
Potassium	Min.	82	66	90	72	71	75	100	101	75	48	84
	Max.	118	113	111	110	110	102	148	124	105	94	108
	Mean	96	89	95	94	94	91	126	110	93	77	88
Calcium	Min.	470	490	470	450	460	555	184	245	600	61	540
	Max.	705	795	690	690	720	795	390	270	770	100	950
	Mean	563	666	548	557	560	664	298	253	680	73	736
Phosphorus	Min.	365	390	350	365	370	405	205	245	405	140	370
	Max.	450	505	505	435	445	500	345	280	500	170	505
	Mean	397	451	396	397	400	449	293	262	448	152	454
Total carbohydrates (mainly lactose)	Min.	44	28	36	51	35	34	62	79	44	1250	47
	Max.	258	280	294	278	211	116	167	107	93	2660	238
	Mean	98	92	122	109	88	69	102	93	68	1950	80

* Number of samples in survey. From Florence *et al.* (1984).

The content of the B complex vitamins and vitamin C in retail cheeses have been reported by Scott and Bishop (1988a).

Analysis of cheese

The routine analysis of cheese should include the determinations of water and fat and hence the calculation of the fat in the dry matter. The rendered or extracted fat may require examination for the presence of fats other than butterfat. Also, particularly in view of legal requirements, it may be advisable to examine samples of cheese for added ingredients, e.g. salt, colouring matter (often annatto), emulsifiers and preservatives, in particular sorbic acid, nitrate, nitrite and possibly nisin and hexamine. Other possible determinations are those of ash, proteins, lactose, acidity and the pH value. Some cheeses have been found to contain copper and lead. Flavoured cheeses should be examined microscopically for the appropriate spice, e.g. celery, sage and if coloured, chemically for the presence of colouring matter.

The presence of alkaline phosphatase activity is indicative of the use of raw or poorly pasteurised milk in the production of the cheese. A screening method using phenolphthalein monophosphate as substrate, and a more quantitative method employing disodium phenyl phosphate and Gibbs reagent is described in the AOAC methods book (see also IDF Standard 53: 1969).

Methods of sampling are included in BS 809 (ISO 707) and some methods of analysis in BS 770. The examination of cheese for nisin is described in BS 4020 and de Ruig (1987) has reported the results of an international collaborative study for natamycin in cheese and rind, where an ultraviolet spectrophotometric method was unsatisfactory and a RP-HPLC method had poor reproducibility.

A collaborative study of three methods for chloride in cheese has been reported by Poortvliet and Horwitz (1982), and recommended the potentiometric method as reference method. Johnson and Olson (1985) compared Mohr, Volhard, ISE and Chloride analyser methods for chloride in seven cheeses and found Volhard the most accurate.

A collaborative study of a molybdenum blue method for phosphorus in twelve samples of processed cheese showed poor reproducibility (Steiger *et al.*, 1985).

Sampling. As cheese tends to dry out quickly on the outside care must be taken to ensure that the sample used for analysis is representative of the whole material. A grater should be used for hard cheeses prior to grinding in a mortar. Soft cheese should be ground in a mortar and all samples should be kept in tightly stoppered jars. The analysis should then be commenced as soon as possible.

Water. Weigh out 3 g sample in a dried dish containing sand and a flat-ended rod. Add water to assist with the spreading of the sample by means of the rod before drying on a boiling-water bath and then overnight in an oven at 100 °C. The reference method in BS 770 prescribes a drying temperature of 102 °C.

Fat. The fat can be determined by the Werner–Schmid process:

Weigh an amount of sample containing about 0·3–0·7 g fat (≡ approx. 1–2 g) into a 100-ml beaker and, using a rod, macerate thoroughly with a few drops of 0·88 ammonia. Then add 10 ml diluted hydrochloric acid (7 vol. conc. HCl + 3 vol. water), cover the beaker with a clock glass and heat carefully while stirring over a small flame or in a boiling-brine bath. When all the particles have dissolved, cool slightly, add 10 ml alcohol, mix, then cool and transfer the liquid down the rod into an extraction tube or separator rinsing in with several small volumes of ether (total 25 ml). Shake the tube vigorously for 1 min, then add 25 ml light petroleum, shake again and transfer the separated extract into a weighed flask. Repeat the extraction three more times, combining the extracts and finally remove the solvent and weigh the residual fat in the usual way.

The Gerber process can also be employed:

To a cheese butyrometer add 10 ml of Gerber sulphuric acid followed by a 6 mm layer of warm water (30–40 °C) and 3 g of sample. Then add 1 ml of amyl alcohol and sufficient of the warm water so that the tube is filled to the shoulder below the neck. Stopper, shake and invert the tube and place in a water bath at 65 °C for 3–10 min. Make sure that no solid particles remain and centrifuge at 1100 r.p.m. for 5 min. Return to the 65 °C bath and read off the percentage of fat directly on the scale. The method is given in detail in Part 2 of BS 696, which also describes the use of milk butyrometers using 0·9–1·1 g sample. Butyrometers designed for cream may also be employed using 5 g of sample.

Also calculate the fat in the dry matter.

Ash. Ignite 5 g at as low a temperature as possible. The ash may be used for the determination of salt, calcium or phosphorus.

Salt. Weigh 2 g sample into a 200–500 ml conical flask and add 25 ml of 0·05 M silver nitrate solution and 10 ml water. Disperse the cheese by swirling and warming to 80 °C, add 10 ml conc. nitric acid and boil gently until the precipitate is granular (about 10 min). To the hot solution add 0·5 g urea, mix, then cool and add 1 ml nitrobenzene and 50 ml water and titrate the unused silver nitrate with 0·05 M potassium thiocyanate solution using ferric alum as indicator.

$$1 \text{ ml } 0{\cdot}05 \text{ M } AgNO_3 \equiv 0{\cdot}002\,922 \text{ g NaCl}$$

NB Triacetin is a preferred coating agent since, unlike nitrobenzene, it is non-toxic and odourless.

The reference potentiometric method (BS 770: Part 4: 1989; ISO 5943–1988) is as follows. Weigh 2–5 g of prepared sample in a beaker, add 30 ml water (55 °C). Mix with blender, rinse blender with 10 ml water. Add 2–3 ml 4 M HNO_3, cool and insert silver electrode and a reference electrode. Titrate with standardised 0·1 M $AgNO_3$ until end-point nearly reached then cautiously titrate to end-point (V_1), i.e. when maximum potential difference between successive drop (0·05 ml) additions. Carry out blank titration (V_0).

$$\text{Per cent NaCl} = \frac{(V_1 - V_0) \times 0{\cdot}5844}{\text{wt of sample}}$$

Protein. Determine the total nitrogen on 1–2 g sample and calculate the crude protein as N × 6·38.

Acidity. Macerate 20 g with warm water (at 40 °C) to produce a total volume of 210 ml and filter. Titrate 25 ml filtrate (≡ 2·5 g) with 0·1 M sodium hydroxide using phenolphthalein and calculate the acidity as lactic acid (1 ml 0·1 M ≡ 0·0090 g lactic acid) or as ml 0·1 M alkali/100 g.

Emulsifying salts. Processed cheese and cheese spread are permitted to contain 'emulsifying salts', namely the sodium, potassium, calcium or ammonium salts of citric, phosphoric and tartaric acid. The Codex code recommends a maximum total level of 4 per cent, but restricts phosphates to 3 per cent.

Citrate and tartrate can be detected after shaking the cheese with hot water (50 °C), then adding a little dilute sulphuric acid and shaking again. Then after adding phosphotungstic acid solution, shaking and filtering, the filtrate can be examined for citrate and tartrate. Both qualitative and quantitative methods are given in detail in the AOAC methods book.

Phosphates can be detected and determined on the ash (see Codex code CAC/M1-1973, Standard No. B-12; IDF Standard 33B: 1982; or Chapter 2, p. 36). In considering the figure obtained from processed cheese as a measure of added emulsifying agent allowance should be made for the natural presence of phosphate in cheese. In this connection, hard-pressed cheeses contain about 0·5 per cent of phosphorus.

Occasionally, crystals of phosphates, citrates and tartrates have been observed in processed cheeses and these have sometimes been mistaken for glass (Leather, 1949).

Nitrate and nitrite can be determined using the Griess–Ilosvay method coupled with a cadmium reduction process (McKay, 1974). (See also p. 78). IDF Standard 84A: 1984 employs sulphanilimide and *N*-1-naphthalene-diamine HCl after cadmium reduction.

Type of fat. In order to detemine the Reichert value by the semi-micro method or for GLC analysis of the fatty acids and sterols it is necessary to extract some fat from the sample. Softer cheeses should be partially dried at 100 °C, cooled and well macerated with ether. With hard cheeses any rind present should be removed before cutting the sample into small pieces and shaking or macerating with ether. Fat obtained from the acid Werner–Schmid process tends to give a slightly lower Reichert value then when this is determined after rendering or ether extraction. Reichert values as low as 20 have been reported for fat in Roquefort, Danish Blue and Stilton cheeses, due presumably to loss of butyric acid during ripening. Up to 1 per cent vegetable oil appears to be permissible in Parmesan cheese due to impregnation of the wrapping cloths used.

ICE-CREAM

Manufacture and composition

Ice-cream consists basically of a mixture of milk, fat, sugars, stabiliser and emulsifier, flavour and colour. The milk solids (MSNF) or 'serum solids' is derived from liquid or skimmed milk or from the condensed or powdered forms supplemented by whey solids. The fat is either a blend of unhydrogenated and hydrogenated vegetable oil (e.g. palm or palm kernel oil) with a slip point of 28–37 °C, or, as in the case of dairy ice-cream, milk fat from cream, butter, anhydrous milk fat or butter oil. The sugars used are now mainly a blend of sucrose and 42 DE glucose syrup giving in the final product about 15 per cent sucrose and 4 per cent dried glucose syrup solids. Egg and lecithin may be employed for their emulsifying properties but GMS with α-monoglyceride content in the range 40–60 per cent, finds widest use. Stabilisers thicken the aqueous phase of ice-cream to give the desired viscosity and also appear to prevent the formation of ice crystals larger than 55 μm which are detectable on the palate. The main stabilisers in common use are guar gum, locust bean gum, carrageenan, sodium alginate, sodium carboxymethyl cellulose, pectin and xanthan gum. Apart from synthetic flavours some types of ice-cream include fruit, cocoa or coffee solids.

In the manufacture the mixture, in which the balance of MSNF and fat is critical, is usually pasteurised by one of the statutory methods of heat treatment. After cooling the mixture may be matured for a few hours and the texture further improved and the volume increased by whipping during freezing. 'Choc ices' are prepared by coating the frozen bricks with chocolate couverture in an enrober. For further details of the manufacture and technology of ice-cream the reader should refer to Hyde and Rothwell (1973), Rothwell (1985) and Arbuckle (1977).

Much ice-cream sold in the UK gives analytical figures which fall within the following ranges:

Constituents	% m/m
Total solids	35–40
Fat	6–12
Sugars (except lactose)	13–18
Milk solids not fat	7·5–11·5
Stabiliser, emulsifiers	0·3–1·0

The overrun (p. 605) is usually between 70 and 110 per cent (by volume). 'Soft-serve' ice-cream normally contains less solids and fat, less ice crystals and gives a lower overrun (40–80 per cent) than conventional ice-cream.

Crowhurst (1963b, c, 1964) has discussed the production of milk ice, water ices and mousse. Mousse has a more gelatinous texture than ice-cream and contains milk, sugar, glucose, syrup, oil, fruit, fruit juice or cocoa, gelatine, partial glycerol esters, carob gum, hydrolysed protein, flavour and colour. It is usually manufactured by producers of conventional frozen foods. Deep-freeze cabinets are now common domestic appliances and at their usual operating temperature of −18 °C, normal ice-cream is frozen hard and portions for meals and snacks are removed from bulk freezer packs with difficulty. This has led to the development of 'soft-scoop' ice-cream, which contains added constituents to lower the freezing point, usually glycerol or dextrose. Ice-cream is normally stored below −18 °C at the factory and below −2 °C in the retailer's cabinet.

The content of the B complex vitamins and vitamin C in ice-creams are reported by Scott and Bishop (1988b).

Legislation and standards

Following the FSC Report on the Ice-Cream Standard in 1957, compositional standards for ice-cream were prescribed in The Food Standards (Ice-Cream) Regulations, 1959, which were subsequently replaced by The Ice-Cream Regulations, 1967 (SI 1967 No. 1866).

The compositional standard in the 1967 Regulations include the general requirements that most ice-cream must contain at least 5 per cent fat and 7·5 per cent MSNF. Ice-cream containing fruit, fruit pulp, purée or juice must either conform to these general standards or the fat + MSNF must be at least 12·5 per cent, which must include not less than 7·5 per cent fat and 2 per cent MSNF. Any ice-cream described as 'dairy ice-cream', 'dairy cream ice' or 'cream ice' must conform to the above standards, whichever is appropriate, and contain no fat other than milk fat or any fat present by reason of the use as an ingredient of any egg, flavouring substance or emulsifying or stabilising agent. 'Milk ice' (including milk ice containing fruit) must contain at least 2·5 per cent milk fat (with some exceptions as above) and at least 7 per cent MSNF. 'Parev ice' (including 'Kosher ice') must contain (1) not less than 10 per cent fat and (2) no milk fat or other derivatives of milk. No ice-cream shall contain any artificial sweetener. All percentages in these regulations are calculated by weight.

If fat other than milk fat is present, the ice-cream shall be labelled with the words 'Contains non-milk fat', or 'Contains vegetable fat' if it contains only vegetable fat. Also labels and advertisements shall not bear any pictorial device that is suggestive of butter, cream or milk or of anything connected with the dairy interest unless the ice-cream contains no fat other than milk fat. It may however bear a statement to the effect that the ice-cream contains skimmed milk solids.

The methods of processing ice-cream are controlled by The Ice-Cream (Heat Treatment) Regulations, 1959, SI No. 734, as amended. For hot mix ice-cream the mixture shall not be kept for more than 1 h above 7 °C before pasteurisation or sterilisation by one of the following methods:

	1. at least 66 °C for at least 30 min.
Pasteurisation alternatives	2. at least 71 °C for at least 10 min.
	3. at least 79 °C for at least 15 s.
Sterilisation	at least 149 °C for at least 2 s.

After the heat treatment the mixture shall (with certain provisos) be reduced to not more than 7 °C within 1½ h and shall be kept at such a temperature until the freezing process is begun. Water ices and ice lollies, whether containing milk solids or not, that are mixed before freezing and have a pH value of 4·5 or less are exempted from the heat-treatment provisions as it is unlikely that pathogenic organisms will multiply in such acid mixtures. In addition to heat treatment, ice-cream must not be sold unless it has been kept at −2 °C or less since freezing. If the temperature rises above −2 °C after freezing, the mixture must be heat treated again and refrozen for sale. Where a complete cold mix is used the reconstituted mixture shall be converted into ice-cream within 1 h of reconstitution.

Analysis of ice-cream

Quality control in the larger manufacturers for the mix composition is now carried out by near infra-red reflectance spectroscopy or infra-red transmission spectroscopy.

The routine analysis of ice-cream should include the determination of fat, the examination of the type of fat present (especially if the product is labelled as being of the dairy type) the assessment of the MSNF and the microbiological purity. Some attention should also be given to the detection of artificial sweeteners, added colours, stabilisers and the determination of trace elements. A more detailed analysis would include the determinations of water, sugars, nitrogen, ash and acidity and an examination for starch, cocoa (from the total alkaloids in chocolate types), fruit and preservatives. Chocolate coatings should be examined for cocoa butter and milk fat. It should be borne in mind that fat from the ice-cream tends to migrate into the coating. Industrially, the overrun is of special interest. Methods of sampling are included in BS 809 (ISO 707) and some methods of analysis in BS 2472: 1986 (see also AOAC methods book).

Sampling. The sample should preferably be analysed immediately after receipt. Otherwise place the sample in a screw-capped jar at −15 °C or below. Prior to the analysis, melt the sample in a water bath at a temperature not exceeding 45 °C, shake and cool to room temperature.

Water. Weigh out 5 g sample into a dried dish containing sand and a flat-ended rod. Add 5 ml water to assist with the spreading of the sample by means of the rod before drying on a boiling-water bath and then for at least 2 h in an oven at 100 °C.

Fat. The fat can be determined by the Rose–Gottlieb process:

Weigh out about 5 g sample into a Rose–Gottlieb or Mojonnier extraction tube, add 1·5 ml of 0·88 ammonia, mix, add 7 ml warm water and re-mix. Maintain at 65 °C for 15 min, add 10 ml alcohol, mix again, cool and extract the fat by shaking with 25 ml ether. Then add 25 ml light petroleum, shake and continue as for milk.

With some types of ice-cream containing added food constituents the Werner–Schmid method may give a more efficient extraction of the fat:

Weigh out about 5 g sample into the extraction tube, add 2 ml hot water, 10 ml conc. hydrochloric acid and place in hot water (50 °C) for 30 min with occasional shaking. Then cool, add 10 ml alcohol and shake with 25 ml ether and then 25 ml light petroleum as for the Rose–Gottlieb method.

The Gerber and Macdonald rapid methods can also be employed using a milk, cheese or cream butyrometer (Hyde and Rothwell, 1973). The Gerber sulphuric acid can be conveniently replaced by a reagent containing equal volumes of acetic acid and perchloric acid (Pearson, 1973).

Butterfat. Products described as 'dairy' have to contain at least 5 per cent milk fat. The type of fat present in the extracted fat can be deduced from the methods described for butter and margarine. For the preparation of fat for examination, heat the ice-cream sample with ammonia/alcohol just to boiling and cool prior to extraction with solvents, or use chloroform/methanol (see Chapter 2).

Sugars. Determine the sugars before and after inversion after clearing with zinc ferrocyanide using Lane and Eynon's method. Calculate the sucrose content from the difference in titrations (p. 197). If the reducing sugars calculated as lactose equals approx. one-half of the MSNF (as assessed by the methods below) it can be assumed for routine purposes that no other reducing sugars are present. The use of TLC will give information on the sugar species present. In the presence of glucose syrup, full analysis for sugars will require HPLC or the use of a specific method described in Chapter 6. Evans *et al.* (1956) have shown that storage at room temperature may cause loss of sucrose due to the action of the diplococcus *Leuconostoc mesenteroides*, which converts it to dextran.

Protein. Determine on 5 g by the macro-Kjeldahl method and calculate the milk protein as N × 6·38.

Milk solids other than fat. The amount of MSNF can be calculated approximately from individual constituents such as the protein, casein, calcium and lactose. The subtraction for the constituent when present from other sources (e.g. calcium alginate, added lactose) is however usually difficult to assess.

The method of Crowhurst (1956) employing the formol titration is one of the simplest for assessing the MSNF content:

To 10 g of ice-cream in a porcelain dish add 1 ml of phenolphthalein solution and titrate the acidity with 0·1 M sodium hydroxide solution. Calculate the acidity as lactic acid (per cent m/m). Add exactly 3 ml of formalin to the neutralised mixture, stir thoroughly and titrate to neutrality again with 0·1 M sodium hydroxide (titration = x ml). Titrate separately 3 ml of the formalin solution (titration = y ml). Then:

$$\text{MSNF (per cent)} = 5{\cdot}67\,(x - y).$$

Note that this calculation assumes that the MSNF are those of skimmed milk in their natural proportions.

Crowhurst showed that the result is not affected by the presence of wheat flour and gelatine. Hill and Stone (1964) titrate the milk protein potentiometrically to an end-point of pH 8.5.

A method for the determination of casein in ice-cream is described in BS 2472. The MSNF can be obtained by multiplying the casein figure by 3 on the assumption that the MSNF is SMP.

For the determination of calcium, ash 10 g sample, take up the ash in hydrochloric acid, precipitate as oxalate and titrate with potassium permanganate. The dry fat-free solids of powdered milk contains about 1·95 per cent (as CaO).

Lactose can be determined by any of the specific methods described in Chapter 6.

Additives. Preservatives are seldom met with in ice-cream. Colouring matters can be detected by the methods described in Chapter 4. As the presence of artifical sweeteners is at present prohibited in ice-cream, suspect samples should be examined for likely additions.

Tests for several stabilisers are given in the AOAC methods book (see also Chapter 4). Glycerol may be determined by oxidation with dichromate (see Chapter 12 and the AOAC methods book) or enzymatically (BCL, 1979).

Overrun. When ice-cream is manufactured, air is whipped in to improve the texture and in consequence the volume of the solidified product is greater than the volume of the original liquid mix. The percentage increase in volume is referred to as the overrun, which can be calculated from the formula

$$\text{Per cent overrun} = \frac{\text{Volume of ice-cream} - \text{Volume of mix}}{\text{Volume of mix}} \times 100 = \frac{y - x}{y} \times 100$$

where x = sp. gr. of the solid ice-cream before melting and y = sp. gr. of the remelted ice-cream after removal of air.

Ice-cream is sold by volume and the overrun may be anything from 20–130 per cent. The FSC discussed the possibility of extending the legal provisions, which are calculated on a m/m basis, so that they included a standard such as a minimum weight of total solids per gallon. Such a provision is included in ice-cream standards in several overseas countries. In view of sampling and other difficulties they decided not to make a recommendation for its enforcement however. For plant control it is convenient to prepare a chart that shows the weight of a 'standard' vessel when filled with ice-cream at various known percentages of overrun (England, 1952). In the AOAC methods book the weight per unit volume of ice-cream is determined on the frozen sample, which is immersed in an overflow vessel filled with aqueous detergent solution and the volume is determined by measuring the amount displaced.

Methylene blue test and bacteriological examination

At the time of publication of the Ice-Cream (Heat Treatment, etc.) Regulations, 1959, the Ministry of Health (Circular 8/59) stated that no bacteriological test for ice-cream was precise enough for statutory purposes and recommended that local authorities should use the methylene blue test for the provisional measurement of bacterial cleanliness. The test has been described and discussed in various Reports of the Public Health Laboratory Service (1947, 1948, 1949, 1950). The main features of the methylene blue test are as follows (cf. Supplement No. 1 (1970) to BS 4285):

SAMPLING

For pre-wrapped ice-cream the sample should consist of one unopened original unit-block or carton. For loose or bulk ice-cream, it should be taken from the surface with the retailer's server. At least 2 fl oz should be put into a sterile wide-mouthed container leaving the minimum amount of air-space. The sampling should be arranged so that the ice-cream reaches the laboratory in not more than 6 h and the test should preferably be commenced at 5 p.m. on the same day. If necessary the sample should be packed in ice and if it has to be melted at room temperature this should not occupy more than 1 h. Alternatively, the sample, provided it has not melted, can be stored at −10 °C for up to 4 days before testing.

PROCEDURE

To a sterile reductase tube (marked at 10 ml) add 7 ml of ¼ strength Ringer solution, 1 ml of methylene blue solution (as used for milk, p. 000) and, using a pipette with a wide-bore tip, 2 ml of the melted sample. Insert a sterile rubber bung, invert the tube once and make up to the 10-ml mark with more of the ice-cream sample. Also prepare two control tubes:

(i) *Ice-cream colour.* To 8 ml of ¼ strength Ringer solution add 2 ml (or more) of the ice-cream so that the volume is 10 ml after inverting.

(ii) *Methylene blue colour.* Heat about 5 ml of ice-cream in a boiling-water bath for 15 min and prepare another tube as in the test but using this sterile sample.

Then incubate the test and control tubes for 17 h in a water bath at 20 °C. Note any decolorisation (at 0 h) and incubate in a water bath at 37 °C, reading and inverting once every ½ h. Note the time taken for complete decolorisation of the methylene blue and assign to it one of the following grades:

Provisional grade	Time taken to reduce methylene blue
1	Fails to reduce in 4 h
2	2½–4 h
3	½–2 h
4	0

The majority (at least 80 per cent) of samples from any one producer (preferably considered over a period) should fall into Grade 1 or 2. Although thermoduric organisms derived from ingredients such as milk powder may tend to put the ice-cream into a lower grade they are seldom responsible for repeated results at Grades 3 and 4 levels. The test is proposed for routine grading and is to be used essentially for indicating where further investigations into manufacturing practices are advisable. Crowhurst (1963a) has drawn attention to the poor correlation that exists between bacteriological counts and the methylene blue test. The same author (1966) also drew attention to the anomalies that could arise if the test is applied to liquid mixes for soft-serve ice-cream (see also Robinson, 1965).

The methylene blue test may be inapplicable to certain samples that contain added colours that interfere with the reading of the tubes, and also if constituents of the flavour tend to reduce the dye. Such effect should normally be revealed by inspection of the control tubes.

Confirmatory evidence of a low grading in the methylene blue test should be sought by carrying out plate counts and tests for coliforms. A maximum plate count of 50 000 organisms/ml seems a suitable general limit to adopt. Also if many samples of ice-cream from the same manufacturer fall into Grade 2 or below, the prevalent organisms should be identified. The bacteriological control of ice-cream manufacture has been discussed by Lloyd (1969a–c), who also comments on the interpretation of the various tests applied to the final product.

REFERENCES

Anselmi, S., Boniforti, L. & Monaccelli, R. (1960) *Bolletino dei Laboratori Clinici Provincali*, **11**, 317.
Arbuckle, W. S. (1977) *Ice Cream*, 3rd edn. Westport: AVI.

Bassette R. & Keeney, M. (1956) *Journal of the Association of Official Agricultural Chemists*, **39**, 469.
BCL (1979) *Methods of Enzymatic Food Analysis.* Lewes: Boehringer Corporation (London) Ltd.
Boley, N. P. & Colwell, R. K. (1987) *Journal of Chromatography*, **410**, 190.
Bolton, E. R., Richmond, H. D. & Revis, C. (1912) *Analyst*, **37**, 183.
Bosin, W. A. & Marmor, R. A. (1968) *Journal of the American Oil Chemists' Society*, **43**, 335.
Budowski, P. & Bondi, A. (1957) *Analyst*, **82**, 751.
Chambers, T. L. & Staruszkiewicz, W. F. (1987) *Journal of the Association of Official Analytical Chemists*, **61**, 1092.
Christie, W. W., Connor, K., Noble, R. C. & Shand, J. H. (1987) *Journal of Chromatography*, **390**, 444.
Christopherson, S. W. & Glass, R. L. (1969) *Journal of Dairy Science*, **52**, 1289.
Crowhurst, B. (1956) *Analyst*, **81**, 123.
Crowhurst, B. (1963a) *Dairy Industries*, **28**, 105.
Crowhurst, B. (1963b, c, 1964) *Ice Cream Industry*, **52**, 92, 290; **54**, 33.
Crowhurst, B. (1966) *Public Health Inspector*, **74**, 476.
Davis, J. G. (1965) *Cheese*, Vol. 1. London: Churchill.
Davis, J. G. & Macdonald, F. J. (1953) *Richmond's Dairy Chemistry*, 5th edn. London: Griffin.
Deeth, H. C., Fitzgerald, C. H. & Snow, A. J. (1983) *New Zealand Journal of Dairy Science and Technology*, **18**, 13.
Dickes, G. J. & Nicholas, P. V. (1972) *Journal of the Association of Public Analysts*, **10**, 87.
Druckrey, F., Hoy, C. E. & Holmer, G. (1985) *Felte Siefen Anstrichmittel*, **87**, 350.
Eisner, J., Wong, N. P., Firestone, D. & Bond, J. (1962) *Journal of the Association of Official Agricultural Chemists*, **45**, 337.
Elsdon, G. D. & Smith, P. (1925) *Analyst*, **50**, 53.
England, C. W. (1952) *Ice Cream Review*, **35**, 40.
Evans, H. J., Kwantes, W., Jenkins, D. C. & Phillips, J. I. (1956) *Analyst*, **81**, 204.
Florence, E., Milner, D. F. & Harris, W. M. (1984) *Journal of the Society of Dairy Technology*, **37**, 13.
Fox, M. & Minchinton, I. R. (1972) *Food Technology in Australia*, **24**, 70.
Gildenberg, L. & Firestone, D. (1985) *Journal of the Association of Official Analytical Chemists*, **68**, 46.
Grosjean, M. H. & Fouassin, A. (1968) *Revue des Fermentations et des Industries Alimentaires*, **23**, 57.
Hall, A. J. (1970) *Dairy Industries*, January, 20.
Hill, R. L. & Stone, W. K. (1964) *Journal of Dairy Science*, **47**, 1014.
Hyde, K. A. & Rothwell, J. (1973) *Ice Cream.* Edinburgh: Churchill Livingstone.
IDF (1971) *Annual Bulletin Part IV, Preliminary Catalogue of Cheese Varieties.* Brussels: International Dairy Federation.
IUPAC (1987) *Standard Methods for the Analysis of Oils, Fats and Derivatives*, 7th edn. Oxford: Blackwell Scientific Publications.
Iverson, J. L. & Sheppard, A. J. (1977) *Journal of the Association of Official Analytical Chemists*, **60**, 284.
Iverson, J. L. & Sheppard, A. J. (1986) *Food Chemistry*, **21**, 223.
Iyer, M., Richardson, T., Amandson, C. H. & Boudreau, A. (1967) *Journal of Dairy Science*, **50**, 285.
Johnson, M. E. & Olson, N. F. (1985) *Journal of Dairy Science*, **48**, 1020.
Kirby, C. J., Brooker, B. E. & Law, B. A. (1987) *International Journal of Food Science and Technology*, **22**, 355.
Kuksis, A. & McCarthy, M. J. (1964) *Journal of the American Oil Chemists' Society*, **41**, 17.
Kuzdzal-Savoie, K. & Kuzdzal, K. (1968) *Le Lait*, No. 475-6, 255.
La Croix, D. E. (1970) *Journal of the Association of Official Analytical Chemists*, **53**, 535.
Lampert, L. M. (1970) *Modern Dairy Products.* London: Food Trade Press.
Laruelle, L., Van Dijck, M. & Daenens, P. (1976) *Journal of Dairy Research*, **43**, 137.
Law, B. A. & Wigmore, A. S. (1983) *Journal of Dairy Research*, **50**, 519.
Leather, A. N. (1949) *Analyst*, **74**, 51.
Lloyd, T. P. (1969a, b, c) *Dairy Industries*, **34**, 199, 271, 363.
Luddy, F. E., Barford, R. A., Herb, S. F. & Magidman, P. (1968) *Journal of the American Oil Chemists' Society*, **45**, 549.
Mahon, J. H. & Chapman, R. A. (1954) *Analytical Chemistry*, **26**, 1195.
Marschke, R., Snow, A. & Tommerup, J. (1979) *Australian Journal of Dairy Technology*, **34**, 36.
McKay, A. J. (1974) *Australian Journal of Dairy Technology*, **29**, 34.
Muuse, B. G., Werdmuller, G. A., Geerts, J. P. & de Knegt, R. J. (1986) *Netherlands Milk and Dairy Journal*, **40**, 189.
Muys, G. T. (1969) *Process Biochemistry*, **4**, 31.
Opfer, W. B. (1978) *Chemistry and Industry*, 681.
Owades, J. L. & Jakovac, J. A. (1963) *Proceedings of the American Society of Brewing Chemists*, p. 2.
Parodi, P. W. (1965) *Australian Journal of Dairy Technology*, **20**, 59.
Paul, A. A. & Southgate, D. A. T. (1978) *The Composition of Foods.* London: HMSO.

Pearson, D. (1973) *Laboratory Techniques in Food Analysis.* London: Butterworth.
Pearson, D. (1976) *The Chemical Analysis of Foods.* Edinburgh: Churchill Livingstone.
Phillips, A. R. & Sanders, B. J. (1968) *Journal of the Association of Public Analysts*, **6**, 89.
Piracox, E. F. (1967) *Chemical Abstracts*, **67**, 89851g.
Pocklington, W. D. & Hautfenne, A. (1986) *Pure and Applied Chemistry*, **58**, 1419.
Pont, E. G. (1960) *Journal of Dairy Research*, **27**, 140.
Pont, E. G. (1964) *Australian Journal of Dairy Technology*, **19**, 108.
Poortvliet, L. J. & Horwitz, W. (1982) *Journal of the Association of Official Analytical Chemists*, **65**, 1350.
Prill, E. A. & Hammer, B. W. (1938) *Iowa State College Journal of Science*, **12**, 385.
Public Health Laboratory Service (1947, 1948, 1949, 1950) *Monthly Bulletin of the Ministry of Health*, **6**, 60; **7**, 84; **8**, 155; **9**, 228.
Pyle, C. A., Holland, P. T. & Payne, E. (1976) *Journal of the Science of Food and Agriculture*, **27**, 219.
Reed, A. W., Delth, M. C. & Clegg, D. E. (1984) *Journal of the Association of Official Analytical Chemists*, **67**, 718.
Robinson, S. V. (1965) *Public Health Inspector*, **73**, 271.
Rothwell, J. (1966) *Process Biochemistry*, **1**, 207.
Rothwell, J. (1985) *Ice Cream Making.* Reading: J. Rothwell.
de Ruig, W. G. (1987) *Journal of the Association of Official Analytical Chemists*, **70**, 949.
Scott, K. J. & Bishop, D. R. (1988a, b) *Journal of the Science of Food and Agriculture*, **43**, 187, 193.
Scott, R. (1967a, b, 1972) *Process Biochemistry*, **2**, 5; **2**, 23; **7**, 33.
Sheppard, A. J., Waltking, A. E., Zmachinski, H. & Jones, S. T. (1978) *Journal of the Association of Official Analytical Chemists*, **61**, 1419.
Slover, H. T. & Lanza, E. (1979) *Journal of the American Oil Chemists' Society*, **56**, 933.
Slover, H. T., Thompson, R. H., Davis, C. S. & Merola, G. V. (1985) *Journal of the American Oil Chemists' Society*, **62**, 775.
Smith, L. M. (1961) *Journal of Dairy Science*, **44**, 607.
Smith, L. M., Dunkley, W. L., Franke, A. & Dairiki, T. (1978) *Journal of the American Oil Chemists' Society*, **55**, 257.
SPA (1936) *Analyst*, **61**, 404.
Steiger, G., Horwitz, W., Poortvliet, L. J. & Soderheim, P. (1985) *Journal of the Association of Official Analytical Chemists*, **68**, 102.
Thorpe, G. W. (1970) *Journal of the Association of Official Analytical Chemists*, **53**, 623.
Timms, R. E. (1980) *Journal of Dairy Research*, **47**, 295.
Veringa, H. A., Verburg, E. H & Stadhouders, J. (1984) *Netherlands Milk and Dairy Journal*, **38**, 251.
Walsh, B. & Cogan, T. M. (1974) *Journal of Dairy Research*, **41**, 31.
Walstra, P. (1965) *Nederlandsch melk-en zuiveltijdschrift*, **19**, 1.
Williams, G. R. (1984) *Journal of the Society of Dairy Technology*, **37**, 66.
Williams, K. A. (1949) *Analyst*, **74**, 508.
Wills, R. B. H., Myers, P. R. & Greenfield, H. (1982) *Food Technology Australia*, **34**, 240.
Withington, D. F. (1967) *Analyst*, **92**, 705.
Wong, N. P., LaCroix, D. E., Vestal, J. H. & Alford, J. A. (1977) *Journal of Dairy Science*, **60**, 1522.

16

Oils and fats

Oils and fats are important parts of the human diet and more than 90 per cent of the world production from vegetable, animal and marine sources is used as food or as an ingredient in food products. Oils and fats serve as a rich source of dietary energy. They contain certain fatty acid components that are essential nutrients (FAO, 1978) and their functional and textural characteristics contribute to the flavour and palatability of many natural and prepared foods.

Modern advances in oils and fats technology and the science of nutrition have led to the need for greater awareness of the composition and structure of dietary lipids and many new and advanced test methods and analytical procedures have been introduced. This chapter describes the general methods used to examine oils and fats for their physical properties and more basic analytical criteria, new procedures to give detailed information on their fatty acid structure and composition, and methods for assessing their quality criteria. Information is also given on the more commonly used dietary oils and fats on methods of analysis for the determination of oil-soluble vitamins. The determination of antioxidants is described in Chapter 4.

Gunstone *et al.* (1986) have produced a comprehensive handbook on lipids and aspects on health, diet and analysis are described by Perkins and Visek (1983). Industrial oils and fats technology is covered by Swern (1979, 1982), Hamilton and Bhati (1980) and Hamilton (1987), and the chemistry of refining by Swoboda (1985).

SOURCES, UTILISATION AND COMPOSITION

LARD

Codex Alimentarius has recommended international standards for lard (Codex Stan 28–1981) and rendered pork fat (Codex Stan 29–1981). Lard is defined as the fat rendered from fresh, sound fatty tissues from swine (*Sus scrofa*). With lard the source of fatty tissue is restricted, but rendered pork fat may be derived from many parts of the animal including clean bones, detached skin, head skin, ears and tails. The main Codex analytical standards are summarised in Table 16.1. In addition there is control over antioxidants, antioxidant synergists, and the amounts of impurities and soaps.

Regulation 565/75/EEC (OJ No. L60, 6.3.75, p. 7) on the characteristics of lard and other pig fat for correct tariff classification requires the following:

Bömer index, 72 minimum.

Table 16.1 Recommended Codex standards for animal fats

	Lard	Rendered Pork Fat	Dripping	Premier Jus
Relative density (40 °C/water at 20 °C)	0·896–0·904	0·894–0·906	0·893–0·904	0·893–0·898
Refractive index at 40 °C	1·448–1·460	1·448–1·461	1·448–1·460	1·448–1·460
Titre (°C)	32–45	32–45	40–49	42·5–47
Saponification value (mg KOH/g)	192–203	192–203	190–202	190–200
Iodine value (Wijs)	45–70	45–70	32–50	32–47
Unsaponifiable matter (g/kg) (max.)	10	12	12	10
Acid value (mg KOH/g) (max.)	1·3	2·5	2·5	2
Peroxide value (mEq/kg) (max.)	10	16	16	10

For all four fats: matter volatile at 105 °C, max., 0·3 per cent; and mg/kg max., iron 1·5, copper 0·4, lead 0·1, arsenic 0·1.

$$\text{Fatty acid ratios, } \frac{\text{total C14} + \text{total C15}}{\text{C16:0}} \times 100\text{, not more than 10.}$$

$$\frac{\text{C18:3}}{\text{C16:0}}\text{, not more than 8.}$$

where total C14 and total C15 represent the total content of acids with 14 and 15 carbon atoms respectively as determined by GLC.

Regulation 171/78/EEC (OJ No. L25, 31.1.78, p. 21) on the conditions for granting export refunds requires lard and other pig fat to have the following:

Bömer index, 73 min.
Peroxide value, 6 mEq/kg max.
FFA, 1·5 per cent as oleic acid max.
Water and dirt content, 0·5 per cent max.

Lard consists essentially of the glycerides of the following fatty acids: oleic, linolenic, lauric, myristic, palmitic and stearic (see Table 16.3, p. 620). The melting point varies from 33 to 45 °C and may give some indication of the origin of the lard. The cholesterol level of normal lard is about 100 mg/100 g but rendered pig carcass fat contains up to 600 mg/100 g.

Analysis of lard

In the examination of lard, the determination of the constants and values shown in Table 16.1 will normally suffice to confirm the identity and quality of the sample but to test for the presence of adulterants, Bömer index, fatty acid ratios and possibly sterol analysis should also be carried out.

Lard should also be examined for the presence of antioxidants and for rancidity values and FFA.

Water. If a sample of lard appears cloudy on melting the presence of water is suggested and the amount should be determined (p. 624).

Iodine value (p. 626). The iodine value of lard varies according to the part of the animal from which the fat was obtained. It has also been shown that the iodine value of lard from young hogs is higher than that of mature animals and oils assimilated by the pig are transferred to its fat with little change in the fatty acids. Figures in excess of 90 have been reported for the iodine value of lard obtained from pigs fed on soya beans (iodine value of soya oil = 128). Sutton *et al.* (1940) have presented a useful survey of the constants of lard. All but 5 of the 113 samples that they examined had iodine values within the range 57·0–72·9. In view of the above findings, samples of lard with iodine values in excess of the Codex maximum of 70 cannot be regarded as necessarily adulterated with vegetable oil.

Bömer index. This is a useful test for the presence of beef fat based on the difference in the melting points of the glycerides and fatty acids.

PROCEDURE

Shake 5 g filtered lard in a cylinder with 20 ml warm (30 °C) acetone and hold at 30 °C overnight. Not more than 3 ml crystals should form. Decant off the supernatant acetone. Add 3 × 5 ml portions warm acetone and decant off the first two portions. Shake the third portion vigorously and rapidly transfer to a filter paper. Wash through with five successive small portions of warm acetone, air-dry the crystals and determine the m.p. (A). If the m.p. of the glycerides is below 63·6 °C the presence of beef fat or other fat containing tristearin is suspected.

For confirmation transfer the crystallised glycerides to a 50 ml beaker and saponify by heating on a steam bath with 25 ml of 0·5 M alcoholic potassium hydroxide. Pour into a separator containing 200 ml water, acidify with dilute hydrochloric acid and shake with 75 ml ether. Wash the upper ether layer at least three times with water. Transfer the solution of extracted acids to a beaker, remove the solvent and dry at 100 °C. Allow the acid to remain at room temperature for about 2 h before determining the m.p. (B).

$$\text{Bömer index} = A + 2(A - B).$$

If the Bömer index is less than 73 the lard is regarded as being adulterated. The interpretation of Bömer index values has been discussed by Sutton *et al.* (1940).

Adulterants. The presence of vegetable fats is readily confirmed by sterol analysis (p. 634). Modern lard substitutes and cooking fats are manufactured from hydrogenated oils or fractionated fats of animal and vegetable origin. Their presence should be identifiable by GLC fatty acid and sterol analysis. Resulting from processing, hydrogenated or hardened oils contain in their triglycerides a proportion of iso-oleic acid (*trans*-octadecenoic acid, elaidic acid). Therefore, the detection of significant amounts of iso-oleic acid in lard is a definite indication of the presence of hydrogenated oils. An argentation TLC/GLC procedure (2.208) for the determination of *trans*-octadecenoic acids is described by IUPAC (1987). *Trans*-acid can also be detected by infra-red spectrophotometry (p. 633). Lard from North American pigs tends to be rather softer than European lard, and blending with hydrogenated fats has been used in the past to make a firmer product. The presence of excessive amounts of foreign or hydrogenated fats will affect the normal constants of lard such as its iodine value, refractive index and melting point.

Beef or mutton fat when present in larger than low percentage amounts can be detected from the GLC fatty acid profile (p. 629) since ruminant fats contain small but significantly greater amounts of branched and odd carbon number fatty acids than normal pig fats (see Table 16.3, p. 620; Paul and Southgate, 1978, 1980) However, Bastijns (1970) has shown that significant changes in lard branched chain fatty acids occur when pigs are fed with diets containing beef fat. He has also pointed out that the antioxidant BHT has the same chromatographic retention time as the

C14 branched chain acid. The fatty acids in the 2-position of the triglycerides of pork fat are significantly different in proportions to those of bovine fat (see under 'Dripping'). Triglyceride profiles by HPLC can also be indicative of adulteration (see p. 635).

The Bömer index and fatty acid ratios (see above) give satisfactory indications whether lard or other pig fat contains more than only small quantities of other fats.

DRIPPING

Dripping (edible tallow) is the rendered fat from various tissue and bones of bovine animals (*Bos taurus*) and/or sheep (*Ovis aries*). Standards are given in Codex Stan 30–1981 and are listed in Table 16.1. The fatty acid profile of edible tallow is given in Table 16.3. In the kitchen the fat that separates from any cooked meat is also referred to as dripping. Regulation 1352/76/EEC on tariff classifications requires that tallows of bovine cattle, sheep or goats that contain small quantities of fats from parts of the animal carcass other than those used for obtaining tallow, should have a titre of 40 °C or higher (p. 625) and an FAC standard colour index of 21 or less (p. 625). The presence of 5 per cent or more of pork fat in beef or mutton fat can be detected and quantified by analysis of the fatty acids in the 2-position of the triglycerides (see p. 633). The ratios C16:0/C18:1ω9 for lard is about 5 whereas for edible tallow it is about 0·4. It is not possible to differentiate analytically between beef and mutton fats. The Antioxidants in Food Regulations 1978 permit the addition of antioxidants to dripping.

PREMIER JUS, SUET

Suet is the fatty tissue situated about the loins and kidneys of oxen and sheep. It consists of microscopic fat particles each contained in a membranous bag. Its composition is about 1 per cent of cellular tissue with some 4 per cent of water and 95 per cent of fat. Premier jus (oleo stock, block suet) is defined in Codex Stan 30–1981 as the product obtained by rendering at low heat the fresh fat (killing fat) of heart, caul, kidney and mesentery collected at the time of slaughter of bovine animals. Standards are listed in Table 16.1 and fatty acid composition shown in Table 16.3.

Because of its hardness and fibrous nature, suet can be shredded or grated but proprietary shredded suet is made from block suet and the particles coated with rice or wheat flour to prevent coagulation of the shreds. Typical analytical values are moisture 1·5–2·7 per cent, dry flour 12–15 per cent and beef fat 83–86 per cent. The Food Standards (Suet) Order 1952 SI No. 2203 requires that block suet shall be free from fibrous tissue and shall contain not less than 99 per cent of beef fat, and that shredded suet shall also be free from fibrous tissue, and contain not less than 83 per cent of beef fat. Since both are rendered fats and therefore almost anhydrous, the Antioxidants in Food Regulations 1978 apply (cf. FSC, 1970).

Premier jus can be fractionated under pressure to produce hard beef stearine and a soft buttery fraction (oleo oil) used for margarine manufacture (p. 589). Suets have largely been replaced for domestic use by shortenings and cooking fats.

Analysis of shredded suet

Sampling. As some of the cereal tends to separate it is especially important to mix the whole sample to ensure uniformity.

Water. Dry 5 g at 100 °C for 2 to 3 h or use Codex method (p. 624).

Fat. Filter a weighed quantity on a tared filter in the 100 °C oven until most of the fat has run through into a Soxhlet flask. Then complete the extraction in a Soxhlet apparatus using light petroleum. Then determine the iodine value and other constants or fatty acid profile to confirm the presence of beef fat and also the FFA and peroxide value to assess the edibility. It is not yet possible to differentiate analytically between beef and mutton fat.

Starch. The dry cereal filler can be obtained by difference or by weighing the residue from the fat extraction. Identify the starch microscopically.

Antioxidants. Test by method decribed in Chapter 4.

COOKING FATS, SHORTENINGS

'Cooking fats' were defined in The Oils and Fats Order, 1949, no longer extant, as 'any fat or mixtures of oils and fats fit for human food or for use in manufacture, preparation, or treatment of human food, having a melting-point which when determined by the method specified in the Fourth Schedule to this Order is not less than 30 °C; but does not include dripping, premier jus, lard, margarine, butter, vegetarian butter, cocoa butter, horse fat, technical tallow, or any unrendered animal fat.'

Shortenings are the same as cooking fats but as their name implies, they are intended for pastry and flour confectionery making. Cooking fats are blends of oils and/or fats chilled and texturized similarly to margarine. Hydrogenated and fractionated animal and vegetable fats are also used. The product is manufactured to a soft texture suitable for easy mixing with other food ingredients. This creamability may be increased by whipping in during manufacture about 10 per cent of an inert gas before chilling the fat. Cooking fats should be examined for moisture, acidity, rancidity and antioxidants and GLC fatty acid and sterol analysis may be used to assist in the identification of the origin of the component fats but this may be difficult because of the extent of hydrogenation. Sheppard *et al.* (1978a) list the fatty acid composition of several different types of household and commercial shortenings.

GHEE, BUTTER OIL

Ghee

Ghee is almost anhydrous milk fat and because of its long shelf-life it is the most widely used milk product in the Indian sub-continent. It is a valued culinary oil and is now widely sold in the UK to the Asiatic community although most manufactured in the UK is exported (Hayes, 1985). In the UK, ghee is made from butter or from butter oil (anhydrous milk fat) not up to the standard needed for ice-cream manufacture. Butter is liquefied at 70 °C to obtain partial separation of water. The separated oil is alkali refined to remove free fatty acids and peroxides before centrifugation and vacuum evaporation. The butter oil is crystallised in a water cooled scraped surface heat exchanger, then canned at 40 °C. In India it is prepared traditionally from buffalo milk, which being low in carotenoids, produces white ghee, whereas the presence of carotenoids in cows milk produces a golden yellow coloured ghee. It is made by a number of slightly differing procedures, the essential step being the removal of water from butter or cream by the application of heat at temperatures between 110–140 °C (Ganguli and Jain, 1973). At the lower temperature the ghee produced is mild in flavour but at the higher temperatures a cooked flavour becomes more pronounced. The molten ghee is clarified by decanting from the residual non-fatty material.

Ghee should have the following characteristics: melting range 28–44 °C; butyro-refractometer reading 40–45 at 40 °C; saponification value not less than 220; iodine value 26–38 and Reichert value 26–29 (cow), less than 26 (goat) or 32 (sheep and buffalo). IDF Standard 68A: 1977 for ghee specifies milk fat 99·6 per cent min., moisture 0·3 per cent max., FFA as oleic acid 0·3 per cent max., copper 0·05 mg/kg max., iron 0·2 mg/kg max., peroxide value not greater than 1·0 mEq/kg, neutralising agents, traces only, and permitted antioxidants 200 mg/kg (gallates 100 mg/kg) max. The standard describes the structure of ghee as a mixture of higher softening point fats in crystal form dispersed through lower softening point fats in liquid form. The fatty acid profile of buffalo and cow ghee has been described by Ramamurthy and Narayanan (1971).

Substitute ghees are ghees manufactured from non-butyric animal body fats or vegetable fats. Substitute ghees based purely on vegetable fats are termed vanaspati. The draft Codex standard for vegetable ghee requires fat 99·5 per cent min., acid value 0·6 mg KOH/g max., peroxide value 10 mEq/kg max., slip point 31–44 °C. In India, vanaspati ghee is required to contain 5 per cent sesame oil in order to aid identification by the application of Baudouin's test. Substantial quantities of a substitute ghee based on fractionated palm oil are manufactured in Malaysia. Genuine ghee may be adulterated with animal body fat or vegetable fats. Vegetable fats are eaily detected by the presence of phytosterols, in particular β-sitosterol, using GLC techniques (p. 634). The presence of animal body fat, unless in high proportion, is very difficult to prove. Lambert *et al.* (1980) detected goat body fat by differential thermal analysis. The Reichert value or the percentage of lower fatty acids by GLC will be lowered by the presence of body fat.

Butteroil

Butteroil or butterfat is required by IDF Standard 68A: 1977 to contain milk fat 99·3 per cent min., moisture 0·5 per cent max., peroxide value not greater than 0·8 mEq/kg and the other limits shown above for ghee. Its structure should be in the form of a smooth, fine grain paste and it is required to be packed in airtight containers flushed with an inert gas. If not packed in this way it must be transported and stored at 10 °C or below. Anhydrous butteroil or anhydrous butterfat is required to contain milk fat 99·9 per cent min., moisture 0·1 per cent max., peroxide value not greater than 0·3 mEq/kg and the other limits as for ghee. Anhydrous milk fat is required to give a peroxide value not greater than 0·2 mEq/kg, to contain no neutralising agents and to conform to the other standards for anhydrous butteroil shown above.

Regulation 349/73/EEC as amended, permits the sale of intervention butter at a reduced price as concentrated butter for cooking, suitably labelled. The product is required to contain at least 95 per cent butyric fats which means that some or all of the SNF (non-fatty milk solids) may be present. Identification markers must be present, namely 2 per cent of C16 and/or C18 fatty acid monoglycerides or 1 per cent of pelargonic acid triglyceride (C9:0 triglyceride) in the butter used. If the concentrated butter contains at least 9 per cent butyric fats, 0·15 per cent stigmasterol of at least 95 per cent purity or 0·17 per cent of at least 85 per cent purity is required in the product.

FRYING OILS

Vegetable oils (maize, soya, groundnut, rape (low erucic), safflower, cottonseed and sunflower) are now used widely for deep frying. Because of the association between heart disease and saturated fats (FAO, 1978), there is an increased tendency for the more unsaturated oils to be marketed, but the cheaper products without specific labelling will probably consist of blends of the more abundant oils such as groundnut, soya bean and rapeseed oils. The smoke point (AOCS, 1978; BS 684: Section 1.8: 1983) of frying oils should be not less than 215 °C. Many commercial frying oils contain small amounts (about 10 mg/kg) of dimethylpolysiloxane anti-foaming agents. These are claimed to reduce lipid oxidation by preventing oxidative foaming and therefore give the frying oil a longer working life. The anti-foaming agent is not truly soluble in vegetable oils and functions by forming monolayers on the surface of the boiling oil. At present there are no satisfactory methods for the determination of dimethylpolysiloxane in frying oils but Doeden *et al.* (1980) have reported on a direct aspiration AAS method with a claimed detection limit of 1 mg/kg. Frying oils may be required to be analysed for antioxidants, rancidity values and authenticity by GLC fatty acid analysis.

During use at deep frying temperatures, frying oils are very prone to oxidative and thermal degradation with the formation of volatile and non-volatile decomposition products some of which in excessive amounts may be harmful to human health. Products include volatiles and polymers (Chang *et al.*, 1978), dimers (Veazey, 1986), oxysterols (Bascoul *et al.*, 1986), cyclic fatty acids (Frankel *et al.*, 1984), and alkaline compounds, soaps (Blumenthal and Stockler, 1986). Stevenson *et al.* (1984) reviewed the chemical and physical changes that occur in frying fats, affecting fat life and product quality.

Pei-fen Wu and Nawar (1986) evluated nine indices for assessing the quality of used frying oil, i.e. viscosity, polymers, dielectric constant (Foodoil Sensor), polar compounds, dimers, FFA, smoke point, carbonyls and cyclic monomers. They found that the ratio of polymers to Foodoil Sensor readings was best and was unaffected by replenishment with fresh oil. Croon *et al.* (1986) compared four rapid methods (Foodoil Sensor, RAU-Test, Fritest and Spot test) with the polar compounds method and found good correlation for all four. Smith *et al.* (1986) similarly found that frying times correlate with increase in dielectric constant, polar compounds and FFA.

Wessels (1983) has reported on the collaborative study of the IUPAC column chromatography method 2·507 for polar compounds (essentially polymerised and oxidised triglycerides), which is also described in the AOAC methods book. Sebedio *et al.* (1986) simplified the method by using Sep-Pak cartridges, as follows:

Inject 80–90 mg of frying oil or fat onto a Sep-Pak cartridge (Waters) or equivalent. Elute non-polar fraction (intact triglyceride) with 20 or 30 ml of light petroleum : diethyl ether (1:1), then the polar fraction with 30 ml methanol. Evaporate solvents from each fraction and weigh.

SALAD OILS

Formerly 'salad oil' was usually olive oil. Nowadays most products contain other oils such as arachis, cottonseed and maize oils.

The examination of salad oil should include the determination of the usual constants, many of which are given in Table 16.2 from Codex standards. As the constants for several of the possible oils are somewhat similar it is necessary to apply special tests for the presence of arachis, cottonseed, sesame and teaseed oil (Pearson, 1976). GLC fatty acid analysis is also desirable. Codex also restricts the additives to be used, e.g. antioxidants, anti-foaming agents, crystallisation inhibitor (oxystearin) and also contaminants, e.g. volatile matter, soap content, Fe, Cu, Pb, As (see footnote to Table 16.2).

COCONUT, PALM KERNEL AND BABASSU OILS

The dried meat (copra) of the fruit of the coconut palm *Cocos nucifera* L. contains from 57 to 75 per cent of coconut oil. Palm-kernel oil is obtained from the hard nut inside the fruit of the palm tree, *Elaeis guineensis* Jacq. Babassu oil is derived from the kernel of the fruit of several varieties of the palm *Attalea funifera*. These three so-called lauric fats (they are solid at room temperature) contain about 50 per cent lauric acid, C12:0, in their triglycerides. Their fatty acid profiles are very similar (Table 16.3, p. 620). Their analytical constants, as recommended in Codex Stan 124–1981, 126–1981 and 128–1981 respectively, are listed in Table 16.2. The following values are also recommended (cf. p. 583):

	Reichert	Polenske
Coconut	6–8·5	13–18
Palm kernel	4–7	8–12
Babassu	4·5–6·5	8–10

Coconut and palm-kernel oils are used extensively in the manufacture of cooking fats, high quality confectionery fats and in filled milk type products. Blends of hydrogenated palm-kernel oil and randomised (interesterified) palm-kernel oil are used as cocoa butter substitutes. So also are partially hydrogenated stearine fractions of palm kernel oil. Babassu oil is finding increasing use.

COTTONSEED OIL

Cottonseed oil is obtained from the seeds of various cultivated species of *Gossypium*. Some of the principal constants in Codex Stan 22–1981 are included in Table 16.2 and the fatty acid profile is shown in Table 16.3, p. 620.

Cottonseed oil is readily identified by Halphen's colour test (Pearson, 1976 and the AOAC methods book). Also the titre is unusually high, generally above 32 °C whereas olive oil gives about 23 °C. The iodine value is also higher than that of olive oil. At a temperature below 12 °C particles of solid fat begin to separate and the oil congeals at a temperature between 0 and −5 °C.

After soya bean oil, cottonseed oil is the most used dietary vegetable oil in the USA, mainly in salad and frying oils, shortenings and margarine.

Table 16.2 Recommended Codex standards for edible vegetable oils

Oil	Relative density 20 °C/water at 20 °C	Refractive index at 40 °C	Saponification value (mg KOH/g)	Iodine value (Wijs)	Unsaponifiable matter (g/kg) max.	Acid value (mg KOH/g) (max.)	Peroxide value (mEq/kg) (max.)
Babassu	0·914–0·917 (25 °C/20 °C)	1·448–1·451	245–256	10–18	12	0·6	10
Coconut	0·908–0·921 (40 °C/20 °C)	1·448–1·450	248–265	6–11	15	VO, 4; NVO, 0·6	10
Cottonseed	0·918–0·926	1·458–1·466	189–198	99–119	15	0·6	10
Grapeseed	0·923–0·926	1·473–1·477	188–194	130–138	20	0·6	10
Groundnut	0·914–0·917	1·460–1·465	187–196	80–106	10	VO, 4; NVO, 0·6	10
Maize	0·917–0·925	1·465–1·468	187–195	103–128	28	VO, 4; NVO, 0·6	10
Mustard seed	0·910–0·921	1·461–1·469	170–184	92–125	15	VO, 4; NVO, 0·6	10
Olive (virgin)	0·910–0·916	1·4677–1·4705	184–196	75–94	15	6·6	20
Olive (refined)	0·910–0·916	1·4677–1·4705	184–196	75–94	15	0·6	10
Olive pomace (refined)	0·910–0·916	1·4680–1·4707	182–193	75–92	30	0·6	10
Palm	0·891–0·899 (50 °C/20 °C)	1·449–1·455 (50 °C)	190–209	50–55	12	VO, 10; NVO, 0·6	10
Palm kernel	0·899–0·914 (40 °C/20 °C)	1·448–1·452	230–254	13–23	10	NVO, 0·6	10
Rape	0·910–0·920	1·465–1·469	168–187	94–120	20	VO, 4; NVO, 0·6	10
Rape (LEAR)	0·914–0·920	1·465–1·467	182–193	110–126	20	0·6	10
Safflower	0·922–0·927	1·467–1·470	186–198	135–150	15	0·6	10
Sesame	0·915–0·923	1·465–1·469	187–197	104–120	20	VO, 4; NVO, 0·6	10
Soya	0·919–0·925	1·466–1·470	189–195	120–143	15	0·6	10
Sunflower	0·918–0·923	1·467–1·469	188–194	110–143	15	VO, 4; NVO, 0·6	10

VO = virgin oil. NVO = non-virgin oil.
For all oils: matter volatile at 105 °C, max. 0·2 per cent (except olive NVO, max. 0·1 per cent); iron mg/kg max. VO, 5; NVO, 1·5; copper mg/kg max. VO, 0·4; NVO, 0·1; lead mg/kg max. 0·1; arsenic mg/kg max. 0·1.

GRAPESEED OIL

Grapeseed oil is obtained from the seeds of the grape *Vitis vinifera*. Constants from Codex Stan 127–1981 are listed in Table 16.2 and the fatty acid composition in Table 16.3 shows close resemblance to that of sunflower and safflower oils.

Apart from olive oil (q.v.) grapeseed oil is different from other edible oils in that it contains significant quantities of the triterpene alcohol erythrodiol. The Codex standard cites a minimum of 2 per cent of the total sterol content of the oil. Gracian Tous *et al.* (1986) have reported on the results of a collaborative study of the IUPAC method 2.431 for erythrodiol in oils. A sample of grapeseed oil used in the study was found to contain about 0·015 per cent erythrodiol in the oil.

The oil is excellent for frying and as a salad oil.

GROUNDNUT OIL

Groundnut oil (arachis oil, nut oil, peanut oil) is obtained from the seeds of *Arachis hypogaea* L., a legume closely related to peas and beans. Most of the world production is grown in Africa, India and China. In the USA, only low grade nuts are used for oil production; the majority are grown for consumption as the nut and as peanut butter. Some of the principal constants in Codex Stan 21–1981 are included in Table 16.2 and the fatty acid profile is shown in Table 16.3 (p. 620). Fatty acid composition is very variable for a number of reasons such as genotype, geography and seasonal weather (Holaday and Pearson, 1974). The Codex also includes a minimum for arachidic C20:0 and higher fatty acids of 48g/kg. Singleton and Patter (1987) have identified 23 triacylglycerols by RP-HPLC and GC/MS in peanut oil. Samples should be examined for the clouding point (Bellier test—see olive oil), which is approx. 39 °C, i.e. much higher than that obtained with other oils.

As with olive oil, a deposit tends to form when arachis oil is stored at low temperatures but the oil is difficult to winterise (removal of the solid fractions formed on chilling) and is therefore not used in salad oils and dressings. Groundnut oil is used in margarines, cooking fats and frying oils.

MAIZE OIL

Maize oil (corn oil) is derived from maize germ, the embyos of *Zea mays* L. and is obtained as a by-product in the manufacture of corn starch. Some of the principal constants in Codex Stan 25–1981 are included in Table 16.2. Maize oil is high in polyunsaturated fatty acids (Table 16.3, p. 620) and therefore finds wide use in frying oils and speciality margarines. The fatty acid composition and triglyceride structure of corn oil, hydrogenated corn oil and corn oil margarine is reported by Strocchi (1981).

OLIVE OIL

Olive oil is defined in Codex Stan 33–1981 as the oil obtained from the fruit of *Olea europaea* L. Standards prescribed for virgin olive oil, refined olive oil and refined olive-residue oil have been revised in 1987 and the revisions appear in this chapter.

Olive residue oil has been renamed olive pomace oil. Various constants are included in Table 16.2. The standard identifies olive oil by its fatty acid profile (Table 16.3) and by maximum specific extinctions, $E^{1\%}_{1\text{cm}}$, in the ultraviolet as follows:

	232 nm	270 nm
Virgin olive oil	3·50	0·30
Refined olive oil	—	1·10
Refined olive pomace oil	6·00	2·00

Also, all three olive oils are required to contain at least 93 per cent of sitosterol, not more than 4·0 per cent of campesterol and not more than 0·5 per cent of cholesterol expressed as percentages of the sum of sitosterol, campesterol and stigmasterol present. Also, the saturated fatty acids (C16:0 + C18:0) at position 2 of the triglycerides shall be for virgin oil, 1·5 per cent max.; for refined olive oil, 1·8 per cent max.; and for refined olive pomace oil, 2·2 per cent max., expressed as the percentage of total fatty acids at position 2.

Regulation 1058/77/EEC (OJ No. L 128, 24.5.77, p. 6) lays down standards for Customs tariff purposes for various qualities of olive oil. Virgin olive oil is required to have FFA content not greater than 3 per cent as oleic acid; $E^{1\%}_{1\text{cm}}$ at 270 nm not greater than 0·25, and after a specified treatment with alumina, not greater than 0·11 as defined by absorption measurements at ± 4 nm of the absorption maximum. The Bellier test for the presence of olive-residue oil is required to be negative.

PROCEDURE

Saponify 1 g of the oil by boiling for 10 min with 5 ml aqueous alcoholic KOH (42·5 g KOH in 72 ml water, bulked to 500 ml with 95 per cent ethanol). After cooling add 1·5 ml aqueous acetic acid (1 + 2 by volume such that 1·5 ml exactly neutralises 5 ml of the aqueous alcoholic KOH) and 50 ml of 70 per cent ethanol warmed to 50 °C. Mix, insert a thermometer and allow to cool. If a precipitate forms above 40 °C, the test for the presence of olive-residue oil is positive.

The Modified Vizern Test for the presence of olive-residue oil is also prescribed. This test is based on appearance of a flocculent precipitate rather than cloudiness when the unsaponifiable matter, having been treated with peroxide, is dissolved in 85 per cent ethanol. The Regulation states that if the palmitic acid, C16:0, content of the fatty acids at the 2-position (p. 633) is higher than 2 per cent, the olive oil is considered as containing re-esterified oil. Also, the β-sitosterol content of the six specified sterols in olive oil must be less than 93 per cent of the total sterol percentage. Methods for the determination of fatty acids at the 2-position and for sterols are included in the Regulation.

The triterpene alcohol erythrodiol is a significant component of olive residue (pomace) oil and to a much less extent is present in genuine olive oil. Apart from grapeseed oil (q.v.), erythrodiol is not found in other edible vegetable oils. Italian legislation allows the presence of erythrodiol up to 5 per cent of the total sterol content of olive oils. More than 5 per cent is regarded as indicating the presence of olive residue oil (which contains up to 20 per cent in its unsaponifiable matter). Itoh *et al.* (1981) have listed the triterpene alcohols and sterols of virgin, refined and residue olive oils.

Complaints occasionally arise in cold weather in relation to a deposit that forms at the bottom of the bottle. This is quite normal and is merely due to the deposition of stearine at temperatures below 10 °C.

Table 16.3 Fatty acid composition of dietary fats and oils (per cent m/m of total fatty acids)

Fatty acid	Ground nut	Cotton seed	Lard and rendered pork fat	Maize	Premier Jus and edible tallow	Safflower	Sesame	Soyabean	Sunflower	Olive	Rapeseed	Low erucic acid rape-seed	Coconut	Palm	Palm kernel	Babassu	Grape-seed
C 6:0													0·4–0·6		<0·8		
C 8:0													5–10		2·4–6·2	2·6–7·3	
C10:0													4·5–8		2·6–5·0	1·2–7·6	
C12:0				<0·3									43–51	<0·4	41–55	40–55	<0·5
C<14	<0·4	<0·1	<0·5		<2·5	<0·1	<0·1	<0·1	<0·4								
C14:0	<0·6	0·4–2·0	0·5–2·5	<0·3	1·4–7·8	<1·0	<0·5	<0·5	<0·5	<0·1	<0·2	<0·2	16–21	0·5–2·0	14–18	11–27	<0·3
C14:iso					<0·3												
C14:1			<0·2		0·5–1·5												
C15:0			<0·1		0·5–1·0												
C15:iso			<0·1		} <1·5												
C15:anti iso																	
C16:0	6·0–16	17–31	20–32	9–14	17–37	2·0–10	7·0–12	7·0–14	3·0–10	7·5–20	1·5–6·4	2·5–6·0	7·5–10	41–47	6·5–10	5·2–11	5·5–11
C16:1	<1·0	0·5–2·0	1·7–5·0	<0·5	0·7–8·8	<0·5	0·5	0·5	<1·0	0·3–3.5	<3·0	<0·6		<0·6			<1·2
C16:2					<1·0												
C16:iso			<0·1		<0·5												
C17:0	<0·1		<0·5		0·5–2·0					<0·5							
C17:1	<0·1		<0·5		<1·0					<0·6							
C17:iso					} <1·5												
C17:anti iso																	
C18:0	1·3–6·5	1·0–4·0	5·0–24	0·5–4·0	6·0–40	1·0–10	3·5–6·0	3–5·5	1·0–10	0·5–5	0·5–3·1	0·8–2·5	2–4	3·5–6·0	1·3–3·0	1·8–7·4	3–6
C18:1	35–72	13–44	35–62	24–42	26–50	7·0–42	35–50	18–26	14–35	55–83	8–60	50–66	5–10	36–44	12–19	9–20	12–28
C18:2	13–45	33–59	3·0–16	34–62	0·5–5·0	55–81	35–50	50–57	55–75	3·5–21	11–23	18–28	1–2·5	6·5–12	1·0–3·5	1·4–6·6	58–78
C18:3	<0·3	0·1–2·1	<1·5	<2·0	<2·5	<1·0	<1·0	5·5–10	<0·3	<1·5	5–13	6·0–14		<0·5			<1·0
C20:0	1·0–3·0	<0·7	<1·0	<1·0	<0·5	<0·5	<1·0	<0·6	<1·5	<0·8	<3·0	0·1–1·2		<1·0			<1·0
C20:1	0·5–2·1	<0·5	<1·0	<0·5	<0·5	<0·5	<0·5	<0·5	<0·5		3–15	0·1–4·3					
C20:2			<1·0								<1·0						
C20:4			<1·0		<0·5												
C22:0	1·0–5·0	<0·5	<0·1	<0·5		<0·5	<0·5	<0·5	<1·0	<0·2	<2·0	<0·6					<0·3
C22:1	<0·3	<0·5							<0·5		5–60	<2·0					
C22:2											<2·0						
C24:0	0·5–3·0	<0·5		<0·5				<0·5	<0·5	<1·0	<2·0	<0·2					
C24:1									<0·5		<3·0	<0·2					<0·1

PALM OIL

Palm oil is obtained from the flesh or mesocarp of the fruit of the oil palm tree *Elaeis guineensis* Jacq. Unbleached palm oil is a deep orange coloured semi-solid fat with a soft buttery consistency and an odour reminiscent of violets. The refining of palm oil is described by Swoboda (1985) and the refined oil is defined in Codex Stan 125–1981. The composition and purity characteristics of the oil are given in Table 16.2; its fatty acid profile is very different from the kernel oil (see Table 16.3).

In Malaya a fractionation process is used to produce a more liquid fraction, i.e. palm olein, and a higher melting palm stearin. A double fractionation process using acetone is also used to prepare a palm oil mid fraction. The composition and purity characteristics of palm oil and its fractions are given by Rossell *et al.* (1985).

Palm oil at 20 °C has 22–25 per cent solid fat content and is therefore suitable for bakery fat shortening formulations. The oil and its fractions are finding wide use in the food industry (Berger, 1986), e.g. in margarine, ice-cream, cooking fats.

Palm oil is unique among edible vegetable oils in that most of the tocopherols are present as tocotrienols (p. 650).

RAPESEED OIL

Rapeseed oil (Colza oil, Sarson oil, Toria oil) is derived from species of *Brassica* (Codex Stan 24–1981). Reference to Table 16.2 shows that the saponification value is lower than that of other vegetable oils. Rape oil has an unusually high viscosity. It also is characterised by the presence of a high proportion of erucic acid (p. 636 and Table 16.3). High erucic rapeseed oil is used in some countries, e.g. Poland, as a dietary oil. However, varieties of rape containing very low levels of erucic acid are widely grown in western Europe and Canada. Low erucic rapeseed oil (LEAR) is defined in Codex Stan 123–1981 and its composition is shown in Tables 16.2 and 16.3. The newer low-erucic acid oils are finding wider use in blended frying oils, cooking fats and margarine.

SESAME OIL

Sesame oil (Gingelly oil, Benne oil, Till oil) is derived from the seed of *Sesamum indicum* L., which is grown in many tropical, sub-tropical and temperate countries. Some of the principal constants in Codex Stan 26–1981 are included in Table 16.2 and its fatty acids are shown in Table 16.3. The fatty acid composition and physicochemical characteristics of sesame oil (and poppy seed oil) are described by Raie and Salma (1985). The oil does not solidify when cooled to 0 °C therefore it is used as a substitute for olive oil as a salad oil. Sesame oil can be identified by Baudouin's test (Pearson, 1976), or the Villavecchia test (AOAC methods book), which are based on the presence of a phenolic substance, sesamol. This substance is a natural antioxidant and makes sesame oil more stable than other oils. It is a high quality oil and is used in cooking fats, margarine and salad oils.

SOYA BEAN OIL

Soya bean oil is derived from the seed of *Glycine max* L., which is a native of China but has been grown extensively since 1940 in the USA. It is the world's major oil seed crop. The principal constants in Codex Stan 20–1981 are shown in Table 16.2. Because of its high linolenic acid, C18:3, content (Table 16.3) the oil is prone to autoxidation and the development of rancidity. This problem has been reduced by genetic manipulation and by selective hydrogenation, bleaching and deodorisation. Soya bean oil is now available that contains less than 1 per cent linolenic acid, about 35 per cent linoleic acid C18:2 and 50 per cent oleic acid C18:1. Such oils treated by hydrogenation may contain small amounts of *trans*-isomers.

Refined and deodorised soya bean is used extensively in frying oils, shortening, salad oils, margarine, fish canning and in many other foods.

SAFFLOWER OIL

Safflower oil (Codex Stan 27–1981, see Table 16.2) is derived from the seeds of *Carthamus tinctorius* L., which is cultivated in the tropical and temperate zones, e.g. India, Mexico, USA. The fatty acid composition is very similar to sunflower oil (Table 16.3) and consists of up to 80 per cent linoleic acid, C18:2, and about 12 per cent oleic acid, C18:1. Plant geneticists have developed cultivars that have these fatty acid concentrations reversed, so there are now available two commercial safflower oils. Safflower oil is used as a frying oil, salad oil and as a constituent of mayonnaise and speciality margarines.

SUNFLOWER OIL

Sunflower oil is derived from the seed of *Helianthus annuus* L. This oilseed is becoming more important because it grows well both in the tropics and in temperate zones, and is resistant to drought. Also, the oil is high in polyunsaturates (Table 16.3). The fatty acid composition alters considerably with variety, soil and climatic conditions (Robertson, 1972). Selective plant breeding in the USA has resulted in the production of high oleic sunflower seed giving 80–86 per cent oleic acid and only 4–8 per cent of linoleic acid in the oil, and independent of climatic and soil conditions (Purdy, 1986). This oil resembles olive oil and gives substantial improvement in oxidative stability. Sunflower oil is used principally as a frying oil and in speciality margarines. The principal constants in Codex Stan 23–1981 are included in Table 16.2. Analytically, it is difficult to distinguish normal sunflower oil from safflower oil (Fedeli *et al.*, 1972).

OTHER VEGETABLE OILS

Mustard seed oil (Codex Stan 34–1981, Tables 16.2 and 16.3) closely resembles rapeseed oil. Since both these oils contain high levels of erucic acid they cannot be used for food (see p. 636).

Almond oil is obtained by expression (without the application of heat) from the seeds of *Prunus amygdalus* var. *dulcis* or *amara*. The colour is described as pale yellow and the taste as bland and nutty. Almond oil should remain clear after keeping at −10 °C for 3 h and should not congeal until the temperature has been reduced to about −18 °C.

Peach-kernel oil and apricot-kernel oil have been sold as adulterants of or substitutes for almond oil. The fatty acid composition of almond, butternut, brazil nut, cashew nut, chestnut, filbert (hazelnut), hickory nut, macadamia, pecan, pignolia, pistachio, and walnut is given by Beuchat and Worthington (1978).

The fatty acid composition of many minor seed oils, and of varieties of common dietary seed oils are listed by Hamilton (1987).

MARINE OILS

Marine oils include fish oils and marine mammal oils, e.g. whale, dolphin and seal oils. Whale oils were once used extensively, after processing and hydrogenation, in margarine and cooking fat but most countries have now banned all whale oil imports for conservation reasons. Fish oils are a by-product of the manufacture of fish meal for animal feed. Several species of fish are trawled for this purpose. Partially hydrogenated fish oils are used extensively in the manufacture of margarines in the UK. Marine oils are notable for their content of long chain highly unsaturated fatty acids. For example herring oil contains about 1·2 per cent C18:3, 1·8 per cent C18:4, 0·6 per cent C20:4, 7 per cent C20:5, 1·1 per cent C22·5 and 6·5 per cent C22:6 (Paul and Southgate, 1978). Whale oils contain variable amounts of these acids according to the whale species (Sonntag, 1979). The fatty acid composition of marine oils is influenced considerably by seasonal, age and other factors.

METHODS OF ANALYSIS

GENERAL METHODS

Introduction

The determinations that follow are those which are commonly used in the examination of oils and fats. In routine analysis, until recent times, determination of the iodine value, saponification value, unsaponifiable matter, acid value and peroxide value, coupled with qualitative tests for appropriate adulterants, has been considered sufficient for confirming the identity and edibility of most oils and fats. In the last 20 years, however, except in the smallest of laboratories, gas chromatographic determination of the fatty acid profiles has largely replaced almost all the traditional procedures and colour tests for identifying oils and fats. Further tests may also be required for the presence of antioxidants and emulsifiers (Chapter 4; Smullin, 1974). Reference methods for the sampling and examination of oils and fats are described in BS 627: 1982 (ISO 5555) (sampling), BS 684 (series of parts on test methods), AOCS (1978), IUPAC (1987) and Codex Alimentarius (CAC/RM 9/14–1969).

Numerous methods are reviewed by Hamilton and Rossell (1986) and described by Williams (1966), who also tabulated the chemical constants for a large range of oils and fats, which are especially useful in interpretation, as do the general books referred to on p. 609. Chromatographic methods for lipids have been catalogued by Mangold *et al.* (1984) and reviewed by Christie (1973, 1980) and chromatographic and spectroscopic methods by Perkins (1975). Christie (1987) has issued a practical guide for HPLC techniques. The constitution and analysis of fats based on older data have been discussed by Hilditch and Williams (1964). The full composition of raw, processed and prepared oils and fats is given in the USDA Agricultural Handbook No. 8–4 issued in 1979.

Water. The moisture content of oils and fats containing 1 per cent or more of water may be determined by Dean and Stark's entrainment distillation method using xylene, toluene or heptane. BS 684: Section 1.16: 1981 (ISO 934) (cf. IUPAC, 2.602; AOCS, Ca 2a–45) gives details and also describes a procedure to be used if water forms an emulsion with the solvent. For oils and fats containing water in the range 0·05–1 per cent, BS 684: Section 2.1: 1976 describes a Karl Fischer procedure using *n*-propanol or 2-methoxyethanol (cf. IUPAC, 2.603; AOCS, Ca 2e–84). Drying methods (volatile matter) on a hot plate or in air or vacuum oven (BS 684: Section 1.10; AOCS, Ca 2b–38, Ca 2c–25, Ca 2d–25) or on a sand bath (IUPAC, 2.601) can be used.

Density. BS 684: Section 1.1 describes pyknometer procedures for the determination of relative density (t°/20 °C) and apparent density (g/ml at t °C) for fats that are liquid and do not deposit stearin at t °C, and for fats which may deposit stearin at that temperature. For normal oils and fats of commerce, the apparent density is 0.0018 less than the relative density. Measurements are made at 20 °C but where the fat is not liquid at that temperature, measurements are usually made at 40 °C or 60 °C.

The IUPAC method (apparent density) is 2.102 and the AOCS method at 25°/25 °C or 60°/25 °C is Cc 10a–25.

Refractive index. BS 684: Section 1.2 (ISO 6320); AOCS, Cc 7–25; IUPAC, 2.102. For the determination of the refractive index using the Abbe refractometer and a sodium vapour lamp the following temperatures are employed: 20 °C for oils, 40 °C for solid fats which are fully molten at that temperature, 60 °C for hydrogenated fats and 80 °C for waxes. The AOCS method includes the use of a butyro refractometer.

Melting point. Commercial fats are heterogeneous mixtures of glycerides and do not have a sharp melting point. Samples may be compared by measuring the temperature at which under specified conditions a column of fat of fixed length rises in an open capillary tube under a definite pressure (slip point). BS 684: Section 1.3 gives details for the examination of animal fats, most vegetable fats and hydrogenated oils and fats, and a modified procedure for the determination of slip point of fats that show polymorphic behaviour, e.g. cocoa butter, illipe butter and cocoa butter substitute. The Ubbelohde method, which employs a small glass cup with a central hole, may be used to determine the flow and drop points of both soft and hard fats of animal and vegetable origin. Temperatures are taken at which a hemispherical protuberance of oil appears at a small opening in the cup (flow point) and at which the first drop falls from it (drop point) (see BP, BS 684: Section 1.4 and BS 894, apparatus). AOCS methods are slip point Cc 3–25 and dropping point Cc 18–80.

Solid–liquid ratio. An important industrial test is the determination of the proportion of solid to liquid triglycerides in a fat. This provides information for example on the extent of hydrogenation of an oil or the suitability of a fat for a particular purpose.

The ratio can be measured by dilatometry. This empirical technique is based on the measurement of isothermal expansion of the fat. The sample is melted, put in an enclosed calibrated glass tube known as a dilatometer and solidified under standardised conditions. The temperature of the solidified fat is then raised in 5 °C stages and the volume of the fat measured each time until it is almost completely molten. By plotting the change in volume against temperature a melting dilatation graph is obtained. A formula may be used to calculate the dilatation of the fat, which represents the difference between the volume that the fat would occupy at a given temperature if it were completely liquid (supercooled) and the actual volume of the fat at the same temperature. It thus gives an empirical indication of the amount of crystalline or solid fat in the sample at the given temperature. The procedure is described in IUPAC method 2.141 and in BS 684: Section 1:12. Bowers (1978) has proposed equations for the rapid conversion of dilatometer readings to Solid Fat Index values (AOCS, Cd 10–57). Dilatation determinations have been superseded by the use of NMR. Broad line (low resolution) instruments differentiate between the protons in solid fats or liquid oils and that when calibrated give a direct indication of the solids content of the sample. Haighton *et al.* (1971) have compared results obtained by continuous wave broad line NMR and conventional dilatometry. Instruments based on pulsed low resolution NMR are stated by Templeman *et al.* (1977) to have the advantage of speed, lower instrument cost and ease of automation over other techniques for the determination of solid–liquid ratio. IUPAC method 2.150 describes both NMR methods.

Titre. The titre of an oil or fat is the solidifying point of the mixed component fatty acids. The titre temperature is of value for characterising oils and fats and for assessing the hardness (see under 'Dripping').

In the method the saponified oil is acidified and the melted fatty acids allowed to cool. After solid begins to separate, the temperature rises slightly due to the latent heat liberated and the highest temperature subsequently reached is taken as the titre. The following method uses glycerol–potash for the saponification, which is shorter than Dalican's alcoholic-alkali method (cf. BS 684: Section 1.6; IUPAC, 2.121; AOCS, Cc 12–59).

PROCEDURE

Heat 19–20 g of potassium hydroxide pellets with about 70 ml glycerol in a 400 ml beaker until the temperature of the solution reaches 150 °C (not more). During the heating the liquid can be gently stirred with a thermometer. Add 50 ml of melted fat and maintain at 145–150 °C for 15 min or until no separation occurs, stirring gently during the heating. Partially cool, pour into a litre beaker containing 450 ml hot water and wash in the remaining liquid with about 50 ml more of hot water. Then add 50 ml dil. sulphuric acid (1 vol. conc. + 3 vols water) and boil until the top layer of fatty acids melts and clears. Siphon off and reject the lower aqueous layer using two pieces of glass tubing joined by a piece of rubber tubing fitted with a spring clip. Add 200 ml of hot water to the beaker and siphon off again. Repeat the washing twice more. Decant the fatty acids carefully from any remaining drops of water into a filter paper and filter into a flask in the 100 °C oven for at least 30 min. Pour the dried melted fatty acids into a large test-tube (9 cm × 3 cm) fixed axially by means of a cork into a wide-necked bottle. Suspend an accurately calibrated 0–50 °C thermometer centrally in the tube by means of a piece of cotton. Allow the liquid to cool and when solid appears, stir with the thermometer, first a few times from left to right and them similarly from right to left. Then stir rapidly with a circular movement avoiding touching the tube. The temperature should fall at first, then remain stationary and then rise before falling away again. The titre is the highest temperature reached after the stationary period. Re-melt the fatty acids and repeat the determination until two successive determinations agree to within 0·2 °C. Typical titre figures for some of the common oils are as follows: arachis 30°, cottonseed 33°, olive 23°, sesame 23°, teaseed 14°.

Colour. The colour of fat and oils is usually measured by comparison with standard coloured glasses in a Lovibond Tintometer or Wesson Colorimeter using

½ in (12·7 mm), 1 in (25·4 mm) and 5¼ in (133·4 mm) cells (see BS 684: Section 1.14; AOCS Cc 13b–45). This method is not satisfactory for oils containing green or brown coloration. A procedure based on FAC standard colours may be used for animal fats and all fats and oils too dark to be read on the Lovibond Tintometer (AOCS Cc 13a–43). A photometric method is used industrially in refining and bleaching operations (AOCS Cc 13c–50; IUPAC 2.103). Percentage transmittance is measured against carbon tetrachloride in a suitable cell (5–50 mm) at wavelengths of maximum and minimum absorbance.

Fats should be melted and any oil or fat that is cloudy should be filtered at a temperature not greater than 60 °C. During colour matching or photometric measurement, the sample should be at room temperature for oils or not more than 10 °C above the melting point.

Iodine value. The iodine value of an oil or fat is defined as the weight of iodine absorbed by 100 parts by weight of the sample. The glycerides of the unsaturated fatty acids present (particularly of the oleic acid series) unite with a definite amount of halogen and the iodine value is therefore a measure of the degree of unsaturation. It is constant for a particular oil or fat, but the exact figure obtained depends on the particular technique employed. The ranges of figures for the iodine values of the various groups of oils and fats are as follows:

Group	Examples	Iodine values
Waxes	—	Very small
Animal fats	Butter, dripping, lard	30–70
Non-drying oils	Olive oil, arachis oil, almond oil	80–110
Semi-drying oils	Cottonseed oil, sesame oil, soya oil	80–140
Drying oils	Linseed oil, sunflower oil	125–200

Iodine values for dietary oils and fats are shown in Table 16.2, p. 617.

The iodine value is often the most useful and easily determined figure for identifying an oil or at least placing it into a particular group. It should also be noted that for natural oils and fats the less unsaturated fats with low iodine values are solid at room temperature, or conversely, oils that are more highly unsaturated are liquids (showing there is a relationship between the melting points and the iodine values). A further point of interest is that, in general, the greater the degree of unsaturation (i.e. the higher the iodine value), the greater is the liability of the oil or fat to become rancid by oxidation.

The iodine value is usually determined by Wijs' method (cf. BS 684: Section 2.13 (ISO 3961); IUPAC 2.205; AOCS Cd 1–25).

PROCEDURE

Wijs' solution. Dissolve 8 g iodine trichloride in 200 ml glacial acetic acid. Dissolve 9 g iodine in 300 ml carbon tetrachloride. Mix the two solutions and dilute to 1000 ml with glacial acetic acid. A method for checking the iodine/chlorine ratio of Wijs' solution has been described by Graupner and Aluise (1966).

Pour the oil into a small beaker, add a small rod and weigh out a suitable quantity of the sample by difference into a dry glass-stoppered bottle of about 250 ml capacity. The approximate weight in g of the oil to be taken can be calculated by dividing 20 by the highest expected iodine value. Add 10 ml of carbon tetrachloride to the oil or melted fat and dissolve. Add 20 ml of Wijs' solution, insert the stopper (previously moistened with potassium iodide solution) and allow to stand in the dark for 30 min. Add 15 ml of potassium iodide solution (10 per cent) and 100 ml water, mix and titrate with 0·1 M thiosulphate solution using starch as indicator just before the end-point (titration = a ml). Carry out a blank at the same time commencing with 10 ml of carbon tetrachloride (titration = b ml).

$$\text{Iodine value} = \frac{(b - a) \times 1{\cdot}269}{\text{wt (in g) of sample}}$$

If $(b - a)$ is greater than $b/2$ the test must be repeated using a smaller amount of the sample.

A method for determining the iodine value using *N*-bromophthalimide and *N*-bromosaccharin has been described by Mohana Das and Indrasenan (1987).

Hydroxyl value and acetyl value. BS 684: Section 2.9; IUPAC 2.241; AOCS Cd 4–40. The hydroxyl value is defined as the number of milligrams of potassium hydroxide required to neutralise the acetic acid capable of combining by acetylation with 1 g of the oil or fat. The acetyl value is the number of milligrams of potassium hydroxide required to neutralise the acetic acid obtained when 1 g of acetylated oil is saponified. The process described in BS 684 involves the acetylation of the oil with acetic anhydride in pyridine, decomposing the excess anhydride by boiling with water and, after obtaining a homogeneous solution on addition of *n*-butanol, the acidity is titrated with alcoholic sodium hydroxide. The acetic anhydride available for acetylation is measured by carrying out the test without the sample. A similar test using the pyridine but omitting the acetic anhydride gives a measure of the free fatty acids present. The acetyl and hydroxyl values are calculated from the three figures obtained. Hartman *et al.* (1987) used toluene-*p*-sulphonic acid instead of the more unpleasant pyridine as the catalyst in the determination of hydroxyl value. This allows the determination to be carried out at room temperature in a much shorter time.

Saponification value. BS 684: Section 2.6 (ISO 3657); IUPAC 2.202; AOCS Cd 3–25. The saponification value of an oil or fat is defined as the number of milligrams of potassium hydroxide required to neutralise the fatty acids resulting from the complete hydrolysis of 1 g of the sample.

A soap is formed during the saponification, for example:

$$\underset{\text{Stearin}}{C_3H_5(C_{17}H_{35}COO)_3} + 3KOH = \underset{\text{Glycerol}}{C_3H_5(OH)_3} + \underset{\text{Potassium stearate}}{3C_{17}H_{35}.COOK}$$

The esters of the fatty acids of low molecular weight require the most alkali for saponification, so that the saponification value is inversely proportional to the mean of the molecular weights of the fatty acids in the glycerides present. As many oils have somewhat similar values (see Table 16.2, p. 617), the saponification value is not, in general, as useful for identification purposes as the iodine value. The saponification value is of some use for detecting the presence of coconut oil (SV 248–265), palm-kernel oil (SV 230–254) and butterfat (SV 225), which contain a high proportion of the lower fatty acids. The presence of any paraffin hydrocarbons may be detected as a cloudiness when water is added to the ethanolic saponified oil or fat solution.

Reagent

Dissolve 35–40 g potassium hydroxide in 20 ml water and dilute to 1 litre with alcohol (95 per cent). Allow to stand overnight and decant off the clear liquid.

Procedure

Weigh 2 g of the oil or fat into a conical flask and add exactly 25 ml of the alcoholic potassium hydroxide solution. Attach a reflux condenser and heat the flask in boiling water for 1 h, shaking frequently. Add

1 ml of phenolphthalein (1 per cent) solution and titrate hot the excess alkali with 0·5 M hydrochloric acid (titration = *a* ml). Carry out a blank at the same time (titration = *b* ml).

$$\text{Saponification value} = \frac{(b - a) \times 28{\cdot}05}{\text{wt (in g) of sample}}.$$

The highest accuracy in the titration is most important. The blank can be omitted if aqueous alkali is used (Hartman and Antunes, 1971).

If the unsaponifiable matter is to be determined the titrated liquid in the sample flask should be retained.

Unsaponifiable matter can be defined as the material present in oils and fats that after saponification of the oil or fat by caustic alkali and extraction by a suitable organic solvent, remains non-volatile on drying at 80 °C.

The unsaponifiable matter includes hydrocarbons, higher alcohols, oil-soluble vitamins and sterols (e.g. cholesterol, phytosterols). Most oils and fats of normal purity contain less than 2 per cent of unsaponifiable matter. Adulteration of oils and fats with paraffin hydrocarbons will appear in the unsaponifiable matter.

The full method for determining the unsaponifiable matter in oils in fats is described in BS 684: Section 2.7; IUPAC 2.401; AOCS Ca 6a–40. BS 684 describes three procedures: for fats of marine animals and most other animal and vegetable fats; for fats such as lard and fats such as shea nut oil, which contain a high proportion of unsaponifiable matter; and for fats containing high proportions of wax esters, such as wool fat.

General Procedure

Weigh to the nearest mg, 2·0–2·2 g of oil or fat sample into a 250 ml flask. Add 25 ml alcoholic potassium hydroxide solution (0·5 M in 95 per cent ethanol) and boil gently under reflux for 1 hour. Transfer the saponified sample solution to a separating funnel, using 50 ml water for washing the flask. Extract the solution while just warm 3 times with 50 ml quantities of diethyl ether. Pour each ether extract into another separator containing 20 ml of water. After the third extract has been added, shake the combined ether extracts with the first 20 ml of wash water and then vigorously with two further 20 ml quantities. Wash the ether extract twice with 20 ml of aqueous 0·5 M potassium hydroxide solution and at least twice with 20 ml quantities of water until the wash water is no longer alkaline to phenolphthalein. Pour the ether extract into a weighed flask, evaporate off the solvent, dry the residue at not more than 80 °C and weigh to constant weight. Dissolve the unsaponifiable matter in neutral alcohol and titrate with 0·1 M alkali (not more than 0·1 ml should be required to neutralise any free fatty acid present).

A procedure using mixed ethers for extraction is described fully under oil-soluble vitamins (p. 647).

A rapid quantitative procedure has been described by Maxwell and Schwartz (1979). Saponification is achieved by grinding the oil with KOH pellets followed by a short heating period. The soap mixture is blended with Celite powder, placed in a small column and the unsaponifiable matter is eluted with small volumes of dichloromethane. The procedure is claimed to give similar results to those obtained by official methods.

Trace metals

Small amounts of metals in edible oils have serious deleterious effects on the stability of these oils. Colour, odour and flavour changes can occur. Copper and iron greatly reduce the oxidative stability by catalysing peroxide formation. Trace metal analysis in oils is made difficult by the low levels present and the analytically difficult matrix they are in. Ooms and Van Pee (1983) have reported on the results of five different analytical techniques for Cu, Fe, Na, Ni and Zn in corn oil. Char ashing and MIBK dilution/direct aspiration (AOCS Ca 15–75) gave similar results by atomic absorption spectrophotometry. For trace analysis down to 0·1 p.p.m., AOCS Ca 18–79 describes a graphite furnace AAS method. BS 684: Sections 2.16 and 2.17 describe colorimetric methods for Cu and Fe on acid digested samples.

COMPOSITIONAL METHODS

Introduction

Modern industrial processing methods and concern over the health and safety aspect of new dietary oil and fat products have led to the develoment of analytical techniques and procedures to give detailed information on their nature, character, treatment and composition. The new procedures are also useful for identifying the biological source of the oil or fat and the presence of foreign fats or contaminants. The extraction of lipid material from composite foods for detailed analysis requires special procedures otherwise incorrect data may be obtained (p. 24). The composition of raw, processed and prepared oils and fats is given in the USDA Agricultural Handbook No. 8–4 issued in 1979, and the composition of dietary oils and fats, with analytical procedures is described by Sheppard *et al.* (1978a).

Fatty acid profile. GLC of fatty acid esters

The commercial introduction of gas–liquid chromatography in the late 1950s and early 1960s made feasible the rapid and accurate measurement of the fatty acid composition in oils and fats following conversion of the triglycerides esters into the more volatile methyl esters. The 'finger-print' fatty acid profile so obtained in many cases made virtually obsolete most of the earlier semi-specific colour and turbidity tests for individual oils. The depth of detail into the positional and geometrical isomeric composition of the triglyceride fatty acids has more recently been extended by newer stationary phases, improved gas chromatographic equipment and especially by the use of capillary columns of considerable chromatographic resolving power.

For most general purposes, however, relatively simple GLC procedures based on packed columns using a flame ionisation detector (or with less sensitivity a thermal conductivity detector) can be used to obtain sufficiently detailed fatty acid profiles. General guidance on suitable techniques has been published in BS 684: Section 2.34: 1980 and by IUPAC 2.302 and AOCS Cd 1–62. For rapid quantitative results an internal standard procedure using for example C15:0 or C17:0 fatty acid methyl ester may be preferred to internal normalisation, which assumes that the total peak areas on the chromatogram represent all of the sample constituents. In either case the use of a modern computer/integrator is invaluable.

The results of a collaborative study on the IUPAC general procedure, listing many suitable columns, stationary phases and operating parameters are reported by Firestone and Horwitz (1979).

Suitable columns for fatty acid profile analysis include:

1. 2 m × 4 mm i.d. glass packed with 12–15 per cent diethylene glycol succinate (DEGS) on 100/120 mesh Gas-Chrom P or equivalent operated at 170–190 °C.
2. 2 m × 4 mm i.d. glass, packed with 10 per cent butan-1,4-diol sucinate (BDS) on Celite or Chromosorb W, 80/100 mesh at 175 °C.
3. 1·8 m × 4 mm i.d., packed with 3 per cent SE–30 on Chromosorb G, silanised, 100/120 mesh, at 190 °C.
4. 2 m × 4 mm i.d. glass, packed with 10 per cent Silar–10C on Gas-Chrom Q, 100/120 mesh, from 130–240 °C at 4 °C/min.
5. 2 m × 4 mm i.d. glass, packed with 10 per cent SP 2330 on Chromosorb W–AW at 200 °C.

The fatty acid profile of the more common dietary oils and fats are shown in Table 16.3 (p. 620), and have also been compiled by Paul and Southgate (1980). A graphical procedure for confirming the identity of oils and fats from fatty acid profile data is described by Spencer *et al.* (1979).

Within the last 12 years, capillary gas–liquid chromatography has become more widely used to investigate the full isomeric composition of oils and fats. Fine open tubular glass and quartz capillaries, internally wall coated (WCOT) with strongly polar cyanopropylsiloxane, e.g. CP-Sil 88, or phenylpropyl siloxane, e.g. Silar 5, Silar 10c, stationary phases, with lengths of up to 100 m, more usually 50 m, are capable of separating double bond positional and geometric (*cis/trans*) isomers of fatty acids. Capillary GLC requires specialised apparatus and techniques and is more expensive to operate than packed column GLC. Accounts of the use of capillary GLC in the analysis of dietary oils and fats have been given by Slover and Lanza (1979), Flanzy *et al.* (1976), and Ojanpera (1978) and Traitler (1987) have reviewed recent advances. Van Niekerk and Burger (1985) have reported on the estimation of the composition of edible oil mixtures containing sunflower, soya, cottonseed, groundnut, maize, palm and olive oils using fatty acid and sterol analysis by capillary GLC and tocopherols by HPLC.

Derivatisation

In oils and fats the component fatty acids are esterified to glycerol at the three hydroxyl group positions. In order to determine the fatty acid distribution (profile) in the oil or fat by gas chromatography, the fatty acids must first be made volatile by quantitative conversion into esters of short chain aliphatic alcohols. In most cases fatty acid methyl esters (FAME) are prepared. The available procedures are reviewed by Sheppard and Iverson (1975), and recommended methods published by IUPAC 2.301; AOCS Ce 2–66; in BS 684: Section 2.34: 1980, and in the AOAC methods book.

The FAME from oils and fats containing short chain fatty acids, e.g. milk fat, may be prepared simply and rapidly by the method of Christopherson and Glass (1969). The procedure is described in BS 684: Section 2.34 and IUPAC 2.301 and briefly is as follows:

PROCEDURE

Weigh about 1 g of the oil or fat or lipid extract into a 20 ml stoppered test tube and add 10 ml heptane. Add 0·5 ml of 2 M methanolic KOH, stopper and shake the contents for 20 s after which time the solution should be clear. Shortly after this time a turbidity forms due to the separation of glycerol. Decant the upper heptane layer containing the methyl esters into a small vial or take a portion of the heptane layer direct by syringe for GLC injection.

The FAME of medium and longer chain fatty acids of dietary oils and fats (more than 6 carbon atoms) are best prepared by the boron trifluoride method (BS 684: Section 2.34; IUPAC 2.301; AOCS Ce 2–66 and the AOAC methods book). The method is suitable for methylating free fatty acids and all lipid classes. The BF_3 method is a good general purpose and reliable procedure but care is needed because the reagent is poisonous.

PROCEDURE

Weigh about 350 mg of the lipid sample into a 50 ml flask. Add 6 ml of approximately 0·5 M methanolic NaOH and a boiling granule, and boil under reflux for 5–10 min or until the droplets of fat disappear. Through the condenser by means of a graduated or automatic pipette add 7 ml of commercially prepared 14 per cent BF_3 in methanol and continue boiling for a further 2 min. Add through the condenser about 2–5 ml of heptane and continue boiling for 1 min. Cool and add saturated NaCl solution with swirling. Transfer about 1 ml of the heptane layer into a small stoppered test-tube or vial and add a little anhydrous Na_2SO_4. This solution will contain about 100 mg/ml of methyl esters suitable for the GLC determination of fatty acid profle.

In order to overcome the volatility and slight water solubility of milk FAME, Iverson and Sheppard (1977) prepared butyl esters by substituting *n*-butanol in the BF_3 method. A satisfactory procedure involving no toxic reagents in which hydrochloric acid is prepared *in situ* was developed by Hartman and Lago (1973).

PROCEDURE

Saponify 200–500 mg of the lipid sample by boiling under reflux for 3–5 min with 5 ml of 0·5 M methanolic NaOH or KOH. Add through the condenser 15 ml of esterification reagent (NH_4Cl, 2 g; methanol, 60 ml; conc. H_2SO_4, 3 ml) and boil for 15 min. Cool, add 50 ml water and 25 ml light petroleum, hexane or ethyl ether, mix and isolate the ethereal layer by means of a separating funnel. Wash the solvent layer twice with 25 ml water and carefully evaporate the solvent preferably under reduced pressure with an inert gas bleed.

Other chromatographic methods in fatty acid analysis

Thin-layer and paper chromatography methods used for the separation of fatty acids according to chain length, number of double bonds and their geometric configuration, have been reviewed by Mangold (1961). Morris (1962) with TLC, and de Vries (1962) with column chromatography first reported the use of silver to facilitate the separation of *cis*- and *trans*-isomers of fatty acids. The complete structural analysis of fatty acids, including the separation of positional isomers, i.e. unsaturated fatty acids of the same chain length but with the double bond in different positions, was reported by Bergelson *et al.* (1964) using silica gel TLC plates impregnated with silver nitrate. The low temperature argentation TLC separation of positionally isomeric *cis*- and *trans*-octadecenoic (C18:1) fatty acids has been described by Morris *et al.* (1967), and a modification of the method used to determine erucic acid (Kirk *et al.*, 1978 and p. 636). Bandyapadhaya and Dutta (1975) have described an argentation TLC procedure for separating the esters of fatty acids containing 1–6 double bonds, and Shukla and Srivastava (1978) have reported on improved TLC of polyunsaturated fatty acids using ammonia-complexed silver ions. The separation of fatty acids and FAME by alumina argentation TLC is described by Breuer *et al.* (1987). The TLC/FID Iatroscan instrument can be used to quantitatively separate FAME on silver nitrate impregnated rods (Sebedio *et al.*, 1985).

The usefulness of HPLC in lipid analysis has been limited because of the lack of suitable detectors. Differential refractometers may be used but are relatively insensitive. Triglycerides do not absorb at analytical ultraviolet wavelengths and therefore derivatisation of the fatty acids into esters containing ultraviolet chromophores is necessary. Durst *et al.* (1975) have described the formation of phenacyl esters of fatty acids by the use of crown ether catalysts, and their separation by reversed-phase HPLC has been reported by Borch (1975). Quantitative separation of fatty acid isomers has been claimed as phenacyl and naphthacyl derivatives (Wood and Lee, 1983) and para-bromophenacyl esters (Chaytor, 1987).

For the separation of fatty acid esters, reversed-phase and argentation HPLC have been used. Silver-loaded aluminosilicate or ion-exchange columns fractionate fatty acid ester mixtures according to the number of olefinic double bonds but in chosen conditions also separate saturated fatty acids and geometric fatty acid isomers (Scholfield, 1979). For example, silver-loaded Nucleosil columns chilled to −15 °C have been used to separate the methyl esters of complex fatty acid mixtures from hydrogenated food fats into total saturated, total *cis*-isomers and total *trans*-isomers (Anon, 1979; Kirk, 1980), prior to capillary GLC of the fractions. Similar pre-fractionation treatment of cod liver oil FAME has been reported by Ozcimder and Hammers (1980) who found that reversed-phase HPLC on LiChrosorb RP–18 with acetonitrile as the mobile phase was more effective than argentation HPLC carried out at room temperature.

Polyunsaturated fatty acids

Polyunsaturated fatty acids (PUFAs) cover the range of acids of 18, 20 and 22 carbon chain length with each of 2–6 double bonds in *cis*-configuration and separated by a methylene group. Khan and Scheinmann (1978) have reviewed modern instrumental methods for their analysis and characterisation.

Linoleic acid (C18:2, ω6) and α-linolenic acid (C18:3, ω3) are PUFAs necessary for the normal growth and function of animal and human tissues, and are termed essential fatty acids. Other longer chain essential fatty acids are arachidonic acid (C20:4, ω6) and docosahexaenoic acid (C22:6, ω3) (FAO, 1978).

Essential fatty acids, namely *cis*, *cis*-methylene interrupted PUFAs, may be determined in oils, fats and lipid extracts from food by saponification to their potassium salts followed by conjugation and oxidation to hydroperoxides by atmospheric oxygen in the presence of the enzyme lipoxidase. The concentration of total *cis*, *cis*-methylene interrupted PUFAs in g/100 g sample is calculated from the absorbance of the conjugated diene hydroperoxide at 234 nm. A centrifugal step was introduced by Prosser *et al.* (1977) to remove palmitates and stearates, which cause interference with absorbance readings, thereby making the method suitable for measuring relatively low levels of PUFA in extracted food lipids. Sheppard *et al.* (1978b) have reported on the results of a collaborative study on two procedures. The method is prescribed in IUPAC 2.209, AOCS Cd 15–78 and the AOAC methods book.

Individual PUFAs may be identified and determined by gas chromatography of their methyl esters. Whether or not the PUFAs are resolved without peak overlap will depend on the complexity of the lipid sample and on the column efficiency. The use of 50 m WCOT columns containing highly polar phases are likely to resolve most

PUFAs but in some seed oils a 2–3 m packed column containing 10 per cent BDS on Chromosorb W may be adequate for the principal essential fatty acids.

Trans-isomers of fatty acids

The double bonds in most fatty acids of natural vegetable oils and fats are in *cis* configuration and are isolated, i.e. they are not conjugated with other double bonds. During partial hydrogenation or oxidation the configuration may change to the *trans* form. Animal and marine fats may however contain small amounts of natural *trans*-isomers. The concentration of isolated-*trans* fatty acid isomers in triglycerides may be determined by measurement of the infra-red absorption at about 10·3 nm (970 cm^{-1}) due to deformation of the C–H bonds adjacent to the isolated-*trans* double bond (Szonyi *et al.*, 1962).

In order to reduce spectral interference, the triglycerides are preferably first converted to the methyl esters of the component fatty acids before infra-red measurement. A calibration curve is prepared using methyl elaidate (C18:1 *trans*) and the result expressed as percentage of methyl elaidate in the methyl esters. Because heavily hydrogenated fats, e.g. in some margarines and shortenings, contain a complex mixture of *trans*-isomers, the calibration of the infra-red method with methyl elaidate will not yield an accurate figure for *trans*-isomers at levels above 15–20 per cent of the total fatty acids. The method is prescribed in IUPAC 2.207, AOCS Cd 14–61 and the AOAC methods book. The infra-red method is not sufficiently sensitive for levels less than about 5 per cent.

The *trans*-isomer content of oils, fats and fatty foods can also be estimated by GLC. Lanza and Slova (1981) first separated *cis* and *trans* FAME by argentation TLC, then chromatographed the *trans* fraction on a 60 m SP2340 WCOT column. Gildenburg and Firestone (1985) reported the results of a collaborativce study on four margarines using 6 m × 2 mm i.d. glass columns packed with 15 per cent OV275 on 100–120 mesh Chromosorb P AW-DMCS at 220 °C. The method studied (not suitable for hydrogenated marine oils) is AOCS Cd 17–85, and is the 1st supplement (1985) to the AOAC method book.

Fatty acids in the 2-position

Natural fats and oils may vary considerably in their physical properties because of the specific distribution of fatty acids in their component triglycerides. The fatty acid profile on the central (β or 2) hydroxyl position of the glycerol molecule may not be the same as the fatty acid profile overall because the component fatty acids in the triglycerides of natural fats are not distributed in a completely random or ordered manner. In nature, the more unsaturated fatty acids usually tend to occupy the central position, but during processing for edible oil manufacture (such as fractionation and re-esterification) the natural distribution of fatty acids in the triglycerides may be made more random although the overall composition or profile remain unaltered. By use of analytical methods based on specific enzymic hydrolysis (Mattson and Volpenheim, 1961) it is possible to determine the fatty acid distribution at the central 2-position of the triglycerides thus indicating by comparison with the overall fatty acid profile whether the fat or oil has been subjected to industrial processing.

A procedure for determining the fatty acid profile at the 2-position is described in

BS 684: Section 2.39 (ISO 6800) and IUPAC 2.210. This procedure is not applicable to fats melting above 45 °C, to oils or fats containing short chain (C12 or less) fatty acids such as coconut and palm kernel oils and butterfat, nor to natural fish or marine oils containing highly unsaturated long chain acids. The oil sample in hexane is purified by passing it through an activated alumina column. A portion of the recovered purified sample is then hydrolysed with pancreatic lipase. The resultant monoglycerides are separated from other products on a silica TLC plate using a hexane/ether/formic acid mixture as the mobile phase, the monoglyceride band scraped off and the component fatty acids from the original 2-position of the triglycerides converted to methyl esters for GLC. The same procedure is described in Regulation 1058/77/EEC (OJ No. L128, 24.5.77, p. 6) to test for the presence of re-esterified oils in olive oil.

A knowledge of the fatty acid distribution in the triglyceride molecules of oils and fats can be very diagnostic. Christie and Moore (1970) demonstrated that the triglycerides in all pig tissues, except the liver, contain mainly palmitic acid C16:0 at the 2-position, and much of the remaining saturated fatty acids in 1-position. Youssef *et al.* (1988), using the palmitic acid enrichment factor (ratio of per cent C16:0 at 2-position to per cent total C16:0, greater than 0·8), detected 5 per cent or more of pork in various meat products.

The composition of the triglycerides of natural unblended oils and fats can be obtained by interpretation of the fatty acid profile on the central 2-position using the random distribution theories expounded by Van de Waal (1964) and Coleman and Fulton (1961). Litchfield (1972) has comprehensively reviewed the subject. (See also under triglyceride analysis, p. 636.)

Sterols

Sterols occur in small amounts in fats and oils in the free form, as esters of fatty acids and as glucosides. The predominant sterol of animal fats and marine oils is cholesterol. Wool grease (lanolin) is a mixture of cholestryl palmitate, stearate and oleate. Other minor zoosterols include cholestanol and coprosterol. Sterols from plant sources are known as phytosterols and include β-sitosterol, stigmasterol, campesterol and brassicasterol. Cholesterol is also found in low concentrations in the sterol fractions of certain vegetable oils and fats. The determination of cholesterol in food and biological lipids is of physiological importance and the identification of the sterols present in an oil or fat is useful for indicating the origin of the oil or fat sample. The presence of animal fat or marine oils in vegetable fats, or vice versa, may also be established by sterol analysis (Table 15.5, p. 592). The cholesterol content of foods is listed by Paul and Southgate (1978) and Sweeney and Weihrauch (1976). The sterol content of numerous plant oils is given by Weihrauch and Gardner (1978) and in seafood by Krzynowek (1985).

Following saponification of oil or fat samples, sterols are present in the solvent extracted unsaponifiable matter. Sterols may be categorised into phytosterols and cholesterol or a mixture of both by precipitation of their digitonides from which the acetates are prepared and melting points determined. Phytosterol acetates melt at 126–127 °C whereas cholesterol acetate melts at 113–115 °C. The method is described in IUPAC 2.402 and is in the AOAC methods book.

Naudet and Hautfenne (1985) have reported the collaborative study of the enzymatic IUPAC method 2.404 for total sterols.

Identification and determination of sterols by GLC has superseded the older procedures. Sterols can be chromatogrphed direct or as volatile derivatives. IUPAC 2.403 describes a procedure based on isolation of the sterols in the extracted unsaponifiable matter by TLC, removal of the sterol band, extraction of the sterols from the removed silica and direct injection of the solvent extract containing the sterols onto a 2 metre glass column containing 3 per cent SE30, OV-17, CPSil 19 or equivalent on 80–100 mesh, acid washed and silanised diatomaceous earth, e.g. Gas-Chrom Q, heated at 230 °C. Sterol standards and normalisation is recommended for quantitative analysis. Regulation 1058/77/EEC on the characteristics of olive oil (OJ No. L128, 24.5.77, p. 6) describe a similar unsap/TLC/GLC procedure but includes the preparation of trimethylsilylether derivatives of the extracted sterols. The Regulation recommends the use of SE30, which does not resolve β-sitosterol and Δ^5-avenosterol, but these sterols give separate peaks on columns containing OV-17 or CPSil 19. BS 684: Section 2.38 (ISO 6799) and the AOAC methods book describe a GLC procedure for β-sitosterol in butteroil following precipitation of the sterols with digitonin, a procedure for the GLC of sterol acetates and an unsap/TLC/ direct GLC procedure using cholestane as an internal standard for the detection of animal fats in vegetable fats and oils. Betulin is now considered by some workers to be superior as an internal standard to cholestane. Slover *et al.* (1983) determined sterols (and tocopherols) in fat, oils and extracted food lipids by TLC/capillary column GLC (Dexsil 400) using 5,7-dimethyltocol as internal standard.

Various procedures have been developed to avoid the lengthy unsap/TLC steps. Thurlow and Palmer (1987) at LGC have successfully eliminated TLC by using Sep-Pak chromatography cartridges. In the normal GLC of FAME of oils, fats and lipid extracts of food, the free sterols are present in the FAME solution and with larger injections and higher column temperature they may be detected. However, the GLC of the sterols may suffer from peak interference from longer chain C20 and C22 FAME in many oils and fats, but this simple method is suitable for samples such as cocoa butter, which contain insignificant amounts of longer chain acids. In such a procedure, betulin is more suitable than cholestane as an internal standard because it has a retention index greater than all the sterols. Sheppard *et al.* (1978a) and AOCS Ce 3–74 describe the preparation of sterol butyrates from a portion of the FAME solution followed by injection onto a 2 m × 4 mm i.d. glass column containing 1 per cent SE-30 on 100/120 mesh Gas-Chrom Q heated to 250–265 °C. Calibration is achieved by running sterol butyrate standards. This derivatisation procedure gives good resolution because solvent, FAME and tocopherol butyrates clear the column before the appearance of the sterol butyrates.

Reversed-phase HPLC has been used to separate the food lipid sterols (Rees *et al.*, 1976).

Triglyceride analysis

Silver ion adsorption column chromatography was introduced by de Vries (1962) to fractionate triglycerides of fats on the basis of the unsaturation of their component fatty acids. At about the same time methods employing enzymic hydrolysis of triglycerides appeared giving information on the distribution of fatty acids in triglycerides (p. 633) and the new technique of gas–liquid chromatography was applied to the direct analysis of triglyceride mixtures (Huebner, 1961).

GLC on silcone columns, e.g. OV-1, OV-17, SE-30, SE-52 and Dexsil 300 on treated silica such as Gas-Chrom Q, separates triglyceride molecules according to their molecular weight, effectively equal to the chain length or number of carbon atoms, in the component fatty acids. For example, oleodipalmitin contains $18 + (2 \times 16) = 50$ carbon atoms in the three fatty acid chains and may be referred to in GLC as the C_{50} peak. Because of the possible combinations of fatty acids giving the same total carbon count, each chromatographic peak may be composed of several structurally different triglycerides. The technique is comprehensively reviewed by Litchfield (1972) and Hamilton (1975). It has been used to characterise many natural oils and fats because each gives a characteristic chromatographic 'finger print'. Official packed-column GLC methods are IUPAC 2.323, AOCS Ce 5–86 and in the AOAC methods book, 2nd supplement 1986.

Padley and Timms (1978) have interpreted GLC data to determine the concenration of cocoa butter substitutes in chocolate. Capillary column GLC gives greater resolution of triglycerides. Hinshaw and Seferovic (1986) used a temperature programmed capillary GC method (Alltech RSL300, a polar high temperature phase based on methyl phenyl silicone) to resolve triglycerides in food lipids. Triglycerides are perhaps the least volatile substances that can be examined by GLC and therefore the high column temperatures required may lead to short column life and some thermal degradation of the injected sample, especially in the presence of traces of oxygen.

Separation of triglycerides by chain length and degree of unsaturation can be achieved by reverse phase HPLC (Dong and Dicesare, 1983; Padley, 1984; Takahashi *et al.*, 1984) and methods have been fully reviewed by Barron and Santa-Maria (1987). Smith *et al.* (1980) have reported on the potential of argentation HPLC for the analysis of triglycerides. The difficulties of the HPLC detection of triglycerides and other lipid classes have been overcome by use of the light scattering (mass) detector (Robinson *et al.*, 1985; Christie, 1986). This detector permits gradient solvent elution to improve separations (Herslof and Kindmark, 1985), and Palmer and Palmer (1989) using improved solvent programming to give a 30 min full separation (see Fig. 16.1) has shown the value of RP-HPLC of triglycerides in the detection of adulteration of oils and fats. Quantitative triglyceride analysis by argentation TLC is decribed by Chobanov *et al.* (1976) and Zadernowski and Sosulski (1979) have used a combination of argentation TLC, GLC and 2-position fatty acid analysis to investigate the triglyceride structures in rapeseed oils. Sebedio *et al.* (1985) used the Iatroscan TLC/FID instrument to separate triglycerides.

Erucic acid

Erucic acid C22:1 ω9 (*cis*-13-docosenoic acid) is a major component fatty acid of *Brassica* seed oils, occurring naturally at levels of 20–45 per cent of the total fatty acids. Rape, one of the most commercially important members of the *Brassica* family, was shown during the Second World War to be an economic source of dietary oil that could be grown in the northern temperate climatic regions (Kirk, 1980). However, it was reported in the 1950s that erucic acid when fed to test animals as rapeseed oil or tri-erucine, caused lipidosis and muscle lesions in the heart (FAO,

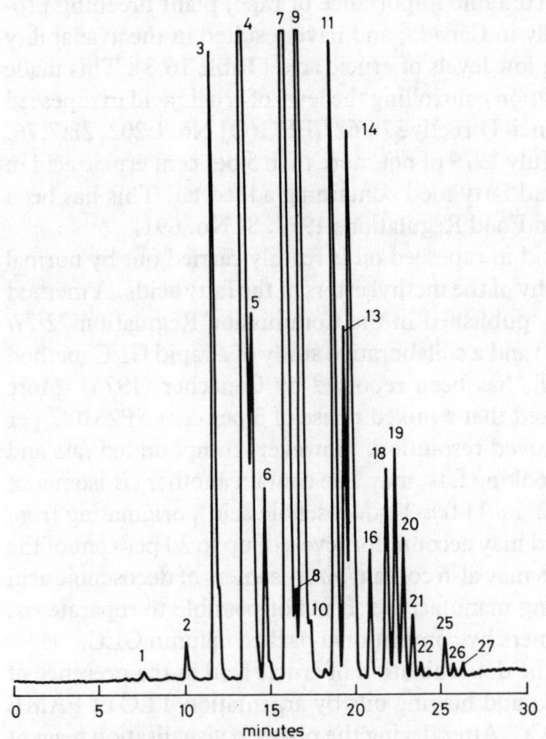

Fig. 16.1 Triglyceride profile of soyabean oil. For peak identification, see Table 16.4. (Figure courtesy of A. J. Palmer & F. J. Palmer.)

Table 16.4 Peak identification table of triglyceride peaks for Fig. 16.1

Peak No.	Triglyceride	Peak No.	Triglyceride
1	LnLnLn	18	OOO
2	LLnLn	19	OLS
3	LLnL	20	OOP
4	LLL	21	PLS
5	LLnO	22	POP
6	LLnP	23	PPaS
7	LLO	24	PPP
8	OLnO	25	OOS
9	LLP	26	SLS
10	OLnP	27	POS
11	OLO	28	PPS
12	OPaO	29	SOS
13	LLS	30	SPS
14	OLP	31	SOA
15	OPaP	32	SSS
16	PLP	33	SSA
17	SLnP		

P = palmitic acid, Pa = palmitoleic acid, S = stearic acid, O = oleic acid, L = linoleic acid, Ln = linolenic acid, A = arachidic acid.

1978). Because of the potential economic importance of rape, plant breeding programmes were undertaken, mainly in Canada, and have resulted in the availability since 1963 of cultivars containing low levels of erucic acid (Table 16.3). This made possible the introduction of legislation controlling the level of erucic acid in rapeseed oil and composite foodstuffs. Council Directive 76/621/EEC (OJ No. L202, 28.7.76, p. 35) set a limit effective from 1 July 1979 of not more than 5 per cent erucic acid in the total fatty acids of oils, fats and fatty food containing added fat. This has been incorporated in the Erucic Acid in Food Regulations 1977, SI No. 691.

The determination of erucic acid in rapeseed oil is readily carried out by normal packed column gas chromatography of the methyl esters of the fatty acids. A method based on an IUPAC procedure is published in the Commission Regulation 72/77/EEC (OJ No. L12; 15.1.77, p. 11) and a collaborative study of a rapid GLC method for docosenoic acid in fat and oils has been reported by Conacher (1975). More recently Daun *et al.* (1983) reported that a mixed phase of 3 per cent SP2310/2 per cent SP2300 (Supelco) gave improved resolution. However, compounded fats and oils, for example margarine and cooking fats, may also contain another *cis*-isomer of docosenoic acid, cetoleic acid C22:1 ω11 (*cis*-11-docosenoic acid), originating from marine oils, such as herring oil and may account for levels of up to 20 per cent of the total fatty acids. Compounded fats may also contain *trans*-isomers of decosenoic acid as a result of hydrogenation during manufacture. It is not possible to separate *cis*, *trans* and positional fatty acid isomers by conventional packed column GLC.

Kirk *et al.* (1978) reported on the determination of erucic acid in the presence of cetoleic acid in blends of rapeseed and herring oils by argentation TLC of FAME using double development at −20 °C. After drying the plates, a visualisation reagent is applied and the methyl erucate streak is scraped off and determined by packed column GLC. This developed into the official EC method in Directive 80/891/EEC of 25 July 1980 (OJ No. L254; 27.9.80, p. 35) for use when samples by packed column GLC show an apparent docosenoic content greater than 5 per cent. It has been collaboratively studied (Wessels, 1984) and appears as BS 684: Section 2.41 (ISO 8209), IUPAC 2.311, AOCS Ce 4–86 and is in the AOAC methods book, 1st supplement (1985).

Ackman *et al.* (1983) have reported on a collaborative study of a rapid capillary column (Silar SCP) GLC procedure and the above TLC/GLC procedure for erucic acid in edible oils and fats containing mixed C22:1 isomers. The methods were equivalent but the capillary column method gave better reproducibility and was much more rapid.

DEGRADATION AND RANCIDITY

Introduction

Oils and fats commence to decompose from the moment they are isolated from their natural living environment. The presence of free fatty acids is an indication of lipase activity or other hydrolytic action. Changes occur during storage that result in the production of an unpleasant taste and odour. Such oils and fats are referred to as having become rancid. The unpleasant organoleptic characteristics are in part caused by the presence of free fatty acids but the major development of rancidity is brought about by atmospheric oxidation (autoxidation). Oxidative rancidity is accelerated by exposure to heat and light, by moisture and by the presence of traces of transition metals (e.g. copper, nickel and iron) and residual natural dyes and pigments. The degradation of frying oils in use is discussed on p. 615.

Oxygen is taken up by the fat with the formation of hydroperoxides (ROOH). These are usually referred to as peroxides. The presence of natural antioxidants, e.g. tocopherols, or synthetic antioxidants (Chapter 4) inhibit the formation of peroxides. The chemical mechanisms in the formation of hydroperoxides and other oxidation products are discussed by Paquette *et al.* (1985) and Frankel (1980).

In general, the greater the degree of unsaturation (the higher the iodine value) the greater is the liability of the fat to oxidative rancidity. When the concentration of peroxides reaches a certain level, complex chemical changes occur and volatile products are formed that are mainly responsible for the rancid taste and odour. For example, the peroxides may suffer further oxidation to form diperoxides, which lead to polymer formation, fission reactions forming aldehydes, semi-aldehydes, aldehydo-glycerides, hydroxyl compounds and subsequently organic acids; dehydration reactions leading to keto-glycerides, and oxidation of other double bonds to form epoxides, hydroxy-glycerides and dihydroxy-glycerides. It is compounds such as these that produce rancidity off-flavours.

The initial oxidation of a fat is usually slow and at a relatively uniform rate. This is known as the induction period. At the end of the induction period, when the amount of peroxide formation reaches a certain level, the rate of oxidation accelerates very rapidly. At this point or soon after, the fat begins to smell or taste rancid.

With most oils and fats, the free acidity increases during storage but, with refined oils particularly, the FFA figure is not necessarily related to the extent to which rancidity has progressed. On the other hand, although the 'peroxides' are possibly not directly responsible for the taste and odour of rancid fats, the concentration of them as represented by the peroxide value is also useful for assessing the extent to which spoilage has advanced (cf. Swoboda, 1973).

Allen and Hamilton (1989) have prepared a comprehensive text on the chemistry and measurement of rancidity, on antioxidants and on the rancidity in various foods.

MEASUREMENT OF RANCIDITY

As rancidity is a complex phenomenon it is advisable to carry out as many tests as possible on all doubtful samples. In routine work, apart from the FFA, the analysis should include the determination of the peroxide value and the application of the Kreis reaction. Pearson (1965, 1974, 1975) has shown that several rancidity tests can be performed on a common chloroformic extract of the fat. A survey of the literature on oxidative rancidity is given by Kochar and Meara (1975) and a full account of reactions in the fatty acid chain has been reported (Sonntag, 1979). Several photometric tests for the measurement of oxidative rancidity have been compared by Tsoukalas and Grosch (1977). They considered that anisidine value and Kreis test were inadequte in being insensitive and unspecific, whereas the ferrous isothiocyanate test and the diene ultraviolet absorption method were most sensitive.

Acid value. Free fatty acids. The acid value of an oil or fat is defined as the number of milligrams of potassium hydroxide required to neutralise the free acidity in 1 g of the sample. Ths result is often expressed as the percentage of free fatty acids (FFA), especially in the UK.

The acid value is a measure of the extent to which the glycerides in the oil have been decomposed by lipase or other action. The decomposition is accelerated by heat and light. As rancidity is usually accompanied by free fatty acid formation, the determination is often used as a general indication of the condition and edibility of oils. There are a number of variations in the titration procedure.

PROCEDURE

Mix 25 ml diethyl ether with 25 ml alcohol and 1 ml of phenolphthalein solution (1 per cent) and carefully neutralise with 0·1 M sodium hydroxide. Dissolve 1–10 g of the oil or melted fat in the mixed neutral solvent and titrate with aqueous 0·1 M sodium hydroxide shaking constantly until a pink colour that persists for 15 s is obtained. The titration should preferably not exceed about 10 ml or otherwise two phases are liable to separate. This does not occur however if hot neutral alcohol is used as solvent (BS 684: Section 2.10: method 1; AOCS 5a–40) or if the acidity is titrated with alcoholic alkali (BS 684: Section 2.10: method 3; ISO 660; IUPAC 2.201).

$$\text{Acid value} = \frac{\text{Titration (ml)} \times 5{\cdot}61}{\text{wt of sample used}}.$$

AOCS Standard 3a–63 uses 10–20 g dissolved in isopropanol/toluene (1:1) and titrated with 0·1 M aqueous KOH against phenolphthalein, or electrometrically for highly coloured oils.

The FFA figure is usually calculated as oleic acid (1 ml 0·1 M sodium hydroxide ≡ 0·0282 g oleic acid), in which case the acid value = 2 × FFA.

BS 684 however recommends that the FFA in the following oils should be calculated as the acid which is more appropriate to that present:

Palm oil as palmitic acid (1 ml 0·1 M NaOH ≡ 0·0256 g).
Palm-kernel, coconut and similar oils as lauric acid (1 ml 0·1 M NaOH ≡ 0·0200 g).
IUPAC 2.201 also includes a potentiometric titration in isopropanol.

With most oils acidity begins to be noticeable to the palate when the FFA calculated as oleic acid is about 0·5–1·5 per cent. Occasionally, however, some quite rancid oils show only minimal acidities (see below).

Baker (1964) and Bains *et al.* (1964) have shown that the FFA of vegetable oils can be determined colorimetrically by shaking a benzene extract with copper acetate solution. The fatty acids react to form copper salts, the blue colour of which in the organic layer can be measured at 640–690 nm and compared with the results obtained using solutions containing known amounts of oleic acid. A rapid colorimetric method is described by Lowry and Tinsley (1974). The free fatty acids in oils and fats may also be identified and determined by GLC after a modified BCl_3 or BF_3/methanol esterification procedure with methyl urea (Chapman, 1979). The GLC of underivatised fatty acids is described by Sakodynsky *et al.* (1979).

Peroxide value. The peroxide value is a measure of the peroxides contained in the oil. During storage, peroxide formation is slow at first during an induction period that may vary from a few weeks to several months according to the particular oil or fat, the temperature, etc., and this must be borne in mind when interpreting quantitative results.

The peroxide value is usually determined volumetrically by methods which have been largely developed by Lea. These depend on the reaction of potassium iodide in acid solution with the bound oxygen followed by titration of the liberated iodine with sodium thiosulphate (see Mehlenbacher, 1960). Chloroform is normally used as solvent. The following method gives rapid results.

PROCEDURE
The test should preferably be carried out in subdued daylight. Weigh out 1 g (or less) of oil or fat into a clean dry boiling tube and while still liquid add 1 g powdered potassium iodide and 20 ml of solvent mixture (2 vol. glacial acetic acid + 1 vol. chloroform). Place the tube in boiling water so that the liquid boils within 30 s and allow to boil vigorously for not more than 30 s. Pour the contents quickly into a flask containing 20 ml of potassium iodide solution (5 per cent), wash out the tube twice with 25 ml water and titrate with 0·002 M sodium thiosulphate solution using starch. A blank should be performed at the same time. The peroxide value is often reported as the number of ml of 0·002 M sodium thiosulphate per g of sample. If the value so obtained is multiplied by 2, the figure then equals milliequivalents of peroxide oxygen per kg of sample (mEq/kg), which has greater international recognition.

The following method is that prescribed in BS 684: Section 2.14; ISO 3960 and IUPAC 2.501. AOCS Cd 8–53 is similar.

PROCEDURE
Weigh accurately (M g) about 1 to 4 g, or less if the sample is clearly rancid, into a dry 250 ml stoppered conical flask flushed with inert gas. Add 10 ml chloroform, dissolve the fat by swirling, add 15 ml of glacial acetic acid and 1 ml fresh saturated aqueous potassium iodide solution. Stopper the flask, shake for 1 min and place the flask for exactly one further minute in the dark (IUPAC specifies 5 min). Add about 75 ml water, mix and titrate (V ml) the freed iodine with 0·002 M or 0·01 M sodium thiosulphate solution using soluble starch solution (1 per cent) as an indicator. Carry out a reagent blank determination (V_0), which should not exceed 0·5 ml of 0·01 M thiosulphate solution.

Where T is the exact molarity of the thiosulphate solution:

$$\text{Peroxide value} = \frac{(V - V_0)T}{M} \times 10^3 \text{ mEq/kg}$$

Fresh oils usually have peroxide values well below 10 mEq/kg. A rancid taste often begins to be noticeable when the peroxide value is between 20 and 40 mEq/kg. In interpreting such figures, however, it is necessary to take into account the particular oil or fat involved.

Colorimetric methods for determining peroxide values have been described by Driver *et al.* (1963), Swoboda and Lea (1958) and Barthel and Grosch (1974).

A reference method for anhydrous milk fat based on the oxidation of Fe(II) to Fe(III) and its measurement by the production of the red thiocyanate colour is described in ISO 3976. The method is very sensitive.

PROCEDURE
Weigh accurately about 0·3 g (W) of lipid sample in a glass cuvette having a volume of at least 15 ml and an optical path length of at least 15 mm. Add 9·60 ml of chloroform/methanol, 70/30 v/v, and mix to dissolve the sample. Add 0·05 ml of ammonium thiocyanate solution (30 per cent m/v), mix and measure the absorbance at 500 nm (E_0) against a blank of the chloroform/methanol mixture. Add 0·05 ml of ferrous chloride solution (0·35 per cent m/v containing 2 per cent of 10 M hydrochloric acid). Mix and after exactly 5 min measure the absorbance again (E_2). Simultaneously, carry out a reagent blank determination (E_1). Prepare a calibration graph of absorbance against μg Fe using 0·25–2·0 ml of ferric chloride solution containing 10 mg Fe/l, and 9·65–7·9 ml of chloroform/methanol mixture, 0·05 ml ammonium thiocyanate solution and 0·05 ml hydrochloride acid solution, 0·2 M. Then where m is the number of μg Fe $\equiv E_2 - (E_0 + E_1)$

$$\text{Peroxide value} = \frac{m}{55{\cdot}84 \times W} \text{ mEq/kg}$$

Kreis test (Rancidity index). The Kreis reaction involves the production of a red colour when phloroglucinol reacts with the oxidised fat in acid solution. The colour formed appears to be related to the increasing production of either epihydrin aldehyde or malonaldehyde (Patton *et al.*, 1951).

The various methods that have been proposed for carrying out the Kreis reaction have been reviewed by Mehlenbacher (1960). Due to its high sensitivity the Kreis test tends to give misleading results as colours are often produced by comparatively fresh oils.

Qualitative test. Shake 10 ml of the oil or melted fat vigorously with 10 ml of 0·1 per cent phloroglucinol solution in ether and 10 ml of conc. hydrochloric acid for 20 s. A pink colour indicates incipient rancidity. If the oil is diluted 1 in 20 with heptane and the test is still positive the rancidity in the sample will probably be evident to the taste and smell.

Quantitative procedure (BS 684: Section 2.32).

Weigh 0·8–10·2 g of oil or fat into 100 ml beaker. Melt sample of fat and add slowly with stirring 20 ml of phloroglucinol (0·1 g/100 ml in diethyl ether, freshly prepared) until sample dissolved. Transfer solution to a separating funnel, add 10 ml conc. HCl, shake well and allow to separate. Run off acid layer into a 1 inch (2·54 mm) Lovibond cell and match the colour using red, yellow and blue glasses. Express the result as red Lovibond units. Up to 3 red units indicates incipient rancidity; between 3–8 units indicates the end of the induction period; over 8 units indicates definite rancidity. The BS lists equivalent CIE colour co-ordinates to measured red, yellow and blue units.

Thiobarbituric acid number

The increase in the amount of red pigment formed in the reaction between 2-thiobarbituric acid (TBA) and oxidised lipids as oxidative rancidity advances has been applied to a wide variety of fatty foods. Working with rancid salmon oil, Sinnhuber *et al.* (1958) have shown that malonaldehyde (an end-product of oxidative decomposition) is probably involved in the reaction. The TBA test appears to measure deterioration in both extractable and non-extractable lipids and therefore has been more frequently applied to compound fatty foods, particularly flesh foods, rather than to 'pure' oils and fats. Earlier workers (Schwartz and Watts, 1957) heated a slurry of the flesh with the TBA reagent and extracted the pigment with amyl alcohol–pyridine mixture. Vyncke (1970) applied the reaction directly to TCA extracts of fish. Many workers have tended to employ a method due to Tarladgis *et al.* (1960) in which the reaction is applied to a distillate produced under standardised conditions from an acidified macerate in order to isolate TBA active interferences from carbohydrates and proteins. The results are expressed as malonaldehyde by reference to a standard graph prepared by using 1,1,3,3-tetra-ethoxypropane, which yields malonaldehyde by acid hydrolysis:

PROCEDURE

Macerate 10 g fatty food with 50 ml water for 2 min and wash into a distillation flask with 47·5 ml water. Add 2·5 ml of 4 M hydrochloric acid to bring the pH to 1·5, followed by an anti-foaming preparation and a few glass beads. Heat the flask by means of an electric mantle so that 50 ml distillate is collected in 10 min from the time boiling commences. Pipette 5 ml distillate into a glass-stoppered tube, add 5 ml TBA reagent (0·2883 g/100 ml of 90 per cent glacial acetic acid), stopper, shake and heat in boiling water for 35 min. Prepare a blank similarly using 5 ml water with 5 ml reagent. Then cool the tubes in water for 10 min and measure the absorbance (D) against the blank at 538 nm using 1 cm cells.

$$\text{TBA no. (as mg malonaldehyde per kg sample)} = 7{\cdot}8\,D$$

Working with distillates from beef, Pearson (1968) found a steady increase in the TBA number as spoilage progressed for individual meat but found that fresh samples gave rather variable figures. The thiobarbituric acid test is reviewed by Gray (1978).

The direct determination (without distillation) of 2-thiobarbituric acid value in oils and fats is described in IUPAC method 2.531; see also Pokorny and Dieffenbacher (1989) for collaborative study results and the method, which essentially is as follows:

PROCEDURE
Weigh 50–200 mg of sample (m) into a 25 ml volumetric flask. Dissolve in small volume of 1-butanol, make to volume with 1-butanol, and mix. Pipette 5·0 ml into a dry stoppered test tube and pipette 5·0 ml of TBA reagent (200 mg of 2-thiobarbituric acid in 100 ml 1-butanol, filtered, store at 4 °C not more than 7 days). Stopper, mix and place in water bath at 95 °C for 120 min. Cool, measure absorbance (A_S) at 530 nm in 10 mm cell against water. Run a reagent blank, absorbance (A_B) not to exceed 0·1.

$$\text{TBA value} = \frac{50 \times (A_S - A_B)}{m}$$

Refined oils in good condition should have TBA value of 0·02–0·08, crude oils or badly stored oil, 0·1–0·2.

p-Anisidine value

The extent of oxidation in oils and fats can be determined by the measurement of the formation of carbonyl compounds. The reaction of aldehydes with benzidine acetate (Holm *et al.*, 1957) has been replaced for safety reasons by a similar method using anisidine (*p*-methoxyaniline) (BS 684: Section 2.24; IUPAC 2.504). List *et al.* (1974) reported good correlation between the *p*-anisidine value (*p*-AV) of salad oils and their organoleptic scores.

PROCEDURE
Weigh accurately 0·5–4·0 g (M) of dry sample into a 25 ml volumetric flask. Dissolve and dilute to the mark with iso-octane (or *n*-hexane). Measure the absorbance (A_1) of the solution at 350 nm in a 10 mm cell against a blank iso-octane. Pipette 5 ml of the sample solution into a 10 ml stoppered test tube and add exactly 1 ml of *p*-anisidine solution (2·5 g/l in glacial acetic acid containing less than 0·1 per cent moisture). Similarly, carry out a reagent blank determination. After exactly 10 min measure the absorbance (A_2) as before against the reagent blank determination.

$$p\text{-AV} = \frac{25 \times (1{\cdot}2A_2 - A_1)}{M}$$

Ultraviolet absorption

Oxidised fatty acids containing conjugated double bonds absorb ultraviolet strongly between 230 and 375 nm, dienes absorbing at about 234 nm and trienes at about 268 nm. Conjugated trienes may be formed by industrial processing, e.g. decolorising with bleaching earths. A secondary absorption by trienes occurs at about 278 nm. In the early stages of oxidation the ultraviolet absorption increases somewhat proportionately to the uptake of oxygen and the formation of peroxides. The ultraviolet absorption curve is said to form a plateau just before the end of the induction period (Privett and Blank, 1962). The magnitude of ultraviolet absorbance is not readily related to the amount of oxidation so the method is best applicable to detecting relative changes in oxidation of an oil in comparison experiments or stability tests.

PROCEDURE (BS 684: Section 1.15; ISO 3656; IUPAC 2.505)
Weigh accurately into a 25 ml volumetric flask, an amount of the oil sample so that the absorbance of its solution in iso-octane in a 10 mm quartz cell lies between 0·2 and 0·8. Trace the absorption curve against iso-octane between 220 and 320 nm and select the wavelenghs (λ) of maximum absorption near 230, 268 and 278 nm, and the absorbance (A) at these points.

$$\text{The specific absorbance } E_{1\,\text{cm}}^{1\%}(\lambda) = \frac{A_\lambda}{c \times d}$$

where c is the concentration of the sample solution (g/100 ml)
d is the cell length in cm.

The method has been extended by Parr and Swoboda (1976) by reduction and dehydration of the hydroperoxides of polyenes in stored foods into further conjugated chromophores.

Stability tests. Accelerated shelf-life testing

Industrially, it is frequently important to find out whether a particular oil or fat is likely to be able to withstand storage over a period. The probable course of the rancidity can be predicted by methods in which the oxidation is accelerated by passing air through the sample or by raising the temperature or both. Samples can be taken off periodically and examined for FFA and peroxide value or other indices. In the Swift Stability Test (IUPAC 2.506; BS 684: Section 2.25; AOCS Ca 12–57), the sample is aerated at 98 °C and the peroxide value is determined at hourly intervals. A peroxide value of 20 indicates the end of the induction period for lard and beef fat. Oxygen absorption methods assess the oxygen uptake of the heated fat by weighing or by measuring the fall in pressure in a closed flask. The most modern and effective method for estimating the shelf-life stability of oils and fats to oxidative rancidity is by use of the Metrohm Rancimat instrument, described by Frank *et al.* (1982). The sample is subjected to a stream of air at temperatures of 100–150 °C and the volatile products, e.g. formic acid (formed during the oxidation phase in which peroxide value and oxygen absorption rapidly increases), are trapped in water and automatically detected conductometrically. Woestenburg and Zaalberg (1988) have reported on interlaboratory tests of the Rancimat method and Läubli and Bruttel (1986) found good correlation between the AOCS active oxygen method and the much labour-saving Rancimat method.

Methods used in accelerated shelf-life testing for oxidative rancidity have been reviewed by Ragnarsson and Labuza (1977) and by Rossell and Hudson (1989).

FAT-SOLUBLE VITAMINS

Introduction

Vitamins A and D are found in animal and fish liver oils, and dairy products but little in animal and vegetable oils and fats, meat and cereals. Conversely, the richest sources of vitamin E are oils and fats of vegetable origin, cereals and eggs. Little comes from animal fats and meat, fruit and vegetables. Vitamin K, which is associated with blood clotting, exists in nature in two forms. Vitamin K_1 occurs mainly in green vegetables and liver, whereas the series of vitamin K_2 are synthesised by bacteria, e.g. in the gut.

In the past, methods of analysis for the fat-soluble vitamins have not always been very successful because each of the vitamins consist of several closely related compounds. However, in recent years improved extraction, clean-up, column materials and detection systems have permitted the development of HPLC techniques that can provide full data on vitamin activity in oils, fats and lipid extracts of foods. For example, Mulholland (1986) reported very sensitive HPLC determination of vitamins A, D and E using narrow bore (100 × 2·1 mm) columns and diode array detection.

Papers published on the applications of HPLC to the determination of fat-soluble vitamins in foods and feeds over the period 1977–88 have been thoroughly reviewed and listed by Ball (1988) and Van Niekerk (1988).

VITAMIN A (RETINOL)

Vitamin A activity in food is mainly due to the all-*trans* isomer of retinol, which is the most abundant and biologically active member of the vitamin A group. The other important isomer is 13-*cis*-retinol, which has about 75 per cent of the activity of the all-*trans* isomer. In animal sources, vitamin A occurs as mixed esters of long chain fatty acids, mainly palmitate, except in egg where it is principally unesterified. In the diet, β-carotene and other carotenoids provide most of vitamin A. The carotenoids are now being used extensively as natural colouring matters. There are more than 80 naturally occurring carotenoids of which about 10 have provitamin A activity. Of these α-carotene, β-carotene, γ-carotene and cryptoxanthin are most important nutritionally. Retinyl acetate or palmitate is used to supplement a variety of food products, e.g. margarine.

The Labelling of Food Regulations require that vitamin A is calculated as micrograms of retinol or retinol equivalent on the basis that 6 μg of β-carotene or 12 μg of other biologically active carotenoids equal 1 μg of retinol equivalent.

Vitamin A is very sensitive to actinic light. Its determination should be carried out in subdued or amber artificial light, using amber glassware.

Vitamin A in oils may be determined by the rapid Carr–Price method, in which the blue colour formed with antimony trichloride is measured, or by measurement of the ultraviolet absorbance in an organic solvent. When the expected concentration of vitamin A in the oil is high, the blue value or the absorption spectrum can be determined directly on a solution of the sample, but with most samples the determination should be made on a solution of the unsaponifiable matter.

Carr–Price blue value

ANTIMONY TRICHLORIDE REAGENT (BP 1932).

Wash chloroform two or three times with its own volume of water and dry it over anhydrous potassium carbonate. Pour off and distil, rejecting the first 10 per cent of distillate. During drying and distillation protect the chloroform from light. Wash antimony trichloride with the pure dry chloroform until the washings are clear. Prepare a solution (saturated at 20 °C) of the washed antimony trichloride with the pure dry chloroform. The solution contains 21–23 per cent m/v of $SbCl_3$ and should be kept in an amber-coloured well-stoppered bottle. To assay mix 1 ml with a solution of 2 g of sodium potassium tartrate in 20 ml of water. Swirl the mixture, add 2 g of sodium bicarbonate and titrate with 0·1 M iodine (1 ml 0·1 M iodine = 0·011 41 g $SbCl_3$).

PROCEDURE USING A TINTOMETER

Pipette 0·2 ml of the prepared solution (either 20 per cent m/v of the oil in chloroform or the solution of the unsaponifiable matter in chloroform) into a 10 mm cell placed on the platform of a Lovibond Tintometer. Add 2 ml of the antimony trichloride reagent rapidly and observe the maximum intensity produced against the blue glass. For maximum accuracy the tintometer reading should lie between 4·0 and 6·0 blue units. Calculate the value of 1 g of sample and multiply the figure obtained by 1·28 to give i.u. per g.

PROCEDURE USING A SPECTROPHOTOMETER
Using a spectrophotometer fitted with a test-tube attachment, pipette 4 ml antimony trichloride into the reference tube, and add 1 ml chloroform. Measure 2 ml reagent into each of the other tubes. With the reference tube in the light beam set the instrument to zero at 620 nm. Move one of the other tubes into the beam, set the instrument to 'Test' and add to it 0·5 ml of the solution containing vitamin A, mix rapidly and measure the optical density. Repeat with the other tubes and take the average optical density. Compare the reading against a standard graph prepared from vitamin A acetate solutions over the range 0–15 i.u. per ml. For the most accurate results all readings should be made 5 s after adding the vitamin A solutions. For other spectrophotometers, using 10 mm cells, scale down quantities accordingly.

The BP spectrophotometric method

The vitamin A content of margarine (p. 593) and liver oils can be determined spectrophotometrically although HPLC is now the method of choice (see p. 647). The BP expresses the result in terms of the Unit (i.e. the international unit) which is 0·344 μg of all-*trans*-vitamin A acetate ≡ 0·3 μg all-*trans*-vitamin A alcohol. Under certain conditions irrelevant absorption of non-vitamin A substances can be allowed for by applying corrections.

Vitamin A in ester form

Samples that are not totally soluble in cyclohexane are fractionated by extraction or other means not involving saponification. Otherwise the sample is dealt with as described under 'Other Vitamin A'.

PROCEDURE
Dissolve the sample or a prepared fraction in cyclohexane such that it contains 9–15 Units/ml and obtain the wavelength of maximum absorption. Measure the extinctions at the wavelengths in Table 16.5 and calculate as fractions relative to that at 328 nm. Also calculate the $E^{1\%}_{1\,\mathrm{cm}}$ figure at 328 nm. If the wavelength of maximum absorption is between 326–329 nm and the observed relative extinctions are within 0·02 of those in Table 16.5.

Table 16.5 Relative extinctions of vitamin A in cyclohexane at various wavelengths (BP 1973)

Wavelength (nm)	Relative extinction
300	0·555
316	0·907
328	1·000
340	0·811
360	0·299

$$\text{Potency (Units/g)} = 1900 \times E^{1\%}_{1\,\mathrm{cm}} \text{ at 328 nm.}$$

If the maximum lies in the same range but the relative extinctions are not within 0·02 of those in Table 16.5 the following correction can be applied:

$$E_{328}\,(\text{corrected}) = 3{\cdot}52(2E_{328} - E_{316} - E_{340}).$$

This is used if the corrected extinction is within −15 per cent and −3 per cent of the uncorrected figure, but if within ±3·0 per cent the potency can be calculated from the uncorrected extinction. If however the corrected extinction is outside −15 per cent to +3 per cent of the uncorrected extinction or the wavelength of maximum absorption is not between 326–329 nm the sample should be dealt with as described under 'Other Vitmin A'.

Other vitamin A

Mix a weighed quantity of sample containing not more than 1 g fat and at least 500 Units of vitamin A with 30 ml absolute alcohol and 3 ml of 50 per cent potassium hydroxide and boil gently under reflux for 30 min in a stream of oxygen-free nitrogen. Cool rapidly, add 30 ml water, transfer to a separator, wash in with 3 × 50 ml ether and extract the vitamin A by shaking for 1 min. After complete separation discard the lower layer and wash the extract with 4 × 50 ml water, mixing especially cautiously during the first two washes to avoid emulsion formation. Evaporate the washed extract down to about 5 ml and remove the remaining ether in a stream of nitrogen at room temperature. Then dissolve the residue in sufficient isopropyl alcohol to give a solution containing 9–15 Units/ml and measure the extinctions at 300, 310, 325 and 334 nm and the wavelength of maximum absorption. If the latter is between 323–327 nm and the extinction at 300 nm relative to that at 325 nm does not exceed 0·73:

$$E_{325}\text{ (corrected)} = 6{\cdot}815\,E_{325} - 2{\cdot}555\,E_{310} - 4{\cdot}260\,E_{334}$$
$$\text{Potency (Units/g)} = 1830 \times E^{1\%}_{1\,\text{cm}} \text{ at 325 nm (corrected).}$$

If the corrected extinction is within ±3·0 per cent of the uncorrected figure the potency is calculated from the uncorrected extinction. If however the wavelength of maximum absorption is not between 323–327 nm or the relative extinction at 300 nm exceeds 0·73 the unsaponifiable fraction of the sample must be chromatographed.

The BP requires that cod-liver oil shall contain at least 600 Units of vitamin A activity per g and halibut-liver oil at least 30 000 Units per g. Halibuit-liver oil capsules contain, in each, 0·12–0·18 ml oil and 4000–5250 Units vitamin A.

Chromatographic methods

The preferred technique for the accurate determination of vitamin A and carotenoids is HPLC. Unsupplemented foods mostly contain retinol and mixed retinyl esters and are saponified, as are supplemented foods, to convert the esters to the free alcohol and to release the vitamin from the food matrix. Published HPLC procedures are reviewed by Ball (1988) and Van Niekerk (1988).

A method for added vitamin A in animal feeds (AMC, 1985) involves saponification of the sample with ethanolic KOH and extraction of the vitamin into light petroleum. The solvent is removed by evaporation and the residue dissolved in propan-2-ol. This solution is analysed by reversed phase HPLC, 5–10 μm particle size, mobile phase methanol/water (97:3), with ultraviolet detection at 325 nm. These conditions result in a single peak that includes all retinol isomers.

The method for all-*trans* and 13-*cis* retinol in foods used at LGC is basically as follows:

HPLC: 25 cm × 4·6 mm Partisil 10 ODS2 or equivalent, ultraviolet 325 nm, mobile phase methanol/water (90:10).

PROCEDURE

Weigh about 50 g sample, or 20 g for a fatty food, to the nearest 0·1 g, into a 500 ml saponification flask fitted with a side-arm and delivery tube. Add 1 g approx. of pyrogallol as antioxidant, and 150 ml ethanolic potassium hydroxide solution (28 per cent m/v). Saponify for 30 min under reflux on a boiling-water bath whilst bubbling a stream of nitrogen through the solution via the side-arm of the flask. Quantitatively transfer the solution, whilst still warm, to a 1 litre separating funnel using 150 ml water to aid the transfer, cool under running water. Add 500 ml mixed ethers, using a portion to rinse the saponification flask, and shake the separating funnel plus contents vigorously for 3 min. Allow to stand until the layers separate and run off the bottom layer into a second 1 litre separating funnel. Extract the solution with a further 500 ml portion of mixed ethers, shaking vigorously for 3 min, allow to stand until the layers separate, and run the bottom layer off to waste.

Some samples such as high cereal foods leave a large solid residue on saponification that will block the separating funnels. A modified extraction technique is used for such samples.

After saponification, allow the saponification flask and contents to cool, and decant the supernatant liquid into the separating funnel.

Add 150 ml water to the separating funnel and shake vigorously for 3 min. Allow to stand until the layers separate, and run off the bottom layer into a second 1 litre separating funnel. Extract the solid residue in the saponification flask with several aliquots from a second 500 ml mixed ethers, as previously, and decant into the second separating funnel. Shake the separating funnel and contents vigorously for 3 min, allow to stand until the layers separate, and run the bottom layer off to waste.

Wash the separate extracts with 150 ml water, inverting the separating funnels gently several times to avoid the formation of emulsions. Allow to stand until the layers separate and run the bottom layer off to waste. Wash the mixed ether extracts with further 150 ml portions of water using progressively more vigorous shaking until the wash solutions are neutral to phenolphthalein.

Transfer the washed extracts to a 1 litre amber rotary evaporation flask, using a few ml ethanol to rinse the separating funnels. Add 2 ml BHT solution 1 mg/ml as antioxidant and evaporate to dryness on a rotary evaporator at 40 °C. Remove any remaining traces of water by adding more ethanol and re-evaporating. Immediately stopper the flask to preserve the inert atmoshere, cool to room temperature, and dissolve the residue in 10 ml methanol. Dilute as necessary to give a concentration of all-*trans*-retinol in the range 1 to 10 μg/ml. Transfer the sample solution to a specimen tube, for chromatography, and immediately protect from light by enclosing in aluminium foil.

Inject 50 μl aliquots of fresh all-*trans*-retinol calibration solution (10 μg/ml in methanol, whose true concentration is calculated from its absorbance at 325 nm in a 10 mm cell, using $E^{1\%}_{1\,\mathrm{cm}} = 1832$) onto the HPLC column until duplicate injections are reproducible to within 1 per cent. Measure the peak area and retention distance of the all-*trans*-retinol peak, and calculate the theoretical retention distance of 13-*cis* retinol using the relative retention distance data obtained from a chromatographic run of the system suitability test solution, namely 5 ml of all-*trans*-retinol + 1 ml 13-*cis* retinol (obtained by saponification and extraction of 13-*cis* retinyl acetate), both 10 μg/ml in methanol; $E^{1\%}_{1\,\mathrm{cm}}$ for 13-*cis* retinol = 1686.

Chromatograph the sample extract under the same conditions, identify the 13-*cis* and all-*trans*-retinol peaks from their retention distances, and measure their peak areas.

CALCULATION

$$\frac{\textit{trans}\text{-retinol}}{\mu\text{g}/100\text{ g}} = \frac{\text{Area } \textit{trans} \text{ sample}}{\text{Area } \textit{trans} \text{ standard}} \times \frac{\textit{trans} \text{ standard}}{\text{conc. } \mu\text{g/ml}} \times V \times \frac{100}{W}$$

$$\frac{13\text{-}\textit{cis}\text{ retinol}}{\mu\text{g}/100\text{ g}} = \frac{\text{Area 13-}\textit{cis} \text{ sample}}{\text{Area } \textit{trans} \text{ standard}} \times \frac{\textit{trans} \text{ standard}}{\text{conc. } \mu\text{g/ml}} \times \frac{1832}{1686} \times V \times \frac{10}{W}$$

where, V = volume of solution containing sample weight
W = weight of sample taken.

Woolard and Woolard (1988) determined retinyl esters in oil based foods without saponification or extraction using unmodified silica HPLC columns. They recommend fluorescence detection (325 nm excitation and 465 nm emission) since interference from other lipid-soluble substances is eliminated. An ester not present in the sample was used for calibration. Collins and Chow (1984) also used fluorescence detection for down to 1 ng of vitamin A with RP-C_{18} column and methanol as mobile phase.

Reversed phase HPLC is very suitable for the determination of β- and α-carotene, the carotenoids of main interest as vitamin A precursors (lycopene, the red carotenoid in tomatoes, does not give rise to vitamin A). The procedure used at LGC involves the identical saponification and extraction as used for vitamin A (see above), with the residue from the rotary evaporator being dissolved in 10 ml of mobile phase (acetonitrile/chloroform, 92:8). The HPLC is 25 cm × 4·6 mm i.d. Spherisorb 5 ODS2 or equivalent, with diode array detection at 450 nm. Calibration solution: 5 ml each of α- and β-carotene solutions 5 μg/ml in hexane, evaporate to dryness under nitrogen, redissolve residue in 5 ml mobile phase.

For foods not containing lipid, various solvent extraction systems can be used, such as acetone, tetrahydrofuran and methanol for wet foods and chloroform or hexane for dry foods.

VITAMIN D

The two nutritionally important forms of vitamin D are vitamin D_2 (ergocalciferol) and D_3 (cholecalciferol). Vitamin D_3 occurs naturally in animal products. It may arise either from the animal's diet or by the action of sunlight on the animal's skin. Vitamin D_2 appears to be as effecive as D_3 in man and is manufactured from plant materials. It is required by law in the UK to be added to margarine. Foods, except for fish liver oils, contain only traces of naturally ocurring vitamin D. Main dietary sources (μg/100 g) are eggs (1·75), butter (0·75), liver (0·75), cheese (0·25) and milk (0·03).

Chemically, vitamin D_2 and D_3 are sterols and may be found along with other sterols such as phytosterols and cholesterol but at much lower concentrations in the unsaponifiable matter of oils, fats and food lipid extracts. Vitamin D_2 and vitamin D_3 in solution are in equilibrium with their provitamins. It follows therefore that provitamins provide potential vitamin D activity.

The World Health Organisation (1949–50) defined the International Unit of vitamin D as the activity of 0·025 μg of vitamin D_3 (activated 7-dehydrocholesterol). The $E^{1\%}_{1\,cm}$ at 265 nm for recrystallised vitamin D_3 in ethanol is given at 490, which corresponds to a molecular extinction coefficient of 18 800. In respect of claims the Labelling of Food Regulations requires the vitamin to be calculated as micrograms of cholecalciferol or micrograms of ergocalciferol. Halibut liver oil may contain up to 300 units of vitamin D per g.

Analysis

The BP requires that the anti-rachitic activity (vitamin D) of cod-liver oil (not less than 85 units) is estimated by biological assay. The sample and a standard preparation of the vitamin are given to separate groups of rats that have been previously subjected to rachitogenic diet. After 10–14 days the extent to which the rickets has been cured is then estimated by an X-ray examination of the bones.

If acetyl chloride is mixed with the Carr–Price reagent, the blue colour so formed quickly fades and an orange colour is developed, which is suitable for semi-quantitative measurements of vitamin D. The reagent, which must be fresh and free from alcohol, contains 20 per cent m/v of antimony trichloride in chloroform with 4 per cent pure acetyl chloride added. For the method, 9 vol. of this reagent are added to 1 vol. of the unsaponifiable matter in chloroform and the value for $E^{1\%}_{1\,cm}$ at 500 nm is measured. This extinction × 1800 represents an approximate measure of vitamin D per gram. Cox (1950) suggested that this method is suitable for control purposes provided that the amount of vitamin A is not more than five times that of the vitamin D. If possible a suitable correction should be made for absorption due to other sterols.

Rogers (1954) reduced the absorption due to other substances present by using as reagent a solution of 20 per cent antimony trichloride in ethylene dichloride with 2 per cent acetyl chloride added. The extinction at 500 nm is recorded every ½ min for 6 min and the resultant graph is extrapolated to zero time.

Chromatographic methods

Because of the very low levels of vitamin D in foods, even in vitamin fortified foods, chromatographic interference from sterols, vitamin E, carotenoids and other substances necessitates extensive clean-up and extraction procedures for HPLC. Saponification of the samples necessary to release the vitamin, followed by ethers extraction. Sterols can be precipitated out and other interferences removed by column chromatography (AOAC methods book), TLC (e.g. Jackson *et al.*, 1982) or clean-up cartridges (e.g. Indyk and Woolard, 1985). Further clean-up by semi-preparative HPLC can be used prior to final HPLC (e.g. Van Niekerk and Smit, 1980). Reversed phase HPLC that separates D_2 and D_3 is preferable so that one or the other can be added as an internal standard. Agarwal (1988) converted the extracted vitamin D from fortified milks following saponification to their isotachysterols with antimony trichloride. These were determined by normal phase HPLC (Spherisorb 3 μm) using hexane/ethyl acetate/methanol (97:2·5:0·05), ultraviolet 301 nm at which wavelength interferences are avoided.

The procedure used at LGC is saponification and extraction as for vitamin A with added D_2 as internal standard as follows. For milk powders, take up evaporated residue in hexane and use silica clean-up cartridge, or for margarine/cereals, take up in ethanol and use RP-C_{18} clean-up cartridge. Evaporate eluant and take up in hexane, inject into semi-preparative HPLC (Partisil 5 PAC, ultraviolet 265 nm, hexane/amyl alcohol; 99:1), collect D_2/D_3 fraction, evaporate and dissolve residue in final mobile phase. Inject into analytical HPLC (Zorbax ODS, ultraviolet 265 nm, acetonitrile/methanol; 90:10) and compare with 10 μg/ml standards.

Published HPLC procedures have been reviewed by Ball (1988) and Van Niekerk (1988).

VITAMIN E

Vitamin E activity in foods is due to four methyl-substituted derivatives of tocol (α-, β-, γ- and δ-tocopherol) and four corresponding α-, β-, γ- and δ-tocotrienols. They have different biological potency, α-tocopherol, the principal tocol in animal tissues, being the most active. In seed oils other tocols are present. For example, β-tocopherol has 30 per cent, γ-tocopherol 15 per cent and α-tocotrienol (ζ-tocopherol) 21 per cent of the activity of α-tocopherol. Other forms have less than 5 per cent of the α-form activity. The tocols are natural antioxidants, blocking free radical chain reactions of lipid oxidation. Vitamin E is found at high concentrations in vegetable seed oils, cereal grains, e.g. wheat germ oil contains 2000 μg/g of total vitamin E, and nuts. Eggs, butter, cheese and liver are also rich in vitamin E. Palm oil contains significant amounts of tocotrienols.

For the determination of vitamin E the method of Emmerie and Engel (1939) is useful for routine purposes. This depends upon the oxidation of the tocopherol in alcoholic solution by ferric chloride and the subsequent measurement (absoptiometrically) of the red colour produced when the resultant ferrous iron reacts with αα′-dipyridyl. The AOAC methods book variation uses bathophenanthroline. Materials such as cholesterol and carotenoids interfere. The Vitamin E Panel of the Analytical Methods Commitee of the SAC (1959) reviewed the methods that have been proposed for the separation and determination of tocopherols and described a method that can be applied to oils, foods and feedingstuffs. Dickes (1966) separated vitamin E using TLC and after applying the Emmerie–Engel reaction determined the α- and β-tocopherols spectrophotometrically.

The Furter–Meyer method makes use of the oxidation of the tocopherols by nitric acid to the red tocoquinone. It is more specific than the Emmerei–Engel method, but suffers from the practical disadvantage that it is more time-consuming and requires the use of a larger sample. The tocopherols are oxidised rapidly in the presence of alkali but are stable in acid so the initial saponification is carried out with alcoholic sulphuric acid.

PROCEDURE

Place a suitable weight of oil (e.g. 1 g) into a 100 ml flask fitted with a reflux condenser. Then add 10 ml absolute alcohol and 20 ml M alcoholic sulphuric acid. Wrap the condenser and flask in aluminium foil, reflux for 45 min and cool. Add 50 ml water, transfer to a separating funnel of low actinic glass (or an ordinary separator covered with aluminiumn foil) with the aid of a further 50 ml of water. Extract the unsaponifiable matter with 5 × 30 ml diethyl ether. Wash the combined ether extracts free from acid and dry over anhydrous sodium sulphate. Evaporate the extract at a low temperature whilst protecting it from light, the final traces of solvent being removed in a stream of nitrogen. Then dissolve the residue immediately in 10 ml absolute alcohol.

Transfer aliquots of solutions of the sample and standards (0·3–3·0 mg vitamin E) to a 20 ml volumetric flask. Add 5 ml absolute alcohol, followed by 1 ml conc. nitric acid (*CARE*: add dropwise with constant swirling). Place the flask on a water bath at 90 °C for exactly 3 min from the time the alcohol begins to boil. Cool rapidly under running water and adjust to volume with absolute alcohol. Measure the absorbance at 470 nm against a blank containing 5 ml absolute alcohol and 1 ml conc. nitric acid treated in a similar manner (cf. Association of Vitamin Chemists, 1966).

Chromatographic methods

Slover *et al.* (1983) used a capillary GLC procedure. About 100 mg of oil, fat or extracted lipid together with 5,7-dimethyl tocol as internal standard is saponified with aqueous KOH in the presence of BHA and pyrogallol. The residual unsap. (extracted with hexane) is silanised to TMS ethers and injected into a 50 m × 0·25 mm glass capillary coated with Dexil 400.

The method of choice for vitamin E is HPLC. Methods published have been reviewed by Ball (1988) and Van Niekerk (1988). The IUPAC standardised method, 2.432, with collaborative study results are reported by Pocklington and Dieffenbacher (1988). In the method used at LGC for α- and γ-tocopherol, γ- and β-tocopherols are not separated but β-tocopherol is very rarely present in significant amounts. The sample is saponified and extracted with mixed ethers as described for vitamin A above. The ethers are evaporated and the residual unsap. is dissolved in methanol, diluted to give α-tocopherol about 20 μg/ml and γ-tocopherol about 10 μg/ml, injected and compared with standards. The column uses Spherisorb 5 ODS 2, methanol/water (95:5), and fluorescence detection with excitation at 290 nm, and emission at 330 nm. Ultraviolet absorbance at 292 nm can be used but with less sensitivity and selectivity.

REFERENCES

Ackman, R. G., Barlow, S. M., Duthie, I. F. & Smith, G. L. (1983) *Journal of Chromatographic Science*, **21**, 87.

Agarwal, V. K. (1988) *Journal of the Association of Official Analytical Chemists*, **71**, 19.

Allen, J. C. & Hamilton, R. J. (eds) (1989) *Rancidity in Foods*, 2nd edn. London: Elsevier Applied Science.

AMC (1985) *Analyst*, **110**, 1019.

Anon (1979) *Report of the Government Chemist.* London: HMSO.

AOCS (1978) *Official and Tentative Methods of the American Oil Chemists' Society*, 3rd edn. Champaign, Illinois: AOCS.
Association of Vitamin Chemists (1966) *Methods of Vitamin Assay*, 3rd edn. USA: Interscience.
Bains, G. S., Rao, S. V. & Bhatia, D. S. (1964) *Journal of the American Oil Chemists' Society*, **41**, 831.
Baker, D. (1964) *Journal of the American Oil Chemists' Society*, **41**, 21.
Ball, G. F. M. (1988) *Journal of Micronutrient Analysis*, **4**, 255.
Bandyapadhaya, G. K. & Dutta, J. (1975) *Journal of Chromatography*, **114**, 280.
Barron, J. C. R. & Santa-Maria, G. (1987) *Chromatographia*, **23**, 209.
Barthel, G. & Grosch, W. (1974) *Journal of the American Oil Chemists' Society*, **51**, 540.
Bascoul, J., Domergue, N., Olle, M. & Crastes de Paulet, A. (1986) *Lipids*, **21**, 283.
Bastijns, L. J. (1970) *Journal of the Science of Food and Agriculture*, **21**, 576.
Bergelson, L. D., Dyatlovitskaya, E. V. & Voronkova, V. V. (1964) *Journal of Chromatography*, **15**, 191.
Berger, K. (1986) *Food Technology*, **40**, 72.
Beuchat, L. R. & Worthington, R. E. (1978) *Journal of Food Technology*, **13**, 355.
Blumenthal, M. M. & Stockler, J. R. (1986) *Journal of the American Oil Chemists' Society*, **63**, 687.
Borch, R. F. (1975) *Analytical Chemistry*, **47**, 2437.
Bowers, R. H. (1978) *Journal of the American Oil Chemists' Society*, 55, 350.
Breuer, B., Stuhlfauth, T. & Fock, H. P. (1987) *Journal of Chromatographic Science*, **25**, 302.
Chang, S. S., Peterson, R. J. & Chi-tang Ho (1978) *Journal of the American Oil Chemists' Society*, **55**, 718.
Chapman, G. W. (1979) *Journal of the American Oil Chemists' Society*, **56**, 77.
Chaytor, J. P. (1987) *Food Chemistry*, **23**, 19.
Chobanov, D., Tarandjiska, R. & Chobanova, R. (1976) *Journal of the American Oil Chemists' Society*, **53**, 48.
Christie, W. W. (1973) *Lipid Analysis*. Oxford: Pergamon.
Christie, W. W. (1980). In *Fats and Oils: Chemistry and Technology*, eds R. J. Hamilton & A. Bhati. London: Applied Science Publishers.
Christie, W. W. (1986) *Journal of Chromatography*, **361**, 396.
Christie, W. W. (1987) *HPLC and Lipids—A Practical Guide*. Oxford: Pergamon.
Christie, W. W. & Moore, J. H. (1970) *Biochimica et Biophysica Acta*, **210**, 46.
Christopherson, S. W. & Glass, R. L. (1969) *Journal of Dairy Science*, **52**, 1289.
Coleman, M. H. & Fulton, W. C. (1961). In *Enzymes for Lipid Metabolism*, ed. P. Desnuelle. Pergamon: Oxford.
Collins, C. A. & Chow, C. K. (1984) *Journal of Chromatography*, **317**, 349.
Conacher, H. B. S. (1975) *Journal of the Association of Official Analytical Chemists*, **58**, 488.
Cox, H. E. (1950) *Analyst*, **75**, 521.
Croon, L. B., Rogstad, A., Leth, T. & Kiutamo, T. (1986) *Fette Seifen Anstrichmittel*, **88**, 87.
Daun, J. K., Mazur, P. B. & Marek, C. J. (1983) *Journal of the American Oil Chemists' Society*, **60**, 1751.
Dickes, G. J. (1966) *Journal of the Association of Public Analysts*, **4**, 50.
Doeden, W. G., Kushibab, E. M. & Ingala, A. C. (1980) *Journal of the American Oil Chemists' Society*, **57**, 73.
Dong, M. W. & Dicesare, J. L. (1983) *Journal of the American Oil Chemists' Society*, **60**, 788.
Driver, M. G., Koch, R. B. & Salwin, H. (1963) *Journal of the American Oil Chemists' Society*, **40**, 504.
Durst, H. D., Milano, M., Kikta, E. J., Connelly, S. A. & Grushka, E. (1975) *Analytical Chemistry*, **47**, 1797.
Emmerie, A. & Engel, C. (1939) *Recueil des travaux chimiques des Pays-Bas et de la Belgique*, **58**, 283.
FAO (1978) *Dietary Fats and Oils in Human Nutrition*. Food and Nutrition Paper, Rome: FAO.
Fedeli, E., Cortesi, N., Camurati, F. & Jacini, G. (1972) *Journal of the American Oil Chemists' Society*, **49**, 233.
Firestone, D. & Horwitz, W. (1979) *Journal of the Association of Official Analytical Chemists*, **62**, 709.
Flanzy, J., Boudon, M., Leger, C. & Pihet, J. (1976) *Journal of Chromatographic Science*, **14**, 17.
Frank, J., Geil, J. V. & Freaso, R. (1982) *Food Technology*, **36**, 71.
Frankel, E. N. (1980) *Progress in Lipid Research*, **19**, 1.
Frankel, E. N., Smith, L. M., Hamblin, C. L., Cheveling, R. K. & Clifford, A. J. (1984) *Journal of the American Oil Chemists' Society*, **61**, 87.
FSC (1970) *Food Standards Committee Report on the Pre-1955 Compositional Orders*. London: HMSO.
Ganguli, N. C. & Jain, M. K. (1973) *Journal of Dairy Science*, **56**, 19.
Gildenberg, L. & Firestone, D. (1985) *Journal of the Association of Official Analytical Chemists*, **68**, 46.
Gracian Tous, J., Pocklington, W. D. & Hautfenne, A. (1986) *Pure and Applied Chemistry*, **58**, 1023.
Graupner, A. J. & Aluise, V. A. (1966) *Journal of the American Oil Chemists' Society*, **43**, 81.
Gray, J. I. (1978) *Journal of the American Oil Chemists' Society*, 55, 539.
Gunstone, F. G., Harwood, J. L. & Padley, F. B. (eds) (1986) *The Lipid Handbook*. London: Chapman & Hall.

Haighton, A. J., Vermass, L. F. & den Hollander, C. (1971) *Journal of the American Oil Chemists' Society*, **48**, 7.
Hamilton, R. J. (1975) *Journal of Chromatographic Science*, **13**, 474.
Hamilton, R. J. (1987). In *Recent Advances in Chemistry and Technology of Fats and Oils*, eds R. J. Hamilton & A. Bhati. London: Elsevier Applied Science.
Hamilton, R. J. & Bhati, A. (eds) (1980) *Fats and Oils: Chemistry and Technology*. London: Applied Science Publishers.
Hamilton, R. J. & Rossell, J. B. (eds) (1986) *Analysis of Oils and Fats*. London: Elsevier Applied Science.
Hartman, L. & Antunes, A. J. (1971) *Laboratory Practice*, **20**, 481.
Hartman, L. & Lago, R. C. A. (1973) *Laboratory Practice*, **22**, 475.
Hartman, L., Lago, R. C. A., Azeredo, L. C. & Azeredo, M. A. A. (1987) *Analyst*, **112**, 145.
Hayes, G. (1985) *Food Trade Review*, 162.
Herslof, B. & Kindmark, G. (1985) *Lipids*, **20**, 783.
Hilditch, T. P. & Williams, P. N. (1964) *The Chemical Constitution of Natural Fats*, 4th edn. London: Chapman and Hall.
Hinshaw, J. V. & Seferovic, W. (1986) *Journal of High Resolution Chromatography and Chromatography Communications*, **9**, 73.
Holaday, C. E. & Pearson, J. L. (1974) *Food Science*, **39**, 1206.
Holm, U., Ekbom, K. & Wobe, G. (1957) *Journal of the American Oil Chemists' Society*, **34**, 606.
Huebner, V. R. (1961) *Journal of the American Chemical Society*, **38**, 628.
Indyk, H. & Woolard, D. C. (1985) *Journal of Micronutrient Analysis*, **1**, 121.
Itoh, T., Yoshida, K., Yatsu, T. & Matsumoto, T. (1981) *Journal of the American Oil Chemists' Society*, **58**, 545.
IUPAC (1987) *Standard Methods for the Analysis of Oils, Fat and Derivatives*, 7th edn. Oxford: Blackwell Scientific Publications.
Iverson, J. L. & Sheppard, A. J. (1977) *Journal of the Association of Official Analytical Chemists*, **60**, 284.
Jackson, P. A., Shelton, C. J. & Frier, P. J. (1982) *Analyst*, **107**, 1363.
Khan, G. R. & Scheinmann, F. (1978) *Progress in the Chemistry of Fats and Other Lipids*, **15**, 343.
Kirk, R. S. (1980) In *Food and Health: Science and Technology*, eds G. G. Birch & K. J. Parker. London: Applied Science.
Kirk, R. S., Mortlock, R. E., Pocklington, W. D. & Roper, P. (1978) *Journal of the Science of Food and Agriculture*, **29**, 880.
Kochar, S. P. & Meara, M. L. (1975) *Scientific and Technical Survey*, No. 87. Leatherhead: British Food Manufacturing Industries Research Association.
Krzynowek, J. (1985) *Food Technology*, **39**, 61.
Lambert, P., Singhai, O. P. & Ganguli, N. C. (1980) *Journal of the American Oil Chemists' Society*, **57**, 364.
Lanza, E. & Slova, H. T. (1981) *Lipids*, **16**, 260.
Läubli, M. W. & Bruttel, P. A. (1986) *Journal of the American Oil Chemists' Society*, **63**, 792.
List, G. R., Evans, C. D., Kwolek, W. F., Warner, K., Boundy, B. K. & Cowan, J. C. (1974) *Journal of the American Oil Chemists' Society*, **51**, 17.
Litchfield, C. (1972) *Analysis of Triglycerides*. London: Academic Press.
Lowry, R. R. & Tinsley, L. J. (1976) *Journal of the American Oil Chemists' Society*, **53**, 470.
Mangold, H. K. (1961) *Journal of the American Oil Chemists' Society*, **38**, 708.
Mangold, H. K., Zweig, G. & Sherma, J. (eds) (1984) *Handbook of Chromatography—Lipids*, Vols 1 and 2. Florida: CRC Press.
Mattson, F. H. & Volpenheim, R. A. (1961) *Journal of Biological Chemistry*, **236**, 1891.
Maxwell, R. J. & Schwartz, D. P. (1979) *Journal of the American Oil Chemists' Society*, **56**, 634.
Mehlenbacher, V. C. (1960) *The Analysis of Fats and Oils*. Illinois: Garrard.
Mohana Das, C. & Indrasenan, P. (1987) *International Journal of Food Science and Technology*, **22**, 339.
Morris, L. J. (1962) *Chemistry and Industry*, 1238.
Morris, L. J., Wharry, D. M. & Hammond, E. W. (1967) *Journal of Chromatography*, **31**, 69.
Mulholland, M. (1986) *Analyst*, **111**, 601.
Naudet, M. & Hautfenne, A. (1985) *Pure and Applied Chemistry*, **57**, 899.
Ojanpera, S. H. (1978) *Journal of the American Oil Chemists' Society*, 55, 290.
Ooms, R. & Van Pee, W. (1983) *Journal of the American Oil Chemists' Society*, **60**, 957.
Ozcimder, M. & Hammers, W. E. (1980) *Journal of Chromatography*, **187**, 307.
Padley, F. B. (1984) *Chemistry and Industry*, 788.
Padley, F. B. & Timms, R. E. (1978) *Chemistry and Industry*, 918.
Palmer, A. J. & Palmer, F. J. (1989) *Journal of Chromatography*, **465**, 369.
Paquette, G., Kupranyez, D. B. & Van de Voort, F. R. (1985) *Canadian Institute of Food Science and Technology Journal*, **18**, 112.

Parr, L. J. & Swoboda, P. A. T. (1976) *Journal of Food Technology*, **11**, 1.
Patton, S., Keeney, M. and Kurtz, G. W. (1951) *Journal of the American Oil Chemists' Society*, **28**, 391.
Paul, A. A. & Southgate, D. A. T. (1978) *The Composition of Foods*. London: HMSO.
Paul, A. A. & Southgate, D. A. T. (1980) *The Composition of Foods*, 1st Supplement. London: HMSO.
Pearson, D. (1965, 1974, 1975) *Journal of the Association of Public Analysts*, **3**, 76; **12**, 73; **13**, 39.
Pearson, D. (1968) *Journal of the Science of Food and Agriculture*, **19**, 553.
Pearson, D. (1976) *The Chemical Analysis of Foods*. Edinburgh: Churchill Livingstone.
Pei-fen, Wu & Nawar, W. W. (1986) *Journal of the American Oil Chemists' Society*, **63**, 1363.
Perkins, E. G. (ed.) (1975) *Analysis of Lipids and Lipoproteins*. Champaign, Illinois: AOCS.
Perkins, E. G. & Visek, W. J. (eds) (1983) *Dietary Fats and Health*. Monograph No. 10. Champaign, Illinois: AOCS.
Pocklington, W. D. & Dieffenbacher, A. (1988) *Pure and Applied Chemistry*, **60**, 877.
Pokorny, J. & Dieffenbacher, A. (1989) *Pure and Applied Chemistry*, **61**, 1165.
Privett, O. S. & Blank, M. L. (1962) *Journal of the American Oil Chemists' Society*, **39**, 465.
Prosser, A. R., Sheppard, A. J. & Hulbard, W. D. (1977) *Journal of the Association of Official Analytical Chemists*, **60**, 895.
Purdy, R. H. (1986) *Journal of the American Oil Chemists' Society*, **63**, 1062.
Ragnarsson, J. O. & Labuza, T. P. (1977) *Food Chemistry*, **2**, 291.
Raie, M. Y. & Salma, A. (1985) *Fette Seifen Anstrichmitel*, **87**, 246.
Ramamurthy, M. K. & Narayanan, K. M. (1971) *Milchwissenschaft*, **26**, 693.
Rees, H. H., Donnahey, P. L. & Goodwin, T. W. (1976) *Journal of Chromatography*, **116**, 281.
Robertson, J. A. (1972) *Journal of the American Oil Chemists' Society*, **49**, 239.
Robinson, J. L., Tsimidou, M. & Macrae, R. (1985) *Journal of Chromatography*, **324**, 35.
Rogers, A. R. (1954) *Journal of Pharmacy and Pharmacology*, **6**, 780.
Rossell, J. B. & Hudson, B. J. F. (1989) In *Rancidity in Foods*, 2nd edn, eds J. C. Allen & R. J. Hamilton. London: Elsevier Applied Science.
Rossell, J. B., King, B. & Downes, M. J. (1985) *Journal of the American Oil Chemists' Society*, **62**, 221.
SAC (1959) *Analyst*, **84**, 356.
Sakodynsky, K. I., Smolyaninov, G. A., Zelvensky, V. Y. & Glotova, N. A. (1979) *Journal of Chromatography*, **172**, 93.
Scholfield, C. R. (1979) *Journal of the American Oil Chemists' Society*, **56**, 510.
Schwartz, M. G. & Watts, B. M. (1957) *Food Research*, **22**, 76.
Sebedio, J. L., Farquharson, T. E. & Ackman, R. G. (1985) *Lipids*, **20**, 555.
Sebedio, J. L., Septier Ch. & Grandgirard, A. (1986) *Journal of the American Oil Chemists' Society*, **63**, 1541.
Sheppard, A. J. & Iverson, J. L. (1975) *Journal of Chromatographic Science*, **13**, 448.
Sheppard, A. J., Iverson, J. L. & Weihrauch, J. L. (1978a) In *Handbook of Lipid Research*, Vol. 1. Fatty Acids and Glycerides, ed. A. Kuksis. London: Plenum Press.
Sheppard, A. J. Waltking, A. E., Zmachinski, H. & Jones, S. T. (1978b) *Journal of the Association of Official Analytical Chemists*, **61**, 1419.
Shukla, V. K. S. & Srivastava, K. C. (1978) *Journal of High Resolution Chromatography and Chromatography Communications*, **1**, 214.
Singleton, J. A. & Patter, H. E. (1987) *Journal of the American Oil Chemists' Society*, **64**, 534.
Sinnhuber, R. O., Yu, T. C. & Yu, Te. C. (1958) *Food Research*, **23**, 626.
Slover, H. T. & Lanza, E. (1979) *Journal of the American Oil Chemists' Society*, **56**, 933.
Slover, H. T., Thompson, R. H. & Merola, G. V. (1983) *Journal of the American Oil Chemists' Society*, **60**, 1524.
Smith, E. C., Jones, A. D. & Hammond, E. W. (1980) *Journal of Chromatography*, **188**, 205.
Smith, L. M., Clifford, A. J., Hamblin, C. L. & Creveling, R. K. (1986) *Journal of the American Oil Chemists' Society*, **63**, 1017.
Smullin, C. F. (1974) *Journal of the Association of Official Analytical Chemists*, **57**, 62.
Sonntag, N. O. V. (1979) In *Bailey's Industrial Oil and Fat Products*, Vol. 1, ed. D. Swern. Chichester: John Wiley.
Spencer, G. F., Kwolek, W. F. & Princen, L. H. (1979) *Journal of the American Oil Chemists' Society*, **56**, 972.
Stevenson, S. G., Vaisey-Genser, M. & Eskin, N. A. M. (1984) *Journal of the American Oil Chemists' Society*, **61**, 1102.
Strocchi, A. (1981) *Journal of Food Science*, **47**, 36.
Sutton, R. W., Barraclough, A., Mallinder, R. & Hitchen, O. (1940) *Analyst*, **65**, 623.
Sweeney, J. P. & Weihrauch, J. L. (1976) *Critical Reviews in Food Science and Nutrition*, **8**, 131.
Swern, D. (ed.) (1979) *Bailey's Industrial Oil and Fat Products*, 4th edn, Vol. 1. New York: John Wiley.
Swern, D. (ed.) (1982) *Bailey's Industrial Oil and Fat Products*, 4th edn, Vol. 2. New York: John Wiley.

Swoboda, P. A. T. (1973) *IFST Proceedings*, **6**, 191.
Swoboda, P. A. T. (1985) *Journal of the American Oil Chemists' Society*, **62**, 287.
Swoboda, P. A. T. & Lea, C. H. (1958) *Chemistry and Industry* (33) 1090.
Szonyi, C., Tait, R. S. & Craske, J. D. (1962) *Journal of the American Oil Chemists' Society*, **39**, 276.
Takahashi, K., Hirano, T. & Zama, K. (1984) *Journal of the American Oil Chemists' Society*, **61**, 1226.
Tarladgis, B. G., Watts, B. M., Younathan, M. T. & Dugan, L. (1960) *Journal of the American Oil Chemists' Society*, **37**, 44.
Templeman, G. J., Sholl, J. J. & Labuza, T. P. (1977) *Journal of Food Science*, **42**, 432.
Thurlow, K. & Palmer, A. J. (1987). Personal communication.
Traitler, H. (1987) *Progress in Lipid Research*, **26**, 257.
Tsoukalas, B. & Grosch, W. (1977) *Journal of the American Oil Chemists' Society*, **54**, 590.
Van der Waal, R. J. (1964) *Journal of the American Oil Chemists' Society*, **37**, 18.
Van Niekerk, P. J. (1988) In *HPLC in Food Analysis*, 2nd edn, ed. R. Macrae. London: Academic Press.
Van Niekerk, P. J. & Burger, A. E. C. (1985) *Journal of the American Oil Chemists' Society*, **62**, 531.
Van Niekerk, P. J. & Smit, S. C. C. (1980) *Journal of the American Oil Chemists' Society*, **57**, 417.
Veazey, R. L. (1986) *Journal of the American Oil Chemists' Society*, **63**, 1043.
de Vries, B. (1962) *Chemistry and Industry*, 1049.
Vyncke, W. (1970) *Fette Seifen Anstrichmittel*, **72**, 1084.
Weihrauch, J. L. & Gardner, J. M. (1978) *Journal of the American Dietetic Association*, **73**, 39.
Wessels, H. (1983) *Pure and Applied Chemistry*, **55**, 1381.
Wessels, H. (1984) *Pure and Applied Chemistry*, **56**, 301.
Williams, K. A. (1966) *Oils, Fats and Fatty Foods*, 4th edn. London: Churchill.
Woestenburg, W. J. & Zaalberg, J. (1986) *Fette Seifen Anstrichmittel*, **88**, 53.
Wood, R. & Lee, T. (1983) *Journal of Chromatography*, **254**, 237.
Woolard, D. C. and Woolard, A. D. (1988) *Journal of Micronutrient Analysis*, **4**, 119.
Youssef, M. K. E., Omar, M. B., Skulberg, A. & Rashwan, M. (1988) *Food Chemistry*, **30**, 167.
Zadernowski, R. & Sosulski, F. (1979) *Journal of the American Oil Chemists' Society*, **56**, 1004.

Appendices

APPENDIX 1

Organisations which issue standard analytical methods for food commodities

Initials	Name of organisation	Subject
AACC	American Association of Cereal Chemists	
AFNOR	Association Française de Normalisation	General
AIIBP	Association internationale de l'industrie des bouillons et potages (International Association of the Stock and Soup Industry)	Soups
AMC	Analytical Methods Committee of the Royal Society of Chemistry	Various
AOAC	Association of Official and Analytical Chemists	Food, Agriculture
APA	Association of Public Analysts	
BSI	British Standards Institution	General
CIRF	Corn Industries Research Foundation Inc.	Starch products
DGF	Deutsche Gesellschaft für Fettwissenschaft	Oils and fats
EC	European Community	Various
EEC	European Economic Community	Various
FAO	Food and Agriculture Organization (UN) (Codex)	Food
FOSFA	Federation of Oils, Seeds and Fats Association	Oils and fats
IASC	International Association of Seed Crushers	Vegetable oils and fats
ICC	International Association for Cereal Chemistry	Cereals
ICMSF	International Commission on Microbiological Specification for Food	Food (general)
ICUMSA	International Commission for Uniform Methods of Sugar Analysis	Sugar
IDF	International Dairy Federation	Dairy products
IOB	Institute of Brewing	Beer
IOCCC	International Office of Cocoa, Chocolate and Sugar Confectionery	Cocoa, confectionery
IOFI	International Organisation of the Flavour Industry	Flavour
ISO	International Organization for Standardization	General
IUPAC	International Union of Pure and Applied Chemistry	General
NMKL	Nordisk Metodik-Kommittee för Livsmedel (Nordic Committee on Food Analysis)	Food
NNI	Nederlands Normalisatie-Instituut	General
OIV	Office internationale de la vigne et du vin (International Office of Wine and Vine	Wine

APPENDIX 2

Codex Standards

Codex Alimentarius Commission. Joint FAO/WHO Food Standards Programme

Subject	Codex Standard
Processed Fruits and Vegetables	
Canned Tomatoes	13-1981
Canned Peaches	14-1981
Canned Grapefruit	15-1981
Canned Green Beans and Wax Beans	16-1981
Canned Applesauce	17-1981
Canned Sweet Corn	18-1981
General Standard for Edible Fungi and Fungus products	38-1981
Dried Edible Fungi	39-1981
Fresh Fungus 'Chanterelle'	40-1981
Canned Pineapple	42-1981
Canned Mushrooms	55-1981
Canned Asparagus	56-1981
Processed Tomato Concentrates	57-1981
Canned Green Peas	58-1981
Canned Plums	59-1981
Canned Raspberries	60-1981
Canned Pears	61-1981
Canned Strawberries	62-1981
Table Olives	66-1981
Raisins	67-1981
Canned Mandarin Oranges	68-1981
Canned Fruit Cocktail	78-1981
Jams (Fruit Preserves) and Jellies	79-1981
Citrus Marmalade	80-1981
Canned Mature Processed Peas	81-1981
Canned Tropical Fruit Salad	99-1981
Pickled Cucumbers	115-1981
Canned Carrots	116-1981
Canned Apricots	129-1981
Dried Apricots	130-1981
Unshelled Pistachio Nuts	131-1981
Sugars (Including Honey)	
White sugar	4-1981
Powdered Sugar (Icing Sugar)	5-1981
Soft Sugars	6-1981
Dextrose Anhydrous	7-1981
Dextrose Monohydrate	8-1981
Glucose Syrup	9-1981
Dried Glucose Syrup	10-1981
Lactose	11-1981
Honey (European Regional Standard)	12-1981
Powdered Dextrose (Icing Dextrose)	54-1981
Fructose	102-1981
Amendment to Explanatory Notes of Codex Standards for Sugars	

Subject	Codex Standard
Processed Meat and Poultry Products and Soups and Broths	
Canned Corned Beef	88-1981
Luncheon Meat	89-1981
Cooked Cured Ham	96-1981
Cooked Cured Pork Shoulder	97-1981
Cooked Cured Chopped Meat	98-1981
Bouillons and Consommés	117-1981
Fish and Fishery Products	
Canned Pacific Salmon	3-1981
Quick-Frozen Gutted Pacific Salmon	36-1981
Canned Shrimps or Prawns	37-1981
Quick-Frozen Fillets of Cod and Haddock	50-1981
Quick-Frozen Fillets of Ocean Perch	51-1981
Canned Tuna and Bonito in Water or Oil	70-1981
Canned Crab Meat	90-1981
Quick-Frozen Fillets of Flat Fish	91-1981
Quick-Frozen Shrimps or Prawns	92-1981
Quick-Frozen Fillets of Hake	93-1981
Canned Sardines and Sardine-Type Products	94-1981
Quick-Frozen Lobsters	95-1981
Canned Mackerel and Jack Mackerel	119-1981
Labelling	
General Standard for the Labelling of Prepackaged Foods	1-1981
General Standard for the Labelling of Food Additives when sold as such	107-1981
Codex Guidelines on Labelling	
General Guidelines on Claims	
Guidelines for Date-Marking of Prepackaged Foods for the Use of Codex Committees	
Cocoa Products and Chocolate	
Cocoa Butters	86-1981
Chocolate	87-1981
Cocoa Powders (Cocoa) and Dry Cocoa–Sugar Mixtures	105-1981

Subject	Codex Standard
Cocoa (Cacao) Nib, Cocoa (Cacao) Mass, Cocoa Press Cake and Cocoa Dust (Cocoa Fines), for use in the manufacturing of Cocoa and Chocolate Products	141-1983
Composite and Filled Chocolate	142-1983
Quick Frozen Fruits and Vegetables	
Quick Frozen Peas	41-1981
Quick Frozen Strawberries	52-1981
Quick Frozen Raspberries	69-1981
Quick Frozen Peaches	75-1981
Quick Frozen Bilberries	76-1981
Quick Frozen Spinach	77-1981
Quick Frozen Blueberries	103-1981
Quick Frozen Leek	104-1981
Quick Frozen Broccoli	110-1981
Quick Frozen Cauliflower	111-1981
Quick Frozen Brussels Sprouts	112-1981
Quick Frozen Green and Wax Beans	113-1981
Quick Frozen French Fried Potatoes	114-1981
Quick Frozen Whole Kernel Cob	132-1981
Quick Frozen Corn-on-the-Cob	133-1981
Quick Frozen Carrots	140-1981
Foods for Special Dietary Uses	
Foods with Low-Sodium Content (including Salt Substitutes)	53-1981
Infant Formula	72-1981
Canned Baby Foods	73-1981
Processed Cereal-based Foods for Infants and Children	74-1981
Gluten-Free Foods	118-1981
Advisory Lists of Mineral Salts and Vitamin Compounds for Use in Foods for Infants and Children	
Amendments to Codex Standards 53, 72, 73, 74 and 118 and to the Advisory List of Mineral Salts and Vitamin Compounds	
Fruit Juices	
Apricot, Peach and Pear Nectars	44-1981
Orange Juice	45-1981
Grapefruit Juice	46-1981
Lemon Juice	47-1981
Apple Juice	48-1981
Tomato Juice	49-1981
Concentrated Apple Juice	63-1981
Concentrated Orange Juice	64-1981
Grape Juice	82-1981
Concentrated Grape Juice	83-1981
Sweetened Concentrated Labrusca Type Grape Juice	84-1981
Pineapple Juice	85-1981
Non-Pulpy Blackcurrant Nectar	101-1981
Blackcurrant Juice preserved exclusively by physical means	120-1981

Subject	Codex Standard
Concentrated Blackcurrant Juice preserved exclusively by physical means	121-1981
Pulpy Nectars of Certain Small Fruits preserved exclusively by physical means	122-1981
Nectars of Certain Citrus Fruits	134-1981
Concentrated Pineapple Juice preserved exclusively by physical means	138-1983
Concentrated Pineapple Juice with Preservatives, for Manufacturing	139-1983
Amendments to Codex Standards 44–49, 63–64, 82–85, 101, 120–122 and 134	
Fats and Oils	
General Standard for Fats and Oils not covered by individual standards	19-1981
Edible Soya Bean Oil	20-1981
Edible Arachis Oil	21-1981
Edible Cottonseed Oil	22-1981
Edible Sunflowerseed Oil	23-1981
Edible Rapeseed Oil	24-1981
Edible Maize Oil	25-1981
Edible Sesameseed Oil	26-1981
Edible Safflowerseed Oil	27-1981
Lard	28-1981
Rendered Pork Fat	29-1981
Premier Jus	30-1981
Edible Tallow	31-1981
Margarine	32-1981
Olive Oil	33-1981
Mustardseed Oil	34-1981
Edible Low Erucic Acid Rapeseed Oil	123-1981
Edible Coconut Oil	124-1981
Edible Palm Oil	125-1981
Edible Palm Kernel Oil	126-1981
Edible Grapeseed Oil	127-1981
Edible Babassu Oil	128-1981
Minarine	135-1981
Amendments to Codex Standards 19–34, 123–128 and 135	
Miscellaneous Products	
Natural Mineral Waters (European Regional Standard)	108-1981
Edible Ices	137-1981
Pesticide Residues	
Codex Maximum Limits for Pesticide Residues adopted by the Codex Alimentarius Commission up to the end of its 15th Session (July 1983)	CAC/VOL XIII-Ed. 1 and Corrigendum

Subject	Codex Standard
Guide to Codex Maximum Limits for Pesticide Residues:	
– Part 1 General Notes and Guidelines	CAC/PR 1-1984
– Part 2 Maximum Limits for Pesticide Residues	CAC/PR 2-1985 2nd Preliminary Issue
– Part 3 Guideline Levels for Pesticide Residues	CAC/PR 3-1984
– Part 4 Classification of Foods and Animal Feed	CAC/PR 4-1984
– Part 5 Recommended Method of Sampling for Determination of Pesticide Residues	CAC/PR 5-1984
– Part 6 Portion of Commodities to which Codex Maximum Residue Limits Apply and which is Analysed	CAC/PR 6-1984
– Part 7 Codex Guidelines on Good Practice in Pesticide Residue Analysis	CAC/PR 7-1984
– Part 8 Recommendations for Methods of Analysis of Pesticide Residues	CAC/PR 8-1984
Food Additives	
Food Additives (comprises the following five parts):	CAC/VOL XIV-Ed. 1
– Part I Definitions	
– Part II General Principles for the Use of Food Additives	
– Part III Principle Relating to the Carry-Over of Food Additives into Food	
– Part IV Guidelines for the Establishment of Food Additive Provisions in Commodity Standards	
– Part V Food Additives Permitted for Use in Codex Standards [NB This is not an exclusive list of food additives]	
Irradiated Foods	
General Standard for Irradiated Foods	106-1983
Contaminants	
Contaminants—contains maximum levels for contaminants permitted in Codex standards adopted by the Codex Alimentarius Commission [NB This is not an exclusive list of levels for contaminants permitted in foodstuffs]	CAC/VOL XVII-Ed. 1
Recommended Codes of Practice (RCP)	
General Principles of Food Hygiene	1-1969 Rev. 1 (1979)
Code of Hygienic Practice for Canned Fruit and Vegetable Products	2-1969
Code of Hygienic Practice for Dried Fruits	3-1969
Code of Hygienic Practice for Desiccated Coconut	4-1971
Code of Hygienic Practice for Dehydrated Fruits and Vegetables including Edible Fungi	5-1971
Code of Hygienic Practice for Tree Nuts	6-1972
International System for the Description of Carcases of Bovine and Porcine Species and International Description of Cutting Methods of Commercial Units of Beef, Veal, Lamb and Mutton, and Pork, Moving in International Trade	7-1974
Code of Practice for the Processing and Handling of Quick Frozen Foods	8-1976
– Method of Checking Product Temperature	Annex 1-1978
– Code of Practice for the handling of Quick Frozen Foods during Transport	Annex II-1983
Code of Practice for Fresh Fish	9-1976
Code of Practice for Canned Fish	10-1976
Code of Hygienic Practice for Fresh Meat	11-1976
Code of Ante-Mortem and Post-Mortem Inspection of Slaughter Animals	12-1976
Code of Hygienic Practice for Processed Meat Products	13-1976
Code of Hygienic Practice for Poultry Processing	14-1976
Code of Hygienic Practice for Egg Products	15-1976
– Microbiological Specifications for Pasteurized Egg Products	Annex II-1978
Code of Practice for Frozen Fish	16-1978
Code of Practice for Shrimps or Prawns	17-1978
Code of Hygienic Practice for Molluscan Shellfish	18-1978
Code of Practice for the Operation of Radiation Facilities used for the Treatment of Foods	19-1979
Code of Ethics for International Trade in Food	20-1979
Code of Hygienic Practice for Foods for Infants and Children (including Microbiological Specifications and Methods of Microbiological analysis)	21-1979
Corrigendum to Code of Hygienic Practice for Foods for Infants and Children (Microbiological Specifications)	21-1979
Code of Hygienic Practice for Groundnuts (Peanuts)	22-1979

Subject	Codex Standard
Code of Hygienic Practice for Low-Acid and Acidified Low-Acid Canned Foods	23-1979
Code of Practice for Lobsters	24-1979
Code of Practice for Smoked Fish	25-1979
Code of Practice for Salted Fish	26-1979
Code of Practice for Minced Fish Prepared by Mechanical Separation	27-1983
Code of Practice for Crabs	28-1983
Code of Hygienic Practice for Game	29-1983
Code of Hygienic Practice for the Processing of Frog Legs	30-1983
Code of Practice for Dried Milk	31-1983
Code of Practice for the Production, Storage and Composition of Mechanically Separated Meat and Poultry Meat intended for further Processing	32-1983
Recommended Methods of Analysis (RM)	
Determination of Total Solids Content (Oven-filter aid method)	1-1969
Determination of Loss on Drying at 120 °C for 16 Hours (USP method)	2-1969
Determination of Loss on Drying at 105 °C for 3 Hours (ICUMSA method)	3-1969
Determination of Sulphur Dioxide (Monier-Williams method)	4-1969
Determination of Sulphur Dioxide (Carruthers, Heaney & Oldfield method)	5-1969
Determination of Colour	6-1969
Determination of Polarization (ICUMSA method)	7-1969
Determination of Conductivity Ash	8-1969
Determination of Relative Density at *t*/20 °C (BSI method)	9-1969
Determination of Allyl isothiocyanate Content (Indian Standards Institute method)	10-1969
Arachis Oil Test (Evers) (BSI method)	11-1969
Sesame Oil Test (Baudouin) (BSI method)	12-1969
Determination of Soap Content (BSI method)	13-1969
Determination of Iron Content (BSI method)	145-1969
Estimation of Milk Fat Content	15-1969
Determination of Fat Content	16-1969
Determination of Water Content by Loss of Mass on Drying	17-1969
Determination of Vitamin E (Tocopherols) Content	18-1969
Determination of Sodium Chloride Content	19-1969
Determination of Bellier Index	20-1970

Subject	Codex Standard
Semi-siccative Oils Test	21-1970
Olive-residue Oil Test	22-1970
Cottonseed Oil Test	23-1970
Teaseed Oil Test	24-1970
Sesameseed Oil Test	25-1970
Determination of Specific Extinction in Ultra Violet ($E^{1\%}_{1\ cm}$)	26-1970
Soap Test	27-1970
Determination of Drained Weight (Method A for Liquid Packing Medium)	28-1970
Determination of Net Contents	29-1970
Determination of Size	30-1970
Determination of Water Capacity of the Container	31-1970
Standards Procedure for Thawing of Quick Frozen Fruits and Vegetables	32-1970
Standard Procedure for Cooking of Quick Frozen Fruits and Vegetables	33-1970
Weight of Quick Frozen Fruits and Vegetables	34-1970
Determination of the Alcohol-Insoluble Solids Content (Quick Frozen Peas)	35-1970
Determination of Drained Weight—Method I (AOAC method)	36-1970
Determination of Drained Weight—Method II	37-1970
Determination of Calcium in Canned Vegetables	38-1970
Tough String Test	39-1970
Thawing and Cooking Procedure—Quick Frozen Fish	40-1971
Determination of Net Contents of Products Covered by Glaze—Quick Frozen Fish	41-1971
Sampling Plans for Prepackaged Foods (1969) (AQL 6.5)	42-1971
Determination of Total Soluble Solids Content of Frozen Fruits	43-1971
Determination of Washed Drained Weight	44-1972
Determination of Proper Fill in lieu of Drained Weight	45-1972
Determination of Water Capacity of Containers	46-1972
Determination of Alcohol-Insoluble Solids (AOAC method)	47-1972
Method for Distinguishing Type of Peas	48-1972
Determination of Mineral Impurities (Sand)	49-1972
Determination of Moisture in Raisins—AOAC Electrical Conductance Method	50-1974
Determination of Mineral Impurities (Sand Test) in Raisins	51-1974

Subject	Codex Standard
Determination of Mineral Oil in Raisins	52-1974
Determination of Sorbital in Raisins and other Foods	53-1974
Determination of Mineral Impurities in Quick Frozen Fruits and Vegetables	54-1974
Determination of Fats for all Infant Foods—Method I	55-1976
Milk and Milk Product Standards	
Part I—Code of Principles concerning Milk and Milk Products	
Part II—Standards for Milk Products:	
Butter and Whey Butter	A-1
(i) Butteroil and (ii) Anhydrous Butteroil and Anhydrous Milkfat	A-2
Evaporated Milk and Evaporated Skimmed Milk	A-3
Sweetened Condensed Milk and Skimmed Sweetened Condensed Milk	A-
Whole Milk Powder, Partly Skimmed Milk Powder and Skimmed Milk Powder	A-5
General Standard for Cheese	A-6
General Standard for Whey Cheeses	A-7
General Standard for Named Variety Process(ed) Cheese and Spreadable Process(ed) Cheese	A-8(a)
General Standard for Process(ed) Cheese and Spreadable Process(ed) Cheese	A-8(b)
General Standard for Process(ed) Cheese Preparations	A-8(c)
Cream for Direct Consumption	A-9
Cream Powder, Half Cream Powder and High Fat Milk Powder	A-10
Yoghurt (Yogurt) and Sweetened Yoghurt (Sweetened Yogurt)	A-11(a)
Flavoured Yoghurt (Yogurt) and Products Heat-treated after Fermentation	A-(11)b
Edible Acid Casein	A-12
Edible Caseinates	A-13
Part III – International Cheese Standards:	
Cheddar	C-1
Danablu	C-2
Danbo	C-3
Edam	C-4
Gouda	C-5
Havarti	C-6
Samsoe	C-7
Cheshire	C-8

Subject	Codex Standard
Emmentaler	C-9
Gruyere	C-10
Tilsiter	C-11
Limburger	C-12
Saint-Paulin	C-13
Svecia	C-14
Provolone	C-15
Cottage Cheese incl. Creamed Cottage Cheese	C-16
Butterkäse	C-17
Coulommiers	C-18
Gudbrandsdalsost (whey cheese)	C-19
Harzer Käse	C-20
Herrgårdsost	C-21
Hushållsost	C-22
Norvegia	C-23
Maribo	C-24
Fynbo	C-25
Esrom	C-26
Romadur	C-27
Amsterdam	C-28
Leidse	C-29
Friese	C-30
Cream Cheese	C-31
Certain Blue Veined Cheeses	C-32
Camembert	C-33
Brie	C-34
Extra Hard Grating Cheese	C-35
Methods of Analysis and Sampling for Milk and Milk Products:	
Sampling Methods for Milk and Milk Products – General Instructions – Sampling of Milk and Liquid Milk Products (except Evaporated and Sweetened Condensed Milk) – Sampling of Condensed Milk and Evaporated Milk – Sampling of Dried Milk and Dried Milk Products – Sampling of Butter – Sampling of Cheese	B-1
Determination of the Fat Content of Dried Milk	B-2
Determination of the Fat Content of Cheese and Processed Cheese Products	B-3
Determination of the Acid Value of Fat from Butter	B-4
Determination of the Refractive Index of Fat from Butter	B-5
Determination of the Fat Content of Milk	B-6
Determination of the Fat Content of Evaporated Milks and of Sweetened Condensed Milks	B-7

Subject	Codex Standard
Determination of the Salt (Sodium Chloride) Content of Butter	B-8
Determination of the Fat Content of Whey Cheese	B-10
Determination of the Dry Matter Content in Whey Cheese	B-11
Determination of the Phosphorus Content of Cheese and Processed Cheese Products	B-12
Determination of the Citric Acid Content of Cheese and Processed Cheese Products	B-13
Polarimetric Determination of the Sucrose Content of Sweetened Condensed Milk	B-14
Determination of the Fat Content of Cream	B-15
Milk Fat, Detection of Vegetable Fat by the Phytosteryl Test	B-16
Milk Fat, Detection of Vegetable Fat by Gas–liquid Chromatography of Sterols	B-17
Cheese, Determination of Chloride Content	B-18
Cheese, Determination of Nitrate and Nitrite Contents	B-19
Anhydrous Milk Fat, Determination of the Peroxide Value	B-20
Butter-Water, Solids-non-fat and Fat on the same test portion	B-21
Caseins and Caseinates—Determination of Water Content	B-22
Rennet caseins and caseinates—Determination of Ash	B-23
Caseins—Determination of 'fixed ash'	B-24
Caseins and caseinates—Determination of protein content	B-25
Caseins—Determination of free acidity	B-26
Milk and Milk Products—Determination of Lactose in the presence of other reducing substances	B-27
Dried milk—Determination of titratable acidity	B-28

APPENDIX 3

EC Directives and Regulations

Subject	Number	Amendments, extensions
Additives Safety Evaluation	80/109	
Criteria of purity, Methods of analysis	81/712	
Alcoholic beverages, alcoholic strength labelling	87/250	
Antioxidants	70/357	74/412, 78/143, 85/7, 81/962, 85/17, 87/55
Specific purity criteria	78/664	82/712
Caseins and caseinates, definitions and standards	83/417	
Methods of analysis	85/503	86/424
Cocoa and Chocolate Products	73/241	74/411, 74/644, 75/155, 76/628, 78/609, 78/842, 80/608, 85/17, 85/7
Coffee Extracts and Chicory Extracts	77/436	85/7, 85/573
Methods of analysis	79/1066	
Colouring Matter	1962/2645	65/469, 67/653, 68/419, 70/358, 76/399, 78/144, 81/20, 85/7
Emulsifiers, stabilisers, thickeners	74/329	78/612, 80/597, 85/6, 85/7, 86/102
Specific criteria of purity	78/663	82/504
Erucic acid, levels in oils and fats and foods containing added oils and fats	76/621	
Methods of analysis	80/891	
Fruit Juices and similar products	75/726	79/168, 81/487

Subject	Number	Amendments, extensions
Honey	74/409	
Jams, fruit jams, jellies, marmalades, chestnut purée	79/693	80/11276, 88/593
Labelling, presentation and advertising of foodstuffs	79/112	85/7, 86/197
Designation of certain ingredients	83/463	
Materials and articles		
intended to come into contact with foods	76/893	80/1276, 85/7
ceramic articles intended to come into contact with foodstuffs	84/500	
migration from plastic materials and articles intended to come into contact with foodstuffs	82/711	
list of simulants to be used in the testing migration of plastics	85/572	
regenerated cellulose film intended to come into contact with foodstuffs	83/229	86/388
which contain vinyl chloride monomer and are intended to come into contact with foodstuffs	78/142	
Methods of analysis of VCM level	80/766	
methods of analysis, vinyl chloride released into foods	81/432	
Meat and Meat products		
Meat inspection, etc.	64/433	66/601, 69/349, 75/379, 81/476, 83/90, 85/323, 85/325
On animal health problems affecting intra-community trade	80/215	80/1100, 81/476, 85/321, 85/3768, 87/491, 88/660
On health problems affecting intra-community trade	88/658	
Minced meat, meat in pieces	88/657	
Milk and milk products	1411/71	3358/75, 556/76
Designation	1898/87	
Milk, certain partly or wholly dehydrated preserved milk	76/118	
Methods of analysis	79/1067	
Sampling	87/524	
Milk, heat treated, intra-community trade	85/397	
Nutritional Foods		
Foodstuffs for particular nutritional uses	77/94	85/7
Poultry,		
Water content of frozen, deep frozen chickens, hens and cocks	2967/76	1691/77, 641/79, 2632/80, 2785/80, 2835/80, 3204/83
Implementing regulation	2785/80	3134/81, 3759/85
Poultry Meat, health problems in intra-community trade	71/118	74/387, 75, 379, 75/431, 80/216, 81/476, 81/578, 82/532, 84/186, 84/335, 84/642, 85/324, 85/326, 3805/87
Pesticides		
Animal and fresh meat, examination for residues	86/469	
Maximum levels in cereals	86/362	88/298
Maximum levels in animal products	86/363	
Maximum levels in fruit and vegetables	76/895	80/428, 81/36, 82/528, 88/298
Sampling fruit and vegetables	79/700	
Prescribed quantities for packaged food	80/232	86/96
Preservatives authorised for use in foodstuffs	64/54	65/569, 66/722, 67/4, 68/420, 70/359, 71/160, 72/2, 72/444, 74/62, 74/394, 76/462, 76/629, 78/145, 79/40, 81/214, 83/585, 83/636, 84/86, 84/223, 84/261, 84/458, 85/7, 85/172, 85/585

Subject	Number	Amendments, extensions
Specific purity criteria	65/66	67/428, 76/463, 86/604
Certain preservatives for surface treatment of citrus fruits	67/427	73/146
Radioactivity in produce from third countries	3955/87	
Sampling and analysis for the monitoring of foodstuffs	85/591	
Solvents for food use	88/344	
Spirits, definition, description and presentation	89/1576	
Sugar, certain sugars	73/437	
Methods of analysis	79/796	
Water, Natural mineral waters	80/777	85/7
Quality of water for human consumption	80/778	81/858
Wines		
Methods of analysis	82/1108	
Organisation of market	87/822	
Quality of specific regions	87/823	
Description and presentation	81/997	

APPENDIX 4

Food composition and labelling: UK Reports and Regulations

Commodity: Subject	FSC* Reports	Regulation†	Amendment†
Alcoholic liquors (amendment of enactments relating to strengths and units of measurements)		1979:241	
Alcohol tables	1979:132		
Alcoholic liquors (amendment of units and method of measurement)		1979:1146	
Bakery products and flour			
Bread and flour		1959:42 1960:43 1974:61	1984:1304
Flour (self raising)	1942:3 1944/5:13	1946:157	
Beer	1976:68	1978:893	1978:1186 1979:132
Cider and perry		1976:1207	1979:241
Chocolate-flavour coatings and fillings	1974:E2		
Cocoa and chocolate products		1976:541	1980:1833 1980:1834 1980:1849 1982:17 1984:1305
Coffee and coffee products	1951:26 1952:34	1978:1420	1980:1849 1982:254 1987:1986
Common Agricultural Policy (wine)		1979:1094	
Compositional Orders (baking powder and golden raising powder, edible gelatine, mustard, curry powder, tomato ketchup, fish cakes and suet)	1970:55 1989:5		
Condiments and sauces			
Curry powder (see also Compositional Orders)	1948:16	1949:1816	1956:1166 1979:1254
Mustard (see also Compositional Orders)		1944:275	1946:157 1948:1073

Commodity: Subject	FSC* Reports	Regulation†	Amendment†
Salad cream	1945:12	1966:1051	1967:1582
			1980:1849
Salt	1949:18		
Tomato Ketchup (see also Compositional Orders)	1948:15	1949:1817	1956:1167
Vinegars	1971:58		
Edible gelatin	1949:20		
	1951:28		
Egg, liquid (pasteurisation)		1963:1503	
Fish products			
Fish cakes (see also Compositional Orders)	1951:25	1950:589	
	1951:31		
Fish paste (Spreadable Products)	1950:21	1984:1566	
	1965:49		
Coated and ice-glazed	1987:3		
General provisions		1944:42	1944:654
Infant formulae	1981:73		
Jams and jellies (see Preserves)			
Margarine (see Oils and fats)			
Meat products	1980:72	1984:1566	1986:987
Bovine offal (prohibition)		1989/2061	
Canned	1962:46		
Mince	1983:77		
Offals in meat products	1972:56		
Paste	1965:49		
Pie and sausage roll	1963:47		
Poultry (absorption of water)			
Sausage and other meat products	1943:1		
	1956:37		
Poultry meat (slaughterhouse checks on absorption of water)		1977:1854	
Meat treatment		1964:19	
Slaughterhouses (hygiene)		1977:1805	
Poultry meat (hygiene)		1976:1209	
Food hygiene (general)	1970:1172		
Public health (shellfish)		1934:1342	
Therapeutic substances (preservation of raw fish)		1964:883	
Milk and milk products			
Butter		1966:1074	1973:1340
			1980:1849
Caseins and Caseinates		1985:2026	1989:2321
Channel Island and South Devon		1956:919	1962:1288
Cheese	1949:17	1970:94	1974:1122
	1956:38		1975:1486
	1962:44		1976:2086
	1982:75		1980:1849
			1984:649
Condensed milk	1969:54		
	1973:60		
Condensed and dried milk		1977:928	1980:1849
			1982:1066
			1986:2299
			1989:1959
Cream	1950:23	1970:752	1975:1486
	1951:29		1980:1849
	1951:30		
	1967:51		
	1982:76		
Cream (heat treatment)		1983:1509	1986:721
Dried milk	1962:45		
Drinking milk		1976:1883	

Commodity: Subject	FSC* Reports	Regulation†	Amendment†
General		1959:277	1962:1288 1973:1064 1977:171
Ice-cream	1950:22 1950:24 1957:39 1989:5	1967:1866	1980:1849 1983:1211
Ice-cream (heat treatment)		1959:734	1962:1287 1963:1083
Importation of milk		1988:1803	
Milk-based drinks		1983:1508	1986:720
Semi-skimmed and skimmed		1988:2206	
Skimmed with non-fat milk	1984:1	1960:2331	1962:1287 1976:103 1980:1849 1981:1174
Special designations		1988:2204	
Yoghurt	1975:64		
Novel protein foods	1974:62		
Oils and fats			
Erucic acid		1977:691	1982:264
Margarine	1954:36 1981:74	1967:1867	1980:1849
Suet (see also Compositional Orders)	1943:7A 1952:33 1952:33A	1952:2203	
Preserves	1968:53	1953:691	1953:1307
Jam and marmalade	1952:35 1952:35A		
Jams, jellies, marmalade and chestnut purée	1976:66	1981:1063	1982:1700 1983:1211
Jams and other preserves	1968:53		
Jellies (table)	1959:41	1949:1656	1955:828
Soft drinks	1951:27 1959:40 1975:63 1976:65	1964:760	1969:1818 1970:1597 1976:295 1977:927 1980:1849 1983:1211
Fruit juices and fruit nectars	1975:63	1977:927	1979:1254 1980:1849 1982:1311
Soup	1954:1 1968:52		
Sugar		1976:1832	1980:1849
Honey			
Specified sugar products		1976:509	1980:1849 1982:255
Water in food	1978:70		
Natural Mineral Water		1985:71	
Wine		1987:1843	
Labelling	1964:48 1979:69	1984:1305	1984:1566 1985:71 1985:2026 1986:987 1989:768

Commodity: Subject	FSC* Reports	Regulation†	Amendment†
Fructose in foods for diabetics	1977:69A		
Exemptions from ingredients listing and generic terms	1977:69B		
Claims and misleading descriptions	1966:50		
	1980:71		
Date Marking	1972:59		

* The FSC and FACC were amalgamated to form the Food Advisory Committee (FdAC) in 1984.
† The numbers and dates refer to England and Wales; separate instruments are issued for Scotland and Northern Ireland.

APPENDIX 5

Food additives and contaminants: UK Reports and Regulations

Additives	FACC Report*	Regulation†	Amendment†
Antioxidants	1953:1	1978:105	1980:1831
	1954:2		1983:1211
	1963:2		1984:1304
	1963:3		
	1965:1		
	1966:4		
	1971:11		
	1974:18		
Artificial sweeteners		1969:1817	1988:2122
Cyclamates	1966:3		
	1967:6		
Saccharin	1952:32		
	1953:32A		
Beer—additives and processing aids	1978:26		
Bread and flour			
Azodicarbonamide	1968:8		
Bulking aids	1980:32		
Cheese	1982:36		
Chloroform	1980:36		
Clouding Agents	1985:2		
Colours	1954:3	1973:1340	1974:1119
	1955:4		1975:1488
	1960:8		1976:2086
	1964:4		1978:1787
	1979:29		1979:1254
	1987:4		1987:1987
Cream	1982:37		
Emulsifiers and stabilisers	1956:5		
	1970:9		
	1972:16	1989:876	
Enzymes	1982:35		
Flavours	1965:5		
	1976:22		
	1978:28		
Freezants	1972:15		
	1974:19		
Mineral oil (hydrocarbons)	1962:1		
	1975:20	1966:1073	

Additives	FACC Report*	Regulation†	Amendment†
Miscellaneous additives	1976:25	1980:1834	1980:1849
	1968:7		1982:14
			1983:1211
			1984:1304
Modified starches	1980:31		
Preservatives	1959:6	1989:533	1989:2287
	1960:7		
	1972:14		
Nitrites and nitrates	1978:27		
	1982:36		
Sorbic acid	1977:24		
Sulphur dioxide	1977:23		
Solvents	1965:2		
	1974:17		
	1978:25	1967:1582	1967:1939
			1980:1832
			1983:1211
			1984:1304
Sweeteners	1982:34	1982:1211	1988:2112

Contaminants	FACC Report*	Regulation†	Amendment†
Aldrin and dieldrin residues	1966:5		
Antimony and cadmium	1955:11		
Arsenic	1949:1	1959:831	1960:2261
	1950:2		1962:1287
	1951:5		1963:1435
	1955:13		1966:1203
	1961:2		1972:1391
	1984:39		1973:1052
			1973:1340
			1975:1486
Asbestos	1979:30		
Copper	1951:4		
	1951:7		
	1956:14		
Tomato ketchup		1949:1817	1956:1167
Fluorine	1953:10		
	1957:15		
Lead	1951:3	1979:1254	1985:912
	1951:6		
	1954:12		
	1961:2		
	1975:21		
Curry powder		1949:1816	1956:1166
			1979:1254
Cooking utensils (safety)		1972:1957	
Materials and articles in contact with food		1987:1523	
Pesticides, maximum residual levels		1988:1378	
Tin	1952:9		
Tetrachlorethylene in olive oil		1989:910	
Zinc	1953:8		
Metals in canned foods	1983:38		

* Reports prior to 1965 were by the Food Standards Committee and its sub-committees. From 1984 the FACC was amalgamated with the FSC to form the Food Advisory Committee (FdAC).
† The numbers and dates refer to England and Wales; separate instruments are issued for Scotland and Northern Ireland.

APPENDIX 6

UK Codes of Practice

Food hygiene

Hygiene in the Retail Meat Trade	1959
Hygienic Transport and Handling of Meat	1959
Hygiene in the Retail Fish Trade	1960
Hygienic Transport and Handling of Fish	1960
Poultry Dressing and Handling	1961
Hygiene in the Bakery Trade and Industry	1966
Hygiene in the Operation of Coin Operated Vending Machines	1967
Hygiene in the Meat Trades	1969
Hygiene in Microwave Cooking	1972
The Canning of Low Acid Foods. A Guide to Good Manufacturing Practice	1981

LAJAC (Local Authorities Joint Advisory Committee on Food Standards, see JAPA 1963, **1**, 50)

Use of the word 'chocolate' in flour confectionery (JAPA 1963, **1**, 49)
Min. 3 per cent dry non-fat cocoa solids in the moist crumb (does not apply if called 'chocolate flavoured').
Labelling of Brandy (JAPA 1963, **1**, 101)
Brandy must come from grapes; 'Cognac' and 'Armagnac' must come from those regions.
Crab meat content in Norwegian canned crab products (JAPA 1963, **1**, 105)
Canned fruit and vegetables. (Now replaced by LACOTS Code of Practice)
Canned beans in tomato sauce (JAPA 1965, **3**, 111)
Covers beans and sauce; standards for total solids, tomato solids and sugar (according to can size).
Marzipan, almond paste and almond icing (JAPA 1969, **7**, 40)
Min. 23.5 per cent almond substance (no other nut ingredient) and min. 75 per cent of remainder to be solid carbohydrate sweetening matter.
Bun loaves, chollas and buttercream (JAPA 1971, **9**, 107)
Bun loaf: min. 5 per cent liquid whole egg and/or fat
min. 10 per cent egg and or fat and/or sugar
Chollas: normally same as for bun loaf
Buttercream (not 'butter cream' or 'butter-cream'):
min. 22.5 per cent butterfat and no other added fat.

For detailed information on these codes, see under the reference given to the *Journal of the Association of Public Analysts* (JAPA).

LACOTS (Local Authorities Co-ordinating Body on Trading Standards)

The following codes have been prepared and adopted by associations within the food industry in consultation with LACOTS:
UK Association of Frozen Food Producers: 'Code of Practice for the Breadcrumb Covering of Scampi'; 1980.
Dairy Trade Federation: 'Code of Practice for the Composition and Labelling of Yogurt'; 1983.
Honey Importers and Packers Association: 'Code of Practice for the Importation, Blending, Packaging and Marketing of Honey'; 1986.
British Fruit and Vegetable Canners' Association: 'Code of Practice on Canned Fruit and Vegetables'; 1986 (replaced LAJAC Code No. 4).
Fish and Meat Spreadable Products Association: 'Code of Practice for the Composition of Meat and Fish Pastes and Pates'; 1986.

APPENDIX 7

Food Surveillance Papers

MAFF, published by HMSO, London.

Food Surveillance Paper No.	Title	Date
1	The surveillance of food contamination in the United Kingdom	1978
2	Survey of vinyl chloride content of polyvinyl chloride for food contact and of foods	1978
3	Survey of vinylidene chloride levels in food contact materials and in foods	1980
4	Survey of mycotoxins in the United Kingdom	1980
5	Survey of copper and zinc in food	1981
6	Survey of acrylonitrile and methacrylonitrile levels in food contact materials and in foods	1982
7	Survey of dieldrin residues in food	1982
8	Survey of arsenic in food	1982
9	Report of the Working Party on Pesticide Residues (1977–1981)	1982
10	Survey of lead in food: second supplementary report	1982
11	Survey of styrene levels in food contact materials and in foods	1983
12	Survey of cadmium in food: first supplementary report	1983
13	Polychlorinated biphenyl (PCB) residues in food and human tissues	1983
14	Steering Group on Food Surveillance progress report 1984	1984
15	Survey of aluminium, antimony, chromium, cobalt, indium, nickel, thallium and tin in food	1985
16	Report of the Working Party on Pesticide Residues (1982 to 1985)	1986
17	Survey of mercury in food: second supplementary report	1987
18	Mycotoxins	1987
19	Survey of colour usage in food	1987
20	Nitrate, nitrite and *N*-nitroso compounds in food	1987
21	Survey of plasticiser levels in food contact materials and in foods	1987
22	Anabolic, anthelmintic and antimicrobial agents	1987
23	The British diet: finding the facts	1988

APPENDIX 8

Serial numbers of permitted food additives

E100	Curcumin
E101	Riboflavin (Lactoflavin)
101(a)	Ribloflavin-5′-phosphate
E102	Tartrazine
E104	Quinoline Yellow
107	Yellow 2G
E110	Sunset Yellow FCF (Orange Yellow S)
E120	Cochineal (Carmine of Cochineal or Carminic acid)
E122	Carmoisine (Azorubine)
E123	Amaranth
E124	Ponceau 4R (Cochineal Red A)
E127	Erythrosine BS
128	Red 2G
E131	Patent Blue V
E132	Indigo Carmine (Indigotine)
133	Brilliant Blue FCF
E140	Chlorophyll
E141	Copper complexes of chlorophyll and chlorophyllins
E142	Green S (Acid Brilliant Green BS or Lissamine Green)
E150	Caramel
E151	Black PN (Brilliant Black BN)
E153	Carbon Black (Vegetable Carbon)
154	Brown FK
155	Brown HT (Chocolate Brown HT)
E160(a)	α-Carotene, β-carotene, γ-carotene
E160(b)	Annatto, bixin, norbixin
E160(c)	Capsanthin (Capsorubin)

E160(d) Lycopene
E160(e) β-apo-8′-Carotenal (C30)
E160(f) Ethyl ester of β-apo-8′-carotenoic acid (C30)
E161(a) Flavoxanthin
E161(b) Lutein
E161(c) Cryptoxanthin
E161(d) Rubixanthin
E161(e) Violaxanthin
E161(f) Rhodoxanthin
E161(g) Canthaxanthin
E162 Beetroot Red (Betanin)
E163 Anthocyanins
E170 Calcium carbonate
E171 Titanium dioxide
E172 Iron oxides, iron hydroxides
E173 Aluminium
E174 Silver
E175 Gold
E180 Pigment Rubine (Lithol Rubine BK)
E200 Sorbic acid
E201 Sodium sorbate
E202 Potassium sorbate
E203 Calcium sorbate
E210 Benzoic acid
E211 Sodium benzoate
E212 Potassium benzoate
E213 Calcium benzoate
E214 Ethyl 4-hydroxybenzoate (Ethyl *para*-hydroxybenzoate)
E215 Ethyl 4-hydroxybenzoate, sodium salt (Sodium ethyl *para*-hydroxybenzoate)
E216 Propyl 4-hydroxybenzoate (Propyl *para*-hydroxybenzoate)
E217 Propyl 4-hydroxybenzoate, sodium salt (Sodium propyl *para*-hydroxybenzoate)
E218 Methyl 4-hydroxybenzoate (Methyl *para*-hydroxybenzoate)
E219 Methyl 4-hydroxybenzoate, sodium salt (Sodium methyl *para*-hydroxybenzoate)
E220 Sulphur dioxide
E221 Sodium sulphite
E222 Sodium hydrogen sulphite (Sodium bisulphite)
E223 Sodium metabisulphite
E224 Potassium metabisulphite
E226 Calcium sulphite
E227 Calcium hydrogen sulphite (Calcium bisulphite)
E228 Potassium acid sulphite (Potassium bisulphite)
E230 Biphenyl (Diphenyl)
E231 2-Hydroxybiphenyl (Orthophenylphenol)
E232 Sodium biphenyl-2-yl oxide (Sodium orthophenylphenate)
E233 2-(Thiazol-4-yl) benzimidazole (Thiabendazole)
234 Nisin
E239 Hexamine (Hexamethylenetetramine)
E249 Potassium nitrite
E250 Sodium nitrite
E251 Sodium nitrate
E252 Potassium nitrate
E260 Acetic acid
E261 Potassium acetate
E262 Sodium hydrogen diacetate
262 Sodium acetate
E263 Calcium acetate
E270 Lactic acid
E280 Propionic acid
E281 Sodium propionate
E282 Calcium propionate
E283 Potassium propionate
E290 Carbon dioxide
296 DL-Malic acid, L-Malic acid
297 Fumaric acid
E300 L-Ascorbic acid
E301 Sodium L-ascorbate
E302 Calcium L-ascorbate
E304 6-*O*-Palmitoyl-L-ascorbic acid (Ascorbyl palmitate)
E306 Extracts of natural origin rich in tocopherols
E307 Synthetic α-tocopherol
E308 Synthetic γ-tocopherol
E309 Synthetic δ-tocopherol
E310 Propyl gallate
E311 Octyl gallate
E312 Dodecyl gallate
E320 Butylated hydroxyanisole (BHA)
E321 Butylated hydroxytoluene (BHT)
E322 Lecithins
E325 Sodium lactate
E326 Potassium lactate
E327 Calcium lactate
E330 Citric acid
E331 Sodium dihydrogen citrate (monosodium citrate), disodium citrate, trisodium citrate
E332 Potassium dihydrogen citrate (monopotassium citrate), tripotassium citrate
E333 Monocalcium citrate, dicalcium citrate, tricalcium citrate
E334 L-(+)-Tartaric acid
E335 Monosodium L-(+)-tartrate, disodium L-(+)-tartrate
E336 Monopotassium L-(+)-tartrate (Cream of tartar), dipotassium L-(+)-tartrate
E337 Potassium sodium L-(+)-tartrate
E338 Orthophosphoric acid (Phosphoric acid)
E339 Sodium dihydrogen orthophosphate, disodium hydrogen orthophosphate, trisodium orthophosphate
E340 Potassium dihydrogen orthophosphate, dipotassium hydrogen orthophosphate, tripotassium orthophosphate
E341 Calcium tetrahydrogen diorthophosphate, calcium hydrogen

E341—*cont.* orthophosphate, tricalcium diorthophosphate
350 Sodium malate, sodium hydrogen malate
351 Potassium malate
352 Calcium malate, calcium hydrogen malate
353 Metatartaric acid
355 Adipic acid
363 Succinic acid
370 1,4-Heptonolactone
375 Nicotinic acid
380 Triammonium citrate
381 Ammonium ferric citrate
E385 Calcium disodium ethylenediamine—*N*,*N*,*N'*,*N'* tetra-acetate (Calcium disodium EDTA)
E400 Alginic acid
E401 Sodium alginate
E402 Potassium alginate
E403 Ammonium alginate
E404 Calcium alginate
E405 Propane-1,2-diol alginate (Propylene glycol alginate)
E406 Agar
E407 Carrageenan
E410 Locust bean gum (Carob gum)
E412 Guar gum
E413 Tragacanth
E414 Gum arabic (Acacia)
E415 Xanthan gum
416 Karaya gum
E420 Sorbitol, sorbitol syrup
E421 Mannitol
E422 Glycerol
432 Polyoxyethylene (20) sorbitan monolaurate (Polysorbate 20)
433 Polyoxyethylene (20) sorbitan mono-oleate (Polysorbate 80)
434 Polyoxyethylene (20) sorbitan monopalmitate (Polysorbate 40)
435 Polyoxyethylene (20) sorbitan monostearate (Polysorbate 60)
436 Polyoxyethylene (20) sorbitan tristearate (Polysorbate 65)
E440 Pectins
442 Ammonium phosphatides
E450(a) Disodium dihydrogen diphosphate, trisodium diphosphate, tetrasodium diphosphate, tetrapotassium diphosphate
E450(b) Pentasodium triphosphate, pentapotassium triphosphate
E450(c) Sodium polyphosphates, Potassium polyphosphates
E460 Microcrystalline cellulose, α-cellulose (Powdered cellulose)
E461 Methylcellulose
E463 Hydroxypropylcellulose
E464 Hydroxypropylmethylcellulose
E465 Ethylmethylcellulose
E466 Carboxymethylcellulose, sodium salt (CMC)
E470 Sodium, potassium and calcium salts of fatty acids
E471 Mono- and di-glycerides of fatty acids
E472(a) Acetic acid esters of mono- and di-glycerides of fatty acids
E472(b) Lactic acid esters of mono- and di-glycerides of fatty acids (Lactoglycerides)
E472(c) Citric acid esters of mono- and di-glycerides of fatty acids (Citroglycerides)
E472(e) Mono- and diacetyltartaric acid esters of mono- and di-glycerides of fatty acids
E473 Sucrose esters of fatty acids
E474 Sucroglycerides
E475 Polyglycerol esters of fatty acids
476 Polyglycerol esters of polycondensed fatty acids of castor oil (Polyglycerol polyricinoleate)
E477 Propane-1,2-diol esters of fatty acids
E481 Sodium stearoyl-2-lactylate
E482 Calcium stearoyl-2-lactylate
E483 Stearyl tartrate
491 Sorbitan monostearate
492 Sorbitan tristearate
493 Sorbitan monolaurate
494 Sorbitan mono-oleate
495 Sorbitan monopalmitate
500 Sodium carbonate, Sodium hydrogen carbonate (Bicarbonate of soda), Sodium sesquicarbonate
501 Potassium carbonate, Potassium hydrogen carbonate
503 Ammonium carbonate, Ammonium hydrogen carbonate
504 Magnesium carbonate
507 Hydrochloric acid
508 Potassium chloride
509 Calcium chloride
510 Ammonium chloride
513 Sulphuric acid
514 Sodium sulphate
515 Potassium sulphate
516 Calcium sulphate
518 Magnesium sulphate
524 Sodium hydroxide
525 Potassium hydroxide
526 Calcium hydroxide
527 Ammonium hydroxide
528 Magnesium hydroxide
529 Calcium oxide
530 Magnesium oxide
535 Sodium ferrocyanide (Sodium hexacyanoferrate (II))
536 Potassium ferrocyanide (Potassium hexacyanoferrate (II))
540 Dicalcium diphosphate
541 Sodium aluminium phosphate
542 Edible bone phosphate

544	Calcium polyphosphates
545	Ammonium polyphosphates
551	Silicon dioxide (Silica)
552	Calcium silicate
553(a)	Magnesium silicate synthetic, Magnesium trisilicate
553(b)	Talc
554	Aluminium sodium silicate
556	Aluminium calcium silicate
558	Bentonite
559	Kaolin
570	Stearic acid
572	Magnesium stearate
575	D-Glucono-1,5-lactone (Glucono-δ-lactone)
576	Sodium gluconate
577	Potassium gluconate
578	Calcium gluconate
620	L-Glutamic acid
621	Sodium hydrogen L-glutamate (monosodium glutamate or MSG)
622	Potassium hydrogen L-glutamate (monopotassium glutamate)
623	Calcium dihydrogen di-L-glutamate (Calcium glutamate)
627	Guanosine 5′-(disodium phosphate) (Sodium guanylate)
631	Inosine 5′-(disodium phosphate) (Sodium inosinate)
635	Sodium 5′-ribonucleotide
636	Maltol
637	Ethyl maltol
900	Dimethylpolysiloxane
901	Beeswax
903	Carnauba wax
904	Shellac
905	Mineral hydrocarbons
907	Refined microcrystalline wax
920	L-Cysteine hydrochloride
924	Potassium bromate
925	Chlorine
926	Chlorine dioxide
927	Azodicarbonamide (Azoformamide)
E1400	White or yellow dextrins. Roasted starch
E1401	Acid treated starches
E1402	Alkaline treated starches
E1403	Bleached starches
E1404	Oxidised starches
E1410	Monostarch phosphate
E1411	Distarch phosphate (esterified by sodium trimetaphosphate)
E1412	Distarch phosphate (esterified by 0.1 per cent phosphorus oxychloride)
E1413	Phosphated distarch phosphate
E1414	Acetylated distarch phosphate
E1420	Starch acetate (esterified by acetic anhydride)
E1421	Starch acetate (esterified by vinyl acetate)
E1422	Acetylated distarch adipate
E1423	Acetylated distarch glycerol
E1430	Distarch glycerol
E1440	Hydroxypropyl starch
E1441	Hydroxypropyl distarch glycerol
E1442	Hydroxypropyl distarch phosphate

APPENDIX 9

Selected Atomic Weights

(From *Pure and Applied Chemistry* (1984), **56** (6), 653)

Name	Symbol	Atomic Number	Atomic Weight
Aluminium	Al	13	26.981
Antimony	Sb	51	121.75
Argon	Ar	18	39.948
Arsenic	As	33	74.992
Barium	Ba	56	137.33
Beryllium	Be	4	9.012
Bismuth	Bi	83	208.980
Boron	B	5	10.811
Bromine	Br	35	79.904
Cadmium	Cd	48	112.41
Caesium	Cs	55	132.905
Calcium	Ca	20	40.078
Carbon	C	6	12.011
Cerium	Ce	58	140.12
Chlorine	Cl	17	35.453
Chromium	Cr	24	51.996
Cobalt	Co	27	58.933

Name	Symbol	Atomic Number	Atomic Weight
Copper	Cu	29	63.546
Fluorine	F	9	18.998
Gallium	Ga	31	69.723
Germanium	Ge	32	72.59
Gold	Au	79	196.967
Hafnium	Hf	72	178.49
Helium	He	2	4.003
Hydrogen	H	1	1.008
Iodine	I	53	126.905
Iron	Fe	26	55.847
Krypton	Kr	36	83.80
Lanthanum	La	57	138.906
Lead	Pb	82	207.2
Lithium	Li	3	6.941
Magnesium	Mg	12	24.305
Manganese	Mn	25	54.938
Mercury	Hg	80	200.59
Molybdenum	Mo	42	95.94
Neon	Ne	10	20.179
Nickel	Ni	28	58.69
Niobium	Nb	41	92.906
Nitrogen	N	7	14.007
Osmium	Os	76	190.2
Oxygen	O	8	15.999
Palladium	Pd	46	106.42
Phosphorus	P	15	30.974
Platinum	Pt	78	195.08
Potassium	K	19	39.098
Rhodium	Rh	45	102.906
Rubidium	Rb	37	85.468
Ruthenium	Ru	44	101.07
Scandium	Sc	21	44.956
Selenium	Se	34	78.96
Silicon	Si	14	28.086
Silver	Ag	47	107.868
Sodium	Na	11	22.990
Strontium	Sr	38	87.62
Sulphur	S	16	32.066
Tantalum	Ta	73	180.948
Tellurium	Te	52	127.60
Thallium	Tl	81	204.383
Thorium	Th	90	232.038
Tin	Sn	50	118.710
Titanium	Ti	22	47.88
Tungsten	W	74	183.85
Uranium	U	92	238.029
Vanadium	V	23	50.942
Xenon	Xe	54	131.29
Zinc	Zn	30	65.39
Zirconium	Zr	40	91.224

APPENDIX 10

Titrimetric analysis data

Chemical factors

For acid substances
1 ml 0·1 M sodium hydroxide solution is equivalent to:

0·006 005 g	Acetic acid	CH_3COOH
0·012 210 g	Benzoic acid	$C_6H_5\ COOH$
0·006 184 g	Boric acid	H_3BO_3
0·007 005 g	Citric acid	$C_6H_8O_7, H_2O$
0·003 646 g	Hydrochloric acid	HCl
0·009 008 g	Lactic acid	$C_3H_6O_3$
0·006 706 g	Malic acid	$C_4H_6O_5$
0·028 245 g	Oleic acid	$C_{18}H_{34}O_2$
0·006 303 g	Oxalic acid	$(COOH)_2 \cdot 2H_2O$
0·004 900 g	Phosphoric acid (ortho)	H_3PO_4
0·004 904 g	Sulphuric acid	H_2SO_4
0·007 504 g	Tartaric acid	$C_4H_6O_6$

For alkaline substances
1 ml 0·1 M hydrochloric acid (or 0·5 M sulphuric acid) solution is equivalent to:

0·001 703 g	Ammonia	NH_3
0·001 401 g	Nitrogen	N
0·008 569 g	Barium hydroxide	$Ba(OH)_2$
0·005 004 g	Calcium carbonate	$CaCO_3$
0·002 804 g	Calcium oxide	CaO
0·006 910 g	Potassium carbonate	K_2CO_3
0·005 610 g	Potassium hydroxide	KOH
0·004 710 g	Potassium oxide	K_2O
0·008 401 g	Sodium bicarbonate	$NaHCO_3$
0·019 072 g	Sodium borate	$Na_2B_4O_7 \cdot 10H_2O$
0·005 300 g	Sodium carbonate	Na_2CO_3
0·004 000 g	Sodium hydroxide	NaOH
0·003 100 g	Sodium oxide	Na_2O

For other titrations
1 ml 0·02 M potassium permanganate solution is equivalent to:

0·002 004 g	Calcium	Ca
0·005 004 g	Calcium carbonate	$CaCO_3$
0·002 804 g	Calcium oxide	CaO
0·005 585 g	Ferrous iron	Fe
0·006 303 g	Oxalic acid	$C_2H_2O_4 \cdot 2H_2O$
0·003 450 g	Sodium nitrite	$NaNO_3$

1 ml 0·1 M silver nitrate solution is equivalent to:

0·003 546 g	Chloride	Cl
0·012 690 g	Iodide	I
0·007 456 g	Potassium chloride	KCl
0·016 600 g	Potassium iodide	KI
0·005 844 g	Sodium chloride	NaCl

1 ml 0·1 M iodine solution is equivalent to:

0·012 610 g	Sodium sulphite	$Na_2SO_3 \cdot 7H_2O$
0·003 203 g	Sulphur dioxide	SO_2

Strengths of common acids and aqueous ammonia (approx.)

	Per cent by weight	Weight per ml at 20 °C	Normality	Molarity	ml required to make 1 litre of molar solution
Hydrochloric acid	35	1·18	11·3	11·3	89
Nitric acid	70	1·42	16·0	16·0	63
Sulphuric acid	96	1·84	36·0	18·0	56
Perchloric acid	70	1·66	11·6	11·6	86
Hydrofluoric acid	46	1·15	26·5	26·5	38
Phosphoric acid	85	1·69	41·1	13·7	70
Acetic acid	99·5	1·05	17·4	17·4	58
Ammonia aqueous	27 (NH_3)	0·90	14·3	14·3	71

Preparation of solutions for titrimetric analysis

Ammonium thiocyanate NH_4SCN, 76·12
  0·1 N = 0·1 M = 7·612 g per litre
Hydrochloric acid HCl, 36·46
  0·1 N = 0·1 M = 3·646 g per litre
Iodine I, 126·90
  0·1 N = 0·1 M = 12·69 g I + 18 g KI per litre
Potassium dichromate $K_2Cr_2O_7$, 294·24
  0·1 N = M/60 = 4·903 g per litre
Potassium iodate KIO_3, 214·02
  Normally depends on the reaction employed, but commonly 0·1 N = M/60. If acidity of reaction exceeds 4 N, then 0·1 N iodate = M/40.
Potassium permanganate $KMnO_4$, 158·0
  0·1 N = 0·02 M = 3·161 g per litre
Potassium thiocyanate KSCN, 97·185
  0·1 N = 0·1 M = 9·7185 g per litre
Silver nitrate $AgNO_3$, 169·9
  0·1 N = 0·1 M = 16·99 g per litre
Sodium EDTA $C_{10}H_{14}N_2Na_2O_8 \cdot 2H_2O$, 372·25
  0·05 M = 18·61 g per litre
Sodium hydroxide NaOH, 40·00
  0·1 N = 0·1 M = 4·000 g per litre
Sodium thiosulphate $Na_2S_2O_3 \cdot 5H_2O$, 248·2
  0·1 N = 0·1 M = 24·82 g per litre
Sulphuric acid H_2SO_4, 98·08
  0·1 N = 0·05 M = 4·904 g per litre

Acid and base indicators

	pH range	Colour change
Thymol blue	1·2–2·8	red–yellow
m-Cresol purple	1·2–2·8	red–yellow
4-Dimethylaminoazobenzene	2·9–4·0	red–yellow orange
Bromphenol blue	3·0–4·6	yellow–red violet
Congo red	3·0–5·2	blue violet–yellow orange
Methyl orange	3·1–4·4	red–yellow orange
Bromcresol green	3·8–5·4	yellow–blue
Mixed indicator 5	4·4–5·8	red violet–green
Methyl red	4·4–6·2	violet red–yellow orange
Lacmoid	5·0–8·0	red–blue
Bromcresol purple	5·2–6·8	yellow–purple
Bromphenol red	5·2–6·8	orange yellow–purple
Bromthymol blue	6·0–7·6	yellow–blue
Phenol red	6·4–8·2	yellow–red
Neutral red	6·8–8·0	bluish red–orange yellow

	pH range	Colour change
Cresol red	7·0–8·8	yellow–purple
m-Cresol purple	7·4–9·0	yellow–purple
Thymol blue	8·0–9·6	yellow–blue
Phenolphthalein	8·2–9·8	colourless–red violet
Thymolphthalein	9·3–10·5	colourless–blue
Alizarin yellow GG	10·0–12·1	light yellow–orange red
Epsilon blue	12·0–13·0	orange–violet

Redox indicators

	E_u(volt) pH 7; 20 °C*	Colour change
Neutral red	−0·32	red–colourless
Methylene blue	+0·01	blue–colourless
Thionine	+0·06	violet–colourless
2,6-Dichlorphenoindophenol	+0·23	blue–colourless
Variamine blue	(+0·60)	
Diphenylamine	(+0·76)	blue–colourless
Diphenylaminesulphonic acid	+0·83†	violet–colourless
Ferroin	+1·06††	red–pale blue

* Redox potential at colour change (50 per cent indicator reduced).
† in 1 M sulphuric acid.
†† 0·025 M in 1 M sulphuric acid.

APPENDIX 11

Freezing mixtures, humidities of salt solutions, sieve sizes

Freezing mixtures

Mixtures by weight	decrease of temperature from (°C)	to (°C)
4 water + 1 Potassium chloride	+10	−12
1 water + 1 Ammonium nitrate	+10	−15
1 water + 1 Sodium nitrate + 1 Ammonium chloride	+ 8	−24
3 ice ground + 1 Sodium chloride	0	−21
1·2 ice ground + 2 Calcium chloride ($CaCl_2 \cdot 6H_2O$)	0	−39
1·4 ice ground + 2 Calcium chloride ($CaCl_2 \cdot 6H_2O$)	0	−55
Methanol or Acetone + solid Carbon dioxide	+15	−77

Humidities of salt solutions

Saturated aqueous solution	Relative humidity (%RH) over the solution at 20 °C
Sodium carbonate $Na_2CO_3 \cdot 10H_2O$	92
Ammonium sulphate $(NH_4)_2SO_4$	80
Potassium chloride KCl	86
Sodium chloride NaCl	76
Ammonium nitrate NH_4NO_3	63
Calcium nitrate $Ca(NO_3)_2 \cdot 2H_2O$	55
Potassium carbonate K_2CO_3	45
Calcium chloride $CaCl_2 \cdot 6H_2O$	35

Sieve sizes and numbers, with close equivalents

US Sieve No. ASTM E-11-61	Nominal aperture size (mm)	DIN 4188 (μm)	Tyler Mesh size	BS 410: 1986
	5600			3
	4750			3½
	4000			4
	3350			5
7	2800	2800		6
	2360			7
10	2000	2000	9	8
12	1680	1670	10	10
		1600		
14	1410	1400	12	12
		1250		
16	1190	1200	14	14
18	1000	1000	16	16
20	841	840	20	18
		800		
25	707	710	24	22
30	595		28	25
		600		
35	500	500	32	30
40	420	425	35	36
		400		
45	354	355	42	44
		315		
50	297	300	48	52
60	250	250	60	60
70	210	212	65	72
		200		
80	177	180	80	85
		160		
100	149	150	100	100
		140		
120	125	125	115	120
140	105	106	150	150
		100		
170	88	90	170	170
		80		
200	74	75	200	200
		71		
230	63	63	250	240
		56		
270	53	53		300
		50		
325	45	45		350
		40		
400	38	38		400
	32	32		440

APPENDIX 12

The SI system of units

The Système International d'Unités (SI) was adopted by the General Conference of Weights and Measures and endorsed by the International Organization for Standardization (ISO) and the International Electrotechnical Commission (IEC).

The SI system is based on the following 7 primary units:

Quantity	Name	Symbol
length	metre	m
mass	kilogram	kg
time	second	s
electric current	ampere	A
thermodynamic temperature	kelvin	K
amount of substance	mole	mol
luminous intensity	candela	cd

Decimal multiples and sub-multiples of the SI units are formed with the following prefixes:

Factor	Prefix	Symbol	Factor	Prefix	Symbol
10^{18}	exa	E	10^{-1}	deci	d
10^{15}	peta	P	10^{-2}	centi	c
10^{12}	tera	T	10^{-3}	milli	m
10^{9}	giga	G	10^{-6}	micro	μ
10^{6}	mega	M	10^{-9}	nano	n
10^{3}	kilo	k	10^{-12}	pico	p
10^{2}	hecto	h	10^{-15}	femto	f
10^{1}	deca	da	10^{-18}	atto	a

Conversion of common units to equivalents in SI units

Length

1 ft	=	0·304 m
1 in	=	25·4 mm = 2·54 cm

Area

1 ft^2 (square foot)	=	0·092 903 m^2 = 929·030 cm^3
1 in^2 (square inch)	=	645·16 mm^2 = 6·4516 cm^2

Volume

1 ft^3 (cubic foot)	=	0·028 316 8 m^3 = 28·3168 dm^3
1 in^3 (cubic inch)	=	16·387 1 cm^3

Capacity

1 gal	=	4·546 09 dm^3 = 4·546 litre
1 US gal	=	3·785 41 dm^3 = 3·785 litre
1 pt (pint)	=	0·568 261 dm^3 = 0·568 litre
1 fl oz	=	28·4131 cm^3

Mass

1 ton	=	1016·05 kg = 1·016 05 t
1 cwt	=	50·8023 kg
1 lb	=	0·453 592 kg
1 oz	=	28·3495 g

Pressure/vacuum

1 bar	=	14·5 lb in^{-1}
	=	0·987 atmospheres
	=	29·53 in of Hg
	=	10^5 N m^{-2}
1 atmosphere	=	101·325 kN m^{-2}
1 Torr	=	133·322 N m^{-2}
	=	1 mm of Hg
1 lb in^{-2}	=	6894·76 N m^{-2}

Temperature

°C:	θ	=	$5(t - 32)/9$
K:	T	=	$5(t + 459{\cdot}67)/9$
where	t	=	Temperature on Fahrenheit scale
	T	=	Temperature on Kelvin scale
	θ	=	Temperature on Celsius scale (°C)

The zero on the Celsius scale is the ice-point (273·15 K)

Energy

1 cal$_{15}$(15° calorie)	=	4·1855 J

APPENDIX 13

Composition of foods

From *Manual of Nutrition*, 9th edn, MAFF, 1985, HMSO

Composition per 100 g (raw edible weight except where stated)

No.	Food	Inedible waste (%)	Energy (kcal)	Energy (kJ)	Protein (g)	Fat (g)	Carbo-hydrate (as mono-saccharide) (g)
	Milk						
1	Cream—double	0	447	1 841	1.5	48.2	2.0
2	Cream—single	0	195	806	2.4	19.3	3.2
3	Milk, liquid, whole	0	65	272	3.2	3.9	4.6
4	Milk, liquid, skimmed	0	32	137	3.4	0.1	4.7
5	Milk, condensed whole, sweetened	0	170	709	8.5	10.2	11.7
6	Milk, whole, evaporated	0	149	620	8.4	9.4	8.1
7	Milk, dried, skimmed	0	339	1 442	36.1	0.6	50.4
8	Yogurt, low fat, natural	0	65	276	5.1	0.8	10.0
9	Yogurt, low fat, fruit	0	89	382	4.1	0.7	17.9
	Cheese						
10	Cheddar	0	406	1 682	26.0	33.5	0
11	Cottage	0	96	402	13.6	4.0	1.4
12	Cheese spread	0	283	1 173	18.3	22.9	0.9
13	Feta	0	245	1 017	16.5	19.9	0
14	Brie	0	300	1 246	22.8	23.2	0
	Meat						
15	Bacon, rashers, raw	11	339	1 402	13.9	31.5	0
16	Bacon, rashers, grilled	0	393	1 632	28.1	31.2	0
17	Beef, average, raw	17	313	1 296	16.6	27.4	0
18	Beef, mince, stewed	0	229	955	23.1	15.2	0
19	Beef, stewing steak, raw	4	176	736	20.2	10.6	0
20	Beef, stewing steak, cooked	0	223	932	30.9	11.0	0
21	Black pudding, fried	0	305	1 270	12.9	21.9	15.0
22	Chicken, raw	41	194	809	19.7	12.8	0
23	Chicken, roast, meat and skin	0	213	888	24.4	12.8	0
24	Chicken, roast, meat only	0	148	621	24.8	5.4	0
25	Corned beef	0	202	844	25.9	10.9	0
26	Ham	0	166	690	16.4	11.1	0
27	Kidney, pigs, raw	6	86	363	15.5	2.7	0
28	Kidney, pigs, fried	0	202	848	29.2	9.5	0
29	Lamb, average, raw	23	295	1 223	16.2	25.6	0
30	Lamb, roast	0	266	1 106	26.1	17.9	0
31	Liver, lambs, raw	0	140	587	20.3	6.2	0.8
32	Liver, lambs, fried	0	237	989	30.1	12.9	0
33	Luncheon meat	0	266	1 153	12.9	23.8	3.3
34	Paté, average	0	347	1 436	13.7	31.9	1.4
35	Pork, average, raw	26	297	1 231	16.9	25.5	0
36	Pork chop, cooked	26	332	1 380	28.5	24.2	0
37	Sausage, beef, cooked	0	267	1 114	12.9	17.7	15.0
38	Sausage, pork, cooked	0	317	1 318	13.6	24.5	11.2
39	Steak & kidney pie	0	274	1 146	9.3	17.1	22.2
40	Turkey, roast, meat & skin	0	189	793	26.2	9.4	0

Water (g)	Calcium (mg)	Iron (mg)	Sodium (mg)	Vitamin A (retinol equivalent) (μg)	Thiamin (mg)	Ribo-flavin (mg)	Niacin equivalent (mg)	Vitamin C (mg)	No.
49	50	0.2	30	500	0.02	0.08	0.4	1	1
72	79	0.3	40	155	0.03	0.12	0.8	1	2
88	103	0.1	50	56	0.05	0.17	0.9	1.5	3
91	108	0.1	50	1	0.05	0.18	0.9	1.5	4
30	270	0.2	140	123	0.09	0.46	2.3	4.1	5
69	260	0.3	170	125	0.07	0.42	2.1	1.5	6
3	1 230	0.3	510	550	0.38	0.16	9.5	13.2	7
86	200	0.1	80	12	0.06	0.25	1.2	0.8	8
77	150	0.1	70	12	0.05	0.21	1.2	0.7	9
37	800	0.4	610	363	0.04	0.50	6.2	0	10
79	60	0.1	450	41	0.02	0.19	3.3	0	11
51	510	0.7	1 170	198	0.02	0.24	0.1	0	12
56	384	0.2	1 260	270	0.03	0.11	4.2	0	13
48	380	0.8	1 410	238	0.09	0.60	6.2	0	14
51	7	0.6	1 340	0	0.45	0.14	6.5	0	15
34	14	1.3	2 404	0	0.57	0.27	12.5	0	16
55	7	1.9	70	10	0.05	0.23	6.9	0	17
59	18	3.1	320	0	0.05	0.33	9.3	0	18
69	8	2.1	72	0	0.06	0.23	8.5	0	19
57	15	3.0	360	0	0.03	0.33	10.2	0	20
44	35	20.0	1 210	0	0.09	0.07	3.8	0	21
67	9	0.7	75	0	0.11	0.13	9.6	0	22
62	13	0.5	90	0	0.05	0.19	13.6	0	23
68	9	0.8	81	0	0.08	0.19	12.8	0	24
59	27	2.4	854	0	0	0.20	9.1	0	25
67	4	0.6	1 405	0	0.54	0.20	6.3	0	26
80	10	6.4	200	160	0.56	2.58	11.1	6.5	27
58	12	9.1	220	220	0.41	3.70	20.1	11.9	28
56	7	1.4	71	0	0.09	0.21	7.1	0	29
55	8	2.5	65	0	0.12	0.31	11.0	0	30
70	6	7.5	73	19 900	0.39	4.64	20.7	19.2	31
54	8	10.9	83	30 500	0.38	5.65	24.7	18.6	32
54	39	1.0	913	0	0.06	0.15	3.9	0	33
47	14	8.2	762	8 300	0.14	1.32	4.3	0	34
57	8	0.9	65	0	0.49	0.20	8.9	0	35
46	11	1.2	84	0	0.66	0.20	11.0	0	36
48	68	1.6	1 095	0	0	0.14	9.0	0	37
45	54	1.5	1 075	0	0.01	0.16	7.2	0	38
51	47	1.8	402	0	0.12	0.25	4.9	0	39
63	7	0.9	70	0	0.09	0.16	12.2	0	40

Composition per 100 g (raw edible weight except where stated)

No.	Food	Inedible waste (%)	Energy (kcal)	Energy (kJ)	Protein (g)	Fat (g)	Carbohydrate (as monosaccharide) (g)
	Fish						
41	White fish, filleted	3	77	324	17.1	0.9	0
42	Cod, fried	0	235	982	19.6	14.3	7.5
43	Fish fingers, raw	0	178	749	12.6	7.5	16.1
44	Herrings, whole	46	251	1 040	16.8	20.4	0
45	Mackerel	40	282	1 170	19.0	22.9	0
46	Pilchards, canned in tomato sauce	0	126	531	18.8	5.4	0.7
47	Sardines, canned in oil, fish only	0	217	906	23.7	18.6	0
48	Tuna in oil	0	289	1 202	22.8	22.0	0
49	Prawns, boiled	0	107	451	22.6	1.8	0
	Eggs						
50	Eggs, boiled	12	147	612	12.3	10.9	0
51	Eggs, fried	0	232	961	14.1	19.5	0
	Fats						
52	Butter	0	740	3 041	0.4	82.0	0
53	Lard, cooking fat, dripping	0	892	3 667	0	99.1	0
54	Low fat spread	0	366	1 506	0	40.7	0
55	Margarine, average	0	730	3 000	0.1	81.0	0
56	Cooking and salad oil	0	899	3 696	0	99.9	0
	Preserves, etc.						
57	Chocolate, milk	0	529	2 214	8.4	30.3	59.4
58	Honey	0	288	1 229	0.4	0	76.4
59	Jam	0	262	1 116	0.5	0	69.2
60	Marmalade	0	261	1 114	0.1	0	69.5
61	Sugar, white	0	394	1 680	0	0	105.3
62	Syrup	0	298	1 269	0.3	0	79.0
63	Peppermints	0	392	1 670	0.5	0.7	102.2
	Vegetables						
64	Aubergines	23	14	62	0.7	0	3.1
65	Baked beans	0	81	345	4.8	0.6	15.1
66	Beans, runner, boiled	1	19	83	1.9	0.2	2.7
67	Beans, red kidney, raw	0	272	1 159	22.1	1.7	45.0
68	Beans, soya, boiled	0	141	592	12.4	6.4	9.0
69	Beetroot, boiled	0	44	189	1.8	0	9.9
70	Brussels sprouts, boiled	0	18	75	2.8	0	1.7
71	Cabbage, raw	43	22	92	2.8	0	2.8
72	Cabbage, boiled	0	15	66	1.7	0	2.3
73	Carrots, old	4	23	98	0.7	0	5.4
74	Cauliflower, cooked	0	9	40	1.6	0	0.8
75	Celery	27	8	36	0.9	0	1.3
76	Courgettes, raw	13	29	122	1.6	0.4	5.0
77	Cucumber	23	10	43	0.6	0.1	1.8
78	Lentils, cooked	0	99	420	7.6	0.5	17.0
79	Lettuce	30	12	51	1.0	0.4	1.2
80	Mushrooms	25	13	53	1.8	0.6	0
81	Onion	3	23	99	0.9	0	5.2
82	Parsnips, cooked	0	56	238	1.3	0	13.5
83	Peas, frozen, boiled	0	72	307	6.0	0.9	10.7
84	Peas, canned processed	0	86	366	6.9	0.7	18.9

Water (g)	Calcium (mg)	Iron (mg)	Sodium (mg)	Vitamin A (retinol equivalent) (μg)	Thiamin (mg)	Riboflavin (mg)	Niacin equivalent (mg)	Vitamin C (mg)	No.
82	22	0.5	99	1	0.07	0.09	6.0	0	41
57	80	0.5	100	0	0.06	0.07	4.9	0	42
64	43	0.7	320	0.2	0.09	0.06	3.5	0	43
64	33	0.8	67	46	0	0.18	7.2	0	44
57	24	1.0	130	45	0.09	0.35	11.6	0	45
74	300	2.7	370	8	0.02	0.29	11.1	0	46
58	550	2.9	650	7	0.04	0.36	12.6	0	47
55	7	1.1	420	0	0.04	0.11	17.2	0	48
70	150	1.1	1590	0	0.03	0.03	7.4	0	49
75	52	2.0	140	190	0.09	0.47	3.7	0	50
63	64	2.5	220	140	0.07	0.42	4.2	0	51
15	15	0.2	870	985	0	0	0.1	0	52
1	1	0.1	2	0	0	0	0	0	53
51	0	0	690	900	0	0	0	0	54
16	4	0.3	800	860	0	0	0.1	0	55
0	0	0	0	0	0	0	0	0	56
2	220	1.6	120	6.6	0.10	0.23	1.6	0	57
23	5	0.4	11	0	0	0.05	0.2	0	58
30	18	1.2	14	2	0	0	0	10	59
28	35	0.6	18	8	0	0	0	10	60
0	2	0	0	0	0	0	0	0	61
28	26	1.5	270	0	0	0	0	0	62
0	7	0.2	9	0	0	0	0	0	63
93	10	0.4	3	0	0.05	0.03	1.0	5	64
74	48	1.4	550	12	0.08	0.06	1.3	0	65
91	22	0.7	1	67	0.03	0.07	0.8	5	66
11	140	6.7	40	0	0.54	0.18	5.5	0	67
67	145	2.5	15	0	0.26	0.16	3.4	0	68
83	30	0.4	64	0	0.02	0.04	0.4	5	69
92	25	0.5	2	67	0.06	0.10	0.9	40	70
88	57	0.6	7	50	0.06	0.05	0.8	55	71
93	38	0.4	4	50	0.03	0.03	0.5	20	72
90	48	0.6	95	2000	0.06	0.05	0.7	6	73
95	18	0.4	4	5	0.06	0.06	0.8	20	74
94	52	0.6	140	0	0.03	0.03	0.5	7	75
92	30	1.5	1	58	0.05	0.09	0.6	16	76
96	23	0.3	13	0	0.04	0.04	0.3	8	77
72	13	2.4	12	3	0.11	0.04	1.6	0	78
96	23	0.9	9	167	0.07	0.08	0.4	15	79
92	3	1.0	9	0	0.10	0.40	4.6	3	80
93	31	0.3	10	0	0.03	0.05	0.4	10	81
83	36	0.5	4	0	0.07	0.06	0.9	10	82
78	35	1.6	2	50	0.30	0.09	1.6	12	83
70	33	1.8	380	10	0.10	0.04	1.4	0	84

Composition per 100 g (raw edible weight except where stated)

No.	Food	Inedible waste (%)	Energy (kcal)	Energy (kJ)	Protein (g)	Fat (g)	Carbohydrate (as monosaccharide) (g)
85	Peppers, green	14	12	51	0.9	0	2.2
86	Potatoes	10* 20†	74	315	2.0	0.2	17.1
87	Potatoes, boiled	0	76	322	1.8	0.1	18.0
88	Potato crisps	0	533	2224	6.3	35.9	49.3
89	Potatoes, fried (chips)	0	234	983	3.6	10.2	34.0
90	Potatoes, oven chips	0	162	687	3.2	4.2	29.8
91	Potatoes, roast	0	150	632	3.0	4.5	25.9
92	Spinach, boiled	0	30	128	5.1	0.5	1.4
93	Sweetcorn, canned	0	85	379	2.9	1.2	16.8
94	Sweet potato	14	91	387	1.2	0.6	21.5
95	Tomatoes, fresh	0	14	60	0.9	0	2.8
96	Turnips, cooked	0	14	60	0.7	0.3	2.3
97	Watercress	23	14	61	2.9	0	0.7
98	Yam, boiled	0	119	508	1.6	0.1	29.8
	Fruit						
99	Apples	20	46	196	0.3	0	11.9
100	Apricots, canned in syrup	0	106	452	0.5	0	27.7
101	Apricots, dried	0	182	772	4.8	0	43.4
102	Avocado pear	29	223	922	4.2	22.2	1.8
103	Bananas	40	76	326	1.1	0	19.2
104	Blackcurrants	2	28	121	0.9	0	6.6
105	Cherries	13	47	201	0.6	0	11.9
106	Dates, dried	14	248	1056	2.0	0	63.9
107	Figs, dried	0	213	908	3.6	0	52.9
108	Gooseberries, cooked, unsweetened	0	14	62	0.9	0	2.9
109	Grapes	5	63	268	0.6	0	16.1
110	Grapefruit	50	22	95	0.6	0	5.3
111	Lemon juice	64	7	31	0.3	0	1.6
112	Mango	34	59	253	0.5	0	15.3
113	Melon	40	23	97	0.8	0	5.2
114	Oranges	25	35	150	0.8	0	8.5
115	Orange juice	0	38	161	0.6	0	9.4
116	Peaches	13	37	156	0.6	0	9.1
117	Peaches, canned in syrup	0	87	373	0.4	0	22.9
118	Pears	28	41	175	0.3	0	10.6
119	Pineapple, canned in juice	0	46	194	0.5	0	11.6
120	Plums	8	32	137	0.6	0	7.9
121	Prunes, dried	17	161	686	2.4	0	40.3
122	Raspberries	0	25	105	0.9	0	5.6
123	Rhubarb, cooked with sugar	0	45	191	05	0	11.4
124	Strawberries	3	26	109	0.6	0	6.2
125	Sultanas	0	250	1066	1.8	0	64.7
	Nuts						
126	Almonds	63	565	2336	16.9	53.5	4.3
127	Coconut, dessicated	0	604	2492	5.6	62.0	6.4
128	Peanuts, roasted & salted	0	570	2364	24.3	49.0	8.6
	Cereals						
129	Biscuits, chocolate	0	524	2197	5.7	27.6	67.4
130	Biscuits, plain, digestive	0	471	1978	6.3	20.9	68.6
131	Biscuits, semi-sweet	0	457	1925	6.7	16.6	74.8
132	Bread, brown	0	217	924	8.4	2.0	44.2

* Old potatoes † New potatoes

Water (g)	Calcium (mg)	Iron (mg)	Sodium (mg)	Vitamin A (retinol equivalent) (µg)	Thiamin (mg)	Ribo-flavin (mg)	Niacin equivalent (mg)	Vitamin C (mg)	No.
94	9	0.4	2	33	0.08	0.03	0.9	100	85
79	8	0.4	8	0	0.20	0.02	1.5	8–19	86
80	4	0.4	7	0	0.20	0.02	1.2	5–9	87
3	37	2.1	550	0	0.19	0.07	6.1	17	88
44	14	0.84	41	0	0.2	0.02	1.5	6–14	89
59	1	0.8	53	0	0.1	0.04	3.1	12	90
65	10	0.62	9	0	0.2	0.02	1.3	5–12	91
85	136	4.0	120	1 000	0.07	0.15	1.8	25	92
72	4	0.5	270	4	0.04	0.06	1.8	0	93
70	22	0.7	19	4 000††	0.10	0.06	1.2	25	94
93	13	0.4	3	100	0.06	0.04	0.8	20	95
95	55	0.4	28	0	0.03	0.04	0.6	17	96
91	220	1.6	60	500	0.10	0.10	1.1	60	97
66	9	0.3	17	2	0.05	0.01	0.8	2	98
84	4	0.3	2	5	0.04	0.02	0.1	5	99
68	12	0.7	1	166	0.02	0.01	0.4	2	100
15	92	4.1	56	600	0	0.2	3.8	0	101
69	15	1.5	2	17	0.10	0.10	1.8	15	102
71	7	0.4	1	33	0.04	0.07	0.8	10	103
77	60	1.3	3	33	0.03	0.06	0.4	200	104
82	16	0.4	3	20	0.05	0.07	0.4	5	105
15	68	1.6	5	10	0.07	0.04	2.9	0	106
17	280	4.2	87	8	0.10	0.08	2.2	0	107
90	24	0.3	2	25	0.03	0.03	0.5	31	108
79	19	0.3	2	0	0.04	0.02	0.3	4	109
91	17	0.3	1	0	0.05	0.02	0.3	40	110
91	8	0.1	2	0	0.02	0.01	0.1	50	111
83	10	0.5	7	200	0.03	0.04	0.4	30	112
94	16	0.4	17	175	0.05	0.03	0.3	50	113
86	41	0.3	3	8	0.10	0.03	0.3	50	114
88	12	0.3	2	8	0.08	0.02	0.3	25–45	115
86	5	0.4	3	83	0.02	0.05	1.1	8	116
74	4	0.4	1	41	0.01	0.02	0.6	4	117
83	8	0.2	2	2	0.03	0.03	0.3	3	118
77	12	0.4	1	7	0.08	0.02	0.3	20–40	119
85	12	0.3	2	37	0.05	0.03	0.6	3	120
23	38	2.9	12	160	0.10	0.20	1.9	0	121
83	41	1.2	3	13	0.02	0.03	0.5	25	122
85	84	0.3	2	8	0	0.03	0.4	7	123
89	22	0.7	2	5	0.02	0.03	0.5	60	124
18	52	1.8	53	5	0.10	0.08	0.6	0	125
5	250	4.2	6	0	0.24	0.92	4.7	0	126
2	22	3.6	28	0	0.06	0.04	1.8	0	127
5	61	2.0	440	0	0.23	0.10	21.3	0	128
2.2	110	1.7	160	0	0.03	0.13	2.7	0	129
2.5	92	3.2	600	0	0.14	0.11	2.4	0	130
2.5	120	2.1	410	0	0.13	0.08	2.9	0	131
40	99	2.2	540	0	0.27	0.10	2.3	0	132

†† The vitamin A content of white and yellow varieties may vary between 0 and 12 000 µg.

Composition per 100 g (raw edible weight except where stated)

No.	Food	Inedible waste (%)	Energy (kcal)	Energy (kJ)	Protein (g)	Fat (g)	Carbohydrate (as monosaccharide) (g)
133	Bread, white	0	230	980	8.2	1.7	48.6
134	Bread, wholemeal	0	215	911	9.0	2.5	41.6
	Breakfast cereals						
135	Cornflakes	0	368	1 567	8.6	1.6	85.1
136	Weetabix	0	340	1 444	11.4	3.4	70.3
137	Muesli	0	368	1 556	12.9	7.5	66.2
138	Cream crackers	0	440	1 857	9.5	16.3	68.3
139	Crispbread, rye	0	321	1 367	9.4	2.1	70.6
140	Flour, white	0	337	1 435	9.4	1.3	76.7
141	Flour, wholemeal	0	306	1 302	12.7	2.2	62.8
142	Oats, porridge	0	374	1 582	10.9	9.2	66.0
143	Rice, raw	0	359	1 529	7.0	1.0	85.8
144	Spaghetti, raw	0	342	1 456	12.0	1.8	74.1
	Cakes, etc.						
145	Chocolate cake with butter icing	0	500	2 092	5.8	30.9	53.1
146	Currant buns	0	296	1 250	7.6	7.5	52.7
147	Fruit cake, rich	0	322	1 357	4.9	12.5	50.7
148	Jam tarts	0	368	1 552	3.3	13.0	63.4
149	Plain cake, Madeira	0	393	1 652	5.4	16.9	58.4
	Puddings						
150	Apple pie	0	369	1 554	4.3	15.5	56.7
151	Bread and butter pudding	0	157	661	6.1	7.7	16.9
152	Cheesecake, frozen, fruit topping	0	239	1 005	5.2	10.6	32.8
153	Custard	0	118	496	3.8	4.4	16.7
154	Ice cream, dairy	0	165	691	3.3	8.2	20.7
155	Rice pudding	0	131	552	4.1	4.2	20.4
156	Trifle	0	165	690	2.2	9.2	19.5
	Beverages						
157	Chocolate, drinking	0	366	1 554	5.5	6.0	77.4
158	Cocoa powder	0	312	1 301	18.5	21.7	11.5
159	Coffee, ground, infusion	0	3	12	0.3	0	0.4
160	Coffee, instant powder	0	100	424	14.6	0	11.0
161	Carbonated 'ades	0	38	166	0	0	10.0
162	Tea, dry	0	0	0	0	0	0
163	Squash, undiluted	0	98	418	0	0	26.1
	Alcoholic beverages						
164	Beer, keg bitter	0	37	156	0	0	2.3
165	Spirits	0	222	919	0	0	0
166	Wine, medium white	0	89	371	0	0	2.5
167	Cider, average	0	43	180	0	0	2.9
	Miscellaneous						
168	Curry powder	0	325	1 395	12.7	13.8	41.8
169	Marmite	0	179	759	41.4	0.7	1.8
170	Peanut butter	0	623	2 581	22.6	53.7	13.1
171	Soy sauce	0	56	240	5.2	0.5	8.3
172	Tomato soup	0	55	230	0.8	3.3	5.9
173	Tomato ketchup	0	98	420	2.1	0	24.0
174	Pickle, sweet	0	134	572	0.6	0.3	34.4
175	Salad cream	0	311	1 288	1.9	27.4	15.1

Water (g)	Calcium (mg)	Iron (mg)	Sodium (mg)	Vitamin A (retinol equivalent) (µg)	Thiamin (mg)	Riboflavin (mg)	Niacin equivalent (mg)	Vitamin C (mg)	No.
38	105	1.6	525	0	0.21	0.06	2.3	0	133
38	54	2.7	560	0	0.34	0.09	1.8	0	134
3.0	3	6.7	1 160	0	1.8	1.6	21.9	0	135
1.8	33	7.6	360	0	1.0	1.5	14.3	0	136
5.8	200	4.6	180	0	0.33	0.27	5.7	0	137
4.3	110	1.7	610	0	0.13	0.08	3.4	0	138
6.4	50	3.7	220	0	0.28	0.14	2.9	0	139
14.0	140	2.0	2	0	0.31	0.04	3.5	0	140
14.0	38	3.9	2	0	0.47	0.09	8.3	0	141
8.2	52	3.8	9	0	0.90	0.09	3.3	0	142
11.4	4	0.5	4	0	0.41	0.02	5.8	0	143
9.8	25	2.1	3	0	0.22	0.03	3.1	0	144
8.4	130	1.6	440	298	0.07	0.09	2.0	0	145
27.7	110	1.9	230	0	0.37	0.16	3.1	0	146
20.6	84	3.2	220	0	0.07	0.09	1.3	0	147
14.4	72	1.7	130	0	0.06	0.02	1.2	0	148
20.2	42	1.1	380	0	0.06	0.11	1.6	0	149
22.9	51	1.2	210	0	0.05	0.02	0.4	0	150
67.5	130	0.6	150	78	0.07	0.23	1.8	0	151
44.0	68	0.5	160	0	0.04	0.16	1.7	0	152
74.9	140	0.1	76	38	0.05	0.20	1.0	0	153
65.7	120	0.3	70	0	0.04	0.15	0.9	0	154
71.8	30	0.1	55	33	0.04	0.14	1.1	0	155
68.1	68	0.3	63	50	0.06	0.10	0.6	0	156
2	33	2.4	2	2	0.06	0.04	2.1	0	157
3	130	10.5	7	7	0.16	0.06	7.3	0	158
98	3	0.1	1	0	0	0.01	0.6	0	159
3	140	4.6	81	0	0.04	0.21	27.9	0	160
91	4	0.1	8	0	0	0	0	0	161
0	0	0	0	0	0	0.9	6.0	0	162
72	11	0.1	35	0	0	0.01	0.1	5	163
93	8	0	6	0	0	0.03	0.17	0	164
68	0	0	0	0	0	0	0	0	165
85	10	0.4	1	0	0	0	0.1	0	166
92	5	0.2	7	0	0	0	0	0	167
9	478	29.6	52	99	0.25	0.28	3.5	11	168
25	95	3.7	4 500	0	3.10	11.0	67	0	169
1	37	2.1	350	0	0.17	0.10	15	0	170
71	65	4.8	5 720	0	0.04	0.17	1.8	0	171
84	17	0.4	460	35	0.03	0.02	0.6	0	172
65	25	1.2	1 120	0	0.06	0.05	0.3	0	173
59	19	2.0	1 700	0	0.03	0.01	0.2	0	174
52.7	34	0.8	840	0	0	0	0	0	175

Index